James Drayton Kidner

MCMXCV

THE ENCYCLOPEDIA OF

RUSSIAN AIRCRAFT

1875 - 1995

THE ENCYCLOPEDIA OF

RUSSIAN AIRCRAFT

1875 - 1995

BILL
GUNSTON

Motorbooks International
Publishers & Wholesalers ®

This edition first published in 1995 by Motorbooks International Publishers & Wholesalers, 729 Prospect Avenue, Osceola, WI 54020 USA

Originally published in Great Britian in 1995 by Osprey, an imprint of Reed Consumer Books Limited.

Library of Congress Cataloging-in-Publication Data is available

ISBN 0-7603-0027-5

Printed in Italy

TITLE PAGE
Even in the rigidly controlled Soviet Union in Stalin's day, wonderful photographs like this were by no means uncommon. These were the members of Chyetverikov's team with the second prototype MDR-6B3

Contents

Introduction

In 1981 an editor friend, Dennis Baldry, said to me 'I have been appointed to create an aviation book list at Osprey. What book would you like to write?' That may not seem remarkable, but in fact every one of the 200-odd books I had written at that time had been done to order. Nobody had ever asked me what I would like to write.

I replied 'I've wanted for ages to write the definitive work on Russian aircraft, but of course it can't be done. We don't have the information'. Cutting a long story short, the book was attempted and it emerged in 1983. Called *Aircraft of the Soviet Union*, it was well received, and I was assured by several Russians that they knew nothing else remotely as comprehensive. Incidentally, one of that volume's reviewers said I might have been a little more generous in acknowledging my debt to Soviet designer V. B. Shavrov, instead of merely putting his name at the head of a long list of others who helped. This I gladly do now. Without the two volumes written by Vadim Borisovich Shavrov I doubt if I would ever have attempted that book. Sadly, he died on 23 Dec 1976, and nobody has found the manuscript of his almost complete third volume, taking the story on from 1950.

When it appeared, *Aircraft of the Soviet Union* (ASU) was the best I could do with the available information. It was heavily laced with 'appears to be . . . probably . . . figure must be an error . . . estimate . . . no info.' In many cases all I could do was offer a range of contradictory figures or supposed facts, citing their sources. Never in a million years did I imagine that the most tightly controlled and centrally run country in the world would break up into a fragile 'federation' of argumentative republics populated by entrepreneurs all hungry for money. Most of them are far too busy trying to make money to bother about history.

How times have changed! Many years ago I was honoured several times to meet the great Andrei N. Tupolev. His memory was perfect, and he was prepared to discuss any of his aircraft up to about 1950 — by sheer coincidence, where Shavrov left off. Any attempt to discuss newer aircraft, even civil ones, was met by a courteous refusal. But at the 1992 Farnborough airshow I at last got to grips with the bomber designers from the great bureau that bears Andrei Nikolayevich's name. The conversation went something like:

'Can you tell me a little about the Tu-98?'

'No, that was not very good. We do not talk about it.'

'Ah, then could you tell me about the history of the Tu-22M?'

'No, you do not need to bother with that. We want to tell you about the Tu-22M3, and plans for its future development . . .'

So we live in an upside-down world, where it is a cinch to find out what the Russians are trying to do for the future, but often still difficult to discover what they did in the past. In general, the bureaucrats in Moscow (still a formidable army) have removed from the secret list only the very latest aircraft, which they want to sell. Such people as Sergei Aleksandrovich Yakovlev, who takes a great interest and pride in everything done by the OKB (now called 'Aircraft Corporation') started by his father, are difficult to find. But so many people asked 'When are you going to update *Aircraft of the Soviet Union*?' that I had to have a stab at it. To do so meant two things: rectifying the numerous errors or uncertainties in the original book; and bringing the story up to date.

Thanks to the political upheavals the title had to change. I refused to entitle this book *The Encyclopedia of Aircraft of Russia, the Ukraine and Uzbekistan*, though several Russians, Ukrainians and Uzbeks pressed me to do so, and the three republics are named in the official title of their aircraft-industry association. In turn, the title change automatically meant that the starting date had to be put back from 1918 to 1875 (I almost included Lomonosov, 1754), and after agonising doubts decided to do this early history very briefly (because while a few readers are passionately interested, most are not) and put it separately at the front instead of slotting each designer into his alphabetical place. What makes this harder is that one or two of them, notably Grigorovich but also Khioni, went on designing for the Bolsheviks and appear in the main book.

Now I cannot avoid the crucial question of how best to update a book 11 years later. If the previous book had never happened, there would be no problem. As it is, a lot of Russian aviation readers have it on their shelves. Big books are not cheap, and I wanted this update to avoid duplicating the first to the maximum degree. It could hardly fail to be a lot more authoritative; there is now much less need for the word 'probably'. But I wanted also to use different pictures, or new drawings, or pictures where *ASU* had drawings. Also, to my astonishment, I found that I had to rewrite a very high proportion of the original text, not because that in *ASU* was nonsense but because it is now possible to present the information in an even tighter form, eliminating background and explanatory material and almost all the previous supposition and guesswork. I hope readers will not be irritated by the terse note form of the writing, which even extends to abbreviating many fairly obvious longer words such as nav/bomb for navigator/bomb aimer. They will appreciate that I was terrified of running out of space somewhere in Sukhoi, so that Tupolev and Yakovlev had to be omitted!

In *ASU* the publisher and I were glad to grab anything we could lay our hands on. Many photographs were almost half a page in size. This has proved to be the key to the present work. At a guess I suppose the text is almost twice as long. I never thought we would get it into the maximum number of pages that made financial sense (*ASU*s 414 plus an extra 146), but by leaving out the big pictures we just made it. Few of *ASU*s photographs are duplicated, and almost all the drawings have been redrawn more accurately, which of course costs money. Like designing aircraft, this is a compromise.

Many of the photographs in this book have been supplied from the archives of the Russian Aviation Research Trust, registered charity number 1000748. The registered address of the Trust is *The Fox, Bambers Green, Takeley, Bishop's Stortford, Hertfordshire CM22 6PB.* The Trust archives have been collected over more than 40 years from many sources including the Trust patrons, Charles Cain, William Green, Par-Erik Nordin, Malcolm Passingham, Wim Schoenmaker, George Stevens, John Stroud, John Taylor, Denys Voaden and the Trustees, Jean Alexander, Alex Boyd, Nigel Eastaway, myself, Jon Lake, Robert Ruffle and Harry Woodman. Contributors to the Trust archives from Russia include Guennadi Petrov, plus Andrew Alexandrov, Je Gordon and Leonid Ugryumov. Mort Stanley, Fred Rumsay, Vaclav Němĕcek, Tet Williams, Peter Kostelnik and Wasac Klepacki have also been significant contributors to the Trust archives, as have others too numerous to mention individually.

The object of the Trust is to collect, research, publish and preserve material on Russian aircraft, aviation and aviators, and contributions to the archives are always appreciated. In many cases, particularly in relation to the older aircraft, the same photograph appears from several different sources. Another important source of information has been the *Bulletin* of the Russian Aviation Research Group of Air Britain, first published in the spring of 1963, which is currently edited by Nigel Eastaway.

Throughout the creation of this book Nigel has been a tower of strength. Without his unstinted help this would be a very much poorer book. I am deeply in his debt.

Among other people who helped I must especially acknowledge the assistance given by Jacques Marmain, a dear friend who sadly died earlier in 1994, and Jon Lake, Editor of *World Air Power Journal,* whose knowledge and attention to detail has spared me many errors. I also acknowledge my indebtedness to numerous authors in the Russian Federation and Ukraine, and to over 100 current workers in the reorganised or recently created design, manufacturing and research organisations in the new republics. Especial thanks are due to Yefim Gordon, a dedicated photographer and archivist; Vyacheslav 'Slava' Savin, Deputy Director of Aviafilm, Kharkov; and to Nikolai Tikhonovich Gordyukov, Sukhoi Art Director. I would also like to thank Andrei Innokentyevich Kandalov, Assistant General Director of the mighty Tupolev bureau, who despite being one of the most senior men in the entire Russian hierarchy took time off to tell me about his (and Tupolev's) imprisonment by Stalin and later wrote out longhand the specifications of different Tu-95/142 versions and managed to browbeat the Russian telephone system hard enough to Fax them to me.

I began *ASU* with a long introduction giving a feel for the background to the Soviet industry. There is no point in repeating this, though much could be said about the traumas faced in privatizing, diversifying and (as far as possible) multinationalizing what was the world's biggest and most centrally controlled aerospace industry. Today the famous design bureaux have become corporations or 'technical complexes' named for the founder, while not one of the new upstart joint-stock firms is named for an individual.

What do you do when overnight you lose almost all your business and wake up to find a totally strange situation in which you have to compete in 'the market place', but still have thousands of mouths to feed? The immediate answers are to move into additional product areas and, to keep in aerospace, start drawing a shoal of commercial and light general-aviation projects. Obviously, only a fraction of the latter will succeed in establishing major production runs. This difficult process is sorting out who is really good in the Russian Federation, and who is sufficiently switched on in the West to participate. I also believe surface-effect vehicles (*Ekranoplans*) are worth noting, and these sometimes awesome vehicles will be found at the back of the book.

Another point worth making is that throughout my lifetime we in 'the West' have taken it for granted that our aircraft are technically superior to those of the USSR. We stuck to this cosy belief throughout the period of The Cold War. It enabled us to shrug off the acknowledged numerical superiority of the USSR. In 1959 an air marshal said to me 'Believe me, they would be impotent; we would slice through their formations like a hot knife through butter'. This appeared to be important, because at that time we Brits had officially given up manned military aircraft altogether!

It is clearly difficult for many people to take a detached view of what seemingly distant humans achieve. During World War 2 the Russians and the Japanese fell into the 'distant' category, so their aeroplanes were invariably described by British or American observers as copies of British or American designs. In the preface to *ASU* I commented how, when the Soviet Union unexpectedly became Britain's ally in June 1941, a British magazine reported on all the Russian aircraft it had heard of and described all but one as copies of well-known Western types. The exception, the Bolkhovitinov DB-A, was called the 'ANT-41' and described as 'most original'. As there was no obvious Western aircraft that could have been its inspiration it was clearly an object of derision.

The attitude dies hard. It is alive and well, and stems entirely from ignorance. Even *Flight International,* staffed by experts, finds it hard to imagine that design teams in the former Soviet Union can ever behave like their Western counterparts and simply try to create the best aircraft. When they saw the first pictures of the Su-27 prototype they called it 'a cross between the F-14 and F-15, with elements of the F/A-18 design thrown in for good measure'. Mikhail Simonov was somewhat miffed by this, and said 'If we were forever copying, how could we ever hope to do better? A fighter pilot cannot go into battle in an inferior aircraft.' I need hardly add that when Western pilots were actually able to fly the Su-27 all thoughts of copying vanished. That's what happens when ignorance is replaced by knowledge.

A corollary of this attitude is failure to consider the importance of timing. Again harking back to World War 2, because of our ignorance we Westerners thought the Soviet air forces flew only completely outdated aircraft. Then we sent a Hurricane Wing to Murmansk, and one day they were asked to escort some bombers. The latter turned out to be Pe-2s, and the Hurricanes found it difficult to keep up with them even with their throttles firewalled. To say this was a shock is an understatement. So when we at last get Eurofighter we have to remember that, for example, the MiG-29 was flying 23 years earlier.

Mentioning the MiG-29 reminds me how impressed I was when, before the 88 Farnborough airshow, I was the first Westerner to be given a briefing on its navigation and weapon-aiming systems. Throughout my lifetime we had cosily 'known' that, while the Russians might have quantity, we in the West made up for it by having quality. Thinking the MiG-29 might have an acceptable radar Washington said it had been produced only because spies stole the secrets of the Hughes APG-65 as fitted to the F/A-18. Suddenly I was confronted by a fighter with an all-can-do radar designed by Fazotron specifically to beat Western equivalents, but which it usually didn't even have to switch to 'transmit' because the pilot could use his infra-red system. Then he could use his helmet-mounted sight and hit with the first round from his gun using the laser ranger. (Western fighter pilots don't have such things, and when the Luftwaffe conducted an in-depth evaluation of their inherited MiG-29s the official view included 'it is outstanding compared to its Western counterparts, and the way the weapons can be used at large off-boresight angles was a surprise'). For the next few days I was under great pressure, but kept thinking 'I must do something about it; we've been kidding ourselves, they have the quantity and the quality. When I had a moment I called *The Times* newspaper and suggested they tell the population, but the desk ed's reply was 'Oh no, we're sick of Farnborough!'

I'm still not sure the penny has dropped. Do we really believe we are the technical leaders, whom the Russians keep copying? Even if the true situation is becoming appreciated, it won't cut much ice with the British Treasury, though they are the people who send our fighting men into battle with some of the most pathetically ill-equipped aircraft in the world. Take a quick look at the condensed list of equipment items under the heading Avionics a few pages further on. Then recall that Harriers in the Falklands had chaff bundles jammed under the airbrakes, because no dispenser installation had been funded. As for IRSTs and HMSs (which friend Ivan had in 1978) we may get these on Eurofighter by about 2005, provided they aren't cancelled to save money.

Many Russians gave me confidence to drop such previously universal abbreviations as TsAGI, TsKB and TsIAM and instead use plain English, such as CAHI, CCB and CIAM (which appears on the reverse [English language] side of their business cards). I hope readers will think this an improvement, unfamiliar as it may seem at first.

All gallons in this book are Imperial (= about 1·2 US gal). All Metric figures use Metric punctuation. Thus 3456 means 3,456 mm; 4,7 means 4·7 metres; 2,5t means 2·5 tonnes (2500 kg or 5,511 lb); 3 km means an altitude of 3 kilometres or 9,843 ft; and a weight of 44 500 kg means 44,500 in English. In the aircraft specifications, most figures are those published by the designers, and I am assured that (at least in the case of modern aircraft) values for range or radius take account of typical allowances. Engine powers or thrusts are not repeated for each aircraft because they can be found in the Engines section at the front of the book, along with lists of weapons.

Sharp-eyed readers will notice that many of the specifications, such as those for the 3MS (called 'M-4' in the West) and Il-54, are quite different from those with which Western readers are familiar. This is because the figures given in this book are correct.

Finally, Jon Lake and I had a difference of opinion over the inclusion of ASCC reporting-names. Useful as these were 30 years ago, today they are merely confusing. Now that we know what the various Tu-95 and -142 versions are called, why invent such unmemorable epithets as 'Bear-F Mod 4'! I asked Jon 'If we wrote a book about the Boeing 747-400, would it help if we insisted on calling it Growler-D?' Jon stuck to his guns, informing me that thousands of unfortunate people in the West are much more familiar with 'Floggers' than with MiG-23s and -27s. Eventually I gave in, and I have now come to agree with Jon that many people would be lost without these names, pure rubbish though they are.

I need hardly add that I welcome critical comment and will be indebted to anyone who can rectify an error.

Haslemere, England 1995

Aircraft Designations

SOVIET AIRCRAFT DESIGNATIONS

During the Civil War most aircraft were a motley collection of foreign types with no national designation system. The Soviet authorities laid down about 1923 a scheme for designating aircraft types according to function (R for reconnaissance, B for bombing, I for fighter and U for training, for example), while non-military machines were named or took the initials of the designer(s) or purchasing authority. By 1925 a scheme had been worked out designating all except light aircraft according to function; occasionally, by error, there was duplication eg, MDR-7. In 1940 this was replaced by a new scheme based on OKB General Constructor names, and these have been continued after the death of their founders. In addition types selected for series production receive VVS, GVF or other service designations, originally with odd numbers reserved for fighters. The reader will soon discover that, as before, some designations were used twice for unrelated aircraft.

A, Sukhoi Su-6, MiG-11
A-1, A-2, Kamerton
A-1 to A-15, Antonov, also CAHI (autogyros)
A-5, Artseulov
A-7, Kamov, also Polikarpov
A-10, Mil Mi-24
A-15, Avietka (1989)
A-40, Antonov KT, Beriev
A-50, Beriev
ACh-39, Tu-2 (reference to engines)
Acrobat, MAI-90, Avietka
AIR-1 to AIR-18, Yakovlev
Aist (Stork) Antonov OKA-38 and ROS-Aeroprogress
AK, Kamov/Mil
AK-1, Alexandrov/Kalinin
Akkord, Avia
Alarm clock (Budilnik), Avietka
Alekseyev, Avietka (1924)
An series, Antonov
Anito, Avietka (1935)
ANT series, Tupolev
Antei (Antheus), Antonov An-22
AO, Polikarpov Po-2
AP, Polikarpov Po-2
Ar, Arkhangyelskii
Arab, Tikhonravov
Argo-02, Avietka (1987)
ARK-3, Chyetverikov
ARK-5, Polikarpov R-5
ARK-Z, Zlokazov
ASh, Shavrov Sh-2
ASK, V. Ya. Krylov
AT, Antonov KT, also MiG
AT-1, V. Ya. Krylov, Avietka (1970)
Atlant, Myasishchyev M
Aushza, Avietka (1989)
Aviatka-900, MAI
Aviatourist, OOS
Aviavnito-3, KhAI-3
B, Sukhoi Su-8, also Arkhangyelskii Ar-2
B-1 to B-6, Chyetverikov MDR-6
B-5 to B-11, Bratukhin
Babochka, Belyayev
Baikal, Avietka (1987)
Bakshayev, NIAI
BB-1, Sukhoi Su-2
BB-2, Grushin BB-MAI
BB-22, Yakovlev Yak-4
BB-MAI, Grushin
BDD, Bolkhovitinov DB-A
BDP, Polikarpov
Be series, Beriev
Bekas, Taifun
Bekshta, Avietka (1974-79)
BI, Beresnyak-Isayev
BICh series (except BICh-17), Chyeranovskii
BICh-17, Kurchyevskii
Blokhi, KhAI-2, Avietka (1935)
BOK series, BOK
Boldrev, Avietka
BOP, Avietka
BS, Antonov
BS-2, Kalinin K-12
BSB, Polikarpov SPB
BSh-1, Kochyerigin V-11
BSh-2, Ilyushin Il-2
BSkhS, Romashka
Budilnik, Avietka (1927)
Buichok (Young Bull), Western reported name for Tupolev Tu-91
Buran, Molniya
Burlak, MiG-15, Tu-160
Buryevestnik, Avietka (1924-29)
CCB-1, Kochyerigin LR
CCB-3, Polikarpov 1-15
CCB-4, Kochyerigin TSh-3
CCB-5, Grigorovich LSh-1
CCB-6, Grigorovich TSh-1
CCB-7, Grigorovich I-Z
CCB-8, Grigorovich TB-5
CCB-10, Chyeranovskii BICh-14
CCB-11, Kochyerigin DI-6
CCB-12, Polikarpov I-16
CCB-15, Polikarpov I-17
CCB-18, Polikarpov I-16
CCB-19, Polikarpov I-17
CCB-21, Grigorovich TSh-2
CCB-23, Grigorovich ShON
CCB-24, Kochyerigin TSh-3
CCB-25, Polikarpov I-17
CCB-26, Ilyushin
CCB-27, Kochyerigin SR
CCB-29, Polikarpov I-16
CCB-30, Ilyushin DB-3
CCB-32, Ilyushin I-21
CCB-33, Polikarpov I-17
CCB-38, Kochyerigin DI-6
CCB-43, Polikarpov I-17
CCB-44, Polikarpov VIT-1
CCB-48, Polikarpov VIT-2
CCB-55, Ilyushin BSh-2
CCB-56, Ilyushin DB-4
CCB-57, Il-2
Ch-1, Avietka (1970)
Chaika, Be-12, Polikarpov I-15, Myasishchyev M-17
Chibis, Avietka (1989)
Chkalov, Avietka (1970)
Chye-2, Chyetverikov MDR-6
Chyerednichenko, Avietka (1935)
Combine, see Kombine
D, Bolkhovitinov, also Borovkov/Florov, also Polikarpov SPB
D-2, Polikarpov
D-3, Polikarpov SPB
DAR, Bartini
DB-1, Tupolev ANT-36
DB-2, Tupolev ANT-37
DB-3, Ilyushin
DB-4, Ilyushin
DB-108, Myasishchyev
DB-240, Yermolayev Yer-2
DB-A, Bolkhovitinov
DB-LK, Belyayev
DBSh, Ilyushin Il-2
DDBSh, Sukhoi Su-8
DDI, Polikarpov VIT-2
Debut, Avietka (1989)
Delfine, Myasishchyev
DF-1, Fedorov
DG-52, Grigorovich IP-1
DG-53, Grigorovich IP-4
DG-54, Grigorovich IP-2
DG-55, Grigorovich E-2
DG-56, Grigorovich LK-3
DG-58, Grogorovich PB-1
D.H.9. Interavia
DI, Yakovlev Yak-7
DI-1, Polikarpov 2I-N1
DI-2, Polikarpov D-2
DI-3, Grigorovich/Polikarpov
DI-4, Laville
DI-6, Kochyerigin
DI-8, Tupolev ANT-46

Shirok, LII
ShM, Avietka (1921)
Sh-MAI, Grushin
ShON, Grigorovich
ShS, Antonov OKA-38
Sh-tandem, Grushin
Shyukov, Avietka (1919)
SI, MiG-17
Sidoryenko, Avietka (1936)
Sigma, Moskalyev
SK, Avietka (1925)
SK-1, SK-2, SK-3, Biesnovat
SK-4, SK-7, Korolyev
SKB, Avietka
SKF, Polikarpov Po-2
SKh-1, Antonov An-2, also NIAI-Bedunkovich LIG-10
Skif, Myasishchyev
SL, Myasishchyev
SL-90, Interavia
SM-1, MiG I-340, Mi-1
SM-2 onwards, MiG
SM-92, Alfa
Sokol, ROS-Aeroprogress
SP, Polikarpov U-2, Avietka (1923)
SP-1, MiG-15
SP-2, MiG-17
SP-5, MiG-15
Sparka, Bolkhovitinov S
SPB, Polikarpov I-16
SPB/D, Polikarpov
SPL, Chyetverikov
SPS-89, Maksimov
SR, Kochyerigin
SRB, Golubkov (ANT-54/ANT-57)
SS, BOK-1
SSS, Polikarpov R-5
Stal-2/3/5/11, OOS
Stal-6/7/8, Bartini
Stal-MAI, Grigorovich/Grushin
Stayer, Taifun
STI, Avietka (1927)
Stratosfera, Myasishchyev M-17
Strekoza, Avietka (Kuibyshev, VEK) (1977)
Strela, Moskalyev
Strelka, Lavochkin La-160
Strizh, Avietka (1990)
Su series, Sukhoi
SUVP, Grigorovich
SVB, Polikarpov VIT-1, also MiG (1991)
Sverchok, Avietka (SKB) (1971)
Szmiel, Avietka (1977)
T-1, Tupolev ANT-41, also Avietka (Mustang) (1975)
T-3/4/4/6/7/8/10/37/43/405/431, Sukhoi
T-101/106/401/407/433/501/602/610, ROS-Aeroprogress
T-107/108/117/200, Bartini
T-405, T-431, Sukhoi
Ta-1, Ta-3, Tyrov
TA, TAF, Chyetverikov
Tarzan, Tupolev Tu-91
TB-1, Tupolev ANT-4
TB-2, Polikarpov
TB-3, Tupolev ANT-6
TB-4, Tupolev ANT-16
TB-5, Grigorovich
TB-6, Tupolev ANT-26
TB-7, Tupolev ANT-42, later Pe-8
TI-28, Gribovskii G-28
TIS, Polikarpov
Titan, ROS-Aeroprogress
TK, Lavochkin La-174
TK-1 Bolkhovitinov DB-A
TOM, Richard
Tourist, Avietka (1970)
TRD, Tu-2
Tri Druga, Avietka (1928)
Triton, KKB
Troika, Avietka (1984)
Ts series, Tsybin
TS-1, Schchyerbakov Shchye-2
TSh, MiG-6
TSh-1, Grigorovich
TSh-2, Grigorovich
TSh-3, Kochyerigin
TShB, Tupolev ANT-17
TSK, Avietka (Pissarenko)
Tu series, Tupolev
Turbolyet, Rafaelyants
U-1, Dudakov/Konstantinov
U-2, Polikarpov, Rafaelyants
U-3, U-4, Mikhelson
U-5, Nikitin NV-5
Ukraina, Antonov An-10
ULK, Grokhovskii G-37
UPB, KAI-3
UPO-22, Avietka (1941)
USB, Tupolev ANT-40
UT-1, Yakovlev AIR-14
UT-2, Yakovlev Ya-20
UT-3, Yakovlev Ya-17
UT-15, Yakovlev AIR-15
UTB-2, Sukhoi but included in Tupolev Tu-2
UTI-1, Polikarpov/Grigorovich I-5
UTI-4, Polikarpov I-16UTI
UTI-5, Nikitin NV-2
UTI-6, Nikitin NV-6
UTI-26, Yakovlev Yak-1
Utka, MiG-8
V-2, Mil Mi-2
V-8, Mil Mi-8
V-10, Mil Mi-10
V-11, Kochyerigin
V-12, Mil Mi-12
V-14, Mil Mi-14
Vasilyev, Avietka (1935)
VAT, Avietka (Chkalov)
VB-109, Myasishchyev
VEK, Avietka (1925)
Veneer-1, NIAI LK-1
VI-100, Petlyakov
VIGR-1 (VIHR), Avietka (1965)
Vigulya, Avietka (1920)
Vinogradov, Avietka (1931)
Vintokryl, Kamov Ka-22
VIT-1, VIT-2, Polikarpov
VK-8, Avietka (1989)
Vlasov, Avietka (1937)
VM series, Myasishchyev
VM-T, Myasishchyev 103M
VNP, Bratukhin B-10
Vnuk Oktyabr, Avietka (VAT)
VO-4, Avietka (1936)
VOP, Avietka (1923)
Voyage, ROS-Aeroprogress
VP, VP(K), Polikarpov
VS-2, Kalinin K-12
VSI, Shchyerbakov
VT-2, Yakovlev Ya-20
VT-11, Polikarpov/Grigorovich I-5
VVA-1, Avietka (1934)
VVA-3, Yakovlev AIR-1
Ya series, Yakovlev
Yak series, Yakovlev
Yak No 104, Yakovlev Yak-30
Yakov Alksnis, Grokhovskii G-31
Yamal, Aviaspetstrans
Ye, Sukhoi Su-10
Ye, numbers up to at least 266, experimental MiGs
Yer, Yermolayev
Yer-4, Yermolayev Yer-2
Z, Grigorovich I-Z
Z-1 to Z-7, Vakhmistrov
ZAOR, Avietka (1935)
Zh, MiG I-270
Zhar Pteetsa (Firebird), Kalinin K-12
Zherebtsov, Avietka (1950)
ZIG-1, Laville
Zlatoust, Avietka
Zvyeno, Vakhmistrov
I-42, MiG
1-EA, 2-EA etc, CAHI
2A, 3A, Sukhoi Su-6
2A/2D/3A/4A/5A, MiG
2I-N1, Polikarpov
2U-BZ, Moiseyenko
3B/M, Avietka
3M, 3MYe, 3MS, Myasishchyev
No 4, 5, Khioni
5, Biesnovat
11, Chyetverikov MDR-3
28, Yak I-26
65, MiG
77, Tupolev Tu-12
81, Sukhoi Su-6
101, Tupolev Tu-144
103, Tupolev 58, 59
103M, Myasishchyev
104, Tupolev Tu-2 variant
105, Gorbunov
110, Tomashyevich
150, OKB-2
201-M, Myasishchyev
302, Tikhonravov
330, Su-1
346, OKB-2
360, Su-3
4302, Florov
7211, Borovkov

AIR WEAPONS

The Russian penchant for weapons resulted in the USSR quickly becoming not merely self-sufficient in aircraft armament – in contrast to Britain, which has relied on guns designed in Germany, the USA, Czechoslovakia and France – but a world leader. No other country has put into service anything like as many types of guns, rockets, bombs and guided missiles, almost without exception of outstanding quality.

GUNS

Until 1932 all aircraft guns were derived from the Maxim/Vickers machine guns, firing rimless ammunition in 7,62 mm (0,300 in) calibre. The designation DA stemmed from *Dyegtyaryev Aviatsionnyi* (Vasilii Dyegtyaryev was perhaps the greatest Soviet designer of small arms); this gun was normally used on pivoted hand-aimed mountings. The very similar PV-1 (*Pulyemet Vozdushnyi,* air machine-gun) was developed mainly by Nadashkyevich and used in fixed, synchronized and movable installations. In 1930 work began on large-calibre recoilless guns, discussed later. The first heavy cannon tested after 1920 was the Hispano *moteur canon,* but this never reached service units and instead B. G. Shpital'nyi created two superb new gas-operated guns, one almost a scale-up of the other. The rifle-calibre ShKAS (*Shpital'nyi/Komarnitskii Aviatsionnyi Skorostrel'nyi*) was the neatest, fastest-firing machine-gun of its day; a Hurricane with five would have had greater hitting-power and more strikes per second than with its eight Brownings. The 20 mm ShVAK (*Shpital'nyi/Vladimirov Aviatsionnyi Krupnokaliber*) was lighter and much smaller than the Hispano but fired with almost the same m.v. (muzzle velocity) at a much faster rate. Both guns were used in fixed, synchronized and movable installations. In 1940 they were joined by the first of the outstanding weapons designed by M. Ye. Beresin, the UB (*Universal'nyi Beresin*). With the same calibre as the famed 0·5 Browning, the Russian gun was much more compact, lighter, faster-firing and had a higher m.v.; it was used as the UBS (*Synkhronnom*), often abbreviated to BS, as the UBK (*Kryl'yenom,* in wing) and as the UBT (*Turel'nom,* turret). The VYa (*Volkov/Yartsyev*) had calibre only 3 mm greater than the ShVAK but fired a projectile more than twice as heavy at higher m.v., penetrating at long ranges over 25 mm armour. Even this gun was dwarfed by the NS-37 (*Nudel'man/Suranov*), which had a bigger punch than US or German weapons of this calibre or the British RR or Vickers 40 mm, with almost guaranteed penetration of 40 mm armour at angles to 45°. The same gun was given a slightly shorter barrel of 45 mm calibre which was satisfactorily installed in Yak-9s (armour penetration, 58 mm) but the recoil of the immense 57 mm version proved to be too much. In 1945 Beresin brought in a lightweight 20 mm gun to replace the ShVAK; designated B-20 it was gas-operated from a point well down the barrel and was used in fighters and as the gun in the remotely directed four-turret PV-20 bomber defence system. The NS team replaced the VYa with their NS-23, the NS-23S variant being synchronized in fighters and two sub-types being used in bomber turrets. The later NR-23 (*Nudel'man/Rikhter*) used the same ammunition but had a higher cyclic rate and was used in fighters and also in bomber turrets (system PV-23) in place of the B-20. A. Suranov, V. Nyemenov, A. Rikhter and P. Gribkov designed the N-37 used in early jet fighters, while from the NR-23 a series of original new features resulted in one of the most formidable single-barrel guns ever created, the NR-30, which, though much lighter than the rival Aden and DEFA of the same era, fired a far heavier projectile at the same m.v. For bombers the AM-23 offered a higher rate of fire, compared with the NR-23, and remains in use to this day. With considerably upgraded ammunition 23 mm remains a very important calibre, and the twin-barrel GSh-23 is seen in the GP-9 scabbed-on pack, the SPPU-22-01 pylon-mounted pod with depressable barrels for surface attack and several types of defensive turret or barbette, while where very rapid fire is needed a six-barrel version is in service. However, 23 mm has been displaced by 30 mm (1·181 in) for today's fighters and attack aircraft, including helicopters; an SPPU depressed-barrel pod is in use. The GSh-30/1 (KB of V. P. Gryasev and A. G. Shipunov) is the lightest and fastest-firing single-barrel 30 mm gun in the world.

RECOILLESS GUNS

As far as is known all Soviet guns of this type were derived from the British Davis of 1915 and, like most similar weapons, cancelled out recoil by firing two equal masses in opposite directions, the projectile down the barrel and gun gas and a balance weight to the rear. Development was headed by Leonid Vasil'yevich Kurchyevskii and led to a series of weapons known by APK designations, from *Avtomatichyeskaya pushka Kurchyevskogo* or as DRP, from *Dynamo-reaktivnoi pushkye.* The original basis had been laid earlier (1922-1926) by Prof Boris Sergeyevich Stechkin, who tested guns of from 1 in (25,4 mm) to 12 in (305 mm) calibre, and flew 6 in (152 mm) weapons under the wings of a Junkers F13 in 1923. Kurchyevskii appears to have picked up the work in 1930 with 3 in (76,2 mm) weapons, which in the APK-4 of June 1934 could fire repeatedly, though at a slow rate. At the same time the APK-11 of 45 mm and APK-100 of 4 in (101,6 mm) calibre were on test. Despite many problems APKs continued to be developed for use against air and ground targets until the DI-8 (ANT-46). Kurchyevskii's KB was liquidated in Feb 1936.

ROCKETS

The roots of Soviet aircraft rocket projectiles lay in studies by artillery officers V. A. Artemyev and N. I. Tikhomirov during the Civil War of 1920-1922. Eventually this led to the GDL (gas dynamics lab) in the Peter and Paul Fortress, Leningrad, where basic designs were prepared under B. S. Petropavlovskii for finned, spinning missiles of 75, 82 and 132 mm calibre. Motor firings were in hand when in 1933 the programme was transferred to the newly formed RNII, where under Ivan T. Kleimenov the theory – completely elucidated by mathematician G. Ye. Langemak – was translated into practice. The 75 mm rocket, for use against aerial targets, went into pilot production as RS-75 (*Reaktivnyi Snaryad,* reaction missile) in summer 1937, and immediately underwent large-scale firing trials from I-16 and other aircraft, often against agile target gliders released by R-5s. About October 1937 Kleimenov and Langemak were arrested and never heard of again, but by 1939 all three missiles were in production, together with simple launch rails and electric ignition circuits for the aircraft. I-15 and I-16 fighters were progressively equipped to fire eight RS-75, and the first operational use came on 22 Aug 39 when 22-IAP (I-16 type 10) fired salvos against Japanese Ki-27s (Western accounts often give the rocket size as RS-82). The latter was made in far greater numbers (at least 2·5 million) than the other species, and was apparently a surprise to the Nazis and the Allies when fired against Panzers from Il-2s in June 1941. A simple projectile with ogive nose (with fuze windmill), drum central portion and four-fin conical tail, it was 864 mm long and weighed 6,82 kg. The RS-132 was 935 mm long and weighed 23,1 kg. After the war the Il-40 and Tu-91 were designed to carry the RS-182, TRS-190 and ARS-212. Modern rockets include the mass-produced S-5 (57 mm) series, fired from UB-8, 16, 19 and 32 pods (the designation being the number of launch tubes); less-numerous are the S-8 (80 mm) fired mainly from the 20-tube B-8M1; the S-13 (130 mm) fired from the 6-tube B-13; and the huge S-16, 21, 24, 28 and 32, usually carried singly. A recent weapon is the RM-122, grouping 4 separate tubes for a very high velocity 122 mm rocket; another is the S-25, with a 250 mm motor but much larger warhead.

AIRCRAFT GUNS

	date	C mm	L mm	W kg	R r/min	P g	V m/sec	
DA	1928	7,62	1270	7,5	600	9,6	840	Pivoted, drum-fed
PV	1929	7,62	1300	14,5	750	9,6	870	Fixed, belt-fed
ShKAS	1932	7,62	1280	10,6	1,800	9,6	825	
ShVAK	1936	20	1760	42	800	96	800	
Ultra ShKAS	1937	7,62	1190	10,0	2,700	9,6	830	
UBK	1940	12,7	1365	21,41	1050	48	850	Wing mount
UBS	1940	12,7	1365	21,45	800	48	850	Synchronised
UBT	1940	12,7	1365	21,43	1050	48	850	Turret
VYa	1940	23	2140	68,5	500	200	905	Also called MP-23
NS-37	1942	37	2670	150	250	735	900	
NS-45	1944	45	2520	152	250	1065	850	
OKB-15-57	1944	57	2905	290	180	2780	980	
B-20	1944	20	1380	25	800	96	800	
NS-23	1945	23	1950	37	550	200	690	
N-37	1946	37	2451	103	400	735	900	
NR-23	1949	23	1980	39	950	200	690	
A-12,7	1953	12,7	1349	25,5	1,100	48	820	
AM-23	1954	23	2175	43	1,300	200	690	
NR-30	1954	30	2158	66,5	900	410	780	
PKT	1960	7,62	1173	8,99	650	9,6	825	
GSh-23	1961	23	1537	50,5	3,600	186	735	Two barrels
9A624	1968	12,7	1350	45	5,000	48	820	Four barrels, YakB-12,7
GSh-6-30	1971	30	1995	290	4,000	410	780	Six barrels
AO-17A	1972	30	2260	80	900	410	1,000	Two barrels
GSh-6-23	1973	23	1525	195	*	186	700	Six barrels
2A42	1976	30	2210	49	900	410	1,000	Armoured-vehicle gun
GSh-30/I	1979	30	2000	50	1,800	410	860	Also called 9A4071K
GSh-30/II	1980	30	2400	123	†	410	940	Also called 9A623K
9A622	1980	7,62	800	18	6,000	9,6	800	Four barrels

C calibre; L length of gun; W weight of gun; R max cyclic rate of fire; P mass of projectile; V muzzle velocity; * two choices, 6,000±500 or 8,000; † two choices, 300/400 or 200/2,600.

MISSILES

One of the handicaps facing vast numbers of former Soviet designers was that the USSR lacked anything resembling the gigantic market for modern electronics seen in other countries, where countless products from desktop computers, word processors and computer games to equipments incapable of being realised without incredibly complex integrated circuits supported a microelectronic output measured in millions (in some cases billions) of circuits a day. This made it hard to produce superior guided missiles, especially in smaller sizes for aircraft. So they tried harder, and (only slightly helped 35 years ago by copying Sidewinder) eventually produced a range of AAMs and ASMs that has no rival in any other country (see tables). Thus, today's Russian fighters can carry a greater variety of weapons, with different guidance systems, than any possible opponents.

FREE-FALL STORES

There are more than 1000 known species of Russian bombs, mines, depth charges and torpedoes. Most have a prefix followed by a number giving the drop mass in kg. Typical prefixes include: FAB, GP HE bomb; OFAB, HE frag; ZAB, incendiary; ZAP, box of ZABs; BETAB, retarded/rocket-assisted anti-runway; PTAB, original hollow-charge bomblet; RBK, KMGU, cluster dispensers, PTK, modern hollow-charge bomblet for cluster dispenser; RPK, frag bomblet for cluster; RRAB, incendiary for cluster; AK, mustard/Lewisite bomblet; KhAB or ChAB, large phosgene; VAP, tanks or spray dispensers for HCN or persistent Toxic-B; SOV-AB, new persistent Toxic-B dispense; NOV-AB, non-persist; PLAB, napalm; FAE, fuel/air explosive; N, TN, nuclear; DAP, DV-AB, smoke; FOTAB, photo-flash; PROSAB, parachute flare.

AIR-TO-SURFACE MISSILES

	Kompleks	L	D	S	G	R	W	DOD/ASCC reporting name		Remarks
		mm	mm	mm	method	km	kg			
KS-4	K87	8250	1220	4900	AP+RH	150	3000	AS-1	'Kennel'	Mikoyan, anti-ship
KSR-2P,2M	K-16	858	1220	4750	AP+TH	330	3800	AS-5	'Kelt'	Bereznyak (TH=ARH or IR)
K-10S	RSL-1	995	900	4900	AP+RH	350	4400	AS-2	'Kipper'	Lavochkin, Tu-16K-10
Kh-20		14960	2150	9200	AP+TH	390	8950	AS-3	'Kangaroo'	Mikoyan, Tu-95K-20
Kh-22		11670	910	2990	I+TH	460	6800	AS-4	'Kitchen'	Tu-95K-22, Tu-22/22M
Kh-22N, 22MP		similar to 22			I+RH			AS-4	'Kitchen'	N active, MP passive
Kh-23	66	3630	278	809	RC	10	282	AS-7	'Kerry'	MiG-23/27, Su-17/24, Yak-38
Kh-23M	68M	3591,5	278	809	RC	10	287	AS-7	'Kerry'	Also 23A and A23
Kh-23		3590	278	809	LH	10		AS-7	'Kerry'	MiG-23/27, Su-17/24/25
Kh-25	69	3680	278	809	PRC			AS-10	'Karen'	MiG-23/27, Su-17/24

	Kompleks	L mm	D mm	S mm	G method	R km	W kg	DOD/ASCC reporting name		Remarks
Kh-25M	711	4220	278	809	PRH	40	300	AS-12	'Kegler'	Zvyezda
Kh-25MR	714	3630	278	809	PRH	8	316	AS-10	'Karen'	
Kh-25ML	713	3680	278	809	LH	7	302	AS-10	'Karen'	Also called S-25LD
KSR-5P,5N	K-26	10900	900	2500	I+TH	400	5950	AS-6	'Kingfish'	Mikoyan, Tu-16K-26, Tu-95M-5
KSR-11	K-11				I+TH			AS-6	'Kingfish'	Tu-16K-11-16
Kh-28	Nisan-28	6025	430	1400	PRH	95	725	AS-9	'Kyle'	Zvyezda
Kh-29L	64L	3900	400	1100	LH	10	660	AS-14	'Kedge'	Vympel, also 29D,L,T
Kh-29T	64T	3900	400	1100	TV	12	680	AS-14	'Kedge'	MiG-29K/M, Su-17/24/25T/35
Kh-31P, RE		5540	360	870	PRH	180	605	AS-17	'Krypton'	Zvyezda; L is laser version
Kh-31A		4750	360	800	ARH	60	655	AS-17	'Krypton'	Anti-ship, MiG-29K/M, Su-27K
Kh-35	3M60	4420	420	1390¶	I+ARH	130+	610	SS-N-25	"Harpoonski"	Zvyezda 300 m/s
Kh-41 Moskit	3M80	9385	760	2100¶	I+ARH	250	3960	SS-N-22	'Sunburn'	Raduga 2 800 km/h Su-27K
RKV-15B		8090	c640	3240¶	I+TC	2400	1700	AS-15	'Kent-A'	Raduga Tu-95MS
RKV-500B	Kh-15S	4780	c750	3240¶	I+RH	150	1200	AS-16	'Kickback'	Raduga Tu-160, 95MS
Kh-58A	112A	4800	380	1170	ARH	100	640	AS-11	'Kilter'	Raduga; 58E and 58U training
Kh-59T	Ovod	5130	380	1170	TV	?	630	AS-13	'Kingbolt'	Raduga, also 59E
Kh-59M	Ovod-M				RC+TV	165	920	AS-18	'Kazoo'	Raduga, turbojet, Su-30M
Kh-65SE	AGM-TVC	6410	500	3100	I+RH	280	1250	AS-19	'Koala?'	Raduga, retractable turbojet
Kh-66					SH			AS-7	'Kerry'	
	Alfa	c9000	c500	?	?	?	?	ASM-L?		Vympel? Retractable wing cruise missile under development

AIR-TO-AIR MISSILES

RS-1U	K-5	2500	180	580	BR/RC	5	83.2	AA-1a	'Alkali-A'	OKB-2 *Almaz* (Grushin)
RS-2US	K-51	2500	180	580	BR/RC	6	84	AA-1b	'Alkali-B'	MiG-19PM, Su-9
?	K-6	3500	200	550	BR/RC	?	135	–		MiG SM-9
R-8M	K-8M	3950	280	1220	SH	12	275	AA-3a	'Anab-A'	Biesnovat, I-75, Su-11
R-8T	K-8T	4150	280	1220	IR	8	287	AA-3b	'Anab-B'	Su-11, all versions 40 kg warhead
R-98RM	K-8M2	3930	280	1220	SH	35	292	AA-3c	'Anab-C'	Su-15, Yak-28P
R-98TM	K-8M1	4230	280	1220	IR	18	299	AA-3d	'Anab-D'	Su-11 and -15, Yak-28P
–	K-9R,K-155‡	4450	280	1300	SH	35	c300	AA-4	'Awl'	Mikoyan, K-9T IR version, Ye-152 family
R-3S	K-13T,310A	2838	127	528	IR	12	73,5	AA-2a	'Atoll-A'	Sidewinder, by OKB-134, warhead 9,5 kg
R-3U	K-13S,310S							AA-2b	'Atoll-B'	Training
R-3R	K-13R,320	3350	127	528	SH	18	90	AA-2c	'Atoll-C'	R-3P, training
R-13M	K-13M,380	2870	127	650	SH	17	83	AA-2d	'Atoll-D'	R-3A, training
–	K-15	6900	400	1390	SH		695			Lavochkin, for La-250
R-4R	K-80R	5250	320	1490	ARH	60	580	AA-5	'Ash'	Biesnovat, also RM, Tu-128
R-4T	K-80T	5160	320	1490	IR	30	545	AA-5	'Ash'	Also TM, Tu-128
R-23R	K-23,340	4060	190	920	SH	25	223	AA-7a	'Apex-A'	Vympel, K-23
R-23T	360	3940	190	920	IR	19	217	AA-7b	'Apex-B'	All versions, MiG-23 family
R-24R(R-23MR)	140	4300	230	1000	dual×R	40	250	AA-7c	'Apex-C'	T-140, training
R-24T(R-23MT)	160	4130	230	1000	IR	c9	248	AA-7d	'Apex-D'	T-160, training

	Kompleks	L mm	D mm	S mm	G method	R km	W kg	DOD/ASCC reporting name		Remarks
R-27R,R1	470	4080	230	772	SH	80	253	AA-10a	'Alamo-A'	Vympel, Su-27
R-27T	470T	3795	230	772	IR	72	254	AA-10b	'Alamo-B'	MiG-29, all versions 39 kg warhead
R-27AE,EM	470A,E	4780	260	800	I+SH	130	350	AA-10c	'Alamo-C'	MiG-29, Su-27
R-27TE	470TE	4500	260	800	IR	120	343	AA-10d	'Alamo-D'	MiG-29, Su-27
R-27RE	470RE	4700	260	800	I+SH	170	350	AA-10	'Alamo-'	Also 27ERE, 27ETE
R-33E		4150	380	900	I+SH	125	490	AA-9	'Amos'	Vympel, MiG-31
R-33T		4105	400	1100	IR	c100	466	AA-9	'Amos'	All versions, 47 kg warhead
R-37	K-37									Smaller wings, folding fins
R-38	–							–		Sukhoi OKB, for T-37
R-40R	K-40R	5490	300	1560	ARH	90	685	AA-6	'Acrid'	Vympel, MiG-25
R-40T	K-40T	5950	300	1560	IR	30	675	AA-6	'Acrid'	Also 40TD
–	K-50					150		–		For Su T-58M
R-60T	K-60T,60	2096	120	370	IR	7	43,5	AA-8	'Aphid'	Vympel
R-60R	K-60R	1995	120	370	SH	c10	c45	AA-8	'Aphid'	All versions, 3,5 kg warhead
R-60M	62	1995	120	370	SH	c20	c48	AA-8	'Aphid'	R-60MK export
R-73A	RMD1,72	2890	170	510	IR	30	105	AA-11	'Archer'	Vympel, MiG-29, Su-24/27
R-73M	RMD2	2890	170	510	IR	40	110	AA-11	'Archer'	8 kg warhead; 73E export
R-?	?	3200	170	404	IR	12	115	?		Vympel, R-73 base, fired astern, vectored motor
R-77	K-77	3610	200	530				AA-12	'Amraamski'	Vympel
R-100	K-100	c5000							?	For Ye-155PA
R-170	170	3605	205	700	ARH	90	175			Vympel, 'RVV-AE', laser fuze
R-172	KS-172	7410†	500	–	I+ARH	400	c390		?	Novator, 'RVV-L' anti-missile

ANTI-ARMOUR MISSILES

	Kompleks	L mm	D mm	S mm	G method	R km	W kg	DOD/ASCC reporting name		Remarks
9M14	PTUR-64	860	120	460	WC	3	11,3	AT-3	'Sagger-A'	Called Malyutka
9M14M								AT-3	'Sagger-C'	
3M11	PTUR-62	1160	130	660	RC/LC			AT-2	'Swatter-A'	Called Falanga
?	?							AT-2	'Swatter-B'	Called Skorpion
9M17		1160	130	660	RC/LC	4	29,4	AT-2	'Swatter-C'	Called Fleyta
9M120		2200	130	305¶	RC/LH	5	24,9	AT-6	'Spiral'	(ex-'AS-8') Called Vikhr
9M117	PTUR-72?	1200	127	305¶	L	c5	16,8	AT-9		Called Shturm
9M119	?							AT-11	'Sniper'	Called Svir

L overall length; D body diam; S wing span; R range; W launch weight; AP autopilot; TH terminal homing. I inertial; RC radio command; PRH passive radar homing; LH laser homing. TV television; TC terrain comparison; ARH active radar homing; BR beam riding; SH semi-active radar homing; IR infra-red; dual×R two forms of radar; WC wire command to line of sight; LC laser command option; c approximate estimate; * three versions, PRH, air/air, air/surface; ‡ Mikoyan designation; † length with booster; ¶ folding wings. Note: range varies enormously with cruise profile or target height, figure quoted is max. In the tables of missiles the first column is the Service designation and the second column the experimental (pre-certification) designation or *Izdelye* (product) number.

MATERIALS

Shorthand used in the text needs a brief explanation and some background on Soviet airframe construction.

Metals Russia had no aluminium smelting capability, but in 1920, at the height of the Civil War, the Soviet Union's infant VVF formed a committee to study light alloys under engineer (later professor) Ivan Ivanovich Sidorin. In August 1922 under his direction the first commercial quantities were made of a direct copy of the German alloy Duralumin, named Kolchugalumin from the town (northeast of Moscow) Kol'chugino where, at the Gospromtsvyetmet works, the ingots were produced. Standard fabricated bar, rolled sections, tube and sheet began to appear in October 1922, and small pieces flew in ANT-1 a year later. By 1939 the name had lapsed, and dural specifications were written for alloys D1 to D16 (sometimes written D-1 with hyphen). Mild-steel specifications are prefaced M or MS. Stainless was first produced on a lab scale in October 1928, with the name *Enerzh.* By 1930 six grades existed, but none was in production; price rivalled silver. *Enerzh-6* was used at great cost to build early Stal aircraft. This was a basic 18/8 Ni/Cr steel, and by 1938 it had been improved to various grades of KhGSA (these are the chemical symbols in Russian). KhMA is a Cr/Mo steel, imported until late 1936. KhNZA or Chromansil is a Cr/Mn/Si alloy. Titanium sponge was produced from late 1953, and today Soviet output far exceeds that of Europe, to which it is an exporter. There are several T-series titanium alloys. Armour has received greater attention than in any other country, mainly because of work at VIAM of Sergei Kishkin and Nikolai Slkyarov from circa 1930 towards fabrication methods for AB-1 (*Aviatsionnaya Bronya,* aviation armour), based on Ni/Mo steel, which had to be formed to approximate shape before being hardened, and which after hardening was difficult to cut or to drill. Major problems continued to be distortion after initial shaping and drilling, so that the *Bronyekorpus* (armour body) did not match drawings, and an extreme scarcity of vital alloying elements.

Wood Soviet Union led world in application of wood to advanced airframes, without resorting to imported balsa widely used in British Mosquito. From two-dimensional plywood workers at several GAZ, assisted by VIAM, developed a 3-D technique called *shpon* in which layers (typically 1 mm) of birch or similar wood were glued one above another whilst formed over a male die giving single or double curvature. No heat or pressure was used, and adhesives were local casein or albumen glues. A major development, by L. I. Ryzhkov (director of a laminated propeller and ski factory), was *delta drevesina* (delta wood) in which each laminate was impregnated with an imported resin adhesive prior to bonding under pressure. Imported resin was finally, about December 1940, replaced by phenol-formaldehyde resin, later with trace additions of borax, often inserted in sheet form. After bonding at 150°C the result was called bakelite ply, widely used in the Second World War. Today wood-based laminates use epoxy resins. Most of today's composites, however, use synthetic (man-made) reinforcing fibres.

It was a hard task building the first ANT-42, even without the supercharger in the fuselage.

ENGINES

From 1924 the NKAP assigned a simple M (*Motornyi*) number to each engine, all of which were then based on imported designs. The first indigenous engine to be qualified was Shvetsov's M-11. From 1941, as with aircraft types, General Constructors were permitted to use their own initials in engine designations.

A, A-117 Free-piston gas generator for long-range bombers, GAZ-117 (1949-1955), chief designer Mikhail Aleksandrovich Orlov, 10,000 hp class; complex, and low target s.f.c. never attained.

ACh-30 43 designation of M-40 for designer A. D. Charomskii; ACh-30B and BF, 1,500 hp, 1150 kg.

ACh-39 Uprated ACh-30, 1,800 hp; 38BF, 1,900 hp.

ADG-4 Four-stroke flat-twin, 35 hp at 2400 rpm.

AI-4 Small aircooled flat-four four-stroke by Aleksandr G. Ivchyenko, qualified 1946 at 52 hp and produced in several forms (4B, 4G for helicopters) at 55 hp.

AI-7 Ivchyenko subsonic ramjet tested 56, 220 mm diam, 56 kg thrust.

AI-10 Ivchyenko 5-cylinder radial, qualified 1946 at 80 hp.

AI-14 Ivchyenko 9-cylinder radial, built in many versions in large quantities for aeroplanes and helicopters, today in production only in Poland and China. Cylinders 105×130 (10,16 lit), qualified 1947 at 240 (later 260) hp, 14RF, 300 hp at 2400 rpm; 14VF, 285 hp; 14RA, 260 hp at 2350 rpm. See M-14.

AI-20 Single-shaft turboprop, begun 1947 at Kuibyshyev as NK-4 under N. D. Kuznetsov with mainly German team as rival to VK-2; in 1952 found to have better s.f.c. and adopted, further development and preparation for production transferred to Zaporozhye under Ivchyenko and redesignated AI-20; qualified 1955 at 2800 kW, AI-20K, 2942 kW; AI-20M 3126 kW (4,192 ehp), later 3169 ekW (4,250 ehp); marinized 20DK 3124 ekW (4,190 ehp); AI-20DM and D Srs 5 uprated to 3760 ekW (5,042 ehp); 10-stage compressor, p.r. 9.2, airflow 20,7 kg/s. Now produced by ZMKB Progress, Ukraine.

AI-24 Ivchyenko single-shaft turboprop; qualified 1959 at 1875 kW; 24A, rating maintained hot/high by water; 24T, V (helo) and 24VT, 2074 ekW (2,781 ehp); 10-stage compressor, p.r. 7.85, airflow 14,4 kg/s. AI-24P multi-fuel Ekranoplan turboprop, 1840 kW (2,467 shp). All now produced by ZMKB.

AI-25 Ivchyenko two-shaft turbofan, 3LP + 8HP, p.r. 7.95, BPR 2.2, 1,5 t (3,307 lb); 25TL and TLM, 9HP, 1,72 t (3,792 lb); 25V, cooled blades, 1,75 t (3,860 lb).

AI-26 Ivchyenko 7-cylinder radial, 155,5×155 (20,6 lit), qualified 1946 at 500 hp, subsequently many versions for normal or vertical installation including 26GR, 500 hp; GRF, 550; GRFL, 575; V, 575; later built in Poland, redesignated Lit-3 and developed into PZL-3.

AL-1 Small single-shaft turbojet by Arkhip Mikhailovich Lyul'ka, 1000 kg (9,8 kN).

AL-3 Lyul'ka turbojet of 1947, not flown.

AL-5 Lyul'ka single-shaft turbojet, 7-stage, prototype 1950 at 4500 kg (9,921 lb), qualified 51 4600 kg (10,140 lb); 5F, 5 t (11,022 lb); 5G, 5,5 t (12,125 lb).

AL-7 Lyul'ka turbojet with supersonic LP stages in 9-stage compressor, p.r. .9.0; 114 kg (251 lb)/s; qualified 1954 at 6,5 t (14,330 lb); 7F, afterburner, same dry, 9 t (19,840 lb) max; 7F-1, 6,9 t (15,212 lb) dry, 9,9 t (21,825 lb) max; 7F-2, 10,1 t (22,282 lb) max; 7F-4, 10,7 t (23,590 lb) max; PB, RV, marinized, 6,5 t (14,330 lb).

AL-21 Developed early 1970s from AL-7, 14-stage compressor, p.r. 14.75, airflow 104 kg (229 lb)/s, 3-stage turbine, series AL-21F-3, 7,8 t (17,200 lb) dry, max afterburner 11,25 t (24,800 lb). Produced by NPO Saturn.

AL-31 Service designation R-32, two-shaft augmented turbofan, series production 1984, now by NPO Saturn; 4 fan stages (b.p.r. 0.6), 9-stage compressor, 110 kg (243 lb)/s; overall p.r. 23; max dry 8,1 t (17,857 lb), max 12,5 t (27,560 lb); 31FM, improved sfc, 13,3 t (29,320 lb).

AL-34 Fuel-efficient small turboprop/turboshaft rated at 373-522 kW (500-700 shp); first run 1990. In development at NPO Saturn.

AL-35 Production derivative of AL-31FM, 13,56 t (29,900 lb).

AL-41 See SAT-41.

AM-3 Single-shaft axial turbojet by P. F. Zubets at Mikulin KB as M-209, official designation RD-3; 8 stages, p.r. 6.4, 135 kg (298 lb)/s, qual 1952 at 6750 kg (14,880 lb); AM-3A, qual 1953 at 8,7 t (19,180 lb); RD-3, 150 kg/s, 100 hp turbostarter, in prodn 1954 at 8,75 t (19,285 lb); RD-3M-500, 1957, 9,5 t (20,950 lb); RD-3M-500A, 9,5 t emergency 10,5 (23,150 lb).

AM-5 Single-shaft turbojet, 670 mm diam, p.r. 5.8, wt 440 kg; qual 1950, 1900 kg (4,190 lb); AM-5A, 2000 kg (4,410 lb); AM-5F, 2150 kg (4,740 lb) dry, 2700 kg (5,952 lb) max a/b.

AM-9 Single-shaft turbojet, supersonic 1st stage, 665 mm diam; qual 1953 at 2,6 t (5,732 lb); AM-9B with a/b, 3250 kg (7,165 lb) max; in 1956 redesignated RD-9, which see.

AM-11 Mikulin two-shaft turbojet with a/b, qual 1955 at 3800 kg (8,377 lb) dry, 5100 kg (11,243 lb) max; in 1956 redesignated R-11, which see.

AM-30 Water (later glycol) – cooled V-12 by AA Mikulin using modified BMW VI blocks (160×190 mm, 45,84 lit), HS12 rear wheelcase, RR Buzzard reduction gear and later supercharger based on Allison; qual 32 at 660 hp.

AM-34 Derived 1935 from AM-30; M-34N, RN, both 820 hp at 1,850 rpm; in 1936 stroke increased to 196,77 mm so 46,66 lit, AM-34FRN 900 hp, FRNV 1,200 hp at 2,000 rpm, RNF 1,275 hp; over 16,000 from GAZ-24 (V. A. Dobrynin) by 1941.

AM-35 Derived engine with improved cylinder head and supercharger and many minor changes, qualified 1939 at 1,200 hp at 2,050 rpm; 35A, 1,350 hp.

AM-36 Derived engine with three 6-cyl banks in Y-form, 2,000 hp.

AM-37 Derived engine, 1,380 hp, 1940; 37F, 1,400 hp.

AM-38 Major production engine, usual c.r. 6·8, numerous refinements; qualified 1941 at 1,550 hp, later same year 1,665 hp; 38F, 1,700 hp, 1,720 hp.

AM-39 Further development, qualified 1942 at 1,870 hp; FN-2, 1,850 hp; A, 1,900 hp; FB, 1,800 hp.

AM-41 Prototype only, 1942.

AM-42 Qualified 1944 at 2,000 hp; 42FB, 1,900 hp with or without TK.

AM-43 Prototype 1944, four ratings from 1,640 or 1,950 hp on 100-oct (43V) to 2,200 hp.

AM-44 Further devt, qual 1945 at 1,950 hp or 2,200 with TK-300 turbos.

AM-47 Prototype 1946, ratings 2,700 hp to 3,100 hp (47F).

AMBS-1, AMTKR-1 Prototypes by Mikulin KB to design of B. S. Stechkin.

AN-1 Two-stroke V-12 diesels by Aleksei Dmitriyevich Charomskii, AN-1A tested 1937 at 900 hp, 1040 kg; AN-1R 1,000 hp; AN-1RTK turbos 1938 1,200 hp, 1100 kg, led to M-40.

ASh-2 Lowest numerical designation of piston engine by Arkadiya Dmitriyevich Shvetsov, but last basic ASh type; basically two ASh-82 in tandem, 28-cylinder tandem to make 4-row radial of 3,600 hp, 1950.

ASh-21 Shvetsov 7-cylinder radial with direct injection, ASh-82 cylinders 155,5×155 (20,6 lit); qualified 1947 at 700 hp, later 730 hp, then (1954) 760 hp; 21V, helicopter engine, 575 hp.

ASh-62 Originally M-62, long series of engines derived from Wright R-1820 Cyclone via M-25; qualified 1937 at 840 hp, later 900 hp and 920 hp; main wartime production 1,000 hp, on 100-octane; ASh-62IR standard An-2 at 1,000 hp with TK maintained to 9,5 km; 62M, 58 hp auxiliary drive-shaft for agricultural equipment. IR still produced in Poland (as ASz 621R) and China.

ASh-63 Usually known as M-63, improved 1939 development of M-62, 1,100 hp.

ASh-71 Essentially two M-63 combined to form large 18-cylinder engine, qualified 1941 at 1,700 hp, 1942 at 2,000 hp, 71F, 2,100 hp, later 2,200 hp.

ASh-72 Improved M-71, rated 2,000/2,250 hp in 1943, qualified 1944 at 2,300 hp.

ASh-73 Ultimate 18-cylinder Shvetsov, tested 1944 at 2,300 hp, qualified 1945 at 2,600 hp; 73TK, rated 2,400 hp with TK maintained to 6,5 km; 73FN, 2,400/2,650 hp.

ASh-82 Important Shvetsov 14-cylinder radials originally (as M-82) using cylinder derived from M-62 but with stroke reduced to give engine capacity 41,2 lit instead of 42,7 lit (cylinders 155,5×155), resulting in compact engine around which KB and CIAM achieved some of world's first really good radial installations; qualified 1940 at 1,250 hp on 87-octane and about 71,000 made in 22 variants with different c.r. and fuel grades, inertia/electric/pneumatic starter and in direct-drive oblique helicopter variant; 82 qualified 1,330 hp later 1,400 hp in 1941; 82A, 1,600 hp; 82FN, 1,540 hp, 1,630 hp, 1,850 hp on 100-octane, direct injection; 82FNV, 1,850 hp; NV, 1,700 hp; T, civil, 1,900 hp; V, helicopter, 1,700 hp.

ASh-83 Derived engine, 1,900 hp.

ASh-90 Derived engine with 18 cylinders; qualified 1941 at 1,500 hp.

BAZ-21083 Piston engine by Volzhsky, on test 1992 at 80 hp.

BD-500 Compound engine using same cylinder as M-501 supplying gas to turbine geared to prop; developed 1946-53 at GAZ-500 under S. S. Balandin.

D-1 Series of liquid rocket engines developed at RNII by team led by L. S. Dushkin and V. A. Shtokolov, culminating in D-1A-1100, with HP pump feed of concentrated nitric acid (RFNA) and kerosine, SL thrust 1,100 kg (10,8 kN), about 1,300 kg at height.

D-10 Alt designation for RD-13 subsonic pulsejet.

D-11 Diesel version of M-11 by T. M. Melkumov 1936, 150 hp, 180 kg.

D-15 Soloviev turbojet rated from 1957 at 13 t (12,75 kN) dry and 15- 17,5 t (147/172 kN) with afterburner.

D-18T Three-shaft turbofan with b.p.r. 5·6, single-stage fan (airflow 765 kg, 1,687 lb/s), 7-stage IP, 7-stage HP (p.r. 27·5), qual 1982 at 23,43 t (51,660 lb). Produced by Lotarev KB, now ZMKB Progress, Ukraine. From 91 D-18TM with larger fan, higher temps, 25-30 t (<66,138 lb).

D-20 Two-shaft turbofan by P. A. Soloviev KB, known only in D-20P form; compressors 3+8, p.r. 13, b.p.r. 1·0; airflow 113 kg/s; rated 1962 at 5,4 t (52,96 kN).

D-21 Soloviev (OKB-19) two-shaft bypass jet for Mach 2·5 cruise application, 1959, wt 900 kg, max TO 5 t (11,023 lb) with a/b.

D-25 Soloviev single-shaft free-turbine helicopter engine; 9-stage, p.r. 5·6, rated 1956 at 4101 kW as D-25V, also called TV-2BM; D-25VF rated about 1963 at 4850 kW (all ratings maintained to 3 km or 40° C).

D-27 Advanced propfan based on Type 27 gas generator, developed from Lotarev D-236 after 1989 by ZMKB Progress, Ukraine, in pusher and tractor forms with one 8-blade and one 6-blade 4,5 m transonic propeller, initial rating 10,350 kW (13,880 hp), 11,2 t (24,690 lb) thrust, sfc 0·286.

D-30 Soloviev two-shaft turbofan; compressors 4+10, p.r. 17·4, b.p.r. 1, airflow 125 kg/s; rated circa 1964 at 6,800 kg (14,990 lb); D-30 II reverser; D-30 III 5-stage LP, ratings cooler and flat-rated hot/high; D-30-10V 9 t (19,841 lb); D-30N for VVA-14. Core used for D-30F6 described separately.

D-30F6 Two-shaft reheat bypass jet (low-b.p.r. turbofan) for M 2·83 cruise application; 5-stage LP (airflow 150 kg, 331 lb/s; b.p.r. 0·57), 10-stage HP, overall p.r. 21·15; qual 1979 9,5 t (20,944 lb) dry; 18,98 t (41,843 lb) max. Devt by Reshetnikov/Kuzmenko at AO 'Aviadvigatel', Perm.

D-30K Soloviev two-shaft turbofan not related to D-30; compressors 3+11, p.r. 20, b.p.r. 2·42, airflow 269 kg/s; rated 1966 at 11 t (24,250 lb) as D-30KU; KU-11, 10,8 t (23,830 lb) to 23°C; KP, KP-II and KPV, 12 t (26,460 lb); N for Ekranoplans.

D-36 First engine developed by new team at Zaporozhye under Vladimir Lotarev (successor to Ivchyenko); three-shaft turbofan, p.r. 20, b.p.r. 5·6, rating 6,5 t (14,330 lb). Basis for 136/236/336/436.

D-100, 110, 112 Aviadvigatel (Perm) turbofans/propfans based on PS-90, not yet firm programmes.

D-136 Turboshaft for large helicopters derived from D-36; compressors 6+6, p.r. 18·3; qualified 1980 at 8,500 kW.

D-200 Rotary engine by RPD NTS VAZ, derived from VAZ-430; geared drive to 2,200 or 2,700 rpm, wt 145 kg, max 162 kW (220 hp) on Mogas.

D-227 Propfan, D-27 scaled down to 8-9 t (17,650-19,840 lb) thrust; pusher or tractor.

D-236 Preliminary demonstrator propfan devt by ZMKB based on Lotarev D-36 core; tested 85 at 7,050 ekW (9,450 ehp), uprated 1992 to 8,090 kW (10,850 shp).

D-336 Turboshaft in 9000 kW (12,000 hp) class.

D-436 Growth D-36 by ZMKB Progress from 1987, four versions from D-436K (7,5 t, 16,535 lb) to D-436T3 (9,0 t, 19,840 lb).

DM-1 Auxiliary ramjet, designation from *Dopolnityelnyi Motor* (supplementary motor), by Ivan A. Merkulov, 400-mm diameter, operating on same fuel as existing piston engine, tested 1939.

DM-2 Improved ramjet, 400-mm diameter, flown Jan 1940.

DM-4 Larger ramjet, 500-mm diameter, weight 30 kg, required special tank with ethyl alcohol added to existing piston fuel, flown Oct 1940.

DN-200 Two-stroke 'diesel Novikov' by RKBM Rybinsk, three double-ended cyls, total 1,759 cc (107 in^3), 200 hp, sfc 0·27, cert 1995.

DN-400 Effectively, doubled DN-200 with 6 cyls, 400 hp, cert 1997.

DV-2 Two-shaft turbofan by ZMKB Progress in partnership with Czech PS (formerly ZVL); DV-2S, qual 1989, 2,2 t (4,852 lb); DV-2B, larger fan, 2,54 t (5,610 lb), several jet and shaft derivatives projected.

ED-1 Soviet designation for series of two-stroke diesels based on Junkers Jumo 204 and 206, tested 1935-40 at around 600 hp.

FED-3, FED-24 CIAM (A. A. Bessonov), four 6-cyl banks in X-form, 950/1,000 hp 1930-36.

GAZ M-1 Modified 4-cylinder water-cooled car engine, Ye. V. Agitov at Gorkii works, 56 hp.

GDL Series of large space rockets, not used for aircraft.

GTD-3 Alternative designation (from *Gelikopter Turbo-Dvigatyel*) for TVD-10.

GTD-350 Small helicopter turboshaft developed by Sergei Pietrovich Isotov; compressor 7+1 centrifugal, p.r. 6·05, airflow 2,19 kg/s; rated circa 1964 at 295 kW, later uprated to 322 kW (350P, 331 kW); made only in Poland.

GTE-400 Turboshaft for 300 kW (400 shp) class by MNPK Soyuz.

M-1 Soviet Gnome engines, usually 100 hp Mono.

M-2 Soviet 120 hp Le Rhône 9.

M-5 Soviet 400 hp Liberty 12.

M-6 Soviet 300 hp Hispano-Suiza 8Fb.

M-8 Earliest known Soviet-designed engine, radial by A. D. Shvetsov circa 1925.

M-11 Classic engine derived from M-8, c125,000 built 1927-59; 5-cylinder 125×140 (8,6 lit), qualified 1927 at 100 hp; 11G, 115 hp; D, 115 hp; K, 115 hp; Ye, 150 hp. Developed by Okromechko: F, 145/165 hp; FM, M, 145 hp; FR, 160 hp; FR-1, 165 hp; FN, 200 hp.

M-12 Developed M-11, 190 hp.

M-13 Water-cooled V-12, 1926, 36,03 lit, 880 hp at 2,100 rpm, 800 kg.

M-14 VOKBM refinement by Ivan M. Vedeneyev, later Prof A. G. Bakanov, of AI-14; M-14V-26 (helos) rated 320 hp, fixed-wing M-14PF 395 hp, 14PM pusher 315 hp, 14PM-1 355 hp, geared, cleared on Mogas and other fuels ±60°C, 14PT pusher 355 hp, 14NTK, direct injection, 424 hp.

M-15 9-cyl radial, 1931, 420 kg, 660 hp.

M-15 Soviet (mainly Shvetsov) 9-cylinder radial, run 1928 at 450 hp, no production.

M-16 VOKBM (Bakanov) piston engine, 8 air-cooled cyls in tandem pairs forming X, 296 hp, to be qual 1994.

M-17 Derived from BMW VI water-cooled V-12 46,8 lit; 17, qual 1929 at 680 hp; 17F, 715/730 hp; 17B, 680; total 27,534 prod 30-42, led to M-34. Also Bakanov (VOKBM) flat-4 piston engine, to qualify 1994 at 175 hp, 260 lb.

M-18 Following design of M-13 two-row radial circa 1926, NAMI team under A. A. Mikulin and N. R. Brilling built prototype M-18, two rows each based on Jupiter; passed 100-hour test at 750 hp and later gave 880 hp at 2100 rpm. No production.

M-18 W-18 by A. A. Bessonov, 1927, 35,32 lit, 1,050 hp, 560 kg. Also in 1994 VOKBM 2-stroke 2-cyl, 1940 or 55 hp.

M-19 In 1928 Bessonov V-12, 27,47 lit, 700 hp, 540 kg. In 1994 VOKBM 2-stroke flat-4, 80 hp.

M-22 Soviet engines originally licensed from GR9ASB (Bristol Jupiter VI) and later locally developed; in production 1930 at 480 hp.

M-23 3-cyls at 120°, also called NAMI-65, 65 hp, 1928.

M-25 Soviet engines originally licensed from Wright R-1820 Cyclone (SGR-1820-F3), cylinder 155,5×174,5 (29,8 lit); first series M-25 October 1934 at 700 hp; later rated 710 hp; 25A, 730 hp; 25V, 750 hp; 25Ye, 750 hp; led to M-62.

M-26 Soviet variant of Bristol/GR Titan using M-15 cylinders; run 300 hp early 1929 but unsuccessful.

M-27 Experimental derivative of M-17; improving cooling and larger radiator.

M-30 Original designation of AM-30; also exptl Charomskii diesel, led to M-40.

M-31 Predecessor of MG-31.

M-32 Water-cooled V-16 31 by V. M. Yakovlev, 1921,7 lit, 750 hp, 460 kg.

M-34 See AM-34.

M-35 See AM-35.

M-36 Soviet radial in 550 hp class run circa 1931 as rival to Jupiter in I-5.

M-37 See AM-37.

M-38 F. V. Kontsevich 9-cyl radial, 1930, 30,7 lit, 600 hp, 450 kg.

M-40 V-12 diesel derived by A. D. Charomskii from AN-1, two turbos each bank, cyls 180×200 = 61,04 lit, 1150 kg, 1,500 hp, became ACh-30.

M-41 Unknown engine (600/700 hp class) for I-10.

M-42 See AM-42.

M-44 Huge N. P. Serdyukov water-cooled V-12, 1933, bore/stroke 222/286 (132,8 lit), 2,000 hp, 1700 kg.

M-48 Soviet (Shvetsov?) 7-cylinder radial tested 1934 at 200 hp.

M-51 1931 variant of M-11 with 'central cooling fins', 125 hp, 165 kg.

M-56 Ye. V. Urmin 14-cyl radial, 1933, 25,1 lit, c625 hp, 450 kg.

M-61K Aircooled flat-twin piston engine, qual 1965, 30,5 hp.

M-62 Derived by Shvetsov from M-25V by fitting two-speed supercharger and improving induction system and other parts. Later see ASh-62.

M-63 See ASh-63.

M-71 See ASh-71.

M-72 Motorcycle engine, 35 hp.

M-73 See ASh-73; same designation, 28-kW/37·5 hp motorcycle engine approved Dosaaf.

M-76 Licensed BMW flat-twin motorcycle engine, 45 hp.

M-81 Experimental 14-cylinder radial flown in I-185 at 1200 hp.

M-82 See ASh-82.

M-85 Baseline licensed version of Gnome-Rhône K14 Mistral Major 14-cylinder, 146×165 (38,65 lit), tested 1934 at 760 hp, assigned to Tumanskii within Mikulin KB and developed to 850 hp at 2,400 rpm Jan 1936.

M-86 Tumanskii development to 870/960 hp in four sub-types 1936-37.

M-87 Tumanskii development 1938-39 to 930 hp; 87A, 930 HP; 87B, 950 hp.

M-88 Tumanskii development with direct injection and 100-octane fuel to 950/1,000 hp Oct 1938; 88A. 1,,000 hp; 88B, 1,100 hp; 88R, 1,000 hp.

M-90 See ASh-90.

M-100 to **109** See VK-100.

M-120 Experimental Klimov engine with three M-103 cylinder blocks at 120° (one vertical above crankshaft), tested 1942 at 1,500 hp, later 1,820 hp; could have TK-3 maintaining power to high altitude.

M-209 Original designation of AM-3 (RD-3).

M-250 Piston engine devd MAI 1939 by V. A. Dobrynin and G. S. Skubachyevsky, 6 banks of 4 cyls, bore/stroke 140/138, 2,500 hp at 3,100 rpm, became VD-4.

M-300 Massive 3,000-hp engine by A. A. Bessonov at CIAM, run 1941; six banks of 6 cyls.

M-501 Experimental compound engine developed at GAZ-500 under V. M. Yakovlev (no relation) comprising gas turbine driven by 4-stroke diesel with 7 Al-monobloc banks each of 6 cylinders 160×170 (143,55 lit) with double-sided central compressor driven by exhaust turbine and separate two-speed geared supercharger; passed 200-hour test 1952 at 6,000 hp at 2,400 rpm with s.f.c. 0,22 kg/h/kW; transferred (in Stalin's last resolution, signed 28 Feb 1953) to marine use.

MB-100 Experimental engine developed under Aleksei Mikhailovich Dobrotvorskii using four M-103 cylinder blocks in X formation with superimposed crankshafts driving common gearbox; tested in neat 1,95-m diam cowl at 3,200 hp Jan 1945, single engine flown in Yer-2.

MG-11 Development of M-11 by M. A. Kossov, rated 1937 at 165 hp; 11F, 165 hp, later 180 hp.

MG-21 Basically 7-cylinder Kossov development MG-11, rated 1938 at 200 hp.

MG-31 Basically 9-cylinder development by Kossov of M-11, rated 1938 at 300 hp; 31F, 330 hp.

MG-40 Inverted aircooled 4-inline by A. A. Bessonov, 120×140 (6,33 lit), 145 hp.

MM-1 Inverted aircooled 6-inline by A. A. Bessonov, 125×140 (10,03 lit), 300 hp.

MV-4 Inverted aircooled 4-inline (Renault licence), 120×140 (6,33 lit), 152 hp.

MV-6 Corresponding 6-cyl (9,5 lit), 270 hp.

MV-12 Designation for 350/380 hp Renault inverted V-12, no Soviet production, imported only.

NII-1 Twin-chamber acid/kerosene rocket by V. P. Glushko, total 1,200 kg at SL.

NII-3 Alt designation for RD-1400.

NK-2 First turboprop project at Kuibyshyev under Nikolai Dmitriyevich Kuznetsov (deputy to Klimov at Ufa plant in WW2), run 1951, qualified 3,730 kW (later 3,805 kW) 1954.

NK-4 Single-shaft turboprop by N. D. Kuznetsov, tested 1955 at 4,000 shp, 200 built for Il-18 but lost to AI-20.

NK-6 Kuznetsov 2-shaft turbofan with duct burning for Mach-2 aircraft, tested 1959 at 20 t (44,092 lb).

NK-8 Two-shaft Kuznetsov turbofan, compressors 2F/2LP+6HP, p.r. 10·8, b.p.r. 1·02; qualified with reverser early 1966 and produced as NK-8-4, 10,1 t (22,270 lb), uprated 1970 to 10,5 t (23,150 lb); 8-2, 9,5 t (20,945 lb); 8-2U, 10,5 t (23,150 lb); 8-4K, Ekranoplan cruise, 10 t (22,050 lb).

NK-12 Large single-shaft turboprop created by mainly German team at Kuibyshyev under Austrian Dipl-Ing Ferdinand Brandner, all virtual prisoners at Kuznetsov KB; 14-stage axial, p.r. varies 9 to 13; airflow 65 kg/s; constant 8,300 rpm; run 1951 and qualified as TV-12 in 1954 at 8,000 kW; NK-12 qual 1955 with 65 hp TS-12 gas-turbine starter at 8,948 kW (11,995 shp); 12M, 12MA, 11,185 kW (14,995 shp); 12MV, 11,033 kW (14,795 shp); 12MK Ekranoplan cruise, 9,900 kW (13,270 shp).

NK-16 Turboprop devd from NK-12, greater airflow, qual 1956 at 13,500 kW (18,100 shp).

NK-20 Two-shaft high-compression turboprop devd in 1966, 14,900 kW (20,000 shp) class.

NK-22 Two-shaft augmented turbojet for supersonic applications qual by Kuznetsov 1967 at 16 t (35,275 lb) dry and 22 t (48,500 lb) max.

NK-25 Two-shaft augmented turbofan for supersonic applications qual by Kuznetsov 1976, production by Samara (Trud) at 19 t (41,900 lb) dry, 25 t (55,115 lb) max.

NK-32 High-pressure core, basis for NK-56 and 321.

NK-44 Three-shaft turbofan, 122 in fan, BPR 6, max 40 t (88,185 lb) on LNG fuel.

NK-46 Giant three-shaft turbofan, to be rated 60 t (132,277 lb) on LNG fuel.

NK-56 Three-shaft turbofan, 18 t (39,683 lb).

NK-62 Counter-rotating unducted propfan, in 1982 ran at 25 t with sfc 0·28 (7.92 mg/Ns).

NK-63 Counter-rotating ducted propfan, 30 t (66,138 lb).

NK-64 Three-shaft turbofan, tested 1984 at 16 t (35,273 lb).

NK-86 Derived from NK-8, qual 1980 at 13 t (28,660 lb), 86A 13,3 t (29,321 lb).

NK-87 Ekranoplan variant of NK-8-4K.

NK-88 NK-8-2 mod 79-88 for liquid hydrogen (LH_2) or (NK-89) liquefied natural gas (LNG), 10,5 t (23,148 lb) on either.

NK-92 Military contra-rotating ducted propfan tested 1987 at 20 t (44,092 lb).

NK-93 Civil contra-rotating propfan, 7 LP, 8 LP, shrouded 114 in fans 8-blade front, 10-blade rear, p.r. 37, BPR 16·6, max 18 t (39,683 lb), sfc 0·49, tested by Samara at 20,4 t (44,974 lb).

NK-94 Three-shaft turbofan, to be qual 1996 at 18 t (39,683 lb) on LNG.

NK-110 Counter-rotating unducted geared pusher propfan derived from NK-62.

NK-112, 114 Turbofans derived from NK-93 to be rated at (112) 8,5 t (18,739 lb), (114) 14 t (30,864 lb).

NK-144 Two-shaft augmented turbofan for SST derived from NK-8, tested 1965 at 17,5 t (38,580 lb); 144A, 20 t (44,092 lb); 144B and military 144-22, 22 t (48,501 lb); 144WT, 17,5 t.

NK-321 Three-shaft augmented turbofan, 3-stage LP (365 kg, 805 lb/s, BPR 1·4), 5-stage IP, 7-stage HP, overall p.r. 28·4; max dry 14 t (30,843 lb), max 23,1 t (50,926 lb). Design under Kuznetsov 1977, produced under Orlov by Samara (Trud).

OR-1 Original liquid rocket (OR, *Opytnyi Raketa*) by F. A. Tsander at TsIAM, later MosGIRD; tested on compressed air and various hydrocarbons 31,5 kg thrust.

OR-2 First liquid engine to fly (BICh-11), run on bench by Tsander at GIRD on lox/petrol 18 March 33,50 kg.

ORM-65 Most important design in series of ORM (*Opytnyi reaktivnyi motor*) rocket engines designed and tested at GIRD and RNII under V. P. Glushko, S. P. Korolyev and G. Ye. Langemak; tested from 1936 on RFNA/kerosene at 175 kg and flown in RP-318 and No 212. Led to RD-1 series. Other members of family ORM-1 (N_2O_4/toluene), ORM-50 and large ORM-52 of 300 kg.

P-020 Flat-twin 2-stroke by KKBM Kuibyshyev, 14,9 kW (20 hp).

P-033 Two-stroke by KKBM, 24,6 kW (33 hp).

P-065 Flat-4 2-stroke by KKBM, 49 kW (65 hp).

PRD-1500 Solid-propellant rocket engine by I. I. Kartukov 1945-49, 1,5 t sea-level for 10 s.

PS-30-V12 Two-shaft low-BPR turbofan by AO Aviadvigatel, using parts related to D-30 III and D-30F6, rated at 5 t (11,023 lb).

PS-90 Turbofan by Aviadvigatel (Soloviev); PS-90A, -90AN, fan 33 Ti blades + 2 LP, 13 HP, BPR 4·8, p.r. 35·5; qual 1992 16 t (35,275 lb); PS-90-76, p.r. 29·5, BPR 4·6, 14 t (30,864 lb), sfc 0·594; PS-90AM, p.r. 34·6, BPR 4·1, 16 t, sfc 0·578; PSX-90 to be devd with CFMI (GE+Snecma) to 17,5 t (38,580 lb) with better sfc; PS-90P to be devd with Pratt & Whitney and MTU for cert 1996; last two projects uncertain.

PVRD General abbreviation for ramjet.

R3-26 Single-shaft turbojet with a/b derived from RD-9B by Mikulin subsidiary at Ufa led by V. N. Sorokin, 3500 kg (7,716 lb); R3M-26 based on RD-9BM, 3800 kg (8,377 lb), both qual 1958.

R-11 Two-shaft turbojet, initially Mikulin AM-11, redesignated R-11 by S. K. Tumanskii; diam 665 mm, 3-stage LP, 5-stage HP, p.r. 8·7-8·9, airflow 63,7-64,5 kg/s; first (qual 1953) R-11V-300 ultra-hi-alt, 3,9 t (8,598 lb); R-11-300, 3,9 t (8,600 lb) dry and 5,1 t (11,240 lb) max afterburner, qual 1954; R-11F-300 improved a/b giving 5,740 kg (12,655 lb); R-11AF-300, 5,9 t (12,676 lb) with a/b; R-11F2-300 and R-11AF2-300, 65 kg/s, 3,950 dry/6,120 (13,490 lb) max, 1961; R-11-F2S-300 equipped for SPS flap-blowing, ratings 3,900/6,175 (13,613 lb max); 20,900 made in USSR, now made in China as WP7.

R-13 Devt by V. Gavrilov (Tumanskii KB) of R-11 with improved compressors, 65,6 kg/s, p.r. 9,05; R-13-300 qual 1966 at 4,07/6,6 t (8,973/14,550 lb), now Soyuz, and also produced (as 'Product 95-1') by UMKB, Ulan-Ude.

R-15 Single-shaft turbojet with a/b for Mach 3, devd by Tumanskii (now Soyuz); diam 1,640 mm, wt 2,590 kg, 5-stages 144 kg/s, p.r. 4·75; R-15K for K-15 cruise missiles 1958; R-15-300 qual 1959 at 6,84/10,15 t (15,080/22,380 lb); R-15B-300, 1962, 7,5/10,21 t (16,535/22,510 lb); R-15BD-300, 1963, 8,8/11,2 t (19,400/24,700 lb); R-15BF2-300, 13,5 t (29,760 lb) all on T-6 fuel.

R-21 R-11F mod by N. Metskhvarichvili (now Soyuz) with greater airflow and larger afterburner; R-21F qual 1962 at 4,700/7,200 kg (10,360/15,875 lb).

R-25 Devd from R-13 by S. Gavrilov, rebladed compressors (1st LP 21 ti blades) 67,9 kg/s, p.r. 9·55, R-25-300 qual 1969 at 4,100/7,100 kg (9,040/15,650 lb), prod by Soyuz and by UMKPA as Product 25-11; licensed to HAL India.

R-27 Further devt by K. Khachaturov, diam 1,060 mm, 5-LP, 6-HP, 89 kg/s, modulated a/b, R-27F-300 qual 1967 at 6,5/9,69 t (14,320/21,365 lb); R-27F2M-300 uprated to 95 kg/s with p.r. 10·5, qual 69 at 6,9/10 t (15,210/22,050 lb); R-27V-300 lift/cruise, twin 100° vector nozzles, 100 kg/s, qual 71 6,7 t (14,770 lb); R-27VM, see R-28.

R-28V-300 Refined R-27V-300, qual 1985 6,94 t (15,300 lb).

R-29 Further R-27 devt by Khachaturov, R-29-300 110 kg/s at p.r. 13·0, qual 71 at 8,3/12,5 t (18,300/27,560 lb), replaced by similar R-29PN; R-29B-300 and BS-300 with small afterburner and simple nozzle for lo-level mainly subsonic, 104 kg/s at p.r. 12·4, qual 1969 at 8/11,5 t (17,635/25,250 lb).

R-31 Designation of R-15B-300 for FAI record submissions.

R-35 Khachaturov upgrade of R-29-300 with higher top temp; R-35-300 qual 1980 at 8,55/13 t (18,850/28,660 lb), made by UMKPA as Product 55B.

R-37 Designation of R-11 versions for FAI record submissions.

R-79 V. K. Kobchyenko (Soyuz) 2-shaft low-ratio augmented turbofan; diam 1,716 mm, wt 2,750 kg with var-nozzle vectoring to 63° or 95°; 5-LP, 6-HP, 120 kg/s, b.p.r. 1·0; 11 t (24,250 lb) dry, 15,5 t (34,170 lb) max.; R-79M for STOL with limited vectoring and FADEC control, 18 t (39,680 lb).

R-95 See R-195.

R-123-300 Turbofan by Soyuz, 2-axial + 1 centrif, b.p.r. c6, TO rating 1,4 t (3,100 lb).

R-127-300 Turbofan by Soyuz, 1 axial + 1 centrif, p.r. 18·9, b.p.r. 5, TO rating 850 kg (1,874 lb).

R-195 Gavrilov devt of R-13 as R-95 without a/b for subsonic use; further devd as R-195 to resist 23 mm hits and run after severe damage, max 4,5 t (9,920 lb). Series prodn at reduced rating (4,1 t/9,037 lb) by UMKPA as Product 95-III; R-195FS supersonic version.

RD-1 Used for two *Reaktivnyi Dvigatel.* (1) Liquid rocket developed from ORM-65 by L. S. Dushkin and V. P. Glushko, normal rating on RFNA/petrol 300 kg at SL (more at height); in RD-1/KhZ (chemical ignition) form, extensively tested on fighters/bombers from 1943. (2) First Lyul'ka axial turbojet with form of afterburner, abandoned after 22 Jun 1941.

RD-2 Next major step beyond RD-1 rocket, qualified at 600 kg late 1945; RD-2M with main chamber 1,450-kg, cruise chamber 400.

RD-3 Service designation of engines of AM-3 family.

RD-3 Turbopump-fed three-chamber engine, run at up to 1,450 kg. Later engines by same KB were for missiles and space launcers.

RD-5 Redesignation of AM-5 after blacklisting of Mikulin, 1956.

RD-7 Service designation of engines of VD-7 family.

RD-9 Important family of single-shaft turbojets developed in Mikulin's KB under Tumanskii 1951-54, originally designated AM-9; qualified 1952 at 2,6 t (5,732 lb); AM-9AK (RD-9AK), 2,8 t (6,173 lb); afterburning AM-9B 1955 at 3,25 t (7,165 lb) with a/b; redesignated RD-9B on Mikulin's removal 1956; in 1957 RD-9AF and RD-9BF (43,3 kg/s, p.r. 7·5) rated 2,6/3,3 t, and for mixed-power fighters RD-9Ye at 3,8 t (8,377 lb); in 1959 RD-9BF-811 qualified at 2,790/3,750 kg (6,150/8,267 lb); produced in China as WP6.

RD-10 Soviet Junkers Jumo 004B, further developed at Kazan as RD-10, 900 or 910 kg; 10A, 1000 kg; 10F, 1100 kg.

RD-13 Improved German As 014 pulsejet, 200 kg thrust.

RD-20 Ex-German BMW 003A further developed, series-produced at GAZ-17, rating 800 kg (1,765 lb); 20F (later redesignated RD-21) hotter, 1,000 kg (2,205 lb).

RD-33 Two-shaft a/b bypass jet (low-ratio turbofan) by Isotov team under A. Sarkisov, now Klimov Corp, 4-LP (76,3 kg/s), 9-HP (overall p.r. 21·9), max dry 5,04 t (11,110 lb), max 8,3 t (18,300 lb); RD-33I, unaugmented, 5,2 t (11,465 lb); RD-33K, greater airflow and p.r., hotter, max 8,77 t (19,335 lb), Service designation RD-37; RD-33-300, unaugmented, max 5,22 t (11,500 lb); RD-33-191, longer-life. Produced by Omsk MKB.

RD-36-35 Kolesov, now RKBM Rybinsk, single-shaft lift turbojet, 6 stages, 40,4 kg/s, p.r. 4·4; max 2,35 t (5,180 lb); RD-36-35FVR, 45,3 kg/s, 3,05 t (6,725 lb); RD-36-35PR Ekranoplan lift engine.

RD-36-41 Kolesov afterburning turbojet for Mach-3 cruise, derived from VD-19, 14,5 t (31,966 lb) dry, 16,15 t (35,605 lb) max.

RD-36-51A Kolesov single-shaft turbojet for supersonic cruise, 14-stage compressor, 3-stage turbine, max 20 t (44,090 lb). RD-36-51V, simpler derivative for ultra-high cruise, max 7 t (15,430 lb), by RKBM Rybinsk.

RD-38 Galigusov (Rybinsk) lift turbojet based on 36-35 but increased gas temperature; 3,25 t (7,165 lb).

RD-41 Novikov (Rybinsk) lift turbojet, 7 stages (55,3 kg/s), titanium and composites, max 4,26 t (9,390 lb).

RD-45 Soviet Rolls-Royce Nene, taking number from GAZ-45 production plant in Moscow despite GAZ-117 having swiftly done most of engineering effort and production of drawings (both plants headed by Klimov); series prod Oct 1948 at 2,200 kg, RD-45F, 2,270 kg from June 1949, led to VK-1.

RD-60 Novikov booster turbojet, based on RD-36-35FVR but horiz attitude and long-life lubrication system; rating 2,98 t (6,580 lb). Devd 1989 by Klimov Corp into RD-60A 2-shaft turbofan for trainer and GA aircraft, 2,5 t (5,511 lb).

RD-500 Soviet Rolls-Royce Derwent V, taking number from GAZ-500 under Vladimir Mikhailovich Yakovlev (see M-501 engine); in production Jan 1949 at 1,590 kg.

RD-600S Alt designation for turboshaft version of TVD-1500.

RDA-1 Rocket engine developed by group under L. S. Dushkin and V. P. Glushko at Aviavnito from 1933, mainly on nitric acid and petrol and flown (RDA-1-150 No 1) Feb 1940 in RP-318 at 100 kg.

RDK-300 Soyuz turbofan for unmanned aircraft, diam 315 mm, 300-350 kg (660-772 lb) class, qual 1977.

RMZ-640 Piston engine, two aircooled cyls in line, rated 28 hp, up to 35 hp at 5,500 rpm with reduction belt.

RTWD-14 Lyul'ka (Saturn) turboshaft aux-power engines of *Buran* spacecraft, using LO_2/LH_2.

RU-013 Controllable liquid-propellant rocket engine by D. D. Sevruk, AK-20 (kerosene) and HTP (high-test peroxide), two main thrust levels (SL) 1,300 kg and 3 t (6,614 lb); installed in U-19D, qual 1957.

RU-19 Tumanskii single-shaft turbojet, 7-stage p.r. 4·6, 16 kg/s; 900 kg (1,984 lb), also called TRD-29; later devd as RU-19-300 TO booster (same) and as RU-19A-300, 7,85 kN (1,765 lb) plus elec power.

S-3 L. S. Dushkin liquid-propellant rocket, usually RFNA/kerosene, devd 1959 as S-3-20M5A installed at 3 t (6,614 lb) SL rating in U-21 package.

S-18 Second Lyul'ka turbojet to run, 1,250 kg Aug 1945, led to TR-1.

S-155 L. S. Dushkin liquid-propellant rocket, devd from RD-2, qual 1955 on petrol/RFNA/HTP at 1,3 t (2,866 lb) at SL.

SAT-41 Extremely advanced augmented turbofan by NPO Saturn (Dr Viktor Chyepkin, previously Lyul'ka KB) for 21st-century fighters; few parts, 1,500C, vectoring con/di nozzle, SAT-41F max 20 t (44,090 lb) with a/b.

SPRD-99 Standard a.t.o. solid rocket, 2,5 t for 10-17·8 s.

SPRD-110 Rocket a.t.o. unit, 3 t for 8 seconds.

SU-1500 Standard a.t.o. solid rocket, 3,5 t for 10 s.

TR-1 First Lyul'ka turbojet to fly, 8-stage axial derived from S-18, qualified December 1946 at 1,3 t; TR-1A qualified 1947-48 at 1,5 t.

TR-3 Single-shaft turbojet by Lyul'ka, qual 1950 at 4,600 kg, later TR-3F 5 t, became AL-5, which see.

TR3-117 Single-shaft turbojet for drones/RPVs, 590 kg (1,300 lb).

TR-7 Prototype Lyul'ka supersonic compressor engine, led to AL-7.

TRD-29 Alternative designation of RU-19.

TS-20 Small Lyul'ka turbojet, 50 kg, also important APU.

TS-31 Small Lyul'ka turbojet, 55 kg; production form TS-31M.

TV-2 Two-shaft turboprop derived from Junkers 022; TV-2F rated 1951 at 3,760 kW (5,050 shp), qual 1952 at 4,660 kW (6,250 shp); TV-2M qual 1955 at 7,650 shp; TV-2VM 6,500 shp for helicopters; twinned 2TV-2F with each half driving separate coaxial prop, rated 8,952 kW (12,000 shp), qual 52.

TV2-117 Unrelated to above, free-turbine helo engine, 10-stage, p.r. 6·6, qual by Isotov 1962 as -117A at 1,250 kW (1,677 shp); -117TG qual on Mogas, diesel, LNG, LPG, propane.

TV3-117 Free-turbine turboshaft derived from TV2, 10-stage, greater airflow, p.r. 7·5; prod by Klimov Corp, 117BK electronic control, 1,618 kW (2,170 shp); 117MT, 1,434 kW (1,923 shp); 117V, 1,633 kW (2,190 shp); -117VK, electronic, 1,618 kW (2,170 shp); 117VM, 117VMA, electronic, flat-rated 1,545 kW (2,070 shp) to 3,6 km.

TV-4 Original designation of VK-2.

TV7-117 Klimov modules assembled to form shaft, jet or propeller engines; core has 7 axial plus 1 centrifugal stages, p.r. 16; prodn 1991 of TV7-117S, also called 117-3, turboprop flat-rated to 35C at 1,839 kW (2,466 shp), and planned at 1,866 and 2,090 kW; 117SV turboprop, 1,865 kW (2,500 shp); 117V turboshaft flat-rated 1,753 kW (2,350 hp) with OEI contingency 2,536 kW (3,400 shp); TVD-117E for Ekranoplans, 1,840 kW (2,465 shp).

TV-12 Original designation of NK-12.

TV-116 MNPK Soyuz, core with 2 centrif compressors, TV116-300 turboshaft and turboprop versions to be rated at 805 kW (1,080 shp).

TVA-200 APU-compressor, 250 shp.

TVD-10 Free-turbine engines by Glushyenkov (now OMKB 'Mars', Omsk), 6 axial + 1 centrif, 4,6 kg/s, p.r. 7·4; produced as GTD-3F turboshaft (671 kW, 900 shp) and GTD-3BM (738 kW, 990 shp); TVD-10 turboprop (754 kW, 1,011 shp) produced in Poland as TWD-10B; OMKB is developing 10BA turboprop (790 kW, 1,060 shp).

TVD-20 Turboprop derived from TVD-10B, 6+1 compressor and 2-stage power turbine, 1025-1066 kW (1,375-1,430 shp).

TVD-100 See TV-O-100.

TVD-450 MNPK Soyuz turboprop, core related to GTE-400, 336 kW (450 shp).

TVD-650 Isotov single-shaft turboprop, 600 kW.

TVD-850 Derivative of TVD-650, same rating.

TVD-1500 Novikov (Rybinsk) core with 3 axial + 1 cent, p.r. 14·4, 2-stage turb 1,267°C, + 2-stage free power turb, 956 kW (1,280 shp); 1500A turboshaft; 1500I pusher propfan; 1500S turboprop; 1500SKh turboprop (ag); 1500V helo with 1,156 kW (1,550 shp) contingency.

TV-O-100 Family based on core by Kobchyenko (Soyuz), devd by OMKB 'Mars'; free-turbine turboshaft, 2 axial + 1 centrifugal, p.r. 9·2, rated 537 kW (720 shp); planned devt at 619 kW; TVD-100 turboprop planned.

U-1 First of series-produced a.t.o. rockets for military use, designation from *Uskoritel* (accelerator); U-5 and U-7 also in use.

U-19 Boost package for supersonic aircraft comprising RU-013 rocket engine, tanks, control system and jettison gear; U-19D with provision for multiple restarts.

U-21 Boost package comprising S3-20M rocket engine, tanks and control system.

VAZ-413 Rotary (Wankel-type) devd by RPD NTS VAZ, Moscow; twin liquid-cooled rotors, 103 kW (138 hp), also 4133 version.

VAZ-430 Twin-rotor, qual 93 on Mogas at 164 kW (220 shp); OEI contingency 201,3 kW (270 shp); turboprop version is D-200.

VD-4 Dobrynin devt of M-250, same cyls but refined, two stage supercharge, in VD-4K single giant exhaust turbo, blowdown turbine on each bank, plus intercoolers, 4,300 hp.

VD-5 Single-shaft turbojet by Dobrynin, tested 13 t (28,660 lb) 56 but not qualified.

VD-7 Single-shaft turbojet by Dobrynin, qual 56 for 3M at 11 t (24,250 lb); derated as VD-7B to 9,5 t (20,943 lb) 57; then devd by P. A. Kolesov as VD-7F 14 t (30,865 lb) with afterburner; in 63 redesigned VD-7P (RD-7P) 11,3 t (24,910 lb) and a/b VD-7M 16 t (35,275 lb), used with a/b removed at 11 t in VM-T; final (65) series version VD-7M-2 (RD-7M-2) 16,5 t (36,376 lb).

VD-19 Reduced-diam version of VD-7 with a/b, max rating 10,2 t (22,480 lb).

VD-251 Dobrynin engine using two banks of VD-4 cylinders to give 2,000 hp.

VK-1 Important centrifugal turbojet derived from RD-45 by GAZ-117 under Maj-Gen Vladimir Yakovlyevich Klimov (see VK-100); airflow increased 30% within same overall engine dimensions and advanced (for era) afterburner developed; qualified Dec 48 at 2,7 t (26,5 kN); VK-1F with afterburner 2,6/3,38 t; VK-1A, 2,7/3,45 t, longer life, different accessory-gearbox system. Almost 40,000 delivered from GAZ-16, 19 and 45.

VK-2 Single-shaft axial turboprop developed at GAZ-117 under Klimov 47 to termination 1952; 8-stage, 3,357/3,580 kW at 9,000 rpm.

VK-3 Pioneer 2-shaft bypass turbojet (Klimov, designer S. V. Lyunevich), 2-LP (airflow 98,4 kg/s), 8-HP (p.r. 12·7), run at GAZ-117 52, design thrust 5,730 kg dry, 8,440 kg (18,600 lb) with a/b, abandoned 56.

VK-5 Developed VK-1 (Klimov, designer A. S. Mevius); increased temp; VK-5F rated 52 at 3 t (6,614 lb) dry and 3,850 kg (8,490 lb) with a/b.

VK-7 Enlarged Nene (GAZ-117 under Klimov), greater airflow, p.r. 6·3, ratings 4,2 t dry, 6,270 kg (13,825 lb) with a/b.

VK-100 In search for powerful new fighter engine circa 34 choice fell on Hispano-Suiza 12Y, imported and licensed by Aviatrust and assigned to new KB of V. Ya. Klimov as M-100 (all following engines redesignated with Klimov's initials December 40 and VK used here). Baseline engine for most numerous family of Soviet engines, over 129,000 produced 35/circa 47. V-12, water-cooled, cylinders 148×170 (35,09 lit), single-speed supercharger, qualified 35 at 750 hp; later 100A qualified 840 or 860 hp, 2,400 rpm.

VK-103 Development with two-speed supercharger, 860 hp January 37, later 103U and 103A, 960 hp; qualified 1,100 hp on 100-octane fuel.

VK-105 Chief WW2 liquid-cooled engine, qual Aug 39 on 87 oct, comp ratio 7·1, 1,050 hp at 2,600 rpm; 105P, 0·59 instead of 0·666 reduction, 1,100 hp; 105PA, strengthened, 2,700 rpm; PF, 1,260 hp Dec 41; PF-2 cleared to 1,100 mm boost in lo-gear; PD with V. A. Dollezhal E-100 2-stage supercharger; R, RA similar PA, able to mount cannon up to 57 mm.

VK-107 Revised valve gear (instead of two inlet, one exhaust, fitted with one inlet, two exhaust and one scav, three cams per cylinder, restressed for higher rpm, revised cylinder liners, equi-spaced exhaust outlets; Jan 42, 1,400 hp; 107A, 43, 1,650 hp; 107P provision for 37-mm gun.

VK-108 Experimental engine based on 107, 1,800 hp in 45.

VK-109 Experimental engine, two series 1,530 or 2,073 hp, 45.

VRD-2 First Lyul'ka turbojet to run, 700 kg, early 43.

VRD-3 Original designation of S-18.

VRD-5 Original designation of TR-3.

VRD-430 400-mm subsonic ramjet by M. M. Bondaryuk flown from late 40.

ZhRD General abbreviation for a liquid rocket; also applied to series of two-chamber units run by L. S. Dushkin and A. M. Isayev at 1,1/1,14 t in 45-47 for Florov No 4302.

Working on the M-22 (derived from the Bristol Jupiter) of an I-4 (ANT-5) fighter in the early 1930s.

AVIONICS

As USSR had no private companies all equipment could be designated according to a common system. Below are some designation prefixes and proper names which have come to light. In most cases prefix has been used for a succession of numbered equipment items over many years. For example SAU numbers exceed 40, and first entry is exceptional. 'Complex' denotes a major system made up of separately packaged subsystems, in one case with 29 LRUs (line-replaceable units).

A General prefix *Aviatsiya* for items which also have army or navy versions; thus, A-100 and -402 are recon cameras, A-336Z is guidance for a cruise missile, A-385A is a SLAR, A397Z is for missile guidance and A-711 and -723 are Loran navaid receivers.

AFA Large reconnaissance cameras.

AGD Horizons and derived situation displays.

Aist-M (Stork) Panoramic single-line TV recon systems.

Almaz (Diamond) Ranging radar, used alone or in *Uragan-1* system.

Amur (river) Radio compass, early jet fighters.

AP Autopilots (pre-1960).

APD Digital secure data-links.

APM MAD (magnetic-anomaly detection) 'birds' for helicopters.

APP Automatic control systems for [usually chaff/flare] dispensers.

Argon (Argon) PRS-1 turret control radar (NATO 'Bee Hind').

ARK ADF (auto direction finders), from 'auto radio compass'.

ARL Analog FM secure data-links and beacon receivers.

Arlekin (Harlequin) Civil radios (plus printers) compatible with *Tsifra*-N and comsats.

ARZ Flight-control artificial-feel systems.

ASO Chaff/flare dispensers.

ASP Gunsights incorporating electronic [usually radar] ranging.

ASU Receivers for ground control of interception.

Avtomat Computer-controlled [usually chaff/flare] dispensers.

Baklan Civil VHF radios.

Barii (Barium) IFF transponders.

Baza (Base) Air/ground telemetry systems.

Berez [phonetically, *Beryoza*] (Birch) Passive RWR (radar warning receiver) complexes.

Binom (Binomial) Data-entry keyboards.

BKO Integrated defensive-electronics complexes, eg *Karpat*.

BPS Rapid-reaction transducers in systems to maintain stable engine operation after nearby warhead explosion.

BRS Broadband radio channel for air/ground delivery of multisensor recon data.

Chaika (Seagull) Loran-type navaids (early versions with 'towel-rail' antenna).

Chaika-1M An airborne telecode terminal.

Cifra-N See *Tsifra*-N.

DGMK Gyromagnetic compasses.

DISS Doppler radars for navigation.

DNM Secure command links.

Doob (Oak) RSIU-4 VHF.

DUA Indicators of combined AOA and slip.

DV Beacon transmitters.

Efir-1M (Ether) Pod containing radiation (surface radioactivity) reconnaissance sensors with tape/data-link output.

Ekran (Shield) Mission data recorders.

Emblema (Emblem) Civil weather radar.

Evkalipt (Eucaliptus) Long-range com radios for interceptors.

Fantazmagoriya LO-80 and -81 Multiwavelength all-aspect passive threat receivers.

Fone (Background) Laser ranger and marked-target seeker (not illuminator).

Gardiniya (Gardenia) Internal ECM active jammer system.

Gelii (Helium) Special com radios.

Geran-F (Geranium) L-101, -102 or -102G internal active jammer systems.

GIK Earth-inductor compasses.

Glonass (acronym in English as well as Russian) National Satcom system.

Gorizont (Horizon) Air/ground link for guidance to target or other surface location.

Grad (Hail) SRD-3 radar ranging sight.

Groza (Storm) Civil weather radars.

GTsK Radio (goniometer) compass magnetic sensor unit.

IK Data blocks in flight-trajectory complexes: VSP velocity, VK course/altitude.

Ilem (Elm) Fighter ADF.

ILS Advanced HUDs with multiple inputs (unconnected with Eng-language ILS).

Initsiativa (Initiative) Family of RBP-4 tactical nav/bombing radars (NATO 'Short Horn').

Iskra (Spark, a gleam of hope) Blind-landing receivers for combat aircraft.

Izumrud (Emerald) Early interception radar RP-1 by V. V. Tikhomirov KB with separate search/track scanners (NATO 'Scan Odd').

K General prefix for computers; also guidance complexes for strategic cruise missiles.

Kaira Dual-wavelength (laser + TV) target sensor and ranger.

KAP Simple autopilots giving stabilization in roll only.

Kara (Retribution) Tactical nav/attack system.

Karpat (acronym) Integrated RWR/DF/ECM defence complex.

Kh Guidance transmitters for Kharkov family of attack missiles.

Khod (Motion) FLIR (forward-looking infrared).

Khrizantem (Chrysanthemum) Fighter homing receiver.

Khrom (Chromium) SRO series IFF.

Khrom-Nikel (Chrome-nickel) SRZO series IFF.

Kinzhal (Dagger) Millimetric (usually 8 mm) radar.

KKR Reconnaissance pods containing cameras and radio-electronic sensors.

Klen (Maple) Laser ranger and missile-guidance illuminator.

Klystron (same in English) ILS for use down to 50 m cloudbase.

KN Navigation computers.

Kobal't (Cobalt) Family of heavy bomber attack and (*K*-N) cruise-missile target-illumination and guidance radars (NATO 'Crown Drum').

KOLS Interception complex: SBI + TP + displays.

Komar (Mosquito) Personal survival beacons.

Konus (Cone) SRD-1, -1M, gunsight/bombing radar rangefinder for fighters (NATO 'Scan Fix').

Kopyo (Spear) Retrofit radar by Fazotron for MiG-21bis and similar fighters.

Korshun (Kite, the bird) Early (RP-3) interception radar developed from *Torii*; too difficult for pilot of single-seater to interpret (NATO 'Scan Three').

KSI Tacan-type fighter navaid.

Kub (Cube) Strategic SLAR (side-looking airborne radar).

Kupol (Cupola) Integrated digital flight/nav complex.

Kvant (Quantum) SRD-5M radar ranging sight (NATO 'High Fix').

L General prefix for defensive ECM systems or controllers.

Ladoga (named for lake) Extremely sensitive MAD installation.

Lazur (Azure) Beam/beacon receiver for fighters.

Liana (Liana) High-power pulse-doppler surveillance ('AWACS') radar.

LO IRWRs (infra-red warning receivers) covering upper or lower hemisphere, with D/F.

Lotos (Lotus) Shoran + Tacan fighter navaids.

Mak-UL (Poppy) LO-82 Missile-launch IR warning receiver.

Marshroot (Route of Journey) Omega satellite nav system receiver.

Merkur (Mercury) An LLTV (low-light TV).

Meteorit (Meteorite) Interceptor beacon transmitter.

Mir (World) Multi-waveband radiogram for civil passengers.

MIS Inertial platforms.

MLP Flight recorders for mission parameters.

MRP Marker-beacon receivers.

MS Monitoring systems (flight or cockpit-voice recorders).

N Coherent pulse-doppler radars by Fazotron, incl N006 *Sapfir*-21, N019 *Sapfir*-29 and *Topaz*, and N010 *Zhuk*.

Neon (same in English) Central controllers for defensive-electronic complexes.

NAS Pre-1960 navigation complexes.

NI Doppler-based nav systems.

NK Tactical nav computers.

OEPS Electro-optical pilot HUD-sights fed by TP and sometimes other sensors.

OKA Helicopter dipping sonars.

Orbita (Orbit) SVU-10 family digital nav computers.

Orel [phonetically *Oryol*] (Eagle) RP-11 family (A,B,V,G,D,DM) of radars for large interceptors, single parabolic dish for search, track and target-illumination for K-8R/R-30 missile (NATO 'Skip Spin').

Orion (same in English) Most powerful multimode nav/attack radar for tactical aircraft.

OSP ILS receivers.

P Vertical-acceleration ('**g**') meters and recorders.

Parol (Password) Advanced IFF transponders.

Parus (Sail) Navigation switching systems.

Pastel (same) Advanced radar-warning system.

PBP Reflector gunsights.

Peleng (Bearing) Navigation complex for supersonic single-seat strategic aircraft.

Pion (Peony) Antenna feeders for PVD air-data booms.

PKI Computing gyro gunsights.

PKV Fixed-reticle gunsights (usually for helicopters).

PNK Navigation and attack complexes for single-seat aircraft.

Polyot (Flight) Tactical navigation and landing complexes, typically RSBN+DV+SKV, later plus IAS/Mach/altitude capture and hold.

PPV HUDs (head-up displays).

Prizma (Prism) HF radio for global communications.

PrNK Navigation and attack complexes, typically NI + SKV + twin RV + K, numbered same as aircraft.

PRS Gunfire-direction radars, inc-1 *Argon*, -2/3, -4, -6 (NATO 'Bee Hind', 'Box Tail', 'Fan Tail').

PSB Sights for radar-directed high-level bombing (NATO name PSBN-M 'Mushrooms').

PV Turret fire-control radars.

PVD Air-data booms, with static/pressure heads and sometimes temperature probes and/or AOA/yaw transducers.

R General prefix for communications radio; also for simple export radars.

Radal-M SRD-6 ranging radar for gunsight (usually ASP-4).

Radan Sh-101, latest production interception radar.

Raduga (Rainbow) Command-guidance transmitters (usually -F) for 9M17.

RBP [see RPB] Major family of radars for blind level bombing, most with individual names (NATO -3 'Look Two', -4 'Short Horn').

Rech' (Speech) Guidance transmitters for air-dropped lifeboats.

REB EW complexes combining Elint and jamming.

Relyef (Relief [topographical]) Radars for terrain avoidance and/or terrain following.

RI Combined aural/visual threat-warning displays.

RIS Stick shakers.

RLS Reconnaissance radars; -BO adds sideways-looking.

RP In-service prefixes of interception radars.

RPB Alternative prefix for radars for nav and blind level bombing.

RSBN, RSDN Short-range guidance systems for homing and bad-weather landing.

RSI Communications radios, mainly trad HF.

RSIU Shorter-wavelength (VHF, UHF) radios.

Rubin (Ruby) Family of RBP strategic bomber radars (NATO 'Down Beat').

RUM Pioneer plan-position bomber radars (NATO [RUM-S] 'Shore Walk').

RV Radar altimeters (c25 types).

S General prefix for pulse-doppler radar fire-controls not yet in service.

SAOG Yaw (helo azimuth) stabilizers.

Sapfir (Sapphire) Pulse-doppler interception radars by various OKBs (Volkov, Kunyavskii and Kirpichyev) with wide variation in power and missile guidance but all having single mechanically scanned twist cassegrain antenna (NATO, RP-21 'Spin Scan', -22 family 'Jay Bird', 23D/23ML family 'High Lark', 25 'High Lark IV', 29 [N019] 'Slot Back').

SARP Central fault-warning systems.

SAU Autopilots.

SBI Interception radars with electronic scanning.

SBU Intercom systems.

SDU Electronically signalled (FBW) flight-control systems.

Sevan Precision timebase so that all elements of (eg, recon complex) operate to common time.

SG Prefix for members of *Sirena* family.

Shkval (Squall) An EO (electro-optic) attack complex, combining TV with LRMTS (laser ranger and marked-target seeker).

Shpil (Spire) Laser linescan reconnaissance pods, all providing broadband radio output and stored film processed in flight.

Shtik (Bayonet) Synthetic-aperture SLAR (sideways-looking airborne radar), mounted internally or in pod.

Sirena (Siren) Passive radar-warning receivers.

Skip Loran (long-range air nav) receivers.

SKV Vertical gyros.

Smerch (Whirlwind) RP-5 family of hi-power interception radars by Volkov OKB with mechanical scan ±60° and look-down capability (NATO names, RP-S 'Big Nose', RP-SA 'Fox Fire', RP-SM not known).

SO, SOD ATC (air-traffic control) SIF (selective identification facility) transponders.

Sokol (Falcon) RP-6 family of hi-power radars for larger gun-armed interceptors. Same name also used 20 years later for one PrNK tactical nav/attack complex.

Sorbitsiya (Break off, miscarry) Smart jammers, usually in wingtip pods.

SORT Central alarm displays.

SOS Aircraft-configuration control systems, eg to retract flaps before sweeping back wings.

SOUA Combined AOA indication and warning (but usually not limiting) systems.

SP ILS (instrument landing systems); also OSP.

SPS *Stantsiya Pormekhovykh Signalov* Station for noise signals, wideband jamming installations (internal), followed after hyphen by number up to (1994) 153.

SPS IR (infra-red) jammers and break-lock emitters, usually in pods or scabbed on.

SPU Alternative to SBU for intercoms.

SRD Radar ranging systems, usually part of gunsight.

SRE IFF interrogators.

SRO IFF systems using group of long, medium and short rod antennas.

SRP Fire-control computers for air/air missile armament.

SRS Elint (electronic intelligence) receivers, most large and internally mounted with side-looking antennas of contrasting forms, some for helos, some for Mach 3, numbered to -9 *Virazh* (NATO [A385A] 'Side Kick').

SRZO IFF systems using blade antennas.

SUA AOA warning systems.

SUNA Engine-protection systems to minimise effect on airflow of close warhead explosion, incorporating BPS.

SUO Tactical-aircraft control complex linking *Relyef*+*Kaira*+PPV+*Tekon*.

SUV Tactical-aircraft armament control and weapon-selection systems.

Svod (Arch, vault) Navaid giving R-Θ (range/bearing) guidance towards ground stations.

SVS Multiple-input air-data systems.

Taifun (Typhoon) RP-26 interceptor radar derived from *Orel*, with look-down and target-illumination performance close to *Smerch* but with smaller antenna (NATO 'Twin Scan').

Tangazh (Pitch angle) Elint receiver installation for detection, analysis, position-fixing and recording of pulse (0·8-300 cm) and CW (2·8-3·6 cm) emissions.

Tekon HUD video monitors.

Tester-UZ Automatic flight-data recorders.

Tigr (Tiger) Multimode nav/attack radar, part of PrNK-24 (*Tigr*-NS, in PrNK-24M).

Ton (Tone) Direction-finding radios (later versions, ADF).

Torii (Thorium) Pioneer (RP-2, 2U) interception radar with search, track and ranging facilities, devd by Slepushkin OKB (NATO 'Scan Can').

TP Opto-mechanical sights for turrets; later IRST (infra-red search/track) sensors.

Trassa (Route) Precision dopplers for jet aircraft.

Tropik (Tropic) Advanced Loran-type navaid.

TS Optical periscopes.

TsD Interception radars (experimental designations).

Tsifra-N (Number) National VHF data-link network.

TsP Interception radars (experimental designations).

Uragan (Hurricane) Large family of interception fire-control systems, initially with gun-ranging radar, later (-5 and -5B) with 5V autopilot link, 5D computer and target illumination for K-8 or K-9 missiles.

UUA AOA measuring/indicating systems; UUAP adds vertical accelerometer (**g**-meter).

UVV Input keyboards.

Uzel (Knot) Beacon-homing systems.

Virazh (Turn) SRS-9 series side-looking radar receiver.

Vikhr (Whirlwind) RP series interceptor radars.

VKU Navigation switching units.

Voskhod (Rising) Tac navigation complex combining inertial platform, two digital computers and autopilot.

Yadro (Kernel) Civil UHF radios.

Ye Bomber radars with (from 1955) missile-guidance capability (NATO [YeN] 'Puff Ball').

Yolka (Fir) ECM (electronic countermeasures) installation for Tu-16.

Zaslon (Barrier) S-800 advanced hi-power interception radars with electronic scanning ±70°, antenna diam 1,195, later 1,400 mm (NATO 'Flashdance').

Zhuk (Beetle) N010 lightweight fighter radar with multi-slot type flat plate mechanically scanned antenna, by Fazotron.

Zhuravl (Crane, the bird) Tactical communications radios.

Zima (Winter) IRLS (infra-red linescan) reconnaissance pods with 0·3°C sensitivity, storing recording and simultaneously transmitting via BRS link.

When the world's biggest aeroplane – An-225 – came to Farnborough few observers thought about its avionics!

ORGANISATIONS

ADD Long-range aviation (bomber force) of VVS.

Aeroflot Civil air fleet.

AGOS Dept of aviation, hydro-aviation (seaplanes and hydroplanes) and experimental construction.

AMG *Maksim Gorkii* propaganda squadron.

Amtorg Organization for importing and licensing US products.

ANTK Aviatsionnyi Nauchno-Tekhnichyeskii Kompleks, Aviation Scientific-technical Complex.

A-VDV Aviation of airborne forces of VVS.

Aviaarktika Independent Arctic directorate of Aeroflot.

Aviavnito Strictly AVIAvnito, aviation dept of Vnito, all-union scientific and technical research organization.

Aviatrust Original (Sept 1923) central aviation management organization with direct authority over CCB and external licensing.

AV-MF Naval aviation.

BAP Bomber regiment.

BNK Bureau of new construction.

BNT NKAP bureau of new technology.

BOK Bureau of special design.

CAHI Central Aerodynamics and Hydrodynamics Institute (previously often spelt in English TsAGI).

CCB Central Construction Bureau (previously often spelt in English TsKB).

Cheka Secret police, 1918-22, became OGPU.

CIAM Central Institute of Aviation Motors (previously often spelt in English TsIAM).

DA Long-range (bomber) aviation, became ADD.

Deruluft Original (1921) German/Russian airline company.

Dobrolyet Original (1922) Soviet airline.

Dosaaf Voluntary society for assisting army, AF and navy.

Dosav Voluntary society for assisting AF.

FA Frontovaya Aviatsiya, tactical aviation of VVS.

GAZ State aviation factory.

GDL Gas dynamics lab.

GIRD Group for studying reaction (in practice, rocket) engines.

GKAP State committee on aviation industry.

Glavkoavia Chief administration of aviation.

Glavsevmorput Chief administration of northern sea routes.

GosNII State scientific research institute of GVF.

Gros, GROS Civil experimental aeroplane construction.

GUAP Chief administration of aviation industry.

GUGVF Chief administration of GVF.

GUSMP Chief administration of navy.

GUVVS Chief administration of air force.

GVF Civil air fleet (of which Aeroflot was operating branch).

IAP Fighter regiment.

IA-PVO Manned fighter branch of PVO.

KAI Kazan aviation institute.

KB Constructor bureau ie, design office.

KhAI Kharkov aviation institute.

KIIGA Kiev Institute of Civil Aviation Engineers.

KOMTA Commission for heavy aviation.

Komsomol Young Communists.

KOSOS Dept of experimental aeroplane construction.

LII Flight research institute.

LIIPS Leningrad institute for sail and communications engineers.

MA Naval aviation, became AV-MF.

MAI Moscow aviation institute.

MAP Ministry of aviation industry.

MAT Moscow aviation technical high school.

MGVF Ministry of civil aviation.

MosGird Moscow Gird.

MOS VAO All-union association for experimental marine aircraft.

MVD Ministry for internal affairs 1946-53, became KGB.

MVTU Moscow higher technical school.

NAMI Scientific auto-motor institute.

Narkomavprom Supreme direction of aviation industry, became GUAP.

NII Scientific test institute (more than 30, including many involved with aviation, eg, NII VVS, NII GVF).

NKAP State commissariat for aviation industry.

NKOP State Commissariat for defence.

NKTP State commissariat for heavy industry.

NKVD State commissariat for internal affairs 1934-46, became MVD.

NOA Scientific experimental aerodrome.

NPO Nauchno-Proizvodstvennoye Obyedinyenye, scientific-production union.

NPP Scientific Production enterprise.

NTK Scientific and technical committee (of VVS).

NTU National technical administration.

OAVUK Society for aviation and gliding of Ukraine and Crimea.

ODVF Society of friends of the air fleet.

OGPU Special government political administration, 1922-34, head Menzhinskii, became NKVD.

OKB Experimental construction (ie, design) bureau.

OKO Experimental design section (Kiev).

OMAG Independent naval aviation group.

OMOS Dept of marine experimental aircraft construction.

OOK Dept of special construction.

OOS Dept of experimental aircraft construction.

Osoaviakhim Society for assistance to aviation and chemical industry.

OSK Dept for special construction.

OSS Dept for experimental landplane construction.

Ostekhburo Special technical bureau for military inventions.

OVI Department of war invention of RKKA.

PVO Protective air defence, disbanded 1983.

RKKA Raboch Krest'yanski Krasno Armiya, workers and peasants Red Army.

RNII Reaction-engine scientific research institute.

SKB Student construction bureau; also *syerinom* KB, production bureau.

SNII Aeroplane scientific test institute of GVF.

Stavka Supreme command staff.

TTT Tactical/technical requirements, ie military specification.

TsAGI, TsIAM, TsKB, see CAHI, CIAM, CCB.

UK Training centre.

Ukrvosdukhput See UVP.

UMS Control board of navy.

USR Special work control.

UVP Ukrvosdukhput, Ukrainian airline.

UVVS Administration of the VVS.

VIAM All-Union institute for aviation material.

VT Internal prison.

V-TA Military transport aviation.

VVA Zhukovskii VVS academy.

VVF Military air fleet.

VVIA VVA engineering academy.

VVS Air force.

ZOK Factory for experimental construction, at GVF and CAHI.

Despite the title of this book, Ukraine is not excluded! The author snapped the two leaders of the Antonov ANTK – Balabuyev and Bulanenko – aboard an An-124.

Miscellaneous Data

USAF TYPE NUMBERS

These invented designations were used 1947-55 to identify new types of Soviet military aircraft and missiles. From 1954 they were replaced by reporting names invented by the ASCC, see next table.

Type 1, MiG-9 (OKB 1-300)
Type 2, Yak-15
Type 3, La-150
Type 4, La-152
Type 5, La-156
Type 6, La-160
Type 7, Yak-19
Type 8, Su-9 (OKB Type K)
Type 9, Tu-12 (OKB Tu-17)
Type 10, Il-22
Type 11, MiG I-270
Type 12, Tu-73
Type 13, not used
Type 14, MiG-15 (OKB I-310)
Type 15, La-168
Type 16, Yak-17
Type 17, Su-11 (OKB Type LK)
Type 18, Su-15 (OKB Type P)
Type 19, MiG KS-1 *Sopka* missile
Type 20, MiG-17 (OKB SI, I-330)
Type 21, La-15 (OKB La-174D)
Type 22, Tu Projekt 64
Type 23, Su-12
Type 24, Yak-14
Type 25, Tsybin Ts-25
Type 26, Yak-17UTI
Type 27, Il-28
Type 28, Yak-23
Type 29, MiG-I5UTI (OKB ST)
Type 30, Il-28U
Type 31, Tu-85
Type 32, Mil Mi-1
Type 33, Be-8
Type 34, Be-6
Type 35, Tu-14 (OKB Tu-81)
Type 36, Mil Mi-4
Type 37, M-4, M-6, 3M
Type 38, Yak-24
Type 39, Tu-16 (OKB N, Tu-88)
Type 40, Tu-95

ASCC REPORTING NAMES

In 1954 the Air Standards Co-ordinating Committee (USA, UK, Canada, Australia, New Zealand) followed the practice adopted in 1942 with Japanese aircraft of assigning a supposed unambiguous name, easily identified over a poor voice radio link, to replace the former less-memorable Type Numbers. Names were assigned according to vehicle mission as follow: A, air-to-air missile (see AAM table); B, bombers (single-syllable propeller, two-syllable jet); C, cargo aircraft (ditto); F, fighters (ditto); G. SAMs; H, helicopters; K, air-to-surface missiles (see ASM table); M, miscellaneous aircraft; S, surface to surface missiles.

Allocated only to aircraft believed to be in service (thus not to MiG-29M, Su-35 etc). Early knowledge of true Russian designations may lead to an end to new allocations. Still useful for early types, where real designations of sub types were often unknown. Allocated in order of being noticed by Western intelligence agencies, leading to anomalies, eg Mi-24 prototypes 'Hind-B' while first production series (seen first) 'Hind-A'.

BOMBERS

'Backfin', Tu-98
'Backfire', Tu-22M
'Badger', Tu-16
'Bank', B-25 Mitchell
'Barge', Tu-85
'Bark', Il-2
'Bat', Tu-2
'Beagle', Il-28
'Bear', Tu-95, -142
'Beast', Il-10
'Beauty', Tu-22 (changed to 'Blinder', judged 'too laudatory')
'Bison', M-4, M-6, 3M
'Blackjack', Tu-160
'Blinder', Tu-22
'Blowlamp', Il-54
'Bob', Il-4
'Boot', Tu-91
'Bosun', Tu-14
'Bounder', M-50
'Box', A-20 Havoc
'Brassard', Yak-28 (changed to Brewer, possible confusion with Max Holste Broussard)
'Brawny', Il-40
'Brewer', Yak-28B, -28I etc.
'Buck', Pe-2
'Bull', Tu-4
'Butcher', Il-28 (changed to Beagle, possible confusion with 'Badger')

CARGO (TRANSPORT AIRCRAFT)

'Cab', Li-2
'Camber', Il-86
'Camel', Tu-104
'Camp', An-8
'Candid', Il-76
'Careless', Tu-154
'Cart', Tu-70, -75
'Cash', An-28
'Cat', An-10
'Charger', Tu-144
'Clam', Il-18 (1947)
'Clank', An-30
'Classic', Il-62
'Cleat', Tu-114
'Cline', An-32
'Clobber', Yak-42
'Clod', An-14
'Coach', Il-12
'Coaler', An-72, -74
'Cock', An-22
'Codling', Yak-40
'Coke', An-24
'Colt', An-2
'Condor', An-124
'Cooker', Tu-110
'Cookpot', Tu-124
'Coot', Il-18, -20, -22
'Cork', Yak-16
'Cossack', An-225
'Crate', Il-14
'Creek', Yak-12 (high-wing)
'Crib', Yak-8
'Crow', Yak-12 (low-wing)
'Crusty', Tu-134
'Cub' An-12
'Cuff', Be-30, -32
'Curl', An-26

FIGHTERS

'Faceplate', MiG Ye-2 etc.
'Fagot', MiG-15
'Faithless', MiG-23PD (23-01)
'Falcon', MiG-15 (changed to 'Fagot', judged 'too laudatory')
'Fang', La-11
'Fantail', La-15
'Fantan', Chinese Q-5
'Fargo', MiG-9
'Farmer', MiG-19
'Fearless', assigned in error
'Feather', Yak-15, -17
'Fencer', Su-24
'Ferret', invented 'MiG-37'
'Fiddler', Tu-128
'Fin', La-7
'Finback', Chinese J-8
'Firebar', Yak-28P
'Fishbed', MiG-21

'Fishpot', Su-9, -11
'Fitter', Su-7, -17, -20, -22
'Flagon', Su-15
'Flanker', Su-27 and derivatives
'Flashlight', Yak-25
'Flipper', MiG Ye-152 family
'Flogger', MiG-23, -27
'Flora', Yak-23
'Forger', Yak-38
'Foxbat', MiG-25
'Foxhound', MiG-31
'Frank', Yak-9
'Fred', P-39 Airacobra
'Freehand', Yak-36
'Freestyle', Yak-141
'Fresco', MiG-17
'Fritz', La-9
'Frogfoot', Su-25, -28
'Fulcrum', MiG-29

HELICOPTERS

'Halo', Mi-26
'Hare', Mi-1
'Harke', Mi-10
'Harp', 'Ka-20'
'Hat', Ka-10
'Havoc', Mi-28
'Haze', Mi-14
'Helix', Ka-27, -28, -29, -32
'Hen', Ka-15
'Hermit', Mi-34
'Hind', Mi-24, -25, -35
'Hip', Mi-8, -9, -17, -171
'Hog', Ka-18
'Hokum', Ka-50
'Homer', Mil V-12
'Hoodlum', Ka-26, -126, -226
'Hook', Mi-6
'Hoop', Ka-22
'Hoplite', Mi-2
'Hormone', Ka-25
'Horse', Yak-24
'Hound', Mi-4

MISCELLANEOUS

'Madcap', An-71
'Madge', Be-6
'Maestro', Yak-28U
'Magnet', Yak-17UTI
'Magnum', Yak-30
'Maiden', Su-9U
'Mail', Be-12
'Mainstay', A-50
'Mallow', Be-10
'Mandrake', Yak-25RV
'Mangrove', Yak-25R, -27R
'Mantis', Yak-25R, -27R
'Mare', Yak-14
'Mark', Yak-7U
'Mascot', Il-28U
'Max', Yak-18
Maxdome, Il-86VPK
'May', Il-38
'Maya', Czech L-29
'Mermaid', Be A-40, Be-42
'Midas', Il-78
'Midget', MiG-15UTI
'Mink', Yak UT-2
'Mist', Tsybin Ts-25
'Mole', Be-8
'Mongol', MiG-21U
'Moose', Yak-11
'Mop', Amtorg GST
'Moss', Tu-126
'Mote', Be MDR-2
'Moujik', Su-7U
'Mouse', Yak-18M
'Mug', Chye-2 (MDR-6), and Be-4
'Mule', Po-2
'Mystic', Mya M-17, -55

RAM LETTERS

In about 1970 the US Department of Defense began assigning a special series of designations to new prototype and experimental aircraft identified on satellite imagery on the test airfield at Ramenskoye (now known to be called Zhukovskii).

RAM-A to G, not yet known but includes Su T6-1 and T6-2IG
RAM-H, Tu-144D (and Su T-4?)
RAM-I not assigned
RAM-J, Su T-8 (Su-25)
RAM-K, Su T-10 (Su-27)
RAM-L, MiG-29
RAM-M, Mya M-17
RAM-N, -O, not known (one believed to be Il-102)
RAM-P, Tu-160
RAM-Q not assigned
RAM-R, Buran (RAM-R1 and RAM-R2 allocated)
RAM-S, not known
RAM-T, Yak-141

In addition the prototype Beriev A-40 was called TAG-D when it was spotted at the Taganrog factory. TAG-A, B and C were assigned previously. SH – reporting names were allocated to prototypes sighted at Sary-Shagan.

SOME GAZ NUMBERS

Under the Gosplan (state plan) of the USSR the aircraft industry was at last properly organized from 1 Oct 1927, with a growing series of GAZ (state aviation factories) which initially all occupied previous old numbered factories. Since 1989 the vast edifice has crumbled, and most of those still active are being run as capitalist firms or as part of groups embracing design bureaux and former rocket factories.

1 Moscow Khodinka (Central Aerodrome), 1912 Dooks factory, renamed Aviakhim Jan 1922, home of OPO-1, R-5 prodn, MiG-1 when evac to Kuibyshyev Oct 1941, later MiG jets.

2 Moscow, Gnome-Rhône engines 1912, from 21 Mar 1927 branch of GAZ-24.

3 *Krasnyi Letchik* (Red airman), Leningrad, in old Factory 23, D. P. Grigorovich OKB and OPO-3.

4 Moscow plant of motor company 1911, from 21 Mar 1927 branch of GAZ-24.

5 Leningrad, in old Factory 25.

6 Kiev, occupied Factory 26, became OKB of O. K. Antonov.

7 Moscow, occupied old Factory 22, renumbered GAZ-22.

8 Occupied Factory 28.

9 Zaporozhye, occupied Factory 29, renumbered GAZ-29.

10 Taganrog, occupied Factory 31, renumbered GAZ-31.

11 Occupied Factory 41.

12 Old 'Radio' factory, housed Aviatrust Sept 1926, assigned to electrical systems initially using Robert Bosch GmbH licences.

15 Simferopol, former Anatra works.

16 Occupied Factory 36, Ufa, then Voronezh, where received evacuated GAZ-82.

18 Voronezh, named for K. Y. Voroshilov, 15 June 1934 Kalinin OKB, then S. A. Moskalyev, Il-2 prodn, evac Kuibyshyev (Samara) Oct 1941; later Tu-95/-142, -154 and parts of -144.

19 Perm, named for I. V. Stalin, produced M-25 engine, from 1 Apr 1934 KB of A. D. Shvetsov.

20 Moscow, from 1931-32 centre for water/fuel/oil systems.

21 Gorki (Nizhni-Novgorod), named for Ordzhonikidze, produced I-5, I-16, Polikarpov OKB to 1935, then Tomashyevich, La OKB from Oct 1941.

22 Moscow (Fili), named for Gorbunov, most impt heavy a/c prodn centre in 1920s (TB-1, R-6, TB-3), home of OPO-4 and P. A. Richard, then Arkhangyelskii OKB, evac as 22A to Povolozhye, Kazan (Pe-2).

23 See 3; major prodn fighters (I-2), trainers (U-1, U-2, UT-2) and many light passenger; home of OPO-3, then Nikitin OKB, Myasishchyev OKB from 1950 before move to Zhukovskii.

24 Moscow, named for M. V. Frunze, home of OPO-2, KB of A. D. Shvetsov to 1 Apr 1934, then A. A. Mikulin, AM-38 prodn.

25 Moscow, OSS from late 1925, evac Oct 1941 to R. Irtysh, nr Omsk.

26 Rybinsk, 17 July 1925 'Russki Renault', then KB of V. Ya. Klimov, M-100/103/105 prodn.

27 Kazan, Jan 1932 engine prodn, later M-105 versions.

28 Moscow, airframes.

29 Zaporozhye, named for Baranov, engines, KB of A. S. Nazarov, M-22 and M-85/86/87/88, later KB of S. K. Tumanskii.

30 Moscow (Botkhinsky), named for Dimitrov, a former Dooks factory, Aviatekhnikum, then Il prodn, then MiG (by road to Lukhovitsky), named *Znamya Trud* (Red Banner), now part of MAPO.

31 Taganrog, one of first two prodn in USSR, R-1/MR-1, R-6, TB-3; Aug 1934 Beriev OKB, number transferred to major prodn plant Tbilisi, Yak fighters, Su-25, now private firm.

32 Named for Menzhinskii, head of OGPU.

33 Also Menzhinskii, opened early 1932 to produce fluid-syst hardware to support GAZ-20.

34 Tashkent, Li-2, Il-14, Il-62 prodn, now part of combine.

35 Smolyensk, home of BOK.

36 Moscow, N. N. Polikarpov OKB to 1935.

38 Moscow Khodinka, named for Ordzhonikidze, Il-2 prodn.

39 V. P. Menzhinskii, also named for 10th anniversary October Revolution, housed VT (internal prison) Dec 1929, Kochyerigin, Rafaelyants, Yakovlev; from 1934 built I-15, R-Zet, I-16, then Ilyushin OKB with 6 brigades KOSOS, built DB-3/Il-4.

43 Kiev, named for Maksim Gorkii, OKO, Tairov, R-10 prodn.

45 Sevastopol, also named for Gorkii, Chyetverikov OKB; later giant engine plant in Moscow.

47 Also named for MG, built trainers, UT-1, UT-2.

51 Moscow, Polikarpov prodn from June 1940, evac Novosibirsk Oct 1941, returned 1943.

53 Post-war bomber prodn, incl Il-28.

56 Moscow Central airfield, light aircraft.

64 Voronezh, branch of GAZ-18, later Tupolev prodn.

81 Moscow Tushino, named for Gorkii, Osoaviakhim, MKB (Gribovskii), OOS Stal KB and prodn, DI-6 prodn, Yatsenko OKB, then Yak prodn. from Yak-4.

82 Engine KB/plant run by NKVD, Charomskii, Stechkin, Bessonov; evac Oct 1941 to Kazan.

83 Khabarovsk, various combat a/c from 1939.

84 Moscow Khimki, named for I. V. Stalin, June 1937 OKB of N. N. Polikarpov, Lisunov KB and Li-2 prodn.

89 Gorkii (Nizhni-Novgorod) GVF repair works, ZIG and OKB of Laville.

95 Josef Stalin, opened 1938 for new technologies incl licences from E. W. Bliss Co of Brooklyn.

98 One of the factories building Tu-95/142.

99 Ulan-Ude, named for Stalin, major eqpt works, later Kamov OKB, An and Su prodn.

115 Moscow (Leningradskii Prospekt), named for Stalin, since 15 Jan 1934 Yak OKB.

116 Semenov-Usurrisk, opened 1941 to build UT-2.

117 Leningrad, Klimov engine prodn, now Isotov Corpn.

120 Central casting and foundry plant, June 1935.

124 Kazan, I. F. Nyezval KB, TB-7/Pe-8 prodn, became world's largest a/c plant (Tu-22M, Tu-160).

125 Irkutsk, began with I-14 Nov 1936, 968 SB, 1,020 Pe-2/-3, 919 DB-3F/Il-4, 325 Yer-2, 218 Tu-2, 42 Tu-14, 459 Il-28, now IAPO (Irk A/c Prodn Assoc), all 2-seat Su-27/30, Yak-112, Be-200.

126 Komsomolsk-na-Amur, opened 1931 to build R-6, later DB-3/Il-4, became Su OKB and prodn factory.

132 Moscow, formed out of No 1 UVVS (Promvozdukh Trust) 1932 to produce elec starters, later giant accessory plant.

135 Kharkov, Kalinin prodn, then Nyemen OKB and R-10 prodn.

145, 149 Metal pressings, stampings, Ustanovka Trust.

150 Moscow Oblast, UT-1 prodn, also Hamilton licensed VISh propellers.

153 Novosibirsk, began Nov 1936 with I-14, I-16 in 1937, LaGG-1/-3, Yak-1/-7/-9, MiG-15 etc, former Po/Su/An OKBs.

155 Moscow, Ulansky Allee, MiG, later A. I. Mikoyan, OKB.

156 Moscow, VT special prison relocated from GAZ-39, Su/Pogosski/Nadashkevich brigades, then STO (special tech depts) numbered from 100 because *sto* = 100 in Russian: 100 Petlyakov, 101 Tomashyevich, 102 Myasishchyev and 103 Tupolev; I-153 prodn, evac Oct 1941 Omsk.

161 Moscow, opened 1934.

162 Propeller prodn.

166 Omsk, Tupolev OKB Oct 1941 to late 1943, Tu and Yak-9 prodn.

167 Kuntsyev, Moscow oblast, from 1938.

168 Ryazan, from 1934.

184 Tashkent, Li-2 prodn 1941, later to GAZ-34.

213 Kutuzovskii, instruments/eqpt from 1933.

214 'Metron', Moscow, inst/eqpt 1932. 'Pirometr', Leningrad, inst/eqpt 1933.

219 Independent prodn plant, wheels, ldg gear, 1935.

224 'Metpribor', Leningrad, inst/eqpt, 1932.

230 'Tizpribor', Kutuzovskii, eqpt group.

234 Leningrad, engine prodn.

240 Yermolayev OKB after Bartini's arrest Jan 1938, hence DB-240.

243 Tashkent, 'Valerii Chkalov', Il-76, An-124 wing.

266 Kulomzino, Irtysh.

286 Kamensk-Uralsk, A. S. Yakovlev OKB 1941-43.

290 Lubertsiy, Moscow suburb, Bartini, now Kamov.

292 Saratov, open Feb 1938 to make R-10, major producer Yak fighters, helos, transports.

301 Moscow suburb, Dubrovin KB, Feb 1938 UT-2 and fuselage of BB-22, 15 Sept 1938 LaGG OKB and prodn.

381 Prodn of I-250/MiG-13, later MiG-15.

382 Moscow, 'Tochizmeritel' insts/eqpt from 1932.

387 Small batches of light a/c; took over U-2 from 1940.

400 Moscow Vnukovo, I-153/-16, then Tu-104/-114/-154.

401 Novosibirsk, M.I.Mil OKB (later moved to Moscow), prodn Mil helos.

402 Moscow Bykovo, Il-18 prodn, D-30 engine.

412 Rostov-on-Don, Feb 1944 UT-2M, Po-2, Yak-14, Il-10M, Il-40, Mi-1/-6/-10/-10K, now Rostvertol.

466 *Krasnyi Oktober*, turbojets from 1947.

ABBREVIATIONS (RUSSIAN) AIRCRAFT CATEGORIES

AK Artillery correction (AOP).
ARK Arctic coastal reconnaissance.
ASK Amphibian for northern region.
BB Short-range bomber.
BSh Armoured attacker.
DB Long-range bomber.
DI Two-seat fighter.
F Frontal (tactical) or boosted (see next list).
G Cargo aircraft, later helicopter.
GK Pressure cabin.
I Fighter.
IP Interceptor fighter.
IS Folding fighter.
K Carrier based.
L Limousine, cabin version, or radar-equipped.
LSh Light attack.
LL Flying laboratory.
MBR Sea long-range reconnaissance.
MDR Sea long-range reconnaissance.
MK Cruiser, armoured seaplane.
MP Sea passenger aircraft.
MR Sea reconnaissance.
MTB Sea heavy bomber.
MU Sea trainer.
P Passenger aircraft, or interceptor.
PS Passenger aircraft.
R Reconnaissance.
SB Fast bomber.
Sh Armoured attack.
T Anti-tank.
TB Heavy bomber.
TSh Heavy attack.
U Trainer, or improved version.
UTI Fighter trainer.

PROPULSION

AD Aviation engine.
APD Accessory gearbox.
D Engine
DPD Supplementary lift engine.
F Boosted, increased power or afterburner.
PD Piston engine.
PuVRD pulsejet.
PVRD Ramjet.
R any jet engine.
RD Turbojet.
TK Turbo-supercharger.
TRD Turbojet.
TRDD Two-spool (two-shaft) turbojet.
TV Turboshaft.
TVD Turboprop.
TvRDD Turbofan.
TVVD Propfan.
U A.t.o. rocket motor.
VFSh Variable-pitch propeller.
VISh Constant-speed propeller.
VPSh Fixed-pitch propeller.
ZhRD Liquid-propellant rocket.

ABBREVIATIONS (ENGLISH LANGUAGE)

a/b Afterburner
AC Alternating current
ADF Automatic direction finder
AOA Angle of attack
APU Auxiliary power unit
ASW Anti-submarine warfare
ATC/SIF Air-traffic control, selected identification facility
a.t.o. Assisted takeoff
b.p.r. Bypass ratio
BVR Beyond visual range
CFRP Carbon-fibre reinforced plastics
CG Centre of gravity
Comint Communications intelligence
c.p. Centre of pressure
c/s Constant speed
c-s Centre, or constant, section
CSD Constant-speed drive
ctp Chief test pilot
DC Direct current
dP Cabin max pressure differential
ECCM Electronic counter-countermeasures
ECM Electronic countermeasures
ECS Cabin environmental control system
Elint Electronic intelligence
EM Electromagnetic
EO Electro-optic (ie usually IR or TV)
ESM Electronic support measures
EW Electronic warfare
FBW Fly by wire (electrical signalling)
FLIR Forward-looking IR
FOD Foreign-object damage
FOV Field of view
FR Flight refuelling
GA General aviation
GFRP Glassfibre reinforced plastics
HDD Head-down display
HF High frequency
HMS Helmet-mounted sight
HP High pressure
HSI Horizontal situation indicator
HUD Head-up display
IAS Indicated airspeed
IFF Identification friend or foe
ILS Instrument landing system
INS Inertial navigation system
IR Infra-red
IRCM IR countermeasures
IRLS IR linescan
ISA International Standard Atmosphere
L/D Lift/drag ratio
LE Leading edge
LLTV Low-light TV
LNG Liquefied natural gas
LP Low pressure
LRU Line-replaceable unit
MAC Mean aerodynamic chord
MLG Main landing gear
MLW Max landing weight
MTO Max takeoff weight
OEI One engine inoperative
OGE Out of ground effect
OTH Over the horizon
OWE Operating weight empty
PRF Pulse-recurrence frequency
PD Pulse-doppler
p.r. Pressure ratio
RWR Radar warning receiver
SAR Search and rescue
SL Sea level
SLAR Side-looking aircraft radar
SST Supersonic transport
STOL Short takeoff and landing
TBO Time between overhauls
TE Trailing edge
UHF Ultra-high frequency
VG Variable geometry
VHF Very high frequency
VLF Very low frequency
VOR VHF omni-directional range
V-p Variable pitch
VTOL Vertical takeoff and landing
WSO Weapon-systems officer

A NOTE ON DATA

It is ridiculous to offer precise conversions from rough original figures (one Western publisher translated a ceiling of 9 km as 29,527 ft 6 in), so imperial conversions vary slightly according to the credibility of input. For speeds the best compromise was judged to be km/h and mph; apologies to modern aircrew who think in knots. All gallons are imperial (for US, multiply by 1·2). Commas are often used to denote what in imperial measures are decimal points; thus 42,3 t means 42·3 tonnes or 93,254 lb. Engine powers are (piston) max rating at best height, (jet) SL static.

A TECHNICAL NOTE

Camloc and Dzus fasteners are patented devices widely copied in the Soviet Union; Camloc is a lever-action latch, Dzus a rotary fastening turned by a screwdriver.

Fowler flaps fit under a fixed wing TE and run out on tracks initially to increase wing area, finally rotating down to increase drag.

Hucks dogs were a bayonet-type drive on the end of early propeller shafts, spun by an external power source (invariably a shaft of adjustable height and inclination on a converted motor vehicle) to start the engine. They were widely used throughout the Second World War after they had long been obsolete in other countries. Picture, p. xxiv.

A *moteur canon* was a Hispano-Suiza patented arrangement of a gun (at first 20 mm, in the Soviet Union later much larger) lying between the cylinder blocks of a geared V-type engine to fire through the hub of the propeller.

A NACA cowl is a long-chord cowling over a radial engine to reduce drag and improve cooling.

Scheibe fins are vertical fins, with or without hinged rudders, attached at or near the tips of the wing. In some cases they have been canted inwards (seen from above, ie LEs closer together than TEs) but the purpose is to reduce tip-vortex formation and increase effective aspect ratio.

Schrenk flaps are split flaps, named for the originator, Dr Martin Schrenk, at the DVL, 1932.

A Townend ring was a simple short-chord cowl around a radial engine, much less efficacious than a NACA cowl.

Zapp flaps were invented by Zaparka in order to reduce the effort needed to lower the flaps against the airstream; they were split flaps pivoted to a series of arms well back from the flap LE so that the flap LE in effect slid backwards along the underside of the recess.

In the aircraft specifications dimensions are given either in metres (6,539 means 6·539, while 6 539 would mean 6,539 in English) or in millimetres, with a span perhaps written 20145. No engine information is given, because engine data are grouped in a special section at the front of the book.

RUSSIAN LANGUAGE

The Soviet Union has a standardized Cyrillic alphabet:

capital	small	italic or handwritten	written or sounds like
А	а	*а*	as English
Ь	б	*б*	B
В	в	*в*	V
Г	г	*г*	G as in give
Д	д	*д*	D
Е	е	*е*	Ye as in yen
Ё	ё	*ё*	Yo as in yo-yo
Ж	ж	*ж*	Zh as in measure
З	з	*з*	Z as in doze
И	и	*и*	I as in ink
К	к	*к*	as English
Л	л	*л*	L
М	м	*м*	as English
Н	н	*н*	N
О	о	*о*	O as in yo-yo
П	п	*п*	P
Р	р	*р*	R
С	с	*с*	S as in bus
Т	т	*т*	as English T
У	у	*у*	U as in rude
Ф	ф	*ф*	F
Х	х	*х*	Kh as in Scots loch
Ц	ц	*х*	Ts as in Tsetse fly
Ч	ч	*ч*	Ch as in church
Ш	ш	*ш*	Sh as in push
Щ	щ	*щ*	Shch as in fish-chip
Ы	ы	*ы*	y as in y'know
	ъ	*ъ*	hard sign
	ь	*ь*	soft sign
Э	э	*э*	E as in men
Ю	ю	*ю*	Yu as in few
Я	я	*я*	Ya as in Yarmouth

Though Russian is phonetic (pronounced exactly as written) English is not, and there are countless problems. Many of the accepted usages could be improved upon. Tupolev would be more faithfully written Toopolyev, but this would be hard to accept. I have written Alexander in the more accurate form Aleksandr, and Maxim Gorky is much better in the form of Maksim Gorkii, but the OKB leader we call Tyrov actually begins with Ta and his Type 110 fighter is thus Ta-110 and not Ty-110. In any case Ty looks exactly like the Tupolev abbreviation written in Russian. Names such as Mikoyan inevitably get pronounced with the middle syllable rhyming with boy; it should rhyme with go. In the same way the Russian word for a glider cannot be written in English in any better way than *planer*, and when seen by an English-speaking reader this is automatically rhymed with strainer; the sound is actually more like plunn-air.

In the reverse direction Russian documents often contain letters which look Cyrillic but are actually Western designations; an example is the Clark YHC wing section, which in a Soviet document is indistinguishable from Russian letters U (oo), N and S.

ІНДЕКС

The main content of this book is in strict alphabetical order, the aircraft of each design bureau then being listed numerically or chronologically. For the convenience of readers in the Russian Federation the following is an index in Cyrillic characters:

AEROPLANES OF PRE-REVOLUTIONARY RUSSIA

PRE - 1918

AIS

Aviatsionnaya Ispitatelnaya Stantsiya, Morskaya Vedomstva, naval air test station, founded late 16 at Petrograd Poly Inst, with seaplane base at Krestovsky Island. Engineers P. A. Shishkov and Sushenkov planned many aircraft, two being built. Shishkov produced torpedo carrier of Farman pusher-biplane type, 130 hp Clerget, made several flights of over 1 h Aug 17. Aist (t added to give word meaning stork) two-seat seaplane with 150 hp Sunbeam and one fixed and one pivoted machine gun, complete autumn 17.

Anatra

Aeroplane factory founded in Odessa 13 by banker Artur Antonovich Anatra, based on naval workshop which built 20 to foreign design 09-12. Built numerous aircraft 13-17 to designs of Farman, Morane, Nieuport, Voisin, output 17 two per day.

Anade Design begun summer 15 of this tandem-seat tractor recon biplane. Wood with fabric covering except aluminium cowling and ply forward fuselage. Ailerons on upper wings only. Fuel tank divided into three with valves to seal off section with bullet hole or other damage. Normal engine as data, a few (called Anaclé) with Clerget; alternative name Dekan. Rear cockpit with pivoted gun, bomb load up to 30 kg (66 lb). First flight 19 Dec 15, resulting in numerous major changes prior to delivery of series aircraft from 16 May 16, 170 delivered.
DIMENSIONS Span 11,5 m (37 ft 8¾ in); length 7,7 m (25 ft 3⅛ in); wing area 35 m² (376·75 ft²).
ENGINE 100 hp Gnome Monosoupape.
WEIGHTS Empty 515 kg (1,135 lb); fuel/oil 90 kg+25 kg; loaded 865 kg (1,907 lb).
PERFORMANCE Max speed (SL) 132 km/h (82 mph); time to 3 km, 26 min; service ceiling 4 km (13,125 ft); duration 3·5 h; TO run 60 m, landing 90 m.

Anasal Broadly an Anade with different engine and slight reduction in dimensions. Water-cooled radial with cylinder heads protruding from aluminium cowl, radiator usually in front of centre section of upper wing, sometimes at sides of fuselage, latter being ply-covered throughout. One 0·303-in Vickers with Dekan interrupter gear, a second on pivot for observer. First flight 25 July 16, about 70 built, plus several **Anasal SS** with 160 hp Salmson, 153 km/h (95 mph).
DIMENSIONS Span 11,4 m (37 ft 4¾ in); length 8,1 m (26 ft 6⅞ in) (SS version 8,95 m); wing area 37 m² (398·3 ft²).
ENGINE 150 hp Salmson.
WEIGHTS Empty 814 kg (1,794 lb); fuel/oil 130 kg+23 kg; loaded 1164 kg (2,566 lb).
PERFORMANCE Max speed 144 km/h (89·5 mph); time to 3 km, 24 min; service ceiling 4,3 km (14,100 ft); duration 3·5 h; TO 75 m, landing 70 m.

Anamon Few details survive of this single-seat fighter with monocoque wooden fuselage, braced parasol wing and deflecting plates on propeller instead of synchronizing gear for single Vickers. Tested from 16 June 16, many criticisms and prototype later crashed.

Anatra D Four-seat biplane bomber, intended to fly to target on all three engines and return on nose engine only. Pusher nacelles each had gunner in front. Bombload 400 kg (882 lb). Grossly exceeded design weights, and prototype not repaired following damage on first flight 23 June 16.

Anadis Broadly an Anasal redesigned as wooden semi-monocoque, with 150 hp Hispano engine. Planned as fighter, built as two-seater (it was claimed) to enable French chief designer Dekan, and French pilot Robinet, to escape in event of revolution. Flown from 23 Oct 16, good results. Capt N. A. Makarov left on planned trip to Western Europe Nov 17 but force-landed near Iasi, Romania.

Anadva Also called Type VKh from designer V. N. Khioni; described under designer's entry on page 4.

Bezobrazov

Army officer Aleksandr A. Bezobrazov, assisted by F. E. Moska, built fighter 14 in Moscow with 60 hp Anzani, later 80 hp Gnome. Extremely slender triplane wings with much stagger. Flown by Moska 2 Oct 14, later flown at Sevastopol.

BIS

Simple pusher biplane completed April 10 by Bylinkin/Iordan/Sikorsky; 15 hp Anzani too weak, rebuilt into BIS-2, tractor 25 hp Anzani, flown by Sikorsky 3 June 10, longest 600 m/42 sec.

Anatra DS (Anasal)

Bezobrazov

Bylinkin

Fedor Ivanovich Bylinkin completed own monoplane July 10, Antoinette style, best 200 m; later biplane a failure.

Bylinkin

ChUR

ChUR No 1 got designation from G. G. Chechet, M. K. Ushkov and N. V. Rebikov. Basically good-looking tractor monoplane, but with variable-incidence short upper wing, with ailerons, and strange main wingtips split into upper/lower parts turned up/down with 90° between them. Even tail also had auxiliary pivoted horizontal upper surface. First flight at Komendantsky aerodrome about 10 April 12. Not repaired after heavy landing 5 July 12.

ChUR

Dokuchayev

Moscow flight instructor Aleksandr Yakovlevich Dokuchayev built six aeroplanes. First completed 10 was biplane with pusher 50 hp Anzani. Tested but possibly on ground only. No 2 built 12, sesquiplane version of Farman IV, 50 hp Gnome, said to have flown well. No 3 again Farman type, but equal-span tandem trainer with 80 hp Gnome, flown late 14. No 4 again Farman type, sesquiplane trainer, three rudders, flown late 15 on skis. No 5 neat monoplane, much in common with LYaM, 80 hp Gnome, flown spring 16. No 6 reverted to Farman layout, 60 hp Gnome, flown with and without front elevator.

Dooks

Often written Dux or Duks, large factory owned by Yu. A. Meller (sometimes written Möller) building French designs (licensed or copied). Blériot copy 11. Improved Farman VII in 12, 'Farman IX' built in series, Meller I vaguely like Blériot but Farman XV 2-seat nacelle, 100 hp Gnome and pusher prop behind pilot. Meller II (13) based on Farman XVI. Meller III monoplane (80 hp Salmson, chain drives to two tractor props) failure. Dooks No 2 another pusher monoplane, modified 15, failure. Final design with twin pusher Le Rhône, unfinished 17. Factory's real job was mass-production of successful Western types.

Dooks Blériot copy.

Dybovskii

Lt Viktor V. Dybovskii, famed pilot, built *Delfin* at Riga 13, strong Deperdussin influence, streamlined monoplane with monocoque veneer fuselage, but bad flyer. Same year produced improved Nieuport IV monoplane.

Engels

Yevgenii Robertovich Engels built *Orel* single-seat high-wing monoplane, tractor 80 hp Gnome, at Sevastopol 15. Wing like insect, five spars all met at pivoted root, pilot rotating wings for lateral control. Tested 16 to takeoff speed, but nobody liked to fly it, designer being then instructor at Baku. Fighter flying boat followed 16, with pusher 100 hp Gnome Mono, downswept tips serving as floats. Flown by designer 5 Dec 16. Navy ordered 60 called Engels III, 120 hp Le Rhône, two completed, one flown Aug 17, second tested 20 by S. A. Kochyerigin.

Grizodubov

Stepan Vasilyevich Grizodubov, at Kharkov, saw film of Wrights flying, produced: No 1, Wright copy, chains kept breaking; No 2, similar, modified to No 3; No 4, switched to Blériot XI type, 25 hp Anzani, flew well in 12. Made powered glider 40, no details.

Hackel

Jacob Modestovich Hackel (in Russian, Yakov Gakkel), electrical engineer, built nine aircraft and flew seven. Hackel I, or YaMG, 09, backstagger biplane with belt drive from 25 hp Antoinette to two props. Hackel II, 10, replaced interplane struts by mass of wires, mod into Hackel III which made second flight in Russia (beaten one day by Kudashyev) 24 May 10. Types IV, V, VI, VII, VIII, conventional tractor biplanes. Hackel IX, braced monoplane with aerodynamics based on German Taube. Dooks pilot bribed mechanic to add acid to cooling water to eliminate IX from major contest.

Dooks Meller I (gun may be dummy)

Hackel III

Dybovskii

Hackel IX

Engels III

Hackel VII

Grizodubov 1

Izhorskii

This large factory concentrated on giant bombers. Photo shows tunnel model of Type A biplane, span 40 m, length 18,5 m, payload 3,5 t, 6×220-hp Renault (4 tractor, 2 pusher) with pneu-powered flight controls; abandoned when half-built 25 May 18. Replaced by triplane project with 6 tractor engines on bottom wing, again never finished.

Izhorskii

Karpeka

Aleksandr Danilovich Karpeka built No 1 tractor biplane 10, flew up to 22 min, subsequently modified as 1 bis and 2 bis; Karpeka No 3 (13) was similar, all sound aircraft.

Karpeka 3

Kasyanenko

Brothers Evgenii Ivanovich, Ivan and Andrei Kasyanenko were students at KPI (Kiev Polytechnic Institute), who in 10-21 designed and built six increasingly original aeroplanes. No 1 was pusher biplane of Farman style, which despite Anzani of only 15 hp made several short flights at Cherkassy. Modified as 1 bis, front elevator removed and tail altered. No 3 (next) built at St Petersburg with left/right biplane wings pivoted so that incidence of each side changed independently. Oerlikon 50 hp water-cooled drove two tractor propellers via chains. Flight tested Oct 12. No 4, tested by Nesterov summer 13, deliberately low-powered monoplane, again with independent variable incidence on left/right wings; flown Kiev with 15 hp Anzani. No 5, built at KPI and tested late June 17, was unconventional biplane fighter, with variable-incidence left/right wings mounted on finely streamlined fuselage with nose gun and cockpit and 100 hp Gnome driving pusher 3-blade propeller (first in world?) behind protective cruciform tail.

Kasyanenko 5

Keburiya

Vissarion Savelyevich Keburiya built glider 09, later bought Blériot which he repeatedly rebuilt (with changes) after crashes. Built second monoplane 12, and No 3 in 13 similar to Morane but Nieuport-type undercarriage.

Khioni

Vassili Nikolayevich Khioni was born 1880 and qualified as engineer 05. In the First World War he was a lead designer at Anatra works at Odessa, assisted in several of that company's designs. He died in 30.

VKh Anadva Neat twin-engined twin-fuselage combat aircraft similar structurally to Khioni Anadye series but larger. Also named *Dvukhvostka* (twin-tail) and Anatra-Khioni No 4. Mainly wood, part ply covering especially on fuselages. Five seats, two pilots with dual control, two rear gunners (one nav/bomb, one observer) and third gunner in nacelle on upper wing. Small bombload. Completed 30 June 16 and flown 9 July 16. Next aircraft more powerful, see Anasalya below.

DIMENSIONS Span 19,0 m (62 ft 4 in); length 7,7 m (25 ft 3¼ in); wing area 62,0 m² (667 ft²).

ENGINES Two 100 hp Gnome Mono rotary.

WEIGHTS Empty and loaded not recorded; fuel/oil 160+45 kg; useful load 600 kg (1,323 lb).

PERFORMANCE Endurance 3 h; no other data.

VKh Anasalya Also called Anadva-Salmson, second VKh series with more powerful engines and minor airframe changes. First flight 5 May 17. Order for 50 for EVK (squadron of flying ships, then equipped exclusively with IM-series Sikorskis); October Revolution came before deliveries began.

DIMENSIONS Span 19,1 m (62 ft 8 in); length 8,1 m (26 ft 7 in); wing area 62,0 m² (667 ft²).

ENGINES Two 140 hp Salmson M9 water-cooled radials.

WEIGHTS Empty 1280 kg (2,822 lb); fuel/oil 180+30 kg; loaded 1930 kg (4,255 lb).

VKh Anasalya

PERFORMANCE Max speed 140 km/h (87 mph); climb 7,6 min to 1 km; service ceiling 4 km (13,120 ft); take-off and landing 60 m (200 ft).

Khioni No 4 Construction interrupted by Civil War, but by 21 Khioni had completed this improved VKh (not same aircraft as 'Anatra-Khioni No 4') with more powerful engines (water radiators on upper wing) and considerably greater fuel capacity. Built Odessa and successfully flown to Moscow where based at Central Aerodrome; damaged by taxiing Nieuport early 22 but repaired. From late 22 'several tens' of successful test flights by K. K. Artseulov and P. Stolyarov. In 23 handed to Serpukhov gunnery and bombing school.
DIMENSIONS As VKh Anasalya.
ENGINES Two 160 hp Salmson P9 water-cooled radials.
WEIGHTS Empty 1300 kg (2,866 lb); fuel/oil 410/40 kg; loaded 2160 kg (4,762 lb).
PERFORMANCE Max speed 140 km/h (87 mph); climb 7·7 min to 1 km; service ceiling 4,4 km (14,440 ft).

Khioni No 5 Named *Konyek-Gorbunok* (hunchback hobbyhorse), neat single-engined biplane built Odessa 23, at former Anatra plant redesignated Air Repair Factory 7. Tandem-seat dual trainer, tested at both Odessa and Moscow. Modified with tank of insecticide powder for anti-locust patrol (believed first ag-aircraft in world) and 30 delivered by 1928.
DIMENSIONS Span (upper) 11,46 m (37 ft 7¼ in); length 7,8 m (25 ft 7 in); wing area 37,0 m^2 (398 ft^2).
ENGINE One 100 hp Fiat 6-inline water-cooled.
WEIGHTS Empty 700 kg (1,543 lb); fuel/oil 110 kg; loaded 975 kg (2,149 lb).
PERFORMANCE Max speed 122 km/h (76 mph); climb 9·8 min to 1 km; service ceiling 3,5 km (11,480 ft); range 480 km (298 miles).

Kudashyev 1

Kolpakov

See main section of book.

Kostovich

Ogneslav Stefanovich Kostovich, originally Hungarian, was pioneer of Russian aircraft, demonstrating models of helicopter, aeroplane and even ornithopter as early as 79. Spent most time with dirigible *Rossiya*, and patented forms of plywood, but built 40 hp aeroplane 11, seaplane 14 and 100 hp amphibious monoplane 16, though at least two of these were never flown. He is today remembered chiefly for his remarkable *Rossiya* airship engine, preserved in Monino museum.

Kress

Vasilii Vasilyevich (Wilhelm) Kress worked in Vienna where around 00 he built ambitious tandem triplane twin-float seaplane. On test it capsized, and he was probably right to blame disaster on ponderous weight of 35 hp Daimler engine.

Kudashyev

Aleksandr Sergeyevich Kudashyev, Kiev professor, flew at Nice with M. N. Yefimov February 10, thereafter building four aeroplanes. Kudashyev 1 was at first glance vaguely like Voisin, but 35 hp Anzani drove tractor propeller. Other features: large ailerons on upper

Lebed VIII

Lebed XI

Lebed LM-1

wing and simple (modern type) tail. Made first flight of any Russian aircraft 23 May 10. Kudashyev 2 was larger, with 50 hp Gnome, flown autumn 10. Kudashyev 3 was monoplane of Blériot type, but with pilot seated to look under wing. Flew winter 10-11. Kudashyev 4 was improved 3 with 50 hp Gnome, even more like Blériot with pilot looking over wing; flew 2 April 11.

Lebed

Vladimir Aleksandrovich Lebed learned to fly in France and opened factory (Shavrov comments, twice burned down to collect insurance). Into production with Deperdussin, Voisin, FBA. Copied foreign types called Lebed I to VI. Lebed VII, copy Sopwith Tabloid, modified as VIII; IX was modified captured LVG. Two Lebed X biplane fighters built, followed by about 10 Lebed XI recon biplanes and no fewer than 225 Lebed XII plus 192 after Revolution. LM-1 float version of XI. XIII was 'high-speed biplane', XVI twin-engine recon-bomber, XVII improved recon biplane, XVIII biplane with 300 hp Fiat, XXI another two-seat recon but a monoplane, and XXIV yet another, halted by Allies stopping flow of Fiat and Hispano engines.

Lomonosov

Father-figure who left copious manuscripts and actually built (said to have flown) clockwork model coaxial helo 1754.

LYaM

Blériot-type monoplane by Maksim Germanovich Lerkhe, Georgi Viktorovich Yankovski and Italian F. E. Moska. With 50 hp Gnome set Russian height record 1775 m 14 May 12.

Meller, Möller

See Dooks.

Moscow Tech School

A. N. Tupolev, B. N. Yuryev and A. A. Komarov, pupils of legendary N. Ye. Zhukovskii (Joukovsky), founder of CAHI, collaborated to build ITU Blériot XI type, flown summer 11 (Tupolev exiled for revolutionary activity). V. L. Aleksandrov left monoplane unfinished. Nikolai Rodionovich Lobanov produced L-1 *Ptenets* (Fledgeling), Farman type, about 100 flights 12-13. A. V. Dukhovetskii and other students made and flew D-1 *Liliput* (8 hp) and D-2 *Malyi Muromets* (50 hp) 13.

Moska

Italian Francesca E. Moska became chief designer at Dooks, worked on LYaM and with Bezobrasov; then managed production of 75 Farman IV.

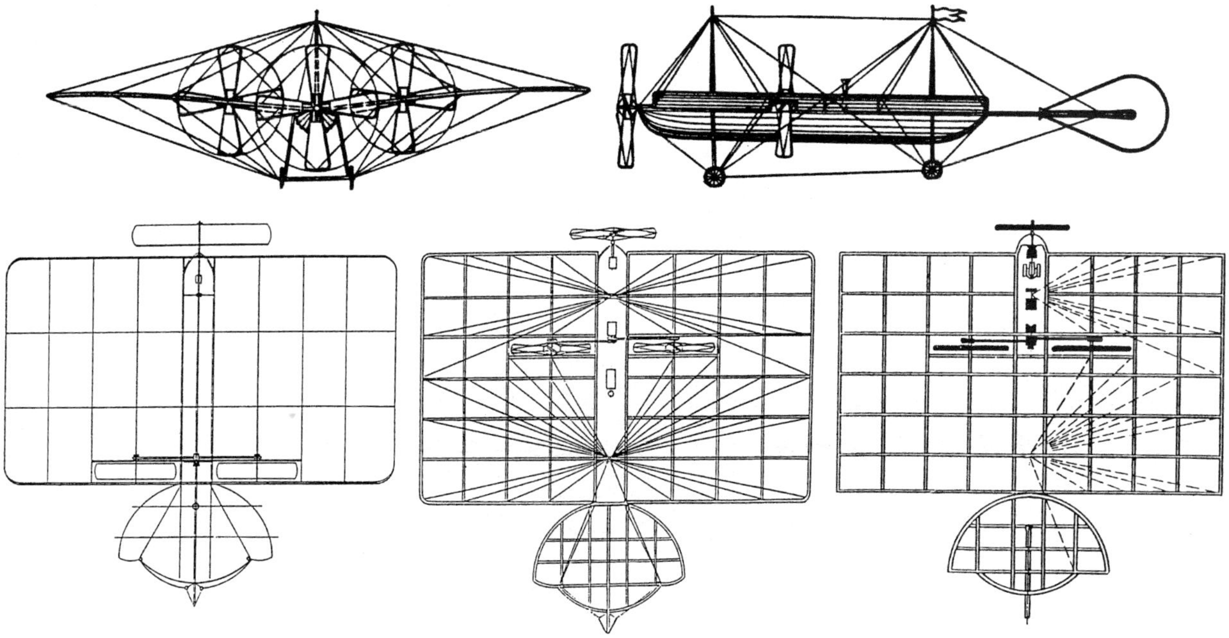

Mozhaiski 1884, left to right, patent drawing, by V. B. Shavrov (also front and side views), CAHI archive.

Mozhaiskii

Aleksandr Fedorovich Mozhaiskii was greatest Russian pioneer constructor of 19th Cent, actually getting as far as attempting ramp takeoff unknown date about Oct 84 (82 date discredited). Large monoplane – span 22,8 m, length 23 m – with 10 hp steam engine driving tractor propeller and 20 hp engine driving pusher screws on trailing edges of rectangular wing; controllable vert/horiz tail but no lateral control. Flew for a second or two but banked and hit lower wing on ground. Second aeroplane abandoned.

Nesterov

Pyetr Nikolayevich Nesterov was most famous Russian pilot pre-14, credited with first loop. Unlike contemporaries, recognised importance of banking in turns and invented (31 Aug 11) complete control system with cams to warp wing. Rebuilt Nieuport IV with this system, rudder being removed.

Olkhovskii

Vladimir Mikhailovich Olkhovskii, commander of 5th Aero Depot at Bryansk 16, built modified Nieuport, Voisin and Morane, *Torpedo* recon monoplane (Morane style) and Voisin-style biplane, incomplete 17. He was Polikarpov's assistant 1923-29.

Porokhovshchikov

Aleksandr Aleksandrovich Porokhovshchikov built monoplane followed by series of Caudron-derived pusher biplanes used as trainers by Imperial Air Service and Red Army and neighbouring countries.

Poro' No 1

No 1 Seeing this monoplane's fuselage was wire-braced bamboo rod, and engine 22 hp, remarkably successful dual trainer, long flights from 25 June 11.

No 2, Bi-Kok 50 hp pusher Gnome-powered sesquiplane with tail carried on twin booms; tandem dual nacelle, good performance Aug-Sept 14.

P-IV First flight 27 Feb 17. Pusher engine, nacelle with pupil seated ahead of instructor. Major production from May 17.
DIMENSIONS Span 10,2 m (lower wing 8,8); length 7,3 m; wing area 33,1 m^2.
ENGINE Usually 50 hp Gnome or 80 hp Le Rhône, but other rotaries also.
WEIGHTS Empty 340 kg typical; loaded around 605 kg.
PERFORMANCE Max speed (80 hp) 120 km/h; climb to 1 km, 9 min; service ceiling 3 km; endurance 4 hr; landing speed 60 km/h.

P-IVbis Similar to P-IV but part-cowled tractor engine and side-by-side seating. Produced in small numbers from February 1920, with major sections made by A. R. Rubenchik.
DIMENSIONS As P-IV.
ENGINE One 80 hp Le Rhône.
WEIGHTS Empty 398 kg; fuel/oil 72+24 kg; loaded 660 kg.
PERFORMANCE Max speed 112 km/h; climb 1 km in 9·5 min; ceiling 3 km; endurance 4·6 h; landing speed 60 km/h.

P-IV 2bis Basically P-IVbis with tandem seating as in P-IV.
DIMENSIONS As before but wing area 32,8 m^2.
ENGINE One 80 hp Le Rhône.
WEIGHTS Empty 400 kg; fuel/oil 72 kg+24 kg; loaded 662 kg.
PERFORMANCE Max speed 113 km/h; otherwise as P-IVbis.

P-V Two-seater based on Nieuport IV, advanced trainer with relatively high wing-loading, ailerons upper wings only, 80 hp Le Rhône.

P-VI Strengthened P-IV type with 110 hp Le Rhône (most common), 80 hp Le Rhône or 120

Poro' Bi-Kok

Poro' P-VI

RBVZ S-7

hp Anzani. Side-by-side seating. No ribs or covering in upper centre-section to give upward view, this series built 21. P-VIbis with covered centre-section, 40 built 23.

DIMENSIONS As before but wing area (VI) 32,0 m², (bis) 33,1.

ENGINE Usually 110 hp Le Rhône.

weights Empty 420 kg; fuel/oil 96 kg; loaded 682 kg.

PERFORMANCE Max speed 120 km/h; climb to 1 km, (VI) poor, (bis) 8·0 min; ceiling (VI) barely 200 m, (bis) 3,5 km; endurance 3·3 h; landing speed 62 km/h.

PTA

First Russian aero firm was Peterburgskii Tovarishchyestva Aviatsii (St Petersburg aviation company), which among other things built PTA No 1, based on Farman IV but made in easily disconnected sections to meet army requirement. Retained 50 hp Gnome, but seat in streamlined nacelle. Completed 26 Jan 11, flown by V. A. Lebed, 'not inferior to Farman'.

RBVZ

Russko-Baltiisky Vagon Zavod (Russo-Baltic wagon factory) was a major industrial combine. When M. V. Shidlovsky became chairman he moved newly formed Aviation Department from Riga to St Petersburg, to be near government contracts, and took on young Sikorsky (which see) as designer. He in turn imported friends from Kiev, including G. P. Adler, K. K. Ergant, M. F. Klimikseyev, A. S. Kudashyev and A. A. Serebrennikov.

S-6B Refined S-6A tractor biplane with pilot geared handle for starting 100 hp Argus, four wheels and other changes. Placed first in 12 military contest, but rejected by War Department.

S-7 July 12 tandem-seat monoplane, 70 hp Gnome, used in Bulgarian war.

S-8 Malyutka Small side-by-side biplane

trainer, 50 hp Gnome; Sikorsky made 90 min night flight to 1500 m 17 Sept 12.

S-9 Kruglyi Spring 13, mid-wing monoplane, 100 hp Gnome, glued monocoque fuselage, two-seat cockpit with single-seat cockpit behind.

S-10 Most in 13, 10 landplanes, 7 seaplanes, all derived from S-6B, some tandem, others side-by-side with 80 hp Gnome, 100 hp Argus or 125 hp Anzani. One seaplane landed normally after being inverted at c300 m.

S-11 Mid-wing side-by-side recon monoplane, 100 hp Gnome.

S-12 Lighter S-11, 80 hp Gnome, looped by G. V. Yankovsky Sept 13 and flown by him to 3860 m.

Russkii Vityaz *Russian Knight* authorised by RBVZ late 12 'for strategic recon', originally called *Grand* or *Bolshoi Baltiiskii.* Biggest aeroplane in world, two 100 hp Argus, two rudders, span upper wing 27 m (88 ft 7 in), loaded 3400 kg (c7,500 lb). Open bow for searchlight (and poss gun), cabin with double doors, glass windows, two pilots, camp stools, sofa, table, water closet (RBVZ passenger coach type), carpet and elec light. Flown by Sikorsky 2 March (15 March, Julian calendar) 13. Modified with two extra 100 hp Argus installed as pushers behind first, modified tail, flown from 10 May 13 (first 4-eng aircraft). Despite fear of asymmetry, Sikorsky again altered to have all four engines along wing as tractors, adding two more rudders. Gross weight 4200 kg, speed 90 km/h. Renamed *Russkii Vityaz,* flown from 23 July 13. On 2 Aug Sikorsky flew 7 pax 1 h 54 min.

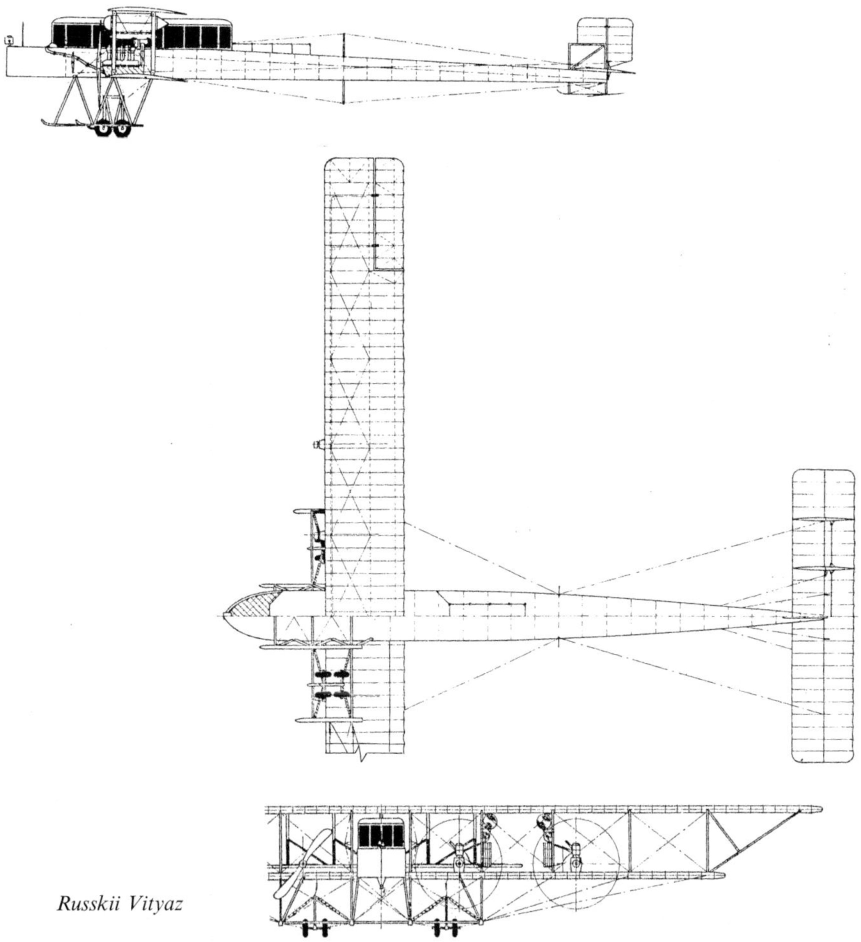

Russkii Vityaz

RBVZ Russkii Vityaz *(4 eng)*

RBVZ S-9

RBVZ S-XVI

RBVZ S-XIX

S-XVI Intended escort fighter for large bombers (see later), side-by-side biplane with synch gun, sometimes observer had aft-firing pivoted gun. RBVZ built 18 by March 16 and more later.

S-XVII Two-seat recon biplane, 150 hp Sunbeam, two sent to front 16.

S-XVIII Fighter biplane, two pusher Sunbeam, gunner in nose; engines said to be poor, two 80 hp Gnome added in tandem.

S-XIX Biplane with two 150 hp Sunbeam on centreline, cockpit in front of each tail boom.

S-XX Neat single-seat fighter, 100 hp Gnome Mono, later 120 hp Le Rhône, excellent but only five built from Sept 16.

Ilya Muromets Success of giant aeroplane led to July 13 contract for 10, all slightly different. Name from legendary hero of Kiev. Greatly refined, upper wing span 120 ft, wing area increased from 120 m^2 to 182 m^2, with same 400 hp (4 Argus) weight of prototype (RBVZ No 107) increased to 5100 kg, 5500 kg and ultimately 6300 kg. Main difference redesigned fuselage, world's first aircraft with enclosed cabin nose to tail. Very short glazed nose, open 'promenade deck' with handrail above rear fuselage. First flights with extra tandem wing on rear fuselage unsuccessful; with this wing removed start of proper testing 12 Dec 13; 12 men and dog carried 12 Feb 14. Flown with wheels, skis and (from spring 14) floats. Second IM, No 128, two 140 hp Argus, two 125 hp Argus, made epic flight to Kiev and back, one stop each direction, total time 30 h 30 min, overloaded with 4 crew, 940 kg fuel, 260 kg oil, 150 kg spares/luggage. Production continued and eventually totalled 80, at first for recon but very soon becoming strategic bombers bristling with guns. Dozens of variations, but basically Type B had Salmson radial engines, Type V (about 30) had Sunbeam (150 hp or 225 hp), Type G had any of six makes of 6-inline water-cooled and reaching up to 135 km/h, Type D

IM No 107 as built with tandem wing

had humpbacked shape with the four tanks faired in above fuselage, and best and final version was Type E (see data). Crew numbered up to six. Most bombs were small, but sizes to 410 kg were dropped. Defensive armament included up to six various Maxim, Lewis, Madsen, Browning and even infantry carbines. Used by EVK 'Squadron of Flying Ships' commanded by Shidlovsky who was made Maj-Gen; first combat mission 15 Feb 15. Only two shot down, one on 6 July 15 which force-landed and fell to pieces and the other on 12 Sept 16 after shooting down three German fighters. Survivors continued into 20s as Bolshevik trainers and transports. Data IM-E2 (Ye-2):

DIMENSIONS Span (upper wing) 34,5 m (113 ft 2¼ in); length 18,8 m (61 ft 8 in); wing area 220 m^2 (2,368 ft^2).

ENGINES Four 220 hp Renault.

WEIGHTS Empty 5 t (11,000 lb); fuel/oil 920 kg+130 kg; loaded 7460 kg (16,446 lb).

PERFORMANCE Max speed 137 km/h (85 mph); service ceiling 3,2 km (10,500 ft); range 560 km (348 miles); TO 450 m; landing 300 m.

Rudlitsky

Georgii Valeryevich Rudlitsky built biplane with forward elevator and pusher 45 hp Anzani, flew well 11. Possibly built further aircraft, with butterfly tail.

Savelyev

Vladimir Fedorovich Savelyev, mechanic at 2nd Aviapark 16, assisted by V. Zalevsky, used fuselage of Morane G in building quadruplane. Span increased progressively from bottom to top wing, flew well from 19 April 16 with 80 hp Gnome, later 100 hp Clerget. Did not build planned eight-wing passenger airliner, but did build and fly two-seat reconnaissance quadruplane 23, with 120 hp Le Rhône; reached 164 km/h (102 mph) on skis.

Shchyetinin

This large factory was managed by D. P. Grigorovich, whose career appears in main part of book under his name. Also see next.

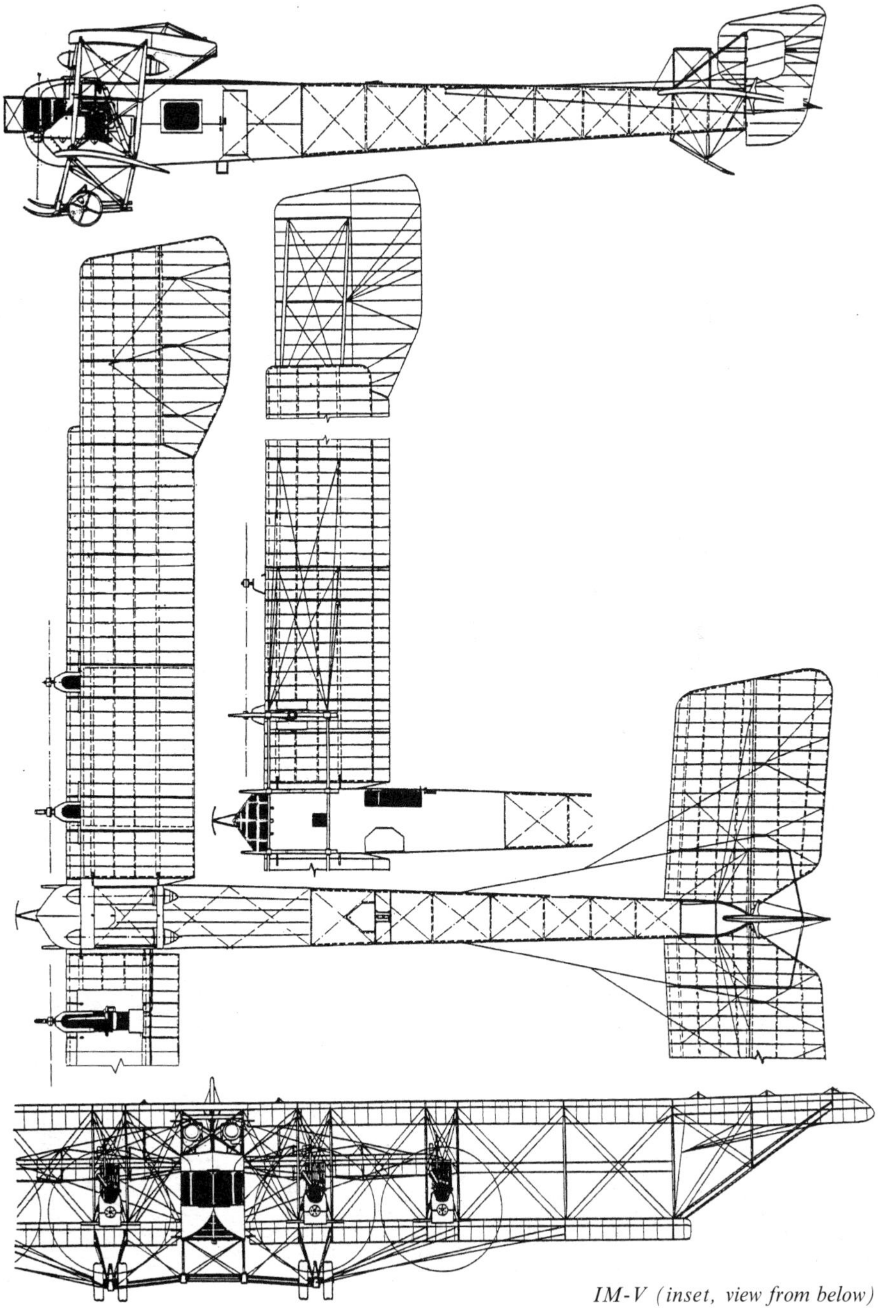

IM-V (inset, view from below)

First IM, No 107 rebuilt on skis

Savelyev

Shishmaryev

Mikhail Mikhailovich Shishmaryev headed construction bureau at Shchyetinin factory. In 16 began design of **GASN**, special-purpose hydroplane. Twin-float seaplane designed for torpedo attack. Observer gunner in nose, side-by-side pilots (one with torpedo sight) and gunner behind wings. First flight 24 Aug 17, three years of tinkering but nine others left incomplete. Two 220 hp Renault, 150 m^2 wing area, 1450 kg useful load, speed 110 km/h (68 mph).

Shkolin

Luka Vasilyevich Shkolin ran out of money when his 09 monoplane was almost finished. Anzani 25 hp had chain drive to two tractor propellers with inflight variable pitch.

Shiukov

Aleksei Vladimirovich Shiukov built and flew gliders 08-10 when student at Tiflis (Tbilisi), today capital of Georgia. Built powered Utka (canard, duck) Sept 12, tail-first monoplane with 50 hp Gnome, repeatedly modified, eventually flew well. Utka 2 (Kanar-2) 80 hp Gnome left incomplete 14 because of war.

Sikorsky

Igor Ivanovich Sikorsky was born in Kiev 89; naval academy St Petersburg, Kiev Poly Inst and engineering student Paris, returning to Kiev with 25 hp Anzani. May-July 09 built this into helicopter **S-1** with co-ax two-blade rotors; lifted but toppled over. **S-2** Improved helo with two three-blade rotors, also unsuccessful. Then followed two BIS (which see).

S-3 Simple tractor biplane derived from BIS-2, 35 hp Anzani. Short flights without turning from Nov 10.

S-4 Similar but 50 hp Anzani, built in one month (Nov 10) for Kiev customer, damaged in forced landing.

S-5 Greatly improved by heavier but more reliable 50 hp Argus, second seat added. Flown April 11, circuits from 17 May, soon set national records for speed (125 km/h), height (500 m) and distance (85 km).

S-5a Twin-float seaplane (unrelated to S-5) built for Navy late 12, 60 hp Gnome, very slender rear fuselage. Inadequate power, modified with central float and 80 hp Gnome, flew 13, accepted by Navy.

S-6 Tractor biplane, 100 hp Argus, flown Nov 11, two pax (Sikorsky once carried three) ahead of pilot.

S-6A Open lattice structure replaced by enclosed fuselage, flown March 12, outstanding aircraft, carried four adult pax at 106 km/h. Subsequent aircraft, see RBVZ.

Slesarev

Vasilii Andrianovich Slesarev, graduate engineer, worked with Zhukovskii, moving St Petersburg 13 to head Poly Inst aerodynamics lab. Improved Nieuport IV and Farman XVI, going on to build **Svyatogor** biplane with two

Shiukov

Sikorsky S-1

Sikorsky S-4 and S-5

Sikorsky S-6

Slesarev Svyatogor in 1916

Steglau 2

engines (300 hp Mercedes not available, 220 hp Renault installed, inadequate) in fuselage driving large pusher screws behind wing. Giant wheels, span 36 m, abandoned after working 14-23.

Slyusarenko

Vladimir Viktorovich Slyusarenko and his wife were both pioneer pilots. In World War I they organised factory at St Petersburg, building French types (total c170) and also making various modifications. Original designs: small fighter with monocoque fuselage. 130 hp Clerget, 17, and uncompleted biplane (G. P. Adler's scheme) with 3-seat nacelle and two Isotta-Fraschini engines (Italian, 100 hp each) in fronts of tail booms.

Steglau

Ivan Ivanovich Steglau built three aircraft at St Petersburg. No 1, biplane with open truss fuselage, 50 hp. No 2, great improvement, normal fuselage, 100 hp Argus, 130 km/h with two aboard. No 3 was high-wing monoplane, two tractor 80 hp Argus, side-by-side seats in front, cabin (unknown seating) filling space under wing; flown 14.

Sveshnikov

Aleksandr Nikolayevich Sveshnikov built monoplane in France, returned to Kiev and built No 2, another Blériot type with 40 hp Anzani, flown 13. He then brought his Sveshnikov-Vandom from France, reassembled it and flew it 14, another monoplane, same engine. No 3 reached 104 km/h (65 mph), considered a good performance on 75 hp.

Tereshchyenko

Fedor Fedorovich Tereshchyenko owned sugar factory in Kiev, built Blériot XI look-alike 09.

Tereshchyenko No 7

This led to four more single-seat monoplanes: Tereshchyenko-Zembinsky, 50 hp Gnome; and Nos 5, 5 bis and 6, all constructed by Austro-French A. de Pischoff, with 60 Gnome, 80 Gnome and 60 Le Rhône, all successful. In 16, with V. P. Grigoryev, built excellent No 7, 2-seat side-by-side recon biplane, 100 hp Gnome Mono.

Ufimstev

Anatolii Georgyevich Ufimstev was convinced wings should be circular discs. Spheroplan No 1, 09, also had circular tailplane, powered by 20 hp engine of Ufimstev design. No 2 was larger, with inventor's own 60 hp flat-6 engine, but equally unsuccessful. Several of his unconventional engines are preserved in Monino museum.

Vegener

Capt A. N. Vegener, director of Gatchina flying school, built training aircraft summer 11 based on Farman but with streamlined nacelle and with ability for instructor to engage or disengage pupil's controls, believed to be first such in world.

Villish

Aleksandr Yustoosovich Villish built *Severenaya lastochka* (northern swallow) in Revel (Tallin) 13; two-seat based on Taube, 60 hp Kalep, forced landing in sea 15 July 13. Two years later built **VM-1** small flying boat for Navy, 100 hp Gnome Mono, variable-incidence biplane wings, hit floating log after second flight and sank 24 Nov 15. Second flying boat, **VM-2**, 2-seat recon with same engine, built by F. Meltser's furniture works and also, with agreement, copied by V. A. Lebedev. Accepted by Navy, which showed more interest in **VM-4**, 110 hp Le Rhône, used at Baku as trainer. **VM-5** sesquiplane fighter flying boat, 130 hp Clerget, reached 170 km/h. **VM-6** was 'counter-fighter', streamlined landplane (single-seat monoplane, 200 hp Hispano) designed to land on water; construction started after Revolution, not completed.

Yur'yev

Boris Nikolayevich Yur'yev (or Yuriev) was pioneer of helicopter. Gold Medal 12 for his deep theoretical analyses, but first actual machine crippled by vibrating and unreliable Anzani engine. For subsequent work, see CAHI in main section.

Aircraft since 1918

AeroProgress

See ROS-AeroProgress.

Aerorik (Aero-RIK)

Privately financed Science and Prodn Enterprise at Nizhny-Novgorod (in 1920-90 called Gorkii, centre of defence production and closed city forbidden to visitors). Complete spectrum of research and design engineers led by Dr Viktor Morozov based at 'Sokol' plant which built MiG-29 and -31. Funding and partners sought.

Dingo So named because at home in outback regions, multirole amphibian with air-cushion landing gear designed to operate from any surface whatsoever, crossing 1 m (40 in) ditch or 0,3 m (12 in) obstacle. Configuration as 3-view, simple airframe except cushion contained by side bladders and front/rear flaps hinged up in flight. Body houses 2-seat cockpit (1 or 2 pilots), cabin for 7 pax or 844 kg (1,860 lb) cargo and rear compartment for cushion-fan engine and propulsion engine driving 3-blade prop. Mockup 91, first flight hoped Mar 94.
DIMENSIONS Span 14,5 m (47 ft 7 in); length 12,8 m (42 ft 0 in); wing area 24,3 m² (262 ft²).
ENGINES Propulsion, one 850-hp PT6A; cushion, one 250-shp TVA-200.
WEIGHTS Empty 2,16 t (4,760 lb); fuel 600 kg (1,323 lb); max 3,6 t (7,937 lb).
PERFORMANCE Max speed 309 km/h (192 mph); cruise 274 km/h (170 mph); service ceiling 3,35 km (11,000 ft); range with max payload 850 km (528 miles); landing 115 km/h (71 mph).

Aerorik Dingo

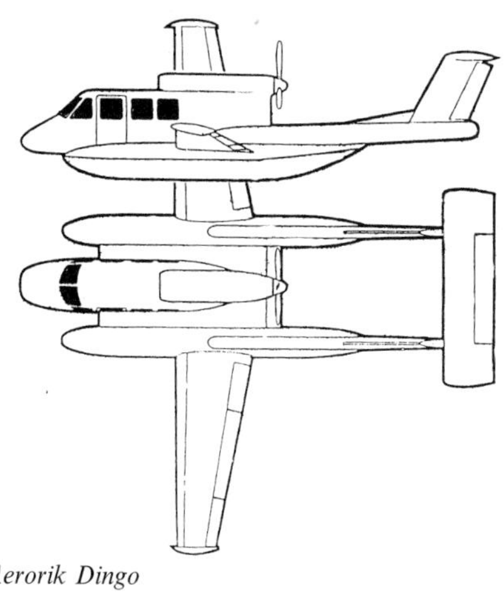

Aerorik Dingo

Alexandrov-Kalinin

AK-1 Apart from *Komta* the first transport aircraft to be designed in Soviet Union. Its need was obvious, especially after failure of *Komta* in 22. With full state backing, design was led at CAHI by V. L. Alexandrov and V. V. Kalinin, assisted by A. M. Cheremukhin. Rather unusually for period, design job was done properly with at least one model tunnel-tested, others dropped from high towers and every part of primary structure stressed. Engine was from wartime imported Voisin LBS, and structure was mixed, with a 12%-thick wooden wing above steel-tube fuselage. Engine completely enclosed, with curved exhaust pipe ahead of open cockpit for pilot and mechanic, and a Lamblin radiator on each side. Up to three passenger seats could be installed in cabin, door being on right. Designated from initials of designers, in style not officially adopted until 41, AK-1 was built at GAZ-5 (and was thus a rival to an aircraft designated for that factory and built there at same time) and first flown by A. I. Tomashevsky on 8 March 24. Results were encouraging, and on 15 June AK-1 was taken on strength of Dobrolyet and assigned to route Moscow-Novgorod-Kazan. Last heard of when it took part in flight to Pekin 10 June to 17 July 25.
ENGINE One 160 hp Salmson (Canton-Unné) RB 9 nine-cylinder water-cooled radial.
DIMENSIONS Span 14,9 m (48 ft 10½ in); length 11,0 m (36 ft 1 in); wing area 37 m² (398 ft²).
WEIGHTS Empty 1145 kg (2,524 lb); fuel/oil 197 + 24 kg (434 + 53 lb); maximum 1685 kg (3,715 lb).
PERFORMANCE Max cruising speed 147 km/h (91 mph); time to 1000 m, 16 min; ceiling 2200 m (7,200 ft).

Alekseyev

Semyon Mikhailovich Alekseyev was lead designer in Lavochkin OKB from 1942 onwards, and was chiefly responsible for La-5 and La-7 detail design. After the Second World War he transferred from Lavochkin bureau and in 1946 was successful in establishing his own OKB to concentrate on high-performance combat aircraft. See OKB-1. He is unrelated to Dr R. Y. Alekseyev.

I-211 First assignment of OKB to produce fighter powered by two TR-1 turbojets (first significant product of Lyul'ka). Ordered by minister A. I. Shakhurin, OKB assisted by S. A. Christianovich and G. P. Svishchyev of CAHI. Prime requirement, heavy armament of three N-37 each with 30 rounds. Alternative four, preferably six, NS-23. Provision in another variant for two 57 mm 'or even 75 mm'. In addition 1 tonne of bombs, cameras or other mission equipment. Unswept wing 1-S10-11 profile, slotted flaps inboard and outboard of nacelles, 20° takeoff, 45° landing, slotted outboard ailerons. High tailplane with elevators dividing rudder into upper/lower sections. Tricycle landing gear each single-strut leg having

twin wheels, main wheels housed in fuselage. On each side of rear fuselage hydraulic air-brake, primarily for dive bombing. Engines centred on wings with spar continuity by banjo rings. Pilot could open sliding canopy manually before landing. OKB chief pilot A. A. Popov began test programme of red prototype 13 October 47. By this time effort switched to derivatives with more powerful engines.
DIMENSIONS Span 12,25 m (40 ft 2¼ in); length 11,54 m (37 ft 10¼ in); wing area 25 m² (269·1 ft²).
ENGINES Two TR-1, replaced before first flight by TR-1A.
WEIGHTS Empty 4360 kg (9,612 lb); fuel 2 t (4,409 lb); disposable 3040 kg (6,702 lb); normal loaded 7400 kg (16,314 lb).
PERFORMANCE Maximum speed 950 km/h (590 mph) at SL, 910 km/h (565 mph) at 4 km; climb to 5 km in 3 min; service ceiling 13,6 km (44,620 ft); range 1550 km (963 miles, with supplementary fuel 1800 km (1,118 miles); TO 800 m (2,625 ft); landing 900 m (2,953 ft).

AK-1

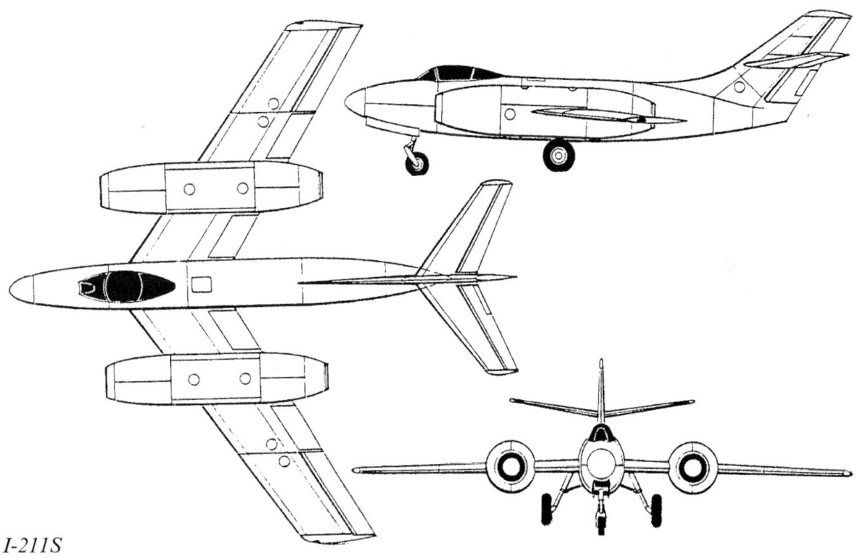

I-211S

I-212 During 47 OKB-21 worked three shifts on totally new and larger fighter powered by two Nene engines. Requirement was for night and all-weather interceptor with crew of two; also sent to Antonov, Lavochkin and Yakovlev. Most challenging, especially demand for long range (2300 km normal, 3100 km with drop tanks). All basic features as 211, apart from engines, size and equipment. Engines ahead of wing box, jetpipe passing through relatively small banjo rings. Wing centre section sharply tapered, outer panels increased in span and ailerons divided into inner/outer sections. Flaps said to be in six sections, angles as on 211. Tail unit enlarged and swept at 45°, with surfaces power boosted. Landing gear stronger version of 211 but with single nosewheel and pneumatic retraction. In nose, Torii-1 interception radar, managed by backseater, who could turn to face rear and manage rear barbette with twin B-20 (L+K '23 mm' incorrect) aimed within 70° cone. Front guns, one NS-37 with 75 rounds plus two NS-23 each with 200. A. A. Popov opened taxi test programme on single prototype 30 June 48, by this time with RD-45 engines. Data for this. According to Sultanov VK-1A engines subsequently fitted. Consensus is that 212 never flew. **I-214** with tail turret removed and replaced by tail radar (increasing fuselage length from 11 570 to 12 150 mm) and with heavier front armament, believed not built. Reason for lack of progress unknown, though Yakovlev's partisan description of 212 as 'another copy of the Me 262' may have hastened closure of OKB.
DIMENSIONS Span 16,2 m (53 ft 1¾ in); length 13,08 m (42 ft 11 in); wing area 33,0 m² (355 ft²).
ENGINES Two 2230 kg (4,916 lb) RD-45.
WEIGHTS Empty 5130 kg (11,310 lb); fuel 4100 kg (9,039 lb): normal loaded 9250 kg, maximum 10 500 kg (23,148 lb).
PERFORMANCE Maximum speed 1000 km/h (621 mph) at SL, 956 at 8 km; time to 5 km 2,8 min; service ceiling 14,8 km (48,560 ft); range (normal) 2300 km, (max with drop tanks) 3100 km (1,926 miles); TO run 910 m (2,990 ft); landing 565 m (1,854 ft).

I-215 Immaturity of TR-1A engine prompted switch to RR Derwent, made available late 1947 as RD-500. Aircraft little changed from I-211, except for new engines and addition of radar (Torii-1 intended but not fitted) and armament choices of: three NS-37 with 30 rounds, two NS-57 with 35 rounds or two 113P with 35 rounds. Aircraft planned to be capable of quick conversion as bomber (dive bomber was intended) using 211-type fuselage airbrakes. Unlike 211, fitted with ejection seat and pressurized cockpit. Two prototypes, first with conventional landing gear and second **I-215D**, with bicycle gear as on Yak-25. Flight testing began Dec 47, initially by Popov, later joined by S. Anokhin, M. Gallai, I. Ivashchyenko and I. Fedorov; Shavrov adds A. A. Yeffimov.
DIMENSIONS Span 12,25 m (40 ft 2¼ in); length 11,54 m (37 ft 10⅜ in); wing area 25 m² (296 ft²).
ENGINES Two imported RR Derwent 5.
WEIGHTS Empty 3995 kg (8,807 lb); loaded 6890 kg (15,190 lb).
PERFORMANCE Max speed (SL) 970 km/h (603 mph), (6 km) 960 km/h (597 mph); time to 5 km 2·7 min; range 2300 km (1,429 miles).

218 Sometimes (apparently in error) called I-218, this was last prototype built. *Shturmovik*, a heavily armed and armoured attacker for close support and anti-armour use in a land war. Pusher AV-28 contraprop with two three-blade units of 3,6 m (142 in) diameter. Armoured cockpit with flat glass slabs giving good view, while backseater faced rear and aimed two defensive barbettes mounted on tail booms, with arc of fire 25° above and below horizontal and out to 50° laterally. These were to be NR-23s with 120 rounds each, and there were various schemes for heavy forward-firing armament such as four NR-23 (150 rounds each), two N-37 (40 each) and two N-57 (30 each). Wing racks were to be provided for 16 rockets, and pylons under fuselage were to carry up to 1500 kg (3,307 lb) of bombs, presumably aimed in dive attacks. Highly stressed or armoured parts were of a new nickel-steel, 30 KhGSNA. Prototype was flown 1948, but no further history known.
DIMENSIONS Span 16,43 m (53 ft 10¾ in); length 13,88 m (45 ft 6½ in); wing area 45 m² (484·4 ft²).
ENGINE One Dobrynin VD-251.
WEIGHTS Empty not known; loaded 9000 kg (19,840 lb).
PERFORMANCE Maximum speed 520 km/h (323 mph) at sea level, 530 km/h (329 mph) at 2 km;

I-215

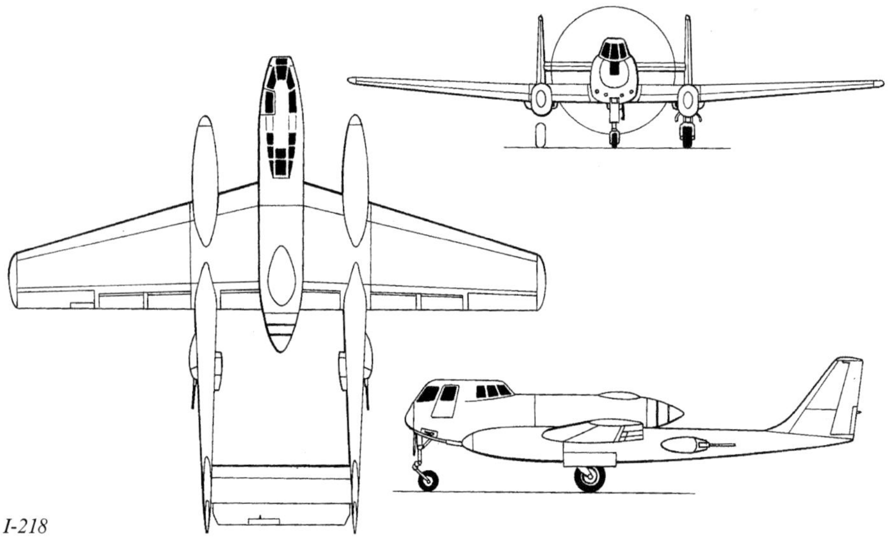

I-218

Alfa

Joint-stock firm formed in Moscow 92, English rendering 'Alpha Airlines'. Project announced at Aero Salon Sept 93.

SM-92 Finist Transliteration of Russian, = Finest. Multirole utility, seats 1 + 6 or 600 kg cargo, wheel/ski option and every imaginable role eqpt; Mühlbauer assisted devt of 3-blade prop; production to start at Smolyensk (former GAZ-35) late 93. Photo in Addenda first page.
DIMENSIONS Span 14,6 m (47 ft 10¾ in); length 9,130 m (29 ft 11½ in).
ENGINE One M-14P.
WEIGHTS Empty 1350 kg (2,976 lb); normal loaded 1950 kg (4,299 lb), MTO 2,2 t (4,850 lb).
PERFORMANCE (est) Max speed 290 km/h (180 mph); cruise 230 km/h (143 mph) at up to 3 km; range (max payload) 820 km (510 miles), max fuel 1000 km; field length 400 m.

time to 5000 m, 5·0 min; ceiling 6600 m (21,650 ft); range 1200 km (746 miles).

Projects Final design studies included: **I-211S**, swept wings and tail, short forward-retracting nose gear; **I-213**, two Nene, heavier (10,95 t) long-range I-212; **I-214**, same but tail barbette replaced by radar, in nose two N-37 (40 rounds) and one NS-57 (20 rounds); **I-216**, I-211 with two Charnko 76 mm guns (used on many armoured vehicles) each 15 rounds; **I-217**, new design, two TR-2 engines, swept-forward wing 14,2 m, length 13,85 m, area 31 m², 8550 kg, three N-37 each 35 rounds; **I-219/I-218Ib**, enlarged and refined I-218; **I-220/I-218-III**, new design, VD-4 engine, span 18,4 m, length 17 m, 55 m², 12 t, 2 × NS-23 (120 rds) nose, 8 × OPO-132 rockets, 2 × NS-23 (150 rds) firing aft; **I-221/I-218-II**, two TR-3 jets, span 17,4 m, length 16,5 m, 50 m², 13 t, 735 km/h, armed as 220 except one gun firing aft (from cockpit).

GST

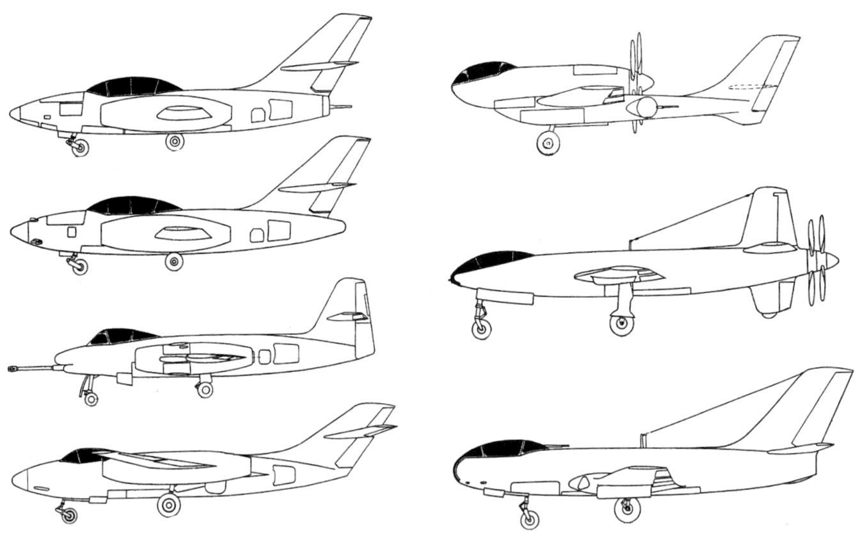

From the top: I-213, I-214, I-216, I-217, (second column) I-219, I-220, I-221.

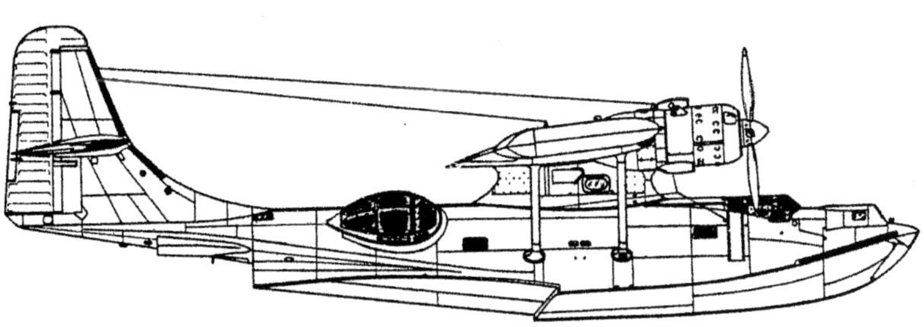

KM-2

Amtorg

Throughout 1930s Soviet Union's vast coastal, river and water areas required good long-range flying boats, but except for small MBR-2 all Soviet marine aircraft were disappointing. By 1936 it was clear the best foreign buy was a Consolidated 28 (PBY). This was released for export in 1938, and a manufacturing license was concluded by Amtorg, Soviet state department responsible for licensing and importing US products. Three complete aircraft imported to approximately PBY-1 standard with 880 hp R-1830-64 engines but without radio or armament. Great difficulty in producing Soviet version **GST** (hydro aircraft transport), though main customer was MA naval aviation. Surprisingly short production run ending late 40, most with 2/3 UBT and underwing racks for up to3 t max. Glavsevmorput purchased roughly equal number as **MP-7** with later engine, up to 20 (some, 25) passengers or cargo. Small number completed as **KM-2** with much greater power and capability. USSR also received 138 PBN-1 and 48 PBY-6A under Lend-Lease. ASCC name 'Mop'.

DIMENSIONS Span 31,72 m (104 ft 1 in); length 20,68 m (67 ft 10¼ in); wing area 130,14 m^2 (1,401 ft^2).

ENGINES (GST) two M-87 or 87A, (MP-7) M-62IR, (KM-2) ASh-82FN.

WEIGHTS Empty (GST) 5580 kg (12,300 lb), (MP) 5130 kg; max (GST) 11,8 t (26,015 lb), (MP) 12,25 t (27,006 lb), (KM) 17,17 t (37,853 lb).

PERFORMANCE Max speed at 3 km (GST) 329 km/h (204 mph), (MP) 277 km/h (172 mph); range with payload (GST) 2660 km (1,653 miles), (MP) 2800 km (1740 miles).

Antonov

Oleg Konstantinovich Antonov was born on 7 Feb 06 in Moscow. In 1921 he helped organize a branch in Saratov of the Moscow Glider Group, and two years later submitted a design in a light-aeroplane competition organized by K. K. Artseulov. In 1924 he built his first aircraft, the *Golub* (Pigeon) glider, which took part in the second national championships at Koktebel that year. Antonov became a student at the Leningrad Polytechnic Institute in 1926, and he continued glider design with the OKA-2 of which several were built at Saratov in 26-27.

In 1930 Antonov graduated, and was appointed first chief engineer and then chief designer at the newly founded Moscow Glider Factory. Many successful gliders were built here, as well as Antonov's first powered aircraft, the OKA-33 (LEM-2). In 1938 Antonov's application to the Jukovsky academy was refused. He then joined Yakovlev to assist on light aircraft. In 1939 he was detached to Saratov again and assigned the development of a STOL aircraft in the class of the German Fi 156. At about the same time his group began the design of one of the first troop-carrying gliders, the A-7. In 1943 Antonov had to rejoin the Yak OKB, as first deputy designer, and saw the war out on Yak fighters at Novosibirsk.

In 1945 he began recruiting people for his own OKB. This was established at Novosibirsk on 31 May 46. He then moved to Kiev. This large establishment, built from scratch, became one of the best-equipped OKBs in the country, and from here came the An-2 and an impressive series of high-wing transports. Production of the latter was centred in eastern Siberia at Ulan-Ude, and even more distant Arsenyev, though limited production took place at Kiev. The Saratov group continued as the leading organization on sailplanes. Many were RF (Red Front) series, last of 36 types being all-metal A-15, 65 (for details see *ASU*).

O. K. Antonov, unusual in bygone years in eagerness to assist Western writers, died 4 April 84. OKB now ANTKI (aviation-scientific tech complex named for) O. K. Antonov, led by General Constructor Pyotr V. Balabuyev, deputy Anatoli G. Bulanenko and asst chief Ye. D. Goloborojko. An-2 produced in China by Harbin and Nanchang, An-2 and -28 at PZL Mielec, Poland, An-32 at Kiev, An-72/74 at Kharkov, An-124 at Ulyanovsk. OKB at Kiev working on An-38, -70/77, -140, -218 and planned -418. Association formed to build 72/74 at Arsenyev/Omsk/Kharkov, with Zaporozhye engine factory. Second association links OKB with giant Yuzhtyazhmash former rocket plant at Dniepropetrovsk to build An-218.

A-7 Also designated RF-8, because it was effectively a greatly scaled-up member of the RF family, the A-7 was the winner in Dec 40 of a competition for a military assault glider. All-wood, with a fabric-covered wing (aft of the main spar) and control surfaces, it was remarkably refined, with the wing set into the top of the streamlined fuselsage, glazed nose cockpit, two sets of double doors (forward on the left, aft on the right), main wheels retracted into recesses by handcranks, and ground-adjustable trim tabs on the rudder and both elevators. Normal load was six soldiers. Production from May 41, 400 built. Some saw action behind the German lines, towed by SB-2s.

DIMENSIONS Span 19,0 m (62 ft 4 in); length 11,5 m (37 ft 8¾ in).

WEIGHT Laden 1100 kg (2,425 lb).

SPEED On tow, 200 km/h (124 mph).

KT *Kr'lya Tanka* (flying tank). First considered 1932, task was to see if it was feasible to fit wings to a light tank (the T-60 was the choice, though the T-26 was more numerous) and tow it like a glider, casting off at its destination for a free landing. The tank's own tracks were, if possible, to serve as the undercarriage. The T-60 was the first tank designed specifically for airborne forces, and it was hoped to tow them to partisan units behind enemy lines. (There was even a design for doing this with the T-34, using two ANT-20bis as tugs!) Antonov's creation, with bureau designation A-40, and also known as the A-T, comprised biplane wings and a twin-boomed tail, made of wood with ply and fabric covering. Both wings had ailerons and 45° slotted flaps, and the flight controls

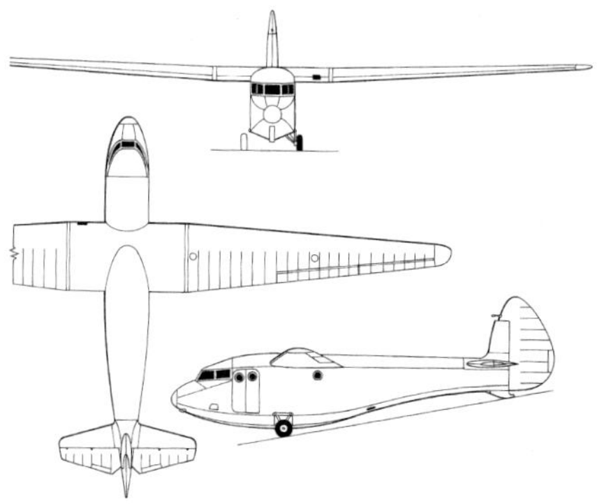

A-7

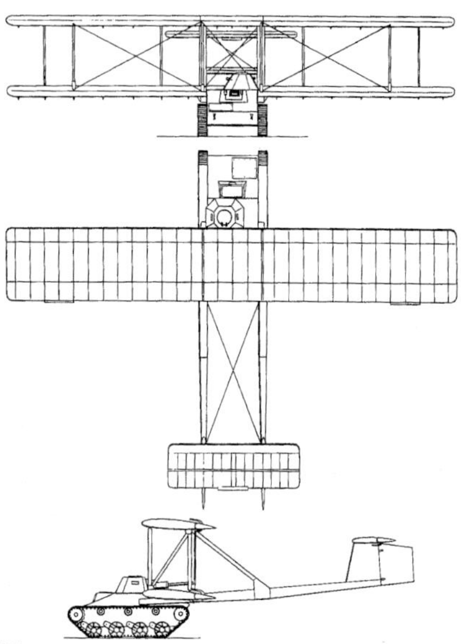

KT

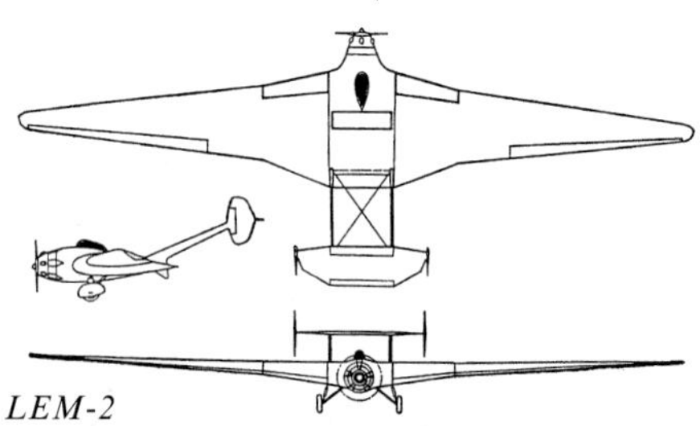

LEM-2

were worked from the driving position of the tank. The wing/tail unit was to be jettisoned, by one lever, immediately on touchdown when the wings were still lifting. As for the tracks, there was argument about whether to declutch them or give the tank a special overdrive top gear to accelerate to take-off speed 160 km/h (99 mph). S. N. Anokhin took quick course in tank driving and made first flight towed by TB-3 early 42. Rough landing but tank driven away afterwards. No production.

DIMENSIONS Span 18,0 m (59 ft 0¾ in); length of air portion 12,06 m (39 ft 6¾ in); wing area 85,8 m^2 (923·5 ft^2).

WEIGHTS Air portion 2004 kg (4,418 lb); T-60 tank 5800 kg (12,787 lb); combination 7804 kg (17,205 lb).

LEM-2 With bureau designation OKA-33, this was Antonov's first powered aircraft. It was built with Osoaviakhim funds in 1936 to create an efficient light transport for isolated communities. Called a 'powered glider', it was a completely new design, almost an all-wing aircraft but with two short booms carrying the tail. The key to efficiency was the wing, using Prof V. N. Belyaev's aerodynamics to obtain a larger inner section (centreline chord 6,7 m, 22 ft) and slender outer panels. Aerofoil section was PZ-2, derived from CAHI R-11, construction being all-wood with ply skin. The engine had a long-chord NACA cowl and drove a U-2 (Po-2) propeller. Up to 1280 kg (2,822 lb) of cargo could be loaded into the interior of the inner wing. N. I. Ferosyev made the first flight on 20 April 37. Though described as first efficient Soviet aeroplane, soon abandoned.

DIMENSIONS Span 27,6 m (90 ft 6½ in); length 10,6 m (34 ft 9⅓ in); wing area 81,4 m^2 (876 ft^2).
ENGINE One 100 hp M-11 radial.
WEIGHTS Empty 1640 kg (3,616 lb); loaded 2920 kg (6,437 lb).
PERFORMANCE Max speed 117 km/h (73 mph); cruising speed 100 km/h (62 mph); service ceiling 1500 m (4,920 ft).

OKA-38 After non-aggression pact with Germany in summer 39 several German aircraft were supplied. The Fieseler Fi 156C Storch light STOL aircraft was obviously relevant to Soviet requirements. Antonov was assigned the task of producing a Soviet equivalent, or making an exact copy. Antonov chose the latter, and though the welded steel-tube fuselage was unfamilar to his bureau it was copied exactly, the wooden wings presenting no problem. Dural slats, slotted ailerons drooping 15° and flaps to 40°. German engine replaced by French Renault copy, much longer, and wing fuel tanks enlarged. Reqt was for two versions, **SS** 3-seat liaison and **N-2** casevac with two stretchers loaded via door on left 2,3 × 1,1, plus seat for medic. Prototype built Kaunas, Estonia, flown Nov 40. Heavier than Fi 156, nearly three times field length needed, but longer range and 42 kg more load. Series production beginning as Germans overran factory.

DIMENSIONS Span 14,28 m (46 ft 10¼ in); length 10,3 m (33 ft 9½ in); wing area 26,0 m^2 (279·9 ft^2).
ENGINE One MV-6.
WEIGHTS Empty 980 kg (2,160 lb); loaded 1343 kg (2,961 lb).
PERFORMANCE Max speed 173 km/h (108 mph); climb to 1 km, 5·5 min; service ceiling 4,4 km (14,435 ft); range 514 km (319 miles); TO 144 m; ldg 63 km/h, 160 m.

SKh, An-2 Appearance of biplane in 47 widely misread; instead of being obsolete anachronism it led to record post-war prodn exceeding 18,500 so far! First task assigned to re-formed OKB was to replace Po-2, reqt issued May 46 by Ministry of Agriculture and Forestry. Studies began with a larger version of the OKA-38 with ASh-21 engine, but the compactness of the biplane carried the day. In the Soviet environment it did not occur to Antonov that others might think a biplane passé; it was the smallest and lightest solution to the task, which as a result of long discussion grew to a machine much larger and more capable than Po-2.

High-tensile (30KhGSA) steel welded forward fuselage carrying engine and landing gear, rest of the fuselage and fin stressed-skin, to give greater capacity and flexibility in providing door, window and other openings. Ruling interior section was fixed at 1,6 by 1,8 m (63 in wide, 70·9 in high), so that in the *Passarzhirski* role there is room for four rows of three seats, with an offset aisle. The wings, of R-11 14% profile, light alloy with I-spars at 15 and 60% chord, skin being stressed light alloy back to the front spar and fabric elsewhere. Upper wings have full-span slats. Both upper and lower wings electrically driven slotted flaps 45°, upper wings also having drooping ailerons. Flight controls manual, rudder, left aileron and elevator electric trim-tabs.

Two prototypes designated **SKh-1** (rural economy 1), first having the ASh-21 and second ASh-62IR. In each case the propeller was the V-509A-D7 with four scimitar-shaped blades of 3,6 m (141·7 in) diameter and large round spinner. Six fabric tanks in the upper wing housed 1200 lit (264 gal) of fuel (91 octane). The engine was fitted with a 27V generator and an air compressor for charging a bottle in the rear fuselage at 49 bars (711 lb/in^2) for pressurizing the main landing-gear oleos according to terrain, and for operating the tailwheel lock; reduced to 9·8 bars (142 lb/in^2) it energizes the wheel brakes. Normal equipment includes blind-flying instruments, gyrocompass, radio compass, two radios, radioaltimeter and intercom. The side-by-side cockpit can have dual control, and its side windows oversail the fuselage to give view downwards. The centre and left windscreens are de-iced electrically, and the right panel uses engine-heated air which also heats the cabin, there is no airframe ice protection.

The first SKh-1, with ASh-21 engine, was flown on 31 Aug 47 by N. P. Volodin. With this engine the payload was limited to 1300 kg (2,866 lb), but with the ASh-621R it could be raised to 2140 kg (4,718 lb) in the first production version, the **An-2T** (*Transportny*), available from October 1948. The 2T remained in production at Kiev as a standard utility machine for Aeroflot and armed forces, supplemented from 1949 by the **An-2TP** for Aeroflot scheduled routes, with comfortable furnishing for 12 passengers, the **An-2P** with up to 14 lightweight seats in a soundproofed cabin, the **An-2S** ambulance version with six stretchers (three super-imposed on each side of the aisle and an attendant) and the **An-2TD** for parachute training and operational parachute missions seating 12 and with special equipment. The first agricultural version was the **An-2Skh** with a 1400 lit (308 gal) copper tank in the fuselage for liquids or solids distributed via a spray-bar system on six pylons under the lower wing or a 300 mm (11·8 in) diameter centreline duct. The eight-blade fan driving the pump was later replaced by a VD-10 with four variable-pitch blades. Operating at 155 km/h (96 mph) with swath width from 60 to 100 m (200-330 ft) the first SKh evaluated by G. I. Lysenko in 1948 on a collective farm near Kiev completed the monthly work quota in three days. In 1948 aerial pesticides were put down on 2·1 million

OKA-38

An-2T

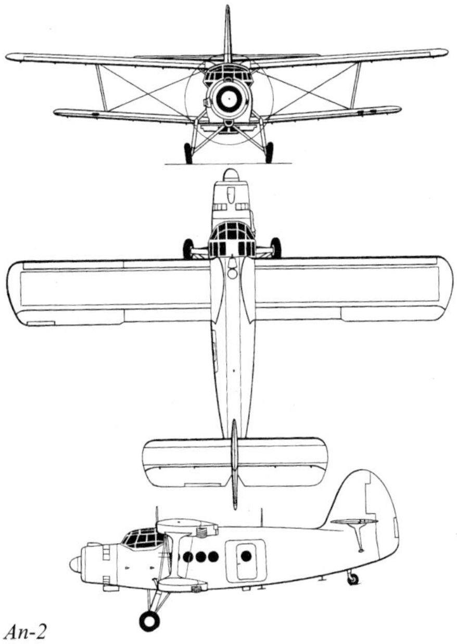

An-2

hectares, but by 1963 the An-2SKh was annually doing nearly 26 million. A hectare, 10,000 m^2, is 2·47 acres.

Normally the SKh variant has a long-stroke landing gear, and like other versions can have two types of ski. In 1949 a seaplane, the **An-4**, was flown and given service designation **An-2V**. The dural floats reduce speed but according to OKB data do not change the empty weight. Most 2Vs are utility transports but in 1961 a water-bomber variant, the **An-2PP** (forest protection), was developed with pilot-activated sluices in the floats through which 1260 litres (277 gal) can be rammed in five seconds' taxiing, and then dumped over a fire. An earlier firefighting model, the **An-2L**, was a landplane carrying water in a fuselage drum or extinguishing chemicals dropped in frangible underwing bottles.

Two unusual early models were the **An-6** and **An-2F**. The former was a high-altitude model with ASh-62IR supercharged by a TK-19 turbo to maintain 850 hp to 9500 m (31,170 ft). This was flown by V. A. Didyenko on 21 March 48 in the *Meteo* form subsequently designated **An-2ZA** (atmosphere sampling) with a heated compartment for a scientific observer faired into the fin. The **An-2F** *Fedya* was completely new aft of the wings, with a glazed rear fuselage tapering to a narrow boom precariously carrying a twin-finned tail. A dorsal cupola mounted a 12·7-mm UBT or 23-mm NS-23 cannon, aimed by hand. A small batch was evaluated in 1955 against unknown Su and Yak rivals, but the requirement was cancelled. An alternative designation was **An-2NRK** (night artillery correction), shortened to **An-2K**. Another oddball not given a designation was the special-chassis An-2 with six low-pressure tyres on the main gears for use from marsh, snow and sand. A field kit added the pair of extra sprung wheels ahead of and behind each existing main wheel, while the tailwheel was given a ski-like surround.

In 1959 general-aviation aircraft were assigned to Poland, and the last of about 5450 An-2s built in the Soviet Union was completed in Aug 60. Subsequently only a few An-2s have been made by the Antonov OKB, the most important being the **An-2M** (modified), flown on 20 May 64. This incorporates 290 engineering changes aimed at improving utility. The

An-2ZA

An-3

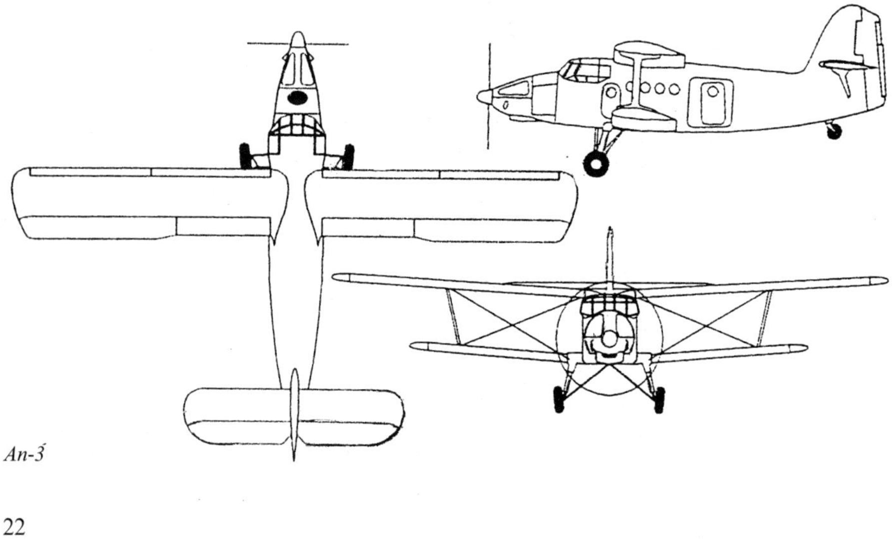
An-3

agricultural hopper is of GRP (glass-reinforced plastics) and enlarged to 1960 lit (431 gal), and a 50 hp power take-off from the engine to a new pump almost doubles dispensing rate from the new spraying and dusting systems. The aircraft is equipped for one-man operation, instead of needing two crew, and the cockpit has air-conditioning and often a separate door instead of access via the hinged left half of the roof (Polish An-2s usually have access from the cabin). The structure incorporates metal bonding and spot-welding, brakes are pedal-operated, main wheels are further forward to preclude nosing over, quick-change arrangements allow for rapid conversion between roles, and the enlarged tail has increased tailplane span and a more rectangular vertical surface. Only a limited number of M-models have been built in Poland.

Since 1959 production of the An-2 has been centred at WSK-PZL-Mielec, where over 11,950 built (10,427 for USSR). Polish designations often differ from the Soviet ones. In the late 1950s about 100, often with rectangular cabin windows, were produced at the DDR plant at Dresden. In 1957, China started building An-2s at Nanchang as the **Y-5**, 727 delivered by 68. Prodn then transferred to Shijiazhuang, where over 288 built by early 92, followed by **Y-5B** with corrosion-resistant structure and special role eqpt of which 150+ by mid-93 to bring grand total beyond 18,565. ASCC reporting name 'Colt'.

DIMENSIONS Span 18,18 m (59 ft 8½ in); length (tail down, most) 12,4 m (40 ft 8¼ in), (tail up) 12,74 m (41 ft 9½ in), (An-2V) 14,0 m (45 ft 11¼ in); height (tail down, most) 4,0 m (13 ft 1½ in); wing area (upper) 43,55 m^2 (468·8 ft^2), (lower 27,96 m^2 (300·96 ft^2).

ENGINE One ASh-62IR.

WEIGHTS Empty (An-2T, An-2V) 3360 kg (7,407 lb), (An-2SKh) 3440 kg (7,584 lb), (An-2M) 3430 kg (7,562 lb); loaded (An-2T and An-2M) 5500 kg (12,125 lb), (most other variants) 5250 kg (11,574 lb).

PERFORMANCE (typical): cruising speed 200 km/h (124 mph); max speed 250 km/h (155 mph); max take-off or landing run on soft ground 200 m (656 ft); service ceiling 4350 m (14,270 ft); range with max (700 kg) fuel 900 km (560 miles).

An-3 In 1959 Antonov studied versions of An-2 with a turboprop. Studies in 1960s were for enlarged aircraft with 1500 hp TV-2-117, but final outcome was same-size project which can be produced by modifying gigantic existing An-2 fleet in Warsaw Pact countries. Antonov OKB built single prototype (SSSR-30576) designated An-3 in 1982 with 940 shp Glushenkov TVD-10B, in production in Poland for An-28. This drives large slow-turning propeller potimised for operating speeds around 150 km/h (93 mph). Fuselage plugs ahead of an behind wings, cockpit sealed and air-conditioned with large blister on right and airtight door to cabin or (ag version) external door on left, new electrical and fuel system and instruments. Spray tank enlarged to 2200 lit (484 gal). Conversion of An-2s at Omsk (Russian Fedn) and Vinnitsa (Ukraine). Deliveries from 93.

DIMENSIONS Span 18,19 m (59 ft 8¼ in); length 13,65 m (44 ft 9½ in); wing area 73,5 m^2 (770 ft^2).

An-8

An-10 prototype

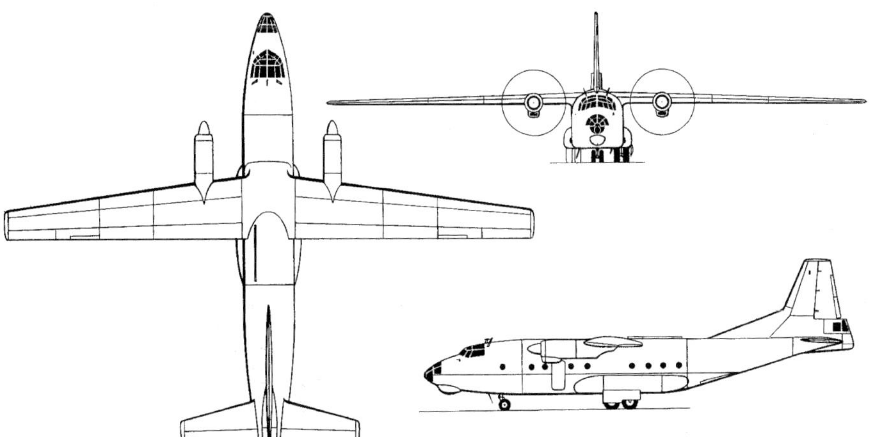

An-8

ENGINE One TVD-20.
WEIGHTS Empty 3,1 t (6,834 lb); payload 2 t (4,409 lb); MTO 5,8 t (12,787 lb).
PERFORMANCE Cruise 220 km/h (137 mph); climb at MTOW 240 m (785 ft)/min; range with 1,5 t 770 km (479 miles).

An-8 After the Second World War, the Soviet Union had a need for civil and military transport aircraft more capable than existing types such as the Li-2. 1952 joint requirement by the V-VS and Aeroflot was won by the Antonov OKB with a totally new design. It was a gigantic step for the bureau, which had no experience of large, highly stressed all-metal structures. In almost all respects the An-8 was a masterpiece, achieved at a time when the configuration that has since become standard was almost unknown (the C-130 had not appeared). Among basic features were an unpressurized hold 2,5 m (98·4-in) square in section, a full-width rear ramp/door, tandem twin-wheel main gears for unpaved airstrips retracting hydraulically into fuselage fairings, hydraulically driven double-slotted area-increasing flaps, manual flight controls, electro-thermal anticing of wings and tail, a glazed nose for contact navigation, and a chin radar for ground-mapping in bad weather and to assist accurate paradrops. Almost all An-8s were originally built with a hydraulic rear turret with a single NR-23 cannon. Crew comprised captain, co-pilot, navigator (nose), radio/radar operator (behind captain at lower level), engineer (behind co-pilot) and tail gunner. The wing, with aspect ratio of 11·7, was one of the most efficient in the world for subsonic transport in the first half of the 1950s, but the NK-6 engines were immature when the prototype flew in 1955. Like many rear-loading transports the An-8 also suffered from rear-fuselage aerodynamic problems and eventually featured long strakes from the wing to just below the tailplane.

Batch of 100 delivered, with engine changed (see data), performance being little impaired and cowlings unchanged. Propellers were four-blade AV-68 with electric anticing. About 11000 lit (2420 gal) of fuel was housed in bag tanks in wing box. Normally the interior was arranged for cargo, with folding seats for handlers, but up to 48 troop seats could be installed. A few An-8s went to Aeroflot, both these and at least half V-VS aircraft later having turret replaced by fairing. In 1966 a military An-8 was monitoring Chinese nuclear-weapon fallout, and type continued to be seen occasionally in the 1970s one at Monino. Nickname *Keet* (Whale). ASCC name "Camp".
DIMENSIONS Span 37,0 m (121 ft 4¾ in); length (no guns) 30,74 m (100 ft 10¼ in); height 9,7 m (31 ft 10 in); wing area 117,2 m² (1,262 ft²).
ENGINES Two AI-20D.
WEIGHTS Empty 21250 kg (46,847 lb); max loaded 38000 kg (83,774 lb).
PERFORMANCE Max speed 600 km/h (373 mph); long-range cruise 480 km/h (298 mph); field length 1000 m (ground run 700) (3,280 ft); range with max payload 2280 km (1,417 miles), with max fuel 3500 km (2,175 miles).

An-10 Ukraina In 1955 Aeroflot issued specification calling for 75 passenger seats in a pressurized fuselage. Antonov succeeded in a minimum change revision of the An-8, main alteration apart from streamlined circular-section body being doubling number of engines. Wing modifications were minor, though 22 bag tanks were housed in main box to raise fuel capacity to 12,710 lit (2,796 gal). Engines, four NK-4, driving AV-68 reversing propellers of 4,5 m (177 in) diameter (at first reverse-pitch was not permitted). The 2,5 m square cabin section had to be maintained, so circular profile drawn around outside to give 222 m³ (7,769 ft³) of clear volume above floor. Cabin dP set at 0,5 kg/cm² (7·11 lb/in²), though glazed nose retained. Cabin conditioning systems housed in main-gear fairings, of semicircular section unlike flat boxes of the An-8; left fairing also housed the APU, a new feature. Each main gear was a four-wheel bogie. Prototype, SSSR-U1957, flown by Ya. I. Vernikov and V. P. Vazin at Kiev on 7 March 57. Shortly beforehand, tunnel testing had shown need for outer-wing anhedral, thus setting a fashion in Soviet high wing transports. Flight testing revealed a need for improved stability and reduced rear-fuselage drag; palliatives included enlarging

An-12BP

dorsal fin, adding rear ventral strakes and extra endplate fins on tailplane, and increasing outer-wing anhedral. Another major change was switch to Ivchenko's AI-20, installed well below wing to give thrust-lines some 15 cm (6 in) lower. Another post-flight change was option of retractable skis with electric heating.

As originally built U1957 had 84 passenger seats in 14 triple-seat rows with a central aisle, with a 'children's playroom' at the rear. By the time Aeroflot services began in July 1959 the rear room was occupied by six revenue seats, and Antonov said the Ukraina (a name applied to all An-10s, not just U1957) was the most economic transport in the world. A year later the **An-10A** was disclosed, with a 2 m (78¾ in) fuselage plug allowing for two more seat rows and with normal accommodation for 100 passengers. The fuel capacity was increased (see data) and propellers were at last cleared for reversing. The 10A entered service in Feb 60, and subsequently set the following world class records: 2000 km with 15000 kg payload (723 km/h, 449¼ mph); 900-km round trip (760 km/h, 472¼ mph); and 500-km circuit (730·6 km/h, 454 mph). The An-10 and 10A were withdrawn in 1973 after a series of structural failures. ASCC name 'cat'.

A single high-density aircraft, with OKB designation **An-16** and service designation An-10V, was flown in 1963. This was shown in a brochure by Antonov in 1958, the fuselage having a further stretch of 3 m (118 in).

DIMENSIONS Span (38,0 m (124 ft 8 in); length (An-10) 32,0 m (105 ft 0 in), (An-10A) 34,0 m (111 ft 6½ in); height 9,83 m (32 ft 3 in); wing area 119,5 m^2 (1286·3 ft^2).

ENGINES Four AI-20K.

WEIGHTS Empty (10) 28900 kg (63,713 lb), (10A) 29800 kg (65,697 lb); max fuel (10) 12,710 lit=9787 kg (21,576 lb), (10A) 13,900 lit=10700 kg (23,600 lb); max payload 14500 kg (31,966 lb); max take-off (10) 54 t (119,048 lb), (10A) 55100 kg (121,473 lb).

PERFORMANCE Normal high cruising speed 680 km/h (423 mph); long range cruise 630 km/h (391 mph); typical take-off (paved surface) 800 m (2,625 ft); range with max payload (10A) 1200 km (745 miles), (max fuel) 4075 km (2,532 miles).

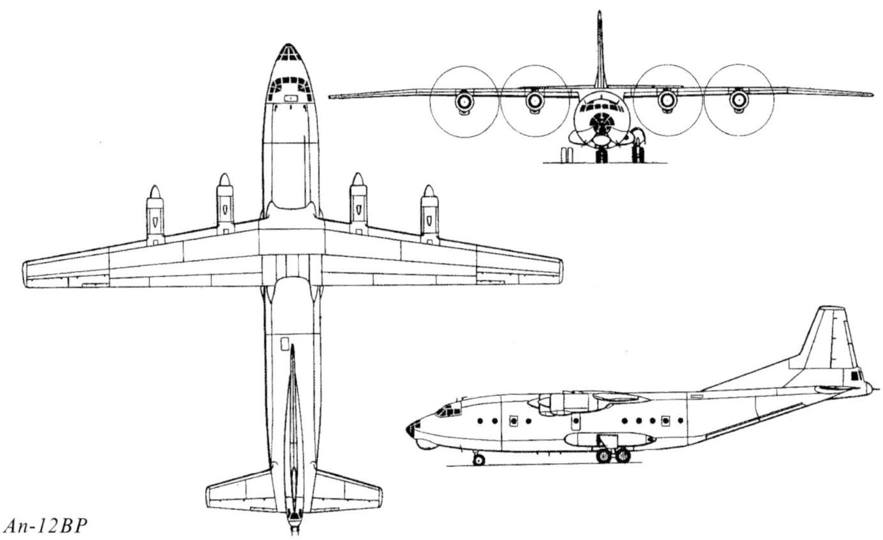

An-12BP

An-12 It was natural that Antonov should produce a cargo aircraft with full-section rear doors using basic airframe of An-10. Developed in parallel with the passenger machine, the NK-4-powered prototype An-12 flew in 1958 (date unknown). Antonov admitted several surprising defeats, one being to abandon pressurization of the cargo hold. Another was to use an integral rear ramp in only a few An-12s, standard model having left and right main doors hydraulically folded upwards, with rear door hinged up from rear edge in usual manner. Thus, though suitable for loading items from trucks, standard An-12 cannot handle vehicles without a separate ramp.

Airframe as an A-10A but restressed to higher weights and with increased fuel in 26 wing tanks and three under floor, total 27900 lit (6,135 gal). FR probe cleared but not adopted. Engines upgraded to 20M. Main-gear track 4,5 m (177 in), nitrogen struts to 4-wheel bogie with tyres 1050×300, pressure 5,6-6,7 kg/cm^2 (80-95 lb/in^2), hyd anti-skid brakes. Twin steerable nosewheels, tyres 900×300, hyd retraction to rear, all doors closed when extended. Left hyd system 2,200 lb/in^2, right system 2,130. Hot-air deicing of wing, electrothermal of tail. Fixed tailplane, flight controls manual with tabs. Twin-NR-23 turret (almost identical with that of Tu-16) was supplied in batches to An-12 production line at Ulan-Ude. Usual direction radar for this turret is not fitted, but tail-warning radar is standard, while under-nose radar was updated to I-band set. Flight crew remained as before, with five forward and sixth in pressurized turret, and in some aircraft a 14-seat pressurized compartment was added behind flight deck for crews of AFVs, handling teams and other passengers. The 2,5 m square section

An-14

allows such AFVs as the ZSU-23/4, ASU-57 and ASU-85, PT-76, and all the ICVs, APCs and scout cars. At Domodyedovo in 1967 a demonstration was put on by An-12s fitted with integral hinged ramps for delivery of such vehicles. Standard **An-12BP** had no ramp and no pressurized passenger cabin, and with 100 paratroops aboard restricted to 5000 m (about 16,400 ft) altitude, drop being made in under a minute with main doors raised on each side. Titanium floor stressed to 1500 kg/cm² point load, same weight /m² as distributed load, cargo positioned by 2,3 t (5,071 lb) elec travelling crane coving whole floor. Total production 1,247 ending 73; ASCC reporting name 'Cub'. Licence to China 1959 led to Shaanxi Aircraft Co producing modified **Y-8**, over 45 in 12 versions by mid-93. At peak, V-TA force of 740 active BP could transport two equipped airborne divisions 1200 km. Export to nine AFs, Egypt using one as E-300 turbojet testbed and India equipping several as makeshift bombers with 16 t (35,273 lb) of palletized bombs pushed out over Pakistan by handlers.

An-12BP entered service 1959, and within 18 months a civil version was supplied to Aeroflot. In 1965 a refined all-civil **An-12B** was exhibited with neat fairing in place of rear turret which had previously either been locally modified (in some cases as a toilet, not usually provided in the BP) or merely by having the guns and rear-fuselage magazine removed. From 1961 the An-12B was Aeroflot's main cargo transport, used by almost every directorate including Polar. Option advanced skis fitted with braking and heating. Aeroflot received about 200 An-12Bs, and at least 12 civil examples were exported. They avoided structural problem of An-10 and were only gradually withdrawn from 1980.

Large fleet of V-TA aircraft progressively replaced by Il-76, whittled down from 1972 by conversion of aircraft to other duties, mainly for EW (electronic warfare). An ECM variant ('Cub-C') has been in use (and supplied to Egypt) since 1973, with several tons of electric-generation, distribution and control gear above the floor and the underfloor area devoted to palletized jammers for at least five wavebands. This was one of the first dedicated ECM platforms with central energy management. Dispensed payloads may be emitted through at least two aperatures. A less-common Elint (electronic intelligence) version ('Cub-B') retains tail turret and uses underfloor region for from five to nine receiver aerials, main cabin being occupied by passive receivers, analysers, communications and recording equipment. Some of these EW versions, and all V-TA transports, were given better radar in larger nose bulge. Dedicated maritime reconnaissance version with different sensors first seen 1982.

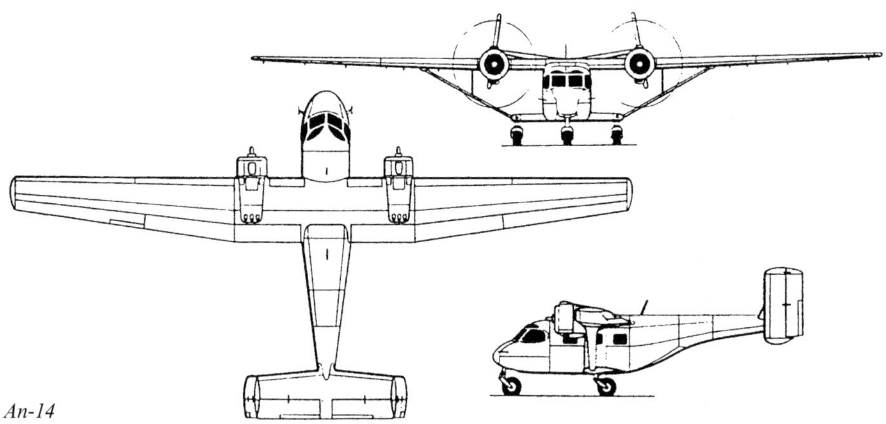

An-14

DIMENSIONS Span 38,0 m (124 ft 8 in); length (BP) 34,05 m (111 ft 8⅝ in); wing area 121,7 m² (1,310 ft²).
ENGINES Four AI-20M.
WEIGHTS OWE (BP) 34580 kg (76,235 lb); max fuel 22300 (49,162); payload 20 t (44,090 lb); MTO 61 t (134,480 lb).
PERFORMANCE Max speed limited to 620 km/h (385 mph); cruise 570 km/h (354 mph); range (max payload) 1450 km (900 miles), (max fuel) 6800 km (4,225 miles); take-off 785 m (2,575 ft); landing run (reverse pitch) 230 km/h, 804 m.

An-14 Named *Pchyelka* (Little Bee), the An-14 was planned around 1955 to meet an Aeroflot need for a light STOL utility aircraft smaller than the An-2 but more capable than the Yak-12 which it wished to replace. The answer was a stressed-skin high-wing monoplane with two radial engines, flown at Kiev by V. Izgeyim on 15 March 58. There was a blaze of publicity; but over the subsequent eight years almost every part of the An-14 was redesigned, numerous ideas being tested on prototypes and rejected until eventually production An-14s became available from Progress Works at Arsenyev, near Vladivostok, entering service in the surrounding Far East Directorate and with the armed forces.

The original prototype (L-1958) had an untapered wing of 19,8 m (64 ft 11½ in) span with full-span trailing surfaces hinged to brackets below the wing, the outer sections serving as slotted ailerons and the inner sections carrying hinged rear portions to form double-slotted flaps. Automatic slats were fitted outboard of the engines, which were 260 hp AI-14Rs. Tapered fins were mounted on the tips of a

An-22M

horizontal tailplane, and the cabin (about 3,1 m long by 1,53 wide and 1,6 m high, 122×50×63 in) was entered via rear ventral doors divided at the centreline and allowed to hang open on each side, with a folding central steps.

The production aircraft has a slender tapered wing, with powered slats inboard of more powerful engines driving V-530 constant-speed propellers usually with three blades. Rectangular fins are mounted on a dihedral tailplane, the nose is longer and, after trying other methods, the original rear door was made standard. A pneumatic system at 50 kg/cm² (711 lb/in²) drives flaps, wheel brakes, inboard slats, engine starting and cabin heating. The nose gear freely castors 70° on each side. Alternative landing gears include skis and twin floats. Engine-heated air is supplied on demand to warm the cabin and the leading edges of the tail, the latter also having electric heater elements.

Dual control can be fitted, but usually the right cockpit seat is taken by a passenger, and the cabin can be furnished for six or eight passengers, six stretcher patients, 720 kg (1,587 lb) cargo or 1000 lit (220 gal) of chemicals distributed via spraybars running along the aileron/flap brackets, down the struts and below the stub wings. There are several other special-role kits. About 200 An-14s are thought to have been built. **An-14M**, see An-28. ASCC name 'Clod'.

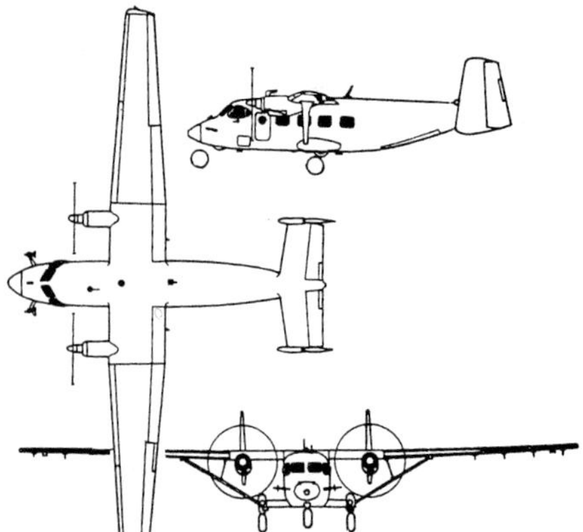

An-14M

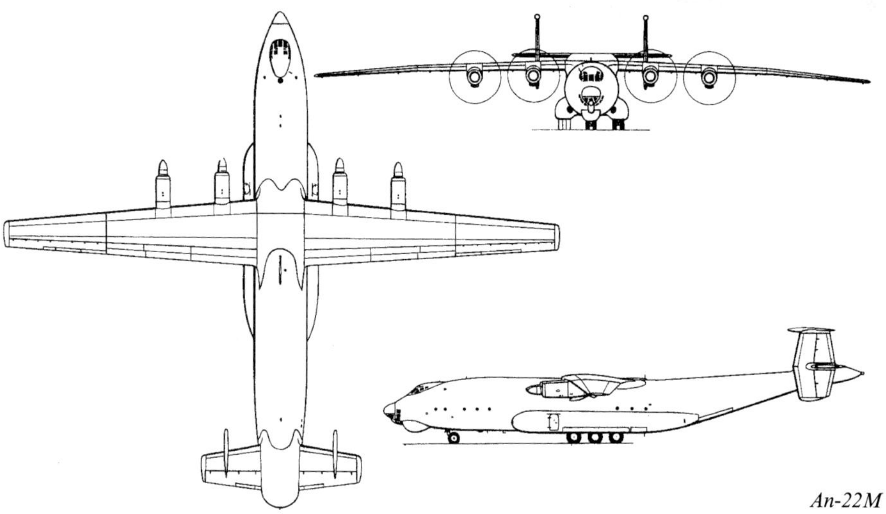

An-22M

DIMENSIONS Span 21,99 m (72 ft 1¾ in); length 11,44 m (37 ft 6½ in); height 4,63 m (15 ft 2¼ in); wing area 39,72 m² (427·5 ft²).

ENGINES Two AI-14RF.

WEIGHTS Empty (typical) 2000 kg (4,409 lb); max take-off 3600 kg (7,935 lb).

PERFORMANCE Max speed 222 km/h (138 mph); cruise 180 km/h (112 mph); landing speed 85 km/h (53 mph); take-off (concrete) 100 m (328 ft); range with max fuel (383 lit, 84 gal) 800 km (497 miles); range with max payload 650 km (404 miles).

An-22 Antei Monster freighter named for giant son of Poseidon and Gaia. Begun early 1962 against a joint V-TA/Aeroflot requirement. Cargo-hold cross-section was selected at 4,4 m (14 ft 5¼ in) square, but again Antonov jibbed at attempting a pressurized fuselage, settling for a pressurized nose and forward fuselage accommodating 29 passengers only. This is separated by a pressure bulkhead with left and right doors from the main cargo hold, which has a floor of titanium with a non-slip surface and cargo locks. There are two rear doors, the forward section hydraulically locking at any height down to ground level (but it is not the patented swinging-link type) and the rear aft-hinged door containing on its underside (the outer skin) two rails which extend the travelling-crane tracks to the rear to assist picking up heavy cargo. Most An-22s have two travelling cranes and two winches each of 2500 kg (5,511 lb) capacity. As in the An-12 two retractable jacks can support the fuselage near the main door sill. Available hold length is 33,0 m (108 ft 3 in). The rear doors can be opened for air-drops.

Normal crew access is via a door in the fairing for each main landing gear, stairs to the main deck, a pressure door to the passenger compartment, and stairs to the flight deck at the upper level. Normal crew is five, the navigator's glazed nose being retained. The first prototype, SSSR-46191, flown (by I. J. Davidov) on 27 Feb 65, was powered by NK-12MV engines driving AV-60N propellers each engine

An-22 mod for An-124 wing carried externally

exhausted through a single central stack containing a large heat exchanger for heating air rammed in through an inlet and subsequently ducted along the wing and tail leading edges. The wing, loaded to an exceptional 0,725 tonnes/m² (148·5 lb/ft²), is of typical gull-wing form. The main box forms integral tanks from tip to tip, housing 43 tonnes (about 55800 lit, 12,285 gal). Structural parts in this wing and the main fuselage frames were made on a 75,000-tonne press and are of impressive size. Three frames lead down to the three tandem sets of levered suspension main gears, 12 wheels in all, hydraulically raised into large fairings which also house cabin-air systems, APUs and the pressure-fuelling couplings. The tyre pressures can be adjusted in flight and the nose gear is steerable. The double-slotted flaps are made in inner and outer sections of equal area, and all flight controls are manual, driven by Flettner tabs. The twin fins are mounted ahead of the tailplane; the lower edges have no tail bumpers though they could touch the ground, and large anti-flutter fairings adorn their tops. The prototype had no nose equipment beyond UHF and VHF blades, the radar being in the underside of the right main-gear fairing. This An-22 set 12 world load/height records in 1966, beaten by another An-22 in 67 with 100 445 kg (221,440 lb) lifted to 7848 m (25,748 ft).

Cleared for prodn 71 as **An-22M,** NK-12MA engines driving largest mechanically coupled props in use (6,2 m, 20 ft 4 in). Nav/mapping radar moved to chin, with weather radar above in nose, retaining glazed nav station. Prodn halted at 66 in 74; half to V-TA with larger nav radar and ability to carry main battle tank or mobile missile system. Rest to Aeroflot mainly for strategic support in Siberia. Early aircraft (64460) modified with central fin and dorsal mountings to carry An-124 wings from Chkalov (Tashkent) plant to Ulyanovsk. Fleet reduced by 93 to 45. ASCC reporting name 'Cock'.

DIMENSIONS Span 64,4 m (211 ft 3½ in); length 57,9 m (189 ft 11½ in); height 12,53 m (41 ft 1½ in); wing area 345 m² (3,713·6 ft²).

ENGINES Four NK-12MA.

WEIGHTS Empty about 114 t (251,325 lb); max take-off 250 t (551,146 lb).

PERFORMANCE Max speed 740 km/h (460 mph); normal cruise 560/640 km/h (368/398 mph); take-off (concrete) 1300 m (4,265 ft); range with max payload 5000 km (3,107 miles); range with max fuel (and 45 t payload) 10 950 km (6,800 miles).

An-24 From 1955 Aeroflot recognised the need to replace the mass of Li-2s, Il-12s and Il-14s which handled its short-range routes, and in 1957 issued a specification for a 32-seater for operation in harsh conditions from primitive airstrips. Antonov's OKB was assigned the project in December 1957, and the first prototype An-24, L1959, was flown at Kiev by Y. Kurlin and G. Lysenko on 20 Dec 59. A conventional high-wing twin-turboprop, it exemplifies Russian disregard for operating economy, and concentration on a tough vehicle for a tough environment. Empty weight of a typical An-24 is more than 3 tonnes, or 25%, greater than that of its Western counterparts, though the latter carry greater payloads further, with about 10% less power. The important point is that the An-24 met the need, and has been most successful aviation export.

Ruling body section is 2,76 m wide by 1,91 m high (108·7 by 75·2 in), and pressure differential of 0,3 kg/cm² (4·27 lb/in²) is retained throughout the hold of cargo versions, unlike earlier An designs. Body section is formed from three circular arcs meeting at the dorsal centreline and at a chine low along each side. Extensive bonding and spot-welding is used throughout, and for the first time great attention paid to corrosion. The body was tank-tested, and in 1961 a life of 30,000 h was stated. The wing has 2° anhedral on the outer panels, and the large machined skins are spot-welded, the outer wing

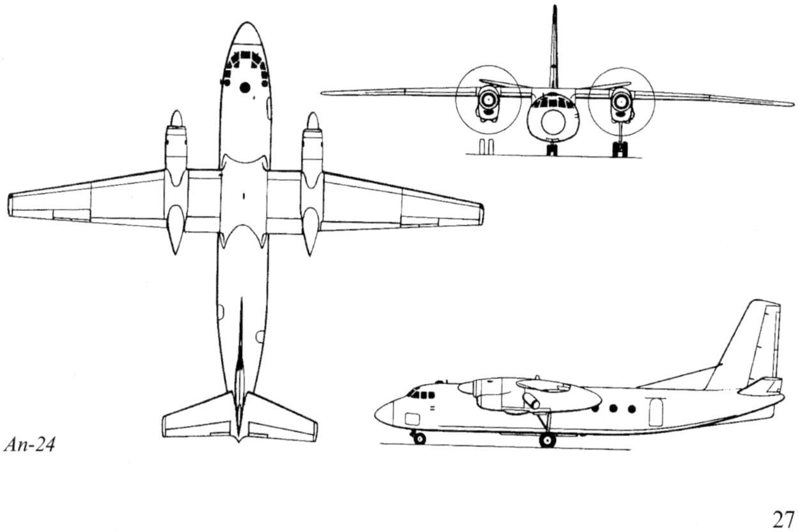

An-24

boxes forming integral tanks. Four centre-section cells bring normal capacity to 5240 lit (1,153 gal), supplemented by four extra c-s cells to total 6000 lit (1,320 gal, 4970 kg). A 155 kg/cm² (2,200 lb/in²) hydraulic system drives the flaps, landing gear, nosewheel steering, wipers, feathering and, on cargo models, freight doors. The prototype had one track-mounted flap on each wing, double-slotted inboard of the nacelle whose rear portion travelled with the flap. Production aircraft have four flaps divided by longer nacelles. Flight controls are manual, with glassfibre servo and trim tabs, that on the rudder being a spring tab. Leading edges are deiced by 10th-stage bleed air, which also drives air-cycle machines in each nacelle for cabin air. Windshield, inlet and propeller anticing is by raw AC. Twin-wheel landing gears have tyres variably inflatable to a maximum of 3·5 kg/cm² (49·8 lb/in²) and can free-fall.

Ivchyenko produced conservatively designed engine. An-24V Series I had AI-24 with short nacelle; standard engine AI-24A with water injection from 68 kg tank in nacelle extended behind wing. AV-72 prop with four blades (feather but no reverse), diam 3,9 m (12 ft/ 9½ in), elec deicing, but engine inlet heated by bleed.

Apart from the longer nacelles, production An-24s have a ventral fin, larger dorsal fin and longer nose. Cargo service began in July 62, and 44-seat passenger service on 1 Dec 63. By this time new AN-24s were of the **An-24V** type seating 50 passengers in a fuselage having an extended (9·69 m, 381·5 in) cabin with nine instead of eight windows each side. The AI-24BV engine introduced water injection, a TG-16 APU was fitted in the right nacelle and there were many detail improvements. By 1967 the inability of this aircraft fully to meet hot/high take-off demands had led to the An-24V Series II, with inboard flaps extended in chord (see wing area) and the AI-24T engine. Another customer option, available as a field modification, was to replace the TG-16 by the RU-19-300 turbojet, nominally rated at 1985 lb thrust but instead giving all the aircraft electric power on take-off plus about 480 lb thrust, and thus putting more power into the propellers. Jet-assisted aircraft are designated **An-24RV**. A third 1967 disclosure was the **An-24TV** (later An-24T) with the passenger door replaced by a rear ventral door hinging up along its rear edge for cargo, loaded by a 1500 kg hoist and floor conveyor on to a reinforced floor 15,68 m (51 ft 5¼ in) long. This of necessity had twin canted ventral fins, also seen on some passenger An-24s. With the auxiliary jet the T becomes the An-24RT, and cargo versions could be equipped for rocket-assisted take-off; many 24Ts have an astrodome!

There are many special role kits including convertible schemes, executive furnishing, casevac with 24 stretchers and, in the **An-24P** (*Pozharny*), chemical packs, equipment and parachutists with 24 stretchers and, in the **An-24P** (*Protivopozharnyi*), comprehensive fire-fighting eqpt including a 3,5 t water tank strapped externally on each side of the fuselage (3 rapid-dump doors), chemical tanks, pylon-mounted dispensers for 384 cloud-seeding payloads and provision for parachuting firefighters. **An-24VMF** dedicated version for naval exploration and geophys survey in Arctic, notable for giant magnetometer blisters each side of fuselage and 12 vertical rod antennas under outer wings. Export **An-24B**, small differences, 126 delivered. Total production 1,172 An-24 (various) plus 165 An-24T, ended 78. Chinese Xian Aircraft Mfg Co produced derived **Y-7-100** and **-200**. ASCC reporting name 'Coke'.

An-24 VMF

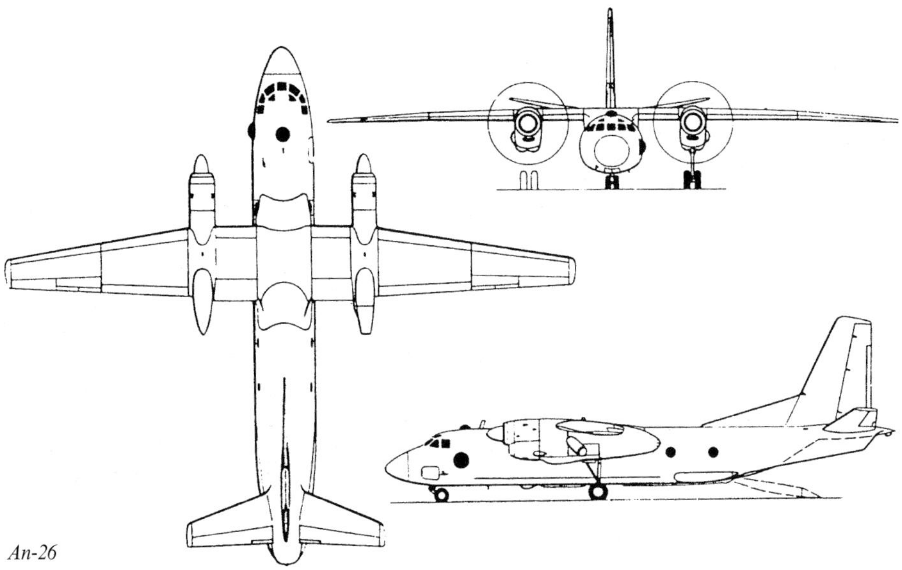

An-26

DIMENSIONS Span 29,2 m (95 ft 9½ in); length 23,53 m (77 ft 2½ in); height 8,32 m (27 ft 3½ in); wing area (Srs I) 72,46 m² (779·9 ft²), (Srs II) 74,98 m² (807·1 ft²).

ENGINES Two AI-24 (prototypes and Srs I); (main production) AI-24VT.

WEIGHTS Empty (24V) 13 300 kg (29,320 lb), (24T) 14 060 kg (30,997 lb); basic operating (24T) 14 698 kg (32,404 lb); max payload (V,RV) 5500 kg (12,125 lb), (T) 4612 kg (10,168 lb), (RT) 5700 kg (12,566 lb); max take-off 21 t (46,296 lb), (RV, RT) 21 800 kg (48,060 lb) to ISA+30°C.

PERFORMANCE Max cruise 498 km/h (310 mph); normal cruise 450 km/h (280 mph); take-off run (concrete) typically 640 m (2,100 ft); initial climb about 7·7 m/s (1,515 ft/min); range with max payload (V, RV) 550 km (342 miles), with max fuel 2400 km (1,491 miles).

An-26 First seen in 1969, this dedicated cargo derivative of An-24 was first Soviet aircraft to have pressurized cargo hold, with full-section rear door. It also introduced a novel door, patented by Antonov himself, which in addition to hinging down as a ramp can also be swung down on two beams and two parallel links to lie just below rear fuselage. Second position is for loading from trucks or for air-dropping. Door operation is hydraulic, if necessary by handpump which also builds up pressure in main system. Underside of rear fuselage sweeps sharply up to tailplane, but side profile is masked by deep glassfibre strakes on each side. Smaller strakes project along each side where door is suspended in air-drop position. Standard engine is VT, auxiliary RU-19A-300, derated to 800 kg (1,765 lb) thrust, always installed, there are two additional fuel cells giving capacity of 7100 lit (1562 gal), structure is restressed for operation at greater weights, main tyres are increased in size to 1050×400, inflated to 6 kg/cm² (85 lb/in²), fuselage under-skin is Bimetal (bonded ply of dural faced with abrasion-resistant titanium), and standard role equipment includes 2000 kg electric hoist, 4500 kg handling system flush with floor, 38 or 40 tip-up seats along walls, and retractable static lines for paratroops or cargo. Conversion to 24 stretchers takes 30 minutes. Most An-26s have bulged observation blister on left side of flight

An-26

deck for contact navigation and precision drop guidance. In 1981 **An-26B** emerged, able to carry three standard cargo pallets, total 5,5 t (12,125 lb), with removable Rollgangs for positioning pallets. All versions can have role eqpt including liquid cabin-heat system, static-line attachments and retrac devices, OPB-1R sight for airdrops, casevac and medical eqpt, external pannier (as on 24RV), chaff/flare dispensers on centre-fuselage pylons and (Angola/Mozambique) racks for bombs or other dropped stores to 1 t (2,205 lb) unit mass on each side of fuselage under TE. Several in CIS regts converted while in service as Sigint platforms with 'farm' of ventral blade and strake antennas ('Curl-B').

Kiev built 1,398 of all versions before switching to An-32 in 83, just over half for military operators, including at least 27 foreign air forces. Civil operators include Aeroflot (over 215), and Afghanistan, China, Cuba, Mongolia, Nicaragua, Romania, Syria, Vietnam and Yemen. Derived **Y7H-500** built in China by XAC. ASCC reporting name for all 'Curl'.

DIMENSIONS As An-24 except length 23,8 m (78 ft 1 in).

ENGINES Two AI-24VT.

WEIGHTS Empty 15 020 kg (33,113 lb); normal payload 4500 kg (9921 lb); normal take-off 23 t (50,706 lb); max overload 24 t (52,911 lb).

PERFORMANCE Cruising speed 440 km/h (273 mph); take-off (concrete) 780 m (2559 ft); initial climb 480 m (1575 ft)/min; range with max payload (no reserves) 1100 km (683 miles), (max fuel) 2550 km (1584 miles).

An-14M

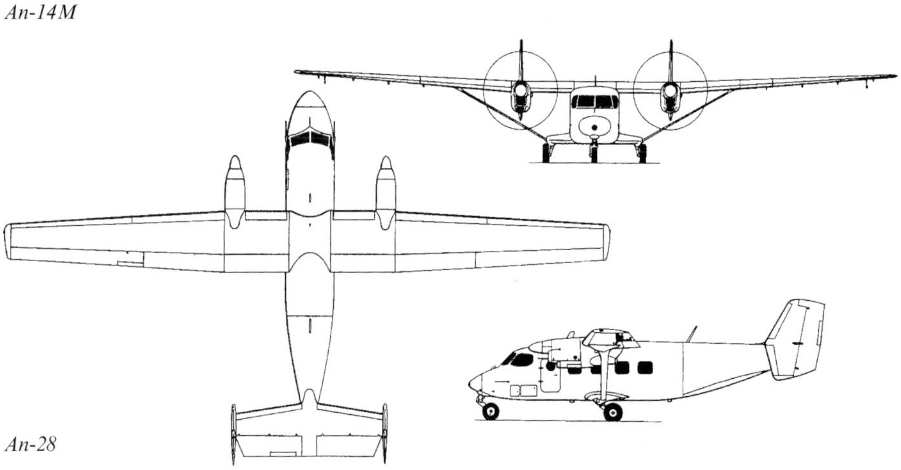

An-28

An-28 STOL twin-turboprop originally designated An-14M. Prototype No 1968 flown by V. Tersky Sept 69 with TVD-850 engines. Vertical tails redesigned, more powerful engine selected and other changes, mod prototype flying April 75. Selected over Be-30 but long delays resulted in major orders for Czech L-410 family.

As finally defined the An-28 has a fuselage larger than that of the An-14, with cabin 5,26 m (17 ft 3 in) long, 1,74 m (68·5 in) wide and 1,6 m (63 in) high. No pressurization or air-conditioning, but cabin heated. Standard seating is for two in cockpit, with dual controls and bulged side windows, plus 17 passengers in six seats to left of 345 mm (13·5 in) aisle and one single and five double seats to right. Seats fold against wall for quick cargo conversion. Entire underside of rear fuselage formed by left/right doors for passenger or cargo, latter positioned by 500 kg (1,102 lb) travelling hoist.

Airframe mainly of duralumin, with some welded or bonded joints. Wing of 14% thickness has full-span automatic slats. Large manual ailerons fabric skinned and droop with hydraulic double-slotted flaps which have CFRP skins. CFRP plate spoilers mounted ahead of flaps and ailerons extended hydraulically as roll augmentors, airbrakes and lift dumpers. Should an engine fail, an automatic spoiler extends ahead of the opposite aileron to restrict wing-drop and yaw. Should the pilot haul back too hard at low speeds the An-28 cannot stall, because of the wing slats combined with the horizontal tail of inverted section with a leading-edge automatic slat (which in addition collects ice rather than the tailplane should the de-icing fail, which is judged an advantage). Leading edges are heated by hot air from engine heat exchangers mixed with bleed air.

Extruded steel strut braces wing to stub wing carrying fixed main landing gear with Russian

An-28

wheels with multi-disc anti-skid brakes and 720×320 mm tyres on levered arms sprung by PZL shock strut. Track 3,4 m (11 ft 2 in). Nosewheel with Stomil (Polish) tyre 595×280 mm, hydraulically steered ±50°. All tyres 3,59 kg/cm^2 (51 lb/sq in). Heated ski gear optional. Four integral wing tanks house 1,960 lit (431 Imp gal) fuel, with overwing fuelling into each tank. Cockpit windows, engine inlets and AV-24AN reversing propellers (three blades, 2,8 m [110·25 in] diameter) deiced electrically. A simple 3-axis autopilot is standard, and Cat-III certification is planned. A wide range of role equipment is available including a 6/7-seat layout with four folding tables, military-style or freight interior with static line for six paratroops, fire-fighting models, casevac with six stretchers and five seated plus attendant, geological or photo survey, and agricultural with 800 kg (1764 lb) tank or hopper and centreline dispenser for solids or tip-to-tip spraybar.

With assignment of GA aircraft to Poland, PZL-Mielec was entrusted with production of the An-28 in Feb 78. It was then expected that deliveries would begin in 1980, but no Mielec-built aircraft flew until 22 July 84. Aeroflot orders reached 163 by end of 90, plus about as many again for other customers. Plan to offer PT6A-65B as alternative engine awaits customer. ASCC reporting name 'Cash'.

DIMENSIONS Span 22,063 m (72 ft 4·5 in); length 13,1 m (42 ft 11·8 in); height 4,9 m (16 ft 1 in); wing area 39,72 m^2 (427·5 sq ft).

ENGINES Two TVD-10B, made by PZL-Rzeszow as TWD-10B.

WEIGHTS Equipped empty 3,9 t (8,598 lb); max fuel 1529 kg (3,371 lb); max payload 2 t (4,409 lb); max take-off and landing 6,5 t (14,330 lb).

PERFORMANCE Max cruise 350 km/h (217 mph), normal cruise 335 km/h (208 mph); range, max payload, no reserve, 560 km (348 miles), max fuel and 1 t payload, 30 min reserve, 1365 km (848 miles); take-off run 260 m (853 ft); landing run 170 m (558 ft).

An-30 First flown in 1974, the An-30 is one of the few purpose-designed surveying and mapping aircraft. As well as optical cameras, it can be equipped with other sensors for all kinds of mapping. Compared with the An-24T the An-30 has a new forward fuselage with an extended and glazed nose and flight deck raised by nearly 1 m (39 in) to give a capacious nose compartment for the navigator with access under the flight-deck floor. The pressurized main cabin houses special navigation equipment (not normally including doppler or inertial, but with VOR/DME), a flight-path computer, radio altimeter and various recording, display and communications sets. In the cartographic configuration there are five major cameras in floor installations, with at least one gyrostabilized. There is a wide range of cameras and mountings, usually with remote control of doors that open below each camera. Normal flight crew numbers five, plus up to four photographer/surveyor/sensor operators. Darkroom and film magazines in main cabin, with crew rest, buffet and toilet. Option of radar in chin fairing, plus various other sensors. Fuel capacity increased to 6200 lit (1,364 Imp gal). Cabin can be converted for cargo. **An-30M** Popularly called Sky Cleaner, cabin contains eight 130 kg (286 lb) modular containers of solid CO_2 dispensed in controlled manner to induce precipitation, plus external pod on each side of fuselage for 384 meteorological cartridges fired into clouds. Chin radar standard.

Total An-30 production was 123, for Bulgaria, Hungary, Romania, Russia and Vietnam. Long-term plan to wet-lease to other users not activated. ASCC reporting name 'Clank'.

DIMENSIONS As An-26 except length 24,26 m (79 ft 7 in); height 8,32 m (27 ft 3·5 in).

ENGINES As An-26.

WEIGHTS Basic operating (without role equipment) 15 590 kg (34,370 lb); typical camera installation 650 kg (1,433 lb); max take-off 23 t (50,705 lb).

PERFORMANCE Max speed 540 km/h (335 mph); typical cruise at 6 km (19,685 ft) 430 km/h (267 mph); service ceiling 8,3 km (27,230 ft) (without APU operating, 7,3 km 23,950 ft); range with max fuel, no reserve, 2630 km (1634 miles) (oxygen for eight hours).

An-32 From the start of design of the An-24 O. K. Antonov has striven to improve performance from hot/high airfields. With this aircraft, announced May 1977 two years after first flight, propulsive power was doubled and maximum lift coefficient raised 60% by automatic leading-edge slats and large triple-slotted trailing-edge flaps. Enlarged tailplane with full-span inverted slat, and much larger canted ventral fins. Different engines drive AV-68 series four-blade reversing propellers with diameter increased to 4,7 m (15 ft 5 in). Thrust lines raised 1,5 m (60 in) above wing mainly to avoid FOD on rough strips, but requiring giant nacelles to house main gears. Airframe restressed for increased power and weight, and whole aircraft cleared for operation without ground support at airfields to 4,5 km (14,750 ft) at ISA+25°C. Thus, auxiliary turbojet replaced by TG-16M APU in right main-gear fairing, improved air-conditioning, electrics, deicing and landing-gear retraction.

Normal crew two pilots and navigator. Interior normally arranged for cargo, aft loading as An-26 except hoist uprated to 3 t (6,614 lb). Floor guides and rollers assist loading and also extraction of loads by parachute. Typical loads 12 pallets, 42 paratroops, 50 passengers or 24 casualties and three attendants. **An-32P** fire-fighting version has two external water/retardant tanks for total 8 t (17,635 lb), cleared to MTO weight 29,7 t (65,476 lb). Other versions agricultural, fisheries surveillance and mobile hospital. Version with extended-chord outer wings, with dogtooth leading edge, and spoilers/lift dumpers ahead of outer flaps, remained project.

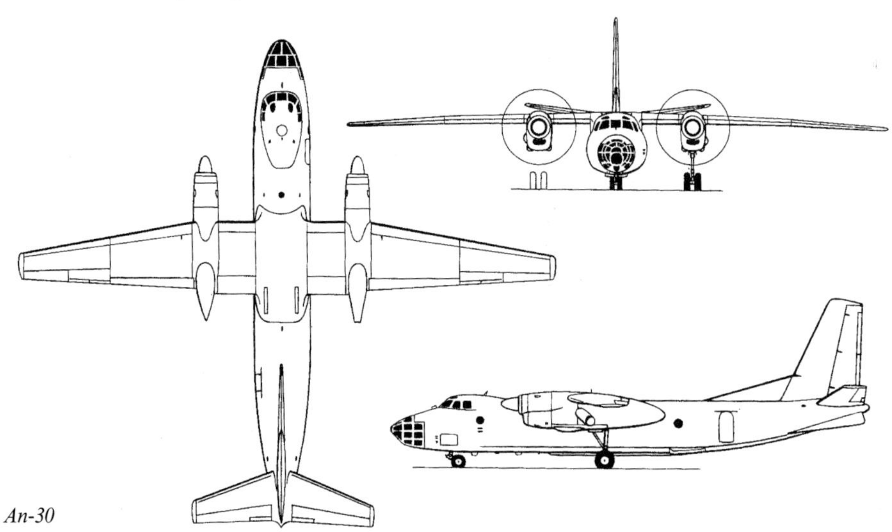

An-30

An-30

An-32P

Replaced An-26 in production at Kiev late 83. Total 231, for CIS and Ukraine forces, plus India (123, named Sutlej) and eight other customers. Peru specified four bomb racks, two on each lower side of fuselage. ASCC reporting name 'Cline'.

DIMENSIONS As An-26 except height 8,75 m (28 ft 8·5 in).

ENGINES Two AI-20D Series 5.

WEIGHTS Equipped empty 17 308 kg (38,158 lb); max payload 6,7 t (14,770 lb); max fuel 5445 kg (12,004 lb); max take-off 27 t (59,525 lb) (see An-32P).

PERFORMANCE Cruise (max) 530 km/h, normal 470 (292 mph); range (max payload) 960 km (597 miles), (max fuel, 5,5 t payload) 2500 km (1,553 miles); TO 760 m; landing 185 km/h, 470 m (1,542 ft).

An-38 Stretched successor to An-28 disclosed May 91. Wing aerodynamically as An-28 but stronger, full-span slats, double-slotted hyd flaps, drooping ailerons augmcnted by hyd spoilers ahead of flaps and outer ends ailerons. Spoilers auto counter roll due to engine failure and also linked to ILS for landings into 650 m strips from –50°/+45°C. Fuselage longer, typically furnished for 26 pax 2+1 (front row 1+1) boarding via rear clamshells with airstairs. Crew door at left front plus new emergency doors both sides at rear. Baggage stowage at rear. With seats folded floor takes 2,5 t (5,511 lb) cargo plus seats for two handlers. Weather radar standard; strengthened An-28 gears exchanged for skis or floats. More powerful engines (see data) driving quiet Stupino AV-36 reversing propellers with six composite blades,

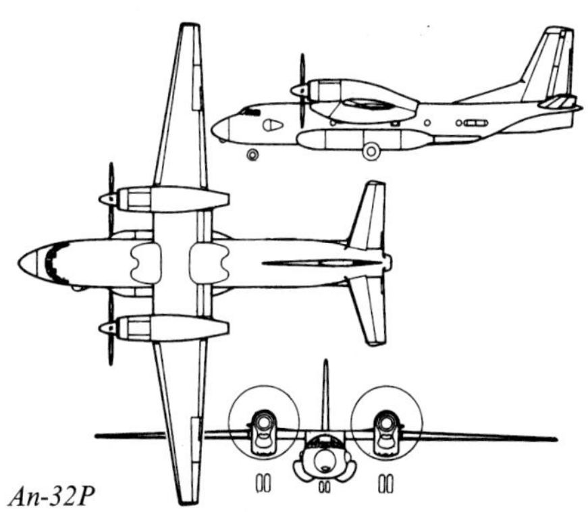

An-32P

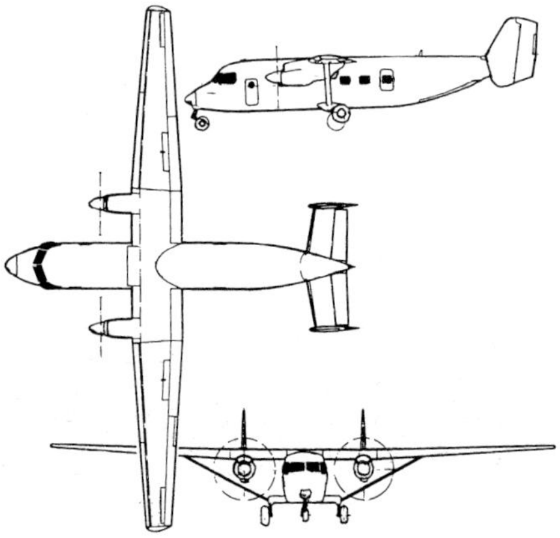

An-38

An-38

diam 2,49 m (98 in); increased fuel capacity. Programme based Novosibirsk. Yakutavia require 100 and India is discussing 50. Western (export) version was first to fly early 94 with 1,960 shp Garrett TPE331-14GR engines flat-rated to 1,500 shp, Hartzell prop, Lucas starter/gen, Amtec panel, Weldon unfeather pump, Hartman power switching and Hydro-Elec pressure switches. Cargo **An-38K** carries four LD-3. No ASCC reporting name.

DIMENSIONS Span 22 063 m (72 ft 4½ in); length 15 540 m (50 ft 11·8 in); wing area 39,72 m^2 (427·5 ft^2).

ENGINES Two TVD-1500B, export versions Garrett TPE331-14.

WEIGHTS OWE 5,4 t (11,905 lb) MTO 9400 kg (20,720 lb).

PERFORMANCE Cruising speed 350-380 km/h (217-236 mph); range with 45-min reserve, (26 pax) 600 km (373 miles), (17 pax) 1450 km (901 miles), (9 pax) 2200 km (1,367 miles); TO 350 m; landing 270 m.

An-50 Projected jet derivative of An-26, four Al-25, 800 km/h cruise.

An-50 model

An-70 Obvious requirement to replace An-12 resulted in VVS contract in 87 for design, prototype manufacture and static-test specimen. Retaining same configuration, An-70 has been designed with careful blend of traditional and new technologies, with much more powerful engine of propfan type (funded specifically for this aircraft) and twice the MTO weight. Result is aircraft which may find global market to replace not only An-12 but also C-130. Compared with An-12, initial An-70T carries ×1·5 cargo, cruises ×1·4 faster, has range with 20 t payload ×10 and has ×2·2 fuel efficiency.

Structure mainly aluminium alloy, with substantial areas of honeycomb sandwich and with wing/body fairing and some other parts fibre composite. Wing supercritical section with root thickness 12%, made as straight-tapered left/right panels with slight sweepback joined to centre section same width as fuselage, with 4°30' anhedral from roots. Leading edge with three sections of powered slat. Trailing edge with extremely powerful double-slotted flaps in two sections on each wing run out on six prominent tracks to give $C_L$max of 5·6, more than double that of An-12 or C-130. Conventional outboard ailerons, plus three sections of spoiler ahead of outer flaps serving to augment roll, as airbrakes

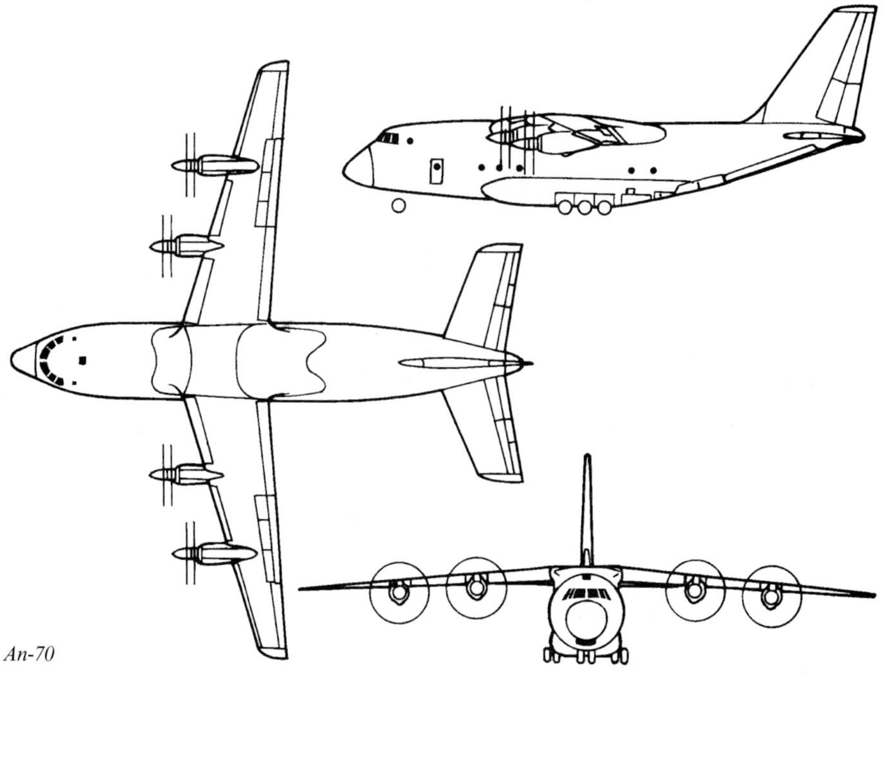

An-70

and as lift dumpers. Tall fin with powered upper and lower double-hinged rudders. Trimming tailplane mounted on fuselage with slight dihedral, carrying inboard and outboard powered double-hinged elevators, outer downstream sections divided into two.

Circular-section fuselage fully pressurized and air-conditioned throughout. Width of hold at floor 4 m (13 ft 1·5 in), increasing to 5·08 m (16 ft 8·5 in) max at shoulder. Height throughout 4·1 m (13 ft 5·5 in). Length of titanium floor 18·6 m (61 ft 0 in), rear hydraulically driven ramp increasing available length to 22·4 m (73 ft 6 in). Rear ramp hinged to sill with height variable by two retractable ram supports to match ground loading systems or trucks. Two hinged extension ramps for loading vehicles, full-width aperture being closed by left/right powered clamshell doors under tail. Comprehensive powered loading systems with computer control for ISO (8ft×8ft) or Yak-series containers, or PA-5,6 rigid pallets or PA-3, 4,0 or 6,8 flex pallets, with max 5 t on rear ramp. Personnel door on each side at front, with two seats for cargo attendants. Forward bulkhead separates flight deck for crew of two pilots plus loadmaster. 'Glass cockpit' with advanced navaids and ILS for Cat II, linked to 1553B digital bus.

Landing gear designed for unpaved strips with bearing ratio σ=0·6 (6 kg/cm^2, 85 lbs/sq in). Each main gear comprises axial row of three units similar to An-124 each with two wheels on single leg retracting inwards into unpressurized fairing which also houses APU, environmental/pressurization, pressure-fuelling control and many other systems components. Steerable twin-wheel nose gear retracting forwards. Hydraulic anti-skid brakes on all wheels. Landing lights on each side of nose immediately behind large radar. Fuel in integral tanks from tip to tip. Engines underslung ahead of leading edge, with annular inlet around spinner and ventral oil radiator and jetpipe. Stupino (Sich) propfans of 4·5 m (14 ft 9 in) diam with six from eight front and six rear deiced reversing composite blades.

Prototype rolled out Jan 94, first flight due Oct 94. Design structure life 45,000 h or 20,000 cycles. Design maintenance burden less than 8 man-h per flight-h (MMH/FH). All systems tailored to global market. No ASCC name known at press time.

DIMENSIONS Span 44,06 (144 ft 6½ in); length 40,25 m (132 ft 0¾ in); wing area 200 m^2 (2,300 ft^2).

ENGINES Four D-27.

WEIGHTS Design empty 57,5 t (126,760 lb); max payload 47 t (103,615 lb); max takeoff 130 t (286,600 lb).

PERFORMANCE Normal cruise 750 km/h (466 mph); cruise ht up to 9,6 km (31,500 ft); range with max payload 5530 km (3,435 miles), with 20 t payload 7250 km (4,505 miles); TO (MTO wt) 1,8 km (5,905 ft); landing 1,9 km (6,235 ft).

An-70T

An-71

An-71 At least three prototypes (one being SSSR-790361) built and/or converted c.84 of AEW&C (airborne early warning and control) version of An-72 to meet requirement of AV-MF for operation from large carriers. Rear fuselage completely redesigned to house DRLD pulse-doppler radar devd by NPO Vega, six operator consoles further forward and special elec power and cooling provisions. Tail also redesigned to carry rotating antenna group above giant forward-swept fin carrying double-hinged rudder. Longit control by slab tailplanes pivoted on rear fuselage. Rejected in favour of Yak-44. ASCC 'Madcap'.

An-72 STOL multirole transport based on USB (upper-surface blowing) from overwing engines to achieve high max lift coefficients. First of two prototypes (SSSR-197774) built at Kiev, flown 31 Aug 77. Eight pre-series also built by OKB, then redesigned as An-72A (ASCC 'Coaler-A') with increased span and length, greater weights and other changes. All production at Kharkov.

Wing of 15·9/14·5% thickness with horizontal centre section and outer panels with 10° anhedral, leading-edge taper 17°. All-metal with three spars, main box forming integral tanks. Outboard of engine, full-span hydraulic leading-edge flap in two sections on prototypes, three in production. Two trailing-edge flaps on each wing, each driven hydraulically out along two tracks. Inboard double-slotted, made of titanium, depressed to max 60° for STOL and low-speed flight with jets from engines attached by Coanda effect. Outboard triple-slotted, again max 60°. Outboard ailerons, short on prototypes, much larger in production and divided into inner/outer, fully powered and inner sections tabbed. Upper-surface spoilers ahead of flaps (five sections in prototypes, four sections ahead of outer flaps only in production), outer pair raised on approach and inner pair triggered by main-gear sensors to dump lift. Circular-section fuselage, fully pressurized and air-conditioned. T-tail with fixed tailplane, manual elevators and double-hinged rudder with tabbed manual lower rear portion used in normal flight, upper rear being brought in in low-speed flight and powered front portion actuated automatically in asymmetric flight. Inverted slat on tailplane opened when flaps lowered. Main landing gear with two levered-suspension oleo-pneumatic legs in tandem on each side, each with mutli-disc brake and 1050×400 mm low-pressure tyre, retracting inwards in side blister. Track 4,15 m (13 ft 7·5 in). Nose gear with twin 720×310 mm tyres, hydraulically steered and tretracting to rear. Wheelbase 8,12 m (26 ft 7·7 in).

Engines mounted in large nacelles ahead of and above wings, blowing aft over titanium upper skin and flaps. Overwing nozzles flattened and incorporate upward-opening target-

An-72P

type reversers. Planned spreaders to diverge jet over whole inboard flap not fitted. APU in right main-gear fairing to provide ground air-conditioning (hot or cold) as well as electric power. Stainless wing and tail leading edges with thermal deicing. Main cabin 10,5 m (34 ft 5·3 in) long, 2,15 m (84·6 in) wide at floor and 2,2 m (86·6 in) high. Underside of rear fuselage has left/right aft clamshell doors and main door with length of 7,1 m (23 ft 3·6 in) which can hinge down hydraulically as ramp or swing forward under fuselage as on An-26. During aft loading rear fuselage supported by telescopic struts hinged down from rear of each MLG fairing. Typical loads four UAK-2,5 containers or PAV-2,5 pallets, moved over rollgangs by 2,5 t (5,511 lb) winch. Up to 7,5 t can be air-dropped by parachute extraction in 2,5 t loads. Folding side seats and removable central seats for 68 passengers; alternative loads 57 paratroops, or 24 stretchers and 12 seated patients plus attendants.

An-72AT ('Coaler-C') is dedicated cargo version able to carry and latch ISO and Yak containers or PA-series flexible or rigid pallets. **An-72P** is maritime patrol version. Main cabin equipped for airdrop of 22 armed paras and equipment, or for 16 casualties plus attendant or 5 t (11,023 lb) of cargo. Enlarged crew compartment for two pilots, engineer, navigator, radio/sensor operator and observer, plus toilet, wardrobe, PSH-6AK life raft, and bulged observation windows on each side. Observer station has oblique camera aimed through port on left. In extreme tail are A-86P vertical/oblique mapping cameras and UA-47 night mapping camera with flash cartridge magazine. Enhanced navigation and radio systems for all forms of self-contained navigation and sea search and air/ship communication. Right MLG fairing houses IPM-10 gas-turbine APU at rear, 38-tube dispenser in centre and searchlight and 1Sh-23L cannon firing straight ahead at front. Left MLG fairing houses optical sight slaved to TV camera. Four FAB-100 (220·5 lb) HE bombs can be hung from rear end of cargo-crane roof rails and dropped through open aft ramp door. Single pylon under each wing for UV-32M (32×57 mm) rocket launcher. Endurance in low-level patrol 5·3 hr. **An-72S** (Salon) ('Coaler-C') is three-compartment executive transport with galley at front, cenre cabin for various arrangements for up to six seats and/or wardrobe, table and other items, and rear cabin for 12 pairs of armchairs. Alternatively 72S can carry cargo and 38 passengers. Also see An-74; 150+ An-72/74 built late 93.

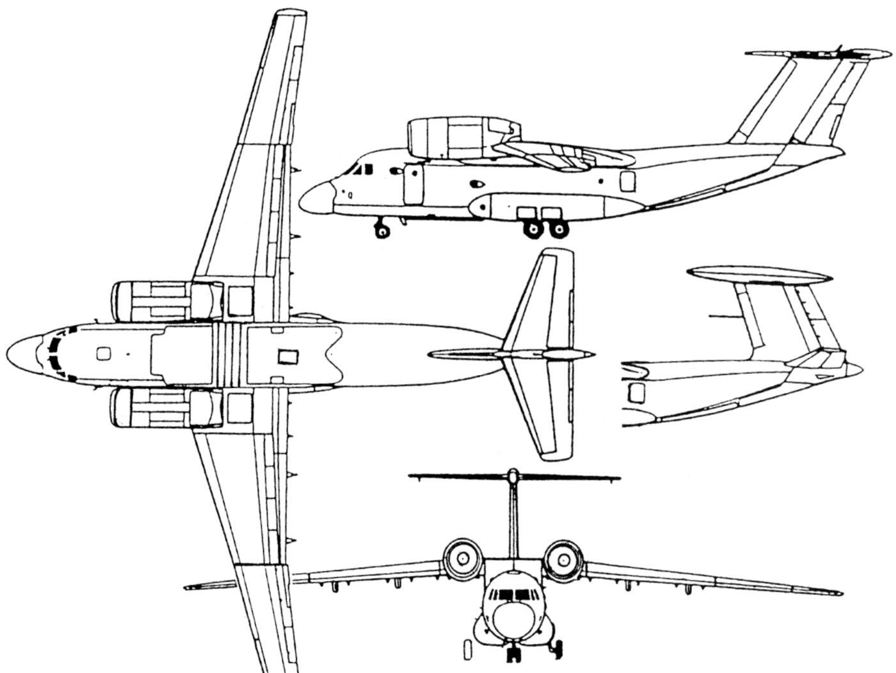

An-72 (inset, tail of An-71)

DIMENSIONS Span (prototypes) 25,83 m (84 ft 9 in), (production) 31,89 m (104 ft 7·5 in); length (prototypes) 26,58 m (87 ft 2·5 in), (production) 28,07 m (92 ft 1 in); height (prototypes) 8,24 m (27 ft 0·5 in), (production) 8,65 m (28 ft 4·5 in); wing area (production) 98,62 m² (1,062 sq ft).

ENGINES (Prototypes and pre-production) two D-36; (production) two D-36 Series 1A.

WEIGHTS Empty (production) 19,05 t (42,000 lb); max payload (prototypes) 7500 kg (16,535 lb), (production) 10 t (22,046 lb); max take-off (prototypes) 30,5 t (67,240 lb), (production) 34,5 t (76,058 lb), reduced according to available STOL run, (72P) 37,5 t (82,670 lb).

PERFORMANCE Max speed (prototypes) 720 km/h (447 mph), (production) 705 km/h (438 mph); cruise 550/600 km/h (342/373 mph); range with 45 min reserve (max payload) 800 km (497 miles), (7500 kg payload) 2000 km (1,240 miles); take-off run (prototypes) 470 m (1,542 ft), (production) 930 m (3,052 ft); landing run (all) 465 m (1,525 ft).

AN-74 Designation of version of An-72 specially equipped for Polar operations. Principal difference is greater fuel capacity. Enhanced avionics (including inertial navigation system) and deicing, optional wheel/ski landing gear,

An-74

different radar causing longer and bulged nose, cockpit with changed instruments and overhead panels for crew of five, with blister window on each side and third blister window on left of main cabin. An-74A ('Coaler-B') was initial production version, An-74T dedicated cargo variant with equipment for loading and locking ISO or YAK containers or PA flex or rigid pallets, An-74T-100 has provision for navigator for day/night flying in Polar regions, An-74TK ('Coaler-B') and TK-100 are corresponding convertible versions for (extremes) from 10 t cargo to 52 passengers, and An-74P-100 a navigator-equipped business version with three lavishly equipped compartments for 16 passengers. An-74-200 versions cleared for higher weights. Airborne early-warning version, see An-71.

First flown 83, an early An-74 demonstrating 300 m landing and 250 m takeoff supplying Polar station SP-27. All variants being refined to use composite structure and D-436 engines. Collapse of procurement system, which was geared to 20 An-72/74 per year, led 92 to formation of special association to continue production using factories at Arsenyev, Kharkov, Omsk; Zaporozhye and other locations. Data as An-72 except:

WEIGHT Max for -100 versions 34·8 t (76,720 lb); -200 36,5 t (80,465 lb).

PERFORMANCE Cruise 700 km/h (435 mph); range (74 with 12 passengers or 1·5 t, 2 h reserve) 4325 km (2,688 miles), 74 T, 5 t 1 h reserve) 3000 km (1,864 miles), (74 T, 10 t) 1000 km (621 miles), 74 TK, 52 passengers, 1 h reserve) 2800 km (1,740 miles), (74P-100) 4500 km (2,796 miles).

An-124 When in 74 Antonov was told An-22 production was being prematurely stopped he was personally commissioned to produce a successor 'in the class of the C-5 Galaxy'. He revealed this project, the An-40, in 77, but designation changed to An-400 before prototype No 680125 was flown by Vladimir Terski on 26 Dec 82. Designation changed again to An-124 before second prototype (No 82002 *Ruslan*) was flown to Paris airshow in May 85. New turbofan-engined airlifter had to meet needs of civil Aeroflot and military V-TA, and it took 75-79 to get agreement on body cross-section, floor strength, loading doors, para-dropping and landing-gear flotation for unpaved airfields. One requirement was to fly 8,000 full-load flights without any fatigue problem. Lotarev KB was assigned task of producing D-18T, first large turbofan in Soviet Union.

Result was in all respects a masterpiece, generally resembling losing Boeing and Douglas C-5 contenders but with superior aerodynamics, structure and engines. Beautiful wing has deep supercritical section, with no dogtooth, fence, vortex generator or winglet, weighing only 145 lb more than troubled C-5A wing yet span 18 ft greater, aspect ratio higher, sweep 32° to 35° and gross aircraft weight nearly 170,000 lb greater. Movable surfaces, all driven hydraulically, comprise full-span leading-edge flaps, almost full-span Fowler flaps in three sections each running out on two tracks, a two-piece outboard aileron, four inboard airbrakes and eight outboard spoilers. Wing box sealed to form ten integral tanks housing 507,063 lb of fuel. Horizontal fixed tailplane mounted on fuselage and carrying four sections of powered elevator and a fin carrying upper and lower powered rudders. All flight controls tabless and have FBW (fly by wire) control. Fuselage, equally beautiful structural exercise, has a pear-shaped section giving a main cargo hold 4,4 m (14 ft 5·3 in) high and 6,4 m (21 ft 0 in) wide, available length (excluding the rear ramp door) being 36 m (118 ft 1 in). Rear ramp door and left/right clamshell doors open hydraulically in

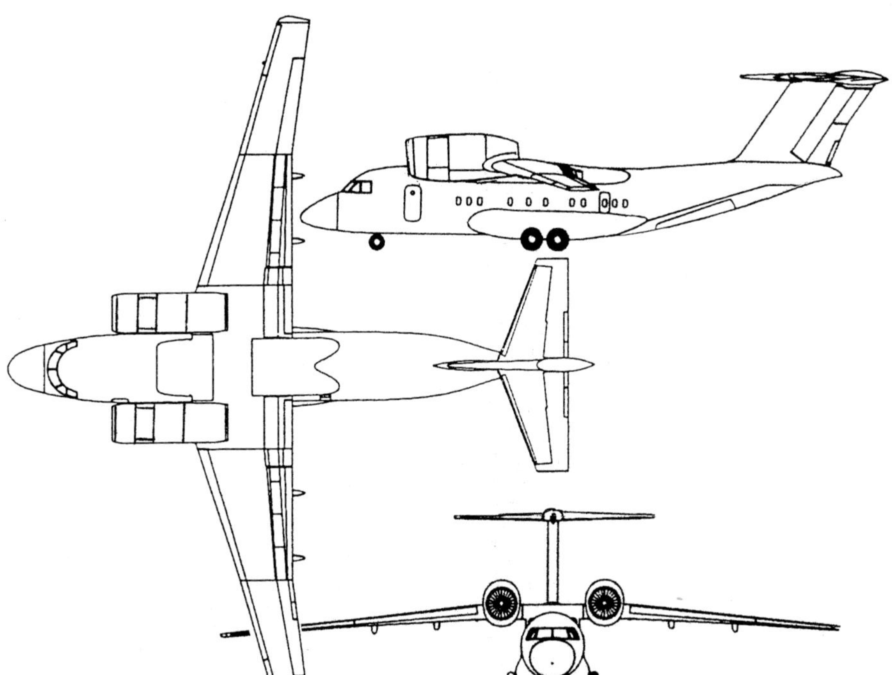

An-74-200

An-124

3 min, but hydraulically powered nose takes 7 min to open fully and extend triple-hinged loading ramp which, like that at rear, can take a 70-ton tank. Cockpit and system pipes and cables are left in place by opening nose, radar cable loom passing in a tube through one hinge. Cargo floor is titanium, attached to fuselage by hinged links. Hold contains two 3 t (6,614 lb) winches and two 10 t (22,049 lb) travelling cranes. Interior is pressurized to 3·55 lb/sq in, upper deck pressurized to 7·8 lb/sq in, this area extending from 6-crew flight deck to rear compartment usually with seats for 88. Over 5,5 t (12,125 lb) of carbon or glass composites. Nacelles and pylon skins almost entirely composite, and incorporate translating-cowl reversers attached to fan case. Deicing of engines and wings by bleed air, but tail uses 'giant pulse' electrical method, claimed to be uniquely energy-efficient. Nose landing gear comprises left and right independent power-steered twin-wheel units, each retracting forwards. A product of Hydromash, each main gear a row of five twin-wheel units, each individually retracted inwards under floor. Each unit has multi-disc carbon brake, front two units on each side having power steering and together with retracting nose units enabling aircraft to kneel on ground to angle of 3·5° with nose resting on two extended feet, or tilt nose-up by collapsing main oleos from front to rear. Nosewheel tyres 1120×450 mm, main tyres 1270×510 mm. Main gears require only shallow low-drag fairing which at rear houses 250 hp APU on each side for ground or inflight hydraulic, electric or pneumatic power (eg for air-conditioning or main-engine starting).

Design payload 150 t, but on 26 July 85 set new record lifting 171 219 kg (377,473 lb) to 10,75 km (35,269 ft). In 87 set closed-circuit record flying 20 151 km (12,521 miles) in 25 h 30 min. Also in 87 deliveries began to V-TA and Aeroflot, which in late 1992 had 24 in service, plus six civil aircraft available on charter to HeaviLift (UK)/Volga Dniepr, one on shuttle of A330/340 parts from Canadair to Toulouse. An outstanding aircraft, world's largest in production. ASCC reporting name 'Condor'.

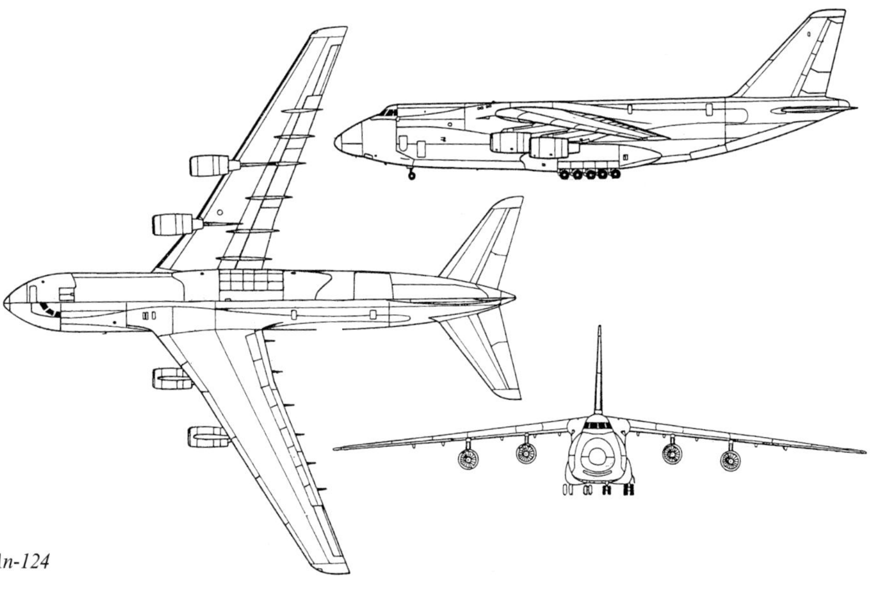

An-124

DIMENSIONS Span 73,3 m (240 ft 5·8 in); length 69,1 m (226 ft 8·6 in); height 20,78 m (68 ft 3 in); wing area 628 m² (6,760 sq ft).

ENGINES Four D-18T.

WEIGHTS Operating empty 175 t (385,800 lb); max payload 150 t (330,693 lb); max take-off 405 t (892,872 lb).

PERFORMANCE Max cruise 865 km/h (537 mph); normal cruise 800 km/h (497 mph); range (max payload) 4500 km (2,795 miles), (max fuel) 16500 km (10,250 miles); balanced field length 3 km (9,850 ft); landing run at MLW 800 m (2,625 ft).

An-140 Intended as a direct replacement for the An-24/26 family, this transport has same configuration but incorporates 30 years of progress in materials, propulsion and systems. Wing centre section with flat top but sharply tapered in thickness on underside, and outer panels tapered in plan and with anhedral. Centre section recessed with large composite fairing on top of metal fuselage with full circular section of 2,93 m (115 in) diameter. Tail conventional with low fixed tailplane, and wing has advanced but simple high-lift system. Engines underslung at ends of centre section with ventral inlet and dorsal oil cooler, driving four-blade reversing propellers of 3,73 m (12 ft 3 in) diameter. Clearance between propeller and fuselage is 1025 mm (40·4 in). Cabin 10,5 m (14,3 with aft toilet/galley/baggage/6,6 m³ cargo), normal seating 52 2+2 or 20 plus 3,4 t cargo, 3,4 m³ underfloor cargo, part composite

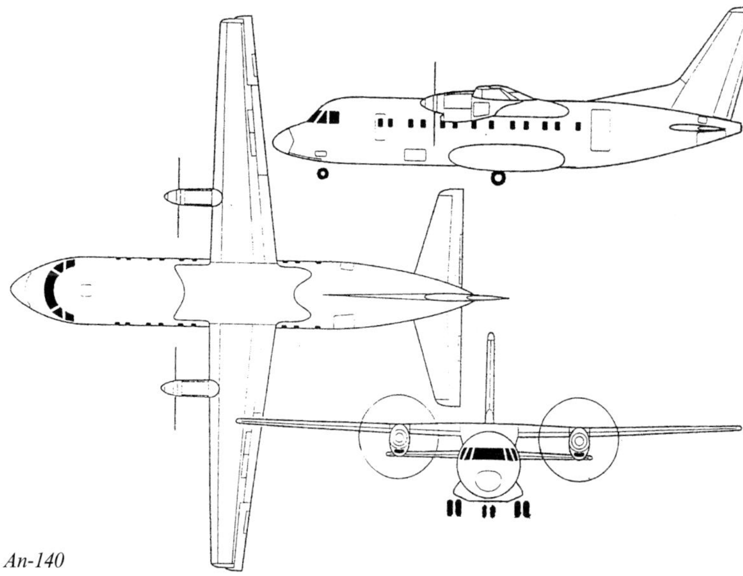

An-140

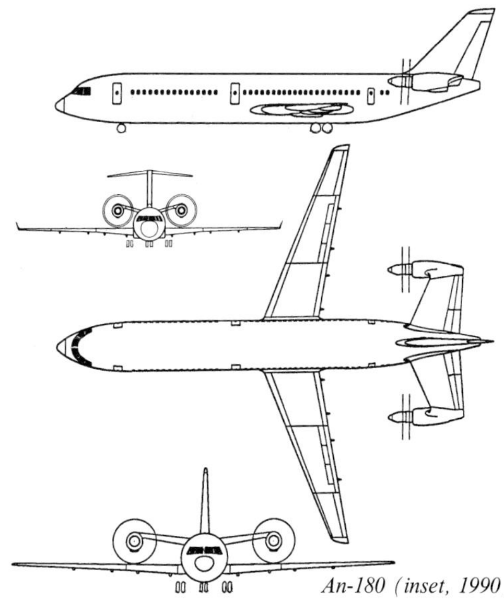

An-180 (inset, 1990 form)

structure, twin-wheel gears track 3180, wheelbase 8150. Normal engine/prop as Il-114, but export option of PW127A with HamStan 247F propeller. Certification to AP-25 and FAR-25 due 96.
DIMENSIONS Span 24,25 m (79 ft 6¾ in); length 22,46 m (73 ft 82 ft 6¼ in); wing area 55,5 m^2 (597 ft^2).
ENGINES Two TV3-117SB2.
WEIGHTS Max payload 4,9 t (10,800 lb).
PERFORMANCE Max cruise 595 km/h (370 mph), econ 520 (323) at 7,5 km (24,600 ft); range (max payload) 1780 km (1,106 miles), (52 pax) 2000, (max fuel) 3750 (2,330); field length 1100 m.

An-180 World's 'first uncompromised twin-propfan passenger transport'. Begun in 90 as OKB project, a generation later than Tu-204, burning 'half as much fuel, meeting all predicted environmental rules'. Design for ultimate efficiency; Balabuyev 'An airliner for people, seating 163 to 180 2+2+2 with twin aisles. It has been given wonderful reports by the authorizing institutes . . .'

Revealed 91 Paris with low wing of supercrit profile, LE sweep 22°, aspect ratio 10 excluding winglets which were omitted 93. Full-span powered slats, two sections TE flap each running on two tracks, three sections spoiler ahead of outer flap, outboard powered aileron. Fuselage circular 4,55 m (179 in) diam, shortened 1993 with 3 instead of 4 doors each side. Originally T-tail (anhedral) with merely trimming elevators on aft-engine struts. By 1993 changed to engines actually hung under tips of low dihedral tailplane with inbd/outbd double-hinged elevators. Engines as on An-70T: tractor Sich (Stupino) 4,5 m (14 ft 9 in) props with 8/6 scimitar reversing blades with elec deicing. Ldg gears as 3-view. Fuel burn 15,0 g (0·033 lb) per pax-km. Two-man glass cockpit for Cat IIIA. Prototype hoped to be at 1995 Paris.
DIMENSIONS Span (1991) 37,92, (1993) 35,83 (117 ft 7 in); length (1991) c45,0, (1993) 40,9 (134 ft 2 in).
ENGINES Two D-27.
WEIGHTS Max payload 18 t (39,683 lb); max TO (1992) 67,5 t (148,810 lb).
PERFORMANCE (Est, 30°C/730 mm Hg) cruise 800 km/h (497 mph) at 10,1 m (33,135 ft); range with max payload, 3300 km (2,050 miles) from 2,6 km runway, 1800 km (1,120 miles) from 2,2 km; (with 163 pax) 4850 km (3,015 miles) from 2,6 km, 3350 (2,080) from 2,2 km; with max fuel 7700 km (4,785 miles).

An-181 Built as OKB tool to investigate validity of Custer-type 'channel wing' with aerofoil curved round under twin props. Single seat, butterfly tail, fixed gear. Exhibited Sept 90, not then flown.
DIMENSIONS Span 7,3 m (23 ft 11½ in); length 7,31 m (23 ft 11¾ in); wing area (projected) 7,0 m^2 (75 ft^2).
ENGINES Two Czech 210-hp M337.
WEIGHT Loaded 820 kg (1,808 lb), (max) 900 (1,984).
PERFORMANCE Max speed placarded as 800 km/h (nonsense); range 750 km (466 miles).

An-218 Twin-engined widebody projected in 88 to fill major gap in Soviet transport aircraft. Primary objective to replace Il-86, burning half as much fuel. Though there may be dedicated cargo or convertible versions initial aircraft will be all-passenger, seating 294 in three classes (18+74+202), or 316 in two classes (38+278) or 350 in single class seated 2+4+2 at 810 mm (31·9 in) pitch. Underfloor compartments will take standard LD3 containers, 14 in forward hold and 10 at rear plus large volume of bulk cargo. Capacious overhead baggage bins each side and down centre, and fullest provision for inflight entertainment and communications.

Features include a supercritical wing of aspect ratio 10, with tip winglets, powered slats, slotted Fowler flaps in inboard and outboard sections each running on two faired tracks, six sections of multipurpose spoiler on each wing, two-section ailerons, a two-part rudder and single elevators hinged to a trimming tailplane. Prototype will have D-18TM engines with flow mixers and full-length pod cowls. Production aircraft planned to be offered with D-18TP or RB.211-524H4. Great attention being paid to airframe, which will use a high proportion of fibre-reinforced composites and Al-Li alloys, with Airbus-type wet assembly joints and most perfect possible exterior finish. Landing gears conventional, four-wheel main units retracting inwards and nose unit aft. Avionics digital, with a two-pilot 'glass cockpit' and ILS for Cat IIIA. Everything possible done to increase life (target 60,000 h) and improve maintainability (target 9·5 man-h per flt-h). Target fuel

An-181

An-218

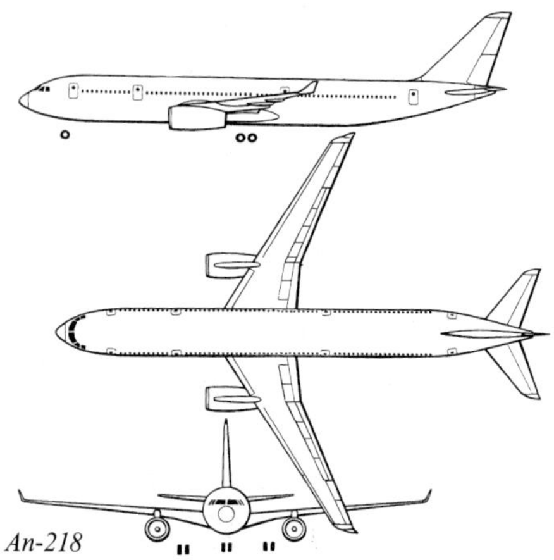

An-218

efficiency 17 g/pax-km (1 oz/pax-mile).

Full go-ahead received late 90, but since then Russian government has withdrawn and funding has to be found by Ukraine. Mock-up review was to take place March 93. First flight scheduled for June 94 (it will slip) and aircraft certificated by ENLG-S were planned for late 96, with JAR/FAR aircraft following in 97. To keep project from collapsing ANTK Antonov has formed joint board of directors with the vast Dniepropetrovsk Yuzhtyazhmash (Southern Heavy Machine Building Plant), which previously built giant rockets, to produce 218 in series. ANTK Antonov plans a short-body 220-seat version to fly 12,000 km (7,457 miles) and a stretched 400-seater. No ASCC reporting name.

DIMENSIONS Span 50 m (154 ft 0 in); length 59,79 m (196 ft 2 in); height 15·6 m (51 ft 2 in); wing area about 270 m² (2,900 sq ft).

ENGINES See text.

WEIGHTS Max payload 42 t (92,593 lb); max takeoff about 200 t (441,000 lb).

PERFORMANCE Cruise 850/870 km/h (528/541 mph) at 10,1-12,1 km (33,140-39,700 ft); range with reserves (max seating) 6300 km (3,915 miles), (3-class) 7000 km (4,350 miles), [production aircraft, max seating, 9100 km, 5,655]; hot (30C) and high (730 mm) runway length on concrete 3200 m (10,500 ft).

An-225 Mriya By a wide margin heaviest and most powerful aircraft ever built, and except for 320 ft span of Hughes Hercules also biggest, yet produced relatively easily by stretching An-124. Requirement was for transporter to carry completed *Buran* spacecraft pick-a-back, a task far beyond capability of Mya VM-T. Could almost be handled by An-124 with new twin-finned tail, but, once studies got under way in June 85, many further uses and pick-a-back payloads became evident, provided fuselage was stretched. Adding new twin-engined centre section followed naturally. Project was given Ukrainian name *Mriya* (= dream).

Wherever possible major sections identical with An-124. Consideration given to replacing entire fuselage by simple girder beam, but this was not worth design effort. In any case internal cargo hold could be useful, though rear ramp door was deleted to ease structural problems caused by greater length and new tail. By far biggest design effort was new centre section, most massive aircraft structure in history. LE swept at same 35° as original inboard wing, but no flap or slat, these surfaces ending at new inboard pylon strut (engines 3 and 4). TE unswept, carries extra section of flap plus four more airbrake/lift dumpers, making total of 32. New centre wing horizontal, and though outer panels retain anhedral pilots have found no difficulty in avoiding scraping tip on ground. Fixed tailplane has dihedral, unlike An-124, and carries three sections of elevator each side, with endplate fins mounted at 90° carrying two-section rudders. All control surfaces retain same FBW signalling and same power units as 124.

Fuselage extended by plugs ahead of and behind wing. Central portion, like aft plug, modified to provide pick-ups for *Buran*, riding on left/right canoe beams and steadied by triangulations of struts at front and rear. Unlike 747 Shuttle carrier An-225 was designed not to need a fairing over enormous flat base of *Buran*. Twelve further attachment points provided to cater for possible future payloads. Internally 43 m (141 ft) cargo hold could accommodate 16 ISO containers in two rows of eight, or 80 family-size cars. Nose gear almost unchanged from An-124, but main gears lengthened by adding two twin-wheel units at front on each side. Steering modified by transferring from front two pairs of main wheels on each side to rear four pairs on each side, thus, including two nose gears, 20 of 32 wheels have power steering.

OKB received funding for single An-225, No 480182 (later changed to 82060), and this ceremonially rolled out at Kiev Nov 88. First flight 'from a 1 km runway' 21 Dec 88, and on 22 March 89 106 records set in single flight, including 2000 km circuit flown at 813 km/h (505 mph) with payload of 156·3 t (344,576 lb), taking off 92 t below maximum and cruising at 12 340 m (40,485 ft). Maximum payload tentatively put at 250 t (551,150 lb). Flights with *Buran* began 13 May 89, and a few days later pilot Alex Galunenko brought combination to Paris airshow, flying tight circuits one-handed at very low level, in pouring rain, at bank angles exceeding 45°! After landing he taxied across sodden grass! At that airshow a scheme jointly proposed with British Aerospace for air-launching Hotol aero-spacecraft from An-225, and there are various proposals for similar launch of Soviet aero-space vehicles. ASCC reporting name 'Cossack'. Now grounded and acting as a 'Christmas tree' spares source for An-124s.

DIMENSIONS Span 88,4 m (290 ft 0 in); length 84,0 m (275 ft 7 in); wing area 905 m² (9,741 sq ft).

ENGINES D-18T.

WEIGHTS Equipped empty 216 t (476,200 lb); max payload 250 t; max fuel 300 t; max take-off 600 t (1,322,750 lb).

PERFORMANCE Cruise (loaded) 700 km/h (435 mph), (clean) 850 km/h (528 mph); range with 200 t internal payload 4500 km (2,795 miles); take-off run with Buran 2,5 km (8,200 ft).

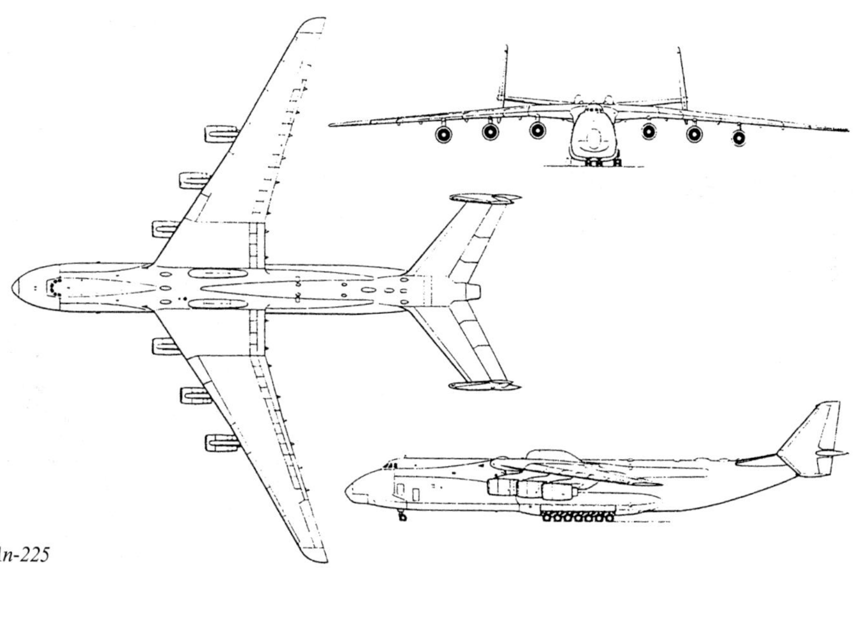

An-225

An-225 with Buran

An-418 ANTK would like to build this projected 'megacarrier', to carry 690 passengers 10 000 km (6,214 miles) or 500/550 passengers 12 000/13 000 km (8,080 miles). It would have to be a new design, though low-mounted wing could be based on An-124, carrying same four D-18T engines. General Designer Balabuyev says 'It would be our 747-400, and would bear comparison with that aircraft'. No ASCC name.

Microlights ANTK has returned to gliders and ultralights with Slavutich and Bravo gliders and a range of microlights, all braced flexwings: M-1 (23 hp Neptun), M-2, two-seat T-2 and faster T-4 (30 hp, with fuselage pod) with 1 h endurance for forestry, ag and transport work.

Ar-2

Arkhangelskii

Aleksandr Aleksandrovich Arkhangelskii was born in December 1892, son of a professor at the University of Kazan. From 1911 he studied at MVTU under Zhukovskii. In the First World War he did tunnel tests and other work on Farmans and the underpowered *Svyatogor*. He joined CAHI on its formation, and for the rest of his life worked closely with Tupolev. He handled much of the design of the ANT-3 and ANT-4, led the ANT-9 team, and was allowed to accompany Tupolev to the USA in 1932. On his return he plunged into the ANT-40 (described under Tupolev). He also designed the vertical tail of the ANT-20. When Tupolev was arrested Arkhangelskii escaped suspicion, and became *de facto* head of the OKB; derivatives of the SB were entirely his creations. (For convenience most are covered in the Tupolev section, but SBB, MMN and Ar-2 are included here.) When new designation system was introduced in 1941 he was allowed privilege of SB-RK becoming Ar-2, but in autumn of that year OKB was evacuated and reverted to Tupolev's name on the latter's release in 1941. Thereafter Arkhangelskii remained First Deputy Designer. He managed the definitive Tu-2S and led the structural team on the Tu-4 and its derivatives. He was effectively chief designer of all early Tupolev jet bombers, turboprop Tu-95, and derived civil transports. In 1945 he was elected to the Moscow Soviet, and in later life devoted much time to research and academic literature.

MMN Spanish civil war showed Tupolev SB still not fast enough. With Tupolev imprisoned, Arkhangelskii in 38 designed new wing of reduced span, NACA-22 high-lift profile, improved slotted flaps, streamlined nose with upper half glazed, nacelles extended to TE almost completely enclosing wheels, otherwise few changes. Three ShKAS (one in turret), bombload 1 t. Prototype 39 recognised as interim aircraft.

DIMENSIONS Span 18,0 m (59 ft 0¾ in); length 12,78 m (41 ft 11⅞ in); wing area 48,2 m^2 (519 ft^2).

ENGINES Two M-105.

WEIGHTS Empty 4820 kg (10,626 lb); fuel/oil 680+70; loaded 6420 kg (14,153 lb).

PERFORMANCE Max speed 360 km/h at SL, 458 (285 mph) at 4,2 km; climb 9·3 min to 5 km; service ceiling 9 km; range 760 km (472 miles); TO 520, ldg 665.

MMN

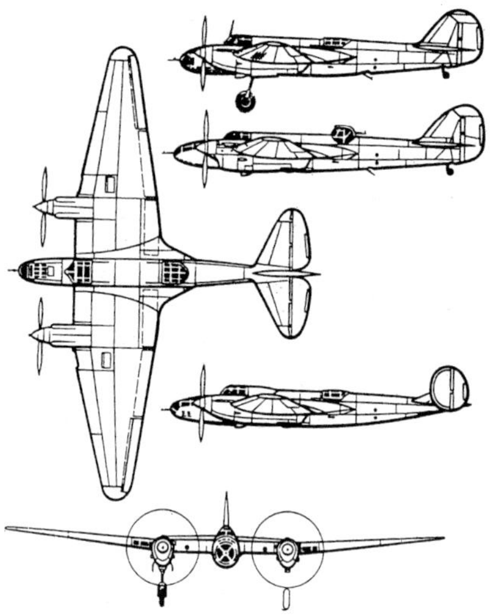
Ar-2 (middle side view MMN-1, bottom SBB-1)

SB-RK, Ar-2 Further 39 devt resulted in improved aircraft. Uprated engines driving VISh-22Ye c/s propellers, in streamlined cowls, radiators being between wing spars fed by LE inlets outboard of engines, outlet in upper surface. Electrically driven multi-slat dive brakes pivoted under outer wings. Fin height slightly reduced, tailplane slightly smaller. Bombload up to 1,5 t (3,307 lb), many combinations. Prototype SB-RK flown early 1940 with original SB twin-gun nose. Nose again redesigned with hemispherical front cupola with single ShKAS; could be two in rear lower and aft dorsal posns (latter replacing turret). Redesignated Ar-2, put into immediate prodn Sept 40, 200 being delivered when GAZ evacuated Oct 41.
DIMENSIONS As MMN except length (SB-RK) 12,27, (Ar) 12,5 m (41 ft 0 in).
ENGINES Two M-105R.
WEIGHTS Empty (RK) 4430 (9,766 lb), (Ar) 5106 kg (11,257 lb); loaded (RK) 6300 (13,889 lb), (Ar) 6600 normal, 8150 (17,967 lb) overload (lightened 1941 aircraft 6,5 t 14,330 lb).
PERFORMANCE Max speed 415 km/h at SL, 475 km/h (295 mph) at 4,7 km, (lightened 1941 aircraft, 512 km/h, 318 mph); climb 7·1 min to 5 km; service ceiling 10 km (1941, 10,5, 34,450 ft); range 1500 km (932 miles); TO 306 m; landing 514.

B, SBB-1 Final devt of SB family, combined engines/props/bombload/aft dorsal cockpit of Ar-2 with nose resembling MMN, new pilot cockpit with aft fairing, revised cooling system (main radiators in wing inboard of engines and oil coolers outboard), new outer wings of even shorter span and without dive brakes and new twin-finned tail, Various structural changes included main spars of multiple glued ply veneers. Flown 40 but considered to offer no advantage over Pe-2.
DIMENSIONS Span 16,0 m (52 ft 5⅞ in); length 12,275 m (40 ft 3¼ in); wing area 46,0 m² (495 ft²).
ENGINES Two M-105R. No record of weights.
PERFORMANCE Max speed 454 at SL, 540 km/h (336 mph) at 4,6 km; climb 6·35 min to 5 km; serv ceiling 10,1 km; range 880 km (547 miles); TO 400m, landing 350 m.

Avia

NPO 'Avia Ltd' joint-stock co formed 92, first product:

Akkord Written ACORD on a/c, multirole private/utility, 5 seats, fuel 300 lit, advanced features include panel with three 8×8 in colour MFDs (MIKBO-43 system). Prototype flew July 92, production expected 95 at $80,000 or (US engines) $150,000.
DIMENSIONS Span 11,3 m (37 ft 0⅞ in); length (landplane) 8,61 m (28 ft 3 in).
ENGINE Two VAZ-4133A or unspecified Lycoming.
WEIGHTS Empty 1014 kg (2,235 lb); payload 500; MTO 1,7 t (3,748 lb).
PERFORMANCE Cruise 65% 260 km/h (162 mph); range (max payload, 45-min reserve) 1000 km (621 miles); TO (land) 175 m, (water) 260.

Aviaprom

AO Aviaprom, 'Aviation Industry', Moscow-based joint-stock company managing several *Konversiya* civil projects with partners.

Eurasia-700 Largest project is high-capacity transport being devd jointly with TAO Sitora and with CAHI help. Blended wing/body, basic wing two spars forming tank 200 m³ (c160 t) with aux tankage available to double capacity. Outer panels CAHI supercrit profile, cruise L/D c20. Fuselage oval 5,8×7,6 m at front, blending into rect mid-section 6,1×12,6, with V-tail at rear. One engine under each wing, one or two above rear. Main gears with 8-wheel (4-axle) bogies for existing airport, retrac inwards, nose unit forwards. Passenger version typically seating 340 in fwd cabin, 440 in aft, all at 780 mm pitch, 4-lane escape slides, various cargo or mixed versions. Fuel burn c.19,3 g/pax-km. Seeking funds.
DIMENSIONS Not yet finalised.
ENGINES Proposed three NK-44.
WEIGHTS OWE c234 t (515,873 lb); MTO 475 t (1,047,178 lb).
PERFORMANCE Cruise M0·82 (c872 km/h, 542 mph); range (max pax) 10000 km (6,215 miles); TO 2240 m; ldg 1980.

Eurasia 18-50 Designation from 18 seats (bizjet) or 50 (regional transport), fuel 25 g/seat-km. Project launched 90 by ANPK MiG, now joint venture with Aviaprom with participation of TAO Sitora and RKIIGA. MiG/MAPO might handle assembly/test. Funds sought.
DIMENSIONS Span 23,3 m (76 ft 5½ in); length 24,9 m (81 ft 8½ in); wing area 64 m² (689 ft²).
ENGINES Two D-36.
WEIGHTS OWE 20,2 t (44,533 lb); MTO 36 t (79,365 lb).
PERFORMANCE Cruise 850 km/h (528 mph); range (50 seats) 8000 km (4,970 miles), (18) 11000-12000 km (<7,455 miles).

Aviaspetstrans

Consortium formed 89 by group of research, engineering and prospecting organizations, bank and Myasishchyev OKB. Objective to promote air transport in remote Arctic/Antarctic, Siberia and far east.

Yamal First aircraft project, multirole amphibian for operation in severest weather with minimal infrastructure. Conventional stressed-skin airframe; wing basic 16% with double-slotted flap with interceptor, 5 screwjacks driven from inside hull; two protected tanks each wing, twin turboshaft engines above hull aft of wing driving high-speed shaft and gearboxes to SV-34 pusher 6-blade prop behind tail; fixed tailplane below prop shaft, rudders above and below prop shaft; main gears retract backwards, tandem wheels somersaulting to lie inverted in blister with side doors, tailwheel retracts upward (nosewheel to be offered later); fixed outer-wing floats; cabin door at right front and (baggage) left rear, 2-seat cockpit under Plexiglas canopy with three metal-framed windscreens, main cabin for 15 pax or various special roles. Digital avionics, option of nose radar. Design objective, every item requiring attention accessible from inside aircraft. Prototypes building by Myasishchyev 93, at least one with PT6 engines donated by P&WC.
DIMENSIONS Span 21,4 m (70 ft 2½ in); length 16825 (55 ft 2½ in); wing area 51,9 m² (558·7 ft²).
ENGINES Two TVD-1500.
WEIGHTS Not stated except payload 2 t (4,409 lb).
PERFORMANCE Max cruise 420 km/h (261 mph); range with max payload 500 km (311 miles), (with 500 kg) over 2500 km (1,553 miles); TO (land/water) 230 m.

Aviatika

Private KB formed 1991 in partnership with MAPO/MiG to produce light aircraft. First designs, 890, 900 Akrobat and 92, see MAI on page 175.

Akkord

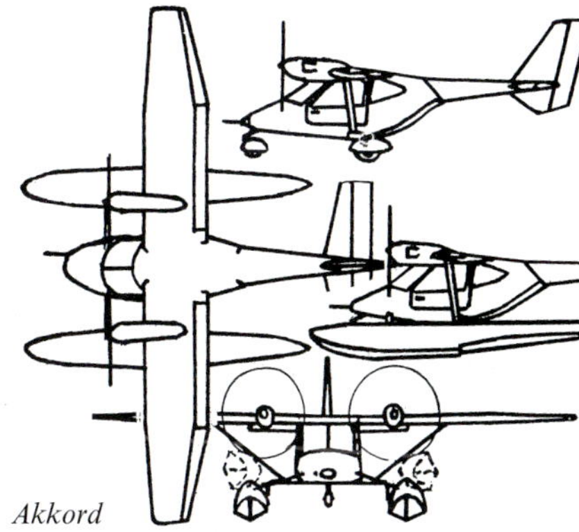
Akkord

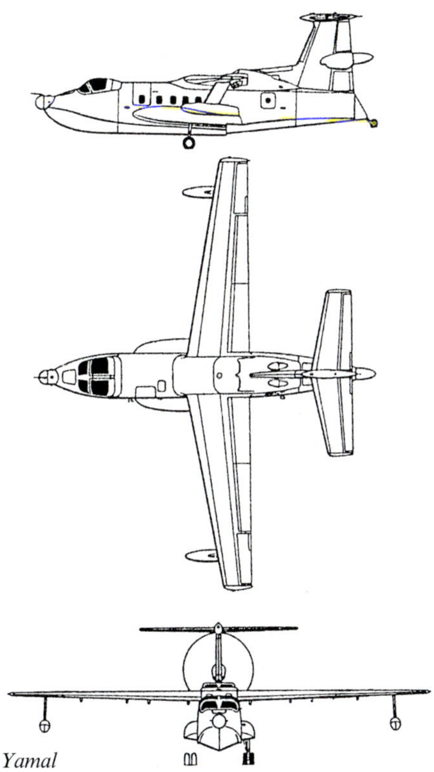
Yamal

Bartini

Roberto L. Bartini (or Bartigni) was born in Italy on 14 May 1897. In 1921 a founder of Italian Communist Party, and organized Milan workers into cells. Two years later Mussolini proscribed the party; Bartini, a qualified engineer, emigrated to the Soviet Union. He spent from 1924-28 as an aviation engineer at the naval aviation experimental centre at Sevastopol, working on seaplanes and flying boats, and then joined the newly formed OSS 'Stal' OKB. By Dec 30 he had shown such talent he was leading his own design group. He continued to show great interest in maritime aircraft, working on Beriev's MBR-2 and Tupolev's MK-1. He suffered political difficulties (apparently because he was of foreign birth) and in Feb 38 he was arrested and sent to Omsk where he formed prison OKB-4, beginning work on supersonic projects in 40! In 42 he was permitted to open special OKB in Siberia (see later) and in 46 he was 'rehabilitated', opening OKB-86, closed 48. After T-117 came string of incredible personal proposals including A/R-55 and A/R-57 strategic attack/recon aircraft of years 55 and 57 both to cruise M2·5+. Later an editor and pioneer of Ekranoplans (which see), receiving Order of Lenin on 70th birthday. Died 6 Dec 74.

Stal-6 Bartini's first 'solo' project was a fighter, not a seaplane or flying boat. In the late 1920s Bartini had devoted great attention to steel construction, and in this experimental aircraft he went all out to create the fastest and most advanced machine possible. He put the proposal to CCB in Oct 30, but it was about 18 months before he obtained permission to build. The key to the design was a finely cowled water-cooled Curtiss Conqueror engine with surface-evaporation radiators in the double skin of the wings. Pilot view was bad. Fuselage: steel tube and ply skin. The wings were of steel, with a double skin of Enerzh-6 back to the front spar to form the radiators. A pilot handcrank worked the large retractable landing wheel on the centre-line and hinged sprung skids under

Stal-6

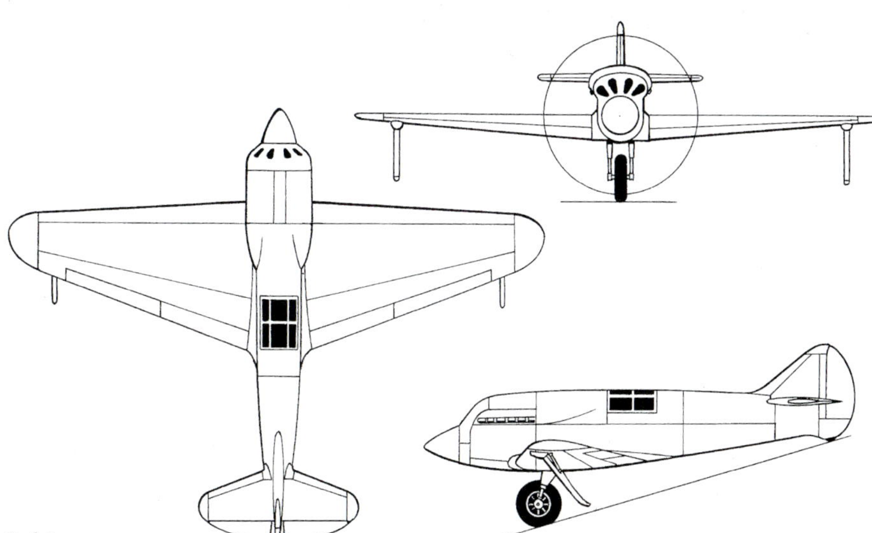
Stal-6

the wingtips. First flight, autumn 33, by A. B. Yumashyev who shared test-flying with P. M. Stefanovskii. First landing was in a cloud of steam, and cooling problems were as intractable as the bad view; but performance was exceptional, setting a Soviet speed record.
DIMENSIONS Span 9,0 m (29 ft 6¾ in); length 6,88 m (22 ft 6¾ in); height 2,23 m (7 ft 4 in); wing area 14,3 m² (154 ft²).
ENGINE 630 hp Curtiss V-1570 Conqueror V-12.
WEIGHTS Empty 850 kg (1874 lb); max 1080 kg (2381 lb).
PERFORMANCE Max speed 420 km/h (261 mph); max climb 21 m/s (4135 ft/min).

Stal-8 Petty jealousies plagued Bartini but he had a true friend in George Ordzhonikidze, who in winter 1932-33 had agreed that the Italian could build a fighter derived from Stal--6, subject to the latter being a success. Stal-8 was almost the same airframe as Stal-6 but stronger for greater weight, and with a proper cockpit, 860 hp M-100A engine (licence-built Hispano-Suiza 12YBR) and two ShKAS in top decking. Design speed was 620 km/h (385 mph), appreciably faster than any other fighter of the day. Probably shortsightedly, VVS rejected the project (on grounds of vulnerability of cooling system) and the prototype remained about 90% finished in Oct 34.

DAR Bartini was assigned in Jan 34 to the ZOK NII GVF to lead the design of the *Dalnii Arkticheskii Razvyedchik* (long-range Arctic reconnaissance, the last word meaning a range of civil rather than military tasks). His answer looked like a copy of the Wal. DAR was entirely of stainless steel, and until a few months before first flight the intention was to shroud the tandem propellers to increase propulsive efficiency. The two engines were to be mounted in tandem in a centreline nacelle carried on struts above the hull, driving via coaxial shafts to two pusher propellers running in a large ring duct. Tests by CAHI confirmed the 'Bartini effect', but at an advanced stage of prototype erection (in the same hangar at the Andre Marti plant in Leningrad as the Sh-5) Bartini reluctantly switched to ordinary tandem engines. Highly stressed structure was Enerzh-6, with spot-welded skins, the trailing edge was hinged tip-to-tip to form slotted flaps, and the ailerons were patented pivoted wingtips made in front and rear sections. DAR was finished in late 1935 and launched at the small-boat port in Leningrad where B. G. Chukhnovskii test-flew it in spring 1936 and was enthusiastic. Later it was fitted with a pair of steel-faced wooden skis, each 5,0×0,32 m, sprung by rubber bags along the sides of the hull, as previously tested on a Wal. Five ordered but no action taken to build; again a matter of personalities.
DIMENSIONS Span 27,4 m (89 ft 10¾ in); length 19,0 m (62 ft 4 in); wing area 100 m² (107·6 ft²).
ENGINES Tandem 860 hp Hispano-Suiza 12Ybrs V-12.
WEIGHTS Empty 4820 kg (10,626 lb); max 7200 kg (15,873 lb).
PERFORMANCE Max speed 240 km/h (149 mph); cruise 229 km/h (142 mph); range 2000 km (1,243 miles); max endurance at reduced speed 20 h.

DAR

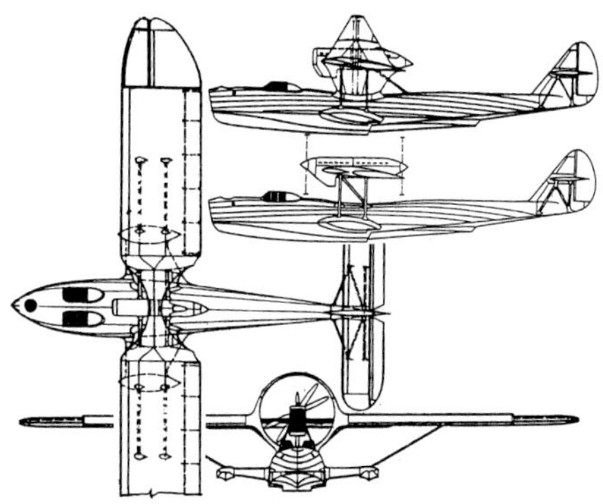
DAR as designed; bottom side view, as built

Stal-7 In 1934 Aeroflot issued a requirement for two modern transport aircraft, the larger to be a twin for 10/12 passengers. Two submissions were accepted for prototype construction, the ZIG and Bartini's Stal-7. The Italian was still heading the NII GVF team, and had already schemed a passenger transport to fly at 400 km/h in 1933. Design began in Oct 34 and though of advanced form was to be a steel truss airframe with fabric covering. Great problems were met with the fuselage, which had 200 primary intersections between tubes of many cross-sections yet deflected excessively – especially in a bomber version which had to be studied in parallel to meet a Stalin decree. By 1935 the Stal-7 had become a light-alloy monocoque aircraft with a somewhat constricted body with sloping sides whose cabin was severely interrupted by the spars of the wing whose inner sections (of cambered profile with concave undersurface) sloped sharply up to pass through the fuselage in the mid-position. The one thing Stal-7 had was efficiency in speed, range and load (though a low wing and convenient cabin would hardly have hurt it). No fewer than 27 tanks of 7400 lit (1628 gal) capacity were installed, and flaps and gear were hydraulic. First flight spring 37 by N. P. Shebanov. Better than predicted performance. But aircraft crashed taking off on a full-load test. This resulted in Bartini's arrest. The aircraft was not repaired, under Yermolayev's direction, until 1939. On 28 Aug 39 Shebanov, V. A. Matveyev and radio operator N. A. Baikuzov flew Moscow, Sverdlovsk, Sebastopol, Moscow (5068 km, 3,149 miles), 405 km/h (252 mph). Performance led to DB-240 long-range bombers (see Yermolayev).
DIMENSIONS Span 23,0 m (75 ft 5½ in); length 16,0 m (52 ft 6 in); wing area 72 m² (775 ft²).
ENGINES Two 760 hp M-100 V-12, later (1939) 860 hp M-103.
WEIGHTS Empty 4800 kg (10,580 lb); fuel/oil 6000 kg (13,230 lb); max loaded 11 000 kg (24,250 lb).
PERFORMANCE Max speed 450 km/h (280 mph) at 3 km (9,850 ft); cruise 360/380 km/h (224/236 mph); ceiling 10 km (32,800 ft); range 5000 km (3,107 miles).

R Around 1942 Bartini was permitted to organise a special OKB in Siberia dedicated chiefly to jet interceptors. Chronologically the first design, never built, was this twin-jet flying wing, remarkably similar to Horten designs but with 35° sweep and with variable span and area achieved by telescopically extending the outer panels.

R-114 Project for vertically launched target-defence interceptor with 33° sweep and aerodynamics based on the R. Amazingly advanced for 1943 (pre-German data); 300 kg (661 lb) Glushko rocket, infra-red target tracker. Construction begun but not completed.

T-107 In 1944-46 Bartini was switched to civil transport. Project for mid-wing double-decker with two ASh-82; recommended for manufacture 1945 but Il-12 chosen instead.

T-108 Unbuilt proposal for high-wing light cargo aircraft with twin 340 hp diesel engines.

T-117 Challenging project for high-capacity transport for main routes, begun in 1944. Schemed in two models, A for passengers, B for cargo. Latter with box fuselage and rear

Stal-7

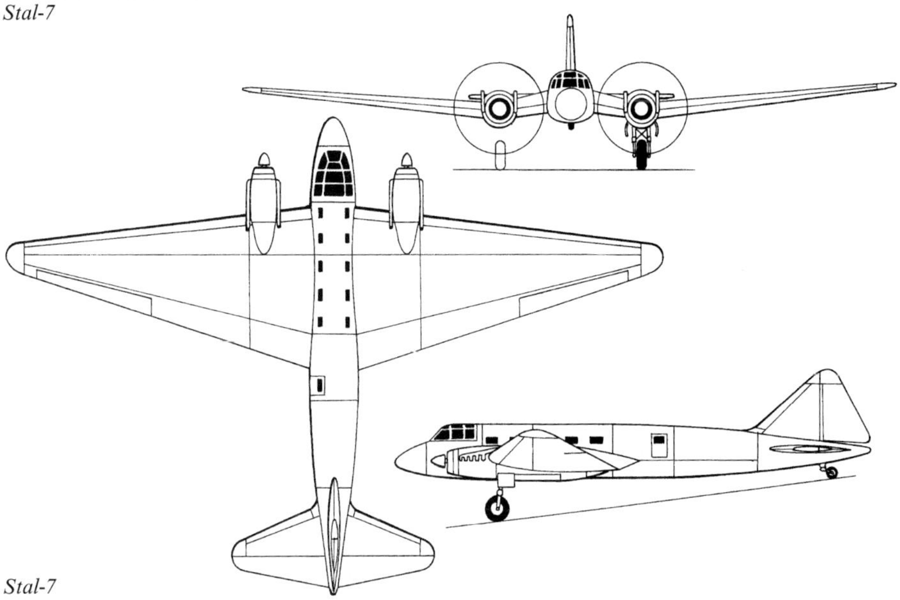

Stal-7

ramp/door, but model A chosen for construction with fully pressurized double-deck, double-bubble fuselage, both upper and lower lobes same size (1,45 m rad) giving body 5 m (16 ft 5 in) high and 2,9 m (114 in) wide. Packed with technically novel features, some even derived from Project R. Discontinued June 48 when 80% complete, mainly because Tu-4 took all available engine production.

DIMENSIONS Span 35,0 m (114 ft 10 in); wing area 128 m^2 (1,378 ft^2).

ENGINES Two 2600 hp ASh-73.

WEIGHTS Empty 11 800 kg (26,015 lb); max loaded (crew 4, 80 passengers) 25 t (55,115 lb).

PERFORMANCE Estimated cruising speed 540 km/h (336 mph); ceiling 12 km (39,400 ft).

T-200 Heavy military transport with main cargo hold 5 m (16 ft 5 in) wide and 3 m (9 ft 11 in) high. Engines, four 2800 hp ASh piston engines (projected) and two 2270 kg RD-45 turbojets. Abounded in interesting features when final design submitted in Oct 47, but Bartini's design OKB-86 was closed the following year. In 1952 Bartini was appointed to strategic planning at the NII. Bartini continued his design work, producing a number of important projects; initially working in Novosibirsk, then in Kamov's Factory 290 near Moscow. The 1955-57 period saw refinement of a giant amphibious bomber carrying a parasite recon aircraft. As the Mach 2 capable A-57 the carriers weighed 75 000 kg empty (90 000 kg max T/O) and was to have been powered by five 26 000 kg Kuznetsov NK-10B or 22 500 kg NK-6 engines. The E-57 was a derived bomber, and the F-57 A 70 seat SST. The carrier aircraft was a massive cranked Delta, with four leading edge sweep angles, inboard sharply swept sections leading forward as LERX/chines. Twin fins were set apart on each side of the boxed in engines, which also acted as a support/launch platform for the parasite aircraft, which resembled the SO9000 Trident in configuration, with trapezoidal wings and tip mounted engines.

The T-200 and -210 were heavy transports with a high-set tapered wing, twin turboprop engines set close-in two the bulky but streamlined fuselage. This tapered to a point in side elevation, thereby acting as a massive aerofoil, with slender booms projecting aft from the wing roots to support the horizontal tail and its endplate fins. The T-217 was of similar configuration, but with four NK-12M engines and was intended for the carriage of ICBMs. These aircraft never progressed beyond model stage, but Bartini's unique concept for an amphibious heavy transport/ASW aircraft using ground effect to allow the carriage of very heavy loads went further. Designated M, and later M-62 the VTO aircraft was intended to fly at very low level, riding on a cushion of air generated by the wings and trapped between them and the land or sea. It was to have an AUW of some 2,500 tonnes – an ambitious target which necessitated an order for three proof-of-concept prototypes. Subsequently Bartini became interested in unconventional anti-submarine systems. These will be found in the *Ekranoplan* section at the back of the book.

Belyayev

Viktor Nikolayevich Belyayev was born in 1896. In 1920 he built a glider, unusual for the day in having wheels, and subsequently was accepted into CAHI. There, among many other tasks, he built his own gliders in 1933, BP-2 and BP-3 (BP, tailless glider), which did well at Koktebel; a BP-2 (CAHI-2) was towed behind an R-5 to Moscow. Subsequently Belyayev worked with Aviavnito and Aeroflot, OMOS, AGOS, KOSOS and the Tupolev OKB, gaining a reputation for versatility and for his technical papers. In 1934 he designed a transport with two tail booms each accommodating ten passengers (M-25 engines).

DB-LK Ideal expression of Belyayev's search for inherently stable wing and superior geometrical efficiency. Initials stood for long-range bomber, flying wing, but it was far from being the latter. Two fuselages of minimum length joined by wing centre section of short span but large chord (5 m) and CAHI MV-6bis profile. Unrelated outer wings of Göttingen 387 profile, aspect ratio 8·2, swept forward 5°42' at leading edge but with backswept tips, taper ratio 7. Frise ailerons with small extra sections on raked tips, slats on outer wings, 45° Zapp flaps. Five-spar light-alloy construction. Large strut-braced fin on centreline carrying near its top a very small (0,85 m^2) tailplane to which were higned relatively enormous (4,8 m^2 total) elevators with large tabs.

Crew of four: pilot in left fuselage, navigator in right, gunners in each tailcone, with access via roof hatches, to manage radio controls and two pairs of ShKAS. Two more ShKAS in leading edge on centreline, firing ahead with remote control within cone of 10°. Total 4500 rounds for six guns. Bomb load carried in fuselage immediately behind single-leg main gears: two FAB-1000 (1000 kg each), four FAB-250, two FAB-500 and two Der-19 or Der-20 containers, or 58 small bombs.

Prototype complete Nov 39. Test pilot M. A. Nyukhtikov, lead engineer T. T. Samarin, test observer N. I. Shavrov. Ready for flight early 40, with Mark Gallai joining team at this time. All-round performance outstanding, but project unable to displace Il-4 even in projected form with ASh-71 engines.

DIMENSIONS Span 21,6 m (70 ft 10½ in); length 9,78 m (32 ft 1 in); height 3,65 m (11 ft 11¾ in); wing area 56,87 m^2 (612 ft^2).

ENGINES Two 1100 hp M-88 in long-chord gilled cowlings driving 3,3 m (130 in) VISh-23D propellers.

WEIGHTS Empty 6004 kg (13,236 lb); fuel (3444 lit, 758 gal) 1048 kg (2310 lb); normal loaded 9061 kg (19,976 lb); max 10 672 kg (23,528 lb).

PERFORMANCE Max speed 395 km/h (245 mph) at sea level, 488 km/h (303 mph) at 5,1 km (16,730 ft); take-off speed 145 km/h (90 mph); landing speed 150 km/h (93 mph); max rate of climb 6,15 m/s (1210 ft/min); ceiling 8,5 km (27,890 ft); range with 1000 kg bombload, 1270 km (789 miles) at normal gross, 2900 km (1,800 miles) at overload.

Babochka Belyayev's curious wing appeared amenable to flexible construction permitting upward deflection under manoeuvre load without pitch problem. Worked with V. I. Yukharin at Kazan KAI on tandem-seat research aircraft to investigate, abandoned at war 41.

DIMENSIONS Span 10,8 m (35 ft 5¼ in); length 6840 (22 ft 5¼ in).

ENGINE One MV-6.

WEIGHTS Empty 680 kg (1,499 lb); loaded 1928 kg (4,250 lb).

PERFORMANCE Max speed 510 km/h (317 mph).

EOI EOI, exptl single-seat fighter, authorized Aug 39. Twin-boom pusher with pilot in completely glazed nose. Few details survive, though recently photo found of mock-up (captioned Secret, Belyayev PBI) which apparently shows a PBI (dive-bomber fighter) version. Prototype and drawings all lost in Oct 41 evacuation and no archive known.

DIMENSIONS Span 11,4 m (37 ft 5 in); wing area 19 m^2 (205 ft^2).

ENGINE One M-105 (prototype), M-106 (production). No other data.

DB-LK

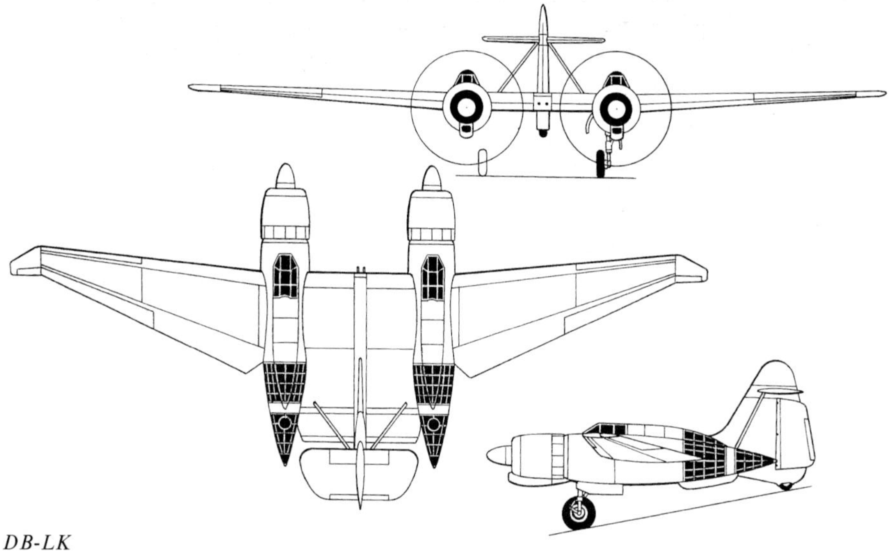

DB-LK

Bereznyak-Isayev

With designation **BI-1**, for the designers, the Soviet Union built the first rocket fighter in the world. Aleksandr Yakovlevich Bereznyak witnessed the tests of Dushkin's definitive rocket engine in 1939, and flight of RP-318 in early 40. He discussed the practicability of a rocket-engined fighter with one of the lead engineers in Dushkin's group, Aleksei Mikhailovich Isayev. The latter had been a designer of RP-318 and was closely involved with Dushkin's next-generation engine, the D-1A. Fired with enthusiasm, Bereznyak went to V. F. Bolkhovitinov and in turn got the mercurial professor so convinced of the rightness of such a project that he took it to GUAP and to Stalin himself.

Kremlin sanctioned five prototypes 9 July 41. Working three shifts, first (unpowered) ready in 35 days. First flight by B. M. Kudrin 10 Sept 41, tug Pe-2, no problems. Factory evacuated 16 Oct to incomplete shed Sverdlovsk. Kudrin (ill) replaced by G. Ya. Bakhchivandzhi. Engine installed Jan 42, but exploded 20 Feb, injuring pilot and technicians. Subsequent management by Prof Puishnov of CAHI, whose T-5 tunnel had tested actual BI-1 (with tip ramjets). Second pilot, K. A. Gruzdyev, appointed, but Bakhchivandzhi made first powered flight 15 May 42.

The BI-1 first flew on skis, attached to the original main legs and retracting inwards by compressed air which also worked the split flaps. The latter were almost the only metal parts of the airframe, structure being a smooth wooden assembly with painted fabric skin over moulded ply. Tailwheel mounted on ventral fin (BI-1 was prone to nose over) and tailplane carried endplate auxiliary fins and on powered prototypes was braced by struts to both upper and lower fins. Nitric acid and kerosene fed from welded steel cylinders in centre fuselage by nitrogen bottles in nose, full-thrust endurance 80 sec. Armament of two ShVAK, each with 45 rounds, in upper nose above gas bottles. Canopy slid to rear, and radio installed.

Prolonged delays caused mainly by acid corrosion, with at least two minor spillages or explosions. By March 43 seven BI-1s had been completed, making numerous towed flights, but powered flights 2 to 6 not accomplished until Feb/March 43. On 21 March height of 3 km (9,843 ft) reached in measured time of 30 seconds. On 27 March, on flight No 7, Bakhchivandzhi made high-speed run at low level and went straight into ground. CAHI had not discovered existence of severe pitch-down at high speeds, and subsequently failed to find a solution. Production batch of 50 BI-1 abandoned, but prototypes continued active, modified with series of later Dushkin engines with large and small (cruise) thrust chambers, with aircraft designations BI-2, BI-3 and BI-7 (originally BI-1 Nos 2, 3 and 7). VVS interest waned, despite excellent flying qualities. Replica at Monino.

DIMENSIONS Span (Nos 1, 2) 6,48 m (21 ft 3 in),

PBI mock-up

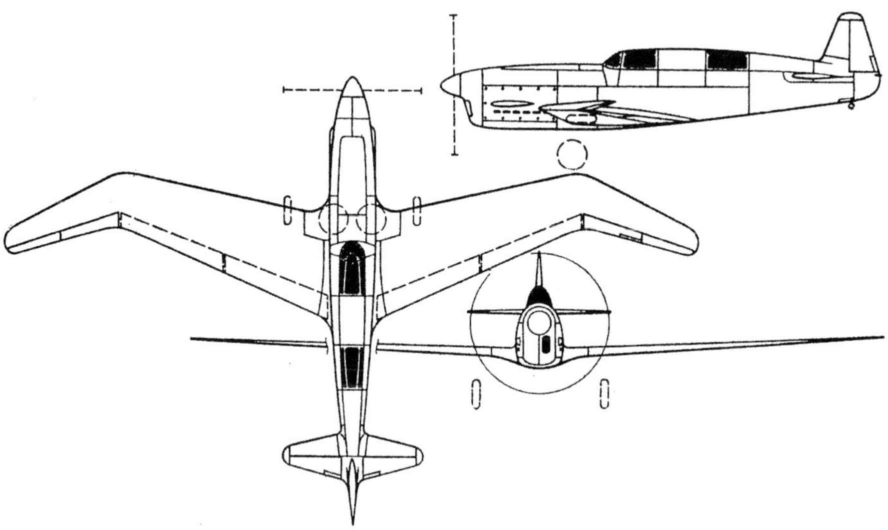

Babochka

(No 3 and probably subsequent) 6,6 m (21 ft 8 in); length (1,2) 6,4 m (21 ft 0 in), (3) 6935 m (22 ft 9 in); wing area (1, 2) 7,0 m² (75·3 ft²).

ENGINE (3) One D-14-1100.

WEIGHTS Empty (1) 462 kg (1,019 lb), (3) 790 kg (1,742 lb), (series) 805 kg (1,775 lb); acid 135 kg (298 lb), kerosene 570 kg (1,257 lb); loaded 1650 kg (3,638 lb); max loaded 1683 kg (3,710 lb).

PERFORMANCE Max speed (est) originally 800 km/h (497 mph), later (early 1943) 900 km/h (559 mph); time to accelerate 800 to 900 km/h, 20s; take-off 400 m (1,310 ft); initial climb 82 m/s (16,400 ft/min); landing speed 143 km/h (89 mph).

Beriev

Few Soviet constructors have had such a straightforward career as Georgii Mikhailovich Beriev. A Georgian, he was born in 1902 and grew up in Tiflis (Tbilisi), and gained acceptance to Leningrad Polytechnic Institute in turbulent 1919. On graduation he joined Aviatrust, and when Richard was appointed to form a design group for marine aircraft, in 1928, Beriev was one of 20 qualified engineers assigned to him. In 1929 he was invited to form a design section for marine aircraft within CCB in Moscow, and he went to Taganrog to undertake improvements to MBR-4. Beriev sought permission to build an improved replacement, and this, MBR-2, launched his own OKB in 1932. Ever since (except for 1942-45) it has been located at Taganrog, and from 1948 has had a monopoly of marine aircraft in the Soviet Union. His last design was a landplane, the Be-30/-32. At this time (68) OKB was, unenthusiastically, assigned to work on Bartini's *Ekranoplan* (see that section). Beriev died 1979,

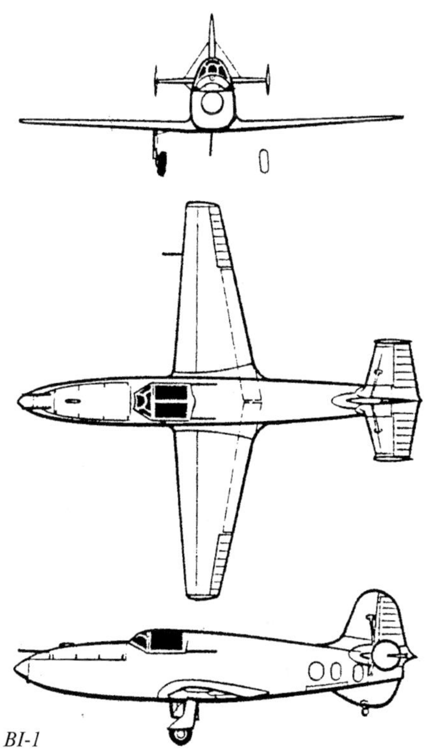

BI-1

successor Aleksandr K. Konstantinov, in 90 expanded as 'Taganrog ANTK named for G. M. Beriev' under General Designer Gennardi S. Panatov.

MBR-4 Though in no way a Beriev design, this flying boat did incorporate his work, undertaken at the Taganrog factory in 1930-31. A single-engined pusher biplane, it was originally the Savoia-Marchetti S.62 bis. After building 22 to the Italian design, a further 29 were built of the MBR-4 version. The specification is the same as for the S.62 bis except for empty and gross weights being 200 kg greater and performance slightly worse.

DIMENSIONS Span 16,6 m (54 ft 5½ in); length 12,26 m (40 ft 2½ in); wing area 69,52 m^2 (748 ft^2).

ENGINE One 750 hp Isotta-Fraschini Asso V-12.

WEIGHTS Empty 2840 kg (6,261 lb); fuel/oil 795+40 kg (1,753+88 lb); loaded 4300 kg (9,480 lb).

PERFORMANCE Max speed 220 km/h (137 mph); cruise 180 km/h (112 mph); range 900 km (560 miles).

MBR-2 This was at once Beriev's first and most important design, being one of a select number of flying boats to achieve a four-figure production total. Designation stood for *Morskoi Blizhnii Razvyedchik*, marine short-range reconnaissance. Beriev began the design at CCB in May 31, assisted by I. V. Ostoslavskii, M. P. Mogilyevskii and A. N. Dobrovolskii, and the prototype was built at the old Menzhinskii works in Moscow in 1932. Straightforward high-wing monoplane with wing thick enough to be cantilever (MOS-27 profile, 18% at root and 10% at tip). Structure entirely of wood except for light-alloy tail unit (based on MBR-4) and ailerons, with fabric covering. Skin mainly 3 mm ply elsewhere except for heavy sections along four hull chines, step and main frames. Hull interior unobstructed for easy access between pilot cockpit (one or two seats side-by-side) and open gun cockpits at bow and in rear fuselage. M-17b engine mounted on pylon chiefly of steel tubes and driving four-blade pusher propeller. Four tanks for 750 lit (165 gal) between wing centre-section spars at 18 and 50% chord. A. K. Belenkov designed a beaching chassis whose main vertical strut on each side clipped into a fixture under the front spar near the root, with a separate auxiliary two-wheel truck placed under the rear step. With small changes this chassis could be used as ski landing gear.

Prototype taken in sections to Sevastopol and first flown there by B. Bukhgolts in Oct 32. At about this time Beriev set up his own OKB at Taganrog, and all subsequent engineering and production was centred there. From the first it was evident MBR-2 was robust and efficient, and first version (about 100) was an armed MA variant. The first delivery took place in spring 34, and the type was to serve for 30 years. Normal mission crew was three, the front gunner doubling as navigator although routine method was to keep coast in sight. Rear gunner, armed with one PV-1 like partner, manned radio which was standard in all MBRs. Max bombload 500 kg (1,102 lb), tested on prototype, but normal load 300 kg (661 lb) carried externally under inner wings. Before end of 1934 Taganrog plant was also in modest production with civil version for Aeroflot, **MP-1** (*Morskoi Passazhirskii*). Central cabin with wall trim, soundproofing and thermal lining, and six seats (not made by OKB) for six passengers, boarding via hatch replacing rear cockpit of military version. Bow cockpit usually deleted and, thanks to better streamlining, speed higher than for original model. Another model was **MP-1T** freighter, identical to original military version but with armament provisions deleted. Later a flat floor was added in Aeroflot workshops.

MBR-2 prototype

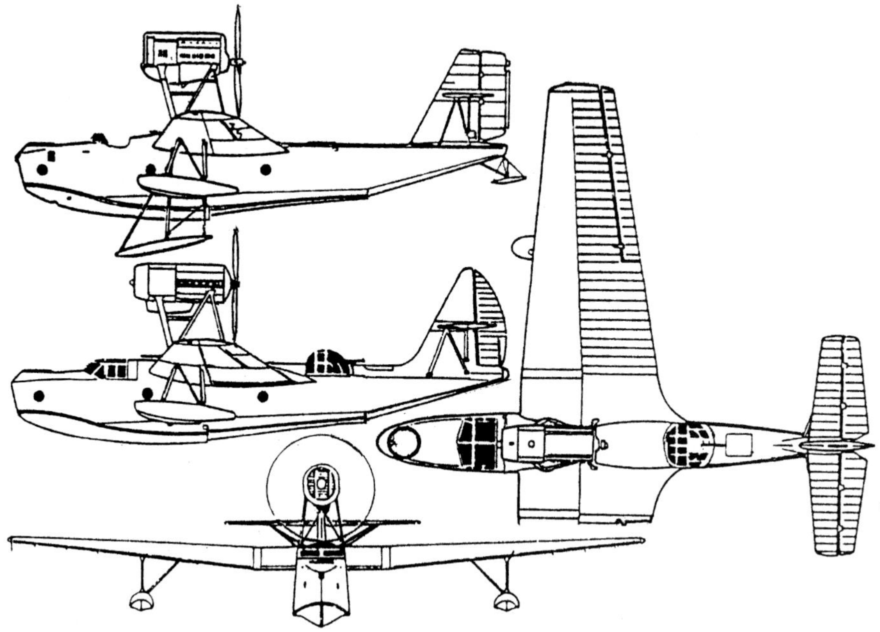

MBR-2/AM-34 (upper side view, MBR-2/M-17b with ski gear)

In 1934 Beriev began improvement programme which resulted in the major production version; no separate designation, though sometimes called MBR-2 AM-34 from the new engine. This was installed in a completely different nacelle with a circular ducted readiator at the front in place of the car-type originally used. New propeller with two or four blades. Tail was redesigned, looking even more like that of S.62 bis, and hull was made deeper, with larger enclosed cockpit and a retractable glazed turret for rear gunner, traversed by hand. Structure strengthened for greater weights and speeds, fuel load varying between 540 lit (original standard), 515 (MP-1), 580 (initial AM-34 batch), 670 (148 gal) and 886 (195 gal). Bomb load of 500 kg routinely carried, often as depth charges, and PV-1s replaced by ShKAS. This variant remained in production until 1942, when Taganrog had to be evacuated. By that time total production had reached 1300 of all variants, which were standard equipment with the 15th Reconnaissance Regt of the Red Banner (Baltic) Fleet, 119th Air Regt of the Black Sea Fleet, the 115th Aviation Regt in the Pacific Fleet and numerous utility, transport and training units. Several dozen M-17b-powered aircraft were rebuilt with new tails, and at least an equal number were converted as **BU** five-seaters for co-operation with fast patrol and torpedo boats.

The civil counterpart was **MP-1bis**, delivered from 1937. They quickly became the most important water-based transports, serving on short hauls from coastal bases and throughout the Siberian river system. In May 37 an MP-1 bis flown by woman-pilot P. D. Osipyenko set up several class records including climbs to 8864 m (29,081 ft), and to 7009 m (22,995 ft) with 1 tonne payload. On 2 July 37 she flew with an all-women crew of three non-stop 2416 km (1,501 miles) over the route Sevastopol, Kiev, Novgorod, Arkhangelsk in 10 h 33 min. The only other major variant was a single prototype flown in 37 with M-103 engine. Significant increase in performance, but no spare engines to support production. ASCC name 'Mote'.

DIMENSIONS Span 19,0 m (62 ft 4 in); length 13,5 m (44 ft 3¾ in); wing area 55 m² (592 ft²).

ENGINE One 500/730 hp BMW VIF (prototype), or 500/730 hp M-17b, or 750/830 hp AM-34B (MP-1 bis) or AM-34NB (military). One aircraft, 850/1000 hp M-103. All V-12 water-cooled.

WEIGHTS Empty (prototype) 2450 kg, (M-17b) 2475 kg (5,456 lb), (MP-1) 2640 kg (5,820 lb), (MP-1T) 2525 kg, (AM-34NB) 2718 kg (5,992 lb), (MP-1 bis) 3119 kg (6,876 lb); normal loaded (prototype) 3700 kg, (most variants) 4100 kg (9,039 lb), (MP-1 overload) 4,500 kg (9,921 lb), (MP-1 bis) 4,640 kg (10,230 lb).

PERFORMANCE Max speed (prototype) 208 km/h, (M-17b) 203 km/h (126 mph), (MP-1) 214 km/h (133 mph), (AM-34NB) 238 km/h (148 mph), (MP-1 bis) 260 km/h (162 mph), (M-103) 295 km/h (183 mph); typical cruise 160 km/h (100 mph); range (M-17b) 650 km (404 miles), (AM-34NB) 800 km (497 miles), MP-1 bis 750 km (466 miles); normal endurance (typical) 5h.

MBR-2/M-17b

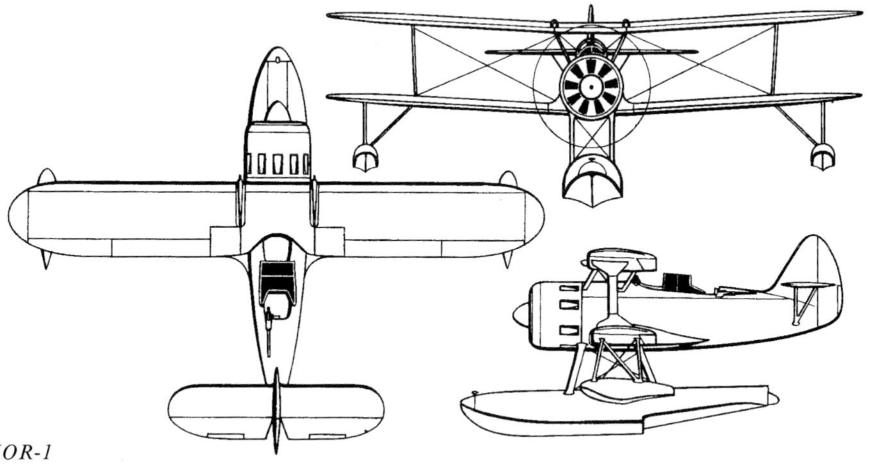

KOR-1

KOR-1 (Be-2) Under Stalin's first Five-Year Plan the Soviet Navy had to be dramatically overhauled and modernised, one provision being installation of (British) catapults on major surface vessels. No time to produce a new aeroplane for shipboard use; the Heinkel HD55 was adopted as the KR-1 (*Korabelnii Razveyedchik*, shipboard reconnaissance). Though adequate, the KR-1 was foreign and could be improved upon in many ways, and Beriev planned a successor as the Be-2 (bureau designation). Simple metal airframe with biplane wings folding to rear, with central float. Fuselage welded from steel tubes, with fabric covering apart from light-alloy engine cowl (with plated front with air cooling apertures ahead of cylinders) and front upper decking. Dural wings with two main spars and Warren-truss ribs, again with fabric covering, and hollow I-struts. Duralumin floats with flush riveting. Ailerons on upper wings, pneumatic landing flaps on lower. Tandem cockpits for pilot and observer/radio/gunner, with dual flight controls. Two windshields or, more often, glazed superstructure around front of large rear cockpit. Normally equipped with two forward-firing ShKAS in fairings above upper centre section with magazines under dural wing-skin covers, single ShKAS aimed by gunner, folding down into recess ahead of fin, and lower-wing racks for two FAB-100 bombs.

First flight, probably April 36; service delivery 37. Considerable problems with seaworthiness, structural deflection afloat and during catapult shots, and also with engine overheating. Though production continued, at least 300 being built by 40, the release for service use in 37, with service designation KOR-1 was only partial. Most were used for shoreline patrol and customs duties. Beriev strove to rectify defects, and in 39 obtained full release for the KOR-1 for naval aviation use without armament and with normal take-offs only. In 39 a landplane version, with no separate designation, was also flown. Though this was ungainly, it was free from structural restrictions. Some were converted seaplanes. Both sea and land versions were still in service in June 41, and the latter was pressed into use as a close-support and attack aircraft on the Romanian and Besarabian fronts.

KOR-1 landplane

DIMENSIONS Span 11,0 m (36 ft 1 in); length

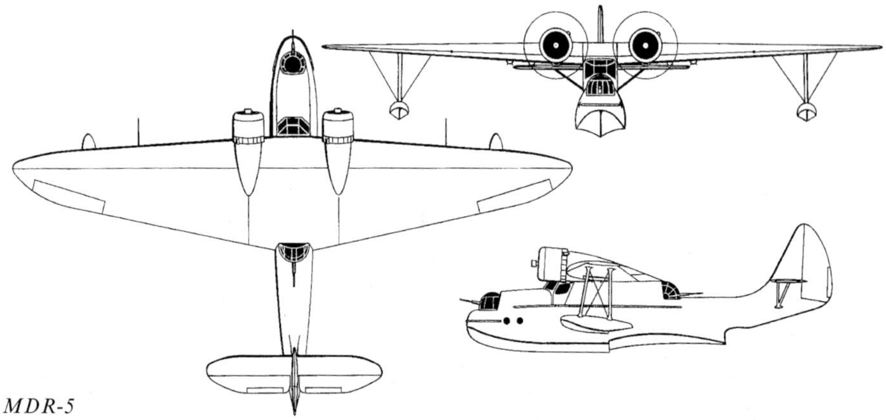

MDR-5

MDR-5

8,67 m (28 ft 5½ in), (landplane 7,2 m, 23 ft 7½ in); wing area 29,3 m^3 (315·4 ft^2).
ENGINE One 635/700 hp M-25 driving three-blade metal (Hamilton) propeller.
WEIGHTS (Seaplane) Empty 1800 kg (3968 lb); fuel/oil 293 kg (646 lb); loaded 2486 kg (5480 lb), overload with bombs 2686 kg (5922 lb).
PERFORMANCE (Seaplane) Max speed 245 km/h (152 mph) at sea level, 277 km/h (172 mph) at 2 km (6,560 ft); climb 1 km (3,280 ft) in 3·2 min; ceiling 6,6 km (21,650 ft); range 530/1000 km (329/621 miles) depending on fuel and overload clearance.

MDR-5 Beriev responded to a 1936 demand for a Russian flying boat for long-range reconnaissance. MDR-5, *Morskoi Dalnii Razvyedchik* = Marine Long-range Reconnaissance, and also called **MS-5** (Marine Aeroplane 5). First MDR-5 provided only for beaching chassis with three twin-wheel units each independently attached to the hull. Structure duralumin, though control surfaces were fabric-covered. Crew of five comprised bow gunner, pilot, navigator, radio operator and observer/gunner. Wing of MOS-2718 profile with four spars and containing eight fuel tanks. Aspect ratio 7·96. Wing mounted on raised part of hull behind side-by-side cockpit with gun turret housing retractable ShKAS (usually a twin installation) at rear of this elevated portion. A second ShKAS in bow turret, traversed by hand. Eight racks under wing at extremity of centre section for total of 1000 kg (2,205 lb) bombs, depth charges or other stores. Engines in long-chord cowlings with gills, driving VISh-3 three-blade constant speed propellers. Special effort made to minimise aileron span to increase size of electrically driven flaps, because wing loading was significantly higher than previous Russian practice.

Two prototypes built, the first flying in 1938. The second was an amphibian, with landing gear similar in geometry to that of the S-43. On test, the MDR-5 proved itself generally satisfactory, but the nose of the stabilizing floats had to be extended 0,3 m (1 ft) forwards to improve seaworthiness, and the amphibian variant was modified by making the wheels completely recessed when retracted. Had it not been for the MDR-6 this design would probably have been produced in quantity.
DIMENSIONS Span 25,0 m (82 ft 0½ in); length 15,88 m (52 ft 1¼ in); wing area 78,5 m^2 (845 ft^2).
ENGINES Two 950 hp M-87A.
WEIGHTS Empty (flying boat) 6083 kg (13,410 lb); fuel either 1917 or 2135 kg (4,226 or 4,707 lb); loaded 8000 kg (17,637 lb), (max fuel and bombs) 9200 kg (20,282 lb).
PERFORMANCE Max speed 283 km/h (176 mph) at sea level, 345 km/h (214 mph) at rated height; landing speed 120 km/h (74·5 mph); ceiling 8150 m (26,740 ft); range (normal fuel) 2415 km (1,500 miles).

MBR-7 Also designated **MS-8** (Marine Aeroplane 8), this was a natural outgrowth of the M-103-engined MBR-2 which had flown in 1937. Though prospects of getting production engines were poor, Beriev redesigned MBR-2 with a smaller and more modern airframe, with a long planing bottom with a step amidships offering reduced drag, and a second step curving down right at the stern. Structure remained all-wood, the hull being covered with glued fabric without application of dope. Bow and tail duralumin, control surfaces with fabric covering, and steel pick-ups for beaching chassis which differed from that of MBR-2. Wing of gull form, with electrically driven flaps. Crew of two only, with glazing over pilot in single-place cockpit, and rear hull decking cranked by screwthread to form shield for observer/gunner with ShKAS. Another ShKAS fixed in nose, fired by pilot. Wing racks for normal bombload of four FAB-100 plus two FAB-50. Prototype built at Taganrog 1938, and flew early 39. Handling at speeds over 200 km/h excellent, but as weight was only slightly less than MBR-2 and wing area had been cut to less than half, pilots found take-off and landing difficult.
DIMENSIONS Span 13,0 m (42 ft 8 in); length 10,6 m (34 ft 9¼ in); wing area 26 m^2 (280 ft^2).
ENGINE One 960 hp M-103, driving large two-

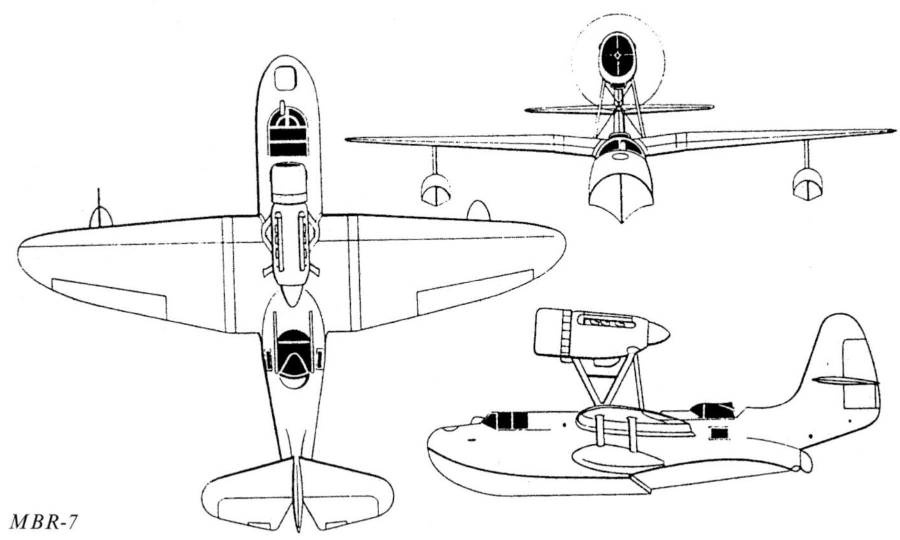

MBR-7

KOR-4 (Be-8)

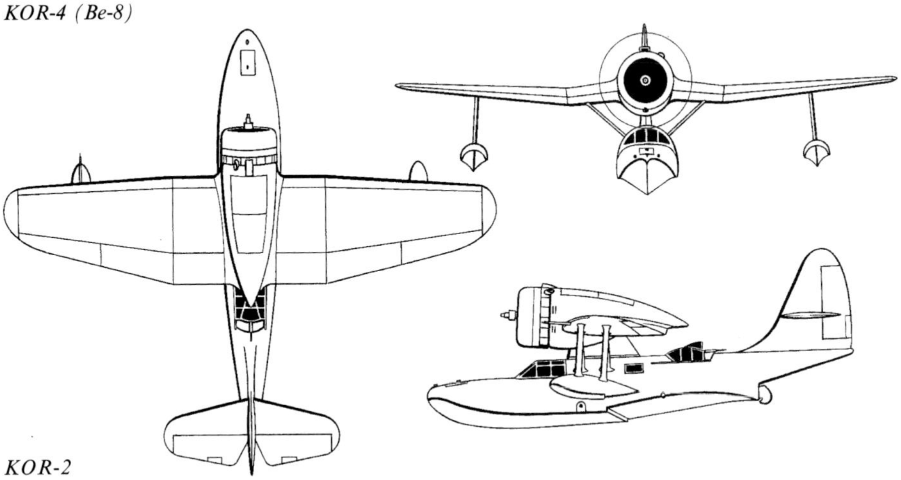

KOR-2

blade VISh-2PT constant-speed propeller.
WEIGHTS Empty 2418 kg (5,331 lb); fuel 398 kg (877 lb); loaded 3,168 kg (6,984 lb), (overload, with bombs) 3600 kg (7,937 lb).
PERFORMANCE Max speed 310 km/h (193 mph) at sea level, 376 km/h (234 mph) at rated height; climb 2·4 min to 1 km, 7·9 min to 2 km ceiling 8500 m (27,890 ft); range 720/1215 km (447/755 miles) depending on fuel and overload.

KOR-2 (Be-4) Distressing at shortcomings of KOR-1, Beriev was determined to create a first-class replacement. Stressed-skin construction decided from outset of planning in August 1939.

Hull duralumin, with efficient planing bottom similar to MBR-7. Side anchorages for beaching chassis. Crew of three: pilot and navigator/radio operator side-by-side (optional dual control) in enclosed cockpit, observer/gunner in glazed rear cockpit with pivoted rear hood forming windshield for gunnery. One ShKAS in rear cockpit, another fixed in bows aimed by pilot. In overload condition (one report states 'overload variant') provision for four FAB-100 or same mass (400 kg, 882 lb) of other stores on racks at extremities of centre section.

Centre section housed tankage and electrically driven split flaps, with additional split flaps on outer panels folded manually on skewed hinges to lie along rear hull with floats outward. Engine set at positive angle of 5°, with all-round cooling gills; nacelle interior layout similar to MDR-5 but extended to behind trailing edge. Fabric-covered control surfaces.

Prototype tested early 41. Tests included catapulting and armament trials, but interrupted by evacuation of Sevastopol in late spring 42. By this time the first two production aircraft were being tested by GUSMP pilot Malkov. Many (about 30) unfinished KOR-2s destroyed at Taganrog during evacuation. In 43 production restarted at Krasnoyarsk, and small number delivered to completion in mid-45. Described as 'average series', possibly 100. They served with the MA, including catapult duties on warships.
DIMENSIONS Span 12,0 m (39 ft 4½ in); length 10,5 m (34 ft 5¼ in); height (keel/fin-tip) 4,04 m (13 ft 3 in); wing area 25,5 m² (274·5 ft²).
ENGINE One 1000 hp M-62 driving VISh-105-62 or AV-24 three-blade c/s propeller.
WEIGHTS Empty 2082 km (4,590 lb); fuel 315 kg (694 lb); loaded 2760 kg (6,085 lb).
PERFORMANCE Max speed 310 km/h (193 mph) at sea level, 356 km/h (221 mph) at 4,7 km (15,420 ft); climb 2 min to 1 km, 12 min to 5 km; ceiling 8100 m (26,575 ft); range 550/1150 km (342/716 miles) depending on use of overload fuel.

KOR-3S Construction abandoned 47 of this multirole three-seat seaplane for catapulting from warships. Roles included recon, gunnery spotting and attack with bombs, rockets or torpedo. A landplane version was also projected.
DIMENSIONS Span 12 m (39 ft 4½ in), folded 4280 (14 ft 0½ in); length 10,58 m (34 ft 8½ in).
ENGINE One M-87A.
No other data.

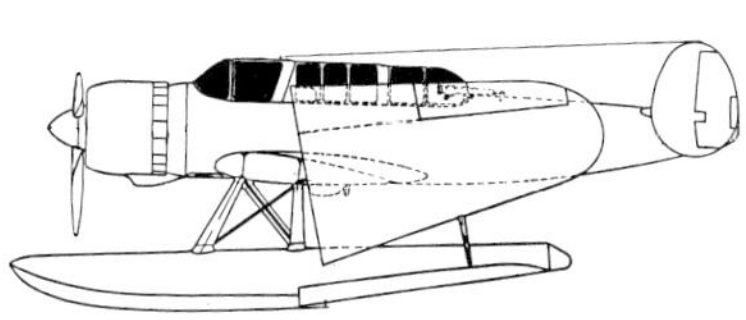

KOR-3S

KOR-9 Another casualty of post-war cutbacks was this more powerful version of **KOR-2**. Wing considerably increased in span, fuel capacity doubled, twin-fin tail mounted on central fin-pylon, increased bombload. Prototypes started Oct 46, stopped near first flight.
DIMENSIONS Span 18,0 m (59 ft 0½ in), folded 4450 (14 ft 7¼ in).
ENGINE One ASh-62.
No other data.

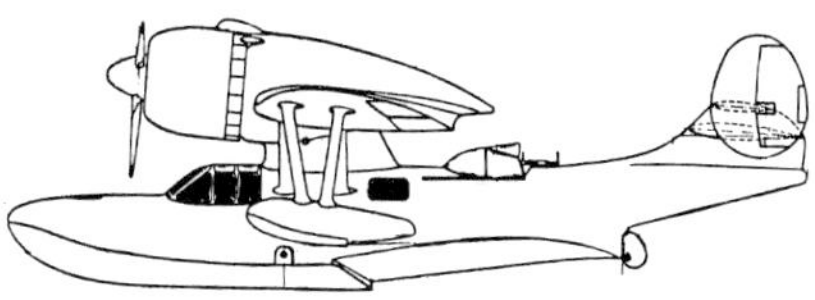

KOR-9

KOR-4 (Be-8) Single-engined amphibian to replace MBR-2 and MP-1 for casevac, photo, pilot training and inter-ship liaison, with ability to taxi to/from land. All-metal stressed-skin. Wing centre-section with four protected tanks, outer panels tapered on LE, slotted flaps, manual flight controls of D16 with fabric covering, all with trimmers. Hull with vertical main step, vee planing bottom continued to tail with water rudder and two hook. Main gear and tailwheel and flaps worked by 27V DC electrics. Side-by-side cockpit with dual controls, cabin with six seats or two seats plus two stretchers. Provision for one fixed and one movable gun. Engine drove prototype VISh c/s three-blade 3 m (10 ft) propeller; exhaust muff heated cabin air supply.

First prototype on taxi trials Nov 47, tail reinforced, new stab floats added, rebalanced prop. First flight by M. Tsepilov 3 Dec. No production order. One prototype (of three) tested several arrangements of hydrofoil(s) mounted on pylon under hull. ASCC reporting name 'Mole'.

DIMENSIONS Span 19,0 m (62 ft 4 in); length 13,0 m (42 ft 8 in); wing area 40 m^2 (430·6 ft^2).
ENGINE One 700 hp ASh-21 driving AV-24 propeller.
WEIGHTS Empty 2815 kg (6,206 lb); fuel 352 kg (776 lb); loaded 3624 kg (7,989 lb).
PERFORMANCE Max speed 266 km/h (165 mph) at 1800 m (5,900 ft); cruise 220 km/h (137 mph); ceiling 5,5 km (18,050 ft); range 1205 km (749 miles).

LL-143 Though Beriev's OKB worked under great difficulties at Krasnoyarsk, immense demands were made and the depleted team of experienced engineers were assigned the most challenging task ever put to Soviet designers of marine aircraft. The overdue need for a modern and capable long-range flying boat was to be met by this OKB, which called the aircraft LL-143 (Flying Boat, Jan 43).

Beriev had sketched the gull-wing layout in 1942, and by April 43 the first engineering drawings were issued. Stressed-skin duralumin structure throughout, except for fabric-covered control surfaces. Large and deep hull with transverse front step and curved V second step with water rudder synchronized with two main air rudders carried vertically on dihedral tailplane. Skin plates flush-riveted to stringers on 44 frames, with planing bottom containing deep keel and transverse bulkheads (each with rubber-sealed hatch) forming eight watertight compartments. Provision for three-unit twin-wheel beaching chassis. Stabilizing floats on two fixed struts, 'loss of float means loss of aircraft', explaining reluctance to adopt retractable floats. Wing built up on strong box with light leading and trailing structures, all riveted or bolted. Section NACA 23020 (root), 23010 (tip). Fuel in 22 flexible cells in wing box both in dihedral centre-section and horizontal outer panels. Slotted flaps in three electrically driven sections each side. Aerodynamically balanced differential ailerons. Dual flight control by cables and push-rods, with AP-5 autopilot driving three power units in control circuits. Comprehensive ice protection: leading edges heated by hot air from combustion heaters, alcohol sprays for navigator's windows and propeller blades, and electric resistance mats for pilots' windscreens. Provision for crew of seven, plus reliefs.

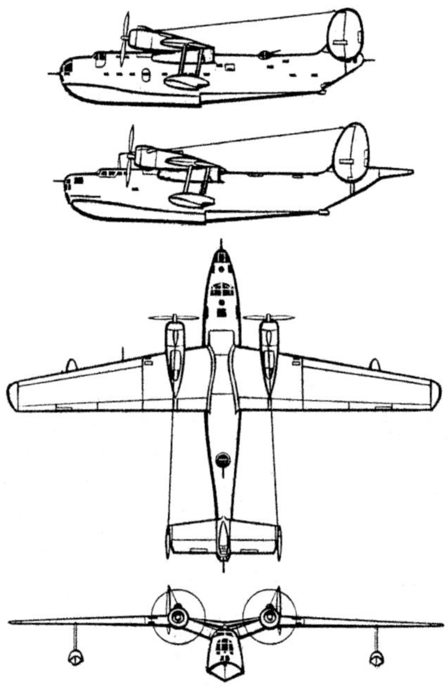

Be-6 (upper side view, LL-143)

LL-143 launched Sept 45 and test-flown at Krasnoyarsk. By this time OKB moving back to Taganrog. While still in Siberia carried full armament: 12·7 mm UBT in nose, left/right beam, upper rear deck, immediately aft of rear step, and extreme stern, six guns in all. Under-wing racks for various bombloads up to 400 kg (normal) or 4000 kg (8,820 lb) overload.

DIMENSIONS Span 33,0 m (108 ft 2½ in); length 23,0 m (75 ft 5½ in); height 7,2 m (23 ft 7½ in); wing area 120 m^2 (1,292 ft^2).
ENGINES Two 2250 hp ASh-72 driving 5056 mm (199 in) V-3BA four-blade propellers.
WEIGHTS Empty 15 104 kg (33,298 lb); normal fuel 4300 kg (9,480 lb) (overload fuel, 8600 kg); normal loaded 21 300 kg (46,958 lb).
PERFORMANCE Max speed 371 km/h (231 mph) at S/L, 401 km/h (249 mph) at 4,3 km; landing speed 140 km/h (87 mph); climb 21·5 min to 5 km; range (normal) 2800 km (1,740 miles), (overload) 5100 km (3,169 miles).

Be-6 Outstanding LL-143 led to production programme at Taganrog. MA (now AV-MF) Be-6 had more powerful engines, slightly modified wing box, redesigned bow with extended nose making room for eighth crew-member, interior provision for complete relief crew, 'balcony' flight-deck windows giving downward view, retractable radar in planing bottom aft of main step, bow spray fences to reduce propeller erosion, provisions for carrying 40 commandos, and different defensive armament: nose, N-2 installation for NR-23 with 200 rounds; top deck, DT-V8 with two NR-23 with 500/550 rounds; tail, Il-K6-53Be with two NR-23 with 225 rounds each. Normal weapon load increased to 16 FAB-100, eight FAB-500, or AMD-500 (actual mass 550 kg) mines, two FAB-1500 or two 1100 kg torpedoes. Normal systems included two GSP-9000 engine-driven generators, M-10B1 APU driving GS-5000 generator, for 28·5 and 115 V systems; B-40 cabin heater, groups of eight KP-19 oxygen bottles; and comprehensive radio fit.

First flight Feb 49 by M. I. Tsepilov, OKB pilot; state trials in same year. Few problems, and service delivery from 50. Over 200 built for service with all MA Fleets, as well as paramilitary and Aeroflot service as utility transports. By 54 modifications in progress, notably replacement of Ilyushin tail barbette by MAD stinger; replacement of nose guns by two different large radars for surveillance and missile guidance in cooperation with warships; and conversion for dedicated transport duties. ASCC reporting name 'Madge'.

DIMENSIONS Span 33,0 m (108 ft 2½ in); length 23,565 m (77'ft 3¾ in); height (keel/fin-tips) 7,64 m (25 ft 0¾ in); wing area 120 m^2 (1,292 ft^2).
ENGINES Two 2400 hp ASh-73 driving V-3BA-5 propellers.
WEIGHTS Empty (as delivered) 18 827 kg (41,506 lb); fuel 7400 kg+500 kg oil (total 17,416 lb); normal loaded 23 456 kg (51,711 lb); max with bomb load 29 t (63,933 lb).
PERFORMANCE Max speed 377 km/h (234 mph) at SL, 414 km/h (257 mph) at 1,8 km; climb 20 min to 5 km; ceiling 6,1 km (20 000 ft); range 4800 km (2983 miles).

R-1 In 1946 Beriev studied LL-143 with two RD-45 turbojets and was sufficiently impressed to request permission to design R-1 (*Reaktivnii* 1, jet 1) to explore possibilities. Much smaller than LL-143, but retaining gull wing to place simple engine nacelles as high as possible. Hull of high length/beam ratio with shallow front step and rear planing bottom tapered to point near stern. Duralumin throughout, including control-surface skins. Wing of NACA 23009 (9%) profile with machined skins and wing box forming integral tankage and Fowler flaps inboard and outboard of engines. Wing-tip

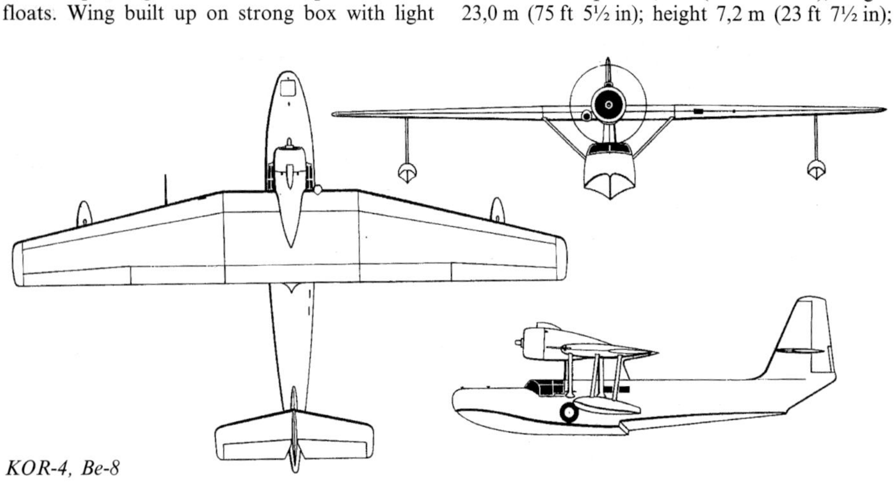

KOR-4, Be-8

Be-6

floats on two struts retracted electrically to form wing-tips. Wing remarkably far back on hull. Fixed tailplane mounted high on large fin with bullet fairing. One-piece rudder and prominent elevator mass-balances. Crew of three: nav/bomb-aimer entering through door in right side of hull to nose compartment with windows both sides, pilot in fighter-type cockpit on left side of hull, and rear gunner in stern compartment sighting through blister on each side to control electrically driven twin-NR-23 barbette. Two NR-23 fixed in bows aimed by pilot. Wing racks for 1000 kg weapon load.

Flown Taganrog by I. M. Sukhomlin 30 May 52. By this time RD-45 engines replaced by VK-1s, set at 5° positive angle. Severe gas/water wake, but well away from rear hull and tailplane.

DIMENSIONS Span 21,4 m (70 ft 2½ in); length (no guns) 19,43 m (63 ft 9 in); wing area 58 m^2 (624 ft^2).

ENGINES Two 2740 kg (6,041 lb) VK-1 turbojets.

WEIGHTS Empty not published; loaded 17 t (37,478 lb).

PERFORMANCE Max speed 760 km/h (472 mph) at SL, 800 km/h (497 mph) at 7 km; ceiling 11,5 km (37,730 ft); range 2000 km (1,243 miles).

Be-10 For over 30 years little was known about this impressive twin-jet flying boat, only aircraft of its class ever to go into operational service. AV-MF requirement of 1953 called for water-based aircraft to replace Be-6 for maritime reconnaissance, anti-ship torpedo attack, bombing of ships and shore installations and various other tasks. Designers to offer proposals for ASW missions; operations to be possible in moderately severe sea states (0,8 m waves, 12 m/sec wind).

Overall design an enlarged sweptback version of R-1 but with engines tucked under wing

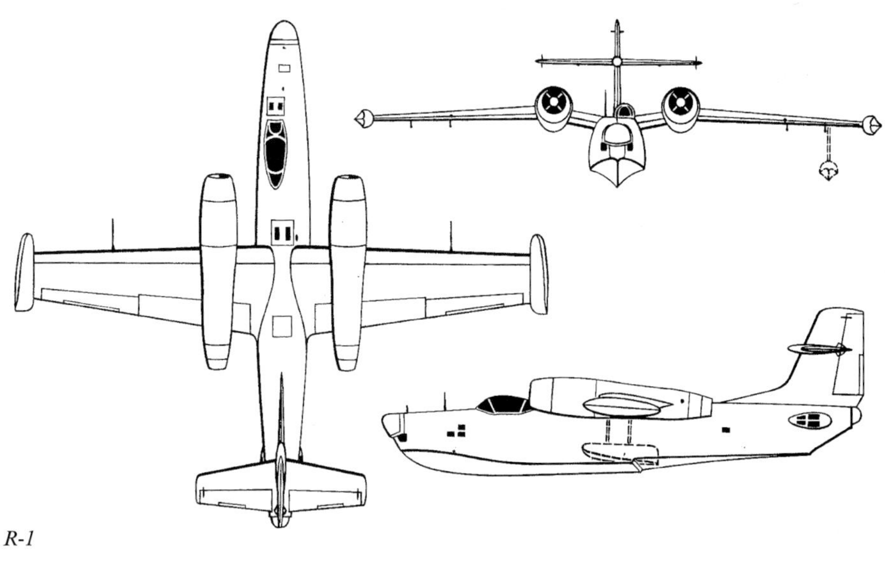
R-1

Be-10

roots; prolonged testing of large spray dams ensured bulk sea-water would not enter inlets. Wing CAHI laminar profile swept 35° on LE, built up from horizontal centre section bolted to sharply anhedralled intermediate and outer panels based on torsion box with two massive spars and 13 ribs each side. During development two fences added to each wing extending round LE to underside. Area-increasing slotted flaps and powered ailerons. Anhedral was means of bringing wingtip floats close to water with only short fixed pylons. Large swept fin and powered rudder able to maintain straight flight in max asymmetric power. Dihedralled trimming tailplane swept at 35° with powered elevators. Necessarily deep fuselage of clad light alloy with 77 frames including five massive bulkheads picking up engines and spars of wing and tail. High length/beam ratio with one deep vertical vee-step amidships, planing bottom continuing to extreme tail. Assembled in three main sections with underfloor area divided into ten watertight compartments. Pressurized compartment at front with door on right for navigator in bow and pilot on centreline under fighter-type canopy, both with upward ejection seat. Tail pressure cabin for radio-operator/gunner with twin NR-23 and downward-ejection seat. Bay immediately aft of bottom step for four torpedoes (3075 kg), up to 3,3 t bombs, mines or depth charges, with doors sealed by pneumatic tube inflated to 3 kg/cm^2 (42·7 lb/in^2). Engines attached to fuselage frames and to front spar, cowl doors forming maintenance platforms. Eight tanks in centre wing, four in each outer wing and two protected tanks in fuselage.

Prototype flown by V. Kuryachi 20 July 1956. In winds below 5 m/sec seaworthiness and flying characteristics generally good, but dangerous porpoising in heavy seas or strong winds. Speed and ceiling double Be-6 and rate of climb three times. Overshoot on one engine no problem at up to 43 t. Concluded type-conversion no problem to (eg) Il-28 or Tu-16 pilots. Four aircraft, reported to FAI as **M-10**, gained world records 1961, including 1000-km circuit with 5 t payload 875,86 km/h and 10 t payload to 12733 m. Equipped two units of Black Sea Fleet. Limited operational usefulness led to early retirement. ASCC reporting name 'Mallow'.

DIMENSIONS Span 28,6 m (93 ft 10 in); length 30,72 m (100 ft 9½ in); wing area 130 m^2 (1,399 ft^2).

ENGINES Two AL-7PB.

WEIGHTS Empty 26,5 t (58,420 lb); normal loaded 45 t (99,206 lb); max 48 t (105,820 lb).

PERFORMANCE Max speed at SL 912 km/h (567 mph); cruise 785 km/h (488 mph); climb to 5 km in 7 min; takeoff 260 km/h in 2,2 km; landing 220 km/h in 1,4 km.

Be-12, M-12 Versatile turboprop amphibian designed from 55 to meet AV-MF reqt for Be-6 replacement for ocean/coastal patrol and reconnaissance, anti-ship, ASW, photo survey, Arctic support, submarine and surface warship co-operation, transport, search/rescue and electronic warfare. Hull essentially derived from Be-6, but stretched in length and with planing bottom nearer to Be-10. Small spray dam along each side of nose, V main step, tapered afterbody to pointed second step under stern with water rudder upstream. Retractable landing gear with main units hingeing up through 150° to lie in hull recesses, covered by door on inner side of leg in extended position; tailwheel retracting rearwards and upwards into diagonal stern compartment closed by twin doors. Tail little changed from Be-6, but wing entirely new with reduced span/area, thinner profile and tracked slotted flaps. Manual flight controls with servo-tabs. Integral tankage along wing box inboard and outboard of engines; none in hull. Stabilizing floats on large-chord single struts.

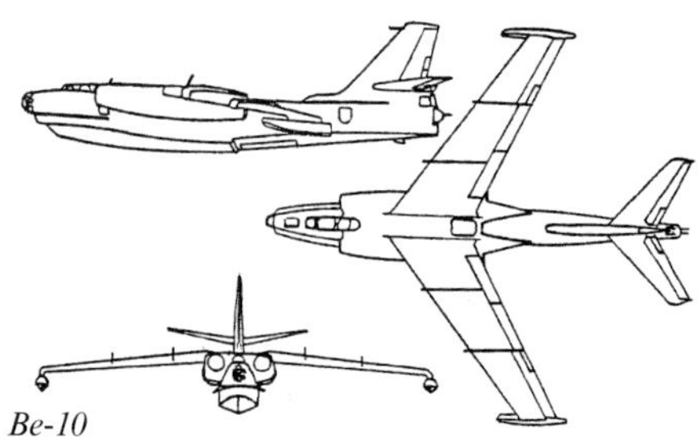

Be-10

Engines carried on welded steel-tube trusses projecting up and ahead from wing box, with cowling panels forming servicing platforms and jetpipes curving across wing to nozzles at trailing edge.

Normal crew five or six; captain, co-pilot/com-radio operator, navigator (in glazed nose), electronics operator, and one or two ASW sensor operators in rear compartment. Interior unpressurized. Internal weapon bay aft of step with watertight doors. Further aft are tubes and retrolauncher for sonobuoys housed in rear compartment. Further aft again is stowage area and, on left side, APU driven by gas-turbine whose exhaust heats leading edges of tail when not discharged through overboard stack. Extreme stern can house MAD stinger. Surveillance radar normally mounted in nose above navigator glazing, a later radar having a radome of wider oval section than drum of original A-304E ASW/nav set. Other emitters designed to include A-321A nav, A-322Z doppler, A-325Z nav, Cross Up beacon and IFF transponder, SRO-2 IFF and SRZ-2 interrogator (also used for rescue and ship homing), and no fewer than four radar altimeters (RV-UM, RV-3, RV-10 and RV-17 high-altitude). Individual aircraft have additional fits including tail-warning and passive receivers, and there is at least one dedicated EW variant.

Be-12

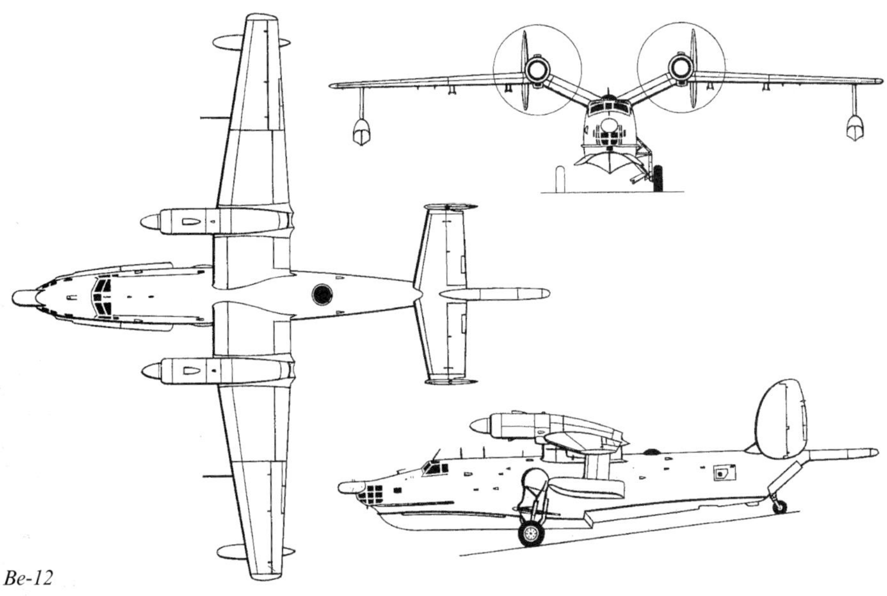

Be-12

Weapons are also carried under the wing well outboard of the propeller discs, there being a minimum of one pylon on each side plus rocket rails.

First flight believed 1960. Displayed publicly 61 and large number (probably 200) delivered from 64. Serving in all fleet commands, number in active inventory in early 1980s being about 75. M-12 holds all 21 FAI turboprop amphibian and all 19 turboprop flying-boat class records; ceiling given below and other figures include payload of 10,1 tonnes (22,266 lb) lifted to 2 km. The flying boat variant averaged 565,347 km/h over 500-km circuit. Always unofficially called *Chaika* (seagull).

New versions promoted from 91, all with MAD and other ASW gear removed, most with stores pylons retained **Be-12P** firebomber with 4500-lit tank in centre of hull, balanced by two of 750 lit further forward, all refuelled in 15s by run across water at 120 km/h; operating 100 km from base and 10 km from water can put down 42 t of extinguishant before needing to refuel. **Be-121** for scientific research, various sensors. **Be-12NKh** for passenger/cargo transport. **Be-12EKO** for ecological reconnaissance. **Be-14** for maritime rescue, carrying life-raft/survival-kit capsules. ASCC reporting name 'Mail'.

DIMENSIONS Span 29,71 m (97 ft 5¾ in); length (with MAD) 30,17 m (99 ft 0 in), (new radar) 30,95 m (101 ft 6½ in); wing area 105 m^2 (1,130 ft^2).

ENGINES Two AI-20DM.

WEIGHTS Empty (estimate) 18 t (39,680 lb); loaded 29 450 kg (64,925 lb).

PERFORMANCE Max speed at optimum height 608 km/h (378 mph); initial climb 912 m (2,990 ft)/min; ceiling 12185 m (39,977 ft); endurance at reduced speed 15 h; normal range 4000 km (2,485 miles).

Be-30 A complete break with tradition, this light airliner was designed in 1965-66 to meet Aeroflot need for multi-role transport to replace An-2. Almost a scaled-down An-24 but with narrow slab-sided unpressurized fuselage. Structure all light alloy with extensive chemical-milling, bonding and spot-welding. Main wing box incorporates four integral tanks; secondary structure mainly honeycomb sandwich, both in wing and tail, though curiously also 'stiffened by stringers'. Prototype flown by M. I. Mikhailov 3 March 67 temporarily fitted with two 740 hp ASh-21. Second aircraft, (possibly first re-engined) SSSR-23166, fitted with imported Astazou XII. Definitive power as in data, driving AV-24AN three-blade reversing propellers. Unusual feature was high-speed shaft linking free-turbine drives of both engines to even out prop torque following failure of one power section. Double-slotted flaps hinged to brackets far below wing, manual flight controls, single-wheel landing gears retracting backwards. Unpressurized fuselage.

Fuselage only 1,5 m (59 in) wide, just sufficient for left/right passenger seats and central aisle. Door at left rear, with integral steps. Normal seating for 14, plus 15th passenger on right of pilot. Alternative schemes for casevac (nine stretchers plus six seated), photo-survey, geo-prospect, offshore/fishery patrol and executive/VIP. Options include autopilot, coupled ILS and roller-map display. Aeroflot service announced for summer 70, but Czech L-410 adopted in preference. ASCC name 'Cuff'.

DIMENSIONS Span 17,0 m (55 ft 9¼ in); length

Be-14 with rescue capsules (MAD removed)

Be-30

15,7 m (51 ft 6 in); wing area 32 m^2 (344·5 ft^2).
ENGINES Two TVD-10B.
WEIGHTS Empty 3360 kg (7,407 lb); fuel 1000 kg (2,205 lb); max payload 1500 kg (3,307 lb); max loaded 5860 kg (12,919 lb).
PERFORMANCE Max cruise 480 km/h (298 mph); normal cruise 460 km/h (285 mph) at 2 km (6,560 ft); take-off 250 m (820 ft); range with 1,25 tonne payload and 30-min reserve, 600 km (373 miles).

Be-32 Final two Be-30 designated Be-32, with TVD-10M engines uprated to 810 kW (1,086 shp), cleared to max wt 6270 kg (13,823 lb) and equipped to carry 18 pax or 1,9 t cargo, one with rear ramp door. Set world climb records March 76, one preserved Monino. In 91 problems with spares and support for Czech L-410 fleets led to resurrection and marketing of updated aircraft with more power, greater weights and other changes. Improved aircraft cleared for rough strips, option skis/floats, various special roles; initial batch of 50 delivered 93 with 17-seat interior to three Russian airlines, further batch probable.
DIMENSIONS As Be-30.
ENGINES Two TVD-10B or PT6A-65.
WEIGHTS OWE 4,76 t (10,495 lb); max 7,3 t (16,090 lb).
PERFORMANCE Max cruise 480 km/h (298 mph), (seaplane 400, 249); range (17 pax) 650 km (404 miles), (7 VIP) 1600 km (994 miles).

A-40, Be-42 In 82 OKB assigned biggest project: design completely new amphibian to replace Be-12 in maritime recon, ASW, minelaying, transport and SAR. Design agreed in outline 83 under Konstantinov as outstanding aircraft with unconventional features. Remarkable similarity to unbuilt Saro P.208.

Wing designed for range efficiency with CAHI supercritical profile, 14·5% root to 11·5% tip, LE sweep 23°13', dihedral 0, twist (wash-in) 4°30', full-span powered LE slats; two double-slotted flaps each wing, inners riding on two tracks, outers on three; outboard ailerons assisted by outermost of four spoilers each wing. Untapered swept vert fin with dorsal fin, carrying two-section rudder entirely above jet level and fairing at top for all-flying horiz tailplane 11870 span with one-piece elevators to increase camber. All flight controls metal-skinned and fully powered with spring feel and elec trim. Oval-section fuselage (hull) of very high length/beam ratio, external width 3,5 m (138 in). Planing bottom extends nearly 38 m from bow to water rudder at stern, initially with normal concave vee profile from which extra chines grow to give three-keel double-vee form as far as shallow V-step amidships, giving improved water stability, easier unsticking (larger upthrust from waves) and claimed reduced vert acceleration on alighting; downstream of step normal vee bottom with (3,5 m aft of step) two patent rectangular wedges with central gap. Entire fuselage pressurized, for combat crew: side-by-side pilots, engineer, radio op, nav/observer and three mission specialists. Outward-opening powered emergency door on left; inward-opening door on right with blast screen upstream for emergency bailout. Wing mounted on very large aux structure (conventional light alloy, integral with fuselage) housing: TA-12 APU, cabin environmental system, main landing gears and sonobuoy bays, and at rear carrying main engines on top and aux booster engines internally. Main engines set at +5° incidence with plain direct inlets and nozzles, no reversers. Large dorsal duct for bleed air. Aux engines fed by direct pipes from main-eng inlets without valves; flow stopped in cruise by closing titanium eyelids over aux nozzles. Tall main landing gears with single legs with nitrogen shock strut, side braces and long forward brace/actuating strut carrying bogie with four wheels with anti-skid multi-disc brakes and tyres 1030×350 (max 142 lb/in^2) retracting rearward into aux underwing structure with truck horizontal. Track 4,96 (16 ft 3¼ in). Short nose gear with twin wheels with tyres 840×290 steering ±55° retracting to rear. Wheelbase 14835 (48 ft 8 in). Two pairs independent hyd systems with AMG-10 non-flam fluid, 210 kg/cm^2 (2,987 lb/in^2), driving slats, flaps, powered flight controls, spoilers, eyelids, water rudder and landing gears. Pneumatic system for brakes and emerg

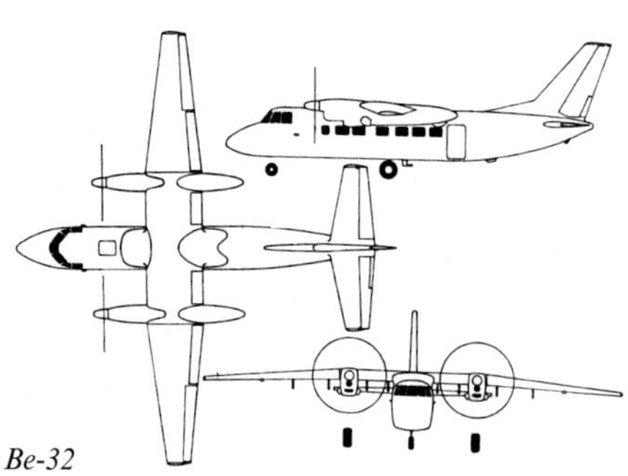

Be-32

Be-32

gear extension, same pressure, and fed at 3 kg/cm² to weapon-door seal. Elec 115/220V from two 60-kVA, static inv for 27V DC. Fuel 35100 lit (7,721 gal) in wing box. Gaseous oxygen, mainly thermal deicing of all LE. All aircraft, HF/VHF/UHF, multimode radar for nav/search/weather/attack/wave measurement, inertial nav/doppler/radar alt.

Two prototypes funded 83, Nos 10 and 20, former making first flight Dec 86 and second in late 88. Both with airframes suitable for **A-40** roles, with FR probe above nose and provision for combined electro-optical (IR+TV) sight system and rails and bay doors for all forms of dropped stores carried along lower part of hull, max 6,5 t (14,330 lb), sonobuoys (eg 48×124 mm) behind main gear bays, MAD boom at tail, *Sorbtsiya* EW (mainly Elint) pods on wingtips and complete AV-MF avionics. Most of this not fitted, and both aircraft used for basic flight devt of type with instrumentation only. In 92 No 20 was preparing for trials with full equipment and weapons to support order

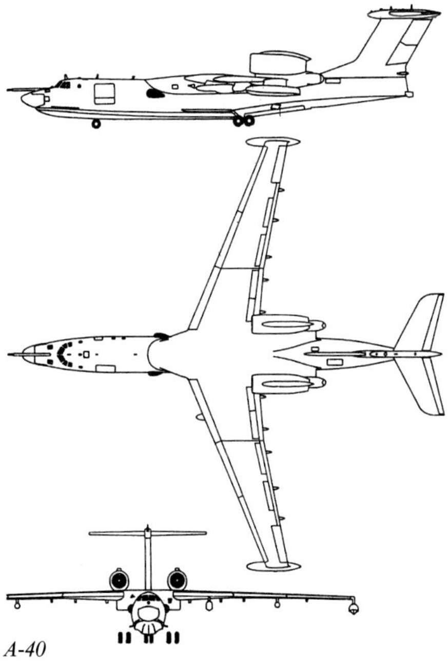

A-40

A-40

A-40

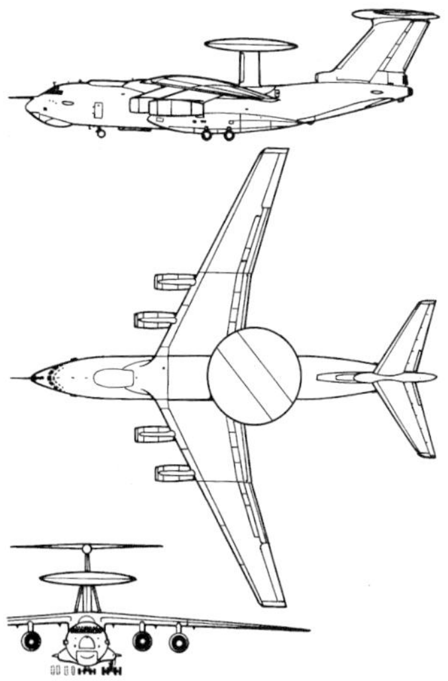

A-50

for 40, deliveries due from 93. ASCC name 'Mermaid'.

Projected variants would delete AV-MF and A-40 mission equipment as well as booster turbojets and tip pods. So far published: **Be-40P** Projected version for 105 passengers (3+2), payload 10 t, range 4000 km; **Be-40PT** mixed 37 or 70 pax, rest (to 10 t) cargo at rear, range 4200 km; **Be-42** SAR version, cleared to 830 km/h, crew nine, comprehensive night/bad weather search sensors, two LPS-6 semirigid motorboats, rescue hoists and special loading for 54 survivors (typically 21 stretchers) with full medical/surgical/anti-hypothermia, range 5500 km. Data for A-40.

DIMENSIONS Span 41,62 m (136 ft 6½ in); length (inc probe) 43,839 m (143 ft 10 in), (fuselage only) 38,915 m (127 ft 8 in); wing area 200,0 m² (2,153 ft²).

ENGINES Two D-30KPV; two R-60K.

WEIGHTS Empty 43,9 t; max TO 86 t (189,594 lb), max landing (land) 73 t, (water) 85 t.

PERFORMANCE Max speed (6 km) 760 km/h (472 mph); cruise 720 km/h (447 mph); service ceiling 9,7 km (31,825 ft); range (max load) 4100 km (2,550 miles), (max fuel) 5500 km (3,420 miles); field length (land/water) 1,8 km (5,900 ft).

A-50, Be-976 In 78 TANTK assigned task of using Il-76 as basis for AWACS-type platform to replace Tu-126. Basic aircraft, see Ilyushin. Partnership with NPO Vega-M (Gen Designer Vladimir Ivanov) who managed entire complex to be fitted by Be into aircraft. Rotodome on twin deiced pylons on new dorsal structure. DRLO radar, based on EKOR-A developed by ISZ 'Kosmos-1870' and cleared for service 87, also used March 91 in spacecraft *Almaz*-1. OTH multimode radar, mainly used as PD elevation-scan at c10 cm, liquid-cooled antenna 9 m×2 m compared with 7,3×1,8 for E-3 radar, giving better definition. Sea-level targets use PDNES (non-elev-scan), with separate processing to refine images. Long horiz plate reflector added along rear sides of main-gear fairings. Hi-power IFF and secure data-link antennas on reverse side of main antenna structural beam. Aft ramp door and other cargo provisions eliminated. Generating plant in rear fuselage, inlet ahead of fin, tail turret replaced by passive receiver installation, forward fuselage lengthened but accomodation for 7 flight crew and 12 mission crew still cramped ('no rest bunks or toilet'), dorsal doghouse over steerable satnav antenna, weather/mapping radars replace glazed navigator station, fixed FR probe centred above nose, numerous added passive receiver and com antennas, threat-warning computers (part of Vega-M complex) control ASO-series 132-shot chaff/flare dispensers scabbed on each side of fuselage. Complex bulkier and heavier than E-3 systems, limiting TO fuel load (FR probe essential), problem not eased by need to do decoding/interfacing on board with duplicate command/IFF/data exchange for different armed forces in various republics. Only advantages (Ivanov) 'better perf against ground clutter and ability to gather and retransmit via satellite.'

Two built new per year 83-85, then five annually to 90, two per year since; first deliveries to AV-MF Chyernomorsk, some to CIS Air Force. Dk grey/white, orange tips; Red 50 has additional antennas. Main force (29 in 92) *Pechora* (Arctic) sending detachments to local needs. Guided MiG-31s to intercept wave-hugging cruise missiles while sending data to subs and controlling interception of naval Tu-22RK 'attackers'. Two operated 24 h per day over Black Sea during Gulf War. Not normally practised, major advantage is ability to fly pattern at 15 km (49,200 ft), double height of E-3. Data similar to Il-76. ASCC name 'Mainstay'.

In addition, six **Be-976** delivered to GVF civil fleet to serve as high-power tracking stations, chiefly in support of missile testing. These use A-50 airframe but with original navigator station in nose, very different processing and com systems, no EW eqpt and large passive receiver pods under wingtips giving accurate triangulation on targets.

Be-103 Light amphibian to fly from any surface in any environment on multiple roles. Advanced structure with much composite and honeycomb, wings NACA-2412M profile, tapered on LE with large root extensions, lie on water at rest obviating need for stab floats; manual flight controls with spring feel on high slab tailplane; mainwheels with 500×150 tyres (3 kg/cm², 43 lb/in²), anti-skid brakes, pneu retrac inwards, nosewheel 400×150 tyre retrac fwd; engines on lateral pylons behind cabin driving tractor 3-blade rev-pitch 1,6 m AV-103 props, 2 wing tanks total 450 lit (99 gal); upward-hinged doors each side to three pairs seats or two stretchers and attendant or 375 kg cargo (disposable load 550 kg of which 320 can be fuel) or role eqpt; digital avionics. Prototype building 93.

DIMENSIONS Span 13,34 m (43 ft 9¼ in); length 10,65 m (34 ft 11¼ in); wing area 25,9 m² (279 ft²).

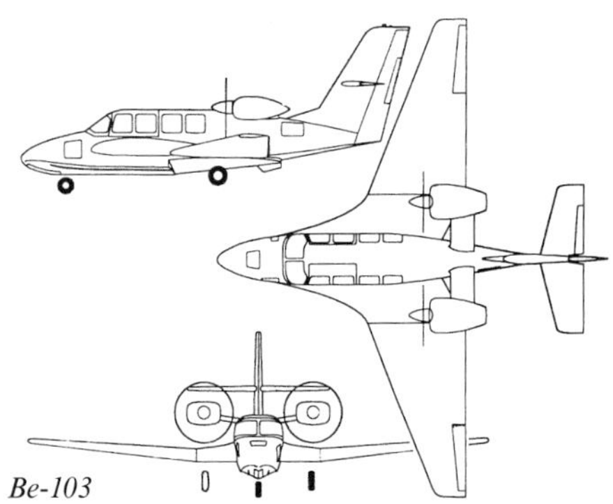

Be-103

A-50

Be-103 model

ENGINES Two M-17.
WEIGHTS Empty 1210 kg (2,668 lb); MTO 1650 kg (3,638 lb).
PERFORMANCE Max cruise 268 km/h (167 mph); range (max payload) 500 km (311 miles), (max fuel) 2800 km (1,740 miles); TO/ldg (water) 560/650 m, and 420/500.

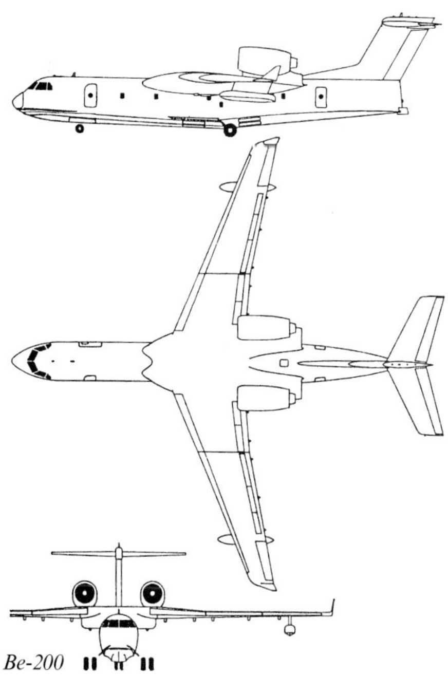
Be-200

Be-200 Scaled-down A-40 to meet wider civil/military markets. Two prototypes funded in partnership ILTA Geneva, first to fly 94. Designed for rather lower speeds, supercrit wing 16/11% with winglets, three spoilers outboard instead of two, modern engines (no boosters), main gears with two 950×300 tyres, nose gear two 620×180, fuel in wing box total 15000 lit (3,300 gal), triple hyd systems. Proposed configurations: 68 pax, 8 t cargo, 30 stretchers + 7 medics, but most important immediate role seen as firefighting ('twice as efficient as CL-215') dumping 12 m³ water + 1,2 m³ chem retardant (can scoop 12 t (26,455 lb) water in waves up to 1,2 m [4 ft]). To be offered with eqpt for mil or paramilitary missions. Certification due 1st quarter 97.
DIMENSIONS Span 31,88 m (104 ft 7¼ in); length 32,049 m (105 ft 1¾ in); wing area 117,44 m² (1,264 ft²).
ENGINE Two D-436T.

Biesnovat

Matus Ruvimovich Biesnovat was born in about 1900 and in the mid-1930s was an engineer in Tairov's OKB. In 1938 he was permitted to form his own OKB, assigned the particular task of building a research aeroplane to test new wing profiles and planforms, new flight controls and new structural forms at the greatest flight speeds attainable. This aircraft was built, and from it was derived a fighter.

SK-1 Designation possibly from *Skorostnii Krylo*, high-speed wing. Smallest possible aircraft to take large V-12 engine, and fitted with smallest possible wing for safe landing on Russian grass or board airstrips. Light-alloy stressed-skin construction, but fabric-covered control surfaces, all with 100% mass-balance. Direct manual flight control but trim tabs on rudder and elevators; no servo assistance. Wing of NACA 23014·5 profile. Single plate-web main spar and sheet skin, finally covered with thin layers of cork dust, marquisette openweave fabric and adhesive, polished off to mirror finish. Slotted split flaps of Vlasov (CAHI) type. Minimum-drag cockpit with flush hood incorporating side windows, hinged up by hydraulic ram together with pilot seat to give view for landing. Minimum-drag engine installation with ejector exhausts, and pressurized (1,1 kg/cm²) coolant system requiring radiator of only 0,17 m² frontal area, said to be half normal for chosen engine. Total fuselage frontal area 0·85 m². Hydraulic actuators for main and tail landing gears, with closure doors throughout.

Flown on skis Jan 39, later on wheels. SK-1 did not frighten test pilot(s) and at high speeds demonstrated excellent handling and manoeuvrability.
DIMENSIONS Span 7,2 m (23 ft 7½ in); length 8,0 m (26 ft 3 in); wing area 9,5 m² (102 ft²).
ENGINE One 1050 hp M-105 driving VISh-52 three-blade propeller.
WEIGHTS Empty 1505 kg (3,318 lb); loaded 2100 kg (4,630 lb).
PERFORMANCE Max speed SL 597 but 710 km/h (421 mph); landing speed 165 km/h (102·5 mph); range 1000 km.

SK-1

SK-2 The success of SK-1 triggered immediate authorization for this fighter derivative. The airframe was the same except for use of a normal light-alloy wing surface, conventional cockpit with fixed seat and sliding canopy, conventional oil cooler and armament. The high-pressure water cooling system was retained, as was the small (unknown) fuel capacity. Never fitted with planned pivoted tray above engine, able to rise 320 mm, carrying two 12·7 mm BS. Flown by G. M. Shiyanov, October 1940.
DIMENSIONS Span 7,3 m (23 ft 11½ in); length 8,28 m (27 ft 2 in); wing area 9,57 m² (103 ft²).
ENGINE As SK-1.
WEIGHTS Empty 1850 kg (4,078 lb); loaded 2300 kg (5,071 lb).
PERFORMANCE Max speed 585 km/h (393 mph)

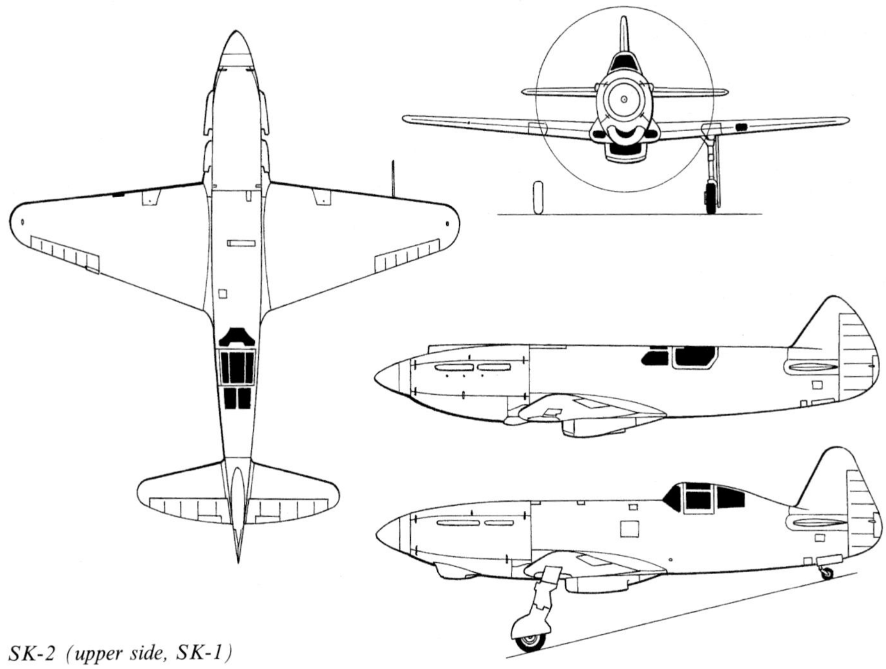

SK-2 (upper side, SK-1)

at SL, 660 km/h (413 mph) at 4·9 km; climb 4 min 20 s to 5 km (16,400 ft); endurance at optimum slow cruise 45 min; take-off 350 m; landing 500 m.

SK-3 Designed at start for combat duty, advanced twin-engined low-wing two-seat fighter. Two AM-37 engines, 33,7 m^2 (363 ft^2) wing, four large-calibre fixed guns. Design speed 555 km/h at SL, 700 km/h (435 mph) at rated height. Project reassessed and terminated Dec 40.

Aircraft '5' To provide Soviet parallel to OKB-2 '346' Biesnovat assigned project for supersonic research aircraft late 45, effectively own OKB, deputy G. Ressing (not to be confused with H. Rössing of OKB-2). Planned in two stages, **5-1** glider and rocket-powered **5-2** differing mainly in increased sweep of vertical tail. Design was of simple aircraft, conventional light-alloy stressed skin. Wing constant chord with 45° sweep, 5° anhedral, profile CAHI 12045 bis root, P2/2M tip, plain inboard flaps, outboard ailerons, two fences each wing from LE to flap. Narrow oval-section fuselage with pressurized jettisonable section ahead of wing for reclining pilot, tandem sections of hinged canopy. 5-1 with ballast in rear fuselage, broad, slightly swept fin/rudder carrying sharply swept (50°) tailplane with mass-balanced elevators half-way up. All control surfs with hydraulic boost, signalled by push/pull rods. Retractable skids just ahead of CG and under tail, retrac outrigger skids at tips. Unusual airbrakes resembling open box hinged out about lower edge each side behind wing. First flight by A. K. Pakhomov 14 July (also reported as 14 June) 48 after drop from parent (Pe-8 with ASh-82FN); wrecked on third flight 5 Sept 48.

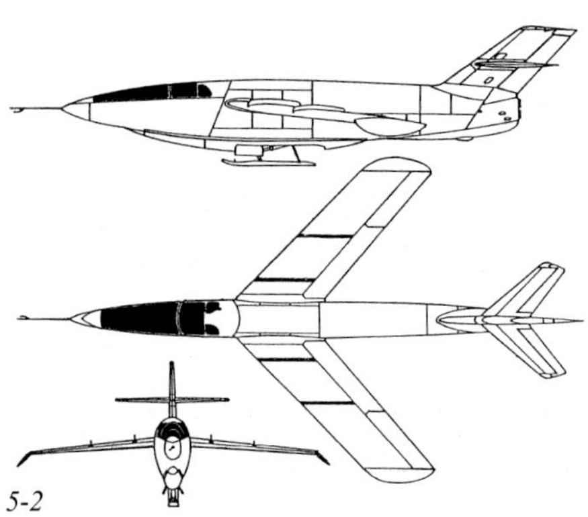

5-2

Aircraft 5-2 fitted with Soviet rocket engine and highly swept vertical tail. First flight by G. M. Shiyanov 26 Jan 49. Five more gliding flights, with modifications including tip skids replaced by steel-edged downturned wingtips, and blister/strake under rear fuselage, but programme cancelled June 49 without having flown under power. Peak Mach reached 0·775.

DIMENSIONS Span (5-1) 6,4 m (21 ft 0 in), (5-2, downturned tips) 6,6 m (21 ft 7⅞ in); length, excl probe (5-1) 9,92 m (32 ft 6½ in), (5-2) 11,2 m (36 ft 9 in; wing area (5-2) 19,87 m^2 (213·9 ft^2).

ENGINE (5-2) One RD-2M3BF.

WEIGHTS (5-2) Empty 1,7 t (3,748 lb); loaded (no propellant) 1,9 t (4,189 lb).

PERFORMANCE Design speed 1200 km/h (M 1·13) at 13 km (42,650 ft).

BOK

BOK *Buro Osovikh Konstruktsii*, OKB for exptl aircraft, ordered by Sov Revolutionary Mil Council Dec 30 as part of CAHI. Vladimir Antonovich Chizhevskii director, N. N. Kashtanov deputy, starting with projects already begun by Skrzhinskii, Kamov, Chyeranovskii, Krichyevskii and Sukhoi. Opened 1 Jan 31, slow progress, transferred to CCB Sept 31 as No 6 Brigade. First ab initio task balloon gondola (see BOK-1). In Feb 33 Chizhevskii transferred to head No 3 Brigade at GAZ-39 Menzhinskii (internal prison), but BOK was restarted June 34, soon transferred to GAZ-35 Smolyensk. Concentrated on tailless and ultra-hi-alt aircraft. Transferred Feb 38 to Podlipky, near Moscow, merged with KB-29; brigades led by Shchyerbakov and Vakhmistrov (see both later). Chizhyevskii arrested Feb 39, BOK merged into Sukhoi OKB in 40. Following is numerical order, not chronological.

BOK-1 Also designated **SS** (*Stratosfernii Samolyet*, stratospheric aeroplane), this was the first pressurized aeroplane to be designed in the world after the Ju 49. BOK was instructed that high-altitude flight was its No 1 priority, and its engineers visited Dessau to talk to the Ju 49 designers. On their return they designed the spherical gondola of the balloon SSSR-1 (exceeded 18 km in 1933), a feature of which was continuous recycling of purified atmosphere, oxygen and water vapour being kept at optimum proportions and carbon dioxide removed. This provided basis for design of 'hermetic cabin' for aeroplane to undertake research at up to 12 km (39,370 ft) to support major 1934 programme of Academy of Sciences.

To save time and effort the airframe was based on the RD (ANT-25), but cabin was completely new unit constructed separately. Cylinder of 1,8 cubic metres volume, constructed of riveted D1 light alloy 1,8-2 mm thick with sealing compound. Convex bulkheads front and rear, with main entry hatch in roof and emergency exit through porthole at rear into fuselage. Small glazed circular portholes, five for pilot and two for nav/radio/observer in rear seat. Design dP about 0,22 km/cm^2 (3·2 lb/in^2), holding interior at 8 km up to ceiling eventually reached of 14,1 km. Controlled leak through dump valve, made good by oxygen from bottles at flow rate keeping oxygen content roughly constant. Engine cooling circuit heated cabin to steady 15/18°C. Four small portholes illuminated unpressurized rear-fuselage.

Span reduced, structure restressed for reduced gross wt, fixed spatted main gears with single wheels. RN engine and 3-blade prop, built GAZ-35, flown by I. F. Petrov summer 36. Later fitted with RNV engine with first one, then two, turbos and 4-blade prop, reaching 14,1 km in 38.

DIMENSIONS Span 30,0 m (98 ft 5 in); length 12,86 m (42 ft 2½ in); wing area 78,8 m^2 (848 ft^2).

ENGINE One 725 hp M-34RN, later 830 hp M-34RNV.

WEIGHTS Empty 3482 kg (7,676 lb) (after engine change 3600 kg, 7,937 lb); fuel 500 kg (1,102 lb) (after engine change 1000 kg); loaded 4162 kg (9,175 lb) (after engine change 4800 kg, 10,582 lb).

PERFORMANCE Max speed 242 km/h (150 mph) at 4 km (after engine change, 260 km/h, 162 mph); endurance given as 4 h for each state. Ceiling 14,1 km (46,260 ft).

BOK-2 Also called RK (*Razreznoye Krylo*, slotted wing), aerodynamic research aircraft. Small wooden single-seater, with M-11 engine, designed by S. S. Krichyevskii in 1934 to test his wing comprising a main front portion and variable-incidence rear portion (too large to be

5-1

5-1 under Pe-8/ASh-82FN

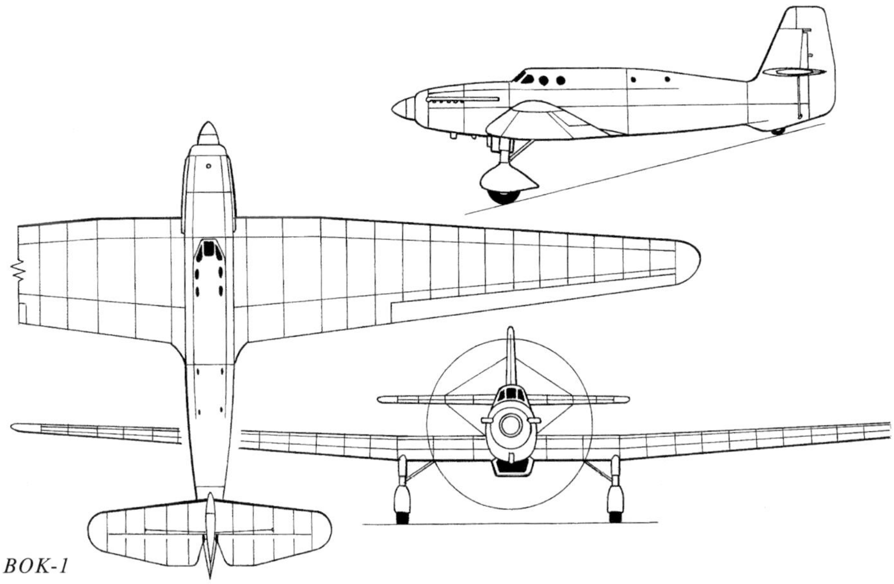

BOK-1

called a flap) with intervening slot. BOK-2 flew successfully in 35, being refined to achieve design objective of automatically adopting most efficient profile, but languished incomplete with designer's death late 35.

BOK-3 Unknown until 91, IS (exec-cttee aircraft), conventional 4-seat liaison aircraft, basically wood, fuselage semi-monocoque, controls dural/fabric, fixed gear, slotted flaps, three tanks total 190 lit. One only, built GAZ-35 and flown in 33. Variants planned including spraying/seeding and mil-trainer with fixed PV-1, turret with DA and FAB-100 or recon camera.
DIMENSIONS Span 11 m (36 ft 1 in); length 7,1 m (23 ft 3½ in).
ENGINE One M-48. No other data.

BOK-5 Simple tailless single-seater. Mainly Dural/fabric (see ASU for details); much devt of TE controls: inboard elevators, then flaps, then ailerons, all to some degree interconnected. Load factor +8. Flown by Stefanovskii and Nyukhtikov from summer 37.
DIMENSIONS Span 9,86 m (32 ft 4¼ in); length 4,365 m (14 ft 4 in); wing area 23,15 m² (249 ft²).
ENGINE One 100 hp M-11.
WEIGHTS Empty 596 kg (1,314 lb); fuel 90 kg (198 lb); loaded 764 kg (1,684 lb).
PERFORMANCE Max speed 174 km/h (108 mph) at SL; landing speed 85 km/h (53 mph); take-off/landing 120/200 m (394/656 ft); ceiling 4850 m (15,900 ft); range 600 km (373 miles), 4 h.

BOK-6 Planned tailless heavy bomber, also called TB, for which BOK-5 provided data; never built.

BOK-7 Devt of BOK-1 with GK (pressure cabin) integral part of fuselage. Again based on RD, using full span. Fuselage slimmer than before, stressed skin, cabin skin thicker, riveted/bonded pressurization to 0,28 kg/cm² by blower driven by main engine. Dorsal domes with portholes, mock-up tested armament for BOK-8.
DIMENSIONS Span 34,0 m (111 ft 6¾ in); length 12,9 m (42 ft 4 in) wing area 87,0 m² (936·5 ft²).
ENGINE One 890 hp M-34FRN with two turbochargers.
WEIGHTS Empty 3900 kg (8,598 lb); no other data survives.

BOK-8 Prototype hi-alt recon-bomber, using second BOK-7 airframe. Dorsal ShKAS barbette with Rezunov optical sight linked electromechanically from second dorsal dome, with synchro link for min lag. Tested on BOK-7 mockup, flown (believed 39) on BOK-8 but records lost.

BOK-10-14 Details lost of these, mainly transport rather than exptl projects. BOK-11 believed to have been built in 38-40, as recon aircraft with ACh-40 diesel in BOK-8 airframe.

BOK-15 Project begun 38, like -11 hi-alt with ACh-40 but min-drag (retrac gear) for 20000-km range to fly round world at 10 km (32,800 ft). Appeals to Stalin led to go-ahead 5 Jan 39, aircraft completed at GAZ-35 June 39 but not flown until 12 March 40, pilot A. B. Yumashyev. Second BOK-15 flown by G. F. Baidyukov.

BOK-3

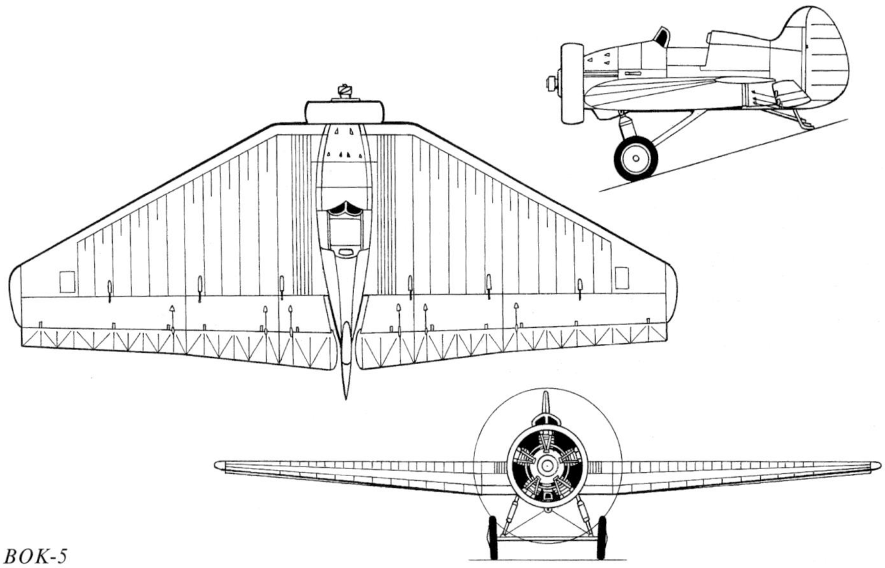

BOK-5

BOK-1 mod

BOK-7

Bolkhovitinov

Viktor Fedorovich Bolkhovitinov was born 1899. An impetuous character, he had a meteoric rise in the VVS. After passing numerous technical courses, appointed a lecturer at the Zhukovskii Academy (VVA) in 32. In 33 became head of VVIA, group of about 20 lecturers and engineers charged with actual design as an OKB, their first task being to plan a replacement for TB-3. About 35 Bolkhovitinov became Professor of Aircraft Design at VVA, continuing in this post until 46, when VVIA was closed. Last hardware programme BI. He died 1970.

DB-A Having planned a successor to TB-3 in 1934 Bolkhovitinov's small staff began serious engineering design at the end of that year. Structural leader M. M. Shishmaryev, aerodynamics Ya. M. Kuritskyes. DB-A (Long-Range Bomber, Academy) was an outstanding accomplishment, fully as advanced for its day as TB-3 had been. Duralumin stressed-skin throughout, except for fabric-covered control surfaces. Fuselage semi-monocoque, with numerous frames and stringers of Z and top-hat section. Main wing spars and ribs built-up trusses like TB-3 but with numerous stringers stiffening skins; aerofoil 20%, dihedral 6°10', 14 wing tanks 14600 lit (3,212 gal), LE hinged down to form servicing platforms each side of engines, deep spars in fuselage filled space above 6×2 m bomb bay with left/right doors. All control surfaces Flettner type, cable operated with trim tabs. Engines drove 3450 (11 ft 4 in) 3-blade v-p props with spinners faired into neat cowls, thin vertical radiators for inner engs down each side of 'trouser', outers simple matrix under short nacelle. Main gears 2000×450 tyres retracting fully into 'trouser' with twin doors, cables also raising tailwheel. Landing gear, split flaps, nose turret and bomb doors all pneumatic. Bomb load 3 t (6,614 lb). Dorsal turret, ShVAK with 250 rounds; nose, tail and rear ventral hatch, one ShKAS with 3,000. Crew 7.

Red prototype flown Khodinka 2 May 35, pilots N. G. Kastanyev and Ya. N. Moseyev completed factory tests March/April 36, NII testing May/June. Good perf, eg sustained 2·5 km on two engines at 19,5 t. Decision to attempt flight over N Pole to USA structurally modified for 34,7 t overload with fuel for 8440 km. Painted red/silver, civil reg N-209, left Moscow Shchyelkovo 12 Aug 37 with crew 6 headed by S. A. Levanevskiy. One eng failed 14 h 32 min later but ETA Fairbanks radioed; no trace of a/c ever found. Despite this, series ordered, see later.

Second aircraft **DB-2A**, increased gross weight, improved airframe, redesigned nose, greater power, APR-3 D/F, crew 8, first flown March 36 on skis. In Nov 36 flown by M. A. Nyukhtikov and M. A. Lipkin to 7032 m with 10 t payload and to 4535 with 13; 14 May 37 Shchyelkovo-Baku and back, 5 t payload, 7 h 2 min. Late 37 order for 16 for DA. Called simply DBA, engines again changed, this time with turbosuperchargers, inner nacelles redesigned with main gears folding back into angled bay with two long doors, inner radiators conventional. Nose again redesigned, cockpit raised, overload bombload 5 t, crew 8 including 5 gunners (nose, dorsal, tail and rear of inner nacelles) with believed twin ShKAS in nacelles, singles nose/tail, ShVAK dorsal turret. Last four not built because of TB-7 (Pe-8). Also remaining on paper **TK-1** (heavy cruiser) with crew 11 and 3 ShVAK (3,000 rounds), 5 ShKAS (11,000 rounds) and underwing rockets.

DIMENSIONS Span 39,5 m (129 ft 7 in); length 24,4 m (80 ft 0⅝ in); wing area 230,0 m^2 (2,476 ft^2).

ENGINES (A) four M-34RN, then RNV, (2A) M-34RNV, (series) M-34FRN.

WEIGHTS Empty (A) 15,4 t, (2A) 16 t, (s) 16,15 t; fuel (2A) 10,2 t; max (A) 21,9 t (48,280 lb), (2A) 24 t (52,910 lb), (s) 22 t (48,500 lb).

PERFORMANCE Max speed at SL (A) 280 km/h, (2A) 305, (s) 300, at ht (A) 330 at 4 km, (2A) 335 (208 mph) at 2, (s) 316 at 6; service ceiling (A) 7220 m, (2A) 6900, (s) 7730 (25,360 ft); range (A [as built], s) 4500 km (2,796 miles), (2A) 4600; TO 400 m; landing 300.

BDD Heavy bomber project Mar 36, four 1,200 hp M-34FRGN, pressurized cabins, span 36,2, length 26, wing area 180, loaded 27 t, max speed 460 (286) at 8 km; plan to build in 38 never materialised.

S An ultra-fast bomber; design had more kinship with speed-record contenders. The designation, said to stand for *Skorostnii* (Speedy), in fact stood for Sparka (Twin, Coupled). Zhukovskii Academy OKB (VVIA) under

DB-A No 1

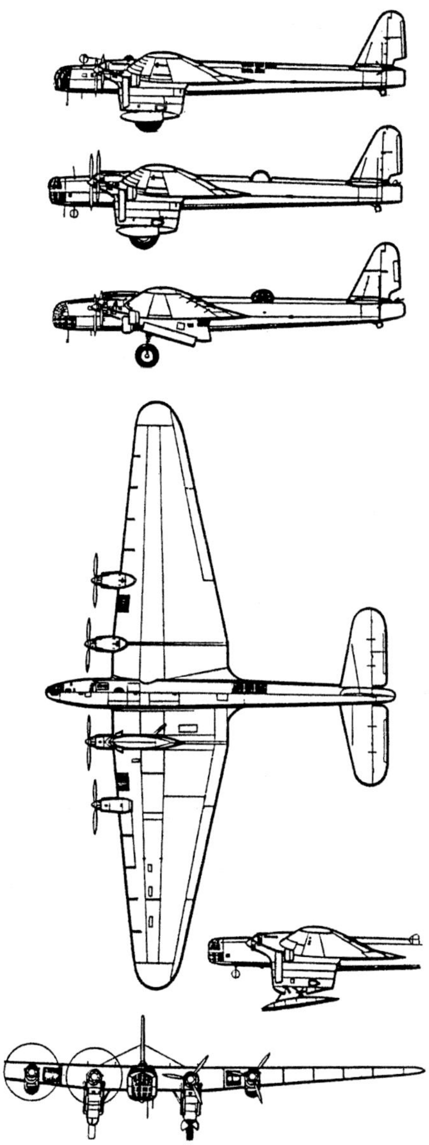

DB-A/M-34RNV (other side views, DB-2A/RNV on skis, DB-A prototype, DB-2A/FRN-TK

DB-2A

S

Bolkhovitinov was enlarged to handle the problems of this tandem-engine system as well as the S aircraft.

The two M-103 engines were mounted in line. The front engine had an extended propeller shaft driving a three-blade propeller. The rear engine was geared to two high-speed (believed crankshaft speed) shafts carried in spherical bearings on the crankcase of the front engine finally driving gears to a second contra-rotating propeller immediately behind the first. The intention was not only that this should confer the highest possible speed but that the aircraft should be able to remain airborne after failure of either engine. It was not intended to take-off on one engine, however, and the wing was made small and heavily loaded in a further attempt to obtain the highest possible speed. It

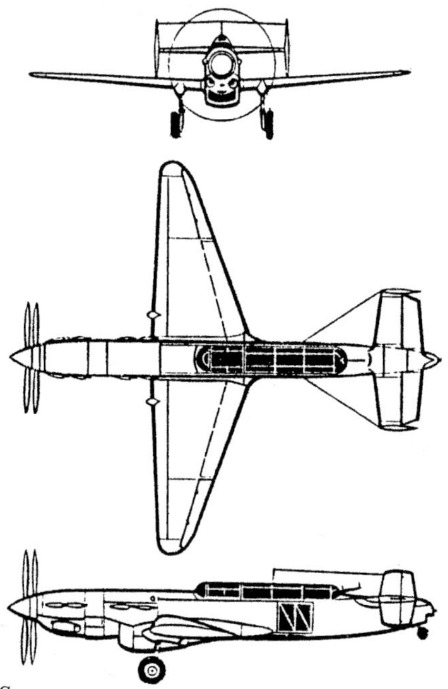

S

was accepted that the exceptional moment of inertia of the tandem engines, and the high wing-loading, would make S far from agile.

Stressed-skin light alloy, including control surface skins. Wing with main two-spar box with heavy upper and lower skins flush-riveted to machined plate spars with lightening holes along neutral axis. Fuselage from top, bottom and two side panels joined at four angle-section longerons. Structure described as original and progressive, and ten years later applied to Il-28. Flaps of Fowler type, driven electrically as was landing gear (wheels turned 90° on backwards-retracting main legs) and numerous other services (29 electric actuators). One actuator drove exit flap on enormous ducted radiator serving both engines. Twin-finned tail, variable-gear drive to elevators, and rudders each with separate trim and servo tabs and inset hinges. Differential ailerons with trimming built into circuit. Pilot and observer in tandem, latter having emergency dual control but seat normally facing to rear with manually aimed ShKAS (typical of S was electric interruption of firing when gun aligned with twin fins, field of fire generally being good). Later ShKAS replaced by much more powerful twin 12·7 mm UBT. Forward-firing guns in original design, not fitted. Internal bay immediately behind rear spar for up to 400 kg of various bombs, with twin electric doors.

Powerplant on test after two years of work in May 38. Detail design of S begun early 37, prototype construction July 38. First flight about September 39 by B. N. Kudrin. Official tests by Kudrin and A. I. Kabanov 40. Performance exceptional (see data) but engine installation ceaseless mechanical failures and a hazard to flight. Nose-up attitude a problem even without simulated bombload (which was aft of CG) and in 1940-41 Z. I. Itskovich, then at VVIA, redesigned wing with modified profile, leading edge and upper surface being covered by ply strips attached by 'secret riveting' and adhesive. Considerable improvement, but inability to solve mechanical problems with powerplant eventually resulted in decision to remove rear engine (probably late 40). In winter 1940-41 testing continued on non-retractable skis. Underpowered with one engine and spare mass could not be used for fuel or bombs. Factory needed for Pe-2 production and development abandoned 41.

DIMENSIONS Span 13,8 m (45 ft 3¼ in); length 13,2 m (43 ft 4 in) (with single engine 13,0 m, 42 ft 8 in); wing area 26 m² (280 ft²).
ENGINE(S) Originally two 960 hp M-103 in tandem; later, one 960 hp M-103P.
WEIGHTS Empty not published; loaded 5652 kg (12,460 lb); (one engine) loaded 3676 kg (8,104 lb).
PERFORMANCE Max speed 570 km/h (354 mph) at 4,6 km; (with one M-103P) 400 km/h (248·5 mph) at 4,4 km; range (two engines) 700 km (435 miles).

I Designation of experimental fighter/dive bomber conceived early 40 and in prototype construction 41. Scheme confirmed by VVS. *Sparka* installation of tandem M-107 engines, though prototype to use M-105 or M-103. Smaller derivative of S but with even more advanced features including Elektron (magnesium alloy) structure, integral tankage, tricycle landing gear and scheme for catapult launch. Lead designer A. M. Isayev. Project stopped at start of war, summer 41.

D Project for impressive four-engined bomber with two *Sparka* tandem M-105 powerplants. Clean spindle-like fuselage with projecting gondola, mid wing and twin-wheel main gears. Max weight 28 t (wing loading 200 kg/m²). Also passenger variant. Abandoned summer 41.

Borovkov-Florov

In 1935 Aleksei Andreyevich Borovkov and Ilya Florentyevich Florov, young engineers at CCB, proposed fighter biplane smaller than any previously attempted except with much less powerful engines. Their studies were so professional that Ya. I. Alksnis at VVS agreed at end 35 to recommend an OKB be started, and this was set up under GUAP authority as OKB-7 early 36.

7211 This beautiful prototype amply fulfilled designers' belief that large engine and heavy firepower could be packaged into significantly smaller airframe, whilst retaining good handling, agility and flight safety. Basis was diminutive fuselage with pilot immediately ahead of fin in part-glazed cockpit. Forward fuselage welded truss of 30KhGSA steel, with dural skin panels; rear fuselage and tail wooden monocoque with ply skin similar to I-16. Identical upper and lower sets of outer wing panels, pinned at front and rear spars to carry-through structure under fuselage and to upper centre section, with considerable stagger. No interplane struts. Wings light-alloy stressed-skin, each with spars and ribs pressed from sheet and skin a sandwich stiffened by thin corrugated light alloy between smooth faces. Three hinged trailing surfaces on each panel, serving as ailerons and drooping as flaps, all worked without interplane connection. Engine assembled from French parts, long-chord cowl with single exhaust at bottom, three-blade prop. Provision for four ShKAS (2,200 rds total) two in sides and two in top of fuselage.

Engine cut on first TO 21 June 37, no blame attached pilot E. Yu. Preman who lost control on essential turn-back. OKB-7 contract March 38 for four follow-on prototypes.

DIMENSIONS Span 6,98 m (22 ft 10¾ in); length 6,35 m (20 ft 10 in); wing area 18,0 m² (194 ft²).
ENGINE One M-85.
WEIGHTS Empty 1321 kg (2,912 lb); loaded 1745 (3,847 lb).
PERFORMANCE Max speed 365 km/h at SL, 416 km/h (285·5 mph) at 4 km; climb 4,6 min to 5 km; service ceiling 13 km; endurance 2 h/ 750 km; time 360° turn 14/15s; take-off 150 m.

I-207 Four prototype and pre-production fighters were built using essentially same airframe as No 7211. *Izdelye 7 No 1* (Product 7, No 1), little changed apart from close-fitting cowl with projecting fairings over cylinder-head rocker-boxes and prominent carb-air inlet at 12 o'clock position. VISh-26 two-blade propeller. Full armament from start of testing. Open cockpit; spatted landing gear without the transverse tie fitted to No 7211. First flight 20 April 39; test pilots included P. M. Stefanovskii, L. M. Maksimov and N. I. Nikolayev.

DIMENSIONS As before except span 7,0.
ENGINE One M-62.
WEIGHTS Empty 1598 kg (3,523 lb); loaded 1950 (4,299).
PERFORMANCE Max speed 387 km/h at SL, 436 (271 mph) at 4 km; climb 6·2 min to 5 km; service ceiling 9150 m (30,000 ft); range 700 km (435 miles); TO 19 s, ldg 115 km/h, 196 m.

I-207/M-63 Three aircraft, *Izd. 7 No 2, Izd. 8 No 3, Izd. 9 No 4*. First three flown 39, direct-drive to VISh-26 2,8-m prop, provision for two FAB-250 or two DM-4 ramjets under lower wings. 7 No 2 flown Ap 39, fixed gear, open cockpit. 8 No 3 Sept or Oct with open cockpit but gear resembling I-153 retrac to rear, wheels turning 90°; 20 flights with ramjets without any emergency. 9 No 4 with geared engine driving 3-blade VISh 3m prop, enclosed cockpit, not flown until spring 41.

DIMENSIONS As before except No 4 length 6,7 (21 ft 11⅞ in).
ENGINE See text.
WEIGHTS Loaded 7/2 1950 kg, 8/3 (no DM-4) 1850 (4,078 lb), 9/4 2,2 t (4,850).
PERFORMANCE Max speed (2) 397 km/h at SL, 423 (263 mph) at 4 km (3, no DM-4) 428/486, (4) est 460/518 (322); climb to 5 km (2) 6·7, (3) 4·6, (4) est 4·5; service ceiling (2) 9,2, (3) -, (4) est 10,5; range (9) est 640 km (400 miles); ldg 115 km/h.

D Experimental single-seat fighter powered by combined 2,000 hp M-71 piston engine in same duct as Merkulov ramjet with rear propulsive nozzle. High monoplane wing of gull form, span 14,8 (48 ft 7 in); leading edge swept (tapered) at 20°. Stressed-skin, tail (at least) having smooth skin stabilized by underlying corrugated structure. Armament two NS-37 and two ShVAK-20 (exceptionally heavy). Active project Jan 41 but abandoned end June 41 (OKB-7 was in extreme Western USSR). Borovkov killed in aircraft accident 1945; Florov continued design (see later).

I-207 (7/1)

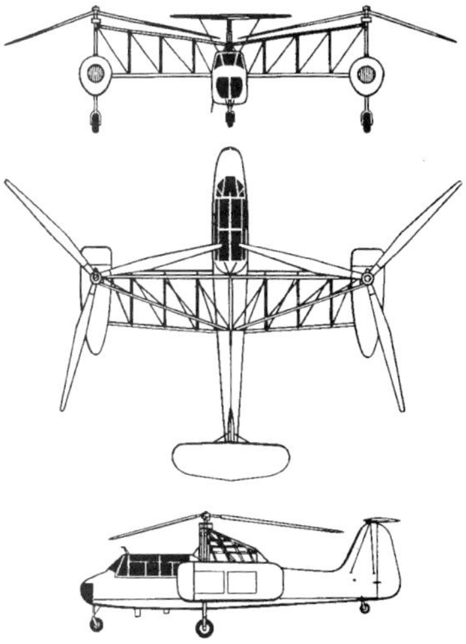

Omega-2MG

I-207 (7/2)

Bratukhin

Ivan Pavlovich Bratukhin was born in 1911. He concentrated entirely on helicopters from his student days, and was one of the pioneers in the CAHI OOK (see later). As early as 33 he was leader of OOK Brigade B, and made significant contributions to CAHI helicopters from 5-EA onwards. During the purges of the late 30s Bratukhin escaped arrest and inevitably moved up to replace those less fortunate, and in Jan 40 was appointed first deputy designer in the newly formed OKB-3 at the MAI, under Yuryev. Only three months later Yuryev was replaced by Bratukhin, who ran the OKB for 11 years. It was closed in 51, after which Bratukhin researched VTOL devices as a member of MAI's staff. Bratukhin was in 80 working as a lecturer at the Mozhaisky Institute in Leningrad.

Omega

Also designated 2MG (twin-engined helicopter), this ambitious project was authorised 27 June 40 and launched Bratukhin as creator of a long series of machines with two engine/rotor groups side-by-side. Steel-tube with fabric covering. Lateral structures carrying engine/rotor units and fixed main landing gears were triangular spaceframes with two lower booms and one at top, with no covering. Fully articulated hubs of steel and light-alloy, carrying three duralumin blades, rotors being handed to rotate in opposite directions. Manual controls, driving swashplates in fixed and rotating parts of head for collective and cyclic pitch control. Much research into optimum method with differential collective to apply initial bank and then T-tail rudder, worked by pedals, to make turns. Emergency control for immediately autorotative setting of both rotors, but single-engine flight intended by connecting shaft between engines with universal joint on aircraft centreline. Final 0·231 gearbox to rotors (577 rpm). No centrifugal clutches. Three mechanical clutches, with overrunning capability, engaged by observer after both engines were running. First drive to associated rotor was clutched-in; then interlink left/right shaft. Fuel tanks behind engines. Observer behind pilot.

Helicopter ready for testing Aug 41. Final design had no wheel or rotor brakes but trimming tailplane with pilot handwheel. Tentative hovering tests, interrupted by engine rough running and overheating. Six-month delay followed evacuation Oct 41. Pilot K. I. Ponomaryov gradually made progress in 1942, discovering structural and control problems but remaining tethered until early 43. Engines seriously overheated.

DIMENSIONS Diameter of each rotor 7,0 m (22 ft 11½ in); span between rotor axes 7,2 m (23 ft 7½ in); length (discounting rotors) 8,2 m (26 ft 11 in); disc area 76,96 m² (828 ft²).
ENGINES Two 220 hp MV-6 inline.

WEIGHTS Empty 1760 kg (3,880 lb); loaded 2050 kg (4,519 lb).
PERFORMANCE Max speed 116 km/h (72 mph) (one report states 186 km/h); range 250 km (155 miles).

Omega II Also sometimes called G-2 (helicopter 2), this was the original design with superior engines in streamlined pods with fan-assisted cooling. Drive ratio 0·32. Rotor masts and outriggers stiffened, and dynamic parts (clutches and gearboxes) redesigned for long life. Test flown by Ponomaryov Sept 44 with good results. Damaged Jan 45, repaired and improved (drive ratio 0·283) and used again from July 45, by this time for research and pilot training. With chief engineer D. T. Matsitskii as observer, gained height of 3 km. Demonstrated Tushino by M. K. Baikalov Aug 46.
DIMENSIONS As 2MG.
ENGINES Two 330 hp MG-31F.
WEIGHTS Empty 1880 kg (4,145 lb); loaded 2300 kg (5,071 lb).
PERFORMANCE Max speed 150 km/h (93 mph); ceiling at least 3 km (9,840 ft).

G-3 (AK) Third Bratukhin helicopter, based on G-2 but planned from outset for operational use by VVS as AK (*Artilleriskii Korrektirovshchik*, artillery correction). No significant differences apart from more powerful engine. Two prototypes ordered 44, flown 45. On completion of State trials in 45 batch of ten AK ordered. Soviet histories mention only original two prototypes as having flown. A single AK delivered to VVS with dual control for training helicopter pilots.
DIMENSIONS As 2MG.
ENGINE Two 450 hp Pratt & Whitney Wasp Junior R-985-AN-1.
WEIGHTS Empty 2195 kg (4,839 lb); loaded 2600 kg (5,732 lb).
PERFORMANCE Max speed 170 km/h (106 mph); ceiling (hovering) 1,1 km (3,608 ft), (forwards flight) 2,5 km (8,200 ft).

G-4 First Soviet helicopter with engine designed for such duty. Cooling fan and front gearbox giving vertical and lateral outputs for rotor and for transverse coupling shaft. Transmission provided with centrifugal clutch; overall drive ratio 0·27 (540 rotor rpm). Rotor diameter slightly increased, inbuilt twist of 6°45' along blade, and blade spar extruded instead of folded from sheet. Two prototypes, first flown Oct 47 by M. K. Baikalov and second a month later. In Jan 48 first G-4 damaged in course of exhaustive autorotational descents and dead-stick landings typically at 12 m/s descent along 15·5-16° glide path. Second G-4 first in Soviet Union to meet stipulated life for dynamic parts (in this case 100 h). Small series constructed 1947-48, according to one account four flying out of ten ordered.
DIMENSIONS Diameter of each rotor 7,7 m (25 ft 3¼ in); span between rotor axes 7,6 m (24 ft 11¼ in); length (discounting rotors) 8,1 m (26 ft 7 in); disc area 93 m² (1,002 ft²).
ENGINES Two 500 hp AI-26GR.
WEIGHTS Empty 2364 kg (5,212 lb); loaded 3002 kg (6,618 lb).
PERFORMANCE Max speed 148 km/h (92 mph); hovering ceiling 2,4 km (7,874 ft); range 233 km (145 miles). (Note: speed low in comparison with reasonable expectation and especially when compared with B-5).

B-5 First helicopter designated for Bratukhin himself. Scaled-up derivative of G-4, with slightly more powerful model of same engines. Transmission and other dynamic parts identical. Same configuration but new airframe with lifting aerofoil wing instead of space-frame outriggers and large passenger fuselage. Latter designed as duralumin semi-monocoque (not steel-tube as has been reported) with level floor and door on right side. Total of eight seats, intended as two crew and three double passenger seats with aisle along right wall. Wings also of light-alloy stressed-skin construction with lifting profile to bear about 25% of weight in cruise. Fixed tricycle gear with bumper tailwheel. Single example only, completed 47; only limited testing because of inadequate wing stiffness.
DIMENSIONS Diameter of each rotor 10,0 m (32 ft 10 in); other dimensions not known; disc area 157 m² (1,690 ft²).
ENGINES Two 550 hp AI-26GRF.
WEIGHTS Empty 2932 kg (6,464 lb); loaded 4032 kg (8,889 lb).
PERFORMANCE Max speed claimed 236 km/h (147 mph); ceiling (hovering) 2280 m (7,480 ft), (forwards flight) 6400 m (21,000 ft); range 595 km (370 miles).

B-9 Ambulance/casevac derivative of B-5, with larger, more rectangular fuselage housing four stretchers in two layers on right and attendant. Wing changed to non-lifting symmetrical section but set at positive incidence. Single example completed 1947 but, because of aeroelastic difficulties with B-5, never being flown.
DIMENSIONS Essentially as B-5.
ENGINES Two AI-26GRF.
No other data.

B-10 Also designated VNP (*Vozdushnii Nabludatyelnii Punkt*, aerial observation point), this could have been AK-2 because its original role was to replace the AK in artillery spotting. In fact by this time the basic Bratukhin helicopter had become rather larger and potentially more capable and VNP was to be a multi-role machine able to fly night recon, tactical supply and even casevac missions. Engines fully boosted AI-26 version, and dynamic parts essentially same as B-9. Wings of same plan and section as B-9 but with pair of bracing struts from bottom of two main-spar frames in fuselage to spar booms at 60% semi-span, and two further bracing struts from upper spar booms at same location to top of rotor masts. Fuselage entirely new: dural monocoque with glazed nose seating pilot on left and navigator on right under large observation dome; observer in tail again with large observation dome. New tail with variable-incidence tailplane mounted on fuselage carrying endplate fins (latter possibly rudders, but not described as such in literature). Usual four-wheel landing gear. Central fuselage available for additional loads: three passengers, or two stretchers (room for more but weight-limited) or 200 kg cargo or various radio or photographic equipment. Entrance door on left.

Single example built and flown 1947. Behaviour satisfactory, and complete performance measurements taken. Later fitted ShKAS at nose and tail. According to Shavrov 1947-48 saw general disillusionment with helicopters and especially with twin lateral rotor configuration.
DIMENSIONS Essentially as B-5.
ENGINES Two 575 hp AI-26GRF.
WEIGHTS Empty 3019 kg (6,656 lb); loaded 3900 kg (8,598 lb).
PERFORMANCE Max speed 218 km/h (135 mph); ceiling as B-5; range 440 km (273 miles).

B-11 Last design authorized from OKB-3, to provide comparison against Mil and Yak single-rotor submissions to VVS requirement for three-seat all-weather communications helicopter. Dynamic parts as before except for hydraulic rotor-hub dampers and faired masts with oil-cooler inlets. Wing of lifting section, set at zero incidence. Fuselage of improved form, with round instead of polyhedral top and bottom. Tail basically as B-5. Pilot in B-5 type nose, main cabin for two seats, with space at rear for freight or two stretchers (in lieu of

G-3 (VVS), pilot Ponomaryov

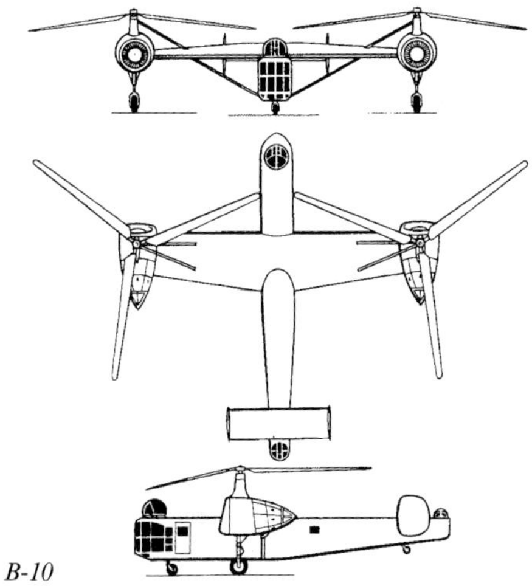

B-10

B-5

passengers) loaded through enlarged door. Improved differential rotor controls with reduced friction.

Two prototypes completed in April 48, respectively flying in June and September. Made good progress with measured performance, sustained (47 min) single-engine flight, autorotative landings and in eradication of various faults stemming mainly from vibration. One problem was lack of wing lift resulting from incidence setting, causing rotors to stall at high forward speeds. Further snag was hydraulic leak, difficult to rectify. On 13 Dec 48 first machine shed blade from right rotor, killing Ponomaryov and I. G. Nilus. Numerous modifications suggested which did not receive support.

DIMENSIONS Rotors as B-5; length of fuselage 9,76 m (32 ft 0¾ in).

ENGINES Two 575 hp AI-26GRF.

WEIGHTS Empty 3398 kg (7,491 lb); loaded 4150 kg (9,149 lb).

PERFORMANCE Max speed 155 km/h (96 mph); range 328 km (204 miles).

B-12 Unbuilt project for single-rotor training helicopter with anti-torque tail rotor in various versions with M-14 (AI-14) or twin M-11 engines. Bratukhin also had studies for a 30-seat assault helicopter powered by two M-82FN engines and, after closure of his OKB, for a ramjet tip-drive helicopter which by 1955 got as far as running a full-scale rig using an engine salvaged from a G-4 to accelerate the test rotor to ramjet self-sustaining speed. This work was done at MAI, where one of the designer's last projects was for a piston-engined (later gas-turbine) convertiplane which formed the conceptual basis of the *Vintokryl* built at the Kamov OKB.

CAHI

CAHI, alternatively TsAGI, Central Aero- and Hydrodynamic Institute, founded Dec 18 by N. Ye. Zhukovskii to centralise national research effort. Aero work immediately branched out beyond aerodynamics into fields of structures, materials and aircraft design. Two major brigades formed for design: AGOS, KOSOS (see Tupolev), and many later General Constructors originally worked in these. In addition, CAHI ZOK (*Zavod Opitnyi Konstruktsii*), factory for exptl construction, built pioneer rotary-wing machines; chief designers were Yur'yev, Chyeremukhin, Bratukhin, Izakson, Kuznetsov, Skrzhinskii, Mil and Kamov. CAHI still makes major input in design of most state-sponsored projects giving recommended configurations, wing sections etc.

Helicopters

1-EA First helicopter to fly in the Soviet Union, brain-child of Prof Boris Nikolayevich Yur'yev, who from 09 had striven to build such machine and under Zhukovskii had been first employee of CAHI and chief architect of its original test facilities. In 25 he organized a *vertolyet* (helicopter) group and in 27 tested a 6-m two-blade rotor driven by 120 hp M-2. In 28, from at least four project studies, design picked for 1-EA (*Eksperimentalnyi Apparat*) and machine built at ZOK under G.Kh. Sabinin. Chief designer Aleksei Mikhailovich Chyeremukhin and Aleksandr Mikhailovich Izakson. Fuselage welded from M1 mild-steel tube forming spaceframe resting on tailwheel landing gear with rubber springing. Twin engines at centre with reduction and bevel gears to vertical shaft to main rotor and side-by-side pairs of 1,8-m anti-torque rotors at nose and tail. Chief development task was finding best structure for main rotor blades, an all-metal blade being rejected in favour of one with dural spar, wood ribs and stringers and ply/fabric skin. Four-blade main rotor with 5° coning angle driven at 153 rpm. Cyclic and collective controls described as similar to those of today, and pedal control of nose/tail rotors from pilot seat just ahead of engines. Chyeremukhin elected to fly 1-EA himself, making first tethered run Aug 30. No disasters and by 32 reliable flights of up to 12 min; 1 Aug 32 same pilot reached 160 m, 3 Aug 230 m, 5 Aug 285 m and on 14 Aug 605 m. Soviet Union not FAI member and did not announce achievement, world record remaining 18 m (d'Ascanio). 1-EA continued to fly until at least 34.

DIMENSIONS Main-rotor diam 11,0 m (36 ft 1 in); length (over nose/tail rotors) 12,8 m (42 ft 0 in); height (main rotor lifting) 3380 mm (11 ft 10¾ in).

ENGINES Two M-2.

WEIGHTS Empty 982 kg (2,165 lb); fuel/oil 70+13 kg; loaded 1145 kg (2,524 lb).

PERFORMANCE Max speed reached 30 km/h; max ht 605 m (1,985 ft); endurance 14 min.

3-EA Second helicopter built 33 for training pilots. Similar to 1-EA and same data. Never flown except on tethers.

5-EA Broadly similar to predecessors, incorporated new rotor proposed by I. P. Bratukhin with three large blades with articulated roots which provided lift, and three smaller blades with rigid attachment which provided control. Some accounts incorrectly state large and small rotors superimposed; all blades were in same plane, large and small alternating. This rotor worked, yet 5-EA appears not to have reached any considerable height or endurance (Western figures 13 miles and 1,200 ft not confirmed in Soviet accounts). 5-EA flew from 33 until late 35 and was instrumental in providing research basis for machine with more powerful and more modern engine.

DIMENSIONS Main-rotor diams (large blades) 12,0 m (39 ft 4½ in), (small) 7,8 m (25 ft 7 in); length of fuselage 11,0 m (4,5 ahead of main axis, 6,5 to rear) (36 ft 10¾ in); length over nose/tail rotors, as 1-EA; height 3105 mm (10 ft 2¼ in).

ENGINES Two M-2.

WEIGHTS Empty 1047 kg (2,308 lb); fuel/oil 70+13 kg; loaded 1210 kg (2,668 lb).

PERFORMANCE Max speed 20 km/h; alt not over 40 m; endurance 8½ min.

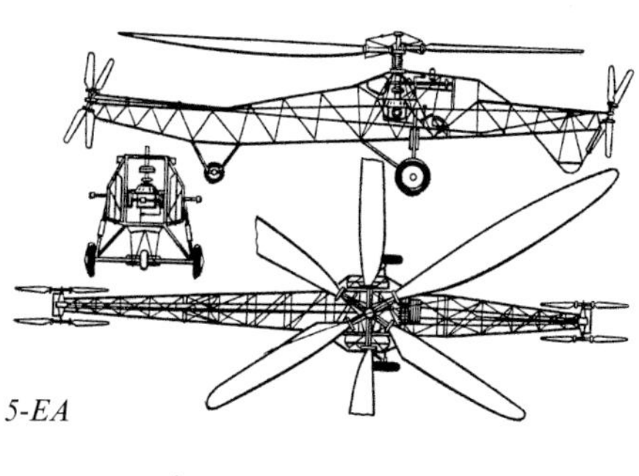

5-EA

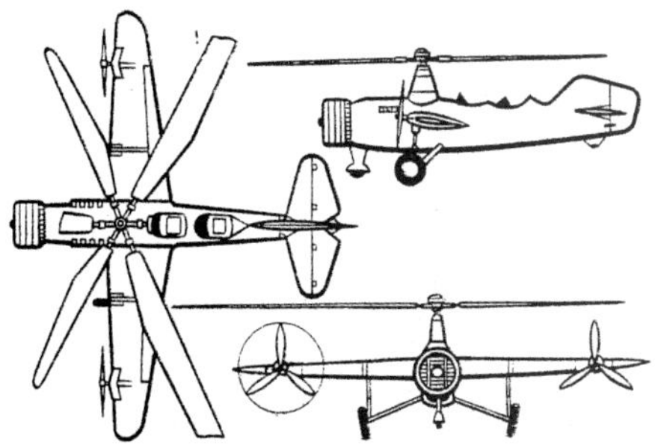

11-EA

11-EA Designed 34, this used Bratukhin's rotor with alternate large articulated blades and small rigid blades but scaled up to absorb power of large US water-cooled V-12 engine. Latter mouned in nose of steel-tube fuselage covered in fabric for streamlining, with tandem cockpits at rear behind main rotor pylons. Engine arranged facing to rear with reduction gear driving second reduction gearbox at foot of rotor shaft. Auxiliary drive from rear wheelcase to three-blade cooling fan drawing air through main radiator on nose. Rotor torque reacted by two 2250-mm three-blade propellers near tips of small fixed wing. Complete aeroplane flight controls – ailerons, elevators, rudder – in addition to improved cyclic/collective controls on main rotor. Intention was to test as helicopter and later arrange for drive to main rotor to be disconnected and aircraft flown as autogyro, with all power used for forward propulsion via propellers, with aeroplane flight control. 11-EA was completed summer 36 and completed tethered testing on a platform, using front cockpit as observer and rear as pilot, though dual controls provided. Considerable difficulties with distribution of power, main blade construction and maintenance of steady height whilst varying push/pull power of propellers. Late 37 wave of arrests removed Chyeremukhin, Izakson and many other helicopter engineers and virtually halted further work through fear of accusation of sabotage.

After much calculation Bratukhin dared 38 to begin alterations and in spring 38 aircraft emerged as **11-EA PV** (*Propulsivnyi Variant*). Main rotor hub fitted with improved blades entirely dural with better profile. Auxiliary propellers replaced by pairs of anti-torque rotors from 5-EA mounted at tips of outriggers of welded steel tube, increasing distance between thrust axes from 8,0 to 11,0 m. Removal of wing improved vertical performance, and smaller screws at greater distance from fuselage reduced power loss in countering torque. Idea of making convertible helicopter/autogyro abandoned. PV ready late 39 and in Chyeremukhin's absence D. I. Savelyev took over flight test, making tethered flights early 40 and free flights from October. Under test director V. P. Lapisov reconstructed machine demonstrated excellent lifting power and good control, but ancient engine no longer ran well (no spares for many years) and programme halted early 41 for new Soviet engine. Never flew again.

DIMENSIONS Main-rotor diams (large blades) 15,4 m (50 ft 6⅓ in), (small blades) 9,2 m (30 ft 2¼ in); span of wing 10,6 m (34 ft 9⅜ in), (PV, 11,2 m, later removed); length (excl rotors) 8,5 m (27 ft 10⅝ in); height 3530 (11 ft 7 in).

ENGINE One 630 hp Curtiss Conqueror.

WEIGHTS Empty not known (PV, 1877 kg); loaded 2,6 t (5,732 lb), (PV) 2250 kg (4,960 lb).

PERFORMANCE PV reached 60 km/h at 50 m.

Autogyros

2-EA Based on tandem-seat Cierva C.19 Mk III, built at OOK under direction of Izakson to design of I. P. Bratukhin and V. A. Kuznetsov and completed 31. First Soviet autogyro for methodical research and development. Fuselage and twin-fin tail welded mild-steel tube with fabric, wing wood and rotor-blade spars and pylon struts welded KhMA. No mechanical drive to rotor. Research complete 33 and 2-EA assigned to *Maksim Gorkii* propaganda squadron.

DIMENSIONS Rotor diam 12,0 m (39 ft 4½ in); span 6,7 m (22 ft 0 in); length 6,5 m (21 ft 3⅞ in).

ENGINE One 225 hp GR Titan (Bristol licence).

WEIGHTS Empty 765 kg (1,687 lb); fuel 90 kg; loaded 1032 kg (2,275 lb).

PERFORMANCE Max speed 160 km/h (99 mph); min speed 58 km/h (36 mph); climb to 3 km, 20 min; ceiling 4,2 km; endurance 1 h; take-off 60 m; landing 0-6 m.

A-4 Designation from *Avtozhir* (autogyro); decree by NII VVS 32 for all-Soviet autogyro, and preparations for production organized forthwith for use as military trainer and observation machine. Design led by N. K. Skrzhinskii using known Cierva technology, regarded as minimal risk. Structure similar to 2-EA but single fin, more power, dual controls. First flight 6 Nov 32; second flight three days later resulted in unexplained rotor vibration and loss of lift, with heavy impact on ground despite maximum power. Production had begun and frantic research with different rotors eventually yielded satisfactory machine, passing NII tests late 33. Ten delivered 34 (possibly a few more later) and served as trainers and in field manoeuvres.

DIMENSIONS Rotor diam 13,0 m (42 ft 8 in); span, as 2-EA; length 7,2 m (23 ft 7½ in).

ENGINE One M-26.

WEIGHTS Empty 1065 kg (2,348 lb); fuel/oil 95+20 kg; loaded 1365 kg (3,009 lb).

PERFORMANCE Max speed 176 km/h (109

A-4

A-8

mph); min speed 50 km/h (31 mph); climb to 3 km, 21 min; ceiling 4,1 km; endurance 1·3 h; range 185 km (115 miles); time 360° turn 15 s; take-off 70-100 m; landing 3-10 m.

A-6 Light autogyro designed at OOK by Kuznetsov. First rotary-wing machine with folding blades, and also folding wings. Construction as before except balloon tyres instead of shock struts. Flown Ts. A. Korsinshchikov early 33 and from start showed outstandingly good flying qualities, far better than A-4 on which all official attention was polarized. Following various research programmes in winter 33, one of which was testing V (butterfly) tail.
DIMENSIONS Main-rotor diam 11,0 m (36 ft 1 in); span, not known but wing area 5,9 m^2 (63·5 ft^2); length 6,3 m (20 ft 8 in); height 3,2 m.
ENGINE One M-11.
WEIGHTS Empty 562 kg (1,239 lb); fuel/oil 67 kg; loaded 815 kg (1,797 lb).
PERFORMANCE Max speed 142 km/h (88 mph); min speed 53 km/h (33 mph); ceiling 2 km; endurance 2·5 h; take-off 50 m; landing 0 m.

A-8 First flown 29 June 34, this was virtually an A-6 with two major alterations: wings had 5° dihedral instead of upturned tips, and as well as ailerons direct rotor control of roll was provided, even during autorotation (and found to be vastly superior to ailerons except at max speed). Other changes included auxiliary fins on tailplane and first air/oil shock struts in Soviet Union.
DIMENSIONS As A-6 except wing area 5,8 m^2 (62 ft^2).
ENGINE One M-11.
WEIGHTS Empty 595 kg (1,312 lb); fuel/oil 67 kg; loaded 837 kg (1,845 lb).
PERFORMANCE As A-6 except min speed only 48 km/h (30 mph) and ceiling 2560 m.

A-7 First of new family developed by 3rd Brigade under Kamov from 31 to meet NII VVS demand for powerful autogyro for front-line duties such as artillery spotting, recon and liaison. Originally designated **7-EA**, broke new ground in many areas. Three-blade rigid rotor with clutch for spin-up to 195 rpm (max in flight 200) for jump start (before achieved by Cierva). Blade spar KhGSA tube, ribs and stringers wood, skin ply over leading portion and *Enerzh-6* stainless sheet to rear. Fuselage welded KhMA steel tube with dural skin, with integral fin. Tail dural with fabric covering, including auxiliary fins below tailplane. Wing wooden, with ply skin. Tricycle landing gear, one of first in world, with steerable nosewheel (believed world first) and spats on all wheels. Townend ring on engine, metal propeller. Slotted elevator and special protection against overturning in gusty conditions or because of short wheelbase on rough ground. Flown by Korsinshchikov 20 Sept 34, but prolonged development needed – interrupted by sensational appearance at Tushino Aug 35 – until released for NII tests Dec 35. Passed NII tests by A. A. Ivanovskii 36 and cleared with small changes for use as observation and spotting platform. Same prototype shipped to Greenland 38 to help in rescue of Papanin expedition, but not needed. Improved **A-7bis** of 36 with modified rotor pylon and other changes led to **A-7-Za** of which five built 40 (see Kamov) and used by VVS. Lighter airframe, improved fuselage, no spats, synchronized PV-1 (later ShKAS) and twin PV-1 in rear cockpit. Active career included participation in 1938-40 Tian-Shan expedition and sporadic reconnaissance missions during the War.
DIMENSIONS Rotor diam 15,2 m (49 ft 10½ in); span 10,4 m (34 ft 1½ in); wing area 14,7 m^2; length not known.
ENGINE One M-22.
WEIGHTS Empty (7) 1300 kg, (bis) 1474 kg, (Za) 1225 kg (2,701 lb); fuel/oil (7) 330 kg; loaded (7) 2056 kg, (bis) 2224 kg, (Za, max) 2300 kg (5,071 lb).
PERFORMANCE Max speed (7) 210 km/h, (bis) 194 km/h, (Za) 221 km/h (137 mph); min speed 53 km/h; ceiling 4,7 km (15,400 ft); climb to 3 km (3 models) 16/22/19 min; endurance 4/4/2½ h; range 600/600/400 km; take-off 60/75/75 m; landing 18/20/18 m.

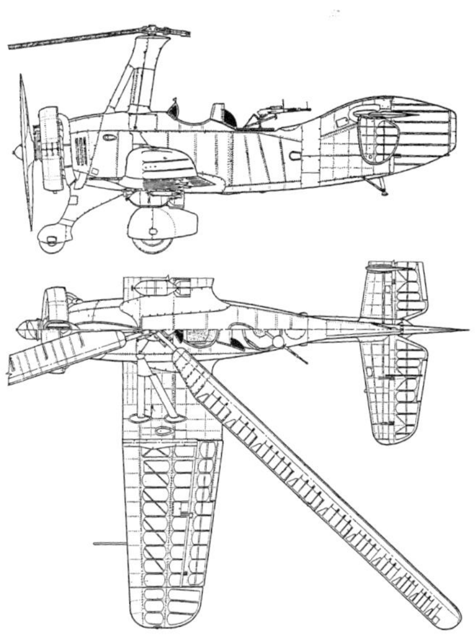

A-7

A-12 OOK 6th Brigade under Skrzhinskii designed A-10 six-seat transport autogyro with M-22 in 34 but dropped this in favour of high-speed wingless machine with fully controllable rotor. Design in parallel with small A-14 which flew first and assisted perfection of A-12 wingless autogyro. Direct-control rotor derived from A-7, with similar engine drive before takeoff. Roll control by rotor only, tail with central rudder (extension of fuselage) and two Scheibe fins on strut-braced tailplane with elevators. Shavrov loosely describes fuselage and engine as I-16 (fighter); in fact steel-tube with fabric, and engine cowl with gills. Two-position Hamilton prop, single-seat cockpit, very strong main gears with oleo struts and wide-track spatted wheels with toe brakes. Cautious taxiing and hops 10 May 36 by A. P. Chyernavskii, eventually 17 h 55 min in 43 flights until rotor blade came off 23 May 37 killing I. Koz'yrev. Cause never found but ascribed fatigue; almost halted autogyro work.
DIMENSIONS Rotor diam 14,0 m (45 ft 11⅛ in); length 6,3 m (20 ft 8 in).

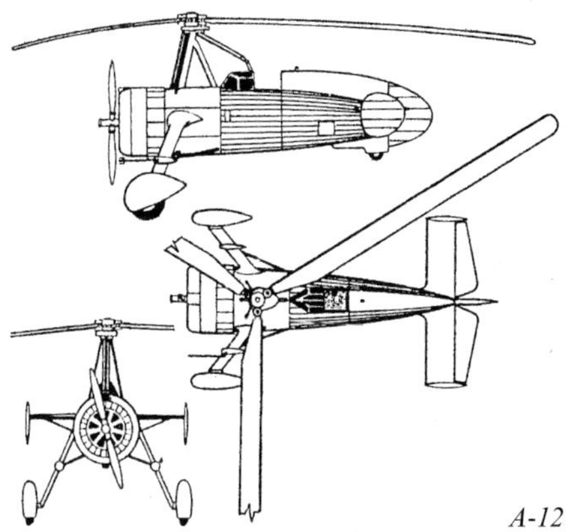

A-12

ENGINE One 670 hp Wright Cyclone (later M-25).
WEIGHTS Empty 1343 kg (2,961 lb); fuel/oil 165+17 kg; loaded 1687 kg (3,719 lb).
PERFORMANCE Max speed 245 km/h (152 mph); min speed 52 km/h (32 mph); ceiling 5570 m (18,275 ft); endurance 1½ h; take-off 25 m; landing 5-10 m.

A-13 Two-seat liaison machine developed from A-6 and A-8. Direct-control inclined rotor with folding blades, folding dihedralled wings and twin-finned tail, internal engine starter and clutch-in drive for rotor before take-off. First flown on skis March 36.
DIMENSIONS Rotor diam 11,5 m (37 ft 8¾ in); span not known but area 5,6 m^2 (60 ft^2).
ENGINE One M-11.
WEIGHTS Empty 540 kg (1,190 lb); fuel/oil 50 kg; loaded 802 kg (1,768 lb).
PERFORMANCE Max speed 151 km/h (94 mph); min speed 45 km/h (28 mph); ceiling 3 km; endurance 2 h; range 250 km; take-off 40 m; landing 0 m.

A-12

A-13

LS

A-14 Final 2nd Brigade (Kuznetsov) autogyro to be flown, derived from A-6 and A-8 without wings and first to be flown in Soviet Union. Trouble-free flight 17 Sept 35 and subsequently major research tool.
DIMENSIONS Rotor diam 11,0 m (36 ft 1 in); length 6,3 m (20 ft 8 in).
ENGINE One M-11.
WEIGHTS Empty 576 kg (1,270 lb); fuel/oil 67 kg; loaded 815 kg (1,797 lb).
PERFORMANCE Max speed 167 km/h (104 mph); min speed 45 km/h (28 mph); endurance 2½ h; take-off 50 m; landing 0 m.

A-15 Intended as fastest autogyro of day, two-seat wingless machine using A-12 and A-14 technology and with usual welded KhMA structure, Scheibe fins on braced tailplane with elevators and direct-control rotor. Main gears with vertical shock struts carried on triangulated outriggers each side of fuselage. Pilot with synchronized ShKAS and observer with twin ShKAS and AFA-13 reconnaissance camera. Chief designer M. L. Mil, construction manager A. A. Kuznetsov. Completed April 37 and taxi tests begun by Chyernavskii and Ivanov, but A-12 crash caused further work to be halted. No reason to doubt airworthiness of A-15 but effectively terminated Soviet autogyro development. Autogyros picked up again at Kharkov, Kuibyshyev and Riga 30 years later.
DIMENSIONS Rotor diam 18,0 m (59 ft 0⅔ in); length 8,6 m (28 ft 2½ in); height 4,1 m (13 ft 5½ in).
ENGINE One M-25V.
WEIGHTS Empty 1695 kg (3,737 lb); fuel/oil 385+40 kg; loaded 2560 kg (5,644 lb).
PERFORMANCE Est max speed 260 km/h (162 mph); min speed 50 km/h; ceiling 6,4 km (21,000 ft); take-off 35-60 m; landing 0 m.

LS Single-seat *Laminarnyi sloi* (laminar layer) research glider built in CAHI workshops from late 41 to investigate achievement of laminar flow over wings. Configuration adopted to enable pusher engine to be installed later whilst retaining non-turbulent wing airflow. Profiles devd from Feb 39 by I. V. Ostoslavsky and K. K. Fedyayevsky, poor lift coefft. First flight 8 Aug 42, towed by DB-3, pilot V. L. Rastorguyev. Stalled on approach, damaged beyond repair. In 43 G. P. Svishchyev devised better laminar profile, but LS not rebuilt.
DIMENSIONS Span 14895 (48 ft 10½ in); length 10390 (34 ft 1 in).
WEIGHT Not discovered.

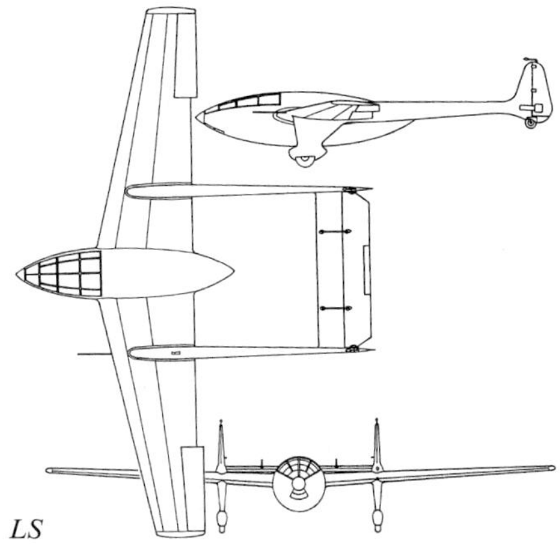
LS

Chyeranovskii

Boris Ivanovich Chyeranovskii was born in 1896 and deeply interested in aeroplanes as a boy. Throughout his 64 years he concentrated on various forms of tailless machine, building more than any other designer in history and in general successful, though often at the expense of ease of handling. He was the pioneer of the so-called Parabola wing with curved leading edge. He was able to enter the Zhukovskii Academy in 22, and in the same year proposed his first published design, a flying wing with aspect ratio 1·5. Many CAHI aerodynamicists were convinced such a wing was useless but tests with model wings of aspect ratio 1·5, 3 and 6 did much to refute their predictions. Via the route of simple gliders in the Crimea, such as the **BICh-1**, Chyeranovskii progressed to aeroplanes. He remained a loner, never really having an OKB, and it is sad that in the jet era when his tailless forms came into their own his failing health prevented construction. He died on 17 Dec 60.

BICh-3 Chyeranovskii's first aeroplane, based on previous gliders. Single-seater with cockpit faired into low vertical tail with rudder. Parabola type wing with ailerons, worked by differential bell-cranks, and elevators, all with inset hinges and aerodynamic balance. Single trousered wheel and wingtip skids. Test-flown

BICh-1

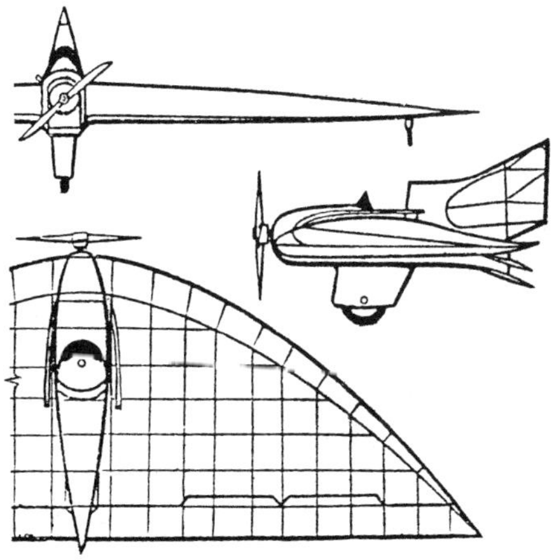

BICh-3

BICh-3

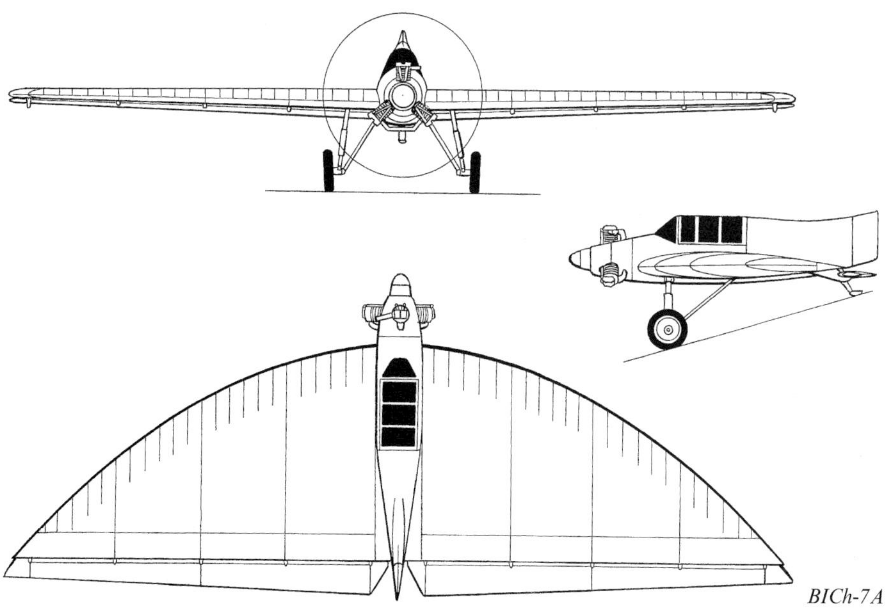

BICh-7A

and demonstrated by B. N. Kudrin at Moscow 26. Rudder powerful because of slipstream, but on the whole BICh-3 not stable. Control improved by minor changes and final assessment was that aircraft safe.
DIMENSIONS Span 9,5 m (31 ft 2 in); length 3,5 m (11 ft 6 in); wing area 20,0 m^2 (215 ft^2).
ENGINE One 18 hp Blackburn Tomtit driving 1400 mm propeller.
WEIGHTS Empty 140 kg (309 lb); fuel 10 kg (22 lb); loaded 230 kg (507 lb).
PERFORMANCE Max speed not known (probably about 140 km/h, 87 mph); landing speed estimated only 40 km/h (25 mph).

BICh-5 Largest of the BICh series, this was to have been aerodynamically clean bomber with two BMW VI engines and retractable main wheels. No data but tested as model in tunnel 28. Much research done on trailing-edge controls which became multi-segment elevons hung below trailing-edge with convex undersides. Work terminated 29.

BICh-6 Not known.

BICh-7 Enlarged development of BICh-3, first flown 29. Scale said to be 1·5 times on area basis, but in fact larger than this. No tail, but small rudders, without fins, on tips of wing. Centreline wheel, with skids under rudders. Two tandem open cockpits. Directional stability and control poor, and take-off almost impossible. Urgent modifications, first being conventional landing gear.
DIMENSIONS Span 12,2 m (40 ft 0½ in); length 4,7 m (15 ft 5 in); wing area 30,0 m^2 (323 ft^2).
ENGINE One 100 hp Bristol Lucifer.
WEIGHTS Empty 612 kg (1,349 lb); fuel 93 kg (205 lb); loaded 865 kg (1,907 lb).
PERFORMANCE Max speed 165 km/h (102·5 mph); landing speed 70 km/h (43·5 mph).

BICh-7A Chyeranovskii completely rebuilt BICh-7 and the resulting BICh 7A did not fly until 32. Enclosed tandem-seat cabin with fin and rudder downstream. No wing-tip surfaces. Normal landing gear with two wheels and tail-skid. Totally transformed, with good directional stability, fine flying qualities and immediate response to pilot demand. Main problem was loss of speed in turn because large elevons caused high pressure on rudder. Test-pilot N. A. Blagin (later collided *Maksim Gorkii*) gradually improved by adding strips to elevons and setting at lower incidence. Control surfaces hung in Junkers double-wing style on five inset hinges below trailing edge of wing. Only other problem was vibration of engine.
DIMENSIONS Span 12,5 m (41 ft 0 in); length 4,95 m (16 ft 3 in); wing area 34,6 m^2 (372 ft^2).
ENGINE One 100 hp Bristol Lucifer.
WEIGHTS Empty 627 kg (1,382 lb); fuel 93 kg (205 lb); loaded 880 kg (1,940 lb).
PERFORMANCE Max speed 165 km/h (102·5 mph); landing speed 70 km/h (43·5 mph); range 350 km (217 miles).

BICh-10 Nothing is known of BICh-8 and BICh-9, and the BICh-10 never flew and led to BICh-14. BICh-10 was first Chyeranovskii twin, with two M-11 engines, tested as tunnel model 33. Few changes led to:

BICh-14 (one modification was addition of Townend ring engine cowls) which was built at Menzhinskii works and received additional designation **CCB-10**. Built to double scale of BICh-7 on area basis, with twin engines on leading edge and from two to five seats in fuselage of same length as wing chord on centreline. Usual wood/fabric construction apart from light alloy in cabin and fin. Centre section of 3,3 m span and outer panels, with total of four spars and 60 ribs. Three surfaces on each side below trailing edge hung on four inset hinges. First flight at end 34 by Yu. I. Piontkovskii. By 36 BICh-14 tested at NII VVS, pilots including P. M. Stefanovskii, M. A. Nyukhtikov, and I.

BICh-14

F. Petrov. General stability and control marginal. Trailing-edge aileron/flaps and inboard elevators of symmetrical section – largely ineffective. Considerable stick force needed to get tail down on landing. Rudder without slipstream and ineffective. Testing discontinued after 37.
DIMENSIONS Span 16,2 m (53 ft 2 in); length 6,0 m (19 ft 9 in); wing area 60 m² (646 ft²).
ENGINES Two 100 hp M-11.
WEIGHTS Empty 1285 kg (2,833 lb); loaded 1900 kg (4,189 lb).
PERFORMANCE Max speed 220 km/h (137 mph); landing speed 70 km/h (43·5 mph); range 370 km (230 miles).

BICh-11 (RP-1) Designed 31 in parallel with BICh-12, this tailless machine was first purpose-designed rocket aeroplane in world. All-wood flying wing, with vestigial central nacelle. Wing-tip rudders, trailing-edge ailerons and elevators. Originally centreline wheel, later changed for normal gear and tailskid. First tests as bungee-launched glider, on skis, early 32. Intended to become RP-1 (*Raketnyii Planer*, rocket glider) with two of F. A. Tsander's OR-2 rocket engines rated at 50 kg (110 lb) thrust each. Ahead of each engine large lagged spherical tank of liquid oxygen and smaller capsule of gasolene. OR-2 never regarded as safe for installation, though run on bench 18 March 33, so BICh-11 flown with small piston engine.
DIMENSIONS Span 12,1 m (39 ft 8½ in); length (rockets) 3,09 m (10 ft 1¾ in), (piston) 3,25 m (10 ft 8 in); wing area 20,0 m² (215 ft²).
ENGINE Intended two Tsander OR-2 rockets, finally one 35 hp ABC Scorpion.

BICh-11

WEIGHTS Only known figure: empty (rockets) 200 kg (441 lb).
PERFORMANCE No data.

BICh-16 Despite attempting ornithopter (flapping-wing) in 21 inventor persisted and in 34 got Osaviakhim funds to build. Birdlike wood/fabric, small horiz surf at rear but no controls other than near-vert actuating rod in front of pilot pivoted by links to about 30% semi-span. Tested Aug 35 to 38, many mods and no success. No data.

BICh-17 Fighter described under Kurchyevskii on p.161.

BICh-18 Named *Muskulyet* (from muscle-power), this was another man-powered device but one with much better chance of success. Again supported by Osoaviakhim and assigned to unfortunate Pishchuchyev to fly, it was basically a high-performance sailplane in style, with nose cockpit and conventional tail. Wings in form of two pairs crossing over in X form on centreline so that lower right plane became upper left and vice versa. Both wings pivoted axially on centreline and driven by pedals and bell-crank linkage. Large portion of each wing freely pivoted to rear of main spar and able to flap up and down. Thus, as pilot pedalled, wings rocked in unison, tips never quite touching, oscillating rear portions giving forward thrust. Wing tips in form of ailerons. Light balsa wood structure. Flown 10 Aug 37 off bungee launch without pedalling, gliding 130 m from release. On launch No 4 Pishchuchyev pedalled and accomplished six wing oscillations. Glide extended to 430 m, pilot reported noticeable forward thrust. Sustained flight admitted to be impossible.
DIMENSIONS Span 8,0 m (26 ft 3 in); length 4,48 m (14 ft 8½ in); wing area 10,0 m² (108 ft²).
ENGINE Pilot's muscle-power.
WEIGHTS Empty 72 kg (159 lb); loaded 130 kg (287 lb).
PERFORMANCE Max range attained claimed to have been 0,45 km.

BICh-20 Pionyer Chyeranovskii's smallest aeroplane, this also marked a shift to a short-span straight-tapered wing of almost delta shape. Trailing-edge ailerons and elevators now of inverted lifting profile. Short fuselage with pilot canopy in effect forming leading edge of fin. Latter broad, to house pilot, and relatively small, but effective because of slipstream. Extensively tested making turns of about 35° in horizontal flight at different heights; stability judged acceptable. Originally flown early 38, on skis, and then re-engined in same year.

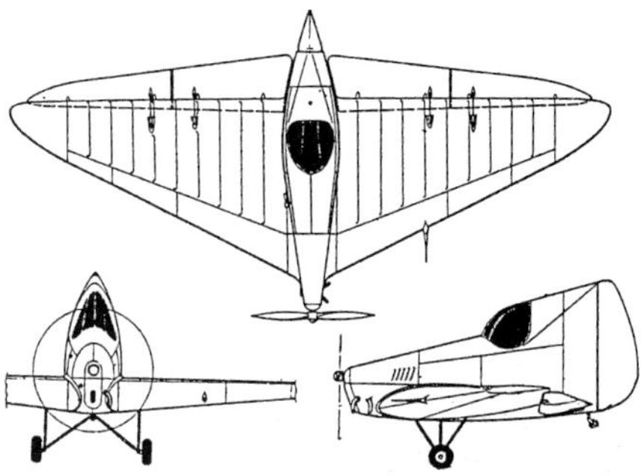

BICh-20

DIMENSIONS Span 6,9 m (22 ft 8 in); length (original) 3,5 m (11 ft 6 in), (re-engined) 3,56 m (11 ft 8¼ in); wing area 9,0 m² (97 ft²).
ENGINE First, one 18 hp Blackburn Tomtit; later, 20 hp, Aubier-Dunne.
WEIGHTS Empty 176 kg (388 lb), (re-engined) 181 kg (399 lb); fuel 24 kg (53 lb), (re-engined) 26 kg (57 lb); loaded 280 kg (617 lb), (re-engined) 287 kg (633 lb).
PERFORMANCE Max speed 160 km/h (99 mph), (re-engined) 166 km/h (103 mph); ceiling 4 km (13,120 ft); range 320 km (199 miles); landing speed 49 km/h (30 mph).

BICh-21 Satisfactory behaviour of BICh-20 prompted Chyeranovskii to design a similar aircraft, but with a much more powerful engine, as an entrant in the Osoaviakhim all-union air race planned for Aug 41. Minimum aircraft capable of taking chosen engine which filled fuselage back to rudder pedals, with hinged canopy forming leading edge of fin. Small fuel tanks in roots of 1,5 m centre section, anhedral with straight leading edge. Bolted-on outer sections with dihedral and sharp taper to round tips. Except for steel-tube and light alloy engine installation, structure all-wood, with polished surface. Ailerons and elevators of usual form below trailing edge. Pneumatically retracted landing gears, folding to rear with wheels partly exposed ahead of fairings under lowest part of wing. Ratier two-blade v-p propeller of 2 m

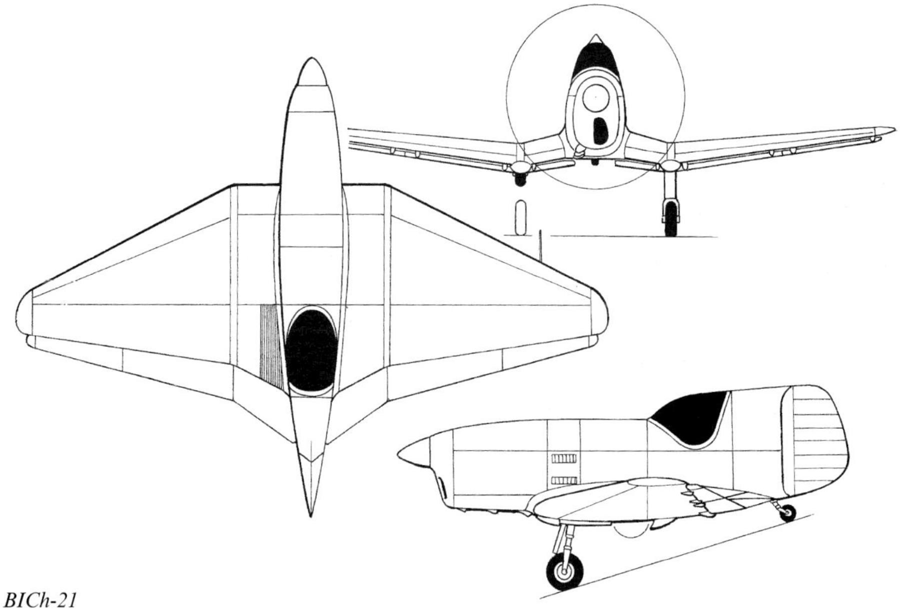

BICh-21

diameter. BICh-21, also styled **SG-1** (*Samolyet Gonochnii*, racing aeroplane), was projected in 38 but not completed until 40. Not flown until June 41. Expected to win, but Nazi attack came first.

DIMENSIONS Span 6,9 m (22 ft 8 in); length 4,74 m (15 ft 6⅝ in); wing area 9,0 m² (97 ft²).

ENGINE One MV-6.

WEIGHTS Empty 526 kg (1,160 lb); fuel/oil 37; loaded 643 kg (1,418 lb).

PERFORMANCE Max speed 385 km/h at SL, 417 (259 mph) at 4 km; ldg 80 km/h.

BICh-26 Between Nov 47 and June 48 Chyeranovskii ran what was in effect an OKB working on advanced jet fighters. BICh-24 and BICh-25 were the first Soviet variable-sweep designs, with plan shapes remarkably like those of modern combat machines, with pivots outboard. These were tunnel-tested but not built, and their place was taken by BICh-26, again with a most modern style of wing and tailless configuration remarkably like 1990s fighter projects. Powered flight controls of typical BICh form, light-alloy structure and pressurized cockpit. Design incomplete.

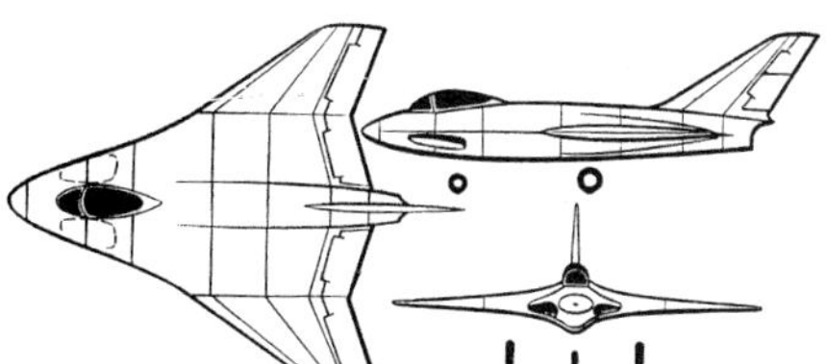

BICh-26

DIMENSIONS Span about 7,0 m (23 ft 0 in); length about 9,0 m (29 ft 7 in); wing area 27 m² (291 ft²).

ENGINE One 2000 kg (4,410 lb) AM-5 turbojet.

WEIGHTS Loaded 4500 kg (9,920 lb).

PERFORMANCE Max speed Mach 1·7 at 7 km (22,960 ft); ceiling 22 km (72,000 ft).

Chyetverikov

Igor Vyacheslavovich Chyetverikov was born in 1904. In 1922 he was admitted to the LIIPS (Leningrad institute of aerial communication engineers) and succeeded in graduating in about 27. He joined the CCB in Moscow and in Jan 31 appointed head of the CCB marine aircraft brigade, located at the Menzhinskii factory. Here he created a major flying-boat which, by all accounts, then had to be designed again by a more experienced team. In 32 he moved to the NII GVF and worked in the OSGA brigade under N. M. Andreyev, where he achieved an original design that did not prove satisfactory. A year later he made a proposal to Glavsevmorput which immediately led to the establishment of his own OKB at Sevastopol in late 34. Here he had another failure, but also one successful design. His only prototype after the Second World War was a failure, and his OKB was closed in 48. He went back to Leningrad to lecture at the Mozhaiskii Academy.

MDR-3 Chyetverikov was appointed 1931 with one overriding commission: design a new long-range flying boat for MA. Chyetverikov's inexperience made him avoid innovation; he sought to make use of existing designs.

In a few weeks he had produced most of the engineering drawings for a fine modern boat that used few parts that were new. Designation: No 11. The hull was almost a direct scale of ROM-2, only major changes being to increase depth and thus allow wing to be mounted direct on hull, and to reduce drag by using curved instead of vertical steps in planing bottom. Complete wing borrowed from TB-5 heavy bomber, with span increased slightly by mating it to wider hull, added centre section having corrugated skin, and with powerplant nacelles attached above fourth rib on each side instead of below. Horizontal tail from same source. There was little that could go wrong with stabilizing floats (again based on existing designs), central fin carrying tailplane, rounded twin fins and rudders (which were original) and pylon-mounted tandem engine installations, each pair of engines sharing common large ducted radiator under front of nacelle.

Provision for crew of six. Dual flight controls in enclosed cockpit which, with position for navigators and gunners, and much on-board equipment, were also based on TB-5. Four pairs of DA guns in two tandem nose, dorsal and tail positions. Racks under centre section for two 250 kg (551 lb) bombs or equivalent.

Swift construction, first No 11 completed Dec 31, and transported from Moscow factory to Sevastopol. First flight, by B. L. Bukholts, 14 Jan 32. Test programme completed 25 March 32. So far story sounds resounding success, and structurally and hydrodynamically No 11, by 32 given service designation MDR-3, was excellent. Trouble lay in all-round poor performance, exemplified by take-off time on smooth water of 36s, time to 2 km of 40 min, and ceiling of 2,2 km. In late 32 project transferred to KOSOS for improvement; it led to ANT-27, MDR-4.

DIMENSIONS Span 32,2 m (105 ft 7¾ in); length 21,9 m (71 ft 10½ in); wing area 153 m² (1,649 ft²).

ENGINES Four 680 hp BMW VI V-12.

WEIGHTS Empty 8928 kg (19,683 lb); fuel 3,3 t (7,275 lb); loaded 13 973 kg (30,805 lb).

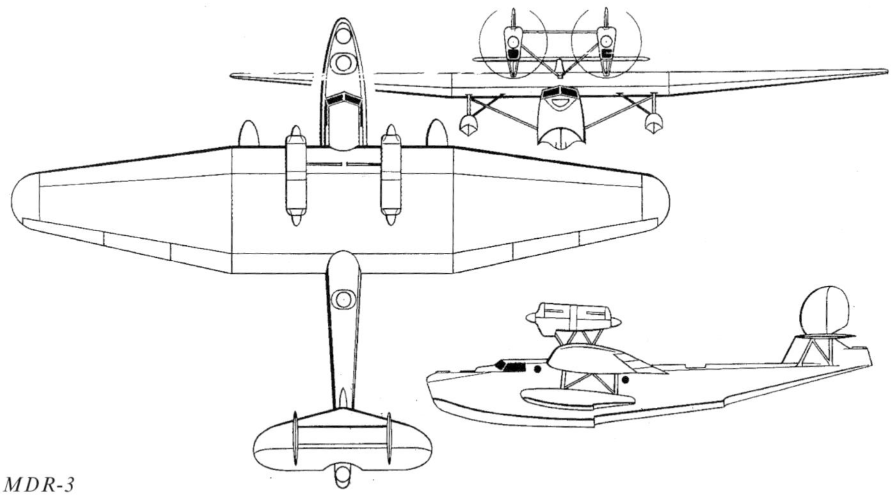

MDR-3

SPL

SPL

PERFORMANCE Max speed 210 km/h (130·5 mph); ceiling 2,2 km (7,218 ft); range 1600 km (1,000 miles); alighting speed 110 km/h (68 mph).

SPL (OSGA-101) British trials convinced Soviet Navy it had to deploy aircraft aboard large cruiser submarines for open-sea recon. Chyetverikov had studied such an aircraft and in early 31 had placed a proposal for an SPL (*Samolyet dlya Podvodnikh Lodok*, aeroplane for submarine boats) on desk of CCB head. No response for two years; then NII suddenly awarded 100,000 rubles for two prototypes of SPL. Chyetverikov decided to make the first a plain non-folding machine with landing gear, and this was designated OSGA-101. After testing, if it was successful, a foldable flying boat would be built as SPL.

OSGA-101 completed spring 34 and first flown by A. V. Krzhizhevskii in July. Main hull and two-spar wing wooden, but tail (with twin fins to reduce height) carried on welded steel tubular booms. Engine, with Townend ring, carried on pylon above wing; two stabilizing floats of wood; landing gear comprised main wheels cranked by hand up under wing (though remaining fully exposed) and castoring tailwheel ahead of the water rudder. Enclosed cockpit seated two side-by-side with a third seat behind, in front of wing spar. Tail construction dural and fabric, flight-control cables running inside the tubular booms. Flight tests were satisfactory.

SPL which followed had almost same airframe but only two front seats, no landing gear, short-span wings folding to lie alongside tail, and an unusual pivoted engine nacelle with bracing struts so that in folded position the nacelle lay between booms with two-blade wooden propeller parallel with longitudinal axis. Folded (specification allowed four minutes for this) SPL fitted watertight cylinder 2,5 m (8 ft 2½ in) diameter and 7,45 m (24 ft 5¼ in) long. Withdrawal and preparation for flight was allowed five minutes. SPL completed in Dec 34, taken to Sevastopol and flown by Krzhizhevskii early 35, testing being completed 29 Aug. MA rejected it on grounds of inadequate seaworthiness, and never did acquire SPL-type aircraft. Prototype rechristened **Gidro-1** and given to Osoaviakhim, later setting up various class records.

DIMENSIONS Span (OSGA) 11,4 m (37 ft 5 in), (SPL) 9,5 m (31 ft 6 in); length (OSGA) 7,6 m (24 ft 11½ in), (SPL) 7,4 m (24 ft 3½ in); wing area (OSGA) 17,0 m² (183 ft²), (SPL) 13,4 m² (144 ft²).

ENGINE One 100 hp M-11.

WEIGHTS Empty (OSGA) 630 kg (1,389 lb), (SPL) 592 kg (1,305 lb).

PERFORMANCE Max speed (OSGA) 170 km/h, (SPL) 186 km/h; cruise (OSGA) 130 km/h (81 mph), (SPL) 183 km/h (114 mph) at 2,5 km; alighting speed (OSGA) 75 km/h, (SPL) 85 km/h); range (both) 400 km (248 miles).

ARK-3 In 1933 Chyetverikov roughed out a compact flying boat with tandem engines, with relatively high speed and long range, but was unable to build it. Off his own initiative he proposed to Glavsyevmorput that this organization, in charge of all northern sea transport, should adopt his design as multi-role vehicle for Arctic. He gained an order for a prototype, designated ARK-3 (*Arktichyeskii*, Arctic, 3). This set him up as constructor, with OKB at Sevastopol.

ARK-3 was straightforward design with dural hull and wooden wings of MOS-27 profile set at 5° incidence, Ailerons and tail, light alloy with fabric covering. Wooden underwing floats well inboard. Dual-control enclosed flight deck and two gunners or observers in bow and rear-hull cockpits. Design configured for minimum dimensions, but structural factor of safety 5·5.

Prototype completed Sevastopol Jan 36 and flight test programme completed September. Modifications needed to bows and floats to give adequate seaworthiness, and engine pylon had to be strengthened, but measured performance

ARK-3-2

good (later, 25 April 37, A. V. Yershov took 1 tonne load to 9190 m) and second prototype authorized as ARK-3-2, original becoming ARK-3-1. Second aircraft more powerful, with longer hull and increased wing chord, and fitted with manual gun turret in bow and rear gun in sliding hatch installation. MA placed small production order for five, speedily implemented, but when first about to fly, on 14 July 37, ARK-3-1 suffered structural failure. Precisely a year later ARK-3-2 also destroyed. Programme abandoned.

DIMENSIONS Span 20,0 m (65 ft 7¼ in); length (1) 14,0 m (45 ft 11 in), (2) 14,6 m (47 ft 11 in); wing area (1) 58,7 m^2 (632 ft^2), (2) 59,5 m^2 (640 ft^2).

ENGINES Two tandem radial driving two-blade Hamilton Standard v-p propellers: (1) 710 hp M-25, (2) 730 hp M-25A.

WEIGHTS Empty (1) 3242 kg (7147 lb), (2) 3642 kg (8029 lb); fuel/oil (1) 870 + 150 kg, (2) 820 + 160 kg; loaded (1) 4787 kg, overload 5800 kg (12,787 lb), (2) 5600 kg (12,346 lb).

PERFORMANCE Max speed at SL (1) 252 km/h (157 mph), (2) 260 km/h (162 mph); at rated height (1) 308 km/h (191 mph), (2) 320 km/h (199 mph); climb to 1 km 3·5 min; endurance 7 h; time 360° turn 45 s; alighting speed 110 km/h (68 mph).

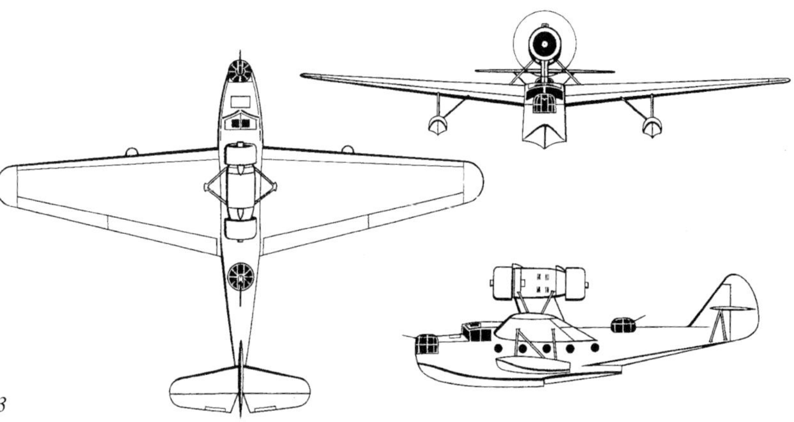

ARK-3

MDR-6 (Chye-2) Designed in 36, this modern long-range military flying boat was designer's only real success. Quite shallow and streamlined hull, with gull wing and high-mounted engines to keep propellers out of spray. Structure light-alloy stressed skin, but fabric-covered flight control surfaces and some other parts. Skin plating of planing bottom and wing primary structure 2 mm, hull mainly 0·8 mm. Generally advanced flush riveting with dimpled holes. No flaps shown or mentioned in most literature, though certainly fitted; one report states full-span, with inset ailerons. Transverse main step, planing bottom tapered to a point at second step. Braced tailplane and fixed stabilizing floats. Fuel in wings. Normal crew only three: pilot, nav/gunner in front turret and radio/observer in rear dorsal turret. Spare seat on right of pilot cockpit. Single ShKAS in nose and dorsal turrets, wing racks for 400 kg bombload, overload 12 stores total 1 t.

Prototype flown July 37 with M-25 engines driving 2-blade Hamilton v-p props, later with M-25Ye and then M-62 with 3-blade VISh. In mid-39 M-63 fitted, and with this engine 20 aircraft (called **N** and also **III**) built at GAZ-31, 13 in 40 and 7 in 41. As **Chye-2** served with Baltic, Black Sea and Far East fleets Oct 41, withdrawn late 42 because Frame 10 weak, props caused hull vibration and fuel system needed remanufacture.

DIMENSIONS Span 21,0 m (68 ft 11 in); length 15,73 m (51 ft 7¼ in); wing area 59,4 m^2 (639 ft^2).

ENGINES See text.

WEIGHTS Empty (37) 4087 kg (9,010 lb), (38) 3940 (8,686), (39) 4,1 (9,039); fuel/oil 1570+110; loaded (37) 5,6/6,45 t, (38) 5,6/6,5, (39) 6,7/7,2 (15,873 lb max).

PERFORMANCE Max speed (37) 338 km/h at 3 km, (38) 350 at 4, (39) 360 (224 mph) at 4; climb to 5 km in (37) 24 min, (38) 10, (39) 9; service ceiling (37) 8,5 km, (38) 10, (39) 9; range (37) 2650 km max, (38) 2700 or 3800 overload fuel, (39) 2650 (1,647 miles); TO 400/400/350; alighting all 100 km/h, 350.

MDR-6A, B1 GAZ-31 evacuated Oct 41, no further prodn. Designer aimed at higher speed with MDR-6A, flown by D. Slobodchikov Dec 40. Total redesign, smaller wing, two spars welded steel 60×60×6, gauge 2 mm in gull-wing centre section, then 0,8 to tip, fabric aft except all-metal elec slotted flaps; streamlined hull 1,9 m wide, twin-finned tail with dihedral, liquid-cooled engines with rads underneath driving c/s 3-blade props, floats pivoted to spars and retrac inwards electrically causing bulge under wing; three hand-aimed ShKAS, four bombs or other stores of 100-250 kg. Completed autumn 39, tested throughout 40 until crashed near Uglich. B1 retrac floats reshaped

MDR-6

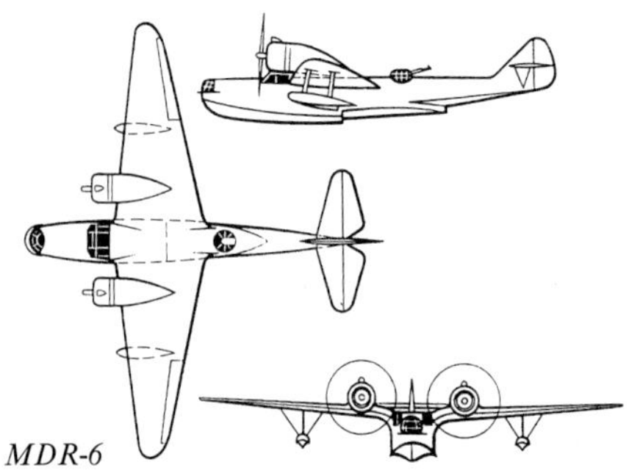

MDR-6

for better hydrodynamic properties, flown spring 41, completed factory test but engine problems. Data for both:

DIMENSIONS Span 16,2 m (53 ft 2 in); length 15,7 m (51 ft 6 in); wing area 48,0 m^2 (517 ft^2).

ENGINES Two M-105.

WEIGHTS Empty 4,2 t (9,259 lb); fuel/oil 1100; loaded 6,9 t (15,212 lb).

PERFORMANCE Max speed 454 km/h (282 mph) at 6,0 km; no other data except TO 400.

MDR-6B2, B3 Fixed floats, engines with flat frontal SB-type rads above spray; B3 with later engines, dorsal blister each side with hand-held UBT so crew 5. Both built 43, B3 first flew early 44.

DIMENSIONS As A, B1.

ENGINES (B2) two M-105, (B3) VK-105PF.

WEIGHTS Empty (B2) 4,3 t (9,480 lb), (B3) 4,7 (10,362); fuel/oil 1100; loaded (B2) 7,2 (15,873), (B3) 7,2 normal, 8,2 (18,078) max.

PERFORMANCE Max speed (B3) 430 km/h (267 mph) at 6 km; no other data except TO (both) 350, alighting (B3) 130 km/h, 400.

MDR-6B4 Near-total redesign of hull, beam increased to 2,2 m, far more capacious, centre section increased span, fixed floats, central fin, more powerful engines with frontal rads, much greater fuel capacity, 5 crew, three UB and 1 t bombload. Flown 44.

DIMENSIONS Span 16,7 m (54 ft 9½ in); length as before; wing area 49,4 m^2 (532 ft^2).

ENGINES Two VK-107.

WEIGHTS Empty 5580 kg (12,300 lb); fuel/oil 3250; loaded 10 t (22,046).

PERFORMANCE No data.

MDR-6B5 Final prototype built 45 with similar airframe to B4 except forward hull 0,5 m longer to put pilot ahead of props, no central fin, engines in streamlined cowls, driving 4-blade props, rads in ducts between spars; hand-held B-20 (200 rds) in bow, SEB elec turret with twin B-20 (300), four FAB-250 or two FAB-500. Flown spring 45 but never passed NII testing. ASCC name for all 'Mug'.

DIMENSIONS As B4 except length 16,2 m (53 ft 1¾ in).

ENGINES Two VK-107A.

WEIGHTS Empty 5610 kg (12,368 lb); fuel/oil 3250; loaded 10 080 kg (22,222 lb).

PERFORMANCE Max speed 380 km/h (236 mph) at SL; range 3000 km (1,864 miles); TO 350 m; alighting 150 km/h, 400 m.

RK-1 Project for tailless swept-wing bomber, May 47, crew 3, 2 Nene, 2 t bombload.

TA Simple dural stressed-skin *Transportnaya Amfibiya* (transport amphibian), designed immediately after war. Spacious hull for crew of two and six to eight passengers (by modern standards room for more) and 1 tonne cargo. Fabric-covered control surfaces. Untapered wing of NACA-23015 15% profile with electric slotted flaps (30° takeoff, 50° landing), mounted on pylon and braced by N-struts. Stabilizing floats fixed near wing-tips. Electrically retracting landing gear, main units raised flush into sides of hull. Completed June 47 and tested from airfield and water until November when main gear broke at take-off and TA landed on keel. Repaired but OKB closed before resumption of testing.

TA-1, second aircraft with semicircular wingtips added and retractable stabilizing floats. Sliding (area-increasing) flaps and

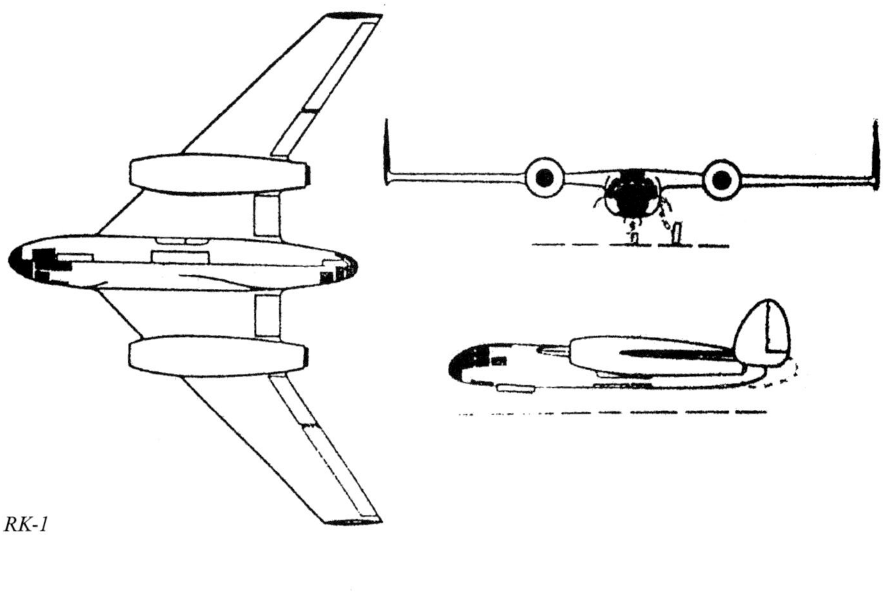

RK-1

MDR-6B4

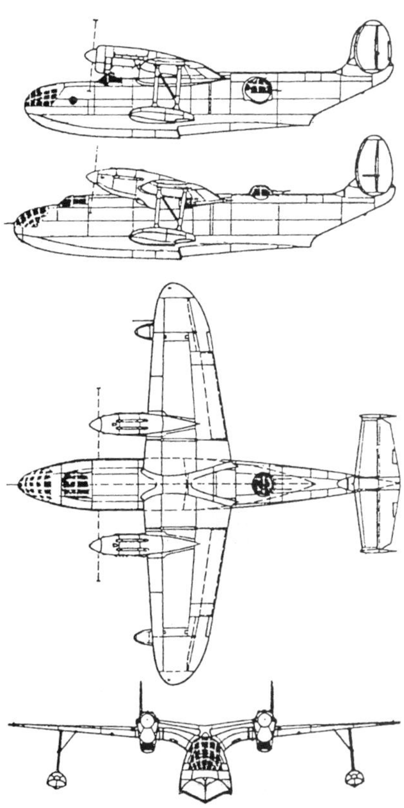

MDR-6B5 (upper side view, B4)

various other changes. Completed May 48. State trial report 20 June but no production decision. **TAF**, *TA Fotografichyeskii*, photo variant with new outer wings of tapered and increased-span design (dimensions not known). Flew successfully late 48 but OKB closed down at end of year.

DIMENSIONS Span 17,2 m (56 ft 5¼ in), (TA-1) 17,8 m (58 ft 5 in), (TAF) over 17,8 m; length 14,0 m (45 ft 11 in); wing area 43,0 m² (463 ft²), (TA-1) 43,6 m² (469 ft²).

ENGINES Two 700 hp ASh-21 driving VISh propellers of 2,8 m diam with three scimitar blades.

WEIGHTS Empty 4658 kg (10,269 lb), (TA-1) 4510 kg (9,943 lb), (TAF) 4268 kg (9,409 lb); fuel/oil 550+55 kg; loaded 6255 kg (13,790 lb),

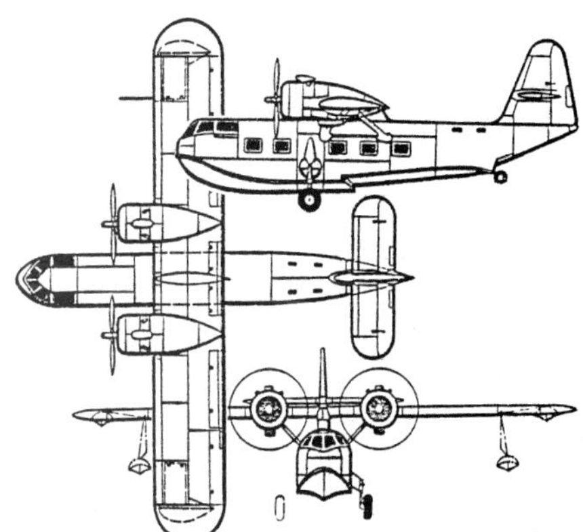

TA-1

TA-1

(TA-1) 6107 kg (13,463 lb), (TAF) 5758 kg (12,694 lb).
PERFORMANCE Max speed 328 km/h (204 mph) at 1,7 km; range 700 km (435 miles), (TAF) 1200 km (746 miles, not explained on same fuel); take-off/landing runs 240/380 m on land.

Dudakov-Konstantinov

These two engineers are the only ones named in connection with the long and important Soviet career of the Avro 504 – apart from Ilyushin who got it started.

U-1 In 1919 an Avro 504K of White (anti-Bolshyevik) forces was shot down and captured almost undamaged. Taken to Moscow by young S. V. Ilyushin, it was soon decided that this famous machine was the best choice as standard trainer for the future Red air force, but it was not until 23 that drawings for U-1 (*Uchyebnii*, training) had been completed at GAZ-1. Manufacture began at GAZ-3, in Petrograd (Leningrad). There was no licence agreement for either the aicraft or its engine, the 110 hp Le Rhône rechristened M-2. A total of 664 U-1 trainers, all similar to 504K, built at GAZ-3 by 31, together with locally developed twin-float variant (73 built 1924-30) designated **MU-1**. In 1927 the production factory completed full static testing and structural calculations for the U-1, the first time this had been done on any aircraft since the revolution.

MU-1 used floats based on the British design but modified in shape and construction. Originally strong but heavy (211 kg the pair) and with heavy steel-tube and wood struts. Later lighter (170 kg). Noteworthy that no satisfactory float version of U-2 (Po-2) was ever developed, and MU-1 remained in use until 34. In 31 a U-1 (land version) was fitted with first assisted take-off powder rockets tested in world. Pilot S. I. Mukhin was airborne in 1·5s. Popular name of U-1 was *Avrushka*. Several modified for special tests, but data basically as for standard Avro 504K.

Ekonomov

Pavel Ivanovich Ekonomov, professional aviation carpenter and joiner, was a prolific inventor and made various propositions while he was working at MAI in 30s. In 36 he constructed primitive but large rotary-wing craft with two contra-rotating sets of two-blade rotors (virtually large wings) of 22 m (72 ft) diameter. Thick wing profile, each with aileron. Called *Zhiroplan*; never completed.

Fedorov

Dmitrii Dmitriyevich Fedorov was born in Kungur in 1875. In 1897 he was arrested for distributing subversive literature and exiled, but managed to become trained aviation engineer in Germany. After 1917 revolution Fedorov returned to his native land and brought with him a design for a two-seat reconnaissance biplane. He managed to continue work at the Anatra factory at Odessa. When this was 'liquidated' he made his way to Factory No 15 at Simferopol (another report states 'aircraft repair plant of 4th Red Army', same location). Here construction began with enthusiasm but difficulties appalling. Constructor became ill before work was finished, and died 22 Dec 22.

DF-1 Tough two-bay biplane, of unusual form with large (5 m span) centre sections on upper and lower planes, 20° swept-back 3,5 m outer panels with transverse-axis ailerons, fuselage mounted on struts midway between wings, and long, shallow vertical tail. Incorporated some Farman and Nieuport parts, but essentially original. Mainly wooden, with truss fuselage and machined wood wing spars. First flown by engineer Pavel Pavlovich Uspasskii, who, with assistance of pilot Ryabovim, carried out test programme in May/June 22. Aircraft proved most satisfactory, with straightforward take-off and good control, small difficulties easily improved. First reconnaissance aircraft built in Soviet Union, but impossible to build in quantity though a second was complete before Factory No 15 was shut.
DIMENSIONS Span 12,0 m (39 ft 4¾ in); length 10,0 m (32 ft 9¾ in); wing area 47,5 m² (511 ft²).
ENGINE 280 hp Maybach 6-inline.
WEIGHTS Empty 1000 kg (2,205 lb); fuel about 500 kg; loaded 1700 kg (3,748 kg).
PERFORMANCE Max speed 170 km/h (106 mph).

Florov

I. F. Florov survived 1941-45 war, unlike co-designer Borovkov on BI-1, and in 43 was designing completely new rocket aircraft. Though it owed much to BI-1, 4302 was more refined and intended not for combat but for aerodynamic research.

No 4302 This aircraft was planned before any German data had become available, and wing and tail were aerodynamically conventional. Wing profile by G. P. Svishchyev, laminar aerofoil of 13% thickness with aspect ratio 5. Trailing edge incorporated three sliding slotted flaps on each side also used for lateral control. Tips turned 30° downwards. Manual tail controls with fixed endplate fins on tailplane. Pressurized nose cockpit with upward-hinged canopy. Jettisonable takeoff trolley and skid/tailwheel alighting gear, but prototype originally built with conventional fixed main landing gears for slow-speed test flying. Tailwheel and strut from La-5. Main tanks in fuselage for propellants in Enerzh 18-8 stainless steel, 3 mm thick, tightly wound with OVS wire (implying gas-pressure feed). Full power for 1 min, normal flight endurance 20 min.

First of three examples built in 44. No 1 with fixed gear (modified La-5FN) solely to develop propulsion. First of 19 gliding flights 46 towed to 5 km by Tu-2. No 2 fitted with engine, wingtips at 45°, one takeoff under power from 3-wheel trolley, Aug 47, pilot A. K. Pakhomov, who shared glider flights with I. F. Yakubov. No 3, later called 4303, had rear-end redesigned for more powerful engine, but latter delayed; never flown, funding transferred Aug 47 to I-270.
DIMENSIONS (Nos 2, 3) Span 6932 (22 ft 9 in); length (2) 7124, (3) 7152 (23 ft 5⅝ in); wing area 8,85 m² (95·26 ft²).
ENGINE (2) Isayev 1,1 t; (2, 3) RD-2M-3.
WEIGHT Empty (1) 970 kg; loaded (1) 1350 kg, (3) 1750 (3,859 lb).

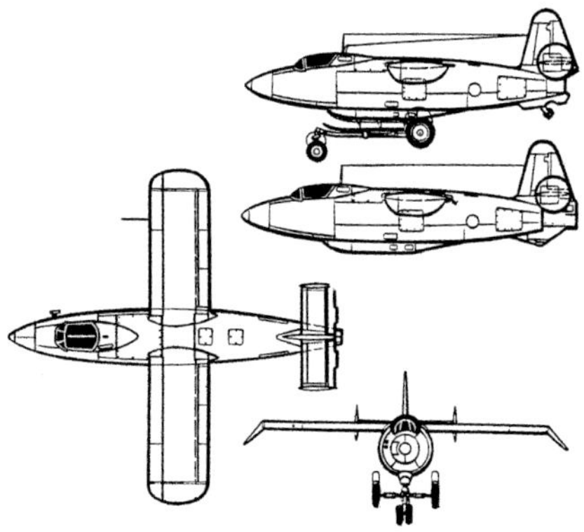

'4302' No 2 (lower side view, No 3)

4302 No 2

PERFORMANCE Max speed (3) 826 km/h (513 mph); landing speed (empty) 125 km/h (78 mph).

Golubkov

From the end of 39 until late 40 A. P. Golubkov was running his own brigade working on the project described below. Main objective was to produce drawings and mock-ups (engineering and systems and for choice of powerplants). Believed this project was passed over from Tupolev OKB with ANT number between 54 and 57. At end of 40 Golubkov was pulled off this project and assigned wartime task of modifying foreign aircraft, including C-47, DB-7, and B-25. After war rejoined Tupolev to work on Tu-104.

SRB Few details available beyond designation *Skorostnoi Razvyedchik-Bombardirovshchik*, high-speed reconnaissance-bomber. Twin-engined, stressed-skin, possibly a cousin of family that led to Tu-2 (speculation).

Gorbunov

Vladimir Petrovich Gorbunov was chief of tech directorate of NKAP when he became partner with Lavochkin in I-22 and subsequent LaGG programme.

GSh This designation has been suggested, but not confirmed, for two-seat armoured attacker (BSh category) with one AM-37 engine. No details, but direct rival to Il-2. Success of latter caused project to be abandoned in 42.

105 In 41 Gorbunov decided to make modifications to LaGG-3 to improve fighting capability. No possibility of switching to more powerful engine, so only possibility was to reduce weight, whilst also improving rear vision. Under designation No 105, also called LaGG-3 *Oblegchennyi* (lightened), work was completed and prototype flown autumn 42. Careful refinement of airframe to reduce weight. Main alterations: cut-down rear fuselage with Yak-1M type canopy sliding to rear, removal of wing slats, aileron mass-balances and main-gear well doors, simplification of fuel and coolant systems, and reduction of armament to one B-20 and one BS. Total weight reduction 300 kg (661 lb), but though 105 proved major improvement, no production ordered in view of equally good rivals such as La-5 already available in quantity.

DIMENSIONS Span 9,8 m (32 ft 2 in); length 8,81 m (28 ft 11 in); wing area 17,5 m² (188 ft²).
ENGINE One 1260 hp M-105PF-1.
WEIGHTS Empty 2400 kg (5,291 lb); fuel/oil 332 kg (732 lb); loaded 2865 kg (6,316 lb).
PERFORMANCE Max speed (SL) 570 km/h (345 mph), (4 km) 623 km/h (387 mph); climb 6 min to 5 km; range 1100 km (684 miles).

105

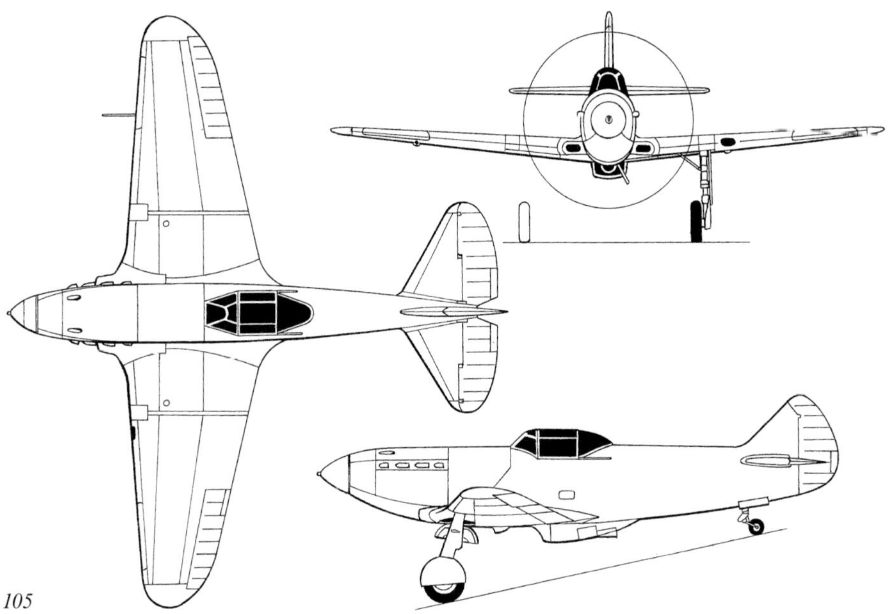

105

Gribovskii

Vladislav Konstantinovich Gribovskii was born 1909 and raised in Gatchina. Passionately interested in aviation, he wrote to first Russian glider pilot, Vyekshin, and pestered other pilots after he came to Petrograd during First World War. In 1919 he tried to get into an aviation school but eventually joined 2nd Petrograd Artillery School. He finally made it to theoretical aviation course in Moscow in Jan 21. In 22 he learned to fly at Sevastopol, and following year was back in Moscow training on Red fighters (mainly Nieuports). Attended first Soviet air firing school, was posted to Kiev, and in 24 attended second all-union glider trials at Koktebel. This set him off designing his own machines, and though he continued in VVS until 30, partly as instructor at air-firing school, he was increasingly involved in designing, helping Osoaviakhim and putting light aviation on map. In 33 he finally became a full-time designer, leading Osoaviakhim Moscow KB (MKB). Altogether he created 14 aeroplane types and 17 gliders.

G-1 Original primary glider, flown Koktebel 25.

G-2 Early (possibly first) Soviet sailplane, jointly with A. B. Yumashyev.

G-3 Project for four-seat transport glider.

G-4 First aeroplane, never flown. Strut-braced high-wing monoplane, 30 hp Bristol Cherub, built Serpukhov 26.

G-5 First aeroplane flown, at Orenburg 28. Neat wooden low-wing monoplane, oval-section monocoque fuselage, wings and tail of high aspect ratio. Single open cockpit. No bracing struts except for landing gear.

SPAN 9,0 m (29 ft 7 in); length 5,1 m (16 ft 9 in); wing area 9,0 m² (97 ft²).
ENGINE One 18 hp Blackburn Tomtit.
WEIGHTS Empty 170 kg (375 lb); fuel 20 kg; loaded 270 kg (595 lb).
PERFORMANCE Max speed 130 km/h (81 mph); ceiling 4,5 km; range 350 km (217 miles)/3 h; landing speed 60 km/h (37 mph).

G-8

G-6 High-performance sailplane, developed with V. A. Stepanchenok. Set record endurance 10 h 22 min.

G-8 Single-seat sporting and training aircraft. Streamlined monocoque fuselage skinned 1·5/2 mm ply, integral with low wing centre-section. Two-spar left/right wings with thin ply skin back to rear spar. Set structural pattern for subsequent designs. Several notable flights.
DIMENSIONS Span 8,0 m (26 ft 3 in); length 5,0 m (16 ft 5 in); wing area 9,0 m² (97 ft²).
ENGINE One 60 hp Walter NZ-60 4-inline.
WEIGHTS Empty 320 kg (705 lb); fuel 70 kg; loaded 483 kg (1,065 lb).
PERFORMANCE Max speed 150 km/h (93 mph); climb 7 min to 1 km, 18 to 2; ceiling 3 km; range 550 km (342 miles)/4 h; take-off 85 m/12 s; landing 100 m/80 km/h/14s.

G-9 Gribovskii's most important training glider, built in large series with numerous variations from 32.

G-10 Another attractive single-seat sporting machine, built at GAZ-1 33. Streamlined monocoque fuselage, two-spar wing mounted on shallow pylon and braced by V-struts, spatted main wheels.
DIMENSIONS Span 8,4 m (27 ft 6¾ in); length 5,6 m (18 ft 4½ in); wing area 11 m² (118 ft²).
ENGINE One M-23; later one 60 hp Walter.
WEIGHTS Empty 335 kg (Walter 330 kg, 726 lb); fuel/oil 75+10 kg: loaded 510 kg (Walter 505 kg, 1,113 lb).
PERFORMANCE Max speed 170 km/h (Walter 165 km/h, 102·5 mph); climb 5 min t km, 11 to 2 km (Walter 13 min); ceiling 5,2 km (Walter 4,8); range 700 km (435 miles) (1300 km overload fuel); take-off 100 m/7s; landing 80 m/67 km/h (42 mph).

G-15

G-12 First Soviet water-based glider.

G-14 Training glider.

G-15 First touring aircraft, two seats side-by-side. Circular monocoque fuselage integral with centre section and fin. Outer panels ply-skinned back to rear spar, slotted flaps and slotted ailerons. Four fuel tanks between spars. Engine in long-chord NACA cowl. Fabric-skinned control surfaces. Trousered main gears. Built at glider works 34; regarded as outstanding in concept and performance. In 35 suffered engine failure when 100 km out from Moscow but made safe forced landing.
DIMENSIONS Span 11,0 m (36 ft 1¼ in); length 6,2 m (20 ft 4¼ in); wing area 14,0 m² (151 ft²).
ENGINE One M-11.

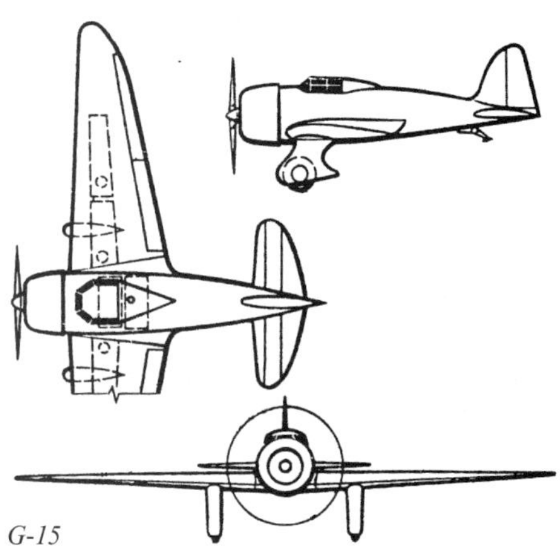

G-15

WEIGHTS Empty 670 kg (1,477 lb); fuel/oil 92+10 kg: loaded 940 kg (2,072 lb).
PERFORMANCE Max speed 185 km/h (115 mph); ceiling 4,5 km (14,760 ft); range 1400 km (870 miles); landing speed 70 km/h (43·5 mph).

G-16 Refined glider flying boat, 35.

G-20 Low-wing strut-braced trainer, with enclosed tandem cockpit. Wing R-II profile, variable chord and thickness with patented manual flaps. Inverted-vee steel struts braced wing and formed apex of crash pylon to protect front-seat occupant. Ground-adjustable braced tailplane. Spatted main wheels. Successful tests but much improved 37 by switch to more powerful engine in same helmeted cowl. Subsequently trained some 70 pilots.
DIMENSIONS Span 9,7 m (31 ft 10 in); length 6,3 m (20 ft 8 in); wing area 13,2 m^2 (142 ft^2).
ENGINE One M-11; from 37 M-11Ye.
WEIGHTS Empty 607 kg (1,338 lb), (M-11Ye) 620 kg (1,367 lb); fuel/oil 69 kg (M-11Ye, 100 kg); loaded 836 kg (1,843 lb), (M-11-Ye) 880 kg (1,940 lb).
PERFORMANCE Max speed 209 km/h (130 mph), (M-11Ye) 235 km/h (146 mph); climb (original engine) 4·6 min to 1 km, 11·1 to 2; ceiling 3870 m (12,700 ft); range 400 km (overload fuel 1017 km, 632 miles); take-off 190 m/14 s; landing 70 km/h (43·5 mph).

G-21 Neat touring, sporting or ambulance aircraft, 36. Low-wing cantilever monoplane, two- or three-seat enclosed cabin in wood monocoque fuselage. Spatted main gears on ends of centre section. Wing geometrically similar to G-15 with skin 1·5 mm ply round leading edge, then 1 mm to rear spar and fabric aft. Automatic slats, joined by steel tubes left/right wings, slotted ailerons. G-21 made several long-range flights with extra fuselage tank.
DIMENSIONS Span 11,0 m (36 ft 1¼ in); length 7,0 m (23 ft 0 in); wing area 14 m^2 (151 ft^2).
ENGINE One M-11Ye.
WEIGHTS Empty 705 kg (1,554 lb); fuel/oil 105+10 kg; loaded 980 kg (2,160 lb).
PERFORMANCE Max speed 220 km/h (137 mph); climb 5·6 min to 1 km; ceiling 4760 m (15,600 ft); range (normal fuel) 500 km (311 miles); landing speed 70 km/h (43·5 mph).

G-22 Extremely clean single-seat sporting aircraft, first flown 36. All wood but fuselage of truss type (vertical sides) instead of monocoque. One-piece wing with single spar, ply ahead and fabric to rear. Frise ailerons hung on door hinges level with upper wing surface. Main gears with spatted balloon tyres. Several notable performances such as 164,94 km/h by Ekaterina Mednikova, women's record for class, 3 June 38. In 39 re-engined with Pobjoy and in 40 with M-23.
DIMENSIONS Span 8,7 m (28 ft 6½ in); length 5,6 m (18 ft 4½ in) (varied slightly with engine); wing area 10,0 m^2 (108 ft^{22}).
ENGINE One 50 hp Walter Mikron inverted 4-inline; (38) 80 hp Pobjoy Niagara 7-cylinder radial; (40) M-23.
WEIGHTS Empty (Walter) 210 kg (463 lb); fuel/oil 35 kg; loaded 325 kg (716 lb).
PERFORMANCE Max speed 165 km/h (102·5 mph), (Pobjoy) 190 km/h (118 mph), (M-23) 170 km/h (106 mph); ceiling (Walter) 3 km (9,842

G-17 canard project

G-20

G-21

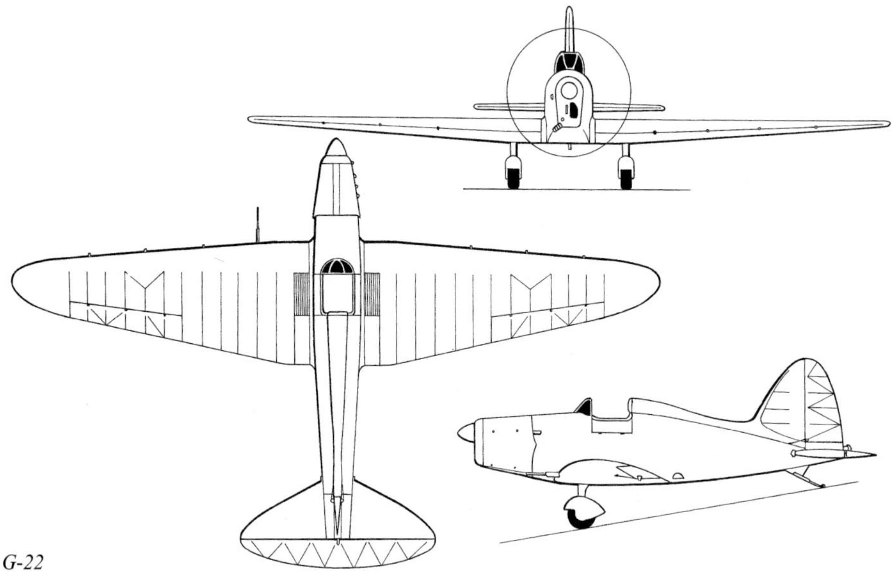

G-22

G-23 (front cockpit replaced by tank)

G-25

ft); range 350 km (217 miles); take-off and landing both 80-85 m/10 s/55 km/h (34 mph).

G-23 Komsomolyets 2 Tandem two-seat low-wing machine completed Feb 37 to evaluate GAZ-M-1 (Avia) car-derived engine. Spatted main gears (most of Gribovskii's aircraft also flew on skis in winter) but simple and no slats or flaps. Tested by NII VVS; speed commendable but climb abysmal. Nevertheless, flown by I. Grodzyanskii 2584 km in 21 h flying time. Second (38) machine, **G-23bis**, totally different animal with M-11Ye radial and startling performance. On 23 July 38 Grodzyanskii reached 7266 m; on 2 Aug N. D. Fedosyeev reached figure given in data. Late 38 fitted with new GAZ-11 as **G-23bis-GAZ.**

DIMENSIONS Span 11,0 m (36 ft 1¼ in); length (M-11-Ye) 10,5 m (34 ft 5½ in); wing area 15,0 m² (161 ft²), (M-11Ye) 14,9 m² (160 ft²).
ENGINE One 56 hp GAZ-M-60; (G-23bis) M-11Ye; (bis GAZ) 85 hp GAZ-11.
WEIGHTS Empty 483 kg (1,065 lb); fuel/oil 40+14 kg; loaded 713 kg (1,572 lb); no figures for G-23bis.
PERFORMANCE Max speed 150 km/h (93 mph), (bis) 179 km/h (111 mph), bis-GAZ 160 km/h (99 mph); climb 19 min to 1 km, 58 to 2 (bis figures, 1·5 and 3·4!); ceiling 2480 m (bis 7985 m, 26,200 ft); range 450 km (280 miles); take-off (GAZ) 150 m/16 s; landing 90 m/10 s/65 km/h (40 mph).

G-24 Not known.

G-25 Possibly Gribovskii's only biplane, this neat machine was built at MKB in 37 to see to what extent a practical tandem trainer could be made smaller and more economical than U-2. Outstanding all round performance and manoeuvrability, but of course could not equal U-2 versatility in utility roles. Engine newly imported from Britain, not same as G-23bis. But when GAZ engine removed from G-23 in 1938 it was installed in G-25, which became G-25bis. Performance unacceptable, but MKB also received newly designed GAZ-11 of 85 hp and with this installed G-25bis flew reasonably well.

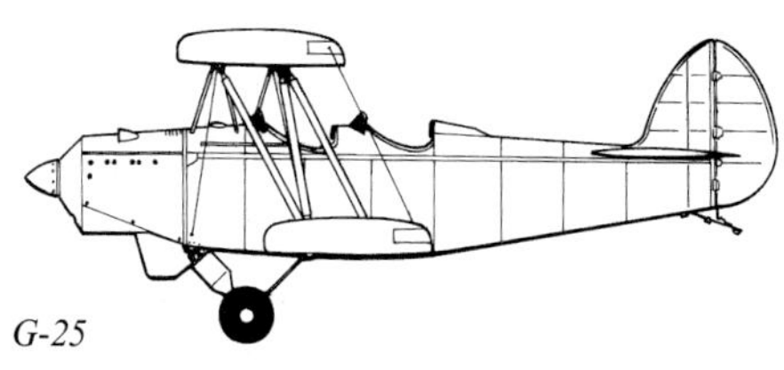

G-25

DIMENSIONS Span 9,0 m (29 ft 6½ in); length 6,4 m (21 ft 0 in); wing area 23,0 m² (248 ft²).
ENGINE (25) one 85 hp GAZ-Avia-11.
WEIGHTS Empty (25) 430 kg (948 kg), (bis) 500 kg (1,102 lb); fuel/oil 40 kg; loaded (25) 630 kg (1,389 lb), (bis) 700 kg (1,543 lb).
PERFORMANCE Max speed 170 km/h (106 mph), (bis) 165 km/h (102·5 mph); ceiling 3,5 km (11,480 ft); range 280 km (174 miles); take-off 40 m (bis, 80); landing 60 m at 55 km/h (bis, 80 m at 65 km/h).

G-26 Attractive low-wing sporting aircraft, structurally almost identical to G-22. Pilot in

open cockpit further aft than in G-22 and design biassed heavily in favour of performance rather than simple training and aerobatics. G-26 not an aircraft for novice. Flew late 1938 on skis. Handling improved during 1939.
DIMENSIONS Span 7,7 m (25 ft 3¼ in); length 6,0 m (19 ft 9 in); wing area 9,0 m^2 (97 ft^2).
ENGINE One MG-40.
WEIGHTS Not known.
PERFORMANCE Max speed 290 km/h (180 mph); range 350 km (217 miles).

G-27 Gribovskii spent much time in 38 designing safe and economical aircraft that could serve as three-seat crew trainer, light transport, ambulance, photo and for various other duties. Structure wooden, with slim monocoque fuselage. Wing with rectangular centre section (1,62 m chord) with small (1,53 m^2) pneumatic split flaps, two-spar all-wood outer panels with slotted ailerons, and tailwheel landing gear. At different times main gears were trousered (first flight), fixed unspatted, fully retractable and skis. Gibovskii designed a floatplane version, not built. Structural provision for nose machine gun and light bombs under centre section. Normal military crew three: nav/bomb in nose, pilot, and observer/radio in rear seat; could have dual control but rear-seat vision not good. Many thought G-27 should have been mass-produced.
DIMENSIONS Span 10,6 m (34 ft 9½ in); length 6,99 m (22 ft 11½ in); wing area 17 m^2 (183 ft^2).
ENGINES Two M-11; in 41 re-engined with M-11Ye.
WEIGHTS Empty 900 kg (1,984 lb); fuel/oil 130 kg; loaded 1300 kg (2,866 lb), (M-11Ye) 1430 kg (3,153 lb).
PERFORMANCE Max speed 240 km/h (149 mph), (M-11Ye) 250 km/h (155 mph); ceiling 4 km (M-11Ye 5 km); landing speed 80 km/h (50 mph).

G-28 Named *Krechyet* (gerfalcon), this single-seat *Trenirovochni Istrebitel* (trainer fighter) received VVS designation **TI-28**. Structure was all-wood, with no fabric anywhere except on flight controls which were duralumin framed. Dural flaps, pneumatic operation. Fin integral with monocoque fuselage, with pilot behind wing under sliding canopy. Two-spar wing, RAF-34 section. Fixed landing gear with single cantilevered oleo struts on front spar, castoring tailwheel. Engine on steel-tube mounts with light-alloy cowl, driving two-blade propeller (described as fixed-pitch, but photos show Ratier v-p). Provision for reflector sight, one synchronized ShKAS with 400 rounds, combat camera in left wing root, and underwing racks for two bombs of 10, 25 or 40 kg. Flown by V. Gavrilov for MKB on 22 May 41, just a month before invasion. NII VVS testing by M. M. Gromov, A. B. Yumashyev and P. M. Stefanovskii. Satisfactory, though Gribovskii wanted more powerful MV-6A engine for production version. Latter was unable to be implemented because of over-riding need to produce G-29 (G-11). Note: there is evidence original design was a racer, with retractable gear, intended for 400 km/h. Not yet explained why first aircraft, completed just before race, was a fixed-gear military prototype.
DIMENSIONS Span 9,0 m (29 ft 6¼ in); length

G-26

G-27

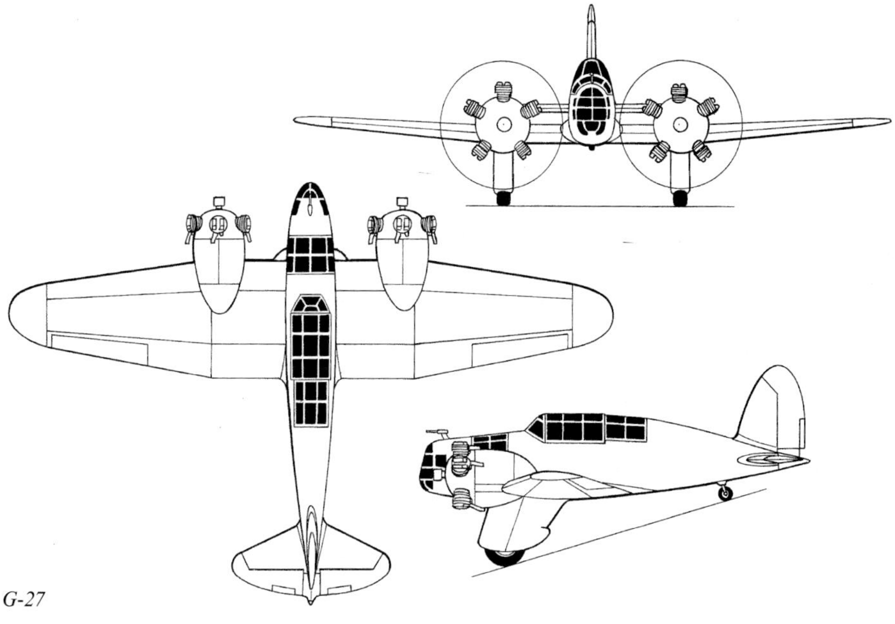
G-27

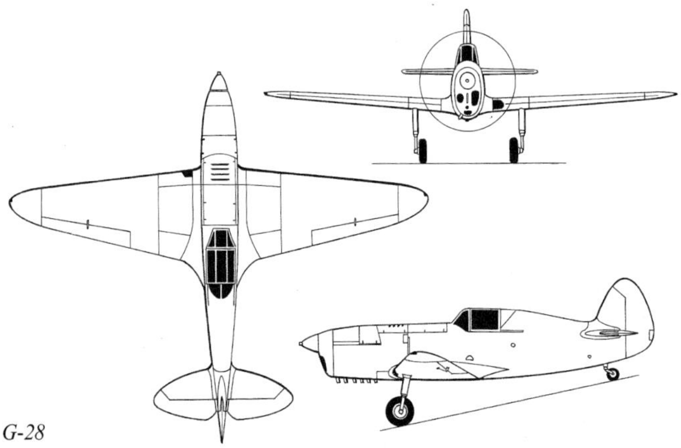

G-28

G-28

7,66 m (25 ft 1⅛ in); wing area 11,6 m² (125 ft²).
ENGINE One MV-6.
WEIGHTS Empty 897 kg (1,978 lb); loaded (no bombs) 1157 kg (2,551 lb).
PERFORMANCE Max speed at SL 275 km/h, at 1,6 km 303 km/h (188 mph); climb 19·4 min to 5 km; service ceiling 6,6 km; range at 90% max speed 500 km (497 miles); time for 360° turn at height 20 s; landing speed with flaps 90,5 km/h (56 mph).

G-29 (G-11) Gribovskii's largest aircraft and one of few to go into production. Designed 40 in parallel with A-7, it was a large assault glider (though small in comparison with later KS-20, BDP and SAM-23) of wooden construction, intended for simple production by quickly trained labour. Frame-truss fuselage with vertical sides, but fish-like streamlined side profile. Two-spar wing with 6,3 m rectangular centre section mounted in high position with spars integral with fuselage frames. Outer panels tapered on leading edge, aspect ratio 10·8. Split landing flaps worked from air bottles with standby hand pump. Landing gear with two main wheels attached to fuselage frame and rear tailskid. Wheels to be jettisoned on operational mission, landing being made on skid on centreline of V-shaped bottom of fuselage. Single-seat cockpit in nose glazed with fixed panes of flat Plexiglas. Main cabin with accommodation for 11 troops or 1,1 tonne cargo, with door on left side and escape hatch in roof. Tow hook well back under nose; usual tug SB, DB-3, Il-4 or Li-2. Tested 41 and NII VVS trials in same year by team of six led by V. Fedorov. Production as G-11 (VVS designation) begun late 41; probably 100 built.
DIMENSIONS Span 18,0 m (59 ft 0¾ in); length 9,8 m (32 ft 1¾ in); wing area 30,0 m² (323 ft²).
WEIGHTS Empty 1200 kg (2,646 lb); loaded 2400 kg (5,291 lb).
PERFORMANCE Max speed on tow or free 280 km/h; normal gliding speed 146 km/h (91 mph), with gliding angle 19·2 and rate of sink 2·2 m (7·2 ft)/s; landing speed 82 km/h (51 mph).

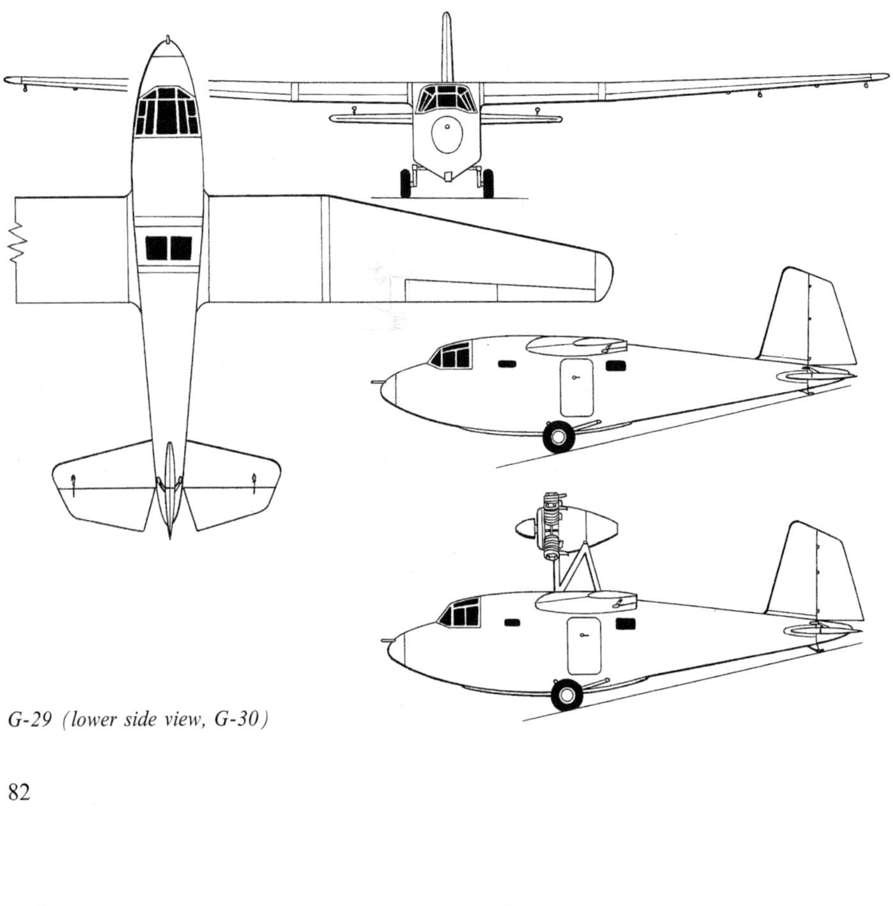

G-29 (lower side view, G-30)

G-30 (G-11M) In 42 Gribovskii decided to build powered G-29; not a true transport; like British Hamilcar X merely intended to return to base after delivering load. Uncowled engine and fuel tank added in nacelle on steel-tube pylon, tractor two-blade fixed-pitch prop. Flown Oct 42 but State trials never took place. Data generally as G-29 except:
ENGINE One M-11.
WEIGHTS Empty 1500 kg (3,307 lb); fuel, 70; loaded 2400.

G-31 In 48, with deliveries of G-11 completed, Gribovskii designed two-seat training glider for instruction of military (G-11) pilots. Resulting G-31 was simple all-wood tandem-seater, with shoulder wing with flaps, tailplane mounted part-way up fin, and enclosed cockpits with hinged canopies. Jettisonable twin wheels and skid landing. Single prototype; no post-war requirement for glider pilots.

Dosav 1950 Gribovskii's last known design was light tourer for 50 Dosav competition (see Avietka section). Most remarkably his offering was light-alloy stressed-skin. Low-wing side-by-side two-seater with enclosed cockpit, dual control, tricycle landing gear and 90 hp engine. No known designation.
DIMENSIONS Span 11,4 m (37 ft 4¾ in); length 7,05 m (23 ft 1½ in); wing area not known.
WEIGHTS Empty 502 kg (1,107 lb); loaded 750 kg (1,653 lb).
PERFORMANCE Max speed 195 km/h (121 mph); ceiling 5,4 km; landing speed 65 km/h (40 mph).

Grigorovich

Dmitrii Pavlovich Grigorovich was the most important constructor in Russia to have stayed in post-revolutionary Soviet Union and become a famed Soviet designer. Born 1883, he was a full-time student at Kiev polytechnic institute and graduated 1910. His first job was as a journalist on an aviation magazine in St Petersburg, but in Jan 13 he was appointed works manager for newly opened aircraft plant of S. S. Shchetinin and M. A. Shcherbakov, second in Russia to build aircraft in series. Batches of Nieuport IV and Farman F.16 were produced, and under Grigorovich's direction the first strength calculations and static structural testing were completed in May/June 13. He set up

M-3

on his own by accident; the accident happened to D. M. Aleksandrov, whose flying boat suffered a crushed nose and other damage. Russo-Baltic Wagon Works (later producer of Sikorskii's IM series) quoted 6500 rubles for repair, and Lebed factory 6000. Grigorovich thought a fair price 400, and did the job. This was so extensive result was first Grigorovich aircraft, designated M-1. Long series of aircraft of many classes followed. He was original head of OMOS and later a renowned lecturer at MAI, but suffered period in wilderness under arrest in Hanger 7 at GAZ-39 in 1930-33. He died in 38, reportedly as result of illness. Many of his team swelled Lavochkin OKB.

M-1 Original Grigorovich flying boat, produced by not only repairing Aleksandrov's French-built machine (1912 Donnet-Lévêcque) but also altering it. Designation from *Morskoi* (marine). Original mixed-construction biplane boat with pusher engine, floats at ends of smaller lower wings and tailplane half-way up fin. Two seats in side-by-side cockpit ahead of wings. Provision for beaching trolley. Grigorovich cut 1 m off nose, redesigned wings closer to Farman F.16 profile and reduced height of step from 200 to 80 mm. Flew as M-1 autumn 13; met requirements with improved flying qualities.
DIMENSIONS Span 9,5 m (31 ft 2 in); length 7,4 m (24 ft 6¾ in); wing area 18,2 m^2 (196 ft^2).
ENGINE One 50 hp Gnome.
WEIGHTS Empty 420 kg (926 lb); loaded 620 kg (1,367 lb).
PERFORMANCE Not known.

M-2 Larger two-seater with more powerful engine, first flown 14. Both wings extended in span, improved planing bottom, modified tail, and wing cellule raised 100 mm above hull. First flying boat in the Soviet Union produced in small series in proper erection berth locating all parts accurately.
DIMENSIONS Span 13,68 m (44 ft 10½ in); length 8,0 m (26 ft 3 in); wing area 33,5 m^2 (361 ft^2).
ENGINE Originally 80 hp Clerget, later 100 hp Gnome Monosoupape.
WEIGHTS Empty, not known; loaded 870 kg (1,918 lb).
PERFORMANCE Not known.

M-3 Also called Shchetinin Third, built autumn 14. Further development to wing profile, giving better flying qualities than M-2, but seaworthiness unacceptable. Data as M-2 with Gnome Monosoupape.

M-4 Developed winter 14 and flown spring 15; further modification to wing profile but main attention to hull. Concave step, small longitudinal angle at keel, stabilizing floats of revised form, tailplane set at positive incidence to lift in slipstream. Built at Sevastopol; four served Black Sea and Baltic. Basically as M-3 and M-5.

M-5 First mass-production marine aircraft in the Soviet Union, also called *Morskoi* 5, Shchetinin 5 or ShchM-5. Lower wing extended, wing cellule raised 150 mm above hull, tail of trapezoidal cross-section with sharper edges, step reduced to 70 mm on centreline and 140 at chines, and other alterations. Set pattern for ongoing family, steadily improved with years in numerous minor variants, many with bow gun and light bombs. About 300 built, saw extensive service in war, revolution and civil war, some remaining in use in late 20s as dual trainer, reconnaissance and utility aircraft.
DIMENSIONS Span 13,62 m (44 ft 8⅓ in); length 8,6 m (28 ft 2½ in); wing area 37,9 m^2 (408 ft^2).
ENGINE One 100 hp Gnome Monosoupape; alternatively 110 hp Le Rhône, 130 hp Clerget (common for armed versions) or other rotaries.
WEIGHTS (typical) Empty 660 kg (1,455 lb); fuel 140 kg; loaded 960 kg (2,116 lb).
PERFORMANCE (typical) Max speed 105 km/h (65 mph); climb 9·6 min to 1 km; ceiling 3,3 km; endurance 4 h.

M-6 Essentially M-5 with 150 hp Sunbeam water-cooled engine and other changes; unsuccessful, especially in poor water take-off.

M-7 Essentially M-5 with rounded hull with larger keel longitudinal angle and raised step. Normal in flight, but water take-off described as heavy.

M-8 Even more rounded hull and higher step (150 mm) with moderately widened keel; water take-off unachievable.

M-9 Apart from MBR-2 and Be-6 this was most important flying boat ever produced by Russian designer. First flown Dec 15, also known as *Morskoi-9, Morskoi devyati* (marine, ninth), Shchetinin M-9, ShchM-9 and ShchS. Basically as M-5 but larger wings and more powerful engine, usually water-cooled with two radiators hung between left and right inboard interplane struts on each side of engine. Hull of different cross section with planing bottom no longer of concave form with side skegs but of modern V-shape with 5° deadrise and outward-sloping lower hull provided with side strakes

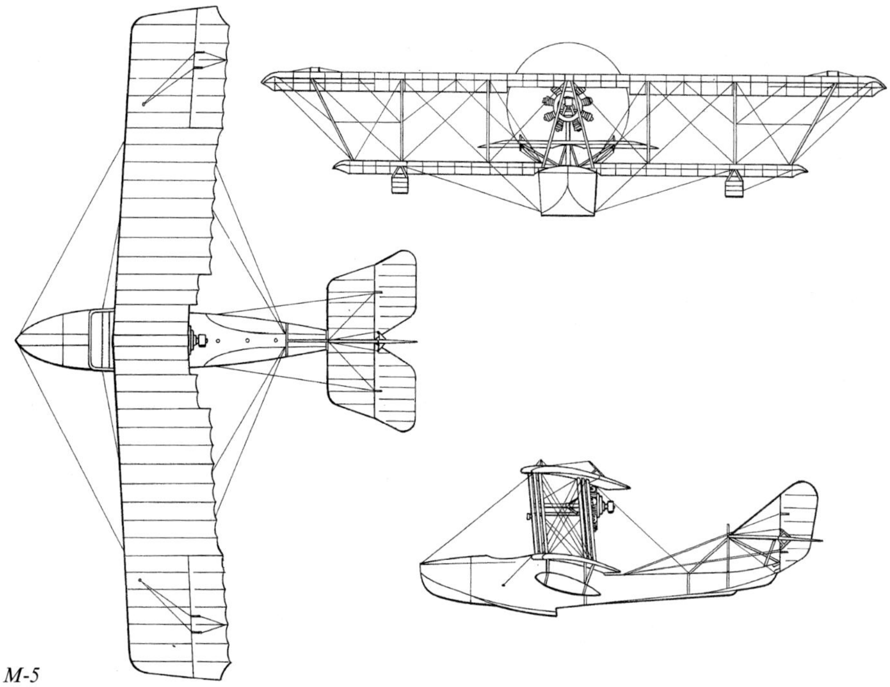

M-5

forward of step 1,5 m long and increasing in width to 100 mm at step. Large horn balance on rudder, usually provision for third crew-member and for gun in bows (7,7 mm Vickers, 7,5 mm Hotchkiss, 20 mm Oerlikon or 37 mm Puteaux), with light bomb racks under lower wings. Rounded top to hull giving egg-shaped appearance. Hull ply planking thicker than before, with several extra frames. Fuel pump driven by windmill. Generally tough and serviceable boat able to operate reliably in waves up to 0,5 m. First M-9 made flight from St Petersburg to Baku 25 Dec 15 to 9 Jan 16. Production machines combat duty start of 17, subsequently used for air combat as well as all other military duties. On 17 Sept 16 made loop, later looped twice with passenger on board. Spirited flying by crews resulted in numerous victories over German Albatros seaplanes and other enemies. Single examples had 140 hp Hispano-Suiza, 220 hp Renault or 225 hp Salmson (Canton-Unné) engines. Total production est 500.
DIMENSIONS Span 16,0 m (52 ft 6 in); length 9,0 m (29 ft 6½ in); wing area 54,8 m² (590 ft²).
ENGINE Usually one 140 or 150 hp Salmson 9-cyl water-cooled radial.
WEIGHTS Empty (typical) 1060 kg (2,337 lb); fuel 220 kg (485 lb); loaded 1610 kg (3,549 lb).
PERFORMANCE Max speed 105 km/h (65 mph); climb 13 min to 1 km, 35 to 2; ceiling 3 km; endurance 5 h.

M-10 Small boat with Gnome-Monosoupape, built at Shchetinin plant without Grigorovich involvement.

M-11 Small flying boat fighter, also called Shch-I (Shchetinin-fighter), first flown summer 16. Two-seater with pivoted bow gun (various but usually rifle-calibre) and also single-seat version with bow gun pivoted or, usually, fixed. Considerable armour round bow cockpit, pilot cockpit and front of engine, thickness up to 6 mm steel. In winter mounted on three skis for operation from snow or ice. At least 100 ordered and 60 delivered, most as single-seaters.
DIMENSIONS Span 8,75 m (28 ft 8½ in); length 7,6 m (24 ft 11¼ in); wing area 26,0 m² (280 ft²).
ENGINE One rotary, (two-seat) 100 hp Gnome-Monosoupape, (single-seat) 110 hp Le Rhône.
WEIGHTS Empty 665 kg (1,466 lb) (single-seat 676 kg); fuel 90 kg (single-seat 106); loaded 915 kg (2,017 lb) (single-seat 926).
PERFORMANCE Max speed 140 km/h (87 mph) (single-seat 148); climb 11 min to 1 km; ceiling 3 km; endurance 2·7 h.

M-12 Single-seat boat distinguished from M-11 by deeper nose and redesigned tail. At least one was fixed-gun fighter, but most had no gun or armour and used for civil utility duties and single-seat training. Data as M-11 except loaded weight 870 kg (1,918 lb) and faster climb.

M-15 Larger tandem-seater designed for long-range reconnaissance, often as landplane with three skis. First flown May 16; about 80 delivered, limiting factor being supply of engines. By 17 used mainly for training; many survived revolution.
DIMENSIONS Span 11,9 m (39 ft 0½ in); length 8,4 m (27 ft 6¾ in); wing area 34,4 m² (370 ft²).

M-9 (Hispano)

M-11 on skis

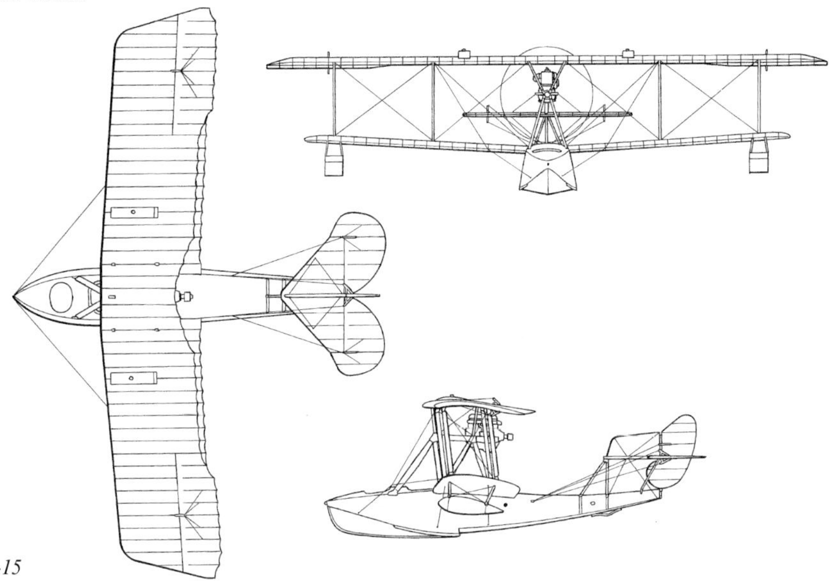

M-15

ENGINE One 140 hp Hispano-Suiza V-8.
WEIGHTS Empty 840 kg (1,852 lb); fuel 184 kg; loaded 1320 kg (2,910 lb).
PERFORMANCE Max speed 125 km/h (78 mph); climb 8·5 min to 1 km; ceiling 3,5 km; endurance 5·5 h.

M-16 Large two-seat reconnaissance aircraft designed for efficient operation on skis (nickname *Zimnyak*, from winter). Nacelle and tail similar to late Farmans. First flown Dec 16 and 40 built 17.
DIMENSIONS Span 18,0 m (59 ft 0¾ in); length 8,6 m (28 ft 2½ in); wing area 61,8 m² (665 ft²).
ENGINE One 150 hp Salmson.
WEIGHTS Empty 1100 kg (2,425 lb); fuel 185 kg; loaded 1450 kg (3,197 lb).
PERFORMANCE Max speed 110 km/h (68 mph); climb 15 min to 1 km; endurance 4 h.

M-20 Variant of M-5 with 120 hp Le Rhône; two-seat reconnaissance 16. M-17 and 19 were related to M-15, no production.

MK-1 Complete break with previous designs, large *Morskoi Kreiser* (sea cruiser), built and tested late 16. Fuselage of Sikorskii IM type mounted on large wooden float and carrying new biplane wings without sweepback (as on M-boats) but with swept-back aileron trailing edges giving large tip chord. Four-seat reconnaissance, with various planned armament. No production.
DIMENSIONS Span 30,0 m (98 ft 5 in); length 16,5 m (54 ft 1⅔); wing area not known (about 165 m² (1,776 ft²).
ENGINES Two 220 hp Renault on inner interplane struts, one 150 hp Salmson (later 140 hp Hispano) on upper wing. No other data.

GASN Twin-float seaplane built 16. Four-seat transport, name *Gidro-Aeroplan Spetsialno Naznachyeniya* (special destination). Mainly wood with fabric skin, struts steel tube with bolted or welded joints. First aircraft in which M. M.

M-16

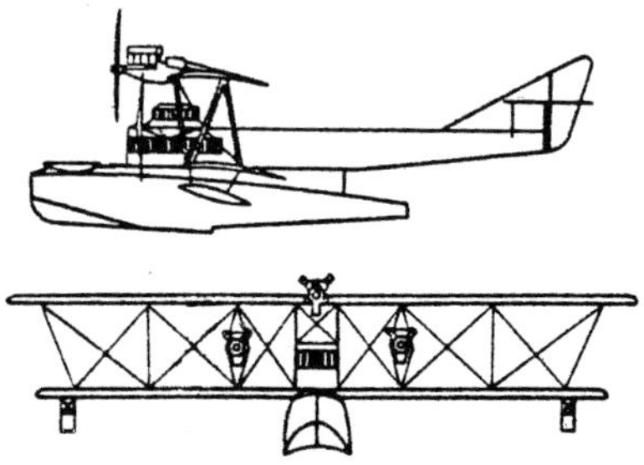

MK-1

Shishmaryev assisted design. Flown A. E. Gruzinov 24 April 17. Production interrupted by revolution.
DIMENSIONS Span 28,0 m (91 ft 10 in); length 14,1 m (46 ft 3½ in); wing area 150 m² (1,615 ft²).
ENGINES Two 220 hp Renault.
WEIGHTS Not known except useful load 1450 kg (3,197 lb).
PERFORMANCE Max speed 110 km/h (68 mph).

M-23bis After five years' interruption because of civil war, Grigorovich was able to resume manufacture in 22 in GAZ-3, completing this flying boat in Sept 23. It was a refined M-9 with smaller wings.
DIMENSIONS Span 12,5 m (41 ft 0¾ in); length 8,7 m (28 ft 6½ in); wing area 45,8 m² (493 ft²).
ENGINE One 280 hp Fiat V-12 water-cooled.
WEIGHTS Empty 1165 kg (2,568 lb); loaded 1615 kg (3,560 lb).
PERFORMANCE Max speed about 165 km/h (102 mph).

M-24 Essentially refined M-9 with more powerful engine, first flown late 23 from GAZ-3. Series of 40 built, but defects. After Grigorovich went to Moscow in mid-23 aircraft was carefully redesigned and emerged mid-24 as **M-24bis**. Series of 20 delivered up to 26 and served with armament on coastal patrol.
DIMENSIONS Span 16,0 m (52 ft 6 in); length 9,0 m (29 ft 6½ in); wing area 55 m² (592 ft²).
ENGINE One 220 hp Renault, (M-24bis) 260 hp Renault.

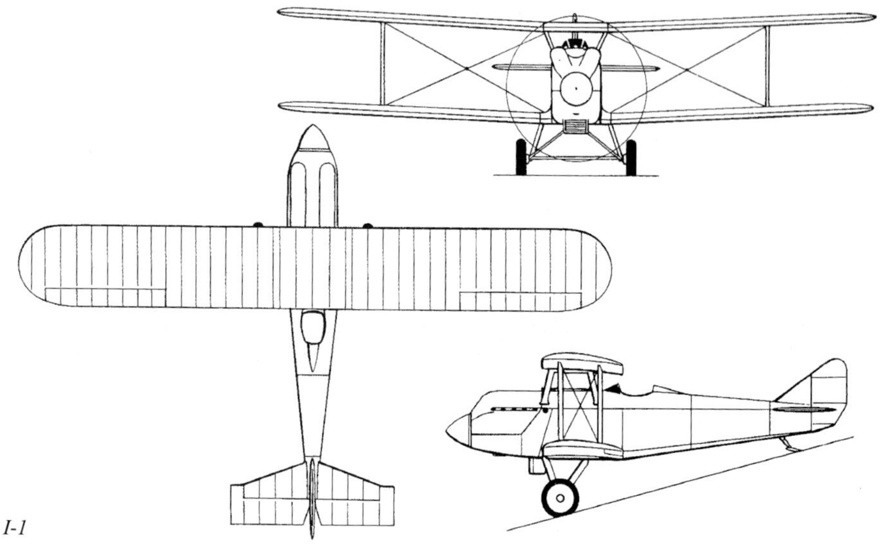

I-1

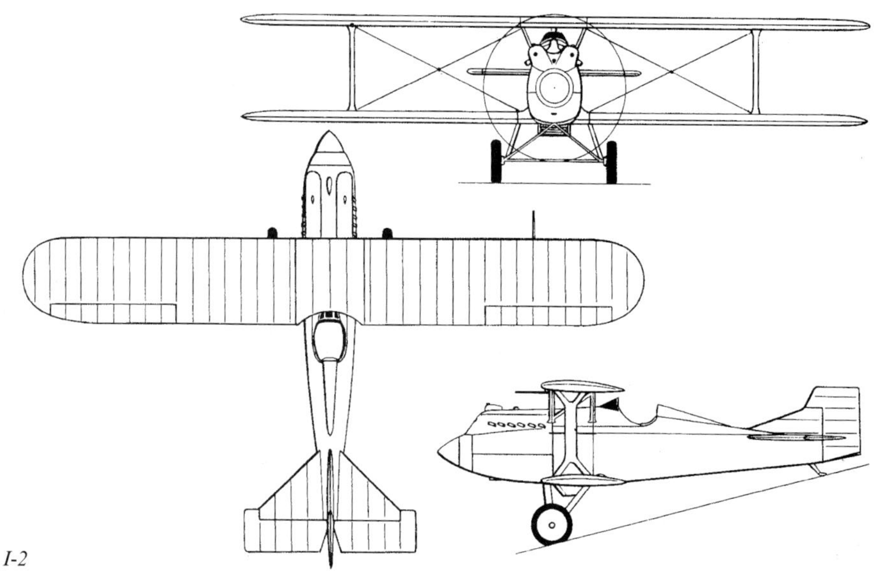

I-2

GASN

WEIGHTS Empty 1200 kg (2,646 lb); loaded 1650 kg (3,638 lb) (M-24bis 1700 kg, 3,748 lb).
PERFORMANCE Max speed 130 km/h (81 mph), (M-24bis 140 km/h, 87 mph); ceiling 3,5 km (bis 4).

I-1 First Soviet *Istrebitel* (fighter), and first design by Grigorovich after transfer to Moscow as technical director of GAZ-1. Design group: A. N. Sedyelnikov, B. L. Korvin, A. A. Krylov, V. V. Kalinin and V. L. Moiseyenko. Neat biplane designed around Liberty engine, wood structure with fabric covering except light-alloy cowl and ply-skinned forward fuselage. Göttingen 436 aerofoil, RAF streamlined bracing wires, wooden interplane struts, radiator hung between landing gears as in water-cooled flying boats between interplane struts. Flown Khodinka early 24. Radiator remounted hung under engine bearers. Stability and climb inadequate. No production.
DIMENSIONS Span 10,8 m (35 ft 5¼ in); length 7,32 m (24 ft 0¼ in); wing area 26,8 m² (288 ft²).
ENGINE One 400 hp Liberty V-12.
WEIGHTS Empty 1090 kg (2,403 lb); fuel 220 kg; loaded 1490 kg (3,285 lb).
PERFORMANCE Max speed 230 km/h (143 mph); ceiling 6 km (19,685 ft); endurance 2·5 h; range 600 km (373 miles).

I-2 Grigorovich refined I-1 to such extent I-2 emerged in autumn 24 as largely new design. Wings of reduced chord without dihedral, cutout in trailing edge, faired I-type interplane struts, more streamlined semi-monocoque fuselage of oval section, long-chord fin and larger rudder with horn balance and angular outline, improved engine cowl, two 7,62 mm PV-1 guns in sides of fuselage. Tested by A. I. Zhukov who appreciated performance and handling but found cockpit narrow and visibility poor. Further major alteration resulted in **I-2bis**, with complete structural redesign of fuselage incorporating welded steel-tube truss with wooden secondary structure and skin. Guns moved to top of fuselage to make breeches accessible to pilot and improve cockpit comfort. Cockpit cut down and seat raised. Ailerons on lower wings only. Flown A. I. Zhukov early 25 and soon cleared for service; first Soviet-designed fighter in production. GAZ-1 built 164, and GAZ-23 another 47 in 1926-29. In 26 airframe completely static-tested at Aviatrust. Several I-2bis modified with twin Lamblin radiators, becoming **I-2′** or **I-2prim**.
DIMENSIONS Span 10,8 m (35 ft 5¼ in); length 7,32 m (24 ft 0¼ in); wing area 23,4 m² (252 ft²).
ENGINE One M-5.
WEIGHTS Empty (2) 1130 kg (2,491 lb), (2bis) 1152 kg (2,540 lb); fuel/oil 236+35 kg; loaded (2) 1530 kg (3,373 lb), (2bis) 1575 kg (3,472 lb).
PERFORMANCE Max speed (2) 242 km/h (150 mph), (2bis) 235 km/h (146 mph); climb (2) 2·1min to 1 km, (2bis) 2·4 min; ceiling (2) 5,8 km, (2bis) 5340 m; range (2) 650 km (404 miles), (2bis) 600 km (373 miles); take-off (2) 120 m/8s, (2bis) 160 m/11 s; landing (2) 190 m/15 s/ 95 km/h (59 mph), (2bis) 210 m/16 s/same.

SUVP (PL-1) Another break with tradition, this high-wing monoplane was a transport built for Ukrainian civil aviation authority, designation thus coming from *Samolyet Ukrvozdukhput*, PL-1 from *Passazhirskii Leningrad*. Almost entirely welded steel-tube fuselage structure, including wing struts and landing gear. Wing (Göttingen profile) and tail of wood, with fabric covering throughout. Door on left side for pilot in enclosed cockpit behind engine, three passengers or 210 kg cargo. Built at GAZ-3 and first flown autumn 25. Successful, and served several years in Ukraine though believed only one built.
DIMENSIONS Span 13,7 m (44 ft 11½ in); length 8,4 m (27 ft 6¾ in); wing area 24,1 m² (259 ft²).
ENGINE One 100 hp Bristol Lucifer 3-cylinder.
WEIGHTS Empty 820 kg (1,808 lb); fuel 120 kg; loaded 1150 kg (2,535 lb).
PERFORMANCE Max speed 139 km/h (86 mph); climb 9 min to 1 km; ceiling 3050 m (10,000 ft); endurance 4·5 h; landing speed 70 km/h (43·5 mph).

I-2

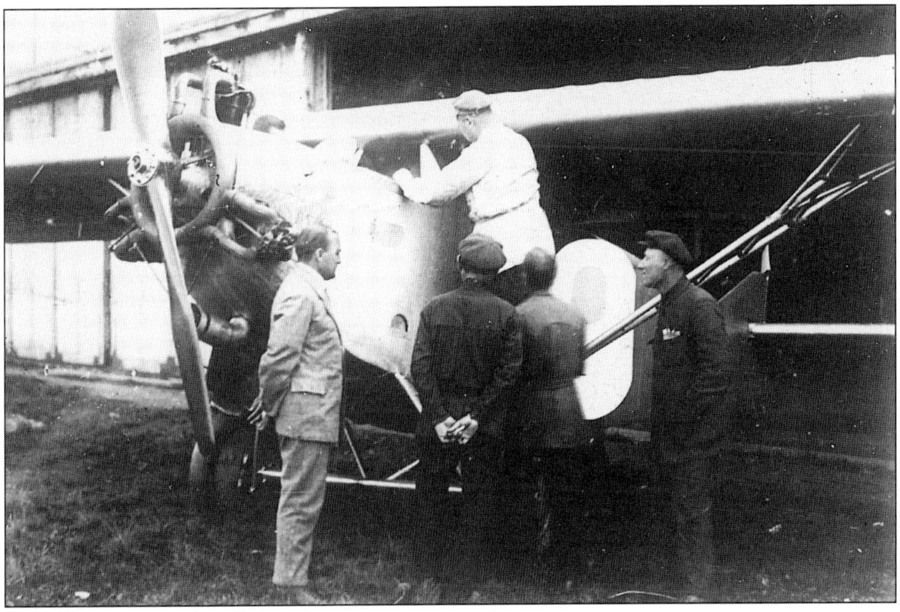
SUVP (before strut faired)

MRL Another single-pusher biplane boat, this was designed in Moscow, though built at GAZ-3. Designation from *Morskoi Razvyedchik Liberty* (sea reconnaissance, Liberty). Single-bay biplane of familiar lines, with shallow V planing bottom extending to tail. Wings Göttingen 436, ply-covered leading edge, N-struts. Hull with layer of fabric between main planks and ply skin. Variable-incidence tailplane. Drawings and many design decisions by P. D. Samsonov and K. A. Vigand. First flown 25. Unimpressive climb and long take-off. Basis for later MR-2.
DIMENSIONS Span 13,2 m (43 ft 3¾ in); length 10,6 m (34 ft 9¼ in); wing area 50,0 m² (538 ft²).
ENGINE One 400 hp Liberty V-12.
WEIGHTS Empty 1660 kg (3,660 lb); fuel/oil 520 kg; loaded 2600 kg (5,732 lb).
PERFORMANCE Max speed 185 km/h (115 mph); climb 11 min to 1 km; ceiling 3050 m (10,000 ft); range 950 km (590 miles); take-off 40s; landing 15 s/95 km/h (59 mph).

MR-2 First aircraft designed at OMOS under Grigorovich's overall direction after its establishment in 25. Improved MRL with later engine, increased-span upper wing, more rounded nose, lower-drag hull planking with aluminium rivets, triple or twin 5 mm steel bracing wires, generally increased dimensions. Latter reduced performance despite extra power, but at least MR-2 could take-off reasonably and climb better. First flight at Leningrad 23 Sept

ROM-1

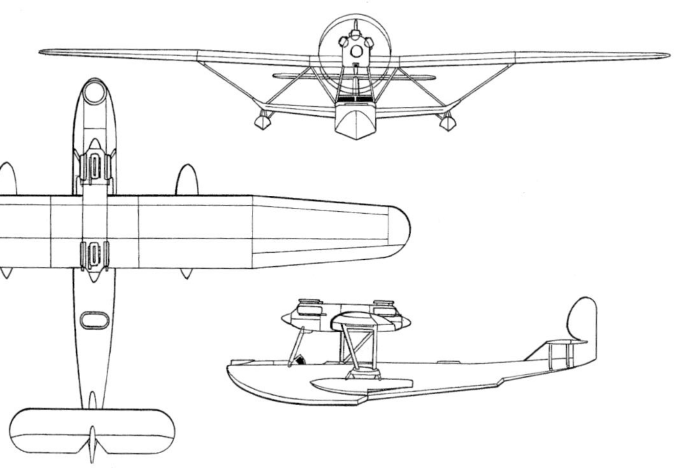

ROM-1

26. NII UVVS pilot F. S. Rastegayev then took over; never flown flying boat before and crashed. No production.
DIMENSIONS Span 15,6 m (51 ft 2¼ in); length 13,6 m (44 ft 7½ in); wing area 56,7 m^2 (610 ft^2).
ENGINE One 450 hp Lorraine-Dietrich V-12.
WEIGHTS Empty 1770 kg (3,902 lb); fuel/oil 560 kg; loaded 2770 kg (6,107 lb).
PERFORMANCE Max speed 179 km/h (111 mph); climb 7 min to 1 km; ceiling 4,2 km; range 900 km (559 miles)/5 h; take-off 25 s; landing 15 s/82 km/h (51 mph).

MUR-1 Traditional-style pusher boat but modernized and refined, though with Soviet-built rotary engine. Intended as dual trainer for all flying-boat pilots. Based on M-5/M-20 but with three-bay wings replaced by single-bay wings of reduced span but deeper aerofoil profile. Tail likewise of thicker aerofoil section and improved overall shape. Designation from *Morskoi Uchyebni Rhône*, designed 25 and flown 26 at Leningrad. Small series, replaced by MU-2.
DIMENSIONS Span 11,5 m (37 ft 8¾ in); length 8,0 m (26 ft 3 in); wing area 33,0 m^2 (355 ft^2).
ENGINE One M-2.
WEIGHTS Empty 700 kg (1,543 lb); fuel/oil 124 kg; loaded 1 t.
PERFORMANCE Max speed 129 km/h (80 mph); climb 8 min to 1 km; ceiling 3,5 km; range 360 km (224 miles); take-off 130 m/12 s; landing 100 m/10 s/70 km/h (43·5 mph).

ROM-1 Grigorovich's most powerful machine to date. Designed 27 after studying Dornier Wal. Intended as long-range sea reconnaissance flying boat; designation from *Razvyedchik Otkrytogo Morya*, reconnaissance open sea. Same configuration as Wal but higher aspect-ratio wooden wing and stabilizing floats on tips of small lower wing with Warren interplane bracing. Hull of Kolchug aluminium fully riveted, with skin thickness 2 mm at step, 1,5 mm over rest of bottom, sides and top 1 mm, fin 0·8 mm. Several AGOS planing-bottom schemes studied, final choice transverse step of 150 mm but angled at 45° instead of vertical. Crew of four; armament four guns (DA, later intended PV-1) in TUR-4 and TUR-5 installations in bow and dorsal, bomb load under lower wing with Sbr-8 sight in bows. First flight winter 27, by L. I. Giksa. From late November moved to Sevastopol for more intensive testing by S. T. Ribalchuk. By autumn 28 established modifications needed, resulting in ROM-2.
DIMENSIONS Span 28,0 m (91 ft 10⅓ in); length 16,0 m (52 ft 5¾ in); wing area 104,6 m^2 (1,126 ft^2).
ENGINES Two 450 hp Lorraine-Dietrich W-12 in tandem pair.
WEIGHTS Empty 4518 kg (9,960 lb); fuel/oil 775+185 kg; loaded 5830 kg (12,853 lb).
PERFORMANCE Max speed 165 km/h (102·5 mph); climb 10·1 min to 1 km; ceiling 3470 m (11,385 ft); range 800 km (497 miles); endurance 5 h; take-off 25 s; landing 13 s/85 km/h (53 mph).

ROM-2 (MDR-1) One fundamental defect of ROM-1 was CG too far aft, and a way to cure this was to replace tandem pair of engines by two larger engines on leading edge. These provided much less power yet ROM-2 managed to have all-round higher performance despite greater fuel capacity and increased gross weight. Important alteration concerned planing bottom, with much sharper V-angle (105° included at keel instead of 155°) curving round to deep downward-pointing chines on each side. Hull longer, steps vertical and second step moved slightly forward. Hull 16,45 m by 1,6 m beam and 2,1 m high. Wings little changed though span slightly reduced; two spars strengthened, ply skin back to rear spar. Lower wing Kolchug alloy, Göttingen 420 profile. Vertical tail redesigned with larger areas. Service designation MDR-1 from *Morskoi Dalnii Razvyedchik*, marine long-range reconnaissance. First flight 29. Behaviour better than ROM-1 but State pilots still not satisfied, and ROM-2 failed structurally during rough landing in 1930.
DIMENSIONS Span 26,8 m (87 ft 11 in); length 17,4 m (57 ft 1 in); wing area 108,2 m^2 (1165 ft^2).
ENGINES Two 680 hp BMW VI (M-17) V-12.
WEIGHTS Empty 4150 kg (9,149 lb); fuel/oil

ROM-2

830+90 kg; loaded 6587 kg (14,522 lb).
PERFORMANCE Max speed 180 km/h (112 mph); climb 7·0 min to 1 km; ceiling 4,5 km; range 900 km (559 miles); endurance 5 h; take-off 250 m/22 s; landing 170 m/15 s/95 km/h (59 mph).

MUR-2 MUR-1 modified at Sevastopol 30 by N. N. Podsyevalov with special planing bottom incorporating membranes, pressure, and force-sensing instrumentation.

MU-2 Improved successor to MUR-1 with much better engine (though less powerful), longer hull of metal construction, of superior form with V instead of concave planing bottom, longer bows, step moved to rear and other changes, slightly swept wings, redesigned vertical tail and longer stabilizing floats set at reduced angle. *Morskoi Uchyebnii* 2, side-by-side dual trainer. Flown Sept 29. No particular faults, but not accepted for production.
DIMENSIONS Span 11,8 m (38 ft 8½ in); length 8,6 m (28 ft 2½ in); wing area 35,6 m² (383 ft²).
ENGINE One M-11.
WEIGHTS Empty 820 kg (1,808 lb); fuel/oil 90 kg; loaded 1086 kg (2,394 lb).
PERFORMANCE Max speed 136 km/h (84·5 mph); climb 10·8 min to 1 km; ceiling 3150 m (10,335 ft); range 380 km (236 miles); endurance 3 h; take-off 35 s; landing 12 s/70 km/h (43·5 mph).

MR-3 Essentially an MR-2 with light-alloy hull and more powerful engine. Same requirements for crew of two/three with bow machine gun and light bomb or stores load. Minor improvements to planing bottom and tail as in other boats. Designed 27, built 28 and flown spring 29. State trials by S. T. Ribalchuk at Taganrog July/Aug 29. Major defects. Intended that MR-3 (alternative designation, MR-5) should be redesigned by P. A. Richard, but task later assigned to CCB at Menzhinskii works and work done by I. V. Chyetverikov. Result was **MR-3bis:** same wings but wooden hull greatly improved hydrodynamic form, and other minor changes including Soviet-built M-17 instead of BMW VI (same engine imported). Tested on Moscow River but, though satisfactory, whole concept considered antiquated and NII UVVS trials not even bothered with. Chyetverikov later used hull as basis for parasol monoplane with M-17 engine, **MR-5bis**, which provided basis for Beriev's MBR-2.
DIMENSIONS Span 15,6 m (51 ft 2¼ in); length 11,4 m (37 ft 5 in); wing area 52,0 m² (570 ft²).
ENGINE One 680 hp BMW VI (bis, M-17).
WEIGHTS Empty 2027 kg (4,469 lb), (bis) 2050 kg (4,519 lb); fuel/oil 440+30 kg; loaded 3082 kg (6,795 lb), (bis) 3100 kg (6,834 lb).
PERFORMANCE Max speed 194 km/h (120·5 mph); climb 7·0 min to 1 km; ceiling 4 km; range 750 km (466 miles); endurance 4 h; take-off 35 s; landing 15 s/85 km/h (53 mph).

MM Project for a *Morskoi Minonosyets*, marine mine carrier. Monoplane with wooden wing (slightly scaled-down ROM-1), twin floats, twin tail booms and tandem Lorraine-Dietrich 450 hp engines above central nacelle. Construction begun 28 but not completed.

MU-2

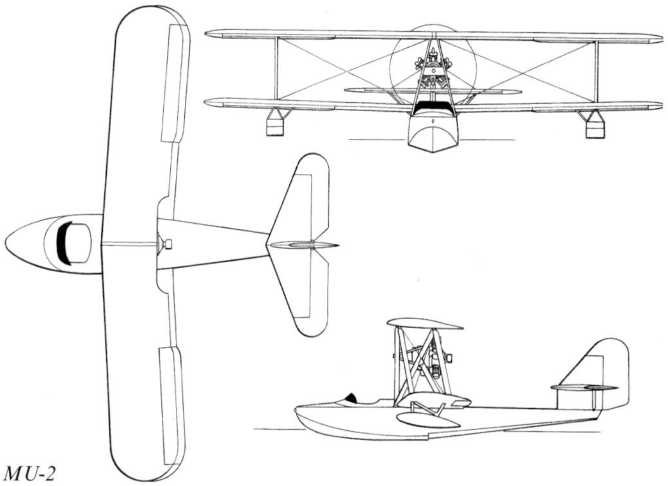

MU-2

MR-3bis

MT Another unbuilt project of 28, this was to be a *Morskoi Torpedonosyets*, marine torpedo carrier. Fuselage of ROM-1 mated to biplane wings with two BMW VI engines (both tractor).

I-5 Though I-5 fighter is accepted as Polikarpov design, it is only fair to note that Grigorovich – oldest and most senior of designers kept in detention at VT at GAZ-39 – himself handled much of design and, according to some reports, was ultimately responsible for all work carried out at VT. Main story appears under Polikarpov.

TB-5 This heavy bomber was largest aircraft produced by internees at VT special detention group at GAZ-39, and it was wholly Grigorovich's design. Bureau number CCB-8. Followed configuration popular at time. Designed for two 800/1,000 hp FED X-24 water-cooled engine being developed by A. A. Bessonov; when this failed to materialise, reliable engines of much lower power were substituted.

Fuselage of welded steel tube with multiple stringers giving polyhedral cross section; fabric covering. Wings with large centre section without flaps, and tapered outer panels with slotted ailerons occupying entire trailing edge. Labour-intensive construction with three spars and many ribs built up as complex trusses from light-alloy tube and sections. Aerofoil R-II, 18% at root and 12% at tip. Light-alloy secondary structure and fabric covering. Tail similar structure, with aerodynamic and mass balance of control surfaces. Normal crew six, with enclosed pilot cockpit, navigator in forward fuselage and engineer/radio in mid-fuselage. Three gunners each with pair of PV-1, front gunner having enclosed turret traversing manually 220° (one of first on any bomber).

TB-5

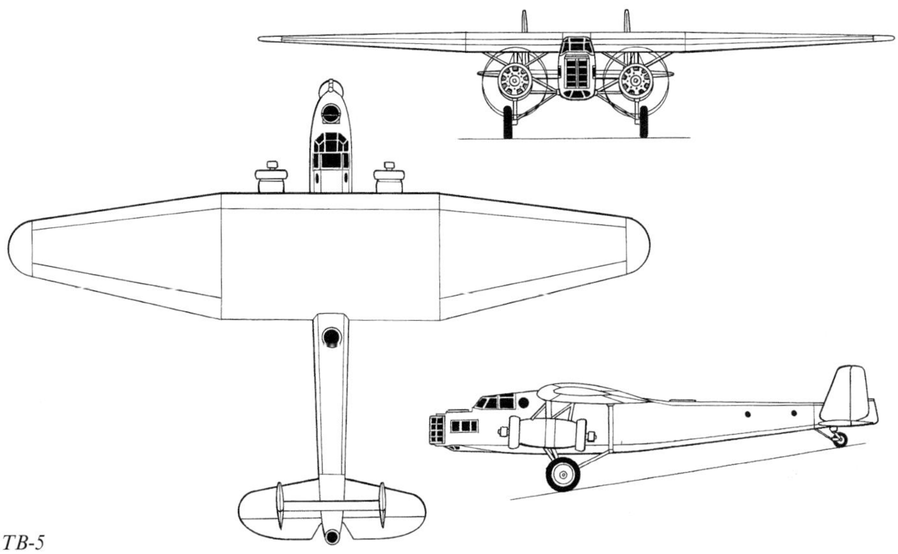

TB-5

Bombs up to total of 2,5 t (5,511 lb) carried in internal bay between wing struts, from 15 to 65% of wing chord.

First flight by B. L. Bukholts, 1 May 31. Performance poor, because of low-powdered engines and apparent inefficiency of tandem arrangement. Special Townend rings added to rear engines, and later also added to front pair. Directional stability marginal. Small improvements made and in late 31 NII VVS trials in hands of M. M. Gromov and test crew. In spring 32, with 12 on board, rear left engine broke away and hung dangling, making control impossible. A. V. Chyesalov escaped by parachute; no other parachutes on board. Gromov managed to get aircraft on ground in severe crash-landing, TB-5 bursting into flames.

DIMENSIONS Span 31,0 m (101 ft 8½ in); length 22,1 m (72 ft 6 in); height 5,8 m (19 ft 0¼ in); track 6483 mm (21 ft 3½ in); wing area 150,0 m^2 (1614·5 ft^2).

ENGINES Four 450 hp (Bristol licence) Jupiter VI.

WEIGHTS Empty 7483 kg (16,497 lb); fuel/oil 3300 kg; loaded 12 535 kg (27,634 lb).

PERFORMANCE Max speed 180 km/h (112 mph) at SL (possibly rather more at best height but not recorded); climb, no figures; ceiling 2,6 km (8,530 ft); range not measured.

DI-3 Another design managed by Grigorovich during his period of detention at VT was this two-seat fighter, which is believed also to have had Bureau number CCB-9. Preparation for production at GAZ-21 went ahead in parallel with design. Conventional except for twin-finned tail, adopted to improve field of fire of observer. Fuselage of welded chrome-molybdenum steel tube with wood secondary fairing structure. Wooden wings, control surfaces duralumin. Fabric covering throughout. Imported engine with underslung fixed radiator. Two PV-1 fixed in top decking, and pivoted DA in rear cockpit with long fairings along sides of fuselage for ammunition drums. First flown Aug 31. Performance satisfactory and no shortcomings, but design rejected on grounds of excessive weight and inadequate manoeuvrability. Skis fitted winter 1931-32.

DI-3

DIMENSIONS Span 11,8 m (38 ft 8½ in); length 7,97 m (26 ft 2 in); wing area 30,1 m^2 (324 ft^2).

ENGINE One 730 hp BMW VI z7,3 V-12.

WEIGHTS Empty 1487 kg (3,278 lb); fuel/oil 200+25 kg; loaded 2122 kg (4,678 lb).

PERFORMANCE Max speed 256 km/h (159 mph); climb 7·4 min to 3 km; ceiling 6,3 km; time for 360° turn 13 s; take-off 150 m/11 s; landing 225 m/17 s/98 km/h (61 mph).

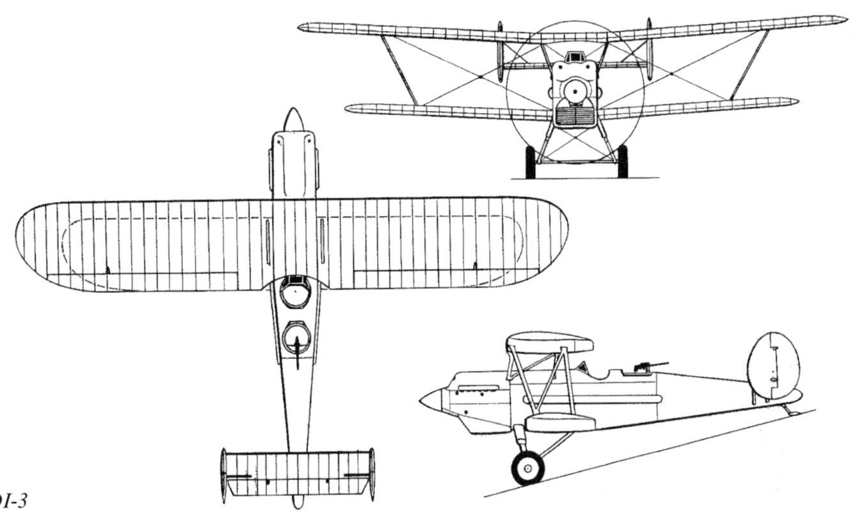

DI-3

I-9 Unbuilt heavy fighter design of 32 with two M-30, M-31, M-32, M-37 or M-38 engines.

I-10 Unbuilt high-speed fighter of Elektron (magnesium alloy) with new M-41 engine.

LSh *Legkii Shturmovik*, light assault aircraft prepared under great pressure during period of detention in 1929-30. Bureau designation CCB-5. To speed project, made use of as many parts as possible of R-5 aircraft, including complete wings, but with considerable local reinforcement. Welded steel armour around engine and tandem cockpits. Twin synchronized PV-1, two aimed by observer from rear cockpit and battery of four under fuselage firing down and ahead at angle set from 30° to 60°. LSh (sometimes reported as LSh-1) completed at end 30, but VVS had decided six months earlier not to buy this type, TSh series offering greater promise.

DIMENSIONS Span 15,45 m (50 ft 8¼ in); length 10,4 m (34 ft 1½ in); wing area 51,2 m^2 (551 ft^2).

ENGINE One M-17.

WEIGHTS Not reliably reported but empty/gross about 2000/3200 kg.

PERFORMANCE Max speed 225 km/h (140 mph).

TSh-1 Again based on R-5 airframe, with much alteration and reinforcement, this was a potent machine in TSh (*Tyazheli Shturmovik*, heavy attacker) category. With bureau number

TSh-1

CCB-6, three prototypes built 30 under Grigorovich direction and all differing in important design features. Much thought given to steel armour in ruling thickness 6 mm and total weight typically 520 kg. In one prototype armour welded, second riveted and third bolted. Armour either mounted on auxiliary duralumin frames or attached to pine and thick ply structure with 35 mm gap. Severe problem accurately matching drawings to armour. Three engine cooling systems: water, fixed radiator; water, retractable; glycol (smaller radiator). Fuselage basically as LSh but lower wings further reinforced and carrying four underwing containers each housing pair of PV-1, and ammunition. Two synchronized PV-1, twin DA in rear cockpit, and in addition *Granatitsa* box from which 300 grenades dispensed. First flight, on skis, Jan 31. Subsequent testing by B. L. Bukholts, Yu. I. Piontkovskii and M. A. Bolkovoinov, as well as NII VVS pilots. Fair behaviour but sluggish above 3,3 t, and chosen water cooling system caused overheating to engine and cockpit because of prolonged high power necessary. VVS rejected TSh-1 in favour of more efficient R-5 derivative, TSh-2.

DIMENSIONS Span 15,5 m (50 ft 10¼ in); length 10,56 m (34 ft 7¾ in); wing area 51,2 m² (551 ft²).

ENGINE One M-17.

WEIGHTS Empty 2495 kg (5,500 lb); loaded 3490 kg (7,694 lb).

PERFORMANCE Max speed 200 km/h (124 mph); range, claimed 650 km (404 miles).

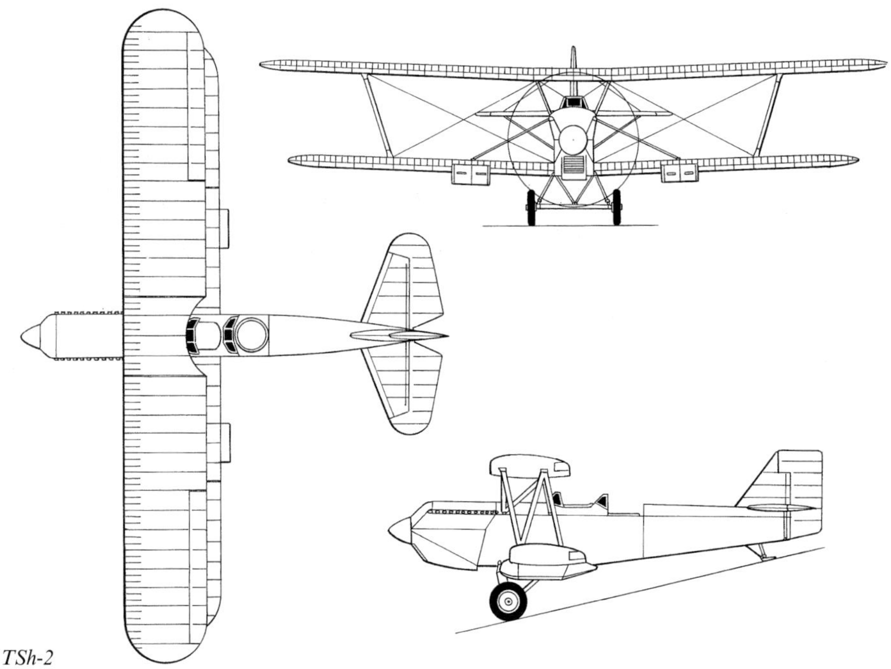

TSh-2

TSh-2 Basically TSh-1 with new lower wings of duralumin construction, greatly thickened immediately outboard of propeller disc to incorporate four PV-1 in each wing, with ammunition boxes all within wing. Twin DA in rear cockpit retained but synchronized PV-1 believed removed. Bureau number CCB-21. First flown late 31, and superior to TSh-1. Pre-series of ten built, but persistent troubles with accurate armour (repeated ten years later with Il-2) and M-17 deliveries prolonged project which was beginning to look dated. In 32 one TSh-2 tested with M-27, of greater power and reduced tendency to overheat, but this engine never cleared for production.

DIMENSIONS, engine as TSh-1.

WEIGHTS Empty not known; loaded 3950 kg (8,708 lb).

PERFORMANCE Max speed 213 km/h (132 mph); ceiling 4220 m (13,845 ft); range 650 km (404 miles).

ShON Prolonged insurrection caused by discontented *Bazmashii* – well armed and organized bands of horsemen in Turkestan – in late 20s brought swift response from Moscow which wished to collectivize all Soviet peoples and crush all dissidents. ShON (*Shturmovik Osobogo Naznachyeniya*, attacker for special purposes) was ordered 30, with designation CCB-23; first example completed in April 31. Basis again R-5, but folding wings to allow transport by rail and road. Much lighter than TSh series, with only light armour around vital areas and no weapons except four twin boxes of PV-1 as in TSh-1. Tested and found satisfactory early 32, but *Bazmashii* eliminated by land forces before ShON reached Turkestan.

DIMENSIONS Span 15,5 m (50 ft 10¼ in); length

ShON

I-Z (series)

10,3 m (33 ft 9½ in); wing area 51,2 m² (551 ft²).
ENGINE One M-17.
WEIGHTS Empty 1610 kg (3,549 lb); loaded 3 t (6,614 lb).
PERFORMANCE Max speed 225 km/h (140 mph); ceiling 6,1 km (20,000 ft); range 730 km (454 miles).

I-Z Yet another of combat landplanes assigned to Grigorovich during detention at VT, this was one of first Soviet monoplane fighters since original IL-400. Created specifically to carry large-calibre DRP (ARK) recoilless cannon designed by L. V. Kurchyevskii, who also participated in aircraft design. Project begun 30 as CCB -7, also called Project Z. To speed project, forward fuselage and engine same as second prototype (*Klim Voroshilov*) I-5; rest of fuselage dural monocoque. New low-mounted wings, of Göttingen-436 profile, mainly of Enerzh-6 stainless-steel lattice construction with torchwelded joints, ribbon bracing wires and underwing bracing by streamlined dural struts to landing-gear structure. Fabric covering throughout wing, rudder and braced horizontal tail mounted well up fin out of way of gun blast. Guns slung under wing attached to both spars. Single PV-1 in left side of fuselage for sighting purposes. First flown late summer 31; bureau pilots B. L. Bukholts and Yu. I. Piontkovskii. Second prototype with modified wing structure flown 32. Small series of 21 aircraft built as I-Z with minor alterations and, instead of helmeted Jupiter, with fully cowled M-22. Further 50 I-Z with wooden wings built at Kharkov 1934-35, but service experience was limied because only at this late time did various shortcomings, notably difficult recovery from developed spin, become appreciated. Most I-Z used in development of later recoilless guns and fast-firing 20 and 37 mm cannon. One used in *Zvyeno* trials (see Vakhmistrov). Led to IP-1, see later.
DIMENSIONS Span 11,5 m (37 ft 8¾ in); length 7,65 m (25 ft 1¼ in); wing area 19,5 m² (210 ft²).
ENGINE One M-22.
WEIGHTS Empty 1180 kg (2,601 lb); fuel/oil 180 kg; loaded 1648 kg (3,633 lb).
PERFORMANCE Max speed 259 km/h (161 mph) at SL, 300 km/h (186 mph) at 3 km; climb 14 min to 5 km; ceiling 7 km (22,970 ft); range 600 km (373 miles); endurance 2·5 h; take-off 110 m/8 s; landing 180 m/15 s/100 km/h (62 mph).

Stal-MAI This advanced structural essay resulted from Grigorovich's role as director and professor at MAI at end of 31, after his release from detention. He organized a group of students to study airframe entirely of Enerzh-6 stainless steel, and investigate structural and manufacturing problems. Group led by P. D. Grushin and included M. M. Pashinin, L. P. Kurbala, V. A. Fyedulov and A. P. Shchyekin. Study concluded aircraft practicable in 32, and became MAI project with Grigorovich supervising. Low-wing cantilever monoplane, tandem enclosed cockpits, fixed landing gear, one 725 hp M-34R V-12 engine. Apart from fabric-skinned control surfaces, entire airframe of various steels, over 90% being Enerzh-6. Thickness remarkably low, typically 0,3 to 0,5 mm, much being sandwich stabilized by corrugations of 80 mm pitch. Corrugations sometimes rounded,

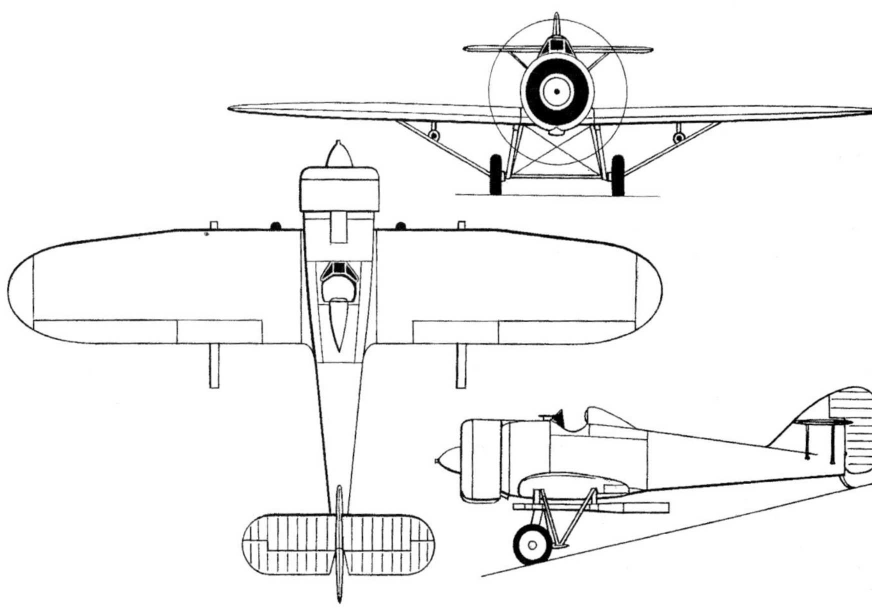

I-Z (guns inboard of struts, Townend-ring cowl)

in some areas trapezoidal. Joints almost entirely welded, by torch and electric arc. Two-spar wing, front spar being at deepest part of profile at 28% chord and with corrugated-sheet web. Semi-monocoque fuselage with integral fin. First aircraft in world with such structure. Completed mid-34 and first flown Sept 34 by I. F. Kozlov and Yu. I. Piontkovskii. Failure of engine mounting on fifth take-off (Piontkovskii flying, Grushin passenger). Stal-MAI repaired but project discontinued. Few numeric details.

Stal-MAI model

IP-1 (DG-52) First aircraft known with alternative designation for Grigorovich's own name. Basic designation from *Istrebitel Pushyechnii*, fighter, cannon. Refined development of I-Z designed in less haste, with much better aerodynamics and structure. All light-alloy stressed-skin construction, elliptical wing with centre section of 55% span carrying APK-4 recoilless cannon at extremities, with bolted wingtips outboard. Two spars built up from chrome-molybdenum tubes, with lattice and sheet ribs and 0,6 to 0,8 mm skin flush-riveted. Slotted flaps and ailerons. Circular-section monocoque fuselage with integral fin skinned in 1,0 to 1,5 mm sheet. Hydraulic flaps and main landing gears folding backwards into underwing blisters. First prototype with Wright Cyclone and Hamilton prop flown 34. Tested with APK-4 with five rounds each, sighted by two synchronized ShKAS. Good aircraft but APK-4 eventually rejected and 200 production IP-1 ordered from Kharkov with armament of two 20 mm ShVAK in wing roots and six ShKAS in boxes at extremities of centre section in place of large recoilless guns, and with M-25. Change of armament moved CG to rear, and spin recovery difficult with full ammunition until large dorsal fin added. Actual production 90, 1936-37.
DIMENSIONS Span 10,97 m (36 ft 0 in); length 7,23 m (23 ft 8¾ in); wing area 19,98 m² (215 ft²).
ENGINE One M-25.
WEIGHTS (production) Empty 1200 kg (2,646 lb); fuel/oil 275+25 kg; loaded 1880 kg (4,145 lb).
PERFORMANCE Max speed 368 km/h (229 mph) at SL, 410 km/h (255 mph) at 3 km; climb 1·3 min to 1 km; ceiling 8,3 km; range 600 km (373 miles) (1000 km overload fuel); take-off 230 m/17 s; landing 175 m/11 s/97 km/h (60 mph).

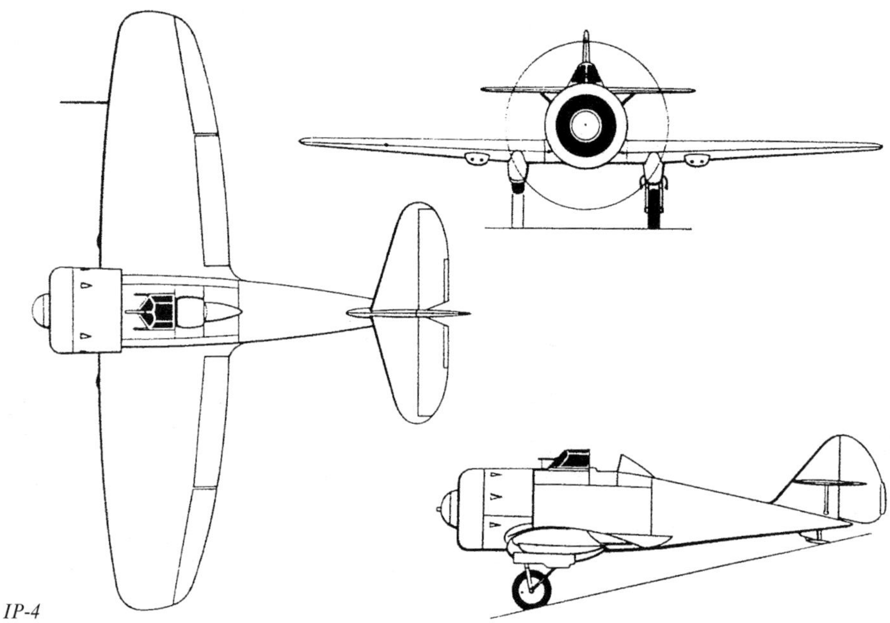
IP-4

Grigorovich with IP-1

IP-4 (DG-53) Family derived from IP-1. Original DG-53, 34, smaller version of DG-52 with four 45 mm APK-11 sighted by two ShKAS. Intended M-25 but flown end 34 with Cyclone. **DG-53bis** with M-25 and armament of two ShVAK and two ShKAS cancelled 35 when nearly complete.
DIMENSIONS Span 9,6 m (31 ft 6 in); length 7,08 m (23 ft 3 in); wing area 16,36 m² (176 ft²).
ENGINE One 640 hp Wright R-1820-F Cyclone.
WEIGHTS Empty 1080 kg (2,381 lb); fuel/oil not known; loaded 1549 kg (3,415 lb).
PERFORMANCE Max speed 382 km/h (237 mph) at SL, 435 km/h (270 mph) at 3 km; climb 1·1 min to 1 km; ceiling 8,3 km; range 600 km (373 miles) (830 km overload fuel); take-off 180 m/8 s; landing 200 m/97 km/h (60 mph).

IP-2 (DG-54) Uncompleted prototype of further development of IP-1 with different engine and devastating armament of either 12 ShKAS or two ShKAS and four ShVAK. Full data have been published including gross weight with 760 hp Hispano-Suiza 12Xbrs of 1952 kg and calculated max speed of 519 km/h (322·5 mph). Cancelled 36.

E-2 (DG-55) Also named *Kometa* (for its design inspiration was obvious) and *Dyevushkovaya Mashina,* Girl's Machine, for there were eight girls in design team at MAI. Intended as an exercise in efficient long-range travel. Simple wooden construction, main wing box being skinned with 8-mm ply. Landing flaps of Zapp type, low hinge-moment allowing manual operation; landing-gear retraction also manual. First flight 35; after test programme, passed to Osoaviakhim and used as courier.
DIMENSIONS Span 11,0 m (36 ft 1¼ in); length 7,9 m (25 ft 11 in); wing area 13,8 m² (148·5 ft²).
ENGINES Two 120 hp Cirrus Hermes inverted 4-inline.
WEIGHTS Empty 1051 kg (2,317 lb); loaded 1546 kg (3,408 lb).

E-2

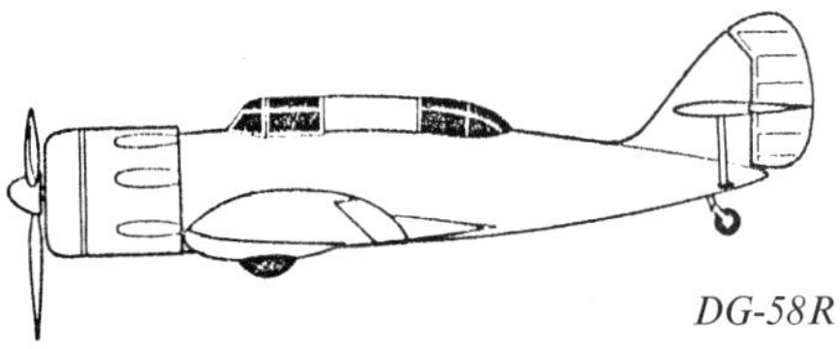
DG-58R

PERFORMANCE Max speed 296 km/h (184 mph); service ceiling 5 km; range 2200 km (1,367 miles); take-off 255 m/15 s; landing 310 m/23 s/102 km/h (63 mph).

LK-3 (DG-56) Project for *Legkii Kreiser* (light cruiser) with armament of eight ShVAK or other impressive alternatives; estimated speed 438 km/h with two 825 hp Hispano-Suiza 12Ybrs. Single-seat prototype begun 35 but not completed; cancelled 36.

PB-1 (DG-58) Last known programme by Grigorovich began as PB-1 (*Pikiriyushchii Bombardirovshchik*, dive bomber) in 36. All metal stressed-skin with M-85 (GR 14K derived) and internal weapon bay. Special hydraulic perforated dive brakes. Main gears semi-retracted into underwing fairings. Crew of two in tandem. Estimated speed with full load 450 km/h (280 mph). In parallel schemed tactical reconnaissance aircraft with bureau designation **DG-58R** (also reported as 58bis). Grigorovich fell ill early in this programme and died 38.

Grokhovskii

Pavel Ignatyevich Grokhovskii born Vyazma 18 March 99. Served with Bolshevik navy, later figher pilot, test pilot and parachutist, ultimately allowed assistants, forming at Leningrad 33 RKKA, this later formalized as Red Army Special Design Team for Aviation Forces, Grokhovskii director. Work centred on methods and equipment for paratroops and air-landed assault troops. Perfected large paradrop containers and carriage of light tank by G-2. Department abruptly closed 36, Grokhovskii dying 2 Oct 46.

Inflatable Throughout his brief design career Grokhovskii was working on inventions that many years later were to be picked up by workers in other countries. In 1950s ML Aviation (UK) and Goodyear (US) claimed originality in building ultralight aeroplanes made of rubberized fabric, unpacked from box and inflated under gentle air pressure. This is what Grokhovskii did in 34. His first *Naduvatsya* (Inflator) was a glider, weighing 77 kg (170 lb) and packing into a box 1×1×0·5 m. Tested from Tushino Jan 35, towed behind U-2. Inventor's intention was to build refined powered version.

Kukuracha Experimental lightplane with wing swept at 35°, slightly swept trailing edge, wingtip rudders and no horizontal tail. Pusher M-11, speed 160 km/h (100 mph). Wood, ply and fabric, with single spar. Built by S. G. Kozlov, V. F. Bolkhovitinov and A. Ye. Kaminov. Structural failure on take-off.

Kukuracha

G-31/M-25 (see overleaf)

G-37

G-26 High-speed aircraft in class of Stal-6. One 860 hp Hispano-Suiza 12Ybrs engine, all-metal stressed-skin construction, cantilever low wing, single retractable landing wheel, single-seat enclosed cockpit. Designer B. D. Urlapov. Built 1935-36 but never quite completed: politically afraid of failure.

G-31 Powered derivative of G-63. Almost identical airframe, initially fitted with uncowled M-11 in nose, later with M-17 and finally 700 hp M-25 with ring cowl for research into flight at extreme altitude. First flown 5 Sept 36. Dubbed *Strekoza*, Dragonfly.
DIMENSIONS Span 28,0 m (91 ft 10¼ in); length 13,9 m (45 ft 7¼ in); wing area 70,5 ² (759 ft²).
ENGINE See above.
WEIGHTS No data.
PERFORMANCE Cruising speed (M-25) 240 km/h (149 mph); service ceiling (calc) 19,0 km (62,300 ft).

G-37 Probably Grokhovskii's best-known achievement, also called ULK (*Universalnoye Letayushchyeye Krylo*, universal flying wing). Though original objective was versatile transport for assault landings, its potential as a commercial carrier was obvious. Idea was that G-37 should pick up a pre-loaded container shaped to conform to rest of aircraft, deliver it and then collect a fresh container. (Same idea seen later in Fi 333, Miles M.68 and Fairchild XC-120.) To reduce time and cost Grokhovskii used ANT-9 wing. On to this he built engines (different installation from PS-9), stressed-skin tail booms and deep trousered landing gear. Nacelle for pilot and flight engineer, with special attachments on underside for payload pod. Tail dural with fabric skin on rudders and elevators. Twin castoring tailwheels. Construction in hands of V. F. Rentel (see Krylov) at Grebno port, Leningrad. Completed 34 (this organization did not number its projects in a consecutive manner). Test-pilot V. P. Chkalov satisfied, and G-37 later flew to Moscow in 2 hr 50 min, average 250 km/h, then almost a record. No production.
DIMENSIONS Span 23,7 m (77 ft 9¼ in); length 16,0 m (52 ft 6 in); wing area 84,0 m² (904 ft²).
ENGINES Two M-17.
WEIGHTS Not known but probably about 3500/6000 kg.
PERFORMANCE Max speed 285 km/h (177 mph); ceiling 5,5 km (18,050 ft).

G-38 (LK-2) This fighter-bomber may have been influenced by contemporary (34) French *multiplaces de combat*. Twin tail booms, central nacelle for 2 crew and guns, and retractable landing gear. Mainly wood construction, except for stressed-skin metal nacelle and fabric-covered light-alloy control surfaces. Greatest possible combination of speed and firepower. Four ShKAS and two ShVAK firing ahead, two ShKAS aimed by gunner at rear of nacelle, various bombs hung under nacelle. Project begun Sept 34. Grokhovskii led design, construction under P. A. Ivyensen and V. I. Korovin, static testing under M. V. Orlov and A. F. Yepishyev from GVF, with consultant professors A. K. Martinov and V. N. Belyaev. LK signified *Leningradskii Kombinat* (Leningrad Combine). G-38 virtually complete when institute closed 36.
DIMENSIONS Span 13,4 m (43 ft 11½ in); length 8,8 (28 ft 10½ in); wing area 32 m² (344 ft²).
ENGINES Two 900 hp Gnome-Rhône K14.
WEIGHTS Empty 2200 kg; loaded 4100 (9,039 lb).
PERFORMANCE (Est) max speed 550 km/h (342 mph); climb 12 min to 8 km; service ceiling 9,5 km (31,160 ft).

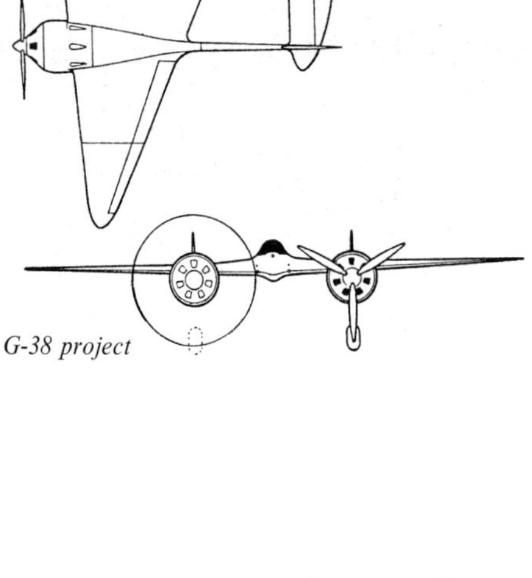

G-38 project

G-61 Not an aircraft but a standard 'people pod' attached under lower wing of R-5. Each container was flat streamlined wooden box of 4,3 cubic metres volume divided into four compartments each accommodating two occupants lying face-down and head-first. Loaded via hinged rounded leading edge. Complete aircraft load 16 persons (troops, casualties or rescuees), total aircraft weight 3800 kg. On 8 Dec 36 Grokhovskii tested R-5 with full load: airborne 400 m/30 s; cruise 201 km/h (125 mph), ceiling 2,8 km. Each container weighed 200 kg (441 lb) empty. Later similar containers were used to rescue crew of icebound vessel *Chelyushkin*.

G-63 When 19-year-old Boris Dmitrievich Urlapov asked for job in 31 he was told by Grokhovskii to design glider towed by R-5 but able to carry 10 armed paras (1 t). Urlapov did this, but was then told 'Now carry 16 troops, 1700 kg'. Only solution was to carry them lying inside thickened wing, reducing fuselage to slim monocoque tube carrying tail. Wood construction, spatted main wheels. First flight by Grokhovskii himself Sept 32. Data for this. Passed **NII** testing, built in small series. In 34 **G-63bis** built with tandem dual cockpits for training cargo glider pilots, named for *Yakov* (Jacob) *Alksnis*. Powered derivative, see **G-31**.
DIMENSIONS Span 28,0 m (91 ft 10⅜ in); length 13,6 m (44 ft 7½ in); wing area 70,0 m² (753·5 ft²).
WEIGHTS Empty 1,4 t (3,086 lb); loaded 3,2 t (7,055 lb).
PERFORMANCE Glide ratio 28·6; normal tow (R-5) 135 km/h (84 mph).

GN-4 High-wing glider designed by Yuri

G-63bis

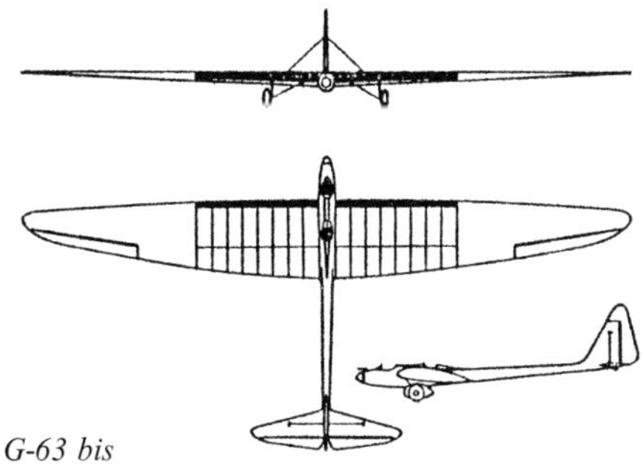

G-63 bis

Groshyev Jan 36 to carry maintenance staff when fighter regiments were moved. Pilot and four passengers or 500 kg.

GN-8 Low-wing glider with trousered main gears, pilot and four passengers in a row under hinged canopies, towed by I-5 or I-16 fighter.
DIMENSIONS Span 13,6 m (44 ft 7½ in); length 7,5 m (24 ft 7¼ in).
WEIGHTS Empty 460 kg (1,014 lb); loaded 900 kg (1,984 lb).

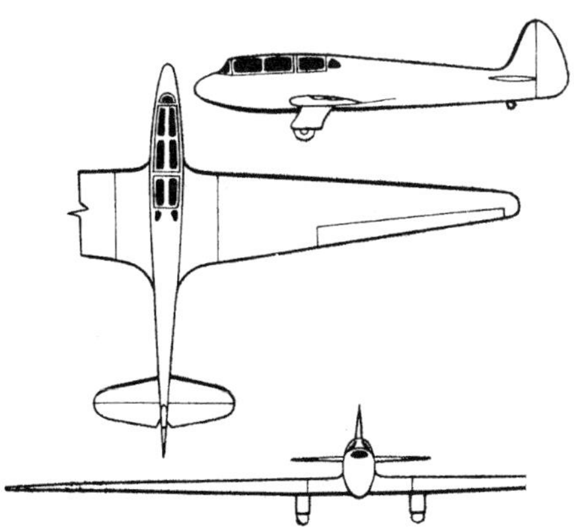

GN-8

Groppius

Ye. E. Groppius is one of least-known Soviet constructors. In 23 he was working on a canard (tail-first) machine, but this was never completed. His only product was a single moderately successful passenger biplane, designated for factory in which it was built.

GAZ-5 Two-bay biplane with a deep fuselage filling space between wings. Light-alloy nose covering engine, with Groppius cooling radiator underneath with regulator flaps. Wooden mid-fuselage with door on right and seats for four passengers. Behind cabin, open cockpit for pilot and mechanic side-by-side. Wings and tail fabric-covered and wire braced. Large horn balances on control surfaces, ailerons on all four wings. Designed 23, construction begun early 24, completed autumn 24. Test-pilot Yefremov unimpressed.
DIMENSIONS Span 11,2 m (36 ft 9 in); length 8,5 m (27 ft 10¾ in); wing area 35,0 m² (377 ft²).
ENGINE One 300 hp Hispano-Suiza 8fb V-8.
WEIGHTS Empty 1118 kg (2,465 lb); loaded 1849 kg (4,076 lb).
PERFORMANCE Max speed 165 km/h (102·5 mph); ceiling 300 m (984 ft).

GAZ-5

Grushin

Pyetr Dmitriyevich Grushin came into prominence working at MAI on construction of Stal-MAI under Grigorovich. This took from 31 until 34, by which time he was designing his own ultralight, named *Oktyabryonok*. He also produced a kind of simplified Link Trainer for blind-flying instruction, and an experimental high-pressure steam engine for aircraft propulsion which was tested in a U-2. In late 41 he had to close his own KB and joined MiG OKB to design DIS.

Sh-tandem This typically bold Soviet prototype got its designation from *Shturmovik-tandem* (tandem attacker); other designations included Tandem-MAI and MAI-3, from institute where it was designed and built. Basic concept was excellent one of making aircraft close-coupled, with tanden wings and rear gun turret with unobstructed field of fire. Wing profile of R-11 series on both main wing and quasi-elliptical rear surface which could be regarded as tailplane or rear wing with 45% of area of main surface. Much argument over controls; final scheme put ailerons on main wing and elevators on rear surface, but alternative (intended to be flown) was two front ailerons and two rear elevons. Twin fins/rudders on underside of rear surface at 50% semi-span, leaving clear field of fire for turret through 250° azimuth. Pilot in enclosed cockpit with sliding canopy. Gunner in MV power-traversed turret (probably electrical) with ShKAS. Four ShKAS fixed in wings firing ahead. Bomb load 200 kg (441 lb). Basically wooden construction, fuselage being monocoque of *delta* (birch plies baked with Bakelite-type polymer). Main gears retracted into wing (poor-quality photos suggest inwards, but could have been to rear, with wheels turned 90° and standing proud of undersurface). Fixed tailwheel.

Prototype constructed in MAI production training school. First flown 5 Dec 37,

Sh-Tandem (with upper fins)

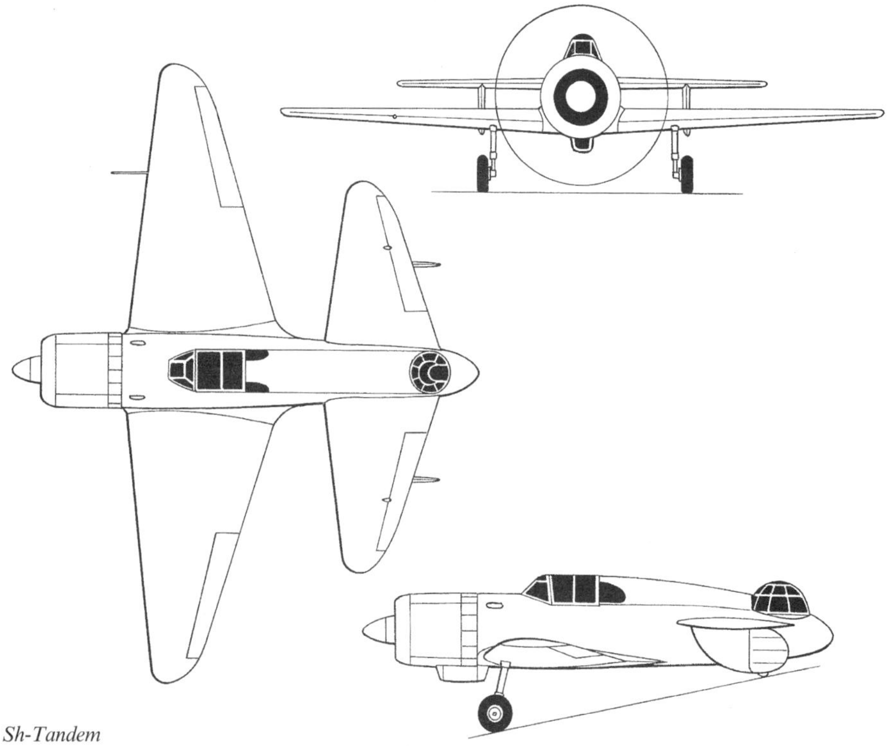

Sh-Tandem

Gr-1 model

subsequently subjected to prolonged test-flight programme by P. M. Stefanovskii.

DIMENSIONS Span 11,0 m (36 ft 1¼ in); span of rear wing 7,0 m (23 ft 0 in); length 8,5 m (27 ft 10¾ in); wing area (combined) 30,4 m^2 (327 ft^2).

ENGINE One M-87.

WEIGHTS Only figure recorded is loaded, given variously as 2560 and 3088 kg (5,644 or 6,808 lb).

PERFORMANCE Max speed 406 km/h (252 mph) at SL, 488 km/h (303 mph) at 4,2 km; no other data.

BB-MAI Handsome tandem-seat short-range armoured attack aircraft. Designed for factor of 12, mainly *delta* bakelite ply (25 mm thick skins of main wing box) bonded with VIAM V-3 adhesive. Two-section split flaps, two-section ailerons, fully retrac tricycle landing gear, castoring nose unit between twin ducted rads. ShVAK thru prop hub, two ShKAS above engine, UB in rear cockpit, bombs or rockets not finalised. Prototype flown summer 41 by A. N. Grinchik. Evacuated with Gr-1, but both destroyed by air attack on train.

DIMENSIONS Span 10,0 m (32 ft 9⅝); length 9,6 m (31 ft 6 in); wing area 16,8 m^2 (181 ft^2).

ENGINE One M-105.

WEIGHTS Empty 2965 kg (6,537 lb); fuel 255+20 kg; loaded 3490 kg (7,694 lb).

PERFORMANCE Max speed at SL 508 km/h (317 mph), at 5 km 550 km/h (342 mph); climb 9·2 min to 5 km; service ceiling 9 km (29,530 ft); range 500 km (311 miles); TO/ldg c500 m.

Gr-1 (IS) Grushin's last known design, this was a break with past in configuration and structure, being of conventional layout and all-metal construction. Main wing box with two sheet webs and spars with steel angle booms, with intermeshing-comb (piano-hinge) joint

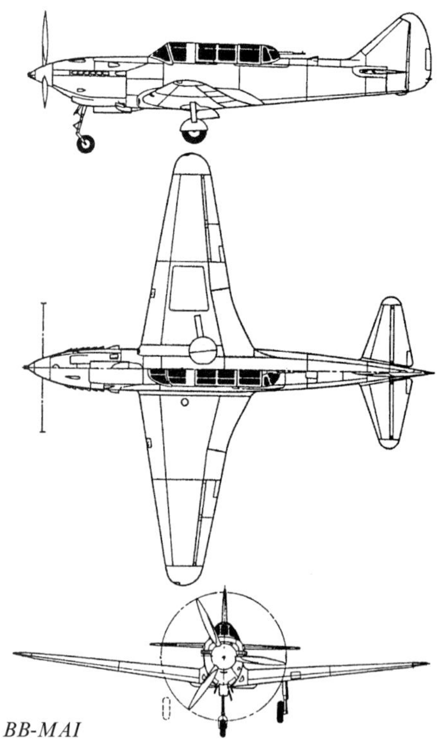

BB-MAI

between centre section and outer panels. Armoured cockpit for single pilot in nose, thinner armour over engines. Designation IS from *Istrebitel Soprovozhdyeniya*, escort fighter. Heavy armament of four ShKAS and two ShVAK, plus eight RS-82 or RS-132 rockets all firing ahead, plus bomb load under fuselage up to 500 kg (1102 lb). No information on landing gear. Prototype construction for nine months, with static testing complete by spring 41. Evacuated October 1941 but destroyed by air attack on train.

DIMENSIONS Span 15,8 m (51 ft 10 in); length, not recorded; wing area 42,0 m² (452 ft²).

ENGINES Two AM-37.

WEIGHTS Empty, about 4900 kg; loaded (normal) 7250 kg (15,983 lb), (max fuel and bombs) 7650 kg (16,865 lb).

PERFORMANCE (estimated) Max speed 480 km/h (298 mph) at SL, 645 km/h (401 mph) at 7,2 km; climb 9·8 min to 8 km; ceiling 11,7 km(38,385 ft); range 1380 km (857 miles), (max fuel 1890 km, 1,174 miles); landing speed 120 km/h (74·5 mph).

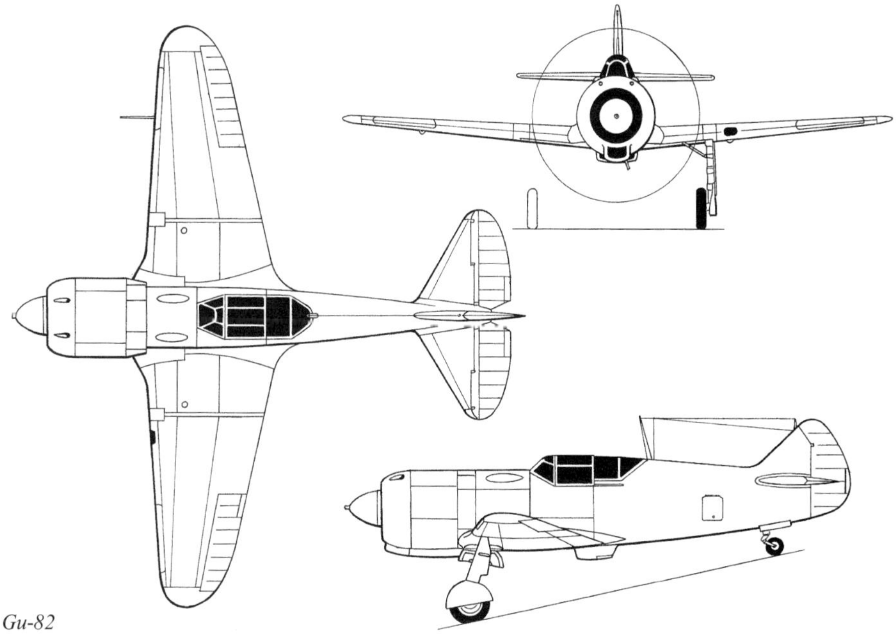
Gu-82

Gudkov

Mikhail Ivanovich Gudkov was a GAZ manager in Sept 38 when he became one of three partners who created LaGG-3 (see Lavochkin). When LaGG OKB was evacuated Gudkov remained in Moscow.

K-37 Also known as **LaGG-3K-37**, this was first of Gudkov's several planned improvements to LaGG's to materialise. Intended primarily for anti-armour attack, it replaced usual 20 or 23 mm engine-mounted cannon by 37 mm weapon. Gun chosen was ShK-37, eventually rejected in favour of NS-37. This big gun rested between cylinder banks, but was so long that firewall, and thus oil tank and pilot's cockpit, had to be moved 35 cm to rear. First prototype ready Aug 41 and successfully tested. According to Soviet records three were completed and were shipped to front, no more being produced. Data as LaGG-3.

Gu-82 LaGG-3 with ASh-82 engine, created hurriedly at GAZ-31, then evacuated to Tbilisi by taking production LaGG-3 and mating it with existing ASh-82 installation complete with engine bearers and cowling from an Su-2. Trailing-edge cooling ring thermostatically adjustable. Oil cooler under fuselage at trailing edge. Venetian-blind shutters retained at front, rear modified by light sheet to conform to slimmer fuselage. Minimum-change lash-up, though superficially adequate. Two ShVAK above engine. One only; second prototype never completed because of Oct 41 evacuation.

DIMENSIONS Span 9,8 m (32 ft 2 in); length 8,75 m (28 ft 8½ in); wing area 17,51 m² (188·5 ft²).

ENGINE One 1330 hp M-82 (ASh-82).

No other data.

Gu-1 Gu-37 Gudkov had been convinced American Bell Airacobra was a sound concept, and in early 40 studied a similar fighter with engine on CG, shaft drive to tractor propeller, and cockpit and guns in nose. Construction mainly wood, except forward fuselage of welded steel tube with removable dural skin panels, steel and dural centre-section spars and dural control surfaces. Wing profile 1V-10 Type V-2, automatic slats, hydraulic flaps and tricycle landing gear. Engine above wing with main radiators in wings, oil cooler under centre section. Drive via 120 mm (4¾ in) diameter shaft, single 37 mm Taubin cannon with 81 rounds (large capacity for this calibre) firing through reduction gear. Prolonged problems with engine and drive, and flight-testing did not begin until summer 1943. Mark Gallai quotes test pilot A. I. Nikashin as being unhappy after taxi trials, saying 'it seems glued to the ground'. First flight 12 June 43; long run, aircraft became airborne, reached about 200 m and then fell off in sideslip and crashed, killing Nikashin. General opinion was wing too small. Gudkov's bureau disbanded.

DIMENSIONS Span 10,0 m (32 ft 9⅝ in); length 10,68 m (35 ft 4¾ in); wing area 20,0 m² (215 ft²).

ENGINE One AM-37.

WEIGHTS Empty 3742 kg (8,250 lb); loaded 4610 kg (10,163 lb).

PERFORMANCE Not measured.

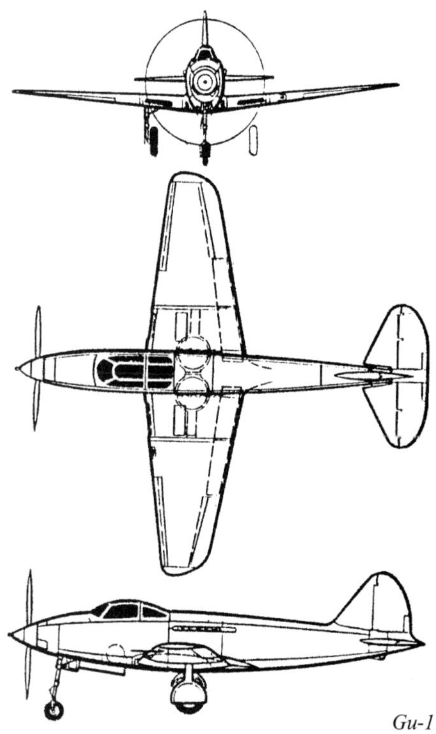
Gu-1

Ilyushin

Sergei Vladimirovich Ilyushin was one of greatest Soviet designers, achieving success with contrasting aircraft plus ability to command respect of Stalin and Politburo. Born 31 March 94 into peasant family in Vologda district, he became a mechanic at Komandantskii airfield, St Petersburg, in 16, working on IM bombers. Called into army and qualified as pilot 17. An enthusiastic revolutionary, he managed to join Red Army and among other contributions realised Avro 504 would fill need for good standard trainer and got captured example to GAZ-1 so that it could be copied (as U-1). In 22 he entered Zhukovskii Academy, and became leader of infant gliding movement, bringing designs to first six All-Union meetings at Koktebel. On graduation 26 appointed to NTK-UVVS, becoming chief. In 32 took civilian appointment with CAHI, being appointed Tupolev's deputy at KOSOS. On formation of CCB, appointed to run its brigade of method studies; after political vicissitudes given charge of long-range bomber brigade in CCB – where one of first projects was a fighter (I-21). In 35 he appointed Vladimir Kokkinaki brigade test pilot, and worked with him for 35 years. In late 37 he was appointed to top job in aviation industry, Director of GUAP; he judged that at his KB he could do more work and probably live longer, and he succeeded in returning there in Feb 38.

On 21 April 38 forced landing, commuting between CCB and GAZ-18 in his AIR-11, left his forehead scarred for life. Some 30 years later he started to relinquish control to his deputy, Genrikh V. Novozhilov, who took over entirely in 76 but retained founder's name. Awarded Order of Lenin and third Hero medal on 80th birthday, and died on 9 Feb 77.

CCB-16 Project for flying-wing bomber with two M-34, assigned to department of V. V. Nikitin; replaced by conventional low-wing development with two M-85. This led to CCB-26 and CCB-30, see later.

CCB-32, I-21 One of greater puzzles: almost every published description and drawing grossly inaccurate, aircraft often said to be I-1 or Il-1, engine fitted with geared drive to *increase* prop rpm from 1,800 to 2,400, dimensions invariably wildly incorrect. OKB drawing reproduced here, no mention in OKB of any other designation but CCB-32, though VVS list does show it as I-21. Single-seat fighter designed 35-36, stressed-skin, designed for speed, unusual features: engine specially built so that deep recesss in crankcase could accomodate front spar of wing; evaporative cooling, with wing surface incorporating stainless-steel condensers; cockpit behind TE. Special attention given to external finish, tailwheel retrac, split flaps, two ShVAK with 200 rounds beyond prop disc. Two prototypes and 20 series aircraft ordered, first flight 'end of 36' by Kokkinaki; programme abandoned early 37 after three flights on grounds of ineffective cooling, excessive field length and dangerously high wing loading.

DIMENSIONS Span 9,2 m (30 ft 2¼ in); length 7,0 m (22 ft 11⅝ in); wing area 13,8 m² (148·5 ft²).

ENGINE One AM-34RNF.

WEIGHTS Empty 1,4 t (3,086 lb); loaded 2125 kg (4,685 lb).

PERFORMANCE Speed reached on test 520 km/h (323 mph).

CCB-26 Precursor of most important Sov long-range bomber of Gt Patriotic War, response to 33 demand for DB (long-range bomber). Objective: equal range of Tu/Su aircraft (ANT-37/DB-2) whilst flying faster. CAHI profile wing with spar truss from 4 mm angles and U-sections in KhMA, ribs trusses from U or top-hat, horizontal c/s and dihedral outer panels joined at spar booms by single KhMA bolts, flush-riveted skin. Fuselage wood monocoque, narrow oval section, glued *shpon* with doped fabric 5 mm at front and 2,5 towards tail. Flight controls D16/fabric. Pneu aft-retracting twin-strut main gears and split flaps. Pilot open cockpit on centreline, struc provision for nav/bomb in nose with gun and radio/gunner aft of wing, external carriage of bombs to total 1 t.

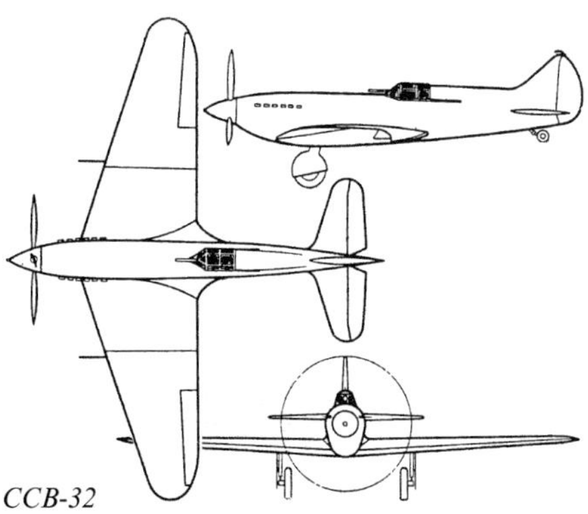

CCB-32

First flight V. K. Kokkinaki June 35, looped by him over Red Square May 36; in Aug 36 set USSR records 12816 m with 500 kg, 12101 with 1 t, 11005 with 2 t, also made 5018 km flight with 1 t in 16 h. Led to CCB-30.

DIMENSIONS Span 21,4 m (70 ft 2¼ in); length 13,7 m (44 ft 11⅜ in); wing area 65,5 m² (705 ft²).

ENGINES Two 765 hp Gnome-Rhône K14 Mistral Major.

WEIGHTS Empty 4,1 t (9,039 lb); fuel/oil 800 kg; loaded 6 t (13,228 lb).

PERFORMANCE Max speed 330 km/h at SL, 390 km/h (242 mph) at 3250 m; climb 15·1 min to 5 km; service ceiling 10 km (reached 13,1 km, 43,000 ft); range 4000 km (2,486 miles).

CCB-30 Prototype bomber based on previous but new metal fuselage with much larger nose and provision for bomb bay and armament. Longerons KhMA with rivets to 6 mm, dural skin mainly 0,6 mm with D1 rivets 2,6 or 3 mm, all external being flush. Completed without armament fitted May 36, painted red, named *Moskva*, flown with many extra tanks (12,6 t takeoff) by V. K. Kokkinaki and A. M. Bryandinskii 27/28 June 38 to Spassk-Dalnyi (nr Vladivostok) 7580 km in 24 h 36 min; by VKK and M. Kh. Gordiyenko 28/29 Ap 39 to Miscou, New Brunswick, various problems preventing New York but still c8000 km in 22 h 56 min. Competed against DB-2 and selected as DB-3 Mar 36; story that Stalin excited by CCB-26 loop, over-ruled decision in favour of DB-2 and ordered Ilyushin is without foundation. Photo shows 12-man assault cabin.

DIMENSIONS Span 21,44 m (70 ft 4⅛ in); length 14,22 m (46 ft 7¾ in); wing area 65,6 m² (706 ft²).

ENGINES Two M-85.

WEIGHTS Empty 4,2 t (9,259 lb); fuel/oil (normal) 800 kg; loaded 6,25 t (13,779 lb).

PERFORMANCE Max speed 335 km/h at SL, 409 km/h (254 mph) at 4 km; service ceiling 10 km; range 4000 km, (max fuel) c8000 kg; TO (max fuel) 2,6 km; ldg 105 km/h, 500 m.

DB-3 Selected for prodn 36; some documents give designation even of initial block as DB-3B. Pilot on centreline with canopy sliding back over fairing, 9 mm back armour; nav/bomb in nose with small windows all around and manual turret (ShKAS, 1,100) and with second ShKAS in ventral posn, sometimes a fourth ShKAS fixed in tail firing aft (300). Bay for ten FAB-100 and three external racks for FAB-500, max 2,5 t with reduced fuel. Six unprotected fuel cells, 1150 lit (253 gal), with only 1 t bombs provision for up to 2860 lit. M-85 engines in long-chord cowls without gills, twin pipes on underside, VISh-23 v-p props, provision for semi-retrac skis secured around wheels.

Deliveries from May 37 from three factories, though complex structure reduced output; total to end-39, GAZ-18 12/204/555, GAZ-39 33/165/279, GAZ-126 [from 38] 30/125, total of **DB-3B** versions thus 1,403 plus further 125 before switch to DB-3F in Dec 39. In Mar 38 **CCB-54** tested revised armament with power turret, better ventral posn and improved sights. In 39 one a/c tested A. I. Privalov D-20

CCB-26

CCB-30 with DK 12

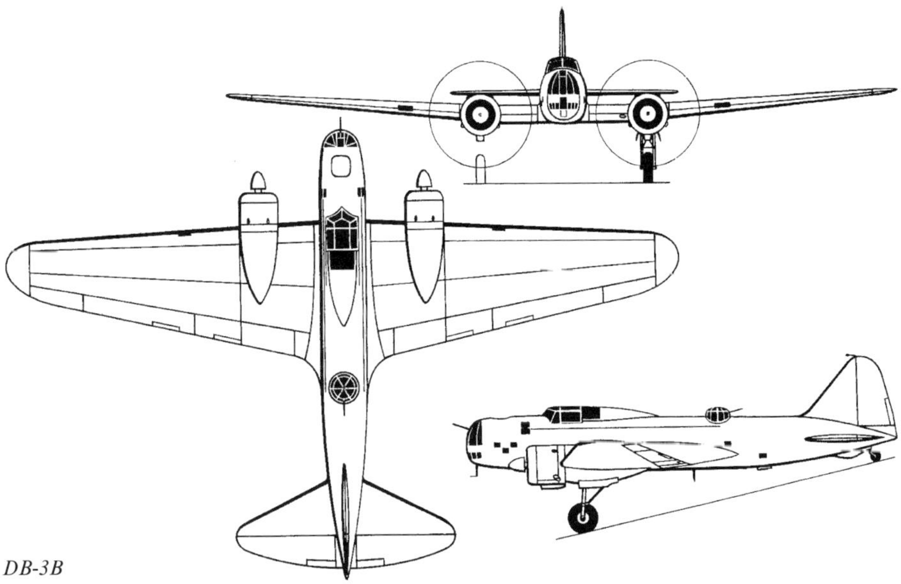

DB-3B

releasable cabin for 10 paratroops (1580 kg). In late 38 engine changed to M-86 and following year to M-87A. Two blocks built 37-38 of **DB-3T** equipped to carry single 450 mm 940 kg torpedo externally, either 45-36 AN for low drop or 45-36 AV for high, with external compressed-air tank behind for running up engine before release; 81st Mine/Torp Regt saw much action summer 41. Single example of **DB-3TP** torpedo-seaplane, as T but two Zh-type floats. Single **DB-3GK** unarmed hi-alt recon aircraft with redesigned pressurized forward cabin and extended wingtips. Single **DB-3UPS** with redesigned wings with slotted flaps/ailerons with boundary-layer suction from 116 hp ZIS-101A engine/blower in bomb bay.

DIMENSIONS As before except TP length 15,083 m (49 ft 5¾ in).

ENGINES See text.

WEIGHTS Empty (36) 4778 kg, (T) 4298 kg, (TP) 5630 kg, (38) 4712 kg, (39) 5030 kg; fuel/oil (36, T) 810 kg, (TP) 502 kg, (38, 39) 810+120 kg; loaded (36) 7 t (15,430 lb), (T) 6,494 t, (TP) 7,550 t, (38) 7,079 t, (39) 7,445 t (16,413 lb).

PERFORMANCE Max speed (36) 327 km/h at SL, 400 at 4,5 km, (T) 320, 395/4, (TP) 292, 343/4, (38) 331, 395/4, (39) 345, 439 (273 mph)/4,9; climb (36) 15·1 min to 5 km, (T) 2·6 min to 1 km, 13 min to 5 km, (TP) 3 min to 1 km, 18·2 min to 5 km, (38) 2·4 min to 1 km, 34·4 min to 7 km, (39)2·4 min to 1 km, 12·1 min to 5 km; service ceiling (36) 8,4 km, (T) 7,8, (TP) 7570, (38) 8,3, (39) 9,6 (31,500 ft); range (36) 4000 km, (T) 1800 km, (TP) 1400 km, (38) 4000 km, (39) 3800 km (2,361 miles); TO (36, T) 200 m, (TP) 600 m, (38, 39) 345 m; ldg (36) 120 km/h, 300 m, (T) 110 km/h, 300 m, (TP) 135 km/h, 600 m, (38) 115 km/h, 390 m, (39) 120 km/h, 450 m.

DB-3M Major structural redesign in 38-39 resulted in DB-3M (modernized), needing about half as many man-hours to produce airframe. Replaced 3B in prodn at all three plants late 39, with M-87A or 87B engine. Data as (39) above.

DB-3F Major series aircraft, distinguished by redesigning forward fuselage with lower drag, more room, full glazing and hand-aimed gun at tip. Further modernization of structure, almost entirely standard dural stock: sheet spar webs, rolled-angle booms, rubber-press sheet ribs,

DB-3M

DB-3GK

Il-4

simpler fuselage, but exterior hardly changed. In prodn Jan 40 with M-87B engines in cowls with gills, M-88 engine in late 40 and M-88B in 41, when four LE tanks added for total 2860 lit. No significant change in armament.

DIMENSIONS As before except length 14,76 m (48 ft 5⅛ in).

ENGINES See text.

WEIGHTS Empty (39) 5373 kg, (40) 5641 kg, (41) 6320 kg; fuel/oil (39) 750+125 kg, (40) 870+125 kg, (41) 1410 kg or 2060+140 kg; loaded (39) 7638 kg (16,839 lb), (40) 8033 kg (17,709 lb), (41) 9470 kg (20,877 lb).

PERFORMANCE Max speed (39) 354 km/h at SL, 445 km/h (277 mph) at 5,4 km, (40) 350 km/h, 435/6,8, (41) 345 km/h, 422/6,8; climb to 5 km (39) 13·6 min, (40) 10·5 min, (41) 14·6 min; service ceiling (39) 9,0 km, (40) 10,0 km, (41) 8,9 km; range (39) 3500 km, (40) 3300 km, (41) 3800 km (2,361 miles); TO (39) 390 m, (40) 400 m, (41) 480 m; ldg (39) 115 km/h, 450 m, (40) 120 km/h, 500 m, (41) 130 km/h, 500 m.

Il-4 Designation of DB-3F from Mar 42 until termination early 45, total (excl DB-3B)

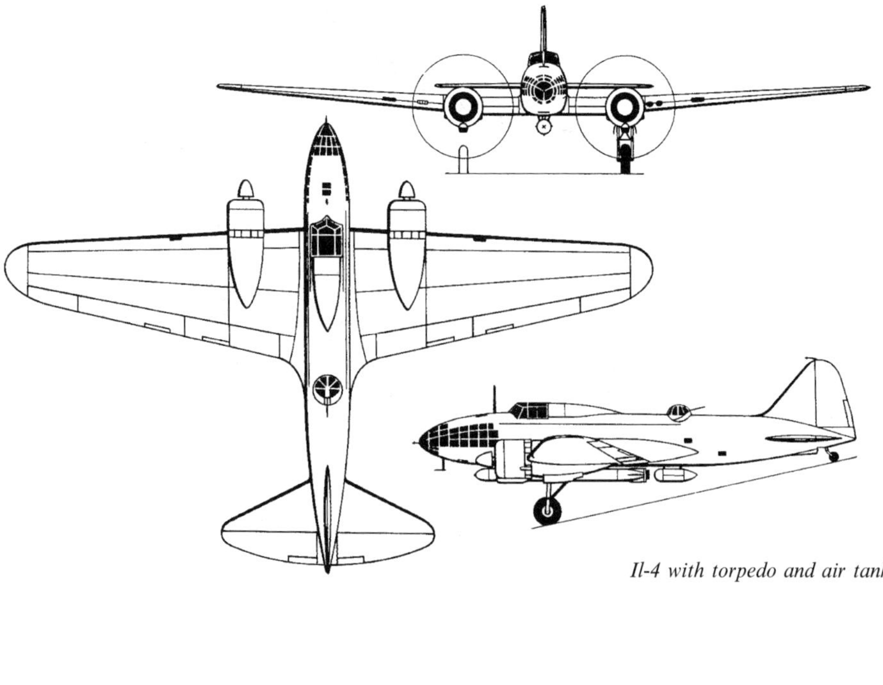
Il-4 with torpedo and air tank

believed 5,256. Progressive changes included defensive armament, initially three ShKAS, later a BS (200 rds) or ShVAK (120) in nose and a UBT (500) or ShVAK (240) in Ilyushin power turret; usual fuel capacity 3855 lit, all in protected cells, instead of 2860 unprotected; all crew 6-9 mm armour; props changed 41 to feathering UF-61-IF and 42 to AV-5FZ-158A; provision for up to 2,7 t bombs or mines or AN or AV torpedoes; most fitted with cleat for towing A-7 or G-11 glider and with mounts for AFA-33 or similar recon camera in bomb bay. In 42 need to conserve alloy resulted in redesign by OKB with wooden nose, fuselage deck and outer wings, incorporated in various ways on production until late 43. After 45 many converted for cargo or mapping/geophys survey. ASCC name 'Bob'.

DIMENSIONS As before 21,44/14,76/66,7.

ENGINES Two M-88B.

WEIGHTS Empty 6421 kg (14,156 lb); fuel/oil 1885 kg or 2750+190 kg; loaded 10055 kg (22,167 lb).

PERFORMANCE Max speed 332 km/h at SL, 404 km/h (251 mph) at 6,65 km; climb to 5 km 19·0 min; service ceiling 8,3 km, (27,230 ft); range (2860 lit) 3585 km (2,228 miles); TO 530 m; ldg 130 km/h, 575 m.

DB-4 model

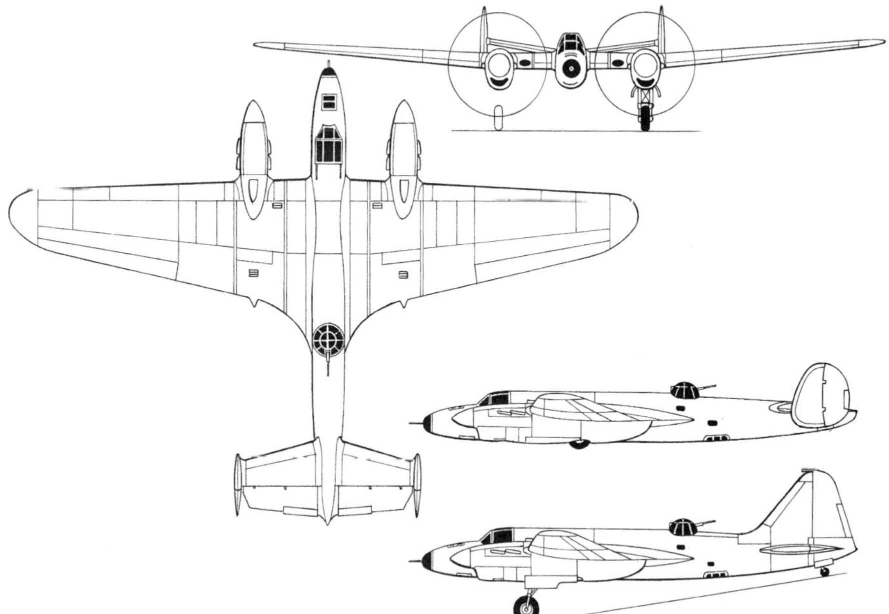

DB-4 (lower side view, second aircraft)

DB-4 (CCB-56) In parallel with DB-3F Ilyushin's brigade planned a superior long-range bomber which, while using a slightly modified form of same wing, would use more powerful engines and have larger bomb bay. Fully retractable tailwheel with doors. Twin-finned tail, four-section split flaps. Crew four; in war either pilot or nav could fly aircraft, other two being turret gunner and radio/ventral gunner. ShKAS or ShVAK in nose, twin ShKAS or one ShVAK in dorsal turret, ShKAS in DB-3F type lower rear position. Deep bomb bay with two doors, normal capacity 1600 kg, maximum 2000 kg. Designed for M-120 engines, but delays with these resulted in temporary use of AM-37, in long nacelles with glycol radiators under engines and oil coolers in centre section inboard of engines. Factory tests started by Kokkinaki Oct 40, and following NII trials given designation DB-4 early March 41. Second prototype built with single fin and planned for M-82 air-cooled radials, but this too had to be powered by AM-37. Early testing encouraging, but Ilyushin KB had to concentrate on Il-2 and Il-4, and production of even interim Mikulin V-12 engines would not suffice for production programme as well as meet needs of MiG-3 and Il-2. Test flying continued only to improve technology of engine installations (for example, for Il-6) and assist Il-4 programmed modifications.

DIMENSIONS Span 25,0 m (82 ft 0¼ in); length 17,85 m (58 ft 6¾ in); wing area 83,0 m^2 (893 ft^2).

ENGINES Two AM-37.

WEIGHTS Empty 7561 kg (16,669 lb); fuel/oil 1250+150 kg; loaded 10,86 t (23,942 lb).

PERFORMANCE Max speed at SL 415 km/h (258 mph), at 6 km 500 km/h (311 mph); service ceiling 10 km (32,800 ft); range 4000 km (2,486 miles); TO 350 m, ldg 360 m.

Il-6 Possibly first aircraft to have borne a designation for Ilyushin from outset, this was second major attempt to develop a successor to Il-4. Work began Feb 42, objective being to use as many Il-4 parts as possible whilst greatly increasing range with 2 t bombload. Early in project decision taken to use completely new wing; prior to start ACh-30B diesel had been selected because of promise of significantly lower fuel consumption and demonstrated efficiency (albeit with poor reliability) in Yer-2. During 1942 design inevitably moved away from Il-4 until final drawings showed noticeably larger and heavier aircraft. Stressed-skin, flush-riveted externally, with fabric-skinned controls. Split flaps divided by large nacelles into four sections; operation of flaps and large main landing gear derived from that of Il-4 but with hydraulic power. Fuselage similar in layout to Il-4 but only rear portion actually same. Cross-section slightly increased to ease access between nose and rear, because of six crew two had to move aft to man guns in combat. Design armament five ShVAK: nose, dorsal turret, two beam windows and rear aft position (different from Il-4 with cannon faired largely external and with restricted cone of fire). Starting with Il-4 as basis kept wing in low position with restricted bomb bay, internal limit being six 500 kg or one 2 t bomb; overload bomb load 3 t (6,614 lb), mainly external. Two torpedoes alternative. Wing totally new with excellent structure (DC-3 influence with multiple spars), aspect ratio 7·98, thickness/chord 16% at root and 12 at tip. Effort made to incorporate thermal de-icing for wings and tail, and to minimize pilot effort with accurate aileron and rudder trimmers. First Il-6 flown by Kokkinaki 7 Aug 43 and put through NII trials by A. Grinchik later 1943. Altogether four examples flown, but programme terminated early 44.

DIMENSIONS Span 26,07 m (85 ft 6⅜ in); length 17,38 m (57 ft 0¼ in); wing area 84,8 m^2 (913 ft^2).

ENGINES Two ACh-30B, later 30BF.

WEIGHTS Empty (No 1) 11,69 t, (No 2) 11,93 t (26,301 lb); fuel 2 t; loaded (1) 15,6 t, (2) 16,1 t (35,494 lb).

PERFORMANCE Max speed at SL (1) 382 km/h, (2) 400 km/h (249 mph), at 6,6 km (1) 445 km/h, (2) 464 (288 mph); climb (1) 15·7 min to 5 km; service ceiling (1) 8 km, (2) 7 km (23,000 ft); range 5450 km (3,387 miles); TO/ldg 600 m.

BSh-2 Also designated **CCB-55** and **DBSh**,

Il-6

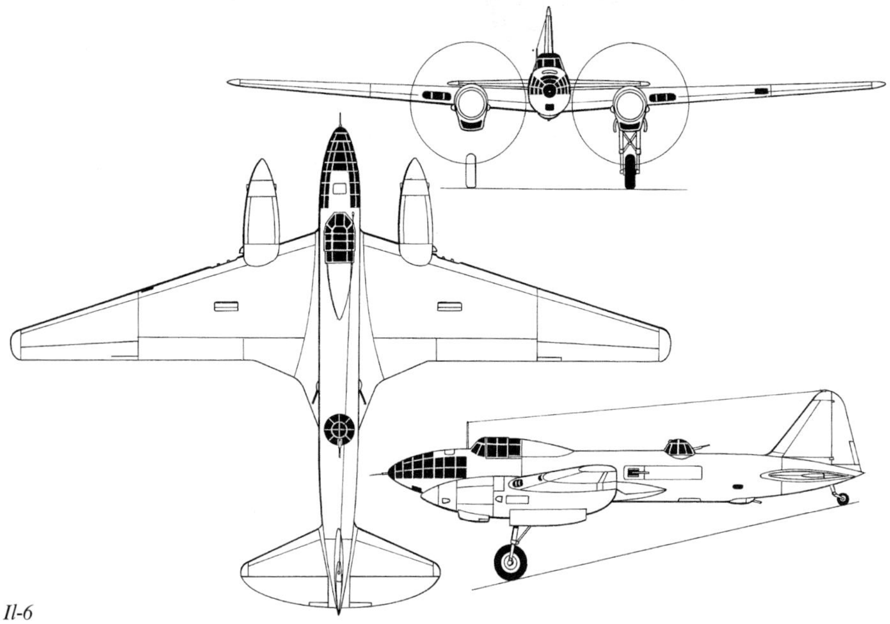

Il-6

BSh-2 (*Bronirovannii Shturmovik*, armoured assaulter) was a single-engined armoured attack aircraft whose design was begun by Ilyushin's KB in 38. Ilyushin took it upon himself to write appreciation of need to produce modern armoured attack aircraft, less vulnerable than existing biplanes and twin-engined bombers, and sent copies to Stalin and various officials on 27 Jan 38. At this time he was still fighting to get out of his official post as Director of GUAP, which he contrived to do a month later. This made Stalin angry (he had torn up an earlier request), and Ilyushin was in a most insecure position when he returned to his KB in Feb 38. Less than two months later he had Politburo go-ahead on BSh-2, and had agreed to fly a prototype by Nov 38. Within days he was seriously hurt in his crash of AIR-11, and programme foundered. He tried to direct his staff from hospital, and returned to GAZ-39 earlier than he should. Basic design was conventional low-wing machine with single engine and tandem seats for pilot and radio/gunner (hence designation DBSh, *Dvukhmestnii BSh*, two-seat armoured assaulter). Armament forward-firing guns in wing, bombs in cells in inner wing, defensive gun in rear cockpit. Landing gears retracting backwards into fairings under wing. Choice of engine a severe compromise: final selection was AM-35, which offered just about adequate power but was large and heavy, and rated for best power at heights far above BSh levels. Main concern was armour, and like contemporary Su and MiG rivals basic philsophy was to fight crippling payload problems by making armour integral with airframe and bear flight loads. Unfortunately this pious hope difficult to achieve because main stresses elsewhere, and heaviest armour not called upon to carry severe loads except underside of engine cowl. From start severe problems with making accurate armour according to drawings, so that bolt and rivet holes matched up. This (far from new) problem resulted in decision to build first prototype as non-flying engineering mock-up. AB-1 (see note on armour in section on materials) used on engine cowl, 4 mm around front and 5 mm on lower sides and underside; also used 7 mm thick for rear bulkhead behind second crew-member. Carefully shaped before case-hardening and then held by 5 mm or 6 mm steel rivets to underlying dural airframe. Top of engine cowl, 5 mm dural sheet. Rough estimate of extra weight of protection, 700 kg. Rest of aircraft mixed construction. Rear fuselage and fin, wood monocoque, mostly from birch *shpon* (laminates 0,8 mm by 100 mm, glued into curved panels 5 mm thick). Wings and tail light alloy, flight controls being fabric-covered. Second BSh-2 made first flight of type on 30 Dec 39. Kokkinaki pleased, but performance marginal. Armament four ShKAS in wings (later, about May 40, increased to two ShVAK with 210 rounds each and two ShKAS with 750 each) plus one ShKAS at rear, with bombload of 400 kg and possible overload of 200 kg carried externally under fuselage. KB tests completed 26 March 40 and NII testing 1/19 April. After years of failure there seemed here to be germ of successful machine, despite unimpressive takeoff, climb, speed, range and longitudinal stability. Ilyushin sensed VVS would ask for second crew-member to be removed to restore performance nearer specification, and began CCB-57 as single-seater. Heated correspondence ensued, finally settled by Politburo in Nov 40 accepting VVS position and noting VVS belief that single-seat BSh-2 would be protected by friendly fighters.

DIMENSIONS Span 14,6 m (47 ft $10\frac{3}{4}$ in); length 11,6 m (38 ft $0\frac{3}{4}$ in); wing area 38,5 m^2 (414 ft^2).

ENGINE One AM-35.

WEIGHTS Empty 3615 kg (7,970 lb); fuel/oil 315+30 kg; loaded 4725 kg (10,417 lb).

PERFORMANCE Max speed 362 km/h at SL, 422 km/h (262 mph) at 5 km; climb 2·3 min to 1 km, 6·3 min to 3 km, 11·5 min to 5 km; service ceiling c9 km; range 618 km (384 miles); TO 340 m, ldg 140 km/h, 270 m.

CCB-57 Single-seat prototype with extra tank replacing rear cockpit (wooden fairing retained) and new engine. Few changes except to make lateral control lighter. First flight by V. K. Kokkinaki 12 Oct 40.

DIMENSIONS As before.

ENGINE One AM-38.

WEIGHTS Empty 3792 kg (8,360 lb); fuel/oil 500+40 kg; loaded 4,988 km (10,996 lb).

PERFORMANCE Max speed 423 km/h at SL, 437 km/h (272 mph) at 2,8 km; climb 1·7 min to 1 km, 5·3 min to 3 km, 10·0 min to 5 km; service ceiling 8,5 km (27,890 ft); range 850 km (528 miles); TO 250 m, ldg 140 km/h, 260 m.

CCB-55P, Il-2 Refined single-seater, bulkhead behind pilot increased to 12 mm, fairing behind cockpit removed, fuel in four cells slightly reduced to 653 lit, production engine 175 mm shorter, driving 3,4 m VISh-23 prop with Hucks dogs, longit stab improved by 'sweeping'

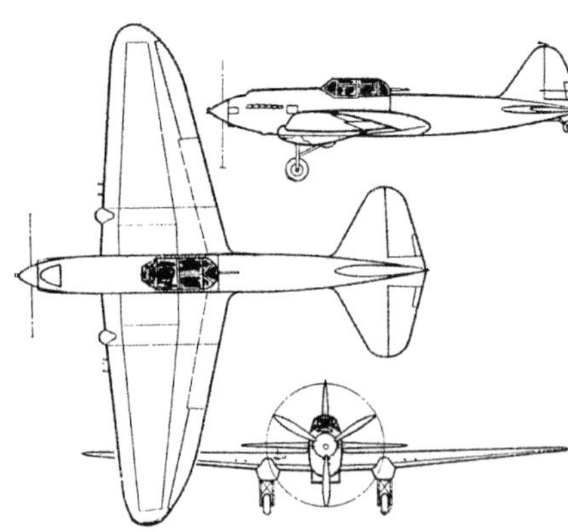

CCB-55

CCB-55

CCB-57

outer wings 5° (CG at 30·2%) and adding 3·1% to tailplane area. Centre section two spars 30KhGSA booms and D16 plate webs, inter-spar box forming two bomb cells each wing with twin doors pulled open by pilot against springs, one FAB-100 each. Six bolts joined all-dural outer wing. Split flaps on c/s and outer wing. Main gears twin shock struts, main tyres 888×260 (2,9 at, 42·6 lb/in²) for soft fields, retrac to rear by same 35-50 at (515-735 lb/in²) pneu system as flaps, wheel protruding for belly landing; a few locally mod for skis. Two ShVAK (420 rds) and two ShKAS (1,500), rails for six (later eight) RS-82.

Flown by VKK 29 Dec 40, NII tests 28 Feb to 20 Mar 41. Numerous small mods, enlarged main/oil radiators with totally revised flow from inlet above cowl to ventral exit, raised seat, rear fairing transparent, 65 mm windscreen and 13 mm back armour, skin removed round tailwheel for better access. Designation changed April 41

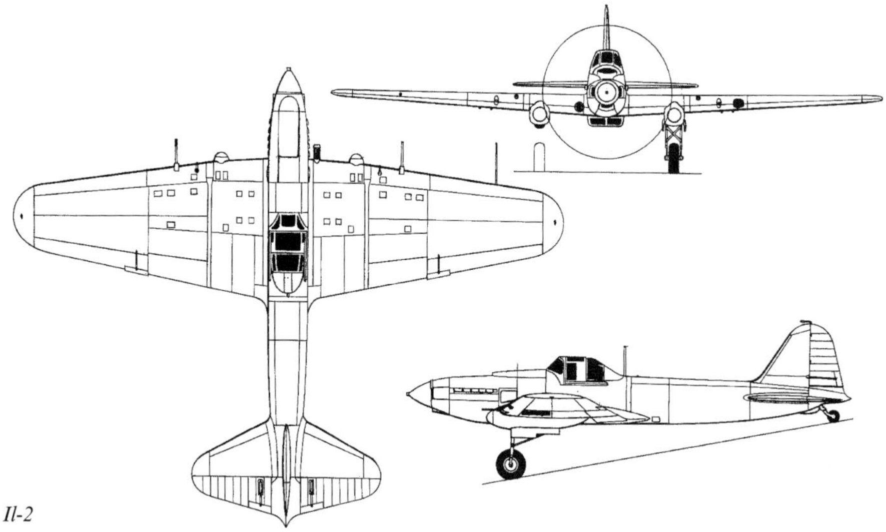

Il-2

to Il-2, by which time already in prodn at GAZ-18, Ilyushin awarded Stalin Prize 2nd class. Stalin still angry, and when prodn failed to exceed one per day by Dec 41 sent famous 'last warning' telegram to designer and GAZ-18 manager. First deliveries 4th Light Bomber Polk, many crashes by pilots converting from R-Zet biplane. Attrition from fighters from 22 June 41 severe, rear defence urgently needed, but effectiveness never doubted.
DIMENSIONS, ENGINE As before.
WEIGHTS Empty 3990 kg (8,796 lb); fuel/oil 470+35 kg; loaded 5310 kg (11,706 lb).
PERFORMANCE Max speed 433 km/h (269 mph) at SL, 450 (280) at 2460 m; climb 1·6 min to 1 km, 9·2 min to 5 km; service ceiling 7,8 km (25,590 ft); range 638 km (396 miles); TO 450 m, ldg 140 km/h, 400 m.
Il-2 Apart from need for rear gunner shortcomings included failure of glued joints and failures of rear fuselage, latter cured by repair engineer A. K. Belyenkov adding four steel angle sections which became standard on production. Many built with wooden outer wings, some with wood rear fuselage and fixed tail surfs, but armour and fuel capacity increased. Firepower against armour greatly increased from mid-41 by replacing ShVAKs by two VYa-23, and introduction of RS-132 rocket. Data for typical 1942 prodn.
DIMENSIONS, ENGINE As before.
WEIGHTS Empty 4261 kg (9,394 lb); fuel/oil 535+65 kg; loaded 5788 kg (12,760 lb).
PERFORMANCE Max speed 396 km/h (246 mph) at SL, 426 km/h at 2,5 km; climb 2·2 min to 1 km, 7·4 min to 3 km, 14·7 min to 5 km; service ceiling 6,2 km (20,340 ft); range 740 km; TO 420 m, ldg 140 km/h, 400 m.
Il-2M Many a/c converted by field units with backseater, but by May 42 permission given to introduce to prodn despite temp loss in output. Gunner isolated behind tank, at least five variations in canopy, but rear gun invariably UBT. Academician Ye. O. Paton solved armour welding problem and typical armour now 990 kg (2,183 lb) in 4/5/6/8/12 mm thickness.
DIMENSIONS, ENGINE As before.
WEIGHTS Empty 4525 kg (9,976 lb); fuel/oil 535+65 kg; loaded 6060 kg (13,360 lb).
PERFORMANCE Max speed 370 km/h (230 mph) at SL, 411 km/h at 1 km; climb 2·4 min to 1 km, 7·8 min to 3 km, 17·8 min to 5 km; service ceiling 6,0 km; range 685 km (426 miles); TO 400 m, ldg 145 km/h, 500 m.
Il-2/AM-38F More powerful engine introduced July 42, designation often **Il-2 Tip 3**. Option of two-tier installation for 32 RS-82. Detail drag reduction followed by decision to fit new *strelkoi* (swept) outer wings with 15° taper on LE, NII test 12 Dec 42, in action Stalingrad Jan 43.
DIMENSIONS As before.
ENGINE One AM-38F.
WEIGHTS Empty 4360 kg (9,612 lb); fuel/oil 535+50 kg; loaded 6160 kg (13,580 lb).
PERFORMANCE Max speed 403 km/h (250 mph) at SL, 414 km/h at 1 km; climb 20 min to 5 km; service ceiling 5500 m (18,050 ft); range 685 km; TO 380 m, ldg 145 km/h, 500 m.
Il-2M3 Alternative designation for final series models. From 43 a proportion had two NS-OKB-16 guns of 37 mm calibire (50 rds), able to penetrate PzKW VI Tiger, as well as two ShVAK and one UBT (no rockets/bombs). In final year (44) structure restored to all-dural. Total deliveries by Nov 44 36,163, greatest for single type of a/c. ASCC name 'Bark'.
DIMENSIONS, ENGINE As before.
WEIGHTS Empty 4525 kg (NS-37, 4625 kg); fuel/oil 535+50 kg or 65 kg; loaded 6360 kg (14,021 lb) (NS-37 6160 kg).
PERFORMANCE Max speed 390 km/h (242 mph) at SL, 405-410 km/h at 1,2 km; climb 2·2 min to 1 km, 6·9 min to 3 km, 15·0 min to 5 km; service ceiling 6 km (19,685 ft); range 765 km (475 m); TO 395 m, ldg 145 km/h, 535 m.
Il-2U Also called **UIl-2** dual trainer with reduced armour/fuel and just two ShKAS, two RS-82, two FAB-100. **Il-2T** tested with 450 mm torpedo. **Il-2/M-82** tested Sept 41 with

Il-2

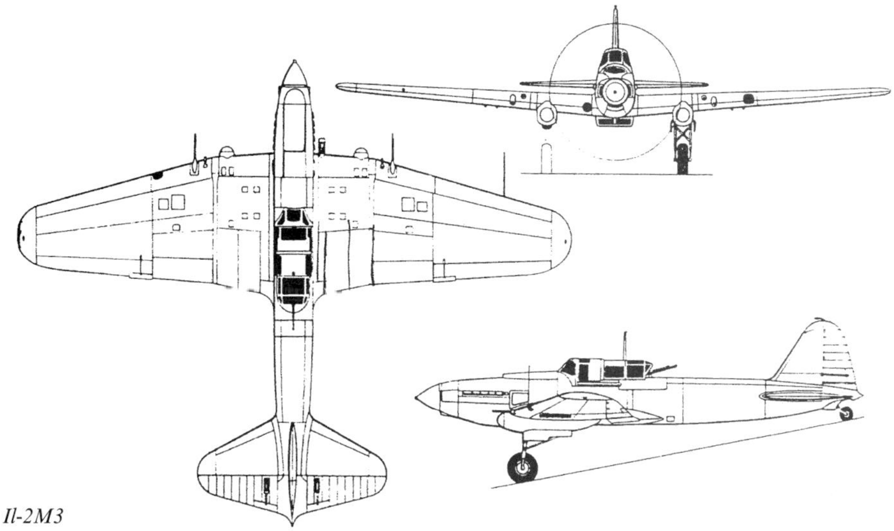

Il-2M3

DIMENSIONS As Il-2M.
ENGINE One AM-38F.
WEIGHTS Empty 4397 kg (9,694 lb); loaded 5383 kg (11,867 lb).
PERFORMANCE Max speed 401 km/h (249 mph) at SL, 415 km/h (258 mph) at 1,3 km; service ceiling 6,5 km (21,325 ft); climb 2 min to 1 km; TO 275 m.

Il-1 Like previous supposed Il-1 (CCB-32) this single-seat fighter is the subject of confusion, especially regarding the plan shape of the wing; based loosely on Il-2 but redesigned airframe, more powerful engine, radiator in wing roots with oil coolers immediately outboard, single-place cockpit with metal canopy in two parts each with two small windows, considerable armour, landing gear as Il-8, two VYa-23 each with 150 rounds outside disc of 3,6 m AV-5L-24 prop, 200 kg bombload or dispenser of 10 AG-2 bomblets. First flight by Kokkinaki 19 May 44; reasonable at low levels but hopeless

Il-2M3

radial engine in poor installn. **Il-2I** described next.
DIMENSIONS As before except M-82 length 11,653 m (38 ft 2¾ in).
WEIGHTS Empty (U) 4300 kg, (M-82) 3935 kg; fuel/oil (U) 470+35 kg, (M) 550+50 kg; loaded (U) 5091 kg, (M) 5655 kg.
PERFORMANCE Max speed (U) 396/414, (M) 365/396; climb (U) 2/6·3/-, (M) 1·8/10·6; service ceiling (U) 7 km, (M) 7,5 km; range (U) -, (M) 700 km; TO (U) 385 m, (M) 275 m; ldg 600 m, 515 m.
Il-2I Uninspired single-seat fighter using essentially same airframe as Il-2, designed July 43, in some documents called **IB** (fighter/bomber). Two VYa-23 each with 150 rds, two bombs of 250 kg. Much lighter than Il-2 but too slow to catch enemy a/c.

Il-2/M-82

Il-1

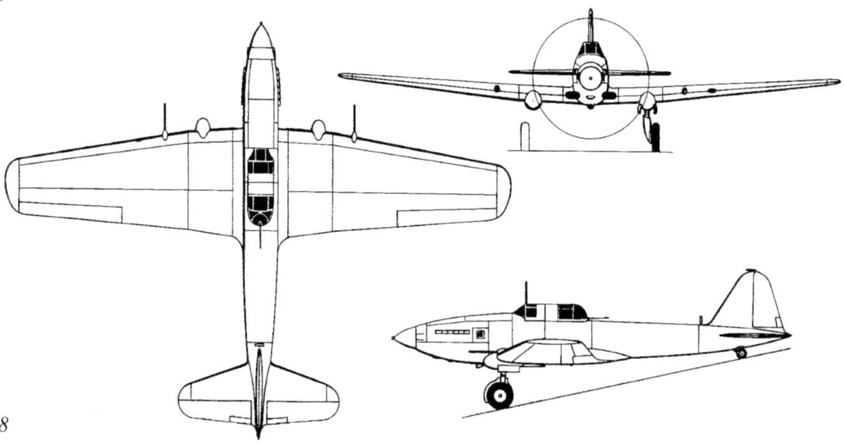

Il-8

Il-8

at height. One a/c only.

DIMENSIONS Span 13,4 m (43 ft 11⅝ in); length 11,12 m (36 ft 5¾ in); wing area 30,0 m^2 (323 ft^2).

ENGINE One AM-42.

WEIGHTS Empty 4285 kg (9,447 lb); loaded 5320 kg (11,728 lb).

PERFORMANCE Max speed 525 km/h (326 mph) at SL, 580 km/h (360 mph) at 3260 m; service ceiling 8,6 km (28,220 ft); climb 1·6 min to 1 km.

Il-8 Kremlin conference of Jan 42 launched several Il-2 offshoots, including Il-1, Il-8, Il-10 and Il-16, all of them major redesigns. Il-8 was least modified, chief alteration being more powerful engine with cleaned-up installation identical to that of Il-1, four-blade prop with anticing, oil cooler radiators in wing roots, improved belly radiator, and new landing gear with single air/oil shock strut retracting rearwards carrying large low-pressure tyred wheel turning 90° to lie inside wing. Armour similar to late Il-2, but extended rearwards to enclose gunner. Outer-wings and tail of duralumin. Two-seat cockpit further forward than in Il-2, improved aerodynamic form, and radio mast on windscreen frame. Armament two VYa-23 and two ShKAS in wings and one UBT aimed by backseater; bombload up to 1 t. Flown by Kokkinaki April 44 and subjected to NII trials, but Il-10 a superior aircraft.

DIMENSIONS Span 14,6 m (47 ft 10¾ in); length 12,93 m (42 ft 5 in); wing area 39,0 m^2 (420 ft^2).

ENGINE One AM-42.

WEIGHTS Empty 5245 kg (11,563 lb); fuel/oil 740+85 kg; loaded 7250 kg (15,983 lb).

PERFORMANCE Max speed at SL 435 km/h (270 mph), at 2,24 km 470 km/h (292 mph); climb 6·2 min to 3 km; service ceiling 6,8 km (22,310 ft); range 1180 km (733 miles); TO 318 m; ldg speed 132 km/h.

Il-10 This second-generation *Shturmovik* used airframe of Il-1 and thus had little in common with Il-2 except basic layout. AM-42 engine with three-blade AV-5L-24 prop with Hucks dogs. Light-alloy construction throughout, except for armour which was identical with Il-8. Modern pressed-sheet and rolled sections in centre section, with improved weapon bays and rearwards-retracting landing gear with pneumatic actuation as in Il-8, with single shock struts. Low-pressure tyres 900 × 300 mm. Flaps extended beneath fuselage. Stressed-skin outer wings of reduced span and area, and with spars swept slightly back to move centre of pressure aft. One-piece Frise aileron. Improved tail, light-alloy structure with fabric-covered movable surfaces, rudder having horn (as on Il-1) instead of external mass balance, slightly greater height of vertical tail, horizontal tail of reduced chord but much greater span and area. Retractable tailwheel. Same crew compartment as Il-8. Original armament two VYa and two ShKAS on first test flights in April 44, changed during KB testing to two NR-23 and two ShKAS, with further changes to two NS-OKB-16 (37 mm) and two ShKAS or four NR-23. Large access doors for speedy replenishment of magazines, unlike Il-2. Normal bomb load 400 to 600 kg, with up to 100 kg size in each centre-section cell, plus eight RS-82. Excellent gunner's cupola with B-20EN cannon with 150 rounds and improved protection, headroom and visibility. Factory tests followed by NII trials by A. K. Dolgov from 9 June 44. Outstanding aircraft in all respects, superior to Il-8 or Su-8, and ordered into production in place of Il-2 on 23 Aug. First Il-10 with troops in October. Enthusiastic reception, and from start availability and effectiveness much higher than Il-2. Unlike Il-2 also produced as dual trainer variant **Il-10U** or **UIl-10** from outset, usually with no armament except two ShKAS and almost same canopy as Il-2U. Production at Rostov ended 1949, tooling to Czech Avia works 1950 where 1,200 built as B-33 and BS-33 trainer. ASCC reporting name 'Beast'.

DIMENSIONS Span 13,4 m (43 ft 11½ in); length 11,12 m (36 ft 5¾ in); wing area 30,0 m^2 (323 ft^2).

ENGINE One AM-42.

WEIGHTS Empty 4650 kg (10,251 lb); fuel/oil 535+65 kg; loaded 6300 kg (13,889 lb).

PERFORMANCE Max speed at SL 507 km/h (315 mph), at 2,3 km 551 km/h (342 mph);

B-33 (Il-10)

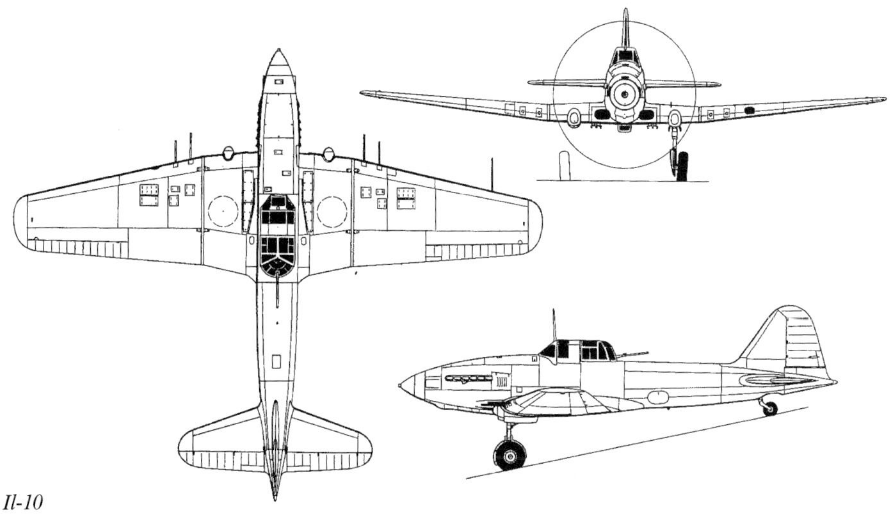

Il-10

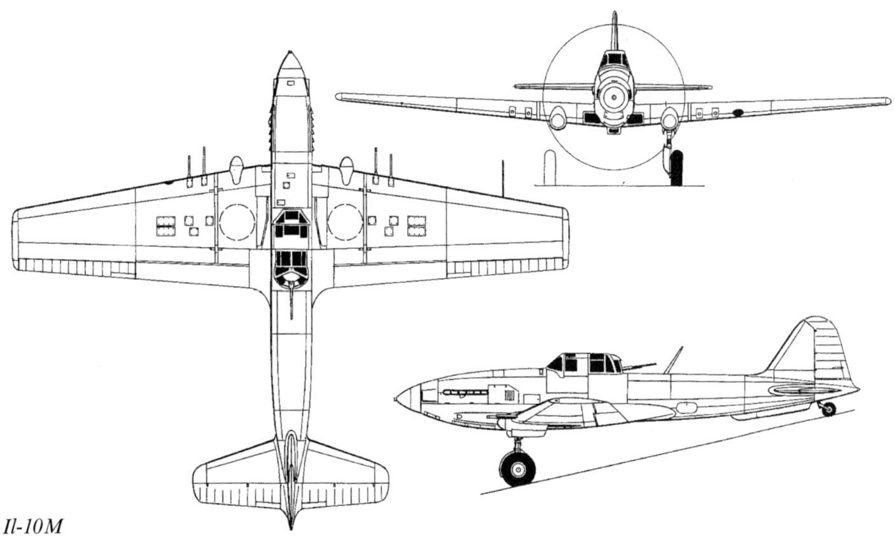

Il-10M

climb 5·0 min to 3 km; service ceiling 7250 m (23,790 ft); range 800 km (497 miles); TO 475 m, ldg 460 m.

Il-10M Major redesign, esp completely different wing. Single structure from tip to tip, with structural joint between centre section and outer panels but no change in taper or dihedral and with one-piece flaps from aileron to centreline. Clark-YH profile, thickness 18% at root and 12% at broad square tip. Span and area slightly greater than Il-10. Ailerons and horizontal tail redesigned, aerodynamics of wing and horizontal tail being scale (analog) of Il-20, then already abandoned. Slightly longer fuselage and improved vertical tail. Normal armament four NR-23 (600 rds) and remote-control turret with B-20EN (150 rds). At least one aircraft with RD-1X3 rocket under rudder plus ventral strake under rear fuselage; this aircraft preserved at Monino. In production 1951-55, one role being to train Il-28 gunners. Total (10 and 10M) 4,966. ASCC reporting name again 'Beast'.

DIMENSIONS Span 14,0 m (45 ft 11¼ in); length 11,87 m (38 ft 11⅜ in); wing 33,0 m² (355 ft²),

ENGINE One AM-42.

WEIGHTS Empty 5570 kg (12,280 lb); fuel/oil 640+65 kg; loaded 7100 kg (15,653 lb).

PERFORMANCE Max speed at SL 476 km/h (296 mph), at 2650 m 512 km/h (318 mph); climb 6·4 min to 3 km; service ceiling 7 km (23,000 ft); range 1070 km (665 miles); TO 440 m, ldg 500 m.

Il-16 Last of direct descendants of Il-2, this two-seat *Shturmovik* was designed to use AM-43 engine, with four-blade prop slightly larger than that of Il-1 and Il-8. Considerably smaller than relatives, so higher performance, but various problems (prop torque, structural weakness) combined with unavailability of engine to halt factory test in Aug 45.

DIMENSIONS Span 12,5 m (41 ft 0 in); length 10,69 m (35 ft 0⅝ in); wing area 24,0 m² (258 ft²).

ENGINE One AM-43NV.

WEIGHTS Empty 4315 kg (9,513 lb); fuel/oil

500+65 kg; loaded 5780 kg (12,743 lb).
PERFORMANCE Max speed at SL 529 km/h (329 mph), at 2,7 km 576 km/h (358 mph); range 800 km (497 miles); TO 400 m, ldg speed 175 km/h (109 mph).

MSh Project of 42 for single-seat *Shturmovik* with engine amidships giving superb forward view. Layout as diagram, but never built.

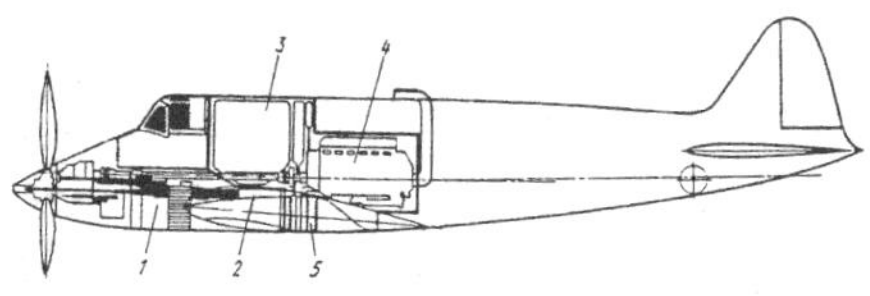

MSh (1, gun; 2, prop shaft; 3, tank; 4, AM-38F; 5, oil cooler)

Il-12 In 43 Kremlin decreed to build modern passenger aircraft to replace US-derived Li-2. Ilyushin received instruction together with Bartini (T-117) and Sukhoi (Yermolayev 2ON). By end 1943 Ilyushin had settled on bluff (hemispherical) nosed pressurized fuselage 2800 mm diam, low wing with Li-2 features (eg Clark YH profile, three spars) and four M-88B engines. This was soon changed to unpressurized fuselage and two ACh-31 diesel engines with coolant radiators in extended inboard LE and nacelles extending back above and below wing behind TE, with large fillets at root and each side of nacelles. Latter restricted size of slotted flaps added on outer wing where profile changed to K-4. Dihedral 2° from root. Simple flight controls, D4 structure with fabric covering, tabs on all except left aileron. Bold choice of tricycle landing gear, main wheels outboard of single legs projecting slightly under rear of nacelles. Castoring levered-suspension nose gear with tyre 770×330, leg leaning back when extended. Flaps and landing gear hydraulic, three-blade 4,4 m propellers, four centre-section tanks (total 4170 lit), LE deiced by exhaust gas cooled by ram air, tail by Goodrich-copy rubber boots, AP-42A autopilot, external riveting mainly flush. First flight by V. K. Kokkinaki 15 Aug 45, but by this time decision taken to change engine. New radial engine driving 4,1 m four-blade AV-9E-91 (Hamilton) autofeathering propeller. Main gears redesigned with two smaller (900×300 tyre) wheels on single leg, retracting forwards, this making possible smaller nacelle which in turn enabled much larger flaps to extend from aileron to fuselage, section outboard of wing centre-section being fabric-covered. Extended inboard LE retained (though of course no radiator). LE deicing changed to hot air from exhaust heat exchanger using air rammed in LE inboard of nacelle, option of long-range tanks in outer wings (total 6020 lit, 1,324 gal). First flight of redesigned aircraft 9 Jan 46. Originally produced for 27 pax (1+2, nine windows each side), or for 32 seated 2+2, but problems required addition of dorsal fin, improved deicing, strut to prevent tipping on to tail during loading (though main door at front, on right) and other modifications added 260 kg and reduced speed 15-20 km/h, engine-out behaviour becoming unacceptable. Seating restricted to 21, but normally only 18 fitted with six windows each side. Aeroflot service from Aug 47, flight crew four plus stewardess. This **Il-12B** was inevitably uneconomic (empty wt similar to 48-seat CV-240); exports to China, Czechoslovakia, Poland. Several Arctic/Antarctic versions built with skis, extra ice protection and special eqpt. **Il-12T** cargo version for 3,5 t, load, 2,2 m wide door on left. **Il-12D** military assault version with optional dorsal turret with UBT, observation blister on left rear of flight deck, 37 canvas seats, glider tow cleat, 3 t load or 16 stretchers. In production 46-49, total 663. ASCC reporting name 'Coach'.

DIMENSIONS Span 31,7 m (104 ft 0 in); length 21,31 m (69 ft 11 in); wing area 103 m² (1,109 ft²).
ENGINES Two ASh-82FN; some later ASh-82T.
WEIGHTS Empty (civil) 11 045 kg (24,350 lb); fuel/oil 3130 kg; loaded (if unrestricted) 17 250 kg (38,029 lb).
PERFORMANCE (nominal) Max speed 366 km/h at SL, 407 km/h (253 mph) at 2,5 km; climb 15 min to 5 km; service ceiling 6,5 km (21,325 ft); range 1500 km (932 miles); endurance 4,5 h; take-off 475 m; landing 563 m/128 km/h (79·5 mph).

Il-14 Ilyushin was determined to rectify underlying deficiencies of Il-12, and this turned out to be major redesign. Only parts left were fuselage and landing gear, and even these were substantially modified. Wing was entirely new, same span but straight taper root to tip entirely on trailing edge without broad centre section; aerofoil SR-5, slightly deeper section than on Il-12, aerodynamically cleaned up, ASh-82T (*Transportnii*) standard, with slightly reduced fuel consumption and 500 h overhaul period to reduce failure-rate. AV-50 props, with feathering time reduced from 18 to 4 or 5 s. Fuel confined to four outer-wing cells. Exhaust system copied from CV-240 with twin stacks taken across wing to ejector nozzles above trailing edge. Improved thermal anti-icing (again based on CV-240) making reduced demand on engine power and, with other changes, greatly improving performance in adverse conditions and especially with one engine out. Large wing-root fillet extended forwards around leading edge.

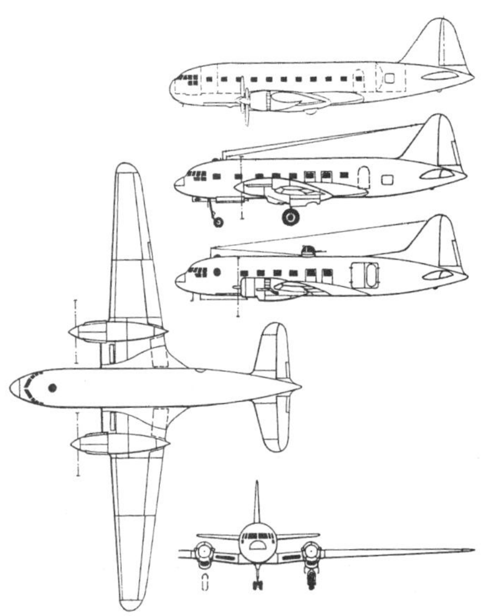

Il-12 prototype (upper side view, 4×M-88 project; lower side view, Il-12D)

Flaps of slotted type. As in practice eight rows of passenger seats never fitted to Il-12, cabin windows reduced to seven each side. First prototype flown by V. K. Kokkinaki 15 July 50, uprated (110 atm, 1,617 lb/in²) hyd syst, improved hot-air heating/deicing. Latter faulty, hence second prototype flown 1 Oct 50 with redesigned hot-air syst, and square-topped fin/rudder of 17% greater area. Cleared for prodn at GAZ-30 June 53, later Chkalov Tashkent also, total 839, plus licence prodn see later.
Il-14P Basic 18-seat version with wt limitation; once cleared to 17 t MTO most mod (by 407 repair works Minsk, backed by 400 Moscow Bykovo) to **P-24**, **P-28** and **P-32**, suffix denoting passenger seating with closer pitch, galley/radio cabin deleted and smaller coat space. **PS** VIP version, 5-8 seats. **S** ***Salon*** VIP version with dividing bulkhead moved forward, oxygen, improved avionics/electrics. **SI** long-range VIP, outer wing tanks added, improved electrics/avionics, better cabin insulation. **SO** as SI but 18 regular pax seats. **D-30** military version based on P, stripped interior, 30 folding paratroop benches, extra search/signal lights plus searchlight on wing. **M** fuselage lengthened 1 m (39·4 in), one extra window each side, strengthened structure, 24 seats, new galley moved to rear, from Feb 57 revised elec syst. **M-14** long-range, outboard wing tanks, 14 seats, oxygen, new avionics, basis for **M-28**, **M-32** and ultimately small number of **M-36**, latter with lighter furnishing, no galley or front baggage. **Iceberg Patrol** (words sound same in Russian), 34 a/c mod at Minsk with extra tanks, a.t.o. rockets, wheel/ski, APU for ground power, gas cooker, ONS-UP Omega nav, AP-6E autopilot, KS-6 gyrocompass, ROZ-1 radar, AB-52 sight, S-1 siren and message release, signal markers, Ladoga photo-fax, Neon transceiver, no replacement a/c in sight. **FK** and stretched **FKM** photo-survey and mapping versions, special navaids including DISS-013FK drift doppler, AFA-TE-55 and -100 plus AFA-33N-20, darkroom. **T** basic mil transport, M airframe but original body length, stripped interior, inward door on R, large cargo door on L with inset para door, windows with blackout screens and shooting apertures, nav cupola, two container racks, provision for towing, OPB-1R sight, Proton-M radio, **T-TD** para version, cargo hoist, P-63 for releasing PDMM-47 or PDTZh-120 containers, 20 folding seats, rubber floor, static lines. **TS** ambulance, 18 stretchers, full oxygen, hot water etc. **TB** glider (Yak-14) tug, tailcone replaced by cleat. **TG** all-cargo, cargo ramp and crane, floor lashing points. **G** civil cargo, as TG but no inset para door, R door larger, 10 folding seats. **RR** fishery surveillance, Groza-40 radar, siren, loudspeaker, various sights/cameras, mail/cargo release, extra fuel in outer wings or fuselage. **Polyarniy** (Polar), retrac skis, 4000-km range, 8 built. **Patrol** versions with comprehensive navaids/sensors for KGB border search, now used for Pacific coast surveillance. **LIK-1**, **-2** flight-research complex calibration, checking accuracy of land navaids and ILS. **Meteo** lab for 22 tasks in weather research, cloud seeding and air/surface pollution measures.

In 55 Avia began prodn at Prague of models which soon included local changes, total 203.

Il-12

Il-14LL

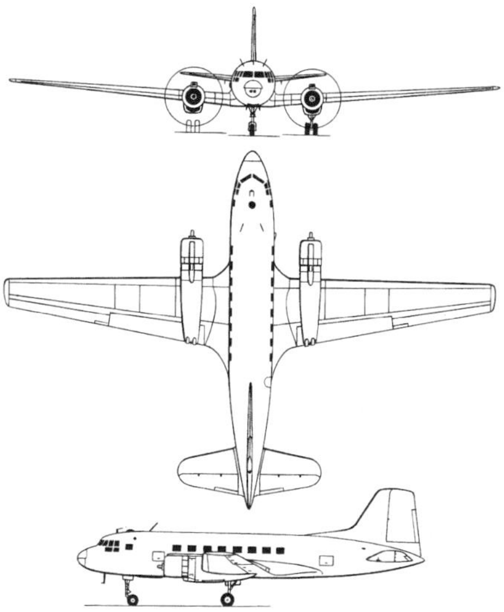

Il-14M

VEB at Dresden built 80, again local mods inc Elint/Comint/jammer. Il-14 was first Soviet aircraft widely exported, to 31 countries. ASCC reporting name 'Crate'.

DIMENSIONS Span 31,7 m (104 ft 0 in); length 21,31 m (69 ft 11 in), (14M, 22·31 m, 73 ft 2¼ in); wing area 100 m^2 (1076 ft^2).

ENGINES Two ASh-82T.

WEIGHTS Empty 12 080 kg (26,631 lb), (14M, 12,7 t, 28,000 lb); fuel/oil 2760 kg; loaded 17 250 kg (38,030 lb), (14M, 14T, 17,5 t, 38,580 lb).

PERFORMANCE Max speed 393 km/h at SL, 430 km/h (267 mph) at 2,4 km; climb 8·5 min to 5 km; service ceiling 7,4 km (24,278 ft) range, typically 1500 km (932 miles) with max payload of 3,3 t; endurance 4·5 h; take-off 485 m; landing 480 m/135 km/h (84 mph).

Il-18 Also called **SPD**, transport for 66 passengers and 900 kg baggage. Physically largest task attempted by Ilyushin, and rivalling anything tackled in USSR in 45, this was a crash programme undertaken by Ilyushin's now very large and capable OKB within CCB, as soon as war over. Designed to cruise 450 km/h at 7,5 km, clear influence DC-4 and -6. Tubular fuselage designed for (not operative) pressurization, 3,5 m diam, seating 60 in 3+2 plus three pairs of seats at rear on left opposite wardrobe on right. Three-spar wing with hyd Fowler flaps, structural joint just inboard of flap/aileron junction. Twin-wheel landing gears, nose unit retracting to rear and main units forwards. Hemispherical nose as planned for original Il-12. Manual flight controls, ailerons in two linked sections. Fuel tanks in inter-spar wing box, total (with oil) 14,01 t. Designed for four ACh-72 diesels, but these never materialised so regrettably switched to petrol and to almost identical engine installation as Tu-70, without turbos, driving 4,8 m AV-16-NM-95 four-blade propellers. This dramatically reduced range from intended 5000 km. First flight by V. K. Kokkinaki 17 Aug 46. When Stalin saw it he asked how many passengers; Il said '66 Comrade Stalin, and planned assault version will carry 90 troops'. Stalin said 'Think what would happen if it crashed. You had better forget about this aircraft'. ASCC assigned name 'Clam'.

DIMENSIONS Span 41,1 m (134 ft 10 in); length 29,855 m (97 ft 11⅜ in); wing area 140 m^2 (1,507 ft^2).

ENGINES Four ASh-73.

WEIGHTS Empty 28 490 kg (62,809 lb); normal payload 5760 kg; loaded 42,5 t (93,695 lb), max 47,5 t (104,718 lb).
PERFORMANCE Max speed 565 km/h (351 mph) at 9 km, cruise 450 km/h (280 mph) at 7 km; range (normal gross wt) 2800 km (1,740 miles); TO/landing 746 m.

Il-16 After Stalin's death OKB was free to build large transport to May 55 order for four-jet (unstated type 3750 kg thrust) 80/85-seater to fly 3000 km at 800 km/h. Cancelled by Aeroflot when 80% drawings issued to shops; replaced by second Il-18.

Il-20 One of ugliest post-war aircraft, this *Shturmovik* was a multi-purpose attacker for use against all surface targets including shipping, though it carried no sensors. Aerodynamically it was direct scale-up of Il-10M. All light-alloy stressed-skin, Clark-YH wing profile, landing gears similar to Il-8/10/16, retractable tailwheel with doors. Most powerful Russian V-12 engine, with large but shallow belly radiator and leading-edge oil coolers. Pilot's cockpit directly above engine with 6 to 9 mm armour covering entire nose; engine also helped protect pilot. Sliding and jettisonable armoured canopy; poor view to rear but superb view ahead, immediately behind four-blade propeller. Second crew-member radio/gunner with remote electrical control of power-driven turret with two 23 mm guns. Turret and ammunition filled fuselage cross-section immediately to rear of gunner; armour unknown. Internal cells for four 100 kg bombs or equivalent; max load an additional pair of 500 kg or equivalent hung externally. Underwing racks for eight RS-82. As far as known, only one prototype, NII trials 1948. Dubbed *Gorbun* (Gorboon), hunchback. Engine undeveloped and general performance, except range, inferior to Il-10.
DIMENSIONS Span 17,0 m (55 ft 9 in); length 12,59 m (41 ft $3\frac{2}{3}$ in); wing area 44,0 m^2 (474 ft^2).
ENGINE One M-47F.
WEIGHTS Empty 7500 kg (16,534 lb); fuel 800 kg; loaded 9800 kg (21,605 lb).
PERFORMANCE Max speed 515 km/h (320 mph) at SL, falling off with height; climb 8 min to 5 km; service ceiling 7750 m (25,430 ft); range 1680 km (1044 miles).

Il-22 Though, like most other high-speed aircraft of its day, a traditional airframe with jet engines, this light bomber is notable on several counts. It was first four-jet aircraft after

Il-14LIK-1

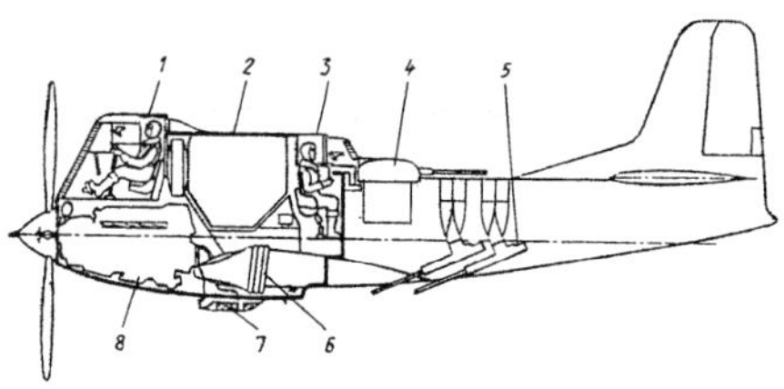

Il-20 (1, pilot; 2, tank; 3, radio/gunner; 4, turret; 5, attack guns; 6, oil cooler; 7, carb inlet; 8 M-47F)

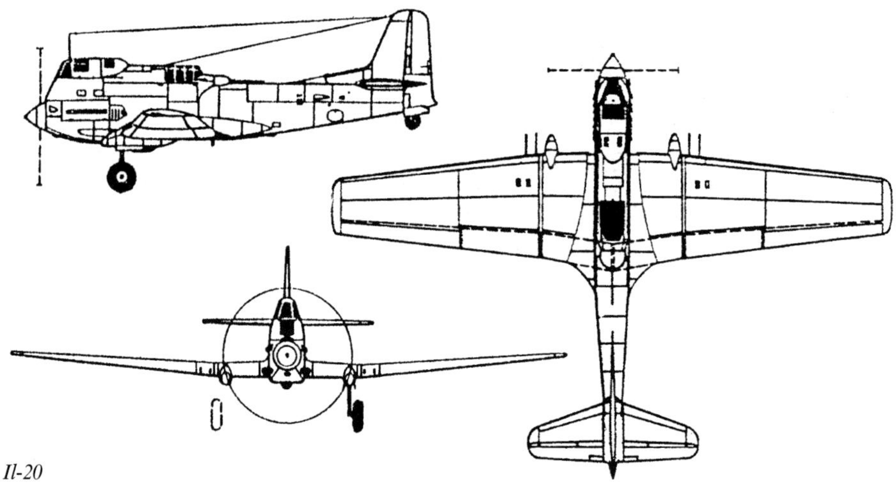

Il-20

Il-18, SPD

Il-20

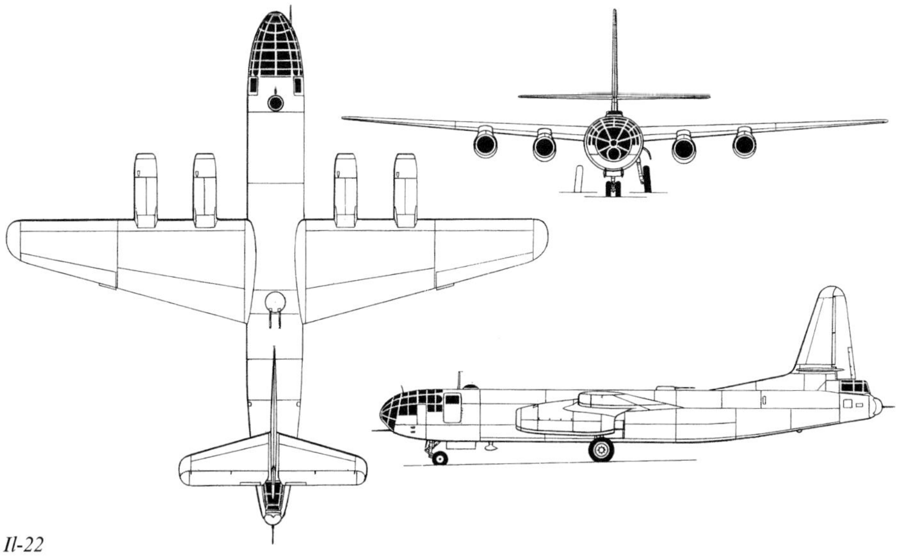

Il-22

wartime Ar 234C to fly in Europe. It was first jet bomber in Soviet Union, four years earlier than Britain's Canberra. It was almost first aircraft to fly with engines designed by A. M. Lyul'ka's KB (just pipped to post by Su-11), and started a partnership between Ilyushin and engine designer that endured through several generations of later bombers.

Design 46 to meet VVS requirement for aircraft powered by four TR-1 engines, to reach Mach 0·75+ and carry 2 t bombload 1250 km at 750 km/h. Rival was Su-10, which used 45° sweep on fin, but Ilyushin decided to ignore sweep. Wing straight taper on TE only, symmetric 1A-10 profile, thickness 12% root, 10% tip. Fixed LE, TE with two sections Fowler flap each side and metal-skinned aileron. Engines in four widely spaced pods almost entirely ahead of LE on sharply swept pylon. Fuselage very wide oval, partly to provide width for main gear each side of bomb bay and partly to obtain adequate track. Pressurized forward section for nav/bomb in nose, pilot and dorsal gunner seated facing aft under Plexiglas dome with remote elec control of VDB-5 turret behind wing with two B-20E each with 800 rounds. Pressurized Il Ku-3 tail turret for gunner with single NS-23 with 225 rounds. Large bomb bay with elec outward-opening doors for max load 3 t (6,614 lb). Simple landing gears, twin-wheel levered-suspension castoring nose unit retracting backwards into bay with three doors, main units swinging forwards to stow large single wheel obliquely in top of fuselage. Conventional tail, tailplane mounted well above level of jets, tabbed rudder and elevators, all flight controls manual. Three self-sealing fuselage fuel cells, total 9300 kg (11 600 lit, 2,552 gal).

Prototype flown 24 July 47, and on 3 Aug took part in Tushino parade (no photo appearing in West). Both Kokkinaki brothers were satisfied with handling, but this obviously interim machine was never considered for production. On 7 Feb 48 V. K. Kokkinaki made takeoff with two SR-2 a.t.o. rockets.

DIMENSIONS Span 23,06 m (75 ft 8 in); length 21,05 m (69 ft 0¾ in); wing area 74,5 m² (802 ft²).
ENGINES Four TR-1.

Il-22

WEIGHTS Empty 14 950 kg (32,959 lb); fuel 6160 kg; loaded 24 t (52,910 lb).
PERFORMANCE Max speed 656 km/h at SL, 718 km/h (446 mph) at 7 km; climb estimate 8·6 min to 5 km; service ceiling 11 km; range estimate 865 km (537 miles); endurance 1·4 h; take-off 2 km; landing speed 190 km/h (118 mph).

Il-24 In 1947-48 Ilyushin produced drawings for this improved Il-22 with four RD-500, and main landing gears, in twin underwing pods.

Il-28 Tactical bomber, launched Dec 47, starting later than rival Tu-73/78. Key to design was twin Nene engines in smaller aircraft, though even more severely penalised by requirement to fit tail turret and 500 kg armour.

Wing superior to Il-22 with SR-5S profile, ruling t/c ratio 12%, set at 3° incidence but only 0°38' dihedral. All taper on trailing edge, aspect ratio 7·55. Two-spar construction in D-16T, skin up to 4 mm. Leading edge (and tailplane) hot-air deicing. Entire airframe structurally divided into upper/lower (or left/right) halves to speed manufacture and installation of systems and equipment, completed halves then being joined, usually by bolting via small hand-apertures with flush-screwed doors; weight penalty 1·5% but saved time. Slotted flaps inboard and outboard of nacelles, hydraulic actuation to 50° by AK-150 pump on each engine. Landing gears retracted pneumatically (55 kg/cm²), track 7,4 m, wheelbase 6,67 m; main tyres 1150×350, folding forward and turning 90° to lie flat under jetpipe. Nose gear twin steerable tyres 600×180, later (51) 600×155V, retracted rearwards. Small fuselage, no access between crew. Pressurized nose and tail compartment, system operated above 2 km to 0·4 ata (5·8 lb/in²) with filtered main bleed air; ATIM-X and ANZM insulation for heat and noise.

Il-28U

Fuselage frames and stringers pressed or rolled, skin 1,0 or 1,5 mm, flushriveted. Fixed tailplane with 7° dihedral mounted on fin, all swept 35° to avoid control problem at max dive speed. Flight controls manual, light-alloy skin, trimmers cast Elektron with elec drive. Nav/bomb on upward-ejection seat in nose under jettisonable roof hatch. Pilot immediately behind on centreline, canopy hinged to right, upward-ejection seat, two fixed NS-23 (100 rds ea). Radio/gunner in turret designed at OKB, reducing weight with Mg structure to only 375 kg without guns but incl armoured ammo boxes and feeds (635 kg with two NR-23 and 450 rds, guns selectable); elec-hyd drive to outstanding firing arcs, became standard in mod forms on Tu-16, Tu-95 and 3M. Normal bombload 1 t (2,205 lb) in internal bay with twin doors opening pneu into airstream and KD-3-Il-28 racks; overload one FAB-3000 or nuclear TN. PSBN-M-8 mapping radar (initially aft of bay) moved in front, OPB-5S optical sight to right of centre in nose plus AB-52 drift sight. Three fuselage fuel cells ahead of wing and two behind for 7908 lit (1,740 gal), all protected and inerted by CO_2. Provision two PSR-1500-15 a.t.o. rockets on bulkheads 27, 30. PB-28 7 m ribbon parachute ahead of turret. Seats armoured 6/10/32 mm, 454 kg; glazing 102 mm windscreen and 68 mm to sides. AP-5 autopilot, RSIU-3M radio, ARK-5 ADF, MRP-48P mkr, Matyerik ILS (GRP-2 g/s rcvr), KRP-1 VOR, SD-1 DME, RV-2 fine and RV-10 coarse radar alts, S-2 (*Sirena*) warning, SRO IFF, PKI-1 gyro sight, S-13 gun camera, blue signalling lamps and three (multicolor) flare chutes.

First flight by V. K. Kokkinaki 8 July 48, fine except elevator immovable at M 0·78 followed by buffet and pitch-down. Critical evaluation against Tu-78 following Oct; Marshal Vershinin ordered three randomly picked crews to fly both; all picked Ilyushin. Stalin demanded 25 to fly May Day 50; V. P.

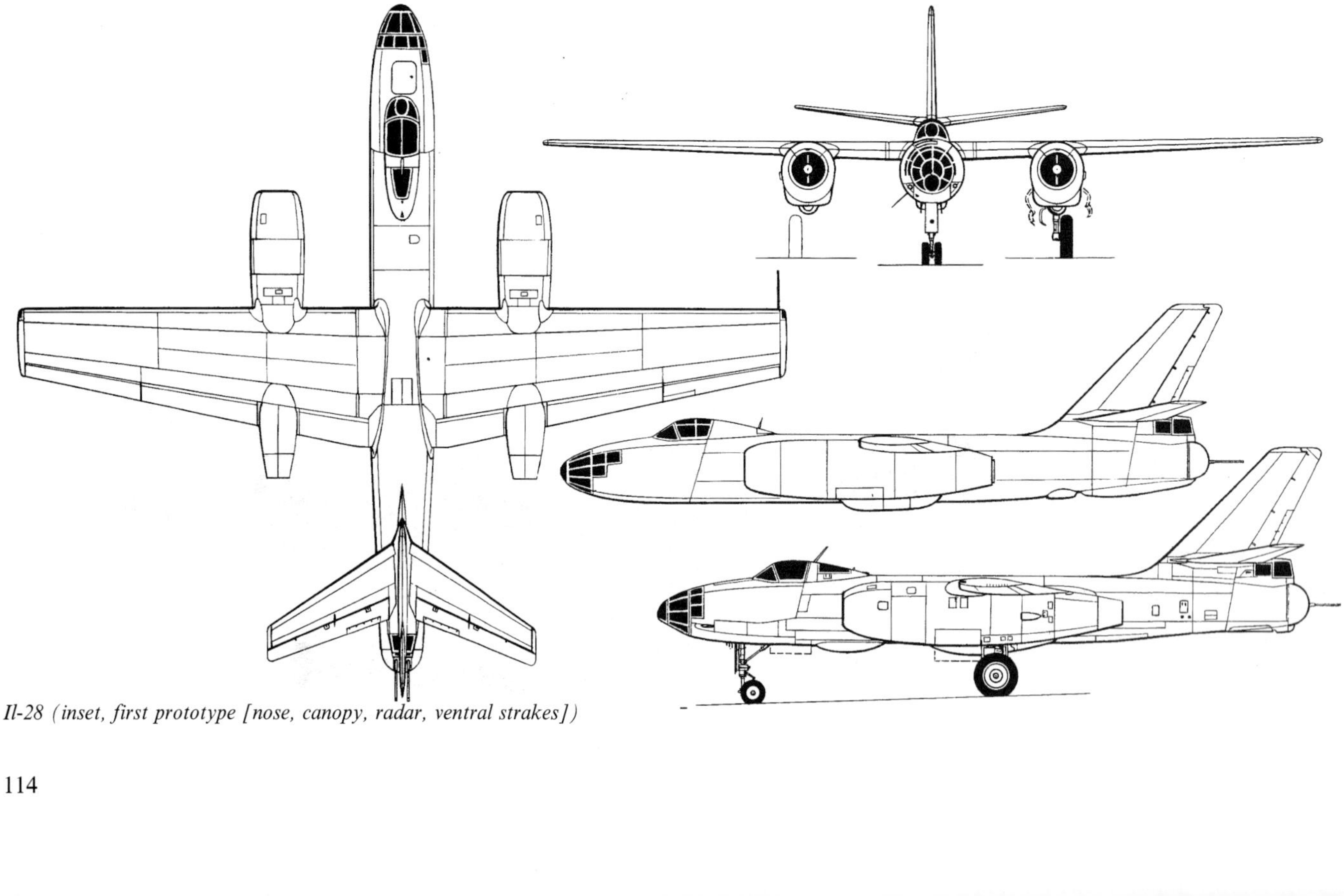
Il-28 (inset, first prototype [nose, canopy, radar, ventral strakes])

Il-28LL (seat testing)

Yatsyenko's management and simplicity of aircraft made this possible though it disrupted devt of aircraft and engine. First polk Sept 50.

Over 1,500 Il-28 built GAZ-53 in 50-55, plus 459 from GAZ-125 in 53-55. **Il-28T** torpedo/bomber/minelayer for AV-MF, flown Jan 50; different radar and sight, weapon bay 2,2 m longer for two 553 mm torpedoes with parabrake and stability ring; strike camera in tail; enlarged bay plus LAS-3 dinghy resulted in fixed 333 lit wingtip tanks being standard. **Il-28U** trainer flew 18 March 50; no radar or armament, redesigned ahead of bulkhead 6 as cockpit for instructor with emerg ability to disconnect pupil controls; tail station instructed radio opr.; 1st U trained pilots for 25 Il-28 for May Day 50, every subsequent unit had 1 or 2. **Il-28R** recon version flown 19 Ap 50, bay redesigned for aux fuel plus heated section for (usually 4) vert/pivoted AFA linear/stereo cameras plus flash/illum FOTAB-50-35 magazine, oblique cameras in rear fuselage; aux fuel plus tip tanks increased wt so main tyres 1150×355 (later standard on bomber), spun-up before landing; landing gear made hyd for faster retraction; right front gun not fitted, LAS-3 dinghy installed. **Il-28D** long-range nuclear, front guns and turret removed (streamlined tailcone), tip tanks and rear-fuse fuel, bay mod with central BD-4 rack and KD-3-Il-28D aft racks, 0,4 m longer. **Il-28RT** only new-build version for EW/Elint with passive receivers and cameras.

Many Il-28 converted for new roles: **Il-28RTR** dedicated electronic reconnaissance; **Il-28REB** standard electronic-warfare jammer with antennas in tip 'tanks' and under fuselage; **Il-28Sh** locally mod for attack with multiple racks under outer wings; **Il-28B** demilitarised for target-towing; **Il-28ZA** demilitarised for weather recon; **Il-28P** demilitarised as transport (see **Il-20**, next). Many aircraft testbeds for radars, jet engines, flight refuelling, upward and (from tail) downward-ejection seats, *Vostok* spacecraft seat, avionic systems, missiles and RPVs/targets. Over 1,000 exported to Afghanistan, Algeria, Bulgaria, China (500), Cuba, Czechslovakia (called **B-228**, later mod as **Il-2B**), Egypt, Finland, E Germany (DDR), Hungary, Indonesia, Iraq, N. Korea, Morocco, Poland, Romania, Somalia, Syria, Vietnam and N. and S. Yemen (YAR and PDRY). In China put into production at Harbin as **H-5**, several hundred exported from this source. Even today est 400 still flying. Swept-wing **Il-28S** never completed and scrapped. USAF designator 'Type 27', ASCC name 'Butcher' later changed to 'Beagle'; 28U name 'Mascot'.

DIMENSIONS Span 21,45 m (70 ft 4½ in), (28R) 22,65 m (74 ft 3¾ in); length (excl guns) 17,65 m (57 ft 10⅞ in); wing area 60,8 m² (654 ft²).

ENGINES Two VK-1.

WEIGHTS Empty 12 890 kg (28,417 lb), (28R) 13 510 kg (29,784 lb), (28U) 11 760 kg (25,926 lb); fuel/oil 6400+200 kg (R, 7831 kg total); loaded 18,4 t (40,564 lb), (R) 19,5 t (42,989 lb), (U) 17,56 t (38,713 lb); max 23,2 t (51,146 lb), (U, 21 t).

PERFORMANCE Max speed (SL) 800 km/h (497 mph), at 4,5 km 900 km/h (559 mph), (R) 876 km/h (544 mph), (U) 895 km/h (556 mph); climb to 5 km 6·5 min (U, 5·5 min); service ceiling 12,3 km (40,355 ft), (U, 13,25 km); range 2180 km (1,355 miles), (R, 2780 km; U, 2260 km); TO 875-965 m (U, 600 m); ldg 185 km/h, 960-1170 m (U, 180 km/h, 700 m).

Il-20 Designation re-used for demilitarised Il-28 bombers, used from Jan 55 by Aeroflot as jet crew trainers, nav-system developers and, esp., as high-speed transports carrying urgent cargo, notably matrices for printing *Pravda* and *Izvestiya* in Sverdlovsk and Novosibirsk. Data as Il-28.

Il-30 In mid-48 OKB assigned challenging task of frontal bomber to carry 2 t bombload 3500 km at 1000 km/h. This demanded swept wings, and only available engine appeared new Lyulka axial. Surprisingly, conventional SR-12s wing, swept 35°, gave prolonged problems. Vicious circles led to Il-30, of Il-28 size, weighing almost twice as much. Structure had to be stiffer and stronger, and another burden was heavy defensive armament. Wing of low aspect ratio, fixed LE, tracked slotted flaps inboard/outboard of nacelles, conventional ailerons, eventually eight fences! Fuselage and tail superficially like Il-28, but *velociped* (bicycle) landing gear. Twin-wheel front and rear main units as close to bomb bay as possible, castoring front unit retracting forwards and tall enough for wing to have AOA close to desired TO value of 9°30' at all weights. Twin-wheel outrigger gears retracted backwards into outer edge of nacelles. Engines hung well ahead of wing, but jetpipe projecting behind TE. Over 10,5 t fuel in wings as well as fuselage, all in protected tanks. Tail based on Il-28 but tailplane higher, greater span and no dihedral. Nav/bomb seat in nose, pilot under fighter-type canopy, radio/gunner seated facing aft and separate capsule for tail gunner. Two gunners shared control of complex sighting and aiming system for six NR-23 cannon, two each in dorsal and ventral Il-V12 turrets and two in Il-K6 tail turret. Bomb bay eventually developed to carry double specified load, ie 4 t (8,818 lb). Pilot had upward ejection seat, others had to escape through ventral hatches forming windbreak.

Prototype completed spring 49 but not flown (by V. K. Kokkinaki) until 9 Sept. By this time door airbrakes added to rear fuselage, though this was intended location for radar. Problems with pressurization and buffet. Official OKB history includes no illustration, nor figures for TO/landing, but comments that Il-30 data assisted Il-46 and 54. No ASCC reporting name.

DIMENSIONS Span 16,5 m (54 ft 2 in); length 18,7 m (61 ft 4 in); wing area 100,0 m² (1,076 ft²).

ENGINES Two TR-3.

WEIGHTS Empty 22 967 kg (50,633 lb); loaded 32 552 kg (71,764 lb), (max) 37 552 kg (82,787 lb).

PERFORMANCE Max speed (SL) 900 km/h, (5 km) 1000 km/h (621 mph); range claimed 3500 km (2,175 miles) with 2 t bombload.

Il-32 About two years after end of the Second World War, Aviation of Airborne Troops issued requirement for transport glider larger than any it then possessed. Specified payload

Il-30

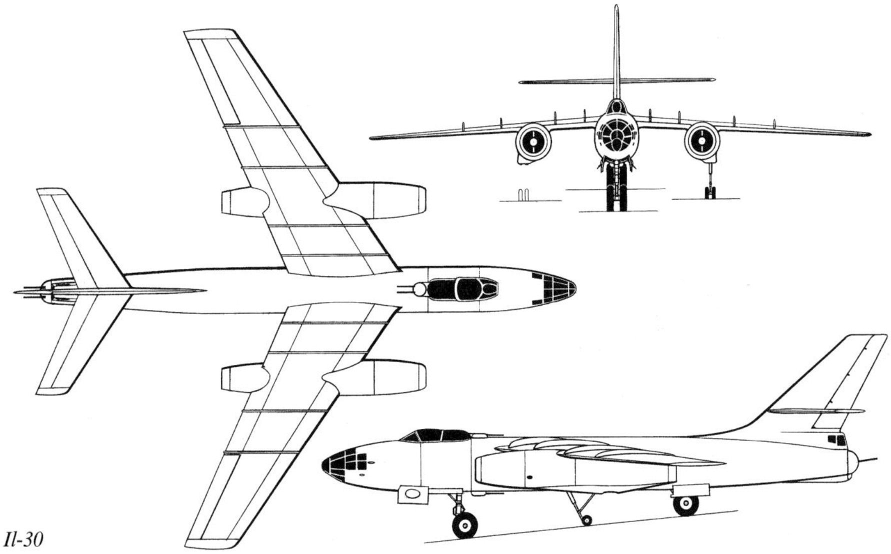

Il-30

3,5 t (7,716 lb), light vehicles and artillery to be readily loaded and disembarked. In competition with Yak-14, Ilyushin OKB produced Il-32. Typical high-wing glider with large box-like fuselage, but most unusual feature was light-alloy stressed-skin construction. Compressed-air system worked spoilers above wings and complete nose and rear fuselage unlocked and hinged to right for both-ends access. Interior 2,6 m high and 2,8 m wide, 7 t cargo or 60 troops. Twin nosewheels and single-wheel jettisonable main gears. To be towed by Tu-4 or similar. Prototype flown 48, small series (believed 10) built 50 at Rostov. No ASCC name.

DIMENSIONS Span 35,8 m (117 ft $5\frac{1}{2}$ in); length 24,84 m (81 ft 6 in); wing area 159,5 m^2 (1,717 ft^2).

WEIGHTS Empty 9,6 t (21,164 lb); max 16,6 t (36,596 lb).

Il-34 At least one Il-32 was converted into a powered aircraft. Two piston engines were installed in wing nacelles; type believed to be 730 hp ASh-21 with fixed-pitch two-blade props.

Il-40 Last of armoured *Shturmoviks*, this prototype of 53 was a strange amalgam of old and new parts. Among old, but highly developed, parts were armoured tandem-seat crew compartment, and weapons comprising devastating forward-firing guns, rear guns for defence and internal bomb cells. Wing design close analogue on reduced scale of Il-30, though with much thicker profile (root t/c 17%). Tail likewise derived from that of Il-28, and twin-NR-23 tail barbette related to bureau's K6, though much further from controlling radio/gunner. Engine installation unique: two small axials, widely separated on each side of fuselage, fed by short ducts. Fuel between and under crew and in tapering nose and rear fuselage. Single-wheel nose gear retracting to rear and main units retracting forwards into bulged compartments, leg being faired beneath main wing box with wheels rotated 90°. Four NR-23 in pallet under fuselage, able to be hinged down for firing on ground targets. Up to 1,5 t bombs in four wing bays and six underwing racks. Two NR-23 in Il-K10 remotely sighted tail turret. First prototype flown 7 March 53. Inlets extended to nose to avoid gun-gas ingestion. Pre-series batch of five at Rostov scrapped before completion because of Khrushchyev's antagonism and VVS belief such aircraft obsolete. Prototype displayed Kubinka June 56, so ASCC bestowed name 'Brawny'.

DIMENSIONS Span 16,9 m (55 ft $5\frac{3}{8}$ in); length 17,0 m (55 ft $9\frac{1}{4}$ in); wing area 47,6 m^2 (512 ft^2).

ENGINES Two AM-5F.

WEIGHTS Empty c8,5 t (18,750 lb); loaded c15 t (33,000 lb).

PERFORMANCE Max speed 964 km/h (599 mph); service ceiling 11,6 km (38,000 ft); range (hi) 1000 km (620 miles).

Il-46 Kremlin demand to improve Il-30 to carry 5 t bombs 5000 km was great challenge, helped by uprated Lyulka engine. Range demand dictated unswept wing, design going ahead Oct 51. Parallel Il-46s study with 35° wing posed problems of bicycle landing gear, structural strength at root, fuel/CG relation and long nacelles with inlets beside pilot. Straight wing therefore adopted, though with proven swept tail. Twin-wheel nose unit hyd steerable and retracted to rear. Four separate single-leg main gears, two retracting forwards and two to rear, wheels rotating 90° to lie under 4,9 m jetpipes. Nav in glazed nose facing aft to manage radar or ahead to aim bombs, with escape through entrance hatch. Pilot on upper level had door on right and ejection seat under jettisonable multi-pane canopy with sliding side windows and front sight for two fixed NR-23 each 640 rounds. Tail gunner/radio operator in Il-K8 turret with twin NR-23 each 320 rounds, rapidly aimed 105° azimuth and from +58° to –39°. Total 880 kg armour. Overload fuel 25 000 lit (5,500 gal), simple fixed LE, slotted flaps, manual controls.

First flight by V. K. Kokkinaki 3 March 52. Excellent aircraft (though marginal on one engine), but could not rival Tu-88. Later fitted with pod for dummy front/rear radar warning system above fin. Previous Western guesses at data were underestimates. No ASCC name.

DIMENSIONS Span 29,0 m (95 ft 2 in); length 24,5 m (80 ft 5 in); wing area 105 m^2 (1,130 ft^2).

ENGINES Two AL-5.

WEIGHTS Empty 26,3 t (57,981 lb); loaded 41 840 kg (92,240 lb); max 52 425 kg (115,575 lb).

Il-32

Il-40-I

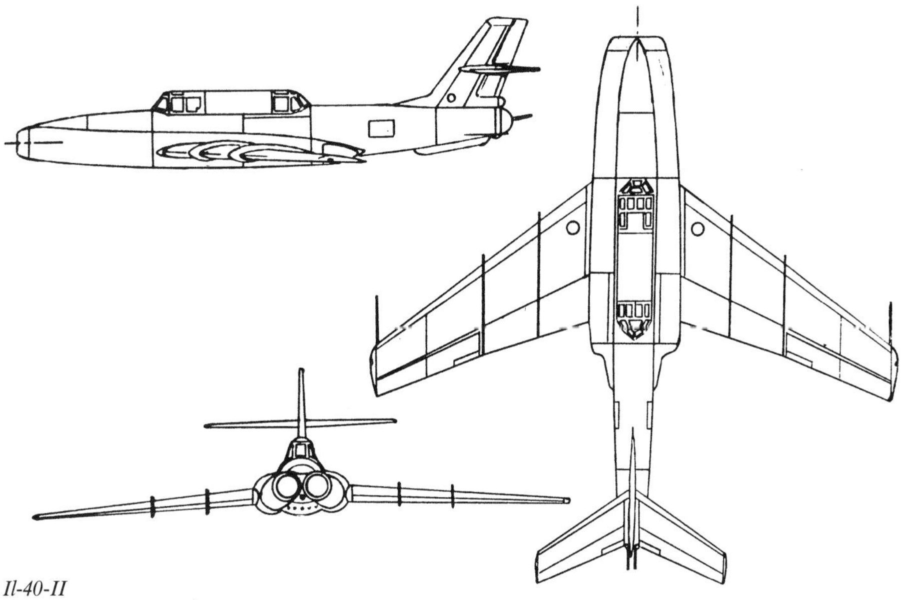

Il-40-II

Il-46

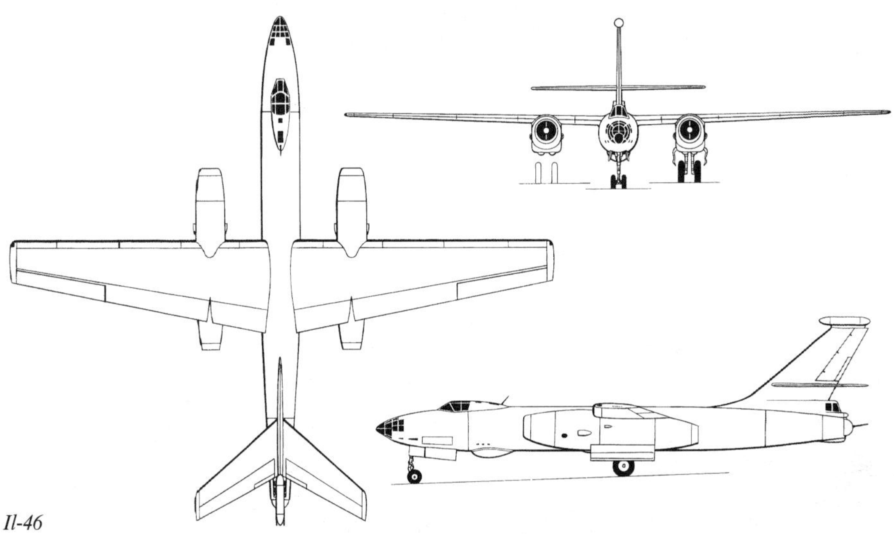

Il-46

PERFORMANCE Max speed (SL) 800 km/h (497 mph), (5 km) 928 km/h (577 mph); service ceiling 12,7 km (41,670 ft); range (3 t bombs, 700 km/h) 4970 km (3,090 miles); TO 1335 m; landing 202 km/h, 673 m.

Il-54 In 52 VVS ordered prototypes of supersonic tactical bombers from Ilyushin and Tupolev. Both used AL-7 engine, but more ambitious Tu-98 needed afterburning engine from outset. Demands included M 1·15 at 4750 m and practical range with 3 t bombload of 2400-2750 km, equating to combat radius under 1000 km (621 miles). After study against usual 35°, Il selected sweep of 55° early 53. First drawing showed engines on each side of fuselage as in Tu-88, with pronounced waisting over wings centred on engines, single main wheel retracting forwards to lie flat in wing, T-tail and manned tail turret. Reluctantly this was rejected and replaced by high wing passing over bomb bay, engines in pods hung close in ahead of and below wing, conventional tail and bicycle landing gear. Major problem was required AOA on takeoff. Despite high power of pivoted dihedral tailplanes (elevators merely adding camber) it appeared pilot would not be able to rotate aircraft nose-up on TO. Only answer was to use extra-extensible nose leg and 'kneeling' rear gear. At rest AOA was 5°45 for min drag; at liftoff speed pilot triggered main gears to rotate aircraft to required 10° (idea repeated in M-50). Penalty was wheelbase of 15,25 m, requiring massive fuselage bending strength around bomb bay. Outrigger gears retracted forwards into tip pods. Fixed LE, Fowler flaps, conventional ailerons, all surfaces fully powered, deep full-chord fence at flap/aileron junction and shallow fence inboard of pylon. Pressure cabin with door on right for pilot in upward ejection seat, with sight for twin 23 mm guns, nav/bomb in pointed glazed nose with downward ejecting seat and radio operator/gunner with downward ejecting seat in turret with twin AM-23. Max bombload 5 t (11,023 lb). LAS-5M dinghy carried, though ditching seemed perilous. Sole prototype completed as frontal bomber and flown by V. K. Kokkinaki 3 April 55. Comment 'landing gear complicated TO and landing'; not recorded whether landing could be made with nose gear unextended. In late 55 AL-7F engines fitted; keel area at rear later increased by adding sloping underfins. NII testing completed but no production; Il-54U trainer and Il-54T torpedo carrier never built. Again, Western estimates grossly wrong despite aircraft being displayed to Western delegation Kubinka June 56. ASCC name 'Blowlamp'.

DIMENSIONS Span 17,65 m (57 ft 11 in); length 28,963 m (95 ft 0 in); wing area 84,6 m^2 (911 ft^2).

ENGINES Two AL-7, later replaced by 7F.

WEIGHTS (54) Empty 23 560 kg (51,940 lb); loaded 36 820 kg (81,173 lb); max 38 t (83,774 lb); (55) empty 24 t (52,910 lb); loaded 40 660 kg (89,638 lb); max 41,6 t (91,711 lb).

PERFORMANCE (54) Max speed (SL) est 1050 km/h, (5 km) 1170 km/h (727 mph); climb 4 min to 5 km; service ceiling 13 km; range with 3 t bombs at 910 km/h, 2200 km (1,367 miles); TO 1075 m; landing 243 km/h, 1150 m; (55) max speed (SL) est 1155 km/h, (5 km) 1250 km/h (777 mph, M 1·09); climb 1·1 min to 5 km; service ceiling 14 km (45,930 ft); range (as before) 2500 km (1,553 miles); TO/ldg as before.

Il-54

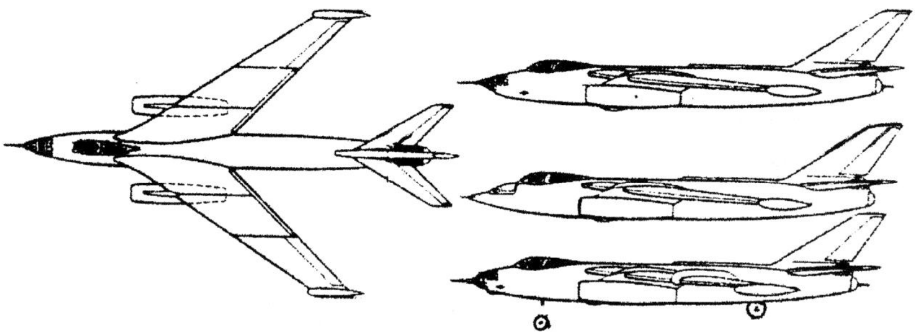

Il-54 (middle side view, Il-54U; bottom, Il-54T)

Il-18 Designation re-used for large modern passenger aircraft made possible by development of NK-4 and VK-2 turboprops and death of Stalin; designation deliberately repeated, project studied from 54. Requirement: 75 passengers, primitive airfields, much better fuel efficiency than Tu-104, bad-weather landing and long structure life.

Low equi-tapered wing, 3° dihedral from root, CAHI SPS profile, 15% root, 13% tip, three spars, skins machined for thickness only. One-piece double-slotted flap each wing, hyd drive via ballscrews to 30°. Fuselage diam 3,5, max dP 0,5 kg/cm^2 (7·1 lb/in^2) with nose fairing riveted on with discontinuity at joint. Flight controls manual/cable but hyd boosters on ailerons/elevators, elec trim on all surfs plus spring tab on rudder. Tricycle ldg gear, all retrac forwards hydraulically into bays with twin doors; nose twin 700×250 tyres steering ±45°, main 4-wheel bogies 930×305 tyres 8 kg/cm^2 (114 lb/in^2), hyd anti-skid brakes, bogie stowed inverted. Hyd syst 210 kg/cm^2 (2,987 lb/in^2), one-shot dry N_2 emergency. Engines drive AV-68I 4,5-m reversing props, axially-hinged cowls, ventral oil coolers, single jetpipes across wing to TE. Fuel in 20 bags from root to aileron, 16400 lit (3,607 gal) plus aux integral outer wings add 7300 lit (1,606), total 23 700 lit (5,213 gal). Electrothermal ice protection, main starter/generators STG-12 two per engine, total 96 kW. Weather radar *Emblema* in unpress nose, Decca Flight Log, VOR/DME, ILS.

Programme based at GAZ-30. Prototype L-5811 named *Moskva* with NK-4 engines flown 4 July 57. Followed by five 75-seat pre-production; flat emerg pressure-bulkhead added behind cockpit, locked in flight, normal flight crew two pilots/nav/radio/engineer. Small batch **Il-18**, Al-20 engines at 3,755 ehp, 75 seats, two small pressurized underfloor holds plus unpress rear hold access external only, MTO 57 t (125,661 lb).

Il-18A July 58, cured vibration and aerodynamic buffet problems, engine uprated to 4,000 ehp, interior rearranged for up to 89 pax with three wardrobes, three toilets and galley, payload 12 t, MTO 58 t (127,866 lb).

Il-18B 58-59, MTO 61,2 t (134,921 lb),

Il-18LL Pushkin

Il-18 Tsyklon

payload 14 t; wing and main gears strengthened, pressure fuelling. Standard at entry to service 20 April 59. One B modified with 31 000 lit fuel and special eqpt, in Feb 81 flew Moscow-Antarctic 15 950 km 25 hr.

Il-18V 60, three cabins seating 20/55/14 (all 3+2 except final row), repositioned doors, props aligned with vestibule instead of seats; in 63 nav/ILS upgraded.

Il-18Ye (Cyrillic, Il-18E) 65, 61,4 t, refined interior up to 120 seats at expense of wardrobe space; alleged greater comfort.

Il-18D 65-69, major series, AI-20M engine, 64 t, four centre-section bags total fuel 30 000 lit (6,599 gal), upgraded avionics incl autoland, data for this.

Total 565 completed Oct 79, incl 121 exported to 16 countries. In Aeroflot over 260 m pax carried in c15 m hr, 5·9 m sectors flown. Served as basis for Il-38. Small number to VVS and AV-MF, mainly as VIP transports, but since 74 growing number of conversions, many as Aeroflot freighters. From 77 30+ rebuilt as subtypes of **Il-20** for recon, **22** for airborne command post ('Coot-B') and **24** (all designations used previously), notably **Il-20DSR** for multisensor recon, in most versions mainly Elint and Sigint but with large SLAR and 12-15 extra antennas, usual crew 4/5 + 20 mission specialists. At least 20 converted as carry-trials aircraft and for meteorological or electronic research. All versions, ASCC reporting name 'Coot'.

DIMENSIONS Span 37,4 m (122 ft 8½ in); length 35,9 m (117 ft 9 in); wing area 140 m² (1,507 ft²).

ENGINES Four AI-20K or (most, from V) 20M.

WEIGHTS Empty (V) 26,57 t (58,576 lb), (D) 27,98 t (61,684 lb); fuel (D, some V upgraded) 23,7-24,54 t; payload 13 or 13,5 t; normal loaded 45,3 t; max (most) 61,2 t (134,921 lb), (D) 64 t (141,093 lb).

PERFORMANCE Max speed, 640 km/h at SL, 685 km/h at height; max cruise 674 km/h (419 mph), econ 625 km/h (388 mph); max climb

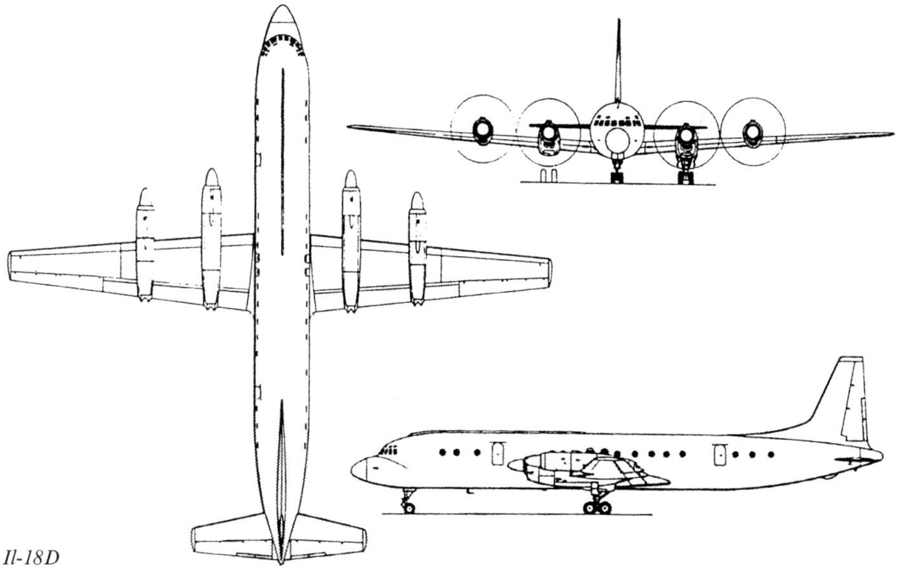
Il-18D

Il-18LL Joukovskii

Il-20 (Elint/SLAR)

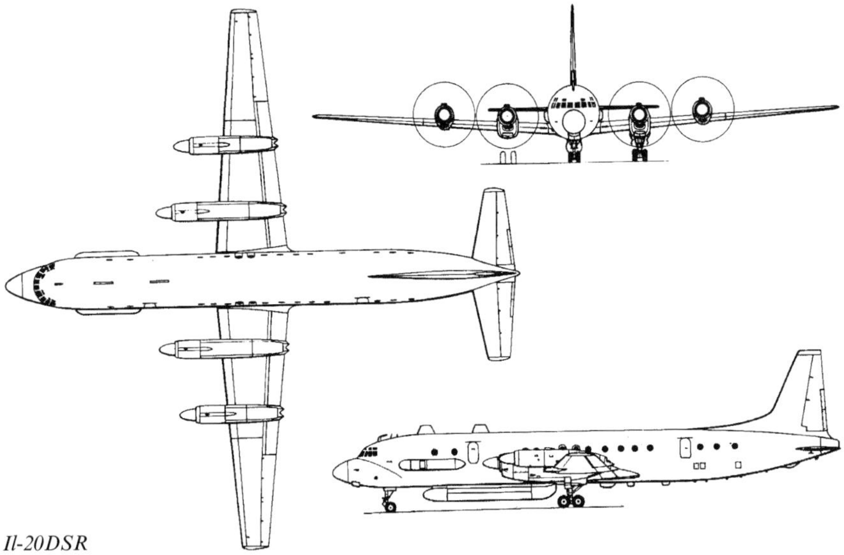

Il-20DSR

10,5 m/s (2,070 ft/min); service ceiling 10 km (32,800 ft); range (D, max payload) 3700 km (2,300 miles), (max fuel) 6500 km (4,040 miles); TO 1,2 km; ldg 190 km/h, 850 m.

Il-38 Patrol/ASW derivative of Il-18 to meet 59 AV-MF demand. Major elements of structure unchanged, but wing moved forward, passenger windows replaced by occasional portholes, press bulkhead retained behind flight deck (three flight crew) but entire intrior can be pressurized for transit to search area; 7/8 tactical crew, small bulged observation window each side flight deck, external cable fairing left side to main search radar under floor, spine fairing from mid-cabin to fin, door at rear on right, MAD sensor tailcone. Shallow bay with twin doors ahead of and behind wing box for 3,0 t sonobuoys and weapons; no external pylons.

Prototype flown 27 Sept 61, production (believed 100) 65-68, five to Indian Navy 75, 315 Sqn. About 59 active AV-MF in 93. ASCC reporting name 'May'.

DIMENSIONS, ENGINES As Il-18 except length 39,6 m (129 ft 10 in).

WEIGHTS Empty 34,5 t (76,060 lb); fuel 24,05 t; MTO 63,5 t (140,000 lb).

PERFORMANCE Max speed at 6,4 km 722 km/h (448 mph); patrol speed (lo) 400 km/h (249 mph); range 7200 km (4,474 miles), 12 h endurance.

Il-62 First large intercontinental jet transport built in the Soviet Union. Basic airframe conventional, but largest use of integrally stiffened panels and large forged frames at time of design (60). Wing swept 35° at ¼-chord, primary box with four spar webs to aileron, two to tip. Built as centre section with trailing edge at 90° to fuselage, and two outer panels. Single flap on centre section and another on outer panel; single-slotted type, track-mounted, driven electrically (DC early, raw AC on 62M). Ahead of outer-wing flap, two spoiler sections, 211 kg/cm^2 (3,000 lb/in^2) hydraulic, for letdown/lift-dump only. Outboard, three-section manual aileron

Il-20DSR

Il-38

with spring tabs and electric trim. T-tail with electrically driven tailplane, manual elevators with separate manual and autopilot tabs, two-section manual rudder with spring and trim tabs and with yaw damper (series type). Fuselage almost circular section (external width 3,75 m, height 4,1) with fail-safe rip-stop structure designed for 25,000 h. Max dP 0,63 kg/cm² (8·96 lb/in²), vertical-elliptical windows almost identical to VC10 structure, floors mainly metal/foam sandwich. Landing gear hydraulic, steerable twin-wheel nose unit retracting forwards and main bogies with four 1450×450 mm tyres (9,5 kg/cm², 135·1 lb/in²) and anti-skid hydraulic brakes inwards under passenger floor. Emergency braking parachute. Twin-wheel vertical rear support strut extended hydraulically to apron after parking. Electrics, originally eight 18-kW generators, most (including all 62M) now generated as raw AC, four 40-kVA. Fuel in seven integral tanks (three in centre section), total 100k lit (21,998 gal), four underwing sockets and eight overwing gravity fillers. All leading edges and engine inlets de-iced by hot air from bleed system; electric windscreens. Cabin environmental inlets at wing roots. TA-6 gas-turbine APU in tailcone.

First prototype No 06156 flown 3 Jan 61. Engines (temporary) Lyul'ka AL-7 turbojets; dorsal spine from fin to wing leading edge, second dorsal service fairing from roof of flight deck half-way to wing and shorter but fatter axial fairing along nose on each side of nose-gear compartment. Second prototype and three pre-production, latter without nose fairings but initially with AL-7 engines. Numerous aerodynamic and systems changes, including major increase in span and chord of drooped outer leading edge, addition of six small fences and small wingtip fairings (incorporating fuel vents and deicing air exhaust) and progressive introduction of NK-8-4 turbofan engine with cascade-type reversers on outboard engines. Prolonged development, first Aeroflot service 10 March 67 with production Il-62 built at Kazan. Basically good long-range aircraft, with exports

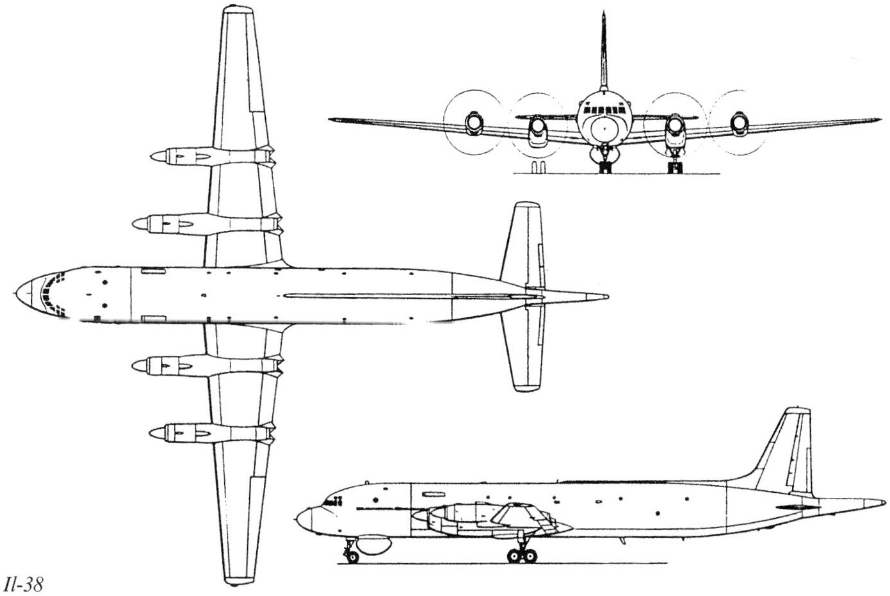

Il-38

Il-62 prototype

Il-62M

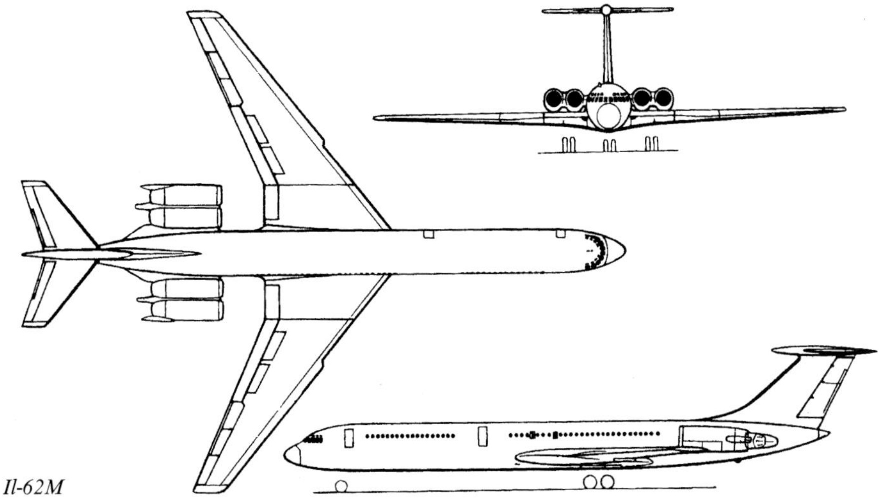
Il-62M

to CSA, LOT, Interflug, Tarom, Cubana and CAAC. Flight crew five, normal seating for up to 186 (usually 163) in triple units with baggage/cargo under floor ahead of and behind wing and (unpressurized) rear fuselage, total 48 m³ (1,694 ft³). In 1970 improved Il-62M-200, later called just **Il-62M**, with different engines, clamshell reversers on outers only, 5000 lit (1,100 gal) fin tank, no fences or tip fairings, differential spoilers for roll augmentation, containerized baggage/freight, modern pilot yokes instead of large ram's horns with 360° nose-gear steering wheels, generally updated avionics and emergency equipment. Entered service early 74. Followed 78 by **Il-62MK** with longer-life airframe cleared to higher weights, wider bogies (lower-pressure tyres, improved brakes), auto spoiler on touchdown, better interior arrangement giving 195 seats yet wider aisles for service carts. Total production 245 by 90, with 25 more to keep factory in employment to 95, one completed 93 as VIP for Russian President. ASCC name 'Classic'.

DIMENSIONS Span 43,2 m (141 ft 9 in), (with tip fairings 43,3/142 ft 0¾ in); length (fuselage) 49,0 m (160 ft 9 in), (overall) 53,12 m (174 ft 3½ in); wing area 279,6 m² (3,010 ft²).

ENGINES (initial production) four NK-8-4, (62M) D-30KU.

WEIGHTS Empty (equipped) (early) 67,8 t (149,470 lb), (production) 69,4 t (152,998 lb), (62M) 71,6 t (157,848 lb); max payload 23 t (50,700 lb), (62M) 23 t, (62MK) 25 t (55,115 lb); max take-off (early) 157,5 t (347,222 lb), (production) 162 t (357,143 lb), (62M) 165 ft (363,757 lb), (62MK) 167 t (368,166 lb).

PERFORMANCE Normal cruise 820/900 km/h (510/560 mph, max M 0·846); range (max payload, 1 h reserve) 6700 km (4,160 miles), (62M) 7800 km (4,846 miles); max climb (62, SL) 1080 m (3,540 ft)/min; FAR take-off field length (62, ISA, SL) 3250 m (10,660 ft).

Il-76 Possibly first aircraft from OKB in which Ilyushin played no direct part, this impressive freighter was designed mid-60s as replacement for An-12 for V-TA and Aeroflot, basic requirement being 40 t cargo carried 5000 km in 6 h.

Wing in five sections, centre section being width of fuselage and structurally part of it, rest of wing –4° anhedral and 25° sweep at ¼-chord with joint outboard of outer engines. T/c ratio 13% at root, 10% at tip. Main box multi-spar integral tank with integrally stiffened skins. Each wing has five sections of hydraulically-powered slat, inboard and outboard hydraulically-powered triple-slotted flap, eight sections of hydraulically-powered spoiler (four ahead of each flap) used for roll augmentation, letdown and lift-dump, and one-piece ailerons with hydraulic boost. Powered tailplane, with aerodynamically balanced tabbed elevators. Single aerodynamically balanced tabbed rudder. All flight controls hydraulically boosted with emergency manual reversion. Fuselage circular, diameter 4,8 m (189 in), pressurized throughout cockpit and cargo hold to 0,5 bar (7·25 lb/in²). Crew door each side at front, hydraulically powered for escape at high IAS. Rear door large left/right halves, hinged outward hydraulically, with upward-hinged ramp able to lift 30 t load placed on it. Hold 3,46 m (136·2 in) wide, 3,4 m (133·9 in) high and 20 m (65 ft 7½ in) long (24,5 m, 80 ft 4½ in, including ramp). Titanium floor, optional or folding roller conveyor panels. Two emergency escape windows each side. Flight compartment for two pilots, radio and nav, plus supernumerary and two freight handlers. Nose glazed for contact nav, heavy dropping and similar tasks. Very comprehensive avionics including weather radar in nose and nav/mapping radar under nose. High-flotation landing gears: steerable nose unit with left/right pair of wheels retracts forward; four main gears, each with left/right pairs of wheels, retract inwards with rotation 90° so that all four axles lie fore/aft, all hydraulic actuation. Nosewheel steering ±50°; nosewheel tyres (4) 1100×330 mm, main tyres (16) 1300×480 mm, tyre pressure variable in flight 2,5/5 kg-cm² (36/72 lb-in²), all doors closed with gear extended to keep wells clean on snow, water, sand. All engines fitted with lateral clamshell reversers. Entire wing box forms six integral tanks, total 81 830 lit (18,000 gal). APU in left landing-gear fairing; environmental-system inlets in wing roots, and hot-air leading-edge de-icing throughout, with electro-thermal glazed panes.

First prototype No 86712 flown by Eduard Kuznetsov from Khodinka 25 March 71. Five pre-production, one civil and remainder military with most features of 76M. Baseline version **Il-76T** with additional centre-section fuel. Two 3 t (6,614 lb) winches at front of hold, and two travelling cranes with total of four hoists each rated at 2,5 t (5,511 lb). Typical loads include large vehicles/armor/earthmovers, ISO containers (8 ft × 8 ft) with lengths to 40 ft, 12 2,5 t containers 1,46 × 2,44 × 1,9 m, six 5670 kg (12,500 lb) pallets 2,99 × 2,44 m or 12 2,5 t

Il-76 prototype

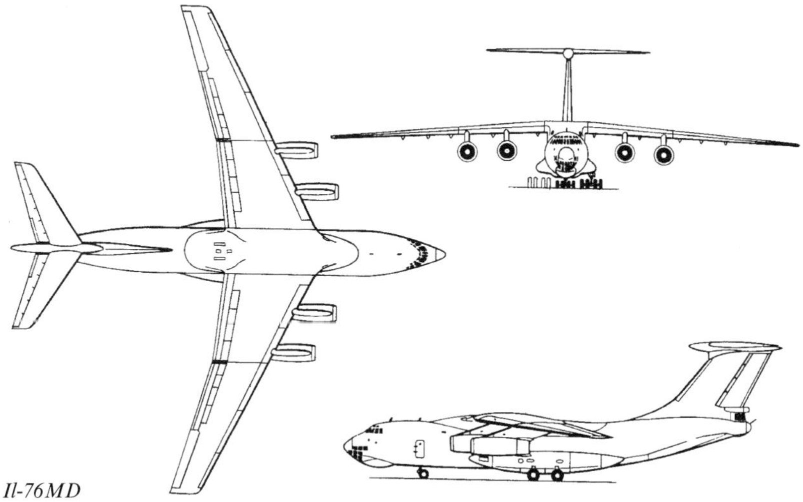

Il-76MD

Il-76VPK

Il-78M

pallets 1,46 × 2,44 m, or three quick-fit modules (20 ft ISO size) each with 2+2 seats for 36 passengers or equipped for litter patients. Analog avionics for Cat II operation includes nose weather radar, chin nav/mapping radar, computer and optional HUD for flight conttrol and autoland. **Il-76M** is baseline military version with Il-type tail turret with twin GSh-23 guns, ECM receiver/jammer fairings on sides of nose and rear fuselage, plus dispensers for 96 flares/chaff of 50 mm calibre on each main-gear fairing and each side of rear fuselage; can carry 140 equipped troops or 125 paratroops. **Il-76TD** replaced 76T in production 81, D-30KP-1 engines flat-rated to ISA+23C permitting increased weights; equivalent military version is **Il-76MD**. All versions, design factor 2·9 to 100 t, 2·0 to 190 t. ASCC names. T, TD, 'Candid-A', M, MD, 'Candid-B'.

In 75 set 25 records including 70 t to 11 875 m, 60 t round 2000-km at 875 km/h and group para jump from 15 386 m (50,479 ft). Tashkent built 760 by 92, then output reduced. Military V-TA and AV-MF (c500), Algeria, Czech/Slovak, India (24 *Gajaraj*), Iraq and Poland. Iraq includes tanker (drogue), *Baghdad-1* with Thomson-CSF *Tigre* radar inverted under tail and *Adnan-1* with rotodome and large strakes. Commercial: former Aeroflot (120), Cuba, Iraq, Italy (Metro Cargo), Libya, NZ (Pacific Express), N. Korea, Switzerland (Tl) and Syria. Total exports 133 by May 93.

Several rebuilt as **Il-76MA** for trials, four available from Gromov Flight Research Inst. Engines tested, NK-86, PS-90A, D-18T, D-236 and NK-93. One ex-VTA mod as **Il-76MDK** for zero-g Cosmonaut research. Two ex-Aeroflot (76450/76451) **Il-76VPK** command posts with 'doghouse' over satcom/IR dishes, ventral canoe radome and strakes, 14 blade antennas, HF probes under outer wings, VLF trailing wire and new APU in left MLG bay, **Il-76DMP** firebomber dumps 44 t (97,000 lb) water/retardant over precise area 500 m × 1000 m in 6 s, reload in 15 min, or drop 40 para-fighters.

A-50 AWACS-type, see under Beriev. **Il-78** tanker first tested as converted MD #78782 77, refined over ten years, entered service 87. Wing fuel 90 t, fuselage 28 t including cylindrical tanks in hold totalling 64 000 lit (14,080 gal). Initial service aircraft with single UPAZ-1A hose-drum unit scabbed on left side of rear fuselage. Crew 7, nose navigator and radars almost same as 76MD but tail turret replaced by observer station. Lights and ranging radar in underside of ramp door. From 89 UPAZ-1A added under each wing, becoming **Il-78M**. Limits for contact 2-9 km (6,560-29,530 ft) at 400-600 km/h (249-373 mph). Est 15 delivered 93, ASCC name 'Midas'. Upgrades described separately. Most VVS aircraft taken over by Ukraine.

DIMENSIONS Span 50,5 m (165 ft 8 in); length 46,59 m (152 ft 10¼ in); wing area 300,0 m^2 (3,229 ft^2).

ENGINES Early, four D-30KP, most D-30KP-2.

WEIGHTS (MD) Empty 97,9 t (215,830 lb); fuel, normal 65,5 t, max 118 t; payload 50 t; max loaded 190 t (418,878 lb); max ldg 151,5 t (333,995 lb).

PERFORMANCE Max speed 850 km/h (528 mph); cruise 750-800 km/h (466-497 mph) at 12 km; service ceiling 14,9 km (48,900 ft); range (50 t payload) 3650 km (2,268 miles), (20 t) 7300

Il-86

km (4,536 miles); TO 1750 m, ldg 900 m (STOL technique) 450 m.

Il-76 upgrades To complete with An-70 in CIS and foreign markets **Il-76MF** launched 92, max effort to fly mid-94. New engines, fuselage stretched by front/rear plugs by 6,6 m, hold 3,45 × 3,4 × 31,14 (26,6 excl rear ramp door), taking 4 ISO or Yak-10 containers or 9 PA-5,6 pallets. New avionics and cockpit including *Kupol* flt/nav complex. Payload vol ×150%, fuel η × 112%, TO run ×50%. Data for this version. **Il-76CFM**, project awaiting customer; original airframe re-engined with CFM56-5C2 (A340 pod) gives range advantage, eg with 50 t payload at 210 t MTO = 6400 KM (3,980 miles), with 30 t = 9400 km. Alternatively, IAE V2500A5 engines in A321 pods.

DIMENSIONS As Il-76 except length 53,19 m (174 ft 6 in).

ENGINES Four PS-90A.

WEIGHTS OWE 101 t (222,663 lb); payload 52 t; loaded 200 t (440,920 lb); MTO 210 t (462,963 lb).

PERFORMANCE Cruise 750-780 km/h at 9-12 km as before; range (40 t payload, full reserves) 5200 km (3,231 miles); TO run 1 km; ldg field length 2,55 km.

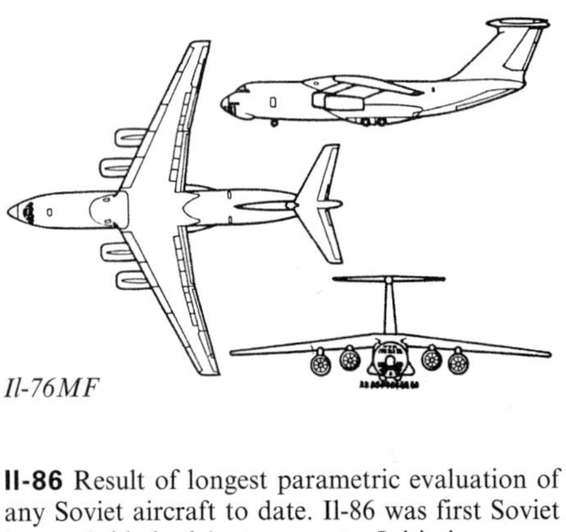
Il-76MF

Il-86 Result of longest parametric evaluation of any Soviet aircraft to date. Il-86 was first Soviet large (wide-body) transport. Initiative came from OKB, Ilyushin and Novozhilov discussing such aircraft with Ministry of Civil Aviation from 1966 as result of launch of 747. Aeroflot stated it had no requirement for such aircraft, which in any case was judged incompatible with

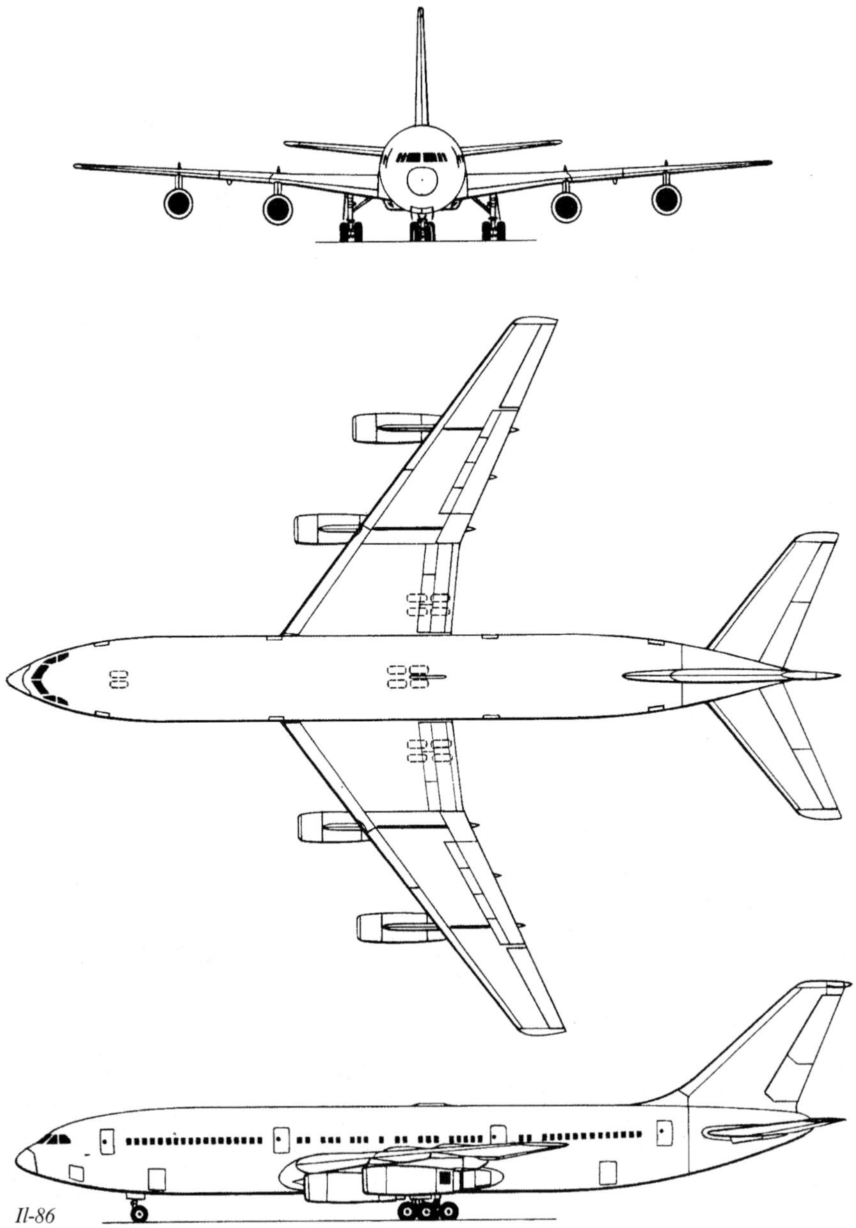
Il-86

almost all Soviet airports. No funding allocated, but OKB continued studies, whilst in parallel working on smaller (150/250-seat) Il-70, 72 and 74. Novozhilov convinced there was need for 350-seater, and obtained interest from military V-TA when he showed such aircraft could operate into airstrip with no facilities whatsoever. Studies continued, centring on body cross-section; superimposed tubes each 3,75 m diameter, side-by-side tubes each 3,6 m, and simple tubes of 5,64, 6,08 and 6,58 m. Final choice fell on 6,08 m (239·37 in), with underfloor hold height 1·95 m, passengers boarding and leaving at this level. Il-86 revealed 71 ability to use airfields with poor runway and few terminal facilities. Original Il-70, 72 and 74 projects disclosed 1967-70, growing from 150 to 250 and then to 350 seats. In 71 Novozhilov revealed Il-86 as medium-range aircraft with large circular-section fuselage, T-tail, low wing and four rear-mounted D-30 engines, to fly 76. Ilyushin OKB brochure mid-72 showed changed layout with underwing engines of different type.

Wing similar to Il-76 but swept 35° at ¼-chord. Two-part hydraulic slat on each leading edge (notch to clear inner pylon). Small upper-surface fences aligned with pylons. Hydraulic double-slotted flaps on centre section and outer panels; inboard end of outer flap made separately and max angle reduced to avoid inner jet. Hydraulic spoilers, two ahead of inboard flap and four ahead of outer, for roll augmentation, letdown and lift-dump. Hydraulically boosted outer-wing ailerons. Flight controls, no manual reversion. Low-mounted powered tailplane with slight dihedral; elevators and rudders in two parts each boosted by separate hydraulic circuits (quad hydraulics 211 kg/cm², 3,000 lb/in², each energized by separate engine-driven pump). Circular fuselage barrel sections of 6,08 m (239·37 in) diameter; selected configuration has upper deck for passengers, with floor of CFRP/honeycomb sandwich (various specimens imported during development). Lower deck 8 to 16 LD-3 containers plus carry-on baggage and coats of passengers boarding or leaving via three powered stairways. Three further internal stairs to main deck, typically furnished with 5,7 m interior width for up to 350 passengers in three triple seats with 550 mm aisles. Enclosed baggage lockers. Four Type A ICAO doors each side at upper level for emergency escape with inflatable slides (and to mate with loading bridge where available). Ten principal doors at lower level, three for passengers, three for cargo and four to systems compartments. Steerable twin-wheel nose gear retracts forwards; DC-10-30 arrangement of main gears but all being four-wheel bogies (tyres 1300×480 mm), two retracting inwards and centreline gear forwards. All landing gears made at Kuibyshyev, hydraulic operation. Intergral tankage in wing box, five compartments, total 114 000 lit (25,077 gal). All engines hung on pylon struts aligned with fences and flap tracks, and equipped with left-right clamshell reversers also said to serve as noise-attenuators. APU in tailcone (occupied by parachute in prototype). First aircraft fitted with novel airframe de-icing (developed on other Ilyushin aircraft since 63) using flex leading-edge skin periodically pulsed by giant electrical currents for a few millisec. Flight deck for two pilots and engineers, with nav if required. Weather radar, doppler, VOR/ILS and computer for Cat IIIA.

Prototype No 86000 flown by E. Kuznetsov from 1820 m runway at Khodinka 22 Dec 76. 86002, flown 24 Oct 77, described as first production Il-86. Prototypes assembled from parts made elsewhere (eg, tail at Kiev). Production wing movable surfaces, tail surfaces, engine pods and pylons all made by WSK-Mielec, Poland. Hoped to begin scheduled services in time for Moscow 1980 Olympics, but in fact start was delayed to 26 Dec 80. Optional variant without three ground-level stairways, with 25 extra (total 375) seats and 7% better economics, but this was directly counter to Aeroflot need for fewer than 350 passengers, not greater. Studies continued of military versions, notably national command centre and AWACS derivatives, but no funding or orders beyond original 100, assembled at Voronezh. Not wholly satisfactory aircraft, suffering from old-technology engines with BPR 1·1 giving poor fuel economy and noise far in excess of FAR 36 Pt II, let alone Pt III. Decision taken Jan 91 to halt production at No 89 and complete remainder (plus possible additional aircraft) with substantial improvements including CFM56-5C2 engines each of 31,200 lb st in A340-type nacelles. This solves noise problem and will increase full-payload range to over 6,000 km. Production due to restart Feb 94, but money for this upgrade is not yet available. ASCC name 'Camber'.

DIMENSIONS Span 48,06 m (157 ft 8¼ in); length 60,21 m (197 ft 6½ in); wing area 320 m² (3,444 ft²).

ENGINES Four NK-86.

WEIGHTS Empty, not disclosed; max fuel 86 t; max payload 42 t (not with max fuel); max take-off 208 t (458,560 lb).

PERFORMANCE Normal cruise 900/950 km/h (560/590 mph); range (max payload) 3600 km (2,237 miles), (max fuel) 4600 km (2,858 miles); published field length 2,6 km (8,530 ft); approach speed 260 km/h (162 mph).

Il-90 In competition with An and Tu, preliminary studies funded 81 for passenger transport powered by propfans. Known internally as DMS (long-haul aircraft), configuration similar to A310 with high-aspect-ratio wings with winglets carrying twin underslung engines. Circular fuselage of 4·22 m diameter seating up to 220 passengers 2+2+2. Project not funded in late 1992, and no present intention to build.

DIMENSIONS Span 47,7 m (156 ft 6 in).

ENGINES Two Samara NK-93 propfans, de-rated to about 17 t (37,480 lb).

WEIGHTS Empty (equipped) 64 t (141,095 lb); max payload 24 t (52,910 lb); max take-off 125 t (275,575 lb).

PERFORMANCE Normal cruise 850/870 km/h (528/540 mph); max range (200 passengers) 12 000 km (7,460 miles); balanced field length 2,65 km (8,695 ft).

Il-96 Gradual recognition that Il-86 was flawed in basic respects led to critical decision in 76 (before first flight of Il-86) to commit large funds to almost total redesign. Il-96 had as objective rectification of all shortcomings, with new wing and tail surfaces, new engines, shorter body and complete revision of structure, materials and systems. Design life 60,000 h, 12,000 landings.

Wing of new supercritical profile, reduced sweep (30° at quarter-chord) and span increased to give aspect ratio of 9·5, plus winglets. Three spars inboard, two outboard, three upper-surface machined skins with integral stiffeners and four on underside, all new hi-purity alloy. Hydraulic double-slotted flap inboard and two sections of single-slotted flap outboard, all track-mounted and of composite material. Small aileron between inboard/outboard flaps. Full-span leading-edge slat in seven sections on each wing. Three airbrakes ahead of inboard flap and six spoilers ahead of outboard flap-pair, two inboard spoilers assisting lateral control and remaining four serving any mixture of roll-control or airbrake. Outboard ailerons used solely in unison to alleviate gusts. Fuselage diameter 6,08 m as before but shortened to 51,15 m (167 ft 9·7 in) by reducing parallel mid-section to frames 40 to 67 only. Main floor and cargo floors of CFRP and honeycomb. Pressurized volume 950 m³, of which 350 m³ is passenger cabin. Redesigned to eliminate unconventional lower-deck boarding, entire underfloor area now being cargo, landing gear and systems. Three passenger doors in upper deck on left, two ahead of wing, all with escape slide but no airstair. Corresponding emergency doors opposite on right. Lower deck cargo compartments for six ABK-1·5 (LD3) containers or

Il-96 (initial version called 96-300)

Il-96M

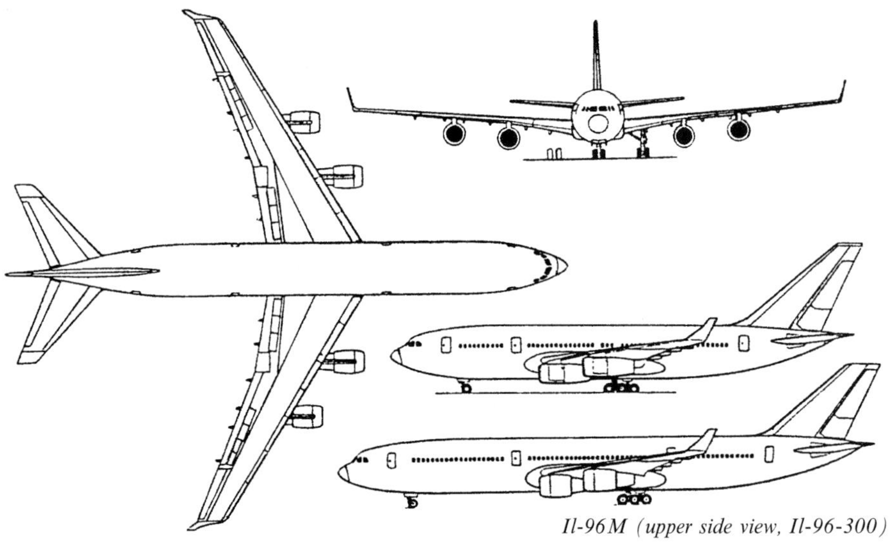

Il-96M (upper side view, Il-96-300)

Igloos at front and ten ABK-1·5 at rear, plus tapering aft bay for general cargo. Omsk VSU-10 54 hp APU in tailcone. Tail surfaces redesigned for greater area and power. Horizontal span unchanged at 20,57 m (67 ft 6 in) but sweep increased to 37°30' at 0·25-chord giving area 96·5 m^2 (1,039 sq ft). Powered tailplane with two-part untabbed elevators. Vertical tail extended to 9,91 m (32 ft 6 in) and swept at 45° at 0·25-chord, with two-section rudder. All flight controls triplex FBW, with manual reversion, to drive fully powered surfaces. Landing gears generally as before, with three main units made at Kuibyshev with KT-204 wheels and 1300×480 tubeless tyres. Track reduced to 10,4 m (34 ft 1·5 in); centreline bogie pivots up 20° before leg retraction. Nose gear tyres enlarged to 1260×460, wheelbase reduced to 20,065 m (65 ft 10 in). Unit retracts forward, two of the six doors remaining open on ground. All tyres 11,5 atm (169 lb/sq in).

New engines of high BPR designed for this aircraft and Tu-204. Nacelles and pylons improved aerodynamically over Il-86. Fan duct reversers incorporate blocker doors and peripheral fixed cascade grills. Inboard inlet ground clearance 1,24 m (48·8 in). Five integral tanks in wing box for 152 620 lit (40,318 USG, 33,572 Imp gal). Four independent 207-bar (3,000 lb/sq in) hydraulic systems. Triplex INS, Omega and Glonass navigation, plus HUD, for automatic guidance in flight and to runway in Cat IIIA. Each pilot has two colour displays and trad standbys, autothrottle with IAS input but no AOA protection. Engineer station in prototypes only. Cabin typically 262 pax in 3 classes 3+3+3, 2+2/2+2 and 2+2+2, or 300 max tourist. One static, one fatigue and three flight articles funded, first flight Khodinka 28 Sept 88, Russian ENLGS cert'n achieved 29 Dec 92, initial 100 in prodn Voronezh (several for CSA), delivered from early 93.

Il-96M, orig (87) called 96-350, stretched 6,05 ahead of wing and 3,3 aft, giving 49,13 cabin for 311 3-class or 386 tourist. Extra length allows smaller fin. Input from 18 US suppliers incl engine pods similar to 757. Rockwell-Collins avionics: EFIS/EICAS/AFCS/FMS and integrated nav (GPS/ADS/IRS), GPWS and blind landing, cockpit 8 (8 in × 8 in) displays, ARINC-700, SAT-900 satcom with Ball conformal antennas. First flight 6 Ap 93, orders placed June 93 by Russian Int'l and Partnairs (Neth), latter selecting **Il-96T** freighter, main deck 23 P1 or P6 pallets, lower deck 9 pallets or 32 LD3 containers, total 74 t for 7500 km. **Il-96MK** projected with 96M airframe and NK-93 propfans giving range 14000 km. **Il-96-500** would be 96K with 2-deck (cottage loaf) fuselage seating 68 upper and 344 in 3-class lower, or 512 all-tourist. No ASCC reporting name.

DIMENSIONS Span over winglets 60,105 m (197 ft 2½ in), bare wing 57,66 m (189 ft 2 in); length (300) 55,345 m (181 ft 7 in), (M) 64,694 m (212 ft 3 in); height (300) 17,57 m (57 ft 7 in), (M) 15,88 m (52 ft 1 in); wing area 391,6 m^2 (4,215 ft^2).

ENGINES (300) Four PS-90AN, (M) four 38,250 lb PW2037.

WEIGHTS OWE (300) 117 t (257,940 lb), (M) 132,4 t (291,887 lb); max payload (300) 40 t, (M) 58 t; MTO (300) 230 t (507,055 lb), (M) 270 t (595,238 lb); ZFW (300) 157 t (346,120 lb); MLW (300) 175 t (385,800 lb).

PERFORMANCE Cruise (300) 850 km/h, (M) 850-870 km/h (528-541 mph); range with max payload (300) 9500 km (5,900 miles), (M) 12 500 km (7,767 miles); balanced field length (300) 2,7 km (8,860 ft), (M) 3350 m (11,000 ft).

Il-102 Revived of interest in *Shturmovik* aircraft in 66 resulted in competition, Su T-8 beating this aircraft devd from Il-40 as private venture. Wing 12% thick, fixed LE 30°, dihedral 0°, powered ailerons, area-increasing slotted flaps, downturned tips. More sharply swept tail with fixed tailplane and powered elevators, balanced rudder with full-length tabs, all surfaces with trim tabs. Two spoilers ahead of each flap, door airbrakes on rear fuselage opening into jets. Two main wing spars, in fuselage bounding main protected tanks, then picking up engines and (rear spar) main landing gear. Engines centred on wing root, plain inlet, access doors above and below wing, jetpipe turned down and out (plan to instal reversers). Twin-wheel main gears with low-pressure tyres retrac forwards into twin-door underwing fairing; steerable nosewheel retrac to rear. Pilot high with excellent view, bulletproof screens and upward-hinged canopy, K-36L seat. Radio operator/gunner over TE with upward-hinged canopy, optical sight and remote control of tail barbette with GSh-23 with superimposed barrels, 600 rounds. Cockpits, underside of engines and tanks protected by titanium/composite armour. Steeply downsloping nose without sensors. GSh-30/II (500 rounds) external under centreline. Three internal bomb cells in each outer wing, further out three pylons. Max weapon load 7200 kg (15,873 lb) including FAB-1000, UV-32-57 launchers, KMG-U dispenser, R-73 and R-60M defence missiles and drop tanks under wing roots.

Single flight article plus static-test specimen. First flight 78. NII considered excellent aircraft, despite losing to Su rival. AKI Ilyushin actively marketing.

DIMENSIONS Span 16,9 m (55 ft 5⅜ in); length 17,75 m (58 ft 2⅞ in); wing area 63,5 m^2 (683·5 ft^2).

ENGINES Two RD-33I (unaugmented).

WEIGHTS Empty 13 t (28,000 lb); fuel 4 t; loaded 18 t (39,683 lb); max 22 t (48,500 lb).

Il-102

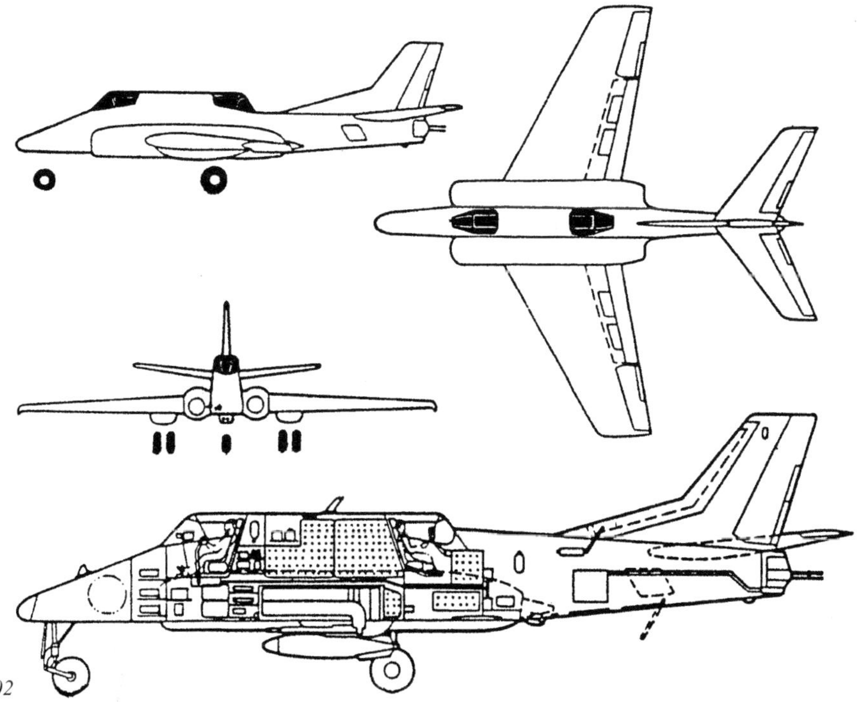

Il-102

PERFORMANCE Max speed 950 km/h (590 mph); combat radius 400-500 km (345-435 miles); ferry range 3000 km (1,864 miles); TO 150 km/h, 290 m; ldg 180 km/h, 300-350 m (with reverse).

Il-103 First lightplane, originally for DOSAAF as trainer, became multirole to FAR-23 with Bendix-King avionics. Simple light alloy, upward 'gull wing' doors, fixed gear but hyd flaps; Mod 1, **Il-103SPO** = 2 seats, dual stick control, +6/−3g, optional tow hook; Mod 2, **Il-103** = wheel control, <5 seats or 440 kg (970 lb) cargo. First flight 17 May 94, deliveries due late 95.

DIMENSIONS Span 10,6 m (34 ft 9 in); length 7,955 m (26 ft 1 in); wing area 14,71 m^2 (158 ft^2).

ENGINE One 210 hp TCM IO-360-ES.

WEIGHT Empty (1) 720 kg, (2) 765 kg; max (1) 965 kg (2,127 lb), (2) 1310 kg (2,888 lb).

PERFORMANCE Max speed (1) 265 km/h; cruise (1) 235 km/h (146 mph), (2) 225 km/h (140); range (2, with reserve) 1240 km (770 miles); TO (1) 160 m, ldg (1) 165 m.

Il-106 Project for new heavy transport in C-17 class to replace Ukrainian An-124. Design payload 80 t (176,370 lb), carried in hold 34 m (111 ft 6 in) long, 6 m (19 ft 8 in) wide and 4,6 m (15 ft) high. Swept wings with winglets and advanced high-lift systems, swept tail with low tailplane, propfans for best speed/economy. Funding sought.

DIMENSIONS Span 58,5 m (191 ft 11 in); length 57,6 m (188 ft 11¾ in); wing area c370 m^2 (3,982 ft^2).

ENGINES Four NK-92.

WEIGHTS Empty c135 t (300,000 lb); loaded 280 t (617,300 lb).

PERFORMANCE Cruise 820-850 km/h (509-528 mph); range with reserves 5000 km (3,730 miles); TO 1550 m; ldg 1400 m.

Il-108 Shown as model in 90, another project initiated at OKB to maintain employment. Planned as business jet, third-level transport or for paramilitary/surveillance tasks, aft-mounted

Il-103

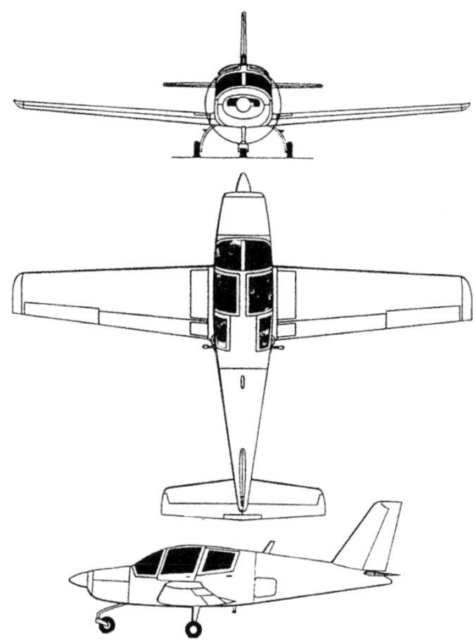

Il-103

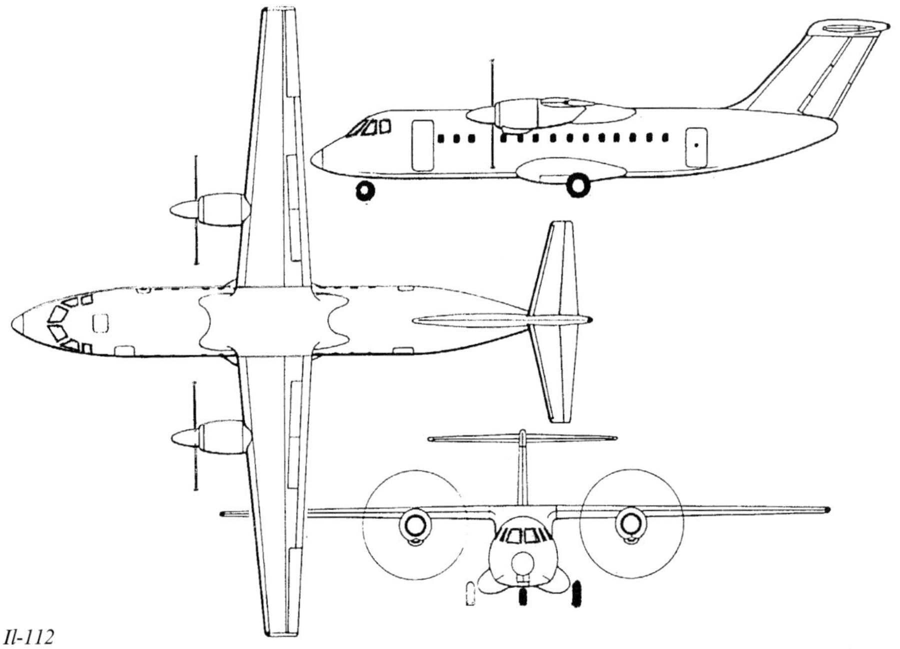

Il-112

engines, T-tail, manual flight controls, integral tanks, twin-wheel main and nose gears and pressurized fuselage of 2,35 m (92·5 in) diameter whose length could be adjusted to suit mission. Initial configuration two crew and nine business passengers or 15 airline passengers seated 2+1, in each case with toilet, wardrobe and front/rear baggage. No funding.

DIMENSIONS Span over winglets 15,0 m (49 ft 2·5 in); length 15,85 m (52 ft 0 in); height 5,5 m (18 ft 0·5 in).

ENGINES Intended two 2,2 t (4,852 lb) DV-2 turbofans.

WEIGHTS Max payload 1,5 t (3,307 lb); max take-off 14,3 t (31,526 lb).

PERFORMANCE Cruise 800 km/h (497 mph); range, 15 passengers 4850 km (2,796 miles), nine passengers 6000 km (3,728 miles); balanced take-off runway 1,8 km (5,905 ft).

Il-114

Il-112 Projected twin-turboprop, pressurized cabin for 32 pax (2+1); cargo version studied.

DIMENSIONS Span 21,0 m (68 ft 10¾ in); length 20,0 m (65 ft 7 in).

ENGINES Two TV7-117.

WEIGHTS OWE 7,6 t (16,755 lb); max 12,36 t (27,249 lb).

PERFORMANCE Cruise (max) 600 (373), (econ) 500 (311); range (econ) 1900 (1,180); TO 380; ldg 350.

Il-114 OKB awarded design contract 83 for future standard regional twin-turboprop to replace An-24/26 with Aeroflot. Design completed early 86, outwardly almost identical to British ATP. Prototype SSSR-54000 flew at Zhukovskii test centre 29 March 90.

Basically conventional low-wing structure, though two-spar wing has Al-Li skins with separate bonded/riveted Z-stringers, many parts liable to corrosion are magnesium alloy and about 8% of airframe is of GFRP/CFRP composites. Production aircraft have over 11% by using CFRP structural boxes for fin and tailplane. Design airframe life 30,000 h. One-piece hydraulic double-slotted flaps mounted on three tracks. Ahead of flaps are airbrakes (one inboard and one outboard of nacelle) and, after wide gap, a spoiler outboard to supplement large manual ailerons with trim and servo tabs. Spoilers used solely following failure of one engine on take-off. Leading edge fixed but removable. Manual elevators with inset hinges and balance tabs, and manual rudder with inset hinges and trim and servo tabs. Fuselage circular, diameter 2,86 m (112·6 in), cabin length 22,24 m (72 ft 11·5 in) with dP 0,38 bar (5·5 lb/sq in). Twin-wheel landing gears retract forwards hydraulically. Tyres 880×305 mm (main), nose 620×180 mm (not 80 as widely reported); tyres 5,9 bar (85 lb/sq in); nosewheels steer ±55°. Engines designed for this aircraft carried on steel-tube frames attached to steel firewall which is separated ahead of wing by main-gear bay, with removable metal panels over engine and overwing jetpipe. Stupino low-noise SV-34 propeller with six CFRP blades, diameter 3,6 m (142 in). Integral wing tanks out to rib 15 total 8125 lit (1,787 Imp gal). APU (production model not yet decided) in tailcone. Main engines drive 207 bar (3,000 lb/sq in) hydraulic pumps and 40 kVA alternators. Four

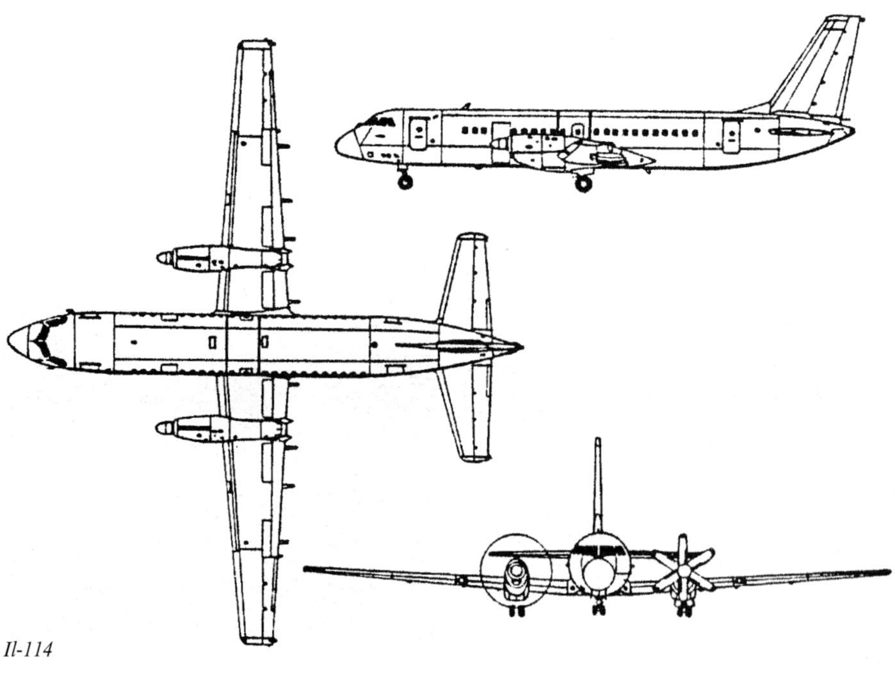

Il-114

separate electrothermal cyclic deicing systems for wings, windscreens and propellers, hot air for engine inlets.

Normal Aeroflot crew two pilots, one steward. Standard cabin 60 seats 2+2, aisle height 1920 mm and width 450 mm, with baggage at front and rear (plus overhead open racks) and aft bar (right) and toilet (left). Outward-opening airstair door at left front with escape-slide service door opposite; at rear a door on left (for airport steps or bridges) with emergency door opposite. Digital avionics for Cat II, two colour displays for each pilot plus central engines/systems displays, all subsystems with suffix -85MVL. Production aircraft to incorporate various Western avionics, with further changes for export.

Three flight prototypes plus static and fatigue airframes, certification complete 93. Initial Aeroflot order for 500, with guaranteed fuel burn 19 g/pax-km, half that of An-24. Production 5 per month by 14 main factories in Russia, Ukraine, Poland, Romania and Bulgaria, with assembly/test at Tashkent and (ex-MiG-29) Moscow. Igor Katyrev predicts orders for further 500 before 2000, including stretched 70/75-seat version and export model with PW127 engines. He predicts output 100 per year by 96. No ASCC reporting name allocated.

DIMENSIONS Span 30,0 m (98 ft 5·3 in); length 26,88 m (88 ft 2 in); height 9,32 m (30 ft 7 in); wing area 77 m^2 (829 sq ft).

ENGINES Two 1,865 kW (2,500 shp) TV7-117-3 turboprops.

WEIGHTS Basic operating 15,0 t (33,070 lb); max payload and max fuel both 6,5 t (14,330 lb); max take-off 22,7 t (50,045 lb).

PERFORMANCE Nominal cruise 500 km/h (311 mph); range, 64 passengers, 1000 km (621 miles), with 3,6 t (7,935 lb) payload, 2850 km (1,770 miles); take-off run on unpaved strip 1,4 km (4,600 ft); approach speed 185 km/h (115 mph); landing run 1,3 km (4,265 ft).

Il-X From 89 OKB studied business or commuter transport with two 1,550 shp TVD-1500 turboprops at rear driving twin-four-blade contraprops. Configuration generally resembling Il-108, with circular fuselage of 2,4 m (94·5 in) diameter (2 in larger than 108!) configured for eight business or 19 airline passengers. No funding by late 92; not judged worth detailed entry apart from intended MTO weight 9180 kg (20,238 lb) and max cruise 620 km/h (385 mph).

Interavia

Joint-stock firm formed in Moscow 91 to build GA aircraft in Lukhovitskii (MAPO) located at the factory airfield for Plant 30, sole production source for single-seat MiG-29. Projects announced Sept 93.

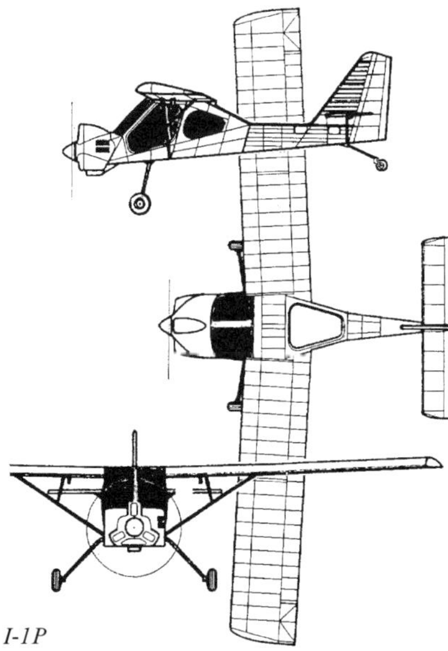

I-1P

I-1P

I-1P Multirole 3/4-seater, aerodynamically resembling KhAI-60-Let with forward-swept high wing. Simple stressed-skin, cantilever legs, manual flaps, two doors in wholly glazed cabin. Prototype flown 93.

DIMENSIONS Span 11,6 m (38 ft 0⅝ in); length 6,4 m (21 ft 0 in); wing area 14,48 m^2 (156 ft^2).

ENGINE One M-3.

WEIGHTS Empty 525 kg (1,157 lb); payload 180 kg; MTO 750 lb (1,653).

PERFORMANCE Max speed 230 km/h (143 mph); cruise 150 km/h (93 mph); range 700 km (435 miles); field length 300 m.

I-3 Sport/aerobatic single- and two-seater variants, factors +10/-8. Both versions displayed statically Moscow 93.

DIMENSIONS Span 8,1 m (26 ft 6⅞ in); length 6,7 m (21 ft11¾ in); wing area 11,54 m^2 (124 ft^2).

ENGINE One M-14PM.

WEIGHTS Loaded (solo) 881 kg (1,942 lb), (dual) 961 kg, (MTO) 1063 kg (2,343 lb).

PERFORMANCE Max speed 350 km/h (217 mph); dive limit 450 km/h; range 700 km (435 miles); climb 1,14 km (3,740 ft)/min.

I-5 High-wing multirole utility, 2+8 seats. Mockup 93.

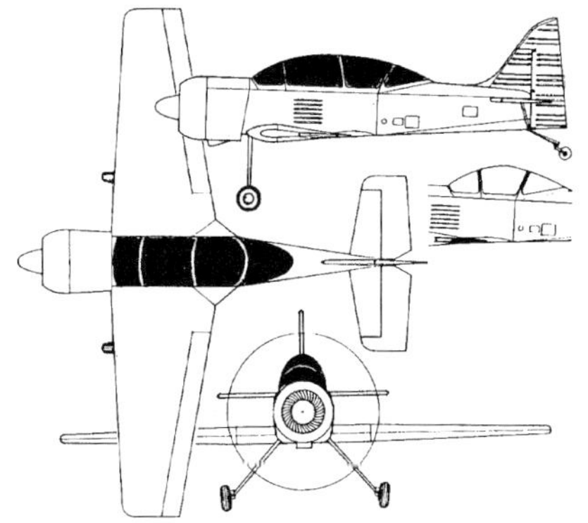

I-3 (inset, single-seat version)

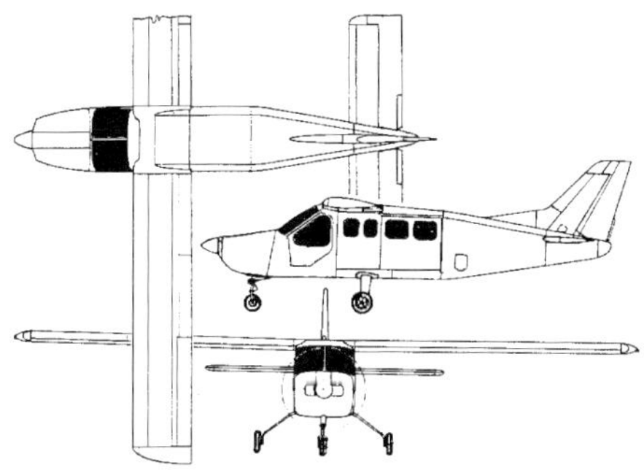

I-5

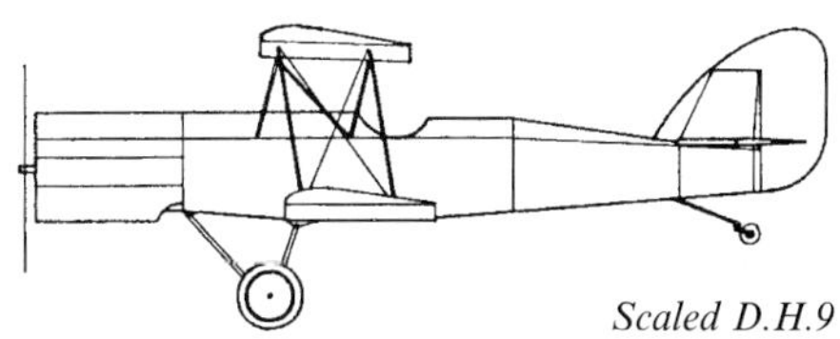

Scaled D.H.9

I-3 (two-seat version)

DIMENSIONS Span 14,6 m (47 ft 10¾ in); length 9,95 m (32 ft 7¾ in); wing area 20,44 m² (220 ft²).
ENGINE One 400-hp Textron Lycoming O-720.
WEIGHTS Empty 1375 kg; fuel 250 or 350; payload 700; MTO 2300 kg (5,071 lb).
PERFORMANCE (est) Max speed 300 km/h (186 mph); cruise 270 km/h (168 mph); climb 330 m (1,083 ft)/min; range 1800 km (1,118 miles).

D.H.9 Replica of famed British a/c, scaled down to 10 m span, with 200-hp Czech M-337.

Isacco⁰

Vittorio Isacco was Italian designer who worked on a helicopter in USSR 1932-36. In 36 he was arrested and worked in a Special KB, possibly at GAZ-39.

I-4 Also called **Isacco-4** and *Gyelikogyr,* this unusual helicopter was only Isacco design actually to be completed. Fuselage of KhMA welded steel tube with fabric covering. Tail, pilot-operated vertical and horizontal surfaces of light alloy, fabric covering. Tailskid and wide-track fixed main wheels. Cabin amidships for pilot and five passengers. On nose, main propulsion engine. Four-blade lifting rotor, each blade having constant profile with light-alloy ribs (welded from 12×10 mm elliptical tubing) located at intervals along light-alloy box spar with two main webs and upper/lower booms, fabric covering overall. Blades supported at rest by bracing ties from central cabane pyramid. Pilot controls to ailerons on pairs of arms behind outer trailing edge for cyclic/collective pitch control. Rotor driven by separate engines on tip of each blade. I-4 begun late 32 and built at ZOK NII GVF. Prof. B. N. Yuryev acted as consultant. Designer's calculations found unreliable, delaying completion until 35. Ground tests in that year caused deformed dural trunnion on engine, remanufactured in steel, but went on to discover severe blade flutter resulting in departure of one of tip engines and severe straining of whole machine. Never flew, and final conclusion was that tip-mounted engine idea was not practical.
DIMENSIONS Diameter of four-blade rotor 24,4 m (80 ft 0⅔ in).
ENGINES One 300 hp Wright J-5 radial driving four-blade propulsion propeller; four 120 hp D. H. Gipsy III each driving small four-blade propeller.
WEIGHTS Loaded 3 t (6,614 lb): also reported (Nemecek) 3,5 t (7,716 lb).
PERFORMANCE Did not fly.

Isacco 4

Junkers

Though strictly German designs, aircraft built at Fili by Junkers company 1921-27 qualify for brief inclusion as they were designed in Soviet Union and had no counterparts outside that country. Part of design work done at AB Flygindustri at Linhamm, Junkers' other foreign subsidiary.

Ju 20 Basically same as **A20** but BMW instead of Mercedes engine. Duralumin low-wing monoplane, pilot and one/two passengers. Fili-built aircraft all military, 20 twin-float seaplanes used by MA (Baltic and Black Sea Fleets) to 1930, then transferred to White Sea; also 20 landplanes for Red Army. Many

Ju 20

famous flights including first to Soviet Arctic (B. G. Chukhnovskii, 1924). In 25 K. A. Vigand installed 310 hp Junkers L-5 in one. Standard military type to 30 re-engined with M-6, aircraft redesignated **R-2.**

DIMENSIONS Span 17,8 m (58 ft 4¾ in); length 8,3 m (27 ft 2¾ in); wing area 32 m² (344 ft²).
ENGINE One 185 hp BMW IIIa 6-inline, later M-6.
WEIGHTS (landplane) Empty 1113 kg (2,454 lb); loaded 1593 kg (3,512 lb).
PERFORMANCE (landplane) Max speed 181 km/h (112·5 mph); climb 10·7 min to 2 km, 43 min to 5 km; range 570 km (354 miles).

Ju 21 Though today known thus in Soviet Union, correct designation **H 21**. Cantilever parasol monoplane, military reconnaissance, pilot and observer/gunner in tandem cockpits. Typical corrugated skin except over fuselage-side fuel tanks which formed large bulges. First flown late 22, about 100 built at Fili by late 25. Fixed Vickers and movable DA guns, light bombs (usually splinter anti-personnel) and fixed and hand-held cameras. Widely used in suppressing peasants resisting collectivization, especially in Turkestan.

DIMENSIONS Span 13,3 m (43 ft 7⅔ in); length 7,8 m (25 ft 7 in); wing area 21,7 m² (233·6 ft²).
ENGINE One 185 hp BMW IIIb.
WEIGHTS Empty 913 kg (2,013 lb); fuel 180 kg; loaded 1350 kg (2,976 lb).
PERFORMANCE Max speed 179 km/h (111 mph); climb 4·5 min to 1 km, 9·7 to 2 km; service ceiling 5,6 km; endurance 2·5 h.

Ju 22 Correct designation **H 22**; single-seat fighter derived from H 21; only three built.

Data as H 21 except much lighter, loaded weight 850 kg (1,874 lb); speed 250 km/h (155 mph); two fixed guns.

Ju 13 Soviet designation for large number of F 13 transports, also used on floats or skis in military roles.

PS-4 Soviet designation of W 33 transport. At least 70 built at Fili with Junkers L-5 engine, widely used in Siberia and Arctic on floats and skis to 41, seven with Dobrolyet, 10 with TsARB and 25 military.

JuG-1 Military variant of G 24 trimotor transport (also built at Fili), this was landplane bomber called K 30C by Junkers and **R-42** by VVS. According to Němeček, also known as **TB-2** despite confusion with Polikarpov bomber. Between 30 and 80 built, in Sweden and at Fili, plus smaller number of twin-float seaplanes designated **R-42M** (*Morskoi*) and **R-42T** (*Torpedonosyets*) used by MA for recon, torpedo dropping and utility transport. One, armament removed, used by Chukhnovskii to direct icebreaker *Krasin* to wreck of Nobile's airship *Norge* north of Spitzbergen, 1928. Others, armament removed and with civil radio, operated by Aeroflot 1930-31 from Lena, Yenesei and other Siberian rivers. Similar float lengthened used on first TB-1 seaplanes.

DIMENSIONS Span 29,9 m (98 ft 1 in); length (landplane) 15,2 m (49 ft 10½ in), (seaplane) 15,5 m (50 ft 10¼ in); wing area 94,6 m² (1,018 ft²).
ENGINES Three 310 hp Junkers L-5.

Ju 21

Ju PS-4

G 24

R-42M

R-42

WEIGHTS Empty (land) 3860 kg (8,510 lb), (sea) 4400 kg (9,700 lb); fuel/oil 1035 kg; loaded 6,5 t (14,330 lb).
PERFORMANCE Max speed (land) 190 km/h (118 mph), (sea) 175 km/h (109 mph); climb (land) 7·5 min to 1 km; service ceiling (land) 4,5 km; endurance 5 h.

Kalinin

Konstantin Alekseyevich Kalinin was born 1890. In the First World War he enlisted in army and trained as pilot, seeing action from 16 on German front. He fought with Reds in civil war, both as pilot and plant engineer, and in 23 appointed director of *Remvosdukhozavod* 6 (Aircraft Repair Factory 6) at Kiev. Here he had plenty of good aircraft-quality wood, and large amounts of steel tube from wartime Voisins. From 25 he designed aircraft, setting up GROSS at Kharkov and becoming one of first well-known Soviet designers. He had the pick of graduates of Kharkov Aviation Institute, and his team soon included I. G. Nyeman, Z. I. Itskovich, A. Ya. Shchyerbakov and V. Ya. Krylov, all of whom achieved fame on own account. His aircraft began with series of civil transports characterized by high wing of elliptical planform and wooden construction, mounted on welded steel-tube fuselage. Later he designed two remarkable bombers, but a third was never finished. Shavrov's semi-official account blandly states Ilyushin's DB-3 was better, so work on last bomber, K-13, was discontinued. He omits to state that in spring 38 Kalinin was arrested on fraudulent charge of conspiracy and spying and executed 24 April 40.

K-1 Also designated **RVZ-6** or *Remvosdukhozavod*-6 (from factory), K-1 set pattern for most of Kalinin's subsequent designs. Given that he had to use available materials and ancient (ex-Voisin) engine, hard to see how result could have been improved upon, and from mid-20s Kalinin was rightly regarded as source of most economic transports in USSR. Strut-braced high-wing monoplane. Wing of elliptical shape, aspect ratio 7, a basic outline Kalinin adhered to rigorously in successive variations. Centre section same width as fuselage and integral with it in welded steel tube. Left and right wings of wood built up on two spars with truss ribs and fabric covering. Two struts each side of steel tube with ply fairing. Fuselage and fin welded from steel tube, first Soviet metal truss fuselage. Engine cowl removable aluminium panels. Skin covering aluminium back to rear of cabin, fabric thereafter. Single door on left for pilot (enclosed cockpit with glazed front and side windows in line with wing leading edge) and four passengers. Main gear, wire-spoked wheels with fabric or metal covering, sprung on steel struts by rubber cords. Engine with ventral water radiator, tall exhaust stack and fuel tank under cockpit floor. K-1 built under Kalinin's direction by RVZ-6 workers led by D. L. Tomashyevich, A. N. Gratsianskii and A. T. Rudyenko. First flight 20 April 25, successful. Sept 25 flew to Moscow to conclude series of State tests. Acknowledged suitable for civil air fleet after being re-engined (see data). Decision to adopt Kalinin designs laid foundation for his own OKB at Kharkov initially named GROSS or GROS (*Grazhdanskoye Opytnoye Samolyetstroenie*, civil experimental aeroplane works), later a production factory.

DIMENSIONS Span 16,76 m (54 ft 11¾ in); length 10,72 m (35 ft 2 in); wing area 40,0 m^2 (430·5 ft^2).
ENGINE Built with 170 hp Salmson RB-9 (data for this engine); later fitted with 240 hp BMW IV.
WEIGHTS Empty 1452 kg (3,201 lb); fuel/oil 170+20 kg; loaded 1972 kg (4,347 lb).
PERFORMANCE Max speed 161 km/h (100 mph); cruise 130 km/h (81 mph); climb 12·3 min to 1 km; service ceiling 3 km; range 600 km (373 miles)/4 h; take-off 120 m; landing 180 m/70 km/h (43·5 mph).

K-2 With support initially from Ukraine government, GROSS was established and developed this refined K-1 with BMW engine. Solitary Kalinin all-metal design of this period, using welded-steel truss structure with skin of corrugated *Kolchug* aluminium, except for fabric flight control surfaces. First flight 26. Four passengers and pilot. Successful performer, and several built and used by Ukrovzdukhput. But this construction was labour-intensive and

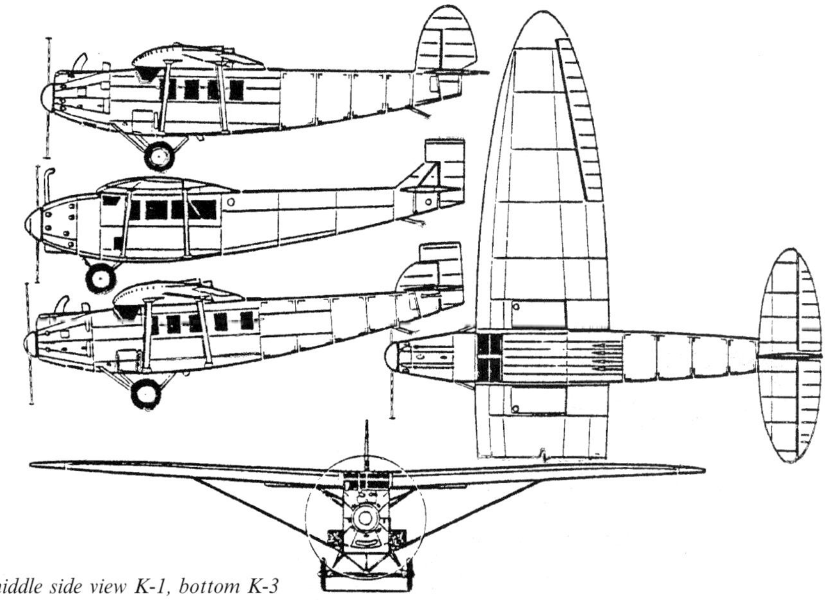
K-2, middle side view K-1, bottom K-3

loaded (passenger) 2350 kg (5,181 lb), (photo) 2040 kg, (ambulance) 2240 kg.
PERFORMANCE Max speed (passenger) 173 km/h (107·5 mph), (others) 175 km/h; climb (passenger) 7 min to 1 km (others 5·4 min); service ceiling 4,5/5,5 km; range 1040 km (646 miles); takeoff (passenger) 270 m, (others) 160 m; landing 200 m/73 km/h (45 mph).

K-5 Growing demand for air transport in Ukraine encouraged Kalinin to build larger, resulting in K-5, most popular and important of all his 13 designs. Consistent low price and operating economy enabled German equipment to be replaced not only in Ukraine but throughout Aeroflot. Basic requirement to carry eight passengers 800 km at over 150 km/h. Wing exact scale of earlier designs. Fuselage welded KhMA steel tube, dural skin from nose to rear of cabin, then fabric at rear and on tail. Enclosed cockpit for pilot and flight mechanic, one set of flight instruments but dual controls. Spoked main wheels (usually with light disc fairing), 1100×250 mm tyres; alternative skis but not floats. Tailplane adjustable in flight ±5° from datum. Compressed-air starting system for main engine. First flight 29 with M-15 engine. Insufficiently powerful, and early forced landing spurred switch to M-22, with which large-scale production organised, 260 examples being delivered by 34. Most passenger aircraft with eight dural seats, with superimposed upholstery. Some had carpet and curtains. Experiments with Townend rings 33, followed by decision to fit more powerful watercooled M-17F. First K-5 with this engine crashed following in-flight failure of nose rib. Prolonged problem with CG

heavy, and Kalinin never again departed from simple and cheap wooden wing for his civil transports.
DIMENSIONS Span 16,7 m (54 ft 9½ in); length 11,17 m (36 ft 7¾ in); wing area 40,0 m² (430·5 ft²).
ENGINE One 240 hp BMW IV.
WEIGHTS Empty 1600 kg (3,527 lb); fuel/oil 200+20 kg; loaded 2236 kg (4,929 lb).
PERFORMANCE Max speed 170 km/h (105·6 mph); cruise 140 km/h (87 mph); climb 12 min to 1 km; service ceiling 3,5 km; range 650 km (404 miles); take-off 220 m; landing 200 m/75 km/h (46·6 mph).

K-3 Variant of K-2 specifically built for ambulance duty. Cabin for doctor and either two stretcher patients or four seated. Long door on left side for loading stretcher. Equipment called 'very complete', and special hanging-strut suspension for superimposed stretchers devised by Dr A. F. Linhart (in Russian, Lingart), Czech-born 'father of Soviet aerial ambulances'.
DIMENSIONS, engine As K-2.
WEIGHTS Empty 1558 kg (3,435 lb); fuel/oil 265 kg; loaded 2235 kg (4,927 lb).
PERFORMANCE As K-2 except range 680 km (423 miles).

K-4 With this 28 machine Kalinin improved same basic design into versatile aircraft which at last weaned away Ukrvozdukhput from Dornier equipment and also sold in three versions, respectively for passengers, photography/survey and ambulance. Refined structure was lighter than K-3, so that even with BMW engine aircraft could carry four passengers, baggage and mail. Photo version with two or three operators in cabin to manage two fixed cameras (normally vertical) installed above hatches in floor. This model used for survey by Ukrainian government and Ukrvozdukhput. Ambulance version almost same as K-3 except for more powerful engine not fitted to other variants. First of this version exhibited at ILA Berlin 28 (note: Shavrov gives this date, but elsewhere dates ambulance model as 29, possibly indicating had not flown when exhibited). Total of 22 of three models built, all sold. One, *Heart of Ukraine*, made record flight Aug 28 Kharkov, Moscow, Irkutsk, Moscow, Kharkov; pilot M. A. Chyegirev, nav I. T. Spirin.
DIMENSIONS Span 16,75 m (54 ft 11½ in); length 11,35 m (37 ft 3 in); wing area 40,0 m² (430·5 ft²).
ENGINE (most) one 240 hp BMW IV; (ambulance) M-6.
WEIGHTS Empty (passenger) 1540 kg (3,395 lb), (others) 1400 kg (3,086 lb); fuel/oil 320 kg;

K-2

K-4

K-5 M-17

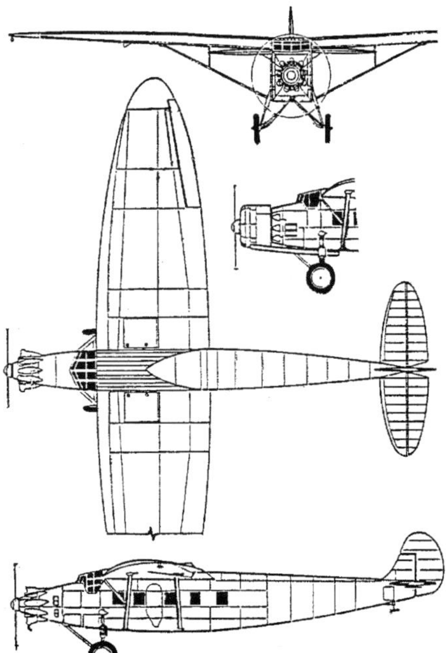

K-5, inset NACA-cowled M-22

and weight of M-17F, and either range or payload had to be cut. Original M-22 aircraft continued to serve in large numbers throughout USSR until 41, together with 'several tens' with M-17F and reduced payload.

DIMENSIONS Span 20,5 m (67 ft 3 in); length (M-15) 15,87 m, (M-22) 15,7 m (51 ft 6 in), (M-17F) about 16,5 m (54 ft 1½ in); wing area 66,0 m² (710 ft²).

ENGINE (prototype) one M-15, (most) M-22, (rebuilds) M-17F.

WEIGHTS (prototype) empty 2275 kg; fuel/oil 500+50 kg; loaded 3750 kg; (M-22) empty 2400 kg (5,291 lb); fuel/oil 400+40 kg; loaded 3925 kg (8,653 lb); (M-17F) empty 2740 kg (6,041 lb); fuel/oil 550+50 kg; loaded 4030 kg (8,884 lb).

PERFORMANCE Max speed (M-15) 190 km/h, (M-22) 185 km/h (115 mph), (M-17F) 209 km/h (130 mph); cruise (M-22, M-17F) 169 km/h (105 mph); climb 8 min to 1 km; service ceiling 4,5/5 km; range 820/960 km (500/600 miles); take-off about 200 m/20 s; landing 250 m/19 s/75 km/h (47 mph).

K-6 Derivative of K-5 for urgent dense cargo, notably mail and matrices for printing of national newspapers in places distant from Moscow. Wing basically that of K-5 but left and right panels joined to new small centre section held on two quad cabane struts above slim fuselage. Latter had modified structure, with thin seam welds and rectangular section just adequate for two pilots and box of freight, with same tankage as M-17F K-5. Main landing-gear shock struts attached further up sides of fuselage. Front cabin two pilots, rear cargo floor 4 m (157·5 in) long to adjust CG. First flight Aug 30.

DIMENSIONS Span 20,0 m (65 ft 7½ in); length 15,0 m (49 ft 2½ in); wing area 64,0 m² (689 ft²).

ENGINE One 420 hp GR9 (Bristol Jupiter IV).

WEIGHTS Empty 1720 kg (3,792 lb); fuel/oil 500+50 kg; loaded 2820 kg (6,217 lb).

PERFORMANCE Max speed 210 km/h (130 mph); cruise 170 km/h (105·6 mph); climb 5 min to 1 km; service ceiling 5,6 km; range 1250 km (777 miles); landing speed 70 km/h (43·5 mph).

K-7 Nearest thing to true 'flying fortress' ever flown. Basic concept was ultimate extrapolation of elliptical wing to very largest size possible. At this size wing depth became large enough for 120 passengers to occupy cabins in wings without affecting space required for fuel. By early 30 he had roughed out remarkable aircraft with six tractor engines and one pusher, twin tail booms and ability to carry 120 passengers or 7 t of cargo. Prolonged discussions ensued, reaching Kremlin, and in 31 Kalinin received permission to build K-7, but as heavy bomber. Even more unusual features then appeared.

Wing of elliptical shape, aspect ratio 6·2, R-II profile with mean thickness of 19% rising to 22% at centreline, giving depth of 2,33 m (91·7 in) on chord of 10,6 m. Two main and two subsidiary spars built up from KhMA steel tube. Lattice girder ribs, multiple spanwise stringers. Rectangular centre section skinned in D1 dural, rest in fabric. Small front nacelle on centreline, all dural construction. Twin tail booms of triangular cross-section (flat top, apex down) of KhMA steel tube, fabric covering. Distance between boom centrelines 11 m (36 ft 1¼ in). Carried monoplane tailplane on which were two fins/rudders 7 m (22 ft 11½ in) apart, fins braced by small auxiliary tailplane at higher level. All flight controls driven manually by large servo surfaces carried downstream of trailing edge on pairs of struts. Six tractor engines on leading edge of wing, one pusher on trailing edge at centreline. Each engine with ventral radiator, two-blade propeller. Aircraft supported by large columns (rear vertical, front sloping) aligned with fourth/fifth rib on each side which also carried tail booms, outboard of engines 3 and 5. These columns linked aircraft to gondolas of all light-alloy construction housing gunners, bombs, entry doors and landing gear comprising one front wheel and side-by-side rear wheels in each gondola, with pneumatic brakes. Stairway in front column, ladder in rear column, to hatches in wing. Inflight access to all engines. Fuel 9130 lit (2,008 gal) in metal tanks in wing. Minimum crew 11: two pilots, nav/bomb, radio/gunner (nose 20 mm gun), engineer, four gunners (each twin DA) in cockpits at front and rear of gondolas and two gunners (each 20 mm) at tips of tail booms. Maximum-bomb load 14,6 t (32,187 lb); overload 19 t (41,887 lb), evenly divided between bays in centre of each gondola, with twin doors. Complete intercom throughout aircraft.

K-7 completed at Kharkov summer 33. First engine run 29 June; severe vibration, especially of booms and tail, at particular engine speeds. Though problem clearly resonance, only solution was local reinforcement, achieved by welding on extra steel angle sections to reduce motion to minimum. Taxi tests 9 Aug. First flight 11 Aug, satisfactory but pilots would have preferred geared engines and larger propellers. Flying continued, causing immense interest around city. On ninth flight, 21 Nov 33, during speed tests at 100 m, sudden vibration in tail broke right tail boom; aircraft went straight into ground and burned. Pilot M. A. Snyegirev and 13 crew and one passenger killed; five crew survived.

This put blight on twin booms and on Kalinin. Latter was sent to new GAZ at Voronezh where at end 33 plan organised under P. I. Baranov to build two more K-7s with various structural changes including stressed-skin booms of rectangular section. In 35 work halted when first was 60% complete; basic concept no longer competitive.

DIMENSIONS Span 53,0 m (173 ft 10⅔ in); length 28,0 m (91 ft 10½ in); wing area 454 m² (4,887 ft²).

ENGINES Seven M-34F.

WEIGHTS Empty 24,4 t (53,792 lb); fuel/oil 6500+600 kg; normal loaded 38 t (83,774 lb), (max bombload) 42,4 t (93,474 lb), (overload) 46,5 t (102,513 lb).

PERFORMANCE Max speed 225 km/h (140 mph); long-range cruise 180 km/h (112 mph); service ceiling 4 km (13,123 ft); normal range 1600 km (994 miles).

K-9 Tandem-seat sporting and liaison aircraft. Parasol monoplane with usual construction for fuselage, but elliptical wing and tail of mainly metal construction and arranged to fold for convenience. Cockpits separated by large chord

K-7

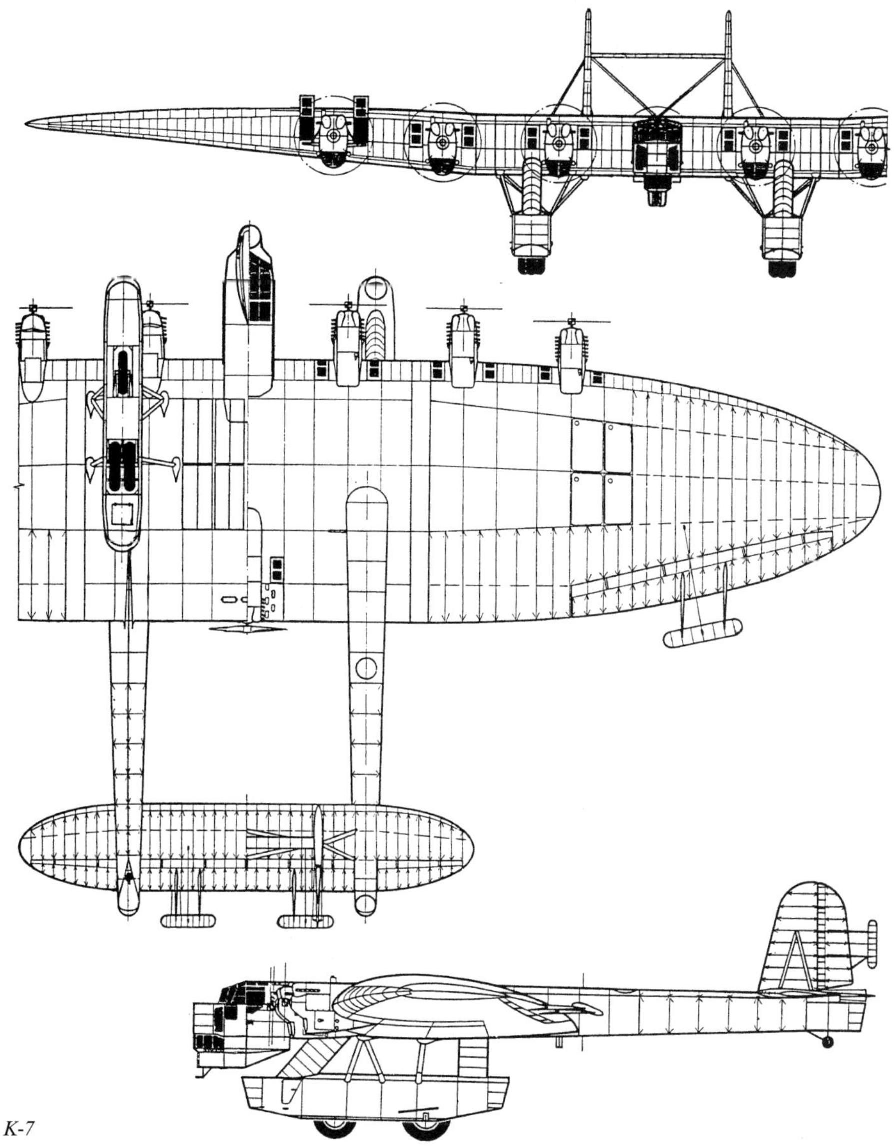

K-7

of wing. Welded steel-tube main gears, with vertical rubber-sprung shock struts well away from fuselage, braced by diagonal to top fuselage longeron.
DIMENSIONS Span 9,8 m (32 ft 1⅞ in); length 7,1 m (23 ft 11¼ in); wing area 18,1 m² (195 ft²).
ENGINE One 60 hp Walter NZ-60 five-cylinder radial.
WEIGHTS Empty 550 kg (1,213 lb); fuel/oil 60 kg; loaded 770 kg (1,698 lb).
PERFORMANCE Max speed 120 km/h (74·5 mph); endurance 3 h; landing speed 60 km/h (37 mph).

K-10 This attractive cantilever low-wing machine used many wing, fuselage and tail parts almost identical to those of K-9, but result was dramatically superior because of greater power. Welded truss fuselage made integral with centre section carrying cantilever landing gears with spatted wheels. Left and right wings arranged to fold manually in same way as on K-9, with skewed hinge at top of rear spar and with trailing edge inboard of aileron previously folded up across wing to provide clearance. Engine fully cowled in long-chord cowl, rare for M-11 in 32, driving 2,3 m two-blade propeller with spinner. Though built as civil sport aircraft, entered as military trainer in competition with AIR-9 (winner) and Gribovskii G-20. After this contest, proposed as agricultural machine for seeding and pollination (presumably also spraying/dusting); with ventral tank. Popular machine, flown Sept 32, but no production.
DIMENSIONS Span 10,7 m (35 ft 1¼ in); length 7,03 m (23 ft 0¾ in); wing area 18,0 m² (194 ft²).
ENGINE One M-11.
WEIGHTS Empty 700 kg (1,543 lb); loaded 1035 kg (2,282 lb).
PERFORMANCE Max speed 175 km/h (109 mph), cruise 145 km/h (90 mph); climb 7 min to 1 km; service ceiling 3,8 km (12,500 ft); range 480 km (298 miles); ldg speed 85 km/h.

K-11 Believed designation of tailless glider of 9 m (29 ft 7 in) span built in 34 as first stage of three-stage development of ultra-fast tailless bomber (see K-12). Completed 100 flights and confirmed stability and control of chosen configuration would be satisfactory.

K-12 Again a radical break with past, planned 33 as **VS-2** (*Vystablyat samolyet*, demo aircraft) as bomber with very short fuselage with large horizontal tail and twin fins with poor moment arm, wing with endplates and superimposed ailerons. By 34 became 11-passenger airliner with no conventional tail. Finally built as intermediate stage in devt of K-14. Wing of R-II profile, unswept with straight main spar and multiple auxiliary spars, stringers and built-up truss ribs. Structure welded KhMA steel throughout (ruling section elliptical, 10× 8,5 mm) with fabric covering. Welded steel-tube fuselage in three sections, main joints and joints between wings and centre section being bolted. Scheibe-type rudders on vertical hinges at wingtips. Double-wing trailing edge with servo-operated ailerons outboard and elevators inboard, all of inverted R-II profile. Crew of three, pilot in enclosed cockpit and two gunners each with ShKAS in manually rotated turret at

K-9

K-10

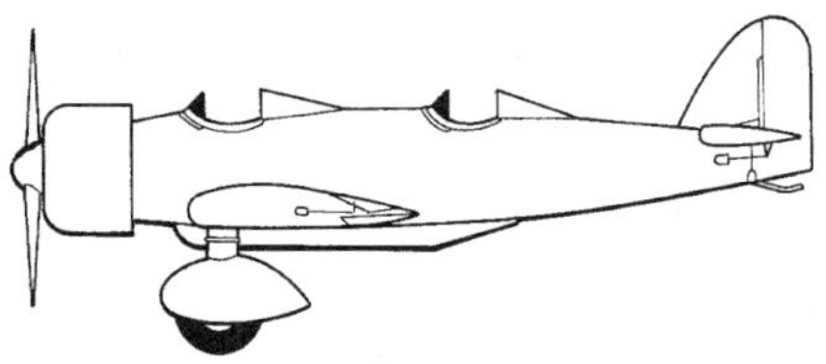

K-10

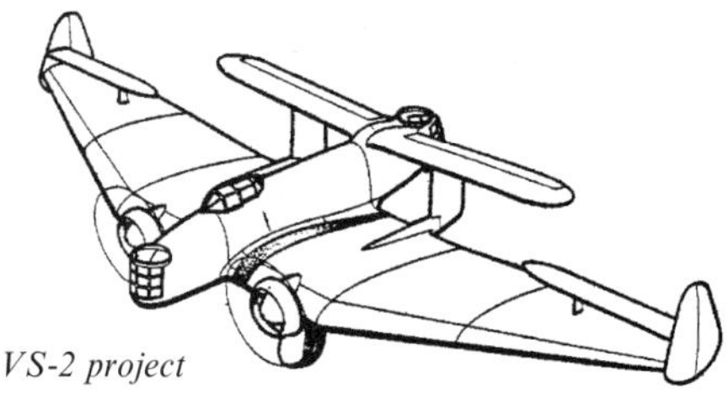

VS-2 project

nose and tail. Internal bay for 500 kg bombload. Main gears retracted backwards into engine nacelles. No attempt at large bombload or long range: two small wing tanks totalled 700 lit (154 gal).

First drawings late 34. First flight Oct 36 by P. M. Stefanovskii. Long take-off and in air demonstrated poor stability and control (much worse than glider), and rudders zero or negative effectiveness at low speeds. Prolonged attempts to improve behaviour yielded slow results. K-12 painted bright red/yellow feathers and named *Zhar Ptitsa*, Firebird (Phoenix) and participated in Tushino airshow 18 Aug 37. On 12 Dec 37 VVS decided to order small series with M-25 engine and tip rudders replaced by more conventional tail, but April 38 stopped at tenth aircraft, few actually flying.

DIMENSIONS Span 20,95 m (68 ft 8¾ in); length 10,32 m (33 ft 10¼ in); wing area 72,75 m² (783 ft²).

ENGINES Two M-22.

WEIGHTS Empty 3070 kg (6,768 lb); loaded 4,2 t (9,259 lb).

PERFORMANCE Max speed 219 km/h (136 mph); service ceiling 7,1 km (23,300 ft); range 700 km (435 miles); TO 700 m.

K-13 Twin-engined medium bomber, wing D16 stressed skin, split flaps, fuselage welded steel tube and fabric, twin-fin biplane tail, nose/tail turrets, crew 3, max bombload 1,4 t. Ordered early 35, prototype flown before mid-36. Planned **K-13M** seaplane with torpedo option not built.

DIMENSIONS Span 20,3 m (66 ft 7 in); length (both) 14,23 m (46 ft 8 in).

ENGINES (13) two M-34RN, (13M) two M-85.

WEIGHTS (M) Empty 4,7 t (10,362 lb); max 8 t (17,637 lb).

PERFORMANCE Max speed (13) 365 km/h (227 mph), (M, est) 332 km/h (206 mph); service ceiling (M, est) 7350 m (24,100 ft); range (M, est) 1250 km with 1 t bombload, 1750 with 400 kg.

K-14 Projected high-speed passenger aircraft, halted by Kalinin's arrest See Addenda.

K-12

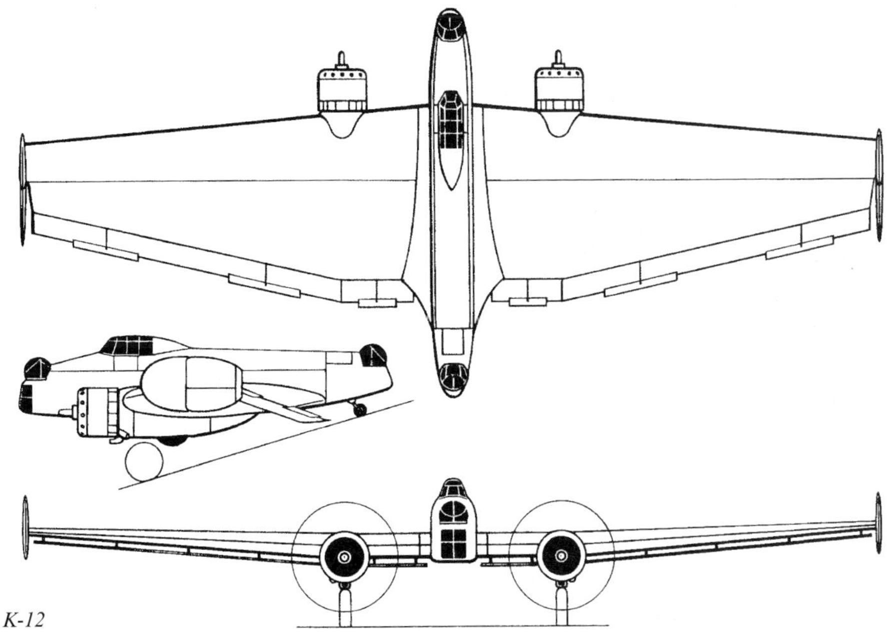

K-12

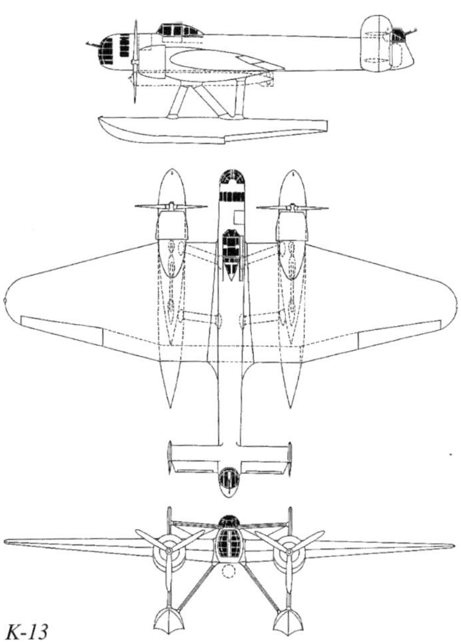

K-13

Kamerton

Developer of light autogyros at NII VVS (V. P. Chkalov) with support of groups 'K & Ko' and 'Room Service'.

A1 Ultralight single-seat, enclosed cockpit with two doors, V-tail. Three flown by late 93. Service life 2,000 h (7 years).
DIMENSIONS Rotor diam 7 m (22 ft 11½ in); length excl rotor) 3,6 m (11 ft 9¾ in).
ENGINE One 38 hp, Mogas.
WEIGHTS Empty 177 kg (390 lb); loaded 260 kg (573 lb), (max) 280 kg (617 lb).
PERFORMANCE Max speed 130 km/h (81 mph); cruise 95 km/h (59 mph); min 43 km/h (27 mph); max climb 192 m (630 ft)/min; range 200 km (124 miles).

A2 Side-by-side 2-seater, two engines driving two-blade rotor and contra-rotating props inside annular duct, cabin with gull-wing doors. Service life 2,000 h. Prototype flying and exhib 93.
ENGINES Two M-18 rated at 55 hp.
WEIGHT Normal/max loaded 500 kg (1,102 lb).
PERFORMANCE Max speed 160 km/h (99 mph); cruise 120 km/h (74·6 mph); range 250 km (155 miles).

Kamov

Nikolai Ilyich Kamov was born 1902 in Siberia. Graduated from Tomsk Technical Institute 26 as locomotive engineer, but two years later joined OMOS, qualified as pilot, and worked on ROM-1 and other flying boats under Grigorovich. In 29 joined forces with another engineer to produce first Soviet autogyro (KaSkr-1, below). In 31 joined OOK at CAHI where his greatest achievement was leader of 3rd Brigade which created A-7. Throughout second half of 30s worked at BOK. In 40 set up A-7/A-7bis production works at Smolensk, but his AK

A1

A2

design had to be abandoned because of war. In 45 set up KB (OKB from 48) with new co-axial lightweight helicopter, Ka-8. From this stemmed major family of co-axial helicopters, initially under Vladimir Barshyevskii and from 60 turbine-engined with greatly expanded design team at Ulan-Ude. Chief deputy M. A. Kupfer, but when Kamov died 24 Nov 73 his place was taken by Dr Sergei Viktorovich Mikheyev.

KaSkr-1 Designation from Kamov and N. K. Skrzhinskii, autogyro *Krasnyi Inzhyener* (Red Engineer), based on Cierva C.8 and likewise using Avro 504 (U-1) fuselage, landing gear and most of flight controls. Original rotor (four braced blades), small wings and other parts. First ground running test 1 Sept 29. Rotor began to spin, but control system ineffective and subjected to major changes. U-1 rudder changed for larger surface; then torque of engine/propeller almost overturned machine, rectified by hanging 8 kg weight under right wingtip. Test pilot I. V. Mikheyev (father of present KB chief) unable to fly aircraft because of lack of power. Rebuilt as KaSkr-2.
DIMENSIONS Diameter of four-blade rotor 12,0 m (39 ft 4¾ in); length 8,8 m (28 ft 10½ in); solidity 0·106 (later 0·127); rotor speed 115 rpm.
ENGINE One M-2.
WEIGHTS Empty 750 kg (1,653 lb); loaded 950 kg (2,094 lb).
PERFORMANCE Achieved speed 90 km/h (56 mph); design endurance 1·5 h.

KaSkr-2 Major reconstruction of KaSkr-1 with much more powerful engine in helmeted cowl forming part of new light-alloy front end. Other minor changes including rudder of better shape. Flown 30; one photograph shows it on skis in winter 1930-31. Pilot D. A. Koshits flew KaSkr-2 on 90 occasions by late 31, reaching figures given below. Pioneer Soviet autogyro, leading quickly to 2-EA and A-4 at CAHI.
DIMENSIONS Diameter of four-blade rotor 12 m (39 ft 4¾ in); length 9·0 m (29 ft 6¼ in); solidity 0·127; rotor speed 135 rpm.
ENGINE One 225 hp Gnome-Rhône Titan 5-cylinder radial.
WEIGHTS Empty 865 kg (1,907 lb); loaded 1100 kg (2,425 lb).
PERFORMANCE Speed 110 km/h (68 mph); altitude 450 m (1,476 ft); endurance 28 min.

AK Discouraged by atmosphere of fear at CAHI which stultified design, Kamov took job of chief designer in factory at Smolyensk opened 39 to make production autogyros (beginning with A-7-Za). Obtained permission to undertake side-by-side observation machine with side doors. Tricycle gear, single fin/rudder on struts from rotor head and main gears, no wings. Delayed by evacuation July 41 but work restarted at new facility in Lake Baikal region 42. Abandoned 43 because of termination of interest by VVS.
DIMENSIONS Rotor diam 13,5 m (44 ft 3½ in); length not known.
ENGINE One MV-6.
WEIGHTS Empty 1026 kg (2,262 lb); loaded 1317 kg (2,903 lb).
PERFORMANCE (est) Max speed 176 km/h (109 mph); ceiling 4,7 km (15,400 ft).

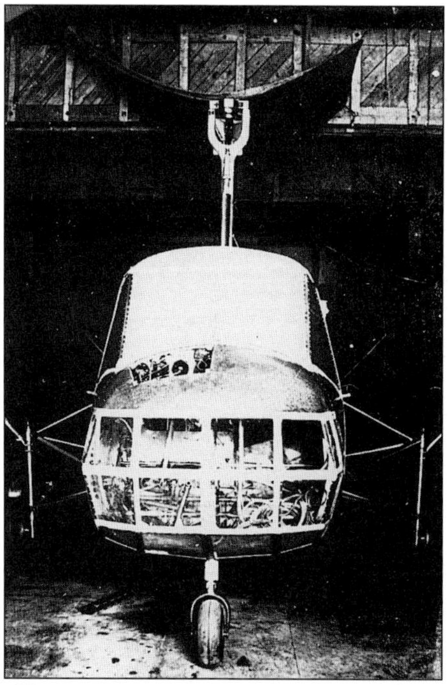

AK

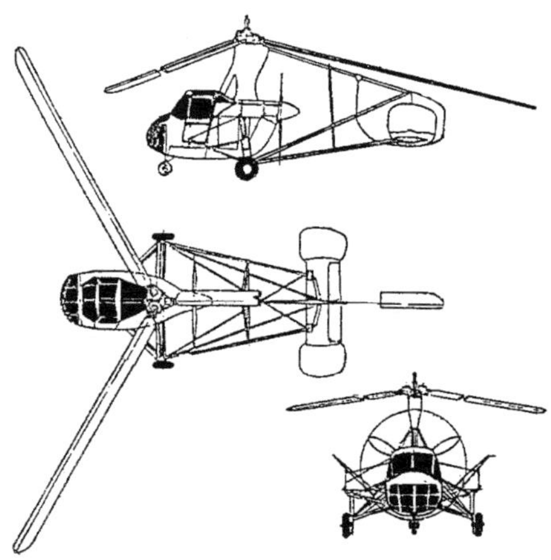

AK

Ka-8 Official disinterest in autogyro made Kamov switch to helicopters, and he decided to build one-man *Vozdushnii Mototsikl* (flying motorcycle) for civil or military use. Main design effort on coaxial rotors, each with three built-up wooden (mainly spruce) blades of NACA-230 profile with glued construction and fabric covering. Metal root held in hub with drag and flapping hinges driven by superimposed swashplates moved directly by pilot. Rest of airframe welded steel tube, with pilot and fixed fin at rear and engine and fuel tank at front, resting on two pontoons of rubberized fabric. Reported to have taken 18 months before permission granted to organise small informal group and build Ka-8. First flown 47, pilot Mikhail Gurov. Handlebar flight control replaced by vertical collective and cyclic levers, pontoons tapered front to rear, and fin changed to rudder driven by pedals. At least three built, one displayed from truck at 48 Tushino show.
DIMENSIONS Rotor diameter 5,6 m (18 ft 4½ in); length (ignoring rotor) 3,62 m (11 ft 10½ in); solidity (each rotor) 0·04; rotor speed 475 rpm.
ENGINE Originally 27 hp BMW flat-twin, modified into 45 hp M-76.
WEIGHTS Empty 183 kg (403 lb); loaded 275 kg (606 lb).
PERFORMANCE Speed 80 km/h (50 mph); ceiling (hover) 50 m; (in forward flight) 250 m (820 ft).

Ka-10 Success of Ka-8 enabled Kamov to open his own OKB in 48, by which time he had completed drawings for this- improved and enlarged (but still single-seat) machine. Structure similar but refined. Engine completely new, designed by Ivchyenko for this application with two stages of reduction gearing, freewheel with centrifugal clutch and improved power split between rotors. Electric started and engine-driven cooling fan. First hover by D. K. Yefremov September 49. Four built, tested by AV-MF, first Sov deck landing 7 Dec 50, minor improvements in **Ka-10M**, 12 built for Coast Guard 54. ASCC reporting name 'Hat'.
DIMENSIONS Diameter of two three-blade rotors 6,12 m (20 ft 1 in); length (ignoring rotors) 3,70 m (12 ft 1¾ in); height 2,5 m (8 ft 2½ in); solidity (each rotor) 0·037; rotor speed 410 rpm.
ENGINE One 55 hp AI-4V.
WEIGHTS Empty 234 kg (516 lb); loaded 375 kg (827 lb).
PERFORMANCE Speed 90 km/h (56 mph); ceiling (hovering) 500 m, (in forward flight) 1 km (3,300 ft); range 95 km (59 miles).

Ka-15 In 50 AV-MF wrote outline requirement for larger two-seat helicopter with enclosed cabin and much greater endurance. Rotor hubs and blades scaled up but compared with Ka-10M only significant change was low-density foam filling of blades between ribs and taper of blade towards tip. Basis of airframe welded steel tube, with covering of ply or dural removable panels and thin plastics glazing and sliding side doors round side-by-side cockpit ahead of rotor, with engine behind. Latter mounted with crankshaft horizontal, driving cooling fan, 90° angle drive to rotor via over-running clutch, and DC generator/starter. From engine to tail, stressed-skin dural construction,

Ka-10M

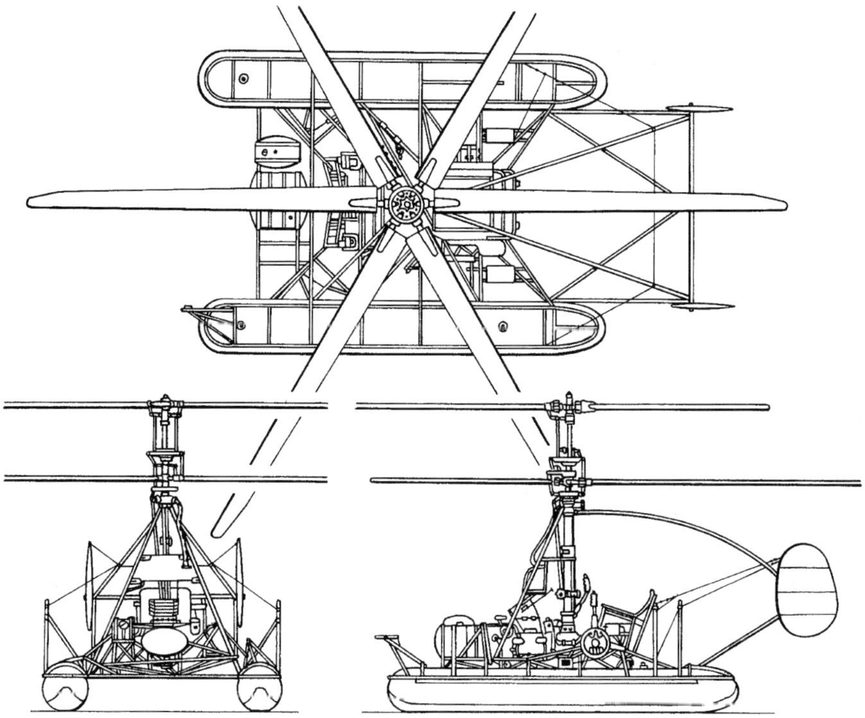

Ka-10M

Ka-15

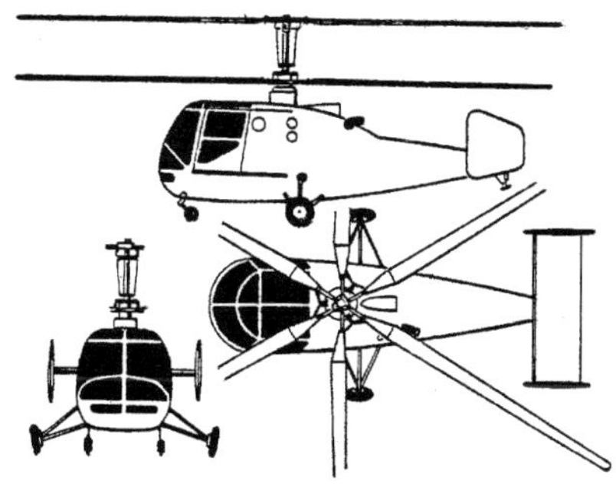

Ka-15

Ka-8

with fixed tail comprising braced tailplane and two endplate fins toed in at front (parallel on some Ka-15s), with pedal-driven rudders for yaw control (poor in hover). First flight by V. V. Vinittskii early 52. Rather slow development, but probably in production 55 and subsequently some hundreds built in several versions. Most fitted with two main and two nose landing wheels, plus tail bumper; nose wheels castoring and main wheels braked. At least one with pontoons and one with three skis. AV-MF used substantial number for liaison, ship-based recon and dual **UKa-15** for training. Ship trials 54 including dipping sonar, but unable to carry eqpt needed for ASW mission. From 58 civil **Ka-15M** used for many roles including ag-spraying; **Ka-15S** equipped to carry two external stretchers. Set records, best (5 May 59) 170,455 km/h over 500 km. Small number exported. ASCC reporting name 'Hen'.

DIMENSIONS Diameter of three-blade rotors 9,96 m (32 ft 8 in); length (ignoring rotors) 5,95 m (19 ft 6¼ in); rotor speed 330 rpm.

ENGINE One AI-14V.

WEIGHTS Empty (15M) 968 kg (2,134 lb); loaded (15M) 1370 kg (3,020 lb).

PERFORMANCE Max speed 150 km/h (93 mph); ceiling (hover) 800 m, (forward flight) 3 km (9,842 ft); range (15M, normal) 290 km (180 miles).

Ka-18 Stretched all-weather Ka-15; up to 4 seats or internal stretcher and doctor or two rear passengers; alternatively 200 kg cargo. Nose no longer transparent; doors hinged at rear; 176 lit (38·7 gal) fuel in floor tanks; extended rear fuselage and fins/rudders of much greater chord; whole aircraft cleaned up aerodynamically. Alcohol de-icing of blades and windscreen, exhaust heater muff feeds hot air on demand to cabin. First flown early 57. State trials 58, produced with blind-flying insts and option of 70-lit ferry tanks for AV-MF, VVS and Aeroflot, **UKa-18** dual control, role eqpt for survey and vineyard spraying. ASCC reporting name 'Hog'.

DIMENSIONS Diam of 3-blade rotors 9,96 m (32 ft 8 in); length (ignoring rotors) 7,03 m (23 ft 0¾ in).

ENGINE (prototypes) as Ka-15, (production) one AI-14VF.

WEIGHTS Empty (production, typical) 1032 kg (2,275 lb); loaded 1502 kg (3,311 lb).

PERFORMANCE Max speed (prototypes) 145

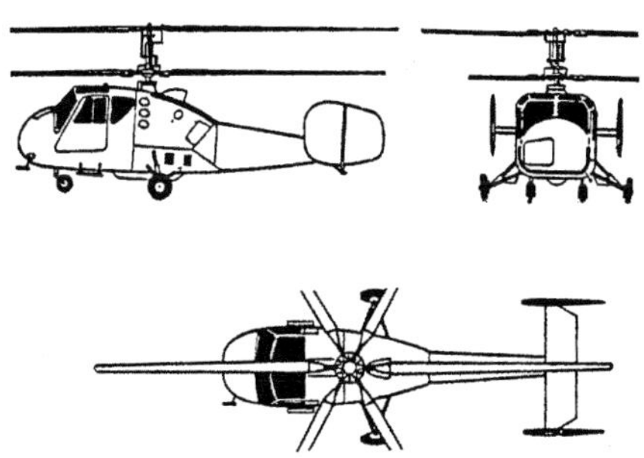

Ka-18

Ka-22

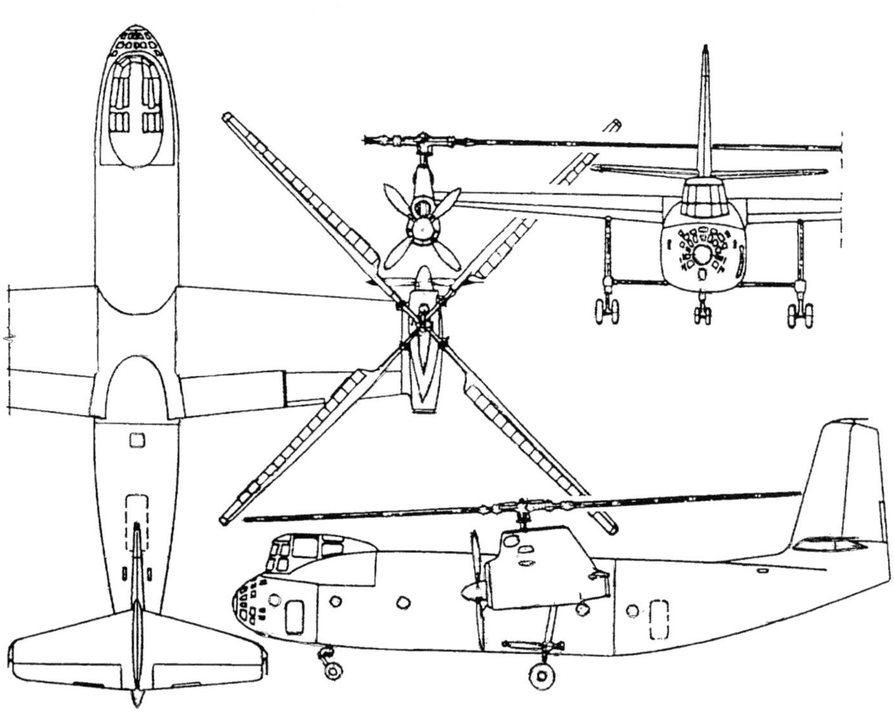

Ka-22

km/h, (VF engine) 160 km/h (99 mph); ceiling (forward flight) 3,5 km (11,483 ft); range (three passengers, 130 km/h cruise) 300 km (186 miles).

Ka-20 See Ka-25.

Ka-22 *Vintokryl* (screw-wing) compound helicopter, project 55, flown (D. Yefremov and crew of 5) 20 April 60. Stressed-skin airframe with engine/propeller/rotor group on each tip of large wing with flaps depressed 90° in helo mode; auto control linking fuel flow with prop pitch in cruise. Wing high on fuselage with interior 17,9×3,1×2,8 for 16,5 t cargo or 80 seats (not fitted). Glazed nose for nav, high flight deck for two pilots and radio/engineer. High-mounted wing with slight taper, outboard tabbed ailerons and inboard flaps which were not lowered in hovering displays. Conventional aeroplane tail, used only in aeroplane mode. Power group comprising turboshaft engine with drive capable of being progressively clutched to either four-blade propeller or four-blade rotor, latter being used for vertical flight and propellers for cruise, with lift shared between wing and windmilling rotors. Selected engine normally has rear drive from free turbine; installation has rear jetpipe and air from inlet ducted round underside of cowling, driving at rear to high-speed shaft to reduction gearbox under rotor shaft, from which front drive goes to propeller. Upper circular inlet feeds fan-assisted oil cooler. Fixed tricycle landing gear with twin main wheels. Superficially rotors and hubs resembled those of Mi-4 and Yak-24, but with trailing-edge tabs inboard; handed, left rotor being clockwise seen from above. Public display (confident) Tushino 9 July 61, D. K. Yefremov. Same pilot, assisted by V. V. Gromov, set world 15/25 km record 7 Oct 61 at 356,3 km/h (221·4 mph), with spatted landing gears, followed by record lift on 24 Nov of 16 485 kg (36,343 lb) to 2 km. OKB testing complete early 64. ASCC name 'Hoop'.

DIMENSIONS Diam of 4-blade rotors 22,8 m (74 ft 9⅜ in); distance between rotor centres 23,53 m (77 ft 2⅜ in); length of fuselage 27,0 m (88 ft 7 in).

ENGINES Two D-25VK.

WEIGHTS Empty 28,2 t (62,169 lb); normal loaded 37,5 t (82,672 lb), (max) 42,5 (93,695).

PERFORMANCE Max speed 375 km/h (233 mph); dynamic ceiling 5,5 km (18,050 ft); TO 300 m, ldg over 25 m 130 m.

Ka-25 Urgent demand for ASW shipboard helo in 58 won easily by compact Kamov coaxial design, with many other advantages. First prototype called **Ka-20** merely test vehicle for turboshaft engines mounted above fuselage, gearbox and rotors, shown Tushino 61 with dummy missiles (ASCC reporting name 'Harp'). Definitive **Ka-25** prototypes incorporated anti-corrosion structure, cabin housing mission eqpt, NII testing 63-69. Rotors with lubricated hinges and aluminium blades with nitrogen-pressure crack warning, hydraulic control, alcohol deicing and auto blade folding. Forward-facing electrically heated inlets, lateral plain (no IR protection) exhausts, and rear drive to rotors and to large cooling fan for oil radiator served by circular aft-facing inlet above rear fuselage. Airframe entirely dural stressed-skin, mainly flush-riveted but incorporating some bonding and sandwich panels. Main fuselage devoted to payload: side-by-side dual control nose cockpit with sliding door on each side, main cabin 1,5 m wide, 1,25 m high and 3,95 m long with sliding door on left and access at front to cockpit, and much of underfloor volume occupied by left/right groups of tanks filled by left-side pressure connection. Cable fairing along right side of cabin. Short boom for tailplane with elevators and central fin and toed-in tip fins carrying rudders. Latter used mainly in autorotation, yaw control by pedals applying differential collective; mixer box holds total rotor thrust constant to reduce workload eg landing on pitching deck. Rotors not designed for negative-g. Two castoring front wheels (tyres 400×150) on vertical shock strut with rear brace pivoted to fuselage to swing up and out to rear out of radar FOV. Two sprag-braked main wheels (600×180) each on vertical strut able to swing vertically on parallel V-struts pivoted to fuselage for same reason; landing loads reacted by diagonal shock/retraction strut. Each wheel fitted for rapid-inflation buoyancy collar.

Autopilot tailored to deck operations and hover, twin-gyro platform and doppler, duplicated HF/VHF/UHF, night lighting plus strobe, radio compass, radar altimeter, IFF and four passive RWRs. Cockpit for two pilots, options include 12 tip-up cabin seats, rescue hoist, external fuel drum each side as alternative

Ka-25BSh

to containers for cameras, flares, smoke floats or beacons. Total production 1966-75 c460 in three versions:

Ka-25BSh ASW version. I/J-band search radar with 360° scan in flat-bottom radome under nose, box for three vertical sonobuoys can be clipped aft on right side, Oka-2 dipping sonar at aft end of fuselage on centreline or (seldom fitted) APM-60 MAD sensor in pod on pylon under tail, large ESM receiver drum above boom with optional ADF sense blister immediately to rear, EO viewing port under boom, upgraded EW suite. Weapon bay 0,9 m wide under centreline, initially with two bulged doors, later as largely external rectangular box, tailored mainly to two AS torpedoes (originally 450 mm calibre) with wire reel on left side of fuselage; alternatively nuclear or conventional depth charges or other stores, max 1,9 t (4,190 lb). Replaced in CIS Navy by Ka-27PL, but serves with India, former Jugoslavia, Syria and Vietnam. ASCC 'Hormone'-A.

Ka-25K OTH targeting version for feeding guidance data to cruise missiles launched by surface warships and submarines. ASW and ESM equipment and weapon bay omitted, internal fuel increased, OTH targeting and cruise-missile guidance radar with large elliptical (instead of rectangular) scanner reflector in bulged radome, secure data link to surface fleet including small antenna in vertical cylinder under rear centreline of fuselage. ASCC 'Hormone-B'.

Ka-25K Unrelated to previous, single civil prototype (SSSR-21110) 1966 with gondola under lengthened nose for controlling 2 t slung load; elec-deiced blades, option 12 passenger seats. No ASCC name.

Ka-25PS SAR and transport version, no weapon bay, radar as BSh, normal equipment includes winch, 12 seats, provision for stretchers and aux tanks; options include nose quad Yagi antenna for homing receiver, ESM, searchlight and loudspeaker. Painted white/red. Replaced by Ka-27PS. ASCC 'Hormone-C'.

DIMENSIONS Diameter of three-blade rotors 15,74 m (51 ft 7¾ in); length (BSh, excl rotors) 9,75 m (31 ft 11¾ in); height 5,37 m (17 ft 7½ in).

ENGINES Originally two GTD-3F, later two GTD-3BM.

WEIGHTS (BSh) Empty c4765 kg (10,505 lb); loaded 7500 kg (16,535 lb).

PERFORMANCE Max speed 209 km/h (130 mph); cruise 193 km/h (120 mph); service ceiling 3350 m (11,000 ft); range 400 km (250 miles), (aux tanks) 650 km (405 miles).

Ka-26 One of few successful Soviet aviation exports, this versatile utility helicopter is refined and cheap to buy. Announced 64, first flight 65. Coaxial rotors with mast inclined 6° forwards and GRP blades (first in Soviet Union and among first in world) with alcohol deicing. Hydraulic control, manual reversion. Light-alloy stressed-skin airframe with external ribbing for skin stability (used on lesser scale on Ka-25). Piston engines mounted with crankshafts horizontal on tips of stub wings attached to roof structure linking nose cockpit and tail booms. Fully glazed cockpit with sliding door on each side equipped for pilot on left; optional passenger or second pilot on right. Engines cooled by fans and driving via right-angle boxes and overrunning clutches giving twin-engine safety. Two castoring nose wheels and two

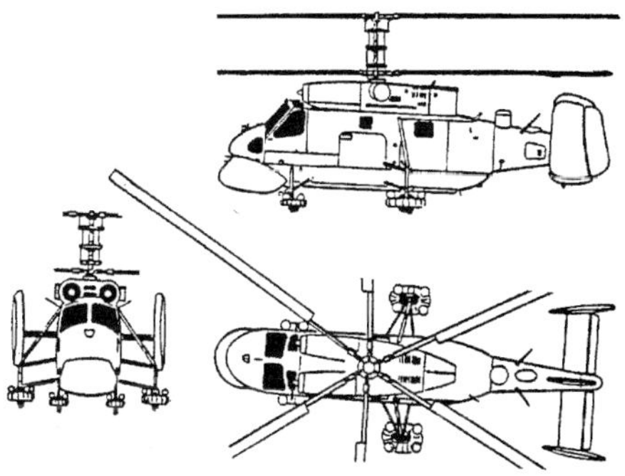

Ka-25PS (with buoyancy collars)

braked main wheels, tyres 595×185 mm. Twin tail booms mainly GRP construction carrying aeroplane-type tail with two vertical surfaces toed in 15°. Payload attached externally in space under rotor. Options include six-seat cabin (warmed like cockpit by combustion heater); same-size compartment for two stretchers, two seated casualties and attendant, with medical equipment; fire-protection model with six firemen and/or chemicals; 900 kg ag-hopper dispensing 1·2 to 12 lit or kg per second via spraybars or dust-spreader; cabin equipped for aerial survey (AFA-31 camera and operator); or geophysical compartment with EM-pulse generator feeding large ring emitter and with towed receiver 'bird' whose cable is sheared if overloaded. Wide use on wheels, skis and pontoons from land and ship, one task to land demolition teams to clear river ice in spring. Exports to 15 countries, total 890 by 93, prodn continuing by Kumertaou Avn Prodn Assoc. ASCC name 'Hoodlum-A'. See Ka-126.

Ka-26

DIMENSIONS Diameter of each 3-blade rotor 13,0 m (42 ft 8 in); length (excl rotors) 7,75 m (25 ft 5 in); height 4,05 m (13 ft 3½ in).
ENGINE One M-14V-26.
WEIGHTS Empty (basic) 1950 kg (4,299 lb); (with cabin) 2,1 t (4,630 lb); loaded 3,25 t (7,165 lb).
PERFORMANCE Max speed 170 km/h (106 mph); cruise 110 km/h (68 mph); hover OGE 800 m (2,625 ft); service ceiling 3 km (9,843 ft); range (7 pax, 30-min reserve) 450 km (280 miles), (max, aux tanks) 740 km (460 miles).

Ka-27 Improved helicopter to fit same hangars, deck lifts and operating platforms as Ka-25 but heavier, more powerful and with expanded capabilities. Later structure with significant titanium and composites, fuselage enlarged to fill size envelope established by folded rotors. Engine power more than doubled but rotor size and number of blades unchanged; power absorbed by blades of new profile with glass/carbon D-spar and 13 trailing pockets of Kevlar-type composite with Nomex-type filling, ground-adjustable tab on trailing edge, all blades elec deicing, upper blades tip lights, lower blades with twin pendulum vibration dampers, further anti-vibration devices in fuselage, manual folding and application of locking struts. Flight controls as Ka-25 but spring-stick trim, full power without manual reversion, increased autopilot authority and new stab-augmentation system for hands-off flight in rough air over wide CG range. Tailplane with bracing struts, no central fins but outer fins toed in 25° with fixed slat on inner side of LE to prevent stall, tail skins composite. New engines with elec-deiced inlets, auto synch and power sharing, starting and control systems vary (see later), gearbox drives oil cooling fan of doubled Ka-25 capacity, AI-9V APU mounted transversely at rear with cabin-air heat exchanger. Normal fuel 3680 kg in 12 underfloor tanks pressure-fuelled from left rear. Provision for aux fuel box high on each side with flat top with gravity filler and maintenance walkway. Landing gears very similar to Ka-25 but designed for doubled loadings and nose gear retracting vertically, leg into fuselage. Provision for attaching large box low on each side housing rapid-inflation flotation bag, better water stability than Ka-25. Design begun 1969, first flight Dec 74. Production versions include Ka-28, 29 and 32.

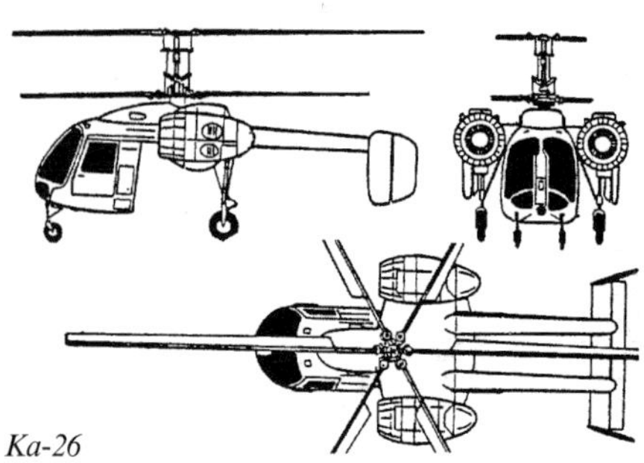
Ka-26

Ka-27PL ASW version, priority task to replace Ka-25BSh. Revised cockpit moved forward ahead of nose gears over radar, much larger windows including more bulged windows in wider right door. Swivelling right seat, third seat behind on right, door to cabin with station for systems operator. Flotation gear optional, aux tanks not fitted and cabin windows near main gears. Autopilot and flight computer for expected functions including auto capture and hover at preselected height in any weather or at night. Lo-IAS system and precision air-data, new doppler (precision hover data) ahead of twin-gyro box under boom, anti-collision and station-keeping beacon, duplicated HSI, new RWR with miniature receivers each side of nose and looking aft from above tailplanes, improved ESM 'flowerpot', new AS radar with rectangular antenna, radome on tailcone, otherwise avionics as Ka-25BSh. Sonobuoys in cabin, dispensed through tube. New centreline weapon bay tailored to various loads including four APR-2E torpedoes; alternatives four groups of S3V guided AS bombs or PA2 torpedo practise weapons. Operational mid-81, about 120 delivered, often operated in pairs though each has full hunter/killer capability. ASCC 'Helix-A'.

Ka-27PS SAR version, standard plane-guard in CIS navy. Radar, ESM, IFF, navaids and autopilot unchanged, AS equipment deleted. Weapon bay replaced by droppable dinghy packs, flotation gear and aux tanks standard, 300 kg hydraulic hoist, searchlight and floodlight, box with tandem hemispherical projections under tailcone, racks for marker floats, cabin windows moved forward. Operational 82, some painted white/red. 'Helix-D'.

Ka-28 Export Ka-27PL, different avionic fit; in service India, former Jugoslavia. 'Helix-A'.

Ka-29TB Assault transport, developed for AV-MF but tailored to operations from land bases. Redesigned forward fuselage, cockpit seating three side-by-side with titanium and composite armour, three flat front windows proof against rifle fire, armoured engine installation, cabin for 16 troops and their equipment, main door outward-opening upper-lower halves; radar replaced by EO sensor on right and millimetric radar on left for terrain following and incorporating missile-guidance link; full-width front of nose formed by radome over main targeting radar and window for FLIR, with sensitive multichannel air-data system carried on triangulated struts on left. New ESM 'flowerpot', IRCM pulsed jammer at aft end of pylon, com antennas changed and augmented. Fixed landing gear with reduced tyre pressure. Aux fuel and flotation gear replaced by four lateral pylons for 32-57 or 20-80 launchers or 9M114 missiles in quad groups; 2A42 fixed firing ahead above left pylon with 250-round box in cabin; 9A622 internal on right of nose, with articulated door over muzzle and ejector chute for cases/links below. To have inlet particle filters and IRCM jetpipes. CIS Navy service 87, disruptive pattern blue/grey camouflage. 'Helix-B'.

Ka-29RDL EW/surveillance platform. Two evaluated 90 aboard *Kuznetsov* (then *Tbilisi*) with Ka-29 features but weapons and combat sensors deleted, air-data sensors moved to centreline, cockpit width maintained back to sudden discontinuity at main gears, new door on right behind cockpit (outward-opening upper/lower halves, large box on top portion), very large boxes both sides ahead of and behind main gears, large pannier running full length under fuselage, unidentified equipment under root of boom, APU installed axially with projecting upturned jetpipe, various new antennas, one with two-tone grey camouflage but Aeroflot titles on both.

Ka-32S Basic civil transport, no naval equipment or avionics but retaining full flight-control autohover and height-hold functions and bad-weather nav, optional radar/flotation gear/aux fuel/winch plus provision for slung load to 5 t (11,023 lb) with floodlight. Normal cabin load 4 t (8,818 lb) or up to 16 passengers on folding seats, or stretchers. Over 100 built, mainly on Aeroflot contract, many fully equipped with radar, elevating landing gear and deck equipment for iceberg recon and ship guidance, ship provisioning (360 t per day) and many other maritime/land duties. 'Helix-C'.

Ka-32T Simple utility transport, IFR nav/inst and aux fuel but most other avionics removed and no flotation/hoist/radar; provision for seats/stretchers/slung load.

Ka-32K Projected crane version with retractable gondola for aft-facing cockpit.

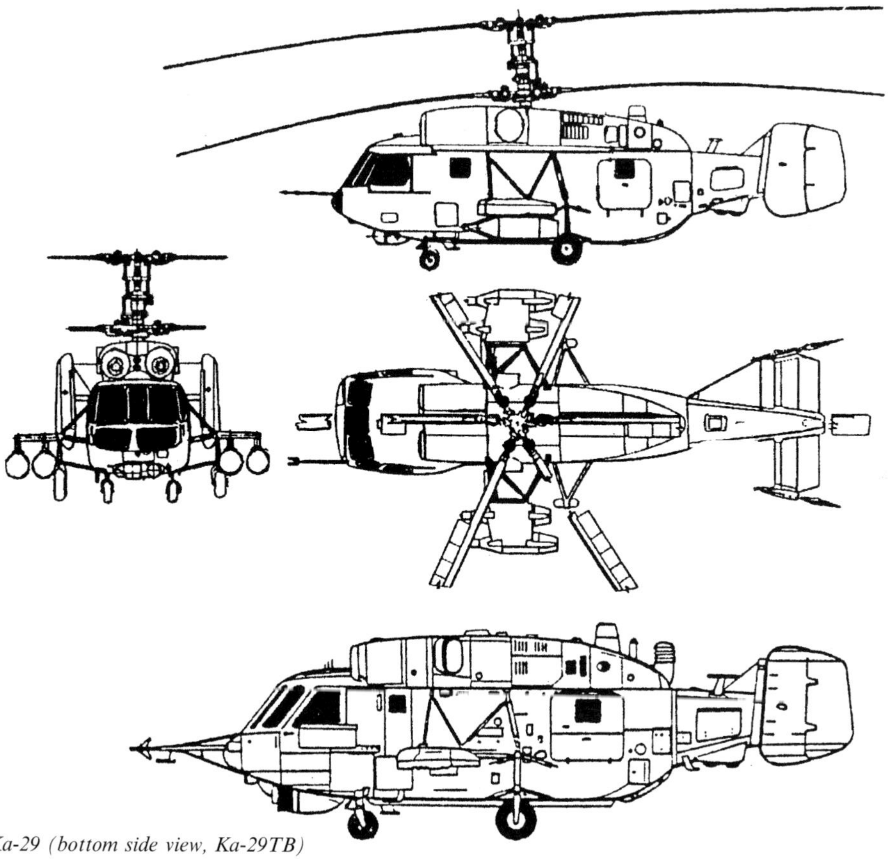

Ka-29 (bottom side view, Ka-29TB)

DIMENSIONS Diameter of three-blade rotors, 15,9 m (52 ft 2 in); length (excl rotors and probes) 11,3 m (37 ft 1 in); track 3,5 m; width over tail 3,65 m; span (29 weapons) 5,82 m; height 5,4 m (17 ft 8½ in); (29 folded) 12,25×3,8×5,44 m.
ENGINES (27PL, 27PS, 28) two TV3-117BK, 2,170 shp, electronic; (29, 32) TV3-117V, 2,225 shp, hydromechanical.
WEIGHTS Empty (27, 32S) 6500 kg (14,330 lb), (29) 5520 kg (12,170 lb); fuel (normal) 3680; loaded 11 t (24,250 lb), (max) 12,6 t (27,775 lb).
PERFORMANCE Max speed (27/28/32) 250 km/h (155 mph), (29) 280 km/h (174 mph); cruise 230 km/h (143 mph); max climb 15 m/s (2,950 ft/min); hover ceiling OGE 3,7 km (12,140 ft); service ceiling 4,3 km (14,100 ft); range (29) 460 km (286 miles), (32) 520 km (322 miles), (aux fuel) 900 km (559 miles).

Ka-37 Unmanned coaxial helicopter devd with Daewoo (S Korea), initially for ag use, later several military roles. Two P-033, 250 kg (551 lb) loaded, 45 min at 110 km/h (68 mph).

Ka-50 Work began 75 under Mikheyev to apply experience to new role of single-seat close-support fighter. Sov Ground Forces required max commonality with rival Mi-28. By 77 Ka were satisfied coaxial formula could yield saving in weight, higher flight performance, wider CG range, greater agility (load factor + 3 g) and reduced vulnerability. BITE diagnostics for 15-day operation away from base without even a ladder. Drag of tall mast minimised by compact semi-rigid hubs with maintenance-

Ka-29TB

Ka-32S

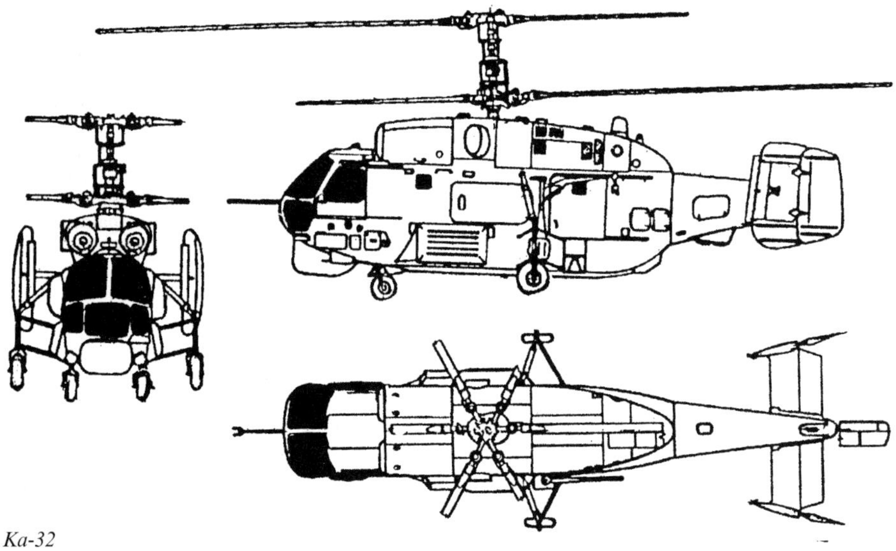

Ka-32

free elastomeric bearings. Blades similar to Ka-27 but designed to survive ground fire, shorter and with tip swept back (as Mi-38, not 28). Four-channel digital autopilot and full hyd power to rotors and negative-camber tailplane with endplate fins and tabbed rudder (but control not lost even if tail shot off). Structural base dural box beam 1 m square housing self-sealing polyurethane-foam protected tanks and carrying large wings with engines above roots. Wings offload rotor at speed (ailerons found unnecessary) and provide four weapon pylons and IRCM/ECM tip pods. Streamlined fuselage around beam formed by largely unloaded CFRP access panels (CFRP 35% of airframe wt). Elektro-Avtomatika designed four-computer management of flight controls, engines, weapons and defence systems, for automatic day-night operation without second crew-member (intention later to integrate with 1553-type digital bus). Engines have inlet particle separators and upswept triple jetpipes (low-drag IRCM mixer added 93) enhancing careful stealth design. Electrothermal anticing of blades/inlets/windscreen. Landing gears hinged to primary beam hydraulically retracting to rear with steerable twin nosewheels and single mainwheels semi-exposed, all tyres low pressure and main oleos 'three times usual energy absorption'. Crashworthy cockpit of alloy/ CFRP honeycomb, entered via door on left, surrounded by two layers of steel armour (arrests 23 mm AP at 90°) and with toughened glass windows (12·7 mm at 90°) with panoramic rear-view mirror. Zero/zero escape system developed by Zvyezda, triggered by pulling seat-pan handles, severs all rotor blades, jettisons cockpit roof, tilts K-37 seat parallel to mast and fires rocket above headrest which pulls via cable. Pilot has flak jacket/survival suit with radio beacon and life raft. VHF/UHF, IFF, RWR, doppler/ twin-gyro nav, beacon, transponder and other routine avionics. Special sensors included (1992) precision air-data down to zero IAS, chin box with separate lasers for ranging/target marking and for guiding missiles. Alongside is LLTV. These feed central display in cockpit, MiG-29 type HUD and pilot's HMD; 93 added FLIR ball on nose. Comprehensive defensive system fed by passive RWR receivers and IR sensor with computer management of Vympel UV-F26 tip pods housing chaff/flare launchers. Primary armament 12 9M117 on outer pylons, leaving inners free for 32-57, 20-80 or 10-130 launchers. Alternative loads, to max 2 t (4,410 lb), four FAB-500 or similar weight PLAB, two Kh-25 or 29 (any variant), four KAB-500Kr, gun pods for 9A622 or 9A624 or various dispensers including smoke or chemical. Single 2A42 gun fixed on right of nose with 500 rounds selectable 50/50 from HE/frag or AP, hydraulically slaved to sight system with +15°/–40° elevation but only 0°/15° laterally (sight aims helo outside this limit).

Design approval May 80, first flight of **V-80** prototype by Nikolair Bezdetnov 27 July 82. At least six prototypes, early aircraft without cannon. Dual trainer and other variants ready for prodn. Export clearance 92, agreement with Marine Services (div of Group Vector, Geneva) for poss license production in Greece, who have yet to find Western rival prepared to accept their challenge to a fly-off. ASCC 'Hokum'. Order for small evaluation batch by US Special Forces reported 93.

DIMENSIONS Diameter of three-blade rotors 14,5 m (47 ft 6⅞ in); length overall 16,0 m (52 ft 6 in), (excl rotors) 15,0 m (49 ft 2½ in); span 7,34 m (24 ft 1 in); height 4,93 m (16 ft 2 in).

ENGINES Two TV3-117VK.

WEIGHTS Empty 4550 kg (10,030 lb); loaded 9800 kg (21,605 lb), (max) 10,8 t (23,810 lb).

PERFORMANCE Max speed (max wt) 310 km/h (193 mph), (limit in shallow dive) 350 km/h (217 mph), (sideways) 70 km/h, (backwards) 90 km/h; vert climb at 2,5 km (8,200 ft) 10 m/sec (1,968 ft/min); hover ceiling OGE 4 km (13,120 ft); combat radius (lo-lo) 250 km (155 miles).

Ka-62 Complete break with coaxial tradition, this fills gap in CIS spectrum with attractive machine in Dauphin class. Single main rotor with four CFRP/Kevlar blades with swept tips, simple composite hub, Kevlar Fenestron tail rotor, fuselage 55% composites, engines close each side of gearbox with slot inlet beside rotor hub, APU low in fuselage behind cabin, 1150 lit (253 gal) fuel in four tanks under cabin, hydraulically retractable tailwheel landing gear, optional electrothermal anticing or inflatable pontoons, cockpit for one or two pilots, cabin (3,3×1,75×1,3) for up to 16 passengers or 2 t cargo, or slung load 2,5 t. Prototype in flight test 94. **Ka-62M** (previously 62R). Export version with RTM322, CT7-2D1 or CTS800 engines and Bendix-King avionics. Data for 62:

DIMENSIONS Diameter of main rotor 13,5 m (44 ft 3½ in); length 15,05 m (49 ft 4½ in), (excl

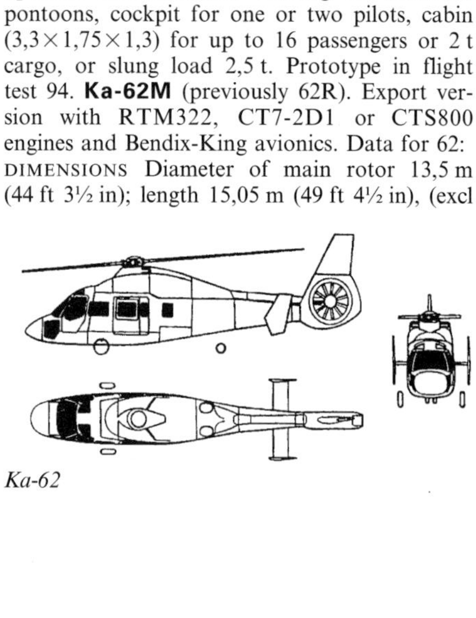

Ka-62

Ka-50 prototype

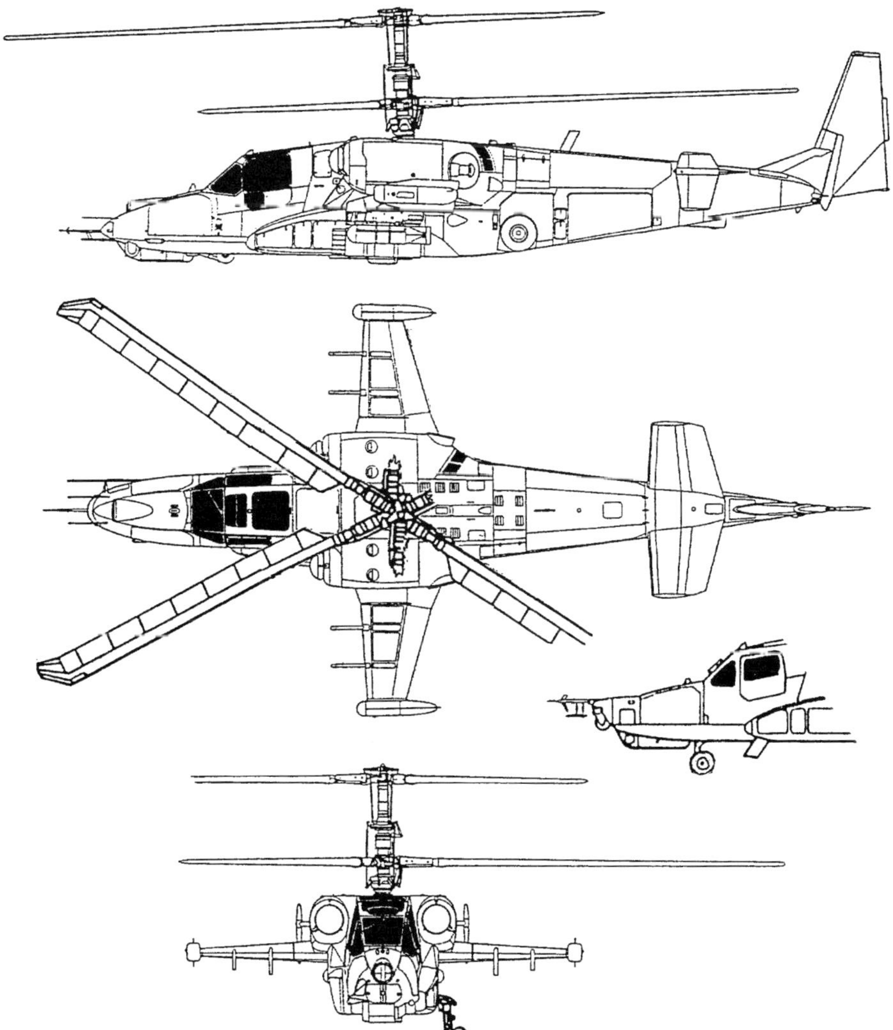
Ka-50 (inset, nose of prototype)

rotor) 13,25 m (43 ft 5⅝ in); height 4,1 m (13 ft 5½ in).
ENGINES Two RD-600.
WEIGHTS Loaded 5,5 t (12,125 lb), (max) 6 t (13,228 lb).
PERFORMANCE Max speed 300 km/h (186 mph); cruise 265 km/h (165 mph); max climb 13 m/s (2,560 ft/min); hover OGE 2,9 km (9,500 ft); service ceiling 5,5 km (18,050 ft); range (standard fuel) 780 km (485 miles), (aux fuel) 1050 (652).

Ka-118 Impressed by McDonnell Douglas NOTAR (no tail rotor), Kamov tested scheme on exptl Ka-26 and now propose this multirole machine for pilot and four passengers. Single main rotor swept tips, V-tail plus upper tailplane, skid gear. Engine not selected late 93 and jet thruster design still in research. Prelim data assume single TVO-100 engine and 700 lit fuel:
DIMENSIONS Diameter of four-blade main rotor 11 m (36 ft); fuselage 10 m (33 ft).
WEIGHTS Payload (internal) 800 kg; loaded 2150 kg (4,740 lb), (max) 2250 kg.
PERFORMANCE Max speed 300 km/h (186 mph); cruise 270 km/h; range 950 km (1500 with aux tanks).

Ka-126 Turbine-engined derivative of Ka-26. Few changes except new engines, revised fuel system (800 lit), cabin pilot + 6 pax; improved blade profile (designer Vyacheslav Savin) and stored-energy system (contra-rotating 24,000 rpm flywheels) for 40 sec flight after total power loss, generally updated eqpt. Mock-up showing two engines 81; actual first flight 86. ASCC name 'Hoodlum'-B.
DIMENSIONS Diameter of each 3-blade rotor 13,0 m (42 ft 7¾ in); length (excl rotors) 7,75 m (25 ft 5 in); height 4,15 m (13 ft 7⅜ in).
ENGINE One TVO-100.
WEIGHTS Empty 1915 kg (4,222 lb); max payload 1 t (2,205 lb); loaded 3 t (6,614 lb).
PERFORMANCE Max speed 190 km/h (118 mph); cruise 170 km/h (106 mph); max climb 8 m/s (1,575 ft/min); hover OGE 1 km (3,280 ft); service ceiling 4,65 km (15,250 ft); range 660 km (410 miles), (aux fuel) 1015 km (630 miles).

Ka-128 Differs from 126 only in engine plus added intermediate gearbox, and Bendix-King avionics. First flight 93.
DIMENSIONS As Ka-126.
ENGINE One French Turbomeca Arriel 1D1, 722 shp.
WEIGHTS Empty 1820 kg (4,013 lb); max 3 t (6,614 lb).
PERFORMANCE Max speed 200 km/h (124 mph); cruise 190 km/h (118 mph); max climb 8 m/s (1,575 ft/min); hover OGE 1,6 km (5,250 ft); service ceiling 5670 m (18,600 ft); range 710 km (441 miles).

Ka-226 Twin-engine derivative of Ka-26/126, few parts common. Improved composite blades and simple elastomeric hubs. Fuselage upper girder housing 870 lit (191 gal) fuel at front above completely new cockpit with enormous curved windscreens, passenger beside pilot, twin booms carry tailplane with inward-toed vertical fins/rudders, four-wheel landing gear as before; various payload pods, esp for six passengers on triple benches facing each other with baggage behind rear bench; Bendix-King avionics. Deal with Allison for 60 engines per year for production at Arsenyev to begin 90, but first flight delayed to 94, FAA certif hoped for 95.
DIMENSIONS Diameter of three-blade rotors 13 m (42 ft 7¾ in); length (excl rotors) 8,1 m (26 ft 7 in); height 4,15 m (13 ft 7⅜ in).
ENGINES Two Allison 250-C20B, each 420 shp.
WEIGHTS Empty 1952 kg (4,303 lb); max payload 1,3 t (2,865 lb); loaded 3100 kg (6,835 lb), (max) 3400 kg (7,496 lb).
PERFORMANCE (est) Max speed 205 km/h (127 mph); cruise 185 km/h (115 mph); vert climb at SL 168 m (550 ft)/min; hover OGE 1280 m (4,200 ft); range (max payload) 37 km (23 miles), (max with aux fuel) 873 km (542 miles).

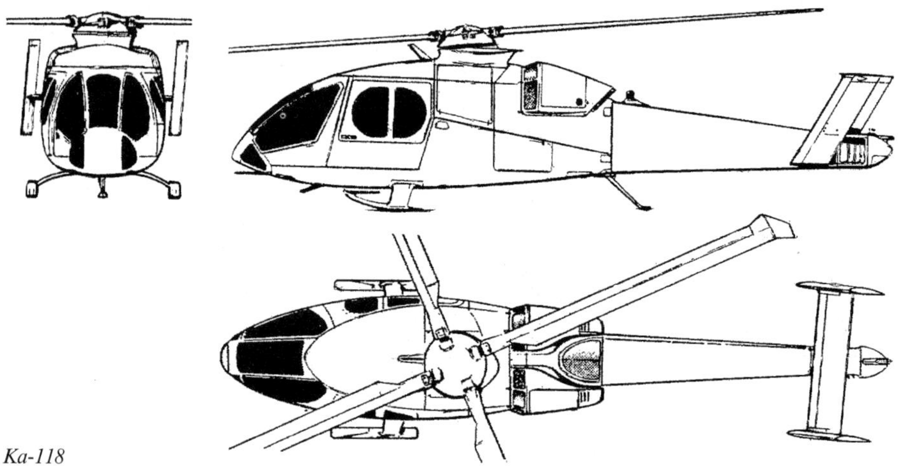

Ka-118

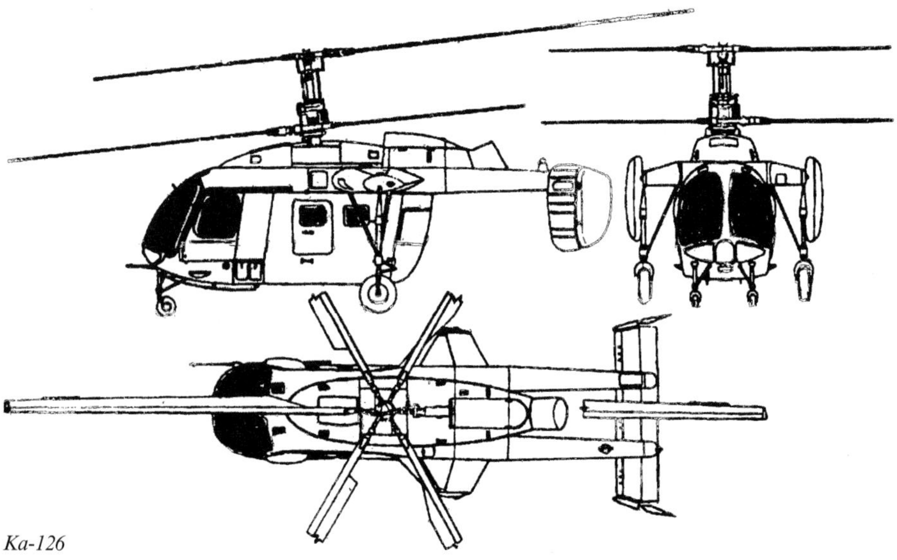

Ka-126

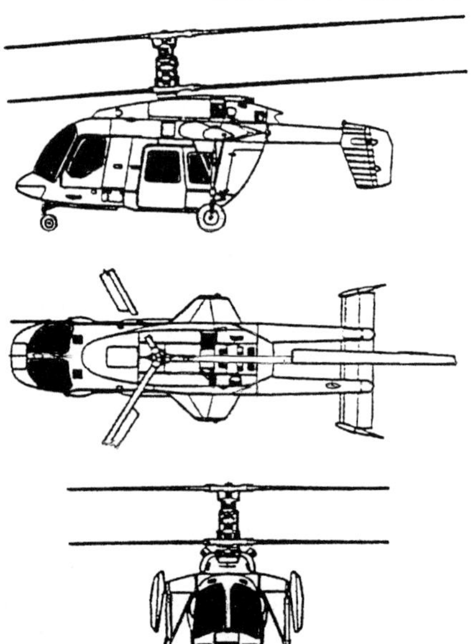

Ka-226

Kazan KAI

Kazan Aviation Institute has since 34 been one of most important SKBs. In 1934-37 KAI designs were created under supervision of S. P. Gudzik by team led by Zyeliman Isaakovich Itskovich, who had been one of Kalinin's designers in 1927-30 and then worked at KhAI. After 37 work lapsed until in 56 a fresh SKB was formed under M. P. Simonov (today head of Sukhoi) to create light aircraft.

KAI-1 First of Kazan designs, this light twin was a clean cantilever monoplane intended for touring, with pilot and five passengers. Glazed nose for fifth pssenger because of intention also to make military version with nav/bomb in nose. Other three passengers in cockpits beside and behind pilot (2 by 2) with two sliding canopies. Structure wooden. Two-spar wing of R-II profile, 14% at root but only 6% at tip, attached to fuselage by four steel fittings. Ailerons of symmetric profile mounted just above level of trailing edge. Zapp-type landing flaps, pulled down against air load by cables; one of first flap installations in country. Monocoque fuselage, elliptical section 1,2×0,94 m, ply skin 1,5 to 2,5 mm thick. Tail with tailplane incidence adjustable in flight. Engines with Townend ring cowls on first KAI-1, individual helmeted cylinder fairings on second. Trousered main gears. Second aircraft had additional features for use as bomber crew trainer, smaller wing area (not explained), normal ailerons and only three (occasional four) seats. First KAI-1 flown Nov 34 by B. N. Kudrin, who was favourably impressed. Recommendation for production of VVS trainer version followed, not implemented.

DIMENSIONS Span 12,6 m (41 ft 4 in); length 7,9 m (25 ft 11 in); wing area (No 1) 22,3 m^2 (240 ft^2), (No 2) 20,6 m^2 (222 ft^2).
ENGINES Two M-11.
WEIGHTS Empty (No 1) 780 kg (1,720 lb), (No 2) 855 kg (1,885 lb); fuel/oil 200+20 kg; loaded (No 1) 1310 kg (2,888 lb), (No 2) 1395 kg (3,075 lb).
PERFORMANCE Max speed 218 km/h (135·5 mph); climb 3·5 min to 1 km; service ceiling (No 1) 4,8 km (15,750 ft); range (No 1) 1000 km (621 miles)/5 hours; landing speed 70 km/h (43·5 mph).

UPB Third KAI-1 was given this designation from *Uchyebno Perekhodnoi Bombardirovshchik*, training transitional bomber. Almost identical in most respects to KAI-1 No 2 but equipped as dedicated military trainer. Three seats: pilot, nav/bomb in nose and radio/gunner in rear cockpit with hand-aimed DA with 200 rounds. Internal cells in fuselage between wing spars for up to 160 kg of various light bombs hung vertically. First UPB tested by B. N. Kudrin from 11 May 35; second by A. A. Ivanovskii from 23 July. Good results, and decision to build initial series of 25; but no spare capacity could be found. Third UPB had retractable landing gear.
DIMENSIONS Span 12,65 m (41 ft 6 in); length 7,98 m (26 ft 2¼ in); wing area (No 1) 20,6 m^2 (222 ft^2), (No 2) 20,8 m^2 (224 ft^2).
ENGINES Two M-11.
WEIGHTS Empty (No 1) 825 kg (1,819 lb), (No 2) 922 kg (2,033 lb); fuel/oil (No 2) 130+16 kg; loaded (No 1) 1250 kg (2,756 lb), (No 2) 1400 kg (3,086 lb).
PERFORMANCE Max speed (No 1) 245 km/h (152 mph), (No 2) 232 km/h (144 mph); climb to 1 km (No 1) 1·6 min, (No 2) 3·0 min; service ceiling (No 1) 5,7 km, (No 2) 5,3 km; range (No 2) 1000 km (621 miles); endurance 5 h; take-off 160 m/15 s; landing 120 m/12 s/73 km/h (45 mph).

KAI-2 Attractive dural stressed-skin trainer for combat pilots, with full aerobatic and inverted-flight capability. Low wing with main gears retracting inwards into fuselage immediately behind radial engine cowl. Single seat in enclosed cockpit. Engine failed to arrive and aircraft never flew.
DIMENSIONS Span 9,0 m (29 ft 6¼ in); length 6,5 m (21 ft 4 in); wing area 13,0 m^2 (140 ft^2).
ENGINE One MG-21.
WEIGHTS Empty 750 kg (1,653 lb); fuel/oil 80+10 kg; loaded 960 kg (2,116 lb).
PERFORMANCE Not measured, but calculated endurance 2 h.

DKL Last of Itskovich's designs at KAI, this

KAI-1, first aircraft, before cowlings fitted

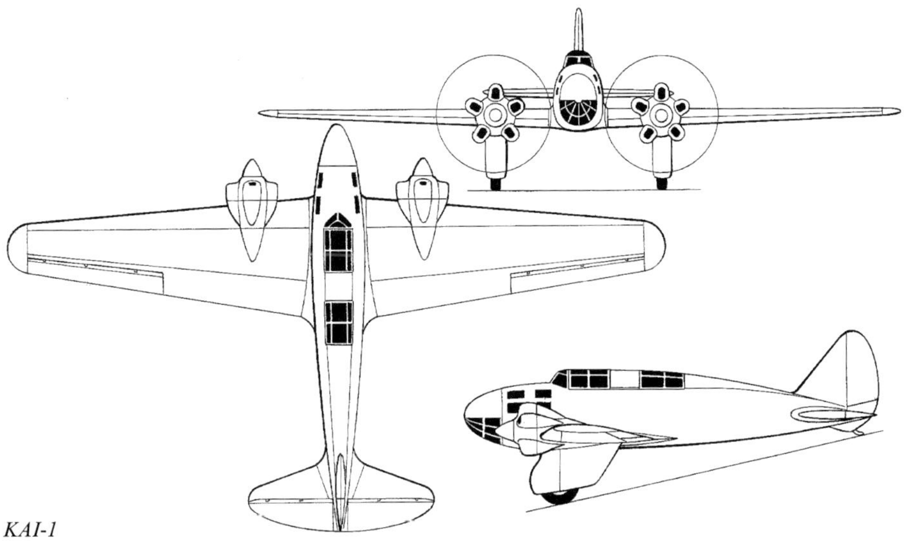

KAI-1

UPB

DKP (V)

10-seat passenger transport got its name from *Dvukhmotornyi Krayevoi Linyeinyi*, twin-engined regional liner. Generally similar to RAF-11, and using identical wheel or ski landing gear, but with low/mid wing to leave room for bomb bay (500 kg of various bombs) in projected military version. Cargo hatch 2,2 × 1,2 m in roof of fuselage. All-metal stressed-skin construction, pneumatic-actuated flaps. Prototype flight-tested at KAI from Aug 37, with improving stability and control resulting from minor changes. Lack of interest, because too close to RAF-11 and Yak UT-3. **DKP(V)** was projected bomber crew trainer.

DIMENSIONS Span 15,05 m (49 ft 4½ in); length 9,1 m (29 ft 10⅓ in); wing area 33,0 m^2 (355 ft^2).
ENGINES Two MG-31.
WEIGHTS Empty, not recorded; loaded 3 t (6,614 lb).
PERFORMANCE Max speed 360 km/h (224 mph); service ceiling 6,5 km (21,325 ft); range with 10 passengers 790 km (491 miles).

KAI-6/7/8/9/11/12/14/17/18/19 Competition sailplanes or training gliders.

KAI-13 Named *Letayushchaya Koltso* (flying platform); single piston engine, tested 1958.

KAI-15 Tailless ultralight; 1960.

KAI-16 Ultralight single-seat aerobatic or tandem trainer.

VTOL Undesignated full-scale model of experimental single-seat VTOL aircraft exhibited 67. One 210 hp M-337 (Czech inverted 6-inline) engine driving two four-blade lift/thrust ducted rotors, pivoting 90° on each side of short fuselage, and ducted four-blade propeller at tail (no other tail surfaces).

KAI-50, 502 Single-seat parasol training glider of 87 dev'd into tandem-seat KAI-502 89. Span 11 m, length 5 m; area 13,2 m^2, loaded 110 kg.

KB-2

Konstruktor Byuro 2 was active in Leningrad 1934-38. Full title was KB-2 OVI RKKA (*Otdel Voyennikh Izobretennii* RKKA, department of air invention of RKKA). Formed by three designers, G. M. Zaslavskii, A. S. Bas-Dubov and G. M. Syemenov.

PI Designation from *Pushechnii Istrebitel*, gun-fighter. Twin-boom pusher with single-seat enclosed cockpit in nose together with two PV-1. Unusual side radiators with engine-driven cooling fan(s). In each wing root, special drum for firing six RS (RS82?) rockets. Prototype construction discontinued 36.

DIMENSIONS Span close to 11 m; wing area 17 m^2.
ENGINE One M-34R.
WEIGHTS Loaded about 2,5 t (5,511 lb).

MPI Designation from *Morskoi Poplavkovyi Istrebitel*, sea floatplane fighter. Conventional configuration with monoplane wing, single seat and central float retracted against fuselage in flight. Project design late 35, prototype construction Aug 37, but never completed.

KB-7

This *Konstruktor Byuro* was organized 36 from a previous design brigade supervised by A. I. Polyarn in order to create aircraft and rocket vehicles to fly in stratosphere. Funding partly Osoaviakhim. Head of KB, L. K. Korneyev, his deputy Polyarn. So far ahead was its thinking, it produced mainly paper projects. These included five types of liquid (mainly Lox/alcohol) rockets, **R-03** to **R-07**, with several variations. Another family of engines included ANIR-5 and ANIR-6 and ENIR-7, and there were several variations of these. Vehicles included winged aircraft and ballistic vehicles, among latter being two-stage **R-10c** to reach 100 km height. **R-05v** was a fighter to climb at 40/50 m/s to a height of 50 km with a cruise engine and two solid boost motors for launch. On a more practical note KB-7 also designed, and apparently built, research rockets for Soviet geophysical institute, programme directed by academician O. Yu. Schmidt.

Kharkov KhAI

Indisputably, *Kharkovskii Aviatsionni Institut* established it147self as leading aviation institute in the Soviet Union for aircraft design. Its KB was opened in 1930 under Iosif Grigoryevich Nyeman, and until 41 produced several major types of advanced conception. In 58 KhAI reopened its KB as student bureau (SKB) confined to ultralight aeroplanes and helicopters.

KhAI-1 With this trim passenger transport Nyeman's design team hit headlines from start. Clean low-wing monoplane, entirely of wooden construction with whole surface fabric-covered and finally doped (silver/red). Monocoque fuselage of elliptical section made chiefly from glued *shpon* in 0,5 mm thickness in five or more layers. Integral centre-section with bolted joints to two-spar wings with 1,5 and 2 mm ply skin. Thin ply skin also on ailerons and tail surfaces. Engine in Townend-ring cowl driving two-blade aluminium propeller. Main gears pulled inwards by cables from cockpit hand-wheel until latched under fuselage. Pilot in open cockpit (retouched to look enclosed in some photos); six passenger seats on right of cabin, each with window; central aisle, five windows and door on left. First flight of prototype L-1351 by B. N. Kudrin 8 Oct 32. Permission for series production given for use on routes to Ukraine and Crimea from Moscow. Production KhAI-1 reached 324 km/h when tested (state trials) Nov 34 by S. I. Taborovskii. Ultimately 43 KhAI-1 delivered 1934-37, all with enclosed cockpit and sliding hood; used on various routes beginning Moscow/Simferopol. In 35 military variant **KhAI-1B** (also called KhAI-1VV, from VVS) produced for bomber training; two-seater with 200 kg load of practice bombs hung vertically in fuselage and one fixed and one movable machine guns, two examples converted from production passenger aircraft.
DIMENSIONS Span (prototype and 36 series) 14,4 m (47 ft 2⅞ in), (34) 14,85 m, (1B) 14,8 m; length (p) 10,4 m, (34 and 1B) 10,21 m (33 ft 6 in), (36) 10,41 m; wing area (p) 34,5 m^2, (rest) 33,2 m^2 (357 ft^2).

KhAI-1

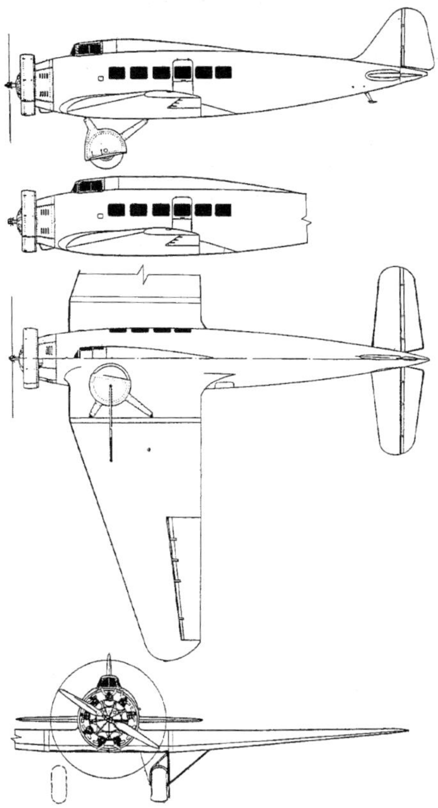

KhAI-1 (inset, 1936 Kiev type)

ENGINE One M-22.
WEIGHTS Empty (p) 1725 kg, (34) 1630 kg, (36) 1830 kg, (1B) 1724 kg; loaded (p, 34 and 36) 2,6 t (5,732 lb), (1B) 2,7 t.
PERFORMANCE Max speed (p) 292 km/h, (34 and 1B) 324 km/h (201 mph), (36) 320 km/h; climb (p) 5 min to 1 km, 23 min to 5 km, (36) 35·4 min to 5 km; service ceiling (36) 6550 m, (rest) 7200 km (23,620 ft); range (p) 820 km, (34, 36) 1130 km (702 miles), (1B) 900 km; TO 12-15 s; ldg (p) 98 km/h, (rest) 85 km/h, 240 m.

Aviavnito-2 A complete contrast, this ultralight trainer was named *Blokha* (Flea), but had nothing in common with Mignet Fleas. Designed under military-glider engineer P. I. Shishov. Tandem open cockpits under parasol wing, mainly wood construction, twin main wheels with nose and tail skids, nose skid being used for bungee launch. Power in cruise from small engine by local engineer P. Labur. Built at Kharkov Young Pioneers Palace in 36. Successful, but no official interest in production.
DIMENSIONS Span 13,0 m (42 ft 8 in); length 6,2 m (20 ft 4 in); wing area 16,0 m^2 (172 ft^2).
ENGINE One 14 hp Labur 2-inline aircooled.
WEIGHTS Empty 200 kg (440 lb); fuel/oil 10 kg; loaded 370 kg (816 lb).
PERFORMANCE Max speed 110 km/h (68 mph); ldg 50 km/h.

Aviavnito-3, KhAI-3 Named for *Sergei Kirov*, second product of Aviavnito brigade under supervision of Aleksandr Alekseyevich Lazarev. It was one of numerous *Planerlyet* motor-glider transports of period (36) intended to reduce costs. Basic concept: all-wing mchine with two axial gondolas each with six seats in line, first seat in right unit being for pilot. Engine mounted on truss pylon above wing on centreline in first drawings, but finally placed on leading edge. All-wing idea adhered to closely, with no other parts except fin/rudder and landing gear. Tapered only on leading edge, giving effective 16° sweep on ¼-chord. Profile V-106 throughout, thickness 14% on centreline and 7% at tip. Rectangular centre section 5 m chord and 4 m span made of welded KhMA steel tube; four spars and six ribs, with covering of 0,5-mm dural. Outer wings, each 9,2 m root to tip, all-wood structures with 8° washout carrying inner and outer trailing-edge controls all driven by single pairs of cables to operate as ailerons or elevators. Flat turns in gliding flight by large rectangular interceptors (operating as airbrakes) near wingtips, opened differentially by pedals; pedals always operated rudder on centreline, with area (inc fin) of 2 m^2. Engine on steel tubes off front spar. Four pairs of wing-root attachments. Tailwheel landing gear, main tyres 800×150 mm, ground angle 15°. Four fuel tanks between gondolas for 8 h endurance. First flight 14 Sept 36; pilots V. A. Borodin and E. I. Schwartz, prototype having dual control from front of both gondolas. Difficulty in getting good turn no matter how elevons, rudder and interceptors were used or not used. Prolonged effort improved control considerably.
DIMENSIONS Span 22,4 m (73 ft 6 in); length 6,8 m (22 ft 3¾ in); wing area 78,6 m^2 (846 ft^2).
ENGINE One M-11.
WEIGHTS Empty 1,440 kg (3,175 lb); fuel/oil 200 kg; loaded 2200 kg (4,850 lb).
PERFORMANCE Max speed 135 km/h (84 mph); cruise 115 km/h (71·5 mph); climb 25 min to 1 km; ceiling 2 km; range 850 km (528 miles); take-off 210 m; landing speed 60 km/h (37 mph).

KhAI-4 Institute's numbering was not consecutive; this small experimental machine ante-dated

KhAI-3

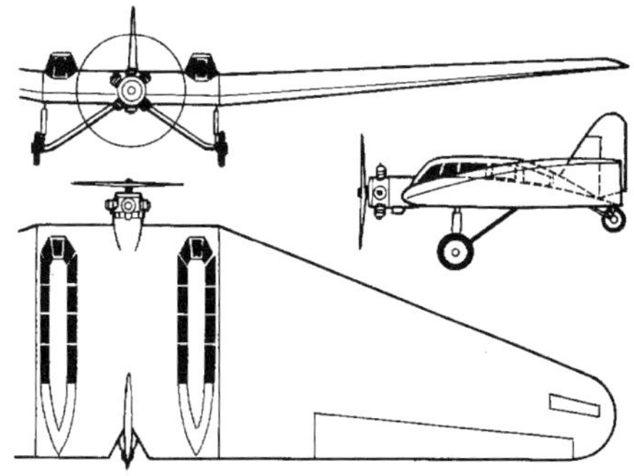
KhAI-3

KhAI-4

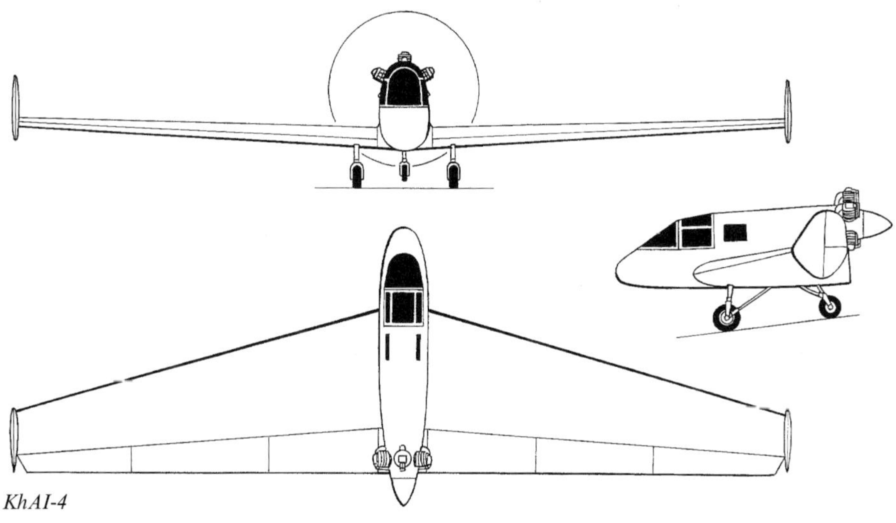
KhAI-4

KhAI-3. Single example, named *Iskra* (spark, flash of light) and also *Osoaviakhimovyets Ukrainy* for local branch of Osoaviakhim which helped finance KhAI-4. All-wood tailless tandem-seat ultra-light, with relatively high power (same engine as 12-seat KhAI-3). Designers: P. G. Benning, A. A. Lazarev and A. A. Krol, determined despite great difficulties to master difficult layout. Wing tapered on leading edge carrying normal ailerons outboard (65% semi-span) with elevons inboard. All surfaces driven together by hand-wheel as in KhAI-3; fore/aft pilot input moved inner surfaces only. Fins and rudders on wing-tips, with separate pedal control. Streamlined but stumpy central nacelle with enclosed tandem cockpits and pusher engine, uncowled. Short landing gears with balloon tyres. First flight Oct 34 by B. N. Kudrin; inadequate elevon moment to lift nose to take-off angle of attack. Left ground at about 180 km/h; trajectory in undulating flight, Kudrin explaining difficulty controlling in longitudinal plane. Acknowledged dangerous, and after third flight KhAI-4 grounded.

DIMENSIONS Span 12,0 m (39 ft 4½ in); length 4,2 m (13 ft 9⅔ in); wing area 21,25 m^2 (229 ft^2).

ENGINE One M-11.

WEIGHTS Empty 550 kg (1,213 lb); fuel/oil 120 kg; loaded 850 kg (1,874 lb).

PERFORMANCE Max speed at SL 180 km/h (112 mph); est service ceiling 3,25 km; design range 600 km (373 miles); landing speed 100 km/h (62 mph).

KhAI-5 Natural devt of KhAI-1 for VVS use, parallel devt to KhAI-1B with more powerful engine driving two-blade Hamilton two-pitch prop. Long-chord cowl with radial frontal air shutters, strengthened wooden airframe (wing skins 1,5-2,0 mm), landing gears modified from KhAI-1, split flaps added, pilot above LE with reverse-slope windscreen and optical sight for twin ShKAS in top decking, gunner with ShKAS in manual turret by I. V. Vyenevidov/G. M. Mozharovskii, internal bay for six FAB-50. Prototype by GAZ-43 Dec 34, nine more in 35 for NII tests 36, satisfactory results.

DIMENSIONS Span 12,2 m (40 ft 0⅜ in); length 9,3 m (30 ft 6⅛ in); wing area 26,8 m^2 (288 ft^2).

ENGINE One 712-hp Wright Cyclone F3.

WEIGHTS Empty 1823 kg (4,019 lb); fuel/oil 260+30 kg; loaded 2515 kg (5,545 lb).

PERFORMANCE Max speed 350 km/h at SL, 388 km/h (241 mph) at 2,5 km; climb 2·4 min to 1 km, 12 min to 5 km; service ceiling 7,7 km (25,265 ft); range 1450 km (900 miles); TO 250 m; ldg 125 km/h, 230 m.

R-10 Despite emergence of later and better aircraft KhAI-5 accepted as basis of prodn a/c to meet 2-seat reconnaissance, light bomber and attack demand. Soviet engine, VISh-6 prop with spinner, chord of fin/rudder increased, tailwheel instead of skid, pilot with normal-slope windscreen, 8 mm back armour and heat from exhaust muff. GAZ-43 built 11 in 36, 17 in 37; Kharkov built 26/100/229 in 37-39 and GAZ-292 15/102/18 in 38/40, total with prototypes 528. Action in Winter War on skis, 60+ several demilitarized (1994 kg empty) to Aeroflot summer 40 as **PS-5** transports (3 pax, cargo, mail).

DIMENSIONS As KhAI-5 except length 9,4 m (30 ft 10 in).

ENGINE One M-25B.

WEIGHTS Empty 2197 kg (4,843 lb); fuel/oil 260+30 kg; loaded 2880 kg (6,349 lb).

PERFORMANCE Max speed 340 km/h at SL, 368 km/h (229 mph) at 2,9 km; climb 2·4 min to 1 km, 14·4 min to 5 km; service ceiling 6,7 km (22,000 ft); range 1300 km (808 miles); TO 350 m, ldg 135 km/h, 240 m.

KhAI-6 Smaller aircraft produced in parallel with KhAI-5 to meet same demands with higher flight perf. Flaps of Zapp type and fabric-skinned control surfs made of light alloy. No weapons; radio operator in rear fuselage managed 2/3 AFA-13 cameras able to photograph up to 50 km from track. First flight Dec 35 by B. N. Kudrin, fast but no order.

DIMENSIONS Span 10,0 m (32 ft 9¾ in); length 6,5 m (21 ft 3⅞ in); wing area 14,0 m^2 (151 ft^2).

ENGINE One 712-hp Cyclone F3.

WEIGHTS Empty 1,2 t (2,646 lb); loaded 1730 kg (3,814 lb).

PERFORMANCE Only fig recorded is max speed 429 km/h (267 mph).

Aviavnito-8 In 37 group under A. A. Lazarev designed 11-passenger improved Av-3: span

KhAI-5

PS-5

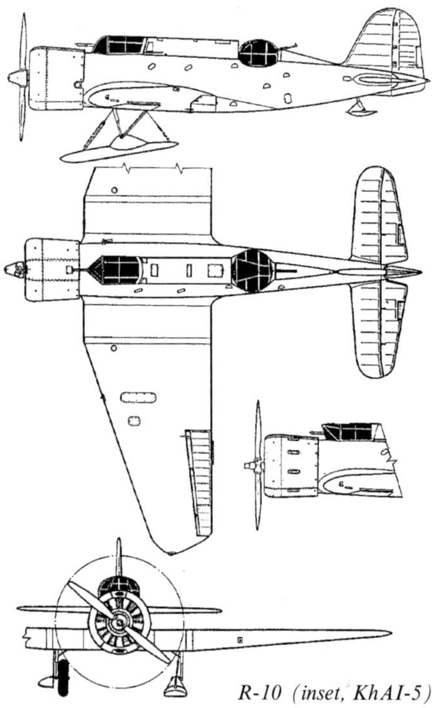

R-10 (inset, KhAI-5)

22,4 m, length 6,8 m, area 78,6 m²; one M-11. Never flown.

Aviavnito-9 Fast passenger aircraft of 34, one M-34, weights 4280/7700, speed 417, unbuilt.

KhAI-51, -52 Last pre-war design, 2-seat recon devd from KhAI-5. Wing spar webs D-16 plate

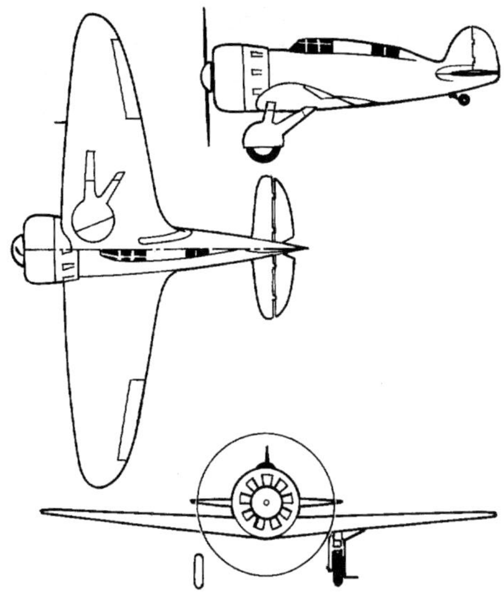

KhAI-6

but still ply skin; six fixed and one movable ShKAS, 400 kg bombload. Kh-51 flown early 39, followed late 39 by short series KhAI-52.
DIMENSIONS Span (51) 12,2 m (40 ft 0⅔ in), (52) 13,42 m (44 ft 0⅓ in); length (51) 9,4 m (30 ft 10 in), (52) 9,6 m (31 ft 6 in); wing area (51) 26,8 m² (288 ft 2 in), (52) 26,73 m² (287·7 ft²).
ENGINE (51) One M-62, (52) M-63.
WEIGHTS Empty (51) 2380 kg (5,247 lb), (52) 2546 kg (5,613 lb); fuel/oil (52) 520+53 kg; loaded (51) 3220 kg (7,099 lb); (52) 3376 kg (7,443 lb).
PERFORMANCE Max speed (51) 358 km/h (222 mph), (52) 410 km/h (255 mph) at 4 km; service ceiling (51) 8,8 km (28,870 ft); range (51) 1000 km (621 miles), no other data.

KhAI-12 Named *Start*, first post-war (49) project, 1-place sporting, never flown. One M-14, wing area 10,125 m², weights 653/830 kg, max 340 km/h, 510 km.

KhAI-17 First post-war to fly, neat wooden low-wing pusher ultralight, slotted flaps, completed (supervisor O. K. Antonov) 59; reflown with geared 4-blade prop 67. **KhAI-18** unbuilt 2-seat version 62.
DIMENSIONS Span 8,6 m; length 5,4 m; wing area 10,5 m².
ENGINE One M-61K.
WEIGHTS 270/350 kg.
PERFORMANCE Max 148 km/h (92 mph); s.c. 2,3 km (7,550 ft).

KhAI-19 Neat tractor ultralight, smaller than pushers yet slower (62). Dim's Span 7,5 m; length 5,2 m; wing area 9,55m²; one M-61K geared to 4-blade, wts 210/312 kg; 140 km/h.

KhAI-20 Much more powerful devt under V. P. Lyushnin, flown 67, dim's 10,44/6,8 m; 140 hp Czech M-332, 500 kg, 270 km/h, 8,6 km (28,220 ft).

KhAI-22, -22A Light helos (68) with 36 hp M-62, 220 kg, 140 km/h.

KhAI-24 Tandem-seat cabin autogyro, functioning mockup 66, 140 hp Czech M-332.

KhAI-25/29 Ultralight a/c, helos, hovercraft.

KhAI-30 Neat flying boat *Professor Nyeman* flown 77, side-by-side enclosed cockpit, tractor engines side-by-side on pylon.
DIMENSIONS Span 9 m (29 ft 6 in); length 7,22 m (23 ft 8¼ in); wing area 13,97 m² (150 ft²).

KhAI-20

KhAI-30

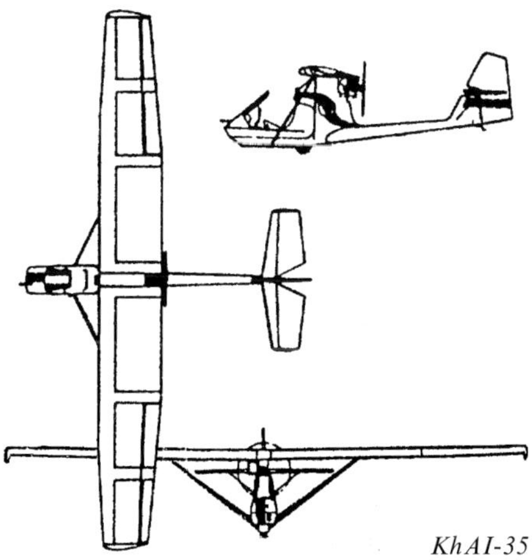

KhAI-35

ENGINES Two 40-hp Vulcan.
WEIGHTS Empty 403 kg; MTO 585 k (1,290 lb).
PERFORMANCE Max speed 125 km/h (78 mph); range 150 km; TO 13 sec, ldg 83 km/h.

KhAI-32 Autogyro, 36 hp MT-9, 220 kg.

KhAI-33 Ultralight seaplane (79), 36 hp MT-9 (Span 9,0 m; length 5,8 m; wing area 12,4 m²; 215/300 kg; 90 km/h); devd 80 into **33M** *50 years KhAI*, 40-hp Vulcan, (Span 8,8 m; length 4,75 m; wing area 9,5 m²; 186/276 kg; 86 km/h).

KhAI-35 Following 18 hp KhAI-34, diploma project by H. Lavrov, pusher ultralight *Entuziast,* wood/GFRP. Flown 81, modified 86 as 35M.
DIMENSIONS Span 10,0 m (32 ft 9⅝ in); length 5,4 m (17 ft 8⅝ in); wing area 11,0 m² (118 ft²).
ENGINE M-61K.
WEIGHTS Empty 200 kg (441 lb); loaded 290 kg (639 lb).
PERFORMANCE Max speed 100 km/h (62 mph); climb 2 m/s (394 ft/min); ldg 60 km/h.

KhAI-36 Two-seat copy of Grigorovich M-9 with 210-hp Czech M-337 engine.

60-let KhAI Following numerous other light and sporting types, this celebrated 60th anniversary, in 90: 3-seat trainer, high forward-swept wing.
DIMENSIONS Span 10 m (32 ft 9⅝ in); length 6,2 m (20 ft 4 in); wing area 12 m² (129 ft²).
ENGINE One 115-hp M-3.
WEIGHTS Empty 406 kg (895 lb); MTO 680 kg (1,499 lb).
PERFORMANCE Max speed 240 km/h (149 mph); range 670 km (416 miles); TO 100 m, ldg 80 m, 70 km/h.

KKB

KKB, Kazan Design Bureau, formed as part of ANTK A. N. Tupolev, Tupolev Aviation Corporation. First project announced 93.

Triton Multirole amphibian, seating 2+8 or 800 kg (1,764 lb) cargo or special role gear; twin diesel engines, prototype said to be well advanced.
DIMENSIONS Not announced.
ENGINES Two DN-400.
WEIGHT MTO 5,8 t (12,787 lb).
PERFORMANCE (est) Max speed 220 km/h (137 mph); cruise 180 km/h (112 mph); range 800 km (497 miles), ferry range 1500 km; max cruise alt 3 km.

Kochyerigin

Sergei Aleksandrovich Kochyerigin was born in 1893. In 1914-15 he trained as military pilot and fought with Imperial Air Service, almost entirely on seaplanes and flying boats. In 17 appointed test pilot at Moscow Central Aerodrome. This brought him into contact with N. N. Polikarpov who appointed him, with A. A. Krylov, chief deputy for aircraft design at OOS on its formation at GAZ-25 in spring 26. Just four years later he went with Polikarpov to CCB on its formation, being appointed head of Reconnaissance Aircraft brigade. He led design of several combat aircraft in pre-war era from KB at Moscow Menzhinskii factory, but never succeeded in opening his own OKB and faded from scene after his first fighters failed to be accepted in 1941-42. He died in 53.

LR Though a modification of ubiquitous R-5 this was such a major redesign it deserves separate inclusion here. LR (sometimes called LR-1) from *Legkii Razvyedchik*, light reconnaissance. Also designated **CCB-1**. Requirement was R-5 replacement with more engine power and reduced weight, able to manoeuvre like contemporary fighter at speed as close as possible to 300 km/h. Authorized Jan 32 together with A. N. Tupolev rival which was never built. Basic design similar to R-5 and used many common parts, but significantly smaller. Wooden wings, Göttingen 436 section, 11% thick. Single I-type interplane struts. Fuselage truss welded Cr-Mo steel tube, dural sheet and panels forward, from cockpit to tail light dural sheet and fabric. Tail and ailerons, dural with fabric covering. Two-spar braced tailplane unusually high on fin, adjustable angle in flight.

LR

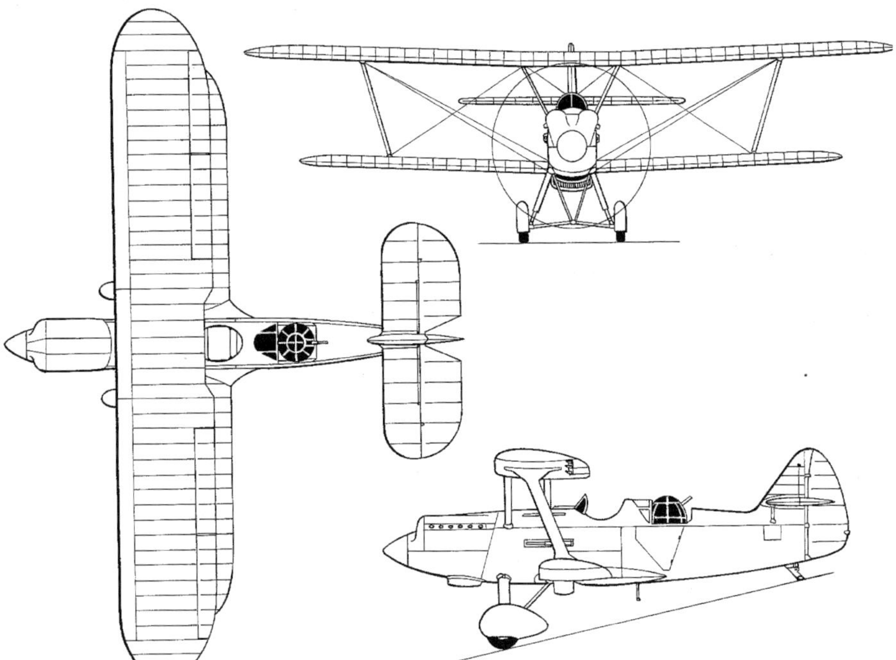

LR

Landing gear as R-5 but air/oil shock struts filled with 90/10 mix glycerine/alcohol. State trials on skis, but wheels had disc brakes, first in USSR. Pilot cockpit open, with left wall folding down for access. Observer/gunner with light-weight glazed turret rotating 360° (one ShKAS) with jettisonable door on left side. Single synchronized PV-1 fired ahead by pilot with optical sight. Bombload 200 kg, usually 2×100 kg, 4×50 kg or 8×25 kg. First flight June 33 and NII State tests early 34. Two built. Production decision in favour of R-Z.

DIMENSIONS Span (upper) 13,0 m (42 ft 7¾ in); length 8,64 m (28 ft 4¼ in); wing area 36,52 m^2 (393 ft^2).

ENGINE One M-34, (No 2 prototype) M-34N.

WEIGHTS Empty 1734 kg (3,823 lb), (No 2) 1812 kg; fuel/oil 275 kg; (No 2) 300 kg; loaded 2426 kg (5,348 lb), (No 2) 2626 kg.

PERFORMANCE Max speed 271 km/h (168 mph) at SL, (No 2) 314 km/h (195 mph) at height; climb 2 min to 1 km; service ceiling 7,4 km (24,278 ft), (No 2) 9,1 km; range 800 km (497 miles); take-off 210 m/13 s; landing 250 m/18 s/ 94 km/h (58 mph).

TSh-3 Also designated **CCB-4**, this was one of first armoured *Shturmoviks* (designation from *Tyazhelyi Shturmovik*, heavy armoured attacker) and was unusual for its day (32) in being a monoplane. Designed to meet requirement of VVS by Kochyerigin staff led by M. I. Guryevich (later partner of Mikoyan), with overall support of S. V. Ilyushin. Remarkably well armed and protected, but thus heavy and rather unwieldy. Construction mainly metal and fabric. Two-spar low-mounted wing tapered from second main rib each side inwards to root in plan chord and thickness as well as towards tip. At second rib, two compression struts bracing wing to top of fuselage. Duralumin truss/plate spars and ribs, fabric covering. Struts, steel tubes, faired with thin dural. Long-span two-part slotted ailerons drooping neutral point 15° to serve as flaps on landing. Basic fuselage and tail structure of welded steel tube, fabric covering, with duralumin keel girder and bulkheads, especially supporting structure for armour. Forward fuselage enclosed by box of bolted armour sections. Frontal thickness 8 mm, complete underside and side 6 mm, top 5 mm across fuel tank. Water radiator and piping protected by 6 mm armour (radiator retractable). Trousered main gears, wheels 900×200 mm, rubber shock-absorption, disc brakes. Crew of two in tandem, both fully armoured. Ten ShKAS installed in wing, firing from leading edge in two batteries of five immediately beyond propeller disc. Inner wing also contained six cassettes (three each side) each containing six frag bombs; additional hardpoints under outer wings for four 100 kg or (eventually) eight RS-82 when available. First flight by V. K. Kokkinaki spring 34.

DIMENSIONS Span 16,5 m (54 ft 1½ in); length 10,75 m (35 ft 3¼ in); wing area 45,04 m^2 (485 ft^2).

ENGINE M-34F.

WEIGHTS Empty 2665 kg (5,875 lb); fuel/oil 230+30 kg; loaded 3557 kg (7,842 lb).

PERFORMANCE Max speed 247 km/h (153·5 mph) at SL, about same at height; climb 10·7 min to 3 km; service ceiling 5,8 km (19,030 ft); range 470 km (292 miles); endurance 2·5 h; landing speed 95 km/h (59 mph).

LR-2 Projected further development of LR begun late 33, for flight late 34. AM-34RN engine, variable-pitch propeller, fully enclosed crew compartment and cantilever main gears with spatted wheels. Prototype never completed.

DI-6, CCB-11 Two-seat fighter designed jointly with Yatsyenko (which see) from Nov 34. Wire-braced biplane with N cabane struts but I interplane struts. Fuselage welded KhMA steel tube, with integral cabane struts and upper centre section. Wings otherwise wooden, two-spar, with ply leading edge and fabric covering. Slotted ailerons with inset hinges. Interplane struts D6 dural with central spar and riveted aerofoil. All tail and control surfaces D6 dural with fabric covering. Fuselage covered in D6 removable panels back to cockpit, fabric aft. Welding by hydrogen, first Soviet use of technique. Engine of prototype imported Wright SR-1820-F3 in NACA cowl with Eclipse-Pioneer electric starter and driving adjustable metal prop (later, two-position Hamilton). Main gears with welded KhGSA struts hinged to both spars of lower wing, retracted inwards by cable pull on unlatched lateral bracing struts whose upper ends slid up vertical guide. Wheels 750×125 mm, Dowty internally sprung. Single fuel tank of riveted D1. Pilot and observer back-to-back, former in open cockpit with optical telescope sight and observer with side and top glazing. One ShKAS with 750 rounds in dural fairing under each lower wing, one ShKAS with 750

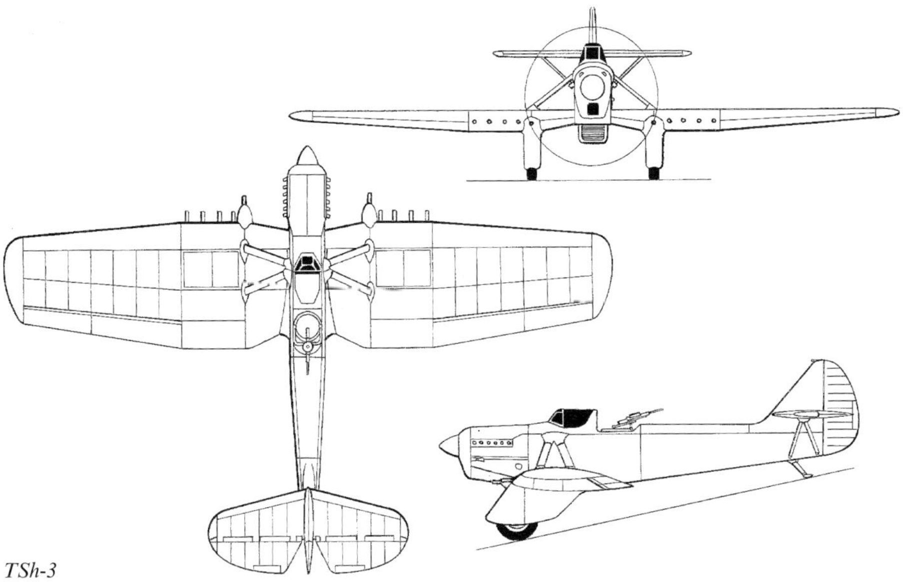

TSh-3

TSh-3

rounds in rear cockpit. Four bomb beams in fuselage (Der-32, Esbr-3) each for one or two bombs of 8 or 10 kg. State trials by Fedorov and Stepanchenok 27 May to 21 Nov 35, ordered as **DI-6** (*Dvukhmestnyi Istrebitel*, 2-seater ftr). Production 10 GAZ-39 in 1936, 61 GAZ-1 and 51 GAZ-81 in 37 and 100 GAZ-81 in 38, total 222. Action in Khalkin-Gol etc 38-40 but few saw action in Winter or Gt Patriotic wars.

Second (*dubler*) **CCB-11Sh**, later **CCB-38**, *Shturmovik* with 8 mm armour, two pairs PV-1 under lower wings, 60 produced 36-38 as **DI-6Sh**. **DI-6MMSh** (*Malya Mod Shturmovik*) with exptl FED-24 engine, never flown. **DI-6bis**, also called *Samolyet 21* (aeroplane 21), intermediate trainer with simple fixed landing gear (speed reduced 25/30 km/h) and dual flight controls (obtained, with instruments, from damaged UTI-4s); some dozens thus modified 1940-41. **DI-6OS** (*Opyshyennyi Stabilizator*, trimmed stabilizer) single aircraft with flight-adjustable tailplane. **DI-6DU** (*Dvoinyi Upravlenii,* war control) with no rear gun and dual control (not same as DI-6bis) for air fighting and bombing instruction; small series of rebuilds.

DIMENSIONS Span 10,0 m (32 ft 9¾ in); length 7,0 m (22 ft 11½ in); wing area 25,15 m² (271 ft²)

ENGINE One M-25; (6bis) M-25V.

WEIGHTS Empty (series) 1360 kg (2,998 lb), (6Sh) 1434 kg (3,161 lb); fuel/oil 172+20 kg (6Sh, 162+20 kg); loaded 1955 kg (4,310 lb), (6Sh) 2115 kg (4,663 lb).

PERFORMANCE Max speed 324 km/h (201 mph) at SL, 372 km/h (231 mph) at 3 km; climb 4·9 min to 3 km; service ceiling 7,7 km (25,260 ft); range 500 km (311 miles); time for 360° circle 12 s; take-off 200 m/10 s; landing 250 m/25 s/ 95 km/h (59 mph).

CCB-27, SR Prototype CCB-27 built 35 to meet Jan 34 **SR** (*Skorostnoi Razvyedchik*, fast recon) reqt., also calling for fighting/bombing.

DI-6

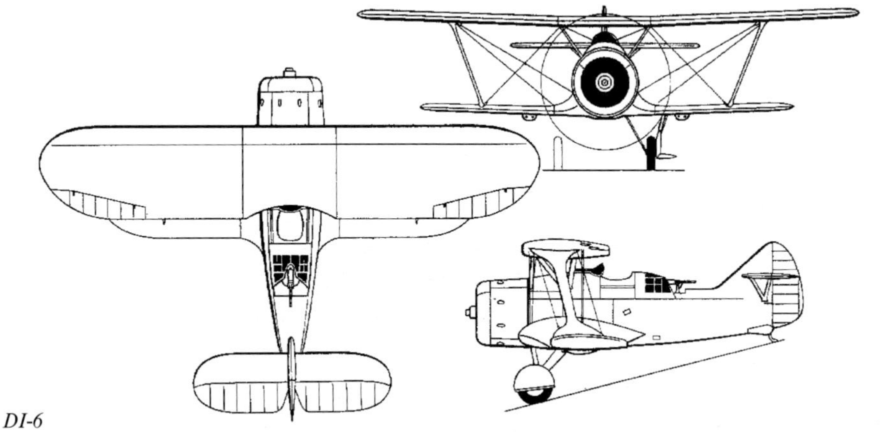

DI-6

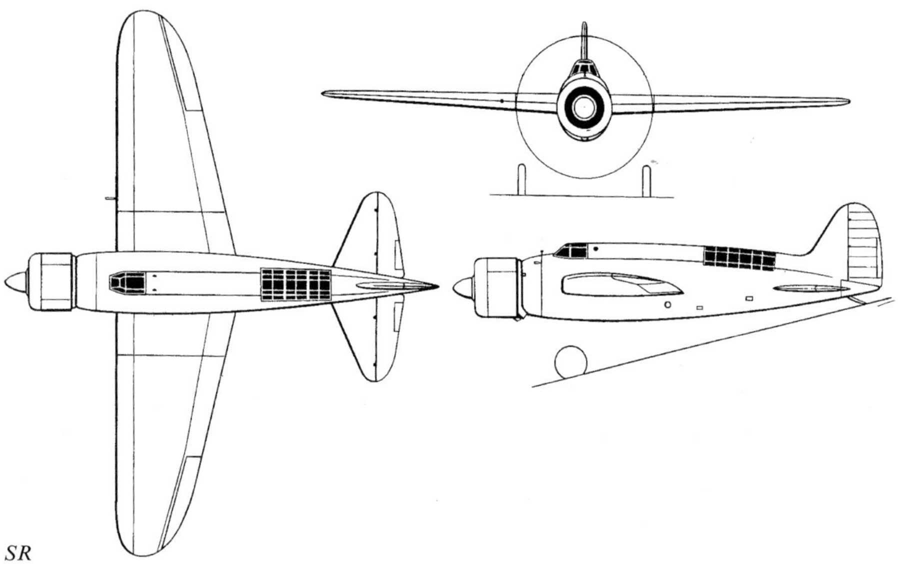

SR

Clean two-seat mid-wing monoplane of mixed construction. Wooden semi-monocoque fuselage. Wings with two KhMA-tube truss spars, Dural ribs and D6 skin 0,6-1,5 mm, flush-riveted. D6 Zapp flaps, all control surfs D6/fabric, rudder/elevators with tabs. Main gears with DI-6 type shock struts pivoted to front spar and pulled in by hyd jack (system energized by pneu bottle recharged by engine compressor), wheels with 800×150 tyres housed in fuselage. Imported engine, ground-adjustable 3-blade prop with spinner. Two ShKAS in rear cockpit, internal bay for up to 400 kg (882 lb) bombs. Troublesome gear usually left locked down, cutting speed to 360 km/h.

DIMENSIONS Span 12,0 m (39 ft 4½ in); length 9,9 m (32 ft 5¾ in); wing area 24,15 m² (260 ft²).

ENGINE One 780 hp Gnome-Rhône 14Krsd Mistral Major.

WEIGHTS Empty 1862 kg (4,105 lb); fuel/oil 260+30 kg; loaded 2649 kg (5,840 lb).

PERFORMANCE Max speed 380 km/h at SL, 460 km/h (286 mph) at 5 km; climb 2·2 min to 1 km, 7·9 min to 5 km; service ceiling 9 km (29,530 ft); range 840 km (522 miles); TO 350 m, ldg 104 km/h, 380 m.

R-9 Two similar aircraft built 36 with Russian engine and v-p prop, many refinements, but rejected in NII testing.

DIMENSIONS As SR except length 10,0 m (32 ft 9¾ in).

ENGINE One M-85.

WEIGHTS Empty 1940 kg; fuel/oil 260+30 kg; loaded 2730 kg.

PERFORMANCE Max speed 366 km/h at SL, 447 km/h (278 mph) at 5 km; climb 8·7 min to 5 km; service ceiling 8350 m (27,400 ft); range 1300 km; TO 360 m; ldg 105/400 m.

Sh, LBSh Major redesign as *Legkii Bronirovannyi Shturmovik*, light armoured attacker. More powerful engine, VISh-22Ye prop. New wooden fuselage with fixed spatted

SR, cowl panels removed

LBSh

PS-43

cantilever landing gears (tyres 750×250), long greenhouse over tankage to fair in MV-3 turret with single ShKAS (500 rds); in wings outside prop disc two ShVAK (150 rds each) and two ShKAS (900 ea); bay for four FAB-100 and two more under wings (Shavrov 'bombload 200, overload 400'). Flown early 38. With six bombs and heavy armour wt 3672, sluggish; so wing bombs omitted. In any case, R-10 won contract. **Sh-2** (*dubler*) and **MMSh** with M-81 never built.
DIMENSIONS As R-9.
ENGINE One M-88.
WEIGHTS Empty 2806 kg (6,186 lb); loaded held to 3450 kg (7,606 lb).
PERFORMANCE Max speed 372 km/h (231 mph) at SL, 446 km/h (277 mph) at 5450 m; climb 2·3 min to 1 km, 9·5 to 5 km; service ceiling 9820 m (32,000 ft); range 1300 km; TO 420 m; ldg 115 km/h, 425 m.

BSh-1, PS-43 Vultee V-11 attack bomber imported 36, Kochyerigin told to adapt this advanced stressed-skin aircraft for licence-prodn. Five assembled from Vultee parts, 31 built from scratch, when discovered not only poorer perf than R-9/R-10 but strength below NII reqts. All BSh-1 and parts handed to Aeroflot for conversion or completion as **PS-43** mailplanes. A few survived war, one to at least 48.
DIMENSIONS Span 15,25 m (50 ft 0½ in); length 11,4 m (37 ft 4¾ in); wing area 35,07 m² (378 ft²).
ENGINE One M-62IR.
WEIGHTS Empty 2911 kg (6,418 lb); fuel/oil 495+50 kg; loaded 4056 kg (8,942 lb).
PERFORMANCE Max speed 318 km/h at SL, 339 km/h (211 mph) at 1,5 km; climb 6·4 min to 3 km; service ceiling 7,2 km (23,620 ft); range 1700 km (1,056 miles); TO 390 m, ldg 280.

OPB *Odnomestnyi Pikiryushchyi Bombardirovshchik*, single-seat dive bomber. Intention to survive in air combat. Forward fuselage duralumin monocoque and two wing spars having plate webs and rolled T or L booms. Rear fuselage wood monocoque. Inverted gull wing with anhedral centre-section leaving room for wide but shallow bomb bay beneath. Root section NACA-230, 14% thick. Main gears hydraulically retracted to rear in centre-section, legs faired but wheels (tyres 750×250 mm) fully housed after rotating 90° and largely enclosed by plate on leg. Retractable tailwheel. Outer wings with full-span slats, D6-skinned ailerons and upper and lower dive brakes (90°) immediately ahead of hydraulic slotted flaps. Fin and tailplane wood, rudder/elevators D6 with fabric. Large engine with ejector exhausts, elec gills, ventral oil rad, driving 3,1 m (122 in) 3-blade prop with large ducted spinner. Cockpit with sliding canopy and good rear view. Two BS with 440 rounds (upper) and two ShKAS with 1700 rounds in sides, firing through front of cowl. Normal bombload one 500 kg (1,102 lb), carried on special mechanism to ensure clean separation without hitting propeller. First prototype flown 41, while estimates made of 584 km/h with ASh-82A and 593 km/h with ASh-71 which were more fully-developed engines. This aircraft often called just OPB, and in some Russian literature it appears as **OPB-5**.
DIMENSIONS Span 10,4 m (34 ft 1½ in); length 8,28 m (27 ft 2 in); wing area 18,0 m² (194 ft²).
ENGINE One M-90.
WEIGHTS Empty 2806 kg (6,186 lb); loaded 3842 kg (8,470 lb).
PERFORMANCE Max speed (500 kg bomb) 600 km/h (373 mph) at 5,7 km (18,700 ft); service ceiling 9,9 km (32,480 ft); range (500 kg bomb) 660 km (410 miles); landing speed (full flaps and slats) 123 km/h (76·5 mph).

OPB

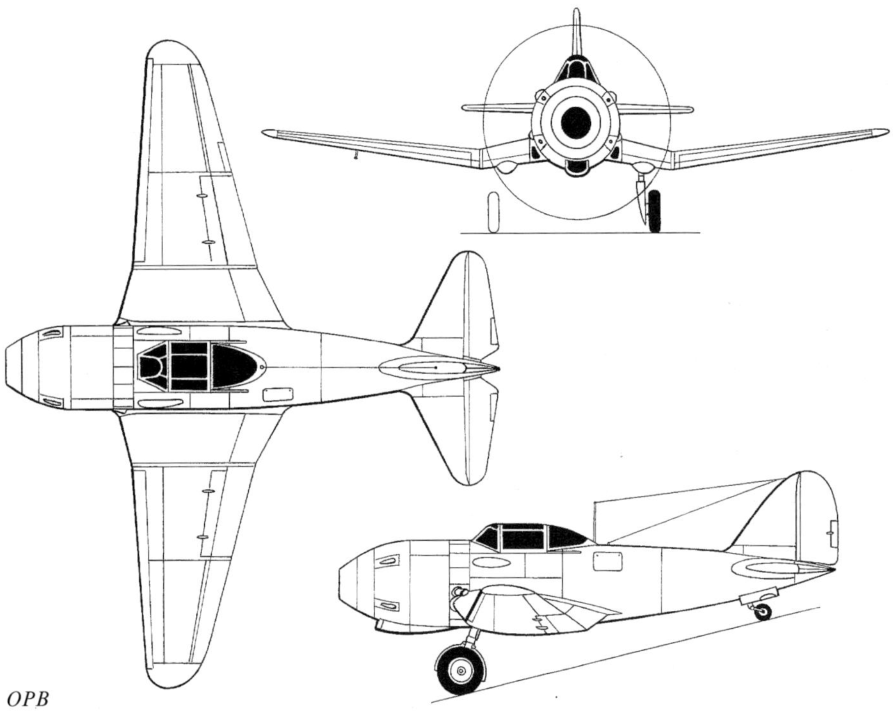
OPB

Kolpakov-Miroshnichyenko

Leonid Dyementyevich Kolpakov-Miroshnichyenko qualified as engineer before turn of century, and in First World War worked at Lebed company where he led design of several aircraft incorporating numerous innovations. Chief among latter was complete installation for machine-gun in cabin aircraft, for which he received royalty of 50R per ten sets supplied to Lebed production. Own designs follow.

K-1 In full, Kolpakov-1, outwardly conventional tandem two-seat reconnaissance biplane with 110 hp Austro-Daimler 6-inline, flown 1916. Unusual in that entire wing cellule variable incidence 7° pivoting about lower front spar; manual screwthread.

Lebed XIV Also called Lebed-Grand, three-bay biplane called *Bolshoi Istrebitel*, large fighter. Two 150 hp Salmson, streamlined near-circular section fuselage with gunners in nose and aft of unequal-span wings, pilot in line with leading edge. Three guns in all, speed 140 km/h (87 mph). Ordered 16, flown first half 17.

LB-2LD Designation from *Lyegkii Bombardirovshchik* (light bomber) with two Lorraine-Dietrich engines. Development of Lebed-Grand, but unable to be drawn, far less built, until after civil war. Two-bay biplane with crew of four. All wooden construction except for struts which were steel tube. Fabric covering, except for occupied part of fuselage which was oval-section plywood semi-monocoque. Design load four guns and 500 kg (1,102 lb) of bombs. Intended engines never installed, and those fitted were only half as powerful. Prolonged difficulties with structural strength and deflection in flight, during test programme begun spring 26 by K. K. Artseulov and Ya. G. Paul. Careful static test showed need for detailed reinforcement, and as R-1 and TB-1 formed adequate team there was no need for production, and rectification never carried out.
DIMENSIONS Span 23,0 m (75 ft 5½ in); length 16,0 m (52 ft 6 in); wing area 140 m² (1,507 ft²).
ENGINES Two 240 hp Fiat A-12.
WEIGHTS Empty 3200 kg (7,057 lb); loaded approx 6200 kg (13,670 lb).
PERFORMANCE No figures except max speed 150 km/h (93 mph).

Kolyesnikov

Dmitri Nikolayevich Kolyesnikov was involved in design of several gliders from about 1930, and in RMK-1 powered aircraft (with Romyeiko-Gurko). Later designed aircraft below with P. V. Tsybin.2

KTs-20 Photograph only recently discovered of this 41 glider, designed in partnership with Tsybin to carry 16 troops. Mainly wood, cantilever wing ply-skinned back to main spar, flaps for landing, fuselage max cross section 1,75 m (69 in) square, dual cockpit in nose manually hinged to right for loading small artillery or vehicles, troop door on left, twin fins, main wheels jettisonable for skid landing. Built in small series. Tsybin's Ts-25 unrelated.
DIMENSIONS (approx) span 22 m (72 ft); length 15 m (49 ft); wing area 57 m² (615 ft²).
WEIGHTS Empty 1,8 t (3,970 lb); loaded 3,5 t (7,720 lb).
PERFORMANCE Landing speed 85 km/h (53 mph).

KOMTA

After 1917 Revolution, only air services in Soviet Union were operated by IM-type bombers converted for transport operations by DVK (*Divisionye Vozdushnii Korablei*, flying ships division, successor to Tsarist EVK squadron of flying ships which had operated same aircraft as bombers). Clear need for transport aircraft led to formation in spring 20 of KOMTA (or KOMPTA), *Kommissii po Tyazheloi Aviatsii*, commission for heavy aviation. Chairman Prof N. Ye. Zhukovskii; members B. N. Yuryev, V. P. Vyetchinkin, V. A. Arkhangyelskii, V. L. Moiseyenko, V. L. Aleksandrov, A. A. Baikov, M. V. Nosov, A. M. Chyeremukhin, K. K. Baulin and A. N. Chyeremukhin, K. K. Baulin and A. N. Tupolev, DVK engineers under Aleksandrov submitted proposal for transport based on IM but KOMTA decided outdated and planned its own aircraft.

KOMTA Planned as triplane with engines under middle wing driving 2-blade propellers. Structure wood joined by small pressed and/or welded mild-steel fittings which also anchored bracing wires. Fabric covering overall. Wing profile better than IM as result of tunnel tests, but underside still concave. Nine ailerons, biplane tail with two tailplanes and elevators (large horn balances like ailerons) and four rudders (no fins). Tailskid and either four main wheels or two skis. Internally wire-braced fuse-

LB-2LD

LB-2LD

lage with up to 10 seats, normally two pilots and six/eight passengers, all boarding via door on left. Parts made by DVK at Sarapul and assembled at Aviarabotnik, Moscow. Complete March 22. Test flown by DVK commander V. M. Remezyuk. Could hardly fly, partly lack of power but mainly CG too far aft. Major reconstruction followed, in course of which engines were moved forward 1 m (39 in). Second series of flights by A. M. Chyeremukhin autumn 23,

KTs-20

KOMTA

and by A. I. Tomashyevskii and B. N. Kudrin spring 24. Inadequate performance. Parked (gate guardian) at Serpukhov air-firing school.

DIMENSIONS Span 15,0 m (49 ft 2½ in); length 9,7 m (31 ft 10 in); wing area 91,0 m² (979·5 ft²).

ENGINES Two 240 hp Fiat A-12.

WEIGHTS Empty 2650 kg (5,842 lb); loaded 2550 kg (7,826 lb).

PERFORMANCE Max speed 130 km/h (81 mph); ceiling 600 m (1,968 ft).

Kondratyev

Experimental OKB formed from members of the disbanded Sukhoi OKB during 51-53 and working at Factory No 51 at Khodinka. The OKB's task was to copy the F-86 Sabre, whose appearance over Korea had come as an unpleasant shock, revealing shortcomings in the MiG-15 and provoking panic measures. Realisation that Sabre superior only in some respects and OKB concentrated on copying US production methods, materials and some systems. Absorbed by Sukhoi OKB late 53.

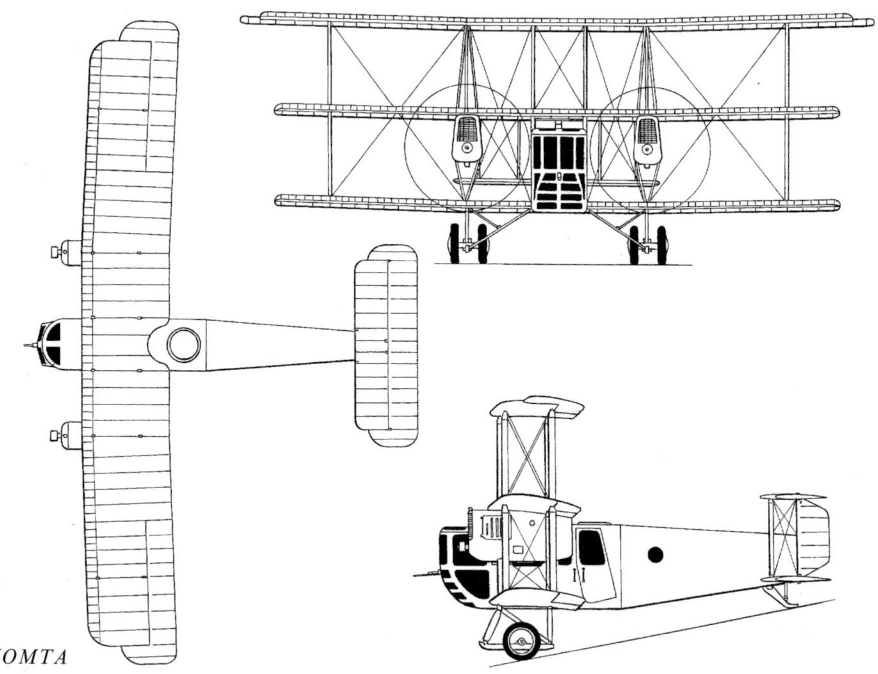

KOMTA

Korolyev

Sergei Pavlovich Korolyev was born 30 Dec 06 at Zhitomir. After basic technical-school training he entered aircraft design in 27, gaining admission to both top aviation schools in Moscow: Bauman Higher Technical School and Moscow School of Aviation. He designed and built light aircraft which flew 30, in which year he graduated from both schools. By 30 also qualified pilot with national reputation on gliders. Never lost interest in aeroplanes and gliders (eg, SK-7) but was so excited by Tsiolkovskii's writings that he began studying rocketry and was appointed head of MosGIRD (Moscow group for study of rocket propulsion). When Marshal M. N. Tukhachyevskii ordered setting up of RNII on 21 Sept 33 Korolyev was appointed deputy chief, and six months later became first head of its winged-aircraft department. In 42 deputy chief designer at GDL-OKB. He flew as flight engineer and twice piloted such aircraft as Pe-2R and La-120R. In 45 appointed faculty member at Kazan University, and at Bauman school, and from this time onwards became not only chief designer of GDL-OKB's large rocket engines but leading technical architect of Soviet Union's immense ICBM and space programmes. He died, laden with honours, on 14 Jan 66.

SK-4 Dubbed *Dalnyego Deistviya*, long-range action, this was intended to be efficient cruising tandem two-seater. Wood construction, monocoque fuselage with 1,5 mm ply skin, and entire braced high wing with aspect ratio 8·17 was also covered with thin ply. Dual control. Six small tanks for 12 hour endurance. First flown 30; programme halted after pilot-error crash.
DIMENSIONS Span 12,2 m (40 ft 0⅓ in); length 7,1 m (23 ft 3½ in); wing area 15,3 m^2 (165 ft^2).
ENGINE One 60 hp Walter.
WEIGHTS Empty 335 kg (739 lb); fuel/oil 190 kg; loaded 690 kg (1,521 lb).
PERFORMANCE Max speed 150 km/h (93 mph).

SK-7 Korolyev could not resist becoming involved with rash of motor-glider transports of 36. This, joint project with P. V. Flerov, had cantilever low wing of span smaller than most. Conventional layout with pilot ahead of five passenger seats, trousered main gears and separate belly cargo compartment. Built at OSGA SNI GVF, work discontinued late 37 when almost complete.
DIMENSIONS Span 20,7 m (67 ft 11 in); length 9,0 m (29 ft 6⅓ in); wing area 39,2 m^2 (422 ft^2).
ENGINE One M-11.
WEIGHTS Empty 780 kg (1,720 lb); fuel/oil 180 kg; loaded 1660 kg (3,660 lb).
PERFORMANCE Max speed calculated 128 km/h (80 mph).

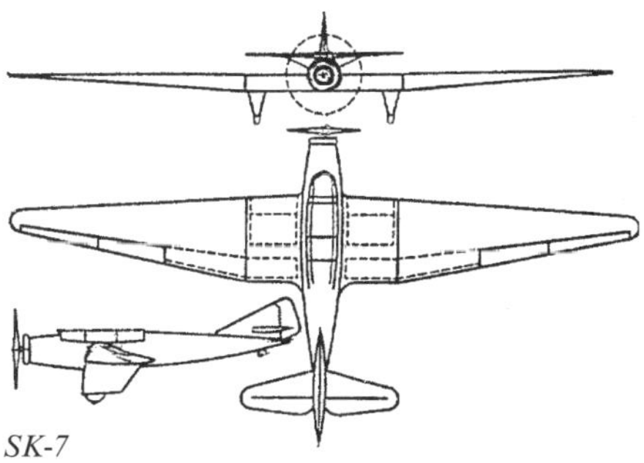

SK-7

RP-318 In amongst SK-4 and SK-7 were Korolyev's gliders, and one of these, SK-9, formed basis for one of world's first rocket aircraft, designed 36 to test ORM-65 engine. This engine was developed under V. P. Glushko, burning RFNA (red fuming nitric acid) and kerosene, SK-9 was two-seat glider originally flown 35. To fly rocket engine it was converted as single-seater, with quite complicated propellant and control system: vertical Dl tank behind seat for 10 kg kerosene, with two vertical stainless-steel tanks total 40 kg of RFNA between wing spars. Thrust chamber at extreme tail, below cut-back rudder. Further details of ORM-65 in Engines section. No 1 engine fired 20 times on bench, nine in Korolyev's Type 212 winged missile and 21 times in RP-318, latter from Feb 37. RP-318 flown as towed glider several times, but no attempt to fly on rocket power until modified ORM-65-2 with improved 'man-rated' installation, had been cleared in static testing. First rocket flight 28 Feb 40, pilot V. P. Fyedorov. Towed by R-5 pilot Fikson, passengers Pallo and Shchyerbakov (later head of own OKB). After 31 min cast off at 2600 m at 80 km/h; after 5-6 seconds speed 140 km/h and pilot then climbed at 120 km/h for 110 seconds gaining 300 metres. Subsequently RP-318 flown nine times, finally with propellants for 30 min.
DIMENSIONS Span 17,0 m (55 ft 9¼ in); length 7,44 m (24 ft 5 in); wing area 22,0 m^2 (237 ft^2).
ENGINE One RDA-1 (ORM-65 derived) rocket, thrust 70-140 kg, normal 100 kg (220 lb).
WEIGHTS Empty 570 kg (1,257 lb); propellants and gas feed 75 kg; loaded 700 kg (1,543 lb).
PERFORMANCE Speed held to 140 km/h (airframe limit).

RP-218 Unbuilt project of 38 with 2-seat pressure cabin, fixed spatted main landing gear.
DIMENSIONS Span 6,5 m (21 ft 3⅞ in); length 8,72 m (28 ft 7¼ in).
ENGINES Three RD-1 rockets. No other data.

Kozlov

Sergei Grigoryevich Kozlov was professor at Zhukovskii VVA throughout first half of 30s. He supervised design of three challenging aircraft, though none was completed.

Gigant First of truly giant bombers, begun as full-scale Academy project Dec 31. Structure basically welded KhMA steel, fabric covering, with some duralumin skin. Mainly enormous wing, with spars so long and deep (3 m) that hinges were inserted to give pin instead of rigid joints. Engines strung along front spar inside wing, in nose-to-nose pairs, each driving gearbox to vertical shaft to tractor propeller on vertical pylon, three propellers above each wing. Short central nacelle for crew of 18-22 with 15 guns and large bombload. Tail carried on twin booms. Static testing of twin-engine group (called *Sparki*, twin) suddenly broke gearbox and front drive-shaft, money running out thereafter in 33.
DIMENSIONS Span greater than 60 m (197 ft); wing area approx 600 m^2 (6,500 ft^2).
ENGINES Six pairs of coupled M-17F.
WEIGHTS Loaded, more than 40 t.

PS Designation not explained, because full name was *Nyevidimyi Samolyet*, invisible aircraft. Full Academy project of 35. AIR-4 used as basis, entire surface covered with Rodoid, patented French organic-glass material. Apart from opaque structure, engine, occupants and other parts, difficulty found in getting effective result. As far as possible opaque parts painted silver-white. Rodoid glass sheet then attached by eyelets and aluminium rivets. First flight 36. Eventually results considered to have achieved 'measure of importance'. Seen from ground, flying PS not seen at first glance as were other aircraft, and often noticed only when observer knew just where to look (height, and accuracy of binaural hearing direction, are in quantified records).

EI Sometimes called **EOI** (*Eksperimentalyni Odnomyestnyi Istrebitel*, experimental single-seat fighter). Project by Zhukovskii academy 39, Kozlov assisted by M. M. Shishmaryev (landing gear), D. O. Gurayev (deputy chief constructor), V. S. Chulkov (wing) and S. N. Kan and I. A. Sverdlov (calculations). Main feature that low wing of 9,2 m span had variable incidence, driven by irreversible Acme-thread on rear spar about ball-bearing trunnions on front spar roots. Spars mixed steel booms and duralumin webs with glued birch *shpon* covering. Fuselage oval-section duralumin monocoque. Engine 1650 hp M-107. Prototype built at outside factory. Severe technical difficulties, project ran late and on 16 Oct 41 prototype and drawings destroyed and factory evacuated.

A. A. Krylov

Aleksei Aleksandrovich Krylov is one of least-known constructors. No relation to V. Ya. Krylov.

R-II Also known as **P-1** (*Perekhodnyi Pyervyi*, interim No 1), this pedestrian two-bay biplane was strictly First World War technology and used wartime engine which at start of design in 23 were all that were available. R-II for *Razvyedchik* (reconnaissance) No 2, intended as replacement for R-1 which itself confusingly had variant designated R-II. Mostly wood/fabric, but metal-covered nose. Two seats in tandem, intended to carry free machine-gun and synchronized gun firing from left side. Interest lapsed after first flight in 25.
DIMENSIONS Span 11,5 m (37 ft 8¾ in).
ENGINE One 260 hp Maybach Mb IVA 6-inline.
WEIGHT Loaded 1650 kg (3,638 lb).
PERFORMANCE No data.

V. Ya. Krylov

This designer worked at GVF Leningrad in 30s, and collaborated with other constructors on two successful light marine aircraft.

AT-1 Joint project with I. N. Vinogradov for three-seat sporting and light transport aircraft 35. Low-wing cantilever monoplane with trousered main gears and fully cowled engine driving propeller with spinner. Mixed construction. Fuselage built on Warren-braced truss of welded steel tube with metal and plywood covering. Centre-section welded steel tube, duralumin covering. Outer wing panels of R-II-C profile 16% at root, 10% at tip, three (probably wooden) spars, ply covering. Split flaps between ailerons including underside of fuselage. Prototype professionally built at Moscow aviation technical secondary school. Completed late 35 and made more than 30 flights.
DIMENSIONS Span 11,8 m (38 ft 8½ in); length 7,0 m (22 ft 11½ in); wing area 22,0 m^2 (237 ft^2).
ENGINE One M-11.
WEIGHTS Empty 750 kg (1,653 lb); fuel/oil 72+12 kg; loaded 1070 kg (2,359 lb).
PERFORMANCE Max speed 180 km/h (112 mph); service ceiling 4 km; range 450 km (280 miles); landing speed 60 km/h (37 mph)

ASK Designation from *Amfibiya Severnogo Kraya* (amphibian for northern territory), sponsored autumn 35 by Glavsevmorput. Cantilever monoplane with twin boat hulls and tractor engine on pylon above wing at centreline. Design in partnership with I. M. Zharnylskii, G. I. Bakshayev and L. S. Vildegrub. All-wooden. Two-spar wing with ply-covered centre-section and fabric outer panels. Hulls stoutly built

R-II

with multiple frames and bulkheads but without internal stringers, though strong keel. Two steps on concave planing bottom giving weak wake. Ply skin basically 3 mm thick, increased to 6 mm over planing bottom and 8 mm at steps. Twin booms carrying three-finned tail. Pilot in open cockpit in small nacelle ahead of leading edge on centreline. Each hull equipped for three passengers, front seat having direct view ahead, or 600 kg cargo. Landing wheels on inner sides of hulls, raised backwards except when needed. Skid at rear of each planing bottom. ASK flew spring 35, test pilots B. V. Glagolyev and P. P. Skarandayev. Satisfactory, but one landing gear broke in Oct 36 and ASK beset by several other troubles culminating in complete write-off in flood of autumn 37.

DIMENSIONS Span 20,8 m (68 ft 3 in) (not 17,4 m); length 12,9 m (42 ft 4 in); wing area 66,4 m^2 (715 ft^2) (not 46,4 m^2).
ENGINE One M-22.
WEIGHTS Empty 2450 kg (5,401 lb); fuel/oil 400 kg; loaded 3450 kg (7,606 lb) (not 4 t).
PERFORMANCE Max speed 215 km/h (133·5 mph); climb 7 min to 1 km; service ceiling 4,1 km (13,450 ft); range 700 km (435 miles); endurance 4 h; take-off 300 m/20 s; landing 170 m/9 s/95 km/h (59 mph).

MA-1 Designation from *Mestnaya Amfibiya*, regional amphibian, this was GVF project undertaken from 38 jointly by Krylov and Vladimir Fedorovich Rentel. Straightforward transport with echoes of ASK in hull shape and landing gear. Cantilever high wing with 30° dihedral gull centre-section integral with hull and outer panels with spars at 15 and 60% chord of wood construction with ply skin 2 mm to rib 9 on each side and then 1,5 mm to rib 20 at tip. Ailerons and tail control surfaces light wood/fabric built up on ply-sheathed nose box. Fin, tailplane and flaps ply skinned, latter driven pneumatically to 60° on long brackets. Hull skin 3-4 mm ply, 4-5 mm near step, covered in doped fabric. Landing gear pneu-retraction back against fairings on sides of hull. First flown Jan 40 on skis, without stab floats, then floats added and tested from Grebno Port. Flown to Moscow Sept 40 and evaluated at Khimki river against Sh-7. Slightly superior, but damaged on takeoff, showing need for hull reinforcement. Tests stopped by war.

DIMENSIONS Span 14,0 m (45 ft 11¼ in); length 11,82 m (38 ft 9⅜ in); wing area 29,6 m^2 (319 ft^2).
ENGINE One MG-31F.
WEIGHTS Empty 1450 kg (3,197 lb); fuel/oil 220 kg; loaded 2200 kg (4,850 lb).
PERFORMANCE Max speed 210 km/h (130 mph) at SL; climb 4 min to 1 km; service ceiling 4,3 km (14,100 ft); range 700/1200 km (435/746 miles); endurance given as 4 h (low); take-off 215 m/11·5 s; landing 107 m/10·2 s/84 km/h (52 mph).

ASK

Kurchyevskii

L. B. Kurchyevskii was designer of APK series of large-calibre cannon. In 35 Marshal Tukhachyevskii got him to organize a USP (control of special work) within NKTP for construction of fighters armed with these guns. He issued invitations to aircraft constructors to join USP: B. I. Chyeranovskii, S. A. Lavochkin and V. B. Shavrov. All were at work by spring 35, but a year later Kurchyevskii was arrested and USP closed.

IL Designated from *Istrebitel Lavochkina*, (Lavochkin's fighter); joint project with S. N. Lyushin (deputy) and also called **LL** from both designers. Appearance similar to IP series but with pilot low in fuselage seeing through periscope. Seat and hinged windscreen raised for landing or for better view in emergency. Mock-up inspected 12 Jan 36 by Gen Ya. I. Alksnis. Eventually abandoned.

BICh-17 Chyeranovskii's fighter (see BICh) followed typical *parabola* lines, with stumpy fuselage with 480 hp M-22 on nose. Gull wing with 5° dihedral to mid-span, then anhedral to tip. Inboard of highest point on each wing was 80 mm APK, just beyond propeller disc and with rear barrel dividing flaps into inner and outer sections. Wooden construction, ply/*shpon* skins, twin-wheel main gears retracting inwards. Enclosed cockpit faired into squat fin. Discontinued when 60% complete when USP terminated Feb 36. By that time it had several other projects including BICh-type fighter with DRP guns and another APK-armed fighter begun by P. D. Grushin at MAI.

DI-4

Laville

André Laville was one of three French designers invited to work in USSR by Aviatrust in 28. First employment was with fellow-countryman P. A. Richard at OPO-4 where he drew plans for two-seat fighter based to some degree on Nieuport-Delage designs on which Laville had worked a year previously. When Richard found political troubles in 30 Laville left OPO-4 and in August began to organize his own Buro Novyikh Konstruktsii (bureau of new designs), BNK, in Moscow. BNK folded 33, Laville going to factory work for NII GVF. In 35 he became Russian correspondent for Paris newspaper, and returned to France Jan 39. He is seldom mentioned in accounts of his aircraft.

ZIG-1

DI-4 Two-seat fighter derived from design prepared at OPO-4. Parasol wing strut-braced to main gears, twin lateral Lamblin radiators and twin-finned tail for good field of fire to rear. Wing profile MOS-27, gull-winged and increasing in thickness from fuselage to standard 16% over horizontal outer portions. Excellent pilot view, first fuselage-joined gull wing in USSR. All-duralumin construction, ruling skin thickness 0,5-0,6 mm. Finely streamlined monocoque fuselage and cantilever tail, with special features ensuring stability of Y-strut under each wing and smoothness of tail skins. Four PV-1, two synchronized and two aimed from rear cockpit. Prototype completed early 33. Factory test-flying by B. L. Bukholts and Yu. P. Piontkovskii, State trials by K. K. Popov. Altogether DI-4 did quite well, only adverse comment being marginal yaw stability and effectiveness of tail.
DIMENSIONS Span 13,3 m (43 ft 7⅔ in); length 8,5 m (27 ft 10⅔ in); wing area 23,9 m² (257 ft²).
ENGINE One 600 hp Curtiss Conqueror V-12.
WEIGHTS Empty 1448 kg (3,192 lb); fuel/oil 200+30 kg; loaded 1949 kg (4,297 lb).
PERFORMANCE Max speed 266 km/h (165 mph) at SL, 238 km/h (148 mph) at 5 km; climb 7·4 min to 3 km; service ceiling 6440 m (21,130 ft); range 500 km (311 miles); endurance 2 h.

PS-89 (ZIG-1) Passenger airliner, major project of NII GVF in May 33. Cantilever low-wing monoplane, all-duralumin construction with flush-riveted stressed-skin covering except for fabric flight-control surfaces. Performance surpassing that of contemporary fighters, yet with maximum security (eg, well-established engines and fixed landing gear). Wing of aspect ratio nearly 11, with three spars, multiple stringers and power-driven landing flaps (shown as split type in drawings). Well-streamlined fuselage with three long but shallow windows each side, squat vertical tail and tailplane braced by double struts. Well-cowled engines with underslung radiators, faired into trousered main gears. Equipped for two pilots and 12 passengers in six pairs facing each other across small tables. Prototype construction assigned to GAZ-89, major aircraft-repair works known as *Zavod Imyennyi Goltsman* (works named for Goltsman), or ZIG, hence designation. Ready for trials spring 35, after final inspection and refinement by A. V. Kulyev. Factory tests began satisfactorily, but on 27 Nov 35 tailplane broke away while gliding in to land (there were rumours of sabotage) and aircraft dived in, killing pilot Ablyazovskii, Kulyev and crew of four engineers. Remarkably, work was allowed to continue, and Aeroflot bought six ZIG-1 with designation PS-89 (*Passazhirskii samolyet*, factory 89). Production aircraft supervised by Pyetr Ivanovich Eberzin. All delivered by late 38 and four proving flights made on Moscow-Simferopol route. Results described as perfect, and PS-89 thereafter established excellent record in several thousand hours' regular line operation.
DIMENSIONS Span 23,1 m (75 ft 9½ in); length 16,245 m (53 ft 3⅝ in); wing area 72,0 m² (775 ft²).
ENGINES Two M-17F.
WEIGHTS Empty 4,9 t (10,802 lb); fuel/oil 800+60 kg; loaded 7,1 t (15,653 lb).
PERFORMANCE Max speed 284 km/h (176 mph); cruise 244 km/h (152 mph); climb 5·7 min to 1 km; service ceiling 4,4 km (14,436 ft); range 1300 km (808 miles); TO 23 s; ldg 95 km/h.

Lavochkin

Syemyen Alekseyevich Lavochkin was born Smolyensk 11 Sept 00. After revolution joined Red Army, but in 20 gained admission to Moscow technical high school to study engineering, later specializing in aviation and assisting in stressing ANT-3 and ANT-4 at AGOS. Graduation 29, joined Richard at CCB, moving to BOK in 31. In late 34 collaborated with BRIZ of Glavsevmorput on design of boats and other craft and became increasingly involved with *Kaplyurit*, invention of BRIZ's O. F. Kaplyur, plastic-impregnated plywood reinforced with internal steel gauze. Within a few months assigned to Kurchyevskii's USP to build IL (LL); on closure of USP in 36 assigned to GUAP as senior engineer. Intimate contact several programmes while spent much time at factory of Leontii Iovich Ryzhkov (products, propeller blades and skis) studying *delta-drevesiny* developed mainly by Ryzhkov from 30 as tough fire-resistant material made from compressed plastic-impregnated laminates of birch (similar to UK's Jablo and Rotol densified woods). By Aug 38 Lavochkin convinced such materials could be of great strategic importance for military aircraft, reducing drain on steels and light alloys without seriously compromising aircraft performance. Teamed up with V. P. Gorbunov and M. I. Gudkov (both mentioned previously) to build *delta* fighter, and received permission by M. M. Kaganovich, GUAP Commissar, to open OKB. Effective 15 Sept 38, in premises vacated by liquidated Silvanskii OKB. Lavochkin attended Kremlin conference Jan 39 called to discuss strategy for future combat aircraft and received permission to build prototype designated I-22. At same time production base granted, and this was found in GAZ-301, former Moscow furniture works used as OKB-301 by A. A. Dubrovin, developing Caudron fighters, and also for manufacture of Yak BB-22 fuselages. Dubrovin went to improve R-10 at Kharkov, deputy A. G. Brunov joined new MiG bureau and LaGG (Lavochkin, Gorbunov, Gudkov) OKB-301 expanded mainly by hiring design engineers from BNK (Laville) and Grigorovich bureau. Success of La-5 and derivatives brought Lavochkin three Orders of Lenin, but from 45 his large and famous OKB – moved early 46, former facilities to S. M. Alekseyev – consistently failed to beat rivals from start of jet era. From late 53 devoted increasingly to SAM (surface/air missile) design, becoming national leader, later adding jet targets. Founder died 9 June 60, bureau then divided and renamed.

I-22/LaGG-1 First LaGG fighter, originally called *Frontovaya Istrebitel* (frontal fighter) to fly tactical missions against air and ground targets at medium heights. Conventional in all respects except materials, and exceptionally tough and simple. Ruling material not *delta* but birch ply and *shpon*, though similar to *delta* in processing and called bakelite-ply. Fuselage narrow oval section, 13 main frames, triangular-section sringers, integral fin and centre-section of 3,17 m span. Engine mounting welded steel tube. Skin thickness 9,5 mm forward tapering to 3,0 mm at fin and tailpiece. Wings with two box spars, close-spaced ribs and bakelite-ply skin mainly 3 mm thick. Duralumin flight-control surfaces with fabric covering. Duralumin split flaps driven hydraulically to 15° or 50° (landing). Main landing gears with 650×200 mm tyres retracted inwards hydraulically, wheels housed in lee of engine air ducts in roots. Retractable tailwheel, twin doors. Engine with Hucks starter dogs, also compressed-air starter in production aircraft. VISh-61P propeller with three blades 3 m (118 in), 58/230°. Ventral oil cooler, main radiator well aft under fuselage at trailing edge with manual shutter. Fuel 340 lit (74·8 gal) in three inter-spar tanks of AMTs (aluminium/magnesium alloy) overlain by four layers of phenol-formaldehyde resin-impregnated fabric and with NG inert atmosphere (prolonged use softened metal of tanks). Spartan cockpit with no gyro instruments or radio but 10 mm seat armour. Plexiglas glazed canopy manually sliding to rear. Simple telescopic sight, later replaced by PBP-1a reflector sight. Armament of first of seven prototypes one Taubin 23 mm (later a VYa) with 80 rounds and two UBS each with 220. First flight by A. I. Nikashin 30 Mar 40. Speed fair but range inadequate and handling tricky to point of being dangerous. Batch of 100 had been laid down at GAZ-301 and these completed, most with one ShVAK and two ShKAS. Crash programme to rectify faults.

DIMENSIONS Span 9,8 m (32 ft 1¾ in); length 8,82 m (28 ft 11¼ in); wing area 17,62 m² (189·7 ft²).

ENGINE M-105P.

WEIGHTS Empty 2478 kg (5,463 lb); fuel/oil 250 kg; max 2,968 kg (6,543 lb).

PERFORMANCE Max speed 515 km/h (320 mph) at SL, 605 km/h (376 mph) at 4950 m; climb 5·85 min to 5 km; service ceiling 9,6 km (31,500 ft); range 556 km (345 miles); TO 23 s, ldg 143 km/h.

I-301

LaGG-3 (1st series, 1941)

LaGG-3 Crash programme produced prototypes called **I-301**, named for plant. Detail analysis to pare 200 kg off structure thwarted by extra eqpt and two extra tanks, total 452 lit (99·4 gal). Outer wings incorporated fixed slots, replaced early 41 by auto slats; armament standardized as one ShVAK with 120 rounds and two ShKAS each with 325; provision for attachment beneath each wing of three RS-82 rockets or one 100-lit (22-gal) drop tank; system of internal pendulum weights inserted in elevator circuit; upper and lower mass-balance fitted to rudder; and intention to fit PF engine when available. First I-301 flown by Nikashin 14 June 40. Major all-round improvement and, following Nikashin's non-stop flight Moscow/Kursk and back with 15% fuel remaining on 28 July 40, decision taken to organize production as LaGG-3 at four plants: No 301, No 31 at Taganrog (which for some months became centre for fighter development), No 21 at Gorkii and No 153 (former tractor works) at Novosibirsk. Thanks to this plan production built up impressively, 322 LaGG-3s being delivered in first half 41 and 2,141 in second half (far outstripping Yak-1 output). But fatalities and write-offs in training were high, morale of LaGG pilots low (initials said to mean *Lakirovannii Garantirovannii Grob*, varnished guaranteed coffin) and defects still numerous.

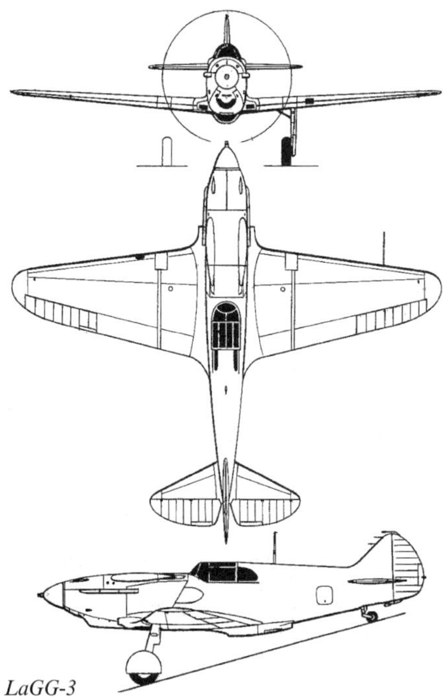

LaGG-3

Radio at last installed, hydraulics made reliable, brakes made controllable, canopy more transparent, rudder lower mass balance removed and later replaced by horn, elevators properly balanced and pendulum masses removed, bulletproof windscreen fitted, and NG tank system made operative. Training eventually removed grosser faults in piloting technique, and various armament schemes fitted, such as VYa or ShVAK plus one/two ShKAS or one/two UB. Field mods resulted in other schemes such as Sh-37 or NS-37 or five UB, two of them in underwing gondolas. Night-flying equipment added to some, and also reconnaissance version with AFA-1 camera built in small numbers. Late 41 main OKB evacuated to GAZ-21 at Gorkii and GAZ-31 evacuated to Tbilisi, but GAZ-153 ordered to switch to Yak-1 instead. LaGG-3 remained inferior combat aircraft, its main asset being toughness. Much of trouble lay in shoddy workmanship by untrained personnel who had had no prior contact with modern technology. Total deliveries 6,528 LaGG-1 and LaGG-3 by August 42, from when all airframes received M-82 engine.

This does not include numerous prototypes from Gorkii and Tbilisi seeking improvements (nor Gorbunov and Gudkov efforts, which see) which included: aircraft with VK-105TK turbocharged engine 40, not flown; aircraft tested 42 with two underwing VRD-430 (boosted speed 30 km/h but drag when inoperative reduced by 50 km/h); **LaGG-3 oblegchyonnyi** (lightened), Series 23, 42, 2865 kg, drastic wt-saving, one MP and one UBS but only 564 km/h. Series 4, 41, burdened by ShVAK/BS/2×ShKAS/6RS-82, max 503 km/h; Srs 23 also RS-82 but no ShKAS, 518 km/h; Srs 28 introduced PF engine Aug 42, 561 km/h; Srs 34 NS-37/BS; fastest Srs 66, May 43, 3023 kg, 580 km/h (360 mph). **LaGG-3F, No 107**, 42 prototype M-107 engine failed on all 33 test flights; another with VK-107 flew 43 but engine still undeveloped; **LaGG-3 dubler** (duplicate), 44, VK-105PF-2, lightweight (2975 kg), one VYa-23 and one UBS, 618 km/h; and growing effort to fit M-82 engine, see next.

DIMENSIONS Span 9,8 m (32 ft 1¾ in); length (excl 23 and 37 mm barrel) 8,82 m (28 ft 11¼ in); wing area 17,62 m^2 (189·7 ft^2).

ENGINE Srs 1-28, M-105PA, 28-66, M-105PF.

WEIGHTS Empty 2620-2890 kg (5,776-6,371 lb); fuel/oil 350 kg; loaded 2990-3363 kg (6,592-7,414 lb).

PERFORMANCE Max speed at SL 498-542 km/h (309-337 mph), at ht see text; climb to 5 km (1943) 5·6 min; service ceiling 9,6 km (31,500 ft); range (1941) 870 km (541 miles), (1943) 650 km (404 miles); TO 20 s, ldg 139 km/h (86 mph).

LaGG-3 (23rd series, 1942)

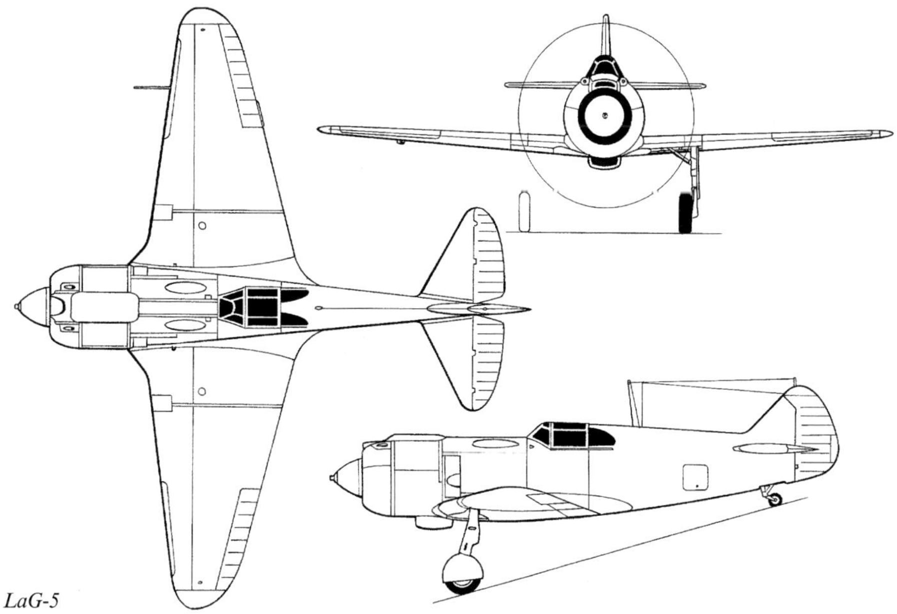

LaG-5

La-5 Within a year installation of M-82 radial engine in LaGG-3 had transformed sadly defective fighter into one of most effective of the Second World War, and also transformed Lavochkin's status. Work began late Sept 41 at Tbilisi. Major problems were: greater weight of engine (about 850 kg compared with 600), compounded by reduction of some 70 kg at rear by removal of water-filled radiator, duct fairing, control flap and piping; greater engine width of 1260 mm compared with 777; and lower thrust-line. On credit side was reduced length of engine which greatly eased problem of CG shift and promised to overcome major cause of poor manoeuvrability. With Shvetsov installation engineer Valedinskii, Lavochkin's designers created outstanding installation with tightly fitting cowling with central fairing over reduction gear, radial controllable cooling-air vanes behind large spinner and exhaust pipes collected in left and right groups of seven discharging past hinged door controlling exit airflow. Inlet to supercharger on top, oil cooler underneath. VISh-105V propeller, three blades, 3,1 m (122 in) diam, hydraulic control and feathering, with Hucks dogs on drive shaft; engine also equipped with compressed-air starter. Auto LE slats retained, with such success Narkomaviprom ordered them fitted to MiG Ye.

First conversion called **LaGG-3M-82** or **LaG-5** (reflecting absence of Gudkov). Compromises including double skin behind engine, both old and new profiles. Seat armour 10 mm, armament two ShVAK each with 200 rounds. Tankage as LaGG-3, five cells totalling 462 or 464 lit (102 gal), rest of aircraft also almost unchanged. Completed in small OKB shop outside main Tbilisi factory Dec 41; thereafter GAZ manager so anxious not to compromise himself he refused to allow it inside a hangar. His cool attitude, combined with severe winter and various teething troubles, delayed first flight until March 42. OKB test pilot G. Mishchyenko not happy with take-off or landing, general control balance or harmonization, but pleased with climb and speed. Initial testing mixed results, and produced numerous detail changes such as ailerons with slightly skewed axes and progressively greater chord on underside only, from tip to inner end, giving more aerodynamic balance. When NII pilots I. Ye Fedorov and A. I. Nikashin flew La-5 in late March they were impressed, and Nikashin told Stalin personally. Result was crash programme of State trials, by NII team led by Col I. V. Frolov and comprising pilots A. Yakimov and A. Kubishkin and engineers V. Saginov and A. Frolov. Begun April 42, and resulted in priority directive in July for maximum-rate production of La-5, with M-82 also fitted to all available LaGG-3 airframes. (When Yakovlev arrived at GAZ-153 in late 42 he found scores of incomplete LaGG aircraft, some mere bumps in snow and many not even discovered until following spring; even these were later turned into La-5s.) Control of GAZ-21 (Gorkii) restored to Lavochkin, previous manager now obsequious, and both this and Tbilisi (GAZ-31) expanded to boost prodn. First ten La-5s completed in three weeks, for service test while flying operationally on war front. Performance far short (up to 50 km/h) of that expected; Lavochkin called to Kremlin after CAHI team had reported reason was chiefly high drag of crude hand-made cowlings. Stalin told designer such things were his responsibility, Malyenkov adding that his 'future performance will be scrutinized closely'. Worse, two wings broke off, one in shallow dive and other on landing approach: in this case cause found to be hammering wing-attach bolts into under-sized holes! Vibration traced to unbalanced propellers, and cannon mounts redesigned to reduce vibration when firing. With these snags overcome special trials regiment formed Sept 42, at Stalingrad; La-5 was universally acclaimed.

Entry over high cockpit sill awkward. Still no gyro instruments and only small inaccurate compass, but RSI-4 single-channel radio standard, with good press/speak button on throttle. No cockpit heating, poor view to rear (or ahead when taxiing) and PBP-1a sight a simple lens with just two deflection rings; for accurate shooting pilot had to be dead astern target. Need to watch cylinder-head temperature constantly; especially in early days heads commonly flew off cylinder when overheated. General flight handling superb, including loop or Immelmann begun at mere 300 km/h (186 mph)! Good recovery from stall or spin, no special limitations and only real problem tendency to bounce on landing (instructors said 'ignore it', but many pilots instinctively added power, causing torque to invert aircraft). Another problem was short range. Five tanks but outers never used in front line, and Lavochkin again called to Kremlin where he reported he could not increase range. Stalin, by now well-disposed to Lavochkin, asked designer to think about it; by early 43 L. A. Zaks at Tbilisi had managed to get full original tank capacity into three tanks in centre section (464 lit), with 160 kg reduction in empty weight. In Dec 42 ASh-82F had replaced original engine, aircraft being designated **La-5F**. At end March 43 important direct-injection ASh-82FN became standard, for first time eliminating engine-cut under negative-g and also giving more power; resulting **La-5FN** distinguished by full-length engine air duct starting at cowl lip. Another important change, introduced soon after completion of last La-5 based on a LaGG airframe (about Nov 42), was improved rear view from cut-down rear fuselage and glazed rear canopy with 75 mm bulletproof glass transversely. Many aircraft had fixed tailwheel for simplicity. Few departures from standard twin-ShVAK armament; some lightened by twin-UBS, and several experimental fits at NII. Underwing loads 200 kg bombs (various) or four/six RS-82. To reduce training losses tandem dual **La-5UTI** produced Aug 43, tested 3 to 30 Sept. (Note: dual two-seaters had been made by field conversion as early as Nov 42, but these poorly engineered.) UTI had tandem sliding canopies, re-routed tail controls and radio mast, relocated battery and, usually, only one ShVAK. By summer 43 UTI had FN engine and interspersed with La-5FN on assembly lines which by this time included not only Tbilisi and Gorkii but also Yaroslavl and a GAZ in Moscow area. Early 43, several **La-5TK** with turbocharger, no series output.

La-5UTI (conversion by 1st Repair Shop, Leningrad)

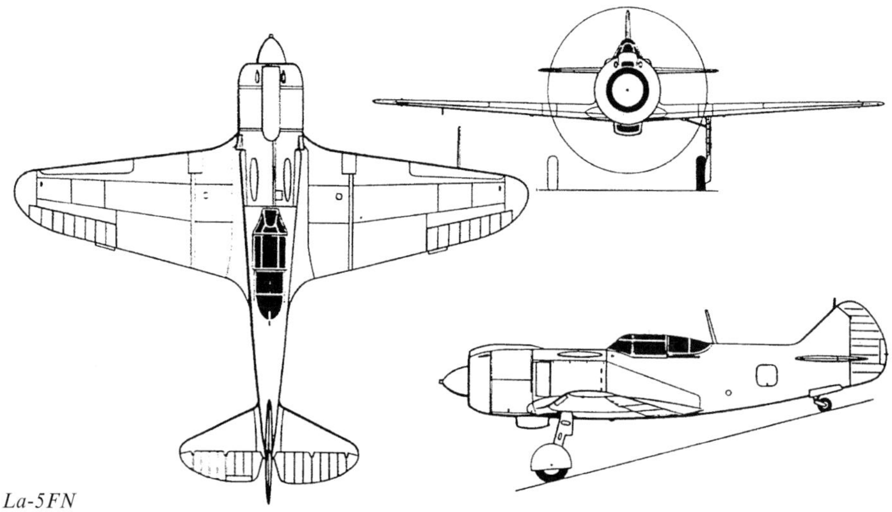

La-5FN

Final change in production was introduction of new wing with D1 duralumin spar webs and 30KhGSA booms in May 44, sub-type called **La-5FN Type 41**; at same time this wing was also fitted to parallel fighter with improved fuselage, La-7. Smaller spar dimensions saved 172 kg structure weight and enabled internal fuel to rise to 560 lit (123 gal). Despite La-7, last La-5FN did not leave factory until late 44. Total production est 9,920. Popular name *Lavochka.*

DIMENSIONS Span 9,8 m (32 ft 1¾ in); length 8,71 m (28 ft 7 in) (FN engine, 8,67 m, 28 ft 5⅓ in; UTI same); wing area (1942) 17,37 m^2, (1943) 17,27 m^2 (186 ft^2), (FN) 17,59 m^2 (189 ft^2).

ENGINE (prototype) one M-82, (1942 production) M-82A, (F) ASh-82F, (FN) ASh-82FN, (TK) ASh-82FNV.

WEIGHTS Empty (1942) 2789 kg (6,149 lb), (FN) 2605 kg (5,743 lb); fuel/oil (typical) 370 kg; loaded (various) 3265/3402 kg (7,198/7,500 lb).

PERFORMANCE Max speed (SL, typical) 562 km/h (349 mph), (altitude) (prototype) 604 km/h, (1942 pre-production) 554 km/h (344 mph) at 6,5 km, (November 1942 production) 603 km/h (375 mph) at 6,5 km, (FN), 648 km/h (403 mph) at 6,5 km (21,325 ft), (UTI) 600 km/h at 3,5 km; climb (early production) 6·2 min to 5 km, (FN) 5·0 min to 5 km; service ceiling (early) 9,6 km (31,500 ft), (FN) 11 km (36,090 ft); range (early) 655 km (407 miles), (FN) 765 km (475 miles); time 360° turn (early) 24 s, (FN) 18 s; landing speed (typical) 148 km/h (92 mph).

La-7 By late 43 it was clear light-alloys would not be in short supply; CAHI meanwhile ran major tunnel programme on La-5FN to find aerodynamic improvements. These were main inputs to Samolyet 120 (also called La-120, first of published designations, begun at 105 and 107, of Lavochkin OKB prototypes). Aircraft No 120 also incorporated many improvements schemed by OKB but ignored through wish not to interfere with production.

Main change was new wing, mentioned at end of La-5, with metal spars. Webs D17-T, booms Chromansil (virtually KhGSA) steel rolled to u.t.s. 130 kg/mm^2. Centre-section joint ribs duralumin, otherwise ribs and skin unchanged. Centre-section leading edge straight taper from root (slightly sharper angle than outer wing). Numerous refinements to fuselage aerodynamics, especially engine installation. Supercharger inlet moved from above cowl to left wing root, with equal-size auxiliary inlet in right root. Oil cooler relocated at trailing edge. New cowling, improved production method,

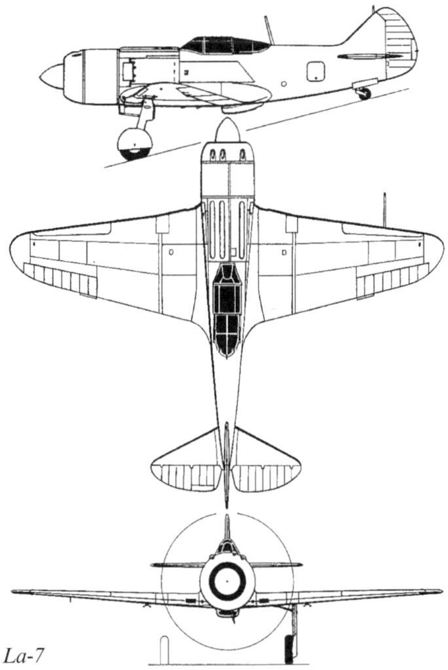

La-7

La-7UTI (one of many field conversions)

La-7ASh-71

with light but stiff left/right panels hinged at upper centreline exposing entire engine when opened and locked by struts. VISh-105V-4 propeller, same size but improved blades and Hucks dogs eliminated. Improved PBP-1B(V) reflector sight and more comfortable cockpit with heating from exhaust muffs, better radio tuning and gyro-horizon. Armament initially three ShVAK; soon changed to B-20 but inadequate supplies so only two guns in Moscow a/c but three in Yaroslavl. Guns in top of fuselage, third gun on left of centreline, all cooled by ram air from small wing-root inlets. Prototype No 120 built at SKB (Yaroslavl) under S. M. Alekseyev and flown 19 Nov 43 by N. V. Adamovich. Outstanding aircraft, hardly any snags, and put through NII trials Jan/April 44, VVS service May 44, used by most of famous IAPs and several top-scoring aces. Dual **La-7UTI** often had one gun and oil cooler under engine. Production terminated early 46 at 5,753.

La-7R

Many modifications and experimental versions, **La-7TK** on factory test July 44 with ASh-82FN/TK-3 (same power but maintained to 8 km) giving better climb and high-altitude performance. Destroyed when one of two turbos disintegrated. **La-7ASh-71** with new 2000 hp engine, weights 2849/3505 kg, factory test 1944. **La-7ASh-83** with new 1900 hp lightweight engine, weights 2522/3140 kg, max speed 725 km/h (450·5 mph) at 7,4 km; built with two NS-23, devastating new gun, completed Dec 44, factory test completed 12 Sept 45. **La-7R** ultra-high-altitude interceptor with V. P. Glushko's RD-1 liquid rocket engine in tail below taller vertical tail with cut-back rudder. Acid (170 lit)/kerosene (90 lit) tanks behind radio bay; armament two B-20. Ground tests 1944, eventually 15 flights by G. M. Shiyanov (first flight, last week of 44) and A. V. Davydov. RD-1KhZ fitted in modified aircraft with ASh-83FN called **La-120R**; Tushino flypast with rocket firing 18 Aug 46. **La-7S** flown by Davydov Sept 46 with two underwing VRD-430 ramjets boosting speed 64 km/h. **La-7PuVRD**, also called **PVRD**, and in some reports described as **La-126PVRD**, with two underwing ramjets also; some reports state VRD-430, others give designation D-10. Speed 800 km/h (497 mph) at 8 km. **La-126** with laminar-section wing with 2° more dihedral, less taper, trailing-edge root further forward; all-metal stressed-skin throughout with several major junctions cast in Elektron magnesium alloy. Many other changes including modified canopy; armament four NS-23 with 290 rounds. Factory test by Davydov, Fedorov and A. A. Popov complete 10 Jan 45; led to La-9. ASCC allocated reporting name 'Fin'.

DIMENSIONS Span 9,8 m (32 ft 1¾ in); length (all except La-126) 8,6 m (28 ft 2½ in), (La-126) 8,64 m (28 ft 4¼ in); wing area (except La-126) 17,59 m^2 (189 ft^2), (La-126) 17,50 m^2.

ENGINE One ASh-82FN (special variants, see text).

WEIGHTS Empty 2,6 t (5,732 lb), (TK, 2711 kg; UTI, 2625 kg); fuel/oil 326 kg (UTI, 443 kg; 7R, 604 kg); loaded 3260 kg (7,187 lb), (TK, 3280 kg; ASh-71, 3505 kg; UTI, 3293 kg; 7R, 3498 kg).

PERFORMANCE Max speed 600 km/h (373 mph) at SL, 680 km/h (423 mph) at 6,8 km (22,300 ft), (TK, 676 km/h at 8 km; La-120, 725 at 7,4; UTI, 648 at 3; -7R, 752, later 795 at 6,3; -120R, 805); climb 4·5 min to 5 km; service ceiling 10,7 km (35,105 ft), (TK, 11,8; -7R, 13);

range 635 km (395 miles), (UTI, 675 km; La-120, 800); time 360° circle 19 s; take-off 340 m; landing 510 m/152 km/h (94 mph).

La-126

La-9

La-9 About mid-44 enlarged OKB at Tbilisi began complete redesign of fuselage, changing not only shapes but also starting afresh with duralumin stressed-skin structure. Many other airframe changes, wing being basically that of La-126 but with new tip of rectilinear form with improved ailerons of almost constant reduced chord and simpler aerodynamic balance. Improved engine installation with better cooling baffles (closer fins on cylinders, no change in engine designation) and exhaust arrangement and with supercharger duct inlet moved from wing roots to top of cowl. Redesigned wider canopy with frameless hood and deeper rear fuselage giving greater comfort and better all-round view. D/F radio behind seat with loop behind rear armoured glass plate. Better tank arrangement making full use of metal wing structure, outer-wing supplementary cells raising total capacity to 825 lit (181·5 gal). Larger reshaped vertical tail and new rectilinear horizontal tail of increased span with spring tabs in rudder and left elevator. Overall structure weight actually slightly less than La-7, allowing for devastating armament of four NR-23 with 300 rounds. Unarmed **La-130** prototype flown 16 June 46, on NII test Oct. Few snags, and production authorized Nov 46; because of end of war and imminent jet fighters numbers small by comparison with previous programmes, and gradually run down from 1947. Some early batches retained armament of three of four ShVAK and many had one of left-side NR-23 removed leaving three (often retaining blister fairing for missing gun). Total production, possibly 1000 including **La-9UTI** dual trainer with one NR-23 first flown Jan 47, NII test June and built in small numbers from Aug 48. Several experimental variants. **La-138** fitted with two PVRD-430 ramjets in same installation as La-7S, complete end 46 and factory test March/April 47; boosted speed 107/112 km/h at expense of high fuel consumption and severe drag when not operating. **La-9RD** fitted with two RD-13 pulsejets attached to same wing hardpoints but with different pylons extending ahead of leading edge. Batch of at least nine aircraft converted, first flying second half 47; formation demonstration at Tushino but vibration severe and soon abandoned despite boost of 127 km/h. ASCC reporting name 'Fritz'.

DIMENSIONS Span 9,8 m (32 ft 1¾ in); length 8,63 m (28 ft 3¾ in); wing area 17,72 m^2 (191 ft^2).
ENGINE One ASh-82FN.
WEIGHTS Empty 2638 kg (5,816 lb), (UTI, 2554 kg; RD, 3150 kg); fuel/oil 595 kg; loaded 3676 kg (8,104 lb), (UTI, 3285 kg; RD, 3815 kg).
PERFORMANCE Max speed 600 km/h (373 mph) at SL (UTI, 558 km/h), 690 km/h (429 mph) at 6250 m (20,500 ft) (UTI, 659 km/h); climb 4·8 min to 5 km (UTI, 5 min); service ceiling 10·8 km (35,433 ft); range 1735 km (1,078 miles) (UTI, 940 km); take-off 345 m; landing 490 m/146 km/h (91 mph).

La-11 Final piston-engine La, begun 46 as long-range escort with provision for two 137,5 lit (30·25 gal) drop tanks on wingtips, giving total fuel 1100 lit, more than three times LaGG-1. Armament three NS-23, lower right position and blister removed; three magazines 225 rounds. Only other significant change relocation of oil cooler in bottom of deeper oval cowling. Prototype **La-140** flown Col A. G. Kochyetkov, NII chief of fighter testing, May 47; NII trials Aug and immediate order for large-scale production (much greater number than La-9). Entered service 48 and subsequently used by virtually all communist air arms well into 50s together with dual **La-11UTI**. ASCC reporting name 'Fang'.

DIMENSIONS Span 9,8 m (32 ft 1¾ in); length 8,63 m (28 ft 3¾ in); wing area (no tip tanks) 17,72 m^2 (191 ft^2).
ENGINE One ASh-82FN.
WEIGHTS Empty 2770 kg (6,107 lb); fuel/oil 880 kg; loaded 3996 kg (8,810 lb).
PERFORMANCE Max speed 562 km/h (349 mph) at SL, 674 km/h (419 mph) at 6,2 km; climb 5·0 min to 5 km; service ceiling 10 250 m (33,630 ft); range 2550 km (1,584 miles); endurance 6·3 h; time 360° circle 19·5 s; take-off 345 m; landing 505 m/149 km/h (92·6 mph).

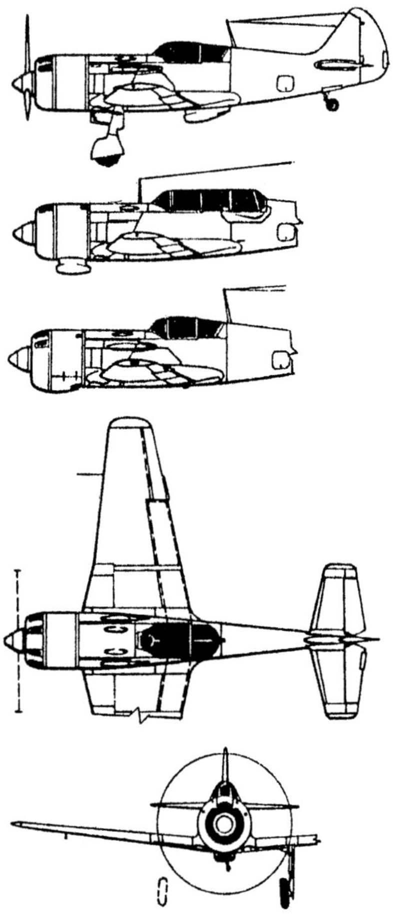
La-9 (centre side view La-9UTI, bottom La-11)

La-150 First OKB jet, result of Kremlin meeting Feb 45 when with MiG, Su and Yak told to use RD-10 engine. Pod/boom layout, wing R-11, 12% root, 9·5% tip, two spars with five fuel cells in inter-spar box 770 lit (169 gal); slotted hydraulic flaps, manual ailerons, fixed leading edge without slats. Wing attached to two strong frames also carrying engine at positive angle 3° and main landing gears, of 1823 mm (5 ft 11¾ in) track, with 825×200 mm tyres, retracted about skewed hinges to lie under front of engine. Nose inlet immediately bifurcated to ducts along sides of fuselage, passing unpressurized cockpit with conventional seat. Nose gear retracted to rear, wheel rotating 90°. Fixed tailplane and left/right elevators mounted quite high on fin entirely ahead of mass/horn-balanced rudder. Entire lower part of fuselage unstressed access panels round engine and armament of two NS-23 with projecting barrels installed below ducts and beside nose-gear compartment. Five prototypes built (for planned 7 Nov 46 parade), first being complete in Sept 46; first flight A. A. Popov, subsequent testing also by I. Ye. Fedorov, M. L. Gallai, N. I. Zvonaryev and A. G. Kochyetkov, continued until April 47. Many shortcomings including high airframe weight, generally sluggish acceleration, poor stability and tail oscillation traced to inadequate stiffness of tail boom. Second prototype completed as **La-150M** (*Modifikatsirovanni*) with wingtips turned down at anhedral of 30° and extended to preserve original span, stronger and stiffer tail-boom, greater fuel capacity, repositioned equipment, low-pressure main tyres, and guns with shorter barrels. Empty and gross weights rose appreciably and performance fell further; first flight Dec 46, testing stopped Sept 47 by which time La-150 no longer active project. Final three prototypes again modified to **La-150F** (*Forsirovannii*) with primitive afterburning engine; on factory test July to Sept 47. First Soviet afterburning aircraft, one of first in world.

DIMENSIONS Span 8,2 m (26 ft 10¾ in); length 9,42 m (30 ft 10¾ in); wing area 12,15 m^2 (131 ft^2).

ENGINE One RD-10, (La-150F) RD-10F.

WEIGHTS Empty (150) 2059 kg (4,539 lb), (150M) 2369 kg (5,223 lb); fuel/oil (150) 553 kg, (150M) 623 kg; loaded (150) 2961 kg (6,528 lb), (150M) 3338 kg (7,359 lb), (150F) 3340 kg.

PERFORMANCE Max speed (150) 840 km/h (522 mph) at SL, 850 km/h at 5 km, (150M) 760 km/h at SL, 805 km/h (500 mph) at 5 km, (150F) 950 km/h (590 mph) at SL; climb to 5 km (150) 4·5 min, (150M) 7·2 min, (150F) 6·0 min; service ceiling (150 and M) 12,5 km (41,000 ft), (F) 13 km (42,650 ft); range (150) 700 km (435 miles); endurance 0·8 h; field length over 1 km.

La-11

La-150 (inset upper, La-150M, lower 150F)

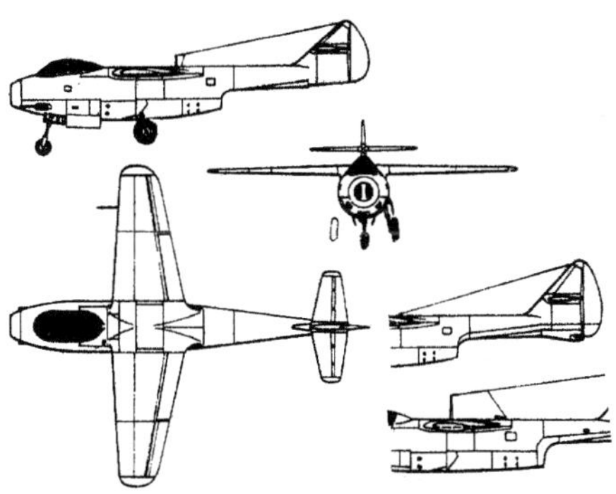

La-150 (upper left, La-13; lower left, La-154)

La-152 Though officially a modification of La-150 this was in effect a different fighter, with totally redesigned fuselage with cockpit above mid-mounted wing and engine in nose. Main aerodynamic change was new wing of mean 9·1% thickness, with improved sealed and internally balanced ailerons and structurally innovative. Rear fuselage of more conventional form, giving no aeroelastic problems. Broad cockpit above wing, later (about Jan 48) with primitive ejection seat, fitted to all subsequent prototypes. Engine in extreme nose with no inlet duct and short tailpipe. Main landing gear not greatly changed from 150 series but folding rearwards immediately beneath wing root with different doors. Deep forward fuselage with NS-23 on each side, engine attached to frames 1 and 4, nose gear to frame 2 and longitudinal walls of compartment, and guns to 3 and 5. Factory test Oct 46 to Aug 47. Undistinguished but no major snag.

DIMENSIONS Span 8,2 m (26 ft 10¾ in); length 9,4 m (30 ft 10 in); wing area 12,15 m^2 (131 ft^2).

ENGINE One RD-10.

WEIGHTS Empty 2310 kg (5,093 lb); fuel/oil 563 kg; loaded 3239 kg (7,141 lb).

PERFORMANCE Max speed 730 km/h at SL, 778 km/h (483 mph) at 5 km; climb 6·5 min to 5 km; service ceiling 12,5 km (41,000 ft); range 500 km (311 miles); endurance 0·9 h.

La-154 Further modification of 152 for research purposes with new wing of different profile but superficially resembling 152. Limited flying from late 46 because of greater potential with afterburning engine in 156.

DIMENSIONS Span 8,5 m (27 ft 10⅔ in); length 9,1 m (29 ft 10¼ in); wing area 13,24 m^2 (142·5 ft^2).

ENGINE One RD-10.

WEIGHTS Empty 2400 kg (5,291 lb); loaded 3500 kg (7,716 lb).

PERFORMANCE Max speed 900 km/h (559 mph) at 5 km; climb 4 min to 5 km; service ceiling 12,5 km (41,000 ft); range 500 km (311 miles).

La-156 Essentially 154 with afterburning engine, installed at angle 5°40' with better secondary airflow than La-150F. Additional NS-23 installed on right side, subsequently retrofitted to La-152. Tail areas; horizontal 2,3 m^2, vertical 2,2. First La with ejection seat from outset. First flight S. F. Mashkovkii Feb 47, subsequently factory tested by I. Ye. Fedorov to late March. NII testing 5 Sept 47 to 31 Jan 48. Best La jet to date, but recognized outdated by swept wing aircraft.

DIMENSIONS Span 8,52 m (27 ft 11½ in); length 9,12 m (29 ft 11 in); wing area 13,32 m^2 (143·4 ft^2).

ENGINE One RD-10F.

La-156

La-174TK

WEIGHTS Empty 2398 kg (5,287 lb); fuel/oil 743 kg; loaded 3521 kg (7,762 lb).
PERFORMANCE Max speed 845 km/h at SL, 905 km/h (562 mph) at 2 km; climb 4·0 min to 5 km; service ceiling 12,7 km (41,670 ft); range 660 km (410 miles); endurance 1·2 h.

La-174TK Confusingly numbered out of sequence, this was basically a 156 with even thinner wing (TK, *Tonkoye Krylo*, thin wing) and at last a different (but still obsolescent) engine. Purely experimental, and 6% wing believed to be thinnest flown in world at that date (certainly on a fighter); but intended as basis for direct rival to Yak-23. Last attempt to use straight wing, prototype being flown Jan 48 more than nine months after swept La-160. Apart from thin wing, main change was bulging forward fuselage to house plenum chamber of UK-imported engine, installed at 6°20' angle and necessitating repositioning three NS-23 in bottom of forward fuselage. Engine right at front, with bullet fairing over forward wheelcase and accessories. Nose gear bay now with single door. Great advance on 156 but slower than less-powerful 160 so straight wings finally abandoned.
DIMENSIONS Span 8,64 m (28 ft 4¼ in); length 9,41 m (30 ft 10½ in); wing area 13,52 m^2 (145·5 ft^2).
ENGINE One 1590 kg (3,500 lb) Rolls-Royce Derwent V.
WEIGHTS Empty 2310 kg (5,093 lb); fuel/oil 945 kg; loaded 3315 kg (7,308 lb).
PERFORMANCE Max speed 970 km/h at SL, 965 km/h at 3 km; climb 2·5 min to 5 km; service ceiling 13 km (42,650 ft); range 960 km (597 miles); endurance 1·5 h.

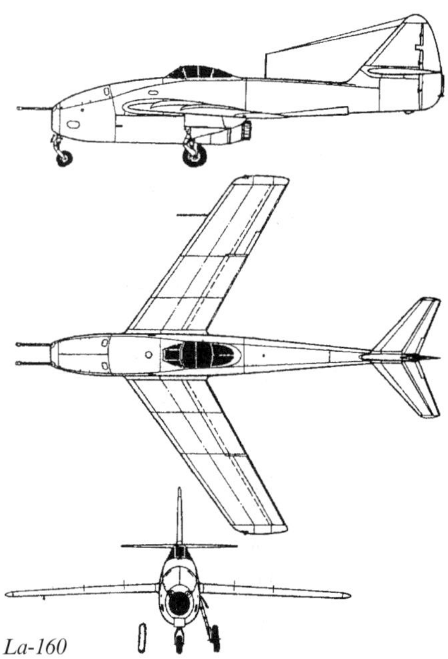

La-160

La-160 Nicknamed *Strelka* (Arrow), this was first Soviet swept-wing aircraft, but otherwise was member of old Jumo-engined family and retained a lengthened version of same fuselage as 154/156 though with reinforced canopy and windscreen for higher indicated airspeeds. Wing swept 35° on leading edge, 9·5% thick making use of German data and with German aerodynamicists involved. About 2° anhedral, two fences on each side extending to trailing edge, believed first on any aircraft. No details of flight controls or flaps. Despite substantial rearwards shift in centre of lift CG was brought forwards by extending nose and fitting armament of two NS-37 as far forward as possible above nose with barrels projecting far ahead of inlet. Yet again a new nose gear, hinged right at front with square door ahead of leg; main door closed on ground. Restressed tail, of reduced thickness and with 35° swept tailplane. Longer fuselage accommodated part of increased fuel capacity of 1528 lit (336 gal). Several La-160s built. First flown 24 June 47; tested four months by team including Fedorov and Mashkovskii in wide programme investigating stability and control. Remarkably few modifications needed, and M 0·92 reached on level; clearly showed swept wing at that time was superior, but La-160 recognized as not definitive fighter because of inadequate engine. Lavochkin OKB built up important team on afterburners and their installation headed by I. A. Merkulov (ramjet engineer, joined OKB April 45) and deputy V. I. Nizhnyego, La-160 being a major vehicle for research.
DIMENSIONS Span 8,95 m (29 ft 4⅓ in); length 10,06 m (33 ft 0 in); wing area 15,9 m^2 (171 ft^2).
ENGINE One RD-10F.
WEIGHTS Empty 2738 kg (6,036 lb); fuel/oil 1100 kg; loaded 4060 kg (8,951 lb).
PERFORMANCE Max speed 900 km/h (559 mph) at SL, 1050 km/h (652 mph) at 6 km; climb not recorded; service ceiling 11 km (36,090 ft); range (Shavrov) 500 km, possibly pessimistic.

La-168 As early as 47 OKB recognized probable superiority of conventional fuselage with nozzle at tail. Resulting 168 was totally new and by far best La jet to date. Wing larger than 160, mounted higher with increased anhedral –4·5° and increased sweep 37°20' at ¼-chord, thickness 12%. Fowler flaps running out to 20° TO and 58° ldg. Fuselage circular section, 9195 from inlet to nozzle, inlet immediately bifurcated (landing light in centre) past forward pressurized cockpit with windscreen and canopy wider than 160. Landing gear with long nose leg extended forward beyond nose (wheelbase 4736) and short main gears (track 1700) retrac forwards into bay with large door. Rear fuselage removable for engine access, swept fin (height 2076 from centreline of nozzle) carrying fixed tailplane same sweep as wing, span 2900, with powered elevators. Door airbrake each side of tailcone hinged about rear edge and opened hyd (like gear and flaps) to 45°. Internal fuel 1230 lit, plus 605 lit conformal belly drop tank. One N-37 on left and two NR-23 on right under nose. Fitted with imported engine and flown by Fedorov 22 April 48. Lighter and in many ways better than MiG-15, but decision to adopt latter already taken. USAF designation 'Type 15'.
DIMENSIONS Span 9,5 m (31 ft 2 in); length 10,56 m (34 ft 7¼ in); wing area 18,08 m^2 (194·6 ft^2).
ENGINE One Nene I.
WEIGHTS Empty 2973 kg (6,554 lb); loaded 4412 kg (9,727 lb).

La-160

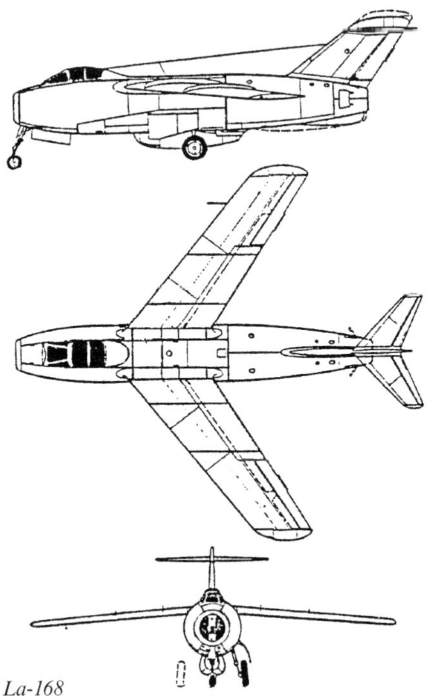

La-168

PERFORMANCE Max speed 1000 km/h (621 mph) at SL, 1084 km/h (674 mph, Mach 0·982) at 2,5 km; climb 2 min to 5 km; service ceiling 14,57 km (47,800 ft); range (no aux tank) 1275 km (792 miles); TO/ldg both 500 m (1,640 ft).

La-174D OKB built similar prototype scaled down to Derwent 5 power. Wing geometrically similar to 168 but smaller, Zapp flaps and anhedral zero. Tail as before so proportionately larger (tp span increased to 3000). Internal fuel 1100 lit, plus 605 lit external conformal. Three NS-23, one left and two right. Tested from late Aug 48 by Fedorov. Excellent aircraft and accepted for production with changes as La-15. USAF 'Type 21'.

DIMENSIONS Span 8,904 m (29 ft 2½ in); length 9,513 m (31 ft 2½ in); wing area 16,16 m² (174 ft²).
ENGINE One Derwent 5.
WEIGHTS Empty 2433 kg (5,364 lb); loaded 3708 kg (8,175 lb).
PERFORMANCE Max speed 900 km/h at SL, 1040 km/h (646 mph) at 3 km; climb to 5 km in 3 min; service ceiling 14,6 km (47,900 ft); range (no aux tank) 1300 km (808 miles).

La-15 Production 174D differed in having –6° anhedral, tailplane lowered, slight increase in length and empty weight, much longer rear ventral strake, canopy shape changed, Sov-built engine and HF wire attached to mast. Single batch (poss only 100) built by GAZ managed by A. K. Belyenkov. USAF 'Type 21', ASCC 'Fantail'.

Two prototypes only of **180**, **La-15UTI**, tandem dual trainer, same dimensions, reduced fuel

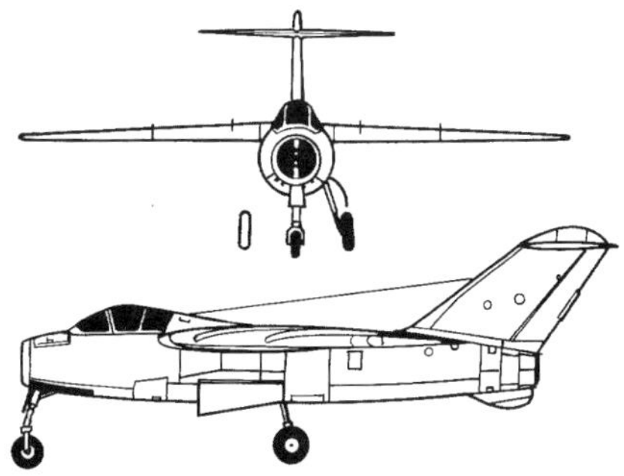

La-174D

(925 kg instead of 1275), single BS with 300 rounds.
DIMENSIONS Span 8,830 m (28 ft 11⅝ in); length 9,560 m (31 ft 4⅜ in); wing area 16,16 m² (174 ft²).
ENGINE One RD-500.

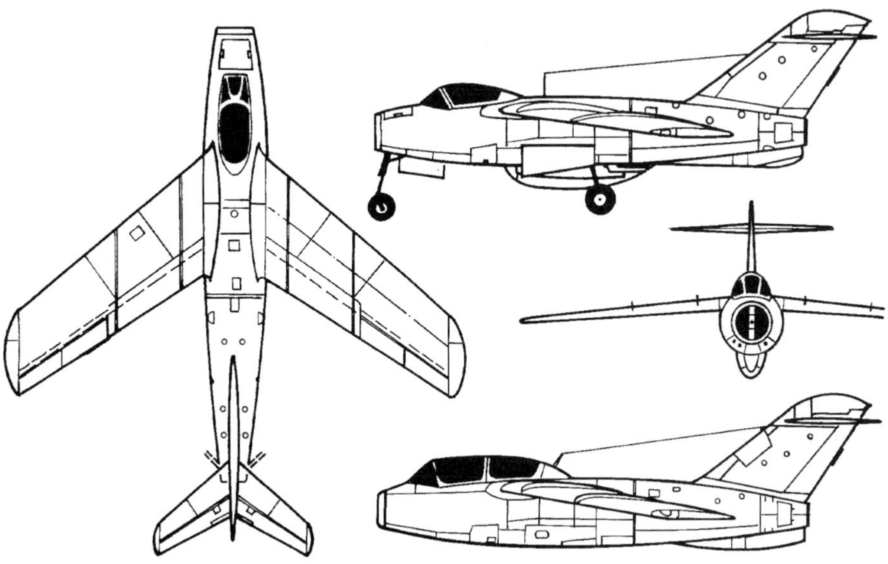

La-15 (lower side view, La-15UTI)

La-15

La-176

WEIGHTS Empty (15) 2575 kg (5,677 lb), (UTI) 2805 kg (6,184 lb); loaded (15) 3850 kg (8,488 lb), (UTI) 3730 kg (8,223 lb).
PERFORMANCE Max speed at SL (15) 900 km/h (559 mph), (UTI) 960 km/h (597 mph); at height (15) 1026 km/h (638 mph) at 3 km, (UTI) 980 km/h at 5 km; climb to 5 km (15) 3·1 min, (UTI) 3·0 min; service ceiling (15) 13,5 km (44,300 ft), (UTI) 12,75 km (41,830 ft); range (15) 1170 km (727 miles), (UTI) 910 km (565 miles); TO (15) 640 m, ldg 630 m.

La-176 Remarkable pace of prototypes continued with this improved La-168, believed first in world with wing swept 45° at ¼-chord; Fowler flaps, anhedral –6°, third fence added ahead of aileron. Tailplane also swept 45° and increased in span to 3 m. Many features borrowed from La-15, incl long aft ventral strake, but instead of lowering tailplane fin was extended at top. Tested by Fedorov and Capt O. V. Sokolovskii from Sept 48, latter diving faster than sound 26 Dec, believed first in USSR. Mach 1 routine in 49 until canopy failed at max speed, Sokolovskii killed. No Western designation.
DIMENSIONS Span 8,59 m (28 ft 2¼ in); length 10,97 m (36 ft 0 in); wing area 18,25 m² (196·4 ft²).
ENGINE One RD-45, later one VK-1.
WEIGHTS Empty 3111 kg (6,858 lb); fuel 1165 kg; loaded 4637 kg (10,223 lb).
PERFORMANCE Max speed 1105 km/h (687 mph) at 7,5 km (Mach 0·992); climb to 5 km in 1·8 min; service ceiling 15 km (49,200 ft); range 1000 km (621 miles).

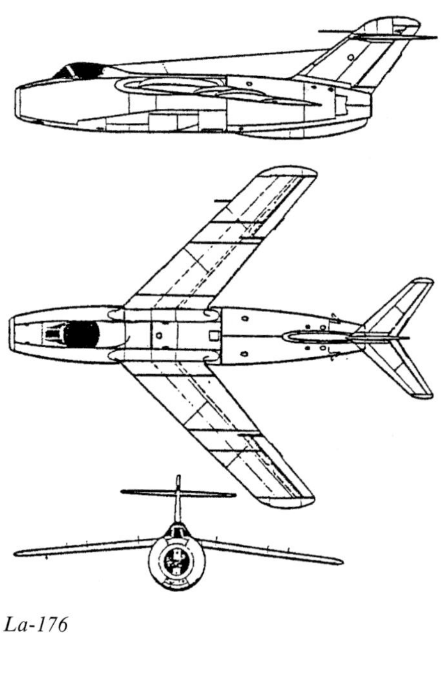

La-176

La-190 NII demand Oct 48 for genuine transonic fighter was modified to require interception radar. Design 49 totally new in all respects with powerful axial engine fed by circular nose inlet modified by presence of radar at top. Impressive wing with sweep 55° and forming integral tanks despite thickness 6·1%; large slot-

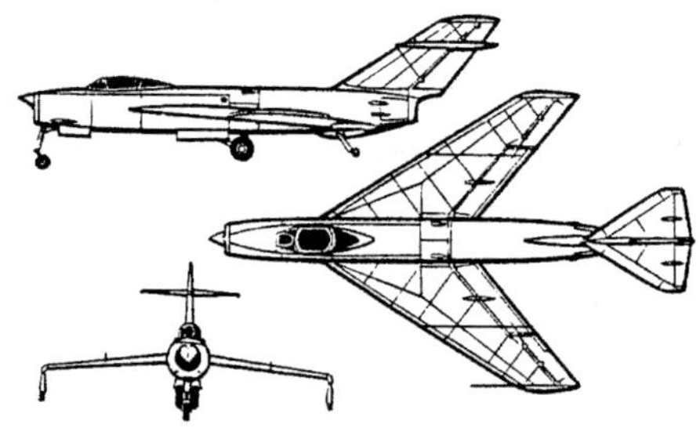

La-190

ted flaps, powered ailerons, single shallow fence across chord near outer end flap. Fuselage circular section, rear fuselage detachable at rear-spar frame, sharply swept fin, rudder above and below remarkable near-delta tailplane with powered elevators. Large airbrakes each side of tail hinged out from front (ie, opposite of previous). Twin mainwheels behind CG on levered leg on centreline, hyd retrac forward; single aft-retrac nosewheel and small stab gears retrac to rear under wingtips. Internal fuel 2100 lit (462 gal). Two N-37 under nose. Flown (without planned *Torii* radar) by Kochyetkov from Feb

La-200-01

La-200-02

La-200B

51; engine never put into production and aircraft made only eight flights. No Western designation.

DIMENSIONS Span 9,9 m (32 ft 5¾ in); length 16,35 m (53 ft 7¾ in); wing area 38,93 m^2 (419 ft^2).

ENGINE One AL-5.

WEIGHTS Empty 7315 kg (16,127 lb); loaded 9257 kg (20,408 lb).

PERFORMANCE Max speed 1190 km/h (739 mph) at 5 km (M 1·03); climb to 5 km in 90 s; service ceiling 15,6 km (51,180 ft); range 1150 km (715 miles).

La-200 NII demand of Jan 48 called for interceptor with crew 2 and fully equipped for night and bad weather. Two prototypes built using known aerodynamics and structures, though larger. Wing 9·5% thick, swept 40°, anhedral –3°, Fowler flaps, ailerons and fences similar previous prototypes, tail equally though tailplane little over half-way up fin with rudder above and below. All flight controls hyd boosted. Airbrakes returned to being hinged at trailing edge and opening out 45°. Unusual main gears similar to previous but with twin wheels on each axle; levered-suspension nosewheel retrac to rear. Engines in tandem as in MiG/Su rivals, likewise fed by circular nose inlet, duct dividing into three, centre past nose gear to forward engine, outers past cockpit to rear engine. Pilot and radar operator side by side with wide flat windscreen and aft-sliding canopy. Internal fuel in two tanks behind cockpit, integral in each wing, total 2800 lit, plus two 1000-lit underwing. Three N-37 (one left, two right) each with 50 rounds. **200-01** designed with *Torii* radar in centre of inlet, flown without radar 9 Sept 49 by S. F. Maskovskii and A. F. Kosaryov. Generally good results, dived to M 1·01. **200-02** completed with *Torii-A* radar in top of nose, small keel underfin, 70 rounds per gun, different underwing tanks and other changes. Flown by I. M. Dzyub and M. L. Baranovskii 51, marginally superior. Devt continued with **La-200B**, single prototype to meet demand for bigger radar and longer range. Major redesign with more powerful engines fed by three separate inlets, lower to front engine, two uppers to rear, entire nose occupied by *Sokol* radar with 1 m mech-steered dish. Small increase in wing span, two inclined rear ventral strakes, single large mainwheels and redesigned nose gear moved right to nose, fin increased in height, finned drop tanks intended 3000-lit each but actually 2650 to bring total fuel to 8110 lit. First flight by Kochyetkov 3 July 52, operative radar from 10 Sept 52, total 109 flights by Kochyetkov and Gallai. Numerous problems, no production. No Western designation.

DIMENSIONS Span (200) 12·92 m (42 ft 4⅝ in), (200B) 12,96 m (42 ft 6¼ in); length (200-1) 16,73 m (54 ft 10⅝ in), (200-2) 16,59 m (54 ft 5⅛ in), (200B) 17,325 m (56 ft 10 in); wing area (200) 40,02 m^2, (200B) 40,0 m^2 (430·6 ft^2).

ENGINES (200) two VK-1, (200B) two VK-1A.

WEIGHTS Empty (200-1) 7090 kg (15,631 lb), (–2) 7675 kg (16,920 lb), (B) 8810 kg (19,422 lb); loaded (–1) 10,375 kg (22,873 lb), (–2) 10,580 (23,325 lb), (B, internal fuel only) 11,56 t (25,485 lb), (B, max) 19,35 t (42,659 lb).

PERFORMANCE Max speed (200) 964 km/h at SL, 1090 km/h (677 mph) at 3,5 km, (B) 1030 km/h (640 mph) at 5 km; climb to 5 km in 2·6 min (200), 2·8 min (B); service ceiling (200) 15,15 km (49,700 ft), (B) 14,125 m (46,350 ft); range (–1) 1300 km, (–2) 1040 km or 2040 km with tanks, (B) 960 km or 2800 km with tanks.

La-250 Culmination of effort on radar-equipped interceptors, biggest fighter of its day, dubbed *Anakonda*, OKB unable to solve problems in time. Launched 53 to meet IA-PVO reqt for long-range two-seat aircraft with armament of large 'second-generation' guided missiles to replace Yak-25. OKB assigned task in parallel with *Burya* long-range cruise missile.

Layout conventional, but innovations sheer size, internal fuel capacity, fuselage length (problem of inertia coupling), fully powered controls, slab tailerons, large radar and even main wheels and tyres. Wing symmetric 4·5/6·5%, almost pure delta, 5°30' anhedral, 57° fixed LE, tiny fence in front of LE (stopping c2% chord) ahead of junction between slotted flaps and powered outboard ailerons, both tapered to align with rear spar. Integral tankage in three-spar wing box. Large fin with powered rudder, mid-high delta tailerons with individual power units, no elevators. Fuselage almost constant section with sharp-edged lateral inlets

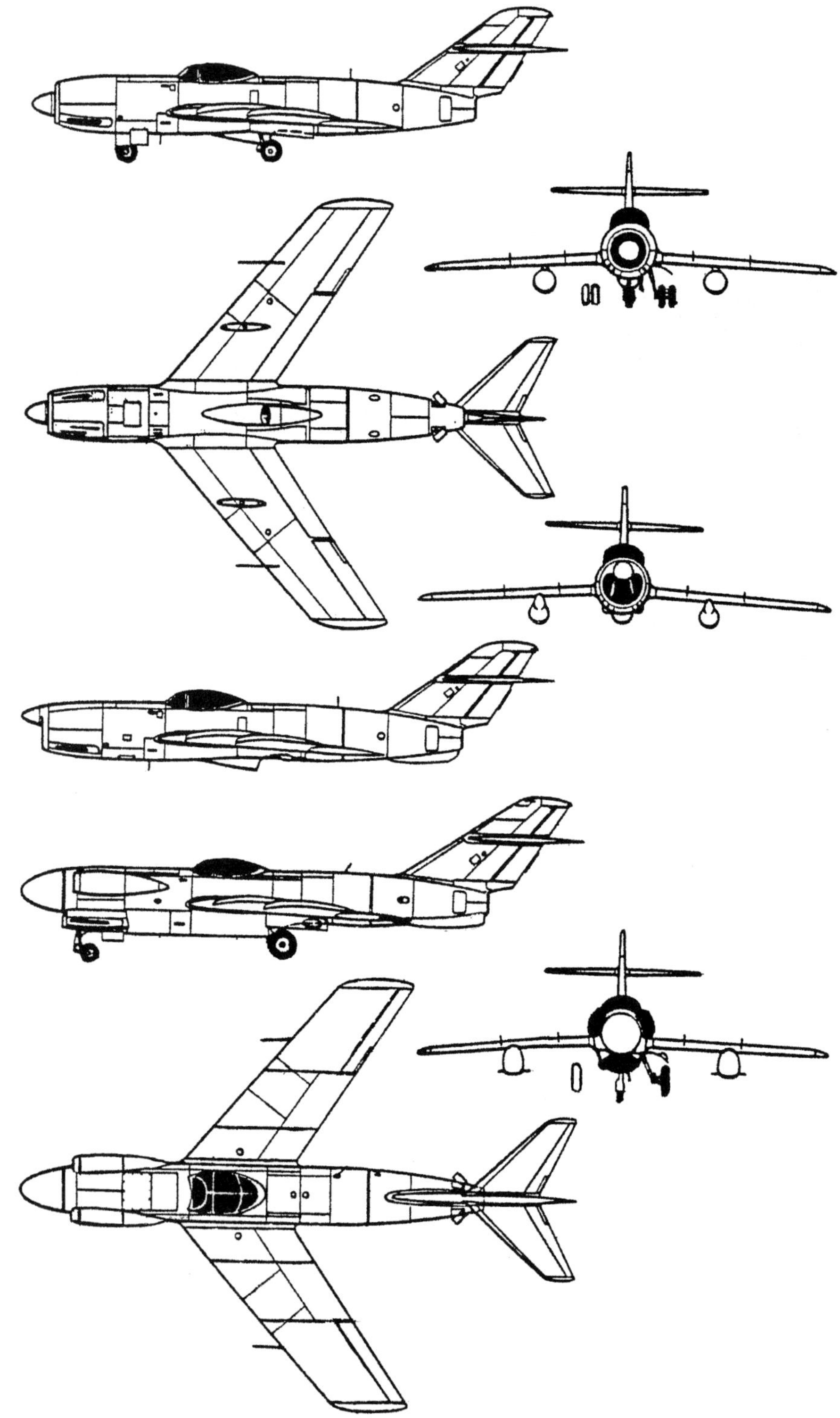

Top, La-200-01; centre (two views only) -02; bottom -03 or La-200B

identical CAHI design as same-engined Tu-98, with translating half-cone; no aux inlets. Pressurized tandem cockpits with KL-2 seats, hinged canopies faired into dorsal spine taking controls and pipes past row of integral tanks between ducts; internal fuel 14500 lit (3,190 gal). Twin-wheel steerable nose gear retrac back hyd to bay with front and side door, single main wheels on legs splayed slightly out and forward, retrac forward into bay with large door with blister. Braking parachute in blister under rear fuselage. Underwing pylons at exactly 50% semi-span for K-15 missiles. Provision for scabbed-on ventral tank.

Detailed mock-up, static-test airframe and four flight articles built Kuibyshyev. First flight by Kochyetkov and observer 16 July 56. Immediate wild excursions in roll, one report 'had to land immediately', another 'crashed but without serious injury'. OKB built simulator on to static airframe, showing unacceptable lag between pilot input and surface power unit response; test pilots G. Shiyanov and Mark Gallai 'succeeded in crashing dozen times in a few minutes'. Flight controls almost completely redesigned, 02 becoming first **La-250A**. This aircraft fitted with underwing pylons for test ballistic bodies. Plan to complete OKB testing at Kuibyshyev, then flying to Zhukovskii, replaced by decision to move 02 by rail/road, causing numerous problems and delays. Finally flown summer 57, Kochyetkov making numerous successful flights punctuated by problems with flight controls, systems and main tyres (large, made treadless without proper moulds, frequent cracks/cuts, changed every third flight). Tyre on fire 28 Nov 57 resulted in short landing, gear collapsed and a/c burned on runway. Further major mods, 03 completed with forward fuselage tilted down 6° to improve pilot view. This assigned to Arkadii P. Bogorodskyi, highly qualified graduate of LII test-pilot school, who 'understood complex aircraft and made number of suggestions', but written off in landing accident 8 Sept 58. Aircraft 04 completed with radar and missiles, Bogorodskyi made good progress marred again by wheel-tyre fire. Programme finally terminated 59, further cause being unreliability of engines. Replaced by much heavier and more capable Tu-128. No 04 at Monino, 02 scrapped at Zhukovskii.

DIMENSIONS Span 13,9 m (45 ft 7¼ in); length (to 03) 25,9 m, (04) 26,5 m (86 ft 11¼ in); wing area 80,0 m^2 (861 ft^2).

ENGINES Two AL-7F.

WEIGHTS Empty (03) 15 t (33,070 lb); loaded 27,5 t (60,626 lb).

PERFORMANCE Max speed (12 km) 1600 km/h (995 mph, M 1·5); service ceiling 17 km (55,775 ft); endurance 2·3 hr.

LII

The LII (experimental flight institute) named for M. M. Gromov is based at Zhukovskii, near Moscow. It has three runways, one 5,5 km long, and operates (93) 70-72 LL (flying lab) research a/c. At the 93 Salon it announced a design of light aircraft.

Shirok Name = wide, meaning many applications; configuration as drawing, with ducted props, inflatable pontoons for amphibian use, up to 3 seats, metal/composite. Prototype due for testing Jan 95.

DIMENSIONS Span 10 m (32 ft 9¾ in); length 5,75 m (18 ft 10⅜ in).

ENGINES Two 64-hp Rotax 582UL.

WEIGHTS Empty 395 kg; normal fuel 32 kg; payload 210 kg; loaded 640 kg; MTO 700 kg (1,543 lb).

PERFORMANCE Max speed 300 km/h (186 mph); cruise 200 km/h (124 mph); practical ceiling 4 km; range (40 lit) 400 km, (max 2000); TO 70 m; ldg 70 km/h.

Lisunov

Boris Pavlovich Lisunov was engineer assigned to supervise technical side of licence agreement

La-250-04

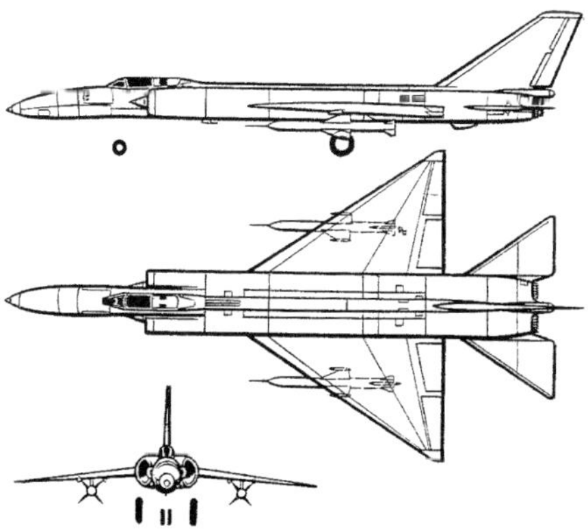
La-250-04

with Douglas Aircraft for DC-3. He was permitted to go to Santa Monica where between Nov 36 and April 39 he went over every part of DC-3 and its tooling and in-service support. Primary reason was use as civil aircraft by Aeroflot; subsequently Lisunov managed development of many other models by V. M. Myasishchyev, I. P. Tolstikh and I. P. Mosolov.

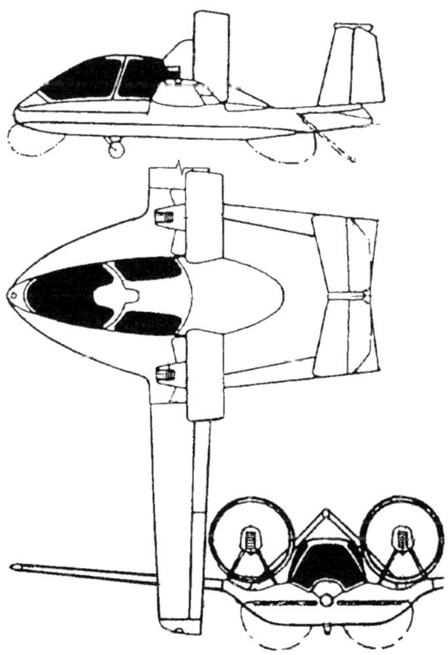
LII Shirok

Li-2 Original Soviet designation **PS-84**, passenger aeroplane 84 (from GAZ-84, Khimki, where production laid down). Despite wish to avoid changes, 1,293 engineering change orders on original Douglas drawings involving part design, dimensions, materials and processes. Engine installation totally different even from Cyclone-powered DC-3; original ASh-62IR driving VISh-21, 1940 to AV-7N (or AV-7NE) later AV-161. Hucks-dogs on all standard models. No cooling gills, many later fitted with front baffle in winter to reduce cooling airflow. Baggage doors on left but main passenger door moved to right (only seven windows). Span slightly reduced, airframe locally reinforced and thicker skin in many vulnerable places adding to empty weight. Main-gear radius arms welded 30KhGSA steel tube, and provision for skis or wheels (1200×450 mm). Performance appreciably below that of R-1820 or R-1830 DC-3. PS-84 in Aeroflot service June 40. In Oct 41 GAZ-84 evacuated, production taken over by GAZ-33 Tashkent. Total production 4,863 completed 45. Variants, often by modification:

Li-2P basic passenger, 14-24 passengers; **Li-2 Salon** VIP 7-14 passengers; **Li-2D** with aux tankage; **Li-2VP** (1,900+) for bombing, troop assault, transport, casevac, minesweeping, glider towing, most with bulged glazed upper part of crew door for observing paradrop DZs, numerous armament options including MV dorsal turret (ShKAS) or VUS (UBT), four FAB-250 and/or 12 RS-82, option fixed 37 mm gun, some with M-88 engines; **Li-2LP** forest patrol and fire-bomber; **Li-2 Polyarnyi** for Arctic; **Li-2RP** fisheries recon; **Li-2RT** radio relay station; **Li-2SKh** for spraying and dusting; **Li-2UT** navigation trainer; **Li-2F** for photography; **Li-2FG** mapping and surveying; and **Li-2M** meteorological. Special conversions included TK-19 turbocharged engines, nav/bombing radar for Tu-4, special laboratory use and S. A. Mostove's caterpillar landing gears. Served in 14 countries. ASCC name 'Cab'.

DIMENSIONS Span 28,81 m (94 ft $6\frac{1}{4}$ in); length 19,65 m (64 ft $5\frac{5}{8}$ in); wing area 91,7 m^2 (987 ft^2).

Li-2D

ENGINES Usually two ASh-62IR.
WEIGHTS Empty (typical) 7,75 t (17,086 lb); payload 2,95 t (6,503 lb); loaded 10,7 t (23,589 lb); max 11,28 t (24,867 lb).
PERFORMANCE Max speed at SL 300 km/h (186 mph); cruise 245 km/h (152 mph); service ceiling 5,6 m (18,370 ft); range 1100-2500 km (684-1,550 miles); landing speed 108 km/h (67 mph).

MAI

Named for Serge Ordzhonikidze, Moscow Aviation Institute is one of largest of former-Sov aviation schools. Except for EMAI, early designs see under P. D. Grushin, who had status of full OKB able to produce hi-power warplanes. In 91 with Kazimir Mikhailovich Zhidovetskiy formed Aviatika joint-stock co to design and market light aircraft, production at Botkhinsky St. MiG-29 factory of MAPO (Moscow A/c Prodn Orgn).

EMAI A bold effort to determine to what extent aircraft could safely be constructed from magnesium. First flown 34, EMAI (Elektron-MAI), also designated **EMAI-1-34** or **E-1**, and named *Sergei Ordzhonikidze*, was major project. Lead designers were Professors S. O. Zonshain, supervising A. L. Gimmelfarb, with full authority of GUAP. Low-wing monoplane with tandem seats in enclosed cockpit, spatted main wheels and tailplane high on fin. Entire airframe Elektron magnesium alloy, mainly welded but some bolted or riveted joints. Fuselage basically hexagonal-section monocoque; cantilever wing based on Stieger's (Monospar) principles. Structure painted, and covered with doped fabric. Judged a complete success; MAI report claimed weight-saving of 42% overall compared with aluminium, wood or steel tube. Widespread opinion this was superior method of construction, despite fire-risk, but shortage of electricity for producing the metal, as much as other factors, prohibited adoption. EMAI flew about 600 times by 40. See Addenda.
DIMENSIONS Span 12,0 m (39 ft 4½ in); length 7,0 m (22 ft 11½ in); wing area 20 m² (215 ft²).
ENGINE One 175 hp Salmson radial.
WEIGHTS Empty 700 kg (1,543 lb); fuel/oil 165 kg (364 lb); loaded 1200 kg (2,646 lb).
PERFORMANCE Max speed 170 km/h (106 mph); range 800 km (500 miles); landing speed 75 km/h (47 mph).

EMAI

MAI-62

E-2 Named for *Klim Voroshilov*, this second magnesium (Elektron) research aircraft was smaller, and had no major backing, though builder M. L. Babad enjoyed active participation of Gimmelfarb and Petrov-Gubish who had worked on EMAI. Tandem-seat mid-wing monoplane, riveted and screwed from heavy (10 mm) sheet, with thinner secondary sections and thin skin over single-spar wing. Flown on 27 hp Anzani by A. I. Zhukov in late 34, 18 hp Blackburn Tomtit then being substituted to improve reliability. Thought that corrosion made structure suspect, and flights ceased early 35. No data.

E-3 Another experimental machine with airframe of Elektron magnesium alloy; monoplane amphibian designed by P.M. Danilov in 36. During construction wing and tail were modified to be covered in tight fabric, hull remaining magnesium. M-11 engine mounted on centreline pylon driving tractor propeller. Work halted 37, apparently because of corrosion.

Boldrev Following natural hiatus in 41-45, next recorded amateur aeroplane is ultralight by Aleksandr Ivanovich Boldrev, senior engineer holding chair of aerodynamics at MAI. Financially assisted by, and skilfully made at MAI in 46-47. Intended as STOL research machine to explore flight control systems. Main feature was a fore-wing looking like a slat but of 286 mm (11¼ in) chord, symmetrical section and mounted on tubular pivots ahead of and above the NACA 23020-section main surface to investigate how its angle and spacing affected lift coefficient. Structure mainly of wood, gondola occupied by pilot being a truss and tail boom a monocoque. Concept does not seem to have been successful, and work ceased at end of 47.
DIMENSIONS Span 6,07 m (19 ft 11 in); length 5,0 m (16 ft 4¾ in); wing area 7,2 m² (77½ ft²).
ENGINE One 22/25 hp M-72.
WEIGHTS Performance, not recorded.

MAI-62 Designed 62, but not flown until later (possibly not at all), this single-seater resembled baby jet fighter. Tailless and virtually all-wing, surface having leading-edge sweep of 45° and centre section of delta shape, joined to swept outer panels with almost square tip ailerons. Cockpit immediately ahead of broad stumpy fin

containing engine, driving pusher propeller. Fixed tricycle landing gear.
DIMENSIONS Span 4,9 m (16 ft 1 in); length 5,25 m (17 ft 2¾ in).
ENGINE One 80 hp M-71.
WEIGHTS Loaded 440 kg (881 lb).
PERFORMANCE Max speed 220 km/h (137 mph); range 800 km (497 miles).

Kvant Named Quantum, and bearing title SKB-EPM-MAI, from student design bureau exptl prodn workshop, this aerobatic single-seater flew late 67. Three-spar wing, steel-tube fuselage, pneumatic gear and flaps, fan-cooled engine driving c/s prop. Load factor +9g/-7 g, made ten flights.

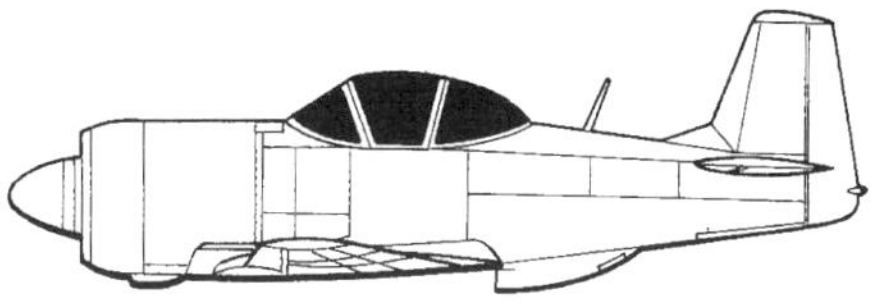

MAI (Aviatika) Kvant

Kvant

DIMENSIONS Span 7,5 m (24 ft 7¼ in); length 5,72 m (18 ft 9¼ in); wing area 8,55 m² (92 ft²).
ENGINE One M-14P
WEIGHT Max 920 kg (2,028 lb).
PERFORMANCE Max speed 381 km/h (237 mph); service ceiling 6550 m (21,500 ft); TO 300 m, ldg 400 m.

Elf Family of metal pusher ultralights with slatted low wing and tricycle gear; Elf-D flown 78, enclosed-cockpit Elf flown 84.
DIMENSIONS Span 5,86 m (D, 5,96 m); length 4,88 m; wing area 6,16 m².

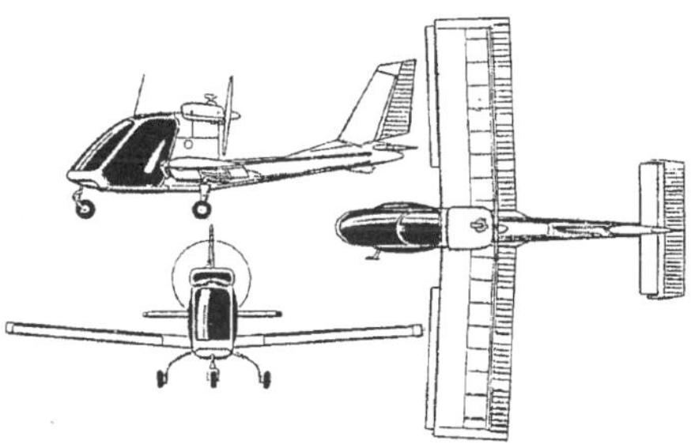

Elf

MAI-89

ENGINE (D) 48 hp, (Elf) 70 hp.
WEIGHT Loaded 380 kg (D, 300 kg).
PERFORMANCE Max speed 100 km/h (D, 80 km/h); climb 3 km/h (164 ft/min).

Foton Remarkable baby jet designed 85 and tested with help of CAHI; object of great interest but so far not taken further.
DIMENSIONS Span 7,32 m (24 ft 0 in); length 8,27 m (27 ft 1⅝ in); wing area 7,32 m² (78 ft²).
ENGINES One turbojet 900 daN (2,050 lb) thrust.
WEIGHT Loaded 2150 kg (4,740 lb).
PERFORMANCE Max speed 640 km/h (398 mph); climb 23,5 km/h (1,290 ft/min); min speed 125 km/h.

Aviatika-890 Based on May 89 **MAI-89** prototype, versatile light 1- or 2-seat biplane with wings replaceable by autogyro unit. Light-alloy

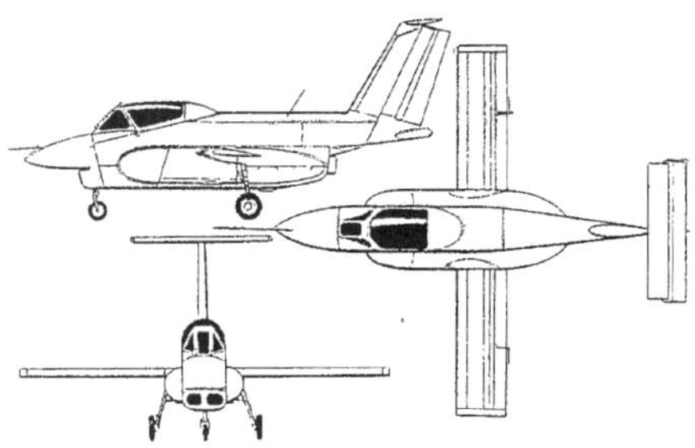

Foton

structure, fabric covering. Fixed tricycle gear, tail carried on tubular boom. Numerous variants, all meeting FAR-23. Chief designer Zhidovetskiy. Flown 90, 28 produced by start of 92. Data for 890U biplane trainer.
DIMENSIONS Span 8,11 m (26 ft 7¼ in); length 5,32 m (17 ft 5½ in), 5,5 m over PVD boom; wing area 14,29 m² (154 ft²).
ENGINE One Rotax (1-seat) 64 hp, (2-seat) 79 hp.

Aviatika-900 Akrobat Competition aerobatic aircraft devd from Kvant. All-metal, symmetric profile, wide leaf-spring ldg gear, claimed record agility, factors ±11, prototype shown early 92.
DIMENSIONS Span 7,15 m (23 ft 5½ in); length 5,7 m (18 ft 8⅜ in); wing area 10 m² (108 ft²).
ENGINE One M-14P.
WEIGHTS Empty 590 kg (1,300 lb); max 715 kg (1,576 lb).
PERFORMANCE Max speed 375 km/h (233 mph); ldg 110 km/h.

Aviatika-92 Ultralight derived from MAI-89, flown 92. High monoplane wing, pusher, single seat, single wheel.

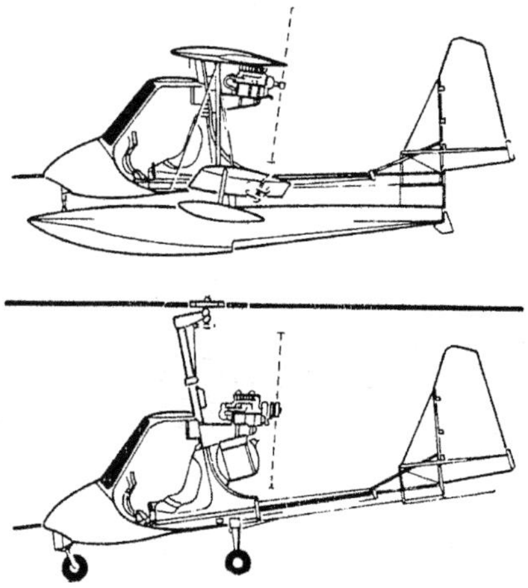

MAI -890 (two of many versions)

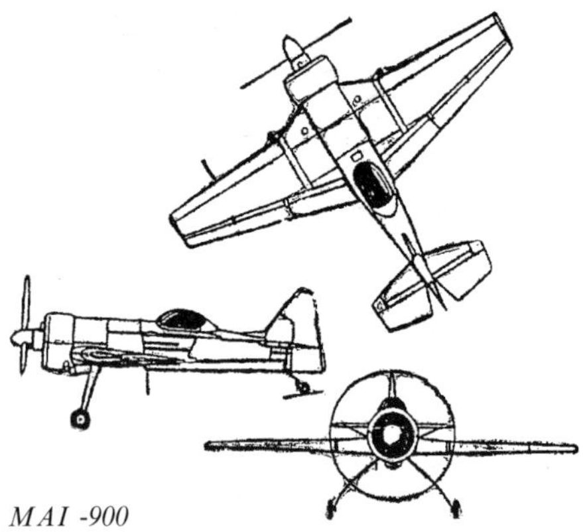

MAI -900

DIMENSIONS Span 10,02 m (32 ft 10½ in); wing area 10,04 m² (108 ft²).
ENGINE Various c60 hp.
WEIGHTS Empty 103 kg (227 lb); max 200 kg (441 lb).
PERFORMANCE Max speed 120 km/h (75 mph); climb 2 m/s (395 ft/min) at 70 km/h.

MAI-900 Akrobat

Maksimov

Dmitri Sergeyevich Maksimov collaborated with aerodynamicist Ivan Ivanovich Drakin working under Bartini on Stal-7. In 1937 they formed own KB, growing by 1940 to 70 persons, at Gorkii Aeroflot repair works to build SPS.

SPS-89 SPS, *Skorostnoi Passazhirskii Samolyet*, fast passenger aeroplane, 89 from GAZ-89 Gorkii where centred. Triggered by study of DC-3 and intended as PS-89 replacement. Slim constant-section fuselage for 17 passengers and crew of three; two M-100 or M-103 on low wing, twin fins, tailwheel landing gear. Begun late 37; in April 40 GAZ handed to Yermolayev for new OKB. SPS not affected, but following further study of project and PS-84 (Li-2) in service, decision to abandon.

MiG

Unlike most of great names in aircraft manufacturing, this OKB has shown little versatility: its products have been almost without exception fighters. It did not exist until immediately before the Second World War and was never eminent until long after that conflict, when its second major type of jet fighter appeared in the Korean war. Since then its designs have been so constantly in news that it has become most famous aircraft-manufacturing organization in world, known to more people than Boeing.

Artyem Ivanovich Mikoyan was son of carpenter in Sanain (now Tumanyan), born 5 Aug 05. Began work as turner at Rostov about 22, then joined Red Army and entered Frunze Military Academy. Moved to Zhukovskii VVA 30, designed *Oktyabrnok* (see Avietka) year before 37 graduation. Joined CCB Polikarpov brigade on I-153. Mikhail Iosifovich Guryevich born into more academic family near Kursk, 12 Jan 1893. Kharkov University, l'Academie de l'Aéronautique (Paris) and Kharkov tech institute; graduated 23. Joined CCB, and in 28 appointed engineer-constructor with Richard's KB, working on TOM programme. In 31 reorganization, appointed deputy chief in Kochyerigin's brigade; led engineering design of TSh-3. In 1936-38 worked at Douglas under Lisunov on DC-3 programme. Polikarpov assigned Kh high-altitude interceptor July 39; this transferred October to detached group OKO-1 led by Mikoyan, with Guryevich, V. A. Romodin, A. G. Brunov, D. N. Kurgouzov and others. Officially independent OKO announced 8 Dec 39, occupying NKAP Factory No 1. Oct 41 evacuated Kuibyshyev. Returned 16 March 42 to Factory 155 (Moscow) becoming OKB-155. Lead designers included Ya. I. Selyetskii, N. I. Adrianov, N. Z. Matyuk, A. A. Andreyev, R. A. Belyakov and P. D. Grushin. Mikoyan promoted General Constructor 20 Dec 56. Guryevich retired 64 but lived to 1976. Mikoyan immobilised by infarctus (blood clot) 27 May 69, dying 9 Dec 70. Succeeded by Rostislav Apollossovich Belyakov.

Kh, I-200 High-altitude interceptor designed with AM-37 engine, but delays and problems forced switch to AM-35A. Chief engineer Brunov, assisted by A. T. Karyev. Prototype transferred to Khodinka airfield 30 March 40 and flown by c.t.p. A. N. Yekatov 5 April.

Very mixed construction but conventional layout. Fuselage back to behind cockpit of welded 30KhGSA steel tube (110 kg/mm²) with duralumin skin, mainly in form of removable panels held by unlicensed Dzus fasteners. Rear fuselage and tail, wooden monocoque with four main longerons and bakelite-ply skin comprising five 0,5 mm layers of *shpon*. Highly tapered single-spar (plus front/rear auxiliary spars) wing of Clark-YH profile, 14% at root, 8% at tip. Centre-section span 2,8 m, no dihedral, duralumin 2 mm webs with heat-treated 30KhGSA booms, 13 dural pressed ribs and five stringers above and five below with flush-riveted skin. Outer wings 6° dihedral, wooden construction with main structure 14 or 15 mm *delta*, main spar 115 mm tapering to 75 mm, auxiliary spars and stringers carrying five layers bakelite-ply 2,5-4 mm applied diagonally and bonded with casein glue. Flying-control surfaces duralumin with AST-100 fabric. All-dural Schrenk-type flaps driven pneumatically to 18° and 50°. Main-gear shock struts with usual 70/30 glycerine/alcohol filling, 270 mm stroke, 600×180 mm tyres, pneumatic retraction inwards with electro-mechanical signalling, bays closed by doors on legs with 90° hinged lower segments to cover retracted wheels. Tailwheel retracting with twin doors. Massive engine resting on 30KhGSA welded mounts, glycol radiator (40-lit system, unpressurized) under fuselage at wing trailing edge, supercharger inlets both wing roots, oil cooler in duct on left of cowl, compressed-air starting, no Hucks dogs on Elektron (Mg-alloy) spinner of 3,0 m VISh-22Ye three-blade propeller. Three main tanks, all AMTs aluminium: 150-lit (33 gal) protected tank behind main spar in left and right centre-section, third tank 110 lit in fuselage between engine and cockpit. Simple cockpit, 13 instruments, no armour. Plexiglas canopy hinged to right (thus always closed in flight), RSI-3 single-channel radio occasionally installed KPA-3bis oxygen, GS-350 generator, 12A-5 battery.

First prototype unarmed. Immediately evident excellent speed, but handling totally unacceptable; high wing-loading, tendency to stall/spin, hopeless longitudinal stability and poor manoeuvrability, and dangerous except to skilled high-speed monoplane pilot. Yekatov soon killed by engine failure on approach, replaced by A. G. Kochyetkov, fellow VVA

student of Mikoyan and later chief test pilot for Lavochkin. Urgent programme to redesign and rectify most faults.
DIMENSIONS Span 10,2 m (33 ft 5½ in); length 8,155 m (26 ft 9 in); wing area 17,44 m² (187·7 ft²).
ENGINE One 1350 hp AM-35A.
WEIGHTS Empty 2475 kg (5,456 lb); fuel 190 kg (419 lb); oil 28 kg (62 lb); loaded 2968 kg (6,543 lb).
PERFORMANCE Max speed 508 km/h (316 mph) at SL, 651 km/h (405 mph) at 7 km (23,000 ft).

I-200

MiG-1 Urgent need for high-altitude fighter resulted in priority production of initial 100 with designation MiG-1. Fitted with armament: one UBS with 300 rounds and two ShKAS each with 375, all above engine. One rack under each wing for FAB-50 or FAB-100. PBP-1 optical sight. From aircraft #9 canopy sliding to rear and openable in flight. Tanks protected by layer of uncured rubber. Second oil cooler added in tunnel on right of cowl, both with thermostatically controlled door at aft end. Wheel doors moved from legs to underside of wing. Radiator duct enlarged and lengthened. Last MiG-1 rolled out GAZ-1 5 Dec 40, in front line April 41.
DIMENSIONS, ENGINE unchanged.
WEIGHTS Empty 2602 kg (5,736 lb); loaded 3099 kg (6,832 lb).
PERFORMANCE Max speed (SL) 486 km/h (302 mph), (7,2 km) 628 km/h (390 mph); climb 5·9 min to 5 km; service ceiling 12 km (39,370 ft); range 580 km (360 miles); TO run 238 m (781 ft); landing speed 141 km/h (88 mph); landing run 400 m (1,312 ft).

MiG-1

MiG-3 Urgent CAHI tunnel programme assisted in rectifying faults. Among major changes, addition of 250-lit (55 gal) tank under cockpit and second oil tank, and all fuel tanks protected by cool exhaust-gas inerting. Engine moved forward 100 mm (4 in) and dihedral of outer

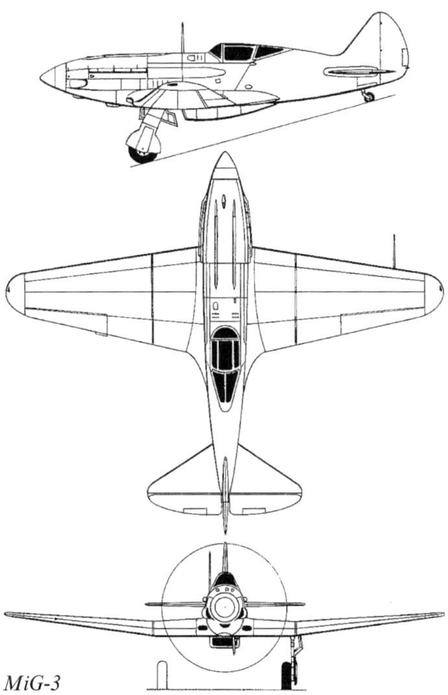
MiG-3

MiG-3

wings doubled to 6·5°. Brakes improved and main tyres enlarged to 650×200 mm for unpaved airfields, some aircraft with fixed tailwheel and no wheel doors. Canopy further improved, and Plexiglas panes to rear enlarged. Armour 8 or 9 mm added behind seat. Propeller VISh-61SH with pitch range increased to 35°. Radiator with revised tube-plate matrix in further improved duct. Improved PBP-1A sight, plus RSI-3 (later RSI-4) radio standard. Total four wing attachments for up to 220 kg of bombs, or eight RS-82 rockets or two VAP-6M or ZAP-6 chemical containers or two BK gondolas with BS gun. Some replaced the two ShKAS by second UBS, but twin-ShVAK installation was tested on two aircraft only. A few aircraft with no armament but one or two reconnaissance cameras. Production at GAZ-1 switched to MiG-3 at 101st aircraft 5 Dec 40, reaching 25 per day by June 41. In Dec 41 Stalin sent telegram to GAZ-1 and GAZ-18 (Il-2) saying latter had priority. To boost AM-38 engine production AM-35A was stopped, halting MiG-3 output at No 3,120 on 23 Dec 41. Still tricky and demanding aircraft, and because air war was mainly below 2 km (6,560 ft) seldom had opportunity to show to advantage at high altitude.

DIMENSIONS Span 10,2 m (33 ft 5½ in); length 8,25 m (27 ft 0⅞ in); wing area 17,44 m^2 (187·7 ft^2).

ENGINE One 1,350 hp AM-35A.

WEIGHTS Empty 2699 kg (5,950 lb); fuel 385 kg (849 lb); oil 55 kg (121 lb); loaded 3350 kg (7,385 lb).

PERFORMANCE Max speed (SL) 505 km/h (314 mph), (7,8 km) 640 km/h (398 mph); climb 10·28 min to 8 km; service ceiling 12 km; range 820 km (510 miles); TO (full power, no flap) 234 m (768 ft).

MiG-3/AM-38 A. I. Zhukov handled test flying May/Sept 41 of one aircraft with AM-37 (at time called MiG-7 by OKB). In Aug 41 many pilots evaluated aircraft with AM-38 (1,600 hp at low levels), giving much higher performance at low altitude, but all engines earmarked for Il-2 so no new production. Some (over 80) damaged aircraft re-engined with AM-38 by repair factory, a few also having twin ShVAK-20 cannon each with 100 rounds. Weights show fuel capacity increased.

DIMENSIONS unchanged.

ENGINE One 1,600 hp AM-38.

WEIGHTS Empty 2582 kg (5,692 lb); fuel 463 kg (1,021 lb); oil 45 kg (99 lb); loaded 3225 kg (7,110 lb).

PERFORMANCE Max speed (SL) 547 km/h (340 mph), (4 km) 592 km/h (368 mph); climb 7·95 min to 5 km; service ceiling 9,5 km (31,168 ft); TO run 380 m (1,247 ft).

MiG-3/M-82 Non-availability of Mikulin engines led Oct 41 to urgent search for alternatives. Five airframes completed with early production M-82A in last nine weeks of year. Team under I. G. Lazarev effected 'crash' installation, not even tunnel tested, and also modified rest of aircraft with: wider and deeper canopy; deeper rear fuselage and larger fin; and automatic slats. Engine installation with internal supercharger inlet, oil cooler on underside and four large hinged cowl panels. Armament three UBS, one above engine, others each side just above wing. Alternative designations **MiG-9**, **IKh** and **I-210**. First flight 2 Jan 42 by V. Ye. Golofastov; poor performance, bad vibration and other faults. According to Belyakov/Marmain, first aircraft given in addition two ShKAS above engine and used operationally on Kalinin Front. Decision to produce properly engineered derivative (**Ye**).

MiG-3/M-82 (I-210, IKh)

I-210

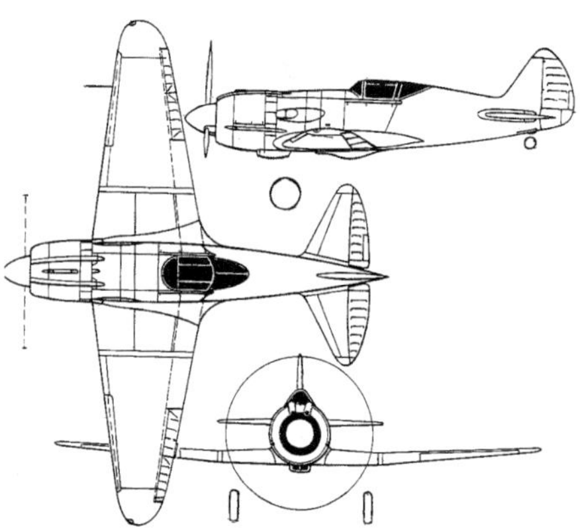

I-210

DIMENSIONS Span 10,2 m (33 ft 5½ in); length 8,078 m (26 ft 6 in); wing area 17,44 m^2 (187·7 ft^2).

ENGINE One 1,700 hp M-82A.

WEIGHTS Empty 2720 kg (5,996 lb); fuel+oil 360 kg (794 lb); loaded 3382 kg (7,456 lb).

PERFORMANCE Max speed (SL) 475 km/h (295 mph), (6150 m) 565 km/h (351 mph); climb 6·7 min to 5 km; range 1070 km (665 miles).

PBSh-1 Armoured assault aircraft authorised May 40, also called **MiG-4**. Large single-seater with low wing of slight inverted-gull form, housing inward-retracting main gears, two tanks and two VYa-23 (96 rounds) plus six ShKAS (750 rounds). Up to 12 RS-82 rockets, or various bombloads: for level bombing, 24 FAB-10, 24 FAB-8, 280 FAB-2,5 or 120 ZAB-1; for dive bombing, 2 FAB-250 or various smaller. Mass of armour 1390 kg (3,064 lb). Mock-up review 24 July 40, but abandoned soon after in favour of PBSh-2.

Dimensions Span 13,5 m (44 ft 3½ in); length 10,145 m (33 ft 3⅓); wing area 30,5 m^2 (328¼ ft^2).

ENGINE One 1,600 hp AM-38.

WEIGHTS Empty 4850 kg (10,692 lb); loaded 6024 kg (13,280 lb).

PERFORMANCE (est) Max speed (SL) 449 km/h (279 mph); service ceiling 7,6 km (24,935 ft); range 900 km (559 miles).

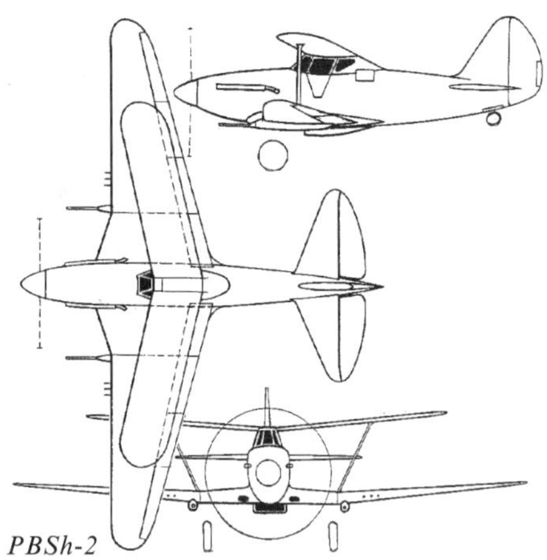

PBSh-2

PBSh-2 Immediately followed PBSh-1 in July 40, also called **MiG-6**. Smaller than predecessor, with main wing with flaps and ailerons raked forward rather than back, and with sharply swept-forward biplane upper wing with no movable surfaces attached above cockpit, with single interplane struts. Guns and wing racks as before, plus cells in inboard wing for small bombs. Armoured cockpit with side door jettisonable in flight. Project abandoned late 40.
DIMENSIONS Span (upper) 8,6 m (28 ft 2·6 in), (lower) 12,4 m (40 ft 8·2 in); length 8,85 m (29 ft 0½ in); wing area (total) 32,4 m² (348·8 ft²).
ENGINE One 1,600 hp AM-38.
WEIGHT Loaded 4828 kg (10,644 lb).
PERFORMANCE Max speed (SL) 426 km/h (265 mph); range 740 km (460 miles).

DIS-200(T)

DIS-200(IT)

DIS, MiG-5 Presence of P. D. Grushin in OKB enabled separate design brigade to be formed for completely different project. Aug 40 decision to produce rival to Polikarpov TIS; launched as DIS (also called **DIS-200**, not explained), *Dvukhmotornyi Istrebitel Soprovozhdyeniya*, twin-engined fighter escort. Capable machine with large wing and consider-

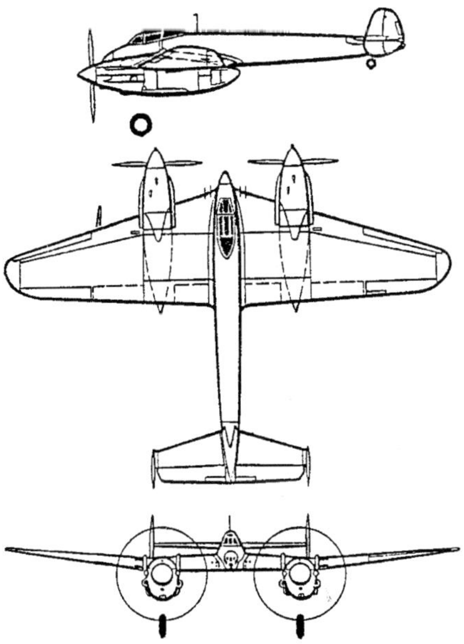
DIS-T, MiG-5

I-230 No 02

able power, but single-seater. Fuselage slim monocoque of *delta drevesina*, eight layers of *shpon* and bakelite-ply skin. Centre-section slight anhedral (1·5°, much less than many Western representations), duralumin spar webs and pressed ribs, 30KhGSA steel booms and main outer-panel joints, dural skin flush-riveted. Wooden outer wings, two main spars tip to tip with multiple stringers. Light-alloy twin-finned tail with all control surfaces fabric-covered. Four protected tanks in centre-section and two in fuselage, total 1920 lit (442·3 gal). Hoped to use ACh diesels, but non-availability led to first of two authorized prototypes (in OKB called **Aircraft T**) having AM-37s. Large underslung nacelles, oil cooler tunnel each side as MiG-3, main radiators with narrow but full-depth inlet on each side of nacelle and exit doors under trailing edge, supercharger inlet in leading edge of wing, exhaust piped over wing. Main gears with single tyres 950×300 mm inboard of single leg, retracting to rear. Tailwheel also retractable, all units (most unusually) having pneumatic actuation. Armament one VYa in underside of nose (magazine behind and below seat) plus two BS and four ShKAS in wing roots (magazines ahead of spar box). Advanced all-glazed canopy with jettisonable sliding mid portion. Armour up to 9 mm in front, rear, sides and underside of seat.

First of two aircraft flown by A. I. Zhukov 15 May 41, also tentatively called **MiG-2**. By this time decision to broaden roles to include heavy bombs (typically two FAB-500, released in dive) or torpedo carried in place of VYa. Good performance but landing too fast and modified with outer-wing slats and ailerons drooping 20° when flaps lowered. Evacuated to Kazan when OKB moved to Kuibyshyev, no available productive capacity, but second aircraft also completed (see next).
DIMENSIONS Span 15,1 m (49 ft 6½ in); length 10,875 m (35 ft 8¼ in); wing area 38,9 m² (418·7 ft²).
ENGINES Two 1,400 hp AM-37.
WEIGHTS Empty 6140 kg (13,536 lb); fuel 1920 kg (4,233 lb); loaded 8060 kg (17,769 lb).
PERFORMANCE Max speed 610 km/h (379 mph) at 6,8 km; climb 5·5 min to 5 km; service ceiling 10,8 km (35,435 ft); range 2280 km (1,417 miles).

DIS-200, IT Second aircraft completed at Kuibyshyev with Perm-built M-82F engines. Split flaps to end of centre section, high-lift slotted flaps thence to aileron. Supercharger intakes in leading edge, oil coolers under

engines. Illustrations show armament unchaged, but written report that two VYa were fitted. Designation believed from *Istrebitel Tyazhelyi,* heavy fighter. Flown by G. M. Chiyakov 22 Jan 42. Later transferred to Kazan, where languished with predecessor. No available factory, and no real requirement, though author told production capacity was sought.
DIMENSIONS as before except length 12,14 m (39 ft 10 in).
WEIGHTS small changes, loaded 8000 kg (17,637 lb).
PERFORMANCE Max speed (5 km) 604 km/h (377 mph); climb 6·3 min to 5 km; service ceiling 9,8 km (32,150 ft); range 2500 km (1,553 miles).

I-230, D Though widely described as specialized high-alt fighter, this was merely next generation after MiG-3, using as many MiG-3 parts as possible for general air fighting at all altitudes. Geometrically similar to MiG-3 but greatly improved aerodynamically, and with steel/dural main spar to tips and mainly light-alloy outer wing generally. Improved balanced ailerons and automatic leading-edge slats. Longer and better-profiled fuselage entirely of wood except forward of firewall. Improved tailplane mounted 200 mm higher. New cockpit with gyro instruments, major advances in equipment, push-button gun/radio/radiator controls and toe brakes. All-round good view from new sliding canopy and rear glazing (except on ground). Revised landing gear and doors, retractable tailwheel. Engine air inlets in roots, water and central oil radiators in wide but shallow group under wing. Intended for AM-39, but this not ready. AM-38F reserved for Il-2 and in any case a low-altitude engine, so prototype fitted with AM-35A. Armament two ShVAK-20 above engine, each with 150 rounds. Alternative designations **I-230**, **MiG-3U** and **D**, D from *Dalnost*, long-range, this

I-231

I-211(Ye)

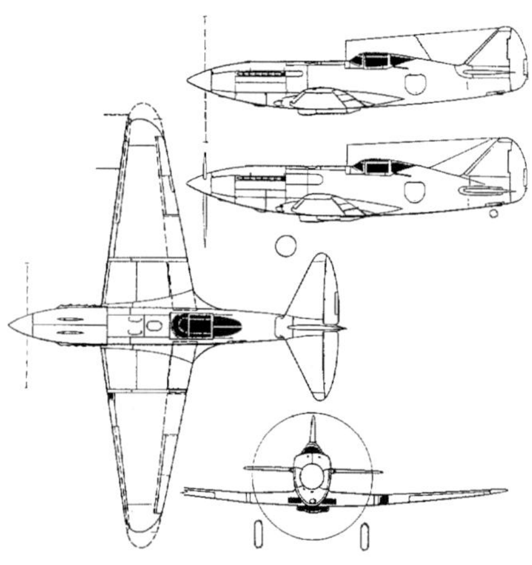

I-230 (lower side view, I-231; wingtips 230-02)

greater range not explained with less fuel. Flown 11 Aug 42. Second aircraft with increased span (11 m, wing area 18 m², 0,5 km higher ceiling not thought worth disruption of manufacture). In any case AM-35A no longer built, and impossible to put this improved fighter into production. All that could be done was find engines for pre-series batch of five aircraft, built at OKB, which fought successfully on Kalinin front.
DIMENSIONS, ENGINE As MiG-3 except length 8,62 m (28 ft 3⅜ in).
WEIGHTS Empty 2612 kg (5,758 lb); fuel/oil 324+56 kg; loaded 3285 kg (7,242 lb).
PERFORMANCE Max speed (SL) 560 km/h (348 mph), (6 km) 660 km/h (410 mph); climb 6·2 min to 5 km; service ceiling 11,5 km (37,730 ft) [I-230 No 02, 12 km] range 1300 km (808 miles).

I-231, 2D Single example built with AM-39A engine. As I-230 except greater power, AV-5A propeller of higher solidity, rear fuselage and tail reverted to MiG-3 (sharp upper taper, lower tailplane), redesigned combined radiator group (which helped efforts to reduce weight) and guns now with 160 rounds. Flown Feb 43 by Yu. A. Antipov, later crashed by P. M. Stefanovskii. No engines, no available factory.
DIMENSIONS As I-230.
ENGINE One AM-39A.
WEIGHTS Empty 2583 kg (5,694 lb); fuel/oil 333+34 kg; loaded 3287 kg (7,246 lb).
PERFORMANCE Max speed (7,1 km) 707 km/h (439 mph); climb 4,5 min to 5 km; service ceiling 11,4 km (37,400 ft); range 1350 km (839 miles).

I-211, Ye Decision taken about Feb 42 to produce properly engineered version of MiG-3/M-82. Full assistance of CAHI and Shvetsov KB. Hermetically sealed engine installation similar to La-5 with cooling exit door on each side, internal supercharger duct and oil coolers in wing roots. Total redesign of fuselage with taper in plan from engine to tail. Cockpit moved 245 mm back and tail moved forward and redesigned as I-230 with higher tailplane. Curiously, slats not fitted (but same slat design said to have been transferred to Lavochkin). Two ShVAK above fuselage each with 150 rounds. Prototype flown by Golofastov 18 Dec 42. Outstanding, said to outperform all contemporary Soviet fighters, but no productive capacity. OKB did manage to build pre-series of ten, used successfully on Kalinin front from Sept 43.
DIMENSIONS As MiG-3 except length 7,954 m (26 ft 1 in).
ENGINE One M-82FN (ASh-82FN).
WEIGHTS Empty 2528 kg (5,573 lb); fuel/oil 385 kg; loaded 3100 kg (6,834 lb).
PERFORMANCE Max speed (7 km) 670 km/h (416 mph); climb 4 min to 5 km; service ceiling 11,3 km; range 1140 km (708 miles).

I-220, A, MiG-11 Ultra-high-altitude reconnaissance of western Soviet Union by Ju 86P

I-220 No 02

from late 1940 triggered early 41 requirement for VP (*Vysotnyi Pyerekhvatchik*, high-altitude interceptor). I-220, Aircraft A, assigned to MiG bureau to result in major programme progressing in stages, little commonality with MiG-3. First experimental aircraft introduced airframe with long-span wing of CAHI laminar profile 20,44 m² area, steel/dural spar, dural extruded stringers and dural skin over centre section, mainly wood outer panels. Two large sections of Schrenk flap each side and outer panels with full-span dural automatic slats. Fuselage longer than MiG-3 but mixed construction; tail all light-alloy to simplify production, with bolted joints. Wide cockpit with sliding canopy. All air inlets in centre-section leading edge. Main cooling radiators rectangular matrices along spar with air exit via variable shutter in upper surface at 52 to 57% chord. Oil coolers inboard with auxiliary inlets at root to serve carburettor and, eventually, pressurized cockpit. Completely new main gears with levered-suspension legs giving 515-mm wheel travel for soft ride on rough surfaces; new-design retractable tailwheel. All fuel tanks soft rubberized-fabric self-sealing cells, between wing spars (4) and in fuselage (2). AM-39 engine selected at start but not then cleared for flight; with a few minor changes, first aircraft completed with AM-38F driving AV-5A propeller (OKB drawing shows four-blade, not seen in any photo). Flown early July 43 by A. P. Yakimov, later joined by P. A. Zhuravlyev. Fitted with two ShVAK, one on each side of crankcase (ie, low on cowling, not above as reported elsewhere), each with 150 rounds. NII testing, with designation **MiG-11**, Sept/Nov. Returned to OKB for re-engining, flown by Yakimov with AM-39 21 or 27 Jan 44. Second aircraft with AM-39 from start, also fitted with vhf whip aerial and heavy armament of four ShVAK, each with 100 rounds, two extra guns being above engine. Flown by I. I. Shelnest Sept 44. Turbocharged engine awaited.

I-220 No 01

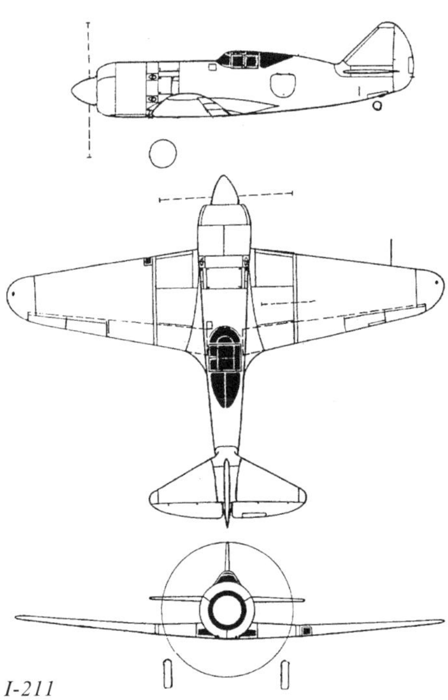

I-211

DIMENSIONS Span 11 m (36 ft 10¾ in); length 9,603 m (31 ft 6 in); wing area (02) 20,38 m² (219·38 ft²).
ENGINE (01) One AM-38F, (02) AM-39.
WEIGHTS Empty (01) 2936 kg (6,473 lb), (02) 3101 kg (6,836 lb); fuel/oil 335+45 kg; loaded (01) 3574 kg (7,879 lb), (02) 3647 kg (8,040 lb).
PERFORMANCE Max speed (SL) 572 km/h (355 mph), (7 km) [01] 630 km/h (391 mph), [02] 697 km/h (433 mph); climb [02] 4·5 min to 6 km; service ceiling [02] 11 km; range [01] 960 km, [02] 630 km (391 miles).

I-221, 2A, MiG-7 Considerably further advanced high-altitude interceptor with new longer-span wing of 22,44 m², NACA-234 laminar profile. Aspect ratio 7·5, thickness 14% root, 10% tip. Duralumin stressed-skin fuselage and wing throughout. Contrary to some reports, cockpit unpressurized, and engine fitted with single TK-2B turbosupercharger on right side. Two ShVAK, one on each side of crankcase, each with 150 rounds. Flown by Zhuravlyev 2 Dec 43, but on an early flight con-rod broke, pilot baling out.
DIMENSIONS Span 13 m (42 ft 7¾ in); length 9,55 m (31 ft 4 in); wing area 22,44 m² (241.5 ft²).
ENGINE One AM-39A (turbo).
WEIGHTS Empty 3179 kg (7,008 lb); fuel/oil 448+40 kg [believed OKB misprint for 348]; loaded 3888 kg (8,571 lb).

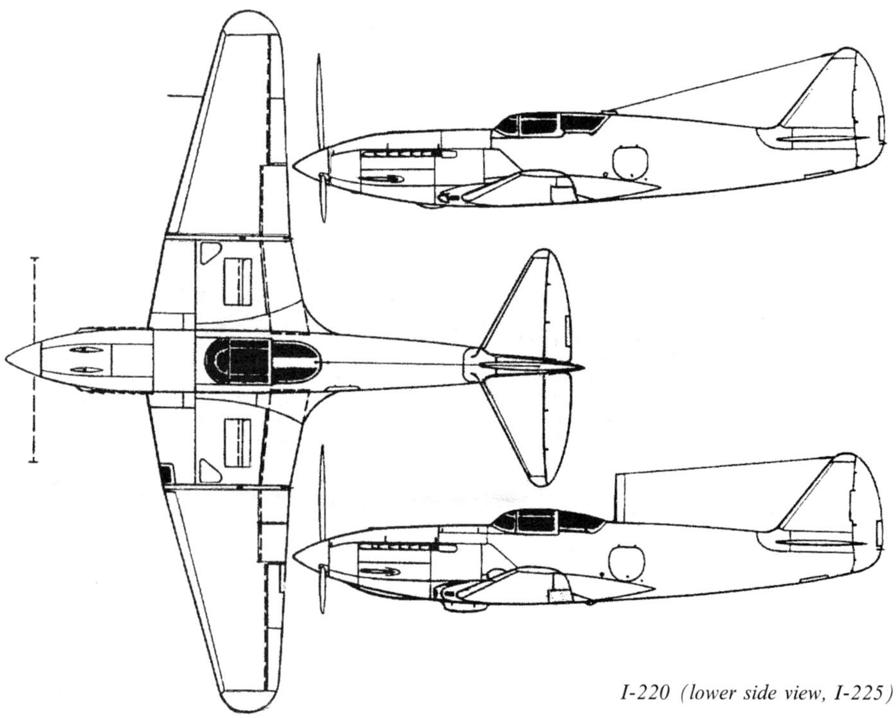
I-220 (lower side view, I-225)

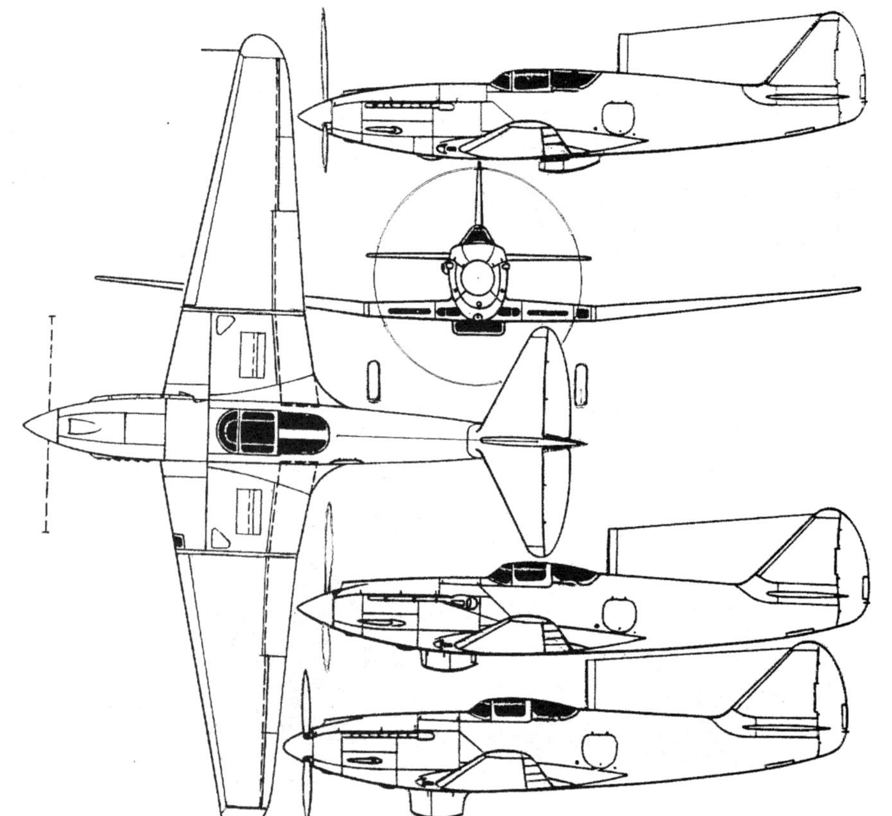
I-221 (middle side view, I-222; bottom, I-224)

PERFORMANCE Max speed (7 km) 689 km/h (428 mph); climb 4·6 min to 5 km; service ceiling 14,5 km (47,570 ft).

I-222, 3A, MiG-7 Fitted with advanced form of Shchyerbakov-type pressure cabin, developed from types fitted to I-153 and I-16. Wooden rear fuselage and outer wings, same geometry as I-221 (2A). Cockpit dural, pressurized from engine turbo and with hot-air de-misting and inflatable seal around sliding canopy of new profile with slightly lower top to rear fuselage. Water and oil radiators of modified design in centre section, large belly inlet for first stage of cooling and engine/cabin air. New tail with greater areas than predecessors. All fuel tanks self-sealing bags in centre section, oil tank armoured in leading edge. Modified engine with experimental high-capacity TK-300B turbo on left side. Cockpit protected by bulletproof windscreen and rear glass, and pressure bulkheads of 8 mm or 9 mm armour. Two ShVAK beside crankcase each with 80 rounds. First flight by A. I. Zhukov 7 May 44 with AV-5A propeller, later replaced by four-blade AV-9L-26V. Development continued, requirement no longer urgent.

DIMENSIONS As 221 except length 9,603 m (31 ft 6 in).

ENGINE One AM-39B-1 (turbo).

WEIGHTS Empty 3167 kg (6,982 lb); fuel/oil 300+40 kg; loaded 3790 kg (8,355 lb).

PERFORMANCE Max speed 691 km/h (429·3 mph) at 13 km; climb 5·5 min to 5 km; service ceiling 14,5 km (47,570 ft); range about 700 km (435 miles); landing speed 169 km/h (105 mph).

I-224, 4A, MiG-7 Last of MiG series of long-span stratospheric interceptors, with more high-altitude thrust from AV-9L-22B propeller of higher solidity (max blade chord 400 mm, 15¾ in) and increased cooling capacity from enlarged centreline heat exchanger. Substantial increase in max fuel capacity. Cockpit constructed in welded aluminium alloy (probably world first) and pressurized to 0,3 kg/cm² (4·3 lb/in²) by new system bleeding air from engine's own supercharger. Main glycol radiator air discharged through four 'sitting dog' pipes over trailing edge. Guns as 222 but 100 rounds each. First flown by Yakimov 20 Oct 44. Adequate, but range 400 km short and ceiling 0,4 km below I-221/222.

DIMENSIONS As 222 except length 9,51 m (31 ft 2⅜ in).

ENGINE As 222.

WEIGHTS Empty 3105 kg (6,845 lb); fuel/oil 476+45 kg; max loaded 3921 kg (8,644 lb).

PERFORMANCE Max speed (SL) 601 km/h, (13,1 km) 693 km/h (431 mph); climb 4·8 min to 5 km; service ceiling 14,1 km (46,260 ft); range 1000 km (621 miles); TO run 230 m; landing speed/run 127 km/h, 440 m.

I-225, 5A Last of wartime MiG fighters, this outstanding aircraft represented a translation from high-altitude series to meet normal medium-altitude demands. Chief engineer Type 5A, A. G. Brunov. Great engine gave such high performance that aircraft 5A not only achieved highest speed of any Soviet piston-only fighter but also satisfactory high-altitude performance. Basically same wing as 2A to 4A, but span reduced back to 11 m. Engine fitted with single TK-300B turbo on right, smaller centreline heat exchanger, underside of cowl deepened and AV-5A-22V three-blade propeller (diam 3,6 m, 142 in) matched to medium altitudes. Pressurized cockpit with improved instruments, armour and vhf radio. Two prototypes, both with four ShVAK with 100 rounds as on I-220-02, which 225 closely resembled. First flight by Yakimov 21 July 44; damaged beyond repair on 15th flight two days later. Aircraft 02, with FB engine, flew 14 March 45. Longest range of MiG with single piston engine.

DIMENSIONS As I-220.
ENGINE (01) One AM-42B, (02) AM-42FB.
WEIGHTS Empty 3010 kg (6,636 lb); fuel/oil 350+41 kg; loaded 3900 kg (8,598 lb).
PERFORMANCE Max speed (SL) 617 km/h, (8,5 km) 720 km/h (447 mph); climb 4·5 min to 5 km; service ceiling 12,6 km (41,340 ft); range 1300 km (808 miles).

I-250, N This unusual little interceptor was, with Su-5, unique example of crash programme for combat aircraft designed for speed at all costs, to meet potential challenge of German jets. Programme begun about March 44 at personal order of Stalin (who criticized industry for not having jets already). Only turbojet, Lyul'ka VRD-2, still under development; alternative was to use most powerful piston engine in conjunction with VRDK (*Vozdushno-Reaktivnyi Dvigatyel Kompressornyi*, air-reaction engine compressor). This developed from about 42 by CIAM team led by K. V. Kholshchyevnikov, and comprised externally driven compressor feeding fresh air through water-cooled radiator of main engine to combustion chamber housing group of seven fuel burners; from here propulsive jet accelerated through variable rear nozzle, capable of being faired over in cruising flight. Basic airframe advanced all-metal stressed-skin construction. Wing of quite large span, two-spars and plate ribs, 10% thickness throughout to preserve aileron control and avoid tip stall. Frise ailerons, CAHI slotted flaps. Fuselage appeared stumpy because of large VRDK air duct from nose, under engine and cockpit to tail. Nearly half-way along was compressor, driven via variable clutch and reduction gear from engine. Immediately behind compressor was entry to upper duct with 180° bend leading to engine supercharger. Normally more than 95% power absorbed by 3,1 m AV-5B propeller. For full-power dash, clutch was completely engaged, transferring power from propeller to compressor, fuel being fed to burners and ignited by plugs, while hydraulic rams opened rear nozzle eyelids. VRDK use limited to 10 min in any flight. Water tank (80 lit) fed cooling radiator behind compressor and also steam cooling system to protect unpressurized cockpit and fuselage structure. At 7 km altitude propeller absorbed 1,450 hp and VRDK added 1,350, a total of 2,800. Fuel (same for both) in 412-litre fuselage tank and two 100-litre wing tanks, all self-sealing rubber. Inlet above spinner served oil cooler with variable exit flaps round top of cowl. Levered-suspension main gears with wheel bays faired by large doors which closed after extending gear. Very small tailwheel retracting into small underfin. First prototype, **N-1**, armed with three B-20, one on each side of crankcase and third firing through propeller hub. Capacity of 160 rounds each, 100 normally loaded. Mockup review 26 Oct 44. Flown by A. P. Dyeyev 3 March 45, VRDK (from Flight 3) adding about 100 km/h. Dyeyev killed 19 May (structural failure under high-g). **N-2**, unarmed, fitted with larger fin but no underfin and thus with fixed tailwheel. Destroyed in forced landing. Urgent order July 45 to have ten I-250 ready for 7 Nov flypast (nine finished but cancelled by weather).
DIMENSIONS Span 9,5 m (31 ft 2 in); length 8,185 m (26 ft 10¼ in); wing area 15 m² (161·5 ft²).

I-222 (3A)

I-224 (4A)

I-225 (5A)

ENGINE One VK-107R with extra drive to compressor of VRDK giving 300 kg (661 lb) thrust at low level.
WEIGHTS Empty 2587 kg (5,703 lb); fuel/oil/water 450+80+75 kg; loaded 3680 kg (8,113 lb).
PERFORMANCE Max speed (SL) 620 km/h (385 mph), (7 km) 825 km/h (513 mph); climb (with VRDK) 3·9 min to 5 km; service ceiling 10,5/11,96 km (with VRDK, 39,240 ft); range (with/without VRDK) 920/1380 km (max 858 miles); TO/landing runs 400/515 m; landing speed 150 km/h (93 mph).

I-250 (N-1)

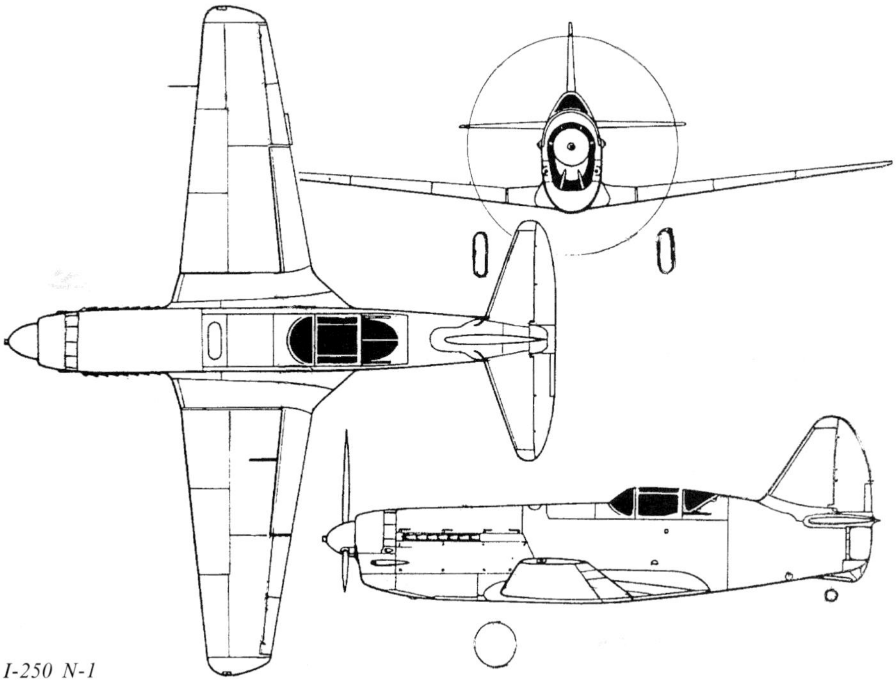

I-250 N-1

MiG-13

MiG-13 Series of 16 additional I-250 aircraft purchased by AV-MF 46. Built by GAZ-381, initially with curved scimitar-blade propellers, used from Riga by Baltic fleet to June 48. Very similar to I-250-02, but with armament as 01, larger wing tanks and radio with wire to mast projecting diagonally ahead from top of windscreen.

DIMENSIONS, ENGINE As I-250.

WEIGHTS Empty 3028 kg (6,675 lb); fuel/oil/water 590+80+78 kg; loaded 3931 kg (8,666 lb).

PERFORMANCE Similar to I-250 except TO/landing at 195/200 km/h (124 mph).

Utka, MiG-8 This unconventional canard (tail-first) aircraft, named *Utka*, Duck, was built in MiG OKB and assigned bureau number which made it sound like production type. Part of design assigned to VVA students under Col (later Prof) G. A. Tokayev; no connection with MiG OKB's normal work, except in important respect that it confirmed low-speed behaviour of slightly swept wings and one of unusual configurations which might be suitable for future jet fighter. High wing with two triangular Scheibe fin/rudder surfaces tested at 55% span and on wingtips. Clark-YH, 12% t/c, constant chord, two spars, strut-braced, sweep 20°, anhedral – 2°, tested wide variety of flap, slat and wingtip configurations. Three-seat cabin with pilot in front. Extended nose carrying fixed tailplane and tabbed elevators. Fixed tricycle gear (nose 300×150, main 500×150 mm), mainwheel spats later removed and new nose gear fitted with same big wheels/tyre as main units. Airframe wood, with wing and control surfaces fabric covered. Pusher M-11F derated to 110 hp; cylinder-head helmets soon removed. Fixed-pitch wood 2,36 m (92·9 in) propeller. Two aluminium fuel tanks in inner wings, total 195 litres. Flown by A. N. Grinchik 19 Nov 45. Few problems but much tinkering, underfins enlarged and slats removed. Soon outstanding stability and refusal to spin. Used as OKB hack for many years.

DIMENSIONS Span 9,5 m (31 ft 2 in); length 6,995 m (22 ft 11⅞ in); wing area 15,0 m² (161·5 ft²).

ENGINE One M-11F.

WEIGHTS Empty 642 kg (1,415 lb); fuel/oil 140+14 kg; loaded (max) 1150 kg (2,535 lb).

PERFORMANCE Max speed 205 km/h (127 mph); range 500 km (311 miles); landing speed 77 km/h (48 mph).

I-270, Zh Aware of Me 163B in early 44, at least two OKBs, including MiG, prepared pre-

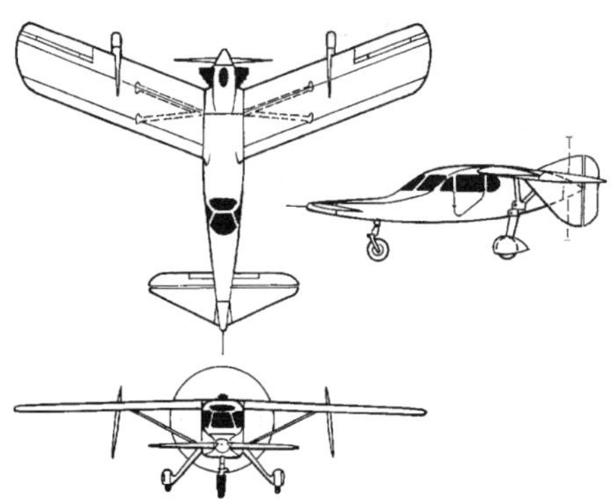

MiG-8 Utka (after modification)

MiG-8 (mod)

I-270 (Zh-1)

lim studies for copy, using Dushkin/Glushko rocket engines, but no prototype construction. After the Second World War considerable Junkers data available on Ju 248 programme (Me 263), including one intact prototype. Decision taken to build Soviet aircraft in same class. OKB assigned task and quickly produced good design. All-metal stressed-skin, fuselage based on Ju 248 but better shape, longer nose and longer pressurized cockpit with hinged canopy, five-spar unswept wings with slotted flaps inboard of ailerons, larger vertical tail with mass-balanced rudder and T-type 30°-swept tailplane with mass-balanced elevators. Engine pump-fed nitric acid and kerosene, with main and upper auxiliary (cruise) chambers respectively giving 4¼ and 9 min endurance under power. (Note: GRD OKB exhibited main chamber only and never publicized this engine, unlike RD-1 family.) Landing gears strong resemblance Ju 248, but aircraft Zh longer, smaller wing, similar empty weight but smaller propellant tankage so lower gross weight. First Zh completed as glider. Radio behind seat with mast/wire on centreline. Round nose, unpainted externally, tested by V. N. Yuganov from early Dec 46 towed by Tu-2. Second, Zh-2, completed with engine. Four tanks of acid, one of kerosene and seven of alcohol/water. Feed by turbopumps, with electric starting pump driven from Me 163B windmill generator on nose. Generally refined airframe. Fitted with planned armament, two NS-23 (40 rounds each) in belly and eight RS-82 rockets underwing. Seat fitted with cartridge ejection. Flown about March 47. High tailplane avoided wing wake at all speeds but manoeuvrability poor because of very high wing loading (until propellants consumed) and endurance with both chambers firing only 255 sec. Nobody very worried when damaged beyond repair in heavy landing, and Zh-1 not repaired after belly landing soon afterwards.

DIMENSIONS Span 7,75 m (25 ft 5 in); length 8,915 m (29 ft 3 in); wing area 12,0 m^2 (129 ft^2).

ENGINE One RD-2M-3V.

WEIGHTS Empty 1546 kg (3,408 lb); acid/kerosene/water 1620+440+60 kg; loaded 4120 kg (9,083 lb).

PERFORMANCE Max speed (SL) 1000 km/h (621 mph), (5 km) 900 km/h, (10 km) 928 km/h, (15 km) 936 km/h; climb 2·37 min to 10 km, 3·03 to 15 km; service ceiling 17 km (55,775 ft); TO run 895 m (2,936 ft); landing speed/run 137 km/h, 493 m.

F, I-300 OKB took no action to build jet until capture of German BMW 003A and Jumo 004B engines, and production factories, in May 45. Then decision taken to build these engines in Soviet Union, and MiG OKB *decided* (author's italics) to use a pair of former. This is at variance with records of La, Su and Yak, which all describe February 45 Kremlin meeting attended by all Fighter OKB leaders at which Stalin instructed crash programme to build fighters using German turbojets, BMW 003 (later called RD-20) and Jumo 004B (RD-10). La and Yak told to use single engine, and fit two NS-23; MiG and Su told to use twin engines and add 57 mm to armament. Three prototypes of each, to fly at earliest date. (After end of the Second World War pressure eased, and despite valuable data from flight-testing of Me 262 original flight expectation of summer 45 slipped by nine months). OKB's Aleksei Timofeyevich Karyev had previously studied Aircraft F to be powered by twin TR-1 engines, and (according to Belyakov/Marmain, with Brunov as chief engineer and Karyev relegated to 'test engineer') this excellent configuration was used as basis but redesigned to take allocated BMW 003A, slimmer and lighter but only 800 kg thrust each. To preserve CG guns fitted far forward, together with unpressurized cockpit with excellent armour and sliding canopy. Two-spar wing of CAHI-1 series profile, ruling thickness

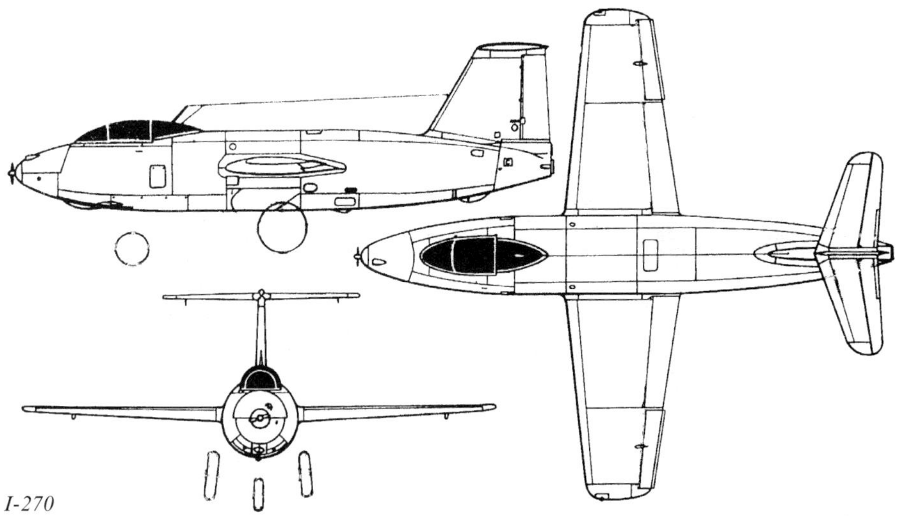

I-270

MiG-9, with tanks

9%, slotted flaps, Frise ailerons. Good fuselage with ample tankage above mid wing and simple ground-level access to engines, guns, radio and other items. Only unusual feature 15-mm air-gap sandwich of stainless steel, with corrugated core, formed into curved panels above jet nozzles. Duralumin flight controls with 0,3/0,5 mm skins, manual system with traditional rod/crank link to elevators and cables to ailerons. Tricycle landing gear with typical MiG levered-suspension throughout, though with lower arms trailing leg instead of leading it as in OKB's previous fighters. Main gears retracted outwards into inter-spar box with bay closed by wing-mounted doors. Castoring nose gear with shimmy-damper, retracting to rear between engine ducts. Hydraulic actuation for all main powered items, single circuit but pump on each engine. Fuel 1635 (also recorded as 1595) litres (351 gal) in four fuselage and six wing tanks, all protected rubber. No aviation kerosene, so T-2 tractor fuel adopted. Small petrol tank for each (German made) Riedel starter engine. Prototype **F-1** armed with giant OKB-16-57 between engine inlet ducts and twin NS-23 (80 rounds each), but monster gun wisely replaced by N-37 (40 rounds) before first flight. No wing hardpoints.

First I-300, aircraft F-1, sent with special test team (OKB pilot A. N. 'Lesha' Grinchik, engineer Karyev and mechanics V. V. Pimyenov and A. V. Fufurin) to NII flight test centre at Chkalovskaya. There met Yak team with Yak-15; tossed coin on 24 April 46 and Grinchik made first Soviet jet flight, unambitious short sortie to feel behaviour. Only real problem (mysteriously came on Flight 8) severe buffet, cured before Flight 12 by reinforcing steel sandwich above nozzles. Grinchik killed on Flight 19, 11 June, when wing-root fillet fairing separated and hit tailplane [for years West has had detailed account of how it was aileron that separated]. Aircraft **F-3** flown by Mark Gallai 9 August and **F-2** by G. M. Shiyanov 11 Aug. F-3 made Aviation Day flypast over Tushino with Yak-15 on 18 Aug. On 28 Aug Mikoyan ordered to have ten aircraft in October Revolution flypast. Mikoyan, Commissar P. Yu Dementyev and factory manager V. Ya Litvinov worked round clock issuing some 60,000 drawings for (ridiculous and delaying) hand-building of these ten aircraft. All ten ready 22 Oct, but flypast cancelled by weather. All similar to prototypes. Despite several hair-raising experiences Gallai said I-300 nice to fly provided 12 to 15 sec always allowed for engines to spool up. High idling thrust led to routine closing HP cocks of one engine on touchdown.

DIMENSIONS Span 10,0 m (32 ft 9¾ in); length [excl guns] 9,75 m (31 ft 11¾ in); wing area 18,2 m² (195·9 ft²).

ENGINES Two BMW 003A, each 800 kg (1,764 lb).

WEIGHTS Empty 3283 kg (7,238 lb); fuel/oil/petrol 1334+35+7 kg; loaded 4860 kg (10,714 lb).

PERFORMANCE Max speed (SL) 864 km/h (537 mph), (4,5 km) 910 km/h (565 mph); climb 4.5 min to 5 km; service ceiling 13 km (42,650 ft); range 800 km (500 miles); TO run 910 m (2,990 ft); landing speed/run 170 km/h, 735 m.

I-301, FS, MiG-9 BMW 003A adapted with Soviet materials and some Soviet accessories and with nozzle bullet eliminated but retaining locally-built Riedel starter as RD-20, produced in limited numbers (believed 500) from mid-46. Production fighter called by OKB FS (F-series), NII designation **I-301** and service designation **MiG-9**. From outset a good and undemanding aircraft needing few changes, but advent of new engines and swept wings made it an unimportant interim type, though order for 500. Among modifications was switch to new multilayer-rubber tank construction, said to enable capacities to be increased though actual capacity slightly reduced. Fin chord considerably increased and small dorsal fin added, wheel brakes and engine-bay cooling improved, longer-life Perlon (German nylon) tyres from Jan 48 and various system improvements. Cockpit unpressurized, seat fixed and agility at high speeds poor. First flight by Gallai 26 Oct 46, 500th delivered December 48. Armament unchanged (N-37 with 40 rounds, two NS-23 with 80) and soon discovered firing guns above 7,5 (24,600 ft) caused engine surge and flameout (same problem with British Hunter 53-56). Cure required major redesign. From Oct 47 all surviving aircraft fitted with airbrakes and with 235-lit (51·7 gal) tanks under wingtips, both feature tested on FT-2 trainer. ASCC reporting name 'Fargo'.

DIMENSIONS As I-300 except length 9,83 m (32 ft 3 in).

ENGINES Two RD-20.

WEIGHTS Empty 3420 kg (7,540 lb); fuel/oil/petrol 1300+35+7kg; loaded 4963 kg (10,941 lb).

PERFORMANCE Almost identical to I-300 (climb/ceiling fractionally better despite greaer weight).

I-301T, FT, MiG-9UTI Need for dual pilot trainer accepted from early 46, mock-up review Oct 46 and prototype FT-1 flown July 47 (not as previously thought Dec 46). BMW 003A engines, one driving German DC generator. Added rear cockpit slightly higher than original cockpit, replacing forward tank. Both cockpits with Machmeter, intercom and later, ejection seat (FT-2). New windscreen and tandem canopies all much taller than on fighter, and sliding window between cockpits. Same guns as fighter. Rejected because of poor view from rear (instructor) cockpit, subsequently used to improve systems and tip tanks. Second trainer,

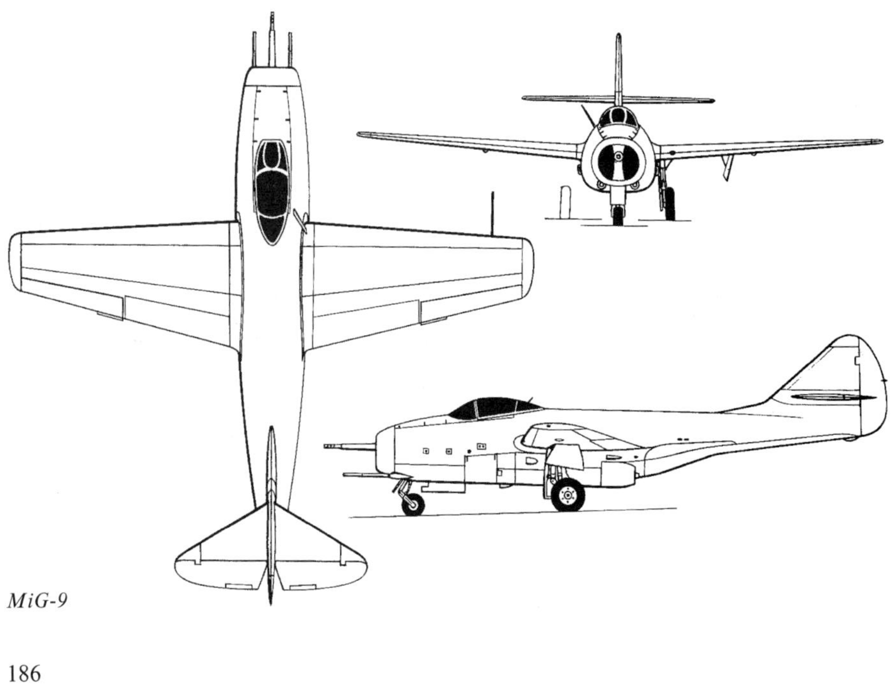

MiG-9

I-301T (FT-1 No 01)

I-308, MiG-9M (FR)

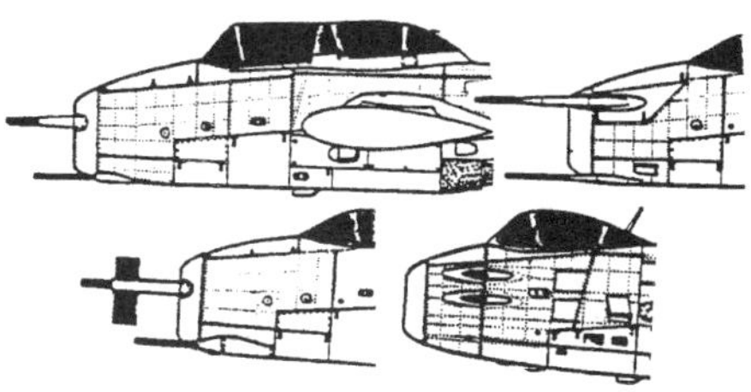

MiG-9 variants: upper L, FT-2; upper R, FP (I-302); lower L, FF (I-307 Babochka); lower R, FR (I-308)

FT-2, flown 25 Aug 47. Redesigned cockpits with rear 0,3 m higher, shallower but wider canopies and two ejection seats designed and made at OKB and sequenced together with canopies by automatic timer. Sliding panel between cockpits removed to improve instructor view further. Fitted with airbrakes of centre-pivot (Vampire) type forming part of each flap. Soviet-made engines with starter petrol capacity doubled. Because of poor endurance, wing plumbed for auxiliary tanks of 235 lit (51·7 gal) capacity under wingtips; these and airbrakes retrofitted to all MiG-9 together with ejection seat. Other changes: addition of S-13 ciné camera, wheel-bay doors closed except during cycling of gear and main wheels/brakes/tyres enlarged. No production, no ASCC name. Data for FT-2.

DIMENSIONS, ENGINES As MiG-9.

WEIGHTS Empty 3460 kg (7,628 lb); fuel/oil/petrol 862+22+14 kg; loaded 4895 kg (10,791 lb).

PERFORMANCE Max speed (SL) 810 km/h (503 mph), (4,5 km) 900 km/h (559 mph); climb 5·3 min to 5 km; service ceiling 12 km (39,370 ft); range (internal fuel) 775 km (482 miles), (tip tanks) 920 km (572 miles).

MiG-9 variants. Improved fighter designated **I-308, FR**. To cure engine flameout when guns fired, entire forward fuselage redesigned with guns higher and further back, all muzzles 1 m (40 in) behind engine inlet. N-37 on right, two NS-23 on left, ammunition unchanged. Cockpit moved forward, requiring sharp bifurcation of ducts, pressurized and fitted with ejection seat, Machmeter and other additions. Uprated non-afterburning engines with electric starters. Fuel capacity unchanged though five wing tanks instead of six. Excellent aircraft, though increased weight forced design load factor to be reduced from previous 6 g to 5·5. Planned to succeed MiG-9 in service as **MiG-9M**, but abandoned because of I-310. Data below for this. Single MiG-9 modified as **I-302, FP**, to see if moving N-37 to new location high on left of nose would avoid flameout problem; not adopted. New airframe built to take single TR-1A engine, with various other changes and all three guns level with centre of nose inlet; designated **I-305, FL** (F-Lyulka), abandoned when almost complete late 47 because of I-310. Version with afterburning RD-20F engines **I-307, FF**, also fitted with extra armour, flown Sept 47. Later tested with large vertical plates fixed to N-37 barrel to duct muzzle gases above and below engine inlet, dubbed *Babotchka* (butterfly). Scheme to fit single RR Nene in **I-320, FN**, abandoned mid-47 [I-320 designation reused later). **MiG-9PB** with large underwing slipper tanks never flown. Single series aircraft rebuilt as **FK** (F-Komet), **MiG-9L** (*Labora-*

MiG-9L (FK)

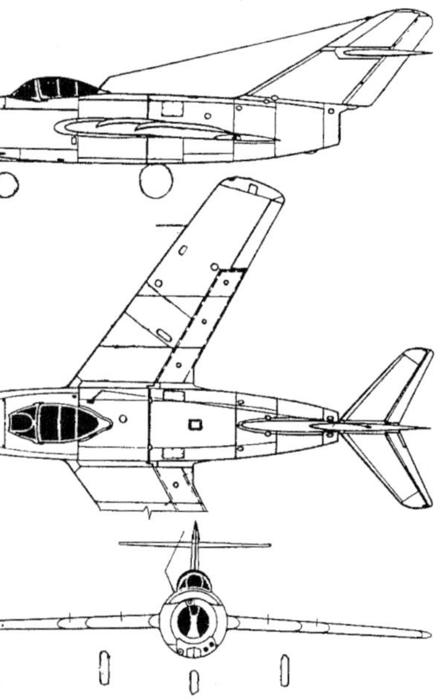

I-310

toriya) to test guidance system of OKB's KS-1 cruise missile for Tu-16. Fuselage stretched to 10,12 m (33 ft 2½ in) with technician cockpit above trailing edge, large target-illuminating radar above nose, receiver antennas in inboard leading edge on each side and transmitter/receiver antennas in pod on top of fin to check on positions of missile and parent aircraft (Tu-4 in early tests). Overall total, all versions, 604. Data for MiG-9M:

DIMENSIONS As MiG-9.

ENGINES Two RD-20F.

WEIGHTS Empty 3356 kg (7,399 lb); fuel 1300 kg; loaded 5069 kg (11,175 lb).

PERFORMANCE Max speed (SL) 950 km/h (590 mph), (5 km) 965 km/h (600 mph); climb 2·7 min to 5 km; service ceiling 13 km (42,650 ft); range 830 km (515 miles); TO run 830 m; landing speed/run 166 km/h, 700 m.

S, I-310 One of most famous aircraft of all time, this simple fighter was not only traumatic shock to Western intelligence when encountered in combat in Korea in Nov 50 but also marked first occasion on which speed of design and development bettered that of any Western rival. Programme begun by Kremlin meeting March 46 attended by all major fighter constructors, charged with design of high-altitude day interceptor able to operate from rough strips, reach Mach 0·9, have good manoeuvrability at high (over 11 km) altitude and flight endurance at least 1 hour. Swept wing taken for granted, and OKBs told to work closely with CAHI where swept wings studied at least from 35 (though not necessarily for high-Mach use). OKB studied forward-swept wing also, as flown by Tsybin. Aircraft S planned on basis of mid-mounted swept wing on shortest circular-section fuselage with large T-tail, configuration studied from late 45 in many German reports and judged to offer best compromise. Engine choice between two competing designs of 2 t (4410 lb) thrust: axial derived from German (mainly BMW) work, similar to first Atar, and large centrifugal based on known external appearance of Rolls-Royce Nene and designated VK-1PO (VK from Maj-Gen V. Ya. Klimov, PO from 'first consignment'). Though latter was crude and generally outside prior Soviet experience MiG OKB told to go ahead, and fuselage sized to 1,45 m diameter in consequence. (No doubt whatever that details and drawings of Nene were soon expected to be obtained by Soviet intelligence.) Design well advanced in Sept 46 when, under terms of trade agreement, UK agreed to export Nene to USSR. Ten engines shipped at once, 15 in March and more later. First engine carefully stripped and issue of Russian production drawings began as early as 30 Oct at No 45 production plant headed by Klimov and to which design staff were specially sent on 21 Sept from Factory 117 (Leningrad), Klimov's main design office. VK-1PO terminated and replaced by RD-45, designated after factory. MiG OKB received accurate installational drawings in Feb 47.

Project authorised same month, headed by deputy general constructor A. G. Brunov and chief engineer A. A. Andreyev, OKB designation **S** (*Strelovidnost*, swept), NII designation **I-310**. Traditional metal stressed-skin design of inspired simplicity, designed to unprecedented load factor of 8 g. Wing swept 35° at ¼-chord, one main spar at ¼-chord with sheet web of D16-T (ruling material throughout airframe) and booms machined from V-95. Two rear spars, 21 stringers and 20 ribs, flush-riveted skin 2 mm between spars, 1 mm elsewhere. Section S-10s at root, high-lift SR-3 at tips, 11% thick throughout; anhedral –2°, MAC 2,12 m, aspect ratio 4·85, taper ratio 1·61, two fences 100 mm high above each wing from leading to trailing edge, sealed and balanced ailerons with rod/crank drive, hydraulic Fowler-type flaps of 2,36 m^2, driven to 20° take-off and 55° landing. Main structural junctions 30KhGSA steel. Fuselage diameter 1,45 m, frontal area 1,16 m^2 (1,35 including canopy), length 8,08 m, fineness 5·57. Pressurized cockpit ahead of wing with rear-sliding canopy and ejection seat. Plain nose inlet to duct bifurcated past cockpit and again above and below unbroken wing centre-section to plenum chamber behind wing. Bold decision to make fuselage in two parts, aft end quickly removed by four bolts at Frame 13 (rear spar) for all-round access to engine. Latter, Nene 1 (4,920 lb) in S-01, carried on 30KhGSA tubular truss off Frame 13 with cantilever jetpipe. Vertical tail 4,0 m^2 swept 56° with main spar 30KhGSA carrying 3,0 m^2 35°– swept horizontal tail 1582 mm above fuselage centreline with incidence ground-adjustable only. Manual tail surfaces with trimmers. Tricycle landing gear with levered-suspension legs throughout, nose unit retracting forwards into lower part of duct bifurcation and main gears with 660×160 mm tyres retracting inwards to lie entirely within wing between spars; actuation, like flaps and airbrakes, hydraulic, with compressed-air standby. Kerosene T-1 fuel in main protected flexible cell behind cockpit between air ducts and in rear fuselage, total 1538 lit (338 gal). Planned armament as FL (I-305) with N-37 in centre of nose and NS-23 at same level each side. Sudden gun-firing engine flameouts with MiG-9 caused rethink and N. I. Volkov engineered outstanding gunpack, winched up and down on four cables, with N-37 and 40-round magazine on right and two NS-37KM each with 80-round box on left, muzzles just sufficiently far from inlet not to disrupt airflow at normal flight speeds.

OKB built two prototype, S-01 and 02, almost identical. S-01 rolled out 27 Nov 47 and flown by Viktor Nikolayevich Yuganov 30 Dec. Only major modifications were to increase tailplane sweep to 40°, slightly modify wing trailing edge and ailerons (all with CAHI

I-310 (S-03)

team), add landing light at top of inlet bifurcation, and cut back rear fuselage to shorten jetpipe by 0,32 m, with aid of CIAM. S-02 fitted with slightly more powerful Nene 2, S-13 cine camera above nose and small rocket added under each wing to assist recovery from spins to which aircraft prone during stall tests. First flown 27 May 48, by Col Gregori Sedov. In March 48 quickly built third prototype S-03 incorporating all required modifications. Wing structure considerably strengthened by using 30KhGSA for main spar web and booms and new light alloy V-95 for skins with ruling thickness increased from 1,5 to 1,8 mm. Flaps increased in chord but reduced in span so that ailerons could be enlarged from 0,96 to 1,17 m², each aileron in two parts linked by universal joint. Door-type airbrakes (initially 0,48 m², later 0,52) installed, swinging out and slightly down on each side of strengthened rear fuselage each driven by hydraulic jack. Tailplane moved 150 mm to rear on modified fin and bobweight added to elevator circuit. Slight reduction in tankage more than made good by provision for detachable (not jettisonable) slipper tank of 496 lit (109 gal) each under wings on hardpoints which could also take FAB-100 bombs. Other changes: removal of gunpack made easier, new canopy locking system fitted, and engine-bay fire-detection and extinguishing system, RSI-6M radio, ASP-1N (British gyro type) gunsight and AGK-47B horizon. S-03 first flown by I. T. Ivashchyenko 17 June 48. OKB testing by him and S. N. Anokhin, reaching M0·934. NII testing at Saki Nov 48. Extremely successful, but still tendency to spin in tight turn, poor behaviour at high AOA and buzz/snaking resulted in decision to redline in service at Mach 0·92 and later airbrakes triggered automatically at about 0·91.

Performance of OKB far outstripped rivals and in Aug 48 design S accepted for mass-production as **MiG-15**, with go-ahead on trainer version at same time. Data for S-03:

DIMENSIONS Span 10,085 m (33 ft 1½ in); length 10,102 m (33 ft 1¾ in); wing area 20,6 m² (221·75 ft²).
ENGINE One Rolls-Royce Nene 2.
WEIGHTS Empty 2955 kg (6,515 lb); fuel/oil 1210+40 kg; loaded 4806 kg (10,595 lb).
PERFORMANCE Max speed (SL) 905 km/h (562 mph), (3 km) 1031 km/h (641 mph), (10 km) 983 km/h (611 mph); climb 2·3 min to 5 km, 7·1 min to 10 km; service ceiling 15,2 km (49,900 ft); range, up to 1530 km (950 miles); TO run 695 m; landing speed/run 160 km/h, 710 m.

SV, MiG-15 Initial series aircraft **SV** (swept, airforce) powered by RD-45. Numerous modifications. Entire airframe reassessed to ensure integrity at higher weights at 8 g. Particular attention to wing spar booms and upper skin, rear-fuselage frames and airbrakes. Latter enlarged to 0,88 m² (9·47 ft²) with geometry revised to eliminate nose-up pitch, and made of El-100N steel for vertical dive held to M0·7. New to OKB, ailerons given hydraulic boost, and trim tab added to left aileron. Outer-wing leading edges given internal 30 kg anti-flutter masses. Overall effort to reduce stick forces at high IAS in both pitch and roll, and second elevator cable circuit added (as far as possible from first). Fuel system revised, tanks pressurizied by engine bleed and provision for slipper tanks (as on S-03), bomb size increased to FAB-250. Much attention to gun pack to improve rearming time further, N-37 replaced by N-37D with new anti-vibration mounts, NS-23 replaced by NR-23 with new mounts, raised hydraulic shock-absorbers and all three guns given larger link chutes (external blisters) to avoid blockage. New main-leg oleo struts. GS-3000 electric starter supplied from higher-capacity battery (12v.30A-hr). Improved cockpit with rear-view periscope, new seat with triggers both sides to jettison one-piece blown canopy with hot bleed-air demisting and then fire seat with different cartridges for summer and winter, enhanced armour protection, ASP-3N gunsight, remote gyro compass (sensing unit near right wingtip), improved IFF (SRO-1) and other items.

Initial production at GAZ-1, delivery to service units (including AV-MF and newly formed PVO) from 8 Oct 48. Ongoing modifications soon included special tank to maintain fuel supply under negative-g and addition of 300-lit (66-gal) underwing drop tanks. Wing further strengthened in torsion, and many pipelines for fuel and hydraulics replaced by steel, with welded joints. Landing light increased in power, RSIU-3 vhf radio and new AGI-51 horizon fitted, with EUP-46 standby, and engine uprated RD-45F. Politburo decision 20 May 49 for production on such scale that plants building La-15, Li-2, Yak-17 and Yak-23 were cleared and switched to MiG-15. Licensed to Czechoslovakia as S-102 (853 built 53-55) and Poland as LIM-1 (54). including licensees, 12,000 built in 17 versions by 56, about half trainers. ASCC reporting name 'Falcon' soon replaced by less complimentary 'Fagot'.

DIMENSIONS As S.
ENGINE One RD-45, later RD-45F.
WEIGHTS Empty 3523 kg (7,767 lb); fuel/oil 1225+40 kg; loaded (normal) 4963 kg (10,941 lb), (max, external fuel) 5405 kg (11,916 lb).
PERFORMANCE Max speed (SL) 1050 km/h (652 mph), (5 km) 1031 km/h (641 mph); climb (clean) 2·5 min to 5 km, 7·1 min to 10 km, (max wt) 3·5/10·5 min; service ceiling 15,2 km (49,900 ft); range 1175 km (730 miles), (external fuel) 1650 km (1,025 miles); TO run 630 m; landing speed/run 160 km/h 720 m.

ST, I-312, MiG-15UTI Need for trainer recognised 47 but project launched by decree 13 April 49, by which time OKB had completed design. Airframe basically identical to MiG-15 apart from forward fuselage. Tandem cockpits each with identical panels and controls and ejection seats with armoured headrest, rear seat fired first. Windscreen taller than on fighter (same as I-301T). Front canopy (pupil) hinged to right, rear canopy (instructor) sliding to rear. Rear cockpit reduced capacity of forward tank (total internal 1120 lit) but provision for 280 or 400-lit underwing tanks. Armament (not always fitted) in first series one UBK-Ye on left side of quickly demountable tray with 150 rounds and one NR-23 on right with 80. Wing racks for two FAB-50 or -100. Front cockpit with ASP-1N sight, S-13 camera above nose, RSI-6 radio. Further armour behind cockpit and gun magazines. Addition from fighter, four signal cartridges (red, green, yellow, white) fired from cockpit.

UTI prototype, **ST**, NII designation **I-312**, built at OKB using fighter components from Kuibyshyev first flown 23 May 49. Prolonged tests to May 50, mainly by user regiment at Kubinka. Then built in very large numbers (c6,500), and also licensed to Czechoslovakia (2,012 built as CS-102), Poland (1,200+ LiM-3) and China (JJ-2). Third block introduced RSIU-3 radio and SRO-1 IFF, fourth block (from No 10444) eliminated NR-23 and replaced sight by ASP-3N, and sixth block installed heavy (84 kg) OSP-48 ILS receiver in place of NR-23 on right side of removable tray. This required further small reduction in capacity of forward tank. This installation tested on prototype **ST-2**, together with KI-11 standby compass, larger UBK-Ye link chute ending in external blister and filter in cockpit pressure bleed line. From 1952 normal to have four UTI with every fighter regiment, and continued to train pilots for MiG-17 and -19. Over 1,960 exported, still used by Algeria, Angola, Congo, Guinea, Mali, Mongolia, Romania, Syria, Vietnam, poss others. ASCC reporting name 'Midget'.

DIMENSIONS as MiG-15.
ENGINE One RD-45FA.
WEIGHTS Empty 3724 kg (8,210 lb); fuel/oil 900+35 kg; loaded (clean) 4850 kg (10,692 lb), (300-lit tanks) 5400 kg (11,905 lb).
PERFORMANCE Max speed at SL 900 km/h (559 mph), at 3 km 1015 km/h (631 mph), (13 km) 950 km/h (590 mph, Mach limit 0·894);

MiG-15UTI

climb 1·6 min to 3 km, 3·2 min to 5 km and 7·8 min to 10 km; service ceiling 14 625 m (47,980 ft); range (clean, 10 km) 950 km (590 miles), (300-lit tanks) 1500 km (932 miles); TO run 570 m; landing speed/run 172 km/h, 740 m.

SD, MiG-15bis Improved engine, similar to original in mountings, connections and weight, though not routinely retrofitted to existing aircraft. Yet again entire structure audited and stiffened and reinforced in 12 places to permit 1076 km/h IAS or Mach 1 in dive. Ailerons redesigned in detail and driven by BU-1 irreversible hydraulic power units (replacing boosters). Airbrakes again redesigned and controlled by stick switch. Elevator and rudder leading edges redesigned and elevator balance increased 22%. Emergency bottle-energised pneumatic system added to drive flaps following engine failure. Wing skins again reinforced and knife-edge stall strips added on each inboard leading edge. Cockpit given improved pressurization and additional hot-air feeds including windscreen, allegedly proof against shells of 64 mm calibre. Armament unchanged though tests carried out with formations dropping showers of 10 kg HE-frag (OFAB) bombs and parachute flares from dispensers against hostile bombers at 12 km altitude all under control of formation leader. Standard equipment RSIU-3 radio, S-13 camera, SRO-1 IFF and ASP-3N sight, but rear-view periscope changed to TS-23 from aircraft 10235, and this in turn replaced by wide-angle TS-25 and finally TS-27A. New pilot connectors including PPK-1 anti-g suit. From 53 a proportion intended to fly at night or in bad weather fitted with OSP-48 ILS (see comments under MiG-15UTI). Many trial installations including 15 m² brake chute for short fields.

Prototype, **SD**, flown at OKB Sept 49 with VK-1 and some of new features, but MiG-15bis did not replace MiG-15 in production until early 52. Licensed to Czechoslovakia (620 built as S-103) and Poland (LIM-2), exported to 12+ countries and still used by Albania (others retain a few as hacks). ASCC name still 'Fagot'.

DIMENSIONS As MiG-15.

ENGINE One VK-1.

WEIGHTS Empty (with OSP-48) 3681 kg (8,115 lb); fuel 1173 kg; loaded (clean) 5055 kg (11,144 lb), (two 600-lit tanks) 6106 kg (13,461 lb).

PERFORMANCE Max speed (SL) 1076 km/h (669 mph), (3 km) 1107 km/h (688 mph), (5 km) 1014 km/h (630 mph); climb 46 m/s, 1·95 min to 5 km, 4·9 min to 10 km; service ceiling 15,5 km (50,850 ft); range (12 km, clean) 1330 km (826 miles), (two 600-lit tanks) 2520 km (1,566 miles); TO run 600 m; landing speed/run 178 km/h, 880 m.

MiG-15 variants To find answers to problems of stability and control at high Mach numbers V. P. Yatsyenko (see his own entry later) assigned two aircraft built on MiG-15 line at GAZ-1 and completed as **SYe** (swept, unique), also called **LL** (flying laboratory). Basically MiG-15 but upper part of vertical tail grotesquely enlarged, chiefly to improve lateral control. Ailerons also enlarged and modified (but without even hydraulic boost) and extra skins added above inner wing. Flown by M. Tyuteryev from July 49, reaching M0·985 on 21 Sept and, after fitting BU-1 aileron power units, 1·01 on 18 Oct. In early 49 OKB designed installation of airborne-interception radar (pioneer *Torii* by A. B. Slepushkin's team), fitted to 102nd aircraft from GAZ-381 and flown by Sedov Dec 49. OKB designation **SP-1**, also called **MiG-15Pbis** (*Perekhvatchik*, interceptor). Pilot-aimed antenna in radome above nose. S-13 camera moving to right side, ARK-5 radio compass and MRP-48 beacon receiver added, 23 mm guns removed, nose gear modified, anhedral of wing increased, airbrakes enlarged and (unique to family) elevator driven by BU-1 power unit. In 51 five **MiG-15bisP** built at GAZ-1 to help develop less-demanding radar/sight system [see SP-2 version of MiG-17]. Much better RP-1 *Izumrud* (Emerald) radar developed by V. V. Tikhomirov's team tested in MiG-15bis from early 50 (not 58 as reported elsewhere). Aircraft designated **SP-5**, tracking receiver in radome in centre of intake and search scanner in upper lip. Guns and camera as SP-1. **MiG-15 Burlak** (barge-hauler) modified 50 with airgun in nose firing harpoon with steel towing cable to engage loop of cable (wing to wing) behind bomber. Fighter then throttles back and towing cable (disconnected

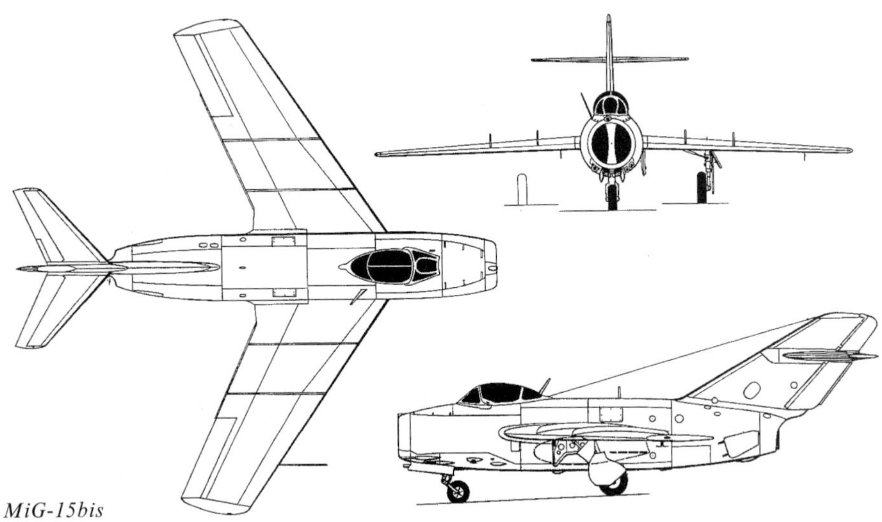

MiG-15bis

MiG-15LL (SYe)

MiG-15Pbis (SP-1)

from one side of towing aircraft) slides until hooks mate in end, fighter then proceeding as glider but ready to defend bomber over distant target. Many difficulties and glider pilot froze. Ciné films show pairs of MiG-15s towing each other. Also in 50, **MiG-15bisS** escort and **MiG-15bisR** reconnaissance versions tested with bigger oxygen supply, 600-lit drop tanks and (for R) one or two AFA-40 cameras. Early 51 **SU** tested new armament; MiG-15 No 109035 (GAZ-1, 35th of 9th block) fitted with V-1-25-Sh-3 system comprising left and right chin gondolas each housing Shpital'nyi Sh-3 cannon with 115 rounds in trunnion mount permitting pilot to pivot both guns from +11° to –7° via switches on throttle (system RUD) and stick (RUS). Tests suggested greater angular range needed, leading to SN version of MiG-17. **SD-21** was MiG-15bis tested early 51 with four wing stores, two 250-lit slipper tanks outboard and two 210 mm S-21 rockets inboard. **ST-7** was MiG-15UTI, also called UTI MiG-15P (trainer interceptor) with RP-1 radar managed by pupil in front cockpit, plus two UBK-Ye guns under nose and S-13 on right. This version put into service in small series. **SD-57** was 803rd GAZ-1 bis equipped to carry a launcher for 12 ARS-57 rockets on each tank pylon; led conversion of 12 as **ISh** (*Shturmovik*) attack aircraft carrying tanks and two types of rocket launcher. In 53 several (at least two) bis used in air refuelling trials with probe on upper left of nose engaging with drogue from Tu-4N. In 55 OKB tested several examples of MiG 15 and 15bis equipped as remotely piloted targets designated **SDK-5**. In same year MiG-15s were tested as **SDK-5s** and 15bis as **SDK-7** cruise missiles carrying 1 t of explosive, using same radio-command guidance system.

I-320, R This was OKB's response to Jan 48

I-320 (R-2)

requirement for fighter to carry radar and intercept day and night at maximum range from defended target. Same requirement also triggered radar developments by Slepuchkin and Tikhomirov and rival prototypes by La, Su and, later, Yak. Specification could not be met by single-engined aircraft, and while adhering as closely as possible to original S (MiG-15) formula OKB adopted novel configuration in which nose inlet (similar to SP prototypes but larger) fed two engines, one at bottom of forward fuselage (rather as MiG-9) and second in MiG-15 type rear end. Inlet divided into three, central portion feeding down to front engine and left/right ducts passing on each side of cockpit and over wing box to plenum chamber of rear engine. Aircraft designed to take off on either engine. Fuel in two fuselage tanks, front 1670 lit and rear 1630, plus 45-lit negative-g supply, total 736 gal. Forward fuselage diam 1,9 m (75 in), with unpressurized cockpit for side-by-side ejection seats for two pilots, each with radar display and able to fly mission unaided (aircraft commander on left). Two bulletproof windscreens and broad sliding canopy. Wing aerodynamics as MiG-15 with same profile and sweep, fences and boosted ailerons, but flap angles increased to 22° TO and 56° for landing. Fin sweep increased to 59°27', tailplane 40°. Tail controls via rods, elevators hydraulically boosted and rudder deflection triggered small electricity driven spoiler under either wing. Airbrakes 1,08 m^2 driven to 45° by same duplicated hydraulics as flaps and landing gear (main tyres 900 × 275, nose 520 × 240).

Three prototypes. **R-1**, flown 16 April 49, with RD-45F engines, no radar and two N-37 guns low in nose, muzzles far back from inlet. Good flying qualities apart from usual wing drop at high IAS or Mach. **R-2** flown Nov 49 with VK-1 engines, third N-37 added high on right, higher windscreen and canopy, with better jettison mechanism, and thermal deicing of inlet and wing/tailplane. Later fitted *Toryii* radar, soon replaced by same KB's much better *Korshun*. Damaged 13 March 50 by explosion of round being fed to N-37 and opportunity taken to increase anhedral, fit third fence on each outer wing, enlarge spoilers and fit automatic Mach-trigger to airbrakes, aircraft flying again 30 March. **R-3**, flown 31 March 50, incorporated R-2 modifications plus 750-lit jettisonable slipper tank under each wing (R-2 plumbed but tanks not fitted). Changed anhedral affected roll/yaw stability and spoilers accordingly linked to ailerons and long quill fairing added at both engine nozzles. No production (Yak-25 chosen later) and R prototypes retained by OKB to help develop ILS systems.

DIMENSIONS Span 14,2 m (46 ft 7 in); length 15,775 m (51 ft 9 in); wing area 41,2 m^2 (443·5 ft^2).

ENGINES (1) Two RD-45F, (2,3) two VK-1.

WEIGHTS Empty (1) 7367 kg (16,241 lb), (2,3) 7460 kg (16,446 lb); fuel 2700 kg; loaded (1) 10 265 kg (22,630 lb), (2,3 max) 12 095 kg (26,664 lb).

PERFORMANCE Max speed, SL (1) 1040 km/h, (2,3) 1090 km/h (677 mph); at 10 km (all) 994 km/h (618 mph); climb (1) 2·3 min to 5 km, 5·65 min to 10 km; service ceiling (1) 15 km, (2,3) 15,5 km (50,850 ft); range 1100 km (2,3 with drop tanks) 1950 km (1,212 miles); TO run (1) 610 m, landing run (1) 770 m.

I-330 (SI-02)

MiG-17, SI MiG OKB decided in January 1949 to transfer substantial part of aerodynamics personnel to plan improved S designated **SI (I-330)** as successor to MiG-15 rectifying faults. Authorization to build SI March 49, funding for SI-1 (static test) and two flight articles, SI-2 and -3, also dubbed **MiG-15bis-45°**. SI-2 completed last week of 49 and flown by I. T. Ivashchyenko 13 Jan 50. Good results, no wing drop and on 1 Feb reached 1114 km/h in level flight at 2,2 km, equivalent at ruling temp to Mach 1·03. On 20 March SI-2 dived into ground, killing Ivashchyenko. Two further prototypes authorized, **SI-01, -02**, SI-3 not flown and SI-02 second to fly, preceding SI-01.

Wing profile S-12s at root thinning to SR-11 at tip, t/c reduced by increasing chord at root from 2,75 to 3,2 m TE sweep zero to second rib) and then sweeping TE 40° to tip with chord increased from 1,5 to 1,7 m, MAC (mean aerodynamic chord) 2,19 m, sweep at $\frac{1}{4}$-chord 45°, LE sweep 49° to mid-span, then 45·5° to tip, anhedral 3°, inner fence moved inboard and

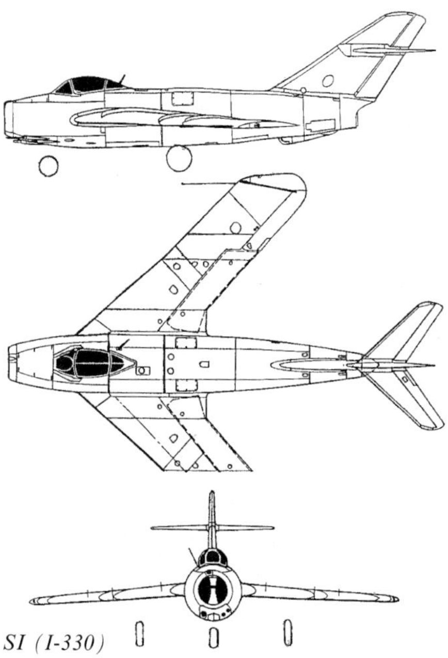

SI (I-330)

third fence added outboard ahead of aileron. From SI-02, pitot boom on both tips instead of inboard on right only. Ailerons improved and enlarged to 0,8 m^2 with tab on right and BU-1U power units. Fowler flaps, again with perforated upper surface, enlarged to 2,86 m^2 and driven to 20° TO and 60° landing. Airbrakes enlarged again to 1,76 m^2. Vertical tail enlarged to 4,26 m^2, horizontal to 3,1 m^2 and swept 45° on LE. Taper of rear fuselage reduced by adding 0,9-m bay behind wing. Fuel capacity unchanged but improved multilayer rubber tank 1250 lit between ducts and metal tank 150 lit under jetpipe. Armament unchanged, underwing attachments for 240-lit tanks or bombs to 250 kg each. RSIU-3 *Klen* vhf, RV-2 *Kristall* radar altimeter with two scales, ARK-5 *Amur* radio-compass, MRP-48 *Xrizantema* beacon receiver, SRO-1 *Barii* IFF, OSP-48 ILS, ASP-4 sight, S-13 ciné camera and EKSR-46 signal pistol.

Production in six GAZ ordered 1 Sept 51. Modifications during early production included change to improved VK-1A engine (same thrust), increase in span (SI-2 was 9,6 m), slight reduction in control-surface movements, addition of ventral strake (underfin) with tail bumper, change to OKB's third pattern of ejection seat with M-B type face blind, further slight enlargement of airbrakes, landing light moved from inlet to under left wing (retractable), introduction of one-piece sliding canopy with no rear frame to obstruct vision, 400-lit pressurized drop tanks and ability to carry tanks and bombs. Production improvements tested by Sedov, Anokhin, Kazmin and Kokkinaki, Mach 1·14 being reached in shallow dive under good control. Service limit initially Mach 1·03, raised April 54 to 1·15. Deliveries from Oct 52, c900 in Poland as **LIM-5** (mainly 5M, see later), most of 1,700+ exports were MiG-17F. Total all versions, excl licensees, c8,900. A few remain active in 13 AFs including USAF. ASCC reporting name 'Fresco'.

DIMENSIONS Span 9,628 m (31 ft 7 in); length 11,264 m (36 ft $11\frac{1}{2}$ in); wing area 22,64 m^2 (243·7 ft^2).

ENGINE One VK-1A.

WEIGHTS Empty 3798 kg (8,373 lb); fuel 1173 kg; loaded 5,2 t (11,464 lb), (max) 5932 kg (13,078 lb).

PERFORMANCE Max speed (SL) 1060 km/h (659

MiG-17 (testbed, two N-37D)

MiG-17PF (of LSK)

mph), (2 km) 1094 km/h (680 mph); climb 47 m/s, 2 min to 5 km, 5·1 min to 10 km; service ceiling 15,6 km (51,180 ft); range (hi, clean) 1295 km (805 miles), (two tanks) 2060 km (1,280 miles); TO run 700 m; landing speed/run 185 km/h, 875 m.

SF, MiG-17F Afterburning engine with large six-petal nozzle, variable 540-624 mm. Initially augmentation limited to 3 min up to 7 km, 10 min at greater altitudes. Modified plenum chamber and jetpipe bay to ensure structural cooling and to fuel system to provide afterburner fuel flow at all heights, with (from May 53) same negative-g supply tank as added to MiG-15. Cleared with 600-lit (132-gal) drop tanks. Further enlargement of airbrakes, with external blisters over actuating rams. Cutback rear fuselage causing longer nib fairing to match rudder of slightly increased chord. VK-1F first fitted to MiG-17 No 850, tested from 29 Sept 51 by A. N. Chernoburov. Deliveries from Feb 53, quickly replacing MiG-17. Early mod addition of negative-g tank (see above) followed Nov 53 by cold-air unit. Polish **LIM-5M**, Chinese **J-5**, Chinese two-seater **JJ-5**, **F-5** and **FT-5** for export. ASCC name 'Fresco-C'.

DIMENSIONS As MiG-17.
ENGINE One VK-1F.
WEIGHTS Empty 3919 kg (8,640 lb); fuel 1173 kg; loaded 5345 kg (11,784 lb), (max) 6075 kg (13,393 lb).
PERFORMANCE Max speed (SL) 1080 km/h (671 mph), (3 km with a/b) 1154 km/h (717 mph), (10 km) 1071 km/h (666 mph); climb (a/b) 65 m/s, 2·1 min to 5 km, 3·7 min to 10 km; service ceiling [prototype, still climbing 3,6 m/sec] 16,6 km (54,462 ft); max range (clean, no a/b) 1240 km, (two 400-lit tanks) 1980 km (1,230 miles); TO run (no a/b) 590 m, landing speed/run 185 km/h, 900 m.

SP-2, MiG-17F Single aircraft fitted with *Korshun* radar, devd by A. B. Slepushkin on basis of *Toryii*, much easier for single pilot to interpret. Other mods included two NR-23 (one 120, one 90 rds), airbrakes auto triggered at M1.03, and aft tank enlarged to 250 lit. Flown by Sedov 11 March 51. NII test Nov-Dec 51.
DIMENSIONS Length increased to 11,59 m (38 ft 0 in).
WEIGHTS Empty 4052 kg (8,933 lb); loaded 5320 kg (11,728 lb).
PERFORMANCE Max speed (3 km) 1109 km/h (689 mph); climb 2 min to 5 km, 5·2 to 10; service ceiling 15,2 km (49,900 ft); range (clean/tanks) 1375/2510 km (854/1,560 miles).

SP-7, MiG-17P Night and all-weather interceptor based on MiG-17 (not 17F). Nose redesigned to accomodate RP-1 *Izumrud* with antennas as SP-5/MiG-15Pbis. Armament revised to three NR-23 each with 100 rounds (two on left, one right). Prototype converted MiG-17 No 209, still with ASP-3N sight. Designated **SP-7**, first flight April 52, very swift development because radar previously flying in several aircraft. One block (believed 100) built for AV-MF, 20 being diverted to PVO. First light radar-equipped jet interceptor in service. ASCC name 'Fresco-D'.
DIMENSIONS, ENGINE As MiG-17 (not 17F) except length 11,68 m (38 ft 3⅞ in).
WEIGHTS Empty 4154 kg (9,158 lb); loaded 5550 kg, (two 400-lit tanks) 6280 kg (13,845 lb).
PERFORMANCE Max speed (SL) 1060 km/h, (3 km) 1115 km/h (693 mph); climb 2·5 min to 5 km, 6·6 to 10; service ceiling 14,5 km (47,575 ft).

SP-7F, MiG-17PF First flown Oct 52, basically MiG-17P powered by VK-1F. Among other modifications were improved afterburner heat shrouds, uprated GSR-6000 generator, Sirena 2 radar warning receiver (first installation) and NI-50B ground-position indicator. Block of 1,000 delivered to PVO units, later (56) all survivors returned to factory for conversion as **PFU** (see next). Also constructed under licence in Poland as **LIM-5P**, in Czechoslovakia as **S-104** and in China as **J-5A**. ASCC name 'Fresco-D'.
DIMENSIONS As MiG-17P.
ENGINE One VK-1F.
WEIGHTS Empty 4182 kg (9,220 lb); loaded 5550 (12,235), max 6280 (13,845).
PERFORMANCE Max speed SL 1050 km/h (652 mph), at 4 km 1121 (697); SL climb 55 m/s (10,830 ft/min); service ceiling 15,85 km (52,000 ft); range 1000 km (621 miles); TO 700 m, ldg 900 at 180 km/h.

SP-6, MiG-17PFU One series PF converted late 53 as **SP-6** with radar modifications and launch control boxes for K-5 (RS-2US) air/air missile. Two APU-4 launch rails under leading

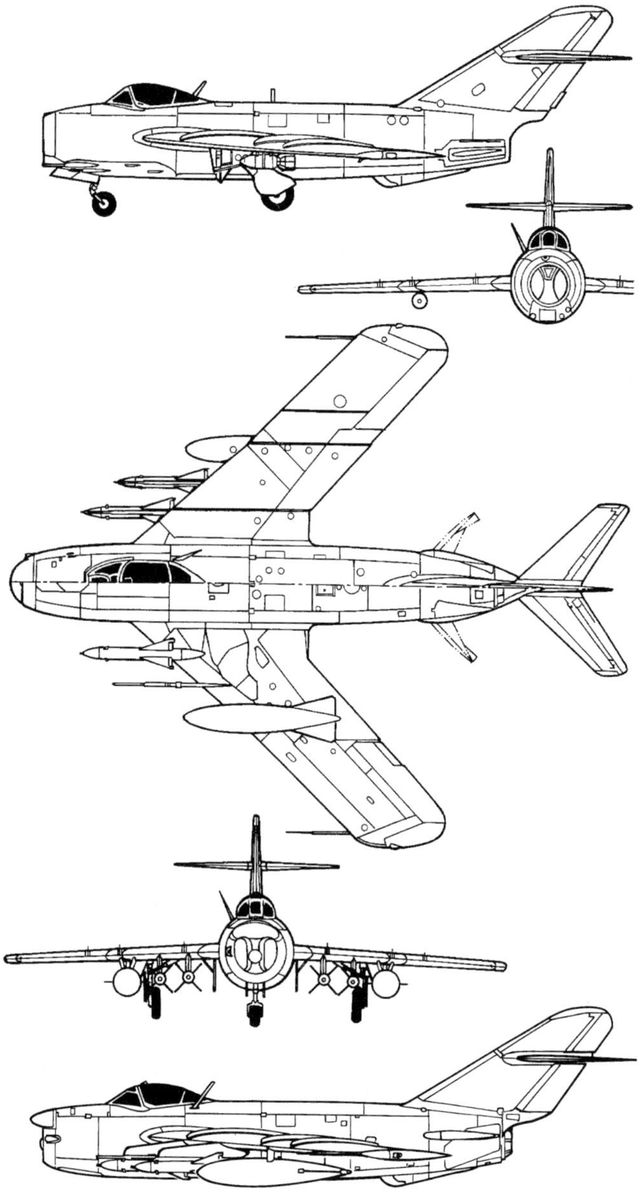

MiG-17PFM (upper side and front views, MiG-17F)

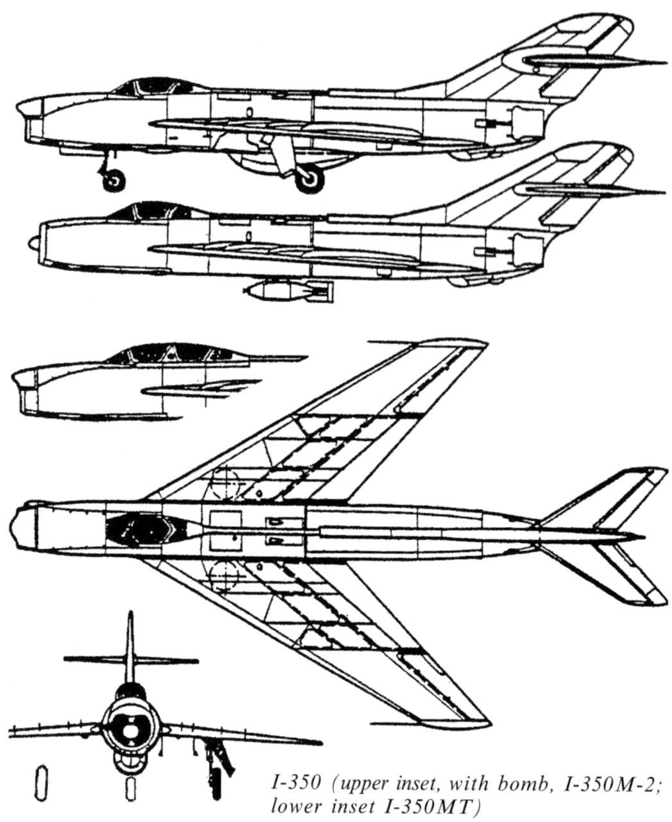

I-350 (upper inset, with bomb, I-350M-2; lower inset I-350MT)

SN

edge of each wing inboard of tanks. This prototype retained one NR-23 on right side, but gun deleted from production **PFU**. First fighter in Europe in production with missile armament, one block being delivered for PVO service from mid-55. APU-4 could equally launch ARS-160 or -212M rockets. All surviving PFs were brought up to same standard. Performance slightly inferior to PF. These aircraft initially equipped front-line regiments, but from 57 served as weapon-system trainers. ASCC name 'Fresco-E'.

MiG-17 variants Several prototype and series aircraft were used to test experimental armament installations. Most visually striking was **SN**, first flown mid-53. Like SU, this was an attempt to instal guns which could pivot in the vertical plane. Unlike its predecessor, SN was totally rebuilt from Frame 13 forward, lengthening fuselage by 1,069 m (42·1 in). Inlet ducts each fed by a plain half-size circular inlet just ahead of wing. This enabled SV-25 armament system to be installed, occupying entire nose ahead of cockpit. On left side were two TKB-495 guns of 23 mm calibre, superimposed on a large rotary mounting which extended across to carry a third TKB-495 on right. This mounting was driven via reduction gear by electric actuator to elevate guns from +27°26' to –9°48'. Three box magazines remained stationary. Mounting and drive weighed 142,4 kg, guns 117 kg, ammunition 139,7 kg and equipment brought total to 469 kg (1,034 lb). SN was also modified in canopy, main landing gear, instrument panel and increased internal fuel. Some 15,000 rounds fired from SV-25, significant combat advantages but also degraded aircraft performance, firing guns (except ahead) altered aircraft trajectory and other problems. Rocket armament trials from **SI-05, 07, 16, 19, 21, 21m** and **91**, and **SP-9** and **11. SP-8** tested *Grad* radar-ranging gunsight. Two prototypes of an experimental twin-barrel 23 mm gun tested under nose of **SP-10**, rebuilt MiG-17PF No 627, in 55. **SI-10**, originally MiG-17 No 214 modified with objective of enhanced manoeuvrability, re-flown Dec 54 with new wing without fences but with essentially full-span automatic slats, modified Fowler flaps and spoilers on undersurface ahead of flaps driven to 55° whenever aileron on same side depressed past 6°; also 'all-flying' tail with original elevators pivoted to tailplane driven by hydraulic power unit from –5°/+3°. As replacement for MiG-9L, **SDK-5** was MiG-17 prototype 007 rebuilt 51 with guns removed and guidance system of K-10 cruise missile installed, with forward-facing radar above nose and aft-facing radar above fin. On 20 Apr 51 permission granted to fit twin AM-5 engines side-by-side in original MiG-15bis45° prototype, which after conversion emerged Dec 51 as **SM-1**, NII designation **I-340**. Few changes except for new rear fuselage retaining almost full width to twin nozzles. Braking parachute (15 m²) added in blister under rear fuselage. Following initial testing by G. A. Sedov SM-1 retrofitted with AM-5F afterburning engines with larger inlet ducts, modified fuselage tanks (forward 1220 litres, aft 330) and reprofiled tail end over larger variable nozzles. Second twin-AM-5F prototype was original SI-2, redesignated **SM-2**, see later. Kremlin order 3 Aug 51 launched project for **MiG-17R** reconnaissance version, powered by single VK-5F; prototype, **SR-2**, fitted with AFA-BA-21s camera on vertical, oblique or gimbal mount, and MAG-9 tape recorder. Armament retained twin NR-23 each with 100 rounds, and hydraulic boost actuation of elevators. Flown June 52, but after two years of testing decision taken to fit same camera/recorder to a number of regular MiG-17F for line service. **SR-2s** was prototype of this aircraft, which in service was still designated MiG-17F. No ASCC reporting names. Data for SM-1:

DIMENSIONS As MiG-17.

ENGINES Two AM-5A (later AM-5F, data with these).

WEIGHTS Empty 3705 kg (8,168 lb); fuel 1240 kg; loaded 5210 kg (11,486 lb).

PERFORMANCE Max speed (1 km) 1193 km/h (741 mph), (5 km) 1154 km/h (717 mph); climb 0·94 min to 5 km, 2·85 to 10, 6·1 to 15; service ceiling 17,5 km (57,420 ft); range (15 km) 1965 km (1,221 miles); TO run 335 m; landing run 568 m.

M, I-350 OKB studied possible supersonic fighters from 48, and Kremlin decree 10 June 50 authorized **I-350**, powered by single TR-3A, a prototype of which was made available same year. OKB designation **M**, totally new wing, and fuselage greatly lengthened over predecessors to take long axial engine even without afterburner. Wing of CAHI form (G. S. Byushgens and team) with profile SR-12s at root reducing to SR-7s at tip, sweep 60° on leading edge, 57° at 25% chord, aspect ratio 8·6, 4° anhedral. Fixed LE, Fowler flaps, outboard ailerons and four full-chord fences on each wing. Remarkable that this high aspect ratio was compatible with normal ailerons, placed at wingtips. All flight controls hydraulically boosted. Nose virtually identical to MiG-17P family, with RP-1 radar and armament of N-37 plus two NR-23 (OKB drawing shows armament of two N-37D only). Tail generally as before but larger, horizontal surface 3,46 m span with LE sweep 55°. Small ventral fin incorporating bumper and compartment for braking parachute, then a novelty. Completely new rectangular airbrakes. Landing gear as MiG-17 but slightly larger and stronger and track increased to 4,47 m. Radar altimeter antenna moved to port wing root. OKB mentions a single drop tank of 800 litres. First flight by G. A. Sedov 16 June 51. Sedov soon returned, and engine failed as airfield was approached. This rendered hydraulics inoperative and Sedov performed miracle in retaining heavy manual control and extending landing gear by emergency air bottle just as wheels hit runway. Flight time nine minutes. Aircraft cancelled after five further short flights because of immaturity of engine. Unbuilt versions included **I-350M-2** with *Korshun* radar and **I-350MT** dual trainer. No ASCC name.

DIMENSIONS Span 9,73 m (31 ft 11 in) [OKB drawing incorrectly shows 9,48]; length 16,652 m (54 ft 7⅝ in); wing area 36 m² (387·5 ft²).

ENGINE One TR-3A (AL-5).

WEIGHTS Empty 6125 kg (13,503 lb); max loaded 8710 kg (19,202 lb).

PERFORMANCE Max speed (SL) 1240 km/h, (10 km) 1266 km/h (787 mph, Mach 1·19); climb 1,1 min to 5 km, 2·6 to 10; service ceiling 16,6 km (54,460 ft); range 1120 km (696 miles), with 800-lit tank 1620 km (1,007 miles).

SM-2, I-360 Several aspects of I-350 were encouraging, notably wing aerodynamics and lateral control. SM-1 (I-340) also confirmed potential of twin AM-5F engines, which suited future SM series supersonic fighter. From July 51 OKB devoted major effort to **SM-2**, headed by Brunov assisted by R. A. Belyakov, with section heads A. A. Shumachyenko (aerodynamics), D. N. Kurguzov (drag), G. Ye. Lozino-Lozinskii (propulsion), V. M. Merzuitov (flight control), N. I. Volkov (armament) and A. V. Minayev (systems). Designed to same Mach (1·19) as I-350 but higher overall performance because greater thrust in smaller airframe. Wing scaled from I-350 with reduced 25% sweep 55°, anhedral increased to –4°30'. Aspect ratio increased, ailerons divided inboard/outboard (inners with tabs), flaps modified, single shallow fence at 50% semi-span plus chordwise stiffener across inboard upper skin. Major effort to provide fully powered control in pitch with variable-incidence tailplane carrying separate elevators all driven by duplicated power units with feedback giving artificial feel. Ailerons and rudder boosted. After CAHI tunnel testing, horizontal tail near top of fin, span 2,78 m, LE 55° as before. Engine installation as I-340, though modified ‘pen nib’ nozzle fairing. Volkov moved two N-37D guns to wing roots, with 50-round magazines along LE, leaving nose free for retractable landing lights and other equipment and avoiding inlet flow distortion. Ventral strake housing braking parachute retained but dorsal spine not needed. Airbrakes as late MiG-17F but longer and of symmetric shape. Track 4,156 m, smaller than I 350; wheelbase 4398 (OKB drawing error 3980). Other effort devoted to improved ejection seat and fuel system. Two SM-2 funded, No 01 flown by Sedov 24 May 52 with AM-5, inadequate for Mach 1 on level until replaced late 52 by AM-5F. Chief problem blanketing of horiz tail by wing at high AOA; also buffet requiring redesign of rear fuselage and later relocation of airbrakes, poor lateral control requiring increase in depth of fences (and later addition of spoilers on MiG-19S), and, after firing guns in 53, adding steel skin near muzzles. Major mod to 01 and 02 in summer 53, move tailplane to top of fuselage, enlarged because of reduced moment arm. With new tail became **SM-2A**, later **SM-2B** after changes to fences, airbrakes and other parts. Failed NII testing March 53, but as SM-2B resubmitted July; reached M 1·33 (1400 km/h) in dive.

DIMENSIONS Span 9,04 m (29 ft 7⅞ in); length

I-340 (SM-1)

I-360 (SM-2 01)

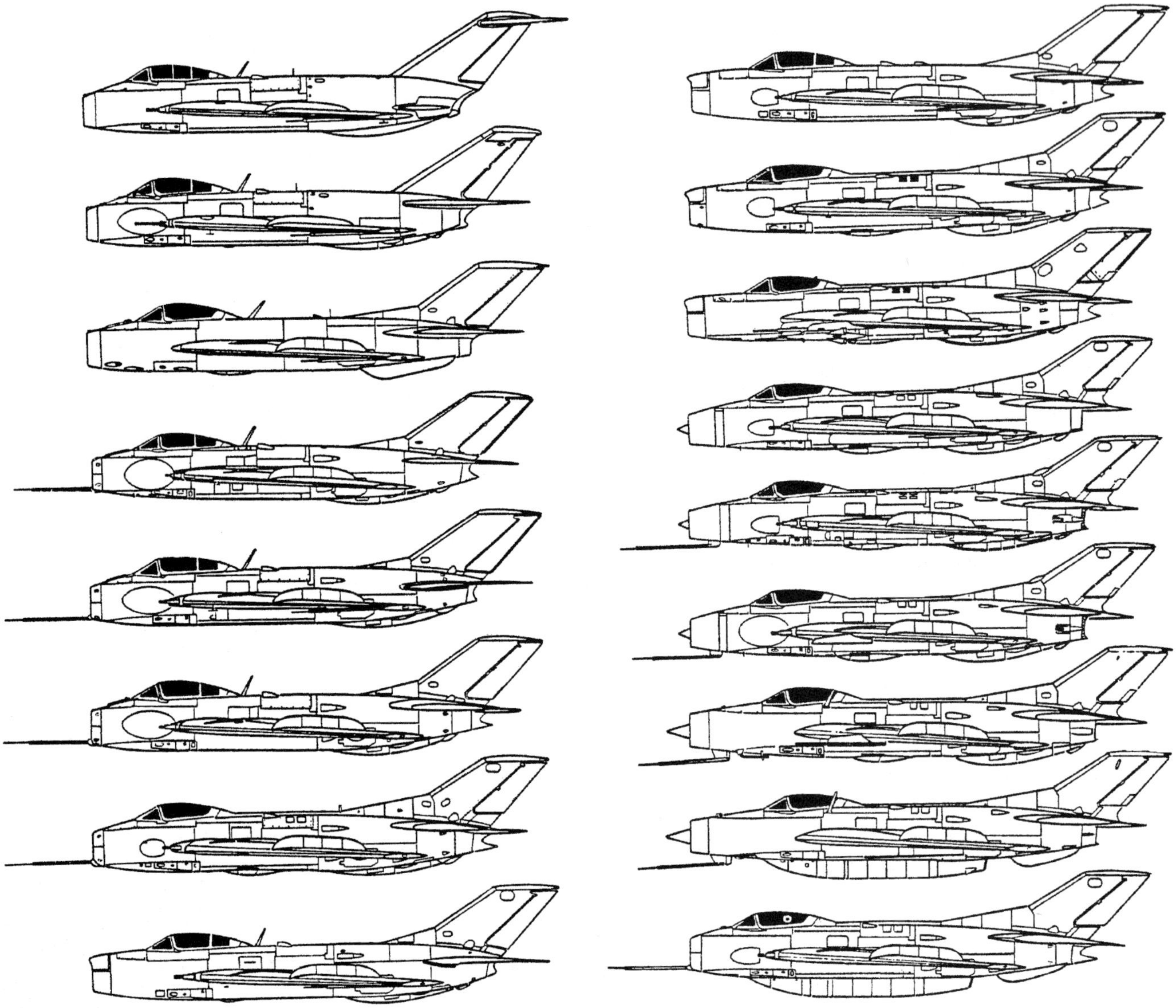

From top left: SM-2, SM-2/2, I-1, MiG-19, SM-9/1, SM-9/2, SM-9/3, SM-7/1, (Col. 2) SM-7/2, MiG-19P, MiG-19PM, 12/1, 12/2, 12/3, 12PM, 12PMU, SM-50

overall 13,9 m (45 ft 7¼ in), fuselage 10,285 m (33 ft 9 in); wing area 25 m² (269 ft²).
ENGINES Two AM-5, later 5F.
WEIGHTS Empty 4929 kg (10,866 lb); loaded 6820 kg (15,035 lb).
PERFORMANCE Max Mach on level, est 1·19.

SM-9/1, MiG-19 When SM-2B was re-engined with AM-9B (later called RD-9B) designation was changed to **SM-9/1**. Other mods included changed armament: two NR-23 in wing roots (replacing N-37D) and a third under right side of nose, all with 120 rounds, plus wing hardpoints for 250 kg bomb or ARS-57 rocket launcher inboard and 800-lit tank or ARS-57 outboard. Other changes improved access to engines and jetpipes, better cooling of engine bay (six, later eight, ram inlets for cold air above and below fuselage) and addition of long pitot boom at bottom of nose (soon made to pivot up on ground to avoid damage).

Wing as before, ruling thickness 8·24%, basic material D16-T with one extremely strong spar at 90° to fuselage axis running from left to right leading edges at mid-span and also carrying main gears. Spar booms and major joints 30KhGSA. Machined skins, max thickness 4,8 mm (more than double MiG-15 or MiG-17). Ribs and aileron hinges at 90° to leading edge. Hydraulic Fowler flaps running aft on two tracks with pneumatic emergency actuation. Hydraulically boosted ailerons with inset hinges and manual reversion, left aileron with trim tab. Fixed leading edge. Fuselage made in front and detachable rear sections with joint immediately behind trailing edge. Inlet duct bifurcated at nose to pass each side of cockpit and above and below aerofoil fairing main spar. Twin hydraulic airbrakes on flanks aft of wing. Entire rear fuselage removable for access to engines, ventilated by four upstream and four downstream ram inlets standing proud of skin. Ventral keel and tail bumper. Vertical and horizontal tail based on MiG-15LL but with increased areas and control power. Wide track (4,15 m) landing gear with levered-suspension legs. Production tyres 660×220 mm (main), usually 883 kPa (129 lb/in²), 500×180 mm (nose). Main gears inwards, non-steerable nose gear forwards, hydraulic with pneumatic emergency. Pneumatic brakes with back-up system. Braking parachute pneumatically released from bay beneath afterburners. Duplicate hydraulic systems, 207 MPa (3000 lb/in²) with main system from right engine serving gear, flaps and engine nozzles and left engine handling flight-control with back-up from right. Production aircraft, two main rubberized-fabric fuel cells aft of cockpit and two shallow curved cells under jet-pipes, total 2170 lit (477 gal). Provision for 800-lit (176 gal) drop tank under each wing. Cockpit with faceblind ejection seat, reflector sight with depressed reticle for air/ground, environmental system packaged in upper mid fuselage behind tanks, fluid windshield anti-icing, aft-sliding canopy later fitted with electro-thermal anti-icing. Standard eqpt

ASP-5M sight, RSIU-4 *Doob* vhf, ILS receiver OSP-48, gyromagnetic compass DGMK-5, radio compass ARK-5, radar altimeter RV-2, IFF SRO-1 and ciné camera S-13.

Flown by Sedov 5 Jan 54. Still imperfect, SM-9/1 clearly worldbeater, with record climb (180 m/s, 35,430 ft/min) and production ordered 17 Feb 54 as **MiG-19**, before NII testing started (30 Sept 54). GAZ-21 and 153 delivered initial 50 aircraft, issued to inventory from March 55, thereafter switching to MiG-19S. ASCC reporting name 'Farmer'.

DIMENSIONS, ENGINES As MiG-19S, see next.

WEIGHTS Empty c5090 kg (11,220 lb); fuel 1800 kg; loaded 7360 kg (16,226 lb).

PERFORMANCE Max speed (SL) 1150 km/h (715 mph), (10 km) 1451 km/h (902 mph, Mach 1·367); climb 1·1 min to 10 km, 3·7 to 15 km; service ceiling 17,5 km (57,415 ft).

SM-9/2, 9/3, MiG-19S Courageous testing by Sedov, Kokkinaki and V. A. Nyefyedov demonstrated danger of longitudinal control, which (e.g.) could augment unwanted pitch excursions and induce violent phugoids. In Jan 54 two new aircraft authorised. **SM-9/2**, fully powered 'slab' tailplane, mounted higher on modified rear fuselage with different pen-nib between nozzles. Together with ailerons (see later) driven by irreversible BU-14MS power units served by main hydraulic system, standby engaged should main system fall below 65 kg/cm². Third fallback, with reduced authority, electric screwjack activated should pressure fall below 50 kg/cm² (711 lb/in²) or by stick button. Devt led by Minayev produced ARU-2 or -2A pitch control to give carefree handling at any dynamic pressure (IAS/altitude) and Mach, with optimised feedback. Each tailplane mounted on pivot at 90° to longit axis outboard of broad fixed root, with anti-flutter mass ahead of tip. To increase roll-rate at high speeds ailerons powered by BU-13M actuators and narrow-chord spoiler added on upper surface ahead of each inboard and outboard aileron, driven by same powered linkage. Airbrakes again changed, with third brake added well forward under ventral centreline causing bulge. Fuel 2155 lit internal plus two 760-lit tanks. Twin pneumatic (air-bottle) rams added to ensure positive ejection of canopy. Chief eqpt addition SRD-3 *Grad* radar-ranging sight with antenna at top of nose next to camera.

SM-9/2 flown by Sedov 16 Sept 54, together with Kokkinaki, Nyefyedov and G. K. Mosolov soon pronouncing it virtually perfect. Second, **SM-9/3**, first flown by Kokkinaki 27 Nov 55. Minor changes to tailplane control, bottom of rudder raised, dorsal fin enlarged and linked to canopy by spine, canopy frameless and armament changed to three devastating NR-30, though steel blast panel on fuselage smaller. Both aircraft dived beyond Mach 1·46, service limit 1·44. With minor changes production as **MiG-19S** (*Stabilizator*) authorized at same plants at Gorkii and Novosibirsk, 2,120 built. ASCC 'Farmer-C'.

DIMENSIONS Span 9,0 m (29 ft 6⅜ in); length overall 14,8 m (48 ft 6⅝ in), (excl air-data boom) 12,54 m (41 ft 1¾ in); wing area 25,16 m² (271 ft²).

ENGINES Two AM-9B (RD-9B).

WEIGHTS Empty 5172 kg (11,402 lb); fuel 1800 kg; loaded 7560 kg (16,667 lb), (max with two tanks and two rocket launchers) 8832 kg (19,471 lb).

PERFORMANCE Max speed at SL 1100 km/h (684 mph), (10 km) 1454 km/h (903 mph, Mach 1·367); climb 115 m/s (22,600 ft/min), 0·4 min to 5 km, 1·1 to 10, 2·6 to 15; service ceiling 17,5 km (57,415 ft); range (14 km) 1390 km (864 miles), (with two tanks) 2200 km (1,367 miles); TO run (with afterburner) 515 m, (dry) 650 m, (dry, two tanks) 900 m; ldg speed/run 235 km/h, 1090 m, (braking on all 3 wheels) 890 m, (plus parachute) 610 m.

SM-7, MiG-19P It was natural to develop radar-equipped SM for service as **MiG-19P**. First prototype, **SM-7/1**, based on SM-9 with original tailplane, with full-depth rudder, small dorsal fin and no spine. Apart from low S 13, nose similar to predecessors with RP-1 *Izumrud* radar, with extra 360 mm (14·2 in) bay ahead of frame 9. Forward tank slightly reduced in size. Armament one NR-23 in each wing root with 120-round belt along LE, four wing hardpoints for tanks or launchers as before. Air-data boom removed from nose and added to both wingtips, that on right serving nav instruments. *Sirena* receiver added at tip of fin, other features unchanged. SM-7/1 flown by Nyefyedov 28 Aug 54. Second, **SM-7/2**, flown 55 with two NR-30 and slab tailplanes, pen-nib and rudder of MiG-19S, but without enlarged dorsal fin or spine. Simiar aircraft but with dorsal fin and spine built in series as **MiG-19P** from late 55 for PVO, FA and AV-MF. ASCC 'Farmer-B'.

DIMENSIONS, ENGINES As MiG-19S except length (excl boom) 12,9 m (42 ft 3⅞ in).

WEIGHTS Empty 5468 kg (12,055 lb); fuel 1700 kg; loaded 7384 kg (16,279 lb), (with two 750-lit tanks) 8738 kg (19,264 lb).

PERFORMANCE Max speed (SL) 1180 km/h (733 mph), (peak, 9,8 km) 1384 km/h (860 mph, Mach 1·3); climb (lighting a/b from 5 km) 1·85 min to 10 km, 3·8 min to 15 km; service ceiling (a/b) 17,6 km (57,745 ft), (dry) 16,15 km; range 1474 km (916 miles), (two tanks) 2318 km (1,440 miles).

SM-7M, MiG-19PM Already tested on other aircraft, K-5 and K-5M missile system was planned for night and all-weather interceptors including single-seaters. First of SM family was **SM-7/M**, authorized Jan 56. Apart from replacing guns by four APU-4 launch rails, major change was RP-2 *Izumrud* 2 radar, search/track/guidance over forward hemisphere to 60° left/right and in elevation +26°/–14°, attack aid ASP-5N optical sight at range below 4 km. Nose deeper to accomodate search and ranging antennas with latter having larger upper dielectric area. Air-data probes on both tips, each able to serve flight and nav instruments giving dual redundancy. Usual compass replaced by GIK-1 Earth-inductor type. Missiles launched by pilot command, using stick button, and guided using PUVS-52 control panel. Production K-5M, service RS-2US, first in Europe cleared for operational use. APU-4 rails could also launch 160 or 212 mm rockets. Fuel 2130 lit internal plus two 760 tanks, but because of weight limitations with four missiles or rockets tanks on outer pylons limited to 400-lit. SM-7/M, with original tail, flown late Jan 56 and built in series for PVO as **MiG-19PM** from mid-57. Second prototype with slab tailplanes, **SM-7/2M**, began testing with K-5M 14 Oct 57; this continued production as **MiG-19PMU** from early 58, total for both versions c400. ASCC 'Farmer-D'.

DIMENSIONS, ENGINES As MiG-19P.

WEIGHTS Empty 5495 kg (12,114 lb); loaded 7730 kg (17,041 lb), (two 760-lit tanks) 9100 kg (20,062 lb).

PERFORMANCE Max speed at SL 1100 km/h, at 10 km 1250 km/h (777 mph, Mach 1·177); climb 105 m/s (20,500 ft/min), to 15 km (a/b) 4·8 min, (dry) 7·2 min; service ceiling a/b 16,7 km (54,790 ft), (dry) 15 km (49,215 ft); range (10 km) 1000 km (621 miles), (two 400-lit tanks) 1415 km (879 miles), (two 760-lit tanks) 1910 km (1,187 miles); TO 900 m, ldg speed/run 235 km/h, 890 m.

SM-9V, MiG-19SV OKB launched programme Aug 55 under A. V. Arkhipov to modify SM (built as MiG-19S) to engage PR Canberra, balloons and other high-flying aircraft then violating Soviet airspace. In parallel reqd sealed pressure suit and helmet, devd as VSS-04A. Crash programme got **SM-9V** in air two months from start, with AM-9BF engines, span increased, wing guns and armour removed and KKO-1 pressure oxygen system. Sedov, Nyefyedov, Kokkinaki and Mosolov quickly completed OKB tests of aircraft and suit, clear-

SM-9/2 with long inlet and short rudder

MiG-19PMU

ing 10° flap with full power at extreme altitude. SM-9V considered easiest of SM family to fly. In NII test on 6 Dec 56 N. I. Korovuchkin recorded 20,74 km (68,045 ft). Suit cleared to 24 km. Limited production as **MiG-19SV**, issued to PVO regiments at strategic points.

DIMENSIONS Span 10,3 m (33 ft 9½ in); length (with boom) 14,64 m (48 ft 0⅜ in); wing area 27 m² (290·6 ft²).

ENGINES Two AM-9BF or BF-2.

WEIGHTS Empty 5058 kg (11,151 lb); loaded 7250 kg (15,983 lb).

PERFORMANCE Max speed (10 km) 1420 km/h (882 mph, Mach 1·336); service ceiling 19 km (62,340 ft).

SM-20

SMR, MiG-19R Four OKB aircraft tested recon installations from 1954, leading to small run of **MiG-19R**; MiG-19S with BA-40 camera and MAG-9 recorder. No ASCC name.

MiG-19UTI OKB designed tandem dual trainer with rear cockpit at same level as front, replacing No 1 tank. Curved top to side-hinged canopies. Retained guns and pylons. Six built, evaluated by VVS but no production, partly because of short endurance and poor instructor view. No ASCC name.

MiG-19 variants Over 50 aircraft (mostly MiG-19S in =400/450 bracket) used, mainly at OKB, in experimental work and trial installations. From late 54 KT-37 main wheels replaced by KT-61 with disc brakes and, from 57 with modified legs, improved KT-87. Nosewheel brake changed to disc; many service aircraft re-equipped. In early 55 work began with Yakovlev OKB on air refuelling, using probe on left wingtip. Trials with **SM-10**, modified #415, completed with dry and wet contacts from Tu-16N in 56, oxygen supply augmented to 18 lit for 10-hr mission with two refuellings. Lack of funds prevented adoption in service. Other 55 trials, S-21 rockets (210 mm calibre) on **SM-21** (#406), and radio nav systems (including *Svod* and *Gorizont*, both using ground stations) on **SM-52P**, modified MiG-19P, which later tested *Almaz* (diamond) radar. Similar tests in 1956 included *Ilem* ADF, trying five antennas, RUP-4 blind landing and NI-501M ground-position indicator. From Oct 56 MiG-19S flew as **SM-20** to develop radio guidance for OKB's K-20 cruise missile, with guidance receiver, whip antenna under nose, position beacon and other eqpt, forged lugs recessed into each side above fuselage at CG for suspension below Tu-95N, with steadying pads behind canopy. Problems with high-altitude engine start led to second aircraft, **SM-20/P**, converted 57 with petrol/carburettor supply to ensure light-up. In 57-58 two more, **SM-K/1** and **SM-K/2**, modified to develop guidance for supersonic K-22. From April 55 Zell (zero-length launch) **SM-30**, original prototype SM-9/1 modified for transport on PU-30 off-road trailer which could elevate to 15°. Under rear fuselage PRD-22 solid rocket, thrust for 2·5 sec peaking at 40 t, launch acceleration reaching 4·5 g. Mods included replacing ventral strake by two strakes at inclined angle, strengthening five fuselage frames, inlet skin and tank No 2, mount for tanks 2 and 3 and wing/fuselage attachment, adding a control lock released after 3 sec of flight (with pilot override) and headrest behind pilot's head. In view of apparent danger first three launches without pilot. NII pilots Anokhin and Chiyanov chosen, latter making first manned firing 13 Ap 57 and former first max-wt launch (760-lit tanks plus rockets) on 3 June. Later system matched with arrester cables, using free parachutes or rotary brakes giving 2 g deceleration within 120 m (394 ft) for off-airfield landing, tested with MiG-19SV.

Another 57 programme to increase speed modified inlet with sharp lip and two-position conical centrebody carried on fixed portion joined to duct bifurcation. This gave increasing gains in pressure-recovery as Mach exceeded unity. First fitted to **SM-12/1** early 57, giving predicted results with RD-9BF-2 engines; loaded wt 7695 kg, max speed 1930 km/h (1,199 mph, M1·817). Two more built mid-57, **SM-12/2** and **12/3**, with R3-26 engines and new flight controls. New engine-bay heat shrouds and redesign of rear fuselage with nozzle petals visible and nib fairing removed. Left air-data boom moved to bracket far back under nose. Tailplane power quadrupled by BU-14MSK power unit (blister fairing at kink of fin LE) with fast-response APS-4MD electric valve drive; likewise roll increased by BU-13MK power units. SM-12/3 had two APU-4 pylons normally carrying ARS-57/8 rocket launchers,

SM-30

and enlarged blast skins beside guns. Fourth aircraft, **SM-12PM**, completed late 57 with APU-4 rails for K-5M (RS-2US) missiles, no guns, magazines or blast skins. To guide missiles inlet centrebody enlarged to house antenna of RP-21 (TsD-30) radar, requiring major enlargement of fuselage ahead of cockpit. High-power tail drive not fitted, but air-data boom mounted with negative incidence on enlarged pylon below nose. Performance well below SR-12/3, max speed 1720 km/h and climb to 10 km taking 4 min. Fifth aircraft **SM-12PMU**, built early 58. Powered by R3M-26, this had in addition U-19D rocket pack as devd with aircraft built at GAZ-21, **SM-50**, **SM-51** and **SM-52**, in programme managed by Yu. N. Korolyev (no relation to S. P. Korolyev). These were basically MiG-19S with original inlets and AM-9BM engines (some had guns removed, fairings remaining). U-19 pack, weighing 900 kg (1,984 lb), scabbed against underside of fuselage, comprised streamlined nose and tail joined by upper beam attached to aircraft. Main section, removable pressurized tanks for propellants (see Engines) with pump feed to RU-013 engine at rear. Subsidiary systems electric start/ignition and final overboard dump and purge. U-19 could be used once only on each flight, but pilot could select either of two thrust levels. At higher level max speed at high altitude, 1800 km/h (1,118 mph, M 1·695), ceiling 24 km (78,740 ft). **SM-51** was MiG-19P with R3M-26 engines and U-19D able to be shut down and restarted five times per flight. **SM-52** was MiG-19P with U-19D and *Almaz* radar. **SM-12PMU** fitted with R3M-26 and U-19D but missile system restricted speed to same (1720 km/h) as PM; used 59 to develop guidance systems, SOD series ATC/SIF and RV/U radar altimeters.

Final test programmes were 59 devt of two air/air missiles. K-13 launched from APU-13 rail attached to universal (outboard bomb/tank) pylon. Test aircraft was SM-9/3 modified as **SM-9/3T** first flown 11 Feb 59 by A. V. Fedotov. Explored safety checks, gas ingestion and

SM-12PM

MiG-19SU (SM-50)

carriage to M 1·245. Second missile testbed was **SM-6**, used 59 to develop K-6 launched from outboard universal pylon with short APU-26 shoe; ciné camera in pod under inboard LE.

Total production of SM family in Soviet Union about 3,700. Polish designation of 19S LIM-7. Similar airframe built by Aero Vodochody, with Soviet engines and equipment, as S-105. Many derivatives later built in China as J-6, from which was developed JJ-6 dual trainer unrelated to OKB version.

MiG-19 derivatives From 55 OKB produced prototypes derived from SM with designations in I-1/2/3 series. These are dealt with next, leaving larger I-7/75 and Ye-150 series to follow MiG-21.

I-1, I-2, I-370 Authorised May 53, I-370, OKB designation **I-1**, matched SM wing with single VK-7 engine. Inlet and ducts smaller

I-1 (I-370)

than MiG-19, but plenum chamber reduced internal fuel to 2025 lit (445·5 gal). Canopy (single forged frame) and three airbrakes linked to seat faceblind, air-data booms on both wingtips, powered tailplane with camber-increasing elevators, fuselage and vertical tail based on SM-9 except around engine, and twin inclined rear ventral strakes. Armament one N-37D and two NR-23 but with ASP-4N sight with *Radal* radar ranging. Initial engine 3525 kg dry and 5235 kg with a/b. First flown by Fyodor I. Burtsyev 16 Feb 55, reaching same 1452 km/h speed as MiG-19S. Later refitted with VK-7 with TET 800C, 4200 kg dry and 6270 with a/b, when reached Mach 1·334.

To go faster, refitted 1956 with wing swept 60° LE, 57° at 25%, basically as I-350 but no fences. Span unaltered and little change in area. Modified **I-2** failed to reach calculated performance.

DIMENSIONS Span 9,0 m (29 ft $6\frac{3}{8}$ in); length overall 12,7 m (41 ft 8 in); wing area 25 m^2 (269·1 ft^2).

ENGINE One VK-7.

WEIGHTS Empty 5086 kg (11,213 lb); loaded 7030 kg (15,498 lb); max with two tanks 8300 kg (18,298 lb).

PERFORMANCE (Original engine) max speed (10,8 km) 1452 km/h (902 mph); climb 1·15 min to 5 km, 3 min to 10 km; service ceiling 17 km; range (hi, two tanks) 2500 km (1,553 miles); TO run 464 m; landing speed/run 180 km/h, 730 m.

I-3, I-380 Front-line fighter ordered by Council of Ministers 3 June 53, hardly any part common to earlier types apart from wing similar to I-2. Sweep 60° at 25%, which this time was allowed to reduce span. Circular inlet with central conical 2-position spike. Large Fowler flaps with kink at inboard TE. Ailerons and slab tailplanes irreversible BU-13MK and BU-14MSK power units, rudder boosted, ram cooling and q-feel inlet ahead of base of fin, single ventral strake with compartment for 15 m^2 braking parachute, two airbrakes on flanks of fuselage behind wing TE. New main gears with wheels pivoted during retraction to remain upright in fuselage. Three NR-30, one left wing root and two sharply staggered on right, each with 65 rounds, breeches and magazines ahead

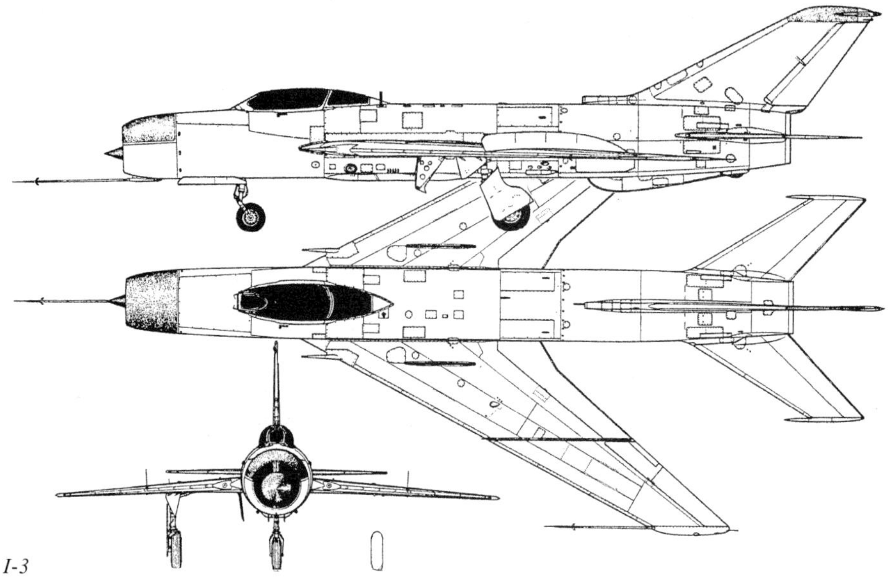

I-3

of front spar. Novel canopy blown Plexiglas with inserted bulletproof front screen all hinged as one unit from front, opening up and forwards by pneumatic ram, jettisoning phased with seat to protect pilot from blast. Engine never fitted; aircraft converted 56 into I-3U.

DIMENSIONS Span 8,978 m (29 ft $5\frac{1}{2}$ in); length overall 14,83 m (48 ft $7\frac{7}{8}$ in); wing area 30 m^2 (322·9 ft^2).

ENGINE Intended one VK-3.

WEIGHTS (Est) empty 5485 kg (12,092 lb); loaded 7600 kg (16,755 lb); max (two tanks) 8954 kg (19,740 lb).

PERFORMANCE (Est) Max speed (SL) 1274 km/h, (5 km) 1311 km/h, (10 km) 1775 km/h (1,103 mph); climb 0·81 min to 5 km, 1·9 min to 10 km; service ceiling 18,8 km (61,680 ft); range 1365 km (848 miles).

I-3P Parallel project for all-weather interceptor using I-3 airframe, to carry *Almaz* radar matched to twin NR-30 guns, also to carry 16 ARS-57 or two TRS-190 rockets. Est empty/gross 6276/9790 kg. Replaced late 54 by I-3U.

I-5, I-3U (NII I-410). Original I-3 fuselage lengthened from 12 275 m to 13 205 m, wheelbase from 5040 m to 5350 m, internal fuel unchaged. Rear fuselage first at OKB to incorporate titanium, mounted two airbrakes able to hold vert dive at full power. Nose redesigned to accomodate *Uragan-1* (hence I-3U) system comprising *Almaz* search radar above circular inlet with tracking dish in centrebody cone well below actual centre; search/track to 17 km, auto ranging of guns. ASP-5M sight, OKB-857 computer and AP-36-118 autopilot. Symmetric armament of two NR-30. Three air-data booms, one close under inlet. Conversion complete 56 but VK-3 never cleared for flight.

DIMENSIONS As I-3 except length (excl boom) 15 785 m (51 ft $9\frac{1}{2}$ in).

WEIGHTS (Est) empty 6447 kg (14,213 lb); loaded 8500 kg (18,739 lb); max (two tanks) 10 028 kg (22,108 lb).

PERFORMANCE (Est) Max speed 1960 km/h; service ceiling 18 km; range 1290 km.

MiG-21 Possibly most famous of all Soviet air-

I-3U/I-5 (I-410)

craft, this light fighter was designed around new engine and with two types of wing, swept and delta. Initially led by Mikoyan, later by Belyakov, 9,000+ aircraft of 32 types in USSR 58-80, also built Czechoslovakia, India and China.

Ye-2 (X-2) Requirement based on Korean experience issued autumn 53 for light fighter to reach M 2 at 20 km whilst carrying limited radar, gun(s) and AAMs with secondary ground-attack weapons and operate under Markham (RSIU) ground control; similar US requirement led to F-104A. First study **Ye-1 (X-1)**, Jan 54, swept-wing aircraft powered by AM-5A, later AM-9B, changing designation to **Ye-2, NII X-2**. Wing aerodynamically related to MiG-19 but scaled down, LE 55°, thickness 6%, two spars, modified two-part ailerons linked with undersurface spoilers, two-part slats over outer LE but no fences. Torque box integral tank with main gears attached to rear face and to aft false spar carrying flaps. Fuselage cross-section dictated by engine duct and seated pilot. Sharp-edged circular nose inlet with conical centrebody, duct bifurcated around cockpit, quickly resumed circular section. No aux cooling inlets but ejector nozzle to induce secondary flow. Front/rear fuselage divided at Frame 28 for engine access or removal. Same frame carries central airbrake, two other airbrakes under fuselage level with LE (around guns). Four tanks around duct and engine separated by firewalls with flame traps. Faceblind seat linked to forward-hinged canopy similar to I-3U. Slab tailplanes, span 3726 with anti-flutter tip masses, mounted high on rear fuselage driven by irreversible BU-44 power unit in fin. Blister each side of fin for compass magnetometer. Rudder merely boosted, all tail controls via push/pull rods in spine linking canopy to fin. Castoring nose gear retracting forwards between ducts, main gears as I-3. Two NR-30 in bottom of forward fuselage fed by 60-round belts around space between duct and skin, cases ejected but links returned to magazine. ASP-5N sight with *Radal* ranging, RSIU-4 vhf, *Uzel* homer, MRP-48 ILS, IFF and radar warning. First flight by Mosolov 14 Feb 55. Satisfactory but underpowered, and programme desultory pending AM-11 engine. ASCC 'Faceplate'.

DIMENSIONS Span 8,109 m (26 ft 7¼ in); length (excl boom) 13,23 m (43 ft 4⅞ in); wing area 21 m² (226 ft²).

ENGINE One AM-9B.

WEIGHTS Empty 3687 kg (8,128 lb); fuel 1360 kg; (loaded 5334 kg (11,759 lb).

PERFORMANCE Max speed (est) 1920 km/h (1,193 mph, Mach 1·8); service ceiling about 17 km (55,775 ft); TO 700m; landing 250 km/h, 800 m.

Ye-4 Enthusiasm of OKB's Pyotr Krasilshchikov and research by CAHI resulted in decision to build two examples of second design with almost unchanged fuselage and tail but wing of delta (triangular) form, called *balalaika*, LE 57°, S-9s profile, anhedral –2°. Balanced one-piece ailerons driven by BU-45 power units, rectangular Fowler flaps, no fences. Main gears modified to retract into triangular space between swept main and transverse rear spars. Few other changes from Ye-2 except later *Konus* ranging radar. First flight by Sedov 16 June 55. Tinkering with wing added two large full-chord fences on underside between flap and aileron, followed (mid-Aug) by replacement of these by three shallower fences on upper surface, outers extending to LE of aileron and others similar length thus extending round LE. Aircraft finally flown with tips removed, see dimensions. Eventually good behaviour to AOA 25+°. ASCC 'Fishbed'.

DIMENSIONS Span (as built) 7,749 m (25 ft 5 in), (later) 7,149 m (23 ft 5½ in); length (excl boom) 13,23 m (43 ft 4⅞ in); length (fuselage only, excl cone) 11,737 m (38 ft 6 in); wing area 23,15 m² (249·2 ft²).

ENGINE One AM-9B.

WEIGHTS Empty 3500 kg (7,716 lb); loaded about 5200 kg (11,464 lb); max (centreline 400-lit tank) 6200 kg (13,228 lb).

PERFORMANCE Max speed 1926 km/h (1,200 mph, Mach 1·81); climb 1·6 min to 5 km; service ceiling 16,4 km (53,800 ft); range (clean) 1120 km (700 miles).

Ye-5, X-5, MiG-21 From outset AM-11 was seen as definitive engine, and since good and

Ye-2 (X-2)

Ye-4 (3 fences)

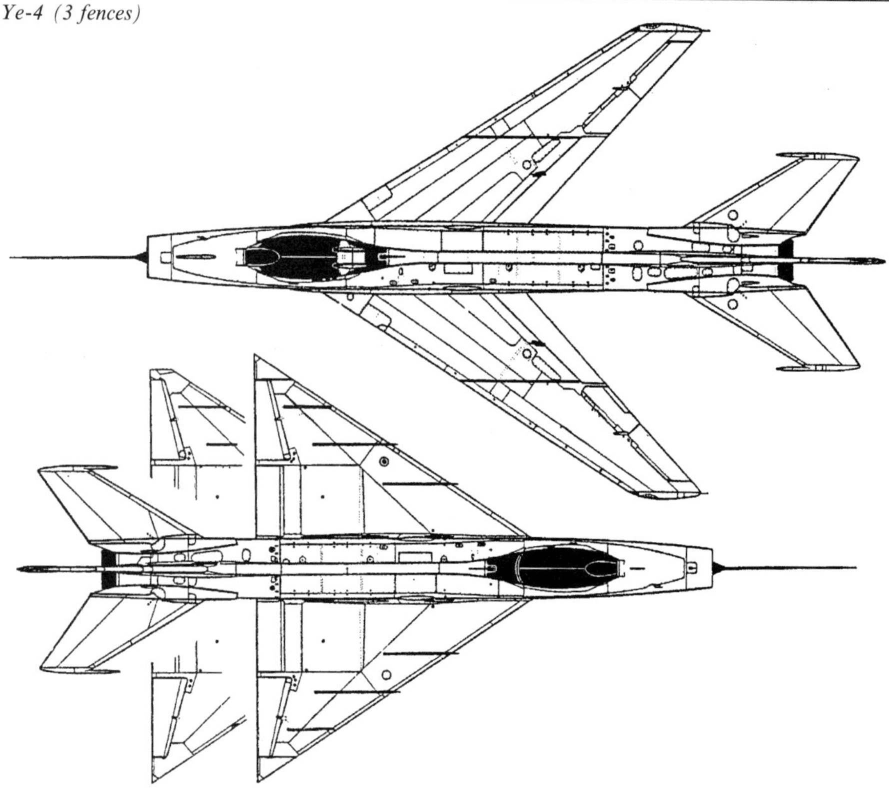

Upper Ye-2A; lower Ye-5 (upper inset wingtip, Ye-4/RD-9I; lower inset, Ye-4/AM-9)

bad features for both types of wing, OKB (GAZ-155) built four further prototypes, two with each wing. First delta, **Ye-5, NII X-5**, had previously been Ye-4/2 and was finished first. Fuselage slimmer, so mainwheel bays caused blisters, but rear fuselage enlarged in diam but shorter. New engine bay with some titanium, improved heat shrouds and ram cooling inlet on each side faired into fixed roots of tailplane, which now oversailed nozzle. Wing three short fences but pointed tips. First flight by Nyefyedov 9 Jan 56. Severe engine fire on ground 20 Feb, eight flights between 26 March and 19 May, then turbine failure. AM-11 (R-11) grounded 18 Oct, fuselage aft of Frame 28 returned to GAZ-155 to be lengthened 0,4 m to shift CG to rear and increase fuel from 1570 to 1810 lit. This helped to counter higher than expected fuel burn. R-11 immature, at least ten being needed to complete contractor flight-test programme. Greater fuel capacity, higher rate of roll, generally better turn radius, lighter structure and marginally better supersonic performance than equivalent swept-wing aircraft, just carried day in favour of delta. GAZ-21 built five further aircraft to this standard, called MiG-21. Some documents call Ye-5 family I-500. ASCC 'Fishbed'.

DIMENSIONS Span 7,749 m (25 ft 5 in); length (as built, excl probe) 13,23 m (43 ft 4⅞ in), (after rebuild, excl probe) 13,46 m (44 ft 2 in); wing area 23,15 m² (249·2 ft²).

ENGINE One AM-11 (R-11).

WEIGHTS Empty 4443 kg (9,795 lb); fuel 1500 kg; loaded 6250 kg (13,779 lb).

PERFORMANCE Max speed 1970 km/h (1,224 mph, Mach 1·85); climb 1·6 min to 5 km; service ceiling 17,65 km (57,900 ft); range 1330 km (826 miles); TO run 730 m; landing run 890 m.

Ye-2A, MiG-23 Corresponding pair of aircraft with swept wing, OKB **Ye-2A**, same rear fuselage as Ye-5. Wing based on Ye-2 but without slats, and with single large fence above wing terminating at flap/aileron LE. Main gears as Ye-2. First flight by Mosolov 22 March 56. GAZ-21 built five further pre-series, designated **MiG-23**. Last of these, Ye-2A/6, used for OKB final test programme by Sedov and Nyefyedov (both HSU for their courage), Mosolov and Kokkinaki. Terminated when delta chosen Dec 56. ASCC 'Faceplate'.

DIMENSIONS As Ye-2.

ENGINE One AM-11 (R-11).

WEIGHTS Empty 4340 kg (9,568 lb); fuel 1450 kg; loaded 6250 kg (13,779 lb).

PERFORMANCE Max speed 1900 km/h (1,181 mph, Mach 1·79); climb 3·3 min to 10 km; service ceiling 18 km (59,050 ft); range uncertain (published 2000 km at variance with smaller fuel capacity than Ye-5).

Ye-50, Ye-50A In search for flight performance, and lacking AM-11 or other suitable turbojet, OKB decided 54 to produce mixed-power prototype with permanent installation of rocket. Result was **Ye-50/1**, **/2** and **/3**. Airframe basically as Ye-2A apart from rocket installation and: lengthened forward fuselage, putting cockpit well ahead of LE; fences slightly further outboard, terminating at inboard LE of aileron; installation of older but uprated AM-9Ye engine; forwards extension of fixed tailplane roots as strakes along fuselage; and (because of increased side area at rear) elimination of twin ventrals and addition of shallow strake on centreline under drain pipes. Vertical tail raised by rocket installation. Tanks for nitric acid, HTP and kerosene occupied extra fuselage bays, with special materials and coatings to protect against acid corrosion or HTP spillage. HP nitrogen for overboard dump or post-burnout purge through two stainless-steel pipes along ventral centreline terminating beyond main-engine nozzle. HTP-driven turbopump and control system in dorsal fin ahead of single thrust chamber above main-engine afterburner. Two NR-30 installed. Ye-50/1 flown by Valentin Mukhin 9 Jan 56 (same day as Ye-5). Tests with rocket fired from 8 June, but on Flight 18 on 14 July aircraft damaged beyond repair in landing short of runway. Ye-50/2 assigned to LII pilot Valentin P. Vasin. Small differences included reduction in height but increase in chord of rudder, and aft extension of base of fin. Max-performance flight 17 June 57 consumed all propellants to reach 25,6 km (83,990 ft) and 2460 km/h. Ye-50/3, flown Sept 57, redesigned nose with central cone of Oswatitsch type to focus two oblique shocks and one normal shock on sharp lip, giving better supersonic pressure recovery. Rear fuselage with extra bay, among other things increasing kerosene tankage (small reduction in HTP). Aircraft destroyed by fire in rocket bay, NII pilot N. A. Korovin killed when ejection seat malfunctioned. Gorki (GAZ-21) produced scheme for **Ye-50A**, with AM-11 engine, S-155 thrust chamber at base of fin but propellants and rocket control system in streamlined demountable pack along underside of fuselage as in rocket-boosted SMs. GAZ-21 authorised

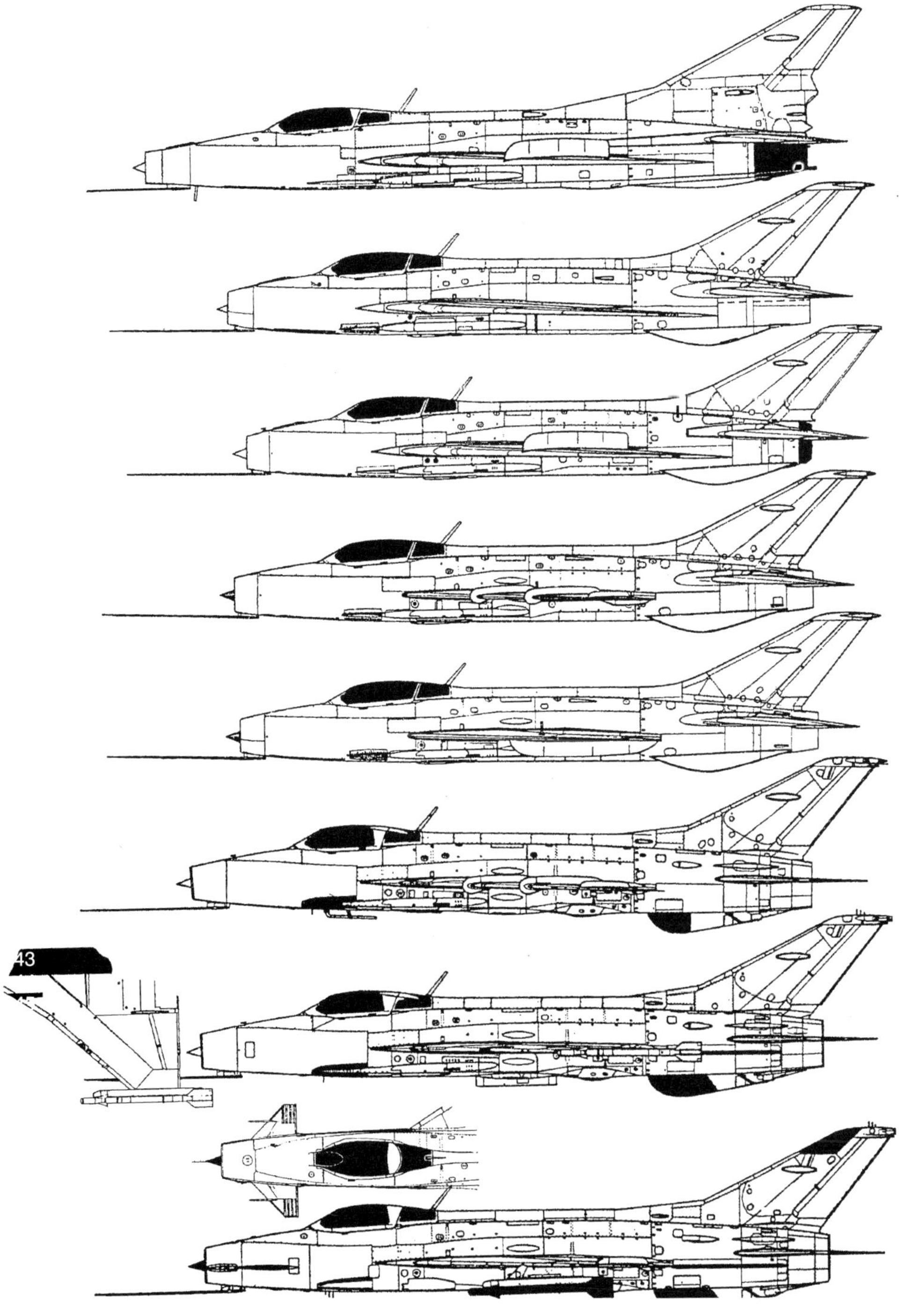

From top left: Ye-50/3, Ye-2, Ye-2A, Ye-4/RD-9I, Ye-4/RD-9Ye, Ye-6/1, Ye-6/2 dorabotnyi (inset plan), Ye-6T/3 (inset plan)

to build 20, in serious attempt to deploy outstanding point-defence interceptor, but closure of Dushkin bureau terminated manufacture of S-155. No Ye-50A completed; published illustrations are artwork. Data for Ye-50.

DIMENSIONS Span 8,109 m (26 ft 7¼ in); length (excl boom) [/1] 13,625 m (44 ft 8½ in), [/3] 14,85 m (48 ft 8⅝ in); wing area 21 m^2 (226 ft^2).

ENGINES One AM-9Ye, one S-155.

WEIGHTS Empty [/1] 4401 kg (9,702 lb); fuel/propellant 1350+2720 kg; loaded 8500 kg (18,739 lb).

PERFORMANCE Max speed 2460 mph (1,529 mph, Mach 2·32); climb unknown; service (not zoom) ceiling 23 km (75,460 ft).

Ye-6 Dec 56 decision in favour of delta enabled GAZ-21 to prepare for production, while GAZ-155 refined design with **Ye-6/1**, **/2** and **/3**, regarded as pre-production aircraft. Built under chief engineer I. I. Rotchik, with help from Minayev's systems team. Ye-6/1 basically as Ye-5 except: later engine; inlet and forward fuselage as Ye-50/3, with centrebody translated forward automatically as Mach rose past 1·4, and automatically retracted on deceleration; modified wing-root attachments with long fairing ahead of and behind wing as on swept-wing prototypes; removal of wingtips (as on Ye-4); tailplanes increased in span and moved down almost to mid position (air inlet left at original location) which in turn demanded relocation of left/right airbrakes under leading edge (profile incorporating gun fairings) and replacement of twin inclined ventrals by single larger underfin; structural redesign of rear fuselage and vertical tail, with slight increase in fin height; addition of dorsal fin (with q-feel inlet) joining fin to spine; extension of rear fuselage well aft of rudder TE (overall length unchanged); and redesign of aft end of canopy (and link to seat to prevent recurrence of failure which killed Korovin) and fixed rear fairing. Assigned Nyefyedov and flown by him Jan 58. On flight 7, 28 May, engine failure and inability to relight caused control to be lost on approach, Nyefyedov dying of injuries. Brunov proposed revision of inadequate electric tailplane drive, but Mikoyan instead accepted Belyakov's insistence on fully duplicated hydraulics. **Ye-6/2** torn apart before first flight and flown by Kokkinaki Sept 58 with redundant system. Major change included new wing with no fences and only 48° LE sweep from mid-span kink to give broad tip for APU-13 launcher for K-13 missile; also combat avionics, as described later. **Ye-6/3** flown Dec 58, generally as 6/2 but normal clipped wingtips, set two world records: absolute speed 31 Oct 59, 2388 km/h, and 100-km circuit 16 Sept 60, 2148,66 km/h, on both occasions with fictitious designation 'Ye-66'. ASCC 'Fishbed-A'.

MiG-21F, Ye-6T, Tip 72 Cleared for production about Oct 58, before first flight of Ye-6/3. Batch of 30 built at GAZ-21 59-60. Following serves as baseline description: airframe designed to ultimate load factor 7 g. Wing profile CAHI S-12 series, MAC 4002 mm (157·6 in), root chord 5970 mm (235 in), thickness 4·2% root, 5% tip, anhedral –2°, main spar 33·3% chord, three aux spars 90° to fuselage, left/right wings attached five fuselage frames. Ruling material D16-T, machined skin up to 2·5 mm, booms V-95 or VM-65, joints 30KhGSA with ML5-T4 at other stress concentrations. Tracked slotted flaps 0,935 m^2 each, hyd drive to 24·5° TO, 44·5° landing. Tabless tapered ailerons each 0,51 m^2, both aerodynamically balanced, left aileron with mass balance, driven by BU-45 power unit, preceded by upper-surface fence with height 7% of MAC 0,67 m from wingtip. Fuselage ruling diameter 1242 mm (48·9 in). Front/rear fuselage disconnected between Frames 28 and 28A, engine removed to rear on internal rails. Sharp-edged circular inlet with centrebody automatically translated to three positions: retracted <M1·5, intermediate M1·5/1·9, fully forward at 1·9 or above. Air-data boom with yaw vanes attached under nose. Duct bifurcated on each side past cockpit, with rectangular spill door each side near nose and almost square aux suck-in inlet each side under LE, then circular duct to engine in fireproof compartment between Frames 29-34, with heat

Ye-5

Ye-2A

Ye-50

shroud and fire-extinguisher bottle. Self-contained lube-oil system with fuel/oil heat exchanger. Electric starting, using petrol from separate small starting tank, plus auxiliary oxygen injection to assist relight at high altitude. Smooth-lightup afterburner with separate fuel pump. Sixth-stage bleed fed to cockpit environmental system and tanks for fuel, starting fuel and hydraulic fluid. Six bladder fuel tanks total 1590 lit, plus 2×175 lit integral in LE plus 2×110 lit integral behind main legs, total 2160 lit (475 gal), later 2280 lit (502 gal). Three door airbrakes, left/right 0,38 m^2 (part of gun blisters) hinged to Frame 11 opening to 25°, centreline 0,47 m^2 hinged to Frame 28 opening 40°. Rear ventral fin 0,352 m^2, door on left of ventral for 16 m^2 multi-ribbon braking parachute. Vertical tail LE angle 60°, 6% CAHI S-11 profile 3,52 m^2, with rudder 0,965 m^2 driven by booster ±25°. Tailplanes symmetric A6A (6%) profile, span 3,74 m including anti-flutter masses on tips, LE 55°, 3,94 m^2, driven +7·5°/–16·5° by BU-44 power unit with ARU-3V artificial feel. All tail control inputs by rigid linkage in dorsal spine. Cockpit pressurized (dP max 25 kPa, c3·6 lb/in²), temp controlled 15±5C. Fixed 62 mm-thick three-ply glass slab ahead of pilot, heavy planks Frames 6-11 and armour behind seat and headrest. One-piece blown-acrylic canopy with bulletproof flat windscreen, hinged at front and raised by pneumatic ram. SK seat fired by cordite cartridges taking rear of canopy with it to serve as protective windbreak, leaving fixed aft fairing. Nose gear with castoring tyre 500 mm×180 mm, wheel with twin-shoe anti-skid brake, retracting forwards between frames 6/11. Main tyres 660 mm×200 mm, inflated to 1,01 MPa (148 lb/in²), anti-skid disc brakes, track 2592 mm (102 in), wheelbase 4806 mm (189·2 in), retracting inwards while wheels pivot on legs 87° to lie almost vertically in narrow space between air duct and skin closed by curved door hinged to fuselage. Two NR-30 each with 30 rounds, two wing pylons under Rib 13 for

UV-16-57 or KARS-57 rocket launchers or single ARS-240 rocket or bombs to FAB-500. Duplicated hyd pumps and 20·59 MPa (nominal 3,000 lb/in²) systems serving flight controls (electric standby), landing gear, flaps, airbrakes, nosecone and nozzle flaps. Main pneumatic system operates brakes, canopy/canopy seal/jettison, pilot suit, gun cocking, tailskid (in ventral fin), main fuel valves, braking parachute and various deice functions. Emergency pneu lowers gear and applies brakes. Main electric power DC from starter/generator (two Ag/Zn batteries guaranteed to start engine in Siberian winter), with inverters giving 400 Hz AC at 115V and 36V. Gaseous oxygen from (usually six) bottles in wing root. Alcohol plus anti-ice additive for deicing windscreen. KAP 2 or -2K autopilot, UKV and RSIU-5V vhf, ARK-10 radio compass under nose with GIK goniometer causing blister each side of fin, MRP-56P beacon receiver, RV-UM radar altimeter, SOD-57M ATC/SIF transponder, SRO-2 IFF and (from early in run) *Sirena* 2 radar warning.

NII testing from Dec 58, and first delivery June 59. Third off line, Ye-6T/3, experimentally fitted with destabilizing powered canards on nose, each with pitot, plus provision for launching two K-13 (R-3S). ASCC 'Fishbed-B'.

MiG-21F-13

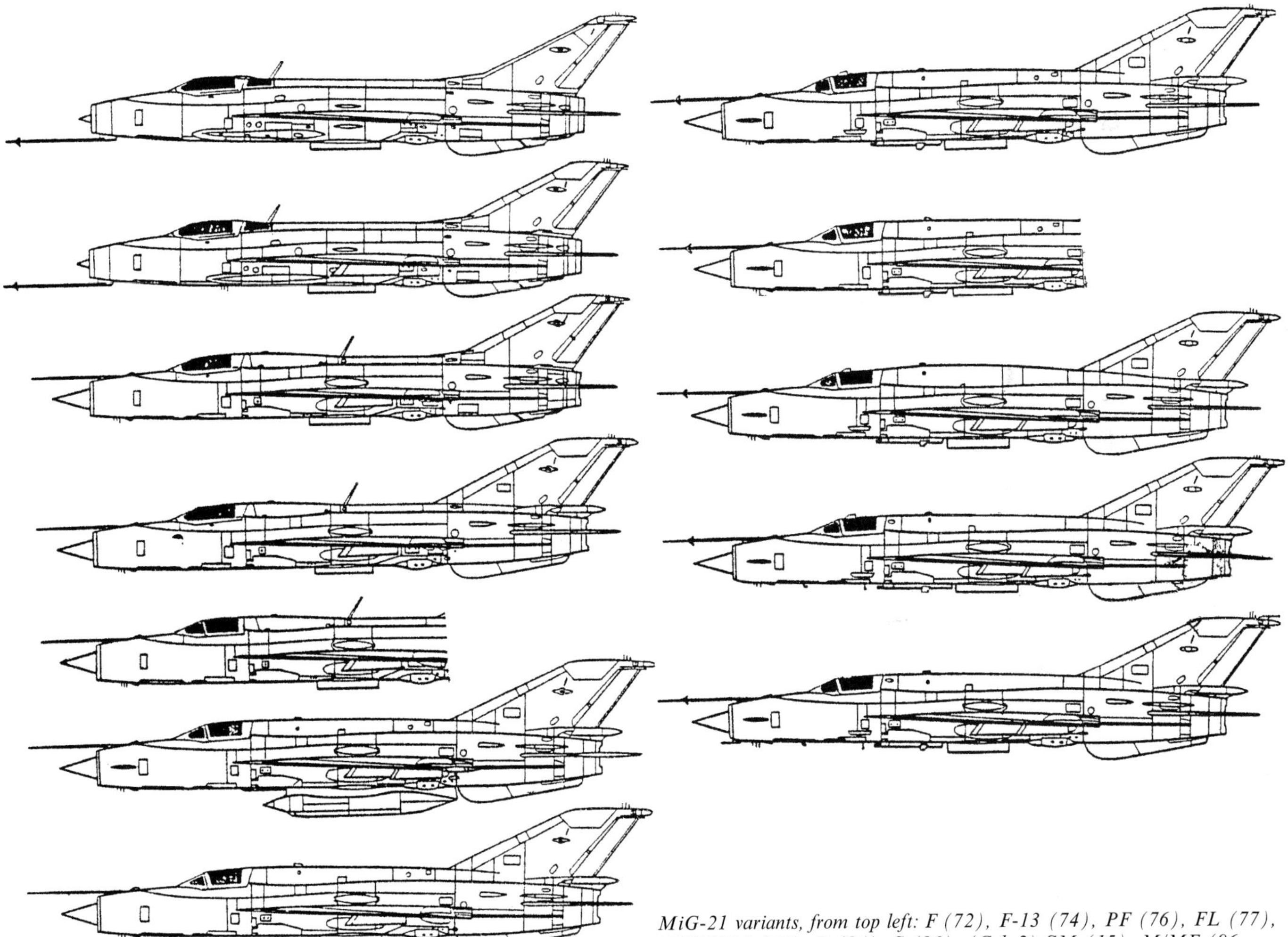

MiG-21 variants, from top left: F (72), F-13 (74), PF (76), FL (77), PFS/PFM (94), R (94), S (96), (Col. 2) SM (15), M/MF (96, 96F), MT (96T) or SMT (50), bis Lazur (75A), bis SAU (75B)

DIMENSIONS Span 7,154 m (23 ft 5⅝ in); length 15,76 m (51 ft 8½ in), (excl boom) 13,46 m (44 ft 1⅞ in); wing area 23,0 m² (247·6 ft²).
ENGINE One R-11F-300.
WEIGHTS Empty 4819 kg (10,624 lb); fuels 1790 kg, later 1820 kg; loaded 6850 kg (15,101 lb), max 8376 kg (18,466 lb).
PERFORMANCE Max speed (SL) 1100 km/h (684 mph), (12,5 km) 2175 km/h (1,352 mph, Mach 2·04); climb 2 min to 5 km, 7,5 min to 18,5 km; service ceiling 19 km (62,340 ft); range 1520 km (945 miles); TO 900 m; landing (with chute) 280 km/h/800 m.

MiG-21F-13, Tip 74 First version for inventory. Designation from K-13 missile, one on APU-13 rail on each wing pylon. Alternative not possible on MiG-21F: two UV-32-57 rocket launchers. Internal fuel increased to 2470 lit, then 2550 lit (561 gal). Centreline pylon for 490 lit or 800-lit four-finned tank cleared to Mach 1·5. ASP-5ND sight, SRD-5M *Kvant* radar ranging and SO-69 *Sirena 2* RWR. Left gun deleted on all but earliest examples. Air-data boom arranged to pivot up on ground. Provision for recon camera, usually AFA-39. Prodn early 60 at GAZ-21. From #115 fin slightly reduced in height but wedge section on LE increased area to 3,8 m². Export **Tip 74** produced at MMZ *Znamya Truda* 62-65. First front-line unit 28th Regt at Odessa, 15 Jan 63. **S-106** produced by Aero Vodochody, Czechoslovakia, 63-66, all of these except first few having all-metal aft (fixed) section of canopy. With small changes licence-built in China from Jan 66 at Shenyang, later at Chengdu, as **J-7**. ASCC 'Fishbed-C'.
DIMENSIONS, ENGINE As 21F.
WEIGHTS Empty 4871 kg (10,739 lb); fuels 2115 kg; loaded (two K-13) 7370 kg (16,248 lb), (max, 490-lit tank and two FAB-500) 8625 kg (19,015 lb).
PERFORMANCE As 21F except climb 13·5 min to 19 km and range slightly shorter, but extended to 1670 km (1,038 miles) with 800-lit tank; TO 900 m (a/b) or 1350 m; ldg 265 km/h, 1250 m, 900 with chute.

Ye-6V Two aircraft for STOL tests. Evaluated original braking-parachute against new location in tube under rudder. Ye-6V/2 made STO (one at Tushino 9 July 61) with two SU-1500 rockets attached under TE with inclined nozzles giving thrust axes through CG. V/2 also tested KL wheel/ski main gears.

Ye-66A Ye-6T/1 rebuilt 61 by GAZ-155 to explore larger flight envelope with more powerful engine and U-21 package housing S-3/20 rocket (canted nozzle), propellant tanks and control system. Vertical tail 4,44 m², ventral fin replaced by two larger inclined ventrals. Canopy faired into spine by metal cover over 170-lit tank. Red '31' on nose. Flown in zoom by Mosolov to 34 714 m (113,891 ft) 28 April 61.

Ye-7, MiG-21P Two prototypes, Ye-7/1 and 7/2, built at GAZ-155, first flown by Pyotr M. Ostapyenko 10 Aug 58, 7/2 following 18 Jan 60. Whereas MiG-21F had been simplest fighter possible, need for radar always realised, and change in policy led to removal of remaining gun and reliance on two K-13 missiles. TsD-30 (later RP-21) radar with antenna scanned conically inside much larger inlet centrebody able to translate to any optimised location, under control of UVD-2M electronics, tip of cone from 200-1213 mm in front of inlet lip. Inlet assisted by SM-12PM; lip diam enlarged from 690 to 870 mm and aft-facing duct exits added above and below nose to suck boundary layer through fixed ring slot behind centrebody. Avionics otherwise as before, except for PKI-1 (ASP-P-21) optical sight, *Lazur* beam/beacon receiver, KSI and single-axis (roll) KAP-1 autopilot. Other changes: airframe restressed to factor 7·8, internal fuel 2750 lit with 170-lit saddle tank in metal fairing behind canopy (RSI vhf and whip antenna moved back along spine), larger main wheels with 800 mm × 200 mm tyres inflated to only 0·785 MPa (115 lb/in²) for unpaved airstrips, two SU-1500 a.t.o. rocket attachments and PVD-5 air-data boom with forward-facing pressure head and three rows of static holes relocated on fixed mount above nose. Considered improved despite loss of gun and rear vision. OKB tests completed 8 May 60 and series **MiG-21P** in production at Gorkii a month later. Data similar to PF. ASCC reporting name 'Fishbed-D'.

MiG-21PF

MiG-21PF, Tip 76 Chief differences, F2-300 engine and first of *Sapfir* radars: Ye-7 tested S-21 radar, for production **MiG-21PF** with RP-22. Replaced 21P in production at GAZ-21 early 62, export *Tip 76* following at *Znamya Truda* early 64. Final block at GAZ-21 and all Tip 76 fitted with relocated braking parachute under rudder as on 6V/1. One aircraft, called *Ye-76*, used to set ladies' records 66-67. ASCC 'Fishbed-D'.
DIMENSIONS As 21F except length (excl boom) 14,7 m (48 ft 2¾ in), length (inlet lip to nozzle) 12,285 m (40 ft 3⅝ in).
ENGINE One R-11F2-300.
WEIGHTS Empty 5256 kg (11,587 lb); fuels 2280 kg; loaded (no tank) 7750 kg (17,086 lb), max 8770 kg (19,334 lb).
PERFORMANCE Max speed (SL) 1100 km/h (684 mph), (13 km) 2175 km/h (1,352 mph, Mach 2·048); climb 8 min to 18,5 km; service ceiling 19 km (62,336 ft); range 1400 km (870 miles), (800-lit tank) 1900 km (1,180 miles); TO run 900 m; landing speed, run 280 km/h, 850 m.

MiG-21FL, Tip 77 Simplified PF for Indian licence construction. Slight increase in fuel to 2900 lit, engine downgraded and radar replaced by R-2L. Produced at *Znamya Truda* 65-68 and (initially by assembly of imported parts) by Hindustan Aeronautics 66-73. ASCC 'Fishbed-D'. Data as PF except:
ENGINE One R-11F-300.
WEIGHTS Fuel 2400 kg; loaded 7830 kg (17,262 lb).
PERFORMANCE As PF except range 1450/1800 km (900/1,120 miles).

MiG-21PFV Export PF for Vietnam; as FL, Tip 77, plus provisions for humid environment.

Ye-8 Visibly striking, two research aircraft with NII designation **MiG-23** under 61 decree seeking basis for interceptor able to engage targets to front or rear in any weather or at night. Based on MiG-21PF with two K-13, but uprated engine fed by ventral inlet, and forward fuselage to house *Sapfir 21* [never actually installed] requiring radome too large to serve as inlet centrebody. There were other advantages in ventral inlet standing away from underside of fuselage with central splitter followed by three variable walls. Duct bifurcated around nose gear with twin suck-in doors each side. Downstream of bifurcation enlarged to 845 mm diam of engine (previously 772). Nose gear redesigned to retract to rear, with wheelbase reduced to 3350 mm (132 in) resulting in tendency to nod on ground. Destabilizing canards (derived from Ye-6T/3) of 2600 mm (102·4 in) span with anti-flutter rods at mid-span mounted horiz below centreline of nose; below Mach 1 these freely weathercocked, above Mach 1 locked in line with longitudinal axis. PVD-7 boom on nose with yaw and pitch vanes; main gears modified MiG-21PF with track increased from 2692 m to 2787 m (109·7 in); improved dorsal spine leading to redesigned vertical tail with rounding top forming vhf/uhf antenna and incorporating other eqpt; tailplanes moved 135 mm below fuselage axis; central underfin with lower portion power-folded to right, triggered by extension of landing gear; and omission of forward airbrakes. **Ye-8/1** (also called **81** and this painted on nose), flown by Mosolov 17 April 62. **Ye-8/2** flown by Fedotov 29 June 62. But Ye-8 project abandoned after 8/1 engine disintegration 11 Sept causing Mosolov serious injury.
DIMENSIONS Span 7,154 m (23 ft 5⅝ in); length (excl boom) 14,9 m (48 ft 10⅝ in); wing area 23,13 m² (249 ft²).
ENGINE One R-21F.
WEIGHTS Normal loaded 6800 kg (14,991 lb), (max) 8200 kg (18,078 lb).
PERFORMANCE Max speed (12 km) 2230 km/h (1,386 mph, Mach 2·1); service ceiling 20 km (65,620 ft).

Ye-7SPS Following CAHI research from 57

OKB introduced SDS (*Sduva pogranichnogo sloya*, blown flaps) to increase lift at low speeds. Each flap simple rectangle, 0,92 m^2 compared with 0,935, hinged to rear false spar and driven by faired actuator at mid-span to 25° TO and 45° landing. Engine of F2S-300 type with manifold around 3rd HP stage, with valve and safety venturi. SPS selected by pilot (not automatic) on TO and/or ldg, introduced to production on Tip 94.

MiG-21PFM, Tip 94 Entered prodn 64, SPS system, KM-1 seat (first of improved standard types, no longer needed canopy as windbreak so fixed windscreen and short jettisonable canopy hinged to right), further increase in fin chord to 5,32 ft^2, PT-21UK cruciform braking chute, RP-21M radar with guidance function for two RS-2US missiles as alternative to R-3S, or for Kh-23 missiles. Previous weapon options available, plus provision for GP-9 pack recessed into belly housing GSh-23L and 200 rounds (self-powered, electrically fired), requiring predictor sight, ASP-PK-21. LSK (E German AF) called this version **SPS-K**. Other avionics *Khrom-Nikel* passive EW receiver antennas on wing LE and each side of fin tip, SRO IFF and Sirena 3M with aft-facing antennas at rear of fin (magnetometer) blisters. Airframe restressed to factor 8·5 (future standard except 21R); provision for SPRD-99 a.t.o. rockets, slight reduction in No 1 tank (overall internal down from 2750 lit to 2650 lit, 583 gal), but both wing pylons plumbed for tanks (centreline tank impossible with gunpack). **MiG-21PFS**, a few supplied to one regiment, F2S-300 engine with two pilot-selectable stages of reheat. Main production **PFM** delivered from GAZ-21 64-65 and export Tip 94 from *Znamya Truda* 66-68. ASCC 'Fishbed-F'.

DIMENSIONS As PF.

ENGINE One R-11F2S-300.

WEIGHTS Empty 5383 kg (11,867 lb); fuels 2200 kg; loaded 7820 kg (17,240 lb); max loaded 9080 kg (20,018 lb).

PERFORMANCE Max speed (SL) 1100 km/h (684 mph), (13 km) 2230 km/h (1,386 mph); climb 130 m/s (25,590 ft/min), 8·5 min to 18,5 km; service ceiling 19 km (62,336 ft); range 1300 km (808 miles), (with 800-lit tank [no GP-9]) 1670 km (1,038 miles); TO 900 m (450 with a.t.o.); landing speed, run (PFM) 275 km/h, 950 m, (SPS) 250 km/h, 550.

MiG-21R, Tip 94R Dedicated reconnaissance version, defined following studies with PF #78 which had four wing hardpoints self-defence missiles and two tanks plus centreline recon container (also had vertical tail of Ye-8 type). **MiG-21R** had vertical tail of Tip 94 (PFM) type, and untapered dorsal spine housing additional 340 lit fuel increasing internal to 2800 lit (616 gal). At last, AOA indicator as requested for 10 years, with sensing cylinder on left of nose with fairing to front and rear. This, plus small pitot on right of nose, fed cockpit instruments (not nose boom). AP-155 three-axis autopilot, SRO-3, but radar still RP-21 and other avionics unchanged. Main payload D-99 pod for day photography, but similar alternatives for night photo, Elint, IR linescan, TV and SLAR. Four wing pylons for two R-3S and two 490-lit tanks; inners could carry rockets or bombs and GP-9 could replace recon payload, wich limited load factor to 6 g. **MiG-21R** and export **Tip 94R** produced at GAZ-21 65-71. ASCC 'Fishbed-N'.

DIMENSIONS, ENGINE As PF.

WEIGHTS Empty (D-99) 5696 kg (12,557 lb); fuels 2320 kg; loaded 8100 kg (17,857 lb).

PERFORMANCE Generally as PFM except climb 8·5 min to 14,6 km; service ceiling 15,1 km (49,540 ft); range inexplicably shorter than PFM at 1130/1600 km.

MiG-21S, Ye-7S, Tip 95 Fighter with five-pylon airframe, tankage and most avionics of 21R, RP-22S multimode radar, ASP-PF sight, improved *Lazur-M* guidance receiver, PVD-7 boom (as Ye-8) and GP-9 gunpack (previously standard only on Indian Tip 94). Wired for tac-nuke bombs on inner pylons in place of R-3S or R-3R missiles. Produced at GAZ-21 for VVS only 65-68. ASCC 'Fishbed-H'. Data as R except:

WEIGHT loaded 8150 kg (17,967 lb).

PERFORMANCE Climb 8·5 min to 17,5 km; service ceiling 18 km (59,055 ft).

MiG-21SM, Tip 15 Similar to S except for new engine and for installation of GSh-23L gun inside belly with only twin faired barrels protruding. Belt of 200 rounds loaded through door at top left to fill space both sides of engine duct. Centreline tank restored, augmenting reduced internal fuel of 2650 lit (583 gal); starting petrol tank no longer needed. ASP-PFD sight and SRO-10 RWR. Produced GAZ-21 68-74. Data as R except:

ENGINE One R-13-300.

WEIGHTS Fuel 2200 kg; loaded 8300 kg (18,298 lb).

PERFORMANCE Climb 9 min to 17,5 km; range 1050/1420 km (652/882 miles).

MiG-21M, Tip 96 Downgraded SM for export, with older engine, RP-21MA radar, PKI (ASP-PF-21) sight and RS-2US missiles. Produced at *Znamya Truda* 1968-71 and by Hindustan Aeronautics 1973-81. ASCC 'Fishbed-J'. Data as SM except:

ENGINE One R-11F2S-300.

WEIGHTS Empty 5950 kg (13,117 lb); loaded 8950 kg (19,731 lb), max 9,4 t (20,723 lb).

PERFORMANCE Max speed at SL 1150 km/h (715 mph), at 11 km 2230 km/h (1,386 mph, M 2·1), climb 160 m/s (31,500 ft/min), 9 min to 16,8 km; service ceiling 18,5 km (60,700 ft); range (3 tanks) 1800 km (1,118 miles); TO 800 m (a/b) or 1350 m; ldg speed/run 270 km/h, 1250 m or 750 with chute.

MiG-2MF, Tip 96F MiG-21M upgraded with engine and radar of SM. New missile option on APU-13 rail, R-60 or 60M. Eqpt as SM except SRZO-2 IFF and (from 72) TS-27AMSh aft-view periscope. Produced at *Znamya Truda* 70-74, followed by GAZ-21 in 75. ASCC 'Fishbed-J'. Data as SM except:

WEIGHTS Empty 5350 kg (11,795 lb); loaded 8212 kg (18,104 lb), max, 2 R-3 + 3 tanks, 9661 kg (21,299 lb).

PERFORMANCE Climb 180 m/s (35,433 ft/min), range (3 tanks) 1800 km (1,118 miles); TO, min 890 m; landing speed/run 270 km/h, 550 m (SPS, chute) or 310 km/h (no blowing).

MiG-21MT, Tip 96T MF with bulged spine, forming unbroken tank fairing extending to drag-chute container, increasing internal fuel to 3250 lit, but for stability reasons aft cell at base of fin deleted and capacity reduced to 2950 (649 gal). Total said to be 15 built at *Znamya Truda* 71. Large number survive as widely dispersed gateguards, perhaps indicating higher total. ASCC probably 'Fishbed-K'. Data as SMT (see next) except:

WEIGHT Normal loaded 8800 kg (19,400 lb).

MiG-21SMT, Tip 50 Developed MT, for inventory; modified fuel and electrical systems and new inlet-control program. Avionics/armament as MF; tanks as MT. 'Fishbed-K'.

DIMENSIONS As PF.

ENGINE One R-13F-300.

WEIGHTS Empty 5700 kg (12,566 lb); fuel 2450

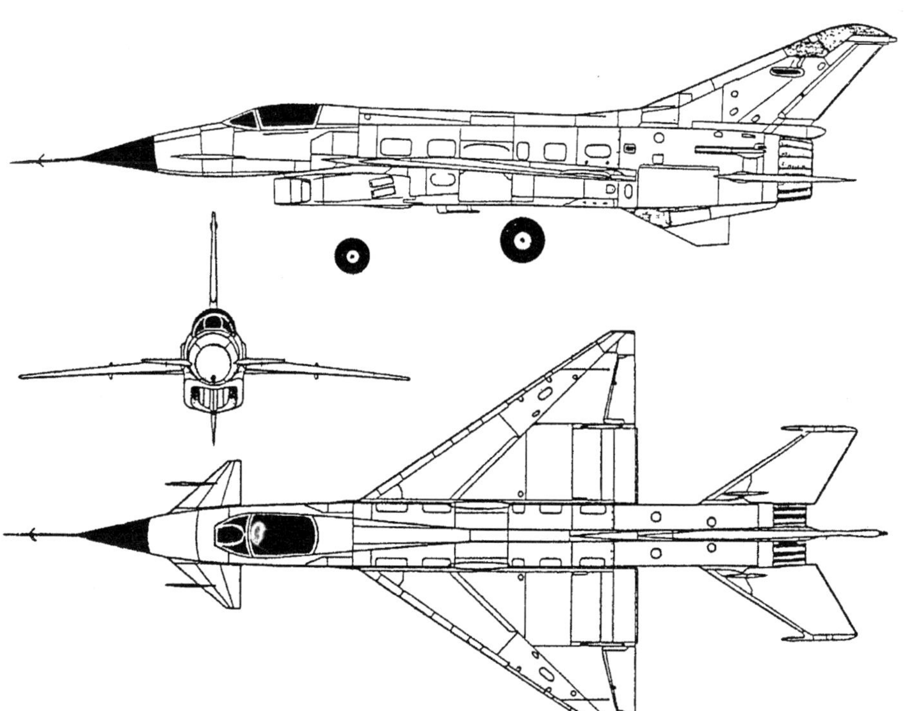

Ye-8 underfin extended)

MiG-21bis

MiG-21UM

kg; loaded 8900 kg (19,621 lb), (max) 10100 kg (22,266 lb).

PERFORMANCE Max speed (SL) 1300 km/h (808 mph), (13 km) 2175 km/h (1,352 mph, Mach 2·05); climb 9 min to 16,8 km; service ceiling 17,3 km (56,760 ft); range 1300 km (808 miles), (one 800-lit tank) 1670 km (1,038 miles); TO run 950 m; landing speed/run (SPS) 250 km/h/550 m.

MiG-21bis, Ye-7bis, Tip 75 Definitive fighter incorporating Vietnam experience so designed for low-alt combat. This required restressing of structure for higher IAS, though SL V_{max} and high-alt Mach unchanged. Design factor 8·5. New engine with revised fuel system and afterburner booster pump. Aerodynamics for min drag, no provision for rear tank and spine tapered off into fin. Internal fuel 2880 lit (634 gal). New options of K-13M (R-3M) and R-55 missiles. Shoran-type navaid and coupled ILS with antennas under nose and above rudder. Automatic monitoring and fault-warning for engine, fuel, hydraulics, electrics and cabin-air systems. Production at Gorkii 72. Licensed to India 74 and **Tip 75** produced by Hindustan 80-86. Variants include **MiG-21bisSAU** with SAU-21 autopilot, **MiG-21bisLazur** with *Lazur* beam/beacon rcvr, and **MiG-21bisN** nuclear strike variant. ASCC 'Fishbed-L' and '–N' (different undernose antennas).

DIMENSIONS As PF except fuselage length 14,1 m (46 ft 3⅛ in).

ENGINE One R-25-300.

WEIGHTS Empty 5895 kg (12,996 lb); fuel 2390 kg; loaded (2 AAM) 8725 kg (19,235 lb), (max) 10,420 kg (22,972 lb).

PERFORMANCE Max speed (SL) 1150 km/h (715 mph), (13 km) 2230 km/h (1,386 mph, Mach 2·1); climb 195 m/s (38,390 ft/min), 8·5 min to 17 km; service ceiling 17,5 km (57,415 ft); range (11 km, clean) 1225 km (761 miles), (2 AAM + 800-lit tank) 1470 km (913 miles); TO run 830 m; landing speed/run (SPS) 270 km/h/750 m.

MiG-21U, Ye-6U, Tip 33 Dual-pilot MiG-21 authorised Nov 59. Based on MiG-21F-13 (second series, 3,8 m^2 fin), with pupil cockpit in original position and instructor cockpit at rear. Original SK seats, but no easy answer to blast protection and each cockpit with canopy hinged to right, braced on ground by stay clipped to windscreen in front of each pilot. Rear cockpit required revision to engine ducts and tankage, but internal fuel only slightly reduced (2300 lit) by removing guns and replacing forward airbrakes by single brake on centreline. KT-50/2 mainwheel/tyre/brake as MiG-21P. PVD-7 boom with pitch/yaw vanes above inlet. Two-position centrebody but no radar. No armament except provision for A-12·7 gun pod on centreline as alternative to 490 or 800-lit tank. KAP-2 autopilot, ASP-5 or 5ND sight, RSIU-5V vhf, SRO-2 IFF. First flight by Ostapyenko 17 Oct 60. Production at Tbilisi for VVS 62-66. Export **Tip 66** at *Znamya Truda* 64-68. One aircraft, '**Ye-33**', used 1965 for ladies' height record (24,336 m). ASCC 'Mongol'.

DIMENSIONS, ENGINE As F-13.

WEIGHTS Empty 5195 kg (11,453 lb); fuels 1950 kg; loaded 7800 kg (17,196 lb).

PERFORMANCE Virtually identical to F-13 except ranges about 100 km shorter.

MiG-21US, Tip 68 Natural development with 5,32 m^2 fin, F2S-300 engine, SPS blown flaps, PT-21UK braking parachute at base of rudder, KM-1M seats, simple two-mirror rear-cockpit retractable periscope and slight increase in internal fuel (2450 lit, though some 2350), but retaining original inlet. No radar or weapons, but wing pylons added for tanks, K-13 family AAMs, rockets or bombs as in fighters. Produced at Tbilisi 66-70 for home and export customers. As **Ye-33** set further ladies' records

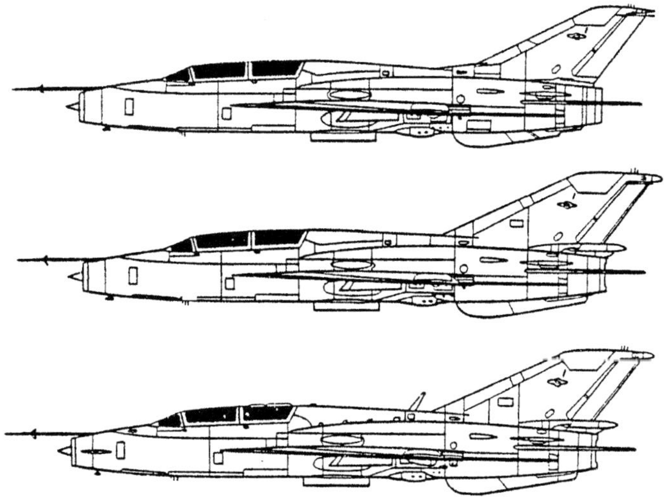

MiG-21 trainers: from the top, U (66), US (68), UM (69)

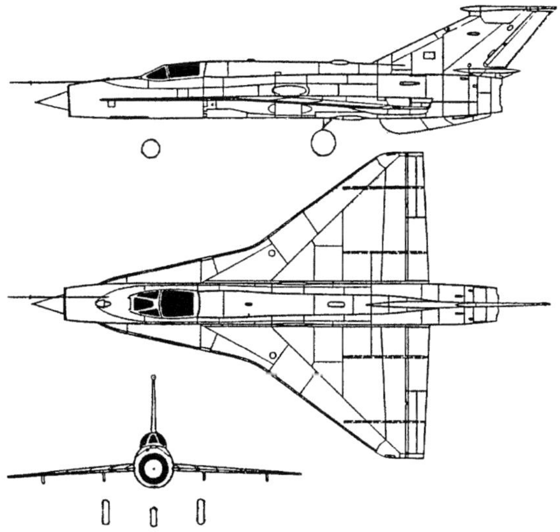

21I/1

June 74, easily beaten five months later by boosted example called **Ye-66B**. ASCC 'Mongol-B'.
DIMENSIONS As PF.
ENGINE One R-11F2S-300.
WEIGHTS Empty 5380 kg (11,861 lb); fuels 2030 kg; loaded (typical) 8000 kg (17,637 lb); max 9,5 t (20,944 lb).
PERFORMANCE Similar to U except max speed at 12 km 2150 km/h (1,336 mph); range 1750 km (1,087 miles); TO 1,1-1,4 km; ldg 900-1250 m.
MiG-21UM, Tip 69 Final and logical dual-pilot version, with AP-155 autopilot, AOA sensor on left of nose, ASP-PDF sight, usually considerably upgraded suite of mission avionics roughly as PFM or MF. Choice of F2S-300 or R-13-300 engine (latter uncommon), and two or four wing pylons. Produced for home and export at Tbilisi 71-72. ASCC 'Mongol-B'.
DIMENSIONS, ENGINE As US.
WEIGHTS, PERFORMANCE As US except max speed at 12 km 2230 km/h (1,386 mph).
ANPK MiG state 'over 10,000 built of 32 MiG-21 versions, excluding foreign manufacture'; actual total c12,500. Western firms have noticed huge market for upgrades, notably two Chinese groups and partnership of IAI (Israel) and Aerostar (Romania). MiG claim no upgrade possible without their co-operation, and offer official packages in partnership with Thomson-CSF (France). Options include *Kopyo* or French radar, HMS, 2-screen cockpit, new avionics throughout plus missiles R-27R1/T1, R-77 Kh-31A/P and Kh-25MP, and S-8 rockets. Command direction (eg from MiG-31) also on offer.
MiG-21-93 Completely refurbished aircraft with new radar, systems, avionics and instruments, available from A. I. Mikoyan OKB. Demo aircraft at Farnborough airshow 94.

MiG-21I, Analog, Tip 21-31 Two aircraft based on 21S completed with R-13-300 and approx scaled wing of Tu-144. Chief engineer I. V. Frumkin, wing led by G. Tseremukhin. **21I/1** inbd LE 78°, outer LE 55°, thickness 2·3/2·5%, inboard flaps and outboard elevons; no horizontal tail. Flown by O. V. Gudkov 18 April 68 (too late to influence Tu-144), crashed a year later. **21I/2** with larger double-ogive wing with four elevons each side, camera fairing at top of fin and on right of canopy fairing, flown 69 by LII's I. Volk, preserved at Monino.
DIMENSIONS Span 8,15 m (26 ft $8\frac{7}{8}$ in); length (excl boom) 14,7 m (48 ft $2\frac{3}{4}$ in); wing area (I/1) 41,1 m^2 (442 ft^2), (I/2) 43 m^2 (463 ft^2).
ENGINE One R-13-300.
WEIGHTS Fuel 2715 kg; loaded 8750 kg (19,290 lb).
PERFORMANCE Tested to M 2·06 up to 19 km, min 212 km/h, ldg 225 km/h.

MiG-21PD, 23-31, Tip 92 Single PFM grossly modified with cockpit moved forward 0,9 m (35·5 in) to provide bay at CG for two lift jets installed with small range of pilot-controlled tilt, drawing air through large rectangular dorsal aperture (in normal flight closed by aft-hinged door with louvres to permit air-start) and with pilot-controlled transverse deflectors under nozzles. Overall length almost unchanged, but centre fuselage lateral bulges to allow main-engine ducts to pass lift bay; this required wings to be attached to new inboard sections. Left/right bleed pipes from main engine to control jets under nose to augment pitch authority at low speeds. Fixed landing gear, track 3400 mm, wheelbase 5150 mm. First flight by Ostapyenko 16 June 66. OKB unimpressed, but went on to build 23-01 (MiG-23PD). ASCC name 'Fishbed-G'.
DIMENSIONS Span 7,765 m (25 ft $5\frac{3}{4}$ in); length (excl boom) 14,72 m (48 ft $3\frac{1}{2}$ in); wing area 26,5 m^2 (285 ft^2).
ENGINES One R-13F-300 and two RD-36-35.
No other data.
MiG-21Ye Remotely piloted target conversions of PF and PFM, designed with assistance of KAI and rebuilt 1966(?)-69 at VVS workshops. Radio receiver and coupled autopilot replaced seat. Limited to subsonic speed.

MiG-21I/1, 21-31, Analog

MiG-21PD, Ye-7PD, Tip 92

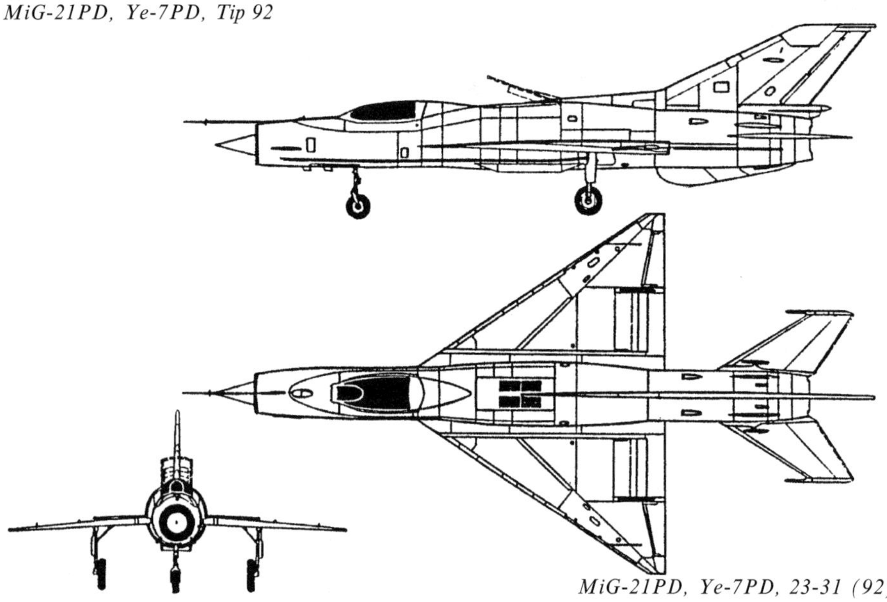

MiG-21PD, Ye-7PD, 23-31 (92)

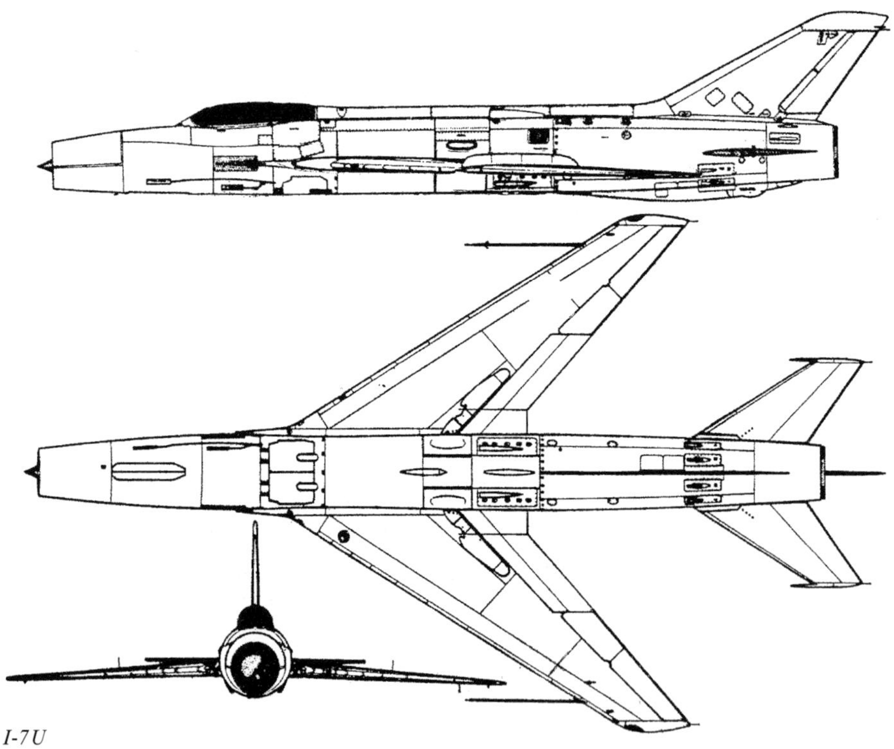

I-7U

I-7U Authorized Aug 56 as next stage in search for large all-weather interceptor. Airframe generally based on I-3U but wing reduced in sweep (55° at 25% chord) which increased span and area. Balanced powered ailerons increased in chord ahead of hinge and, like all control surfaces, with low-density rigid filling instead of ribs. Fowler flaps with inner TE at 90° to fuselage. Latter redesigned with slightly greater diameter, radar antenna in cone centred in inlet to modified duct leading direct to compressor face of axial engine. Airbrakes redesigned and moved forward in line with wing LE. Canopy incorporating ARK-5 antenna along top (as later fitted I-3U) and joined to fin by dorsal spine housing tail control rods; tailplane span reduced from 4457 mm to 4434 mm. Shallow ventral blister with adjacent drag chute. Less wing sweep reduced fuselage TO/landing attitude, enabling main gears to be reduced in height and hinged between integral wing tanks and flaps, with wheels stowed (as MiG-21) in fuselage. Track reduced from 4036 mm to 3242 mm but wheelbase up from 5350 mm to 5965 mm. First OKB fighter designed to load factor 9. Designation U from *Uragan-1* fire-control, unchanged from I-3U. Two NR-30 moved inboard to wing roots with breeches and 65-round magazines in fuselage. Provision for four pylons for 16-57 rocket launchers. Flown by Mosolov 22 April 57; wing damaged landing from Flight 13 on 21 June, grounded to Jan 58. Six more flights, speed 880 km/h below estimate because of instrumentation problems; returned to GAZ-155 12 Feb 58 to be rebuilt as I-75F. I-7K with *Almaz-3* radar, two NR-30 and two K-6 missiles, not built. No ASCC name.

DIMENSIONS Span 9,976 m (32 ft 9½ in); length overall 16,925 m (55 ft 6⅜ in), (fuselage lip to nozzle) 15,692 m (51 ft 5¾ in); wing area 31,9 m^2 (343 ft^2).

ENGINE One AL-7F.

WEIGHTS Empty 7952 kg (17,531 lb); fuel 2 t; loaded 10 200 kg (22,487 lb), (max) 11 540 kg (25,441 lb).

PERFORMANCE Max speed (11 km) 1420 km/h (882 mph, Mach 1·33); climb 0·6 min to 5 km, 1·18 to 10; service ceiling 19,1 km (62,665 ft); range 1505 km (935 miles); TO 570 m; landing speed/run 295 km/h/990 m.

I-75 Authorised 7 March 57, OKB's first gunless fighter, for high-alt interception of supersonic bombers. New fire-control *Uragan-5B* with larger antenna in more capacious Oswatitsch (3 angles) conical centrebody of dielectric material, 5T-1 controller, high-power 5D computer and 5V (AP-39) autopilot. CW guidance function for two K-8 AAMs on outer underwing pylons. Guns removed. Provision for two drop tanks on inner pylons (not fitted). Full mission avionics installed. New vertical tail with LE sweep reduced from 64° to 61·5°, thus height increased. MiG-21F type canopy linked to SK seat. Uprated engine fed by original ducts from new inlet of increased diameter with variable geometry provided by translating ring around throat. Fully powered controls: BU-44 ailerons, BU-44B tailplanes (ARU-3V feel feedback) and BU-45 rudder. Twin brake chutes, one each side of shallow ventral. Instrument boom on right wing only. First of two

I-7U

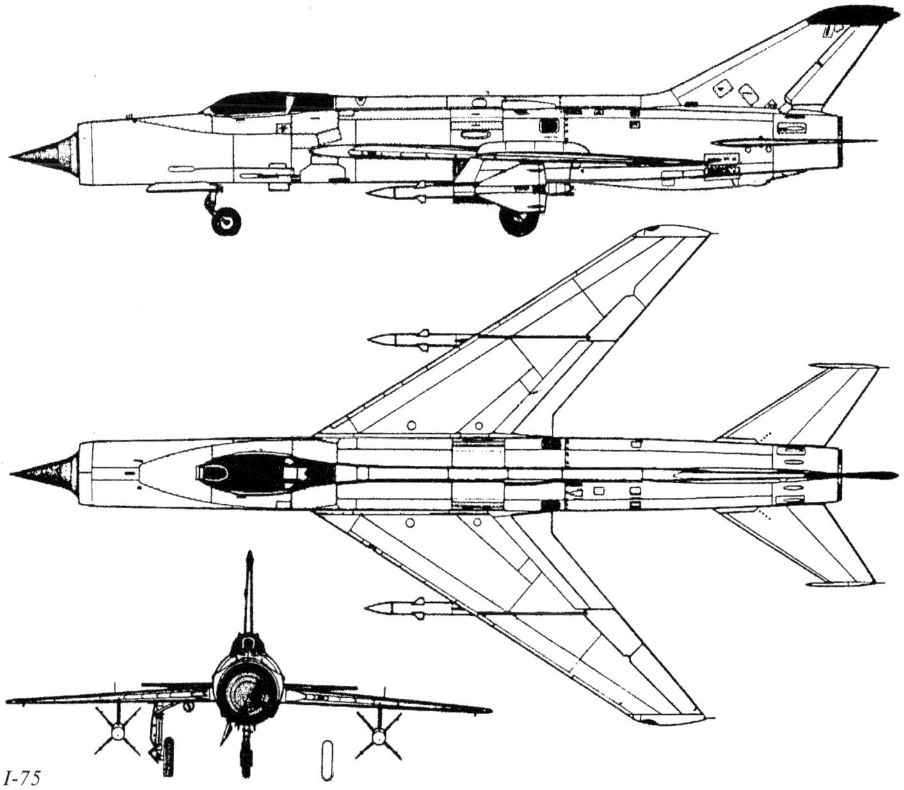

I-75

I-75

prototypes taken without radar to LII field 1 March 58, canopy jettison perfected and flown by Mosolov 28 April. *Uragan-5* installed 25 Dec. Termination (Su-9 selected) 11 May 59.
DIMENSIONS As I-7U except length overall 18,275 m (59 ft 11½ in).
ENGINE One AL-7F.
WEIGHTS Empty 8274 kg (18,241 lb); fuel 2 t; loaded 10 950 kg (24,140 lb), (max, 2 tanks) 11 470 kg (25,287 lb) [last figure suspect, should exceed I-7U].
PERFORMANCE Max speed (11,4 km) 2050 km/h, (12,4 km with AAMs) 1670 km/h (1,038 mph, Mach 1·57); climb 0·93 min to 6 km, 3·05 to 11; service ceiling (clean) 18,7 km (61,350 ft), (AAMs) 16 km; range (12 km, two tanks) 1470 km (913 miles).

I-75F Second prototype was I-7U rebuilt, uprated engine. Flown 58, terminated with I-75. Data as 75 except:
ENGINE One AL-7F-1.
WEIGHT Est max loaded 11,38 t.
PERFORMANCE Est max speed 2360 km/h (1,466 mph, M 2·22) at 18 km; service ceiling 21 km.

Ye-150 Urgent need for interceptor with speed (to catch B-58) and height (to reach U-2); Ye-150 to develop propulsion and also serve as prototype fighter with two K-9 (no guns). Wing S-12, 3·5%, almost scale of Ye-4 but LE 60°, no fence and two-part ailerons extending to slightly clipped tips. Cylindrical fuselage 1600 diam except for experimental ejector nozzle 1650. Inlet as I-75 but *Uragan-5B* not fitted; pitot boom on tip of cone. Five fuselage bladder tanks total 3720 lit plus 245-lit integral tank in each wing. Fully powered controls: tailplanes with anhedral, larger than I-75 (5292 mm span), drive by BU-65 and ailerons and rudder by BU-75. Single ventral fin housing PT-5605-58 mm braking parachute. Landing gears similar to I-7, track 3322 mm, wheelbase 5996 mm. Previous twin airbrakes replaced by single large brake. Single shallow (min-drag) canopy with titanium frame hinged at rear incorporating single moulding of T2-55 heat-resistant glass without antenna; KM-1 seat. Prototype to LII airfield Dec 58, engine delayed first flight by Fedotov to 8 July 60. By end of OKB project on 25 Jan 62 had flown 42 times with three engines and two designs of ejector, but no missiles. No ASCC name.
DIMENSIONS Span 8,488 m (27 ft 10⅛ in); length (excl probe) 18,14 m (59 ft 6⅛ in), (fuselage lip to nozzle) 15,6 m (51 ft 2⅛ in); wing area 34,615 m^2 (372·6 ft^2).
ENGINE One R-15-300.
WEIGHTS Empty 8276 kg (18,245 lb); fuel 3410 kg; loaded 12 435 kg (27,410 lb).
PERFORMANCE Max speed (SL) 1210 km/h (752 mph), (19,1 km) 2890 km/h (1,796 mph, Mach 2·72); climb 80 sec to 5 km, 305 sec to 20 km; service ceiling 23,25 km (76,280 ft); range 1500 km (932 miles); TO 935 m; landing speed/run 285 km/h/1250 m (with chute).

23-01, MiG-23PD In late 60 OKB began investigating ways of reconciling Mach 2-2·3 performance with STOL capability, to avoid missile-targeted airfields. One team assigned with CAHI to VG 'swing wing'. Alternative used lift jets (though no attempt at VTOL). Khachaturov KB produced R-27 engine for

Ye-150

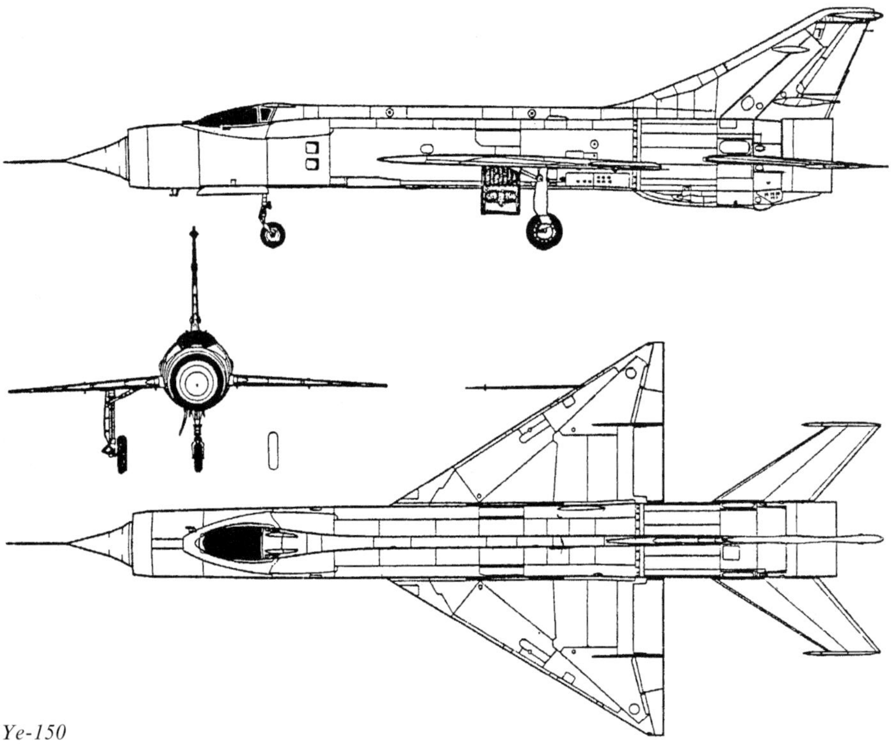

Ye-150

both prototypes, that with lift jets being called **23-01**. Mid-mounted wing scaled from MiG-21, similar powered ailerons and blown flaps (final compressor stage). Slab tailplanes mounted below mid-point of fuselage for pitch only. Sharply tapered vertical tail, powered rudder immediately above container for 21 m^2 cruciform drag chute triggered by long sprung rod hinged to rear of ventral fin. Entire fuselage and vertical tail similar to rival **23-11**, complete with same nose gear, unusually retracting to rear with twin 520 mm × 125 mm tyres and MRK-30 steering ±90°. Semicircular inlets with five-angle translating centrebodies separated from fuselage by wide boundary-layer bleed incorporating two ram inlets each side. Ducts taken straight past lift-jet bay, then curving sharply in past upright retracted KT-133 main wheels with 830 mm × 225 mm tyres; track 3460 mm, wheelbase 6130 mm. Lift bay covered by single louvred door hinged at rear with side panels. Two lift engines in tandem (gross thrust 29·5% of MTO weight) fixed at 85° with multiple vanes across nozzles to give ±25° fore/aft deflection. Armament: GSh-23 gun and two R-23 missiles on underwing pylons. With such modest jet lift no necessity for jet reaction controls. First flight by Ostapyenko 3 April 67. Even at this time scheme considered not worth severe penalties, though unarmed 23-01 displayed impressively at Domodyedovo 9 July 67. Did not fly again. ASCC reporting name 'Faithless'.

DIMENSIONS Span 7,72 m (25 ft 3⅞ in); length (excl boom) 16,8 m (55 ft 1⅜ in); wing area 40 m^2 (430·6 ft^2).

ENGINES One R-27-300, two RD-36-35.

WEIGHTS Empty 12 020 kg (26,500 lb); fuel c3700 kg; loaded 16 t (35,273 lb), (max) 18,5 t (40,785 lb).

PERFORMANCE Not explored except for TO (light) 180-200 m, landing 250 m.

23-11/1 Seemingly superior answer to MiG-23 problem. Wing mounted high (above inlet ducts) with fixed centre section without glove vanes, LE sweep 72°, carrying pivots 1,5 m from centreline on carry-through box bounded by Frames 18 and 20 of welded steel which also formed integral tank. Each outer wing SR-12s profile, 6·5/5·5% thick measured at zero sweep, fitted with four sections LE flap driven hydraulically, synchronized with TE flaps, 17° TO, 19° ldg; four sections TE slotted flap 25° TO, 50° ldg; and two spoilers ahead of two middle flap sections all opened to 45° to kill lift after landing, differentially until 45° sweep, then locked and roll controlled by tailerons, each 5,3 m span, LE sweep 57°, mounted shoulder-high on fixed root with ram-air inlet. Wings positioned by horn on front of front-spar root with universal joint driven by ball-screwjack rotated by main and standby hydraulic motors in SPK-1 drive system; separate circuit to each motor, each able to drive alone at half-speed. Nominal sweep angles 16°, 45°, 72°. Sweep entirely under pilot control, by lever on left of cockpit. Structural problem (eg non-redundant pivots) limited design load factor to 3·1 g. Vertical tail as 23-01, LE angle 65°, rudder driven by irreversible power unit with spring feel. Wing TE flaps, major section of tailerons and rudder aluminium honeycomb. Single lever controlling engine and narrow rectangular inlets with vertical splitter plate standing 55 mm away from fuselage, gap having central ram inlet for cockpit ECS and avionics cooling and boundary-layer channels above and below. Downstream of splitter inner wall of inlet completely perforated and variable in position and angle under control of UVD-23 hydro-electronic system activated at all Mach numbers above 1·15 to adjust pressure ratio between 11 (M1·15) and 4 (peak Mach), simultaneously adjusting taileron neutral angle. Two rectangular auxiliary suck-in doors in outer wall under wing glove, large pressure-relief holes in underside of fuselage between Frames 24/28.

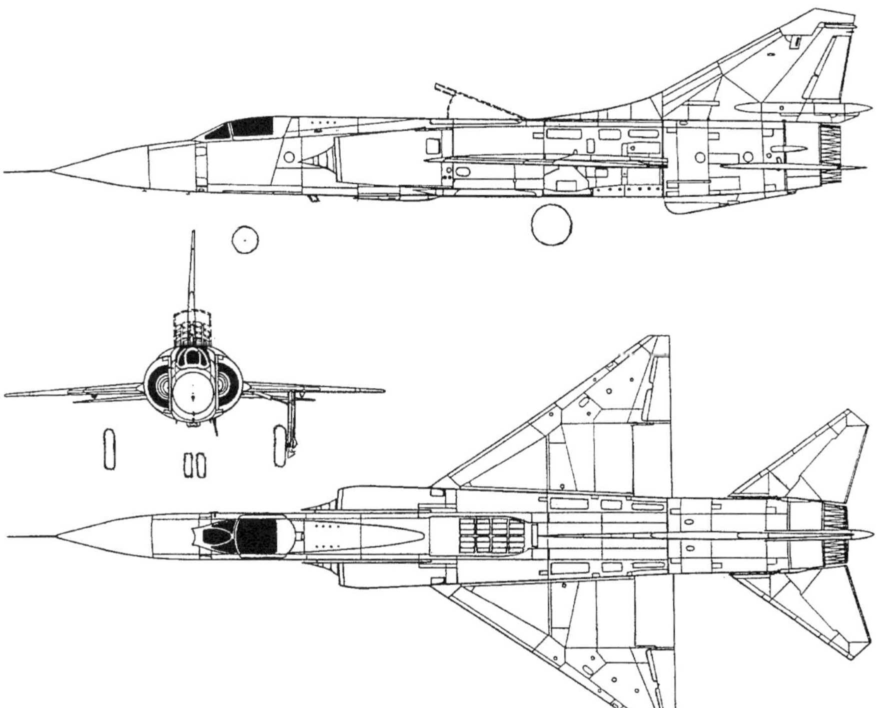

MiG-23PD, 23-01

23-11/1

Same bay incorporates boxes housing inner flap with wing at 72° normally closed by hinged flaps edged by Teflon. Rear fuselage detachable at Frames 28A/B for engine change. Frame 31 carrying taileron bearings, four long-chord door airbrakes, rear spar of fin and main spar of deep ventral fin folded to right by hydraulic ball-screwjack synchronized with landing gear. Nose gear as 23-01. Main gears enlarged MiG-21 type, wheels etc as 23-01, carried on single J-shaped leg horizontal with gear extended (track 2658 mm, wheelbase 5772 mm), pulled by near-vertical jack directly up and in while levered-suspension trailing arm is pulled down to stow wheel diagonally. PT-10370-65 drag chute (21 m^2) as on 23-01. Fuel in three fuselage integral tanks (1920 lit between ducts, 820 lit in wing centre box, 710 lit between Frames 25/28) and six integral wing tanks (62,5 lit, 137,5 lit and 200 lit in each wing, extending close to tip and feeding through pivoting telescopic pipes). Rear-fuselage 28-lit tank for injection water.

Team leader Mikhail R. Waldenberg. Aircraft completed with balancing mass in place of radar, and without armament. First flight by Fedotov 10 June 67, impressive display at Domodyedovo (Flight 14) when wings repeatedly cycled. Despite greater size/power fuel burn per kilometre at max sweep 25% less than MiG-21. In early 68 engine changed at 25 h, later in year replaced by R-27F2-300, AP-155 autopilot installed, APU-13 pylons with AAM rails added under fuselage and wing gloves and inert R-3 and R-23 missiles fired. OKB testing complete 9 July 68, aircraft later moved to Monino museum. Two further prototypes, third (call sign 234) with RP-22 radar and pitot boom offset to upper right of nose. ASCC reporting name 'Flogger'.

DIMENSIONS Span (16°) 13,965 m (45 ft $9\frac{3}{4}$ in), (72°) 7,779 m (25 ft $6\frac{1}{4}$ in); fuselage length (excl probe) 15,795 m (51 ft $9\frac{7}{8}$ in); wing area (16°) 32,1 m^2 (345,5 ft^2), (72°) 29,89 m^2 (321,75 ft^2).

ENGINE One R-27F-300, later F2-300.

WEIGHTS Empty 9290 kg (20,481 lb); fuel 3530 kg; loaded 12 860 kg (28,351 lb), (4 missiles) 13 300 kg (29,321 lb).

PERFORMANCE Limiting IAS 1178 km/h; max speed (16°) normally 800 km/h (497 mph), (72°) 2260 km/h (1,405 mph, Mach 2,13); service ceiling 17,2 km (56,430 ft); range (16°, 2 missiles) 2045 km (1,271 miles); TO speed/run 270 km/h, 320 m; landing speed/run 230 km/h, 440 m (750 m without drag chute).

23-11, MiG-23S Interim series version. Airframe as 23-11/1, imposing 3·1 g manoeuvre limit, but uprated engine. Pulse-doppler S-23 (*Sapfir 23*) radar not ready and low-power RP-22 installed (from 5th aircraft) in shorter radome of new dielectric material, consigning aircraft to operational training role with little to offer over MiG-21. Normal armament one GSh-23L with 250 rounds recessed under fuselage, two fuselage and two glove pylons for R-3R/R-3S missiles. Chief additions: centreline attachment for 800-lit drop tank, windmill-driven emergency hydraulic pump, retractable landing light under each inlet duct, brakes for nosewheels; electrically heated rear-view mirror in fixed fairing in top of canopy. SAU-23 flight-control system, tested in automatic interceptions and coupled ILS landings; SARP-12G centralised fault-warning; ARK-10 radio compass; ASP-PF sight with combined display for radar and added TP-23 IR detector in small gondola under nose. First flight by Fedotov 28 May 69. Prolonged testing of gun, missile firing, radar and automatic flight modes. Series production of about 50 ending late 1970. ASCC name 'Flogger-A'.

DIMENSIONS as 23-11/1.

ENGINE One R-27F2M-300.

WEIGHTS About 400 kg heavier throughout than 23-11/1.

PERFORMANCE Max IAS with missiles 1280 km/h (795 mph); Mach limit 2·27; service ceiling (clean) 18 km (59,055 ft); range (wings 16°, 4 missiles) 1800 km (1,118 miles), (with tank) 2500 km (1,553 miles); TO run 550-700 m; landing run (chute) 450-600 m.

23-51, MiG-23UB Tandem-seat pilot training version authorised 17 Nov 67, two prototypes built at OKB in parallel with 23S. Fuselage modified between Frames 8 and 18 to accomodate rear (instructor) cockpit at higher level. Two KM-1 seats (H=0, V=130-1250 km/h). Front cockpit with blind-flying hood sliding forward on side rails, canopy enlarged, sloping up to hinge on inter-cockpit crash arch with side windows, no rear mirror. Rear canopy pivoted to greatly enlarged dorsal spine, with 460 mm retractable periscope extending from metal centre of roof. Major redesign of tail with tailerons moved aft 0,6 m (24 in) to leave large gap between LE and TE of wing at 72° and leaving jetpipe which appeared not to project so far aft. Vertical tail likewise moved aft 0,6 m, but dorsal fin upstream of kink (Frame 26) unaltered, leaving dorsal fin greatly enlarged and kink moved to rear. Ventral fin not moved

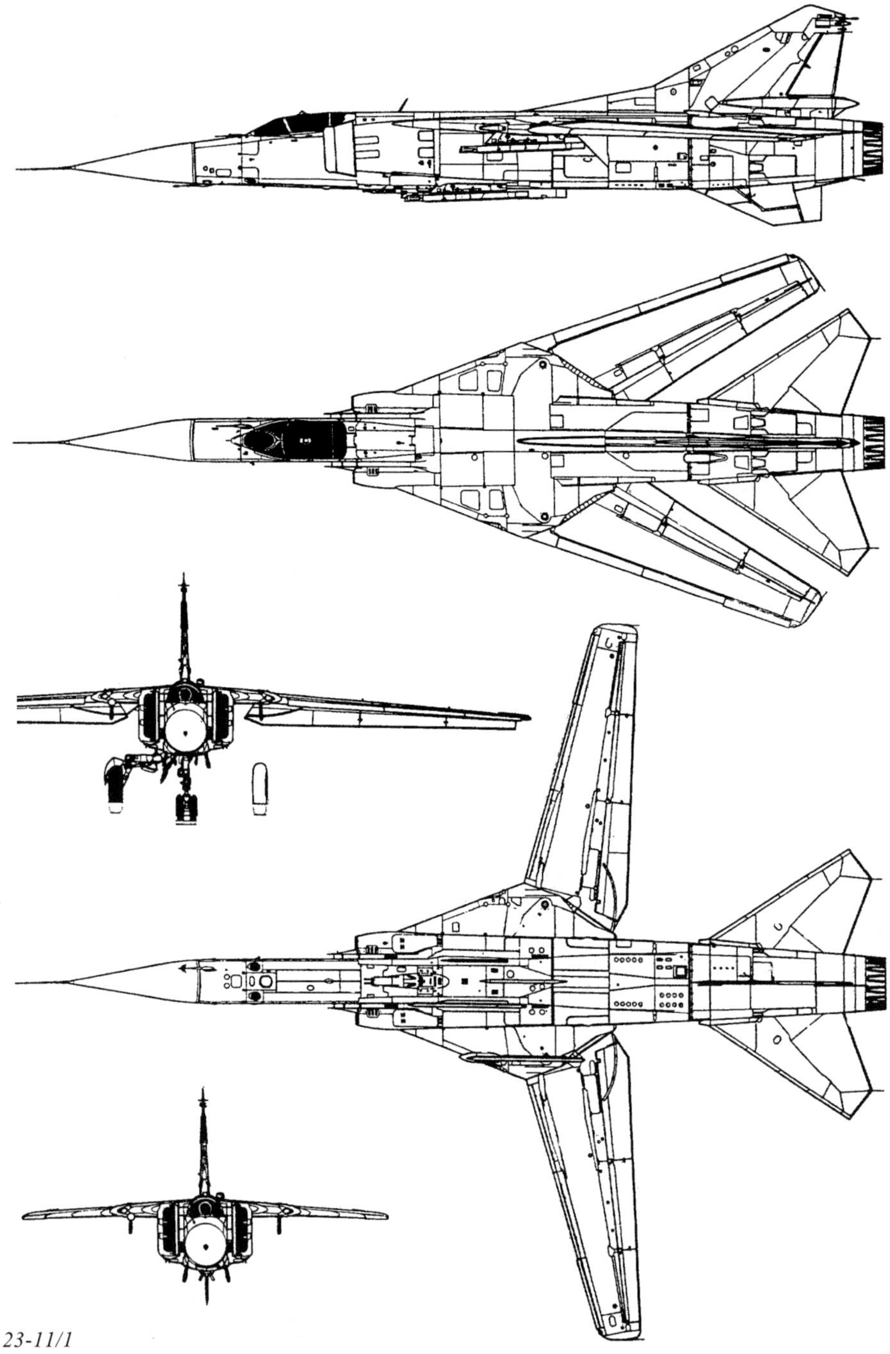

23-11/1

but vhf antenna added ahead of root. TE flaps made double-slotted. Braked nosewheels retained, with larger tyres causing small door bulges. Emergency hydraulic group retained by reducing size of No 1 tank (internal total 4000 lit, 880 gal), but 470-lit tank added in rear fuselage. SAU-23B flight-control provided with ARZ-1A artificial feel on all three axes, responding to airspeed/air density/ wing sweep, and also linked with *Polyot*-11-23 system comprising RSBN-6S homing/landing system, SKV-2N2 vertical gyro and heading system and DV-10 and DV-30 beacon transmitters. As before, ARK-10 radio compass, SRO-2 IFF, beacon receiver, SOD ATC/SIF, plus Sirena 3 RWR system feeding SORT central alarm display, RV-5 radar altimeter and target ident system. In emergency instructor can disconnect pupil's flight controls or eject pupil, though he has no weapon controls. SPU-9 intercom with MS-61 CVR, and comprehensive manoeuvre-limit and stall-warning with either SOUA high-authority AOA (angle of attack) limiter with UUA-1 indicator or SUA-1 AOA-warning system plus RIS stick-shaker. GSh-23L and 200 rds, four APU-13 pylons. Pupil has ASP-PFD-21 sight, without radar ranging (nose ballast only). For training with R-23R missiles guidance pod carried on glove pylon.

Two prototypes, first flown by Komarov 11 May 69. In production at Irkutsk 70-78. From 71 entire structure restressed to achieve design load factor 7. Outer wings redesigned as Type 3, omitting Type 2 (see MiG-23M) with LE extended over entire length to increase LE sweep by 2°40', increase area and provide large dogtooth claws at root to generate vortices at high AOA whilst moving CP forward, greatly improving manoeuvrability (four-section LE flaps retained). This was made possible by redesigned tail, enhanced by welding on plate strips 40 mm wide along entire TE of rudder and inboard 40% of taileron. Airbrakes slightly enlarged and strengthened for unlimited IAS by two deep chordal ribs, and moved back to Frame 32. ASCC name 'Flogger-C'.

DIMENSIONS, ENGINE As 23S, except (almost all) wing area (16°) 37,35 m^2 (402 ft^2), (72°) 34,16 m^2 (367·7 ft^2); tailplane span 5510 mm, area 6,01 m^2; fin/rudder 16,0 m^2; spoilers (2) 2640 mm, 1,2 m^2; flap (2) 5,9 m^2.

WEIGHTS Empty (Sov) 10 950 kg (24,140 lb); fuel 4 t; loaded 15 740 kg (34,700 lb), (max) 18 t (39,683 lb); (export) 8700/14780/17640 kg; max landing 11,8 t.

PERFORMANCE IAS limit (16°) 800 km/h, (72°) 1200 km/h (746 mph) at SL, 1400 km/h at 5 km, 2500 km/h (1,553 mph, M 2·35, clean), M limits 2·0 (R-3S missiles), 0·8 (tanks); climb 1·4 min to 11 km; service ceiling 15,8 km (51,840 ft); max range 3000 km (1,864 miles); TO 1,2 km, ldg 260 km/h, 1,2 km.

23-11, MiG-23M, MF, MS Fighter, most numerous variant. Differed from 23S in wing, engine, radar, tail and equipment. Structure, especially wing centre box, strengthened for combat load factor 8 g at up to Mach 0·85 (wings 45°) and 7 g over 0·85. Wing initially Type 2, with geometric changes of Type 3 but with LE flaps omitted in order to meet 8 g requirement. Tail and airbrakes as UB, but engine replaced by R-29-300, with fully variable afterburning, giving increased flight performance and sustained agility. Attachment for jettisonable a.t.o. rocket each side of Frame 24. Usual RSIU/SRO/SOD/ARK/Sirena etc, but doppler added with quad antenna in blister on left flank of forward fuselage, and S-23D-Sh fire-control at last available, comprising *Sapfir*-23D-Sh radar and TP-23 IRST feeding combined. display on ASP-23D HUD sight. SAU-23A flight-control with ARZ-1A feel. Fuel system as UB but internal 4700 lit (1,034 gal) with Tank 4 at rear; 28-lit water retained, three external tanks, slightly smaller at 790 lit each. Armament as 23S, with wide range of air/ground stores plus ability to guide R-23R and carry two R-60 on each body pylon.

First flight by Fedotov June 72. From late 73 airframe again upgraded by improving quality control of centre box and pivots (earlier aircraft reinforced), second hinge added at centre of rudder and wings replaced by Type 3 (see UB) with large spigot on centre spar and plumbing to carry 800-lit tank on fixed pylon when ferrying with wings at 16°. Subsequently about 2,000 delivered from three plants by 76, many used examples for sale. Export MiG-23MS with no BVR capability: RP-22 radar, no IRST, no R-23 missiles, no doppler and other limitations, but supplied to Angola, Cuba, Egypt, Libya, with customer paint schemes. Some used briefly by Frontal Aviation. Export MiG-23MF differs from MiG-23M only in details and customer fits, supplied to Bulgaria, Czechoslovakia, E Germany, Hungary, India, Poland, Romania, Syria, Vietnam. All sub-types initially suffered high attrition, mainly from failure of wing box or sweep mechanism. Later gained high reputation for reliability and maintainability. Poor showing in unskilled hands caused superior FA versions to be under-rated. ASCC name 'Flogger-B'.

DIMENSIONS Span as 23-11/1; length overall 16,7 m (54 ft 9½ in); length of fuselage (excl probe) 15,88 m (52 ft 1¼ in); wing area (16°) 37,35 m^2 (402 ft^2), (72°) 34,16 m^2 (367·7 ft^2)

ENGINE One R-29-300.

WEIGHTS Empty 8,2 t (18,078 lb); fuel 3800; loaded 15 750 kg (34,722 lb); (max weapons, internal fuel) 18 400 kg, (one tank) 19 130 kg, (2 tanks) 19 940 kg, (3 tanks) 20 670 kg (45,569 lb).

PERFORMANCE IAS and Mach limits as UB, but service ceiling 17,5 km (57,415 ft), climb to 11 km in 1·4 min; range (wings 16°), clean 2200 km (1,370 miles), (3 tanks) 3300 km (2,050 miles); TO 580-600 m, ldg 255-275 km/h, 750-800 m.

23-12, MiG-23ML Improved fighter, designation from *Legkii*, lightened. Smaller dorsal fin (LE kink hardly noticeable) and folding ventral (TE moved forward), made possible by removal of No 4 tank, internal fuel 4300 lit, 28 lit water. Engine similar in size, weight and airflow, but greater thrust and improved sfc. Airframe cleared to 8·5 g to Mach 0·85, 7·5 g at higher speeds. Two inboard TE flap sections made as one, and LE flap made in lighter but stronger sections. Tyre pressure (main) 11±0,5, (nose) 7±0,5. KM-1M seat (0 m, 130-1250 km/h). Upgraded flight-control SAU-23AM, linked to *Polyot* with TAS channel and height capture and hold. Radar S-23ML, IRST TP-23ML and combined sight ASP-17ML, all with improved performance; radar lighter, with added dogfight mode; IRST bigger optics in enlarged fairing. Nose gear strengthened and numerous antennas, AOA vanes and air-data probes added or repositioned. Eight stores pylons, tandem on centreline, flanks of fuselage, gloves and outer wings. Glove pylons equipped to fire UPK-23-250, each housing GSh-23L with 250 rounds. In production 76-81. Team from Kubinka with various items removed visited Finland and France 78, first modern Soviet fighters seen in West. Export sub-types sold to Czechoslovakia, East Germany, Iraq, North Korea, Syria and South Yemen. All in FA service subsequently upgraded to MLD standard. ASCC reporting name 'Flogger-G'.

DIMENSIONS As 23M.

ENGINE One R-35-300 (export usually 29-300).

WEIGHTS Empty 8,2 t (18,078 lb); fuel 3310 kg; loaded 14 800 kg (32,628 lb); (max, 3 tanks) 17 800 kg (39,242 lb).

PERFORMANCE Max speed (SL, 72°) 1350 km/h (839 mph, Mach 1·1), (11 km, 16°) 940 km/h (584 mph), (72°) 2500 km/h (1,553 mph, Mach 2·35); service ceiling 18,5 km (60,700 ft); range (16°, clean) 1950 km (1,212 miles), (3 tanks) 2820 km (1,752 miles); TO 1,2-1,3 km, ldg 255 km/h, 1,2-1,3 or (with chute) 750-800 m.

23-14, MiG-23P Sub-type of ML equipped for automatic interception with PVO. Digital data-link and on-board computer linked to SAU-23AM flight control to steer aircraft on best trajectory to target. Display indicates to pilot when to engage afterburner, fire gun or launch particular missile. ASCC name 'Flogger-G'. Data as ML.

23-18, MiG-23MLD All ML in VVS updated during normal cycling through refurbishment 81 onwards, designation from *dorabotanyi* = 'given finishing touch'. LE root of wing glove cut back to form 'claw' to generate vortex at high AOA; smaller vortices from small sharp-edged plates at root of nose probe. SOS-3-4 system added to retract LE flaps when sweep exceeds 33° or when speed exceeds 900 km/h plus 'wings deployed' warning if sweep is not 16°. R-73A missiles added, plus pivoting outer-wing pylons (replacing fixed), new RWR subsystem governing chaff/flare dispensers added above fuselage on each side. Gun/AAM firing simulation system added. Nosewheel mudguard often removed. Data as ML. ASCC reporting name 'Flogger-K'.

Product 32, MiG-23B In 69 OKB began study of attack aircraft, considering all-new 'jet Shturmoviks' before deciding best answer was to use MiG-23S as basis. After argument, supersonic performance retained, but aircraft tailored to attack mission from front-line strips. Heavier engine for increased power and better fuel economy. All landing gears strengthened and fitted with high-capacity brakes and low-pressure tyres, nose tyres larger diameter causing blisters on doors and main gears with wider 10+0,5 kg/cm^2 tyres causing lateral bulge of centre fuselage. Track 2728 mm, wheelbase 5991 mm. Fuselage tankage enlarged, total internal 5380-5420 lit, all tanks protected by inert-gas system. Comprehensive ECM/IRCM systems (see MiG-27). Armour slabs scabbed on each side of cockpit. Ahead of cockpit *Utkanoz*, ducknose, made possible by removing radar, giving improved view ahead. Main nav-attack system PrNK-23S *Sokol* with twin-gyro platform, doppler, radar altimeters and simple computer for air/ground gunfire or delivery of free-fall bombs or rockets. GSh-23L retained. Six stores pylons, four as on 23S (side-by-side mid-fuselage and under gloves) and one projecting on each side aft of main gear; also centreline for 800-lit drop tank. Loads could include six TN (tactical nuclear) bombs of five yields or FAB-500 or UV-16-57 or 32-57 rocket launchers or 18 FAB-100, or other stores including two UPK-23-250 gun pods. With Type 3 wing using jettisonable fixed pylons three 790 or 800-lit tanks could be carried with other loads subject to overall weight limit; outer wing pylons could carry UV-32-57, for example, but this capability (applicable to later versions) seldom used, neither are heavy stores often carried on rear-fuselage pylons because CG shift then precludes rear-fuselage fuel. These attachments alternatively used for a.t.o. rockets.

First article **Product 32-21, MiG-23B**, callsign 321, flown by Ostapyenko 20 Aug 70. Interim aircraft, 24 built. Allocated ASCC reporting name 'Flogger-F'. Replaced by **32-**

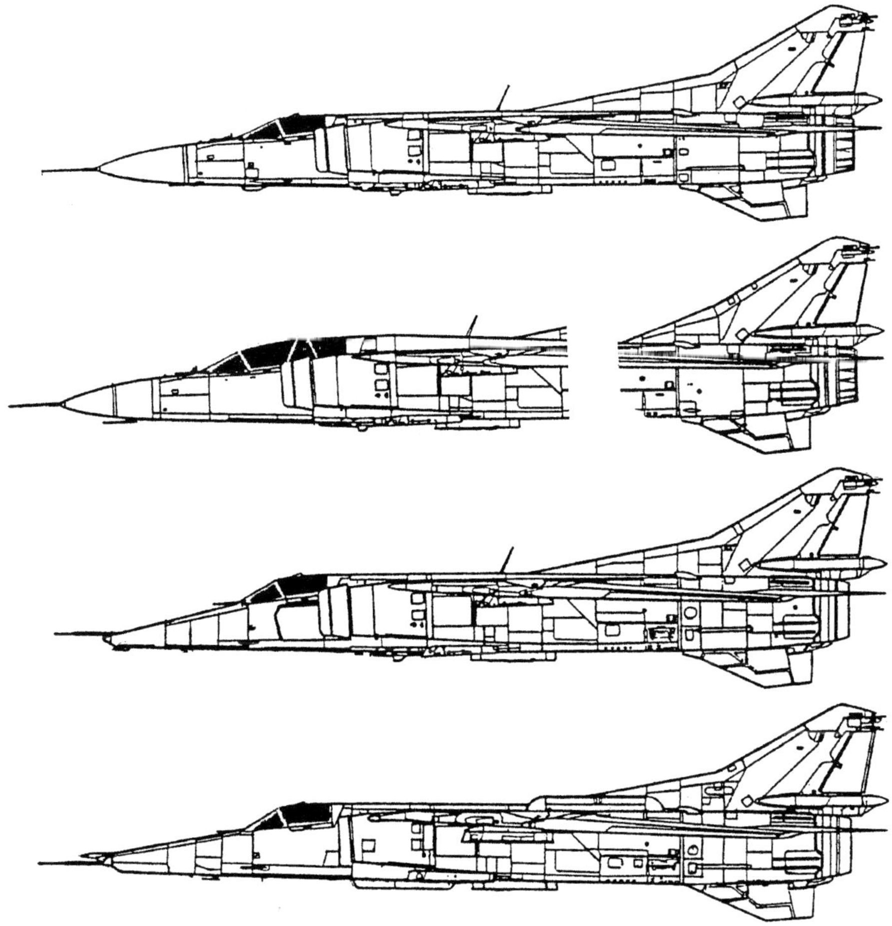

From the top: MiG-23M, MiG-23UB, tail of 23ML, MiG-27BM, MiG-27M

23, MiG-23BN, with Type 2 or (later standard) 3 wing and same fuselage/tail as MiG-23M and R-29B-300 engine with simple afterburner with short two-position nozzle. Nav-attack PrNK-23N, with laser ranger and command guidance for Kh-23 in pod on right glove pylon; some aircraft added TV camera in similar pod on left. Built from 72 in parallel with **32-25, MiG-23BM**, with PrNK-23 controlling trajectory and weapon release via digital computer, and **32-26 MiG-23BK**, with further eqpt changes, notably SG-1 RWR blister on each flank of forward fuselage. These are given the reporting name 'Flogger-H'. Made in large numbers and repeatedly updated, most BN to BK standard. Exports Algeria, Angola, Bulgaria, Cuba, Czechoslovakia, Egypt, Ethiopia, East Germany, India, Iraq (with locally-installed Dassault fixed flight-refuelling probe), Libya and Sudan. East German aircraft known locally as MiG-24.

DIMENSIONS Span (16°) 14,0 m; fuselage length (excl boom) 15,349 m (50 ft 4¼ in), length overall 16,88 m (55 ft 4½ in).

ENGINE (23B) one AL-21F-300, (others) one R-29B-300.

WEIGHTS (BN) Empty 10,7 t (23,589 lb); fuel 4500 kg; loaded (clean) 16,75 t (36,927 lb), (3 tanks + 4 UV-16-57) 18 600 kg (41,005 lb), (6 FAB-500) 18 900 kg (41,667 lb).

PERFORMANCE IAS limit (16°) 800 km/h (M 0·8), (45°) 1100 km/h, (72°) 1350 km/h; max speed at 11 km 1900 km/h (1,181 mph, M 1·78); climb 11 km in 1·7 min; service ceiling 16,8 km (55,120 ft); radius (lo-lo-lo, 5 min over target), 400 km with six FAB-500, 500 km four FAB-500, 600 km four FAB-250; TO 650-700 m, ldg 260 km/h, 800-850 m.

MiG-27D

MiG-27KR

MiG-27 Uncompromised attack aircraft based on 32-25 (MiG-23BM) but with simple fixed-geometry inlets with bulged outer profile and small unperforated splitter plates standing 80 mm away from fuselage. Type 3 wing, with auto rapid-action sweep control to suit CG and IAS. Internal fuel 5400 lit, plus 28-lit water. Long pitot boom above right of nose and q-feel probe on right of windscreen as on 23B, PrNK-23 nav-attack, SAU-1 flight control with SUA-1 AOA warning plus stick-shaker, INS plus KN-23 nav computer, SG-1 RWR and IR warning feeding RI-65 16-channel vocal and display warning, SPS-141 IR jammer, SRZO or SRO-1P IFF, SO-69 ATC/SIF transponder, SP-50 ILS with forward antenna under centre of nose and rear at tip of fin, usual uhf/vhf, RV-5R/RV-10 radar altimeters, *Fone* laser ranger (later aircraft) and (some aircraft) Kh-23 guidance pod. Pylons: centreline for 790-lit tank; under inlets, under gloves and (reduced fuel) on sides of rear fuselage for stores to 500 kg; under outer wings (16° only) for 790-lit ferry tanks. Total weapon load 4 t (8,818 lb), including up to eight TN/FAB-500/PLAB/FAE, UV-32-57 or S-24, or nine FAB-250 or 22 FAB-100 or four KMG-U each dispensing 48 AO-2·5 bomblets, or two Kh-23 or Kh-29 missiles or SPPU-22 pods with GSh-23L gun with depressed barrels. Can carry three-camera recon pod or self-defence R-3/13/60 missiles. Later given twin dorsal KDS chaff/flare dispensers.

First flight by Valery E. Menitsky 73. In production 74 and, together with 27K, in three FA regiments in East Germany 75, when previous 23B family unknown in West. MiG-27K replaced 27 in production 75, with wider and deeper nose housing PrNK-23K *Kara* nav-attack system for automatic trajectory control, gun firing and weapons release with PMS mode for control during manoeuvres and PKS mode for control in cloud or at night using programmed target co-ordinates. SUV weapon selector, computer-managed EW/ECM/IRCM subsystems. *Fone* laser with less-sloping window, later aircraft with designator (see 27D). Scabbed armour often removed or replaced by structural flush armour. GSh-23L replaced by exposed GSh-6-30 with 260 rounds; later given cover over mechanism/feed and gas deflector vanes. Can guide smart bombs and lay smoke-screen. Produced for FA and AV-MF. ASCC reporting name 'Flogger-D' for basic -27 and -27K.

DIMENSIONS As MiG-23M except fuselage length (excl boom) 15,489 m (50 ft 9¾ in), length overall 17,076 m (56 ft 0¼ in).

ENGINE One R-29B-300.

WEIGHTS Empty 11 908 kg (26,252 lb); fuel 4560 kg; normal loaded (unpaved strip) 18 100 kg (39,903 lb); max (27) 20 300 kg (44,750 lb), (27K, 8 FAB-500) 20 670 kg (45,570 lb); 1/2/3 tanks increase TO weight by 750/1530/2280 kg, so in theory MTO weight is 22 950 kg (50,595 lb).

PERFORMANCE Max speed (SL, clean) 1350 km/h (839 mph, Mach 1·1), (8 km) 1885 km/h (1,171 mph, Mach 1·7); combat radius (lo-lo-lo, two Kh-29) 225 km (140 miles), (two Kh-29, three tanks, 7% reserve) 540 km (336 miles); TO run 950 m; landing speed/run 260/270 km/h, 900 m (chute).

MiG-27D Product 32-27 First definitive version, replaced original 27 and 27K from 80. Wing roots extended forward to end at ECM jammer each covering 100° arc behind inlet lip.

Externally identical aircraft (with RSBN navigation system and different construction number prefix – 619 not 837) designated **MiG-27M** (possibly **32-29**) by Soviet air forces. Both types serving (in two separate Regiments) with GSFG until 1993 withdrawal. Indian aircraft (OKB **MiG-27L, Product 32-29L**) externally identical, known locally as MiG-27M and (briefly) as *Bahadur* (Valiant). 165 aircraft 84-93, licence assembled with progressively higher indigenous content by HAL Nasik, leading to local airframe manufacture.

MiG-27KR New nose shape, with 'pimple' radome housing terrain-avoidance (following?) and ranging radar above new broad oval window for new LLTV system, (or, alternatively, for Klen laser rangefinder, target seeker and designator). New undernose fairing (lower, wider and further aft than on 27D/27L) with wide rectangular 'window'. Reportedly houses TV tracker or new *Kaira* laser designator. Very similar to *Kaira-24* fitted to Su-24M. Pitot (stbd) and ILS (port) mounted lower on nose. Picture caption in Belyakov/Marmain book led to erroneous belief that this aircraft was MiG-

27M (32-29) but Soviet air force use MiG-27K designation, and OKB sources indicate MiG-27KR. Configuration not described in Belyakov/Marmain because still classified. ASCC "Flogger-J2".

Ye-151 Unbuilt projects (two) with TKB-495 gun on pivoting mount, one in nose, second aft of cockpit. Led to improved inlet and duct on 152.

Ye-152 Long before Ye-150 was completed plans were prepared for improved versions. Fedotov/Mosolov emphasized great superiority in climb/acceleration of 'heavy fighters' over MiG-21, but R-15 was unreliable, so **Ye-152A** was designed with two MiG-21F engines in area-ruled fuselage widened at rear. Structure restressed to 7 g. Inlet redesigned with fixed three-angle cone but translating lip carried on four slides with jacks driven by air-data system to three positions. Inner ring and four dump doors as before. Wing and tail changed only by wider rear fuselage (tailplane span 5850 mm) and modified fin LE (smaller dorsal fin similar to MiG-21F-13). Canopy and spine as Ye-150, despite poor pilot view. Nine fuel tanks (one between wheel bays and one in each wing) total 4400 lit (968 gal). Airbrakes redesigned with two above fuselage and one below all at TE; two large inclined ventral fins and twin braking parachutes. Fixed rear fuselage, no ejector. TsP-1 radar coupled through autopilot and SRP computer to fire-control for two K-9 AAM on underwing pylons, no guns. RSIU-4V vhf, SRO-2 or 2M IFF, ARK-54N radio compass. *Meteorit* beacon, *Sirena*-2 at fin tip. First flight by Mosolov 10 July 59 (long before Ye-150); 55 flights by 6 Aug 60, last four with pylons and last two with missiles. Displayed Tushino 61, continued as carry-trials testbed, crashed 65 killing OKB pilot A. Kravtsov. OKB tinkered with related prototypes with different wing sizes and R-15 engine. ASCC name 'Flipper'. Misidentified as MiG-23.

Ye-152A (with K-9s)

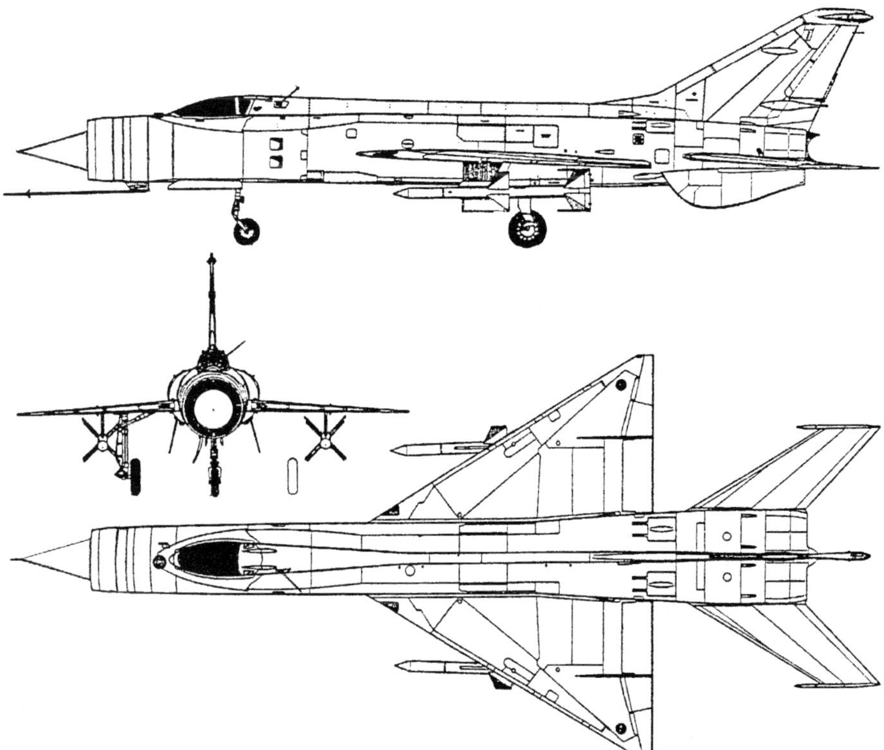
Ye-152A

DIMENSIONS As Ye-150 except length 19,0 m (62 ft 4 in) and wider fuselage reduced wing area to 34,02 m^2 (366·2 ft^2).
ENGINES Two R-11F-300.
WEIGHTS Empty 8740 kg (19,268 lb); fuel 3560 kg; loaded (clean) 12,5 t, (max, AAMs) 13,55 t (29,872 lb), (+600-lit tank) 13,96 t (30,776 lb).
PERFORMANCE Max speed at 20 km (AAMs) 2135 km/h (1,327 mph, M2·0), (clean) 2500 km/h (1,553 mph, M2·35); climb 10 km 1·48 min, 20 km 7·64 min; service ceiling 19,8 km (65,000 ft); TO/ldg (no chute) 1295/1600 m.

Ye-152-1 Basically Ye-152A with bigger wing and single Ye-150 engine. Fixed inlet cone housing *Uragan-5B* radar, with band 300 mm wide downstream of final (normal) shock covered in close perforations, air dumped overboard through aft-facing ejectors above and below forward fuselage, translating outer lip. Major difference insertion of extra fuselage bay and combined increase of span and reduction of LE angle (to 53°47') to increase chord to give greater area and enable AAMs to be attached direct to tip. Greater span increased track to 4200 mm and wheelbase to 6265 mm. Redesigned ailerons with BU-120M power units (overwing fairings) to eliminate flutter. Shallow fence under wing from LE to tip of flap. Vertical tail redesigned with reduced height, rudder moved down and driven by BU-120M but tailplanes unchanged. Extra fuselage bay increased gap between wing and tail and increased internal fuel to 4960 lit (1,091 gal) in 6 fuselage cells (fasteners visible externally) and 4 wing, with provision for 1500-lit drop tank. Single large airbrake under fuselage at wing LE. Entire tail removable for engine change. Considerably enlarged centreline ventral fin, with adjacent box for PT-5605-58 chute. Single PVD-7 instrument boom mounted well back under nose. Arrived LII field 16 Mar 61, first flight by Mosolov 21 April. Set 100-km circuit record 2401 km/h 7 Oct 61; 15/25-km record 2681 km/h 7 June 62, both reported as **'Ye-166'**. By 11 Sept 62 (programme halted incomplete by cancellation of K-9) made 11 flights without AAM rails, next 51 with empty rails and final five with K-9s fitted. ASCC name 'Flipper'.
DIMENSIONS Span 8,793 m (28 ft 10¼ in); length (excl boom) 19 656 m (64 ft 5⅞ in); wing area 40,02 m^2 (430·8 ft^2).
ENGINE One R-15-300.
WEIGHTS Empty 10,9 t (24,030 lb); fuel 4150 kg; loaded 14,35 t (31,636 lb), (max) 14,9 t (32,848 lb).
PERFORMANCE (clean) Max speed (10 km) 2510 km/h, (15,4 km) 3030 km/h (1,883 mph, Mach 2·85); climb to 10 km in 84 s, 20 km in 5·33 min; service ceiling 22,68 km (74,410 ft); range (with tank) 1470 km (913 miles); TO/ldg 1185/1270 m.

Ye-152-2 Second aircraft differed in detail. Centrebody with wider band of perforations,

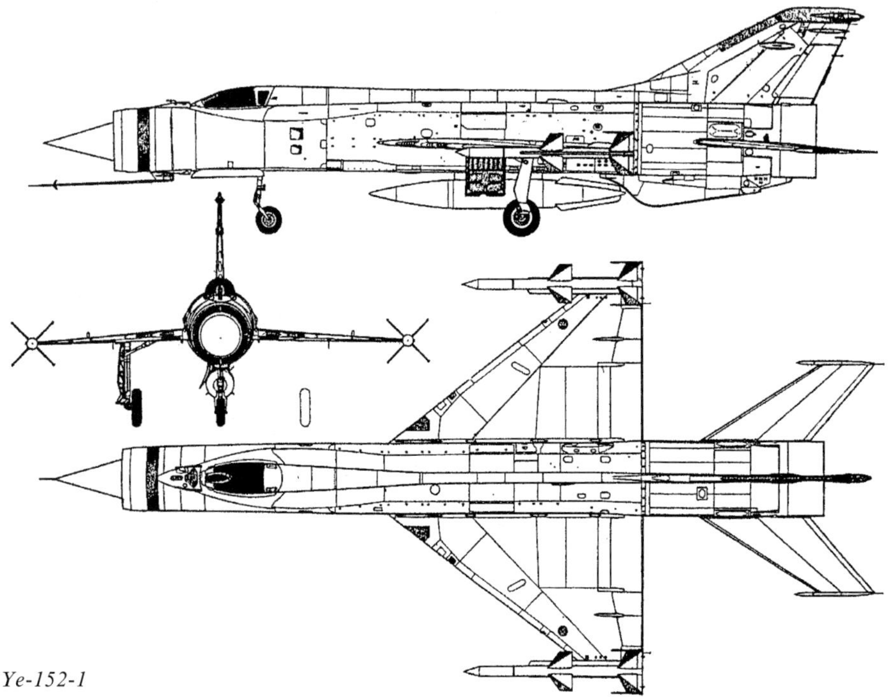

Ye-152-1

mid-span (but canard roots were built-in, showing parallel double-wedge profile with marked AOA and 1872 root chord). Abandoned 63 in favour of Ye-155, but displayed Domodyedovo July 67 with three red stars and title 'Ye-166', but in fact records were set by Ye-152-1. Preserved at Monino. ASCC 'Flipper'.

DIMENSIONS Span as Ye-152 (152M was to have been 10,3 m); length as Ye-152, but fuselage from lip to nozzle shortened by removal of ejector from 16603 to 16350); wing area 42,89 m² (462 ft²) (152M intended 44,5 m² (479 ft²).

ENGINE One R-15B-300.

WEIGHTS Empty 11,44 t (25,220 lb); fuel (with drop tank) 6,8 t; loaded 18469 kg (40,716 lb).

PERFORMANCE Level speed reached 2681 km/h.

designed to house TsP-1 radar as part of *Smerch* fire-control matched to K-80 missiles on wingtips. Tank sequence altered to use No 6 first to improve longitudinal stability. Arrived Zhukovskii 8 Aug 61, Pyotr Ostapyenko starting flight programme 21 Sept; reached 2740 km/h and 22,5 km, but engine troubles halted at 16th flight July 62. Aircraft rebuilt as 152M. Data as 152-1.

Ye-152M, 152P Planned as definitive interceptor with greater internal fuel, uprated engine and 'perfected' missile and avionics. Usual six fuselage tanks (#1 to 6, 550, 1100, 1120, 120, 460 and 380 lit, total 3730) plus four 300-lit wing integral cells, one ahead of main spar and one behind on each side, now added large spine tanks (#1 to 3, 750, 630 and 380 lit, total 1760), giving internal fuel 6690 lit (1,472 gal). Larger dorsal fin with bullet fairing at kink on LE. Uprated engine with fully profiled con/di nozzle, ejector eliminated. In original form **Ye-152P** designed with foreplanes (scaled up from Ye-6T-3) and wingtips with sharp dihedral and deep vertical pylons for K-80 AAMs. Aerodynamic and structural problems led to elimination of upswept tips and return to missiles attached direct to broad tip, as **Ye-152M**. First flown mid-61 without foreplanes but with weighed dummy K-80s; these fluttered and when actual missiles were carried it was clear launch trajectory would be unpredictable. Intention was to modify aircraft as Ye-152M with 754 mm tip extensions which could fold down to form AAM launchers (whilst increasing torsional rigidity of outer wing), lie horizontal to reduce wing loading or hinge diagonally up as winglet. These were never fitted, nor large (3500 span) powered canards with anti-flutter rods at

Ye-155, MiG-25 This programme ran earlier than MiG-23, and followed from Ye-152M. Reason was threat posed by XB-70 and Lockheed A-12. Despite experience with Ye-150/152 up to same peak speed (3000 km/h class) this programme was to pose OKB's greatest challenge, from sustained high speed causing heat soak to 300C, with radar and missiles able to destroy any hostile winged target at any Mach number or height from 25 km to sea level.

CAHI worked with OKB to determine configuration, decided in 61 before programme authorized. Unique layout with thin high wing passing above ducts to two engines separated at front by width of small forward fuselage, with ducts from fully variable inlets of form decided 62. Twin outward-canted vertical tails (62 decision on fixed fins) and outward-canted ventrals. Major 61 decision, by A. I. Mikoyan, to use welded steel for most primary structure, with 11% D19 heat-resistant light alloy (similar to British RR.58) and 8% titanium alloy for hottest parts. Engine existed, flown on cruise missiles and Ye-150/152P. Go-ahead Feb 62 for interceptor MiG-25P and reconnaissance MiG-25R; later authorization for PU and RU trainers. First flight by Ye-155R-1, first prototype of recon variant, 6 March 64, followed by Ye-155P-1 on 9 Sept 64. No attempt to achieve high lift coefficient, nor typical fighter load factor. Latter set at only 4·5 g at over Mach 1, partly because at high speeds aircraft travels in essentially straight trajectory. IAS

Ye-152M (painted 'Ye-166', which it was not)

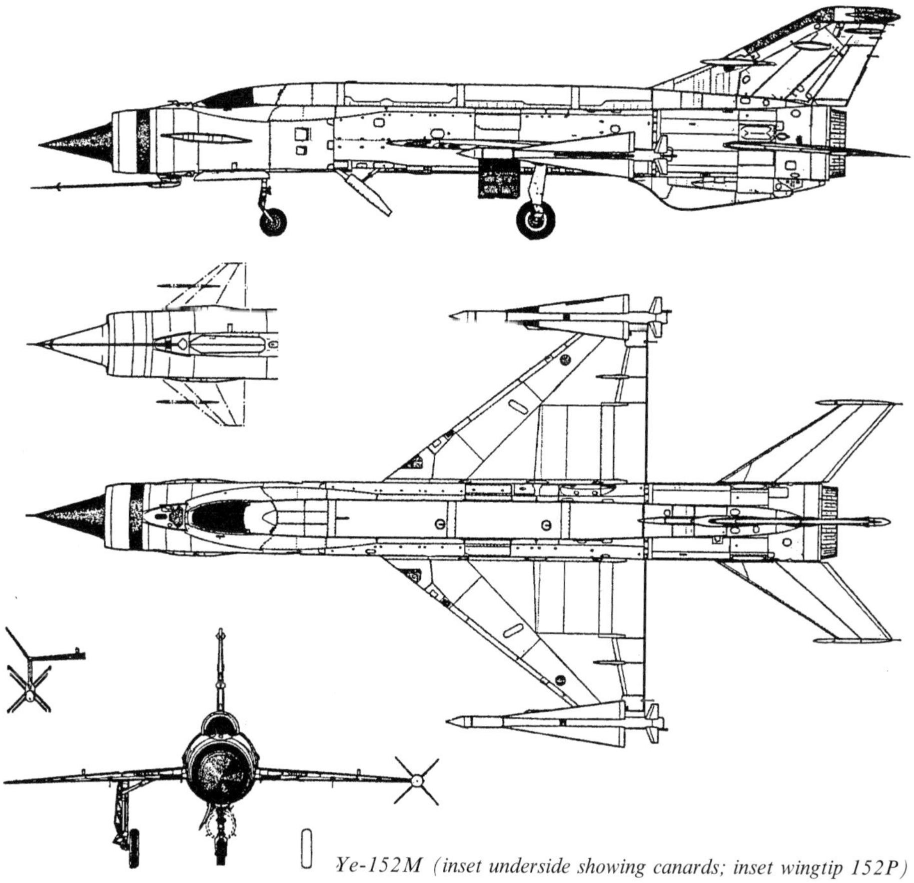
Ye-152M (inset underside showing canards; inset wingtip 152P)

design limit 1300 km/h. Objective: catch any aerial target and destroy from a distance with specially developed AAM. Thus, high wing loading and fast TO/landing.

Wing of SR-12s profile but reduced LE radius, t/c typically 4·4%, LE sweep 42·5° from root to outer AAM pylon (just over half semi-span), then 41° to tip. TE almost square to fuselage, so sweep at 25% chord only 32°. Initial dihedral 0°. Two main spars forming torsion box filled with fuel to LE kink in integral welded container carried within wing and not subject to major flight loads. LE fixed, machined titanium. TE with simple plain flap inboard and plain ailerons in inner/outer halves (total 2,82 m²) stopping well inboard from tip. Tip options: large delta winglet (equal above and below tip); 1200-lit tip tank with outward-canted delta underfin; large anti-flutter mass. Originally a single fence on upper surface from LE to stop just inboard of outer end of flap. Ahead of wing but at same level, large destabilizing canard surface, parallel double-wedge with mid-span anti-flutter body, normal AOA +2°. Fuselage oval profile, disappearing between large air inlets of vertical wedge form; sharp lips, horizontal bottom lip hinged and variable, upper wall with perforated throat also variable, both driven by jacks under electronic control which also opened large square discharge door in roof of duct. Fuselage again welded steel, except for titanium inlet lips and other areas subject to severe kinetic heating. Total 14 primary frames, Nos 6B, picking up wing front spar, 7 and 9 main spars and 10 and 11 rear spar and flap inboard hinge. Engine inlets well outboard of fuselage, with gap above and below for boundary layer, ducts curving in to leave bays outboard for main landing gear and to meet close-together engines in fireproof compartments with inner liners of thin steel with 30-micron coating of silver (max 5 kg silver per aircraft). Engines as on Ye-152P. Four fuselage integral tanks, which with wing tanks gave total 17 660 lit (3,885 gal). Entire nose erosion-resistant dielectric (good for 300°C) covering scanner of *Smerch-A* fire control with TsP radar with limited look-down capability, with 18 avionic LRUs in next two fuselage bays. Main Frame 1 in line with flat bulletproof windscreen, KM-1 seat, canopy hinged to right between upper inlet lips. Spine downstream of canopy ending in tube for twin drag chutes (circular 30 m² each or cruciform 25 m²) with large door hinged upward at front. On each side of this a door-type airbrake synchronized with single ventral brake. Twin fins canted out 11° with machined titanium LE at 60°, each 7,8 m², no dorsal fin, small inset rudder, like ailerons driven by BU-120M power units. Twin ventrals incorporating retractable skid, same lateral spacing as fins but inclined 15°. Slab tailplanes, machined titanium LE at 50°, total 9,81 m². Main landing gears with single legs carrying single wheel outboard with multi-disc antiskid brake and tyre 1300 mm × 360 mm (12+0,5 kg/cm²), retracting forward to house wheel upright between duct and outer skin. Nose gear with hyd steered and braked twin wheels with tyres 700 mm × 200 mm (10+0,5 kg/cm²), retracting forwards. Track 2850, wheelbase 5139 m. Main and booster hyd systems, each serving

Ye-155R-1

one half of twin-chamber flight-control power units; four tanks pressurized by N_2 and four engine-driven pumps. Main: landing gear, air inlets, airbrakes, nosewheel steering and main/emergency brakes; booster serves emergency brakes. Taxi light on nose gear, retractable landing light under each inlet. Variants follow, starting with first to fly:

Ye-155R, Three development prototypes, initially similar to Ye-155P (see next) except for slim conical forward fuselage with interception radar replaced by reconnaissance sensors. These not fitted to #1155 (first aircraft), which merely had nose profile and flat panels. Canards and wing as other prototypes (kinked LE, dihedral 0°) but permanent 1200-lit tank on each wingtip, with outward-canted ventral. Tail fins each forming 350-lit integral tank, top of fin horizontal. First flown by Fedotov 6 March 64, suggesting need for new wing with reduced span but constant 41° sweep to leave area unchanged (5° anhedral as on 155P). On #3155 (Ye-155R No 3) this wing was fitted, tip tanks replaced by anti-flutter bodies. Vertical tails modified as 155P, but integral tanks retained. Overall capacity increased by giant (5300-lit) drop tank. Ventral fins slightly reduced in chord, rear corners cut off at angle. BD engines fitted. Cleared for production as MiG-25R at GAZ-21 Gorkil Feb 69, ahead of interceptor. Sensors, see MiG-25R. ASCC 'Foxbat'.

Ye-155P, Seven development prototypes of interceptor, all initially with canard and deep symmetric fins above and below slim pods on tips of horizontal wing. Nose more capacious accomodating large antenna of *Smerch-A* radar scanning 60° left/right. Associated sight K-10T. Intended IR sensor not ready. *Prizma* RPS hf with shunt antenna in LE of left fin, R-832M uhf/vhf with uhf blades under and above fuselage and vhf flush antennas in ventrals and top of left fin. *Lazur* beam/beacon receiver, RSBN-6S short-range navaid, SO-63 ATC/SIF transponder with antenna forming front top of right fin, MRP-56P beacon receiver, ARK-10 radio compass, SP-50 ILS, RV-UM or RV-4 radar altimeter, SRO-2M/SRZO-2 IFF with antennas ahead of windscreen and sometimes ahead of nose gear, and *Sirena*-3 radar warning with antennas facing out from mid-point of tip pods and aft from top of right fin. PVD boom on nose with twin yaw vanes and horiz ILS antenna; second pitot on right of windscreen. Single pylon under each wing at LE kink to carry R-40 missile.

First flight by Ostapyenko 9 Sept 64. By this time 155R flying had confirmed need for removal of canards; replacement of tip fins by heavy anti-flutter masses carried far ahead of tips of wing with −5° anhedral; addition of shallow fences at LE kink from 15% chord to stop ahead of joint between inner/outer ailerons; increase in fin chord at LE to area 8 m²; modification of tailplane drive system so that above 800 km/h IAS surfaces could be driven in opposition as tailerons; and reduction in depth of ventrals to avoid scraping ground. Later improvements: water/methanol injection to increase inlet air density at over Mach 1·5 with wing tanks feeding external pipe to injection tree projecting down between throat and engine; addition of second pair of pylons inboard (at main fences) to carry two R-40T and two R-40R; addition of CW illuminating transmitters on inner side of nose of each wingtip tube; and mudguard behind nosewheels. Difficulty with roll control and roll/yaw, notably matching of four ailerons and two tailplanes, and asymmetric missile hang-up at high speed. Though series parts made at Gorkii from 69 clearance for service withheld until 13 April 72. During test programme 155R-1, 155R-3 and 155P-1 set 16 world records, including 2981,5 km/h 500-km circuit (Komarov, 5 Oct 67), altitude 36,24 km (118,898 ft) by Fedotov 25 July 73, and climb to 20 km in 2 min 49·8 s and 30 km in 4 min 3·86 s. Aircraft submitted to FAI as 'Ye-266'.

MiG-25P, Product 84 Interceptor, to final standard of pre-series. Cleared for fixed 5300-lit tank but seldom fitted. Operational 73, with engine life 150 h. Despite interim radar, built in larger numbers (about 600) than any other version, and exported to Algeria, Iraq, Libya and Syria. All in VVS/PVO service in 79 rebuilt to PDS standard, see later. ASCC 'Foxbat-A'.

DIMENSIONS Span 14,015 m (45 ft 11¾ in); length 21,42 m (70 ft 3½ in), (excl boom) 19,75 m (64 ft 9⅝ in); wing area 61,4 m² (661 ft²).

ENGINES Two R-15B-300.

WEIGHTS Empty 20 020 kg (44,136 lb); fuel 14 570 kg; loaded (clean) 34 920 kg (76,984 lb), (max) 36 720 kg (80,952 lb).

PERFORMANCE Max speed (SL) 1200 km/h (746 mph), (13 km) 3000 km/h (1,864 mph, Mach 2·82, Mach limit 2·83); climb, see records, but climb to 20 km and acceleration to Mach 2·35 8·9 min; service ceiling 20,7 km (67,915 ft); range on internal fuel (subsonic) 1730 km, (supersonic) 1250 km; TO speed/run 360 km/h, 1250 m; landing speed/run, 290 km/h, 800 m.

MiG-25PD, Product 84D Improved, replaced P in production 78. Engines uprated and cleared for 1,000 h. Electrics with mixed supply, each engine mounting direct-drive 27V DC generator and constant-speed 3-phase alt for 200/115 V for stabilized 400 Hz. Nose lengthened by 250 mm ahead of cockpit for retractable FR probe and improved avionics. At bottom of extra section, IR tracker in small gondola with optics to cockpit. RP-25 *Sapfir*-25 pulse-doppler radar with improved look-down capability. Outer pylons and fire-control modified to launch four close-range R-60 as alternative to outer pair of R-40; later, other AAMs (R-23 and R-73 series) as options. All equipped for 5300-lit tank. Production complete 82. ASCC 'Foxbat-E'. Data as 25P except:

DIMENSIONS Length 21,67 m (71 ft 1⅛ in).

ENGINES Two R-15BD-300.

MiG-25PDS MiG-25P (about 370) rebuilt 1979-84 to PD standard with nose stretch. To remain in service to at least year 2000. ASCC 'Foxbat-E'.

MiG-25R, Product 02 Recon, faster than 25P and less-troubled development. Wing as Ye-155R, variable-freq elec supply from two combined generators at 115 and 36 V, stabilized 400 Hz for completely different avionics. All production recce variants have lengthened jetpipes and slender wingtip anti-flutter fairings by comparison with interceptors, and have a symmetrical brake chute fairing. *Peleng* nav system based on duplicated inertial platform plus DISS doppler and other means to refine inertial output. Linked with flight control derived from SAU-155 of interceptor, to steer in any weather

MiG-25R and RBK

with high precision to predetermined point at start of run. Unarmed, so normally making run at Mach 2·83 at min 20 km. Pointed nose with small radar for mapping and weather avoidance. Remainder of forward fuselage houses sensor pallet with any of three arrangements of optical cameras, typically one 650 mm focal length tilting left/right covering band of terrain width five times aircraft altitude, one 1300 mm swinging L/R covering band 2·5 times altitude and short-FL fixed vertical camera under cockpit. Alt pallet houses three vertical and two fixed oblique, and third with one vertical and four oblique, all with large flat outer windows. Standard Elint (electronic intelligence) receiver SRS-4A with square dielectric window on left of nose. All installations pressurized and heated, with optics prevented from misting or icing. Minor airframe mods to ensure capability of flying mission at Mach 2·35. Parallel devt of **MiG-25RBK**, see later. Total production of MiG-25R fewer than ten in 69, all later brought up to RB standard. ASCC 'Foxbat-B'. Data, see RB.

MiG-25RB Recon, aircraft with secondary high-level bombing capability, replaced R in production 70. Wingtip pods, new threat-warning and countermeasures; Elint changed to SRS-4B (similar dielectric window). Nav sufficiently accurate for free-fall bombs at Mach 2·35 at 20 km without sight of targets whose geog coordinates were known accurately. This required RB to have digital computer(s) and DISS Doppler fairing below nose. Standard bombload six FAB-500 (1,102-lb HE), one each on four wing pylons and two under fuselage, local structure reinforced. Cleared to Mach 2·83 with full bombload, and to 4·5 g at lower speeds. VVS service early 72 and produced (later as sub-types, see next) to 82. Exported to Algeria, Bulgaria, India, Iraq, Libya and Syria and with VVS personnel in Egypt 71. 'Foxbat-B'.

DIMENSIONS Span 13,418 m (44 ft 0¼ in); length 23,20 m (76 ft 1⅜ in), (excl boom) 21,55 m (70 ft 8½ in); wing area 61,4 m^2 (661 ft^2); tailplane 8,74 m; ailerons 2,72 m^2; flaps 4,3; rudders 2,12.

ENGINES Two R-15BD-300.

WEIGHTS Empty 20755 kg (45,756 lb); fuel 15 245 kg (with external tank, 19 615 kg); loaded 37 t (81,570 lb), (max) 41 200 kg (90,829 lb); max landing 24 t.

PERFORMANCE Max speed (SL) 1200 km/h, (13 km) 3000 km/h (Mach 2·82); climb to 19 km (clean) 6·6 min, (with bombs) 8·2 min; service ceiling 21 km (68,900 ft); range (with external tank) 2130 km (1,324 miles) at Mach 2·35, 2400 km subsonic; TO/ldg as 25P except landing 280 km/h.

MiG-25R sub-types In parallel with first versions **MiG-25RBK, Product 02K**, launched 62 as multirole bombing/electronic reconnaissance. Sensors did not normally include cameras, but five or six Elint installations, all flush antennas with RWR antenna fairings on intake sides. Largest, *Kub* SLAR (hence designation RBK) required dielectric panel 1600×930 on left side ahead of cockpit. In production Gorkii 71-80. **MiG-25RBS** devd from 65 with different sensors, mainly Elint, in production 71-77. From 78 Gorkii added **MiG-25RBV** and **RBT**, differing from RB in nav system, optical sensors completely replaced by Elint including *Virazh* SRS-9 side-looking receiver (hence RBV). From 81 RBS upgraded with new sensors and other mods redesignated **MiG-25RBSh**. Surviving RB rebuilt from 81 as **MiG-25RBF, Product 02F**, with new panoramic vertical camera and upgraded ECM with flush square antennas in approx same positions as original oblique camera ports. All differ from RB in new environmental system, FR probe, revised instruments, improved navaids and upgraded ECM/IRCM. Since 90 some RB variants grey/green camouflage. ASCC names: F, K, S, Sh, 'Foxbat-D'; T, V, 'Foxbat B'.

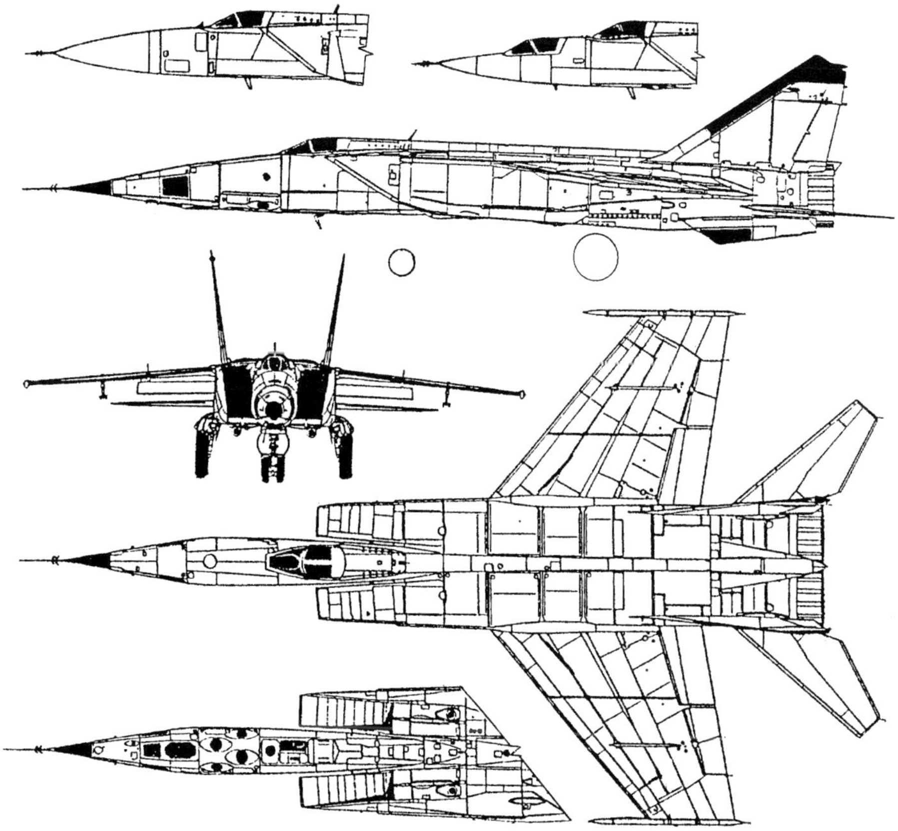

MiG-25RBV, including inset of underside of 5-camera nose (inset top left 25P, top right 25U)

Ye-133, MiG-25PU (Product 39), RU (Product 22) Dual-control pilot trainers for interceptor (PU) and recon-bomber (RU). Unusually, original (rear) cockpit for pupil and new front cockpit at lower level for instructor. Shallow nose with no radar. Drag increased and M 2·65. No radar or weapons (PU) or sensors (RU), but comprehensive simulation for target tracking and missile launch, with pylons fitted (PU) and for optical or Elint sensor operation (RU). In addition, 14 onboard systems or equipment items modified to simulate inflight failure, and instructor trigger to eject pupil in emergency. PU flown 68, RU 72, both produced to 80. One PU called **Ye-133** used by S. Ye. Savitskaya for ladies' records 75-78. RU has lengthened jetpipes, symmetrical brake chute fairing and slim anti flutter fairings associated with recce variants. *Kresla* (armchair) for ejection-seat testing. ASCC (all) 'Foxbat-C'.

MiG-25BM, Product 02M Defence-suppression version devd 72 on basis of RB. Conical nose lengthened by 0,72 m for passive antennas, Elint and ECM tuned to defence systems. Additional antennas in blisters on sides of nose and on front of each outer pylon. Several systems and cockpit modified. Four Kh-58 on wing pylons, 5800-lit tank and probe. In production 82-85. ASCC 'Foxbat-F'. Possible sub-variant or alternative designation is MiG-25RBU. Total of MiG-25 production, 1,200+.

Ye-155M, Product 99 From 72 OKB cleared to investigate improving speed, height, climb and range. First stage, R-15BF2-300 engines; second stage, increase length for greater internal fuel and eliminate D19 alloy and other parts unable to soak at Mach 3. Second stage never realised, but several aircraft re-engined as Ye-155M. First flight 74, and on 17 May 75 set climb records (25 km 164·2 s, 30 km 189·85 s, 35 km 251·7 s) which still stand, followed by alt records to 37 650 m July/August 77 which still stand; FAI told 'Ye-266'. Two further aircraft rebuilt with 19 700 lit internal fuel and D-30F engines whose power and economy gave 21,9 km service ceiling, range 3310 km; leading to MiG-31. During testing of these aircraft it was recognised more could be done with aluminium alloys, and proportions of these were actually increased for Ye-155MP.

Ye-155MP, 83, MiG-25MP During devt of MiG-25 it was clear its technology was obsolescent. So important was strategic interception of all hostile aircraft, particularly cruise missiles, that authorization given to OKB in 71 to develop improved aircraft with digital electronics, *Zaslon* radar and new missiles. Complexity of avionics combined with flight-refuelled mis-

MiG-25PU

sions lasting up to 6 h demanded second crew-member, called navigator but really tactical director. Basic requirements to extend mission radius and detect, track and destroy targets throughout forward hemisphere, down to sea level, and also to rear, and to have unrestricted flight envelope at all heights. Despite high wing loading structure restressed to 5 g at all heights and speeds. Better fuel economy, extra thrust used to enhance climb and acceleration rather than speed. Audit of structure soaked at 3000 km/h enabled arc-welded steel to be reduced to 49%, with D19 aluminium alloy 33% (increased proportion of bonded honeycomb) and titanium alloy 16%; composites 2%. Prototype, OKB Product **83, Ye-155MP**, #831, flown by Fedotov 16 Sept 75. This was one of first flights of D-30F engine, only days after first of two Ye-155M as noted earlier. Longer, heavier and more expensive than R-15, this engine quickly attained higher level of reliability, besides reducing dry sfc from 1·01 to 0·71. Airframe incorporated most changes planned for MiG-31: rear cockpit, redesign of fuselage to increase fuel capacity, enlarged air ducts, relocation of airbrakes on flanks of fuselage ahead of main gears opening diagonally, and installation of new main gears. Latter designed at OKB to permit sustained operations from unpaved airstrips or snow, which MiG-25 would soon render unusable from ruts. Single main leg carrying diagonal levers to front and rear each with own shock strut, front arm carrying KT-175 wheel with 950 mm × 300 mm tyre on inner side, rear carrying similar wheel outboard, so each makes its own (shallower) rut. Gears retract forward to lie with twin-wheel trucks inverted outboard of ducts. Twin steerable KT-176 nosewheels with 660 mm × 200 mm tyres on vertical leg retracting to rear into bay with twin doors plus forward door attached to leg carrying three landing/taxi lights. Wing TE redesigned to increase span of two-part slotted flaps and two-part ailerons, latter extending almost to tip.

MiG-31, Product 01 Several development aircraft MiG-31 cleared for production 79, work delayed by severe fuel-system failure later 79 resulting in Menitskii making dead-stick landing (Fedotov had authority to reject certain items of equipment, notably in fuel system, to guard against repetition). Principal change: redesign wing, modified profile, LE 68° at root then 40° to tip, anhedral reduced to –4°, third main spar, LE full-span flap in four sections with 200 mm fixed portion attached to fence/pylon, anti-flutter tip bodies not required. Airbrakes again redesigned and moved inboard to lie under ducts, opening downwards. FR probe on upper left of nose, extending diagonally up on telescopic pipe from small blister fairing. Cockpits with K-36D seats, rear cockpit with simple flight controls. Canopies metal frames with inserted windows hinged up hyd at rear, front canopy with retractable mirror, rear with 920 mm periscope to enable WSO to land aircraft in emergency. Definitive integral-tank fuel system occupying entire wing boxes, all available fuselage space and fins, total 20 250 lit (4,455 gal). SBI-16 Zaslon radar in pressurized nose, electronically scanned circular antenna 1195 mm diam. KOLS opto-electronic system with Type 8TP IRST (IR search/tracker) in retractable drum hinged down under forward fuselage feeding HUD and HDD together with radar. Digi-

MiG-25 (Kresla)

MiG-31 OKB demonstrator

tal avionics including *Marshroot* Omega and *Tropik* Loran-type navigation and central computer dedicated to managing defensive avionics and countermeasures. Further computer linked with APD-518 secure data transmission for guidance (three modes) by ground station or for leader of finger-4 to be guided by ground net and automatically guide other three aircraft to defend front 900 km wide (outer aircraft 600 km apart). One GSh-6-23 gun with 250 rounds recessed on right of rear fuselage firing ahead. Under fuselage front and rear pairs of AKU-33 parallel-link trapeze launchers for R-33, front pair recessed. Single pylon under each wing for one R-40 (usually 40T) or two R-60 (one of each) or one 2500-lit drop tank.

MiG-31 cleared for production at Gorkii late 79, in PVO service early 83. From 86 retractable FR probe added on left of windscreen and two extra underwing pylons. Total complete 89) 500. ASCC 'Foxhound-A'.

DIMENSIONS Span 13,464 m (44 ft 2 in); length overall 22,688 m (74 ft 5¼ in); wing area 61,6 m^2 (663 ft^2).

ENGINES Two D-30F6.

WEIGHTS Empty 21 825 kg (48,115 lb); internal fuel 16 350 kg; loaded 41 t (90,388 lb), (max with external fuel) 46,2 t (101,852 lb).

PERFORMANCE Max speed (SL) 1500 km/h (932 mph, Mach 1·23), (17,5 km) 3000 km/h (1,864 mph, Mach 2·82); max cruise Mach 2·35; Mach limit 2·83; climb reported incorrectly; service ceiling 20,6 km (67,600 ft); range (external fuel) 3300 km (2,050 miles); combat radius (at M2·35, 4 AAMs) 720 km (450 miles), (at M0·85, 2 tanks) 1400 km (870 miles); TO run (max wt) 1200 m; landing speed/run 280 km/h, 800 m.

MiG-31M, Product 05 Next-generation interceptor, launched 83 under K. Vasilichyenko, incorporating mods introduced to MiG-31 production since 86, notably two additional outboard pylons as on MiG-25P. Redesigned fins mounted on bulged dorsal fairings, rudders of extended chord, no sweep on TE and large gap between rudder and fuselage. Wing with long sharp-edged root extension, fences cut back to start at 20% chord, joints between inboard/outboard flaps and ailerons at 90° to TE, and wingtips either with large flush antennas at front (rounded) and rear or tip bodies with delta fins as on Ye-155 housing two passive receivers and two emitters. New defensive electronics with high-performance passive receivers at top of inlets (replacing those of MiG-31) and in various aft-hemisphere configurations. New radar with 1,4 m electronically scanned antenna, axis tilted 7° down. KOLS system upgraded but IRST no longer retractable but removable. Gun removed, provision for six R-37 missiles on parallel-link trapeze body launchers and four R-77 on wing pylons. Frameless polycarbonate windscreen and wider canopies, pilot all-transparent, backseater no longer a pilot (but emergency sidestick), side windows only and no periscope, three multifunction displays for nav and system management. Capacious (300 lit extra fuel) spine tapered off at rear without cylindrical drag-chute container. FR probe transferred to right and fully retractable. Redesigned nose gear, no front door but two side doors.

Eight prototypes. First to incorporate M features lost 4 April 84, killing Fedotov and observer Valery Zaitsyev. First flight of MiG-31M 89 but this aircraft also lost, 9 Aug 91. All survivors flying 92, but production not yet initiated. ASCC 'Foxhound-B'.

MiG-31D, Product 12 Interim aircraft designed under leadership of A. Belosvyet retaining original radar but compatible with R-37 missile. Intended as new production aircraft. Also poss prodn by conversion of existing MiG-31 as **MiG-31BS**, **Product 01BS** already on test, with original radar but R-37 and -77 missiles, choice of wingtips and dorsal fins, outer pylons plumbed for 2250-lit tanks. An exptl prototype of 87 was **Product 07**, satellite killer with underwing RVV-1 missiles, wingtip bodies with large upper/lower delta fins, larger tail dorsal fins, flat underside to fuselage with no missile recesses and small conical nose with radar replaced by ballast.

DIMENSIONS classified

ENGINES

WEIGHTS Empty c24 t; loaded 46 t (101,411 lb); max 52 t (114,638 lb).

105-11 OKB designation for low-speed test vehicle to explore aerodynamics and control of planned 50-50 EPOS aerospace-plane. Product 105-11 was generally similar to EPOS but used conventional powered controls, available production systems and instruments and a conventional turbojet, without jet deflection, and with no rocket engines. Configuration was a tailless delta, with pressurized cockpit near the nose and a dorsal air inlet. Wheels were fitted to the retractable levered legs for taxi testing and

MiG-31

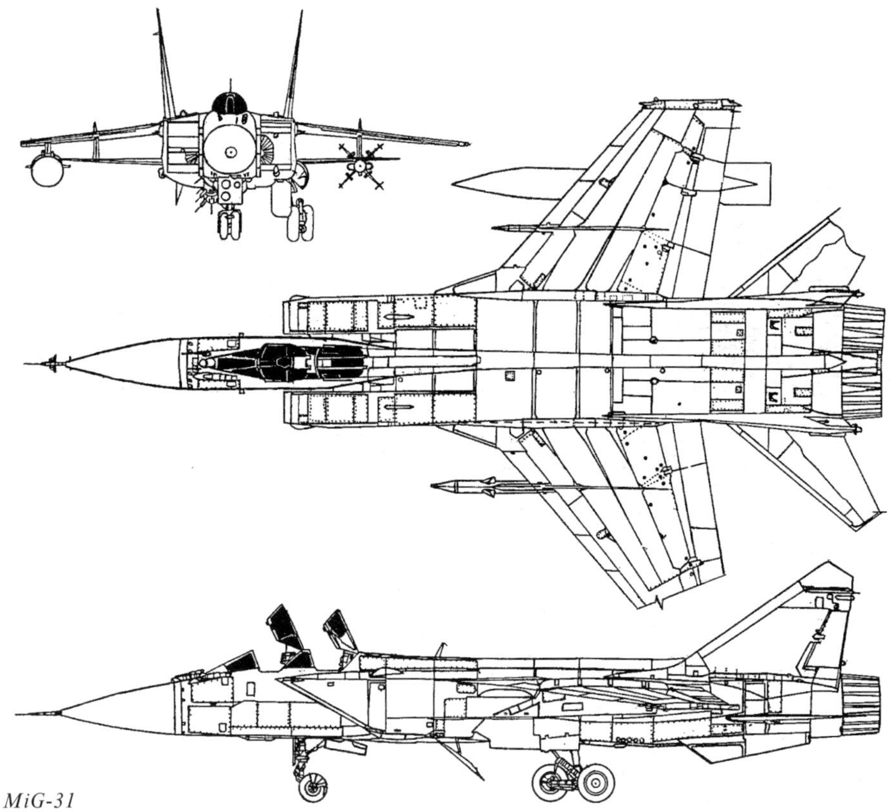

MiG-31

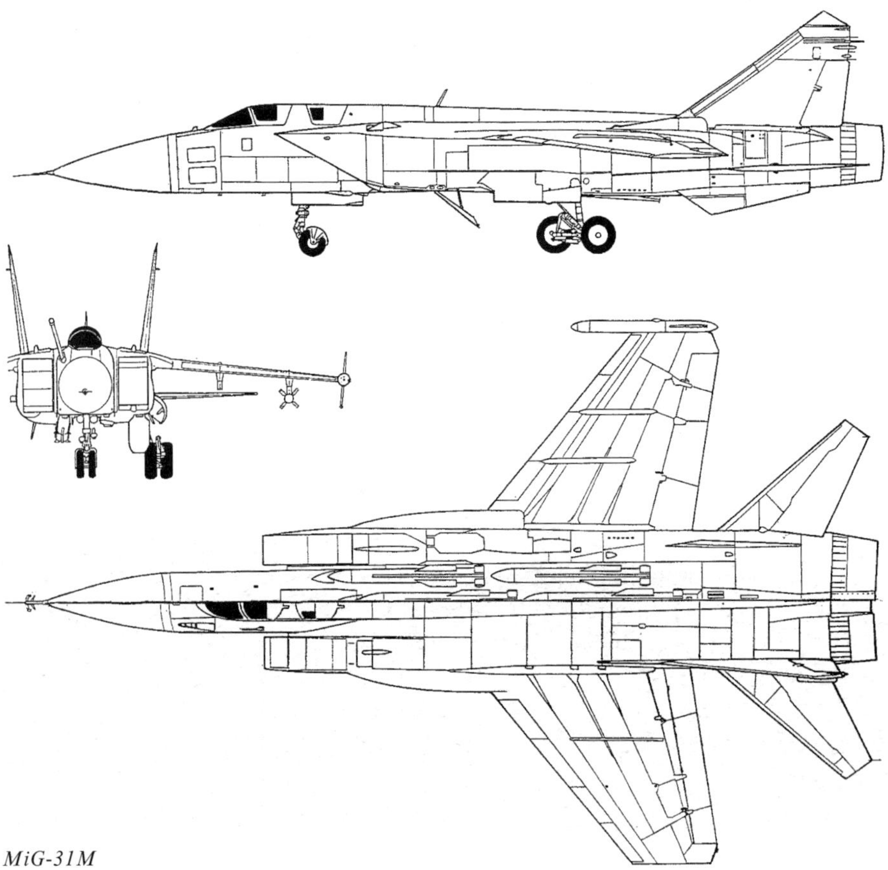

MiG-31M

ground manoeuvring, and on 11 Oct 76 A. G. Fastovets made the first flight, taking off from one Moscow runway and landing at another (Zhukovskii) 19 km directly in front. On 27 Nov 77 he was dropped at 5 km (16,400 ft) from the OKB Tu-95K previously used for K-20 development and followed a prescribed trajectory. The vehicle made six further flights by Sept 78, latterly with skids only, before being retired to Monino.

DIMENSIONS Areas: wing 6,6 m^2, fuselage (plan) 24 m^2.

ENGINE One RD-36-35K.

WEIGHTS Empty 3,5 t (7,716 lb); fuel 500 kg; loaded 4220 kg (9,300 lb).

PERFORMANCE Landing speed 250-270 km/h (155-168 mph).

MiG-29, Product 09 Uncompromised design begun 72 in search for LFI (light frontline fighter). Requirement: autonomous operation from austere site to achieve air superiority over tactical theatre, including escort and surface attack, in sky occupied by latest Western aircraft. Detail design 74 by Belyakov assisted by A. A. Chumachyenko, V. A. Lavrov and M. R. Waldenberg. From outset: 'integral aerodynamics' with lifting fuselage disappearing into large wing, two underslung engines with variable inlets, structure for sustained 9 g, multimode PD radar, comprehensive fire-control and EW avionic systems and gun plus not fewer than six AAMs. Eleven prototypes (9-01/9-11).

Aerodynamically 40% lift from central component comprising fuselage and inner wing between fins, LE 73·5°. Outer wings symmetric 4% profile, LE 42°, aspect ratio 3·5, tip chord 1270 mm, anhedral –2°. Hyd LE flaps computer-linked with plain TE flaps to vary camber throughout flight. Outboard powered ailerons with ground-adjustable TE strip. Twin vertical tails, 1690 from centreline, canted out 6°, LE 47°50', small inset powered rudders, prototypes and first 100 series aircraft with ventral fins outboard of engines. These remain on many aircraft, but further spin testing eliminated them with extended dorsal fin across wing to form BVP-30-26M chaff/flare dispenser, increasing effective keel area. Horizontal taileron, LE 50°, span 7,78 m, limits +15°/–35°. Flight-control system with mechanical links to power units and computer tailoring for 'carefree' handling in symmetric manoeuvres; in rolling flight AOA limit normally 26° with aileron authority progressively phased out, but in combat pilot can override pitch and g-limiters. Typical sustained turn radius 225 m at 450 km/h, 350 m at 800 km/h. Horizontal acceleration at Mach 0·85 11 m/s^2 at SL, 6·5 m/s^2 at 6 km.

Mainly al-alloy, with Al-Li skins over three-spar wing box. No 1 tank ahead of front spar (2550 lit) single component welded aluminium alloy forming top of inlet duct behind louvres and major part of blended inboard wing. Door-type airbrake above and below rear fuselage between jetpipes. Four keel beams, two between engines and two outboard, latter cantilevered at rear to carry tails. Over 7% airframe composite, with CFRP skins over tail, and ailerons, flaps, rudders and aft part of tailerons CFRP honeycomb. Small areas steel or titanium to meet strength or temperature demands. Engines separated to leave deep and wide channel on ventral centreline. Rectangular inlets with fixed sharp

MiG-31M/02 (Izdelye 05 No 7)

105/11

lips sloping 60° well back under wing for good recovery at extreme AOA; 100 mm gap for boundary layer and inlets canted out 8° to match taper of wing thickness. Upper wall three hinged ramps to vary angle and throat area. Front ramp hinged at front and (when load on front oleo) is hyd rotated fully down so that (ignoring 887 perforations) duct is closed at front and all air enters via five spring-loaded transverse louvres in top, avoiding FOD. When engine is shut down loss of hyd pressure allows door to open for maintenance access. Engines started by 98 hp GTDE-117 APU fed by inlet projecting above rear fuselage with ventral exhaust passing through centre drop tank if fitted. As hyd spool up, inlets close, opening again only as pressure comes off nose gear with airspeed exceeding 200 km/h, pilot unaware of transition. Top louvres free to open in flight according to airflow demand; 800 km/h poss with main inlets closed. Downstream are three exits for air from ramp perforations. All fuel in integral tanks in centre section and wings, total 3200 kg/4300 litres (945 gal), to which can be added 1520-lit non-conformal centreline tank and on 40% of Soviet MiG-29s, two 1150-lit underwing tanks, entire system pressure-filled through coupling in left MLG bay. Landing gears designed at OKB, made by Hydromash. Long nose gear with twin wheels with large mudguards on trailing link with 570 mm × 140 mm tyres with hyd steering and rearwards retraction. Replaced on 9-02 by shorter gear with small mudguard ahead of nose wheels formed from part of nose gear door. This replaced by lightweight slotted debris guard behind wheels during production series, moved much further back to reduce wheelbase from 5210 mm to 3645 mm, retracting between inlets. Main gears with single legs hinged under wing root, track 3090 mm, hyd retraction forwards while wheel with 840 mm × 290 mm tyre rotates 90° to lie flat in centre section above leg. All wheels have pneumatic steel antiskid disc brakes. Cruciform 17 m^2 drag chute deployed from tube between upper/lower airbrakes prior to touchdown. Nose PVD probe 1040 mm long, no transducers. Forward fuselage semi-monocoque housing radar, forward avionics bay, cockpit, centre and rear avionics bays, hydraulic centre and then tankage. Cockpit with one-piece frameless polycarbonate windscreen followed by canopy often with three rear-view mirrors filling periphery hinged up hyd at rear. K-36DM seat, inclined back 10°, outstanding all-round view. SUV fire-control complex linking RP-29 (N-019) coherent pulse-doppler radar with drum-like twist cassegrain antenna collimated with IRST and laser precision ranger both using same optics served by glass sphere on right of windscreen protected by fairing. All three can feed HUD, supplemented by radar display on right, giving outstanding target search/acquisition/track under all conditions, in close engagements augmented by HMS to cue IR missiles (especially R-73) on to off-boresight target. Prototype 9-01 fitted with twin barrel GSh-23L or GSh-2-30 at front of left wing root, with magazine inboard and collector box outboard. This was replaced by single-barrel 9A4071 (GSh-30/1) in same installation, with 150 rounds. Waldenberg: 'If we had known MiG-29 would unfailingly hit with first round we'd have made magazine smaller'. Three stores pylons under each wing for various combinations R-60, R-27 and R-73 missiles all cleared to launch at 8 g. Inboard four pylons available for tanks or secondary surface-attack stores

9-01

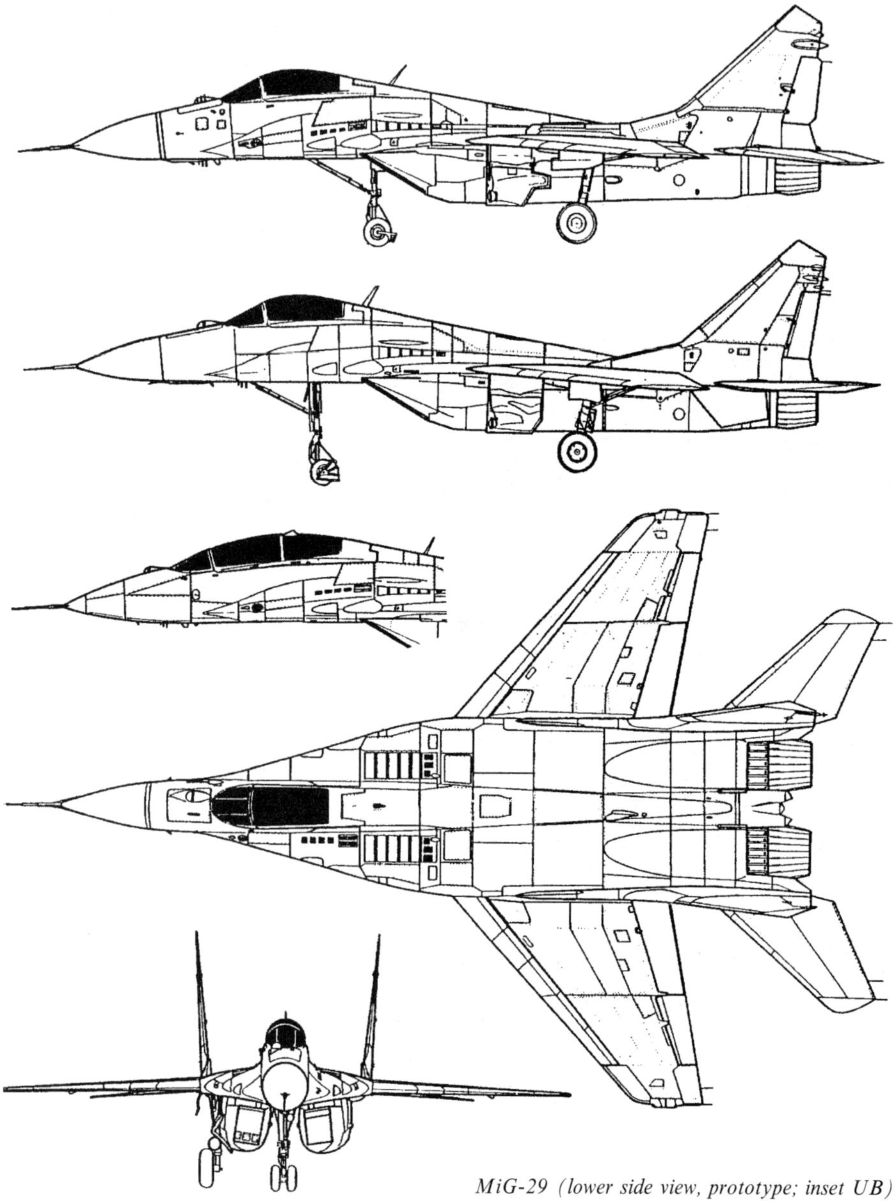

MiG-29 (lower side view, prototype; inset UB)

MiG-29 (2 × R-27, 4 × R-60)

including RN-40 on port inboard pylon perhaps incl other TN- and RN-series, rockets 57/80/240 mm, dispensers, bombs etc to total 2 t, later 3 t (6,614 lb). Other avionics INS, ILS, hf/vhf/uhf, RSBN Tacan, SOD ATC/SIF, SRZO-2 IFF and *Sirena*-3 (SG-15 and -69) threat warning feeding computer-managed ECM/IRCM systems including dispensers.

9-01 flown by Fedotov 6 Oct 77, 9-02 June 78 and 9-11 Jan 79. Very successful but Menitskii insisted on perfection of handling, especially in extreme AOA and stall/spin regimes. Eight pre-production, 9-12/9-19, completed OKB testing and hi-rate production began 82 at Znamya Truda Moscow and of UB at Gorkii (now called Nizhny-Novgorod). By late 82 build-standard deleted ventral fins but added TE strips to rudder and small sharp fairings each side of PVD boom to cause vortices which prevent aileron reversal above 25° AOA. These vortex generators and extended chord rudders even applied to surviving early production aircraft with ventrals. Deliveries to VVS and AV-MF about 1,600, completed 92. Exports, no special designation though many lack certain items: Bulgaria 18, Cuba 7, Czechoslovakia 18, E Germany 20, Hungary 20, India 45+20, Iran 14 (+ Iraq defections), Iraq 36, Jugoslavia 14 (now Serbia), N Korea (about 24), Poland 9, Romania 14, Slovakia 5 (plus 9 transfers from Czech) and Syria (reported 80). Luftwaffe took over former NVA (E German) force and in formal assessment reported 'Aerodynamic performance better than we expected . . . In some areas outstanding compared to Western counterparts. With a simple but very effective HMS, the way the weapons can be used at large off-boresight angles was a surprise'. Luftwaffe to retain 24 aircraft combat-ready to 2003. ASCC 'Fulcrum-A'.

DIMENSIONS Span 11,36 m (37 ft 3¼ in); length overall 17,32 m (56 ft 10 in); length of fuselage (excl boom) 14,72 m (48 ft 3⅜ in); wing area (net) 38 m² (409 ft²), (gross) 43,5 (468 ft²).

ENGINES Two RD-33.

WEIGHTS Empty equipped 10,9 t (24,030 lb); fuel 3623 kg; loaded (clean) 14 670 kg (32,341 lb); max loaded 18 480 kg (40,741 lb).

PERFORMANCE Mach limit 2·35, equivalent to 2500 km/h (1,553 mph) at height; max IAS 1500 km/h, but SL limit 1250 km/h (777 mph, Mach 1·02); initial climb 19,8 km (65,000 ft)/min, time to 6 km 55 s; service ceiling 17,9 km (58,730 ft); range (clean) 1500 km (932 miles), (1500-lit tank) 2100 km; TO 220 km/h, 290 m; landing 270 km/h, 900 m, (600 m with chute).

MiG-29UB, 9-51 UB (trainer, fighting) not flown until #51 (by Fedotov, 29 April 81), but was second major variant in service. Radar removed, nose 100 mm longer, pupil cockpit ahead of and lower than original. Two K-36DM seats, instructor can eject pupil but cannot disconnect his flight controls. Single upward-hinged canopy with panoramic periscope for instructor. Unchanged fuel 4300 lit. Gun (usually 50 rds) and weapons (normal weaps 2,3 t max) retained, plus advanced simulation system generating symbology on pupil's 'radar' and HUD displays, together with simulated emergencies. Produced at Nizhny-Novgorod for VVS, Frontal Aviation and in small numbers for previous export customers. ASCC 'Fulcrum-B'. Data as MiG-29 except:

MiG-29S

MiG-29M

DIMENSIONS Length 17,42 m (57 ft 1⅞ in).
WEIGHTS Empty 10,88 t (23,986 lb); loaded (clean) 15 t (33,070 lb), max 18 t (39,683 lb).
PERFORMANCE Max speed at SL 1350 km/h (839 mph), at 12 km 2400 km/h (1,491 mph).
MiG-29S, 9-13 Aircraft 13, second of pre-production batch, completed with bulged spine (fuselage aft of cockpit) to accomodate avionics displaced by extra 300 kg (240 litres) fuel. No change in designation – though ASCC 'Fulcrum-C' – and dubbed *Gorbatov* (hunchback) serves alongside other MiG-29s including early ventral-fin examples. This was basis for **MiG-29S, 9-13S** built new and by conversion. Refinement of flight-control system with enhanced computer power and greater surface movement to expand operational AOA (30°) and permissible g. Sighting system upgraded and provision for RVV-AE missile. Provision for two 1150-lit underwing and centreline 1520-lit drop tanks may also have 75 litres more internal fuel; up to 4 t (6,614-lb) bombload with inner wing pylons carrying tandem side-by-side pairs of FAB-500. Different shellcase disposal to avoid centreline tank, added simulation mode for pilot training and BITE for avionics test. First of three prototypes flown 84, in production 92 by MAPO, Moscow. **MiG-29SE (9-13SE)** export version, tailored to customer choices, engine certified 2,000 h, first customer Malaysia. Contrary to reports Hungary received standard 'Fulcrum-A'. Further improvements will include optional FR probe, radar mapping mode, compatibility with TV, laser, and radar-homing ASMs.
DIMENSIONS, ENGINE As MiG-29.
WEIGHTS Empty 11,2 t (24,691 lb); loaded 15,3 t (33,730 lb); max 19,7 t (43,430 lb).
PERFORMANCE As MiG-29 except range (3 tanks) 3000 km (1,864 miles).
MiG-29M, 9-15 Second-generation development with major advances. OKB considered most important change FBW (fly-by-wire) flight controls giving 35° α limit, but airframe changes are also significant. Structure re-audited for 2,500 or 4,000 hour service life to sustain 9 g at higher weights, proportion of CFRP increased and forward fuselage redesigned in welded Al-Li alloy to increase space for fuel and equipment. Aircraft now 29% Al-Li 1420, 35% steel, 27% aluminium alloys, 3% titanium and 6% composites by weight. Wing changes: inboard LE sharp (radius c3 mm), with previous pairs of EW antennas merged and reprofiled; modified tips with MiG-31 type fore/aft antennas; aileron span extended to aft tip antenna, flap angles 25°/29°. Fuselage: profiled radome indicating removal of old bulky Cassegrain antenna, new IRST fairing, longer canopy 40 mm higher over raised seat and with embedded ARK antenna, bulged spine with ARK antenna and two integral 60-round chaff/flare dispensers followed by giant CFRP honeycomb door-type airbrake (original brakes removed) hinged on main-spar frame leading to broad 'platypus' tailcone housing twin 13 m^2 drag chutes well forward in box with twin dorsal doors. Propulsion: uprated engine served by inlets with downward-hinged lower lips to handle greater airflow at low speeds. Original shut-off inlet doors replaced by mesh grill which admits full airflow when closed to prevent FOD; accordingly no aux inlets needed above duct (these painted on some M prototypes!). Absence of these inlets allows forward part of inboard wing (apart from EW antennas) to be large 2550 lit integral tank, single complex assembly in welded Al-Li, playing chief role in increasing internal fuel by 1500 kg this tank also replaces outer and middle exits for throat airflow each side. Resulting greater weight, plus heavier attack loads, required strengthening of landing gears. APU inlet moved to right side. Tail modified with dorsal fin extensions (dispensers) removed, lower part of fins extended aft to line up with extended TE of rudder, tailerons strengthened and increased in area with CFRP strip along TE and LE extended outboard terminating in mid-span dogtooth. Agility brought to remarkable level by aft shift of CG and greater control power, especially in pitch/roll. FBW flight-control system with flat multicore cables duplicated each side of aircraft. Pitch channel quadruplex, roll/yaw triplex, but whole system of analog type mainly to avoid susceptibility to EM damage and with mech back-up. Cockpit with two multifunction displays, though retaining numerous dial instruments, HOTAS controls. New wide-angle HUD fed from N-010 *Zhuk* radar with multi-

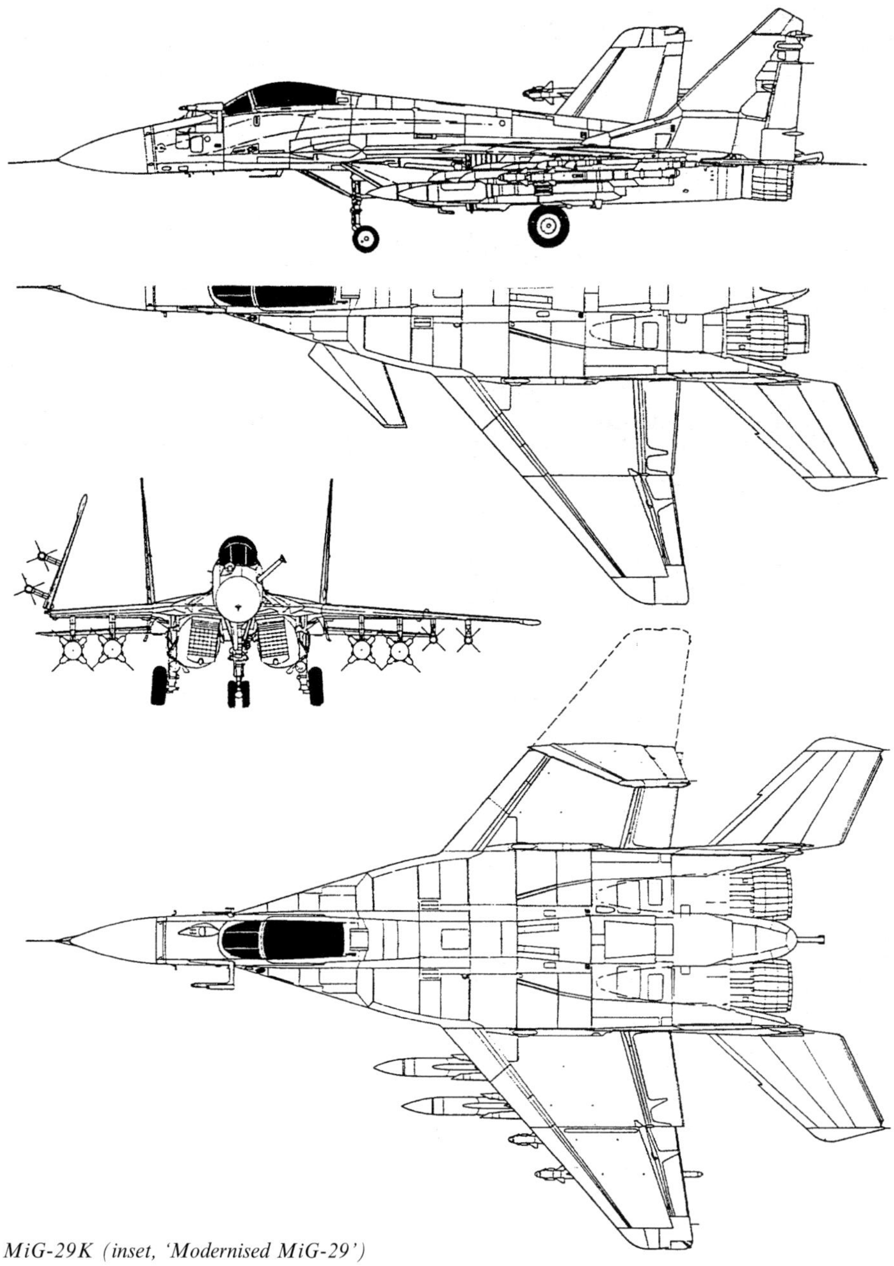

MiG-29K (inset, 'Modernised MiG-29')

target and mapping capabilities and with 25% greater detection range. New optics served by upgraded and better-refrigerated IRST, laser ranger (with smart-weapon guidance) and TV, all collimated together. New L-150 Pastel RWR and Gardeniya 1-FUE active jammer. Eight underwing pylons, four inboard stressed for 1000-kg loads, others to 500 kg, plus centreline. New underwing tanks of deeper section and 2500 lit capacity under development. Stores include laser- TV- and active radar-guided ASMs and RVV-AE. Centreline limit 2200 kg, will also carry 2500 lit tank. External load limit now 4500 kg. Gun magazine reduced to 100 rounds.

First prototype flown 25 April 86. Six prototypes, all built in OKB workshops. First with RD-33K engine flown by Menitskii I Nov 89. Prolonged certification while avionics perfected. Fifth M (155) exhibited Minsk 92. Full funding still sought; production may be launched with foreign partner (other than India, which expects to assemble 30 plus two two-seat Ms under licence). Su-35 reportedly selected instead, despite suitability of cheaper, lighter, more versatile 29M.

DIMENSIONS As MiG-29; aileron (aft of axis) 1,72 m^2; flap 2,92 m^2; fin 10,1 m^2; rudder (aft of hinge) 1,36 m^2; tailplane 8,19 m^2.

ENGINES Two RD-33K.

WEIGHTS Empty c11,5 t (25,350 lb); fuel 4980 kg; loaded c17 t (37,480 lb), (max) c21 t (46,300 lb); max landing 15 560 (34,303 lb).

PERFORMANCE Generally similar to MiG-29 except range (clean) 2250 km (1,400 miles).

MiG-29K, 9-31 ANPK began work on naval version 83, designation from *Korabelnyi*, ship-based. AV-MF wished competitor for Su-27K and India (possibly China) export prospects. Airframe incorporates new materials or coatings to withstand marine environment. Wing as M except for power fold upwards at flap/aileron joint; span increased to broad tip bulged on underside housing warning and jamming systems; chord of LE extended to increase area and preserve tip chord despite increase in span; redesign of flaps as double-slotted with chord extended at TE; ailerons increased in span as M but on K still not reaching near tips. Fuselage generally as M but airbrake not identical and rear fuselage strengthened for greater landing impacts and pull of square-section arrester hook; belly mounted spotlight illuminates hook; for LSO. Drag chutes deleted. Tail and propulsion as M, APU inlet on right. No provision for catapulting, but nose gear strengthened and fitted with 'traffic lights' to assist carrier batsman (auto carrier landing also fitted), nosewheels unbraked. Main gears strengthened as M and extended in stroke, with shortening link to restore length on retraction. All tyre pressures increased. Retractable FR probe installed left of windscreen. Radar of *Zhuk* family, new IRST/TV/laser as M, air/ship radios, individually tailored defensive electronics. K is multirole aircraft with anti-ship attack as important as air combat. Typical armament: gun (100 rounds) plus four Kh-29T or Kh-31 on inner pylons plus (on folding outer wings) four R-73.

Project pilot Takhtar Aubakirov made first flight with #311 (9-31/1, first of two prototypes) 23 June 88. He also made first 20 of 66 flights from *Admiral Kuznetsov* (then *Tbilisi*), running up to full afterburner against ship's restrainers and then taking off in 100 m over ski jump. Approaches under ACLS at 240 km/h at 14° AOA, with single-engine overshoots. Both 311 and 312 exhibited Minsk 92. Carrier trials completed 93.

DIMENSIONS Span 12,0 m (39 ft 4½ in), folded 7,8 m (25 ft 7 in); length overall 17,32 m (56 ft 9⅞ in); wing area 39,6 m^2 (426 ft^2).

ENGINES Two RD-33K with 92,22 kN (20,725 lb) emergency regime.

WEIGHTS Empty c12 t; fuel c5000 kg; loaded 15,24 t (33,598 lb); max 17,7 t (39,021 lb).

PERFORMANCE Generally similar to MiG-29M.

MiG-29KVP Several 9-12 converted to *Korotkaya vzlet posadka* (short landing) standard (poss **9-18**) since 87 for pilot training and to develop techniques on dummy deck with ski jump at Saki, Crimea. KVP has ACLS and hook but retains original airframe and engine; some have ventral fins.

MiG-29KU Should MiG-29K go into production OKB hope to build this dual-pilot trainer, based on K but probably no radar. Rear K-36DM seat significantly higher than front.

Modernised MiG-29 Projected version with new wing, canards and new avionics.

33 OKB designation for light-fighter project 80-86, single RD-33, F-16 configuration.

301, 321 Numbers assigned by Deputy General Designer Anatoliy Bielosvet to extreme-alt hypersonic research/recon platforms.

701 Number assigned to projected replacement for MiG-31. Delta-winged interceptor designed

MiG-29K

to meet CIS air forces requirement for a *Mnogofunktsionalnyi Dalniy Perekhvatchik* (long-range multi-function interceptor) for the PVO, with engines above the fuselage and internal missile bays for long-range AAMs. Tandem cockpits. **701P** with side-by-side cockpit proposed as long range SST biz-jet.

MiG-101 First of OKB's *konversiya* projects to seek civil work, multirole STOL transport twin-turboprop with twin tail booms. Replaced by SVB and MiG-110.

SVB Project for twin turboprop utility transport in 20-tonne class with conventional fuselage with rear ramp door. Studied 90-94.

MiG-110 Pressurized high-wing twin-turboprop passenger/cargo transport. Hi-lift wing with slats and double-slotted flaps, twin tail booms, advanced materials and systems, twin-wheel gears, 6-blade reversing props. Pressurized cabin 7,4 m×2,2 m (2,76 m max)×2,2 m, rear ramp/door for small vehicles, 4×ZAK-1 containers, 5 t cargo or 35 pax. Fuel burn 222 g/t-km. Devt with Klimov Corp (engine), Nizhegorodskii 'Sokol' factory and many suppliers; customer delivery hoped 96.
DIMENSIONS Span 22,12 m (72 ft 6⅞ in); length 18295 (60 ft 0 in).
ENGINES Two TV7-117SV.
WEIGHT Max 15,3 t (33,730 lb).
PERFORMANCE Cruise 500 km/h (311 mph) at 8-11 km (<36,000 ft); range (5 t load) 1550 km (963 miles), (max fuel) 4050 (2,520); runway length 600 m (unpaved, 3,5 t, 470 m).

18-50 Project for large bizjet, see under Aviaprom.

MiG-AT Launched 90 as contender for VVS advanced-trainer requirement. In 93 became training complex (with ground facilities, software etc) with French engines (Turbomeca-SNECMA), hyds/gear (Messier-Bugatti), elecs (Auxilec) and avionics (Thomson-CSF/SFENA/Sextant); wing devd with Daewoo (S Korea). Unswept wing, three panels Al-Li alloy honeycomb, CFRP slats/hi-lift flaps/outboard ailerons. Fuselage 40% of skin CFRP/GRP, rear integral with T-tail, overwing inlets to engines on sides of rear fuselage, tailcone airbrakes 'Glass' cockpit, single canopy, specially designed zero-zero seats. Two fuselage tanks, 850 kg, and one in wing, 200 kg, system pressurized by bleed air. Nose gear 500 mm×150 mm, 5+0,5 kg/cm^2; main gear 660 mm×200 mm, 9+0,5, track 3,8 m, wheelbase 4,56. Electronics MIL-1553B type bus. Seven stores pylons for guided/unguided weapons. Design factor +8/–3. First flight due end-94.
DIMENSIONS Span 10,0 m (32 ft 9¾ in); length 11,3 (37 ft 0⅞ in); wing area 17,67 m^2 (190 ft^2).
ENGINES Two R-35 (DV-2) or Larzac 04-R20, 3,175 lb each.
WEIGHTS Empty c3280 kg (7,230 lb); fuel 850/1650; loaded 4610 (10,163); max 5460 (12,037).
PERFORMANCE Max speed 850 km/h (528 mph); service ceiling 15 km (49,200 ft); range 1200 km (746 miles), ferry 3000 km (1,864); TO 450 m; ldg 480.

ATTA Seeking a partner to complete in USAF JPATS, ANPK MiG picked Belgian Promavia Squalus; Menitskii evaluated it and reported favourably. Objective: keep good features, modify for two engines, fit US avionics, market jointly with Promavia. Two prototypes building for first flight March 94 (missed date).
DIMENSIONS Span 9,73 m (31 ft 11 in); length

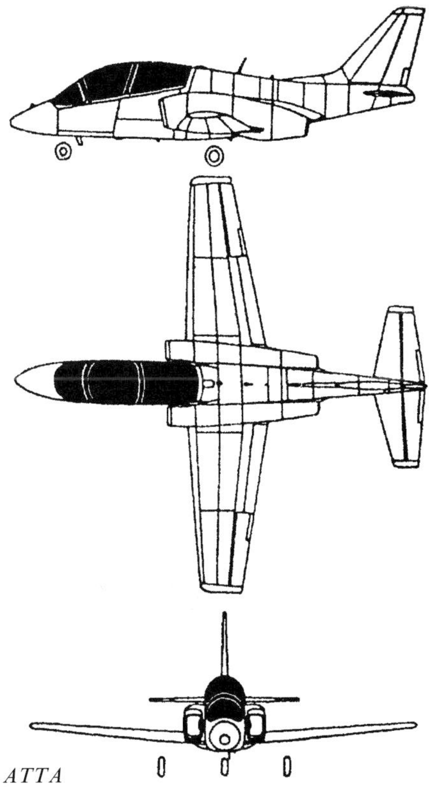

ATTA

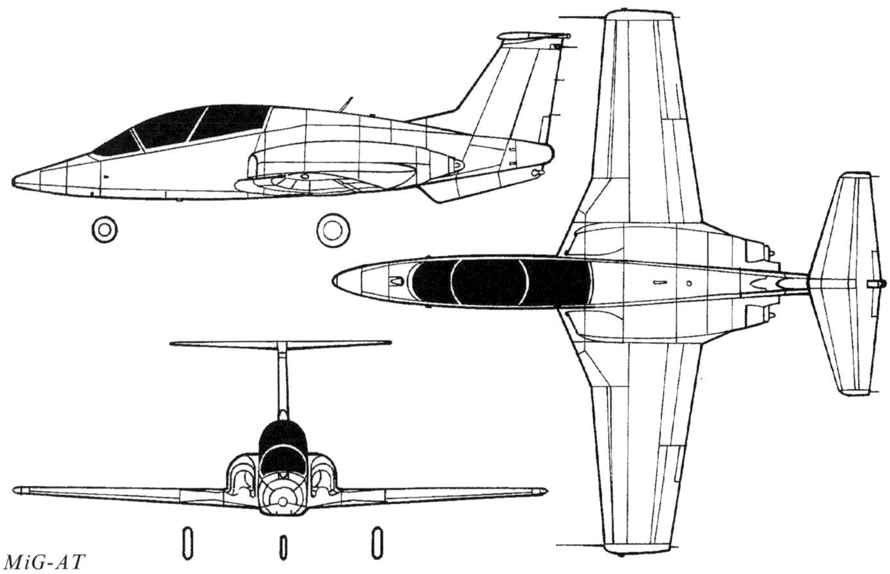

MiG-AT

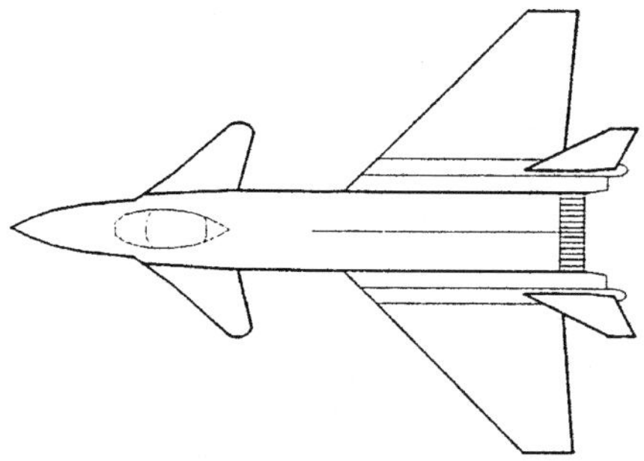

A possible 1-42 configuration

10,42 m (34 ft 2¼ in); wing area 15,86 m² (171 ft²).
ENGINES Two Williams-Rolls FJ44 or Garrett F109, each 1,900 lb.
WEIGHTS Empty 2060 kg (4,541 lb); fuel 780 lit; loaded 3050 kg (6,724 lb); design MTO 7,064 lb.
PERFORMANCE Max speed (SL) 700 km/h (435 mph), (9100 m, 30,000 ft) 800 (497); climb to 5490 m (18,000 ft) 3·5 min; service ceiling 13,72 km (45,000 ft); range 2200 km (1,367 miles); TO 170 km/h, 390 m; ldg 165, 415.

TA-4 Small *Konversiya* project in partnership with Transal AKS to carry 3 passengers and baggage or 300 kg (660 lb) from any surface, using air-cushion landing gear. Configuration as 3-view, advanced corrosion-proof structure, manual flight controls, to be cleared to grass 1 ft, tussocks or furrows 200 mm, snow/ice, waves 250 mm, climb river bank 5% grade. US engines, Bendix-King avionics, deicing. To fly late 94.
DIMENSIONS Span 12,4 m (40 ft 8¼ in); length 9054 (29 ft 8½ in).
ENGINES Propulsion, Teledyne Continental IO-550 (300 hp); cushion, Nelson N-63P (48 hp).
WEIGHT Fuel 130 kg; loaded 1740 kg (3,836 lb).
PERFORMANCE Max speed 270 km/h (168 mph); cruise 180-200 km/h (112-124 mph); ceiling 4 km (13,125 ft); range with reserves 1000 km (621 miles), ferry range 2000 km (1,242 miles); TO 100 km/h, 415 m; ldg 100, 255 (water 305).

New trainer ANPK MiG has published plan view of proposed future airborne part of 'Russian Training Centre'. Very advanced unstable fighter-like twin-jet with novel features.

1-42 OKB designation for **MFI** (multifunction frontal fighter), totally new design of air-combat fighter for foreign sales. Aft delta wing with clipped tips, canards, widely spaced twin verticals, Saturn AL-41F engine with vectoring nozzle, radars fore/aft, internal missile bays, limited stealth design. Design started 83, two prototypes built said by Belyakov Oct 93 to be awaiting engines, first flight expected 94 with interim AL-41 engine. No data.

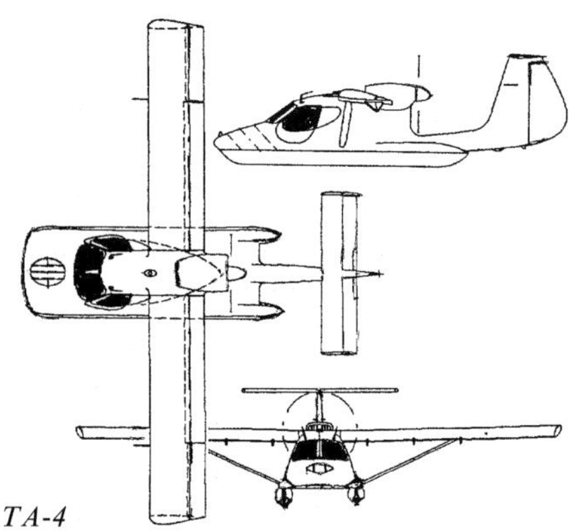

TA-4

MK-1

Mikhelson

N. G. Mikhelson was a First World War designer who joined with several other partners over many years to create diverse types which, though often successfully flown, saw no production. In 1930s he worked at GAZ-3 where, in addition to types described, he produced several light projects including PSN and MU-4 (both see Nikitin).

MK-1 Ryibka Name meaning Little Fish, joint project with Mikhelson responsible for management and drawings, M. M. Shishmaryev for calculations and V. L. Korvin for manufacturing. Begun as project at former Anatra works at Taganrog 1 May 21. Single-seat fighter seaplane, designed for UVVS competition. Single-bay biplane of wooden construction with fabric wings and tail and birch-ply semi-monocoque fuselage of excellent profile with fully cowled engine. Radiator in upper centre-section. Single main spar in each wing and single interplane strut each side. Four-blade propeller with large spinner. Completed as twin-float seaplane spring 23 and made 'fairly good' test flights, major problem being indifferent float design. Later flew better as landplane. No production.
DIMENSIONS Span about 10 m; length about 8 m; wing area in region of 35 m².
ENGINE One 200 hp Hispano-Suiza (judged inadequate for 1920s).
WEIGHTS Probably about 1200 kg loaded.
PERFORMANCE Max speed (landplane) 190 km/h (118 mph).

U-3 Development of U-2 by Mikhelson in partnership with A. I. Morshchikhin. Project started at *Krasnyi Lyetchik* plant (GAZ-3) 1934. Intended to be superior for flying instruction to U-2(AP), with more power and various detail features considered advantages (eg, seven-sided Townend ring cowl made of Enerzh-6 stainless). Flown on factory test 35 but consensus was U-3 offered nothing not available more cheaply from U-2; in any case engine not produced in quantity.

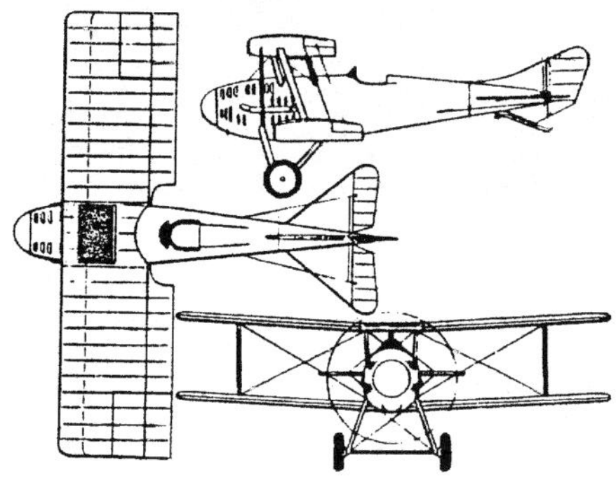

MK-1

DIMENSIONS Span 11,0 m (36 ft 10⅔ in) length 8,4 m (27 ft 6¾ in); wing area believed 30 m² (323 ft²).
ENGINE One M-48.
WEIGHTS No data.
PERFORMANCE max speed 210 km/h (130 mph).

U-4 Second attempt to find superior replacement for U-2. Speed increase obtained by aerodynamic refinement. Slimmer streamlined fuselage, different wings (same as U-3), spats and many other changes including fully cowled engine. Tested A. A. Ivanovskii 36. Satisfactory aircraft, but again offered no advantage over U-2 for instruction and both costly and more fragile.
DIMENSIONS Span 11,0 m (36 ft 10⅔ in); length 8,1 m (26 ft 7 in); wing area 30,0 m² (323 ft²).
ENGINE One M-11.
WEIGHTS Empty 750 kg (1,653 lb); fuel/oil 90 + 10 kg; loaded 1016 kg (2,240 lb).
PERFORMANCE Max speed 170 km/h (105·6 mph); climb 29·6 min to 3 km (U-2, 40-48 min); service ceiling 4 km; range 550 km (342 miles); endurance 4 h; take-off 130 m/11 s; landing speed 70 km/h (43·5 mph).

RV-23 Designation from *Rekord Vysoty*, record height, major project by Mikhelson at GAZ-3 to gain world seaplane height record, with much-modified U-2! Single-seater with greatly extended span, restressed airframe and much more powerful engine. New wing centre-section of 7,1 m span, outer panels 2,8 m longer each side. Forward fuselage redesigned to mate

U-3

with engine installed in I-15 type cowl driving special large-blade propeller. Supplementary tanks ahead of enclosed cabin and beneath upper wing. Wooden floats on faired steel-tube struts. Completed Aug 37 and tested by F. F. Zhyerebchyenko from Moscow reservoir. On 9 Sept reached 11 280 m; 23 Oct reached 11 869 m; finally (no date) 13 430 m. Later taken to Sevastopol for continued testing by Polina (Paulina) Osipyenko, after which decision to build small series.

DIMENSIONS Span 17,0 m (55 ft 9¼ in); length 9,8 m (32 ft 1¾ in); wing area 51,7 m^2 (556·5 ft^2).

ENGINE One 710 hp Wright Cyclone SGR-1820-F3.

WEIGHTS (landplane) empty 994 kg (2,191 lb); fuel/oil 130+20 kg; loaded 1244 kg (2,743 lb).

PERFORMANCE Max speed (landplane) 130 km/h (81 mph); ceiling 14 km (45,930 ft); landing speed 90 km/h (56 mph).

MP

MP Designation from *Morskoi Podvesnoi*, naval suspended; radical high-performance torpedo carrier projected by Mikhelson/Morshchikhin at Leningrad 36. Original idea by N. Valko based on Vakhmistrov (q.v.). Reasoning was that high-performance aircraft able to attack heavily defended ship would lack range; but if carried to target under parent aircraft with just sufficient fuel for return trip it could be smaller and more agile. Moreover, it could be water-based yet carry torpedo under or recessed in keel because it would not have to take off by itself. Krasnyi Lyetchik plant took on construction job to drawings by M/M team, with assistance from Vakhmistrov. Apart from being designed to hang (in flight of two, possibly three, aircraft) under large carrier aircraft, MP was itself unconventional. Cantilever low-wing monoplane single-seater (pilot needed to navigate only on return trip) with powerful water-cooled engine. Mixed construction, mainly welded Cr-Mo tube fuselage with dural covering, wing all-metal but structure unknown. Tough planing bottom from bow (under nose of engine) to second step aft of wing, with large recess to accommodate torpedo (553 mm with twin fins, Type 45-36-AN). Aircraft armed and hung under parent aircraft by single locking attachment above CG in front of inverse-raked windscreen. Near target, engine started; release controlled by MP pilot. After high-speed low-level attack, return to base and water landing, engine being tilted 20° upwards on long mounting beams to raise propeller clear of water. Wings designed to serve as sponsons giving lateral stability. Retractable landing gear also fitted. Ready for flight 1938, but political atmosphere so frightening nobody dared to sanction flight testing in case anything went wrong. See Addenda.

DIMENSIONS Span 8,5 m (27 ft 10⅔ in); length about 8,0 m (26 ft 3 in); wing area 20,0 m^2 (215 ft^2).

ENGINE One 860 hp Hispano-Suiza 12Ybrs.

WEIGHTS Empty about 2200 kg (4,850 lb); loaded 3200 kg (7,055 lb). No other data.

Mil

Mikhail Leontyevich Mil was Siberian, like Kamov, born son of mining engineer at Irkutsk 22 Nov 09. Siberian Tech Inst, Tomsk, from 26; Don Poly Inst, Novochyerkassk, 28, transferring to Novochyerkassk Aviation Inst, graduating 31. In 29 through Osoaviakhim assisted with KaSkr-1. From 31 in Izakson's brigade at CAHI and played leading role in A-15 autogyro. Deputy chief designer to Kamov 36. Dr Mil served at front with A-7 autogyros 1941-43. In 45 head of rotating-wing lab at CAHI; formed own OKB 47. Died 31 Jan 70, succeeded by Marat N. Tishchyenko to 92, then Mark V. Vineberg.

Mi-1, GM-1 First project of OKB, this was (remarkably, in view of widespread prior

Mi-1T

rotary-wing efforts) first Soviet helicopter of now-classic layout with single main rotor and anti-torque tail rotor. Original designation GM-1 (*Gelikopter* Mil); design begun September 47 and prototype completed exactly a year later. Three-blade main rotor, blades based on A-15 and related autogyros, mixed steel/ply/fabric NACA-230 profile, fully articulated hub with friction dampers, normal speed 232 rpm. Fuselage light alloy, except for welded steel-tube basis of mid-section housing engine with crankshaft horizontal and cooling fan, driving through angle box to transmission with centrifugal clutch and rotor brake. Four-seat cabin with left/right hinged doors. Fuel in welded aluminium tank 240 lit (53 gal) behind engine and, from about 40th production, provision for external supplementary tank of 160 lit (35 gal) on left side. Monocoque tail boom and pylon for tail rotor with three wooden blades. Fixed nosewheel-type landing gear with brakes, plus long rear skid to protect tail rotor. First flight M. K. Baikalov (ex-Bratukhin) Sept 48. Both first two GM-1 lost, second killing Baikalov after weld failure in tail-rotor bearing. Project taken over by Mark Gallai and V. V. Vinitskii, followed in summer 49 by NII testing by G. A. Tinyakov and S. G. Brovtsyev, reached height 6800 m and speed 190,5 km/h. Yak-100 delayed so production authorized as Mi-1. Eight took part in 51 Aviation Day display. Civil and military variants including ambulance with left/right stretcher pods externally, agricultural **Mi-1NKh** with two 500 lit (110 gal) hoppers (note: solids only, liquid being weight-limited to smaller capacity) and many specialized models, with optional pontoon landing gear. From about 40th called **Mi-1M** with 0,32 m² adjustable stabilizer (tailplane). From 57 new blades with extruded steel-tube spar. By this time basic model called **Mi-1T**, pilot and two passengers plus radio and fluid de-icing; also dual **Mi-1U** trainer. In 61 Mi-1 *Moskvich* with all-metal blades of almost untapered plan, hydraulic controls and better standard of equipment and soundproofing. Name dropped and improvements (initially at request of Aeroflot) mostly standardized. Variant with four-blade rotor (erroneously dubbed Mi-3 in West) remained prototype. From 54 main blades bonded metal, prodn transferred to Poland as **SM-1**, 'some thousands' built to 65. Last example withdrawn 83. ASCC name 'Hare'.

DIMENSIONS Diameter of three-blade main rotor 14,346 m (47 ft $0\frac{3}{4}$ in); length (rotors turning) 16,95 m (55 ft $7\frac{3}{8}$ in), (ignoring rotors) 12,1 m (39 ft $8\frac{3}{8}$ in); height 3,3 m (10 ft 10 in).

ENGINE One AI-26V.

WEIGHTS Empty (Mi-1M) 1,8 t (3,968 lb), (others, 1,760-1,890 lb); fuel (typical) 360 lit; loaded 2,3 t (5,071 lb), (max) 2,4 t (5,291 lb).

Mi-2

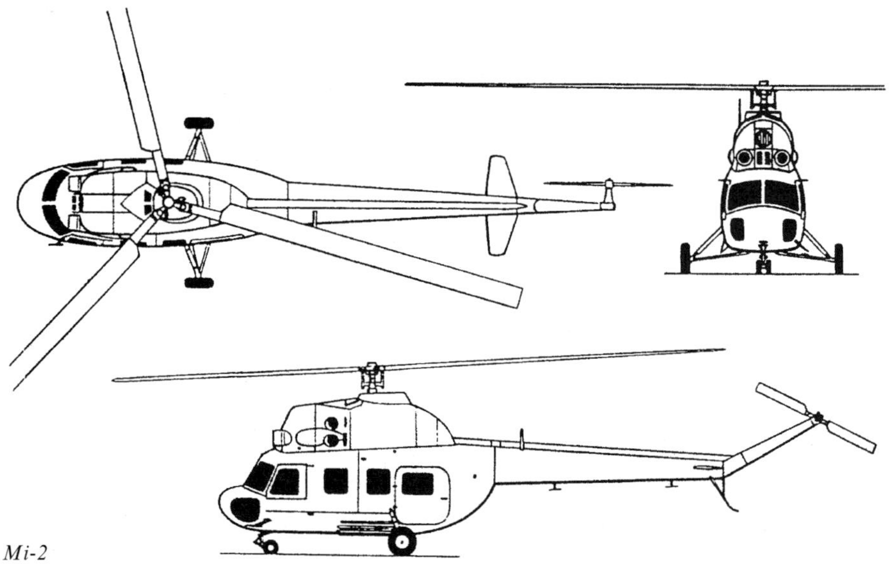

Mi-2

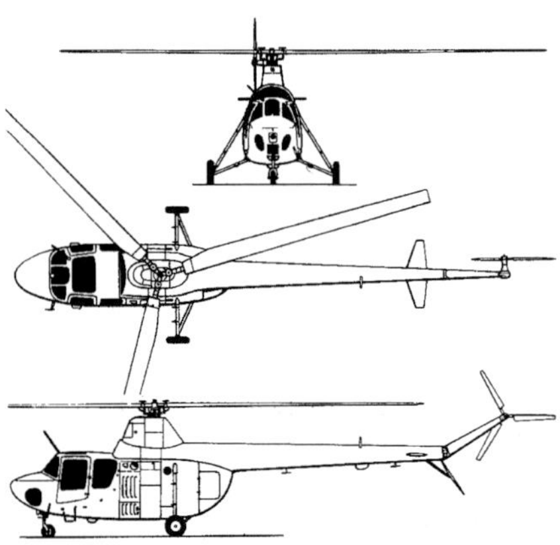

Mi-1T

PERFORMANCE Max speed at SL 170 km/h (106 mph), at 2 km 185 km/h (115 mph); cruise 135 (84); climb 6,5 m/s (1,280 ft/min); service ceiling 3 km (9,842 ft); range 620 km (385 miles).

Mi-2 Chronologically later than Mi-4 and Mi-6, Mi-2 was planned 1958-59 as modernized turbine derivative of Mi-1, just as Mi-8 was to

Mi-4. Originally Mi-1 dynamic parts retained and mated with completely new fuselage, all light-alloy monocoque with steel forgings at concentrated loads, with twin turboshaft engines above cabin ahead of gearbox. Structural basis deep floor box carrying wheel or ski landing gears and housing flexible fuel cell of 600 lit (132 gal). Normal accommodation for pilot and passenger (on right) with main cabin for 700 kg (1,543 lb) cargo or seat unit for three passengers facing forward and three facing aft, with eighth passenger seat on right opposite rear door. Option of four stretchers and attendant, or slung load of 1,2 t (2,645 lb) or two 600 lit ag containers. All versions plumbed for two 250-lit (55-gal) auxiliary tanks on sides. First flight of **V-2** Sept 61. Subsequently developed as production Mi-2 with bonded/welded fuselage, hub with hydraulic instead of friction dampers, bleed-air anticed intakes, tail rotor with bonded-metal honeycomb blades and electro-thermal de-icing on all blades. Whole programme transferred to WSK-Swidnik in Poland in Jan 64. All production and development subsequently at WSK (outside scope of book), involving 12 series versions including SAR, photo, anti-armour and gunship. Total by 93 over 5,320. ASCC name 'Hoplite'.

DIMENSIONS Diameter of main rotor 14,56 m (47 ft 9¼ in); length (rotors turning) 17,42 m (57 ft 2 in), (ignoring rotors) 11,94 m (39 ft 2 in); height 3,75 m (12 ft 3½ in).

ENGINES Two PZL-built Isotov GTD-350 or GTD-350P.

WEIGHTS Empty (passenger) 2402 kg (5,295 lb); loaded 3,55 t (7,826 lb); max (not ag versions) 3,7 t (8,157 lb).

PERFORMANCE Max speed (except ag) 210 km/h (130·5 mph), (ag models) 155 km/h (96·5 mph); cruise 190 km/h (118 mph); SL max climb 4,5 m/s (885 ft/min); hover ceiling OGE 1 km (3,280 ft); service ceiling 4 km (13,125 ft); range (max payload) 170 km (105 miles), (ferry, no reserve) 797 km (495 miles).

Mi-4T

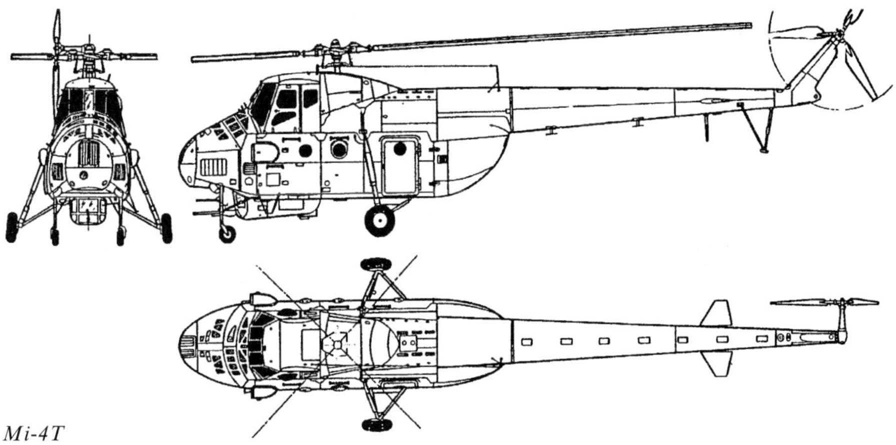

Mi-4T

Mi-4 Second challenge to OKB, this was dramatic leap in capability and resulted from Kremlin meeting of constructors September 1951 at which Stalin insisted on sudden great advance in Soviet helicopters. All backed away except Mil and Yakovlev; on following day these two were given one year to design, build and fly prototypes, Mil's assignment being single-engined 12-passenger machine. Mil had prepared outline design beforehand which he produced at Kremlin; basically scaled-up rotors of Mi-1 with added fourth main-rotor blade, and S-55 configuration. Rotor axis inclined forwards 5°. Fuselage light-alloy semi-monocoque with extensive use of magnesium. Engine installed at 25° in nose accessible through upper/lower hinged nose doors and left/right hinged side doors. Cooling fan and centrifugal clutch immediately to rear of engine, with inclined shaft between pilots to main gearbox. Separate cooling systems for oil radiator and hydraulics for flight control. Straight tail boom with adjustable stabilizers, deep skid/bumper and narrow fin carrying tail rotor on right, with three bakelite-ply blades. Main fuel tank welded aluminium, 960 lit (211 gal), behind gearbox; optional aux tank 500 lit in hold or externally. Quad landing gear with pneumatic-braked mainwheels and castoring nosewheels; optional pontoons for water. Main hold 4,15 m (13 ft 7⅓ in) long, about 1,8 m (71 in) square section. Max internal load 1740 kg (3,836 lb) including small vehicles loaded through left/right rear doors and clip-on ramps. Slung load to 1,3 t (2,866 lb). Alcohol de-icing of blades and windscreens as on Mi-1. First flight delayed several weeks by blade flutter in ground-running from 14 April 52. First flight May, Vinitskii assisted by Brovtsyev. Remarkably few subsequent snags and NII testing completed before end of year. Main production model military **Mi-4T**, with increased-diam main rotor, aluminium cargo floor, bulged circular windows with gun ports, ventral gondola for nav/observer, and tactical avionics. From 54 **Mi-4P** civil variant with large rectangular windows, spatted wheels (often later removed), no gondola, and interior heated, soundproofed and equipped for ten passengers (each 20 kg baggage)/wardrobe and toilet Small batch of **Mi-4L** (*Lyuks*, de luxe) six-seaters. Some civil and mil equipped for casevac with eight stretchers and attendant. **Skh** multi-role ag variant for spraying (1600 lit, 352 gal) with wide spraybars, dusting (1 t, 2,205 lb) or forest fire-fighting. Urgent development of improved metal blades 1954-60 culminating in dural blade with extruded spar and honeycomb box trailing sections. Magnesium fuselage skins replaced by aluminium, and better flight control and avionics. April 56 various records including 500 km circuit at 187,254 km/h, and later 1012 kg lifted to 7575 m (24,850 ft). Main prodn continued with **Mi-4A** assault transport with A-12·7 with 200 rounds, 1740 kg cargo or 14 troops; **Mi-4S** Salon VIP version (7315 kg max) and, first flown 62, **Mi-4PL** ASW version for AV-MF with four 250 kg or six 100 kg A/S bombs, chin radar (also fitted to various military variants), short but deeper gondola, towed MAD at rear and sonobuoys on external rack on right side. High altitude (2-speed supercharged) model 65, **Mi-4M** tactical variant with gun turret and air/surface rockets 68, and EW platform first seen 77 with two pairs of lateral *Yagi* arrays and other aerials mainly for communications jamming. Production ceased 64 at c3,200, of which c700 exported to 6 countries, plus 545 licence-built in China from 59 as Z-5 with many changes. ASCC name 'Hound'.

DIMENSIONS Diameter of four-blade main rotor (initial) 17,22 m (56 ft 6 in), (series) 21,0 m (68 ft 11 in); length (rotors turning) 25,02 m (82 ft 1 in), (ignoring rotors) 16,79 m (55 ft 1 in); height 4,4 m (14 ft 5¼ in).

ENGINE One ASh-82V.

WEIGHTS Empty (A, T) 4,9 t (10,802 lb), (M) 5365 kg (11,828 lb); loaded (A, T) 7,15 t (15,763 lb), (M) 7,29 t (16,071 lb); max (A, T) 7,55 t (16,645 lb), (M) 7575 kg (16,700 lb).

PERFORMANCE Max speed at SL 175 km/h (109 mph), at 500 m (A, T) 200 (124 mph), (M)

Mi-6T

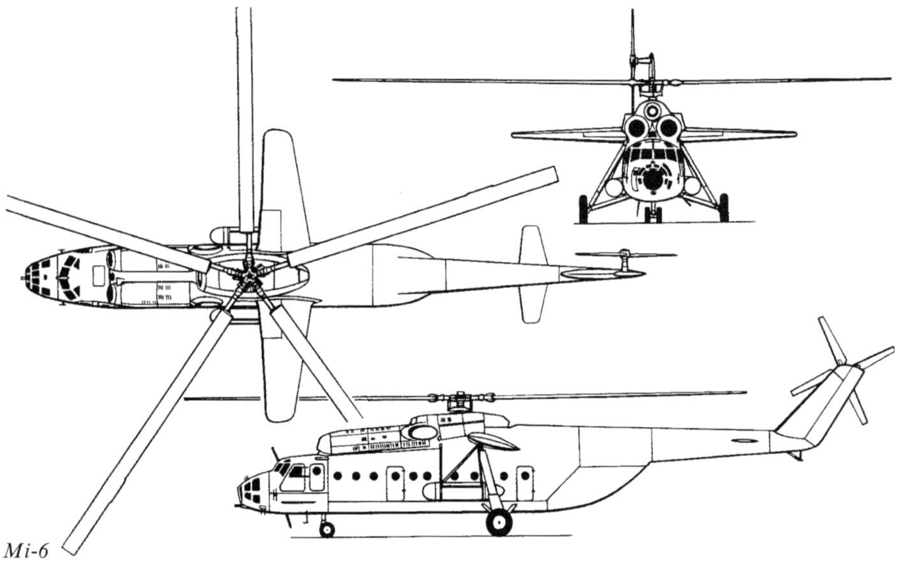

Mi-6

185 (115 mph); SL climb (A, T) 6,5 m/s (1,280 ft/min), (M) 3,9 m/s (768 ft/min); service ceiling (A, T) 5,5 km (18,045 ft), (M) 4 (13,125 ft); range (A, T) 650 km (404 miles), (M) 370 km (230 miles).

Mi-6 Heavy transport helicopter, built to meet combined military/civil requirement of June 54, possibly greatest single advance in history of rotary-wing aircraft. By far largest helicopter of its day, and also fastest; pioneered large turbines, use of lifting wings and (with wings removed) heavy crane role.

Prime original requirement was VVS need for VTOL airlift to complement An-12 and carry similar-size items (two lifts with Mi-6 equal one full load of An-12). Main Aeroflot need was strategic airlift in Siberia and other developing regions, though passenger version also built. Cargo hold cross-section smaller than An-12, width 2,65 m and height 2,5 m at rear reduced over front half to 2,01 m. Other basic factor was engine/gearbox, sized to give twin-engine safety in this application. Solovyev bureau adopted free turbine to give speed flexibility and easy starting, and eliminate clutch. Engines same each side except for handed jetpipes. Trolley APU driven by 100 hp AI-8 turbine carried on board for ground power and to supply engine-mounted 12 kW starter/generators . Each engine also drives 90 kVA alternator feeding 360 V at 400 Hz, partly radio but mainly for heavy electrothermal load anticing engine inlets and main-rotor blades. Left/right engines attached flexibly to sub-structure above forward fuselage with drive shafts at rear to R-7 gearbox weighing 3,2 tonnes not including oil system and housing four sets of gearwheels reducing speed 69·2 times. Box also drives oil cooling fan served by central intake ahead of rotor hub and three hydraulic systems (11,77/15,2 MPa, 1705/2205 lb/in²) for flight control. Rotor axis angled forward 5°. Five blades, final standard based on extruded cold-rolled tubular spar 40KhNMA steel with screwed-on aerofoil sections in short lengths made from 2 mm D19A-M dural giving profile CAHI modified NACA-230 with thickness 11% from root to tip. Normal speed 120 rpm. AV-63B four-blade tail rotor with blades of bakelite ply on steel spar, electrothermal anticing changed to alcohol late in production. Conventional stressed-skin fuselage and tail boom. Wing of P35 profile, 15% root and 12% tip, set at 14°15'/15°45' incidence to provide 20% gross lift in cruise; removed in crane role. Horizontal stabilizer 4,87 m² trimming –5°/+13°. Fixed landing gear, main tyres 1320×480 mm, 0,686 MPa (99·5 lb/in²), pneumatic brakes, dual-pressure legs interconnected across aircraft to damp ground resonance. Nose gear free castoring, twin tyres 720×310 mm. Cargo via left/right full-section rear doors/ramps with hydraulic drive. Hold length 12 m, volume 80 m³ (2,825 ft³), floor stressed to 2 t/m² and electric winch (800 kg pull) multiplied by pulley-blocks. Slung cargo to 8 t, internal overload 12 t (26,450 lb). Flight deck with two jettisonable doors each side for two pilots, nav (glazed nose), radio and engineer; electrothermal window anti-icing, three-channel autopilot. Normally tip-up seats along main hold, max 65; up to 90 passengers with clip-on central seating. Max fuel capacity 17 250 lit (3,794 gal) comprising eleven fuselage tanks total 8250 lit and four-overload (ferry) drum tanks of 2250 lit each, two in cabin and two external. Engine side panels form maintenance walkway, closed hydraulically .

Five development aircraft, first flight R. I. Kaprelyan Sept 57 without wings. Many subsequent records by Brovtsyev, N. Lyeshin, V. Kolosychyenko, B. Galitskii and other pilots, including 20 117 kg useful load lifted to 2738 m by Kaprelyan 61, and speed over 100 km circuit of 340,15 km/h (211·35 mph) with specially prepared aircraft. Pre-series of 30 followed by **Mi-6T** for airborne troops (provision for A-12·7 in nose fired by nav), **Mi-6A** for Aeroflot (by 1990 airlifted 15 m tonnes plus 12 m people), **Mi-6S** casevac (41 stretchers, 2 seats, oxygen), and **Mi-6P** firefighting (water dumper, wings removed). Single **Mi-6Pass** 80-passenger airline version with large rectangular windows. Popular name *Shestyorka* (six); ASCC name 'Hook-A'. Several **Mi-6R** and related versions equipped as com-relay and airborne command posts, with c8 t payload of radio including 12-15 antennas and heat exchanger on right of fuselage, 'Hook-B'; a later com-relay version with totally different antennas is **Mi-22**, ASCC 'Hook-C'. Among export customers air forces of Algeria, Iraq, Peru and Vietnam. Total about 800, complete 1980. All versions 'being grounded', said Vineberg Aug 92.

DIMENSIONS Diameter of main rotor 35,0 m (114 ft 10 in); length (rotors turning) 41,739 m (136 ft 11½ in), (ignoring rotors) 33,179 m (108 ft 10½ in); height 6,71 m (22 ft 0 in), (tail rotor turning) 9,86 m (32 ft 4 in); span 15,3 m (50 ft 2½ in); wing area 35,0 m² (376·7 ft²).

ENGINES Two D-25V.

WEIGHTS Empty (6A) 27 240 kg (60,055 lb); fuel (normal) 6315 kg; loaded (9016 kg payload) 44 t (97,002 lb); limit for normal VTO 42,5 t (93,695 lb).

PERFORMANCE Max speed (normal) 300 km/h (186 mph); max cruise 250 km/h (155 mph); climb at 40,5 t, 9·7 min to 3 km, 20·7 min to 4·5 km; service ceiling 4,5 km (14,750 ft); range with 8 t payload at 1 km econ cruise 600 km (373 miles), (4,5 t and external tanks) 1050 km (652 miles), (ferry) 1450 km (900 miles).

V-7 Few details available of this experimental tip-drive helicopter completed 1959. Pilot and three passengers in stressed-skin nacelle with three doors, tail on tubular boom, skid landing gear and kerosene tank in roof. All-metal blades (related to final Mi-1 but shorter) carrying subsonic tip ramjets. Hub of design unlike other Mil helicopters. Inclined drive shaft to tail rotor. Believed never flew with more than pilot on board, and never publicized.
DIMENSIONS Diameter of two-blade main rotor 11,6 m (38 ft 7 in); length not recorded.
ENGINES Two AI-7.
WEIGHTS Empty 730 kg (1,609 lb); loaded 835 kg (1,841 lb).

V-7

V-7

Mi-8 Great success of Mi-6 spurred development of smaller and more versatile machine, and V-8 (series designation Mi-8) was designed on basis of Mi-4 and retaining existing rotors, transmission (except clutch and fan) and tail boom. Fuselage pod much larger, and because of high installation of engine wholly available for payload. Light-alloy semi-monocoque with nose cockpit for two pilots, with jump seat between for engineer. Main cabin 6,42 m (later 6,36 m, 20 ft 10¼ in) long, 2,2 m (later 2,34 m, 7 ft 8¾ in) wide and 1,82 m (later 1,80 m, 5 ft 10¾ in) high. Engine KB under S. P. Izotov assigned development of new free-turbine engine which, like that for Mi-6, was a conservative design and eschewed performance in favour of robust reliability. Drive via VR-8 two-stage gearbox reducing engine input (max 12,000 rpm) to rotor speed 192 rpm. Engine not flight-cleared when prototype appraching completion, and to begin flight development **V-8** prototype flown with single AI-24V and four-blade main rotor from June 61. Improved rotor then designed with five blades of better structural design than previous Mil rotors. NACA-230 profile, entire leading edge extruded aluminium-alloy spar with rear formed from 21 honeycomb-stabilized sections screwed on. Production spar gas-pressurized to warn of fatigue crack, and balance tab added to trailing edge. New hub scaled from that of Mi-6, with irreversible hydraulic control. Axis inclined forwards 4°30'. New blades of similar construction also designed for three-blade tail rotor. All blades with electrothermal anticing. Before these blades were ready, TV2-117 engine became available, at rating of 1400 shp, installed in second prototype flown Aug 62, still with Mi-4 dynamic parts. New rotor hub and blades first flown 64, but several pre-production Mi-8s flew with old rotor which caused prolonged difficulties. Engine uprated 66 to 1,500 shp as TV2-117A. Small internal fuel bag of 445 lit (98 gal), main supply in external drum tanks housing (L) 745 lit (164 gal), (R) 680 lit (149·5 gal); provision for one or two similar tanks in cabin raising max ferry fuel to 3700 lit (814 gal). Right external tank faired at front into large cabin heater on production aircraft (heater sometimes not installed). Main and standby hydraulics at 4,4/6,38 MPa (640/925 lb/in²), lower than Mi-6 and Mi-10. DC starter/generators, but main anti-icing load AC at four voltages from 7·5 to 208. Comprehensive instruments (duplicated) and avionics including autopilot giving stabilization about all axes and set hover height or IAS. Main gear with tyres 865×280 mm, pneumatic brakes, twin steerable nosewheels locked central in flight.

Production authorized 66. Basic utility variant **Mi-8T**, civil or military, with circular windows, aluminium cargo floor, rear full-width doors and hook-on ramps and hold thereby reduced in length to 5,34 m (17 ft 6 in). Capacity 23 m³ (812 ft³). Optional 24 tip-up seats along walls, 200 kg (441 lb) electric winch with pulley-blocks, sling for external load of 3 t (6,614 lb) and max internal load 4 t (8,820 lb). Optional rescue hoist, electric 150 kg (330 lb), above sliding jettissonable front left door. Normal fuel 1870 lit + 2×500 lit external. Military usually with AP-34B four-axis autopilot linked through SAU-8 flight-control system, R-842 M hf and R-860 vhf, SBU-7 intercom, ARK-9 ADF, RV-3 radar altimeter (aural dangerous-height warning), SRO-2 IFF and often DISS-15 doppler in box under tail boom. Option of fitting strut-braced outriggers for two payloads each side, usually UV-16-57 and later 32-57 rocket launchers or bombs to FAB-250 or many special role fits. Export sub-types with weather radar in nose pod or ventral ASW search radar, engine inlet filters or other equipment. ASCC name 'Hip-C'. When fitted with special command communications in pylon-hung boxes, with large tubular and wire antennas above/below tail boom, **Mi-8 VZPU** (*Vozduzhni'i Zapasnoi Punkt Upravlenya*) Aerial emergency command post 'Hip-D'. **Mi-8SMV** with border-surveillance sensors along left side called 'Hip-J'. **Mi-8TG** conversions to operate on LPG (liquefied petroleum gas), 'Hip-C'.

Mi-8P civil 28-seater (32 seats without toilet/wárdrobe/baggage), convertible to cargo. Same large square windows as prototypes, early examples with spats, fuel 2615 lit. **Mi-8S** (Salon) civil/military VIP, as 8P but luxury interior up to 11 passengers. Both these 'Hip-C'. **Mi-8TB** armed vesion, fuel 1870+915 lit, A-12·7 with 700 rounds on pivot in nose, six pylons for 32-57 or B8V20 launchers or FAB-250 (or mix) plus upper rails for four/six 9M17 missiles, gyrostabilized missile sight and various ECM/IRCM, without weapons 22 troops or 12 stretchers or 3 t cargo; ASCC 'Hip-E'. **Mi-8TBK** (originally DDR, E Germany, Nicaragua because 9M17 not available) similar but rails for six 9M14M missiles; 'Hip-F'. **Mi-9** command relay platform, fuel 2615 lit, cabin filled with HF R-886, UHF R-802 and R-111, ADF R-405, central language/alphanumeric station, GS-24A generator driven by AI-8 APU, two Chipstone 'hockey stick' and one mast antennas, long strakes under fuselage; 'Hip-G'. **Mi-8M, Mi-8TV**, see Mi-17. **Mi-8PPA** (*Patanovchik Pamech Aktivni'i*) comint and jamming platform with six similar X-shaped antennas in frame each side with avionics cooled by six fan-driven radiators under fuselage; 'Hip-K'.

Most successful European helicopter, 8,100 built at Kazan and Ulan Ude by 82 and over 500 since, for 33 air forces and many civil customers. Assisted Mi-24, basis of Mi-14 and developed into Mi-17.
DIMENSIONS Diameter of 5-blade main rotor 21,29 m (69 ft 10¼ in); length (rotors turning) 25,28 m (82 ft 11¼ in); (ignoring rotors) 18,31 m (60 ft 0¾ in); height 5,65 m (18 ft 6½ in).
ENGINES Two TV2-117A.
WEIGHTS Empty (T) 7160 kg (15,785 lb), (S) 7420 kg, (TB) 7422 kg, (9) 7500 kg; loaded (T) 11,1 t (24,471 lb), (S) 10,4 t, (TB) 11 564 kg, (9) 11 t; max (all) 12 t (26,455 lb).
PERFORMANCE Max speed at SL 250 km/h (155 mph), (TB, 245), cruise 225 km/h (140 mph) (TB 205); climb at SL 4,5 m/s (886 ft/min); service ceiling 4,5 km (14,764 ft); range (T) 930 km (578 miles), (TB) 775 km, (S, 9) 480 km.

Mi-10 It was logical to mate dynamic parts of Mi-6 with rest of helicopter tailored to crane role, lifting bulky external load and positioning over short ranges. Start of design Feb 58, **V-10**, prototype of initial **Mi-10R** version, flown 60 in VVS markings. This version with dedicated

Mi-8TB

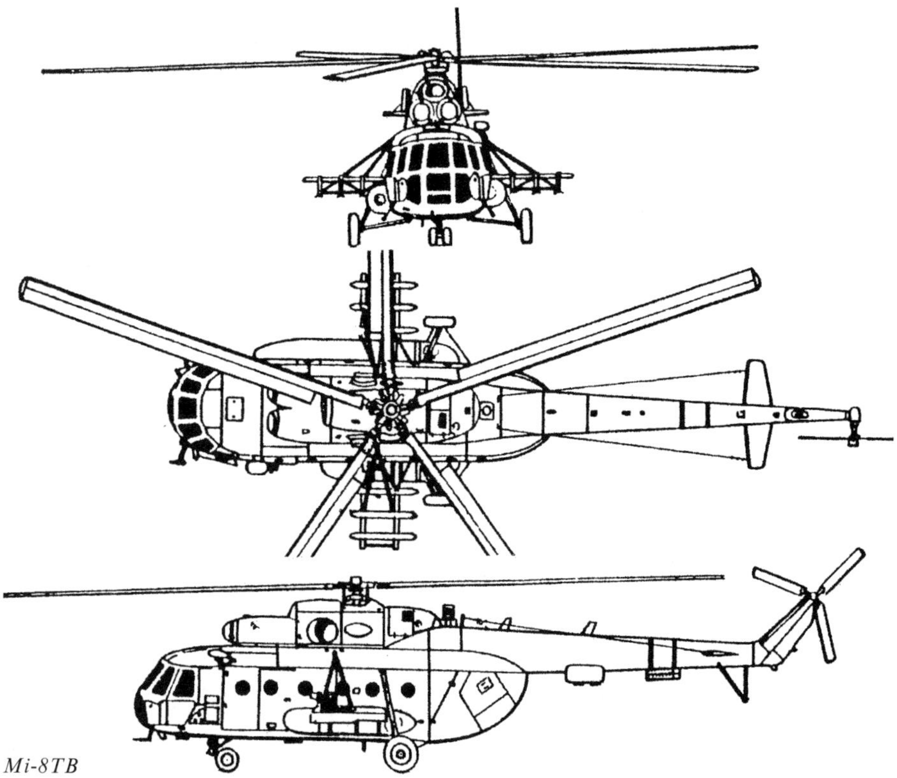

Mi-8TB

hydraulic locks for securing cargo in space 4 m high between four tall legs. Mi-6A dynamic parts, but main-rotor axis inclined forwards only 1°30'. New fuselage with shorter nose, different flight deck with no eye brow windows but deep bulged side windows and single door each side well aft (under engine compressors), two pilots and engineer but no nav so no nose compartment. Slim fuselage with cabin 14,04 m (46 ft 0¾ in) long, 2,5 m (8 ft 2½ in) wide and 1,68 m (5 ft 6 in) high for cargo and with 28 tip-up wall seats. Almost straight bottom line; top of rear fuselage slopes down to bring tail fin and rotor much lower than in Mi-6A. Tall quad landing gears with two pairs 1230×260 mm main tyres and two pairs 950×250 mm tyres on castoring nosewheels, with flight-crew steps in left front leg fairing, in some Mi-10s both front legs. Development aircraft had trapeze wires from above flight deck to front wheels for emergency crew escape in low hover. To counter side-thrust at tail and torque effects right legs 300 mm (11·8 in) shorter than left, crew cabin being canted to keep it laterally level on ground. Fuel in two external pods (3500 lit, 771 gal each) plus internal service tank (731 lit, 161 gal); total 6184 kg (13,633 lb). Optional ferry tanks in hold, total 2400 lit (528 gal, 1920 kg, 4,233 lb). Provision for pressure-fuelling from ground whilst hovering. Internal cargo loaded via rear right door with 200 kg electric hoist. Main load slung externally. Ground clearance 3,75 m (12 ft 3½ in) at full load. Normal procedure to taxi over and straddle load which can measure 20 m by 10 m (actual Mi-10 track and wheelbase, measured from strut centrelines, 6,92 and 8,29 m). Two load-carrying systems: 500-kg winch through floor hatch, and group of four swinging struts picking up corners of load with hydraulic grips controlled from cockpit or remote panel in cabin via TV from two locations. Load can be mounted on special platform 8,53×3,54 m (27 ft 11¾ in×11 ft 7⅓ in) with or without wheels. AI-8 turbine APU permanently installed for electric/hydraulic power without main engines. Anticing as Mi-6A. Can make running take-offs and landings at 100 km/h. Total Mi-10R about 60. ASCC name 'Harke'.

Mi-10K (*Korotkonogii*, short-legged) presaged by special Mi-10 flown 65 with single centreline

Mi-10

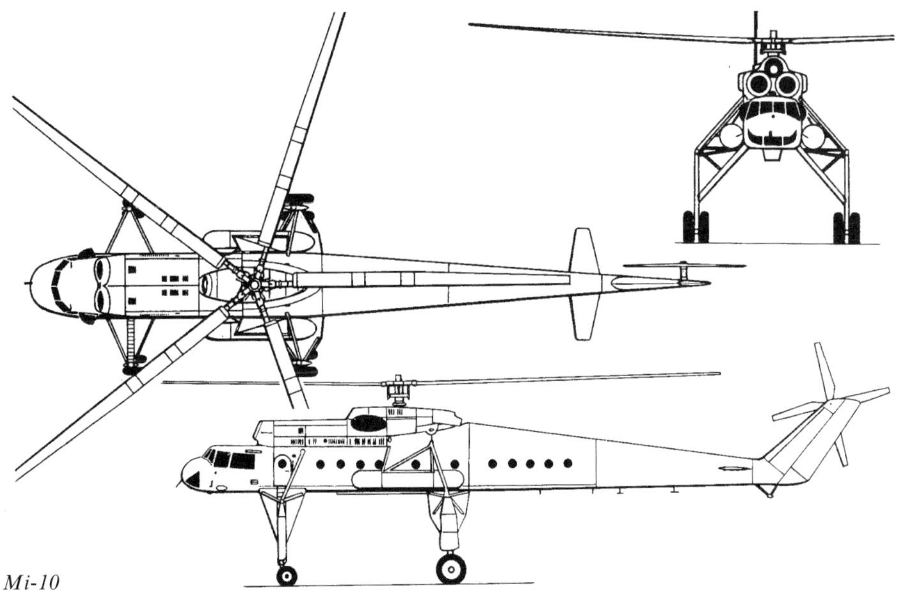

Mi-10

nose gear and single spatted main gears of min length and weight, used to lift 25 105 kg (55,347 lb) to 2840 m (G. Alferov, 28 May 65). This was restored to Mi-10R standard, but in 66 OKB flew **Mi-10K** with four short landing gears reducing door sill height from almost 4 to 1,8 m. New crew compartment with single pilot at original level and second in central all-glazed gondola facing aft with full controls for helicopter and load. Larger internal fuel cells, giving total with two external tanks of 9000 lit (1,980 gal). Tail bumper and much narrower tail fin. Cleared for production October 66, about 20 built. ASCC 'Harke'. All versions being grounded Aug 92.

DIMENSIONS Diameter of main rotor 35,0 m (114 ft 10 in); length (rotors turning) 41,89 m (137 ft 5½ in), (ignoring rotors) 32,86 m (107 ft 9¾ in); height (10R) 9,8 m, (10K) 7,8.

ENGINES Two D-25V.

WEIGHTS Empty (R) 27,3 t (60,185 lb), (K) 24 680 kg (54,410 lb); max payload (R) 15 t internal, 8 t slung, (K) 11 t; loaded (R) 43,7 t (96,340 lb), (K) 38 t (83,775 lb).

PERFORMANCE Max speed (10, max wt) 200 km/h (124 mph); cruise (10, max wt) 180 km/h (112 mph), (K, empty) 250 km/h (155 mph), (K, slung load) 202 km/h (125 mph); authorized height limit 3 km (9,843 ft); range (typical load) 250 km (155 miles), (max ferry) 795 km (494 miles).

V-12 By far largest helicopter ever built, this was unusual extrapolation of Mi-6 a decade later to match greater fixed-wing airlift of An-22 and Il-76. To avoid immense task of developing new set of rotors, reduction gears and transmission, decision taken to double up Mi-6 dynamics and use two sets of Mi-6 engines, gearboxes and lifting rotors side-by-side, left rotor being mirror image, with small overlap. Rotor rpm reduced to 112; gearboxes linked by transverse shafting. Axes inclined forwards 4° 30'. Engine/rotor groups carried on wings of light-alloy stressed-skin construction with 8° dihedral, sharp inverse taper and set at incidence 7° root 14° tip. Braced at root and tip to main landing gears with torque reacted by horizontal bracing to rear fuselage. Inner/outer trailing-edge flaps fixed in up position after flight trials. Fuel in outer wings and two external tanks; optional ferry tanks in cabin. Fixed twin-wheel landing gear with main tyres 1750×730 mm, pneumatic brakes, and steerable nose tyres 1200×450 mm. Large stressed-skin fuselage with crew door each side, three sliding side doors and full-section rear clamshell doors and ramp with left/right twin-wheel ventral bumpers. Aeroplane tail with fin, tabbed rudder, dihedralled tailplane with tabbed elevators, and endplate fins mounted vertically but toed inwards. Flight deck for pilot (left) with engineer behind and co-pilot (right) with elec-syst operator behind. Upper flight deck for nav with radio operator behind. Hydraulic flight control

V-12

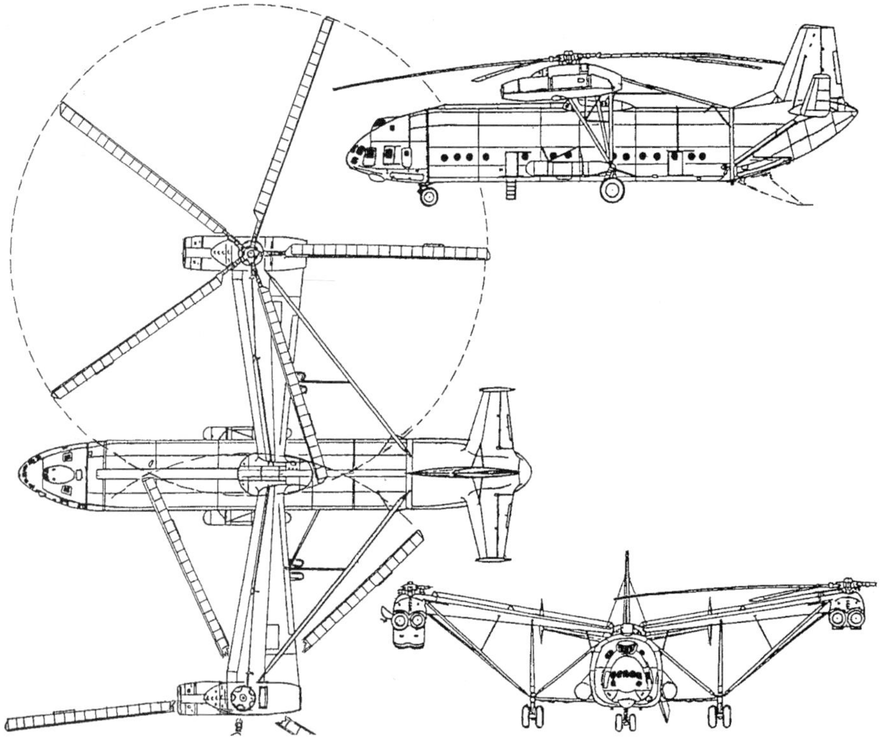

V-12 (starboard engines open for access in front view)

with emergency manual reversion. Autopilot with three-axis autostab; mapping radar under nose, AI-8 turbine APU for ground power and engine start. Main cabin 28,15 m (92 ft 4 in) long, 4,4 m (14 ft 5 in) square. Overhead gantry crane with four 1 t hoists. Tip-up seats along sides (50 to 120). First hover 67 terminated by impact with ground causing severe damage; cause coincidence of primary airframe aeroelastic freq with natural freq of control system, causing uncontrollable vertical oscillations. Second (21142, now at Monino) flown by V. P. Koloshchyenko Aug 69 to 2255 m with payload of 40 204·5 kg; NII tests completed and demos at Paris, but abandoned because Mi-26 far superior. ASCC name 'Homer'.

DIMENSIONS Span over rotors 67,0 m (219 ft 10 in); length 37,0 m (121 ft 4½ in); height 12,5 m (41 ft 0 in) (Soviet figures, close approximations).

ENGINES Two pairs of D-25VF.

WEIGHTS Empty not disclosed; normal payload 25 t (55,115 lb) VTOL, 30 t (66,140 lb) STOL; loaded (normal) 97 t (213,850 lb), (max) 105 t (231,480 lb).

PERFORMANCE Max speed 260 km/h (161 mph); cruise 240 km/h (149 mph); service ceiling 3,5 km (11,480 ft); range with max payload (payload given as 35,4 t, not explained) 500 km (311 miles).

Mi-14, V-14 Maritime helicopter for ASW and other roles, using Mi-8 as basis. Programme begun 66, V-14 flown Sept 69 with Mi-8 engines and dynamic parts. Boat-type lower part of fuselage for amphibious operation, clamshell doors deleted, sponson each side at rear housing flotation bag (4 m^3 inflated), inflatable tail float (0,185 m^3), landing gear with left and right castoring levered-suspension nosewheels hydraulically retracting forwards into watertight hull boxes with doors, left and right braked levered-suspension twin-wheel main units hydraulically retracting aft into sponson with closing doors, fuel 3500 lit plus aux tank 465 lit, watertight axial weapons bay along centreline with powered doors and 12-M search radar under cockpit. ASCC name 'Haze-A'.

Mi-14PL, ASW version, prototype 73, deliveries from 76, fitted with APM-60 towed MAD sensor, twin chutes for sonobuoys/flares/smoke floats. OKA-2 dipping sonar, weapons bay equipped for torpedoes and depth charges, retractable lights each side of nose, extra rotating beacon on boom, WAS-5M-3 liferaft. APM-60 'bird' originally housed at 45° under root of tail boom, later under recessed pylon at rear of fuselage. Crew two pilots with engineer between them and systems operator in cabin. Ten delivered 75-76. Main production (same designation) with more powerful engine in shorter cowl, tail rotor moved to left side, gearbox and other dynamic parts also as Mi-17, increased tankage, landing-gear doors deleted, standard 150-kg rescue hoist above front of main door, avionics upgraded. Production 240+ from 78, including exports to Bulgaria, Cuba, East Germany, Jugoslavia, N Korea, Libya, Poland (first recipient, calls it **Mi-14PW**), Romania and Syria. ASCC name still 'Haze-A'. One Polish PL modified with extra searchlights as **Mi-14PX** SAR trainer. **Mi-14PLM** uprated engines. APM-60 moved to bottom of rear end of fuselage, *Khrom-Nikel* IFF with blade antennas; 'Haze-A'. **Mi-14BT** (*Buksirtralschchik* – towing minesweeper) MCM (mine-countermeasures), prototype 73, deliveries from 76. No ASW sensors or weapons but tows any of three types of mine-activating sled; windows in lower rear of fuselage for sled operator, searchlight box under front of boom, SKW cabin heater in large pod

Mi-14PL

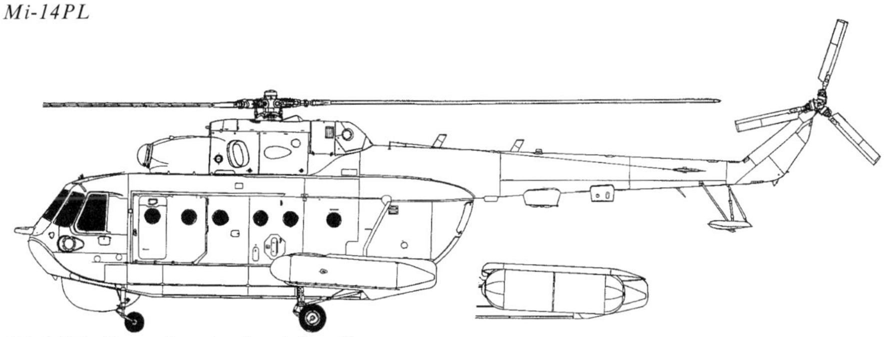

Mi-14PL (inset, flotation bag inflated)

on upper right side, tow-winch hydraulics and heavy electric power for some sleds in fairing along along right side, deliveries from 86, exports to E Germany; 'Haze-B'. **Mi-14PS** SAR (search and rescue) version based on BT but no MCM equipment; instead double-width main door, increased-capacity winch with 3-person basket, two extra searchlights on nose, two additional radios with dorsal blade antennas, carries ten 20-place liferafts and can tow these when filled. Various fits, including vertical searchlight on nose and vertical camera under boom, some modified from BT. Low-rate production to 92, five exported to Poland. ASCC 'Haze-C'.

DIMENSIONS Diam of main rotor 21,29 m (69 ft 10¼ in); length (rotors turning) 25,32 m (83 ft 0⅞ in), (ignoring rotors) 18,37 m (60 ft 3¼ in); height 6,94 m (22 ft 9¼ in).

ENGINES (V-14, early PL) two TV2-117A, (all later) two TV3-117MT.

WEIGHTS Empty (PL) 8902 kg (19,625 lb), (BT) 8,8 t (19,400 lb); loaded (all) 13 t (28,660 lb), max (all) 14 t (30,864 lb).

PERFORMANCE Max speed at SL 230 km/h (143 mph); cruising speed 205 km/h (127 mph); SL climb 7,8 m/s (1,535 ft/min); service ceiling 4 km (13,123 ft); range (max fuel) 1135 km (705 miles); endurance 5·9 h.

Mi-17, Mi-8M Upgrade of Mi-8 incorporating engines, gearbox, left-side tractor tail rotor and other dynamic parts of Mi-14. Forged titanium main-rotor hub, PZU engine-inlet particle deflectors. AI-9V gas-turbine APU providing pneumatic power for engine starting, automatic control of rotor speed and synchronization of engine power, fuel system provides for up to two auxiliary tanks in cabin (see weight data), other detail refinements. First flight 76, marketed by Kazan Helicopters for civil cargo, secondary passenger role. Military versions given Mi-8 designations by CIS forces. Basic **Mi-8MT** has role equipment as Mi-8T plus VMR-2 set for airdrop of special loads (eg mines) and duplicated vhf/uhf. ASO-2 chaff/flare dispenser can be strapped under boom and coded-pulse

Mi-171

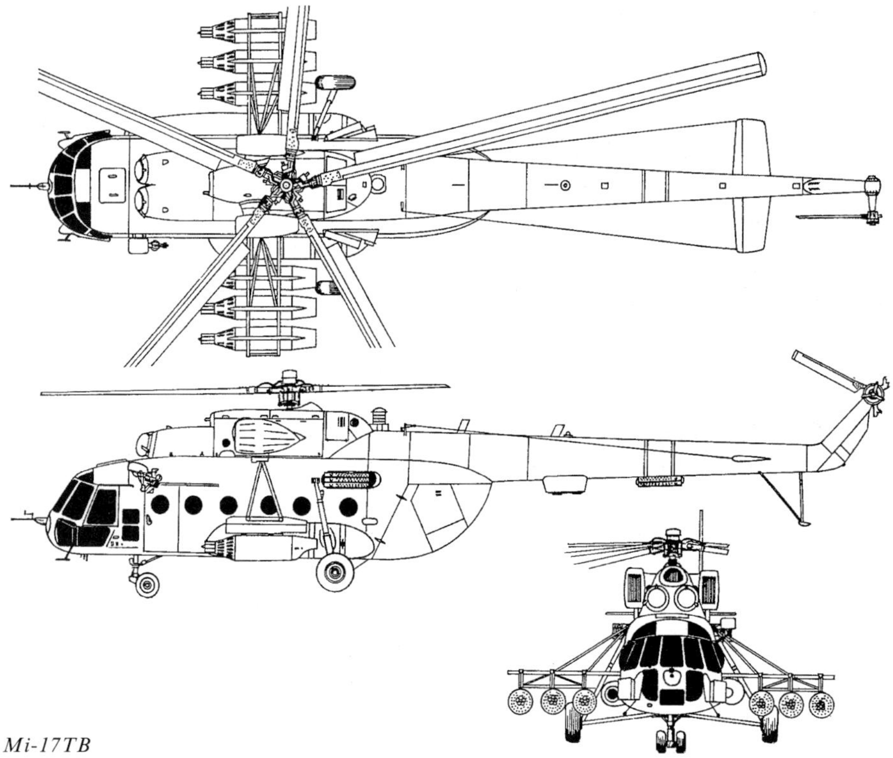

Mi-17TB

IRCM jammer installed above front of boom. Option of EVU exhaust IR-suppressors. **Mi-17TB** adds A-12·7 gun in nose, weapon options as Mi-8TB plus aft-hemisphere PKT gun pack, four B8V20 or FAB-500 or UPK-23-250 gun pods, stabilized missile sight, scabbed-on cockpit armour and option of ASO-3 flare dispenser each side of fuselage, forward-facing combat camera on left pylon and 250-kg hoist over door. ASCC 'Hip-H'. **Mi-17PP** Comint/jammer platform (for Hungary, instead of Mi-8PPA), with similar power units and row of heat exchangers under fuselage but antenna groups comprising primary array on each side of 32 planar spirals (4×8, in flat upright panel) and four more (2×2) on each side of tail boom; Units 1/2/3 cover Bands F/D/B respectively, Elint PTP covers 8 pulse or CW in Bands H/I. ASCC name 'Hip-H EW'. Other eqpt: rapid (10s) buoyancy system, snow skis, sport-para gear, AF-42/50 survey camera, firefighting (20 firemen with SU-R fast-descent system and VSU water-dropping), SAR (winch, boom, cradle, lighting, RSN-20AK raft etc), casevac (20 rescuees, KSO-LS2 oxygen), and LG surgical hospital. Since 89 Ulan Ude Aviation Industrial Assoc has produced **Mi-17-1, Mi-171**, upgraded civil versions with TV3-117VM engines and: two 40kW generators, *Baklan*-20 vhf, *Yadro*-1G1 uhf, ARK-UD and ARK-15M ADF/radio compass, 8A-813 weather radar in chin pod, DISS-32-90 doppler, A-723 long-range (Loran type) navaid, AGK-77 AND -74V horizon dislays, BKK-18 attitude monitor, ZPU-24 course selector, A-037 radar altimeter, DAS-DS star tracker and BUV-8A hover control unit. Cleared to carry 4 t (8,818 lb) cargo to height of 5 km, or 24 passengers in forward-facing airline seats instead of tip-up inward-facing as on Mi-8. ASCC 'Hip-H'. **Mi-17-1BA** is surgical hospital, produced by Kazan (originally with Hungarian industry) to fly to disaster sites; again 'Hip-H'. Czech (not Slovak) air force used locally modified variant with lateral outriggers carrying four large drums housing high-power jammers; no ASCC name allocated. **Mi-172** Kazan all-weather VIP meeting FAR-29. **MK-30** improved version by Daewoo S. Korea. Export customers include Angola, Cuba, Czechoslovakia, Hungary, India, N Korea, Nicaragua, Papua New Guinea, Peru and Poland.

DIMENSIONS As Mi-8 except length (rotors turning) 25352 (83 ft 2 in), (excl rotors) 18424 (60 ft 5⅜ in).

ENGINES Two TV3-117MT, (171) 117VM.

WEIGHTS Empty (17) 7100 kg (15,653 lb), (171) 7055 kg (15,553 lb); fuel 2027 kg, (+1 aux tank) 2737 kg, (+2 tanks) 3447 kg; loaded 11,1 t (24,471 lb), (max) 13 t (28,660 lb).

PERFORMANCE Max speed 250 km/h (155 mph); max cruise 240 km/h (149 mph); hover ceiling OGE 3980 m (13,000 ft); service ceiling (11,1 t) 6 km, (13 t) 4,8; range (no aux fuel) (17) 495 km (308 miles), (aux tanks) 950 (590).

Mi-22 See Mi-6.

Mi-24 After studying Bell 209 in 65 Mil proposed concept of combat helo, but with unique addition of squad of eight troops. Prolonged disbelief by military finally overcome and funding received late 67, Mil supervising basic design before handing over to Tishchyenko. He died Jan 70, a few weeks before first flight of prototype. Based on Mi-8 but airframe had to be totally new. Dynamic parts little changed, but uprated for eventual TV3 engine, though prototypes and first production block had to have TV2. Main rotor designed for higher g-load, aerodynamically similar with blades constant chord, NACA 230 profile 11/12% thick, but totally different construction with extruded D-section titanium spar, honeycomb filling and glassfibre skin, N_2 pressurized with crack indicator in hydraulic reservoir in root cuff; blades shorter to reduce disc diameter, fitted with balance tab on outer TE. Hub of forged titanium, hydraulic lead/lag dampers, elastomeric bearings deliberately avoided. Tail rotor diam unchanged but chord of aluminium blades increased (profile 230M) and TE cut off diagonally at tip. Main and tail blades deiced electrically. Fuselage aluminium semi-monocoque tapered on underside at rear to transport joint with deep boom, losing pod-and-boom appearance. Fin offset 3° and pivoted tailplane both new, with increased area. Tricycle landing gear with steerable levered-suspension nose unit (twin low-pressure tyres 480×200) retracting to rear and simple main units with low-pressure tyre 720×320 retracting aft and up while wheel rotates square-on to leg, bay covered by blistered door. Track 3030, wheelbase 4390. Engine installation generally as Mi-8 with fan-blown oil cooler immediately in front of rotor. Fuel in main tank filling rear fuselage and bag tanks under cabin floor, total 2130 lit; provision for 1800-lit auxiliary tank in cabin or (24D) two 850-lit or (24P) four 425-lit tanks on inner pylons. Cockpit with flat panels and limited side armour. Weapon operator in front (with, from 'Hind-A' A-12·7 gun on universal mount aimed by periscope looking through ventral turret) access via upward-hinged left window. Pilot/copilot behind with engineer in centre, access via sliding door with bulged window on left. Two pitots and one or two air-data probes on nose, retractable landing/taxi light underneath. Cabin 2610 long, 1470 high and 1200 wide, with large door each side opening into interlinked upper/lower portions, upper with two windows opening for troop AK-47 fire, lower incorporating steps. Eight fold-down seats and provision for four stretchers. On each side a horizontal 'wing' set at 19° incidence carrying two pylons for UV-32-57 or B8V20 launchers, bombs to 250 kg or other load. Cleared for production as **Mi-24A** early 73. ASCC name 'Hind-B'.

In 74 armament increased by fitting larger 'wings' with 12° anhedral, extended out to carry third (endplate) pylon horizontally bifurcated at base for rails for two 9M17 missiles. Combat camera at top of left inner pylon. Equipped Sov regt in E Germany in 74. ASCC 'Hind-A'. Limited production at Arsenyev, together with **Mi-24U** training version without gun or missile rails; 'Hind-C'. In 75 trainer with small mods set women's records including speed over 100 km 334,464 km/h and time 2 min 33·5 sec to 3 km and 7 min 43 sec to 6 km. Arsenyev reported this as 'A-10' to FAI. Also in 75, interim **Mi-24D** with redesigned forward fuselage. Front cockpit for weapon operator, with head-down binocular missile sight and vane-compensated reflector sight, duplicate sticks and retractable pedals, bulletproof front screen with wiper, upward hinged canopy with steps up left side. Higher rear cockpit for pilot, again with thick front screen with wiper and PKV fixed-reticle reflector sight, access via hinged door on right. Flight engineer on jump seat behind pilot in passage to cabin. Titanium armour under

Mi-172 (Kazan buoyancy system)

floor, round cockpits and protecting engines, anti-fragment panel between cockpits, both canopies with quadruple NBC seals. Small number built 75-78, some mod as **DU** unarmed dual trainer. 'Hind-D'.

New engine available 76, replaced TV2 in production 78, changes on **Mi-24F** including shorter cowlings, addition of AI-9W APU in rear of rotor-head fairing with inlet on right and exhaust on left, and transfer of tail rotor to left side. Optional additions eventually included inlet particle filters and large IRCM exhaust boxes. Still 'Hind-A'. In 78 new engine fitted to **Mi-24D**, together with upgraded armament and avionics. UBT replaced by USPU-24 turret mounting 9A624 (YakB-12·7) with 1,470 rounds, power traverse ±70°, elev +15°/–60°; a few instead with 9A622. Avionics: box under right side of nose with two front windows (normally closed by powered metal shutters) for low-light TV and laser designator/ranger, traversing squared off *Raduga*-F missile guidance pod under left side, these requiring longer nose gear semi-exposed when retracted; other avionics, hf and duplicated uhf/vhf, ARK-UD/R-852 radio compass and ARK-15M ADF, DISS-15D doppler driving roller-map pilot display, duplicate RW-5 radar altimeters under boom, SRO-2 IFF (antennas usually above front windscreen and under tail), *Sirena*-3M RWR (antennas usually each side of nose and facing aft from tail). Large boom projecting ahead from top of front windscreen later carrying sensitive (low speed) pitch and yaw vanes and pitot, linked to PTV-241 simulator. Cabin 8 troops or 1,5 t (3,307 lb) cargo. Weapons included four 9M17, four 32-57 or B8V20 launchers, up to 1500 kg of conventional, chemical or other bombs, PFM-1 mine dispensers, UPK-23-250 pack containing GSh-23L gun and 250 rounds or GUB pack containing YakB-12·7 plus two GShG-7·62 with total 4,350 rounds or YakB plus 30 mm grenade launcher, or four FAB-250 or two 500 kg or similar bombs, in each case plus four 9M17 (M-17P) missiles. Cleared for production 79, never less than 15 per month at Arsenyev and Rostov. **Mi-24DU** dedicated pilot trainer with turret removed. Permanent dual controls. Late 80 production switched to **Mi-24V**: missiles replaced by tube-launched 9M114, max 12 on six twin pylons, matched with larger fixed guidance radar pod under nose with dielectric domed front. SRO-2 replaced by *Khrom-Nikel* with blade antennas (same locations), coded-pulse IRCM emitter mounted on rear end of rotor-head fairing, colour-coded flare ident system, ASO-2V chaff/flare dispenser under rear of boom later augmented by ASO-3 dispenser on each side of fuselage (total 192 payloads), combat camera moved to left 'wingtip', instrument probe fitted with sensitive yaw/pitch vanes, *Sirena* system augmented with forward oblique receivers on LE of missile pylons, pilot sight replaced by HUD, forward oblique searchlight added under nose, added options of UV-20-76 launchers and twin self-defence R-60 missiles. Change to 9M114 missile altered reporting name to 'Hind-E' (Polish **Mi-24H**). Experience in Afghanistan led 80 to **Mi-24P**, with chin turret replaced by 9A623K (GSh-30/II) with large flash eliminators mounted externally on right side with cabin magazine for 126 or 250 rounds; avionics unchanged. ASCC 'Hind-F'. **Mi-24G** alternative sub-type with large gun on right side. **Mi-24VP** (also known as **PV**) with 23 mm AO-17A in undernose turret. **Mi-24RKh** special NBC version with sensors and missiles removed, several extra avionics and protective systems and 'Clutching hand' devices in place of missile rails. Each with three side-by-side scoops for taking soil samples. Optional marker dispenser on tail bumper. No missile director or EO box flanking gun turret; used at Chernobyl 86 but normal Ground Forces equipment. ASCC 'Hind-G1'. **Mi-24K** recon and fire-correction version with large camera on right side and video/EO device

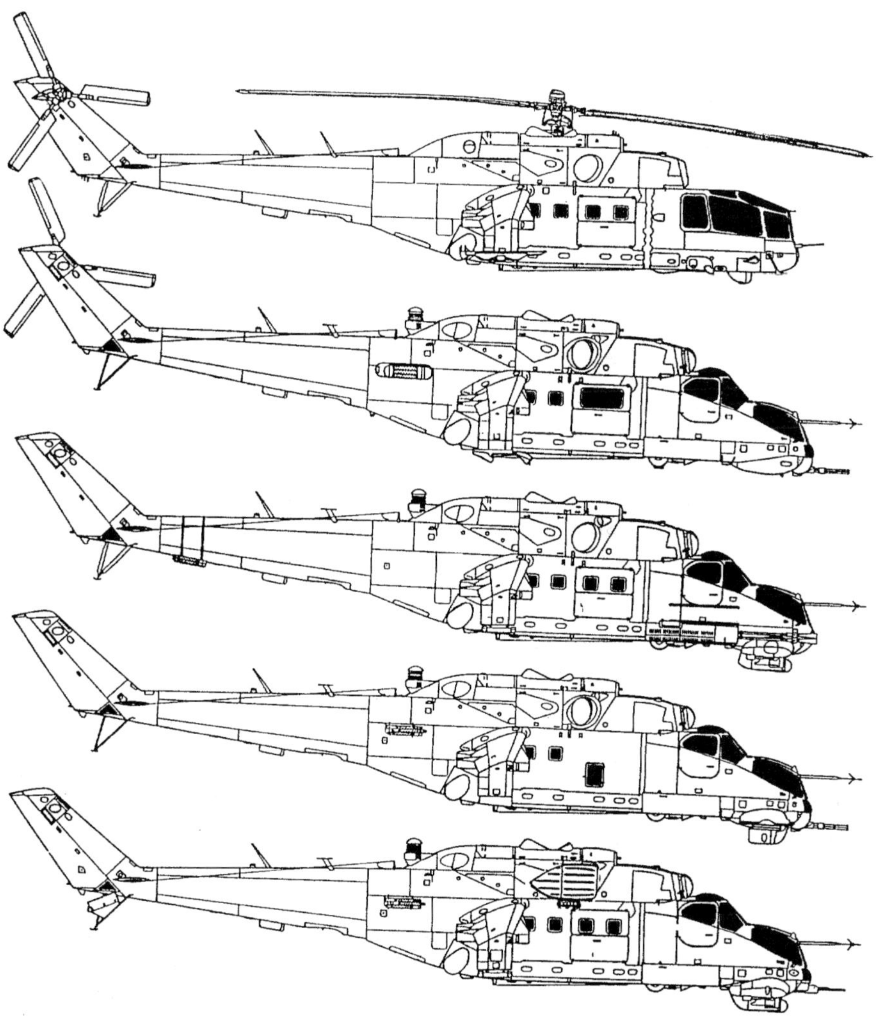

From the top: Mi-24A, Mi-24D, Mi-24P, Mi-24K and Mi-24VP

Mi-24P

to right of nose turret, 'Hind-G2'. Supplied to Afghanistan, Algeria, Angola, Bulgaria, Cuba, Czechoslovakia, E Germany, Hungary, India, Iraq, Libya, Mozambique, N Korea, Nicaragua, Peru, Poland, Vietnam and Yemen. **Mi-25** Sanitised Mi-24D for export. **Mi-35** export version of 24V, **Mi-35P** export version of 24P, **Mi-35D** upgraded with some Mi-28 dynamics. European users generally given full-standard Mi-24D, Mi-24V and Mi-24P. About 2,400 of all versions.

DIMENSIONS Diameter of main rotor 17,3 m (56 ft 9 in); length (rotors turning) 19,19 m (62 ft 11½ in), (excl rotors, gun) 17,506 m (57 ft 5¼ in); height (inc rotors) 6,5 m (21 ft 4 in).

ENGINES (24) two TV3-117MT, (25, 35) two TV3-117VM.

WEIGHTS Empty (D) 8500 kg (18,739 lb), (P) 8550 (18,849); fuel (normal) 1800 kg, (max) 4200; loaded 11,2 t (24,691 lb), max (D) 11,5 t (25,353 lb), (P) 11,8 t (26,014 lb).

PERFORMANCE Max speed at SL 300 km/h (186 mph), at 1 km 330 km/h (205 mph); cruise 260 km/h (162 mph); SL climb 12,5 m/s (2,460 ft/min); hover ceiling OGE 1500 m (4,920 ft); range (D) 750 km (311 miles), (P) 1200 km (746 miles).

Mi-25 See Mi-24.

Mi-26 By 70 it was clear V-12 was major error. Tishchyenko at once decided to start with clean sheet of paper. GUAP agreed this course, despite extra time needed and, esp, need to create completely new dynamic parts on much greater scale in order to achieve desired load-carrying ability with conventional helicopter lifted by single main rotor. Everything possible done to reduce risk, even at cost of greater time or less-competitive final result. From start **V-26** planned for military and airline tasks, but chief duty support of exploitation of Soviet Union's undeveloped regions where environment severest in world and reliability essential in week-long operations away from main base. Lotaryev KB already working on new D-136 free-turbine engine, vital to success and efficiency of new helicopter. Final breakthrough was Mil OKB's reluctant decision to handle design and development of new main gearbox itself, finding no external bidders.

GUGVF laid down severe numerical requirements, one of which was empty-weight fraction of 0·5, figure missed by large margin with V-12. This worked against low-risk approach, yet Tishchyenko's team spent fruitful three years in detail examination of engineering, and result is excellent compromise. V-26, designated **Mi-26** in production, is similar in configuration to Mi-6 but totally different dynamic parts, different structure and modern flight deck and avionics. Main rotor with eight blades surprisingly accomplished within smaller diameter than Mi-6 and Mi-10 despite handling twice engine power. Prolonged research on blades led to compromise with extruded high-tensile steel spar forming nose carrying 26 glass-fibre trailing-edge sections (called *Kolpachok*, cap). This handled power with eight blades, with adequate autorotative inertia and sufficient torsional stiffness. In-flight vibration one-tenth that of Mi-6. No expected corrosion problem. For 7,61-m tail rotor Kamov experience with GRP used to underpin all-GRP blades, five used, again with constant profile.

Main hub machined from titanium forging, largest on any helicopter. Test hub made with elastomeric bearings and may introduce later, but standard production hub has trad articulation in lubricated bearings. Main gearbox handles twice power and twice torque of R-7 box used in previous generation. Max torque is 90 000 kg-m (650,992 lb-ft). Despite this it is smaller, and only slightly heavier at 3,5 t (7716 lb). Engines identical except for handed jetpipes, and have large FOD deflectors upstream of airframe inlets to avoid ingestion of ice and birds. Dust and sand removed by centrifugal separators in each duct. Cowl panels fold down to form work platforms. Ahead of rotor is large oil radiator in duct with powered compressor to increase airflow. Gas-turbine APU under flight-deck floor instead of in cabin, and all fuel in eight integral tanks beneath main cargo floor, instead of in drums internally and externally. Bleed-air inlet anticing, rest of helicopter de-iced by pulsed raw AC electric power.

Stressed-skin pod/boom fuselage with much lower % weight than Mi-6. Conventional flush-riveted construction, some parts Al-Li alloy, frames notched to receive close-pitched stringers. Cross section similar to Il-76T (3,2 m square) with strong titanium floor all at one level over usable length of 15 m. Left/right rear clamshell doors opening downwards, with vehicle access up left/right hinged ramps positioned according to vehicle track. Doors closed hydraulically, with hand-pump for use on ground. Normally interior bare and unlined, but with fold-down wall seats for handling crews. No passenger windows but two Type A doors on left side. Rear internal ladder giving access to tail boom. External step/handholds aft of rear side door for access to top of boom, tail and rotor pylon. Large cambered fin with ground-adjustable tailplane (stabilizer) short way up leading edge, with tail rotor on right. Twin steerable nosewheels, twin main wheels with LP tyres 1120×450 and hyd brakes on trailing links pivoted bottom of fuselage, track 717, wheelbase 8950, long tail bumper. Provision for lifting wings but, though flight-tested, not normally fitted. Single 30-t cargo hook under CG for external slung load; interior cargo handling by two 2,5-t electric hoists running full length of hold on tracks along upper sides (not a gantry).

Spacious flight deck at main floor level with large windows bulged at sides for direct view of slung load or ground below, augmented by TV cameras under tail looking obliquely forwards, at front of hold looking aft and looking vertically down through floor hatch at slung load. Pilot (left) and co-pilot with TV screen, doppler-driven moving-map display giving accurate blind nav and drift in hover, autopilot and stab-augmentation system. Nav to rear (right) handles all basic nav problems, radio and advanced mapping/weather radar in nose. Flight engineer behind on left manages systems. Central jump seat for loadmaster or supernumerary.

First V-26 flown in hover 14 Dec 77 and began full translational flying 21 Feb 78. World class records 82 including payload 25 t (55,116 lb) lifted to 4,1 km (13,450 ft). Deliveries from Rostov 83 to VVS, AV-MF.

Variants (deputy general designer Aleksei Samusenko), **Mi-26T** civil utility, **Mi-26P** (up to 96 passengers), **Mi-26MS** medevac (up to 70 patients), **Mi-26A** with PNK-90 auto nav complex, **Mi-26TZ** tanker (14 040 lit extra fuel for dispensing through 10 hoses), **Mi-26TM** crane (second pilot station looking aft behind nose gear, third plus trainee looking forward from under rear ramp). Role eqpt for mechanized handling of containers (eg Yak-5) or large pipes, firefighting (1,6 km firebreak per release of 15 500-lit extinguishant), SAOG azimuth-stab and USP stab platform for slung loads,

Mi-26T

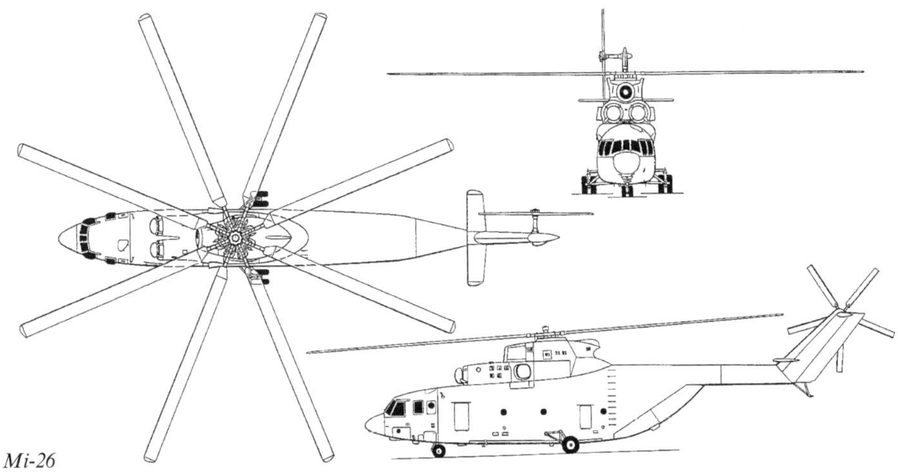

Mi-26

VG seismic survey towing large sensor platform (could be MCM platform), LG surgical hospital and TM aux pilot cabins. Marketed by Rostvertol Helo Mfg Assoc. To fly 94-95, **Mi-26M** with new engines, improved composite rotors, new nav complex, glass cockpit, FBW, 22 t payload. ASCC name 'Halo'.

DIMENSIONS Diameter of main rotor 32,0 m (104 ft 11⅞ in); length (rotors turning) 40,025 m (131 ft 3¾ in), (excl rotors) 33,727 m (110 ft 8 in); height (top of rotor head) 8,145 m (26 ft 8¾ in).

ENGINES Two D-136, (26M) two D-127.

WEIGHTS Empty equipped 28,2 t (62,169 lb); normal payload (excl fuel) 20 t (44,090 lb); normal loaded 49,6 t (109,350 lb); max loaded 56 t (123,457 lb).

PERFORMANCE Max speed at SL 295 km/h (183 mph); normal cruise at max wt 255 km/h (158 mph); hover ceiling OGE at max wt 1,8 km (5,900 ft); normal operating ceiling 4,5 km (14,760 ft), (has lifted 15 t to 5 km); range with 20 t payload 800 km (497 miles).

Mi-28 Slim gunship version of Mi-24 discussed from 72, design started 80, led by Aleksei Ivanov. Commonality almost vanished. Design factor +3·65/–0·5 g. Main-rotor blades of high-lift cambered profile with spirally-wound glassfibre D-spar and 14 rear pockets of Nomex-type honeycomb with Kevlar-type skin. LE with titanium sheath and electric deicing, tip unit with diagonal LE with pitot tube and tracking weight, full-span upswept tab along TE. At root, hydraulic reservoir and N_2 crack indicator, then long non-lifting tie to twin-bolt attachment to elastomeric anchor to new forged titanium hub with auto grease lubrication, hydraulic lag damper and fixed droop stop, agility enhanced by doubled hinge offset. Tail rotor has four glassfibre blades (diam 3840) with elastomeric bearings arranged 35°/145° (ΔH) on right of new tail with high two-position braced tailplane on left side only. Transmission by aluminium shaft inside tube of composite armour. Conventional fuselage with some Al-Li, composite doors and composite and titanium armour. Shallow rotor hub fairing, main VR-28 gearbox recessed into fuselage, driven by long diagonal shafts from rear of engines mounted in separate nacelles well back on each side. Whole transmission designed for 30 min without lubrication. Inlet fitted with particle separator, exhaust with large IRCM mixer (eg, composites box housing triple downward-ejecting pipes mixed with air from rectangular inlet at front). Fan-blown oil cooler and AI-9V APU mounted in tandem behind rotor mast, APU axial not transverse. Internal fuel 1900 lit in multiple self-sealing tanks in metal belly with outer layer composite armour and inert-gas purging, provision for four 500-lit pylon tanks, all tanks filled from single pressure socket. Main landing gears have single trailing leg pivoted to fuselage reacted by 'crashworthy' shock strut, low-pressure tyre 720×320. Castoring tailwheel with tyre 480×200. Track 2290, wheelbase 11 m. Cockpits for weapon operator and (higher, to rear) pilot have comprehensive titanium and ceramic armour and tinted armour-glass windows 35 mm thick (front windows 50 mm). Front cockpit has door on left, pilot door on right, both doors hinged at rear to stay open for emergency escape by inflating chute below door to deflect occupant past mainwheel. Front cockpit without flight controls, dominated by large CRT display fed by nose radar (terrain-avoidance and missile radio command link), laser designator/ranger in nose turret (usually slaved to gun), fixed FLIR (left of nose) and LLTV (right). Double handgrips for laser turret which also aims periscopic sight for completely exposed 2A42 (armoured vehicle type) gun on NPPU-28 mount slewing ±110°, elevation +13°/–45°, GP or AP ammunition selectable from 150-round boxes on each side moving with gun. Pilot has large forward sight and HUD (with HMS sockets) plus duplicate sensor-fed display and roller map driven by INS and doppler; firing controls for gun when fixed central and other stores except missiles. Weapons carried on anhedralled stub wings whose two spars also carry engines. Each wing contains hand winch system for loading stores (limit 500 kg) to two deep pylons. Typically outer pylon can carry eight 9M114 or 9P149, inner a 32-57 or B8V20 launcher, R-60 missile or drop tank. Comprehensive air-data, UHF/VHF/ADF/IFF/RWR/ECM, with twin 32-tube chaff/flare dispenser on each wingtip and provision for IRCM jammer behind rotor; formation-keeping lights along tailboom. Main

Mi-28

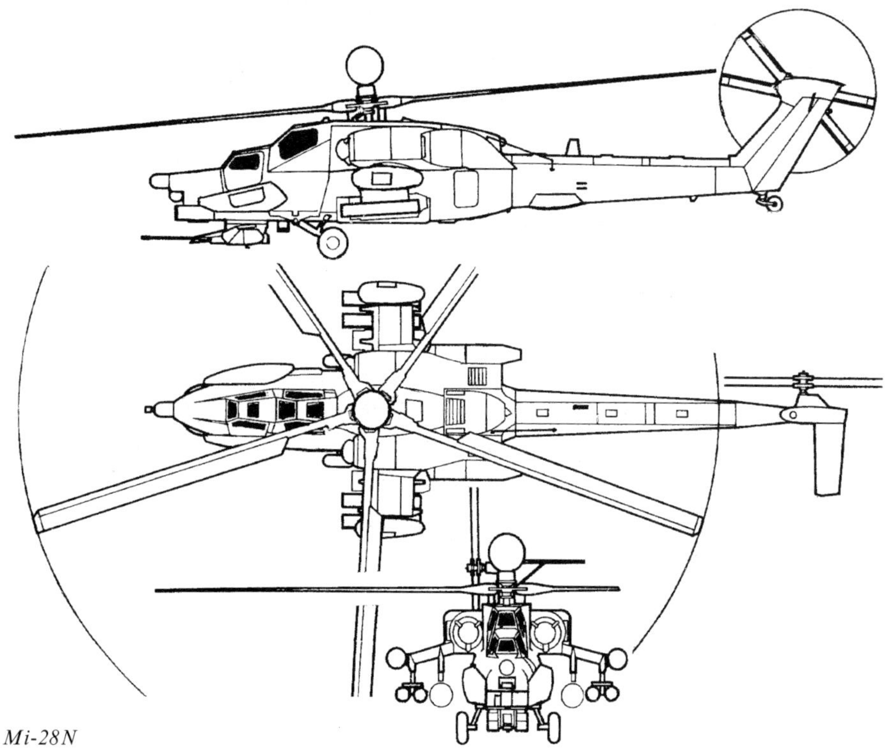

Mi-28N

avionics bay in rear fuselage can in emergency house two rescuees.

First of three prototypes flown by ctp Jurgen P. Karapetyan 10 Nov 82. Deliveries to Russian army by Rostvertol Dec 94. Sikorsky chief project pilot opinion 'Most effective attack helo yet developed'. **Mi-28N** for night and bad weather to fly 96, uprated engines, millimetric mast-mounted radar, IR sensors replaced by FLIR ball and LLTV, other new nav/attack avionics and improved protection; Vineberg expects CIS orders. Projected variant for air combat and naval amphibious assault; airmobile transport version see Mi-40. ASCC name 'Havoc'.

DIMENSIONS Diameter of main rotor 17,2 m (56 ft 5 in); length (rotors turning) 21,7 m (71 ft 3½ in), (ignoring rotors) 17,01 (excl gun) 16,85 m (55 ft 3½ in); height (rotor head) 3,82 m (12 ft 6½ in).

ENGINES (28) two TV3-117VM, (28N) TV7-117V.

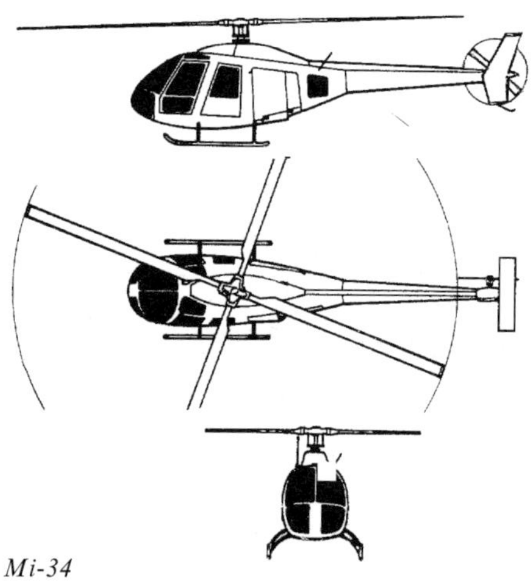

Mi-34

WEIGHTS Empty 8095 kg (17,846 lb); fuel 2200 lit; wing load 2400; loaded (28) 10,4 t (22,925 lb), (N) 10,5 t (23,148 lb), max (N) 11,66 t (25,705 lb).

PERFORMANCE Max speed (28) 300 km/h (186 mph), (N) 324 km/h (201 mph); max cruise 270 km/h (168 mph); hover OGE (28) 3,6 km (11,800 ft), (N) 3,7 km (12,140 ft); service ceiling (both) 5,75 km (18,865 ft); range (28) 470 km (292 miles), (N) 435 km (270 miles); ferry 1100 (683).

Mi-34 Pet project of Tishchyenko was sporting machine for aero clubs, refined into agile trainer also suitable for competition. Also replace Mi-2 in DOSAAF clubs, which wanted same engine as fixed-wing aerobatic aircraft, and M-14V also offered rapid power change and insensitivity to exhaust ingestion. Detail design 84, with cabin 1420 wide for four seats opening up other transport, patrol and observation uses. Design load factor +3·0/–0·5 g. Four-blade main rotor with glass/carbon blades held by laminated flex steel straps to hub with flapping and cyclic hinges but lead/lag flexing. Tail rotor on right, 1480 diam with two similar blades. Fixed fin and tailplane on end of riveted aluminium fuselage. Engine mounted sideways with cooling fan, 170-lit tank below for all-attitude supply, skid landing gear, track 2060. Four doors, pilot on right, optional dual, VHF, ADF, RV, horizon etc. First flight 86; brilliant display by Sergei Barkov 92. Was to have been assigned to Polish industry, now in production Arsenyev. Export **Mi-34P** with AlliedSignal Lycoming TIO-540-J2B, range increased to 450 km. ASCC name 'Hermit'. See next.

DIMENSIONS Diameter of main rotor 10,0 m (32 ft 9¾ in); length (rotors turning) 11,4 m (37 ft 4¾ in), (excl rotors) 8713 m (28 ft 6⅞ in); height (rotor head) 2,8 m (9 ft 2¼ in).

ENGINE One M-14V26.

WEIGHTS Empty 770 kg (1,700 lb); fuel 1220 kg; max payload 240 kg; MTO 1350 kg (2,976 lb), (aerobatic limit 1100).

PERFORMANCE Max speed 220 km/h (137 mph); cruise 160/180 km/h (<112 mph); service ceiling 4,5 km; hover ceiling OGE 1,5 km (4,920 ft); range (165 kg payload) 180 km (112 miles), (140 kg) 360 km (224 miles).

Mi-34VAZ Uprated version made possible by availability of VAZ-430 rotary engine. Engine more compact, start at –25C without warm-up, twin installation much more powerful, enabling fuel capacity to be increased (245 lit Mogas or various other fuels, plus 245 lit optional aux tank) and interior rearranged for greater load or stretcher. New CFRP starplate rotor hub with large apparent offset for great control power. Main-blade and window deicing. Prototype to fly 93; production from 97 by Rostvertol. Export **Mi-34A** with single Allison 250-C20R, range 500 with payload 380. Data for VAZ as 34 except:

ENGINE Two VAZ-430.

WEIGHTS Max loaded 1960 kg (4,321 lb).

PERFORMANCE (Est) max speed 220 km/h as before; cruise 185 km/h (115 mph); hover ceiling OGE 1,6 km (5,250 ft); service ceiling 5,0 km (16,400 ft); range (340 kg load, 30-min reserve) 600 km (373 miles), (max fuel) 980 km (609 miles).

Mi-35 See Mi-24.

Mi-38 Major OKB project for modern transport machine to replace Mi-8. Numerous civil and military versions studied since 87. Main rotor with six blades based on Mi-28, swept-back tips, elastomeric titanium hub. Airframe mainly glass/carbon composites. Detachable rear boom high enough to clear trucks carrying symmetric left/right tailplanes with elevators and tall offset fin carrying ΔH-type tail rotor on right. Large but shallow rotor hub fairing enclosing engines together behind gearbox with lateral filtered air inlets and drive shafts together on rear centreline. Full duplication of fuel system and triplexed electrics and hydraulics. VD-100 APU ahead of rotor (thus, arrangement opposite to Mi-8). Twin-wheel landing gears with low-pressure tyres, track 3300, wheelbase 6610, steerable nose gear retracting forwards, main units forward and up into door-enclosed bay in sponson; optional flotation pontoon bags. Main cabin with engine-bleed air conditioning, 2200-2360 wide, 1800/1850 high, length 6800 to hinge of full-width hydraulic rear ramp door, powered hoists and pallet/container rollers. Cargo load 5 t (11,023 lb) or 30 light airline type seats or folding troop seats or provision for many role fits including cargo sling, rescue hoist, stretchers, surgical hospital, survey cameras. Dual flightdeck but single-pilot cargo operation. Completely new digital avionics with FBW flight controls, five large colour displays in cockpit, computer controlled nav/flight system including Glonass satellite, Omega, doppler, weather radar, autostab, autohover and autoland. Designed to meet FAR 29, all-weather operation –60/+50°C, intended basic cargo version price US$30 m. Deal concluded Jan 93 with Eurocopter to market, Western avionics/furnishings and possibly RTM322 or other Western engine, in competition with EH 101. Four static/flight prototypes by 95, production by Kazan Helicopters from 97-98.

DIMENSIONS Diameter of main rotor 21,1 m (69 ft 2¾ in); length (excl rotors) 19,7 m (64 ft 7½ in); height to main-rotor head 5,13 m (16 ft 10 in).

ENGINES Two TV7-117V.

WEIGHTS Empty c6300 kg (13,900 lb); fuel (standard underfloor bags) 4 t; loaded 13 460 kg (29,672 lb), (max) 14,5 t (31,965 lb).

PERFORMANCE (Est) max speed 275 km/h (171 mph); cruise 250 km/h (155 mph); hover OGE 2,5 km (8,200 ft); range with 30 min reserve (max 5 t payload) 325 km (202 miles), (30 pax/baggage) 530 km (329 miles); (1,8 t payload and external tanks) 1300 km (808 miles).

Mi-40 Stretched version of Mi-28 for multirole use by airmobile units. Schemed 80, then abandoned, resurrected 91 with airmobile interest and incorporating Afghan experience. Maximum Mi-28 input including dynamic parts, but fuselage ahead of boom joint entirely new. Dual side-by-side cockpit with integrated navigation, sensing and sighting *kompleks* including mast-mounted radar, FLIR, laser ranger and optics for day/night all-weather ops at ground level; twin-barrel 23-mm chin turret, 12·7-mm mount for aft-hemisphere protection, various anti-armour and surface attack missiles firing ahead. Cabin derived from Mi-24 for 8 troops plus wall armour. Simple landing gear, track 3000, with 'crashworthy' main and nose wheels and tail ski, every possible form of protection. Pro-

Mi-34

Mi-38

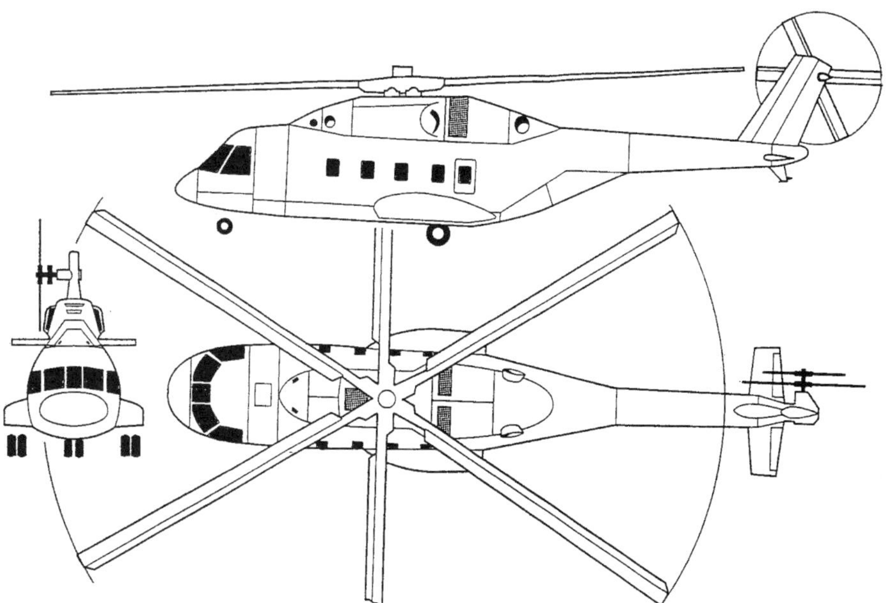

Mi-38

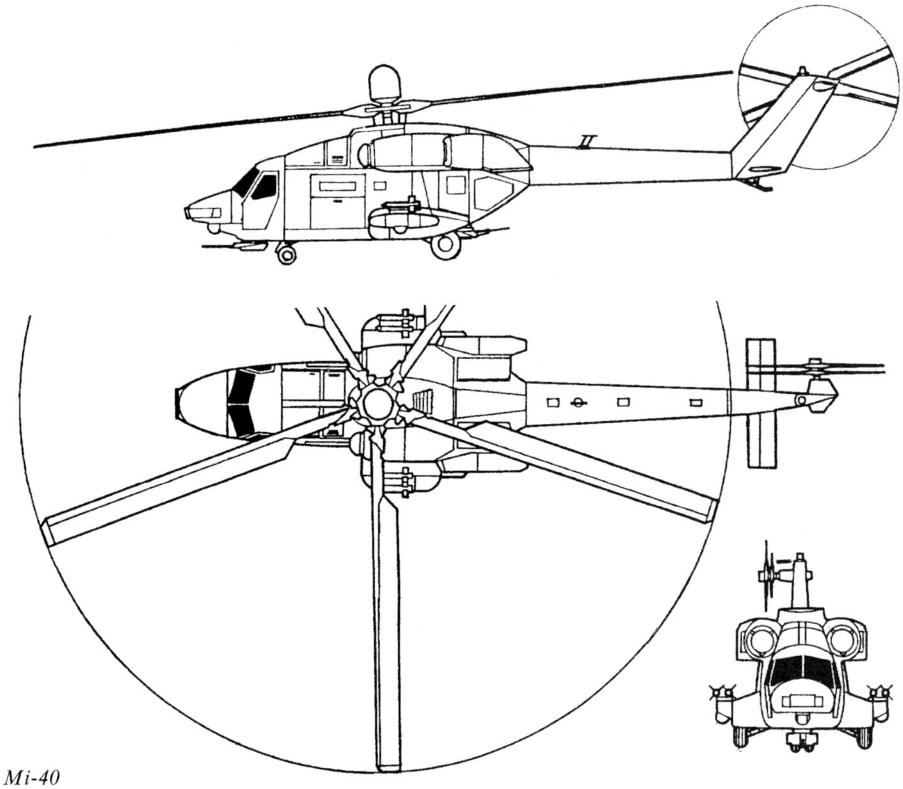

Mi-40

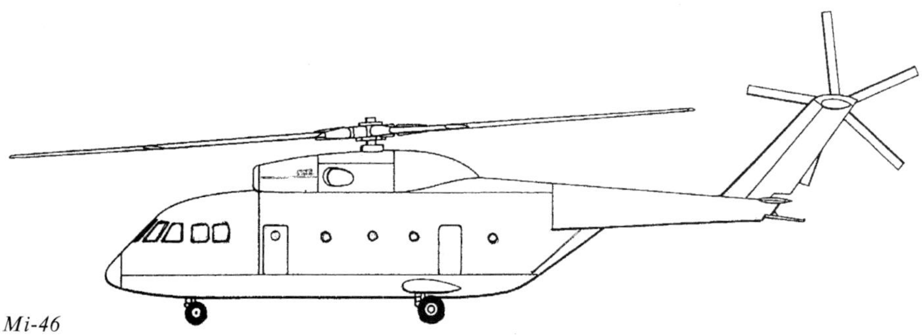

Mi-46

totype '2 years behind Mi-28N'.
DIMENSIONS Diameter of main rotor 17,2 m (56 ft 5⅛ in); length (ignoring rotors) 16,00 m (52 ft 5⅞ in); height (to radar) 4,600 m (15 ft 1 in).
ENGINES Two TV3-117VM.
WEIGHTS Empty 7675 kg (16,920 lb); loaded 11 t (24,250 lb); max 11,9 t (26,235 lb).
PERFORMANCE Max speed 310 km/h (193 mph); cruise 260 km/h (162 mph); service ceiling 5,8 km (19,000 ft); hover OGE 3,3 km (10,827 ft); range 5% reserve 400 km (249 miles), (aux fuel) 960 km (597 miles).

Mi-46 Project to replace Mi-6 (Mi-46T) and Mi-10K (Mi-46K). Modern design, just over half wt of Mi-26, 7-blade rotor, 5-blade tail rotor. **Mi-46T** with layout similar to Mi-26 with unobstructed hold with rear ramp door, twin engines and oil cooler ahead of main gearbox, fixed twin-wheel tricycle landing gear. **Mi-46K** flying crane similar to Mi-10K with stub wings, long braced landing gear, aft-facing pilot position used when picking up or setting down.
DIMENSIONS Not finalised.
ENGINES Two unspecified 8,000 shp.
WEIGHTS (T) Empty 16,2 t (35,715 lb), max 30 t (66,140 lb); (K) empty 19,7 t (43,430 lb), max 36,5 t (80,467 lb).
PERFORMANCE Cruise (T) 270 km/h (168 mph); hovering ceiling, max wt (both) 2,3 km (7,546 ft); range, 30-min reserve (T 10 t load, K 11 t) 400 km (248 miles).

Mi-52 Light utility helicopter to carry pilot and two pax or 250 kg. Advanced 4-blade rotor, T-tail, spatted tricycle gear. Designed for min cost, min maintenance, multi-fuel, all-climate. Prototype 94, to be made (expected large numbers) at Mil Gelikopter Plant.
DIMENSIONS Diameter of main rotor 10,0 m (32 ft 9¼ in); length (excl rotors) 8,71 m (28 ft 7 in).
ENGINE One VAZ-430.
WEIGHT Max payload 320 kg; MTO 1150 kg (2,535 lb).
PERFORMANCE Max speed 215 km/h (134 mph); cruise 170 km/h (106 mph); hover OGE 1,6 km (5,250 ft); service ceiling 5,0 km (16,400 ft); range (250 kg load, 30-min reserve) 400 km (249 miles), (max fuel, 160 kg payload, 30-min reserve) 800 km (497 miles).

Mi-54 Medium utility helicopter with large power reserve; single-engine version studied for Asian markets. Four-blade rotors, low tailplane, fixed tricycle gear, 10/12 pax or up to 1,3 t (2,866 lb) cargo. To be available from 97 in CIS and FAR-cert versions, with many projected derivatives using same rotor system.
DIMENSIONS Diameter of main rotor 13,5 m (44 ft 3½ in); length (excl rotors) 13,2 m (43 ft 3¾ in).
ENGINES Two unspecified turboshaft.
WEIGHT Max 4200 kg (9,259 lb).
PERFORMANCE Max speed 280 km/h, cruise 260 km/h (161 mph); hover OGE 1,8 km (5,900 ft); range, 30-min reserve, 1,2 t 450 km (280 miles), max int fuel 800 km (497 miles).

Molniya

Molniya (Lightning) design bureau organised 76 at MAPO (Moscow Aviation Production Organization) to produce reusable aerospace vehicle inspired by NASA Shuttle. MAPO traditionally a MiG factory, and Mikoyan already responsible for much space-related work, including MiG-105. General Designer Gleb E. Lozino-Lozinskii (ex-MiG). Work in partnership with NPO 'Energiya' Imeni S. P. Korolyev at Kaliningrad, producer of launcher. In 89 became NPO Molniya engaged in civil *Konversiya* projects.

50-50, MAKS Project for two-stage piggyback aerospace plane by Lozino-Lozinskii at MiG OKB. Assembled takeoff weight 140 t (308,650 lb), single-seat 10,3-t (22,700-lb) upper component (called **50**) to separate at Mach 5·5 and fly orbital missions. Unmanned **BOR-4** test vehicle for 50 flown four times on Cosmos launchers 82-84. Continued by small team at Molniya as **MAKS** with model shown at Farnborough 94 together with model of proposed **BOR-5** manned upper-stage test vehicle.

Buran Name 'snowstorm', vehicle similar in size to Shuttle Orbiter, but lighter and simpler because it has no main engines; instead four 590 t (1·3 m lb) LO_2/LH_2 engines are in core of Energiya launcher to which are strapped (instead of two solid rockets) four 726 t (1·6 m lb) LO_2/kerosene boost engines. Thus, with less weight in tail, *Buran* wing is further forward; nose gear is further aft. Payload bay longer than Shuttle at 18,55 m (60 ft 10⅜ in), payload limit normally 31 t (68,342 lb), but two-deck crew compartment larger, accommodating 10 Cosmonauts. Vert liftoff of assembled vehicle weighs c2436 t (5,370,000 lb), though heavier versions possible. Thrust rises with falling atmos press until at boost separation total reaches 4037 t (8·9 m lb). Core engs continue to c8 min, at V=7620 m (25,000 ft)/s. Orbiter continues with main OME (orbital manoeuvring engines) and control thrusters. Latter rotate vehicle tail-first for OME retrofire; re-entry surface temp on 38,000 ceramic tiles up to 1,535 C

Mi-52 model

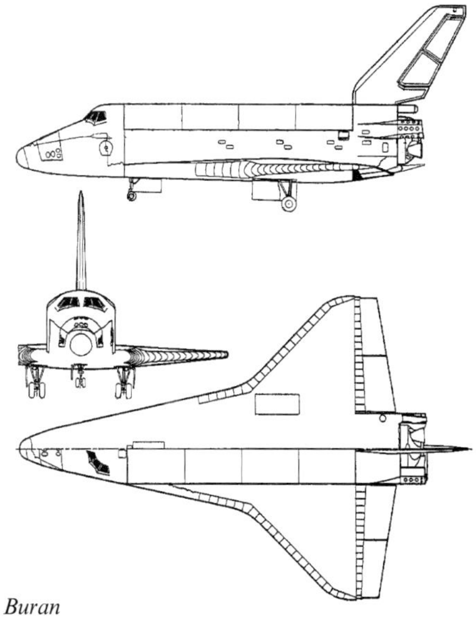
Buran

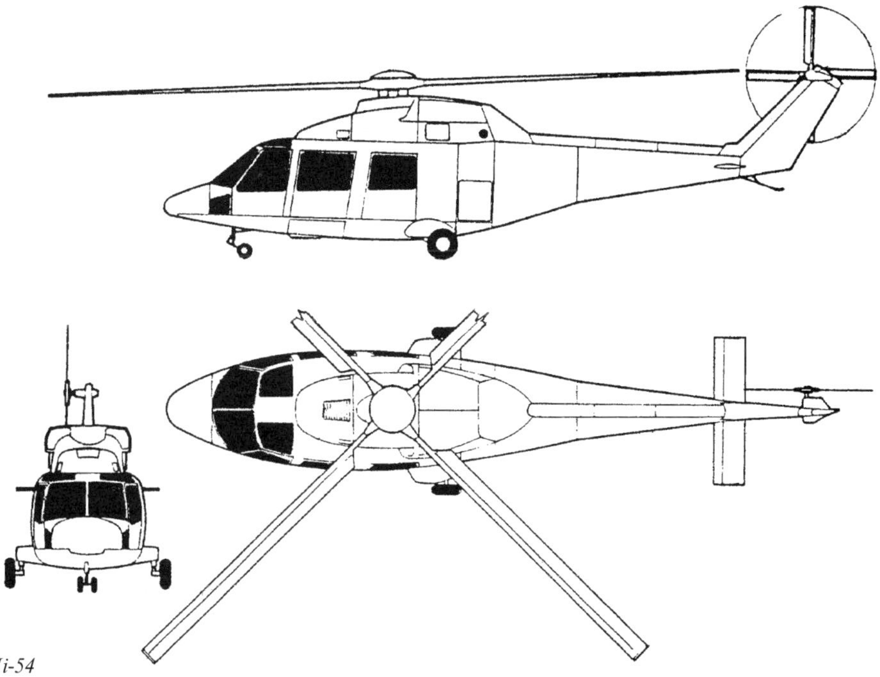
Mi-54

Models of MAKS and (right) BOR-5

(2,800 F). Landing automatically guided, ending with triple cruciform drag chutes. Five built, differing in mission eqpt. Assembled vehicle cannot be carried by VM-T *Atlant* but could be carried and launched from An-225; first flew unmanned. Data for orbiter only.
DIMENSIONS Span 23,9 m (78 ft 4⅞ in); length 36,37 m (119 ft 3⅞ in); wing area 250 m^2 (2,690 ft^2).
ENGINES Two OMEs, 42 control thrusters.
WEIGHTS Loaded, 101-107 t (222,660-235,890 lb).
PERFORMANCE Orbit depends on payload and inclination, typically 250 km (155 miles); re-entry Mach 25 (26 600 km/h, 16,530 mph); landing speed 340 km/h (211 mph).

Molniya-1 Exptl aircraft to explore Lozino-Lozinskii's wish to eliminate stall/spin accidents with inexperienced pilots. Key factor, a lifting foreplane. Light-alloy/GFRP airframe, folding wings (driving as road vehicle intended), pusher piston engine, fixed nosewheel gear or skis/floats, three pairs seats, payload 505 kg (1,113 lb) of which 180 kg fuel, to be flyable by pupil with 8-16 h total dual time. Intention of many extras (avionics, role eqpt). Prototype flown 18 Dec 92.
DIMENSIONS Span 8,5 m (27 ft 10¾ in); length 7,86 m (25 ft 9½ in).
ENGINES One M-14PM-1.
WEIGHTS Empty 1,075 kg (2,370 lb); MTO 1580 kg (3,483 lb).
PERFORMANCE Cruise 360 km/h (224 mph); range (max payload) at 270 km/h (168 mph) 1200 km (746 miles); TO 260 m.

Molniya-022 Devt of -011 with twin engines driving through combing gearbox to single prop. Commonality 85%, first flight late 93, planned production 100 per year at $270,000.
DIMENSIONS As –1 except length 8,2 m (26 ft 10⅞ in).
ENGINES Two (probably US) piston, 210 hp each.
WEIGHTS Fuel 240 kg, payload 425 kg, MTO 1660 kg (3,660 lb).
PERFORMANCE (Est) Cruise 220-330 km/h;

Buran testbed (extra engines)

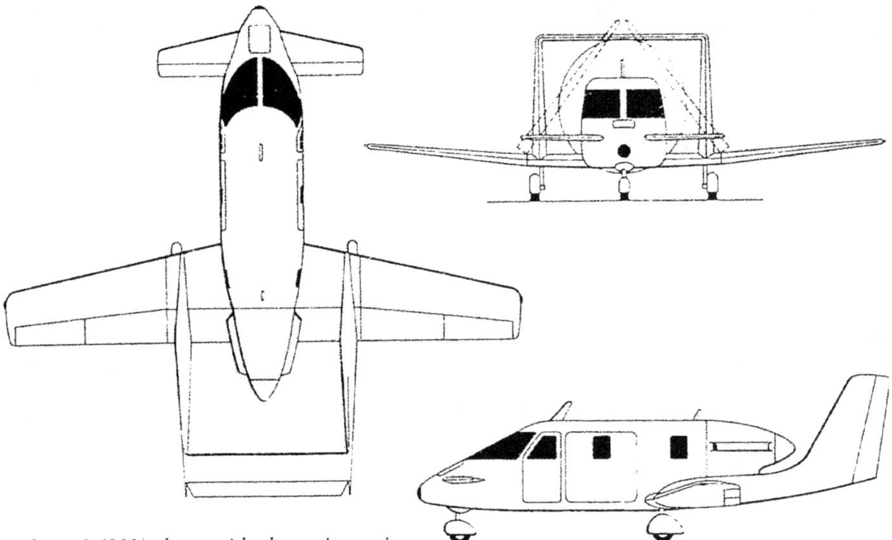

Molniya-1 (011) shown with alternative engine

range (max payload) 500 km (373 miles), (max fuel) 1500; TO 400 m, ldg 360.

Ladoga Named for Russian lake, light multirole amphibian, normal (not 'triplane') layout, 6 seats or cargo or role eqpt. Two unspec. 450-hp pusher piston engines burning mogas. Production due second quarter 95.
DIMENSIONS Span 12,1 m (39 ft 8⅜ in); length 8,7 m (28 ft 6½ in).
ENGINES See text.
WEIGHTS MTO 2190 kg (4,828 lb).
PERFORMANCE (Est) Max speed 310 km/h, cruise 270 km/h (168 mph); opg ht 3,5 km (11,500 ft); range 800-1350 km (to 839 miles); ldg 120 km/h.

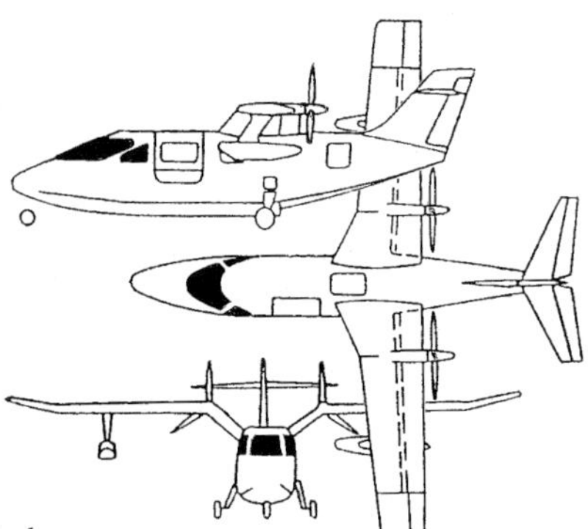

Ladoga

Molniya-100 Projected 'triplane' multirole, basic 2+15 seats, 1,5 t cargo, 4 stretchers or 8 biz pax; deliveries planned mid-95 at $500,000.
DIMENSIONS Span 12,5 m (41 ft 0 in); length 12,3 m (40 ft 4¼ in).
ENGINES Two pusher M-14PM.
WEIGHTS Fuel 750 kg; MTO 4 t (8,818 lb).
PERFORMANCE (Est) Cruise 275-360 km/h (to 224 mph); range (max load) 620 km (385 miles), (max fuel) 2000; TO 420 m, ldg 350 m.

Molniya-300 Projected pressurized multirole 'triplane', fuselage related to –100, new wing and tail.

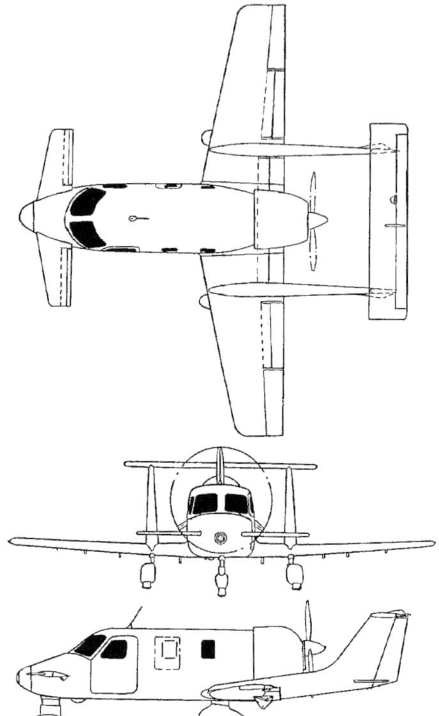

Molniya-022

DIMENSIONS Span 13,5 m (44 ft 3½ in); length 13,2 m (43 ft 3⅝ in).
ENGINES Two AL-34.
WEIGHTS Empty 2,7 t; fuel 1,5 t; payload 1,35 t; MTO 5 t (11,020 lb).
PERFORMANCE (Est) Cruise from 800 km/h (497 mph) with range 3750 km to 670 km/h (416 mph) with range 5000 km; TO 575 m, ldg 465 m.

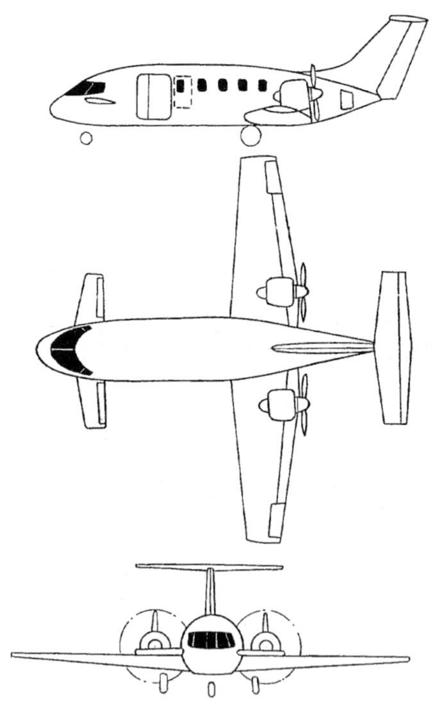

Molniya-100 (now projected with winglets)

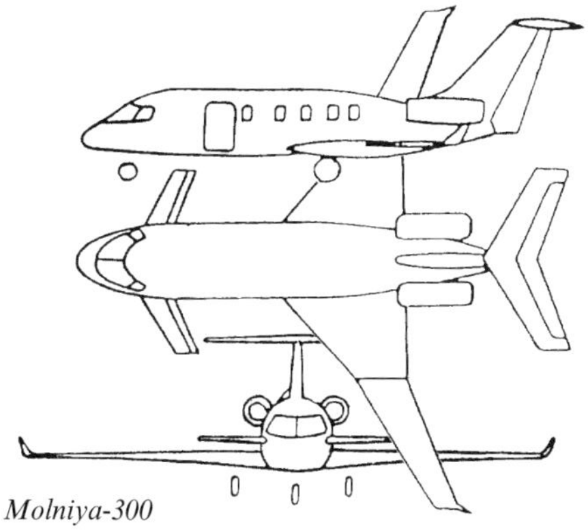

Molniya-300

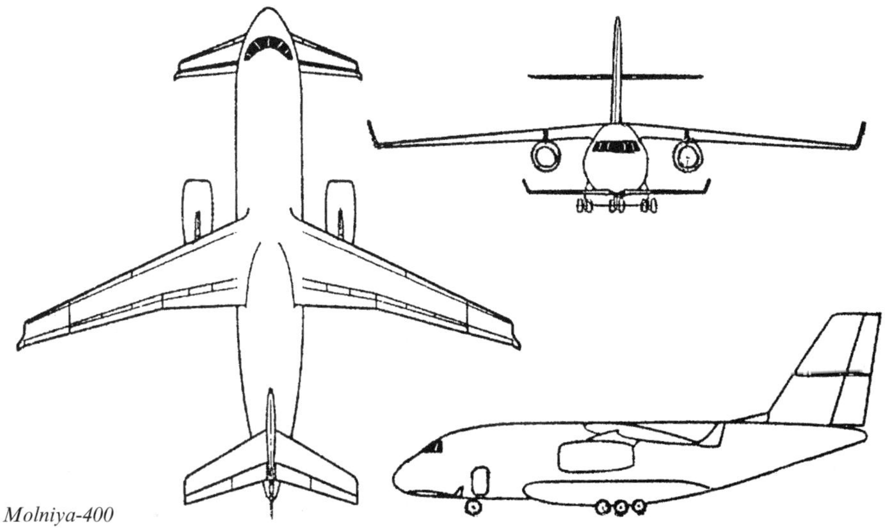

Molniya-400

Molniya-400 Larger 'triplane', flown by professionals. Foreplane unloads main wing, which need be only 67% normal size, and reduces fuselage stresses (to one-third normal) and size of horiz tail needed for longit stab and control. Result is smaller and lighter aircraft, burning less fuel. Molniya-400 seen as multirole, mainly cargo, transport; fully pressurized, full-section rear ramp door, rough-field landing gear (An-70 type), 60,000-h structure; max payload 50 t or 300 pax. Fuel burn 130 g/t-km (0·412 lb/US ton-mile). Funding for prototype sought.
DIMENSIONS Span 42,2 m (138 ft 5½ in); length 41,5 m (136 ft 2 in); height 16,1 m (52 ft 10 in).
ENGINES Two PS-90A.
WEIGHTS Loaded 112-121 t (246,900-266,750 lb).
PERFORMANCE Max speed 850 km/h, cruise 780-800 km/h (<497 mph) at 9,5-12 km; range 2000 km (1,242 miles) with 50 t payload, 8000 km with 27 t.

Gerakl (Heracles) Project for giant 'triplane' freighter, hanging payload or pod under arched centreline between twin slim fuselages carrying lifting foreplane and π-tail. Lozino-Lozinskii 'On same engines we could carry 1·5 × payload of An-225'.
DIMENSIONS Span 90,4 m (295 ft); length 73,4 m (240 ft 9 in).
ENGINES Six D-18T or NK-93.
WEIGHTS Payload 450 t; MTO 900 t (1,984,000 lb).
PERFORMANCE Cruise c800 km/h (497 mph); range (450 t) 3100 km (1,925 miles), (240 t) 10300 (6,400); TO 2,37.

Molniya-1000 Gerakl (art)

Moiseyenko

Born about 1890, Viktor Leonidovich Moiseyenko was cadet in Czarist army 13, and became pilot of IM-class bombers, rising to be flight commander in post-revolutionary DVK 19. He then enjoyed 30 years as a chief regional supplier of aircraft bought-out components and accessories.

2U-B3 Interim training aircraft, built at GAZ-1 25. Many details said to be copied from P-2, built at same time, though Polikarpov transport was much larger. Designation from 2-*Uchyebnii* BMW-III (2nd training type, BMW III engine). Originally designed as equal-span biplane with N struts, upper wing removed and finally completed as steel-braced low-wing monoplane. Engine 185 hp BMW IIIa, with metal cowl. Forward fuselage wire-braced wooden with ply skin. Tandem dual cockpits. Tested late 26 and flew adequately but no production.

Moskalyev

Aleksandr Sergeyevich Moskalyev pronounced Moskalyov was one of indefatigable Soviet constructors who, having no assured finances and no production base, managed to build an impressive run of aircraft. Many were of outstanding merit. By sheer enthusiasm on part of pilots who flew them, some gained production orders; many more Moskalyev aircraft would have served their country had not GUAP Commissar Kaganovich been antagonistic and done all he could to hamper work of this dedicated and good-natured enthusiast. Born 1898, Moskalyev qualified at Univ of Leningrad and then joined *Krasnyi Lyetchik* plant as stressman in Grigorovich KB, being assigned to structure of I-2bis. He showed promise and made major contribution to improved I-2bis radiator installation. Continued post-grad studies at Lensovyet Tech Inst as external student, whilst beginning design of SAM-1 and -2 described below. Successful flight of second of these resulted in its designer being sent in 33 to vast newly built GAZ-18 at Voronezh, where under impossible conditions in makeshift shed he created further designs which included tailless configurations as well as conventional production transports. He also engineered later TB-3 for production, designed installation of Charomskii diesel in DB-1, and made many contributions to other constructors' aircraft. In 36 he was permitted to organize his own OKB, No 31, but with no official support. Following year appointed Director of Voronezh Aviatekhnikum (air commercial school). Partly through aiming too far ahead of available technology OKB was shut 46, Moskalyev becoming lecturer at VVA (facilities went to Chyetverikov). In 80s worked at MAI.

SAM-1 Designation from *Samolyet* A. Moskalyev (aeroplane, A. Moskalyev). Single-seat gull-winged monoplane fighter, design speed 345 km/h. Prototype begun 30 but M-34 engine never made available.

SAM-2, MU-3 Designation from *Morskoi Uchyebnyi*, marine trainer. Simple wooden biplane flying boat with side-by-side open cockpit and pusher engine based on Grigorovich MU-2 and resulting from Moskalyev's work for that KB. Designed using many MU-2 components by group including N. G. Mikhelson and O. N. Rozanov and completed about Feb 31. Improved planing bottom, smaller wings and much lighter. State tests spring 31 resulted in selection of Sh-2 instead.
DIMENSIONS Span about 10,5 m (34 ft 5⅓ in); length about 8,0 m (26 ft 3 in); wing area about 28 m^2 (300 ft^2).
ENGINE One M-11.
WEIGHTS Empty 650 kg (1,433 lb); fuel/oil 90 kg (198 lb); loaded 920 kg (2,028 lb).
PERFORMANCE Max speed SL 132 km/h (82 mph); climb 10 min to 1 km; service ceiling 2,3 km (7,550 ft); range 400 km (249 miles), 4 h; take-off 400 m/30 s; landing 100 m/10 s/65 km/h (40 mph).

SAM-5 (spats removed, with Moskalyev)

SAM-3 Designed 31, this seemingly efficient 12-passenger transport was intended to run shuttle service over 27 km route Leningrad-Kronstadt (island). Never completed.

SAM-4, Sigma First and most radical of Moskalyev's tailless designs, on drawing board at Leningrad early 33. Serious attempt to exceed Mach 1, with Glushko rocket. Pure ogival (Gothic) delta, no protruding surfaces except fins/rudders at tips; elevons at trailing edge, pilot prone. Collaboration with V. P. Glushko at GRD, but obvious that ORM series engines far from yielding sufficient thrust. Decision 34 to shelve and build piston-engined fighter to test aerodynamics (SAM-7). No other data.

SAM-5 -2bis

SAM-5 Most important of SAM series, made Moskalyev famous and led to numerous aircraft in several variants. High-wing monoplane transport, designed in light duralumin construction. First project at Voronezh, built under great difficulty by team with no experience of stressed-skin structures. Fuselage skinned with 0,3 mm sheet stiffened by single beads (longitudinal corrugations) at pitch of 15 cm; wings and tail 0,3 mm corrugated sheet. Accommodation five-seat. Enclosed cockpit under leading edge, entry door on right, cantilever wing but braced tailplane, uncowled engine, spatted main gears attached only to fuselage. Approved by S. P. Korolyev for Osoaviakhim, but no production.

Voronezh team – L. Polukarov, Ye. Serebryanskii, M. Shubin and B. Dyakov – redesigned aircraft in wood. Result was **SAM-5bis**, late 34. Same dimensions but slightly narrower fuselage and two bracing struts under each wing. Wheel spats removed, and other simplifications. Dramatic reduction in empty weight, and disposable load better than 45% of gross weight. Much of skin fabric, but ply over leading edge and front half of fuselage. Spring 35 completed NII testing by P. M. Stefanovskii, who recommended series production. N. D. Fikson and mechanic A. S. Buzunov, 21 Sept 36 flew 1600 km in 12 h; 21/22 Oct flew 3200 km in 25 h 05 min. GAZ-18 given order for 37, almost all ambulance version with stretcher door on left. Carried three patients plus attendant; allegedly often six adults on board. Ambulance had no separate designation; delivered 1937-38 and served into War with GVF.

Final model of immediate series was **SAM-5-2bis**, with reduced span and great attention to drag reduction, all bracing struts being removed from wing, tailplane and main landing gear. Wing profile R-II, thickness 14%. Rubber bungee sprung cantilever main-gear legs of curved 10 mm ply forming rigid streamline section, extremely successful. Fuel tanks between two spars of wing outside cabin area. NII tests early 37 by V. Borodin and leading engineer M. P. Mogilyevskii. On 23/24 Sept 37 A. N. Gusarov and mechanic V. L. Glebov set FAI distance record 3513 km Moscow/Krasnoyarsk non-stop in 19 h 59 min. Moskalyev then fitted MG-21 engine and Borodin reached 8 km (class ht record). In summer 39 supercharged M-11FN installed, driving metal propeller, and B. K. Kondratyev reached 8,9 km. VVS did thorough testing and in 38 ordered 200 SAM-5-2bis ambulances, but Kaganovich countermanded the order.
DIMENSIONS Span (5, 5bis) 12,5 m (41 ft 0 in), (5-2, 5-2bis) 11,49 m (37 ft 8⅓ in); length (5, approx) 8,0 m (26 ft 3 in), (5-2) 8,02 m (26 ft 3¾ in); wing area (5) 24,0 m^2 (258 ft^2), (5-2) 21,8 m^2 (235 ft^2).
ENGINE One M-11; 5-2bis later fitted with MG-21 and M-11FN.
WEIGHTS Empty (5) 626 kg, (bis) 580 kg, (bis-ambulance) 710 kg (1,565 lb), (5-2bis) 656 kg (1,446 lb); fuel/oil (bis amb.) 146+19 kg, (2bis) 90+14; loaded (5) 1106 kg, (bis) 1070 kg, (bis amb.) 1219 kg (2,687 lb), (2bis) 1160 kg (2,557 lb).
PERFORMANCE Max speed (5) 175 km/h, (bis) 185 km/h, (bis amb.) 173 km/h (107·5 mph), (2 bis) 204 km/h (127 mph); cruise 140 km/h (87 mph), (2 bis) 175 km/h (109 mph); climb 7 min (bis amb., 11) to 1 km; service ceiling 3,7 km (bis amb., 2,8); range (5) 1760 km, (bis amb.) 900 km, (2 bis) 515 km; take-off (bis amb.) 13 s/150 m; landing 16 s/250 m/70 km/h.

SAM-6 Small single-seat research aircraft to investigate single-wheel landing gear, planned for tailless fighter. SAM-6 wooden with open cockpit, and conventional tail added for prudence. Scheibe-type vertical surfaces on wingtips with skids providing lateral support on ground. Built 33 but flown on ski early 34 with fair success. Later trousered wheel tested. By end 34 modified into **SAM-6bis** with conventional gear (two trousered wheels) and tandem

SAM-6

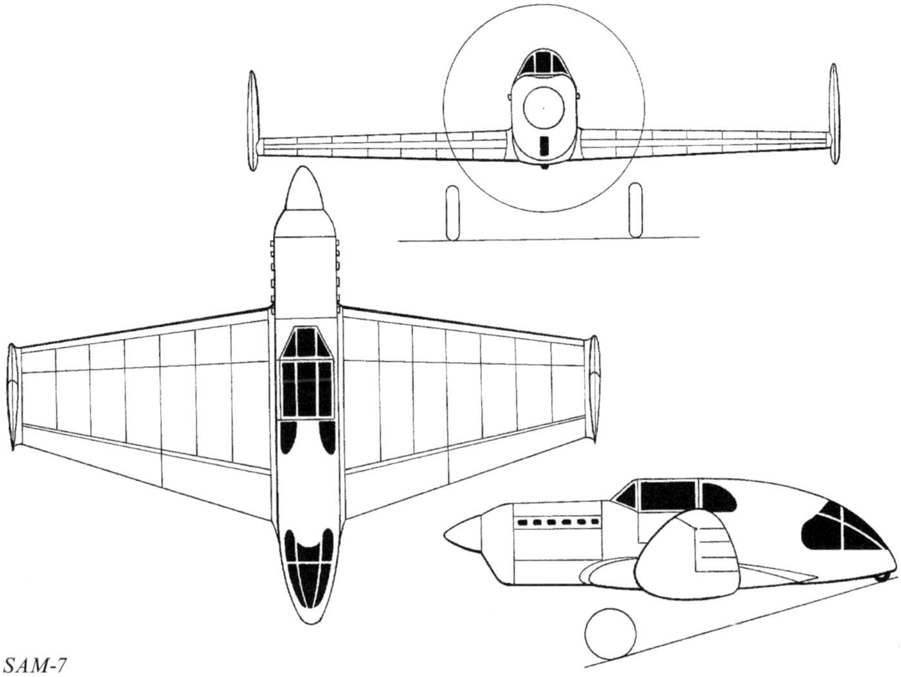
SAM-7

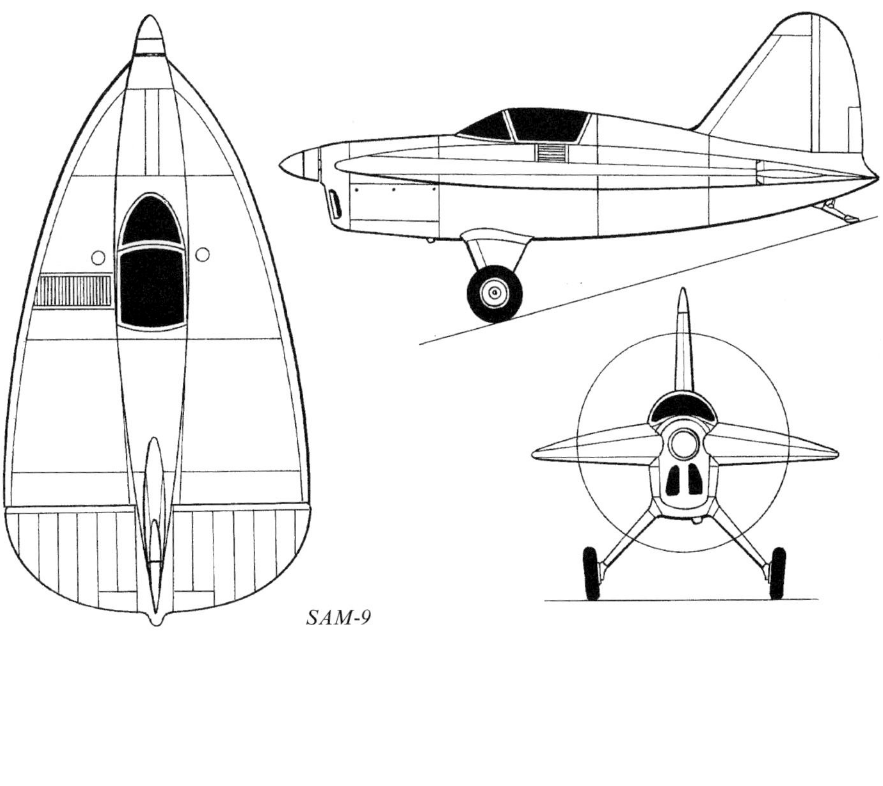
SAM-9

enclosed cockpits for pilot and passenger, plus instrumentation.
DIMENSIONS Span 8,0 m (26 ft 3 in); length 4,5 m (14 ft 9 in); wing area 12 m^2 (129 ft^2).
ENGINE One M-23.
WEIGHTS (bis) Empty 380 kg (838 lb); fuel 50 kg; loaded 500 kg (1,102 lb).
PERFORMANCE Max speed 130 km/h (81 mph); service ceiling 3 km; range 200 km (124 miles); landing speed 55 km/h.

SAM-7 Sigma As interim step to *Sigma* (SAM-4) Moskalyev designed piston-engined two-seat fighter intended to offer superior speed manoeuvrability and rear field of fire. All-metal contruction with two-spar wing R-II profile 12%, with CG 13/15%. Spar webs 2/2,5 mm, fixed leading edge, inboard elevators and outboard ailerons with neat drives from stick via push/pull rods and bell-cranks. Scheibe fin/rudder on each wingtip. M-34 with four-blade wooden propeller from TB-3. Cooling by surface radiators, supplemented at low speeds by retractable radiator. Single -strut main gears pivoted to front spar and retracting inwards, small wheel at extreme tail. Two fixed ShKAS above engine and one (option two) aimed by rear gunner. Though flown Oct 35, fin/rudder shape differs in all six known Soviet drawings. Aircraft judged dangerous, with fast landing and difficult to keep straight on ground. Never reached maximum speed.
DIMENSIONS Span 9,46 m (31 ft 0½ in); length 7,0 m (22 ft 11½ in); wing area 20,0 m^2 (215 ft^2).
ENGINE One M-34.
WEIGHTS Empty 940 kg (2,072 lb); loaded 1480 kg (3,263 lb).
PERFORMANCE Max speed (est) 435 km/h (270 mph) at SL, 500 km/h (311 mph) at height; ceiling est 9,2 km; range est 800 km; landing speed 138 km/h (86 mph).

SAM-9 Strela SAM-8 not known. SAM-9 *Strela* (arrow) was major step towards supersonic gothic delta SAM-4, and in recent years hailed (justifiably) as aerodynamic pioneer of today's SSTs. Again failing to get 760 hp HS12Y engine Moskalyev had to settle for low-powered Renault; and permission to build such radical machine granted only in late 36 after prolonged aerodynamic testing by CAHI which, taking cue from Kaganovich, failed to show interest in concept. Wooden construction with high surface finish, fabric-skinned control surfaces. Wing root extended entire length of aircraft except for spinner; aspect ratio 0·975, aerofoil RAF-38 with Moskalyev's local modifications. Pilot in small enclosed cockpit with linkage to two large trailing-edge elevons. Cantilever fixed main gears as on SAM-5-2, and tailskid. Prototype also known as **Aircraft L**, built in first 70 days of 37 and N. S. Rybko made first flight at Voronezh. Controllable, but demanded intense concentration and 20° attitude on approach then novel and frightening. Flown by A. N. Gusarov and A. P. Chyernavskii. Moskalyev's unofficial status and supposed fantastic ideas warped judgement, and project banned mid 37.
DIMENSIONS Span 3,55 m (11 ft 5¾ in); length 6,15 m (20 ft 2 in); wing area 13,0 m^2 (140 ft^2).
ENGINE One MV-6.
WEIGHTS Empty 470 kg (1,036 lb); fuel/oil 60+10 kg; loaded 630 kg (1,389 lb).

PERFORMANCE Speed reached at SL 310 km/h (193 mph); height reached 1,5 km; take-off 200 m; landing 100 m/102 km/h (63 mph).

SAM-10 Conventional low-wing 5/6-seater developed from SAM-5-2bis and using actual rear fuselage, tail and wing panels from earlier high-wing machine. Wooden, with ply sides to fuselage as far back as rear of cabin, fabric elsewhere. Light-alloy cowl over experimental engine driving two-blade Ratier propeller. Light-alloy cabin roof frames with Celluloid windows. Trousered main gears with 470×200 mm tyres and air/oil shock struts. Two-spar wing with ply skin, panels removed by unscrewing beneath tankage. Outstanding performance on SNII GVF testing 5 June to 9 Aug 38, and ordered as passenger and ambulance aircraft. Thwarted by unavailability of engine. Two **SAM-10bis** built with MV-6 but this likewise failed to get into production.
DIMENSIONS Span 11,49 m (37 ft 8⅓ in); length 8,5 m (27 ft 10⅔ in); wing area 21·86 m² (235 ft²).
ENGINE (10) One MM-1; (10 bis) MV-6.
WEIGHTS Empty (10) 866 kg (1909 lb), (bis) 873 kg (1925 lb); fuel/oil 150+20 kg; loaded (10) 1436 kg (3166 lb), (bis) 1448 kg (3192 lb).
PERFORMANCE Max speed (10) 311 km/h at SL, 336 km/h (209 mph) at 1,7 km, (bis) 262 km/h at SL, 275 km/h (171 mph) at 1,5 km; climb (10) 2·7 min to 1 km, (bis) 3·2 min; service ceiling (10) 7,1 km, (bis) 5910 m; range (10) 1000 km (621 miles); take-off (10) 95 m/7 s, (bis) 100 m/9 s; landing (10) 120 m/9 s/87 km/h (54 mph).

SAM-10

SAM-11 Yet another adaptation of SAM-5-2 bis components, this amphibian was built 39 and used same wings, tail and engine as SAM-10, plus SAM-5-2 bis main landing gears. Wooden boat hull seating pilot and three passengers. Main gears hinged to strong sidewalls and manually cranked up so that wheels lay in recesses in wing. Fixed tailwheel at second step in keel. Tractor engine caused severe turbulence making tail ineffective; pilot lost control and aircraft damaged. Rebuilt as **SAM-11 bis** with MV-6 engine in better nacelle. Test flown Sept/Oct 40 by P.V. Yakovlev and A.N. Gusarov; flew satisfactorily but after NII testing at Sevastopol both VMF and VVS decided payload too small.
DIMENSIONS Span 11,49 m (37 ft 8⅓ in); length 8,74 m (28 ft 8 in); wing area 21·86 m² (235 ft²).
ENGINE (11) One MM-1; (bis) MV-6.
WEIGHTS Empty 1094 kg (2,412 lb), (bis) 1030 kg; fuel/oil 95+10 kg; loaded 1560 kg, (bis) 1510 kg (3,329 lb).
PERFORMANCE Max speed (SL) (11) 225 km/h, (bis) 217 km/h; max speed at 2,4 km (both) 240 km/h (149 mph); service ceiling (11) 4,7 km, (bis) 5,6 km; take-off (bis) 200 m/16 s; landing (bis) 110 m/9 s/85 km/h (53 mph).

SAM-11bis

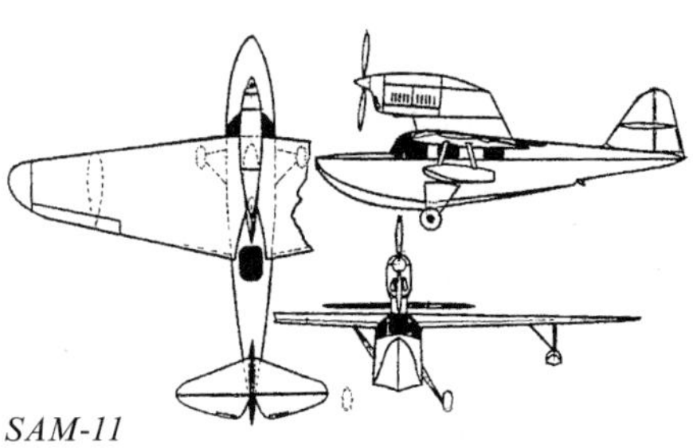
SAM-11

SAM-12 Tandem dual trainer of conventional layout, low wing, tricycle landing gear, wooden construction, enclosed cockpits. Designed for MM-1 but converted to MV-6 and believed completed late 40. Features included Ratier v-p propeller, split flaps and steerable nosewheel, but no record of flight test.

SAM-13

SAM-13 Persisting in MV-6 (Renault) engine, Moskalyev designed this fighter from scheme of 37, inspired by Fokker D.23, but was unable to build until 39. Rather 'hot' design with small wing. Push/pull aircooled engines with pilot between under rear-sliding hood. Twin tail booms 2,38 m apart carrying wide-chord tailplane with vertical tail on centreline. Fully retractable tricycle landing gear, nose gear with trailing levered-suspension and shimmy damper. All-wooden construction except for welded

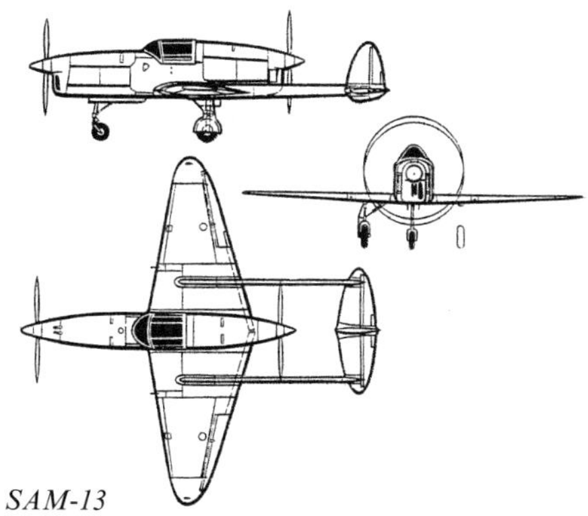

SAM-13

chrome-steel engine mountings, ply skin throughout, two-spar wing set at +2° incidence with split flaps, all flight controls with aerodynamic and static balance. Four ShKAS, two above front engine and two at outer ends of centre section beyond front propeller disc (2,2 m diam). First factory test flight late 40 by N. D. Fikson. Poor handling and, because of error in drawings, nose gear failed to retract fully; but speed still reached 560 km/h. Rapid rectification of faults, and LII testing spring 41 under M. L. Gallai. Promising, and entered for summer-41 high-speed race, but after invasion work halted and SAM-13 destroyed prior to evacuation of OKB to Omsk *oblast* (possibly Irtysh).
DIMENSIONS Span 7,30 m (23 ft 11½ in); length 7,85 m (25 ft 9 in); wing area 9 m^2 (96·9 ft^2).
ENGINES Two MV-6.
WEIGHTS Empty 754 kg (1,662 lb); loaded 1183 kg (2,608 lb).
PERFORMANCE Max speed (est) 463 km/h (288 mph) at SL, 680 km/h (423 mph) at 4 km; service ceiling 10 km; range 850 km (528 miles); landing speed 125 km/h (78 mph).

SAM-14 Substantially modified SAM-5-2bis built for Aeroflot 39. Utility transport retaining simple features but more powerful engine. Wing of R-II section, 14% thick, fabric covering, wood truss fuselage with ply skin from nose to rear of cabin, fabric at rear. Ventilated cabin for pilot and five passengers, two doors on left, Kalman wing flaps, trimmers on control surfaces. NII testing Nov 39 to June 40; GVF order rescinded by Kaganovich.
DIMENSIONS Span 11,49 m (37 ft 8⅓ in); length 8,06 m (26 ft 5⅓ in); wing area 21,86 m^2 (235 ft^2).
ENGINE One MV-4.
WEIGHTS Empty 765 kg (1,687 lb); fuel/oil 95+10 kg; loaded 1280 kg (2,822 lb).
PERFORMANCE Max speed 196 km/h (122 mph) at SL, 170 km/h at height; service ceiling 3360 m; range 550 km (342 miles); take-off 240 m/17 s; landing 190 m/15 s/68 km/h (42 mph) with flaps.

SAM-14

SAM-16 model

AM-14

SAM-16 SAM-15 not known. SAM-16 small twin-engined three-seat wooden reconnaissance flying boat, with engine nacelles central on gull wing. Wing profile NACA-230, 16% at root and 12% at tip, slotted flaps and slotted ailerons. Stabilizing floats on neat fixed single struts. Twin-step hull with pilot cabin ahead of leading edge and two ShKAS in glazed turrets in bow and between wing and twin-finned tail. In erection shop June 41, and destroyed prior to evacuation.
DIMENSIONS Span 15,5 m (50 ft 10¼ in); length 12,03 m (39 ft 5¾ in); wing area 32,15 m^2 (346 ft^2).
ENGINES Two MG-31F.
WEIGHTS Empty 2160 kg (4,762 lb); fuel/oil 390+30 kg; loaded 3 t (6,614 lb).
PERFORMANCE Max speed (est) 362 km/h (225 mph); service ceiling (est) 4,6 km (15,000 ft); alighting speed 100 km/h (62 mph).

SAM-18 SAM-17 not known. SAM-18bis under construction 41 as civil/military amphibian version of SAM-16.

SAM-19 Yet another use for well-proven SAM-5-2bis wing, light attack aircraft (*Shturmovik)*, project dated 42. High wing on central nacelle, pilot cockpit with side door, tricycle landing gear, pusher MG-31F on pylon above wing at centreline, and twin tail booms. Accepted power inadequate.

SAM-23

SAM-25

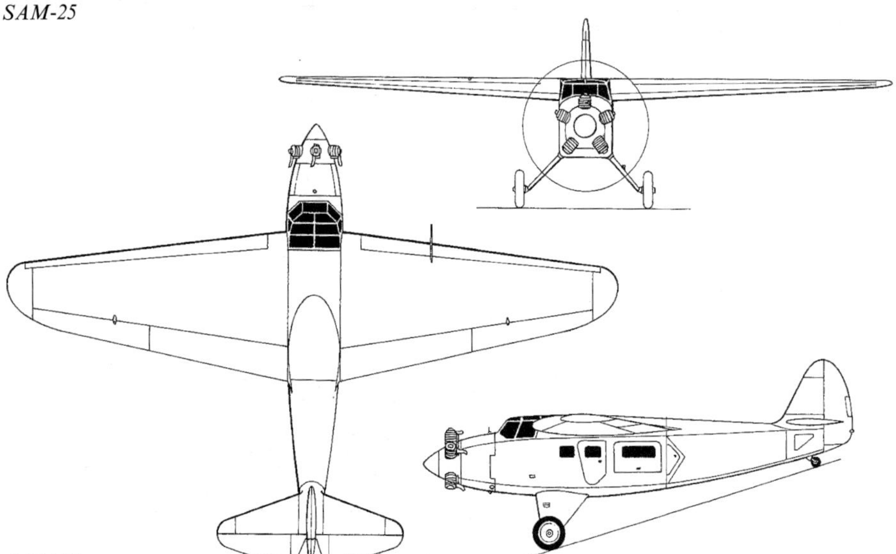

SAM-25

SAM-22 Project for slow economical transport seating 16 passengers, with single pusher 140 hp M-11F on pylon above high wing; 43.

A-2bis Modernized Antonov A-2 glider; 42.

AM-14 Modernized A-7 assault glider, designation from Antonov/Moskalyev; dated 43.

SAM-23 Evacuated OKB in Omsk region concentrated on assault trooping pods (DK, *Desantnyi Kabine*), seating 10 (DK-10) or 16 (DK-16) men, carried by retired DB-3 bombers, and also on large assault gliders. Only completely new aircraft design to fly was SAM-23, inspired by Go 242 assault glider and with same configuration but larger. Wooden construction, with box-section tail booms off high wing, and upward-hinged rear to give full-section access to 16-seat nacelle, with hinged ramp for small vehicles. Max tow 241 km/h (150 mph), gliding speed after cast-off 172 km/h (107 mph). Prototype flown behind DB-3 May 43, but no production.

SAM-24 Derivative of SAM-23 with two M-11F engines; not flown.

SAM-25 This was main powered aircraft built by Moskalyev-led assault-transport group in evacuated OKB-31. In all respects outstanding, updated SAM-5-2bis with cleaned-up airframe designed 42 to meet military/civil need for passenger, cargo, ambulance and assault utility transport. Wooden construction retained but refined and simplified further, but with addition of automatic slats, drooping ailerons and slotted flaps. On left side main entry door, wide stretcher (litter) hatch and rear door/window for photography. Tabbed rudder/elevators. Tested at Omsk-district OKB mid-43, followed by NII testing by A. Dabakhov during which flown in stages to Moscow and then back nonstop. Arguments over production; impossible during war, but recent info suggests series built 46 for GVF.

DIMENSIONS Span 11,49 m (37 ft 8¾ in); length 8,02 m (26 ft 3¼ in); wing area 21·86 m² (236 ft²).

ENGINE One M-11F.

WEIGHTS Empty 720 kg (1,587 lb); fuel/oil 140+10 kg; loaded 1280 kg (2,822 lb).

PERFORMANCE Max speed 200 km/h (124 mph) at SL, less at height; service ceiling 4850 m (15,900 ft); range 1760 km (1,094 miles); take-off 150 m; landing speed 65 km/h.

SAM-26 Largest SAM design, mid-wing assault glider of 42 seating 102 troops on two decks; rear ramp for vehicles. Not completed.

SAM-28 Improved SAM-22 with deeper fuselage and tricycle landing gear; not built.

SAM-29, RM-1 Final SAM project, in detail design 44. Ultimate *Strela* rocket fighter with ogival (gothic) delta wing blended into needle fuselage, no horizontal tail but large vertical tail, tricycle landing gear. Dushkin RD-2M-3V rocket engine of 2 t SL thrust (more at altitude). Despite support of S. P. Korolyev project collapsed through being ahead of its time in immediate post-war era. OKB shut Jan 46, but studies continued to 48.

Mozharovskii-Venyevidov

Georgii Mironovich Mozharovskii and Ivan Vasilyevich Venyevidov were specialists in aircraft armament 1926-50. Their work encompassed design and development of turrets, ammunition supply systems, sights and various dropped stores. From 40 they were also engaged in design of an aircraft.

MV Kombain Named (Combine) because of its versatility, this was to be a single-seat single-engined *Shturmovik* (armoured attacker). Engine, AM-38, driving pusher propeller at rear of central nacelle. All-metal construction, low wing, twin tail booms. Forward-firing armament two or four ShVAK at front of tail booms, projecting ahead of leading edge, two or four ShKAS in nose on pivoted mounts to fire from 0° to –20° (20° downwards). Heavily armoured enclosed cockpit. Small compartments for fragmentation bombs. Tricycle landing gear with all units retracting to rear. Radiator under wing with engine-driven fan in cooling duct. Sponsors collected good team but lacked design, calculation and test skills. Money voted and substantial personnel assigned to project, but design never completed. VVS NII mock-up conference late March and early April 41. Decision to terminate in view of success of BSh-2 (Il-2). No other data.

Myasishchyev

Vladimir Mikhailovich Myasishchyev was born 1902, pupil of Zhukovskii at MVTU in 1918-21. Joined CAHI 21 and from 24 assigned increasingly important roles in ANT designs including flush radiators of R-6, rear fuselage of TB-3, TB-4 and ANT-20 and entire design of DIP. Accomplished designer of metal structures and led move away from traditional Tupolev truss/corrugation to smooth stressed-skin. At Douglas 37 helping translate DC-3 drawings. Arrested 38 and imprisoned as chief of KB-102 brigade. Produced DVB-102, delayed by evacuation Oct 41 to Irtysh, and additionally in late 43 assumed responsibility for Petlyakov programmes in succession to Putilov, and handled testing of Shchyerbakov GKs. Succession of major projects, mainly bombers, did not lead to production until 3M strategic aircraft (OKB-23), winning Lenin Prize 57. M-50 and several unbuilt projects led to major share of Buran aerospacecraft and mod programme to produce pick-a-back carrier of outsize loads. More recently ultra-high recon aircraft and various civil *Konversiya* projects, plus German link in Tec Avia Inc (see MM-1). V. M. Myasishchyev died 14 Oct 78, succeeded by V. Fedotov. OKB now 'Eksperimentalnyi Mashinostroitelnyi Zavod im V. M. Myasishchyev', General Designer Valerii Konstantinovich Novikov.

DVB-102 before first flight

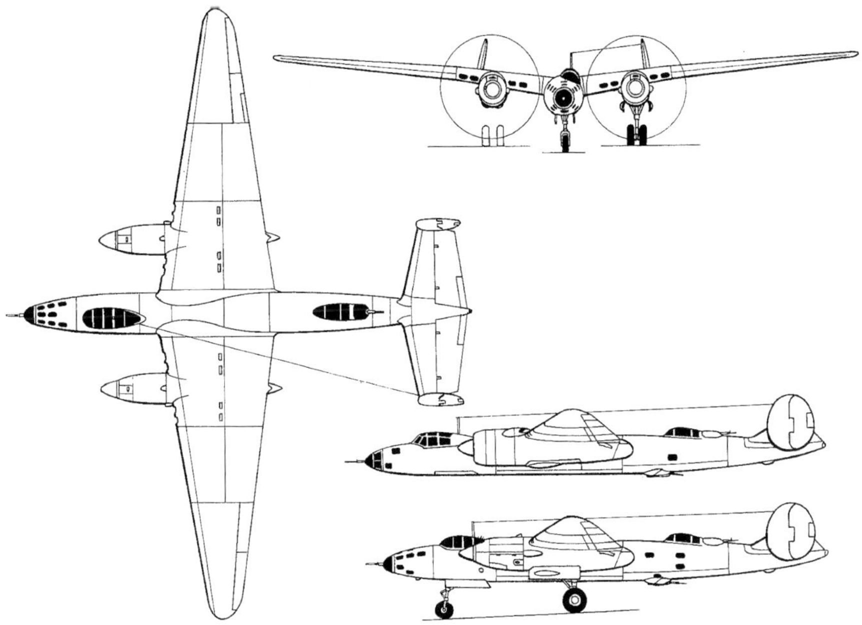

DVB-102 (upper side view 102/ASh-71)

DVB-102 First Myasishchyev design not to bear another OKB's designation, DVB-102 title from *Dalnyi Vysotnyi Bombardirovshchik* (long-range high-altitude bomber) numbered in same series as Petlyakov 100 and Tupolev 103 from NKVD-run Special Brigades. Project authorized to meet VVS need late 39; design started early 40. Prototype construction begun late April 40, with airframe in static test summer 41. Late Aug 41 factory evacuated and prototype seriously delayed, finally being completed with numerous modifications. An advanced and efficient design. All stressed-skin, with high-mounted gull wing of aspect ratio 8·2, thickness 14·5% root and 10% tip, three spars inboard and two to tip, max skin thickness 3 mm, held by countersunk screws for access to large integral wing tanks filling whole inter-spar box. Pioneered construction in upper/lower halves, joined on neutral axis (later used on Il-28). Three flap sections each side, electric actuation along steel tracks inside wing. Circular-section fuselage 1,6-m diameter with pressure cabins in nose and rear for crew of four: nav/bomb (glazed nose with hand-aimed ShVAK), pilot (fully glazed canopy over cockpit offset left of centreline), and two rear gunners with sealed sight blisters and remote-directed barbettes with electric drive. Rear armament two gunners using DUS-1 system to control upper UBK/ShKAS aimed together and lower UBK. Later nose gun exchanged for single NR-23. Weapons bay for 28 combinations of bombs and other stores with individual weight to 2 t; normal load 2 t for long range, overload to 4 t. Bomb doors hydraulically retracted up inside fuselage. Tricycle landing gear with single legs and twin main wheels, hydraulic retraction (nose to rear, main forwards) and pneumatic brakes. Twin fins at 90° to tips of tailplane with 11° dihedral. First aircraft completed with M-120 engines with TK-3 turbochargers, all radiators in four wing ducts, AV-9L-80 three-blade props.

Manufacture of prototypes directed by S. P. Korolyev. First flight, by V. I. Zhdanov, Feb 42. Testing shared with F. F. Opadchii increasingly concentrated on M-2 (see next), and both abandoned 44, development moving to DB-108 and VB-109.

DIMENSIONS Span 25,15 m (82 ft 6⅛ in); length (excl ShVAK) 18,9 m (62 ft 0 in); wing area 78,3 m² (843 ft²).

ENGINES Two M-120TK.

WEIGHTS Empty 10 966 kg (24,175 lb); fuel/oil 2000+150; loaded 14 906 kg (32,862 lb).

PERFORMANCE Max speed 440 km/h (273 mph) at SL, 540 (336) at 6,25 km; climb 13·5 min to 5 km; service ceiling 8,3 km (27,230 ft); range 3340 km (2,075 miles); TO 640 m; ldg 165 km/h, 340 m.

102/M-71 Prototype with ASh-71 engines, retaining wing oil radiators; increased span, shorter nose with different glazing, metal aft

102/M-71

section to pilot canopy, various other changes. First flight autumn 43, flown on mainly hi-alt testing with twin TK-3 turbos Mar 44 to July 45. Several other variants never built, including DVB-102N without pressure cabins and DVB-102DM with MB-102 engines.
DIMENSIONS Span 26,04 m (85 ft 5¼ in); length 18 850 (61 ft 10⅛ in); wing area 78,5 m^2 (845 ft^2).
ENGINES Two M-71TK-3 (ASh-71F).
WEIGHTS Empty 12 173 kg (26,836 lb); fuel/oil 2000+150; loaded 16 037 (35,355 lb).
PERFORMANCE Max speed 430 km/h (267 mph) at SL, 529 km/h (329 mph) at 8,45 km; climb 12 min to 5 km; service ceiling 10,5 km (34,450 ft); range c3600 km (2,237 miles); TO 670 m; ldg 145 km/h, 340 m.

VM-13 See Petlyakov Pe-2.

DB-108 After Petlyakov's death Myasishchyev continued devt of Pe-2VI into further outstanding mid-wing attack aircraft. DB-108 (*Dnyevnoi*, day) used latest Klimov V-12 engine in beautiful cowl with exhaust along sides and top, four-blade VISh-107 props, engine rads inboard LE, oil rads outboard LE. Basic variant **VM-16** from designer's initials, 2-seater with bomb bay for 2 t load or option of two NS-37 or NS-45 in same bay; designed for overload addition of two FAB-1000 under wings. With shortened fins one FAB-2000 possible internally. Nav/bomb could sit beside pilot facing aft and aim remote-control gun (UBK, later ShVAK) in extreme tail. First flown 30 Dec 44, data below for this.
VM-17 Second aircraft added third crew member to manage revised defensive armament of three B-20 and one UBS; prolonged ground testing but aircraft eventually rebuilt as VB-109. **VM-18** lengthened to accommodate fourth crew-member, yet further modified firepower with B-20 fixed firing ahead, UBS at rear of crew compartment and upper/lower remote-control ShVAK as in DVB-102. Flown 46, tests never completed.
DIMENSIONS Span 17,8 m (58 ft 4¾ in), (18) 20,6 m; length 13,47 m (44 ft 2¼ in), (18) 15,02 m; wing area 43,0 m^2 (463 ft^2), (18) 48,0 m^2.
ENGINES Two VK-108.
WEIGHTS Empty 6953 kg (15,328 lb), (17) 7512 kg; fuel/oil 1000+131; loaded 9431 kg (20,791 lb), (17) 9990 kg, (18) 10 530 kg.
PERFORMANCE Max speed at SL 583 km/h (362 mph), (17) 545 km/h, (18) 542 km/h; at 5,95 km 700 km/h (435 mph), (17) 670 km/h, (18) 660 km/h; climb 5 min to 5 km; service ceiling 12 km; range (2 t bombs) 2250 km (1,400 miles); TO 420 m; ldg 140 km/h, 485 m.

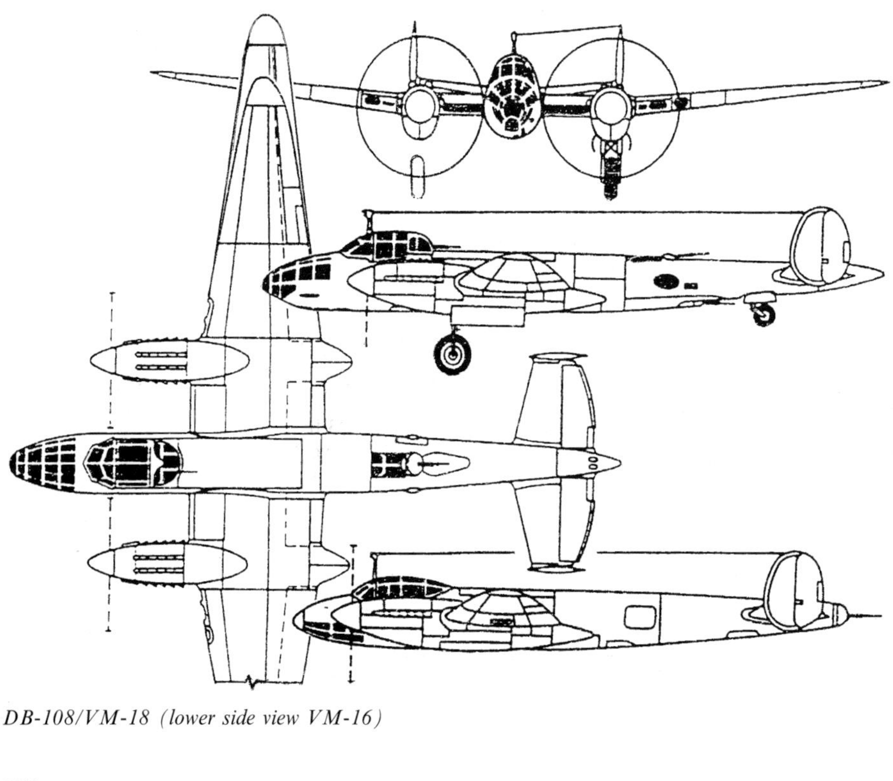

DB-108/VM-18 (lower side view VM-16)

VB-109 Klimov's relentless devt of more powerful engines made possible this simpler yet heavier rebuild of second DB-108. Engines fitted with two-speed two-stage superchargers, rads and inlets again in LE, driving VISh-107L30 3-blade 3,3 m paddle-blade props. Fuel in protected flexible cells in wing and fuselage, normal 900 kg, max 1740 kg (when 1 t bombload taken 2200 km). Normal bombload up to 2 t (4,410 lb), plus overload two underwing containers each for 500 kg. Single fixed B-20 and one remotely controlled in tail as in VM-16. Pilot and radio/gunner in pressure cabin with KPA-12m oxygen and armour up to 10 mm, with 64 mm transparencies, periscope under rear fuselage. First flight late 45 with interim VK-107A engines, then cancelled.
DIMENSIONS Span 17,8 m (58 ft 4¾ in); length (excl B-20) 14,17 m (46 ft 5⅞ in); wing area 43,16 m^2 (464·6 ft^2).
ENGINES Two VK-109.
WEIGHTS Empty 7508 kg (16,552 lb); fuel/oil see text; loaded (normal) 9,9 t (21,825 lb), max 11,9 t (26,235 lb).
PERFORMANCE (est) Max speed 595 km/h (370 mph) at SL, 690 km/h at 5 km, 720 km/h (447 mph) at 9 km; climb 7·2 min to 5 km; service ceiling 12,5 km (41,000 ft); range see text; TO (est) 650 m; ldg 151 km/h, 570 m.

DIS Last of Myasishchyev's devts of Pe-21, *Dalnyi Istrebitel Soprovozhdyeniya* (long-range fighter escort) was, except in firepower, inferior to predecessors. Lower-powered engines and, for its size, enormous fuel capacity. Nav seated beside pilot in pressure cabin, two B-20 in nose, two NS-37 or NS-45 in ventral gondola and one B-20 in dorsal dome remotely sighted by nav; weight of fire per second 12,3 kg. OKB test completed final weeks 45 but taken no further. [This armament and data below totally different from 1st edition of relevant Shavrov.]
DIMENSIONS Span 18,12 m (59 ft 5⅜ in); length 13,8 m (45 ft 3¼ in); wing area 43,8 m^2 (471·5 ft^2).
ENGINES Two VK-107A.
WEIGHTS Empty 7230 kg (15,939 lb); fuel/oil 1675 kg; loaded 10 042 kg (22,138 lb).
PERFORMANCE Max speed (est) 531 km/h (330 mph) at SL, 627 km/h (390 mph) at unstated ht; climb 7·2 min to 5 km; service ceiling est 9,6 km (31,500 ft); range (est) 4000 km (2,485 miles); TO 570 m; ldg 146 km/h, 530.

SDB Prototype *Skorostnyi Dnyevnoi* B, fast day bomber, left incomplete at end of war. Basically a DB-108 with stretched fuselage, greater wing area, max 13,5 t (29,760 lb); speed with VK-108 engines 660 km/h. Work switched to DSB-17.

DVB-202, VM-22a Attempted copy of B-29, rival to Tupolev 64. Three pressure cabins, bomb load (normal) 5 t, (max) 2×4 t internal, same external; four body turrets each 2×B-20 (500 rounds), tail turret 1×B-20 (300) + 1NR-23 (100). Main tyres (2) 1325×485, nose (2)

VB-109

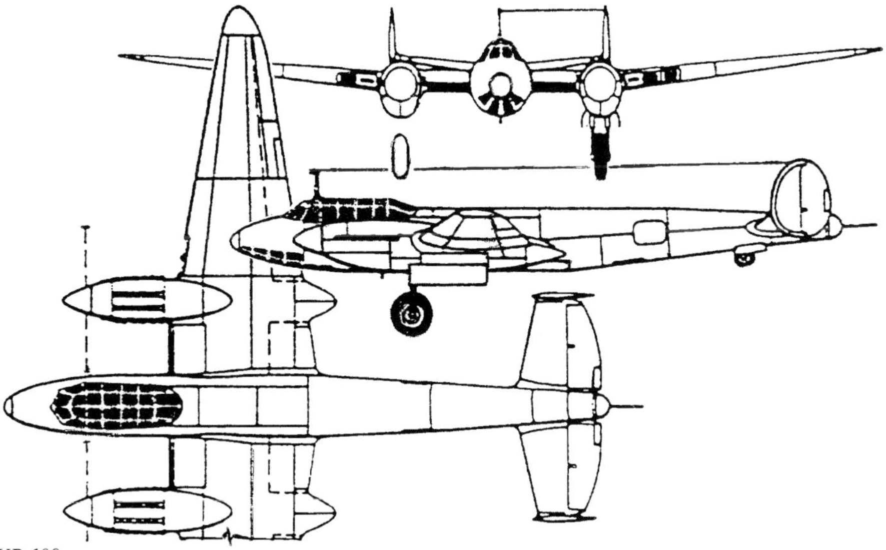

VB-109

700×220. Abandoned Oct 44.
DIMENSIONS Span 42 m; length 29,2; wing area 150 m².
ENGINES Four ASh-72TKM.
WEIGHTS Empty 22,1 t; max 44 t.
PERFORMANCE Max speed (10 km) 650; range (42 t, 5 t bombs) 6000.

DVB-302 Upgraded DVB-202, late 44, not built.

DSB-17, VM-24 DSB = Long-range Mid-wing Bomber; wing scaled down from DVB-302; max bomb load 5 t, single NR-23 in manned tail turret, main gears retracting behind bomb bay. Abandoned 45. See Addenda.
DIMENSIONS Span 20,8 m; length 16 m; tailplane span 6,5 m; wing area 75,6 m².
ENGINES Four RD-10 (drawings say Jumo 004B).
WEIGHT Max 28 t.
PERFORMANCE Range (1 t bombs, 800 km/h) 3000 km.

Bison The following, from Project 25 to M-28, tells the story of the aircraft known to ASCC as 'Bison'.
Project 25, M First Soviet strategic jet bomber, proposed in paper by V. M. Myasishchyev in 50 to have range 11 000-12 000 km to counter threat of B-52/B-60. Stalin quickly ordered go-ahead, to tightest possible schedule. On 24 March 51 Kremlin ordered MAP's OKB-23 to be reopened, headed by Myasishchyev. In June VVS specified range with 5 t bombload not less than 11 000 km at 900 km/h at 9 km. To meet timescale max help from establishments and other OKBs; chief constructor L. I. Balabukh, aerodynamics led by I. Ye. Baslavskii and at CAHI by section under A. I. Makarevskii, initial drawings by L. Selyakov.

Project launched as **25**, planned around four AM-3 turbojets. Alternative **26** schemed with two VD-5, but this impossible in timescale. Wing aerofoil SRs series, nominal 12% at root (modified by engine ducts) and 7% tip. Design factor (initially at 140 t) 2 g. Three-spar centre section, two-spar outer panel to give aspect ratio 8·7 (B-52, 8·5). Ribs 90° to rear spar, outer-wing joint 10,35 m along this spar from centreline. Fixed LE swept 37°30' to joint, thence 34°48' to tip. Inter-spar box machined rectangular skin panels, sandwich skin behind rear spar. Inboard TE 90° to centreline occupied by jetpipe fairings with undersides covered by split flap; TE outboard of kink swept 25°. Enormous double-slotted flap on each wing in two parts, inboard running on one track and one end-rib, outer directly aft on four tracks. Outboard of flap, 2,18 m (85·8 in) fixed portion to second kink, thence 29° sweep, powered aileron in two sections, inner having long tab on both wings. Fence on upper surface 10-85% chord at inner end of aileron. Pivoted trimming tailplane with dihedral carrying powered un-tabbed elevators part-way up fin, inset-hinge tabbed rudder.

Four engines aft of rear spar on beams aligned with centreline, axially staggered because of sweep. To avoid interference, oval inlets projecting ahead of LE for each engine, duct through spar webs, outer fairing tapering off into wing to reappear again around engine. Upper-cowl access doors and aux inlets, lower access doors followed by split flap under jet-pipes. Outer pipes longer, all four angled 4° outwards.

Circular-section fuselage, ruling diameter 3,5 m (138 in), unfaired upper joint with pressure cabin housing nav/bomb-aimer in glazed nose, two pilots, *radist* managing RPB-4 main radar (rear-fuselage location studied, moved to front) and three gunners. Cabin wt reduced to 500 kg. Tail-gunner capsule without communicating tunnel; total crew 8, all with downward-ejecting seats. Pilot sight for fixed cannon (later omitted). Forward gunners TP-1 electro-optical sights at three hemispherical windows, electro-mechanical links to drive dorsal and ventral turrets just behind pressure cabin. Gunner in Ilyushin tail turret able to take control of other turrets in emergency. All three turrets twin NR-23. Bomb bay aft of wing bridge with left/right double-fold doors, max one TN-9000 or FAB-9000 or lesser bombs to design max 24 t. Protected and inerted fuel cells along top of fuselage, full depth front and rear of bomb bay and in wing box, design total 132 390 lit but practical limit 123 600 lit (27,189 gal).

Weight saved by bicycle landing gear. Entire weight supported by main four-wheel bogie at front and rear, tyres 1550×480, all with multi-disc anti-skid brakes and hydraulic retraction forwards, each bogie somersaulting inverted into bay with twin doors. Front bogie hyd steerable ±15°; on takeoff, steering auto-centred and, at rotation speed, strut auto extended to optimise AOA. Outrigger gears with twin wheels on long levered struts retrac forwards

M-4-2

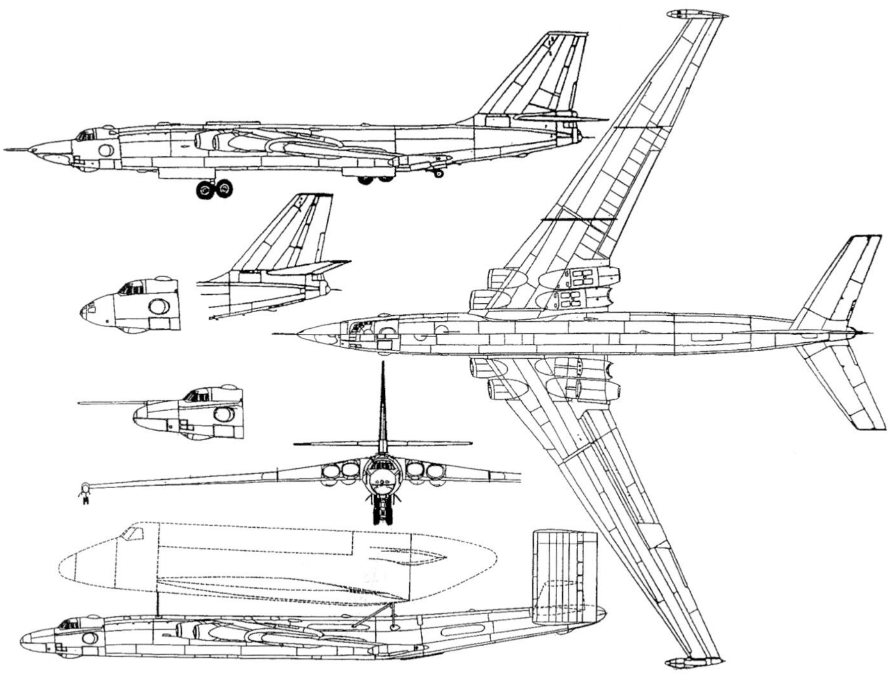

*3MD (side view VM-T/*Buran, *inset nose/tail M-4 and nose 3M)*

into tip pods with four doors. (Landing gear tested on Tu-4.) Box for large braking parachute under rear fuselage. Hydraulic system of twin-duplex (four-pump) type, 210 kg/cm^2 (3,000 lb/sq in). Elec based on four DC 27-volt generators, with rectifiers to provide AC for radar and deicing. Over 40 gaseous-oxygen bottles. APU added later. Deicing mainly by hot air from engine bleed and heat exchangers, electrothermal elements for cockpit and blister glazing.

Prototypes built by two brigades, Fedotov/Zholkovskii, led by Balabukh and then by F. M. Dostoevskii. Recognised shortfall in range made interim-engined aircraft of little value as strategic bomber. Available TO thrust limited wt while poor fuel economy limited range. CAHI veteran N. N. Korchyemkin suggested discuss with Tupolev, leading to improved versions. First aircraft rolled out 30 Dec 52, flown by crew headed by F. F. Opadchi 20 Jan 53. OKB-23 testing complete 15 April 54.

DIMENSIONS Span 50,526 m (165 ft 9¼ in); length 47,6 m (156 ft 2 in); wing area 340,2 m^2 (3,662 ft^2).

ENGINES Four AM-3A.

WEIGHTS Empty c77 t (169,750 lb); max 181,5 t (400,132 lb).

PERFORMANCE Max speed 947 km/h (588·5 mph) at 6,7 km (22,000 ft); service ceiling at 138 t 12,5 km (41,000 ft); range with 5 t bombload (est) 8000 km.

103, M-4 Second (*dubler*) **Product 103** flown late 53, in May Day flypast with four MiG-17 to give scale. Loran rail antenna and defensive avionics, bombload limited to 18 t (39,683 lb), fixed gun omitted, turrets modified (AM-23 guns) and numerous systems changes. OKB-23, LII and NII testing began 4 May 54, qualified for DA as **M-4**, in training role because did not meet reqd range. Limited production at Kazan, first delivery Nov 55. In 56-57 re-engined with more powerful and fuel-efficient RD-3M and RD-3M-500A engines, latter enabling max range to be increased. **M-4A** and **M-4-2** conversions, see below. Popular name *Molot* (hammer), ASCC name 'Bison-A'.

DIMENSIONS Span 50 526 (165 ft 9¼ in); length 47 665 (156 ft 4½ in); wing area 340,2 m^2 (3,662 ft^2).

ENGINES Four RD-3, later RD-3M, RD-3M-500 and RD-3M-500A.

WEIGHTS Empty 79,7 t (175,705 lb); max 184 t (405,644 lb).

PERFORMANCE Max speed 930 km/h (578 mph) at 7,5 km (24,600 ft); range with 5 t bombload 8100 km (5,035 miles).

M-4A One aircraft (#85) retrofitted 59 with inflight-refuelling system by S. M. Alekseyev, fixed probe on centreline above nose, glazed nose retained.

M-4-2 Series aircraft converted as tankers with 48 000 lit bomb-bay tankage and hose/drogue unit. Glazed nose retained. Redundant bombers replaced by 3M converted to this standard 59-63.

28, 2M Unbuilt project of 56-58 intended to confer higher-altitude performance, dubbed *Vysotniyi* (hi-alt climber); main difference wing redesigned to carry four VD-5 in widely spaced underwing pods, outers housing stabilizer landing gears; tail surfaces extended; design ceiling 16,2 km (53,150 ft). Dropped because 3M met requirement.

3M OKB effected major redesign, initially called **M-6**, **Article 201**. Wing increased span, more efficient structure, later materials, integral tanks, outer TE straight, outer panels sharp washout, large fence added bisecting structural joint extended back to flap and around LE. Redesigned installn for VD-7 engines, increased wing fuel plus two aux tanks in bomb bay (reducing bombload to 11 t). Fin chord increased at expense of rudder, tailplane without dihedral. FR probe above lengthened unglazed nose housing RBP-4 (not RPB-4) radar with nav/bomb relocated below cockpit with side windows and ventral gondola, landing gears strengthened, upgraded nav/bombing subsystems improving accuracy at night or in bad weather, ILS and radar warning updated and chaff/flare ejectors added. First flight by M. L. Gallai and crew 27 March 56. Full prodn from late 56 as **M-6**, changed same year to **3M**. Outstanding aircraft, though still not quite able to fly design mission, and utilization poor because of short engine TBO. Two aircraft used in 59 for FAI records: **103M** set 1000 km circuit record with 27 t payload at 1028 km/h; **201M** set 19 world records including carrying 10 t to 15 317 m (50,253 ft) and 55,22 t (121,480 lb) to 2 km. ASCC 'Bison-B'.

DIMENSIONS Span 53,14 m (174 ft 4⅛ in); length 51,7 m (169 ft 7½ in); wing area 350 m^2 (3,767 ft^2).

ENGINES Four VD-7.

WEIGHTS Empty 74 430 kg (164,087 lb); loaded 193 t (425,485 lb), (max, bomb-bay fuel) 202 t (445,326 lb).

PERFORMANCE Max speed 940 km/h (584 mph) at 8,5 km (27,890 ft); service ceiling 14,9 km (48,900 ft); range with 5 t bombload 12 000 km (7,457 miles), (1 refuel) 15 400 km (9,570 miles).

3MS Reluctant acceptance of need to replace VD-7 engine led to major conversion in 58-62 with refined and uprated Mikulin turbojet as each aircraft came up for major overhaul. Approximately half 3M force were re-engined. ASCC name unknown.

DIMENSIONS As 3M.

ENGINES Four RD-3M-500A.

WEIGHTS Empty 75 740 kg (166,976 lb); max 192 t (423,280 lb).
PERFORMANCE Max speed 625 km/h at SL, 925 km/h (575 mph) at 8,5 km; service ceiling 14 km (45,930 ft); range with 5 t bombload 9400 km (5,840 miles); TO 310 km/h, 2950 m; ldg 210/1800.

3MN In late 58 Dobrynin KB achieved clearance with engine which, mainly by considerable derating, achieved similar TBO and reliability to RD-3. This became standard production engine, and was also retrofitted to some earlier bombers instead of RD-3M-500A. It reduced empty weight and increased range by 15% on typical missions. ASCC name unknown.
DIMENSIONS Unchanged.
ENGINES Four VD-7B.
WEIGHTS Empty 74 990 kg (165,322 lb); max 192 t (423,280 lb).
PERFORMANCE Generally as 3MS except range with max bombload 10 200 km (6,340 miles); TO 310 km/h, 2500 m.

3MD Refinements introduced 60 to improve performance with VD-7B engine, notably nose redesigned ahead of front pressure bulkhead as cone with probe at tip; small increase in wing area, span unchanged. Numerous equipment items upgraded. Myasishchyev's final contribution (assigned to CAHI later in 60). 3MD #49 displayed Domodyedovo July 67, so given ASCC name 'Bison-C' (again thought to be a maritime recon platform).
DIMENSIONS As before except length 51,8 m (170 ft 0 in); wing area 351,78 m² (3,787 ft²).
ENGINES Four VD-7B.
WEIGHTS Empty 76 800 kg (169,312 lb); max 192 t (423,280 lb).
PERFORMANCE Max speed 930 km/h (578 mph) at 8,5 km; range with 5 t bombload 10 950 km (6,805 miles), (1 refuel) 13 600 km (8,450 miles).

3MYe Final attempt to produce hi-alt strategic bomber, single example tested 63, basically 3MD re-engined with VD-7P (RD-7P). Design ceiling 17,5 km (57,415 ft).

3MS-2, 3MN-2 In conformity with SALT-2 remaining 3MS and 3MN bombers were scrapped in 80s or converted into single-point tankers, with almost same tank/hose installn as M-4-2 and nose as 3MD but without probe, plus receiver homing avionics. Served ADD as standard tanker. Final 40 were to be replaced by Il-78 in 91-93, but these retained in Ukraine. Still 'Bison-C'.
DIMENSIONS As 3MD.
ENGINES (S) RD-3M-500A, (N) VD-7B.
WEIGHTS, PERF Similar to 3MD; range with 40 t transfer fuel (N) 9440 km (5,866 miles).

3M-T, VM-T In 78 Myasishchyev returned to OKB, and proposed to solve problem of rapid transport of large space launchers and components to Baikonur by modifying 3M. He died later 78, work continued by Fedotov, three 3MN taken to SibNIA for detailed structural audit to plan rebuild of fuselage/wing and grafting of new rear fuselage (7 m longer) and twin-fin tail. Redesignated **3M-T**, zero-lifed airframe. As thrust more impt than TBO, re-engined with VD-7M (ex-Tu-22) with afterburner removed. One static-tested by CAHI, two for flight, one with FR probe, first flown 80. Both aircraft subsequently mod to carry outsize pick-a-back loads. Notably *Buran* airframe or *Energiya* tank (40 t, diam 8 m/26 ft), on front/rear trusses above fuselage, civil regis-

3MS

3M

3M

3MD

VM-T

M-28 model

tered and redesignated **VM-T** *Atlant*. First flight with payload by A. Kurchyerenko and crew 6 Jan 82, subsequently 150+ transport missions to Baikonur.

DIMENSIONS Length 58,7 m (192 ft 7 in).

ENGINES Four unaugmented VD-7M rated at 11 t.

M-28 Project c70 for double-deck military cargo transport on basis of 3M airframe, with provision for loading vehicles and large pallets. Also unbuilt, civil project with double-deck pressurized fuselage for 380 passengers.

M-50 Launch 55 of USAF Weapon System 110 (later B-70) spurred response by GUVVS, who decided there was need to strategic supersonic bomber. Task assigned Myasishchyev OKB. Suitable engine assigned to KB of P. F. Zubets (Zubts), devd with CIAM team led by G. P. Svishchyov. Prelim design handled in partnership with CAHI team led by M. V. Keldysh, A. I. Makarevskyi and G. S. Byushgens; OKB project team led by L. L. Selyakov, L. I. Balabukh, I. B. Baslavsky, L. M. Rodnyansky, V. A. Stopachinsky and V. A. Fedotov. Demand included radius exceeding 3000 km and dash speed Mach 2.

Over 30 configurations investigated, 13 with canard foreplanes and 16 with twin vertical or inclined tails (this ignores engine disposition). Final choice Dec 56 to use delta wing in shoulder (mid-hi) position so that thickest portion almost at top of circular fuselage, with conventional tail. Planned to use four engines, final choice of many arrangements was two under wing at mid-span and two on broad (cropped) tips. Several features preceded Concorde by a decade, including flight control system and longitudinal trim.

Apart from XB-70, wing was largest ever built for supersonic flight. Thickness 3·7% at root, 3·5% at tip, set at marked positive incidence, 2° anhedral, LE sweep 50° from root to inboard engines at 55% semi-span, thence 41°30' to tip; TE at 90°. Four main spars all at 90°, seven full ribs each side, machined skins, fixed LE, TE occupied by rectangular inboard double-slotted flap and tapered outboard flap serving also as aileron. Tail comprised modest delta fin and rudder and swept slab tailplanes with tip anti-flutter masses below level of wing. All flight controls signalled by pioneer fly-by-wire, in first prototype with mechanical backup. Cockpit inputs converted by potentiometer into varying voltage, without computer shaping. Control surfaces powered by quad actuators each in separate (twin-duplex) hydraulic system.

Very large fuselage of almost perfect form.

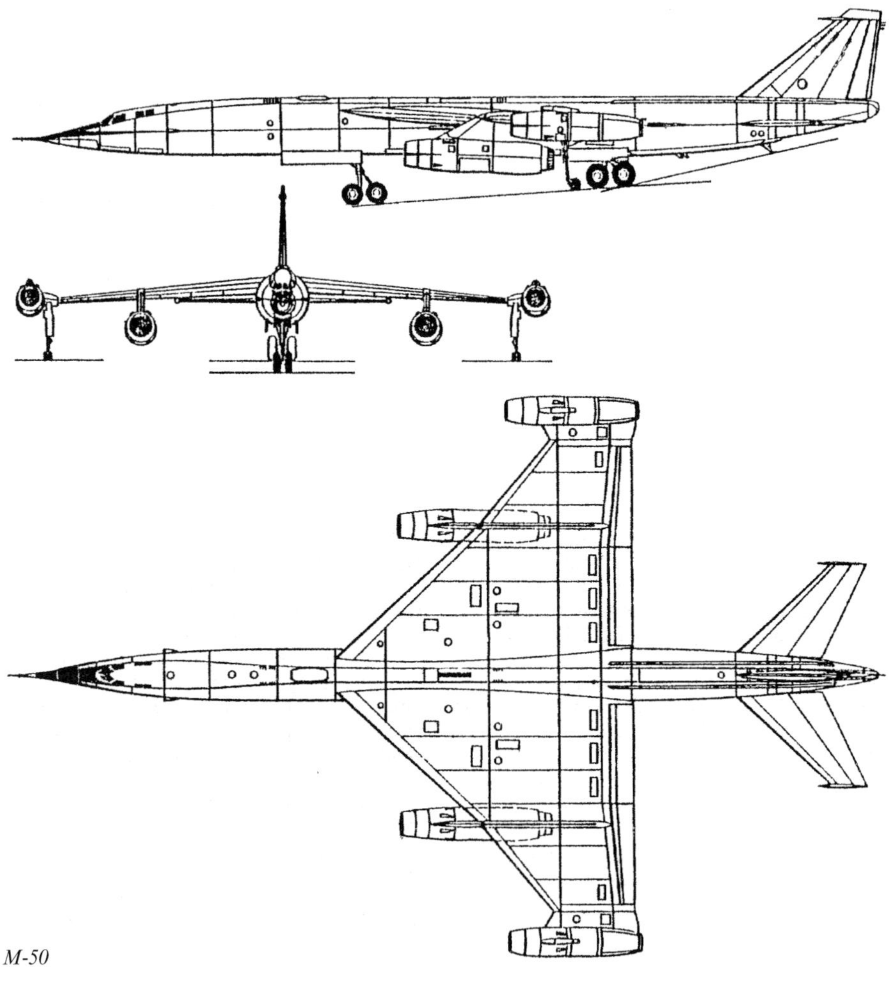

M-50

M-50

Pointed nose with instrumentation boom (to house radar in series aircraft), then unbroken circular profile to tail where hemispherical end replaced turret (required in original study). Crew of two pilots in tandem, backseater normally managing nav, bombing and systems. Each strapped into downwards-ejection seat hanging on four cables below forward-hinged ventral door. After checks by ground crew, hoisted by elec winch into pressurized cockpit drum with outer and inner doors closed. Sharp vee windscreens, side windows but fixed metal roof leading into large control and pipe fairing to extreme tail.

During transonic acceleration this fairing carried fuel from trim tank behind cockpits across tops of integral tanks filling front and rear fuselage to aft trim tank in extreme tail. On transonic deceleration with fuel levels lower fuel was pumped forward from aft trim tank into rear main fuselage tank by pipes faired under rear fuselage. This was first use of fuel transfer to alter longitudinal trim. Further fuel housed in wing integral tanks.

Another problem was how to arrange landing gear, but from outset 3M configuration was chosen, using front and rear bogies generally very similar to those of predecessor. Front bogie steerable, rear bogie (bearing 63% of weight at takeoff) fixed. Both main gears retracted forwards into bays with twin doors. Remaining problem was that, as with 3M, tailplane authority was marginally adequate to rotate aircraft to flight angle of attack on takeoff. Contributory factor was that M-50 was designed primarily as rocket launcher, and 11-m cruise missile had to fit between landing-gear bays, preventing rear bogie from being moved any further forward.

Solution was to repeat previous scheme (also used by Il-54) and use rapid-extension strut on forward gear. Double-extension hydraulic struts were triggered automatically as speed reached 300 km/h on takeoff, forcing up nose of aircraft through 10° rotation, continuing by rotating bogie beam nose-up to give even higher angle of attack for positive climb. Myasishchyev called this his 'galloping bicycle'. Wingtip outrigger gears with vertical oleo struts carrying two small wheels, pivoted to rearmost spar inboard of tip engine and retracting forwards.

Programme involved exceptional research, devt of new analytical methods, flutter models, algorithms which led to first Sov use of finite-element analysis and special computer centre (then unique to OKB). Here pilots N. I. Goryainov and A. S. Lipko learned how M-50 would fly whilst aircraft characteristics themselves were refined. Programme involved static-test specimen and two flight articles.

Planned engine was to be installed in nacelles with efficient supersonic inlets with translating conical centrebodies as in B-58. This engine ran very late in timing and in 58 decision taken to complete first M-50 with substitute engine, less powerful and less efficient and offering only enough power for brief sorties at greatly reduced weight. These were installed in nacelles with plain pitot inlets, inboard pylons being extended back across upper surface as fence terminating 1 m ahead of outboard flap/aileron.

Aircraft finished unpainted except red stars and number 023 on forward fuselage. Successful flight 27 Oct 59, Goryainov/Lipko praising pilot view, comfort, cockpit layout and 'simple yet effective flight controls'. Aircraft subsequently **M-52** with new wingtips in form of reverse-taper pylon housing redesigned outrigger gear retracting rearwards, increasing span, and with prominent ram-air inlet high above each engine nacelle. Contrary to common Western belief, Mach 1 was exceeded on original engines. Myasishchyev always careful about relation of thrust/weight, and four -7F engines would have been too powerful. His decision was two -7 and two afterburning -7F.

Only M-50 was flown. It appeared at 61 Tushino airshow with new inboard nacelles, with larger inlets and increased diam throughout, housing afterburning engines, with auxiliary inlets at front of pylon struts removed, and with various painted decoration and number 12. Photograph also exists showing it without painted decoration and bearing number 05. As '12' it is today parked in flight attitude at Monino. Reasons for termination included major shortfall in combat radius, heavy expenditure on ICBMs, unsuitability of configuration for SST and cancellation of B-70. Data below wholly unconfirmed. ASCC name 'Bounder'.

DIMENSIONS Span 37 m (121 ft 4 in); length 57 m (187 ft); wing area 195 m² (2,099 ft²).

ENGINES (originally) four VD-7, (later) two VD-7, two VD-7F.

WEIGHTS Empty 74,5 t (164,240 lb); max 198 t (436,500 lb).

PERFORMANCE Max speed 1280 km/h (795 mph, M 1·2); range 6000 km (3,730 miles).

Subject 34 OKB awarded contract 67 for single-seat fighter to destroy CIA high-alt recon balloons. Led by V. M. Myasishchyev, after his death by Leonid Sokolov. Section leaders V. N. Arnoldov, A. A. Brook, Yu. A. Gorelov, S. G. Smirnov and A. D. Tokhoonts. Kolesov KB produced RD-36-52 engine; CAHI (Prof Ya. M. Serebrisko) refined P-173-9 supercritical wing profile. Aircraft *Subject 34* dubbed *Chaika* (Gull) from inverted-gull wing. Armament two AAM plus dorsal turret as on 3M bomber but with GSh-23, 600 rds.

M-17 Stratosphera Changed to recon platform. Straight-taper anhedral wing with eight-part flap outboard of booms and very short tip ailerons built at Kumertau, Bashkiri, to be flown by factory pilot Kir V. Chyernobrovki, but finally taken to Zhukovskii and flown by NII pilot Eduard V. Chyeltsov 26 May 82. This aircraft, #17401, later covered with environmentalist slogans and mod to series standard.

Series aircraft with redesigned wing. Stressed-skin airframe designed to load factor 2, wing aspect ratio 11·9 sagging with −2°30' anhedral on ground, zero at 1 g, 3 sections tracked flap, 3 sections airbrake and long tabbed 3,79 m² aileron each side, twin tail booms with added fin/tailplane bullets, oval-section nacelle with plain side inlets, cockpit dP 0,43 atm (6·32 lb/in²), two 10-lit (150 kg/cm²) O_2, upward-hinged canopy, K-36L seat. Ruling material D16 or AL-9, strong central frames OT4-1, heaviest skin 5 mm AK4-1. Engine with air-turbo starter fed from five wing tanks total 1600+2×2650+2×1550 = 10 000 lit (2,200 Imp gal). Manual flight controls with SAU autopilot, hyd syst 210 kg/cm² for gear/flaps/airbrakes, pneu 170-260 kg/cm² for wheel brakes, strut and tank pressn and cabin/canopy control, comprehensive ice protection. Retrac landing lights under front of booms. Avionics inc: PRNK-17 nav complex, PKP-72 flt command, NPP nav-plan, RSBN *Kobalt* radar, ARK rad-compass, DA-200 vert/lateral V, PVD air data; SAS-1

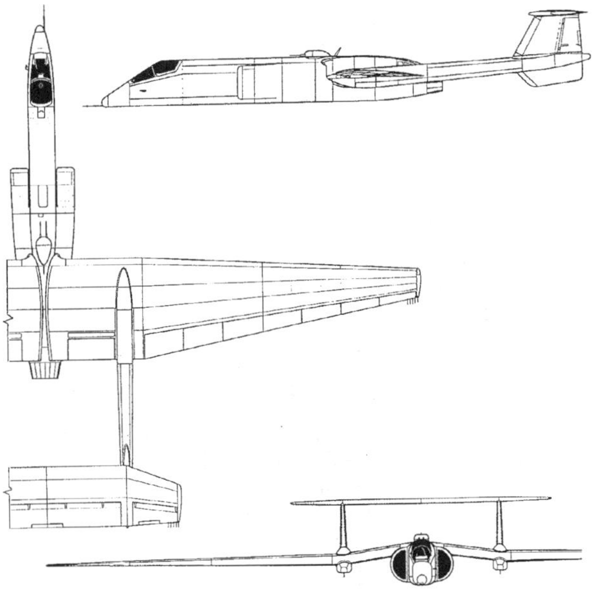

34, Chaika

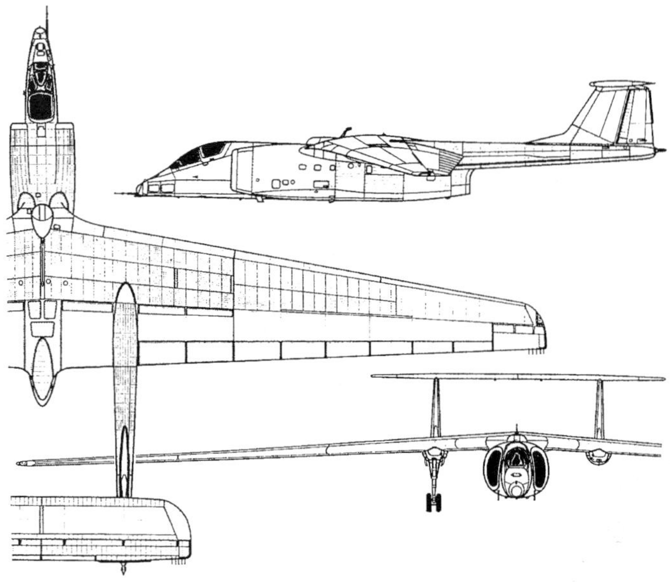

M-17 prototype

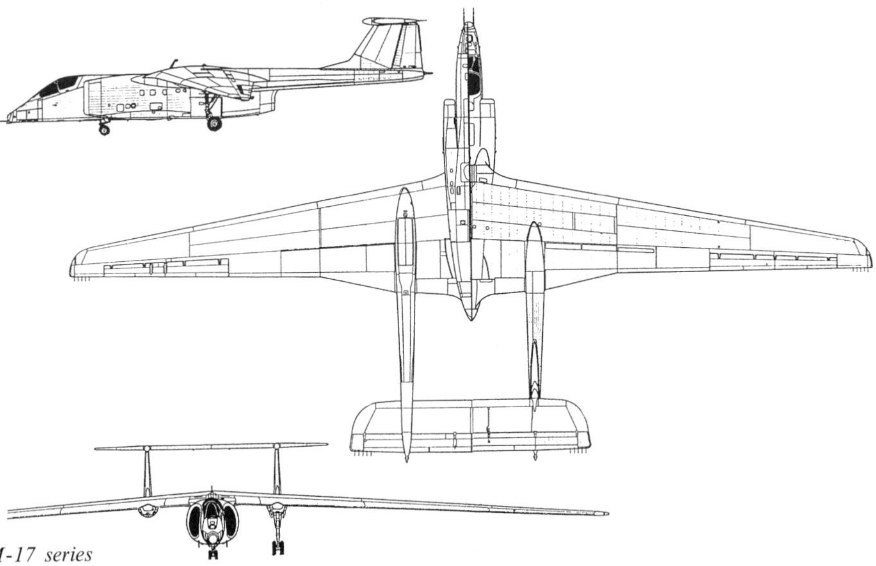

M-17 series

emergency locator, R-863 and 864 radios, provision for 1 t sensor payloads (mainly air-data).

Prototype achieved L/D of 30, set 25 FAI S-11 speed/climb/height records March/May 90 flown by ctp Vladimir Arkhipyenko, Nikolai Generalov and Oleg Smirnov. In 92 investigated Antarctic ozone 'hole'. #17103, stored Monino with Aeroflot titles, long strake antennas under tail booms, dorsal doghouse fairing over LE and several fairings, blade and spiral-planar antennas on underside. Data for series. ASCC name 'Mystic-A'.

DIMENSIONS Span 40,32 m (132 ft 3½ in); length 22,27 m (73 ft 0¾ in); wing area 137,7 m^2 (1,482 ft^2).

ENGINE One RD-36-51V.

WEIGHTS Empty typically 11 900 kg (26,235 lb); normal loaded 18,4 t (40,564 lb); max 19,95 t (43,981 lb); max landing 16,3 t (35,935 lb).

PERFORMANCE Max speed 332 km/h (206 mph) at 5 km, rising to 743 km/h (462 mph) at 20 km; service ceiling at max wt 21,55 km (70,700 ft); range (at 20 km/M0·7, 5% reserve) 1315 km (817 miles); endurance 2·25 h; TO to 10,5 m, 175 km/h, 875 m; landing (16,3 t) 188 km/h, 950 m.

M-17

M-55 Geofizika Environmentally designed aircraft to study Earth's ozone problem. Higher weights, greater fuel capacity (7900 kg, later 8300 kg), higher load factor (reduced span), twin engines with ram inlet under centreline of jetpipes, revised structure with horiz wing centre section, redesigned flaps, more capacious nose, raised cockpit, no tailplane bullets, wheelbase 5735 m (226 in), and many other changes. Prototype #01552 flown by Chyeltsov 16 Aug 88. Three further M-55 (first #55204) retain blade antenna under L inlet but replace strake antenna under L boom by blade under R. No dorsal fairing but radomes in nose and ahead of nose gear, blade-type IFF above nose and under ventral inlet. **M-55UTS** dual trainer, same length, rear cockpit at same level with periscope replac-

ing eqpt moved underfloor, no radars. Together with M-17 used in Earth-resource surveys, anti-hail (convert to rain) missions and other tasks to assist CIS economy. 'Mystic-B'. Version with conventional fuselage, swept tail and engines under wingroots under devt.
DIMENSIONS Span 37,46 m (122 ft 10¾ in); length 22 867 m (75 ft 0¼ in); wing area 131,6 m^2 (1,417 ft^2).
ENGINES Two D-30-10V.
WEIGHTS Empty 13 995 kg (30,853 lb); loaded 23,4 t or 23,8 t (max 52,469 lb).
PERFORMANCE Max speed limits same as M-17 but rising at 20 km to 750 km/h (466 mph); service ceiling 21,5 km (70,540 ft); climb to 21 km, 35 min; endurance at 17 km 6·5 h; range 4965 km (3,085 miles); TO 900, ldg 780.

M-101 Gzhel (Gazelle) Single-engined lightplane projected 90 with 360-hp M-14P piston engine, winglets and butterfly (Vee-type) tail. Devd into business aircraft with Czech turboprop driving five-blade V-510, conventional swept tail, up to four pairs seats, front and rear doors on left, airline avionics, certification due 94, series from Sokol Nizhegorodsky (MiG) plant.
DIMENSIONS Span 13,0 m (42 ft 7¾ in); length 9,975 m (32 ft 8¾ in).
ENGINE Prototype one Czech M-601F (580 kW, 778 shp), production one AL-34.
WEIGHTS Fuel 650 kg; payload 630; MTO 3 t (6,614 lb).
PERFORMANCE Max cruise 500 km/h (311 mph) at 7,6 km (24,900 ft); range (max load) 800 km (497 miles), (max fuel) 2500 km (1,553 miles); TO 350 m; ldg (reverse pitch) 280.

M-103 Skif High-wing 4/5-seater designed 90, work deferred to 95.

Yamal See under Aviaspetstrans.

M-102 Duet/Saras Light transport, launched 90 as **Delfin** with two pusher AL-34 turbo-

M-55

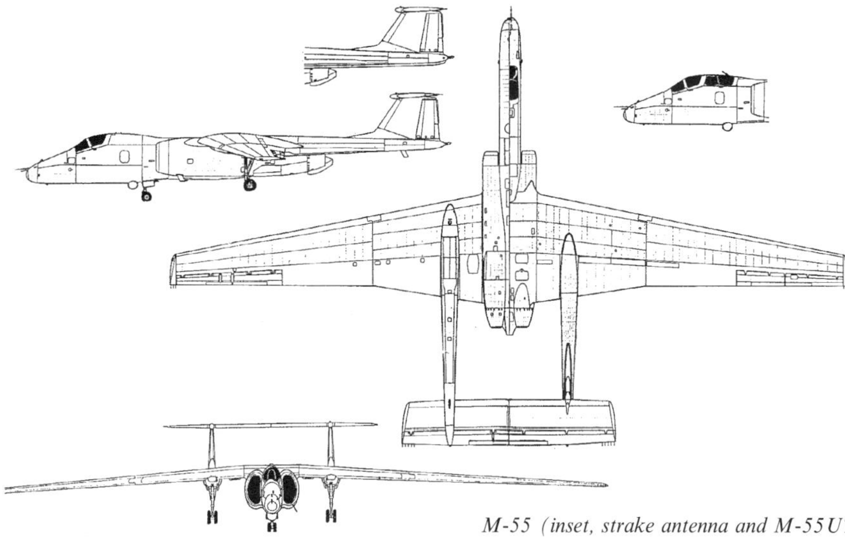
M-55 (inset, strake antenna and M-55UTS)

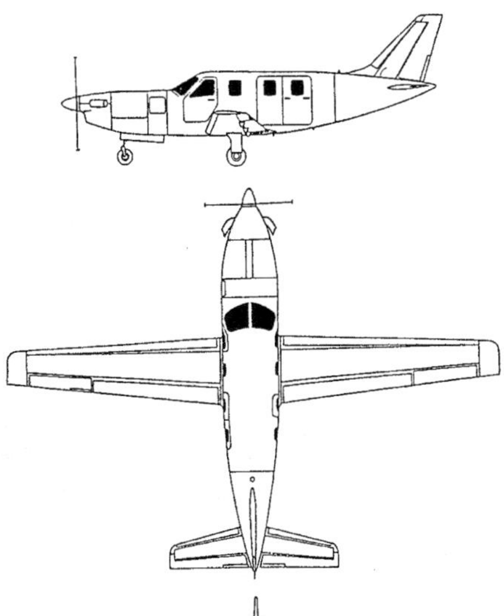
Gzhel

OKB drawing of M-53 supersonic transport project, last signed by V. M. Myasishchyev

Gzehl model

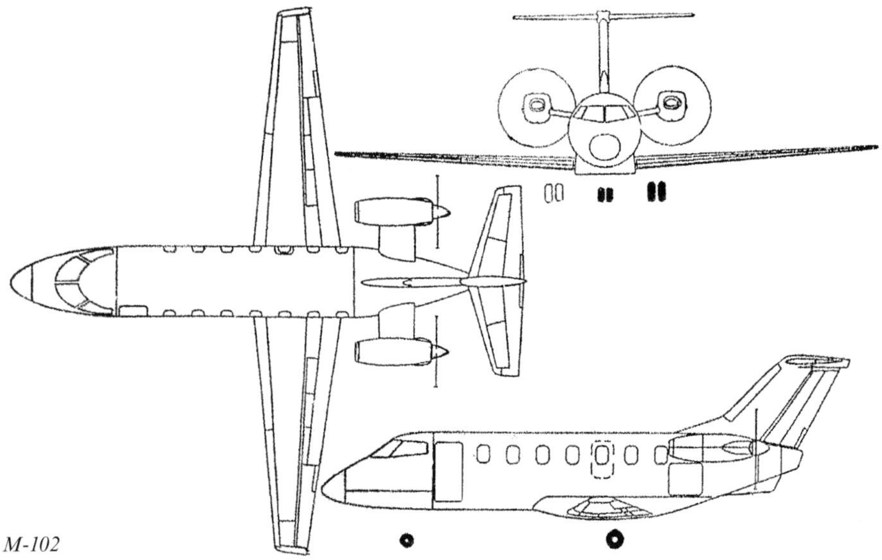

M-102

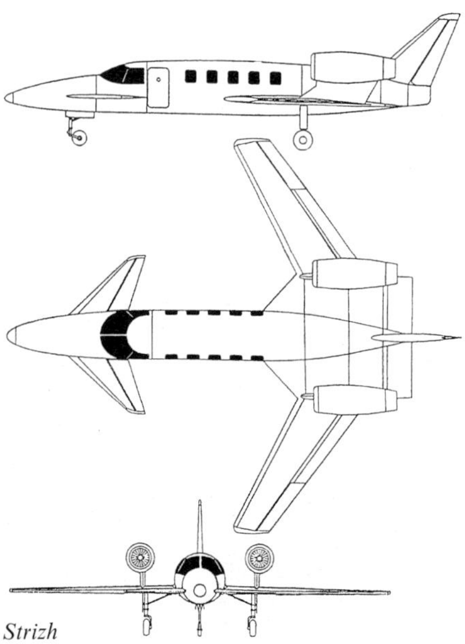

Strizh

props above wing, by 93 engines moved to rear fuselage, driving 6-blade pusher props, tailplane moved to top of swept fin. Cabin 1,77 wide, 1,7 high, dP 0·55bar (8 lb/in²) for up to 14; agreement 93 with NAL of India, Mya fuselage/ldg gear, NAL wings/tail/eng; two prototypes Mya (first Nov 95) and one NAL.

DIMENSIONS span 14,7 m (48 ft 2¾ in); length 14,2 m (46 ft 7 in).

ENGINES Two TVD-20 (export, 850 shp P&WC PT6A-66).

WEIGHTS Max fuel 1,2 t; payload 1,3 t (2,866 lb); MTO 5,7 t (12,566 lb).

PERFORMANCE Max cruise 650 km/h (404 mph); range (LR cruise, 1,3 t) 2400 km (1,491 miles); TO 380 m; ldg 260.

Strizh Projected bizjet, up to 10 pax, forward-swept wing and aft-swept canard, funds sought 93 for prototype. Name = swift.

DIMENSIONS Span (wing) 13,3 m (43 ft 7⅝ in); length 14,0 m (45 ft 11¼ in); wing area 25,0 m² (269 ft²).

ENGINES Two unspec. turbofan 2 t (4,409 lb) each.

WEIGHTS MTO 10 t (22,050 lb).

PERFORMANCE (est) Cruise M 0·95-0·98; range 6000 km (3,730 miles); service ceiling 15,5 m (50,850 ft).

MM-1 Twin-turboprop transport based on German Air-Metal AM-C 111 of 72-75. Design resurrected 89 by Tec Avia inc of Munich/Washington DC, deal struck 93 for final improved design by Myasishchyev and series production from 96 at Smolensk. Wing with six integral tanks for 2500 lit (550 gal), Fowler flaps and flaperons, fuselage interior width 2750 (108 in), dP 5·8 lb/in², rear ramp door 2×2 m (80×80 in), 24 pax plus galley/toilet/wardrobe/attendant or equipped for various roles. Weather radar, IFR with single pilot, manual controls, full deicing. Max cargo 2400 kg (5,291 lb) but announced at Moscow Sept 93 as 3,2 t (7,055 lb); expected price US$5·5 m.

DIMENSIONS Span 19,2 m (62 ft 11⅞ in); length 18,25 m (59 ft 10½ in); wing area 37,7 m² (406 ft²).

ENGINES Two P&WC PT6A-67 rated at 1,657 shp driving 5- or 6-blade props.

WEIGHTS OEW 4,8 t (10,582 lb); MTO 9,2 t (20,282 lb).

PERFORMANCE (Est) Max speed 450 km/h (280 mph); TO 875 m, ldg 620 m.

M-200 Jet trainer begun 91 as UTK-200 to meet UTS reqt; tandem seats, engines under wingroots, adaptive programmable control, factors +8/-3 g.

DIMENSIONS Span 9,4 m (30 ft 10 in); length 10,37 m (34 ft 0¼ in).

ENGINES Two R-35 (DV-2 mod).

WEIGHTS Empty 3805 kg; fuel 1380; max 4700.

PERFORMANCE Max speed at Sl 850 km/h (528 mph); cruise 700 km/h; ceiling 13,1 km (43,000 ft); range 1400 km, max 2200; TO 200; ldg 480.

M-90 Samson Projected giant cargo airlifters as photo with 35 m payload pod; **MGS-6** with six NK-12MV turboprops, span 96 m, MTO 650 t; **MGS-8** with eight, MTO 850 t, payload 400 t.

NIAI

In Sept 30 Leningrad LIIPS formed educational institute, UK GVF (training combine of civil air fleet). In turn this formed NIAI (*Nauchno-Issledovatelskii Aero-Institut*, scientific research aero institute) in 31. This attracted good design engineers, and after appearance of LK-1 it was reorganized in 34 into OKB, O in this case standing for *Osoboye* (personal). Brigades led by Lisichkin, Bedunkovich, Krylov, Domrachyev and, later, Bakshayev. NIAI liquidated 38, but construction of experimental aircraft continued by GVF group at Leningrad to 41. Grigorii Ivanovich Bakshayev was born 18. On leaving school he joined the Leningrad institute of aerial communication (LIIPS), and gained experience in aircraft engineering. His first responsibility was for parts of the ASK amphibian.

LK Bakshayev's first lead assignment was a simple tandem two-seat cabin monoplane, *Leningradskii Komsomolyets* (Leningrad young communist). Mixed construction with welded

MM-1

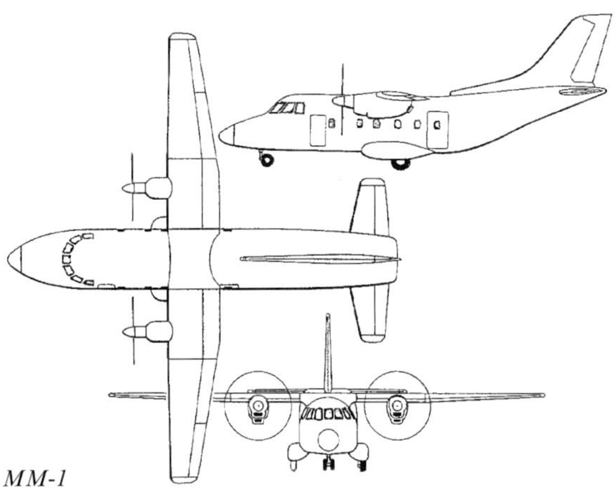

MM-1

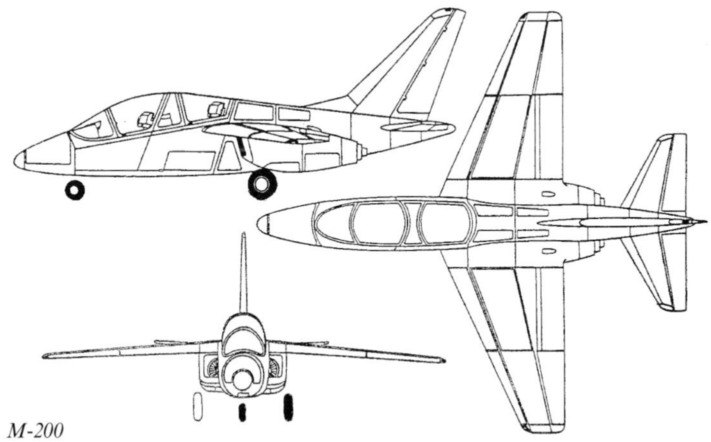

M-200

truss fuselage and two-spar wooden wing with duralumin slats and flaps. Fixed endplates on wings with patented spoiler (interceptor) type ailerons carried on rotating brackets and projecting above the outer wings. Prototype built April to July 36. Following successful flight tests in Aug, it flew to Moscow. One built.
DIMENSIONS Span 10,0 m (32 ft 9¾ in); length 6,5 m (21 ft 4 in); wing area 15,5 m^2 (167 ft^2).
ENGINE M-11 radial.
WEIGHTS Empty 740 kg (1,631 lb); fuel/oil 120 kg (265 lb); max 1100 kg (2,425 lb).
PERFORMANCE Max speed 175 km/h (109 mph); landing 75 km/h (47 mph); range 750 km (470 miles).

MGS-6 model

LIG-8 Development of LK with restressed structure and wider fuselage accommodating five-seat cabin. Test-flown 37 and built in small numbers.
DIMENSIONS Span 10,0 m (32 ft 9¾ in); length 6,6 m (21 ft 8 in); wing area 15,5 m^2 (167 ft^2).
ENGINE One MG-31.
WEIGHTS Empty 1000 kg (2,205 lb); fuel/oil 220 kg (485 lb); max 1600 kg (3,527 lb).
PERFORMANCE Max speed 245 km/h (152 mph); landing 100 km/h (62 mph); range 800 km (497 miles).

RK Also designated **LIG-7**, this research monoplane was Bakshayev's first project at the Leningrad GVF, which he joined in 36. Influenced by the contemporary interest in variable-geometry wings, he sought to prove a different principle from other VG designers. Like the *Matroshka* (set of dolls, each fitting tightly inside the next) he designed a wing around which was a set of telescopic larger wings with which wing area could be reduced for high speed or increased for take-off and landing. The RK (*Razdvizhnoye Krylo*, extending wing) was a simple low-performance tandem-seat cabin machine of mixed construction, with a wire-braced low wing of constant M-6 profile. Around each half-wing were arranged five telescoping ply aerofoils (CAHI-846 profile) normally housed in the fuselage. The observer had a handcrank with which he could pull out the outermost section on each side more than halfway to the tip, where bracing wires were attached. Each section had a span of 460 mm and pulled out the one behind it. Extension or retraction took about 40 seconds. Trials in 37 paved the way for fighter application of RK principle.
DIMENSIONS Span 11,3 m (37 ft 0¾ in); length 7,34 m (24 ft 1 in); wing area variable 16,56-23,85 m^2 (178·25-256·72 ft^2).
ENGINE One M-11.
WEIGHTS Empty 667 kg (1,470 lb); loaded 897 kg (1,978 lb).
PERFORMANCE Max speed 150 km/h (93 mph), landing (small wing) 250 m/18 s/100 km/h, (big) 110 m/9 s/70 km/h.

RK-I Also known as RK-2, this was to be the definitive application of the telescopic wing to a combat aircraft (RK-I from variable-wing fighter). For the first and probably only time it used tandem wings as the supports and guides for telescopic sections which, when extended, multiplied wing area by nearly 2·4. The wings were light alloy and steel, with spot-welded 30KhGSA stainless skins, the rear wing having flaps and ailerons. Fuselage, light-alloy monocoque. There were nine sections of 'large wing' on each side, housed with main gear in fuselage. Large pipes between engine and coolant radia-

LK

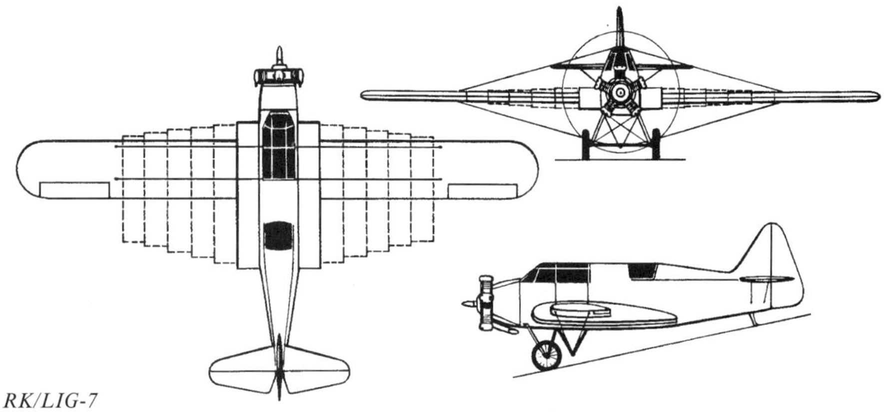
RK/LIG-7

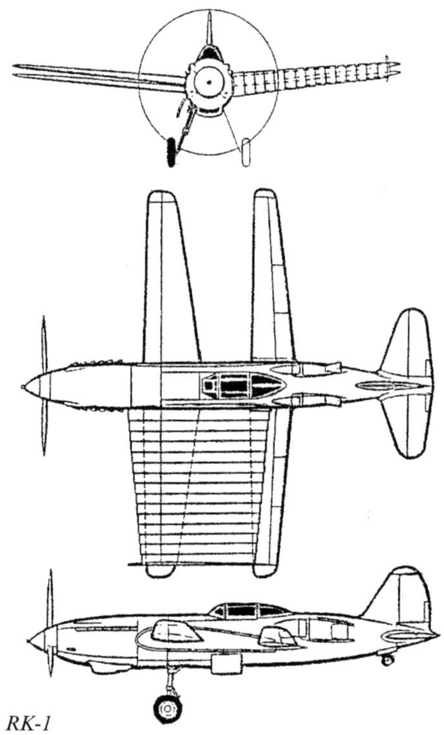
RK-1

tors on each side of rear fuselage for minimum drag. Stalin was enthusiastic and insisted on use of most powerful engine available. Original design of Oct/Dec 38 based on M-105 was replaced by enlarged aircraft with M-106, retaining same armament of two ShVAK and two ShKAS all in nose. In definitive form large wing was extened hydraulically in 14 seconds, first section forming a large endplate at the tip and the rear of each duralumin segment running in tracks ahead of rear-wing ailerons and flaps. Prototype complete early 40, but engine in serious difficulty. Nobody dared to fit any engine other than the type Stalin had suggested, and RK-I languished in erection shop. Bakshayev probably later joined Bartini's 'Special OKB' to work on Project R.

DIMENSIONS Span 8,2 m (26 ft 10¾ in); length 8,8 m (28 ft 10½ in); wing area (high speed) 11,9 m² (128 ft²), (landing) 28,0 m² (301 ft²).
ENGINE One M-106.
WEIGHTS Max loaded 3100 kg (6,834 lb).
PERFORMANCE (Est) max speed 780 km/h (485 mph).

LK-1, NIAI-1 Designated from *Leningradskii Kombinat*, and also called *Fanera-2* (Plywood No 2). Designed by Alexei Ivanovich Lisichkin and Vladimir Fedorovich Rentel. Basically simple all-wood passenger aircraft but with thick inboard wing integrated into lifting fuselage to seek reduced drag. Wing with one main spar, large ailerons, no flaps, spar carried across cabin forming support for two rear passenger seats. Two front seats (pilot on left) with view ahead through sloping Celluloid front windows; rear passengers with three large windows on each side sloping 60° in hinged frames serving as doors. Trad pyramidal main gears and tailskid. Prototype flown May 33, and good behaviour (though obviously pilot had asymmetric view ahead). Autumn 33 A. Ya. Ivanov flew to Moscow where NII trials resulted in order for 20 series examples. Several saw transport service in Arctic with Aeroflot. Some operated on skis, and at least one (**NIAI-1P**) had twin floats. Series aircraft had revised wing root and rear fuselage, and redesigned much taller but shorter-chord vertical tail.

DIMENSIONS Span 12,47 m (40 ft 11 in); length (landplane) 8,87 m (29 ft 1¼ in); wing area 27,6 m² (297 ft²).
ENGINE One M-11.

RK-2 (RK-I)

LK-1 prototype

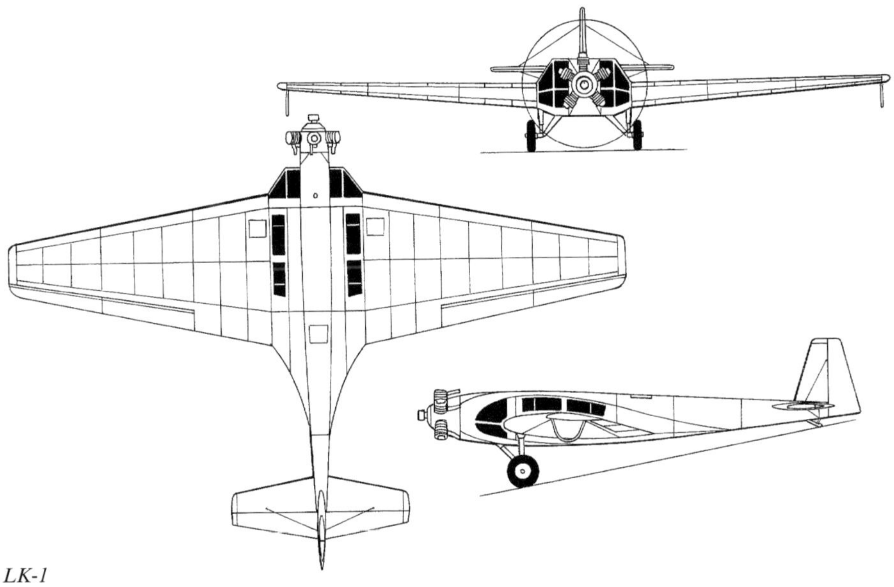

LK-1

LK-1 series

WEIGHTS Empty (prototype) 730 kg, (production) 746 kg (1,645 lb); fuel/oil (production) 170 kg; loaded (prototype) 1035 kg, (production) 1160 kg (2,557 lb).
PERFORMANCE Max speed (prototype) 157 km/h, (production) 154 km/h (96 mph); climb 10 min to 1 km; service ceiling (prototype) 3950 m, (production) 3370 m (11,000 ft); range (prototype) 770 km, (production) 850 km (528 miles); take-off 200 m/7 s; landing 120 m/12 s/65 km/h (40 mph).

LK-4, NIAI-4 Designed by Anatolii Georgievich Bedunkovich, engineer-colonel who wrote many instructional books and among other designs produced AB-55 engine (55 hp, 30) using Anzani cylinders from 10. LK-4 was neat trainer which, as explained, was 'four aircraft in one'. Original machine, flown early 34, had spatted main wheels. Wooden sesquiplane with ply leading edge and forward fuselage, fabric elsewhere. Two tandem cockpits with dual controls, slotted ailerons. Engine in Townend-ring cowl. Later spats removed and new leading edge to upper wing with large plywood slats. Flown in four configurations. **LK-4-I**, slatted upper wing with twin parallel bracing struts each side, cantilever lower wing (upper wing 13 m^2, lower 7 m^2). **LK-4-II**, different upper wing leading edge without slats, upper wing moved aft and lower wing forward. Flew faster but landed faster, spun more easily and simpler to perform aerobatics. **LK-4-III**, parasol monoplane with lower wing removed, faster still but required harsher control. **LK-4-IV**, low-wing monoplane with original lower wing replaced by upper, with twin bracing struts to top of fuselage. Fastest of all, but control described as 'more complicated'. Simple aircraft to make, and propeller and instruments standard U-2. NII testing late 34 resulted in recommendation for start of series production. This never materialized, though in different configurations LK-4 successful in various light-aircraft competitions 1934-36.
DIMENSIONS Span 9,0 m (29 ft 6⅓ in) (lower wing of LK-4-I, LK-4-II) 5,7 m (18 ft 8½ in); length 7,07 m (23 ft 2⅓ in); wing area (LK-4-I, LK-4-II) 20,2 m^2 (217 ft^2), (LK-4-III, LK-4-IV) 13,0 m^2 (140 ft^2).
ENGINE One M-11.
WEIGHTS Empty (I) 565 kg (1,246 lb), (II) 558, (III) 517, (IV) 510; fuel/oil 45 kg; loaded (I) 790 kg (1,742 lb), (II) 783 kg, (III) 742 kg, (IV) 735 kg.
PERFORMANCE Max speed (I) 157 km/h (97·5 mph), (II) 168 km/h, (III) 177 km/h, (IV) 180 km/h; climb to 1 km (I, II) 6 min, (III) 5 min, (IV) 4·5 min; service ceiling (I, II) 3300 M, (III) 4500 m, (IV) 3815 m; range (I) 250 km, (II) 280 km, (III, IV) 300 km; take-off (I) 70 m, (II) 80, (III) 90, (IV) 130; landing (I) 100 m/60 km/h, (II) 150/70, (III) 170/80, (IV) 220/90.

P-3, LIG-5 Success of LK-4 prompted this faster and more powerful machine in LIG (Leningrad Institute GVF) series. Wooden, mainly ply covering fully cowled engine. Three basic configurations, **P-3DP**, **P-3OB** and **P-3ON**. **DP** (*Dvukhmestnyi Polutoplan*), two-seat sesquiplane for training pilot/observer of R-5 and similar aircraft. Full combat equipment including fixed and free ShKAS, radio and recon camera. **OB** (*Odnomestnyi Biplan*), single-seat biplane for training pilots for I-15 and other biplane fighters. Upper centre-section removed together with side cabane struts, interplane slanting struts and bracing wires, leaving diagonal struts to upper plane only. Rear cockpit faired over. **ON** (*Odnomestnyi Nizkoplan*), single-seat low-wing monoplane for training pilots of I-16 and similar low-wing fighters. Cantilever low wing only, and smaller horizontal tail. Aircraft did well in testing from late May 36, and completed factory testing Feb 37. To Moscow for NII tests, with expectation of major production; crashed by pilot error, not repairable.
DIMENSIONS (Span (DP) 10,6 m (34 ft 9⅓ in), (others) 9·0 m (29 ft 6⅓ in); length 7,5 m (24 ft 7¼ in); wing area (DP) 30 m^2 (323 ft^2), (OB) 26,0 m^2 (280 ft^2), (ON) 14,0 m^2 (151 ft^2).
ENGINE One MG-31F.
WEIGHTS Empty (DP) 1100 kg (2,425 lb), (OB) 960 kg, (ON) 850 kg; fuel/oil (DP) 220 kg, (others) 110 kg; loaded (DP) 1560 kg (3,439 lb), (OB) 1260 kg, (ON) 1150 kg.

LK-4

LEM-3

PERFORMANCE Max speed (DP) 210 km/h (130 mph), (OB) 220 km/h, (ON) 280 km/h (174 mph); range (DP) 500 km (373 miles), (OB) 280 km, (ON) 300 km; landing speed (DP) 80 km/h, (OB) 75 km/h, (ON) 100 km/h.

LEM-3, LIG-6 Another of LIG series, this also had designation in LEM series (initials coined from Lev Pavlovich Malinovskii, director of NTU-GVF in 30s) of *motor planer* (powered gliders). No NIAI number known. Designed by Yuri Vladimirovich Domrachyev in partnership with Leonid Sergeyevich Vild'grub. Remarkable use of horsepower: Soviet sources agree accommodation in enclosed cabin for two pilots side-by-side and eight passengers in four double seats behind. Access by hinged glazed roof panels with small windows. In addition, four cargo compartments in wing. All-wooden construction, with fabric-covered control surfaces. Wing braced by struts from main spar to main gears with horizontal strut between legs (drawings show trousered gears, not flown). Townend-ring engine cowl, U-2 propeller. First flight 36, completed factory tests early 37 and flew to Moscow for NII testing but suffered forced landing on arrival.

DIMENSIONS Span 26,0 m (85 ft 3⅔ in); length 13,3 m (43 ft 7⅔ in); wing area 57,0 m^2 (613 ft^2).

ENGINE One M-11.

WEIGHTS Empty 1050 kg (2,315 lb); fuel/oil 170 kg; loaded 2 t (4,409 lb), (overload) 2,240 kg (4,938 lb).

PERFORMANCE Max speed 122 km/h (76 mph); cruise 100 km/h (62 mph); ceiling 2,2 km; range 800 km (497 miles); landing speed 55 km/h (34 mph).

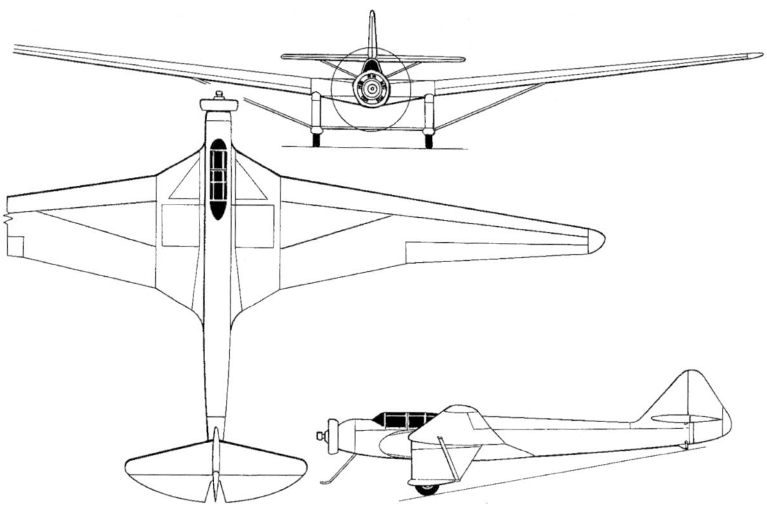

LEM-3

SKh-1, LIG-10 Another of LIG series, this was ancestor of An-2 and designed by Bedunkovich for same wide spread of duties. Designation *Syelskolkhozyaistyennyi*-1, agricultural-1. Two-bay biplane with equal-size wooden/fabric two-spar wings, R-II section 12%, four slotted ailerons and slotted flaps, hinged to centre section and folding to rear. Fuselage welded from mild-steel tube, fabric covering. Tail wood and fabric, with pilot control of tailplane incidence. Large cabin with pilot high up for good view through all-round Celluloid windows. Dural fuel tanks filling upper centre section outboard of fuselage. Accommodation for eight persons (front left seat for pilot, or side-by-side dual control); casevac with four stretchers, 690 kg cargo, or 600 kg ag-chemicals with ventral spreader for solids (no record of liquid spraying). Prototype flown 37; prolonged testing in many roles including full season seed-sowing and casevac role throughout Finnish war 1939-40. Early 41 decision to produce in series, but thwarted by evacuation of chosen factory.

DIMENSIONS Span 12,8 m (42 ft 0 in), (folded 5,85 m); length 10,7 m (35 ft 1¼ in); wing area 41,17 m^2 (443 ft^2).

ENGINE One MG-31F.

WEIGHTS Empty 1215 kg (2,679 lb); fuel/oil 175 kg (later 275 kg, with reduced payload); loaded 2150 kg (4,740 lb).

PERFORMANCE Max speed 182 km/h (113 mph); climb 5·6 min to 1 km; service ceiling 3,8 km (12,500 ft); range 420 km (long-range tanks, much greater); take-off 210 m/19 s; landing 180 m/18 s/65 km/h (40 mph).

Bedunkovich 1947 Col Bedunkovich never had his own OKB, but recently a 3-view has been found of a 47 project for a heavily armed Shturmovik, noted for its twin-boom layout making possible a rear turret. Guns: 2×37 mm, 1×20 mm, 5×7,62 mm. Not explained is the title: *Odnomotornyi* (single-engined).

Nikitin

Vasilii Vasilyevich Nikitin was born 01 and studied at an institute of architecture. No formal engineering training yet learned how to design aircraft and, again without instruction, how to fly them. Began designing details under D. P. Grigorovich 22, assisting N. N. Polikarpov 25-29 and in 30-36 at CCB. Designed and built his own aircraft 33-39, whilst continuing to assist others. At OKB-30 in 38-40. From 41 manager of repair factories, converted captured Ju 52/3 m into flying lab and rebuilt several

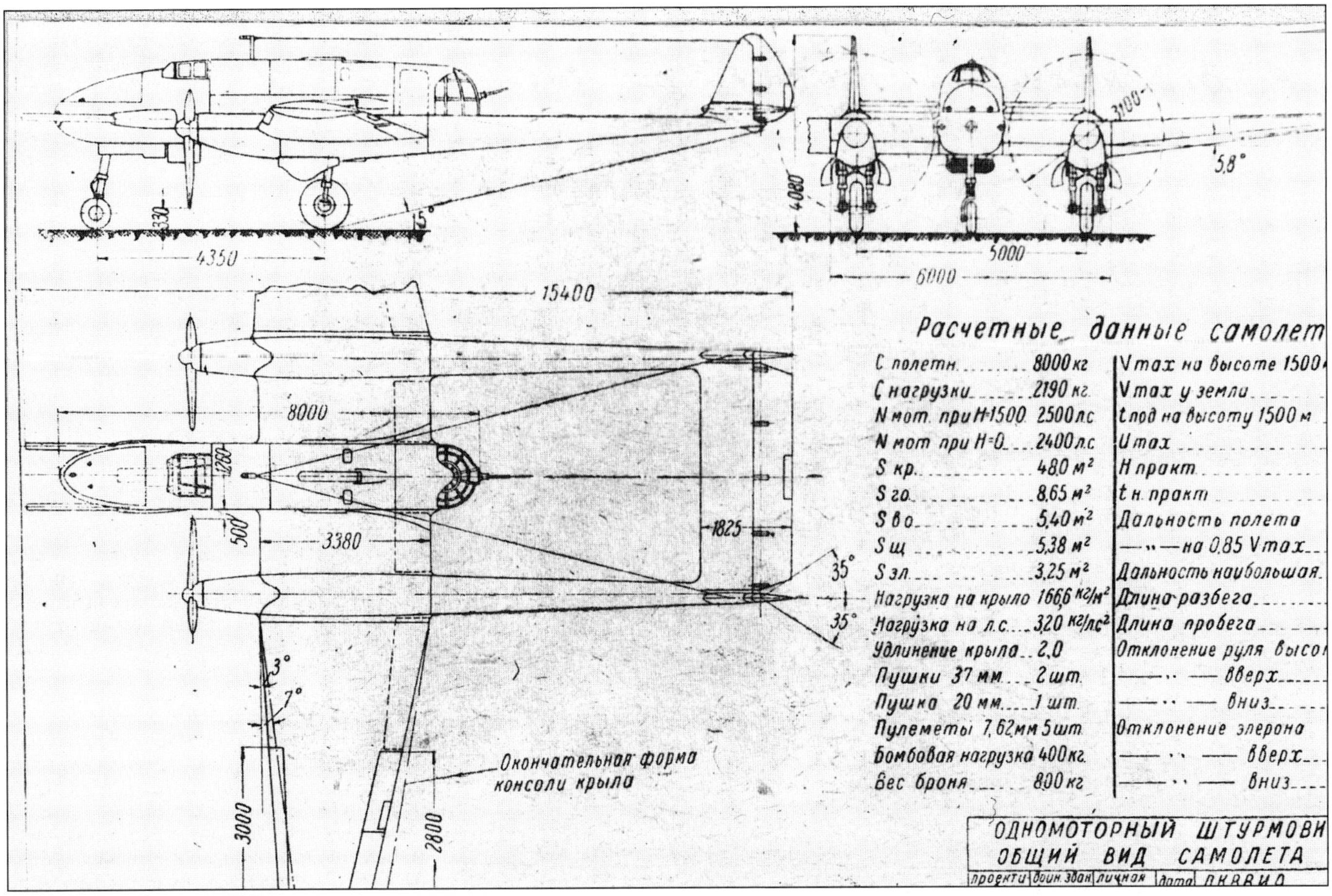

OSh Shturmovik project 1940

LIG-10

NV-1

Douglas Boston/A-20 into six-passenger staff transports. Post-war deputy to N. I. Kamov, dying in this post 55.

NV-1 Single-seat sporting aircraft inspired by US racers. Wooden wing with bracing struts to fuselage and to trousered main gears. Fuselage welded from mild-steel tube, duralumin tail, fabric covering throughout. Control by push/pull tubes. Unsprung main legs except for low-pressure tyres. First flight by V. P. Chkalov Sept 33. Nine flights but difficult to fly and regarded as little practical use.
DIMENSIONS Span 6,4 m (21 ft 0 in); length 4,25 m (13 ft 11⅓ in); wing area 6,85 m² (73·7 ft²).
ENGINE One M-11.
WEIGHTS Empty 350 kg (772 lb); fuel/oil 70 kg; loaded 510 kg (1,124 lb).
PERFORMANCE Max speed 232 km/h (144 mph) at SL; service ceiling 4,8 km (15,750 ft); range 850 km (528 miles); landing speed 95 km/h.

NV-2 Single-seat sporting and training aircraft, again with M-11 in full-length NACA cowl but still driving U-2 propeller. Wooden construction with monocoque fuselage with wall-thickness 2 mm made from glued layers of 0,5 mm birch *shpon*. Wing with 2 mm ply back to rear spar, fabric covering overall. Single-leg main gears retracted backwards into underwing fairings. Built 35 with Osoaviakhim funds at OKB-30 and proved to have excellent flying qualities, partly stemming from superb finish.
DIMENSIONS Span 8,6 m (28 ft 2¼ in); length 6,15 m (20 ft 2 in); wing area 11 m² (118 ft²).
ENGINE One M-11.
WEIGHTS Empty 385 kg (849 lb); fuel/oil 250+25 kg; loaded 750 kg (1,653 lb).
PERFORMANCE Max speed 230 km/h (143 mph) at SL; service ceiling 5,8 km (19,000 ft); max endurance 10 h; landing speed 75 km/h.

NV-2bis, UTI-5 Original NV-2bis built at OKB-30 in 38, essentially an NV-2 with more powerful MG-11 and slight local strengthening.

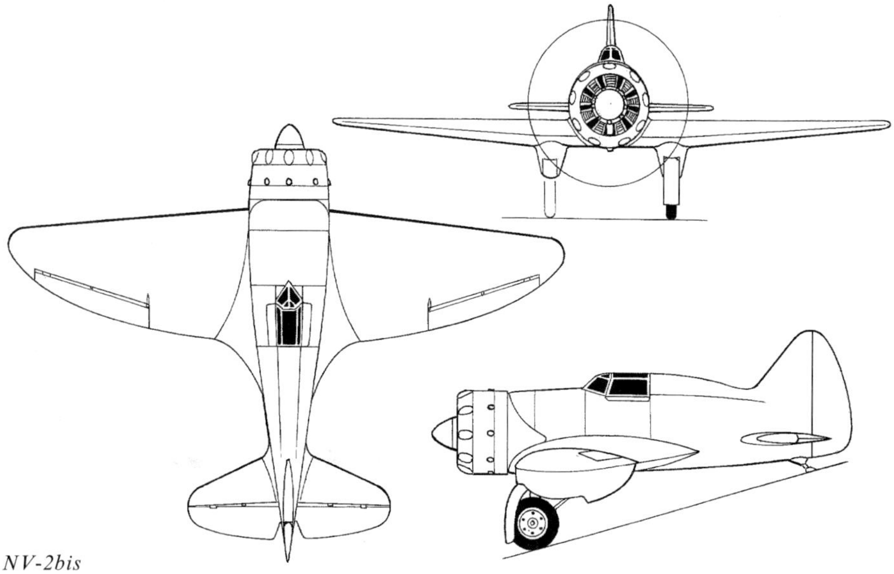

NV-2bis

NV-2bis

Successfully tested; small series of ten ordered by VVS but abandoned when plant assigned to produce UT-2. Nikitin then built further prototype called UTI-5 (NV-2bis/MG-31) to order of UVVS as training fighter. This was an NV-2bis with further increase in power, restressed airframe and other changes, plus synchronized ShKAS with 250 rounds. Hinged glazed canopy, fully retracting main gears and metal propeller with spinner. Flown early 39 and tested by more than 40 pilots including representatives of VVS, NII, GVF and Osoaviakhim. Excellent aircraft, much better finished than I-16; 200 (Shavrov, 20) ordered by VVS but never produced.

DIMENSIONS Span 8,0 m (26 ft 3 in); length 6,15 m (20 ft 2 in), (UTI-5) 6,3 m (20 ft 8 in); wing area 11,0 m^2 (118 ft^2).

ENGINE One MG-11, (UTI-5) one MG-31.

WEIGHTS Empty 435 kg (959 lb), (UTI) 560 kg (1,235 lb); fuel/oil 250+25 (UTI, +30) kg; loaded 800 kg (1,764 lb), (UTI) 950 kg (2,094 lb).

PERFORMANCE Max speed 260 km/h (162 mph), (UTI) 350 km/h (217 mph); service ceiling 7 km (23,000 ft), (UTI) 8 km (26,250 ft); endurance 7 h, (UTI) 4·5 h; range 800 km (497 miles), (UTI) not known; landing speed (both) 75 km/h.

NV-4 Biplane amphibian with two seats in tandem enclosed cockpits. Fuselage circular-section monocoque glued from layers of 0,5 mm *shpon*, with ply and dural panels at front, and integral wooden float carrying main gears retracted by hand-crank. Two-spar wooden wings with ailerons (upper only) and landing flaps (lower only). Wooden stabilizing floats, without step, beneath lower planes in line with N interplane struts. Equipped for night flying. Completed 36 and flown by Nikitin, but had to be shelved through pressure of other work.

NV-4

DIMENSIONS Span 10,5 m (34 ft 5⅓ in); length 8,7 m (28 ft 6½ in); wing area 28,5 m^2 (307 ft^2).
ENGINE One M-11.
WEIGHTS Empty 825 kg (1,819 lb); fuel/oil 90 + 15 kg; loaded 1090 kg (2,403 lb).
PERFORMANCE Max speed 160 km/h (99·5 mph) at SL; ceiling 3,5 km (11,500 ft); endurance 4 h; landing speed 65 km/h.

MU-4 Small training amphibian designed by N. G. Mikhelson working with Nikitin at OKB-30. Mainly wooden, but dural struts and fabric on rear of wings and on control surfaces. Single-bay biplane, side-by-side dual-control cockpit with glazed hinged canopy, tractor engine on upper wing, and air/oil shock struts on main gears manually retracted along sides of hull. First MU-4 completed 36 and flew well, but after prolonged testing crashed because of failure of glued joint.
DIMENSIONS Span 12,0 m (39 ft 4½ in); length not known; wing area 33,0 m^2 (355 ft^2).
ENGINE One MG-11F.
WEIGHTS Empty 900 kg (1,984 lb); fuel/oil 80+ 20 kg; loaded 1200 kg (2,646 lb).
PERFORMANCE Max speed 168 km/h (104 mph); no other data.

PSN Unusual projects of 37 were two species of *Planer Spetsial'no Naznachenaya*, glider for special purposes. Project started by S. F. Valk 33 who proposed glider anti-ship bomb with IR guidance; gradually embraced six labs and depts, by 35 working on **DPT** (long-range glider torpedo), **LTDD** (long-range flying torpedo) and **BMP** (towed mine glider). Work led to **PSN-1**, all assigned to OKB-21 at GAZ-23 Leningrad, designed by Nikitin and Mikhelson (though chief designer also named as Arharov), to be carried to target under TB-3 or TB-7 and there released to deliver bombs or torpedo. Small single-seat flying boat, pilot to evaluate *Kvant* IR guidance/autopilot for operational pilotless version, 10 built, several flown. Span 8 m (26 ft 3 in); weight empty 970 kg (2,138 lb), payload 1 t. Later *Izumrud* radio guidance to be carried by I-16 for a high-speed combat DPT. Totally different **PSN-2** designed at Nikitin OKB by M. M. Yefimov, also single-seat monoplane, twin floats each with fin; carried under parent June 40 and also flown on tow, again with pilot to monitor *Kvant* IR guidance and autopilot. Span 7 m (22 ft 11½ in), length 7,98 m (26 ft 2⅛ in); planned mission (pilotless) 40 km range at 700 km/h (435 mph). All work stopped 19 July 40.

MU-5 Development of MU-4 with MV-6 (Renault) inline engine of 222 hp. Construction abandoned 38.

NV-5, U-5 Simple tandem dual biplane, trainer, smaller in dimensions than U-2, with I-type single interplane struts and small (500 mm) semi-balloon tyres. Various other refinements, such as rubber shock-isolating engine mounts. Mixed construction, fuselage truss welded 30KhGSA steel tube with ply covering, wooden wings and fabric-covered control surfaces. Design prompted by Osoaviakhim/Aviavnito competition in late 34 for safe aeroplane. First NV-5 flown early 37 and won first prize. This machine had MV-4 engine driving an efficient Kuznetsov propeller, plain U-2 type ailerons

MU-4

PSN-1 No 6

PSN-2

NV-5/MV-4

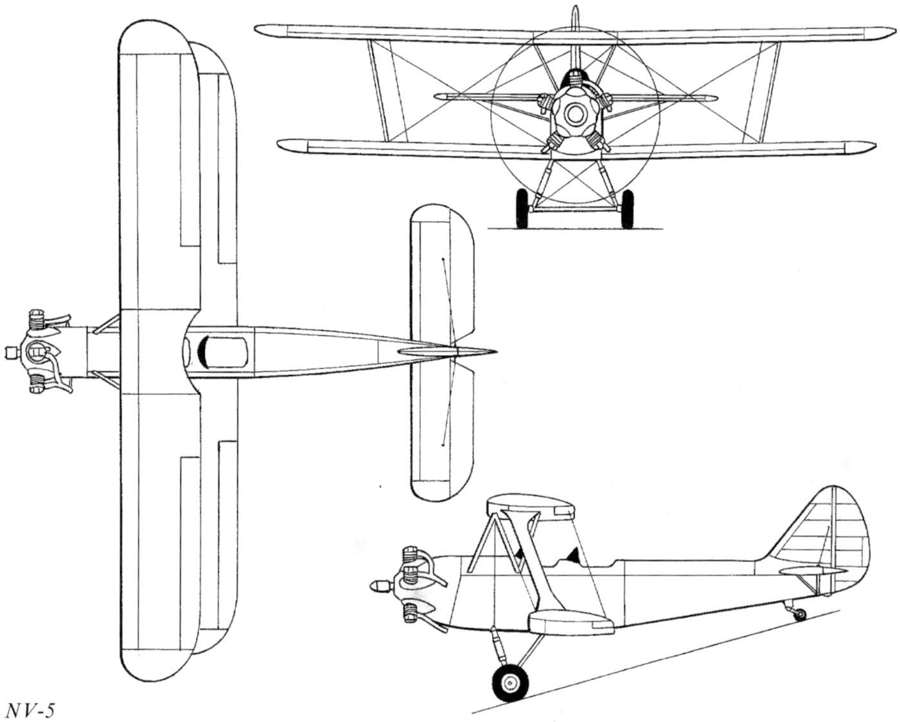

NV-5

NV-5bis/M-11F

and tail skid. Subsequently fitted with slotted ailerons and tailwheel. Made 250 flights in hands of 15 pilots, with wheel and ski landing gear. In 38 first **NV-5bis** flown, with regular production engine. Even better than NV-5 and drawings prepared for production for Osoaviakhim and, later, UVVS. In late 38 **U-5** produced as prototype and four series examples. M-11 and M-11G engines, simplified fuselage, wing with flat underside, control-surface frames of dural, control system by push/pull rods and aluminium propeller. Prolonged testing by 60 pilots, all satisfied. In 39 one U-5 fitted with ShKAS on lower right wing outside propeller disc and four RS-82 rockets under lower wings. Later in 39 one prototype and four series **U-5bis** built with minor strengthening and MG-11F engine. UVVS tested 12 examples of U-5 family.

DIMENSIONS Span (NV-5) 9,82 m, (U-5 family) 9,84 m (32 ft 3⅓ in); length (NV-5) 7,7 m, (U-5) 7,62 m (25 ft 0 in); wing area (NV-5) 25 m^2, (U-5) 25,53 m^2 (275 ft^2).

ENGINE (NV-5) one MV-4, (5bis) MG-11F, (U-5) M-11 or M-11G, (U-5bis) MG-11F.

WEIGHTS Empty (NV-5) 612 kg, (U-5) 700/711 kg, (U-5bis) 773 kg (1,704 lb); fuel/oil (all) 75+12 kg; loaded (NV-5) 850 kg (1,874 lb), (5bis and U-5) not known, (U-5, M-11G) 974 kg, (U-5bis) 1036 kg (2,284 lb).

PERFORMANCE Max speed (NV-5) 202 km/h, (5bis) 220 km/h, (U-5) 170 km/h, (M-11G) 181 km/h, (U-5bis) 205 km/h (127 mph); climb to 1 km, (NV-5, 5bis) 3 min, (U-5) 5 min, (U-5bis) 3·8 min; service ceiling (NV-5) 6 km, (U-5) 3,75 km, (U-5bis) 4,5 km; endurance (NV-5) 5·5 h, (5bis) 4 h, (U-5) 3 h, (U-5bis), 2·5 h; takeoff (NV-5) 120 m, (U-5) 70 m/9 s; (U-5bis) 70 m/7 s; landing (NV-5) 110 m/60 km/h, (U-5) 65 m/10 s/60 km/h, (U-5bis) 120 m/10 s/70 km/h.

LSh, U-5/MG-31

Though a U-5 derivative, this variant sufficiently different for separate treatment. Built 42 at request of Moscow defence HQ in MVO frontline repair shops where Nikitin was chief of design and technology. Much more powerful, structurally strengthened, and new upper wing cannibalized from I-153 but with roots slightly shortened to preserve original span. Large glazed cockpit with pilot in bucket seat and longitudinally arranged plank (*sic*) on which two passengers sat astride at rear. Tough aluminium cockpit able to take weapons and heavy supplies without damage. Outstanding aircraft with STOL and good range and speed. No available capacity for production, but sole LSh (*Legkii Shtabno*, light staff aircraft) became famous from Leningrad to Caucasus, making over 600 operational missions often over ground battles and frequently riddled with holes.

DIMENSIONS Span 9,84 m (32 ft 3⅓ in); length 7,75 m (25 ft 5¼ in); wing area 25,53 m^2 (275 ft^2).

ENGINE One MG-31F.

WEIGHTS Empty 880 kg (1,940 lb); fuel/oil 250+30 kg; loaded 1400 kg (3,086 lb).

PERFORMANCE Max speed 272 km/h (169 mph); cruise 240 km/h (149 mph); range with reserves 1000 km (620 miles) in 4·5 h; take-off 40 m/6 s; landing 130 m/11 s/71 km/h.

NV-6, UTI-6 Last of Nikitin's pre-war sporting machines, this nimble biplane was tailored

LSh

NV-6

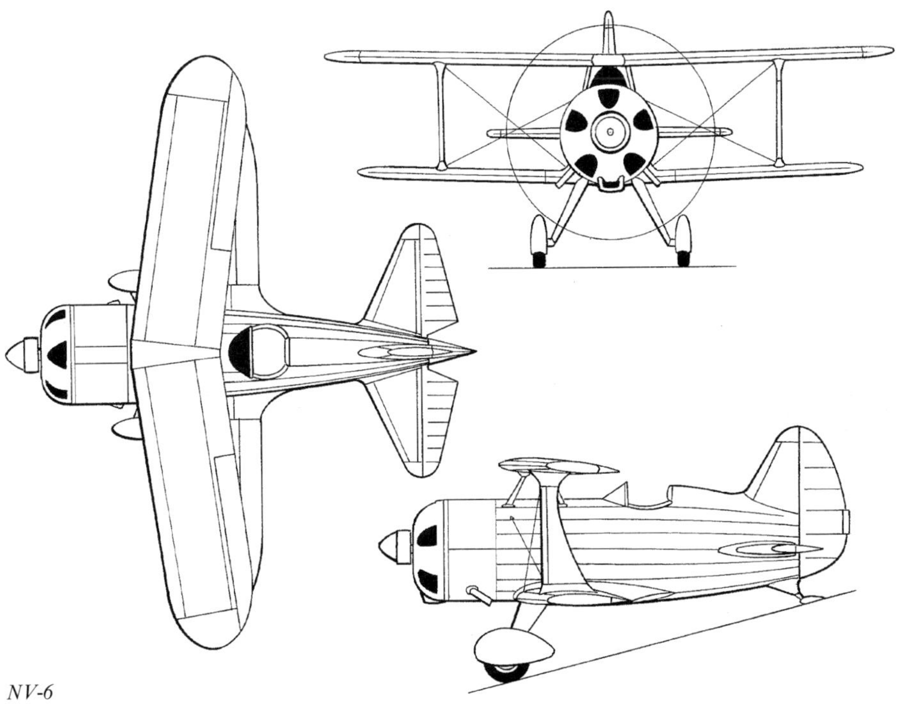

NV-6

for aerobatics, then an unusual design objective (39). Fuselage truss of welded KhMA steel tube taken from NV-1, together with complete wooden lower wing and dural/fabric tail. New upper wing of slightly greater span and 8° sweepback on leading edge, single large-chord-streamlined interplane struts. Cantilever main legs with spatted balloon-tyred wheels. Engine with M. A. Kossov's special carburettor for sustained inverted flight. No engine available until Dec 40, when Nikitin and V. V. Shyevchyenko carried out test flying with ski gear. War prevented further work on this attractive machine.

DIMENSIONS Span 7,0 m (22 ft 11½ in); length 5,8 m (19 ft 0⅓ in); wing area 14 m² (151 ft²).

ENGINE One MG-11F.

WEIGHTS Empty 560 kg (1,235 lb); fuel/oil 80+20 kg; loaded 750 kg (1,653 lb).

PERFORMANCE Max speed 270 km/h (168 mph); service ceiling 4,5 km (14,750 ft); endurance 2·5 h; take-off 50 m; landing 170 m/ 75 km/h.

IS-1 Soviet designers such as Bakshayev and (working in France) I. Makhonine were pioneers of polymporphic (variable-shape) aircraft. In 38 Nikitin's test pilot, Vladimir Vasilyevich Shyevchenko, began to investigate practicability of making a biplane which, after taking off from short field, could retract lower wing and turn into fast monoplane fighter. Nikitin assisted, and in 39 Shyevchyenko built working model at MAT. Crucial factor was relative insensitivity of underside of aerofoil to cavity big enough to receive smaller wing retracted into it from below. Effect further reduced by fact that almost half lower wing could fold into recess in side of fuselage, reducing disruption of upper plane in biplane regime. In mid-39 OKB-30 was organized for design task, with NII sponsorship, engineering strength gradually growing to reach 60. IS-1 (*Istrebitel Skladnoi*, folding fighter) based on I-153 in size, power and style, but with fuselage truss, wing spar booms and basic structure of lower inner wing and landing gears welded from 30KhGSA Cr-Mo steel tube. Rest of structure D16 duralumin, with flush-riveted dural skin except fabric control surfaces, driven by push/pull rods. Main landing gears attached to outer ends of inner section of folding lower wing, with single leg in trunnion on third rib and dural sheet fairing over leg and wheel. After take-off pilot selected single retraction lever to 'chassis up' position. Pneumatic retraction of main gears inwards, legs and front half wheels (700×150 mm tyres) lying ahead of recesses forming inboard lower leading edges. Selection of 'wing fold' position could be made at any time; in theory IS-1 could fight as biplane, gaining in turn radius but slower than monoplane. To fold wing, further pneumatic actuator mounted vertically near CG pushed upwards to break lower-wing bracing strut on each side and pull wings inwards and upwards. Instead of allowing lower wingtips to scrape along underside of upper wing, linkage held lower-wing outer section horizontal, this section being housed in underside of gull-type upper plane. On extension, actuator had to push lower wings down despite lift. Armament four ShKAS with 1000 rounds in inboard part of upper wing. Fuel in centre fuselage. Fabric-covered tail, fixed tailwheel, open cockpit with deep screen and down-folding side door.

Nikitin-Shyevchenko IS-1

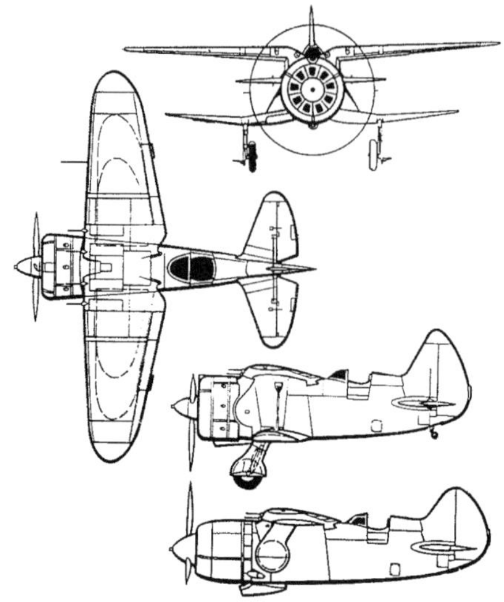

IS-1; lower side view IS-2

Imported three-blade 2,8 m Hamilton propeller. First flight 6 Nov 40, on wheeled landing gear (skis would have precluded wing fold). Satisfactory, but by this time performance poor in comparison with monoplane fighters. Completed factory tests, and G. M. Shiyanov tested IS-1 at LII.

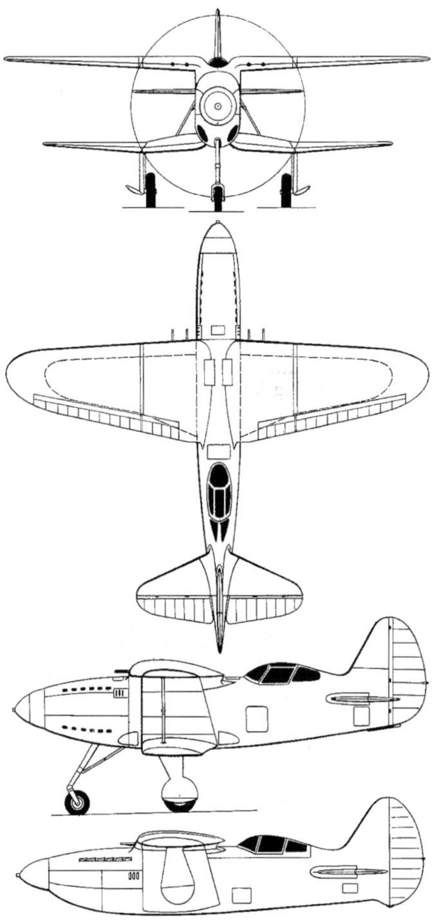

IS-4 (lower side view, as built with AM-37)

DIMENSIONS Span 8,6 m (28 ft 2½ in), (lower, extended) 6,72 m; length 6,79 m (22 ft 3⅓ in); wing area 20,83 m^2 (224 ft^2), (lower 7,83 m^2, upper 13,0 m^2).
ENGINE One M-63.
WEIGHTS Empty 1400 kg (3,086 lb); loaded 2300 kg (5,070 lb).
PERFORMANCE Max speed 453 km/h (281 mph); climb to 5 km, 8·2 min; service ceiling 8,8 km (28,870 ft); range 600 km (373 miles); landing speed 115 km/h.

IS-2 Though Nikitin had reached conclusion IS was not worth while, construction of second airframe had been put in hand parallel with first, and this was completed with various improvements. Engine was changed for M-88 of reduced diameter in long-chord cowl with gills, driving VISh-61 series propeller. Outer ShKAS replaced by two BS. Revised vertical tail of almost circular shape. Retractable tailwheel, and possibly other modifications. First flight early 41, testing discontinued after invasion.

DIMENSIONS Same wings, but length 7,36 m (24 ft 1¾ in).
ENGINE One M-88.
WEIGHTS Empty not recorded; loaded c2810 kg (6,195 lb).
PERFORMANCE (est) Max speed 468 km/h at SL, 588 km/h (365 mph) at 6,7 km; service ceiling 11 km (36,000 ft); endurance 1·2 h.

IS-4 Last two designs, IS-3 and -4 never built, but latter is on record. Natural next stage with more power (M-120 chosen, see drawing, but finally settled for available AM-37), tricycle landing gear and enclosed cockpit. Two BS, two ShKAS. Abandoned 41.
DIMENSIONS As IS-2 except considerably longer, and lower wing same area but extended span to 7,1 m (23 ft 3½ in).
ENGINE One AM-37.
WEIGHTS Loaded (est) 2900 kg (6,393 lb).
PERFORMANCE (est) Max speed 720 km/h (447 mph), (as biplane) 436 km/h (271 mph); service ceiling 12,5 km (41,000 ft); landing 107 km/h.

IS-2

OKB-1

In 46 large numbers of 'potentially useful' German aircraft engineers were transported with their families to prepared secure compound at Podberezye, Oskov Ob., 120 km E of Moscow, where they were housed in quite comfortable 4-storey blocks. Single large office block housed design teams OKB-1 and OKB-2 (q.v.), both officially constituted 22 Oct 46, former headed by P. N. Obrudov, but design leader Dipl-Ing Brunolf Baade. Assigned task: continue devt of former German Junkers EF 126 crude pulsejet 'emergency fighter' and EF 131 jet bomber. Former soon dropped. From Sept 48 all work of both bureaux coordinated by S. M. Alekseyev (q.v.), whose own OKB had been closed.

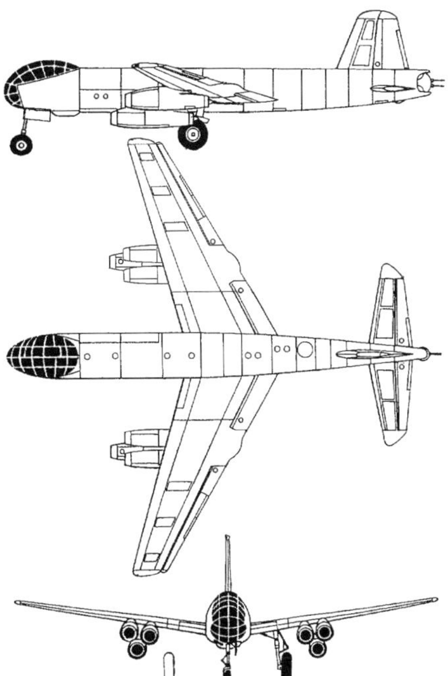

131

Type 140 On 16 Aug 44 Junkers flew a crude prototype, Ju 287 V1. Though powered by four turbojets, it was intended to test at low speeds radical concept of an FSW (forward-swept wing). Junkers planned derived aircraft intended to serve as high-speed bombers, notably EF 131. Improved Ju 287 V2 was frantically built, but captured by Russian troops before first flight. It was taken to Russia, but interest centred on much superior EF 131. This was actually constructed (obviously to very tight timescale) and tested at LII, initially by Flugkapitän Dülgen, in 47, but regarded by captors as inferior to Il-22. Late 47 decision to rebuild as **140** with two VK-1 under direction of G. N. Nazarov abandoned, and sole EF 131 languished derelict until 49. Alekseyev assigned OKB-1 to design completely new and larger bomber, initially given Junkers number EF 150, soon changed to Soviet Type 150. In 53 OKB-1, now redundant, moved to nearby Kimri; later 53 those wishing to leave (incl Baade) were allowed to go to DDR (E Germany), where formed VEB Dresden.

DIMENSIONS (287 V2) Span 20,11 m; length 18,6 m; (131) span 17,88 m; length 17,8 m.
ENGINES (287 V2) Six BMW 003A (RD-20); (131) six Jumo 004B-2 (RD-10).
WEIGHTS (287 V2) Empty 11,93 t; max 21,52 t; (131) empty 11,9; fuel 8680 lit; max 23.
PERFORMANCE (Est. 287 V2) Max speed (no bombs) 819 km/h at SL, 865 km/h (538 mph) at 5 km; service ceiling 12 km; range at 792 km/h with max (4 t) bombload 1585 km; (131) max speed 970 km/h (603 mph) at 9 km; SC 11,1; range c2300.

Type 150 Soviet historian I. Sultanov described *Samolyet* (aeroplane) 150 as having been launched by S. M. Alekseyev OKB using input from CAHI workers V. N. Belyaev, A. N. Makarevskii, G. P. Saishchyev and S. A. Khristianovich. He also made passing mention of Prof Brunolf Baade, but did not comment that Type 150 had originally been planned in Germany as Junkers EF 150 by Baade, Hans Wocke and other Junkers designers who had formed core of OKB-1. While it was wrong to ignore German genesis, it is worth noting that when work began in Soviet Union '150' was nothing but a few drawings.

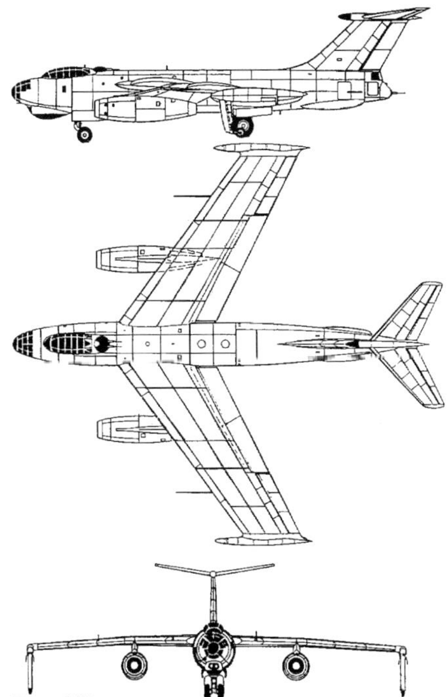

Type 150

These showed bomber with shoulder-high almost untapered swept wing carrying two underslung turbojets. Wing with fixed LE swept 35°, profile S-10s-9 at root changing to SR-3-12 outboard of engines, Dihedral 0° inboard, –1°20' (OKB drawing shows c-4°) beyond engines, skin reinforced by internal corrugated doubler. Two-piece slotted flaps, total 16,4 m², driven electrically to 33° TO as well as landing, three-piece ailerons, two large fences from LE to flap. Fin swept 45° with two-part rudder, boldly carrying tailplane at extreme top with large acorn fairing; tailplane LE 45°, fixed incidence 3°30', dihedral 8°, carrying three-part elevators. All flight-control surfaces driven by high-speed electric screwjacks; dual DC systems augmented by emergency wind-driven generator. Ventral strakes mounted at 45° under rear fuselage with braking-parachute bay between them. Door-type airbrake each side of rear fuselage. Fuselage almost circular section with pressurized compartment at front with door on right side for nav/bomb in glazed nose and at higher level pilot, copilot/commander and radio/gunner facing aft to control remotely sighted dorsal turret with two NR-23. In extreme tail a small capsule for rear gunner again with twin NR-23. RPB-4 main radar filling lower part of nose. Capacious bomb bay 2650 square by 7 m long with two hydraulically driven doors rotating inside fuselage. Design bomb load 6 t (13,228 lb). Ahead of and behind bay were two main gears each with twin braked wheels, forward unit hydraulically steerable, both retracting to rear. For takeoff rear gear shortened to give ground angle +3°30' for optimum wing lift. Rear gear retraction jack pinned to trunnion at extreme top of aft bomb-bay

150

frame to rotate gear up 92°, wheels retracting almost vertically. Long wingtip outrigger gears retracting aft into prominent pod. Track (outriggers) 23,6. To investigate such gear I-215D was built. AM-3 engine desired, smaller engine offered, installed in neat pod 6639 mm long on sharply swept (79°30') pylon 5,1 m from centreline. Eight fuel cells in fuselage; optional bomb-bay fuel.

Single flight article plus static airframe. First flight by Ya. I. Vernikov 14 May 51. Politically disfavoured because of ancestry, Type 150 generally exceeded expectation (required SL speed was 790). Waiting for engines brought it into timescale of Tu-88 which was demonstrably superior. On Flight 16, 9 May 52, aircraft stalled during landing approach, resulting in severe damage which was not repaired. Baade and colleagues returned to DDR where same design was basis for two forms of BB-152 passenger transport by VEB Dresden. No ASCC reporting name.

DIMENSIONS Span 24,1 m (79 ft 1 in); length 26,74 m (87 ft 8¾ in); wing area 125 m^2 (1,346 ft^2).

ENGINES Two AL-5.

WEIGHTS Empty 23 064 kg (50,847 lb); loaded 54 t (119,000 lb).

PERFORMANCE Max speed (SL) 850 km/h (528 mph), (10 km) 930 km/h (578 mph); service ceiling c13 km (42,650 ft). No other data but range required varied with load from 1500 to 4500 km.

346-1 under B-29

OKB-2

Under this designation, unrelated to pre-war KB-2, a large group of German engineers, mainly ex-Siebel, DFS and DVL, was formed 22 Oct 46 at Podberezye, in same building as OKB-1. Director A. Ya. Bereznyak (q.v.), design leader Hans Rösing (in some documents Rössing). From early 48 work co-ord with that of OKB-1 all under S. M. Alekseyev (q.v.) until both OKBs disbanded 53.

346 Assigned task: continue devt of wartime single-seat jet and rocket projects, notably DFS 8-346 supersonic recon aircraft. Mission soon changed to exptl. Using mainly DFS drawings, tunnel models tested at CAHI from 3-47. Final design mid wing NACA-012 profile, sweep 45°, two fences each side, plain flaps, split ailerons with inner sections locked at high speed. T-tail with tailplane pivoted –2°40'/+2° carrying split elevators operated like ailerons. Fuselage with Plexiglas nose around prone pilot, tankage around CG for kerosene and nitric acid, plus HTP to drive turbopump, with retractable skid underneath. Rocket engine in tail with main/aux chambers. Cockpit jettisonable (tested with nose built on B-25). **346P** glider taken to test site at Tyopil Stan, Moscow Ob., and flown after release from B-29 (42-6256) by Wolfgang Ziese. **346-1** high-speed glider version flown 48 after drop from Tu-4, again by Ziese. Powered **346-2**, also called **346D**, flown after drop from B-29 by Ziese 30 Sept 49; damaged on landing (too fast, skid failure), pilot injured. Repaired, 346-2 flown from Oct 50 at Lyukhovitsi by LII pilot P. I. Kasmin. Ziese recovered, flew **346-3** under power 15 Aug 51; flew 2 Sept but third flight on 14 Sept ended in loss of control, Ziese abandoning aircraft, ending programme. Developments said to be **446** and **468** also abandoned; parts donated to MAI. Biesnovat '5' programme was contemporary. In Jan 51 Kasmin also tested **KS-1 Analog**, manned version of Mikoyan KS-1 cruise missile. Data for 346-3:

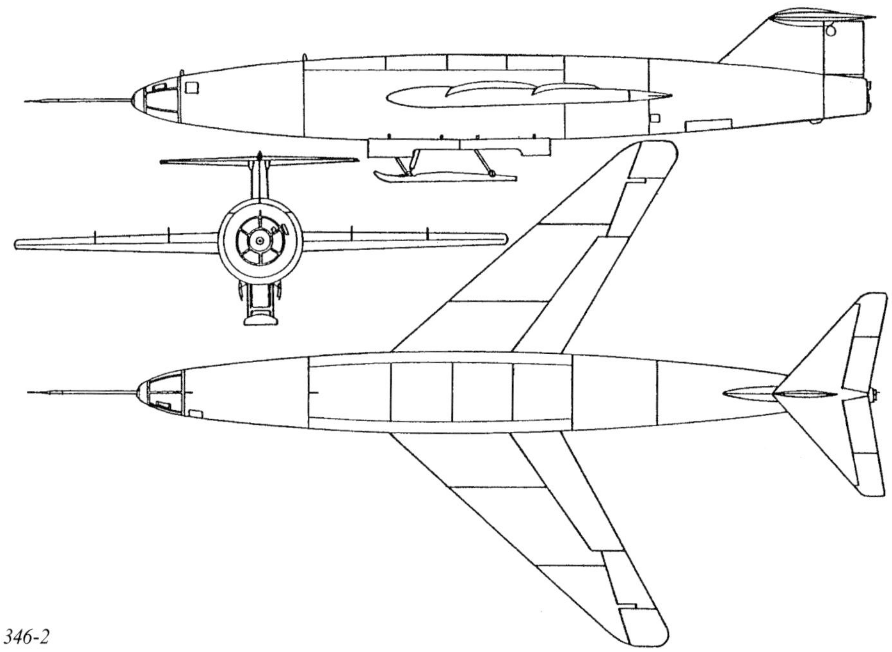

346-2

DIMENSIONS Span 9,0 m (29 ft 6⅜ in); length (excl probe) 13,45 m (44 ft 1½ in); wing area 19,87 m^2 (213·9 ft^2).

ENGINE One Walter HWK 109-509C (called ZhRD 109-510).

WEIGHTS Empty 3180 kg (7,011 lb); propellant 1,9 t; loaded (excl instrumentation) 5230 kg (11,530 lb).

PERFORMANCE Max speed intended Mach 2 during 2-min burn.

OMOS

This organization was founded in Leningrad 1925; *Otdel Morskogo Opytnogo Samolyetostroeniya* (Dept of Experimental Marine Aircraft Design). Headed by D. P. Grigorovich, deputy A. N. Syedelnikov, other engineers included Korvin, Gimmelfarb, Shavrov, Samsonov, Mikhelson and Vigand. Products described under constructors.

KR-1 Only OMOS product not assigned to particular constructor, this reconnaissance flying boat was Heinkel HD 55 assembled under licence and fitted with Soviet engine and equipment. In 30 purchased two Heinkel ship catapults and 30 HD 55 airframes. Wood hull and fin, fabric wings and tail surfaces. Stressed for operation from skis but otherwise not amphibious. Tractor engine between folding biplane wings, pilot under engine, and observer/gunner aft of wings with DA-2. Aircraft assembled at OMOS 1930-31 and served from surface warships and shore stations until 38. Designation from *Korabelnyi Razvyedchik*, fleet reconnaissance.

DIMENSIONS Span 14,0 m (45 ft 11¼ in); length 10,4 m (34 ft 1½ in); wing area 56,9 m² (612 ft²).

ENGINE One M-22.

WEIGHTS Empty 1550 kg (3,417 lb); fuel/oil 450+50; loaded 2200 (4,850).

PERFORMANCE Max speed 194 km/h (120·5 mph); climb to 1 km, 4 min; service ceiling 4,8 km; endurance 5½ h; range 800 km (497 miles); alighting speed 80 km/h (50 mph).

Stal-3

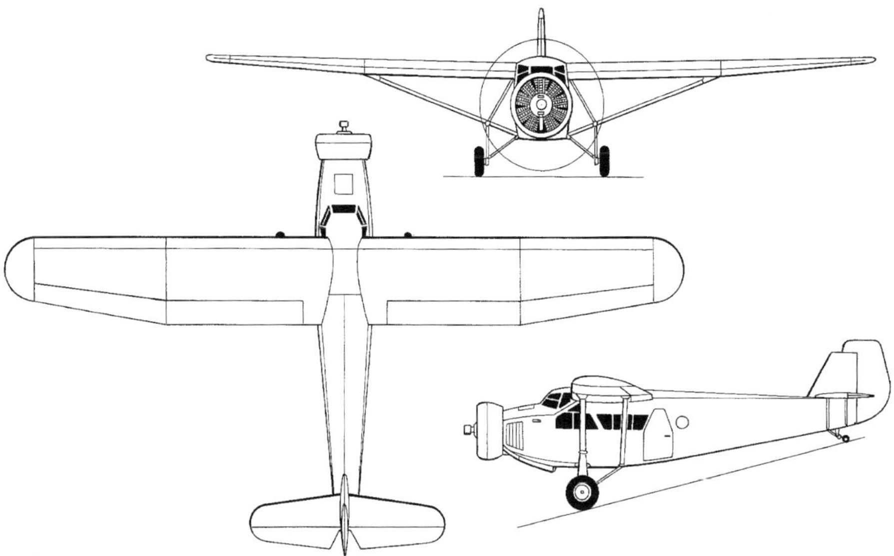

Stal-3

OOS

In 20s national shortage of both aluminium and necessary electric power for smelting prompted growing research into steel as material for primary airframe structure. Work centred at VVA under Zhukovskii, where in 28 special *Stal* (steel) group formed. In 29 this was joined by Aleksandr Ivanovich Putilov from CAHI. With dept manager S. G. Kozlov and VVA lab chief P. N. Lvov, Putilov drafted detailed programme for research into high-tensile and special steels (chiefly *Enerzh*-6), welding methods, and overall plan for future research. In 30 programme took shape under OOS (*Otdel Opytnogo Samolyetostroeniya*, section for experimental aircraft construction). After some months outgrew VVA lab and relocated in former Dobrolyet hangar (preferable, as OOS ostensibly civil, with admin by GVF). Later, probably late 32, expanded into large new OKB offices at GAZ-81, Tushino. Aircraft designated Stal, from material used; but some Stal aircraft not OOS but by other constructors (see Bartini, Grigorovich) and even Putilov's later designs are often ascribed to him rather than to OOS.

Stal-2 First product of OOS (no record of Stal-1). Conventional cabin monoplane seating pilot (plus optional extra passenger) in front and four passengers in main cabin, with door on left. Structure entirely of stainless steel, chiefly *Enerzh*-6, with fabric covering. Sections never thicker than 1 mm and ranging down to 0,1 and 0,15 mm for ribs and other detail components. Joints spot and seam-welded, but wing spar built up from sheet rolled into complex U-sections resembling those of Petlyakov. Upper/lower booms each based on single rolled U-sections riveted inside larger section with two Us, and finally capped by outer three-U section, booms connected by pairs of diagonal struts each stiffened by being pressed to dished section. Supply of *Enerzh*-6 slow and costly (same price as silver) until bulk purchase of similar 18-8 stainless from Sweden 32, and import of Krupps rolling mill same year. Low-rate production begun. First Stal-2 flown successfully 11 Oct 31 and production ordered from GAZ-81. Manufacture still labour-intensive, and first production aircraft delayed until early 34. Shavrov: Wright engine, plus small number also with M-26 or MG-31. Single **Stal-2bis** with Frise ailerons. Total production 111 by late 35, many remaining in use during Second World War.

DIMENSIONS Span 16,2 m (53 ft 1¾ in), length 9,74 m (31 ft 11½ in); wing area 31 m² (333·7 ft²).

ENGINE One radial, uncowled: (prototype) 300 hp Wright J-6, (series) M-26, then (most) MG-31, driving aluminium propeller.

WEIGHTS Empty 1030 kg (2,270 lb), fuel/oil 290

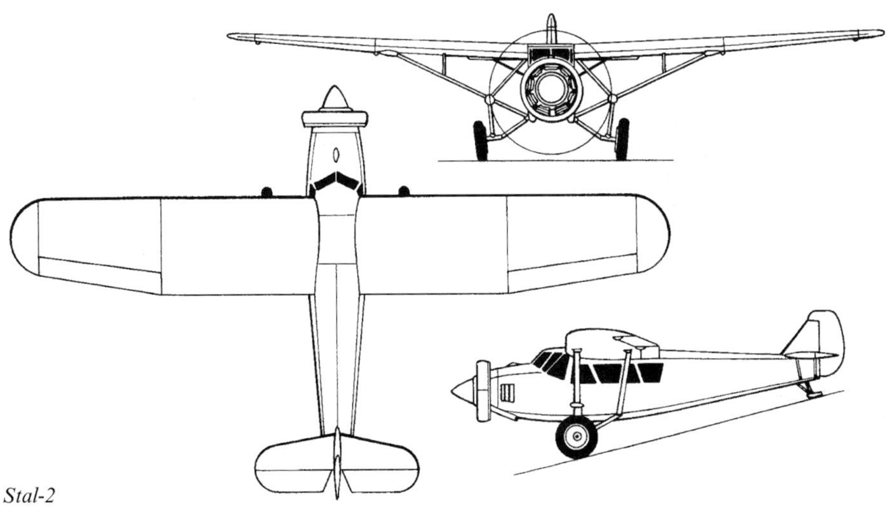

Stal-2

kg; loaded 1800 kg (3,968 lb).
PERFORMANCE (MG-31) Max speed 200 km/h, cruise 170 km/h (106 mph); service ceiling 5 km (16,400 ft); range 750 km (466 miles); TO 130 m; ldg 140 m, 70 km/h.

Stal-3 Enlarged successor or Stal-2 with refined but simplified structure cutting man-hours dramatically. Structure weight reduced from 43 to 41% and design load factor raised. Pilot and (normally) co-pilot with dual control in front, six passengers in main cabin of increased cross-section. Slotted ailerons and manual slotted flaps, engine in Townend-ring cowl, spatted wheels fitted with brakes, or streamlined skis. Prototype tested 33, series production ordered and 79 delivered 1935-36. Several fitted with water-cooled M-17. Important Aeroflot aircraft until summer 41, thereafter continuing on utility duties, some impressed into VVS.
DIMENSIONS Span 17,02 m (55 ft 10 in); length (prototype) 10,8 m (35 ft 5¼ in), (production) 10,68 m (35 ft 0½ in), (M-17, 11,3 m); wing area 34,8 m^2 (374·6 ft^2).
ENGINE One M-22; some M-17.
WEIGHTS Empty (production) 1672 kg (3,686 lb); fuel/oil 440 kg; loaded 2817 kg (6,210 lb).
PERFORMANCE Max speed 237 km/h (147 mph); climb to 1 km, 5 min; service ceiling 5340 m (17,520 ft), (prototype 6550 m); range 940 km (584 miles); take-off 280 m/16 s; landing 230 m/15 s/88 km/h.

Stal-5 Impressive flying-wing schemed by A. I. Putilov in 33, to reach ultimate transport efficiency. Payload (normally 18 passengers) in broad centre section, carrying two long but shallow vertical tails at extremities with hinged rear part of wing forming rectangular elevator between them. Tapered outer wings with slotted flaps and ailerons. Two main and two tail wheels, all retractable. Fully glazed cockpit in leading edge on centreline. Structure entirely *Enerzh*-6 with skin of bakelite-ply on centre section and fabric elsewhere. Also planned **KhB** version (*Khimichyeskii Boyevik*, chemical fighter) for spraying war gases. In 34 complete wing spar made for static test, and flying scale model (span 6,0 m, wing area 15,0 m^2, two 45 hp Salmson) flown late 35 by V. V. Karpov and Ya. G. Paul. Stability and control response poor, and after some tinkering project abandoned.
DIMENSIONS Span 23,0 m (75 ft 5½ in); length 12,5 m (41 ft 0 in); wing area 120 m^2 (1,292 ft^2).
ENGINES Two M-34F.
WEIGHTS (est) Empty 5,5 t (12,125 lb); loaded 8 t (17,640 lb).
PERFORMANCE No data.

Stal-11 High-speed light transport intended primarily to carry pilot and four passengers, mail and baggage, but also planned as military recon aircraft. Mixed construction: fuselage built on truss of *Enerzh*-6 with bakelite-ply skin, wings of high-speed profile with all-wood construction around two spars with sheet webs, built-up ribs, multiple stringers and exceptionally smooth skin. Engine (HS 12Y derived) with detachable metal cowl, belly radiator and two-blade v-p propeller. Main landing gears retracted outwards into wings, with bays sealed by hinged doors on inside of legs. Exceptionally large slotted flaps depressed to 60° with slot progressively closing. Modified Frise ailerons. Prototype completed autumn 36 and tested on wheels and also on fixed skis. In 37 tested as reconnaissance version. Flown approximately 300 times, but no further action. Performance good, but expensive to make and maintain.
DIMENSIONS Span 15,0 m (49 ft 2½ in); length 12,5 (41 ft 0 in); wing area 31,0 m^2 (333·7 ft^2).
ENGINE One M-100A.
WEIGHTS Empty 1830 kg (4,034 lb); fuel/oil 380 kg; loaded 2,7 t (5,952 lb).
PERFORMANCE (wheels) Max speed 430 km/h (267 mph); cruise 370 km/h (230 mph); service ceiling 8 km (26,250 ft); range with full payload 1000 km (621 miles); landing speed 80 km/h (50 mph).

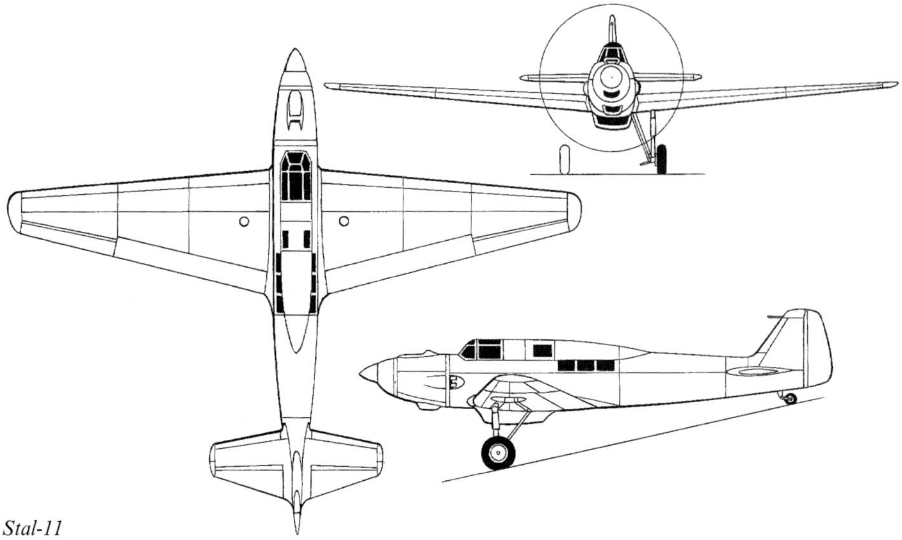

Stal-11

Aviatourist Smaller edition of Stal-11 intended for long-range racing, with D. H. Comet configuration. Two crew in tandem enclosed cockpit, and two 130 hp Gipsy Major engines on wing of 10,18 m^2 area. Structure mainly bakelite ply. Airframe ready late 36, but despite visit to Stag Lane in England engines never received and project eventually abandoned.

Pashinin

Mikhail Mikhailovich Pashinin was deputy to Polikarpov on fighters of 1930s. In 38 combat reports from Spain and Siberian frontier prompted him to plan modern fighter able to dive at highest possible airspeed.

I-21, IP-21 Authorized at Kremlin meeting Jan 39; number assigned by NII even though previously used for defunct Ilyushin CCB-32. Conventional low-wing monoplane, mixed construction. Wing of NACA-0012 (symmetric) profile tapering to 0009 at tip. Wide (3 m) centre section, no dihedral, 30KhGSA steel tube, ribs D16, aluminium-riveted 3/1,75 mm wood-ply skin. Outer panels dihedral and sharp taper, all-wood with ply skin. Forward fuselage welded KhGSA steel tube, removable dural skin panels. Rear fuselage, monocoque of wood *shpon*, 3 mm thick tapering to 1,5 mm over tail surfaces made as one unit. Unarmoured cockpit with hinged canopy. Three retractable landing wheels, each main gear folding to rear with wheel rotating 90° to lie just ahead of split flap. Oil cooler under engine and main radiator under wing on centreline. Carburettor air via wing-root inlets. Intended engine M-107, not developed in time; all known prototypes with M-105 or 105P. Firt prototype fitted with BT-23 firing through propeller hub, two ShKAS above. Construction finished 1 May 40, first flight P. M. Stefanovskii 18 May. Despite severe handling problems NII tests began 5 July and continued with four further prototypes all with modifications to improve handling. Second I-21 had taller vertical tail and greater 'sweep' on outer wings. Factory testing shared with S. Suprun who recommended reduced span and greater dihedral; new wing on third aircraft flown Jan 41, but NII claimed to notice no improvement. Believed there were variations in armament, height of radiator and arrangement of exhaust stacks. Many details exactly as I-16, and consensus of opinion was that I-16 was safer than I-21! Abandoned April 41.
DIMENSIONS Span (Nos 1 and 2 prototypes) 11,0 m, (3rd prototype) 9,43 m (30 ft 11¼ in); length 8,73 m (28 ft 7¾ in); wing area (long span) 15,8 m^2 (170 ft^2).
ENGINE One M-105P.
WEIGHTS No data except first prototype loaded 2670 kg (5,886 lb).
PERFORMANCE Max speed (No 1) 488 km/h at SL, 573 km/h (356 mph) at 5 km, (No 2) 506 km/h at SL, 580 km/h (360 mph) at 4750 m; climb to 5 km (No 1) 6 min; service ceiling (No 1) 10,6 km (34,777 ft); time 360° turn 25/26 s; landing speed (No 1) 165 km/h (102·5 mph).

S-82 Two-seat multi-role attack aircraft with long range and mixed propulsion, schemed 46 and basis of sudden expansion of Pashinin OKB 47 onwards. All-metal stressed-skin, low-wing monoplane with powerful piston engine in nose, with two turbochargers (and intended to drive five-blade propeller), and turbojet in rear fuselage with wing-root inlets. Turbojet used only for take-off and combat. Crew in central pressurized tandem cockpits designed by V. B. Shavrov. Heavy armament (details lacking). Prototype construction 47, but Pashinin did not

I-21

expand manufacturing strength and in 48 work abandoned and OKB closed.

DIMENSIONS Span 21,2 m (69 ft 6⅔ in); length 18,0 m (59 ft 0⅔ in); wing area 61,5 m^2 (662 ft^2).

ENGINES One ASh-73TK (two TK-19) and one RD-45 turbojet.

WEIGHTS Empty about 9 t (19,840 lb); fuel/oil 4400+300 kg; loaded 15 t (33,070 lb).

PERFORMANCE Max speed (est) 870 km/h; endurance on piston engine 10 h.

Petlyakov

Vladimir Mikhailovich Petlyakov was born 27 July 91 at Sambek (now Novoshakhtinsk), near Taganrog, where he attended Tech Sch; under Zhukovskii at MVTU, staying on as draughtsman; to CAHI 21 as lab asst, graduated from MVTU 22 and designed structures from ANT-2 on, deputy to Tupolev, head of heavy aeroplane brigade Oct 31, independent July 34 (ANT-42/Pe-8); head of CAHI ZOK (factory xptl const'n) July 36, and chief constructor 20 July 37 when arrested, concentrated in CCB-29 to form STO (special technical dept) 100 in GAZ-156. Success led to reinstatement July 40, general constructor of own OKB, A. M. Izakson deputy, evac to Kazan Oct 41, but killed 12 Jan 42 when Pe-2 caught fire en route to Moscow. Izakson was soon replaced by A. I. Putilov, in turn replaced 22 June 43 by V. M. Myasishchyev, previously chief designer Pe-2 prodn and modn. OKB closed 46, facilities to Myasishchyev.

ANT-42, TB-7, Pe-8 Only Soviet heavy bomber in Gt Patriotic War. Created to July 34 reqt for precision bombing from 10-11 km at >440 km/h with range 4500 km with 2 t bombload. Task to A. N. Tupolev, long discussions with CIAM and engine KBs leading to ATsN (*Agregat Tsentralnovo Nadduva*) central engine driving large supercharger feeding four propulsion engines. Work assigned to Petlyakov brigade, with co-ordinating design role by Iosif Fomich Nyezval.

Ruling structural material D16 duralumin, general form as SB (ANT-40) with same stress limits and factors but twice linear scale. Main exception, two wing spars of trad Petlyakov truss-girder type but for the first time made from 30KhGSA. Basic method rolled strip re-rolled to U-section and assembled by bolts, rivets and some welding; truss ribs, stringers and secondary structure all-riveted; smooth skin in over 60 long narrow strips each wing; profile CAHI-40, root t/c 19%, tip 15·5. Mean chord 5,35 m, CG at 28·4% empty and up to 34% loaded. Fuselage oval section with main longerons tubular including four of 30 KhGSA; rest conventional D16 riveted structure with all exterior riveting round- or pan-head. First use of large plastic templates in lofting, and tooling more extensive than on any previous Soviet aircraft. Fabric-covered flight-control surfaces, all manual and with trim tabs on rudder and both elevators; long narrow ailerons in three sections pivoted on underside to 8 brackets. Split flaps (one each outer wing, one on c/s) driven hydraulically to 45°. Main gears with pneu brakes, hyd retraction to rear, twin doors with part of 1600×500 tyre showing; fixed tailwheel with tyre 700×300. Fuel in 16 protected AMTs tanks in LE and between spars. Engines on 30KhGSA tube trusses, driving 3,9 m VISh-2 3-blade props, water radiator beneath each engine with controllable side exit door, exhaust from each bank to single pipe above wing.

Fuselage dominated by ATsN installation (see later), absent from prototype but space provision with high deck continued forward to cockpit for two pilots in tandem just to left of centre with sliding canopies; at lower level, radio op to L and engineer to R; further forward nav station and bomb aimer in large chin gondola with nose turret with ShVAK; dorsal gunner at rear of ATsN fairing with sliding hood over pivoted ShVAK; second ShVAK in tail turret, ShKAS in ventral hatch and ShVAK in manual barbette at rear of each inner nacelle with access through wing or via

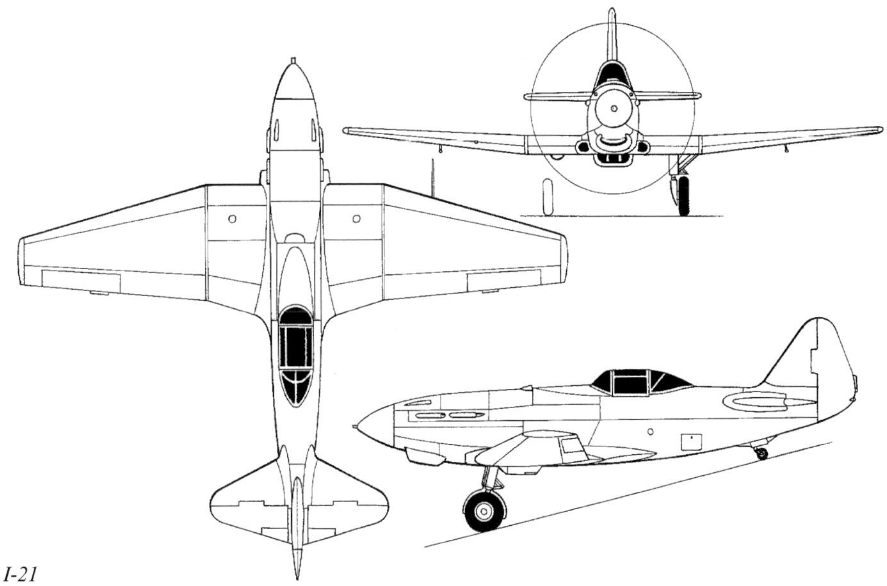

I-21

ANT-42 first prototype

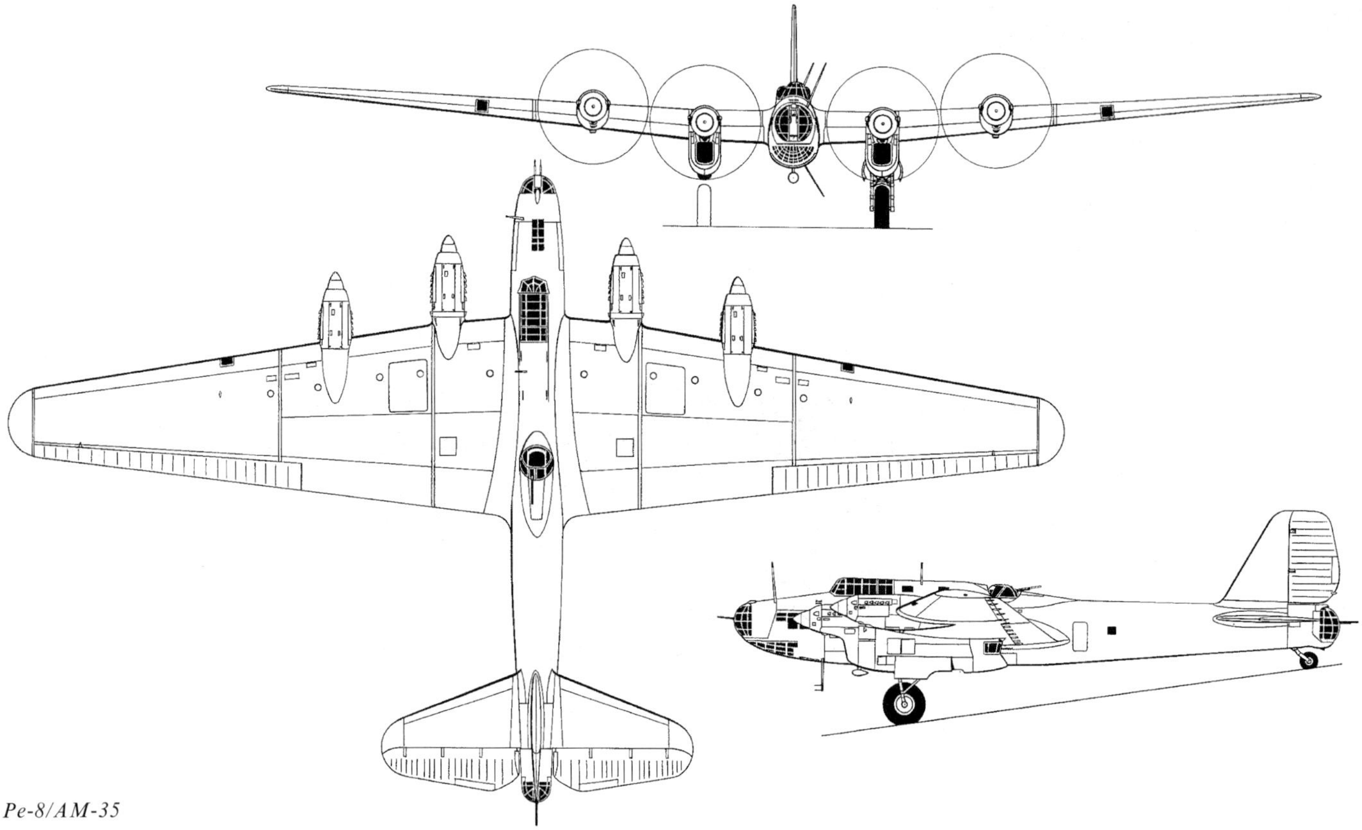

Pe-8/AM-35

trapdoor in upper wing surf. Fuselage cross-section pear-shaped giving large bomb bay with two doors driven hyd on swing-links; racks for up to 4 t (8,818 lb) bombs incl two FAB-500 externally under inbd wing.

Prototype delayed by crash of ANT-20 and Petlyakov/Saukke work on ANT-20bis, completed Khodinka 9 Nov 36, no ATsN or armament, tailplane braced to fin. First flight on wheels 27 Dec 36 by M. M. Gromov and N. S. Rybko with NII engineer M. F. Zhilin. Satisfactory, but outer-nacelle poor aerodynamics; decision to put all rads in inbd nacelles. Rebuilding of Khodinka demanded move to small field outside Moscow, where CAHI pilot Rybushkin damaged landing gear/nacelles/props. Rebuild with inner/outer rads staggered in deep duct under each inner engine with superimposed exit doors; rudder horn balance replaced by deeply inset hinges. During 37 ATsN installed with M-100 in compartment in top decking with blister and exhaust pipe on right, driving via step-up gearbox to 25,000 rpm compressor feeding main engines via aluminium pipes. Once faulty prop corrected 444 km/h reached, so (against KB opinion) armament reduced to ShKAS throughout. Successful tests on skis Nov 37. NII testing at Eupatoria, Crimea, from 6 Mar 38.

DIMENSIONS Span 39,01 m (127 ft 11¾ in); length 23,4 m (76 ft 9¼ in); wing area 188,4 m² (2,028 ft²).

ENGINES Four AM-34FRN.

WEIGHTS Empty 17 885 kg (39,429 lb) [after mods 18 t]; fuel 8250 kg; loaded 23 860 kg normal, 30 t (66,138 lb) max [after mods 24/32 t].

PERFORMANCE Max speed 320 [315] km/h at SL, 403 [430] km/h at 8 [8,6] km; time to 5 km 18·8 [16·3] min; service ceiling 10,8 [11,25] km; range 3000 [3500] km.

ANT-42 dubler ANT-42 accepted for service as **TB-7** 20 Ap 38, pre-series with FRNV engines and improved ATsN-2 laid down at GAZ-124. Second prototype flown 26 July 38 with wider chin gondola, improved engines, two extra centre-section tanks, better armour and tank protection, and defensive a/t installed: bombload 4×FAB-250 or equal with full fuel; overload 4 t (eg 40 FAB-100 or two FAB-2000). Programme ground to standstill 39 because of failure to produce ATsN-2 and suspension of prodn of FRNV engine. Dimensions unchanged.

ENGINES Four AM-34FRNV + ATsN-2 driven by M-100A.

WEIGHTS Empty 18 520 kg (40,829 lb); fuel 10,8; loaded 24 594 kg normal, 32 t (70,547 lb) max.

PERFORMANCE Max speed 305 km/h at SL, 414 km/h (257 mph) at 7,5 km; climb 20 min to 5 km; service ceiling 10,4 km (34,120 ft).

TB-7/AM-35A Unavailable FRNV engine and continuing ATsN tech and supply problems resulted in decision early 39 to switch to AM-35A and abandon ATsN. Resulting TB-7 with 19 fuel tanks, twin ShKAS in nose turret, chin gondola removed, radio masts nose and amidships canted to L, ATsN replaced by new station for commander and radio operator behind pilots (normal crew 11), bomb bay able to take one FAB-5000, rack for one VAP-500 or VAP-1000 under each wing root, deeper rear fuselage with side door, tailplane now shoulder-high and unbraced, VISh-24 props, two HF radios, D/F loop, bcn rcvr and IFF. Supply of AM-34FRN erratic and in autumn 39 ceased, with two series aircraft complete, four more with ATsN-2 and no engines and two completed with AM-35A engines. Whole

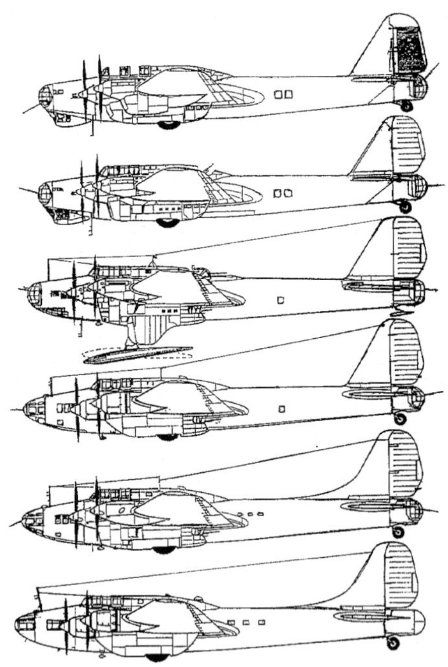

From top: ANT-42 prototype March 37, same May 40, No 4211 1st series, Pe-8/ASh-82, Pe-8ON/ACh-30B, Pe-8ON Polyarnyi

Nacelle turret

programme ended Dec 39, tooling at GAZ-124 dismantled. Sudden reversal April 40, order reinstate prodn 4 May with diesel engines. OKB urgently designed installn for M-30/M-40 diesel engines (similar except for turbos). GAZ-124 given new director M. M. Kaganovich, who as stop-gap obtained some AM-35A engines, enabling prodn to resume, sending by end-40 18 TB-7/AM-35A to regt at Borispol (which quickly ceased flying because no spares and no contact with GAZ, rectified when situation discovered April 41). Unit became 412 Polk (Brig M. V. Vodopyanov) near Leningrad, attacking Berlin 9 Aug 41. During 41 almost all had diesels, but their unreliability led to resumption with AM-35A late 41 and to several a/c re-engined with AM-35A. This was only suitable engine for impt diplomatic missions, notably by No 66 to Washington May/June 42, but supply ceased mid-42, Deputy Commissar P. V. Dement'yev suggesting switch to ASh-82 (see later). Dimensions unchanged.

ENGINES Four AM-35A.

WEIGHTS Empty 19 986 kg (44,061 lb); fuel 13 025 lit; loaded 27 t normal, 35 t (77,160 lb) max.

PERFORMANCE Max speed 347 km/h at SL, 443 km/h (275 mph) at 6360 km; climb to 5 km in 14·6 min; service ceiling 9,3 km (30,500 ft); range 3600 km (2,237 miles).

TB-7/M-40 When interest reawakened May 40 urgent need for greater range recognised, order to fit M-30 or M-40 diesels. One aircraft immediately built with each, first (M-40, later ACh-30) flown spring 41 by G. F. Baidukov. M-30 a/c handed over to LII test June 41, war intervening. Combat missions showed a/t inadequate, so (often as retrofit) MV-6 dorsal turret (one ShVAK, 200 rds) with retrac fairing, turrets behind nacelles (one UBT, 220) and tail turret (one ShVAK, 200), all elec powered. Also, many diesel problems, eg slightly opening throttle could cause engine failure, no restart poss above 3 km and much skill reqd. Greater range nullified by enormous lube-oil consumption, missions impossible from rear areas until aux oil tanks added; several a/c re-engined with AM-35A. By end-41 17 diesel a/c delivered but serviceability poor; only three more (with M-30) built 42. Dimensions unchanged.

ENGINES Four M-40.

WEIGHTS Empty 19 790 kg (43,629 lb); fuel/oil 13 025 kg + 670 kg; loaded 26 t normal, 33,5 t (73,854 lb) max.

PERFORMANCE Max speed 345 km/h at SL, 393 km/h (244 mph) at 5680; climb 16·2 min to 5 km; service ceiling 9·2 km (30,200 ft); range 5,460 km (3,393 miles).

Pe-8/AM-35A

Pe-8/ASh-82FN of Polar Directorate

TB-7/ASh-82 Dement'yev suggested reliable radial engine, OKB designing new mounts and nacelles with complex exhaust syst. Single a/c converted Sept 42, tested Oct by GAZ-124 pilot B. G. Govorov; good except inability to use hi-gear supercharge, problem defeated eng-designer S. K. Tumanskii. ADD commander said range more impt than speed, and from then on eng supply no problem. One more ASh-82 delivery 42; regular prodn 43-44 with hiccup caused by re-emergence of diesel, see later. At end-43 nose redesigned with hand-aimed ShVAK at tip, less drag and weight, increased firepower. Aircraft redesignated Pe-8 during 43.

DIMENSIONS Unchanged except length (new nose) 23,59 m (77 ft 4¾ in).

ENGINES Four ASh-82.

WEIGHTS Empty 18 570 kg (40,939 lb); fuel/oil 13 120 kg + 670 kg; loaded 27,2 t normal, 36 t (79,365 lb) max.

Pe-8 test bed (test engine not identified)

PERFORMANCE Max speed 362 km/h at SL, 422 km/h (262 mph) at 5,6 km; climb 15·0 min to 5 km; service ceiling 9,5 km (31,168 ft); range 5800 km (3,605 miles).

Pe-8ON Direct order from Stalin for increased range led to final effort with diesel engines, I. F. Nyezval preparing 42612 cOct 44, Kazan effected further mods in 42712 March 45. ON, *Osobogo Naznachyeniya*, personal assignment, refined airframe, pointed nose (ShVAK), no dorsal turret, dorsal fin, VISh-61V1 props, new lube-oil and cabin-air systems. Nyezval "No op'l improvement apparent". Total of all versions 93. Dimensions as ASh-82 a/c.

ENGINES Four ACh-30B.

WEIGHTS Empty 22 864 kg (50,406 lb); fuel/oil c13 t + 800 kg; loaded 30 t normal, 35,5 t (78,263 lb) max.

PERFORMANCE Max speed 342 km/h at SL, 390 km/h (242 mph) at 6,0 km; climb 19·5 min to 5 km; service ceiling 8,2 km (26,900 ft); range 5600 km (3,480 miles).

Pe-8 mods In late 45 I. I. Chyerevichny requested *Polyarnyi* version for Arctic use; two mod from ASh-82 bombers, all mil gear removed, extra fuel and much special eqpt, second flown by Zadkov many years, often on skis. Another mod for NII-VVS as testbed for aircooled engines, esp ASh-2 and VD-4K.

'100' STO-100's first assignment March 38 'No 100', also called **VI-100**, high-alt fighter and dive bomber to replace seemingly terminated TB-7. Turbocharged engines, pressure cabins

VI-100

Pe-2 on skis

Pe-2 First series

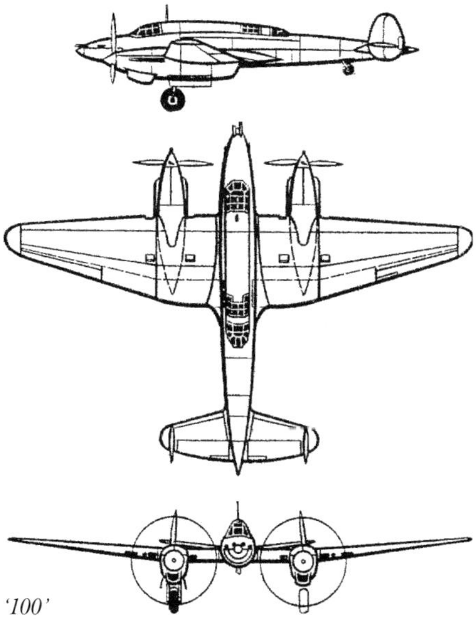

'100'

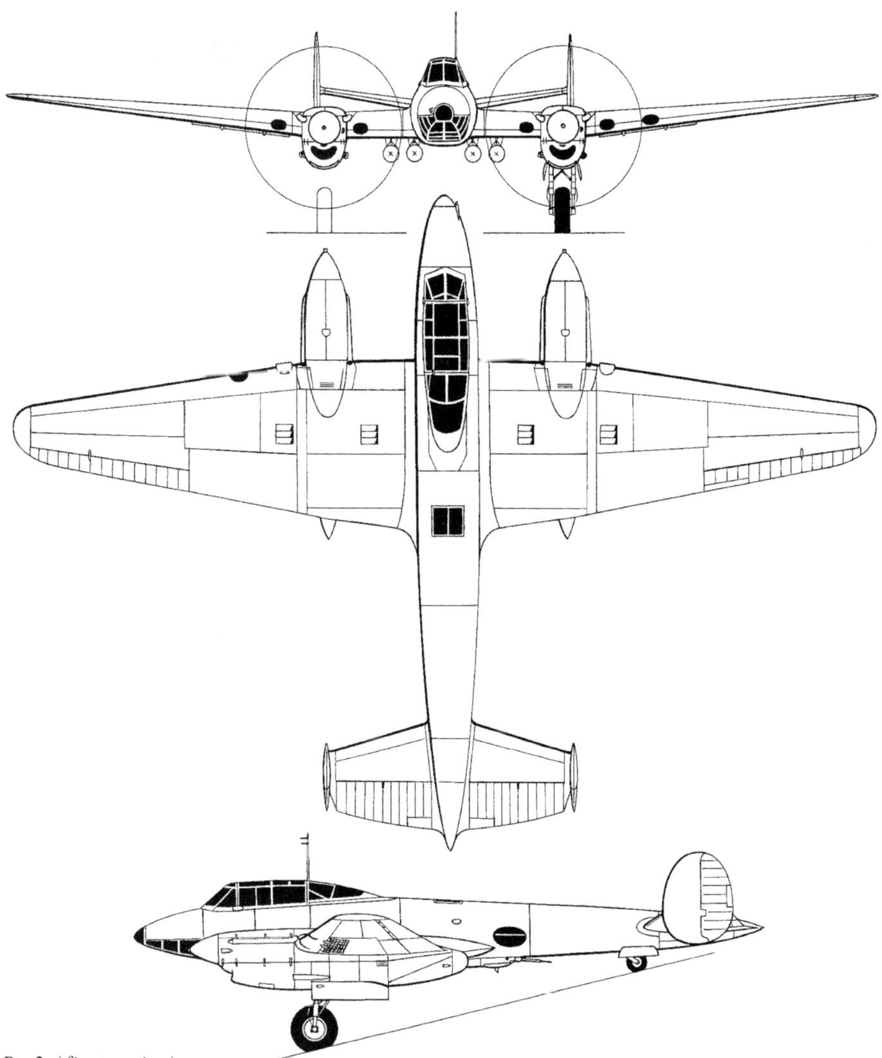

Pe-2 (first series)

by M. N. Petrov's brigade, elec servo-motors in flight controls and auto correction of gunsight according to wing AOA. Flush-riveted skin except fabric control surfs with trim tabs. Wing 'VVS' profile (V at root, VS at tips), two spars, four sections split flap driven like other services by extensive 28V electrics. Radiators in wing inbd and outbd of each engine with LE inlets and upper-surf exits; oil coolers under engines. Separate pressure cabins for pilot and radio/gunner, but latter depressurized for firing single ShKAS (this never installed). Nose occupied by two ShVAK and two ShKAS. Original scheme for offensive load 3-inch gun fed from 24-projectile drum; scheme replaced by external racks for two bombs of 100, 250 or 500 kg. First flight 22 Dec 39 by P. M. Stefanovskii and engineer I. V. Markov; on May Day 40 rolled over Red Sq with gear extended. Similar *dubler* flown Jan 40, but on 25 May it was decided interceptions could be done by MiG-3. Pe team given until 1 June to produce conversion to **PB-100** 3-seat dive bomber, but UVVS rejected pressure cabins and turbos, demanding simpler bomber, see next.

DIMENSIONS Span 17,16 m (56 ft 3⅜ in); length (to rudder TE, excl tailcone) 12,6 m (41 ft 4 in); wing area 40,5 m^2 (436 ft^2).

ENGINES Two M-105 + TK-2.

WEIGHTS Empty 5172 kg (11,402 lb); loaded 7260 kg (16,005 lb).

PERFORMANCE Max speed 455 km/h at SL, 535 km/h (332 mph) at 6 km; climb 6·8 min to 5 km; service ceiling (est) 12,2 km (40,000 ft); range (est) 1400 km (870 miles); ldg 128 km/h.

PB-100 Frantic redesign of VI-100 for new role, borrowing c200 engineers from other OKBs, mock-up passed 1 June 40, two prototypes built. Airframe basically as VI-100 without pressure cabins and with normal M-105R engines. Pilot moved forward, bomb-aimer in glazed nose, nav at rear with ShKAS. Two prototypes, no dates, first force-landed by Stefanovskii, second crashed by A. M. Khripkov and P. I. Perevalov. Despite this, series approved 23 June 40, designation changed late 40 to Pe-2.

Pe-2 Standard tac bomber of VVS in Gt Patriotic War. Wing mod for equal taper on LE/TE, vert tails increased in chord and moved to tips, engines without turbos in redesigned nacelles, landing gear hydraulic but c50 elec motors driving other items, completely new fuselage, airframe simplified for production (eg, no fuselage stringers).

Wing with two spars of D16 sheet with 30KhGSA tube booms, pressed ribs, skin 0,6/0,8 mm flush-riveted, 7° dihedral outer panels only, fabric ailerons (R with tab), split flaps driven to 45°. Welded-steel Venetian-blind dive brake under each wing with AP-1 (*Avtomat Pikirovaniya*) auto dive control [deleted on prodn a/c]. Tail D16 with tabbed fabric control surfs, tailplane adjustable +1°15'/–3°45', elevs +31°/–18°, rudders ±25°. Five rubberized fuel cells inerted by cooled exhaust, prototype 440 lit fuselage, 180 lit each c-s, 143 lit each outer panel, total 1086 lit (239 gal). Pilot left of centre with walkway and floor hatch with integral steps on R; nav/bomb prone posn in nose with sight, swivel seat on R behind pilot with hand-aimed ShKAS; radio/gunner isolated behind tank with roof hatch, two side windows and periscopic sight for ventral ShKAS. Two ShKAS fixed in nose fired by pilot. Normal bombload 4 × FAB-100 in bomb bay, single FAB-100 in each nacelle and 4 × FAB-250 external. Landing gear track 4730 m, main tyres 900 × 300, pneu brakes, retrac tailwheel with twin doors; retrac skis tested in 42 but perf too degraded for adoption. VISh-61 props, ATP horizon, RPK-2 radiocompass.

Prodn drawings released to GAZ-22 23 July 40, first a/c complete 1 Dec 40, tested 3-19 Dec by N. K. Fyodorov and 20 Dec by Stefanovskii. By 22 June 41 (war) 459 delivered, c290 operational. Oct 41 evacuated to Kazan, by this time GAZ-39, -124 and -125 also in prodn. Total 31 Dec 41 1,626. Popular, called *Peshka* (pawn, but in this case 'little Pe'). ASCC reporting name 'Buck'.

DIMENSIONS Span 17,15 m (56 ft 3¼ in); length 12 241 (40 ft 2 in); wing area 40,5 m^2 (436 ft^2).

ENGINES Two M-105.

WEIGHTS Empty 5863 kg (12,925 lb) [skis, 6165 kg], fuel 730 kg; loaded 7536 kg (16,614 lb) [skis 7900 kg].

PERFORMANCE Max speed 452 km/h (281 mph) at SL [skis 405 km/h], 540 km/h (336 mph) at 5 km [skis 474 km/h at 4,8 km]; climb 9·3 min to 5 km [skis 10 min]; service ceiling 8,8 km

(28,900 ft) [skis 8,5 km]; range 1200 km (746 miles); TO 584 m; ldg 230 km/h.

Pe-3 Day/night interceptor, pilot and radio/gunner back-to-back on left side, dive brakes and bomb gear removed, fuel in fuselage/nacelle bomb bays, prototype Feb 41 with two UBS and one ShKAS in nose and two ShKAS aft. GAZ-125 built 220 for defence of Moscow later 41, two UBS and three ShKAS in nose Prototypes only of **Pe-3bis** multirole fighter/interceptor/bomber/photo-recon two-seater, as Pe-3 but nose one ShVAK and three UBS, one ShKAS at rear, bombload 300 kg, wing racks for eight RS-82. Tested by S. I. Sofronov; useful in some situations but no ventral protection.

DIMENSIONS, ENGINES As Pe-2 except 3bis M-105R.

WEIGHTS Empty not recorded; loaded (3) 7880 kg (17,372 lb), (bis) 8040 kg (17,725 lb).

PERFORMANCE Max speed 444 km/h (276 mph) at SL (bis, 448 km/h), 535 km/h (332 mph) at 5 km (bis 530 km/h); climb 9 min to 5 km (bis 10·2 min); service ceiling (bis) 9·1 km (29,860 ft); range (both) 2150 km (1,336 miles).

Pe-2 Series 22, M-105RA driving VISh-61B. **Series 83** new nav cockpit with hand-aimed UB. **Series 105**, nose side glazing much reduced, RPK-10 radiocompass. **Series 110**, also called **Pe-2FZ** (*Frontovoye Zadaniye*) [and in some places **FT**, *Trebovaniye*, front-line request] mid-42 with nav in MVT turret with UBT as well as ShKAS fired through openable beam window (left or right); nose side glazing elimd. **Series 179** lower cowling and oil rad redesigned. **Series 205** larger bomb bay, usually PF eng, new carb inlets. **Series 211** modified underwing racks for range of loads. **Series 249** oil dilution for winter starts. **Series 265** longer-range radio, mast on windscreen. **Series 275** DAG-10 with two ejectors each five AG-2 grenades rear defence. **Series 301** revised dorsal hatch. **Series 354**, also called **Pe-2B**, some (**359**, all) with individual exhaust stubs, no fixed guns but UBT turret and aft ventral ShKAS. **Series 382** improved winter starting. **Series 410/411** drag reduction, eg ventral ShKAS installn refined. All with PF engine and fuselage tank enlarged to 518 lit, 53 lit in centre wing box, plus 107 lit outboard of each outer wing tank, total 1484 lit (326·5 gal).

DIMENSIONS, ENGINE As before, except from late 43 with VK-105PF engine.

WEIGHTS Empty (FZ) 6020 kg, (FT) 6,2 t, (B) 6210 kg (13,690 lb); fuel/oil 1200+150 kg; loaded (FT) 8135 kg, (FZ) 8,3 t, (B) 8580 kg (18,915 lb).

PERFORMANCE Max speed (FT, FZ) 437 km/h at SL, 482 km/h (300 mph) at 3,2 km, (B) 472 km/h at SL, 534 km/h (332 mph) at 3,9 km; climb to 5 km (FT, FZ) 11·1 min, (B) 10·5 min; service ceiling (FT, FZ) 7,8 km (25,600 ft), (B) 9·1 km; range (all) 1300-1400 km.

Pe-2UT Also called **UPe-2**, multirole trainer with No 1 (fuselage) tank replaced by extra pilot station at rear with dual controls; crew 2 to 4; two UBS and two ShKAS firing ahead and 600 kg bombs internal or two FAB-500 external; made in large numbers from autumn 42.

DIMENSIONS, ENGINES As before.

WEIGHTS Empty 5956 kg (13,131 lb); loaded (4 crew) 7344 kg (16,190 lb).

PERFORMANCE Max speed 454 km/h at SL, 508 km/h (316 mph) at 3,6 km; climb 8·5 min to 5 km; service ceiling 8,7 km (28,550 ft); range 890 km.

Pe-2UT

Pe-2M/VK-107A (Myasishchyev VM-13)

Pe-2/ASh-82FN

Pe-2R 3-seat day recon, no bombs or dive brakes, various cameras (AFA-B, AFA-1, AFA-27T1, AFA-3S, NAFA-19), two underwing 290 lit drop tanks, AK-1 auto-heading, small series most on base Series 105 or 359, typically 3 BS, 7603 kg, 1700 km.

Pe-2M Designation used twice. First flight 16 Oct 41 by S. A. Shestakov and ldg engineer A. A. Rozenfeld of a/c with M-105/TK engines, auto slats and internal bay for two FAB-500. Repeated March 45 for prototype (also called Myasishchyev VM-13) 3-seat bomber with refined airframe, hi-power engines, extra fuel, three B-20 with 520 rds. Data for second a/c:

DIMENSIONS Span 17,99 m (59 ft 0¼ in); length 13,57 m (44 ft 6¼ in); wing area 43,5 m² (468 ft²).

ENGINES Two VK-107A.

WEIGHTS Empty 7458 kg (16,442 lb); loaded 10170 kg (22,421 lb).

PERFORMANCE Max speed 545 km/h at SL, 630 km/h (391 mph) at 5,6 km; climb 8·8 min to 5 km; service ceiling 8,5 km (27,900 ft); range with 2 t bombload 2050 km (1,274 miles).

Pe-2/M-82 Single example early 43 with radial engines, minor changes; built in small series (135), to 99 Polk, one with revised wing profile reducing ldg speed to 200 km/h.

DIMENSIONS As FT/FZ.

ENGINES Two M-82 (ASh-82).
WEIGHTS Empty not found; loaded 8125 kg (17,912 lb).
PERFORMANCE Max speed 458 km/h at SL, 547 km/h (340 mph) at 6,2 km; climb 7·9 min to 5 km; service ceiling 9,1 km (29,860 ft); range 1170 km (727 miles).

Pe-2Sh *Shturmovik* version mod from prodn a/c with two ShVAK and a UBS firing obliquely down under fuselage; second version tested with oblique gun batteries at front and rear of former bomb bay each with ShKAS on left and ShVAK on right. A third, **Pe-2/MV-3** tested 42 with gondola under fuselage with two ShVAK and two BS firing ahead. None prod in series.
Pe-2VI *Vysotnyi istrebitel* (hi-alt fighter) with pressure cabin(s) studied from outset of project '100', prototype being built but abandoned at evacuation Oct 41 along with **Pe-2VB** high-alt bomber. Former designation reused for new fighter with pressure cabins by A. I. Putilov (installed 1 April 43) and engines with two-stage (Dollezhal) superchargers; nose guns one ShVAK, two BS and one ShKAS, backseater one BT. NII-tested, formed basis for Myasishchyev projects.
DIMENSIONS As FT/FZ.
ENGINES Two VK-105PD.
WEIGHTS Empty 5790 kg (12,765 lb); loaded 6870 kg (15,146 lb).
PERFORMANCE Max speed (est) 557 km/h (346 mph) at ht; climb to 5 km in (est) 10·4 min; service ceiling 10,5 km (12 km planned).
Pe-2I Major mod under Myasishchyev to produce better dive bomber, crew pilot and

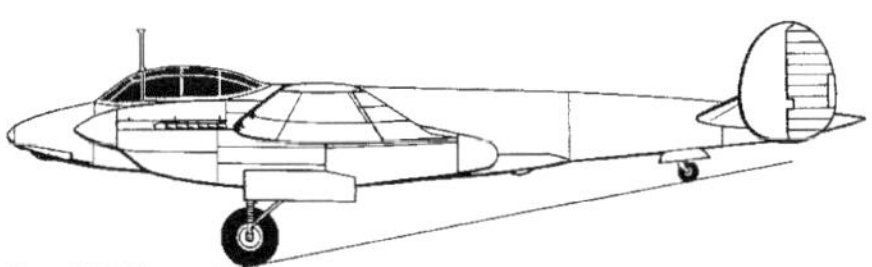

Pe-2VI

nav/gunner; wing NACA 23012 profile mounted in mid posn above bay for 1 t bombload with 1 t bombload external; one UB fired by pilot, second in tailcone aimed remotely by backseater; new engs with rads in wing, oil rads outbd, all exits under wing, fuel capacity increased. State test May-June 44, built in small series.
DIMENSIONS: Span 17,18 m (56 ft $4\frac{3}{8}$ in); length 13,4 m (43 ft 11½ in); wing area 41,8 m^2 (450 ft^2).
ENGINES Two VK-107A.
WEIGHTS Empty 7080 kg (15,608 lb); fuel/oil (max) 3 t; loaded 9058 kg (19,969 lb), (max) 10,4 t.
PERFORMANCE Max speed 556 km/h at SL, 656 km/h (408 mph) at 5,6 km; climb 7·0 min to 5 km; service ceiling 9350 m (30,700 ft); range 2275 km (1,414 miles).
Pe-2 exptls Over 30 sub-types and mods, many post-war; those on OKB record include: **Pe-2RD** 3-seat dive bomber with M-105RA engines and RD-1 (later RD-1X3) rocket in tail; 2 BS + 1 ShKAS, max 1 t bombload; tested by Mark L. Gallai 43, 169 in-flight firings, 8200 kg, 520 km/h without rocket, range 800 km with 500 kg bombload. **Pe-3bis** 2-seat

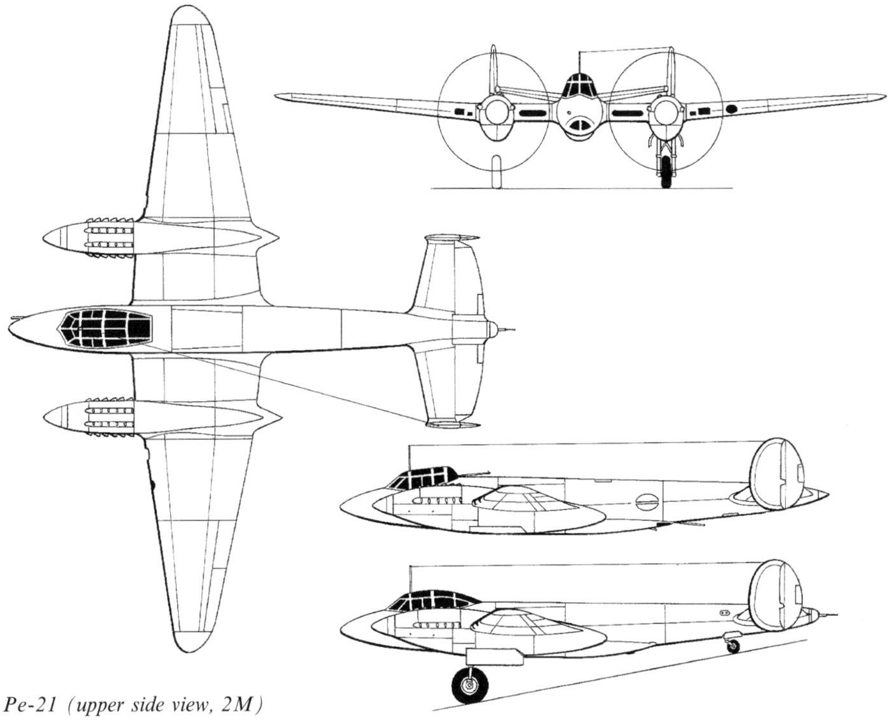

Pe-2I (upper side view, 2M)

Pe-2 (109-014 pulsejet testbed)

fighter with VK-105PFs, 2ShVAK, 3 BS and 2 DAG-10 plus 700 kg bombs, total 8300 kg, 545 km/h at 4 km. **Pe-2D** 3-seat bomber with VK-107As, 2,5 t bombs plus 3 BT (not BS) and 1 DAG-10; 8750 kg, 600 km/h at 5640 m, range 1700 km. **Pe-2 Paravan** dural 5 m nose girder carrying sharp cables leading to wingtips. **Pe-2DV** long-range (2500 km) bomber with bigger wing, ASh-82FNV engs, pressure cabin for 12 km, 1 t bombs but no guns, never completed. Post-war tested ejection seats, guns and German As 014 pulsejet.

Total for regular series 11,427, completed early 45. Post-war to Bulgaria, Czech (as B-32 and trainer CB-32) and Poland. ASCC name 'Buck'.

Phoenix

Joint-stock company Phoenix-Aviatekhnika formed Moscow 91.

LKhS-4 Light utility transport intended for taxi, commuter and other local tasks throughout CIS. Slender 2-spar 15·5% biplane wings, lower with full-span flaps, folding to width 3 m (9 ft 10 in); nacelle with 2 pairs seats and door each side, boom carrying three identical tail surfs, pusher engs driving 1,6 m (63 in) pressed-metal props, two fuselage tanks total 100 lit (22 gal), fixed tricycle gear with leaf-spring legs and 1,5 kg/cm^2 (22 lb/in^2) tyres for soft snow or

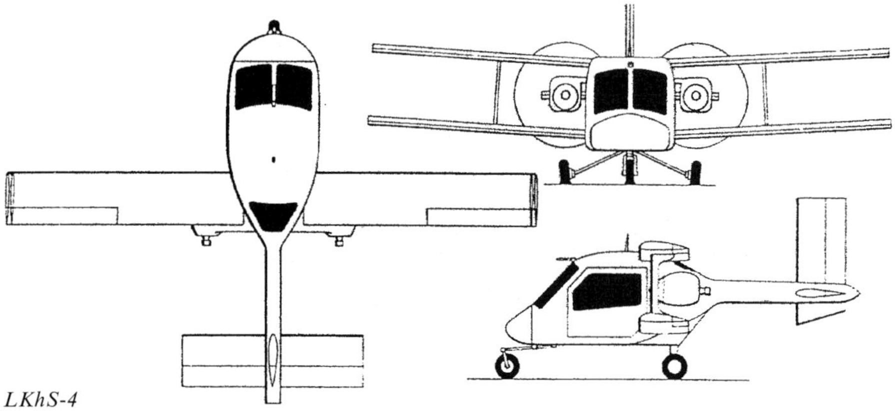

LKhS-4

mud. UHF radio, unspecified deicing. Target cost Rb10m, first flight 93, deliveries planned for 94.

DIMENSIONS Span 8,2 m (26 ft 10⅞ in); length 5,0 m (16 ft 4⅞ in); wing area 13 m² (140 ft²).

ENGINES Two BA3.

WEIGHTS Empty 550 kg (1,213 lb); MTO/MLW 950 kg (2,094 lb).

PERFORMANCE Cruise 200 km/h (124 mph); climb OEI 102 m (335 ft)/min; range (max payload) 500 km (311 miles); TO/ldg 200 m (656 ft).

Polikarpov

Nikolai Nikolayevich Polikarpov was, with Tupolev, most famous Soviet designer before the Second World War. In general he concentrated on small aircraft, while Tupolev built large ones; in particular Polikarpov created roughly 99% of Soviet trainer, fighter and reconnaissance aircraft up to 1941. Born 8 July 92 at Georgyevsk (now Livensk) east of Saratov. Graduated as mech and aero engineer early 16 from St Petersburg Poly Inst, working until 18 at Russo-Baltic Wagon Works (RBVZ) on IMs and other aircraft. In turmoil of civil war only surviving aircraft factory was former Duks plant, subsequently GAZ-1, whence Polikarpov went in 18 and was appointed chief engineer. Supervised Spad S.VII production and then conversion of D.H.9a into important R-1 programme. First original design IL-400/I-1 (22). In 26 appointed Director of OSS, GAZ-25, with V. M. Olkhovskii deputy and including many later famous such as P. F. Fedorov, A. A. Dubrovin, M. M. Shishmaryev, V. V. Nikitin, V. P. Yatsenko, A. G. Brunov, S. O. Zonshain, M. K. Tikhonravov, V. P. Nyevdachin, V. F. Savelyev, I. V. Venyevidov, L. D. Kolpakov-Miroshnichyenko, K. A. Petrov, S. A. Kochyerigin and A. A. Krylov. Wrote *The Manoeuvrability of Powerful Fighters* (27). In 29 arrested and, with his staff, set to work (alongside Grigorovich and his staff) in Hangar 7 at GAZ-39, pioneering VT 'internal prison' type of design office. Created I-5, released 33 as CCB chief landplane designer (Brigade 2, GAZ-39), formed own OKB June 37 at GAZ-84, crash of M. M. Gromov brought retribution to colleagues but Polikarpov enjoyed Stalin's protection and still at desk when collapsed and died 30 July 44; OKB abruptly closed.

R-1, R-2 Possibly fair to regard as first mass-produced aircraft of Soviet Union, derived by Polikarpov from D.H.4 and D.H.9. Several flyable examples of both, captured 1917-21, and greatly admired. Embryonic VVF clamoured for more, and Polikarpov's main task at Duks factory from late 18 was examining D.H.4 and preparing Russian drawings. No Rolls-Royce engines, so designed installation of Fiat A-12 (240 hp) and oversaw production of 20 examples 1920-21. In 22 Polikarpov switched to D.H.9a type airframe, powered by captured Mercedes D.IV(a) (260 hp), receiving designation R-1 (Razvyedchik, recon, type 1); 100 built 1922-23, some dozens using British (Airdisco) airframes purchased as job lot. By 23 Duks factory organized as GAZ-1 and in 1923-24 delivered 130 R-2 (also written R-II) with 220 hp Siddeley Puma. By early 24 Polikarpov had completed design of definitive R-1 with numerous detail improvements, powered by M-5 (copy of Liberty 12) and normally armed with fixed Vickers (PV-1) with 200 rounds, single (sometimes twin) DA in rear cockpit with 500 rounds and up to 400 kg (882 lb) bombs. Data below for this standard R-1, produced by GAZ-1 (by years 26-30) 392/383/445/225/56 and GAZ-31 (29-32) 312/329/303/2, total 2,447. Single example with Lorraine-Dietrich and 4-blade prop, poor. M. M. Gromov and M. A. Volkovoinov flew to Pekin June/July 25, many other long flights. After removal from first-line units 20 refitted with BMW IVA and used as trainers. Polikarpov designed twin-float MR-1 with two wooden floats and taller vet tail, built by GAZ-31 (27-29) 49/43/32, total 124, and single MR-2 in 27 with Munzel floats in Kolchug alloy.

DIMENSIONS Span 14024 (46 ft 0 in); length 9236 (30 ft 3⅜ in); wing area 44,54 m² (479 ft²).

ENGINE One M-5.

WEIGHTS Empty 1450 kg (3,197 lb); fuel/oil 327 kg+40 kg; loaded 2217 kg (4,888 lb).

PERFORMANCE Max speed 185 km/h (115 mph); climb to 1 km 4·5 min; service ceiling 5,8 km (19,000 ft); range 700 km (435 miles); TO 250 m; ldg 90 km/h, 200 m.

IL-400, I-1 Prime need of VVF was for fighter, and especially for national design capability. Project for Soviet single-seat fighter begun early 22 at about time factory organized as GAZ No 1. Polikarpov decided to use Liberty engine (later made as M-5) and unbraced low-wing monoplane configuration. Assistance by I. M. Kostkin, head of technical section at GAZ-1, and A. A. Popov, head of manufacturing. Wooden construction with four main fuselage longerons, secondary fairings giving rounded top/bottom and 3/5-mm ply covering; wing with two box spars, lattice ribs and ply back to front spar, fabric elsewhere. Aerofoil RAF-derived, 20% root and 15% tip, set at −3° and with zero-lift angle of −7°. D.H.9a radiator and propeller. Wire-braced main gears and curved tube skids to protect wingtips. First flight 15 Aug 23; K. K. Artseulov could not prevent aircraft rearing up to about 20 m, with stall and fall in flat attitude. Discovered c.g. much too far aft (52% MAC) and aircraft redesigned as **IL-400B** or **IL-400bis**. New wing of reduced thickness (16% root, 10% tip), areas of vert/horiz tail increased, wing and tail redesigned in Kolchug aluminium with sheet ribs and beaded 0,8-mm skin; cockpit and engine moved forward and radiator of Lamblin type underneath; cutout at wing root TE for pilot view. First flight 18 July 24 by Artseulov, subsequent testing by A. I. Zhukov and A. N. Yekatov until cleared for

R-1 civil demo Iskra

Il-400 bis

production 15 Oct as **I-1**.

Next aircraft still a prototype, metal skins replaced by plywood and static tests carried out to establish safe factor of 10. Lamblin replaced by deep honeycomb radiator obliquely under nose, two PV guns mounted on forward fuselage (location apparently not recorded) and fuel tanks (plural) welded from Kolchug following discovery of suitable flux by alloy's inventors (see materials section). Total 33 I-1 ordered, but much variation; even first 13 had three variants, and welded tankage soon dropped. All 33 sent to NOA where prolonged tests showed recovery from spin often difficult; 23 June 27 M. M. Gromov escaped by parachute first time in USSR, while A. I. Sharapov later in 27 hit ground in flat spin and survived by sheer luck. Thus, though fast and in many respects advanced fighter, I-1 never in VVF service.

DIMENSIONS Span 10,8 m (35 ft $5\frac{1}{4}$ in); length (400) 8,2 m (26 ft $10\frac{7}{8}$ in), (bis) 8,3 m (27 ft $3\frac{3}{4}$ in); wing area 20,0 m^2 (215 ft^2).

ENGINE One Liberty, (I-1) M-5.

WEIGHTS (400bis) empty 1112 kg (2,451 lb); fuel/oil 230 kg; loaded 1510 kg (3,329 lb).

PERFORMANCE Max speed 264 km/h (164

PM-1

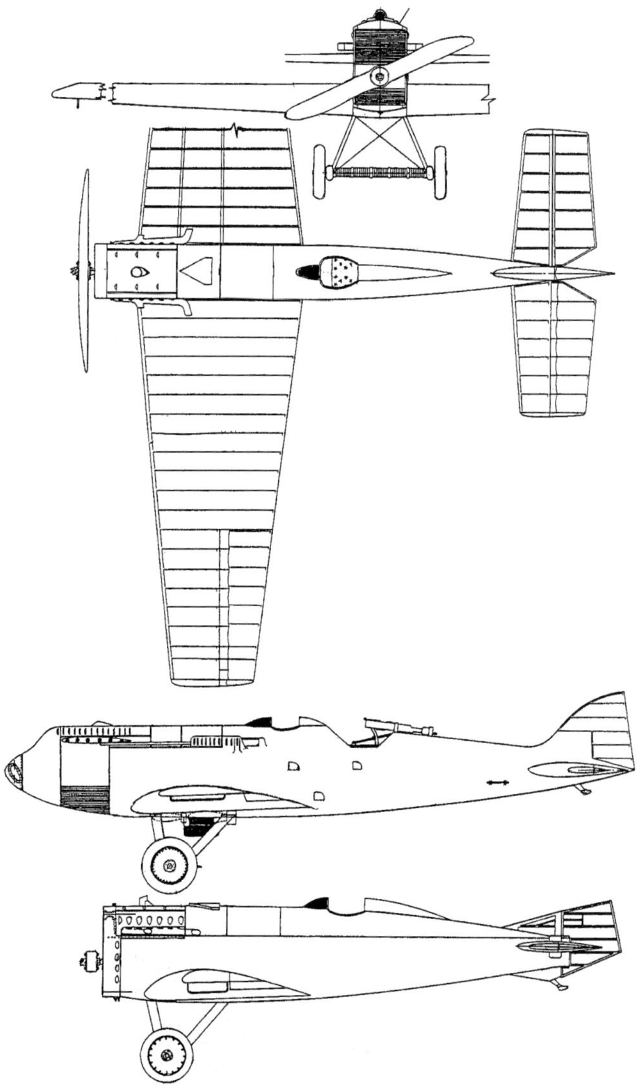
IL-400 (upper side view, projected unbuilt RL-400V)

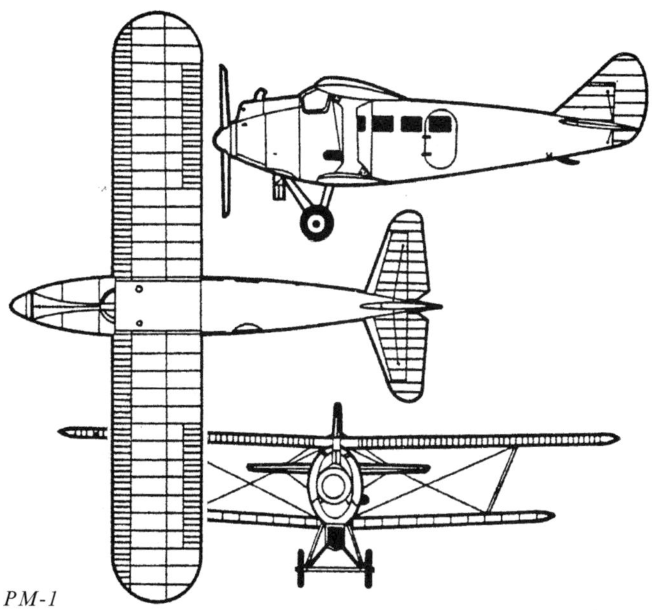
PM-1

mph); climb to 1 km, 1·9 min; service ceiling 6750 m (22,150 ft); range 650 km (404 miles)/2·5 h; take-off 7 s; landing speed 105 km/h.

PM-1, P-2 Need for Soviet transport prompted this beautifully executed sesquiplane, basic design of which was due to Aleksandr Aleksandrovich Syemyenov. All-wood construction; monocoque fuselage of glued *shpon* with integral fin, two-spar wings attached above and below main fuselage frames and joined by single streamlined I-type interplane struts, fabric covering aft of front spars and on movable control surfaces. Open cockpit for pilot (room for copilot but believed seat not installed) and comfortable cabin for five passengers (believed two pairs plus single at rear). Honeycomb water radiator extended by hand wheel and chain on sliding mounts ahead of main gears. Aircraft completed after three months' work and named PM-1 (*Passazhirskii* Maybach). First flight A. I. Zhukov 10 June 25 (same day as start of group flight to Pekin in which PM-1 had hoped to participate). No problems and all tests soon completed. Production of ten begun for Deruluft, and after public demo 26 July first PM-1 flew to Leningrad and in early August began service Moscow-Berlin; on second (Königsberg-Berlin) sector damaged in forced landing, not repaired because engine not in production.

DIMENSIONS Span 15,5 m (50 ft 10¼ in); length 11,0 m (36 ft 1 in); wing area 38,5 m^2 (414 ft^2).
ENGINE One 260 hp Maybach IVa.
WEIGHTS Empty 1380 kg (3,042 lb); fuel/oil 440 kg; loaded 2360 kg (5,203 lb).
PERFORMANCE Max speed 180 km/h (112 mph); climb to 1 km, 8 min; service ceiling 4,1 km (13,450 ft); range 1200 km (746 miles); take-off 22 s; landing 15 s/90 km/h (56 mph).

2I-N1, DI-1 Designations from 2-seat fighter, Napier No 1 and, later, two-seat (*Dvukhmyestnyi*) fighter No 1. Outstanding aircraft in both aerodynamic and structural design, and powered by most impressive available imported engine. Fuselage monocoque of glued *shpon*, oval section with multiple layers built up to thickness 4 mm over forward section tapering to 2 mm at tail. This fuselage by Vladimir Denisov, specialist in wooden monocoques through VT prison period and into late 30s. Sesquiplane wings without internal bracing wires, with two wooden spars, ribs of 3-mm ply with large lightening holes capped by ply strips 23-mm wide, plus multiple stringers 10 × 10 mm, with 1,5-mm ply skin. Single spar in lower wing joined by V-struts to two spars in upper, all struts streamline section in Kolchug alloy, with steel Rafwires for external bracing. Aerofoil to fair main-gear axle; aircraft flown with braced skis. One fixed PV and one DA on ring mount in rear cockpit. First flight 12 January 26, excellent; Polikarpov as observer on flight 4 and flight 8 (ceiling climb). On flight 9, 31 March 26, V. N. Filippov, with test observer V. V. Mikhailov, made speed runs over measured kilometre at Central Airfield. At max speed at 100 m covering ripped off right upper wing, upper surface first, then lower; lower right wing then also collapsed. Both crew killed. Cause eventually judged to be bad glueing of skin, with large areas of rib cap and stringer not glued at all, and many panel pins had missed underlying structure. Omission of bradawl holes

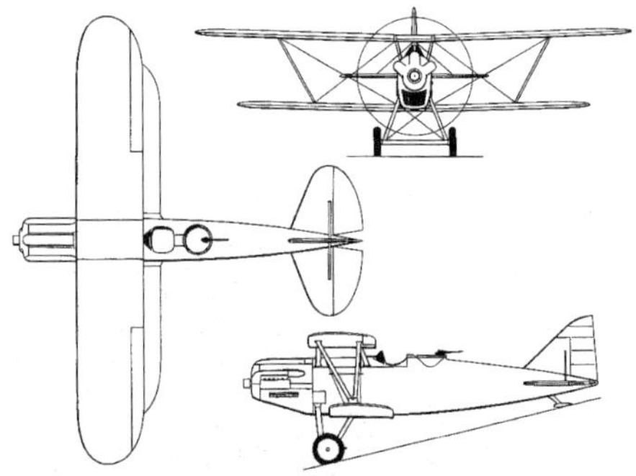

2I-N1

in skin (usual practice to enable internal pressure to balance that outside) had resulted in pressurization on ceiling climb, forcing skin off wing, suction at high speed finally pulling it off. The crash of such advanced aircraft shocked entire industry, six months elapsing before resumption of designing and with many subsequent aircraft (I-3, U-2, DI-2 and even R-5) being over-strength.

DIMENSIONS Span 12,0 m (39 ft 4½ in); length 9,75 m (32 ft 0 in); wing area 27,15 m^2 (292 ft^2).
ENGINE One 450 hp Napier Lion.
WEIGHTS Empty 1153 kg (2,542 lb); fuel/oil 547 kg; loaded 1,7 t (3,748 lb).
PERFORMANCE Max speed 268 km/h (166·5 mph); climb to 1 km, 1·8 min; service ceiling 7,1 km (23,300 ft); range 800 km (500 miles)/3 h; time turn 360°, 12 s; take-off 100 m/7 s; landing 200 m/16 s/92 km/h.

U-2, Po-2 Though not famous outside the Soviet Union, this aircraft was possibly built in greater numbers in more variants than any other in history. Designed as simple biplane trainer, its robust reliability led to adaptation to countless other duties and continued production 1928-51 in USSR with others assembled from spares until at least 59; plus licence prodn Poland 48-55; total factory-built variants c80, total number built est 32,600-33,300.

In 26 VVF used assortment of aircraft as trainers in addition to U-1, and wished to establish single Soviet type with reliable engine around which uniform syllabus could be writ-

U-2 prototype

2I-N1

ten. As newly appointed head of OSS, Polikarpov instructed to build required aircraft. Emphasis on simplicity and cheapness resulted in compromised design in which aerodynamics and performance played little part. Long project definition with NTK UVVS (S. V. Ilyushin) resulted in decision to use new M-11 engine, to eliminate previously important training stage of prolonged taxiing exercises in aircraft with wing skin removed, and to make maximum use of concept of interchangeable parts. Decision to follow U-2 phase with more powerful P-2 intermediate trainer led to stipulated speed not exceeding 120 km/h. Result was ugly single-bay biplane with four identical rectangular wings (upper span greater because of centre section) each 14% thick with constant-section spars, all ribs identical, and same hinged surface used as rudder, elevators and ailerons. This prototype, later called **TPK** (thick wing section), completed about Feb 27. Tough and simple, but performance so poor it became a problem (speed reached 142 km/h, but climb to 1 km took 11·5 min, and to 2 km 29 min), and payload could not exceed 120 kg. Static tests showed excessive factor of 17.

DIMENSIONS Span 11,0 m (36 ft 1 in); length 8,1 m (26 ft 6⅞ in); wing area 34,0 m^2 (366 ft^2).
ENGINE One M-11.
WEIGHTS Empty 758 kg (1,671 lb); fuel/oil 95 kg; loaded 1013 kg (2,233 lb).
PERFORMANCE Max speed 142 km/h (88 mph); climb 11·5 min to 1 km; range 350 km (217 miles); TO 100 m; ldg 65 km/h, 85 m.

Aircraft completely redesigned to have efficient structure without such emphasis on simplicity. Wings with rounded tips, all wings still interchangeable but of 8·1% CAHI-541 profile resembling Clark Y with flat underside. Pine box spars, ply ribs and caps, rectangular strip stringers and leading-edge rib caps, and fabric covering. Fuselage pine truss with wire bracing, front part covered with removable 1 mm ply panels mainly held by screws, rear with fabric. Periphery of wings and tail surfaces dural tube, tail ribs, ply, fabric covering. All struts originally streamlined-section dural tube, later (in 29) replaced for cheapness by round mild-steel tube with downstream fairing of wood attached by doped fabric. Tailplane normally braced by two struts with screwed end-fittings adjustable for length. Rotating Y-forked bolt on fin spar enabled tailplane incidence to be altered on ground. All fittings mild steel, and casein glue used throughout. Engine hung on large ring stamped from dural plate 4-mm thick held on steel pipes with pinned forked end-fittings. In 30 changed to copy of Sh-2 engine mount (made at same plant) of simple welded steel tubing. Pin-jointed main gears with rear struts sprung by 16 m on each leg of 13-mm bungee; spoked wheels with 700×120 mm tyres. Ash tail-skid joined to rudder by twin torque-springs. Skis, main 2200×344×264 mm, tail 500×204 mm. Engine, originally in streamlined cowl leaving cylinders only exposed, in production U-2 completely exposed but with four panels to rear held by cable and spring and released almost instantly for access to rear of engine. Many variations, and crankcase often covered by locally made cowl in winter.

First improved U-2 flown by M. M. Gromov 7 Jan 28. Acclaimed for good qualities, especially positive longitudinal stability and absence of vices. Spun only with difficulty, and immediately recovered by releasing or centring stick (not necessarily good quality in primary aircraft). Prototype on skis completed state tests in ten days; then exhibited at Berlin. Series production at GAZ-23 (by years 29-40) 28/89/283/942/1,381/1,100/327/968/1,782/2,016/1,584/472, switched to GAZ-387 50 in 40, 1,245 in 41, total by 22 June 41 12,267.

DIMENSIONS Span 11,4 m (37 ft 4¾ in); length 8,17 m (26 ft 9⅝ in); wing area 33,15 m^2 (357 ft^2).
ENGINE One M-11.
WEIGHTS Empty (typical) 635 kg (1,400 lb); fuel/oil 71 kg+10 kg; loaded 890 kg (1,962 lb).
PERFORMANCE Max speed 156 km/h (97 mph); climb to 1 km 5·6 min; service ceiling c4 km (13,100 ft); range c400 km (250 miles); TO 70 m; ldg 65 km/h, 120 m.

U-2

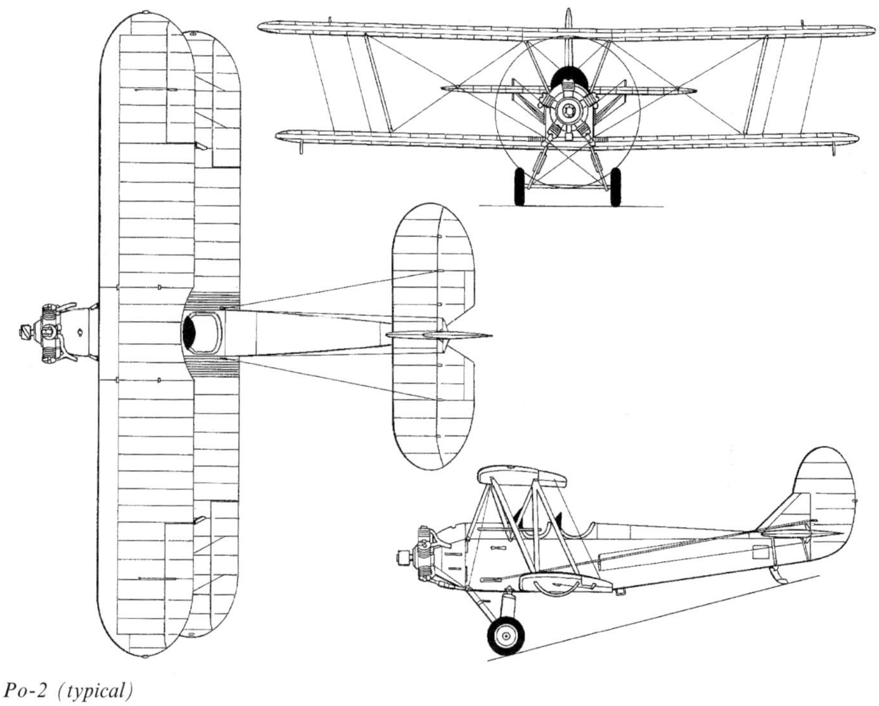
Po-2 (typical)

First main variant **AP** (*Aeropyl*, air dust) or **AO** (*Aeroopylitel*, air duster), flown October 30. First production aircraft designed for ag-aviation, front seat moved 250 mm forward, space behind occupied by tank (usually aluminium but type and capacity varied) for up to 250 kg (551 lb) chemicals dispensed across 10 m swath by dural sheet box-venturi, usually assisted by four-blade windmill and impeller. Production 260 by 32 and 1,235 by 40.

U-2M, MU-2, seaplane variant designed at CCB by S. A. Kochyerigin. MU, *Morskoi Uchyebnii*, marine trainer. Central float wooden, single step, dual tandem steel-tube struts with wire bracing; N-strutted wing-tip floats. Tested

Moscow River spring 31; poor performance and handling. Subsequently floatplane variants produced by V. B. Shavrov, N. G. Mikhelson, A. Ya. Shchyerbakov and S. A. Mostov, none being fully successful despite fitting M-11D engine.

U-2 Siemens Single aircraft re-engined 31 with 125 hp Siemens Sh 14 for Arctic service. Equipped with skis, radio and extra gravity fuel tank on centre section.

U-2KL Designated from name of factory; two aircraft completed with large rear cabin with bulged streamlined hinged canopy with Celluloid windows, flown early 32.

U-2SP, SP Designation from *Spetsprimeneniya,* special purposes. Incorporated structural modifications of AP, but instead of tank had two open rear cockpits (often one large one) without instruments or controls and with easily removable seats. Great success and 861 built 1934-39, most for Aeroflot for utility transport.

U-2S, S-1, S-2 Designation from *Sanitarnyi Samolyet*, ambulance aeroplane. Following proposal by Dr A. F. Lingart, one aircraft completely rebuilt 32 with forward cockpit as on AP and SP, with closed cubicle to rear for doctor and compartment for stretcher patient between that and fin, covered by top decking folded to right (when closed, interior dark except later when Celluloid portholes added). Doctor could speak to patient and touch patient as far as waist. Batch of 100 built 1934-36.

U-2G Designation from Gorzhan, name of pilot who suggested radically different flight-control interface in cockpit. No modification to control surfaces but rudder linked to modified control column with ram's horn spectacles operating in three senses, pedals being removed. Tested by NII GVF summer 34.

U-2 V-tail (no designation known). Single aircraft tested 34 by K. A. Kalilyets and A. A. Ivanovskii fitted with V-tail, or *typa babochki* (butterfly); twin oblique tail surfaces with included angle 140°; ply skin on wood (no mention of dural periphery) with fabric to rear of spar. Described as 'no worse than U-2' but no production. Similar tail later on R-5.

E-23 Designation from *Eksperimentalnyi*, special training and research variant designed 1934-35 by N. G. Mikhelson for sustained inverted flight. Wings of symmetric profile 9% thick, with flaps on lower wing. Modified fuel system and experimental carburettor for aerobatics. Flown May 35, successful but many manoeuvres called for more power. No production.

SPL Designation from SP Limuzin, Sp by 35 also coming to mean *Svyazno Passazhirskii*, liaison passenger. Development of original SP with two passengers in more comfortable cabin with hinged lid, exhaust muff heating, Celluloid windows and wicker chairs. SPL usually also fitted with Townend Ring or NACA cowl and often with spats. Oil company Bashnyeft adopted their own model; features attributed to engineer Barsuk. Shavrov says no production, but Bashnyeft had several.

U-2 Grushin No designation known for beautifully styled improved version produced at MAI under P. D. Grushin in 35. NACA cowl, smoothly faired cockpit canopy, spatted wheels on streamlined struts and two 55-lit supplementary tanks. Whole aircraft brightly doped.

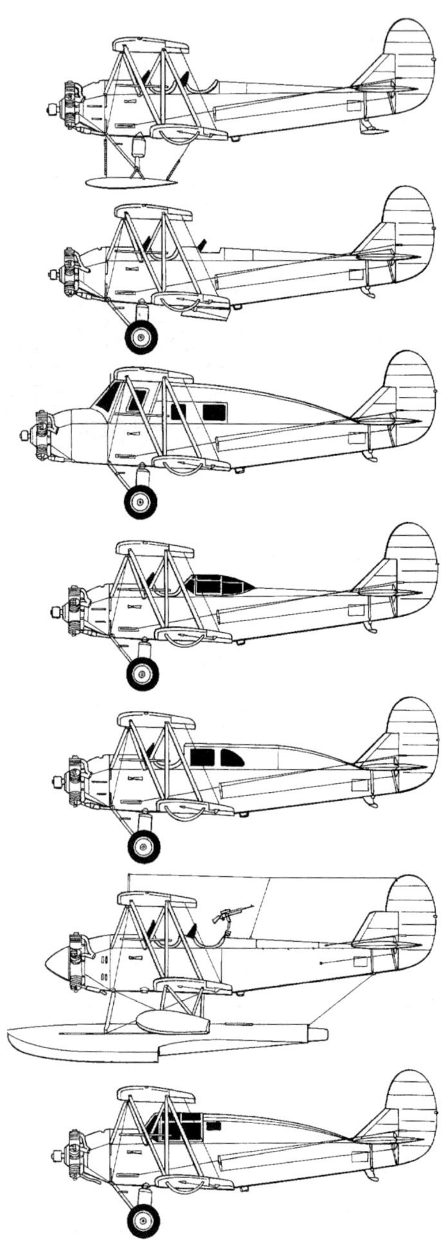

Po-2 variants, from top: Std aircraft on GAZ-25 skis, U-2AP (Po-2A), U-2L (Rafaelyants), U-2KL, Po-2L, U-2M, typical ShS/Po-2S

Weight increase only 27 kg; speed increase 20 km/h.

U-3, U-4, RV-23 These major rebuilds are described under designer, Mikhelson.

U-2UT Designation from *Uchyebno-Trenirovochnyi*, educational training. First change in basic series, in Jan 41, with switch to M-11D as standard. Few UT built, distinguished by individual helmets over cylinders and enclosed canopy over rear cockpit.

DIMENSIONS Standard.

ENGINE One M-11D.

WEIGHTS Empty 705 kg (1,554 lb); fuel/oil 98 kg+12 kg; loaded 964 kg (2,125 lb).

PERFORMANCE Max speed 152 km/h (94 mph), cruise 112 km/h (70 mph); time to 1/1,5 km 8/13 min; service ceiling 4,2 km (13,800 ft); range 500 km (311 miles); TO 77 m, ldg 65 km/h, 95 m.

S-2, S-3 Following S-1 these ambulance variants of 39 and 40 were produced in series. **S-2** with good arrangement for single patient (340 kg limit) built in large numbers; many mods incl clipped-on pods (see **Po-2S**) and S. A. Mostov's dural floats for Arctic/Siberia in 46. **S-3**, also called **SKF** (*kabina Filatov*) devd during Winter War in Finland; ply upper fuselage for two stretchers.

DIMENSIONS Standard.

ENGINE One M-11D.

WEIGHTS Empty (2) 760 kg (1,675 lb), (3) 719 kg (1,585 lb); fuel/oil 90 kg+10 kg; loaded (2) 1,1 t (2,425), (3) 1071 kg (2,361 lb).

PERFORMANCE Max speed (2) 151 km/h (94 mph), (3) 149 km/h; cruise (both) 115 km/h; time to 1/1,5 km (both) 8·5/20 km; service ceiling (2) 3,2 km, (3) 3460 m; range (both) 450 km (280 miles); TO/ldg (2) 125/130 m, (3) 110/100 m.

U-2L, PAM-11 A. N. Rafaelyants rebuild 43 with new fuselage filling space between wings giving cabin 2,6×1,1×1,6 for pilot and 4 (overload 6) pax; handling little changed.

DIMENSIONS According to Shavrov, standard.

ENGINE One M-11D.

WEIGHTS Empty 833 kg (1,836 lb); fuel/oil 105 kg+10 kg; loaded 1350 kg (2,976 lb).

PERFORMANCE Max speed 160 km/h (100 mph), cruise 120 km/h; time to 1/1,5 km, 12/29 min; service ceiling 2,8 km (9,190 ft); range 450 km (720 kg overload fuel).

U-2LPL, LPL Designation from *Lyezhachi Polozheni Lyetchik*, prone-position pilot. Need for prone piloting first argued 35 for ultra-fast aircraft; then decided useful for air-combat manoeuvres. Eventually U-2 rebuilt with new rear cockpit accommodating pilot lying on front in inclined position, with chin-rest and soft pad against forehead. Pilot Bragin did NII VVS testing from 39, results suggesting excellent idea for record-breaking or interceptor-type aircraft. Provided background for Type 346.

Po-2A Following Polikarpov's death in 44 aircraft redesignated Po-2, and Po-2A agricultural variant remained in production throughout war and in Poland to 53, post-war models having M-11K. Improved (V. F. Stepanov AOD-S3) dispenser increased swath width to 30 m, and raised working rate from 17 to 30 hectares/h. Production (with AP) 8,980.

DIMENSIONS Standard.

ENGINE One M-11K.

WEIGHTS Empty 740 kg (1,631 lb); fuel/oil 134 kg+15 kg; loaded 1250 kg (2,756 lb).

PERFORMANCE Max speed 140 km/h (87 mph), cruise 115 km/h (71 mph); time to 1/1,5 km 12/28 min; service ceiling 2,4 km (7,874 ft); range 720 km (447 miles); TO 170 m, ldg 75 km/h, 90 m.

Po-2VS Basic VVS front-line aircraft, VS (*Voiskovoi Seriya*, military series) often impressed ex-GVF SP with three open cockpits. Standard Soviet military liaison aircraft, also built from new and issued to all senior officers of all services, large numbers with prison and labour-camp commandants and local military governors. Inventory of 45 showed over 9,200 in liaison duty. Many locally modified with extra tank on centre section, enclosed transparent canopy (often from other aircraft), slats and/or flaps (variant by M. M. Kulik), various engine cowls and external payload panniers.

Po-2Sh Light attack variant with fixed ShKAS KM-33 above each lower wing, DA in rear cockpit, two FAB-100.

LSh *Lyegkii Shturmovik* (light attacker), some hundreds of existing (mainly pre-41) aircraft converted in various workshops as close-support aircraft; alternative designation **U-2VOM-1**. Normally equipped with ShKAS aimed by observer in rear cockpit, 120 kg of bombs and four RS-82 on underwing rails. Even without armour usually weighed 1,4 t, sluggish performance. Established great reputation; German troops said to believe LSh would look into each window to see if invaders inside houses.

DIMENSIONS Standard.

ENGINE Usually one M-11D.

WEIGHTS Empty 797 kg (1,757 lb); fuel/oil 90 kg+10 kg; loaded (no FAB/RS) 1150 kg (2,535 lb).

PERFORMANCE Max speed 130 km/h (81 mph), cruise 110 km/h (68 mph); time to 1/1,5 km 11/24 min; service ceiling 3,0 km (9,840 ft); range 430 km (267 miles); TO 130 m, ldg 75 km/h, 190 m.

Po-2SP No relation to previous SP, but designation again from *Spetsialno Primeneniya*, special tasks; photography, survey, instrumentation, post-war magnetometer survey etc.

Po-2LNB *Lyegkii Nochnoi Bombardirovshchik*, light night bomber, wartime production version based on emergency conversions in late 41. Most two-seat (some single-seat) with two FAB-100 or similar 200 kg (441 lb) bombload. Invariably dull black overall, silencer on exhaust pipe and if possible with engine idling in target area. Target illuminated by flares or, according to many contemporary reports, searchlight carried by companion LNB. Large number used against Luftwaffe airfields and front-line troops. Countless mods including VMF version with elec generator and large ring antenna for exploding sea mines.

DIMENSIONS Standard.

ENGINE One M-11D.

WEIGHTS Empty 773 kg (1,704 lb); fuel/oil 90 kg+10 kg; loaded 1,4 t (3,086 lb).

PERFORMANCE Max speed 134 km/h (83 mph), cruise 100 km/h (62 mph); time to 1/1,5 km 25/48 min; service ceiling 1,5 km (4,900 ft); range 450 km (280 miles); TO 270 m, ldg 82 km/h, 140 m.

NAK, Po-2NAK *Nochnoartilleriskyi Korrektovovshchik*, night artillery correction, local reconnaissance in dull black with observer equipped with army radio to report on effects and accuracy of gunfire.

Po-2P Designation from *Poplavkii*, floats; substantial number built or converted with central/tip floats to designs of (40) Shavrov (Sh in data) as trainer and (44) Polikarpov (P) for recon with ShKAS or DA in rear cockpit, four FAB-25 (overload FAB-41), most with extended-chord fin; many sub-variants.

DIMENSIONS Standard except length (both) 8,9 m (29 ft 2⅜ in).

ENGINE (Sh) M-11, (P) M-11D.

WEIGHTS Empty (Sh) 880 kg (1,940 lb), (P) 949 kg (2,092 lb); fuel/oil (Sh) 90 kg+10 kg, (P) 105 kg+12 kg; loaded (Sh) 1150 kg (2,535 lb), (P) 1350 kg (2,976 lb).

PERFORMANCE Max speed (Sh) 147 km/h (91 mph), (P) 144 km/h; cruise (both) 120 km/h; time to 1/1,5 km (Sh) 10/25 min, (P) 12·7/30 min; service ceiling (Sh) 3, (P) 2,5 km; range (Sh) 480 km (298 miles), (P) 500 km; TO (Sh) 200 m, (P) 450 m; ldg (Sh) 70 km/h, 110 m, (P) 75 km/h, 115 m.

Po-2SKF (Bakshayev pods)

Po-2L

Po-2ShS Shtabnoi samolyet, staff aeroplane, transport by Polikarpov 44: cockpit 1020 mm wide for pilot and four pax with sliding canopies, larger rudder, elev trimmer and aerody balance, latter also ailerons; GS-35 generator for engine start, lighting and radio; 10-min conversion to ambulance.

DIMENSIONS Standard.

ENGINE One M-11F.

WEIGHTS Empty 822 kg (1,812 lb); fuel/oil 105 kg+10 kg; loaded 1350 kg (2,976 lb).

PERFORMANCE Max speed 163 km/h (101 mph); cruise 120 km/h; time to 1/1,5 km 13/31 min; service ceiling 2,6 km (8,530 ft); range 425 km (715 km aux fuel); TO 255 m; ldg 76 km/h, 410 m.

Po-2S General designation for wartime ambulance variants, both new and rebuilds (see previous U-2S/S-2/3-3). Standard versions had fuselage behind pilot arranged for two passengers or two stretcher patients, but some **U-2S** had provision for four sitting pax and those with patient pods could carry as overload five. Original pod (41) by A. Ya. Shchyerbakov accommodated patient semi-reclining ahead of and below lower leading edge. Improved type (44) by G. I. Bakshayev resembled box-like drop tank into which stretcher slid from end. Most variants could seat third patient or doctor in rear cockpit. Large number of pre-war S-1 and S-2 continued in use, with many variations (rear decking hinged on left, or hinged up from front, or in front/rear sections, latter being fixed and stretcher slid in via front half, thick centre-section fuel tank, and glazed canopy in sections

Po-2Sh (fixed ShKAS KM-33 above lower wings, 7,62 Degtyaryev at rear)

over patient and doctor/pilot). Po-2S remained in production in Poland post-war.

DIMENSIONS Usually standard.

ENGINE (U-2S) M-11G, (others) M-11K.

WEIGHTS Empty (pods typical) 862 kg (1,900 lb), (U-2S) 740 kg (1,631 lb), (Po-2S) 810 kg (1,786 lb); fuel/oil (p) 93 kg+10 kg, (U, P) 145 kg+15 kg; loaded (p) 1415 kg (3,119 lb), (U) 1220 kg (2,690 lb), (P) 1250 kg (2,756 lb).

PERFORMANCE Max speed (all) 140 km/h (87 mph), cruise (p, U) 110 km/h, (P) 115 km/h; time to 1/1,5 km, (p) 12/- min, (U, P) 12/29 min; service ceiling (p) 1,2 km (3,940 ft), (U, P) 2,45 km (8,040 ft); range (p) 450 km, (U, P) 720 km; TO (p) 375 m, (U, P) 170 m; ldg (p) 82 km/h, 305 m, (U, P) 75 km/h, 90 m.

Po-2GN *Golos Nyeba*, voice from sky, psy-war variant of 44 with 80-W amplifier and loudspeakers broadcasting messages usually direct from microphone in rear cockpit; same with silenced exhaust as LNB.

Po-2 About half final batches were simple dual trainers. Following is typical (46):

DIMENSIONS Standard.

ENGINE One M-11K.

WEIGHTS Empty 750 kg (1,653 lb); fuel/oil 100 kg+13 kg; loaded 1023 kg (2,255 lb).

PERFORMANCE Max speed 150 km/h (93 mph), cruise 111 km/h (69 mph); time to 1/1,5 km 8/20 min; service ceiling 3,3 km (10,800 ft); range 410 km (255 miles); TO 150 m; ldg 70 km/h, 170 m.

Po-2LS *Limuzin svyarznoi*, April 45 by GVF engineer M. M. Kulik, 3-seat cabin with Plexiglas canopy, lower wing 1 m shorter with full-span ailerons serving as slotted flaps lowered to 45°, upper wings full-span slats of 0,8 mm Dural, tailplane pivoted and adjusted by slat/flap selection. Weights 800/1150 kg, ldg speed 45 km/h.

Po-2L Standard production passenger version 48, upper centre section reduced from 0,8 to 0,6 m, open pilot cockpit, rear enclosed for two pax or stretchers.

DIMENSIONS Standard.

ENGINE One M-11K.

WEIGHTS Empty 780 kg (1,720 lb); fuel/oil 145 kg+15 kg; loaded 1250 kg (2,755 lb).

PERFORMANCE Max speed 140 km/h (87 mph), cruise 110 km/h; time to 1/1,5 km 12/28 min; service ceiling 2,4 km (7,874 ft); range 720 km (447 miles); TO 170 m; ldg 75 km/h, 110 m.

Popularly called *Kukuruznik* (corn-cutter), many still in use. Variant like LNB used by N. Korea 1950-53, licence prodn in Poland as **CSS-13** and ambulance **CSS-S-13** 1948-55. ASCC name for all 'Mule'.

P-2 Designation from *Perekhodnyi*, intermediate; between U-2 and R-5, to train pilots and observers. Single-bay wing box with four identical wings with 25° stagger, joined by N-struts and long diagonal spanwise strut (tear-drop section 100×40 mm) instead of bracing wires, with objective of eliminating time-consuming rigging by adjusting wire tensions. Construction same as U-2 and R-5, no forward-firing gun but DA on Scarff ring. Water radiator retractable. Four identical interlinked ailerons, braced tailplane, rubber-sprung main gears, aluminium engine cowl and pointed propeller spinner. Prototype completed autumn 27, and flight tests satisfactory, resulting in order for 55, delivered from *Krasnyi Lyetchik* plant 1928-30. It was then discovered diagonal struts serious mistake. Handling was unpredictable, and many aircraft reluctant to recover from spin. Many P-2s modified with conventional wires. Eventually so many adjustments made that wing separated in flight, B. L. Bukholts escaping by parachute. Training schools hardly used P-2, preferring R-1 (esp 240 hp BMW IVa model) with armament removed.

DIMENSIONS Span 10,4 m (34 ft 1½ in); length 7,84 m (25 ft 8⅔ in); wing area 28,0 m² (301 ft²).

ENGINE One M-6.

WEIGHTS Empty 1 t (2,205 lb); fuel/oil 200 kg; loaded 1470 kg (3,241 lb).

PERFORMANCE Max speed 206 km/h (128 mph); climb to 1 km, 3·3 min; service ceiling 5680 m (18,635 ft); range 800 km (500 miles)/4 h; take-off 80 m/7 s; landing 150 m/13 s/85 km/h.

I-3 First Soviet fighter preceded by lengthy planning. Polikarpov had devoted long attention to rival claims of water-cooled engines (long, heavy, but streamlined if one ignored radiator) and air-cooled radials (short, light, but higher drag). OSS decided to build example of each. Water-cooled fighter was I-3, plans being approved 14 May 27. Fuselage with pine truss framework, four longerons and 13 frames with integral fin, with light fairing structure to support semi-monocoque oval-section covering of *shpon*, from 0,5 mm glued veneer built up to thickness 5 mm forward and 3 mm near tail. Wings of Clark Y profile with two spars of pine/ply box type, ply ribs with ply caps and pine stringers, internal wire bracing, ply and fabric covering. All struts dural teardrop section with adjustable screwed end-fittings. All control surfaces dural framework with fabric covering, tailplane adjustable incidence in flight, elevators with large cut-outs though at 0° missed rudder. Elevator push/pull rods, cables to other controls, ailerons differential Frise type. Pin-jointed main gears with rubber shock-absorption, tyres 800×150 mm, skis R-1 type 2,73×0,34 m (area 1,86 m²). CG at 26% MAC. Radiator extended from belly. Two synchronized PV-1 (Vickers in first prototype). Armament by Aleksandr Nadashkevich, responsible for armament of all subsequent Polikarpov aircraft. First flight 4 May 28, second aircraft three months later. Good performance (283 km/h with high-speed prop but long take-off). Fuel consumption 142 lit/h. NII tests late 28 resulted in order 399. Most had imported engine, licensed M-17 later substituted.

DIMENSIONS Span 11,0 m (36 ft 1 in); length 8,01 m (26 ft 3⅓ in); wing area 27,85 m² (300 ft²).

ENGINE One 500/730 hp BMW VI 7,3 z.

WEIGHTS Empty 1,4 t (3086 lb); fuel/oil 210 kg+33 kg; loaded 1846 kg (4,070 lb).

PERFORMANCE Max speed (standard prop) 278 km/h (173 mph) at SL; time to 1 km, 1·8 min; service ceiling 7,2 km (23,620 ft); range 585 km (364 miles); take-off 150 m/12 s; landing 250 m/18 s/100 km/h (62 mph).

D-2, DI-2 Second two-seat fighter merely enlarged I-3; wings joined to extended c-s, fuselage one frame longer and rudder enlarged. Two synchronized PV-1, twin DA on Scarff ring for observer. Completed early 29, but on early flight tailplane fluttered and broke away, A. V. Chyekarev killed. Replaced by DI-3.

P-2

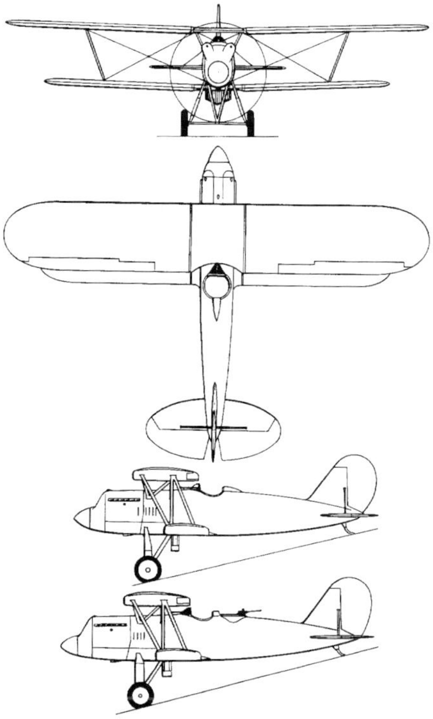

I-3 (lower side view, DI-2)

I-3

DI-2

DIMENSIONS Span 11,8 m (38 ft 8½ in); length 8,2 m (26 ft 10¾ in); wing area 31,8 m² (342 ft²).
ENGINE One 500/730 hp BMW VI.
WEIGHTS Empty 1557 kg (3,433 lb); fuel/oil 210 kg+33 kg; loaded 2122 kg (4,678 lb).
PERFORMANCE Max speed 256 km/h (159 mph); climb to 1 km, 2·2 min; service ceiling 6,3 km (20,670 ft); range 510 km (317 miles); time 360° circle, 14 s; take-off 180 m/14 s; landing 250 m/18 s/100 km/h.

TB-2 Recognizing obsolescence of R-42 and TB-1, OSS instructed 27 to design modern bomber to replace these and Farman F 62. Project begun as **L-2**, Kolpakov-Miroshnichyenko playing leading role. Major changes resulted in delay, but prototype finally appeared spring 30. Sesquiplane upper/lower wings in ratio 2·6:1; profile modified Clark Y, t/c 16%. Wooden construction with two box spars, ply ribs and ply covering back to rear spar. Engine nacelles on lower wing. Upper wing mounted on reverse-N cabane and no other wing struts except single diagonal pair from lower spars just outboard of engines to upper spars well outboard. Struts steel tubes with wooden fairings. Box-like wooden fuselage with ply rounded edges. Simple single tail, control surfaces dural/fabric. Main gears beneath engines with rubber springing and small mudguards. Radiators manually extended upwards behind engines. Crew five, including three gunners (one with radio) each manning pair of DA. Bombload 800 kg (1,764 lb). On test TB-2 flew well, but regarded as obsolescent and decision taken to put all trust in ANT-6.
DIMENSION Span 27,0 m (88 ft 7 in); length 17,6 m (57 ft 9 in); wing area 128 m² (1,378 ft²).
ENGINES Two 500/680 hp BMW VI.
WEIGHTS Empty 4220 kg (9,303 lb); fuel/oil 1 t; loaded 6770 kg (14,925 lb).
PERFORMANCE Max speed 216 km/h (134 mph); climb to 1 km, 3·2 min; service ceiling 6,8 km (22,300 ft); range 1200 km (746 miles) (presumably with bombs); take-off 210 m/18 s; landing 170 m/15 s/86 km/h (53 mph).

TB-2

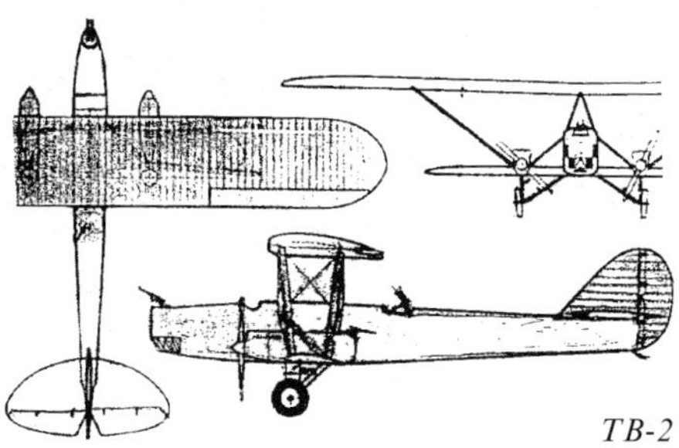

TB-2

R-4 Min-cost improvement of R-1, designed 27. Engine retained but moved 140 mm forward and lowered 29 mm, giving improved pilot view and better CG posn. Prop given spinner faired into smaller header radiator and improved

cowl, ventral retrac radiator added. Wooden landing gear legs replaced by metal tube. Factory test 13 June to 3 July 28. Better than R-1, but not in same class as R-5. Taken no further, but improvements incorporated in last batches of R-1.
DIMENSIONS As R-1, except length 9625 (31 ft 6⅞ in); wing area 43,54 m² (468·7 ft²).
WEIGHTS Loaded 2219 kg (4,892 lb).
PERFORMANCE Not recorded.

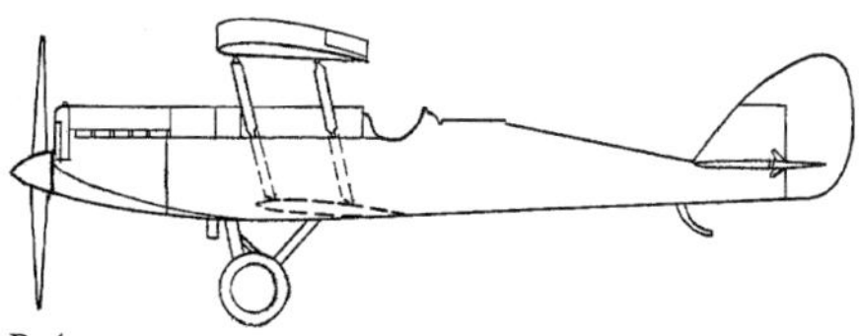

R-4

R-5 This two-seat biplane was one of first Soviet original designs to be made in large numbers; host of variants included VVS standard recon and attack a/c and 1,000+ civil. Design 25-28 included exhaustive statistical research in replacing R-1.

Wings modified Clark Y profile, 10% thick, CG from 24·5% empty to 35·8 max. Two box spars each wing with 5 mm ply webs, 30 mm pine booms and box width 80 mm (upper wing), 55 mm (lower wing). Ply ribs with strip capping and oval lightening holes providing spaces for diagonal internal bracing wires. Multiple pine stringers, ply skin to front spar, fabric behind. All struts dural tube of streamlined section with adjustable machined fork ends. Streamlined external bracing wires (lift, pairs No 12; landing, individual No 11; cabane, Nos 10 and 7) all with 12 mm threaded end-fittings. Wooden fuselage truss with four longerons and 12 frames, diag side bracings and light secondary fairing to give round top, skin of ply panels 3 mm at front and 2 mm at tail. Wooden tail, but periphery of wings and all control surfaces dural, with fabric covering. Dual flight controls by cable, differential ailerons on upper wings only, 32° up and 10°30' down. Engine mounted on dural box sections joined by riveted arch frames below and with main side ties steel/dural pipes with machined end-fittings adjustable for length. Two main fuel tanks of galvanized iron in centre-section feeding via copper pipes. Brass honeycomb radiator well ahead of main gears extended by chain drive from front cockpit. Two-blade wooden 3350 mm propeller with spinner and Hucks dogs. Main gears with pin joints and front strut sprung by rubber laminates, wheels spoked to 33 and disc subsequently, tyres 900 × 200 mm. Four patterns of ski; floats discussed later. Normal armament one synchronized PV-1, one (from 34 usually twin) DA; 250 kg bombs on wing racks, (bomber version) overload up to 400 kg bombs. Throughout construction casein glue used, mild-steel fittings (most with two or three parts flame-welded) and galvanized iron screws and pins.

Prototype flown 28, before October. Satisfactory, but prolonged testing and discussion prior to launch of series production. First R-5 (*Razvyedchik*, reconnaissance) completed GAZ-1 mid 30, deliveries (by year to 37) being 30/336/884/1,572/1,642/450, total 4,914. Other variants brought grand total to 6,726.

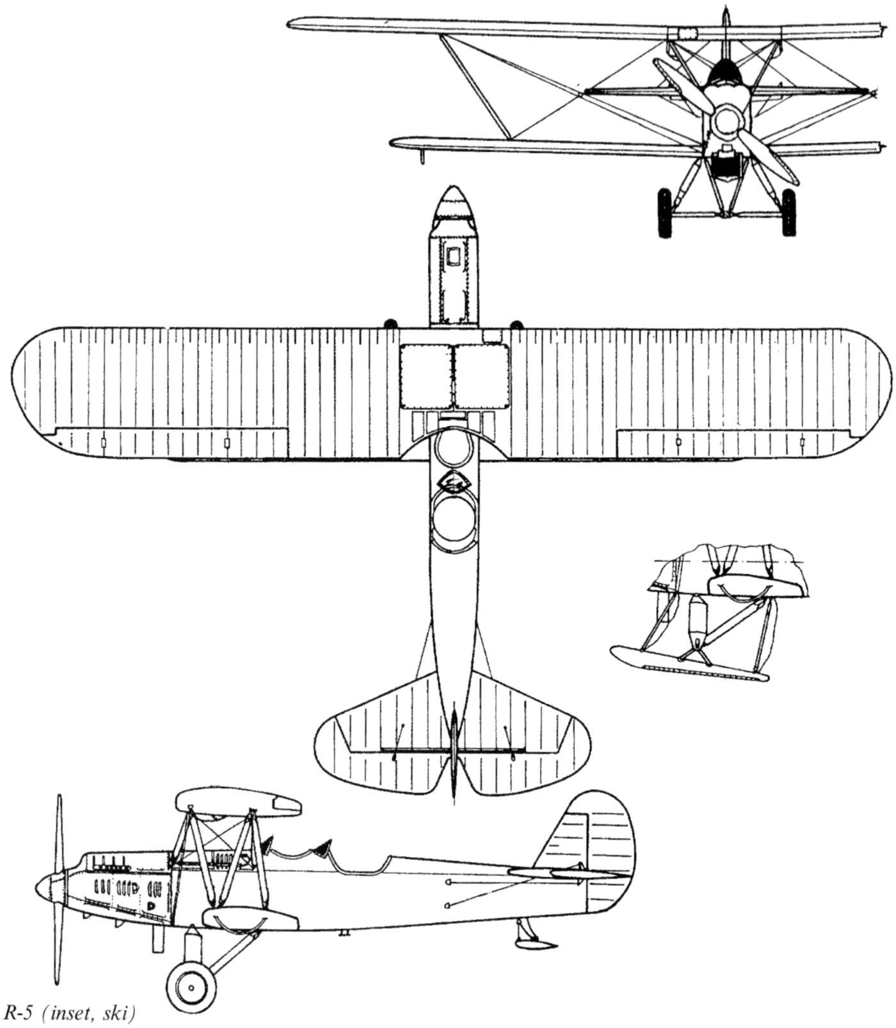

R-5 (inset, ski)

R-5 installation VAP-40 gas dispensers

R-5 reached VVS early 31, subsequently equipped over 100 regts up to close of Second World War and seeing especially intensive duty in Spain, at Khalkin-Gol and in Winter War. In 1941-44, most subjected to various mods including fitting heavier guns, RS-82 launch rails, special communications and various kinds of role eqpt including searchlight, side stepladder, loudspeaker and glider tow hook.

R-5Sh *Shturmovik* (attacker) first major variant, 31. Merely used payload to carry more weapons. First example had two PV-1 in box above each lower wing and Der-5 box for 240/500 kg small bombs under fuselage, retaining existing guns. By 33 in production with M-17B, eight PV-1 in two quads (later eight ShKAS) and minor changes.

R-5a, MR-5 Original designation for *Morskoi Razvyedchik*, marine reconnaissance. Twin-float seaplane; floats wooden with 21 frames and eight bulkheads arranged along two keelsons as on Sh-2 and finally covered with fabric and varnished. Attached by 12 struts of dural tube streamline-section with wire bracing. Enlarged fin, hand-crank starter attachments on each side of engine and other minor changes. Project of V. B. Shavrov and D. A. Mikhailov, prototype flown April 31. Adequate but performance capable of improvement. Often called **MR-5bis** (because of abandoned Chyetverikov MR-5); total 111 built GAZ-1 34-35.

P-5 Designation from *Passazhirskii*, but usually used to carry 400 kg mail/cargo. From about mid-31 R-5 supplied to GVF as standard aircraft off GAZ-1 line but with armament removed. Aeroflot workshops increasingly devised ways of improving this transport version which by 40 numbered over 1,000. At first rear seat and control column removed and cargo not even secured. Later rear cockpit reconstructed without instruments and enlarged into cockpit for two passengers in wicker chairs with baggage. Smaller number rebuilt with enclosed rear cabin for two – possibly more – passengers with at least three types of lid or glazed roof (see *Limuzin* and PR-5). Aeroflot was also only known user of seaplane version, **P-5a**; believed small numbers. Important addition to P-5 was use of underwing *Kasseta* (drum or casket) designed by D. S. Markov, P. I. Grokhovskii (which see) and others. Some had lifting profile and thus could accommodate 1,3-t load in single container. As noted under Grokhovskii, that designer flew a P-5 with 16 adults, including seven in each *Kasseta*. Smaller *Kasseta* model played central role in dramatic rescue of *Chyelyuskin* expedition from Arctic ice in 34, P-5s and other transports being on skis. P-5 important well into the Second World War, outlasting P-Z.

R-5L, Limuzin Developed at CCB by A. N. Rafaelyants and B. L. Bukholts as comfortable transport with pilot in front and cabin for two passengers behind. Dural framed canopy with Celluloid windows and passenger door with airstairs in left side. First factory-built *Limuzin*, flown late 31.

R-5D Designation from *Dalnost*, distance; long-range example of late 31 with large fuselage tankage which cracked open in flight (safe forced landing without fire).

R-5PS Author's shorthand for *Povorotnyim Stoika*, rotating struts (may have been actual designation). Front interplane struts broad surfaces of aerofoil profile, chord 200 mm, mounted in bearings and separately controllable by pilot either to weathercock freely or be pulled almost broadside to airflow. Followed research into spin-recovery. Tested on R-5 and also on I-5 with dire results, causing death of R-5 test-pilot M. A. Bolkovoinov Sept 32.

R-5M-34 Standard R-5 used 32 as flight test-bed for first M-34 engine. No major problem.

R-5RK Author's shorthand for *Razreznyim Krylo,* slotted wing. Upper wing full-span slats 227 mm chord and full-span ailerons, lower wing full-span slotted flaps. Two built 32, 20° AOA held at 70 km/h but no prodn.

R-5T Designation from *Torpedonosyets*, redesign as single-seat torpedo carrier in 32 under V. V. Nikitin. Main modification redesign of main landing gear with two separate units each with diagonal side strut pin-jointed under fuselage longeron and with air/oil shock strut almost vertical from bottom of this strut to spar of wing, which had to have additional diagonal upper bracing strut to fuselage upper longeron. Torpedo crutch geometry resulted in increased landing attitude of 17° with torpedo on board. Idea put forward by A. I. Grebenyev at GAZ-1 in 33 and results were fully satisfactory. Batch of 50 R-5T built 35 serving at Far East MA.

P-5L Another *Limuzin* built at GAZ-1 in 33 similar to R-5L but without airstairs. Several produced from new, called 'first production Limousine'.

SSS, R-5SSS Designation from *Skorostnoi Skoropodyemnyi Skorostrelnyi* (speedster, high-speed lift, fast-shooter). Major revision of R-5 by D. S. Markov and A. A. Skarbov for increased performance. M-17F with improved cowl, air/oil shock strut landing gears with streamlined casings and spats, detail attention to joints and finish, streamlined cockpit, many exterior rivets made flush. Increased armament of two ShKAS or one ShKAS and one PV-1 firing ahead (in *Shturmovik* variant, four additional ShKAS on lower wings). Made vital injection of increased performance (22/30 km/h, extra 2 km alt) leading to introduction of R-Z described separately. SSS built at GAZ-1 35-37 total 221/129/270 = 620.

R-5 V-tail V-type (butterfly) tails studied at VVA since 29 and designs worked out in detail by Yezhi Rudlitski. Two tail surfaces, 140° angle, RAF-30 section, 12%, wooden fixed portions with dural/fabric movable surfaces geared to work together as elevators or differentially as rudders. Flight tests from summer 35 by M. M. Gromov, K. K. Popov and I. F. Kozlov. Handling little different from series R-5, though pedals heavier, qualities poorer at point of landing or in tight turn, and spin characteristics worse.

R-5 Jumo, ED-1 Possibility of using Jumo-4 diesel in VVS derivative of ANT-25 led to import of several engines and modification of

R-5T

an R-5 as flying testbed. Deep cowl, four-blade propeller (as used on G 38) and large radiator under belly. Rear cockpit enlarged, and equipped for two test observers. Aircraft flown Oct 35, designation from *Eksperimental Dvigatyel*. Made over 200 flights, one by A. I. Zhukov lasting 12 h 01 min.

R-5USh Author's shorthand for *Ubirayemy shassi*, retractable gear. Project by Markov and Skarbov, main gears retracting inwards into thickened inboard lower wing as on DI-6.

R-5G Author's shorthand for *Gusenichnyi*, caterpillar tracklaying; R-5 tested tracked main gears designed by Chyechubalin 35, but overall considered clumsy.

PR-5 Final modernized transport version, built 36 by A. N. Rafaelyants. Semi-monocoque fuselage with pilot in high front cockpit with glazed canopy and four-passenger cabin occupying enlarged cross-section below and behind. Pilot access from above via hinged canopy, passengers via door with airstair on left. Tested spring 36, CG adjusted by upper wing moved 100 mm to rear. Result called **PR-5bis**, CG never aft of 36% MAC and no problems. Total 210 built by conversion and used by Aeroflot at least through 41.

ARK-5 Following proposal by polar pilot M. V. Vodopyarnov, two R-5s rebuilt for Arctic use. Streamlined ply containers for payload faired into sides of fuselage and lower wing; enclosed cockpit with heating. On 26 March 36 Vodopyarnov and Makhotkin flew these to Novaya Zemlya and Franz-Josef Land.

PR-12, R-Z, P-Z Described separately.

LSh, TSh-1, SHON Described under Grigorovich.

LR, LR-2, TSh-3
Described under Kochyerigin.

DIMENSIONS Span (upper) 15,5 m (50 ft 10¼ in), (lower) 12,0 m (39 ft 4½ in); length 10,555 m (34 ft 7½ in); wing area 50,2 m² (540 ft²).

ENGINE One BMW V-12: (prototype) 680 hp BMW VI, (main series to 33) M-17B, (post-33) M-17F, except most P-5 still M-17B.

WEIGHTS Empty (1930 reconnaissance) 1969 kg (4,341 lb), (R-5a) 2330 kg, (RK) 2184 kg, (P-5) 2040 kg, (PR) 2118 kg, (PR bis) 2200 kg; fuel/oil (standard) 400 kg+30 kg, (P-5) 500 kg, (PR) 585 kg+50 kg; loaded (prototype) 2955 kg, (1930) 3247 kg (7,158 lb), (Sh) 3410 kg, (5a) 3230 kg, (1934) 2997 kg (6,607 lb), (P-5, PR, bis) up to 3350 kg (7,385 lb).

PERFORMANCE Max speed (1930) 228 km/h (142 mph), (Sh) 202 km/h, (5a) 198 km/h, (RK) 205 km/h, (1934) 244 km/h (152 mph), (P) 215 km/h, (PR, bis) 233 km/h; cruise (P-5) 165 km/h (102·5 mph); climb to 1 km (1930) 2·1 min, (1934) 2·5 min; service ceiling (1930) 6,4 km (21,000 ft), (1934) 5940 m (19,490 ft), (5a) 4350 m, (P-5) 5,8 km; range (1930) 800 km, (1934) 1000 km, (5a) 700 km, (P-5) 1200 km; time 360° circle (reconnaissance, typical) 17 s; takeoff (typical) 300 m/25 s (1934, 22 s); landing 220 m/18 s/95 km/h (59 mph).

PR-5

PR-12 Passenger monoplane based on PR-5 but new low cantilever wing. One only, L-3600, flown on skis early 38.

DIMENSIONS Span c17 m (55 ft 9¼ in); length 10,56 m (34 ft 7¾ in); wing area c45 m² (484 ft²).

ENGINE One M-17F.

WEIGHTS Empty 2100 kg (4,630 lb); fuel/oil 585 kg+50 kg; loaded 3200 kg (7,055 lb).

PERFORMANCE Max speed 250 km/h (155 mph); cruise 228 km/h (142 mph); landing speed 100 km/h (62 mph), no other data.

R-Z Developed at instigation of A. M. Byelinkovich, director of GAZ-1, under chief engineer Ye. P. Shyekunov. Design by Markov and Skarbov, of Tupolev, based on SSS. More powerful engine, more efficient radiator behind landing gear. Redesigned fuselage with modern wood monocoque structure, deeper section with larger cockpits and larger tanks. Glazed sliding canopy over pilot cockpit and fixed glazed fairing over gunner. Redesigned tail of larger area

ARK-5

PR-12

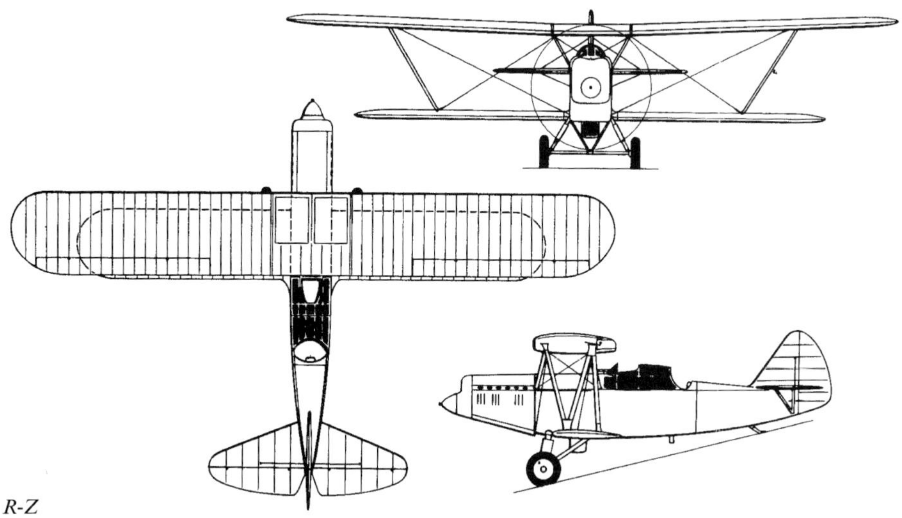

R-Z

with balanced surfaces without horns. Centre section of upper wing without trailing-edge cutout. Low-pressure tyres 800×200 mm, with spats on first R-Z, thereafter left open. Wheel brakes (no details).

First flight Jan 35; handling soon superior to R-5, and accepted for VVS service. Total 1,031 built at GAZ-1 terminating spring 37. Armament usually fixed PV, one or two movable ShKAS, six wing bombs eg two FAB-125 and four FAB-50. Prolonged service in Spain, nicknamed *Rasante* (ground-shaver) and in conflict with Japan. Still numerous in Winter War.

R-ZSh Single *Shturmovik* prototype with four ShKAS on lower wings outboard of propeller disc. Believed later test-bed for heavier and more unconventional armament schemes.

P-Z Commercial variant of R-Z. First in service 36, few changes apart from more powerful engine and removal of armament, payload 275 kg. Used for carriage of mail and (usually two) passengers facing each other. Substantial number in use 1937-39.

PT Attempted reconstruction to improve performance (T, *Transportnyi*). Streamlined rectangular-section cargo box on each side above lower wing, 1,5 m^3 each. Upper wing moved back 50 mm (half as far as PR-5), lower wings fitted with flaps (believed slotted), rear-fuselage top raised, tail extra bracing wires and more powerful radio. Payload 494 kg but poorer flying qualities.

R-ZR, *Rekordnyi* R-Z with armament removed and detail improvements to reduce drag. Built on instigation of pilot V. V. Shyevchyenko and constructor V. V. Nikitin. Flown by Shyevchyenko to 11,1 km on 8 May 37.

R-Z GK *Gruzovymi Kassetami*, various transport R-Z with streamlined payload containers on and beneath lower wings.

DIMENSIONS Span 15,45 m (50 ft 8¼ in); length 9,72 m (31 ft 10⅔ in); wing area 42,52 m^2 (458 ft^2).

ENGINE One M-34N, (PZ, PT) M-34NB.

WEIGHTS Empty 2007 kg (4,425 lb), (PZ) 2154 kg, (PT) 2345 kg; fuel/oil 584 kg+55 kg; loaded 3150 kg (6,944 lb), (PZ) 3161 kg, (PT) 3575 kg.

PERFORMANCE Max speed 275 km/h at SL, 316 km/h (196 mph) at 3,5 km, (PT) 296 km/h at 3,5 km; cruise (PT) 213 km/h (132 mph); climb to 3 km, 6·6 min, (PT) 13·9 min; service ceiling 8,7 km (28,550 ft), (PZ) 9150 m, (PT) 6240 m; range 1000 km (621 miles), (PZ) 1130 km; time 360° turn, 18 s; take-off 270 m/15 s, (PT) 300 m/18 s; landing 220 m/14 s/105 km/h, (PT) 195 m/11 s/97 km/h.

I-6 As soon as I-3 was released to manufacturing, OSS began design of rival single-seat fighter with air-cooled radial. Polikarpov drew first three-view Sept 28 and all brigades assisted with detail design 29. Thanks to short and relatively light engine, overall dimensions appreciably less than I-3, and weight only 69% as much. Nevertheless trad OSS structure retained: two-spar wooden wings with dural periphery and fabric covering, streamlined-section dural tube struts, oval-section monocoque fuselage of glued layers of *shpon*, 2,5 to 5,0-mm thick, dural movable control surfaces with fabric covering, and rubber-sprung landing gear. Original engine mount by I. A. Tavastshyern rejected in favour of

I-6

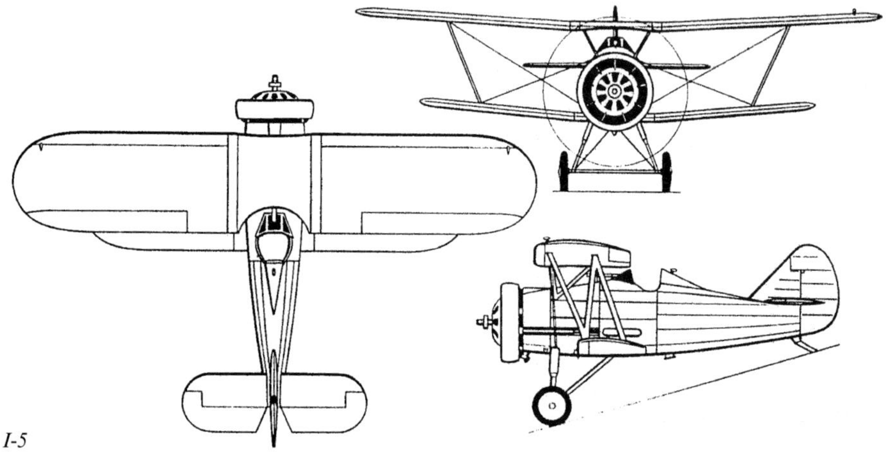

I-5

VT-11 No 2

welded steel tube, with dural outer cowl leaving cylinders open at front but faired by individual part-helmets at rear. Wooden propeller with spinner and Hucks dogs. Though hard to believe, photographs (all retouched) show diagonal interplane struts as on 2I-N1 and P-2. OGPU had constructor arrested Sept 29. By this time prototype almost complete, and in Polikarpov's absence completion delayed to following spring. Two examples built, first flying 30 March 30 and both in May Day parade. On 15 June test-pilot A. D. Shirinkin left I-6 by parachute (Shavrov 'without valid cause'). Eventually (31) about a year spent comparing remaining I-6 with I-5 and latter chosen mainly on account of four guns, fractionally better time for 360° turn and supposed tougher structure. Polikarpov never officially informed of I-6 results until his release in 33.

DIMENSIONS Span 9,7 m (31 ft 10 in) (lower, 7,5 m); length 6,78 m (22 ft 3 in); wing area 20,5 m² (221 ft²).

ENGINE (1) Gnome-Rhône Jupiter VI, (2) M-22.

WEIGHTS Empty 868 kg (1,914 lb); fuel/oil 180 kg; loaded 1280 kg (2,822 lb).

PERFORMANCE Max speed 280 km/h (174 mph); climb to 1 km, 1·5 min; service ceiling 7,5 km (24,600 ft), range 700 km (435 miles); time 360° circle, 15 s; take-off 90 m/7 s; landing 200 m/15 s/95 km/h).

I-5 Chief Soviet fighter of its day (30s). Assigned to AGOS under first five-year plan, drawn up by Aviatrust 27 and accepted 22 June 28. To be mixed-construction, metal-framed fuselage, fighter otherwise similar to OSS-designed I-6 with wood monocoque fuselage, and likewise powered by Jupiter VI for which Aviatrust was then negotiating licence. Tupolev gave priority to ANT-6 and proposed I-5, with AGOS number ANT-12, lagged. Polikarpov and Grigorovich arrested and told to build I-5 quickly. After disorganized autumn special *Vnutrenniya Tyurma* (internal prison) established November at Hangar 7, V. R. Menzhinskii factory (GAZ-39).

Fuselage truss gas-welded from standard steel tubing, as used in contemporary engine mounts, with light dural frames and stringers to give streamline form for fabric covering. Rest of airframe similar, or identical, to I-6, except for shape of fin/rudder, use of normal wing bracing wires, small increase in span upper wing and revision of landing-gear pivots and bracing wires. Design finished in two months and approved 28 March 30. First flight **VT-11** by B. L. Bukholts 29 April 30. This had supercharged Jupiter VII in completely helmeted cowl; doped silver and with VT badge proudly painted on rudder. Second had large faired rubber shock-absorbers on main legs, unblown Jupiter VI and was doped red, with fuselage emblazoned *Klim Voroshilov*. Third fitted with second licensed Jupiter VI (M-15) 'Gift No 16'.

Menzhinskii plant kept building I-5, gradually improving design by introducing Townend-ring cowl, crankcase fairing with cooling holes, dural propeller with pitch adjustable on ground, improved landing gear (some with spats), pilot headrest and fairing, pitot/static head on upper right wing and stronger tailskid. Seven aircraft produced by September 30, and from No 10 armament increased from two synchronized PV-1 with 1,200 rounds to four guns (two each side of fuselage) with 4,000 rounds, or two guns and two 20 kg bombs. Final NII testing of definitive I-5 by V. A. Stepanchenok, A. F. Anisimov, I. F. Petrov, I. F. Kozlov and others. GAZ-1 delivered 66 in 31 and 76 in 32; GAZ-21 delivered 10 in 32, 321 in 33 and 330 in 34, total 803. Standard advanced trainer 1936-41.

Many local modifications including reinforced plate under rear fuselage for Zvyeno parasite experiments, fitting of MF radio with wire aerial stretched from oblique mast ahead of upper centre-section, fitting of generator, battery and nav lights (widely installed), LSh light *Shturmovik* variant with four guns, and three types of skis. Factory-built **UTI-1** trainer with second cockpit (original moved forward) with dual control; 20 built 34 but not used by VVS.

DIMENSIONS Span (upper) 10,24 m (33 ft 7⅛ in), (lower) 7,4 m (24 ft 3⅓ in); length 6,78 m (22 ft 3 in); wing area 21,25 m² (229 ft²).

ENGINE (VT-11) 450 hp GR Jupiter VII, 525 hp Jupiter VI, (I-5) first with M-15, (production) M-22.

WEIGHTS Empty (VT-11 No 1) 919 kg, (production) 934 kg (2,079 lb); fuel/oil 180 kg; loaded (VT) 1331 kg, (production) 1355 kg (2,987 lb).

PERFORMANCE Max speed (VT) 273 km/h (238 at SL), (production) 278 km/h (173 mph) at SL, 252 km/h at 2 km; climb to 1 km, (VT) 1·4 min, (production) 1·6; service ceiling (VT) 7,8 km, (production) 7,5 km (24,600 ft); range 660 km (410 miles); time 360° circle, (VT) 14 s, (production) 10/11 s; take-off 100 m/8 s; landing 200 m/15 s/95 km/h (59 mph).

I-7 VT had largely completed design of I-5 by spring 30 and played major role in studying further fighters. I-9 and I-10 (see Grigorovich) not built, partly owing to absence of engines, and Heinkel HD 37c adopted and built under license as I-7, with existing licensed BMW engine. Few changes apart from PV-1 guns and Russian radio (same aerial installation as I-5). Regarded as stop-gap, yet many survived alongside I-5 and were withdrawn only upon debut of I-15.

DIMENSIONS Span 10,0 m (32 ft 9⅔ in); length 6,95 m (22 ft 9⅔ in); wing area 26,7 m² (287 ft²).

ENGINE One M-17F.

WEIGHTS Empty 1296 kg (2,857 lb); fuel/oil 200 kg; loaded 1729 kg (3,812 lb).

PERFORMANCE Max speed 290 km/h (180 mph) at SL; climb to 1 km, 2 min; service ceiling 7,2 km (23,620 ft); range 700 km (435 miles); time 360° circle, 12 s; take-off 90 m/7 s; landing 160 m/13 s/96 km/h (59·7 mph).

I-11 Unbuilt Polikarpov (VT) monoplane single-seat fighter with emphasis on speed rather than manoeuvrability (in conformity with 32 VVS doctrine of fighter force made up of agile biplanes collaborating with fast monoplanes). One M-34F; metal airframe.

I-13 Unbuilt project of 32 for agile biplane fighter with Wright Cyclone. Led to I-15.

I-15, -3, Chaika Polikarpov's final months in detention were spent examining ways of improving I-5, then (Jan 33) still not in full production. He returned to CCB with detailed three-view of new fighter remarkably similar to definitive I-5 but with gull-type upper wing (hence nickname *Chaika*, seagull) to improve pilot forward view. Desired engine, Wright Cyclone F-series.

I-5

I-15

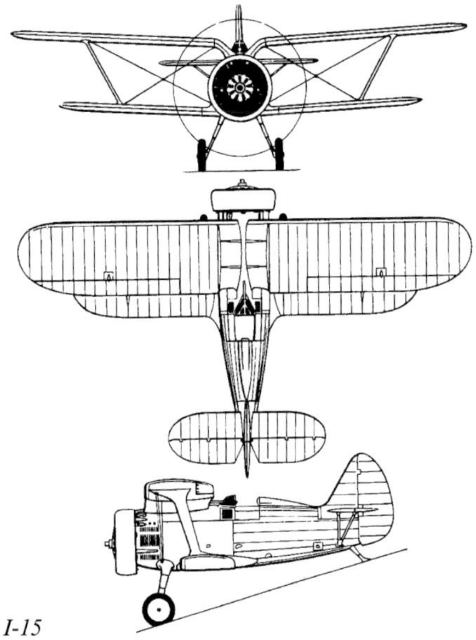

I-15

Design undertaken as -3. Wings almost identical to I-5 except for welded steel-tube basis of inboard sections of upper wings, inclined at 30°. I-type interplane struts of KhMA tube with light dural fairing. Fuselage truss of gas-welded KhMA tube with light secondary structure of rolled dural L-sections to give streamline form. Skin of removable D1 panels back to windscreen, fabric to rear. Large Frise-type ailerons on upper wings only, differential action. All control surfaces D6, with fabric covering. Most outstanding feature was cantilever main legs, each of machined steel tube with lower portion sliding in upper and located by internal slot 1,5×12 mm section with up to 190 mm stroke, with air/oil shock-absorbing. Wheels fitted with disc brakes and tyres 700×100 mm; or skis 1650×370 mm.

Prototype fitted with Cyclone SGR-1820-F3 driving Russian dural two-blade propeller with ground-adjustable pitch. Two synchronized PV-1 with 2,000 rounds, optional overload of two 20-kg bombs. First flight by Valerii P. Chkalov, Oct 33. Excellent handling, set record for 360° circle (just over 8 s). Factory testing took 26 days, and NII tests barely another month, resulting in immediate series production as I-15. License for Cyclone and initiation of production as M-25 took four years (1932-Oct 36) so GAZ-1 went into prodn with imported engine, building 60/273/2 in 34-36 backed up by GAZ-39 34/15/0 total 384. During prodn two extra PV-1 fitted in sides of fuselage, total 3,000 rds, seat given armour, fuel tank reduced from 310 to 210 lit. Licensed to CASA in Spain. Prototype called **CCB-3bis** flown by V. K. Kokkinaki 21 Nov 35 in specially lightened condition to 14575 m (47,818 ft).

I-15 saw extensive service in Spain, China, at Khalkin Gol, Nomonhan and in Winter War, many remaining operational in June 41. Gull wing was not universally liked, and in early 36, 12 I-15 built with normal upper centre section. In 38 about 40 versions with normal guns replaced by two BS. In same year, following long research, A. Ya. Shchyerbakov fitted rubber-sealed pressure cabin to production I-15, with circular pilot windows. Light attack version studied but not accepted.

DIMENSIONS Span 9,75 m (31 ft 11⅞ in); length 6,1 m (20 ft 0 in); wing area 23,55 m² (253·2 ft²).
ENGINE One 710 hp Wright Cyclone F-3.
WEIGHTS Empty (4 guns) 1130 kg (2,491 lb); fuel/oil 168 kg+35 kg; loaded 1390 kg (3,064 lb).
PERFORMANCE Max speed 315 km/h at SL, 367 (228 mph) at 3 km; climb to 5 km in 6·2 min; service ceiling 9,8 km (32,150 ft); range 550 km (342 miles); 360° circle 9·0 s; TO 80 m; ldg 90 km/h, 70 m.

I-15bis, I-152 Gull wing of I-15 pleased many pilots but some complained view worse than before; following VVS tests with 12 un-gulled aircraft Polikarpov brigade, about this time given status of OKB, effected major revision of I-15 as **CCB-3 ter**, later I-152 and more commonly I-15bis. New upper wing with 0,6 m² greater area, normal centre section, simplified and restressed structure and profile of Clark-YH type. Front lift wires duplicated in tandem. Improved and more powerful engine, in totally different long-chord cowling with shuttered front, large removable side panels with apertures for individual exhaust stacks and ducted oil cooler at top. AV-1 propeller usually with spinner. Improved landing gear and brakes, and tyres now 120 mm wide. Pilot given 9-mm back armour. Empty weight 350 kg greater, but performance generally improved. Four PV-1 (often later replaced by ShKAS) with 2600 rounds or two BS, bombload up to 150 kg. First flight late 37, followed I-15 at GAZ-1 which delivered 1,104 in 38 and 1,304 in 39, total 2,408. Together with I-15 total of 550 supplied to Republican Spain, where dubbed Chato (flat-nosed) by Nationalists. Fair results against Nationalist fighters, C.R.32 possibly having edge in manoeuvrability but I-15bis excelling in dive/climb and firepower. Many served in Far East, proving generally inferior to Ki-27, and in Winter War; over 1,000 were still operational in June 41.

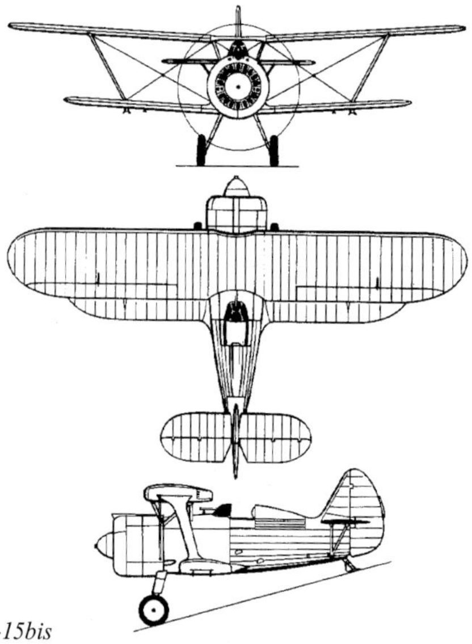
I-15bis

I-15 bis

DIT

I-152TK Single 39 aircraft fitted with two TK-3 turbochargers driven by exhaust (4½ cylinders to each turbo, top pipe being divided) from otherwise standard M-25B. Speed 435 km/h at 6 km. Empty weight increased by 140 kg.

I-152GK Single 39 aircraft rebuilt with improved Shchyerbakov pressure cabin, of rubber-sealed aluminium shaped to fit profile of rear fuselage. Cabin weighed 78 kg, but no details of any in-flight pressurization. Described as first really successful Soviet GK (*Germetichyeskoi Kabine*, hermetic cabin).

DIT Single I-152 rebuilt as two-seat *Shturmovik* with modified armament. Believed only two-seat biplane built by this OKB after 34. Only 10 kg heavier empty and 22 kg heavier loaded, and speed actually higher at 327/379 km/h. No change in length.

DIMENSIONS Span 10,2 m (33 ft 5½ in); length 6,2 m (20 ft 4 in); wing area 22,5 m^2 (242 ft^2).

ENGINE One M-25B.

WEIGHTS Empty 1310 kg (2,888 lb); fuel/oil 177 kg; loaded 1700 kg (3,748 lb).

PERFORMANCE Max speed 327 km/h at SL, 379 km/h (235·5 mph) at 3,5 km; time to 5 km 6·6 min; service ceiling 9,3 km (30,500 ft); range 520 km (323 miles); 360° circle 10·5 s; TO 160 m; ldg 90 m.

I-153 Again dubbed *Chaika*, but hardly ever called **I-15ter**, this final member of Polikarpov biplane family was probably best of all biplane fighters but represented too-prolonged attempt to perpetuate concept of agile if slow fighter in age of more powerful stressed-skin monoplane. Soviet Union had no way of knowing I-153 would be outclassed in nation's hour of dire need. Indeed I-153 stemmed from Kremlin meeting with Stalin and VVS at which results of Spanish war were examined and firm decision made to press ahead with improved biplane because of apparent superiority of Fiat biplane over I-16. This meeting late July 37, when Bf 109B-1 already showing superiority in Spain. In August Polikarpov drew up plans and I-15ter approved 11 Oct 37. Programme assigned to A. Ya. Shchyerbakov.

Remarkably few changes apart from reversion to gull wing, switch to M-62 engine and AV-1 prop and retractable landing gear. Principal air/oil shock strut braced laterally by diagonal strut, both pivoted to fuselage to retract directly to rear, wheel rotating 90° to lie flat in well in lower wing and with tie-rod to doors on ventral centreline hinged outwards to cover legs. Actuation by cockpit handcrank, bevel and screwjack acting on mid-jointed rear bracing struts. Tyres further widened to 150 mm. Instead of skid, fixed tailwheel with solid tyre, castoring within small limits set by rudder position. Engine-driven generator, battery and nav lights standard, and radio in a proportion of series aircraft. Standard armament four ShKAS with 2,600 rounds. Additional pre-production aircraft with four BS or two ShVAK, extremely powerful for 38. Four racks under lower wing each stressed for FAB-50 or various other stores to same 200-kg total such as 25-kg anti-personnel bombs. First flight summer 38, and NII tests with several aircraft completed in autumn. Followed I-15bis on GAZ-1 line, which delivered 1,011/2,362/64 in 39-41, total 3,437.

First deliveries to Far East, where 70 IAP were in action 25 July 39. Though not totally outclassed by lower-powered (650 hp) Ki-27 it was clear I-153 was becoming obsolescent. In absence of alternative, production was continued with improved version, **I-153BS**. Designation from four BS guns, but main advantage in combat was M-62 engine giving substantially more power at altitude and thus better climb, high-alt speed and ceiling. This version available from spring 39, small number also being produced of **I-153P** (*Pushyechnyi*) with two ShVAK in place of four BS. Following service trials I-153 also cleared mid-39 to launch up to eight (six more common) RS-82 from rails beneath lower wing. From early in production of M-62 model lower wing was plumbed to

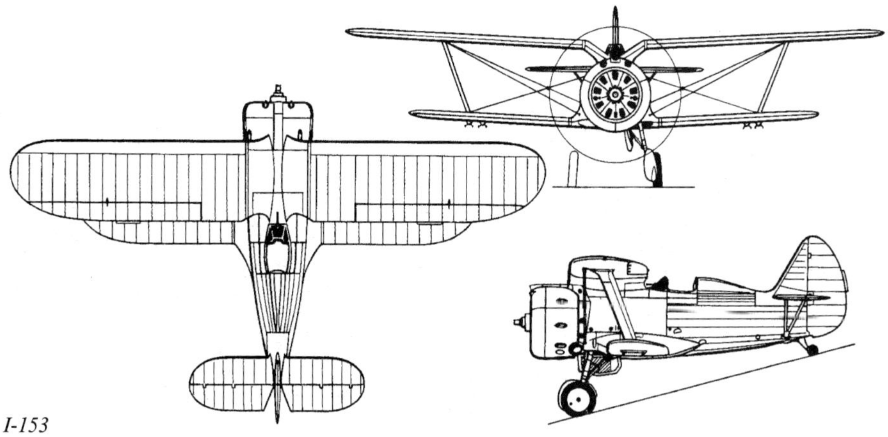
I-153

I-153

I-153/DM-4

allow inboard racks to carry two 100-lit (22-gal) drop tanks. In early 40 engine switched to M-63, again AV-1 prop, some with four BS in place of ShKAS. Many survived in close-support role to 43 (Finland used 14 captured examples, with Browning guns and German radio).

I-153V Single aircraft with definitive Shchyerbakov pressure cabin. Capsule of minimum-leak regenerative type, of welded AMts-L0,8 aluminium with all apertures sealed by rubber. Much-improved glazing with large slabs of curved 6-mm Plexiglas in heavily framed hinged hood giving view not much inferior to that in normal enclosed cockpit. This cabin weighed 45 kg, and empty weight of I-153V was 190 kg heavier at 1538 kg. Designation from *Vysotnyi,* height.

I-153V-TKGK Single aircraft fitted late 39 with M-63 engine with twin TK-3 turbochargers as well as flexible rubberized-fabric GK (pressure cabin) by Polikarpov KB. At 1,9 t this was lighter than 1959 kg of I-153V, though empty weight must have been appreciably more. Shavrov does not explain why speed at best height should have been same as regular 153BS, ceiling same and rate of climb worse, despite highly blown 930 hp engine.

I-153DM Designation from *Dopolnityelnyi Motor*, supplementary motor; aircraft used by Merkulov at Central Airfield Sept 40 to test two 400-mm DM-2 ramjets, and (Oct) two 500-mm DM-4. On 27 Oct 440 km/h reached at 2 km with ramjets burning.

I-190 Described separately.

DIMENSIONS Span 10,0 m (32 ft 9½ in); length 6,17 m (20 ft 2⅞ in); wing area 22,14 m^2 (238 ft^2).

ENGINE (39) One M-62, (40) M-63.

WEIGHTS Empty (4×ShKAS, typical) 1375 kg (3,031 lb); fuel/oil 150 kg+19 kg; loaded (39) 1765 kg (3,891 lb), (40) 1902 kg (4,193 lb).

PERFORMANCE Max speed 365 km/h at SL, 424 km/h (40, 427, 265 mph) at 5,1 km; climb to 5 km 5·7 min; service ceiling 10,7 km (35,100 ft); range 560 km (348 miles) (40, 510 km); 360° circle (39) 13,3 s, (40) 14,5 s; TO 110 m; ldg 105 km/h, 150 m.

I-16, CCB-12 While Polikarpov and aides were in detention in early 32 NII concluded two-year study with strategic decision to use mix of agile (biplane) fighters and fast (possibly monoplane) fighters. Much argument over practicability of monoplane, especially one without bracing struts. When Polikarpov regained freedom in early 33 he had already decided top priority was to make up for lost time on new monoplane fighter to compete with KOSOS ANT-31, then about to fly. CCB-12 project launched about beginning of March 33 and pressed ahead with same intensity as prevailed in VT prison. Resulting design was stumpy close-coupled monoplane of mixed construction, with pilot well aft as in I-15 and facing same problem of delay with licensed Cyclone engine. Though far from faultless it overtook ANT-31 to become leading Soviet fighter of rest of 30s. Unfortunately it was still chief VVS and MA fighter in June 41, when obsolescent.

Wing R-II profile, t/c 16% root, set at +3° in low-mid position with large root fillet. Centre section with two spars, widely separated, of riveted KhMA steel tube (from 36, KhGSA) built up into strong truss, with ribs of truss type built up from riveted dural box sections, with flush-riveted dural D1 skin. Main wing panels with spars of D6 sheet webs riveted to KhMA tube booms and truss ribs riveted from D6 box and angle sections, with skin of flush-riveted 0,5-mm D1 back to front spar and fabric aft. Fuselage of trad multi-layer *shpon*, 4 mm at front and 2,5 mm towards tail, built up with casein glue on carcase made in left and right halves each with four pine longerons and 11 half-frames. Very large slotted ailerons of D6 with fabric covering, with differential push/pull rod actuation and drooping 15° for landing. Entire tail of D6, mainly pressed from sheet with lightening holes, with fabric covering; surfaces unbalanced and driven by cables (rudder) or rods and bell-cranks (elevators). Fixed tailskid but fully retractable main gears, with air/oil sprung legs braced by rear strut both pivoted to fold inwards under pull of cable from cockpit screw-jack (there was prolonged study of pneumatic and hydraulic actuation). Tyres 700×100 mm, wheels with brakes (originally manual, later pneumatic). Only available engine M-22, but for first time installed in long-chord cowl with optimized cooling-air slot at rear. Usual perforated crankcase fairing, standard 2,8 m ground-adjustable duralumin propeller without spinner. Single aluminium fuselage tank of 4251 lit (93·5 gal). Rather narrow cockpit with folding left sidewall and seat adjustable for height only (as in I-15). Glazed aluminium/glass canopy with integral vee-windscreen, sliding forwards assisted by bungee cords to overcome drag in flight. Aldis-derived CP optical sight integral with windshield. Two ShKAS each with 900 rounds in centre section beyond propeller disc.

First prototype airframe 445 kg, engine and equipment 450 kg, gross weight 1311 kg. Doped red overall and flown by Valerii P. Chkalov 31 Dec 33. Adequate handling; performance fair considering low power. Second (*dubler*) prototype. **CCB-12bis**, flown 18 Feb 34 with imported Cyclone F3 driving 9 ft 2 in (2794 mm) Hamilton two-pitch prop. Intensive factory and NII trials quickly established CCB-12 as fit for production as I-16, but not easy to fly. Marginally stable about all axes, no trimmers, and in glide tended to stall, also tending to stall/spin in hard manoeuvres; severe vibration from engine, great difficulty in cranking up landing gear (44 turns, progressively harder) and gear tended to jam when part-extended prior to landing (cable

I-16 tip 29

shears later standard kit in cockpit), shock struts undamped and landing dangerous even on rare smooth field, landing speed high and wing dropped if speed allowed to bleed off, and with M-22 high power needed whenever gear extended, and throttle still well open as flown on to ground. Despite this, considered best available fighter and ordered into production, May 34. Subsequent models given *tip* (pronounced teep,=Type) numbers, reflecting justified belief there would be many future variants.

I-16 Often called **I-16M-22**, this was original pre-production version, flown late 34 and displayed in two flights of five in May Day parade 35. Such pressure was put on introduction of more powerful versions that few Type 1 built. But by late 35 substantial number with M-22 engine delivered as two-seat trainers.

UTI-2 Originally I-16 *tip*-1 with armament removed and fuselage redesigned by A. A. Brovkov to accommodate pupil in rear cockpit moved 450 mm to rear and instructor and modified fuel tank of almost unchanged capacity despite reduced length of tank. Both cockpits open, with separate windscreens.

UTI-3 New-build aircraft with UTI-2 fuselage and simple fixed landing gear almost identical to original but with retraction mechanism removed and wells plated over. From early 36 many fitted with blind-flying canopy of plywood for rear cockpit, manually sliding to rear and with black curtain hanging across at front to pass over pupil's head.

I-16 tip-4 First main production model, closely resembling CCB-12bis and powered by imported engine driving imported Hamilton propeller. Cowl and engine installation similar to tip-1 though cowl extended 405 mm aft. Landing gears fitted with extra wheel doors, with hinged lower portion, to cover wheels in flight. Many hundreds delivered 1934-35.

I-16 tip-5 First variant with licensed M-25 engine driving AV-1 propeller. Seat fitted with 8 mm back armour as on final batch of *tip*-4. Attachment added under each outer wing for FAB-100, total bombload 200 kg (441 lb). Replaced *tip*-4 in production July 35.

UTI-4 Two-seat conversion of *tip*-5 with essentially same fuselage as UTI-3 and often with standard retractable landing gear. By late 35 every fourth I-16 was dual-control, often also known as **I-16UTI** (*Uchyebno-Trenirovochnyi Istrebitel*, training fighter). Small number of dual trainers also built with later engines, or produced by field conversion of subsequent fighter models.

I-16Sh, CCB-18 Single 35 *tip*-1 modified for *Shturmovik* role with four ShKAS (Shavrov, or PV-1) and 100-kg bombload. First Soviet aircraft with bucket seat for seat-pack parachute incorporating rear and bottom armour; further armour ahead of pilot.

I-16P Single *tip*-5 with F3 Cyclone modified in 35 with heavier armament; designation from *Pushyechnyi*, cannon. Two ShKAS in top of fuselage, and at extremities of centre-section two unsynchronized ShVAK with 150 rounds each. Racks under outer wings for six Der-31 bomb containers. First installation of wing cannon, but took two years of testing before production.

I-16 tip-6 Standard 36 production with locally reinforced airframe, adding about 50 kg to structure, and more powerful M-25A engine.

SPB, CCB-29 Designation from *Skorostnoi Pikiruyushchnyi Bombardirovshchik*, fast dive bomber, I-16 *tip*-5 with pneumatic system added for retraction of landing gear and operation of dive brakes added in place of inboard section of aileron on restressed wings which were also provided with racks for two FAB-

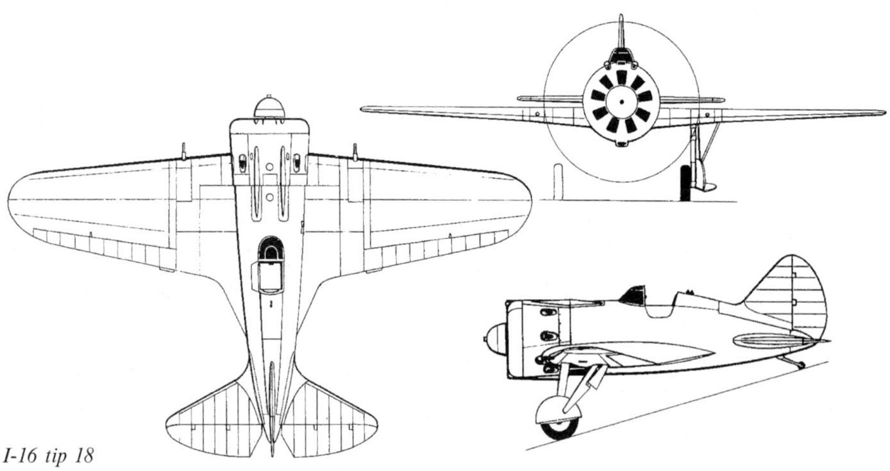

I-16 tip 18

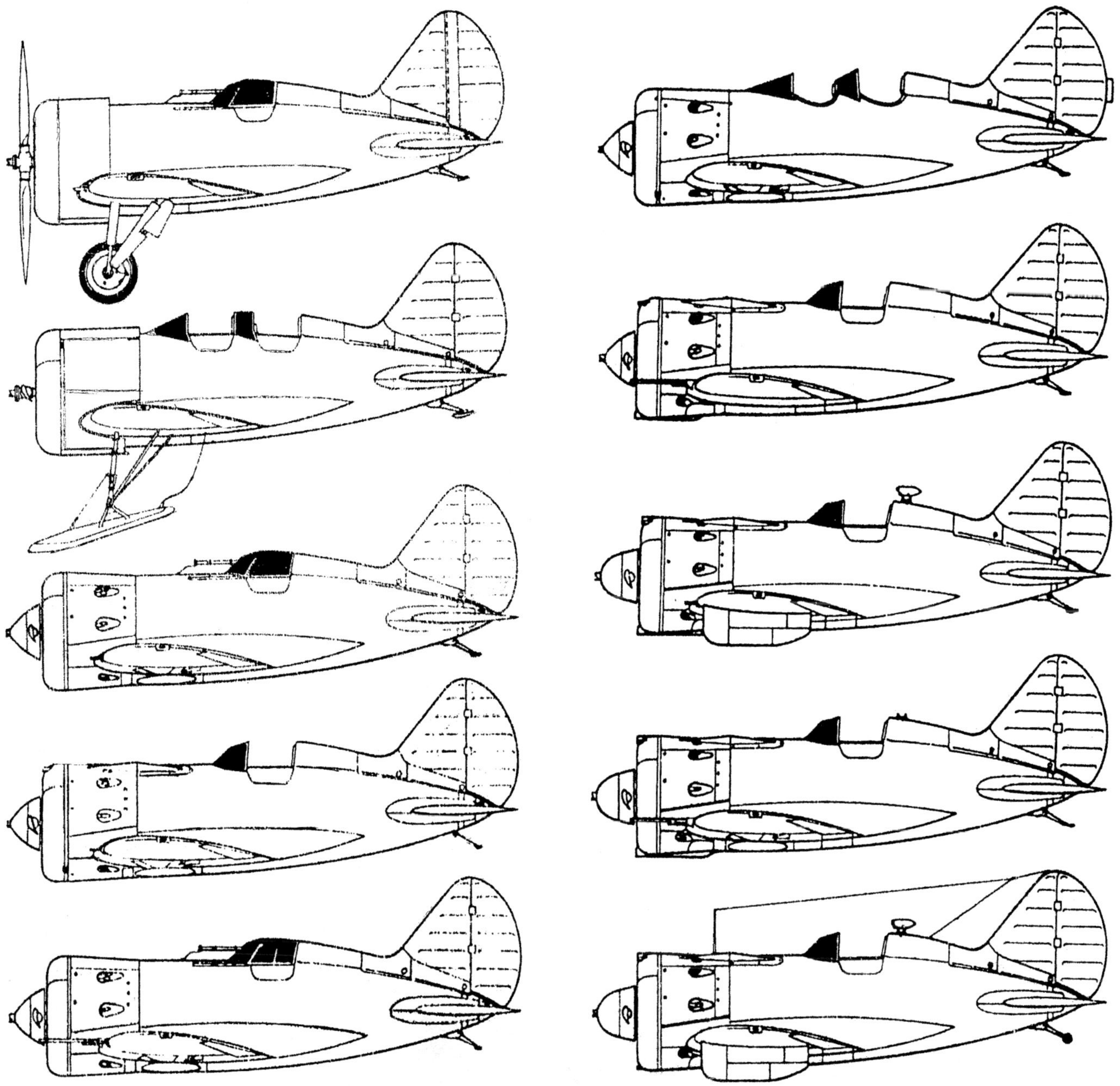

I-16 variants, from top left: CCB-12, UTI-2, Tip 5, Tip 10, Tip 12, (Col. 2) UTI-4, Tip 17, Tip 24, Tip 27, Tip 29

250. Initial tests with two FAB-100 May 36; full bombload used in connection with air-launched *Zvyeno* trials (see Vakhmistrov). Guns, two ShVAK in outer centre section. No production but unknown number of field conversions in 1941-42.

I-16 tip-10 Standard 37 model, produced in larger quantity than any other. M-25V engine, further local strengthening to structure, armament four ShKAS, extra pair being added under fairings in top of fuselage each with 650 rounds. Fixed windscreen, otherwise open cockpit, with simple reflector sight. First variant with retractable skis (previously in winter I-16s had used skis with legs locked down); new I-16 model had skis mounted on hinges on existing landing gears with small tie-rod to keep skis in unchanged attitude during retraction to lie flat under wing with toe in deep recess under engine (legs being flush as usual). According to VVS combat performance almost same as with wheeled gear, whereas fixed skis had reduced max speed 60/80 km/h.

I-16P, CCB-12P Further cannon-armed prototype, 38, based on *tip*-10 with ShVAK each with 150 rounds in outer centre-section. In this installation they were mounted much further forward, most of barrel projecting from leading edge. After refinement accepted as production armament. Same aircraft also tested four ShVAK (possibly two in fuselage in place of ShKAS but no details).

I-16 tip-17 Standard 38 model introducing two fuselage ShKAS and two wing ShVAK, as previously tested. M-25V unchanged, but structure further improved with wing spars of KhGSA stainless steel and tailskid replaced by wheel with solid rubber tyre. Because of greater weight, new main gears introduced two-stage shock struts with steel main legs 80 mm diameter, reflector sight improved and in some aircraft armour 9 mm instead of 8 and of larger area. With improved landing gear, cleared to 1925 kg with extra BS gun in pod on ventral centreline (just cleared by wheels or skis) and two 100-kg bombs. First variant armed with RS-82 rocket, usually six (rarely eight) on underwing racks; RS-82 trials with earlier models 36, operational on *tip*-17 early 39 and in action against Japanese Aug 39.

I-16TK Several, possibly many, I-16s fitted

with CIAM turbochargers. First was this *tip*-10 with two TK-1 turbos, and engine driving AV-1 hydromatic with 30° pitch range; rated height increased from 2,9 to 7,25 km.
I-16 tip-18 Standard 39 aircraft, introducing M-62 engine with two-speed supercharger. Modified exhaust system discharging through three holes on each side, alternately with single and paired pipes for nine cylinders. Propeller AV-1 or VISh-6A, smaller fuselage tank of 255 lit (56 gal) for highest flight performance, made up by plumbing for optional pair of 200-lit drop tanks as on I-153; otherwise as tip-10 with four ShKAS with 3,100 rounds.

I-16 tip-24 Final major production version, second half 39, similar to *tip*-10 or *tip*-18 but with M-63 engine, driving VV-1 or VISh-6 propeller. Wing stiffened by increasing number of ribs, increasing gauge of leading-edge metal skin to 0,6 mm and adding 3 mm ply between spars on both upper and under surfaces, retaining outer layer of fabric. Drop tanks in this model cleared to 254-lit (55·9-gal) size. Diverse weapons fitted, including four ShKAS, two ShKAS and two wing ShVAK, and four ShKAS plus single BS between fuselage guns on upper centreline. Underwing loads six RS-82 or bomb loads up to two FAB-250 (500-kg, 1102 lb). In addition, PBP-1 or RBP-1a reflector sight, RSI-1 or RSI-3 HF radio and oxygen increasingly made standard.

I-16 tip-28 to tip-30 Closely related to *tip*-24 but with direct-drive M-63 engine. Tip-29 was best and most famous version. Production of I-16 comprised: GAZ-39, 50/4/4 in 34-36; GAZ-21, 527/902/1,881/1,070/1,571/2,207/337 in 35-41; and GAZ-153, 6/105/264/503/19 in 37-41, total 9,450, of which c1,800 two-seat. Various popular names, notably *Mosca* (fly), also adopted officially by Republican Spain. Called *Rata* (rat) by Nationalists. Well-documented operational history includes dive-bombing by SPB conversions (see Vakhmistrov) and deliberate ramming attacks. Several preserved.
DIMENSIONS Span 9,0 m (29 ft 6⅓ in); length (CCB-12) 5,9 m, (–1 to –6) 6,0 m (19 ft 8¼ in), (–10, –17) 5,98 m (19 ft 7½ in), (M-62) 6,04 m (19 ft 9¾ in), (M-63) 6,13 m (20 ft 1⅓ in); wing area 14,54 m² (156·5 ft²).
ENGINE One radial, (prototype, –1) M-22, (–4) 710 hp Cyclone F3, (–5) M-25, (–6, –12) M-25A, (–10, –17) M-25V, (–18, some –24) M-62, (–24 and later) M-63.
WEIGHTS Empty (UTI-2) 1030 kg, (–5) 1200 kg (2,646 lb), (–6) 1260 kg (2,778 lb), (–10) 1350 kg (2,976 lb), (–17) 1495 kg (3,296 lb), (–18) 1428 kg (3,148 lb), (–24, M-63) 1490 kg (3,285 lb); fuel/oil (UTI) 155 kg+16 kg, (most) 183 kg+25 kg, (–24) 280 kg, (–24, M-63, internal) 190 kg+30 kg; max loaded (UTI) 1,4 t, (–4) 1420 kg (3,131 lb), (–5) 1460 kg (3,219 lb), (–6) 1660 kg (3,660 lb), (–10) 1715 kg (3,781 lb), (–17) 1810 kg (3,990 lb), (–18) 1830 kg (4,034 lb), (–24, –M-62, normal) 1912 kg (4,215 lb), (–24, M-63, max) 2095 kg (4,619 lb).
PERFORMANCE Max speed (CCB-12) 359 km/h, (UTI, typical) 364 km/h, (M-22, wheels) 360 km/h, (–4, –5) 395 km/h (245 mph) at SL, 455 km/h (283 mph) at 4 km, (–10, –17) about 385 km/h at SL, 440 at 5 km; (–18) 411/464 km/h (255/288 mph), (–24, either engine) 440/489 km/h (273/304 mph), M-25A aircraft (with fixed skis) 354/385 km/h, (SPB) 350 km/h; climb to 1 km (–5) 1·3 min (–18) 1·8 min; service ceiling (typical) 8,6/9,1 km (–18) 9470 m (31,000 ft), (–24, M-63) 9,7 km (31,825 ft); range (typical) 820 km (500 miles), (–24) 700 km; time 360° circle (early) 15 s, (late models) up to 18 s; takeoff (early) 230 m/13 s, (–18) 320 m/15 s, (–24, max wt) 450 m/22 s; landing (typical) 280 m/18 s/100 km/h (early) to 130 km/h (late models).

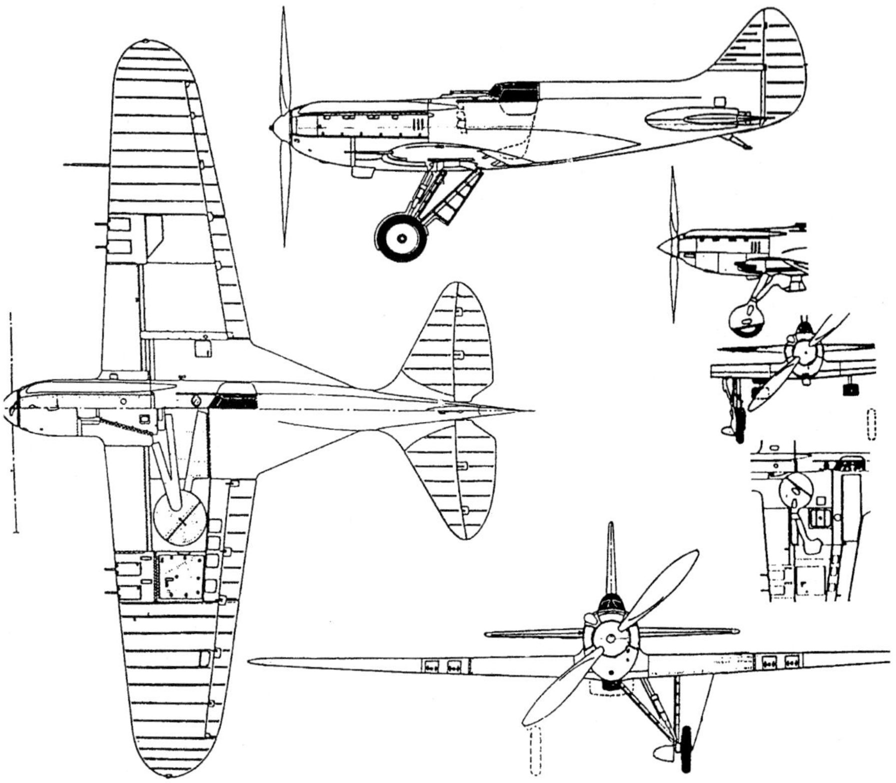
I-17, CCB-15 (insets, CCB-19)

I-17/CCB-19

I-17 To provide advanced fighter with water-cooled engine for comparison with I-16 Polikarpov's CCB collective designed **CCB-15** in first half 34, building prototype very quickly in summer and starting flight testing (V. P. Chkalov) on 1 Sept. Technology as **CCB-12** (I-16) with wooden *shpon* monocoque fuselage and KhMA steel/D6 dural wing with D1 skin back to front spar and fabric to rear. Bulky V-12 engine, then subject of licence negotiation, carried on mounting fabricated from dural box sections, with D1 skin panels. Large metal two-blade prop. Large unducted radiator extended by manual cranking from belly. Main gears similar to I-16 with manual cranking but retracting outwards from pivots near wing roots. Provision for two ShVAK (then still experimental) and two ShKAS in centre section outboard of propeller disc, plus 50-kg bombload (almost certainly 2×25-kg). Forward-sliding canopy with integral windscreen with Aldis sight. Attractive aircraft, with cockpit nearer CG than in other Polikarpov fighters, but Chkalov found cockpit narrow and shock struts stiff in their cylinders, and on demo before VVS delegation one leg failed to extend. Nevertheless, designation I-17 allotted, and second prototype authorized. This was modified, designated **CCB-19**. New wide-track landing gear retracting inwards with pneumatic (30 at, 440 lb/in²) operation. Licensed M-100 engine driving 3,6 m (11 ft 10 in) licensed French propeller with ShVAK firing through hub. Four ShKAS in outer centre-section. Wing of reduced aspect ratio, with shorter ailerons and Zapp-type flaps. Racks under outer wings for two 50-kg bombs. Unique coolant system with two equal-size radiators packaged in short ducts mounted on sliding semi-retractable pylons containing piping. One radiator for water, other half-water, half-oil. Excellent aircraft, faster than any I-16, better handling and good gun platform. Exhibited Paris Salon 36. Third prototype built, **CCB-33**, with surface-evaporation steam cooling using 7 m² (75 ft²) of smooth D1 double skin along wing leading edges (details show intimate knowledge of British R-R Goshawk system). Armament reduced to three ShKAS to keep weight to 1935 kg and cooling system thought vulnerable. Perhaps unfortunately, Chkalov was by late 35 preoccupied with planned RD record flight, improved I-16 was considered adequate and programme lapsed.

I-17Z Development for launching from TB-3 *Aviamatka* (see Vakhmistrov) with only 9 m² wing and no landing gear except emergency skid; three ShVAK or one ShVAK and two ShKAS, (est) speed 575 km/h, to hook back on parent aircraft after combat. Not built.

CCB-25 Projected development of I-17 with M-34RNF or GR14Krsd, studied late 35.

CCB-43 Projected development of CCB-19 with HS12Ybrs and four ShKAS plus 100 kg (two FAB-50) bombs; (est) speed 520 km/h, again Dec 35.

DIMENSIONS Span (1) 10,2 m (33 ft 5½ in), (2) 10,0 m (32 ft 9⅔ in); length 8,8 m (28 ft 10½ in); wing area (1) 17,7 m² (190 ft²), (2) 17,9 m² (193 ft²).

ENGINE (1) One 760 hp Hispano-Suiza 12Ybrs, (2, 3) M-100.

WEIGHTS Empty (2) 1620 kg (3,571 lb); fuel/oil 155 kg+35 kg; loaded (1) 1655 kg (3,649 lb), (2) 1950 kg (4,299 lb).

PERFORMANCE Max speed (1) 424 km/h at SL, 455 km/h (283 mph) at 3 km, (2) 'near 500 km/h'; climb to 1 km, 1·1 min; to 5 km, 6·3 min; service ceiling (1) 9260 m, (2) 9,9 km (32,480 ft); time 360° circle, 16 s; takeoff 240 m/13 s; landing 280 m/24 s/127 km/h.

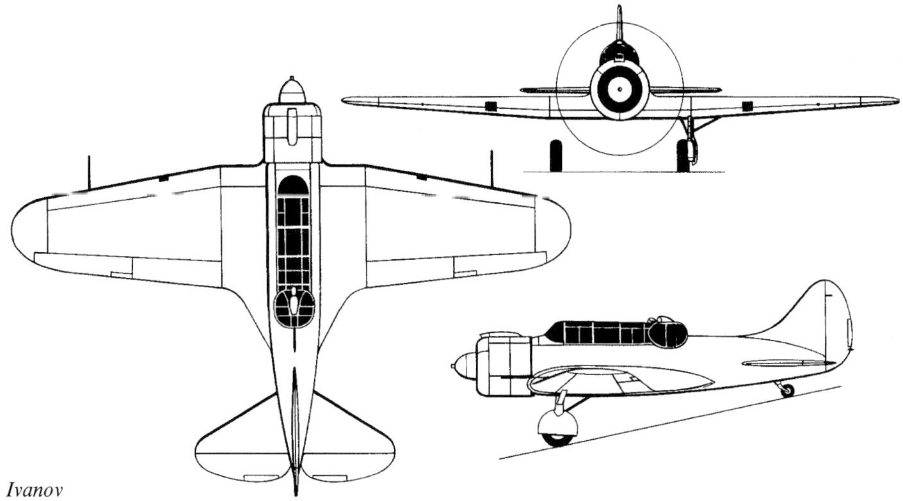

Ivanov

Ivanov

Ivanov Stalin's personal telegraphic code-name was used for 1936 VVS competition for a fast reconnaissance *Shturmovik*. Won by existing KhAI-5 as R-10, but several other later contenders, CCB entrant having no known designation other than name of competition. Exceedingly like later Su-2 but lighter and different systems. Fuselage oval-section wood monocoque put together from left/right halves, *shpon* skin 3 mm front and 2 mm rear. Light-alloy wing first to have T-section booms of 30KhGSA stainless steel; rest of structure mainly D6 with 0,8 mm flush-riveted skin. Control surfaces D6 with fabric, all with trimmers. First Polikarpov levered-suspension landing gear, pneumatic retraction back into wing with wheels rotating 90°; tailwheel also retractable. Engine in long-chord cowl with pilot-operated gills, driving 3,0 m three-blade constant-speed propeller with spinner. First of two prototypes with four synchronized ShKAS in forward fuselage, two BK and one ShKAS in wing (inboard end of outer panels), one BS in manual dorsal turret and one ShKAS in retractable lower rear position. Remarkable bomb load, 400 kg in fuselage bay and 500 kg under wings. Design first quarter 37, start of construction 4 July 37, first flight Feb 38. Better than R-10, but by then too late; flying qualities unimpressive but judged amenable to improvement. Tested until August as reconnaissance/bomber; second *Ivanov* not tested.

DIMENSIONS Span 14,0 m (45 ft 11¼ in); length 9,4 m (30 ft 10 in); wing area 28,07 m² (302 ft²).

ENGINE One M-62.

WEIGHTS Empty 2662 kg (5869 lb); fuel/oil 540 kg; loaded 3930 kg (8664 lb).

PERFORMANCE Max speed 410 km/h at SL, 425 km/h (264 mph) at 3 km; no other data.

VIT-1, SVB, MPI-1 OKB instructed 1936 to prepare project for fast twin-engined aircraft with heavy fixed guns and bombload, designations from *Vozdushnyi Istrebitel Tankov* (aerial tank fighter), *Samolyet Vozdushnogo Boya* (aircraft for air combat) and *Mnogomestnyi Pushyechnyi Istrebitel* (multiseat cannon fighter). Aerodynamically clean, low wing, mixed structure, fuselage left/right halves of *shpon* monocoque, wing aerodynamics as VIT-2 (next) with large root fillets, two truss spars welded KhMA tube, rolled-dural lattice ribs, flush-riveted D1 skin. Single-fin tail similar, pioneer use of metal-skinned control surfs. Split flaps and twin-leg main gears all pneumatic operation, cable to retract tailwheel. Nav/bomb in nose with single ShVAK aimed through ±10°, pilot above LE under sliding canopy with sight for two ShK-37 (exptl Shpitalnyi) cannon with long barrels projecting ahead of wing roots, radio op close behind pilot with single ShKAS in manual turret, two FAB-500 under inner wings plus up to 600 kg under fuselage. First flight summer 37 by Chkalov. Dropped in favour of VIT-2.

DIMENSIONS Span 16,5 m (54 ft 1⅝ in); length

12,7 m (41 ft 8 in); wing area 40,4 m² (435 ft²).
ENGINES Two M-103.
WEIGHTS Empty 4013 kg (8,847 lb); fuel/oil 980 kg+70 kg; loaded 6453 kg (14,226 lb).
PERFORMANCE Max speed 450 km/h (280 mph) at SL, 530 km/h (329 mph) at 3 km; climb 8·4 min to 5 km; service ceiling 8,0 km (26,250 ft); range 1000 km (620 miles); TO 390 m; ldg 460 m.

VIT-2, CCB-48, basically same airframe as VIT-1 but more power and even heavier armament. Wing of Clark-YH profile, 14% root and 6·35% tip, aspect ratio 6·67, improved four-part flaps. Redesigned fuselage with large fuel tank separating pilot from radio/gunner now aft of wing with hand-aimed ShVAK. Larger glazed area in nose, same hand-held ShVAK. In wing roots, two ShK-37 and two ShVAK. Addition of retractable lower rear installation of two ShKAS, also managed by radio operator. Same bombload as VIT-1. Revised twin-finned tail, main gears with large dural fairings forming underside of nacelle when retracted, radiators just outboard of engines under wing ahead of flaps. First flown by Chkalov 11 May 38; soon afterwards fitted with VISh-61 constant-speed 3,3-m propellers. Statically tested to design factor of 10·9 (VIT-1 was designed to excessive factor of 13). Factory testing by B. N. Kudrin and NII testing Sept 38 by P. M. Stefanovskii. Generally excellent aircraft; decision to build in series, but with some armament removed in order to increase speed and ceiling.

OKB had detailed proposal for special long-range aircraft derived from VIT-2, with armament replaced by fuel tankage. May have been designated DDI, though not a fighter; wing span increased and crew two pilots only, in tandem. Range 7900 km (4,900 miles) at 350 km/h, 6200 km at 500 km/h. Never built.
DIMENSIONS Span 16,5 m (54 ft 1⅝ in); length 12,25 m (40 ft 2¼ in); wing area 40,76 m² (439 ft²).
ENGINES Two M-105.
WEIGHTS Empty 4032 kg (8,889 lb); fuel/oil 780 kg+100 kg; loaded 6302 kg (13,893 lb).
PERFORMANCE Max speed 486 km/h (302 mph) at SL, 513 km/h (319 mph) at 4,5 km; climb 6·8 min to 5 km; service ceiling 8,2 km (26,900 ft); range (est) 800 km (500 miles); TO 22 s, 450 m; ldg 400 m.

VIT-2

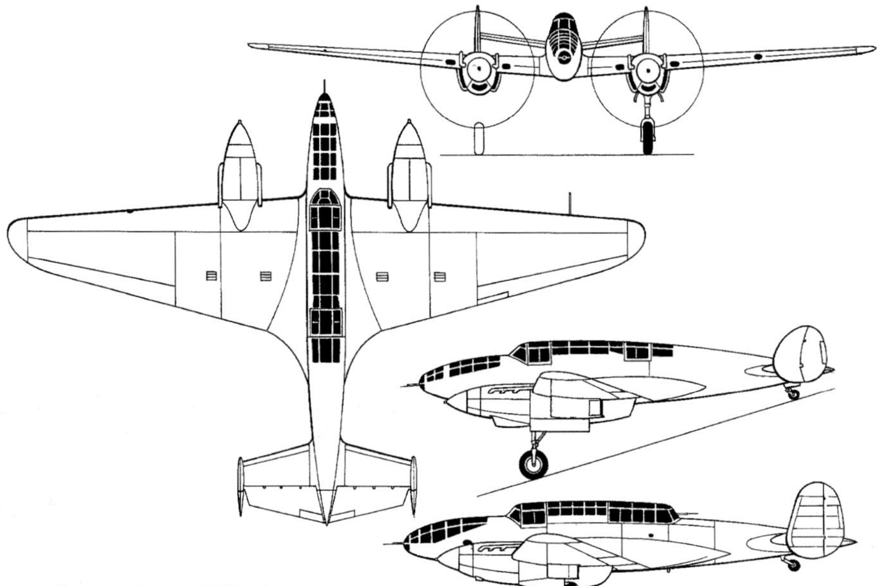
SPB (lower side view, VIT-2)

SPB, D, D-3 Designation from *Skorostnoi Pikiriyuschii Bombardirovshchik* (fast dive bomber) plus D for *Dalnost* (distance). Instead of guns, this aircraft carried fuel, heavier than before. However, by refining aerodynamically, faster than VIT-2. Outwardly looked like predecessor, though fuselage smaller and more streamlined, wing longer span, and low aspect ratio tail surfaces. Structurally modern stressed-skin, with spar booms rolled or extruded, webs plate and ribs and frames pressed sheet. Reluctantly adhered to three-seat layout, with nav/bomb in nose (single UB), and radio/gunner at back with single UB raised from retracted position and aimed from rear cockpit after sliding canopy forward, and periscopic sight to paired ShKAS extended from floor of rear compartment. Bombs: 800 kg in fuselage bay and 700 kg under wings. Improved single-leg main gears retracting into nacelle bays with twin doors, improved pedal-actuated brakes, fully retractable tailwheel, oil coolers in wings each side of engine, and deep but narrow radiator in duct on each side of nacelle. Design structural factor 11, just higher than VIT-2. In conformity with VVS demand OKB built five SPB(D), first flown spring 40 by B. N. Kudrin. On 27 April 40 P. G. Golovin killed when refused to recover from spin. Second disintegrated, killing M. A. Lipkin. Investigation revealed failure to fit leading-edge balance weights to ailerons. Third suffered runaway of rudder trim, but Kudrin managed to land undamaged. Terminated before end 40 because of proven excellence of Petlyakov's No 100. Most KBs would have been shut, but Polikarpov was favoured by Stalin and only chief project engineer, N. A. Zhemchuzin, was imprisoned.

Termination halted work on **SPB(P)**, twin-float version with reversing propellers (first in world) to assist water manoeuvres; speed 435 km/h at 7 t gross. **SPB(BSB)** was 38 project for reconnaissance-bomber version with M-88 engines, 1,5 t bombload, 535 km/h and range 2900 km at 6,5 t wt, with ceiling 11 km. **PB-1** was dive bomber of 1939-40 with two M-71, single ShKAS in nose and BT dorsal. Also **SPBM-71** high-level bomber, 643 km/h (400 mph), 1939-41. **E, E-2**, dive bomber of 40 with two M-37, 3,3-t bombload, 622 km/h.
DIMENSIONS Span 17,0 m (55 ft 9¼ in); length 11,18 m (36 ft 8⅛ in); wing area 42,93 m² (462 ft²).
ENGINES Two M-105.
WEIGHTS Empty 4480 kg (9,877 lb); loaded 6850 kg (15,100 lb).
PERFORMANCE Max speed 490 km/h at SL, 520 km/h (323 mph) at 4,5 km; service ceiling 9,0 km; range 2200 km (1,370 miles); take-off 500 m/25 s; landing 450 m.

I-180 Kremlin meeting spring 38 took decisions on future fighter programme, strongly influenced by Spanish experience. C.R.32 continued to suggest need for agile biplane, but prime need was improved monoplane fighter(s). I-16 *tip*-10 was considered markedly superior to rivals, and situation (wrongly) assessed as not urgent. Decision taken to build new fighters in two stages: first generation with 14-cylinder Tumanskii engine, I-180 monoplane and I-190 biplane; second stage with 18-cylinder Shvetsov or Tumanskii of much greater power, I-185 monoplane and I-195 biplane.

Polikarpov SPB

I-180/2

I-180/3

I-180-1 Selected I-180 on 14 June 38 from original studies but redesigned with traditional I-16 structure with wood monocoque fuselage. Wing Clark-YH profile, built-up KhMA tube spars and D6 lattice ribs, D1 sheet leading edge with adhesive tape edges securing fabric attached over rest of wing. Control surfaces D6 and fabric; three-strut main gears folding inwards pneumatically, split flaps worked by 7-m cables, single protected AMTs tank ahead of cockpit, engine in extended I-16-type cowl driving 2,9-m VISh-23Ye 3-blade prop, two ShKAS above engine and two in inner wings, bombload 40 kg or overload 200 kg. Flown at Central Aerodrome by V. P. Chkalov 15 Dec 38; engine left uncovered all night in –24°C and, after listening to engine, Chkalov rashly decided to fly, suffered engine seizure in circuit and cartwheeled avoiding obstructions. Chkalov, national hero, killed; while body lay in state, many in OKB arrested including Tomashyevich, as well as Usachyev, director GAZ-156, and Belyaikin, director GAP.

DIMENSIONS Span 9,0 m (29 ft 6⅜ in); length 6,79 m (22 ft 3⅜ in); wing area 14,68 m² (158 ft²).

ENGINE One M-88.

WEIGHTS Empty 1710 kg (3,770 lb); fuel/oil 200 kg+25 kg; loaded 2021 kg (4,455 lb), max 2331 kg (5,139 lb).

PERFORMANCE Max speed est 435 km/h (270 mph) at SL, 557 km/h (346 mph) at 5 km; climb (est) 6·3 min to 5 km; service ceiling (est) 10,25 km (33,630 ft); range (est) 800 km (500 miles).

I-180-2 Second aircraft with larger wing, engine with 2-speed supercharger. First flight 19 April 39 by Ye. G. Ulyarkhin; flown by S. P. Suprun in May Day parade, but on 53rd flight crashed killing Tomas P. Suzi, believed due to oil-cooler fire. Considered not to discredit aircraft and series ordered.

DIMENSIONS Span 10,09 m (33 ft 1¼ in); length 6,79 m (22 ft 3⅜ in); wing area 16,11 m² (173 ft²).

ENGINE One M-87B.

WEIGHTS Empty 1847 kg (4,072 lb); fuel/oil 200 kg+30 kg; loaded 2240 kg (4,938 lb) normal, 2459 kg (5,421 lb) max.

PERFORMANCE Max speed 408 km/h (254 mph) at SL, 540 km/h (336 mph) at 5,85 km; climb 6·25 min to 5 km; service ceiling 10,25 km (33,630 ft); range 800 km (500 miles).

I-180-3 Polikarpov got GAP to prevail over GAZ production director and modern stressed-skin construction finally accepted, with simple presswork from sheet. Spar booms machined T-sections in KhGSA. Engine M-88R, more powerful and with geared VISh-23 series propeller, in improved installation with ejector exhausts grouped at rear. Impressive armament, two packages each of one BS and one ShKAS, firing between upper cylinders of engine grouped as one battery; bombload as before. Cockpit enclosed by hinged canopy. First flight 10 Feb 40; factory testing by Ulyarkhin encouraging, series production to start Aug 40, but a few weeks earlier NII's A. G. Proshakov broke aircraft doing incorrect barrel roll (escaped by parachute). Series continued.

DIMENSIONS Span, area, as -2; length 6,88 m (22 ft 6⅞ in).

ENGINE One M-88R.

WEIGHTS Empty 2020 kg (4,453 lb); fuel/oil 200 kg+30 kg; loaded 2429 kg (5,355 lb) normal, 2638 kg (5,816 lb) max.

PERFORMANCE Max speed 455 km/h (283 mph) at SL, 575 km/h (357 mph) at 7 km; climb 5·6 min to 5 km; service ceiling 11,05 km (36,250

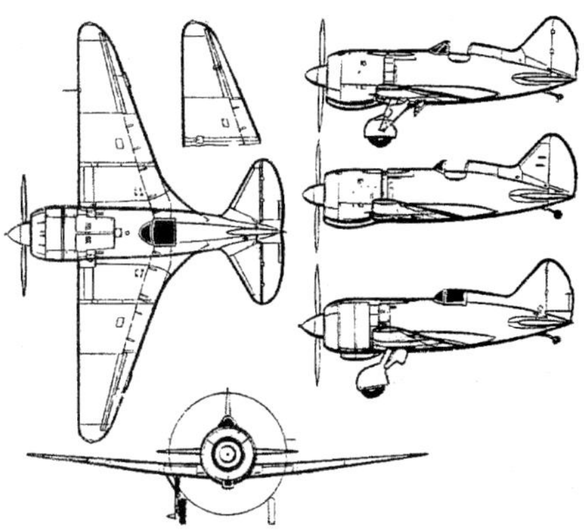

I-180-3 (upper side view and inset wing -1; middle side view -2)

ft); range 900 km (560 miles); TO 240 m; ldg 130 km/h, 200 m.

I-180S Series (*Serinyi*) version basically as -3 (which was completed later) except for open cockpit and insistence by GAZ-156 on restoration of old I-16 type wing structure. Intention was to add underwing tanks for *Shturmovik* role and strengthen landing gears but insufficient room (see I-180Sh). New 4-point pick-up for complete engine and armament pack. First three of initial batch of 10 completed Dec 39, but no more built because judged obsolescent.

I-180Sh Designation not *Shturmovik* but *shassi* (landing gear); this component redesigned with single cantilever shock-strut instead of triangulated pattern (called pyramid-type in Russian) used on all previous I-16 and I-180 fighters. Discontinued Jan 40 when it was recognized I-180 family no longer competitive, and requiring less wing area and more power to meet more stringent demands for performance and firepower. Urgent redesign to second stage of development with new 18-cylinder engine resulted in I-185 series.

DIMENSIONS, ENGINE As -3.

WEIGHTS Empty 2046 kg (4,511 lb); fuel/oil 200 kg+30 kg; loaded 2456 kg (5,414 lb) normal, 2675 kg (5,897 lb) max.

PERFORMANCE Max speed 470 km/h (292 mph) at SL, 585 km/h (364 mph) at 7,15 km; climb 5·0 min to 5 km; no other data.

I-185 Despite accidents Polikarpov OKB had by 40 built fast and simple fighter ready for service; but it had been overtaken by rivals both at home and abroad. To achieve superiority in 41 needed more power, higher wing loading and heavier armament, and though based on I-180 resulting I-185 was virtually fresh design. New wing with NACA-230 profile with thickness 14% on centreline, 12·2% at end of centre section and only 8% at tip. Further increased aspect ratio of 6·18. Improved structure yet still of mixed type, built to factor of 13. Light-alloy stressed-skin wing, large four-section pneumatic split flaps, large automatic leading-edge slats added on outer panels. Dihedral 3° on outer panels only. Flight controls D6 and fabric, fully balanced with trimmers on left aileron, left (later both) elevator and rudder, control by cables from pedals and rods/bell-cranks elsewhere. Fuselage *shpon* monocoque with integral fin, wide cockpit with broad shallow windscreen, central canopy sliding to rear and Plexiglas rear section. Oil cooler in duct under fuselage at trailing edge. Large engine installed as complete unit on welded 30KhGSA tube mounting. Protected tanks between centre-section spars 540 lit (119 gal). Single-leg main gears, track 2,7 m, tyres 700×220, pneu retraction inwards causing small blister above wing root. Tailwheel separate actuating cyl, twin doors.

First aircraft called **R**, or **I-185/M-90**, built amazingly fast, 25 Jan to 10 Mar 40. Notable for ducted spinner forming nose of fuselage with 205-mm inlet for cooling air expelled from gills as in I-180, downdraught carb fed from separate inlet above cowl, exhaust from each cyl row grouped into two tandem pipes each side. Armament as I-180S. Engine never received, partly through conflict over installation.

DIMENSIONS Span 9,79 m (32 ft 1⅜ in); length 7,53 m (24 ft 8½ in); wing area 15,53 m² (167 ft²).

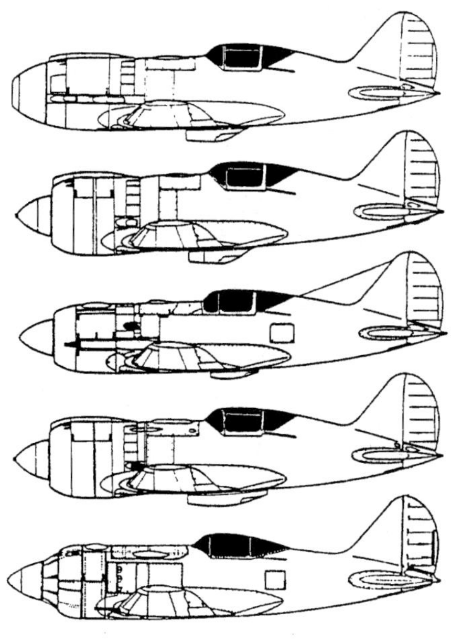

From the top: I-185/M-90, I-185/M-71 (1941), I-185/M-82A, I-185/M-71 (1942), I-187

ENGINE One M-90.

WEIGHTS Empty 2068 kg (4,559 lb); fuel/oil 389 kg+40 kg; loaded 2708 kg (5,970 lb) normal, 3223 kg (7,105 lb) max.

PERFORMANCE (est) Max speed 580 km/h (360 mph) at SL, 701 km/h (436 mph) at 6 km; climb 4·95 min to 5 km; service ceiling 10,25 km (33,630 ft); range 800 km (500 miles); TO 280 m; ldg 355 m.

I-185/M-81, RM, 02 Second aircraft with conventional installation of exptl Shvetsov M-81 engine, tight-fitting cowl, all exhaust grouped into single pipe each side at rear, otherwise no significant change. First flight 11 Jan 41, but engine seriously down on expected power and in May removed; by Dec 41 replaced by M-71 with which factory tests completed.

DIMENSIONS As R, except length 7,74 m (25 ft 4¾ in).

ENGINE One M-81, later M-71 (data for latter).

WEIGHTS Empty 2471 kg (5,448 lb); fuel/oil 389 kg+50 kg; loaded 3119 kg (6,876 lb) normal, 3534 kg (7,791 lb) max.

PERFORMANCE Max speed 505 km/h (314 mph) at SL, 615 km/h (382 mph) at 6,2 km; climb 5·7 min to 5 km; service ceiling 10 km (32,800 ft); range (est) 900 km (560 miles); TO 300 m; ldg 150 km/h, 370 m.

I-185/M-71, 04 Fourth aircraft completed after evacuation to GAZ-51. Wing changed structurally with single spar with sheet web and T-booms machined from 30KhGSA, landing light in left LE, set at +1°30' incidence, smaller (650-mm) main tyres, two UBS (400 rds) and two ShKAS (1,400) plus 4×FAB-100 or 2×FAB-250 or 8×RS-82. Flown c Oct 41 (before re-engined RM), tested on Kalinin front by VVS pilots in company with M-82A aircraft. Later rebuilt with 3×ShVAK as later prototypes.

DIMENSIONS As R except length 7,68 m (25 ft 2⅝ in).

ENGINE One M-71.

WEIGHTS Empty 2846 kg (6,274 lb); fuel/oil 389 kg+56 kg; loaded 3500 kg (7,716 lb) normal, 4015 kg (8,851 lb) max.

PERFORMANCE Max speed 556 km/h (345 mph) at SL, 630 km/h (391 mph) at 6,17 km; climb 5·2 min to 5 km; range (est) 835 km (520 miles); TO 300 m; ldg 148 km/h, 370 m.

I-185/M-82A, I, 06 First I-185 with redesigned forward fuselage housing slimmer 14-cyl engine and totally new armament. NKAP sent instruction to La, MiG and Yak OKBs to fit this outstanding and recently qualified engine, and in I-185 result was significant reduction in drag as well as in weight. Increased tankage, 650 lit (143 gal) in three protected tanks in c-s. Pioneer installation of three ShVAK, one above fuselage and one on each side; recognizing any malfunction of synchronization could destroy aircraft, great care taken with installation by A. G. Rotenberg (Polikarpov) and K. A. Bortnovskii (B. G. Shpitalnyi KB); total 500-540 rounds. Wing racks for 4×FAB-100 or 8×RS-82. Structure weight: fuselage 211 kg, wing 410 kg, tail 49 kg, with poss 30 kg reduction in prodn using thinner material gauges. Completed early 42, air firing tests by Ulyakhin, then NII testing 13 Apr to 5 July, followed by Nov 42 battle testing (with M-71 aircraft 04) with 728 IAP on Kalinin front. Prodn recommended but not actioned because La-5 with same engine used less scarce duralumin.

DIMENSIONS Span 9,8 m (32 ft 1⅞ in); length 8105 m (26 ft 7 in); wing area 15,53 m² (167 ft²).

ENGINE One M-82A.

WEIGHTS Empty 2717 kg (5,990 lb); fuel/oil 360 kg/450 kg+50 kg; loaded 3328 kg (7,337 lb), max 3418 kg (7,535 lb).

PERFORMANCE Max speed 549 km/h (341 mph) at SL, 615 km/h (382 mph) at 6,47 km; climb 6·0 min to 5 km; service ceiling 10,7 km (35,100 ft); range est 1015 km (630 miles); TO 404 m; ldg 155 km/h, 348 m.

I-185 Etalon The *Etalon* (Standard) aircraft was seen as prototype of production I-186, and combined new slimmer fuselage, wing tanks and three ShVAK with bigger engine, cowl diam 1400 mm, length 1360 mm. First flight 10 June 42, factory test to Jan 43, then NII test. Considered fastest fighter in world, plus heavy firepower and 500-kg bomb, but crashed (carburettor blockage) 5 April 43, killing V. A. Stepanchenok.

DIMENSIONS As M-82A except length 8052 m (26 ft 5 in).

ENGINE One M-71.

WEIGHTS Empty 3130 kg (6,900 lb); fuel/oil 360 kg/450 kg+55 kg; loaded 3735 m (8,234 lb), max 3825 m (8,433 lb).

PERFORMANCE Max speed 600 km/h (373 mph) at SL, 680 km/h (423 mph) at 6,1 km; climb 4·7 min to 5 km; no other data.

I-187 Never completed, these were immediate successors to 185 family. Modified engine installation with engine moved back above wing, exhausts grouped into ejector stubs at sides, longer reduction gear and extension shaft and variable-profile inlet shutters from spinner of AV5-118A 3,4 m 3-blade prop; further increased fuel/oil tankage, new armament of two ShVAK wide apart above engine and two at inner end of each outer wing, 640 rounds.

I-185/M-71

I-185/M-82A

Projected early 43, in parallel with M-90 engine, and shortly followed by **I-188** with M-95.
DIMENSIONS As 185E.
ENGINE One M-71F or M-90.
WEIGHTS (M-71F) Empty 2490 kg (5,489 lb); fuel/oil 495 kg+63 kg; loaded 3667 kg (8,084 lb).
PERFORMANCE Not calc.

I-190 Superior biplane fighter derived from I-153, two built 39. Main difference two-row engine of reduced diam. Fin/rudder enlarged, armament two ShVAK or four ShKAS plus 200 kg bombload. First flight 30 Dec 39, by A. I. Zhukov, but crashed 13 Feb 41 and discontinued. Second aircraft, **I-190GK**, fitted with ducted spinner (150-mm diam inlet) as on I-185/M-90(R), turbocharged engine and pressurized cockpit. See Addenda. Data for No 01.
DIMENSIONS Span 10,2 m (33 ft 5⅝ in); length 6,48 m (21 ft 3⅛ in); wing area 24,83 m² (267 ft²).
ENGINE One M-88.
WEIGHTS Empty 1761 kg (3,882 lb); fuel/oil 200 kg+20 kg; loaded 2112 kg (4,656 lb), max 2321 kg (5,117 lb).
PERFORMANCE Max speed 375 km/h (233 mph) at SL, 450 km/h (280 mph) at 7,05 km; climb 5·9 min to 5 km; service ceiling 12,4 km (40,700 ft); max range 720 km (447 miles).

I-195 Ultimate biplane fighter, projected 40. Considerably strengthened I-190 with much more powerful engine cooled via ducted spinner, enclosed (not pressurized) cockpit. Armament as 190. Never completed.
DIMENSIONS Length 7,65 m (25 ft 1 in); wing area 28,0 m² (301 ft²).
ENGINE One M-90.
WEIGHT Loaded 2916 kg (6,429 lb).
PERFORMANCE Max speed at height (est) 580 km/h (360 mph).

ODB Few records of *Odno-dvigatyel Bombardirovshchik* (single-engine bomber) apart from fact that bevel/shafts drove two propellers on wings. Engine AM-37, or AM-38 or M-120. Single-seat aircraft to weigh (est) 5,3 t (11,685 lb), and reach 585 km/h (364 mph) with 2 t bombload carried 880 km (550 miles). Started/abandoned 41.

TIS Designation from *Tyazhelyi Istrebitel Soprovozhdeniya*, heavy fighter escort. Appearance of large twin-engined fighters in other countries caused urgent reappraisal and issue of VVS requirement (according to 80 account, mainly to escort TB-7). Urgent demand could have been met by Yak Ya-22, but this became bomber. Polikarpov received parallel assignment 5 Sept 38, but in this OKB work was much slower, and prototype did not appear until two years after Yak aircraft because engines were not ready.

First aircraft, also called Aircraft **A** or **TIS(A)**, started serious design early 39, as parallel project to VIT/SPB series. Similar aerodynamically, and, though smaller, TIS more powerful and startlingly heavier (empty weight almost equal to loaded weight of heavily armed VIT-2). Heavy structure partly because of design load factor of 12 and large, heavy engines. Wing made as broad centre section and bolted outer panels, profile NACA-230, thickness 14·1% on centreline, 12·85% at joint and 7·8% at tip, dihedral 7° from roots, aspect ratio 6·9, taper ratio 4·06, incidence 1°30'. Modern wing structure with dural plate webs to two spars and ribs pressed from sheet with lightening holes. Light-alloy monocoque fuselage with streamlined form, cockpit well ahead of wing for pilot and aft-facing radio/gunner back-to-back under sliding canopies. Protected tanks in centre fuselage and inner wing for 2430 lit (534·5 gal) fuel. Metal-skinned balanced ailerons, but fabric-covered elevators and rudders all with trimmers. Automatic leading-edge slats, four-section split flaps divided by long nacelles. All three landing gear units fully retractable with doors, pneumatic actuation. Main coolant radiators under engines, oil coolers above. Four ShKAS with 3000 rounds in nose, two ShVAK with 1600 rounds in wing roots and with backseater manning upper and lower ShKAS each with 750 rounds. Weight of fire 5,12 kg/s. Racks under centre section for

TIS-A

TIS-MA

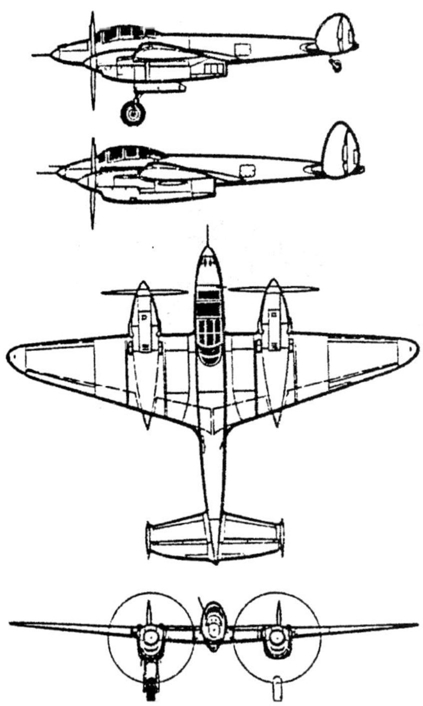

TIS(A) [lower side view TIS(MA)]

two FAB-500 or equivalent bombload.

Aircraft complete 30 April 41 but (possibly lack of engines) did not fly until September, and only one or two flights possible before evacuation. Second aircraft. **TIS (MA)** assembled at Novosibirsk from previously made parts, designation from *Mnogotsyelevoi-A* (multirole A). More powerful engines in shorter nacelles with carb inlet above behind spinner, rads in inner wing with LE inlet and upper-surf exit, oil coolers under engine. Reduced fuel capacity. Vertical tails considerably taller. Two UBS in nose, two Sh-37 in wing roots, one NS-45 in under-fuselage gondola, radio op with hand-aimed ShKAS (no aft lower gun); no bombs. First flight 13 June 44, but discontinued on death of Polikarpov. From 41 OKB studied derived fighter with spring-powered ejection seats.

DIMENSIONS Span 15,5 m (50 ft 10¼ in); length 11,7 m (38 ft 4⅝ in); wing area 34,8 m² (374·6 ft²).

ENGINES (A) two AM-37, (MA) AM-38.

WEIGHTS Empty (A) 5,8 t (12,787 lb), (MA) 6261 kg (13,803 lb); fuel/oil (A) 2420 kg, (MA) 1087 kg; loaded (A) 7840 kg (17,284 lb), (MA) 8280 kg (18,254 lb).

PERFORMANCE Max speed at SL (A) 490 km/h (304 mph), at height (A) 555 km/h (345 mph) at 5,8 km, (MA) 530 km/h (329 mph) at 4 km; climb to 5 km (A) 7·3 min, (MA) 6·4 min; service ceiling (A) 10,25 km (33,630 ft), (MA) 12·5 km (41,000 ft); range (A) max 1720 km (1,070 miles); TO (A) 433 m; ldg (A) 150 km/h, 236 m, (MA) 139 km/h.

ITP, M Polikarpov OKB began work Nov 40 on ITP (*Istrebitel Tyazhelyi Pushechnyi*, fighter with heavy cannon) and to broaden scope of engine experience and acquire direct comparison with I-180 chose liquid-cooled M-107P. Totally fresh design but still mixed construction with wood *shpon* fuselage and integral fin. Ahead of firewall, engine mount structure welded 30KhGSA with dural cowl panels. Wing NACA-230 series, made as centre section 3,22 m span (0° dihedral) and outer panels (5°30' dihedral). Two spars with T-booms of 30KhGSA, structure otherwise dural with flush-riveted stressed skin. All flight control surfaces fabric covered. Same landing gears as I-185, these and tailwheel retracted pneumatically and same system drove split flaps extending in two sections on each side of centreline under fuselage. Automatic leading-edge slats. Neat coolant radiator blocks in wing immediately aft of front spar with leading-edge inlet to duct passing over retracted main legs and through truss section of front spar; exit on underside immediately ahead of rear spar. Inlet under engine for oil cooler and carburettor. Tankage between spars of c-s, total 624 lit. Cockpit, rear fuselage and tail as I-185 except rudder extended to bottom so cutouts in elevators. One ShK-37 firing through hub of prop (Shavrov 40 rounds, Gordon 50), two ShVAK beside engine crankcase with 200 rounds each, wing racks for eight RS-82.

Half-finished (together with static-test second airframe) Oct 41; decided to leave at GAZ-156 and complete 02 as flight article called M-2 (next), so 01 became **M-1**. First flight 23 Feb 42, by A. I. Nikashin. Excellent but bedevilled by engine unreliability; replaced by different engine at end 42, see data.

DIMENSIONS Span 10,0 m (32 ft 9¾ in); length 8,95 m (29 ft 4⅜ in); wing area 16,45 m² (177 ft²).

ENGINE VK-107P, later VK-107A.

WEIGHTS Empty 2960 kg (6,526 lb); fuel/oil 517 kg+30 kg; loaded 3366 kg (7,421 lb).

PERFORMANCE Max speed (est) 568 km/h (353 mph) at SL, 655 km/h (407 mph) at 6,3 km; climb 5·9 min to 5 km; service ceiling 10,4 km (34,120 ft); range 1280 km (795 miles); TO 350 m; ldg 130 km/h, 375 m.

M-2 Second ITP completed as flight article with bigger engine, slightly less fuel and armament of three ShVAK as on later I-185s. First flight 23 Nov 43 by N. V. Gavrilov, engine later replaced by more powerful type (see data). Factory testing to June 44, by which time no question of production.

DIMENSIONS As M-1 except length 9,2 m (30 ft 2¼ in).

ENGINE One AM-37, later AM-39A.

WEIGHTS (AM-39A) Empty 2910 kg (6,415 lb); fuel/oil 440 kg+50 kg; loaded 3570 kg (7,870 lb).

PERFORMANCE (AM-39A) Max speed 600 km/h (373 mph) at SL, 650 km/h (404 mph) at 2,5 km; climb 6·0 min to 5 km; service ceiling (est) 11,5 km (37,730 ft); range 980 km (610 miles); TO 326 m; ldg 140 km/h, 390 m.

VP,K

(*Vysotnyi Perekhvatchik*, high-altitude interceptor) programme began in 42 and embraced aircraft M (Mikoyan) and K (Polikarpov). Aircraft K was being built March 43, using ITP tail and landing gear with different fuselage with pressure cabin and new high aspect ratio wing of greater span but fractionally reduced area. Appreciably lighter, partly because design factor only 10·5, despite heavy engine tur-

ITP

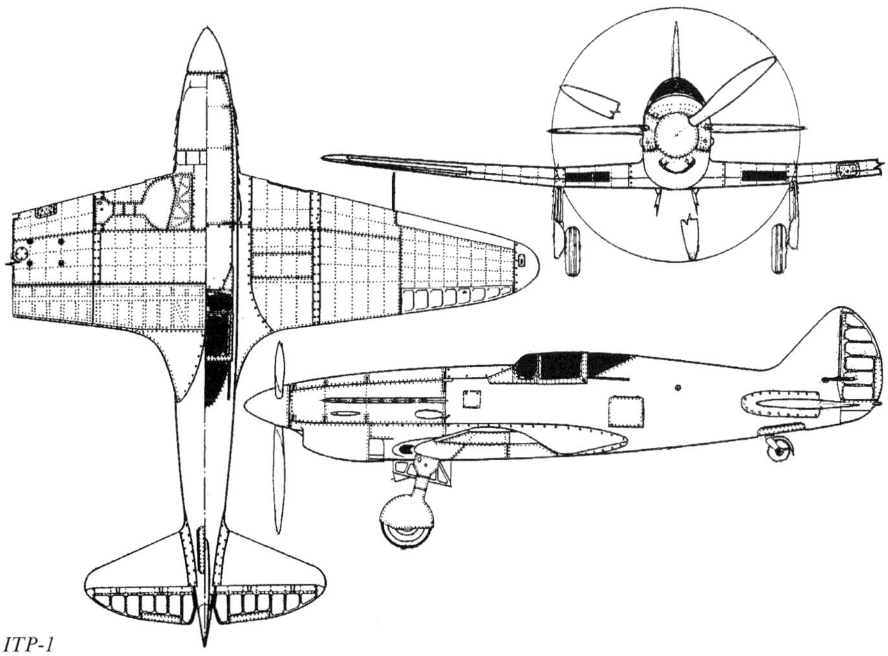

ITP-1

bocharged by two TK-300B and two powerful 23 mm guns, in top of fuselage (intended for new NS-23 but had to be designed for heavy VYa as being available). Prototype left unfinished after Polikarpov's death.

DIMENSIONS Span 11,0 m (36 ft 1 in); length 9,2 m (30 ft 2¼ in); wing area 16,2 m^2 (174 ft^2).

ENGINE One AM-39A with turbos.

WEIGHTS Empty 2727 kg (6,012 lb); fuel/oil 550 kg+55 kg; loaded 3320 kg (7,319 lb) (these are Shavrov's figures, leaving minus 12 kg for ammo and pilot).

PERFORMANCE Max speed 500 km/h at SL, 715 km/h (444 mph) at 14 km; time to 5 km, 3·5 min, to 14 km, 18 min; service ceiling 14,05 km (46,000 ft); landing speed 150 km/h (estimated data).

NB, T Little known about circumstances in which OKB became charged with developing Aircraft T, *Nochnoi Bombardirovshchik* (night bomber). VVS had no known specific requirement before or after 22 June 41 and OKB had no bomber experience since TB-2 and was already heavily committed. Start dated from winter 1941-42, insistence on cavernous internal bomb bay. Immense effort on minimum structure weight, success achieved shown by achieved weight of 2853 kg compared with max loaded weights up to 18 850 kg. This done despite, or because of, complex mixed construction with several materials in close proximity. Wooden fuselage, mainly *shpon* layers built up to 5,5 mm forward and 4 mm at tail, with frames of glued box section with inner/outer rings 17×14 mm and webs 1,5 mm ply. Whole mid-section affected by bomb-bay aperture and cut-out for wing, so stiffened by framework of welded 30KhGSA tube with diagonal struts and even network of bracing wires adjusted for tension. Wing shoulder-high, NACA-230, 16% centreline, 15·1% root, 8% tip, aspect ratio 7·96, taper ratio 3·47. Single, strong main spar with largest sections of 30KhGSA then attempted, machined to form upper/lower T booms joined by double webs of 1,5/2 mm D16. Lighter pressed-plate nose spar and rear spar carrying movable surfaces. Ribs of centre section (joint just outboard of nacelles) pressed from D16 plate; outer-panel ribs built up from wood trusses with ply sheet each side. Centre-section skin flush-riveted D1; outer panels skinned with ply of varying thickness. Area-increasing slotted flaps of all-dural construction divided by long nacelles with electric actuation to 42°. Automatic leading-edge slats on outer panels, all-dural. Most unusually, all control surfaces wooden structure, fabric covering, push-rod actuation. Tailplane with 7° dihedral with twin fins, all dural structure with plywood skin. Landing gears with two shock struts, 460 mm stroke, 1200×450 mm tyres for soft surfaces, pneumatic retraction to rear into bays covered by doors, separate actuator for fully retractable tailwheel. Hydraulic wheel brakes. Six protected fuel tanks in centre section and optional four in outer wings, all linked into system with electric valves to maintain trim and feed either engine from any tank. Engines originally (Jan 42) to be ASh-82A, then ASh-71 chosen and finally ASh-82FNV. Efficient close cowls, fully sealed with electric gills and other cooling valves. Unspecified de-icing for

NB

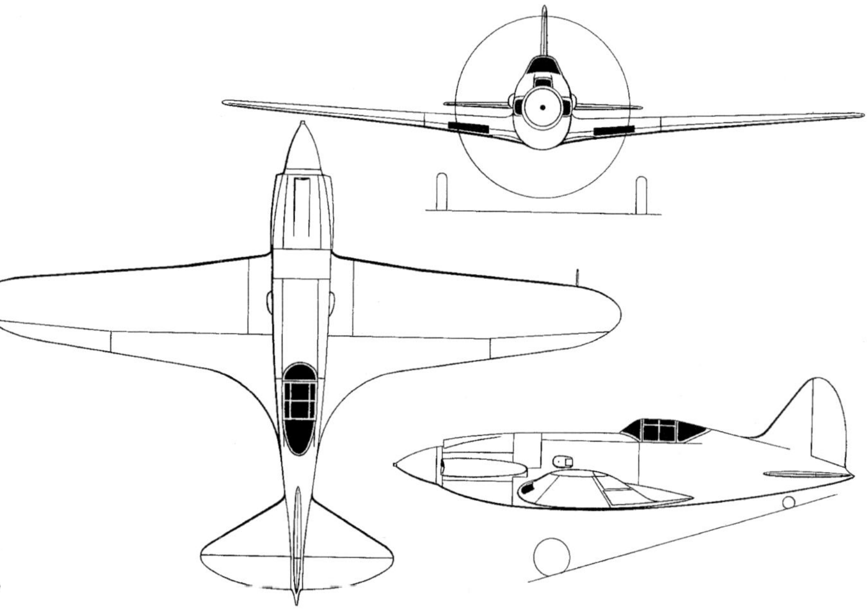

VP

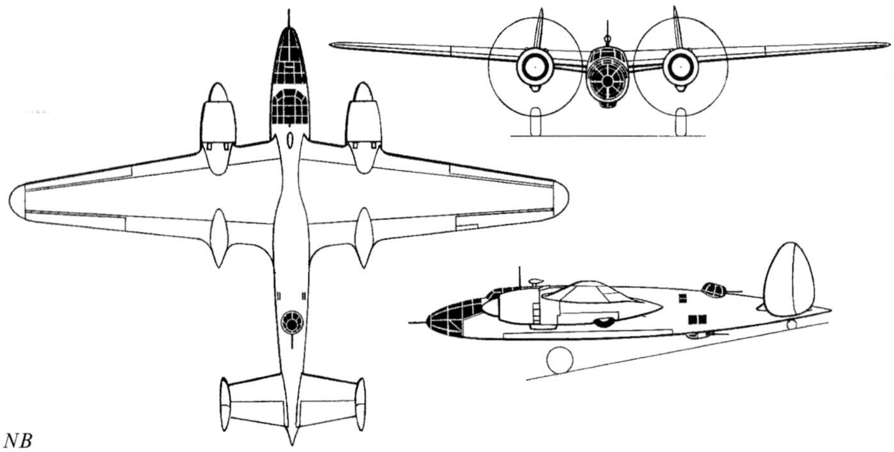

NB

wing/tail leading edges, carb inlet and AV-5-167 propellers. Crew of five in heated cockpits, with pilot reflector sight for fixed UB, dorsal turret for UBT and lower rear UB. Bomb bay with racking for many arrangements of bombs up to 2-t size to total mass 5 t (11,023 lb). Twin doors positioned manually via ratchet mechanism. Advanced flight system with autopilot and pneumatic inputs to command desired heading and attitude, with heading command input by nav/bomb in nose during bombing run. Design load factor 6·18, CG 22·9% with limits 18·6/26%.

FNV engines only installed at last moment, prior to first flight 23 May 44. Promise of outstanding aircraft, see data from measured GAZ testing completed Aug. By this time OKB was being closed.

DIMENSIONS Span 21,52 m (70 ft 7¼ in); length 15,29 m (50 ft 2 in); wing area 58,1 m² (625 ft²).
ENGINES Two ASh-82FN.
WEIGHTS Empty 8843 kg (19,495 lb); fuel/oil 2760 kg; loaded 13,8 t (30,423 lb).
PERFORMANCE Max speed 445 km/h (277 mph) at SL, 510 km/h (317 mph) at 5 km; climb 12 min to 5 km; service ceiling 6150 m (20,200 ft); range with 3 t bombs 3030 km (1,883 miles) [intended to be cleared to 18,85 t, giving 4000 km with 5 t bombs]; TO 660 m; ldg 122 km/h, 570 m.

BDP Polikarpov himself oversaw design in 41 of assault glider to compete with A-7 and G-11, but this BDP (*Boyevoi Desantnyi Planer*, troops assault glider) probably never flown. Same basic design used to meet 42 requirement but considerably modified as **BDP-2**, also called **S-1**, and flown in four weeks. Wooden construction with rounded monocoque fuselage of glued *shpon* and wooden box spar (two in centre-section, one in outer panels) in wing with ply-skinned leading edge, fabric elsewhere. Fully balanced ailerons, Schrenk flaps worked by compressed-air bottle or cockpit handpump. Two pairs of wheels, low-pressure tyres 700 × 220 mm, dropped after take-off (tug, SB-2, Il-4, Li-2) with landing on large left/right skids. Interior cross-section 1000 × 610 mm, max 20 troops and kit, but 12 more common. Glazed nose cockpit for single pilot, roof hatch for defensive gun(s). No provision for large cargo. Evaluation quantity built, no production.

DIMENSIONS Span 20,0 m (65 ft 7⅓ in); length 13,5 m (44 ft 3½ in); wing area 44,7 m² (481 ft²).
WEIGHTS Empty 1923 kg (4,239 lb); loaded 3 t (6,614 lb).
PERFORMANCE No data.

MP Like other constructors Polikarpov fitted prototype BDP-2 with engines in 43. Believed single aircraft converted, designation from *Motor Planer*. Standard M-11F engines with steel-tube mounts on leading edge, aluminium pressed fairings over crankcase and 2,35 m (Po-2) propellers. Aluminium tanks adjacent to engines for 7 h flight. Several windows arranged to open for firing DP machine guns, and 12 sheets of armour for protection against small-arms fire. One reason given for no production is availability of Shchye-2.

DIMENSIONS As BDP-2.
ENGINES Two M-11F.
WEIGHTS Empty 2,3 t (5,071 lb); fuel/oil 365 kg; loaded 3,7 t (8,157 lb).
PERFORMANCE Max speed 172 km/h (107 mph) at SL; practical ceiling 2 km (6,500 ft); range 930 km (580 miles) at 130 km/h (81 mph); landing speed 98 km/h (61 mph).

PB Unbuilt project for *Planer Bombardirovshchik* (glider bomber). Schemed 42 with single pilot and 2 t bombload internally. Not a suicide aircraft and intended for operation at appreciable heights, but no details.

D, Limuzin D was unbuilt wartime project for five-seat liaison aircraft with comfortable cabin; 145 hp M-11F, empty/gross 798/1375 kg, speed

BDP

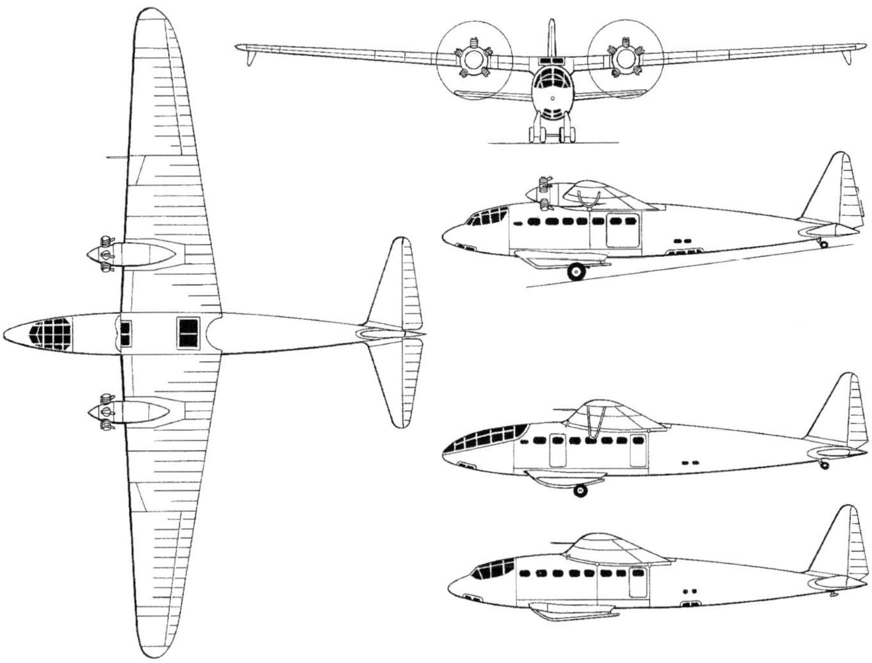

MP (lower side view BDP, middle BDP prototype)

BDP hatch for troops' DPs

Limu'zin (D) mockup

250 km/h, ceiling 3,75 km, climb to 2 km in 17·5 min.

Malyutka Name meaning *Little One*, a simple rocket interceptor in same class as No 302 and BI-1 and overseen by Polikarpov himself from start in 43 at Novosibirsk. Intended for target defence, with endurance under power 8 to 14 min. Clean low-wing aircraft with wood *shpon* fuselage and dural stressed-skin wings and tail. Pilot in pressurized nose cockpit, tricycle landing gear actuated like flaps by air bottles (probably with recharging handpump) and armament of two 23 mm cannon on sides of nose. Construction of prototype begun early 44 but Polikarpov's rapidly failing health delayed completion, and abandoned with his death.

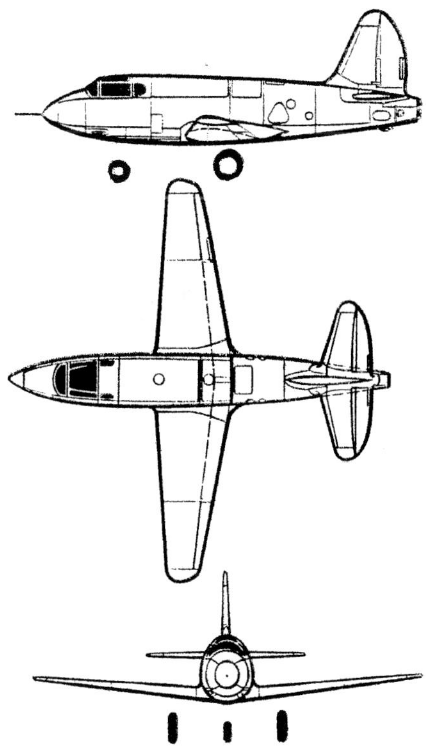

Malyutka

DIMENSIONS Span 7,5 m (24 ft 7¼ in); length 7,3 m (23 ft 11⅜ in); wing area 8,0 m² (86 ft²).
ENGINE One NII-1.
WEIGHTS Empty 1016 kg (2,240 lb); propellant 1,5 t; loaded 2795 kg (6,162 lb).
PERFORMANCE (est) Max speed (est) 890 km/h (553 mph) at SL, 845 km/h (525 mph) at 15 km; climb (est) 1·0 min to 5 km; service ceiling 16 km (52,500 ft).

Rafaelyants

Aram Nazarovich Rafaelyants was born 97 and qualified at VVA 27. For 30 years (about 1929-59) chief engineer of GVF (Aeroflot) aircraft repair and modification shops at Bykovo, on Volga. Designer of several important modifications including improvements to early German transports, PR-5 and other variants derived

RAF-11

Turbolyet, O

from R-5, modified Li-2, and U-2L derived from U-2. Post-45 leader in turbojet test equipment. He died 60.

RAF-1 Ultralight built and flown 25; simple wood/fabric low-wing monoplane of exceptionally clean design. Flew well.
DIMENSIONS Span 9,4 m (30 ft 10 in); length 5,5 m (18 ft 5½ in); wing area 12,6 m^2 (136 ft^2).
ENGINE One 18 hp Blackburn Tomtit.
WEIGHTS Empty 175 kg (386 lb); fuel/oil 20 kg; loaded 273 kg (602 lb).
PERFORMANCE Max speed at SL 105 km/h (66 mph); ceiling reached 3,250 m (10,700 ft); landing speed 47 km/h (29 mph).

RAF-2 Cantilever low-wing monoplane two-seater, built 26. This was a failure, span and drag being too great for engine, flight only just being accomplished. Constructor never fitted a more powerful engine, through lack of both money and time.
DIMENSIONS Span 12,0 m (39 ft 4½ in); length not recorded; wing area 20 m^2 (215 ft^2).
ENGINE One 60 hp Airdisco Cirrus.
WEIGHTS Empty 435 kg; fuel/oil 50 kg; loaded 645 kg.
PERFORMANCE Poor.

RAF-11 By far largest of RAF's aircraft projects, this twin-engine transport was designed mid-30s and built 1937-38. All-wood, with *shpon* monocoque fuselage of fine aerodynamic form and single-spar cantilever wing with ply covering back to spar and fabric elsewhere. Well-engineered, and twin-leg main gears retracted to rear under power of hydraulic system, with pump on each engine; pneumatic (presumably bottle) system provided for emergency extension against air drag. Split flaps, hydraulic actuation with pneumatic standby. Also flown with retractable skis. Flight deck with windows large for period, passenger cabin seating six (three on each side, beside windows) with door at rear on right. Measured performance safe but unimpressive. Rafaelyants spent 18 months refining aircraft, dramatically reducing weight (and slightly reducing fuel capacity, and fitting more powerful engines) and in 40 tested resulting **RAF-11bis**. Outstanding machine, flown with abandon ('like fighter') by many pilots. NII GVF recommended production early 41 for second-class Aeroflot routes, and also as interim civil crew trainer, but this had not been organized when USSR invaded.
DIMENSIONS Span 15,0 m (49 ft 2½ in); length (11) 10,86 m (35 ft 7½ in), (bis) 10,1 m (33 ft 1⅔ in); wing area 30,0 m^2 (323 ft^2).
ENGINES (11) two MG-31, (bis) MG-31F.
WEIGHTS Empty (11) 2,5 t (5,511 lb), (bis) 2097 kg (4,623 lb); fuel/oil (11) 400 kg+56 kg, (bis) 380 kg+40 kg; loaded (11) 3270 kg (7,209 lb), (bis) 3 t (6,614 lb).
PERFORMANCE Max speed (11) 289 km/h (180 mph) at SL, (bis) 294 km/h (183 mph); cruising speed 257 km/h (160 mph); climb to 1 km (bis) 3·5 min; service ceiling (bis) 4,5 km (14,750 ft); range (11) 1000 km, (bis) 930 km (578 miles); take-off (bis) 355 m/18 s; landing (bis) 290 m/18 s/107 km/h (66·5 mph).

Turbolyet, O Inspired by Rolls-Royce TMR (Flying Bedstead) this VTOL was designed in partnership with Prof V. N. Matveyev and team at LII and first hovered under tethers early 57. Publicly flown free Oct 57. Engine single AM-3 (RD-3) mounted vertically with special bearings and oil system, bellmouth inlet pointing upwards and bleed pipes to four reaction control jets on long outrigger arms. Four vertical sprung legs, three fuel tanks around engine and pilot cabin with strong crash-proof roof structure. Most flying done by helicopter pilot Yu. A. Garnayev. Later used for research into jet deflection and engine noise. Played role in development of control systems for Yakovlev VTOL aircraft.

REDA

REDA-MDT Ltd, joint-stock firm established 91 to carry out broad range of R&D, mainly into GA aircraft. First project, team mainly students, was *Priz* (Prize), unconventional amphibian. This has been refined into *Poni*, produced by firm set up for this purpose by REDA-CAHI and NII.

Priz Devd from 90, versatile 4-seat amphibian with unique 3-point hydrodynamics and twin ducted props. First four being built, first flight Nov 94, certification to FAR-23 hoped 96.
DIMENSIONS Span 15,36 m (50 ft 4¾ in); length 11,84 m (38 ft 10⅛ in); wing area 19,52 m^2 (210 ft^2).
ENGINES Two 125-hp TCM IO-240A.
WEIGHTS Empty 1,3 t (2,866 lb); fuel 130 kg; payload 280 kg; MTO 1,7 t (3,748 lb).
PERFORMANCE (est) Max speed 225 km/h (140 mph); cruise 185 km/h (115 mph); climb 180 m (591 ft)/min; range 750 km (466 miles), ferry tanks 1860 (1,155); TO 159 m land, 214 water; ldg 170/150 m.

Poni 2-seat amphibian with same 3-point hydrodynamics as Prize, claimed to make water takeoffs as easy as on land. All-dural structure with some polymer fabric skin on tail and on wing aft of spar. Usable fuel 50 kg, electrical syst, IFR, to FAR-23. Four in build, to fly Sept 93, certification late 94.
DIMENSIONS Span 11,0 m (36 ft 1 in); length 8,5 m (27 ft 10⅝ in); wing area 16,5 m^2 (178 ft^2).
ENGINE One 81-hp Rotax 912A.
WEIGHTS Empty 444 kg (979 lb); MTO 680 kg (1,499 lb).
PERFORMANCE (est) Max speed 150 km/h (93 mph); cruise 135 km/h (84 mph); climb 180 m (591 ft)/min; range 400 km (249 miles); TO 90 m land, 150 m water; ldg 80/120 m.

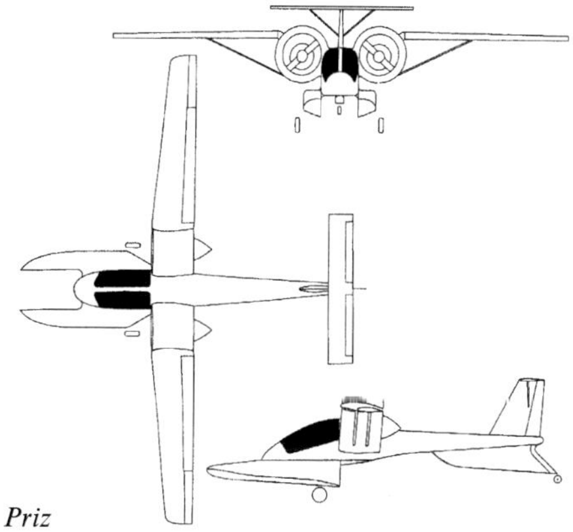

Priz

TOM-1 (but see Addenda)

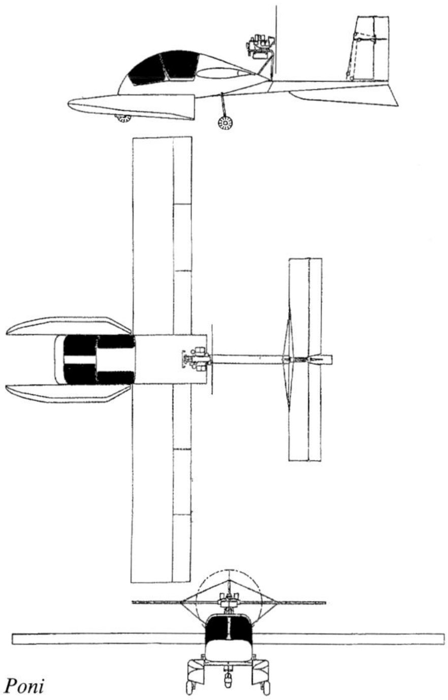

Poni

Richard

Paul Aimé Richard was one of several foreign (in this case French) designers to work in the Soviet Union. Invited to do so by Aviatrust, arriving Aug 28. Worked at MOS VAO at GAZ-28 while organizing his own KB at OPO-4 as part of CCB. Team included I. I. Artamonov (head of manufacturing), D. M. Khomskyi (head of technical section), P. D. Samsonov (projects), S. A. Lavochkin (calculations), fellow-countrymen Augé and Laville, and many other engineers including G. M. Beriev, D. A. Mikhailson, G. M. Mozharovskii, N. K. Skrzhinskii, N. I. Kamov, A. L. Gimmelfarb, M. I. Guryevich, S. P. Korolyev, I. V. Chyetverikov and V. B. Shavrov! Projects included two-seat fighter (later basis for Laville's DI-4), monoplane reconnaissance flying boat (later used by Beriev as starting point for MBR-2), parasol-winged military flying boat with two M-17 engines on wing, four-engined transport flying boat of stressed-skin construction, and studies for flying boats of 100 and 200 tonnes. When TOM failed to win production order he returned to France, late 31.

TOM-1 Designation from *Torpedonosyets Otkrytogo Morya* (torpedo carrier, open sea). Up-to-date and relatively clean design with cantilever mid wing with twin engines neatly faired in, riding on twin floats each with tandem struts with diagonal wire bracing, single fin carrying high tailplane with N-struts and well-streamlined fuselage. Duralumin construction throughout. Wing with St Cyr 60 (French) profile and high aspect ratio (9·075), two built-up truss spars with T-section rolled/machined booms, ribs pressed from sheet and stringers spaced at 100 mm. Fuselage with pressed open-section frames and L stringers at 150-mm spacing. Entire skin smooth dural sheet 0,5-0,6-mm, in places flush-riveted. Slotted flaps power-driven to 40°. Three PV-1 guns in nose and dorsal gunners' cockpits and sliding (retractable) lower rear turret. Torpedo in internal bay under wing with manually operated doors. Prototype

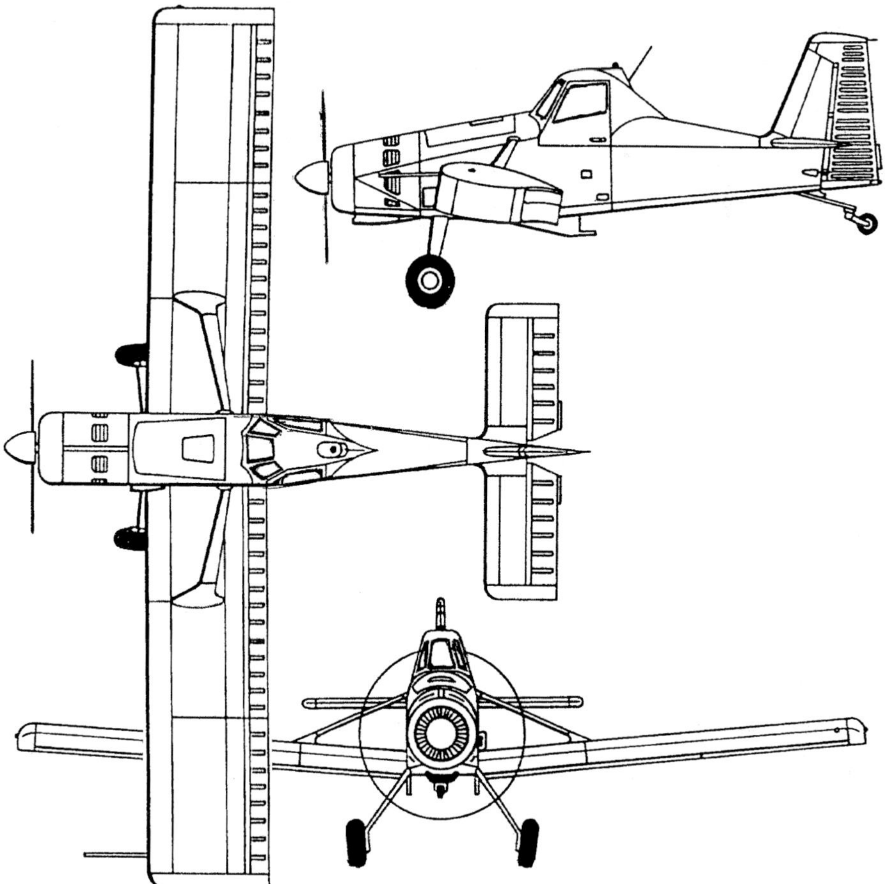

BSKhS (see overleaf)

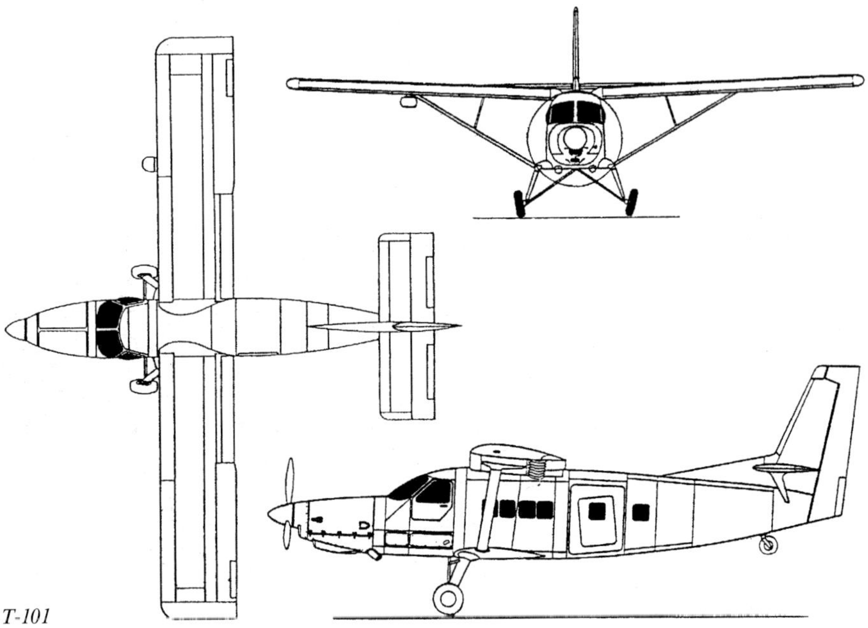

T-101

launched Sevastopol 1 Jan 31 and tested until Aug. Pilots N. I. Kamov and N. A. Kamkin favourably impressed, but series production rejected on grounds that TB-1P was adequate (and, doubtless, that TOM-1 would be expensive and structurally difficult to build and maintain).
DIMENSIONS Span 33,0 m (108 ft 3¼ in); length 19,0 m (62 ft 4 in); wing area 120 m^2 (1,399 ft^2).
ENGINES Two 680 hp BMW VI (in 33 re-engined with M-17).
WEIGHTS Empty believed 4929 kg (10,866 lb); fuel/oil 1300 kg+100 kg; loaded 8030 kg (17,703 lb).
PERFORMANCE Max speed 210 km/h (130·5 mph) at SL; patrol speed 171 km/h (106 mph); service ceiling 5,5 km (18,000 ft); range 1500 km (932 miles); take-off 25 s; alighting speed 80 km/h (50 mph).

Romashka

Romashka Agricultural Co is based at Voronezh.

BSKhS Conventional ag-aircraft, metal/composites, fixed-pitch prop, hopper for <1 t between engine and cockpit feeding spraybars behind wing; mockup shown 93.
DIMENSIONS Span 12,56 m (41 ft 2½ in); length 8,62 m (28 ft 3⅜ in); wing area 22,5 m^2 (242 ft^2).
ENGINE One M-14NTK.
WEIGHTS Empty 1096 kg (2,416 lb); max 2190 kg (4,828 lb).
PERFORMANCE (est) Max speed 240 km/h (149 mph); cruise 170 (106); SL climb 270 m (886 ft)/min; range 510 km (317 miles); TO 166; ldg 144.

Romeiko-Gurko

Daniil Aleksandrovich Romeiko-Gurko (98-47), glider designer/pilot, produced passenger a/c 35 and sport a/c 38.

RG-1 Designation has appeared in West and 'RGP' in East Europe; neither found in Soviet source. Unconventional high-speed passenger/mail aircraft designed (believed in GVF workshops) about 34. Mixed-construction, braced low-wing monoplane with large water-cooled engine on CG driving tractor propeller via extension shaft under pilot's seat. Rather cramped rear cabin for 1 t mail or cargo but room for only four passengers, two at front facing aft and two in tandem on right at rear facing forward; door or left. Trousered main gears, wing with split flap(s). Entered for Aviavnito competition, high-speed section, but judges said 5/6 passengers were stipulated and they criticized passenger accommodation and alleged rear CG position. Not flown, no data.

RMK-1 Second of constructor's designs. One of crop of *planerlyet* powered-gliders intended 36 to achieve lowest-cost air transport. High wing pusher with tail carried on slim lower boom on centreline. Engine installation essentially U-2 in reverse. Welded AMTs fuel tank in nacelle ahead of engine. Fixed spatted main gears braced to fuselage. Well-glazed fuselage nacelle with pilot on centreline at front and six passengers in three rows (or 1 t cargo) with door on left. Shavrov, construction not completed, but recent suggestions RMK-1 flew late 36.
DIMENSIONS Span 28,0 m (91 ft 10⅓ in); length 12,4 m (40 ft 8¼ in); wing area 64,0 m^2 (689 ft^2).
ENGINE One M-11.
WEIGHTS Empty 1170 kg (2,579 lb); fuel/oil 230 kg; loaded 2,5 t (5,511 lb).
PERFORMANCE Max speed (Shavrov) 110 km/h (68 mph); cruising speed 90 km/h (56 mph); practical ceiling 2,5 km (8,200 ft); range (Němeček) 1000 km (620 miles); landing speed 56 km/h (35 mph).

Akula Inspired by Caudron racers, used French engine in odd single-seater completed late 38; name = shark. Engine on steel-tube mounts with dural cowl, otherwise mainly wood; low gull wings, 4 sections split flap, spatted wheels, fin-like fairing behind canopy.
DIMENSIONS Span 7,0 m (22 ft 11⅝ in); length 8350 m (27 ft 4¾ in); wing area 8,6 m^2 (92.6 ft^2).
ENGINES One 220hp Renault Bengali 6.
WEIGHTS Empty c 700 kg; loaded 980 kg (2,160 lb).
PERFORMANCE Max speed 400 km/h (249 mph).

ROS-Aeroprogress

General Designer Evgeniy P. Grunin originally formed Delaero, to build modest General Aviation types. In 90 this was replaced by ROKS-Aero Corp, member of Business Aviation Assoc, other members being Moscow Avn Prodn Assoc, Yak 'Skorost' plant, Myasishchyev OKB/factory, and production plants at Komsomolsk, Luchovitsy, Novosibirsk, Smolyensk and Ulan-Ude. Hope to build wide range, inc Ekranoplans. In 93 renamed ROS-Aeroprogress Corpn, but aircraft designed by ROKS-Aero retain that title.

T-101 Gratch (Rook) Turboprop monoplane to relace An-2. All stressed-skin, wing P-11-14 profile, auto slats and elec slotted flaps (TO 25°, ldg 40°), manual flight controls with fixed tailplane, tyres all 3,5 kg/cm^2 (50 lb/in^2), hyd anti-skid brakes, option skis/floats, AV-24AN reversing 3-blade 2,8 m prop, 6 wing fuel cells total 1200 lit (264 gal), 6-kVA alternator, AP-93 autopilot, adv avionics inc satnav and emergency locator beacon and radio buoy. Cabin 4,2 m×1,6 m wide×1,8 m (13 ft 9 in×63 in×71 in), 2 crew+9 pax, front door and rear 1,46 m (57·5 in) cargo door, payload 1,6 t (3,527 lb). First of five prototypes flew May 93, versions planned with nose gear, AL-34 engine and advanced avionics.
DIMENSIONS Span 18,18 m (59 ft 8 in); length 15 042 m (49 ft 4¼ in); wing area 43,63 m^2 (469·6 ft^2).
ENGINE One TVD-10B.
WEIGHTS Empty 3330 kg (7,342 lb); MTO 5,5 t (12,125 lb).
PERFORMANCE Max cruise (3 km) 298 km/h (185 mph); econ 235 km/h (146 mph); service ceiling 3,6 km (11,800 ft); range (max fuel) 1320 km (820 miles); TO 454 m; ldg from 15 m 370 m.
T-101E Variant with imported engine, Hartzell prop, B-K or Collins avionics, 2 t payload.
DIMENSIONS Span 18,15 m; length 15,3 m; area 42 m.
ENGINE One P&WC PT6A-65AR or Garrett TPE331-14.
WEIGHTS MTO 5670 (12,500 lb).
PERFORMANCE (est) Cruise 320 km/h; range 1 h reserve 1040 km.
T-101V Amphibian, two floats each with front and main wheel hyd retraction to rear, AV-17 reversing 3,6 m prop, fuel 1240 lit (272 gal), 12 pax or 1,5 t (same cabin as 101).
DIMENSIONS Span 18,5 m (60 ft 8½ in); length 15 225 m (49 ft 11½ in); wing area 42,02 m^2 (452·3 ft^2).
ENGINE One TVD-20.
WEIGHTS Empty 3,7 t (8,157 lb); MTO 5715 kg (12,600 lb).
PERFORMANCE Econ cruise 240 km/h (149 mph); service ceiling 7 km (23,000 ft); range

ROS-Aero T-101 (Floatplane)

(max payload) 600 km (373 miles); TO/ldg 350 m.

T-106 Twin devt of T-101, larger wings with 3-section slats and double-slotted flaps braced to stub wings carrying fixed levered main wheels, spoiler ahead of each aileron, nose for baggage or radar, cabin 5,7×1,65×1,85, castoring nose-wheel.
DIMENSIONS Span 19,9 m (65 ft 3½ in); length 15,678 m (51 ft 5¼ in).
ENGINES Two TVD-10B (export option PT6A-65).
WEIGHTS Payload 1,8 t (3,968 lb); fuel 1,6 t; MTO 6,3 t (13,889 lb).
PERFORMANCE Max speed 390 km/h (242 mph); cruise 360 km/h (224 mph); op ceiling 4 km (13,125 ft); range, 30-min reserve (max payload) 450 km (280 miles); max fuel 1250 km (777 miles).

T-108 Zolotoi Orel (Golden Eagle) Twin multirole utility based on T-101, seats 2+19.
DIMENSIONS Span 19,9 m (65 ft 3½ in); length 16,8 m (55 ft 1⅜ in).

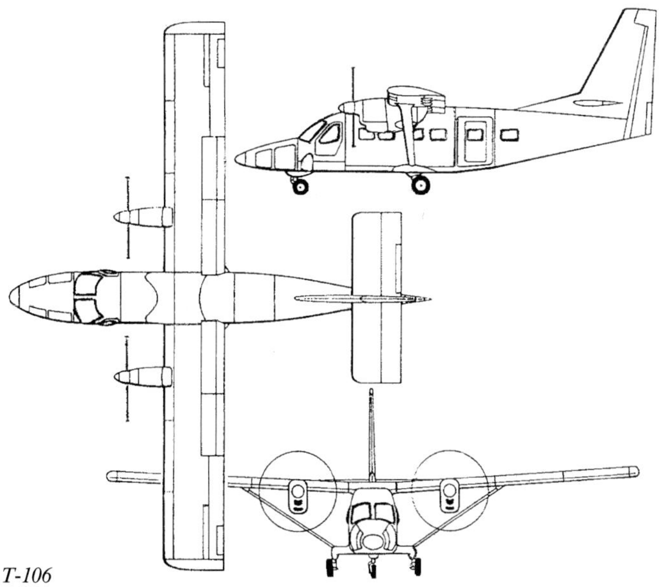

T-106

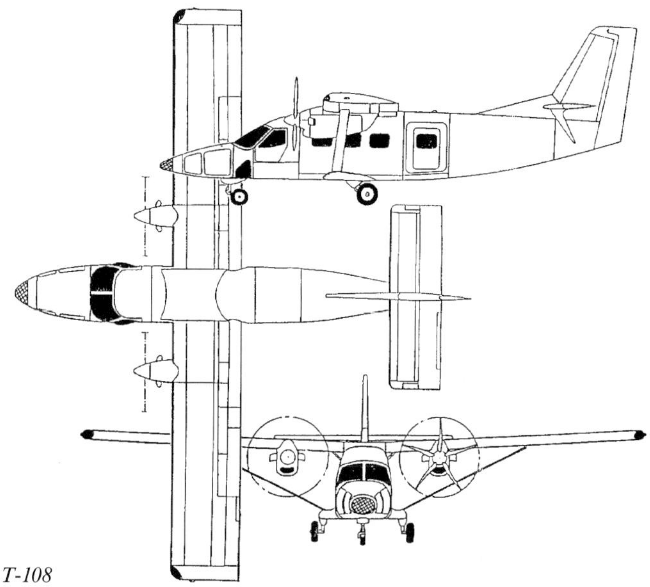

T-108

T-108

ENGINES Two 760 shp Czech M-601E.
WEIGHTS Fuel 1250 kg; payload 2 t; MTO 5950 kg (13,117 lb).
PERFORMANCE (est) Cruise 420 km/h (261 mph); op'g ceiling 6 km; range 1350 km (839 miles); TO 650 m, ldg 600.

T-110 Upgraded T-106, 36 pax, TVD-20 engines; T-112, cabin cross-section increased to 2,1 m square, TVD-20; **T-132** amphibian version of 106, TVD-10B.

T-130 Fregat Utility amphibian; metal airframe, cabin 2,6×1,34×1,3 high for <15 pax, 600 kg or role eqpt, hyd flaps/gear, fuel 1000 lit, bleed-air LE deicing. To fly late 95.
DIMENSIONS Span 12,57 m (41 ft 3 in); length 10,1 m (33 ft 1¾ in); wing area 25,14 m^2 (271 ft^2).
ENGINES Two 751 shp Walter M701E (or PT6A).
WEIGHTS OWE 1,25 t (2,755 lb); max 2,08 t (4,585 lb).
PERFORMANCE Max cruise 228 km/h (141 mph); range (max load, 30 min reserve) 600 km (373 miles), (max fuel) 1320 (820); TO (land) 380; ldg 345.

T-201 Aist STOL utility, refined T-101, no slats but double-slotted flaps, same cabin size, 2 t payload, 1400 lit fuel, hot-air deice, to fly Dec 94.
DIMENSIONS Span 19,9 m (65 ft 3½ in); length 15,1 m (49 ft 6½ in); wing area 43,78 m^2 (471 ft^2).
ENGINE One TVD-20 (export, PT6A-67R).
WEIGHTS OWE 3,4 t (7,496 lb); max 5,7 t (12,566 lb).
PERFORMANCE Max cruise 285 km/h (177 mph); range (max load, 45 min reserve) 180 km (112 miles), (max fuel) 1370 (851).

T-203 Pchel (bee) Low-wing ag-aircraft, based on T-108, fuselage part-composite to avoid corrosion, 14% wing flaps and full-span slats, main tyres 720×320 3,5 kg/cm^2, 1200 lit fuel, 60 lit oil, chem tank 2200 lit, cable-collision warning. Work began May 93.
DIMENSIONS Span 19,9 m (65 ft 3½ in); length 15,1 m (49 ft 6½ in); wing area 47,67 m^2 (513 ft^2).
ENGINE One TVD-20 (export option PT6A-67R 1,424 shp).
WEIGHTS Empty 3370 kg (7,429 lb); max (also max ldg) 5,7 t (12,566 lb).
PERFORMANCE Max speed 315 km/h (196 mph); econ 240 km/h (149 mph); range (45 min reserve) 1370 km (851 miles); TO 295 m; ldg 108 km/h/160 m.

T-274 Titan Projected STOL airlifter, LE and triple-slotted flaps, pressurized fuselage, cabin 10,5m×2,15×2·2 m high for 13 t, hot-air deice, APU, rough-field gear. Predictably, funds sought to build prototype.
DIMENSIONS Span 31,89 m (104 ft 7½ in); length 28,07 m (92 ft 1¼ in); wing area 98,62 m^2 (1,062 ft^2).
ENGINES Four TV7-117S.
WEIGHTS Max 36 t (79,365 lb).
PERFORMANCE Normal cruise 600 km/h (373 mph); range (13 t, 60-min reserve) 900 km (559 miles), (max fuel) 6600 (4,100); TO 750; ldg field length 1150.

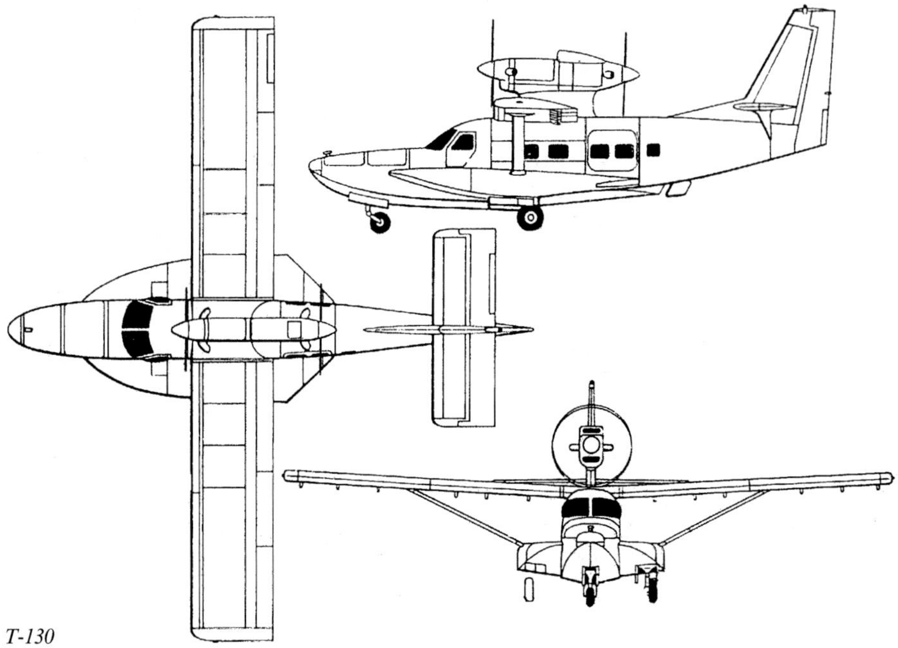

T-130

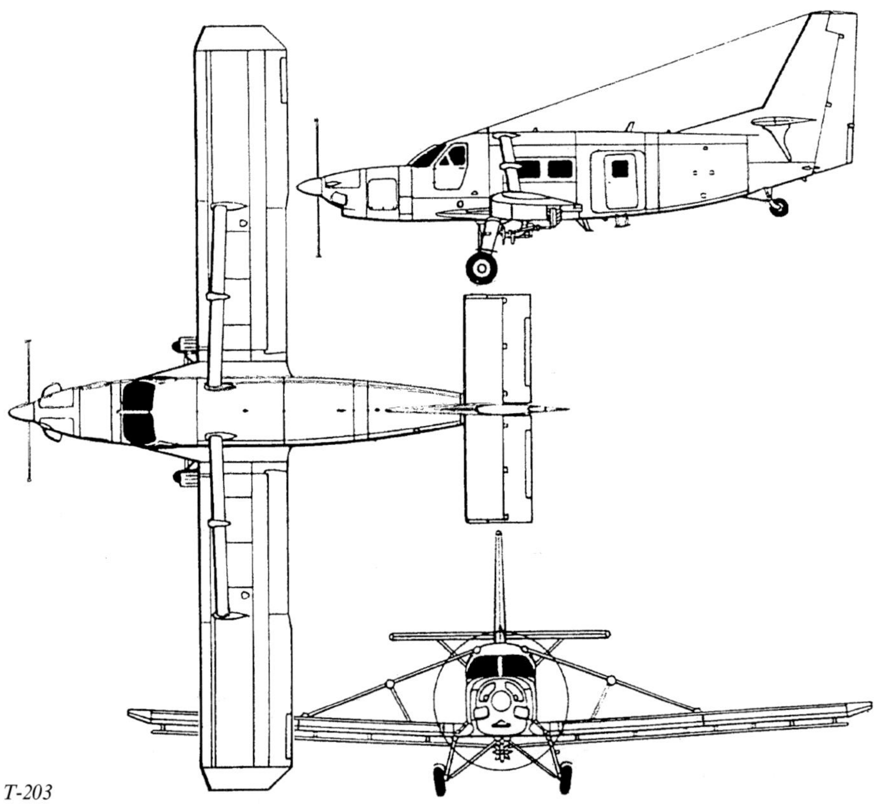

T-203

T-282 Multirole twin-turboprop utility, unpressurized box fuselage, cabin 2,1 m (82·7 in) square, 6,6 m long; 18-24 pax, rear ramp door for car etc, max payload 2 t (2,2 with TVD-20), max fuel 2100 kg.
DIMENSIONS Span 22,06 m (72 ft 4½ in); length 16,4 m (53 ft 9½ in); wing area 39,8 m^2 (428 ft^2).
ENGINES Two TVD-10B (alternative, TVD-20).
WEIGHTS Empty 4,9 t (10,802 lb) (TVD-20, 5,2 t); max 7,5 t (16,535 lb) (TVD-20, 8,3 t, 18,298 lb).
PERFORMANCE Max speed 350 km/h (217 mph) (TVD-20, 400 km/h); econ cruise 275 km/h (TVD-20, 315 km/h); range (45-min reserve) 1700 km (1,056 miles) (TVD-20, 1640 km); TO 390/460 m, ldg 250/280 m.

T-401 Sokol (Falcon) Stressed-skin utility aircraft, two-section flaps, fixed nosewheel gear or

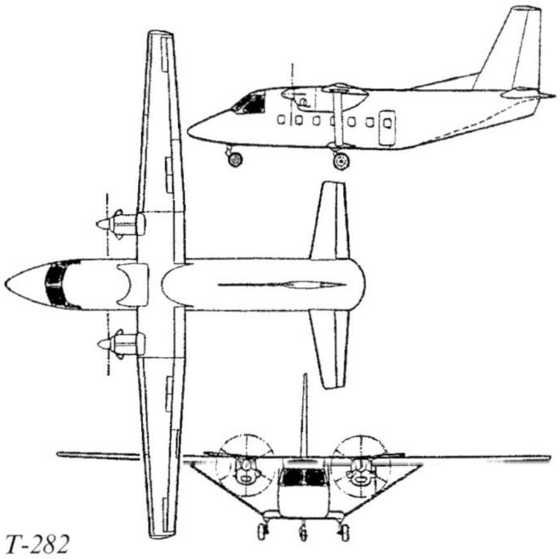

T-282

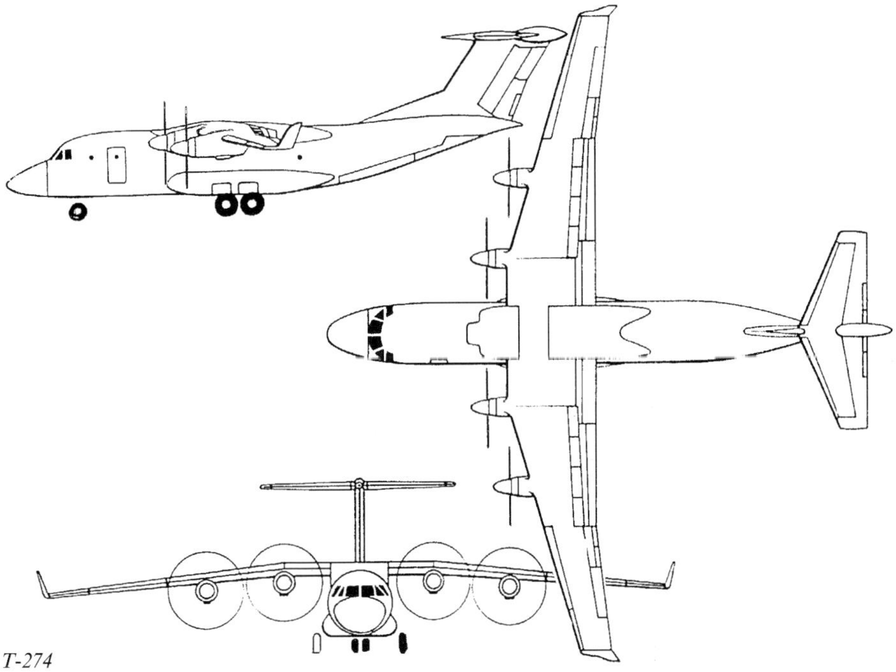

T-274

floats/skis, 3-blade 2,4 m c/s prop, two tanks total 350 lit (77 gal), 3 pairs seats or 500 kg (1,100 lb) or various role eqpt. Prototype to be built 94.
DIMENSIONS Span 13,66 m (44 ft 9¾ in); length 9,88 m (32 ft 5 in); wing area 20 m² (215 ft²).
ENGINE One M-14PR.
WEIGHTS Empty 1430 kg (3,153 lb); MTO 2030 kg (4,475 lb).
PERFORMANCE Max cruise 270 km/h (167 mph), econ 215 km/h (134 mph); service ceiling 3 km (9,850 ft); range (max payload) 800 km (497 miles); field length 600 m.

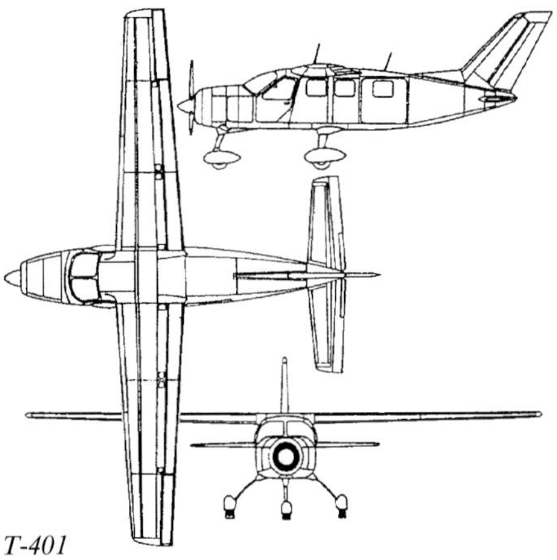

T-401

T-407

T-407 Low-cost utility partner to T-401, square steel-tube truss fuselage, two in cockpit and five in cabin or 600 kg (1,323 lb), 380 lit (83·6 gal), fuel, export option Bendix-King avionics.
DIMENSIONS Span 12,5 m (41 ft 0 in); length 9,8 m (32 ft 1⅞ in); wing area 25,14 m² (270·6 ft²).
ENGINE One M-14P (export option 350-hp TIO-540).
WEIGHTS Empty 1250 kg (2,755 lb); MTO 2080 kg (4,585 lb).
PERFORMANCE Max cruise 215 km/h (133 mph); range (max payload) 580 km (360 miles); TO to 15 m 400 m, ldg from 15 m 550 m.

T-411 Aist-2 Light aircraft, aluminium with synth-fabric, pneu flaps/brakes, 200 lit, pilot + 3, IFR; first flight 10 Nov 93.
DIMENSIONS Span 12,6 m (41 ft 4 in); length 9355 (30 ft 8½ in); wing area 24,05 m² (259 ft²).
ENGINE One M-14P.
WEIGHTS OWE 1,1 t (2,425 lb); payload 363; max 1,6 t (3,527 lb).

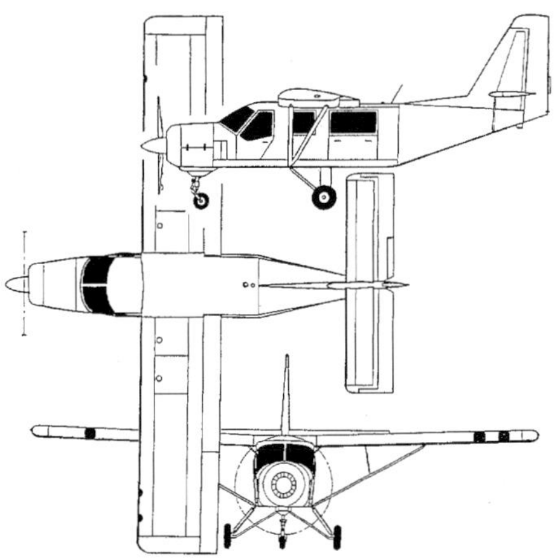

T-407

T-724 Twin-turboprop 19-seat multirole, pressurized cabin 2050 wide and 1870 high, nose radar, tail APU, hi-lift flaps/slats, winglets, manual controls, pneu gear retraction.
DIMENSIONS Span 22,9 (75 ft 1½ in); length 17 (55 ft 9¼ in).
ENGINES Not selected, 1,300 hp each.
WEIGHTS Empty 4590 (10,119); payload 1,7 t; max 7,5 t (16,535).
PERFORMANCE Max cruise 500 (311); service ceiling 8 km (26,250); range 3500 km (f2,175); TO to 10,7 m 500 m; ldg from 9 m 600 m.

T-433 Flamingo Multirole amphibian, mid

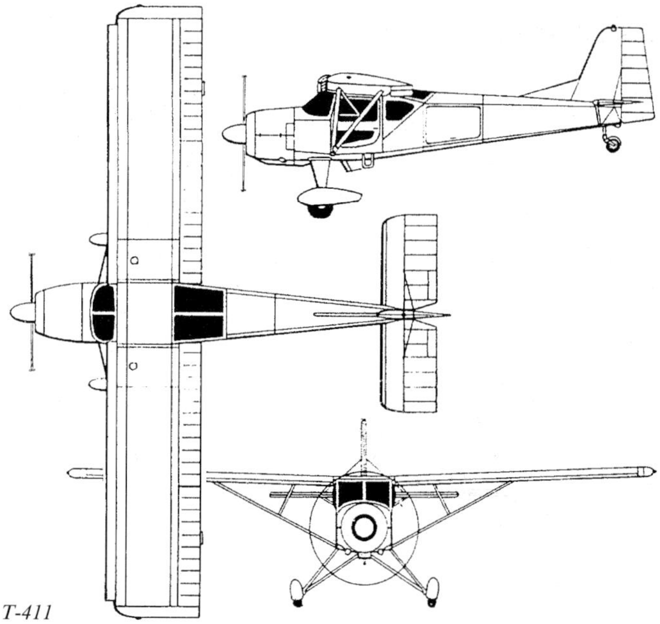

T-411

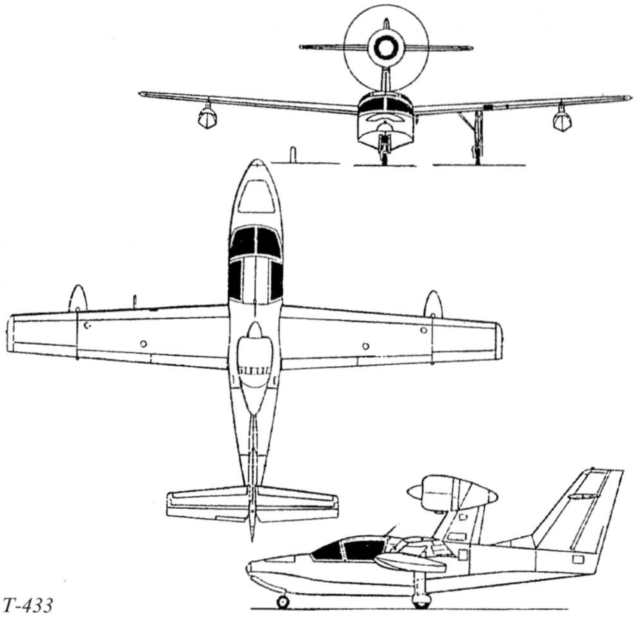

T-433

wing with slotted tracked flaps, main gears hyd retraction inwards to wing, nosewheel into hull, 2-blade c/s prop, two wing tanks total 300 lit (66 gal); upward-hinged canopy doors, two cockpit seats plus 3-seat bench or 370 kg (815 lb) cargo; prototype building 93.
DIMENSIONS Span 14,2 m (46 ft 7 in); length 10,62 m (34 ft 10¼ in); wing area 20,68 m² (222·6 ft²).
ENGINE One M-14P.
WEIGHTS Empty 1470 kg (3,240 lb); MTO 2050 kg (4,520 lb).
PERFORMANCE Max cruise 230 km/h (143 mph); econ 180 km/h (112 mph); range (SL, max payload) 630 km (390 miles): TO/ldg (land) 220 m, (water) c300 m.

T-501 Turboprop trainer, stepped tandem ejection seats, hyd gear/flaps, fuel 500 kg (1,102 lb), prov'n for drop tanks or 500 kg armament. Contract US$22 m for two prototypes plus test airframe placed with Mikoyan April 92.
DIMENSIONS Span 11,0 m (36 ft 1 in); length 9,66 m (31 ft 8½ in); wing 16,5 m² (177·6 ft²).
ENGINE One TVD-10B.
WEIGHTS MTO 2670 kg (5,886 lb).
PERFORMANCE Max speed 530/570 km/h (330/354 mph); SL climb 21 m/s (4,135 ft/min); max range (hi, tanks) 1800 km (1,115 miles).

T-602 Orel (Eagle) Multirole twin, unpressurized 2-seat cockpit plus 9 cabin or various role eqpt or 800 kg (1,764 lb cargo), slotted flaps, nosewheel, (all units retrac forward), V-530A-D35 props, 720 kg fuel, AP-23 autopilot, option A-813 radar. Prototype building 93.
DIMENSIONS Span 13,66 m (44 ft 9¾ in); length 12,12 m (39 ft 9⅛ in); wing 22,5 m² (242 ft²).
ENGINES Two M-14P (export option, TIO-540).
WEIGHTS Empty 1980 kg (4,365 lb); MTO 3,2 t (7,055 lb).
PERFORMANCE Max cruise 320 km/h (199 mph); normal cruise ht 4 km (13,125 ft); range (max payload) 1100 km (685 miles), (max fuel) 2100 km (1,305 miles); TO 380 m; ldg 475 m.

T-610 Voyage Turboprop STOL utility, stressed-skin, Fowler flaps, spoilers ahead of ailerons, leaf-spring main gears 700×200 tyres, hyd brakes, castoring nosewheel, option floats/skis; prototypes Czech M601E engine, prodn options TV-O-100, AL-34 or 600-shp PT6A-114; V-510 reversing 5-blade 2,3 m prop, wing tanks 1300 lit (286 gal), seats arranged 2-3-3-2-1 or 1,1 t (2,425 lb) cargo or role eqpt. Prototype to fly 94.

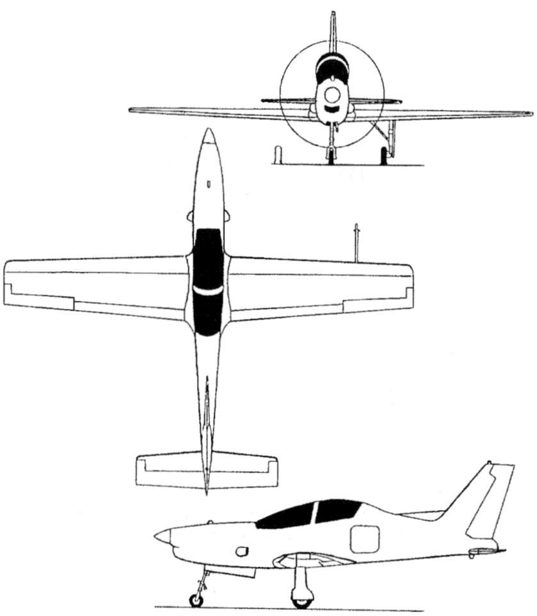

T-501

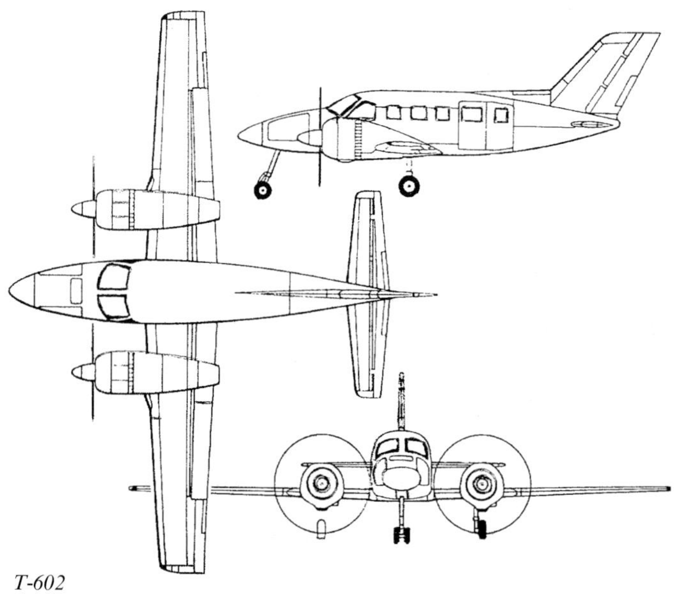

T-602

T-602

DIMENSIONS Span 16,16 m (53 ft 0¼ in); length 12,08 m (39 ft 7½ in); wing area 28,0 m^2 (301 ft^2).
ENGINE See text.
WEIGHTS Empty 1955 kg (4,310 lb); MTO 3850 kg (8,488 lb).
PERFORMANCE Econ cruise 250 km/h (155 mph); range (max payload) 1600 km (995 miles); TO/ldg (reverse prop) c270 m.

T-724 Twin turboprop 19-seat multirole, pressurized cabin 2050 mm wide and 1870 mm high, nose radar, tail APU, hi-lift flaps/slats, winglets, manual controls, pneu gear retraction.
DIMENSIONS Span 22,9 m (75 ft 1½ in); length 17 m (55 ft 9¼ in).
ENGINES Not selected, 1,300 hp each.
WEIGHTS Empty 4590 kg (10,119 lb); payload 1,7; max 7,5 t (16,535 lb).
PERFORMANCE Max cruise 500 km/h (311 mph); service ceiling 8 km (26,250 ft); range 3500 km (2,175 miles); TO to 10,7 m 500 m; ldg from 9 m 600 m.

T-910 Kur'yer (Courier) Pressurized light transport (<10 pax); al-alloy, ails+elec-hyd spoilers, elec trim tailplane, main tyres 800×200, nose 500×180, 4550 lit fuel, elec 115V 400 Hz + 27V DC, Progress AI-9 APU, digital avionics. Work began Jan 93.
DIMENSIONS Span 17,0 m (55 ft 9¼ in); length 15,24 m (50 ft); wing area 32,2 m^2 (347 ft^2).
ENGINES Two AI-25TL.
WEIGHTS Empty 5,7 t (12,566 lb); payload 1,1 t; MTO 9,7 t (21,384 lb).
PERFORMANCE (est) Max speed 780 km/h (485 mph) at 8 km; econ cruise 720 km/h (447 mph) at 12 km; range (1-h reserve) 4500 km (2,796 miles); TO 500 m; ldg 400 m; field length 1,1 km.

Samsonov

Pyetr Dmitriyevich Samsonov was (possibly apprenticed) at Shchyetinin works in the First World War. In 25 appointed joint head, with V. B. Shavrov, of manufacturing at OMOS. With K. A. Vigand handled drawings and calculations for MRL-1 and other Grigorovich boats.

MBR-3 See Addenda.

MBR-5 Funded as possible successor to MBR-2. Clean amphibian of mixed construction. Two-spar wings of MOS-27 profile joined on centreline, structure similar to DB-3B with spars welded from KhMA tube and light ribs built up from riveted dural box, tube and angle sections. Spars joined by diagonal struts of welded KhMA tube braced with double steel tapes. Fabric covering, varnished. Wooden hull with two-step planing bottom without inner keelson, wide-spaced frames and close-pitch stringers carrying ply skin and varnished fabric covering. Control surfaces dural with fabric covering, actuation by cables throughout, pilot control of tailplane incidence via screwjack in wood fin integral with hull, wings with slotted flaps. Main landing gears with compressed-air shock struts also arranged when required to unlatch sprung locks and raise gear into well in side of hull. Retractable tailskid. Engine in NACA cowl carried on N-strut on centreline with V side-brace struts on left only. Oil cooled in nine-tube ring radiator round nacelle. DA gun in bow and dorsal turrets (manual drive), four FAB-50 in enclosed underwing fairings, radio and reconnaissance camera. First flight Aug 35; good performance but caught fire when moored. NII tests at Sevastopol, during which met with accident; abandoned Sept 36.
DIMENSIONS Span 15,4 m (50 ft 6⅓ in); length 11,2 m (36 ft 9 in); wing area 32,5 m^2 (350 ft^2).

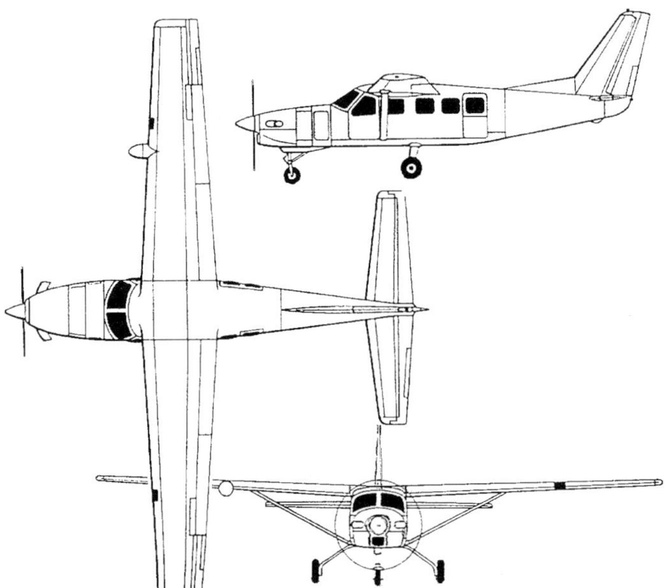

T-610

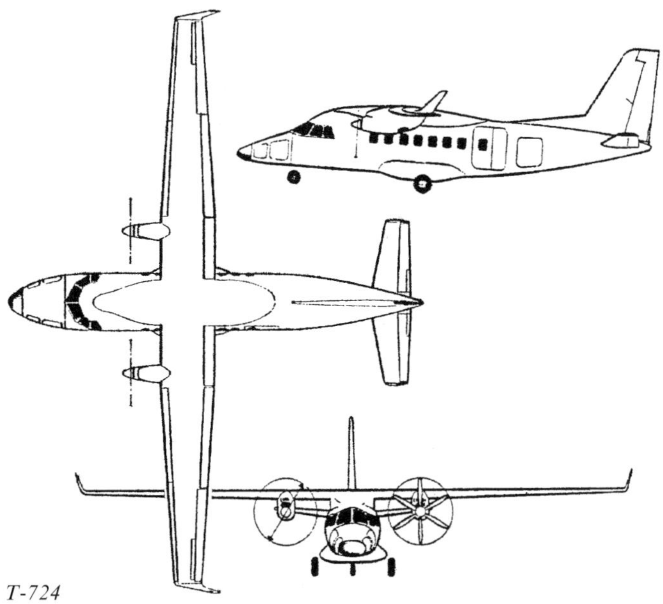

T-724

MBR-5

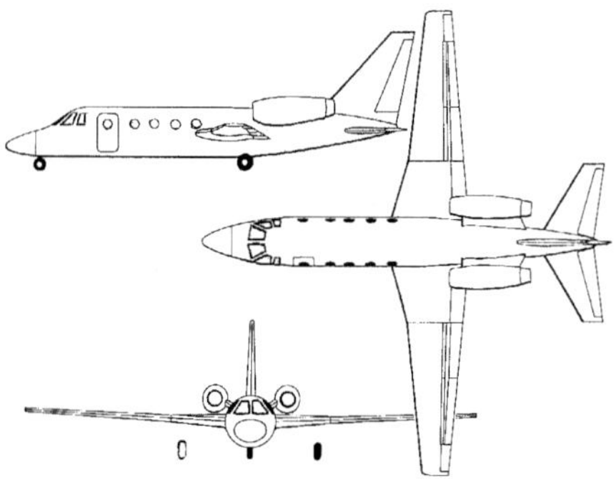

T-910 *(see p.321)*

ENGINE One 710 hp Wright Cyclone SGR-1820-F3.
WEIGHTS Empty 2060 kg (4,541 lb); fuel/oil 441 kg; loaded 3100 kg (6,834 lb).
PERFORMANCE Max speed 256 km/h (159 mph) at SL, 306 km/h (190 mph) at 1,8 km; climb, 8 min to 3 km, 18 min to 5 km; service ceiling 7,5 km (24,600 ft); range 750 km (466 miles); alighting speed 110 km/h.

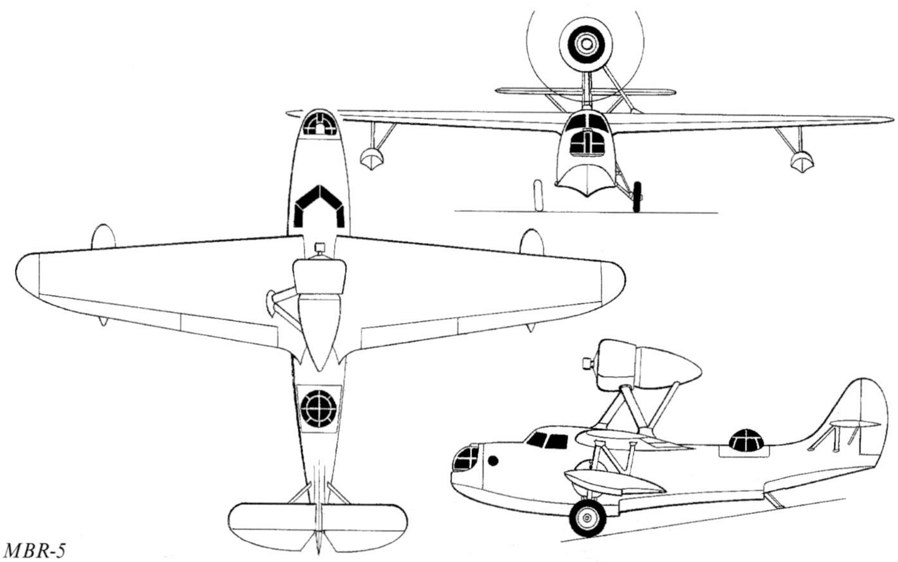

MBR-5

MDR-7 Recon amphibian, design accepted by MA 14 Oct 38. Shavrov built flying boat with same designation; slight evidence Samsonov was later allocated 'MDR-8' to remove confusion. Clean cantilever monoplane, all-metal stressed-skin construction, excessive wing-loading (188 kg/m^2) to obtain speed much higher than normal for class. Long-life (DC-3 inspired) wing with five sheet-web spars root to tip, made as centre section and tapering outer panels. Upper surface strengthened by corrugated inner skin between second and third spars to act as tank bay. Skin thickness 0,6/1,2 mm. Automatic slats ahead of slotted ailerons, slotted flaps and pneumatically retracted wing-tip floats folding into outer side of streamlined engine nacelles. Two-step planing bottom to streamlined hull, first step V in plan on first of three prototypes, transverse on others, and rear step tapering to knife edge. Keel of twin dural webs, angle at bottom 30° and not concave as in MBR-2 and many other boats. Relatively thin hull skin spot-welded to numerous stringers. Cantilever tailplane half-way up fin. Engines in low-drag cowls driving VISh-22-3 three-blade two-pitch propellers. Novel riveted aluminium tanks with varnished fabric tape along joints. Crew 3 or 4, flight controls cable-operated, armament two BS in dorsal turret, two ShKAS (location not stated) and underwing racks for ten FAB-100. Prototype flown by T. V. Ryabenko 25 July 40. Turning in to land aircraft slipped in steeply and broke up on impact. Second aircraft flown by I. M. Shyevnin spring 41; concluded inadequate wing. Third aircraft with increased span and fixed floats never flown.
DIMENSIONS Span 18,0 m (59 ft 0⅔ in); length 14,0 m (45 ft 11 in); wing area 40 m^2 (430·5 ft^2).
ENGINES Two M-88.
WEIGHTS Epty 5100 kg (11,243 lb); fuel/oil 1100 kg+140 kg; loaded 7500 kg (16,534 lb).
PERFORMANCE Max speed at SL 395 km/h (245 mph); intended range 1000 km (621 miles); take-off 27 s; alighting 20 s/150 km/h/93 mph.

Shavrov

Vadim Borisovich Shavrov was born 7 Nov 98, served in Red Army after revolution and qualified as engineer 23. In 25 joined OMOS on its formation, being assigned joint responsibility with P. D. Samsonov for manufacturing. Had idea for smaller and simpler machine, a U-2 style amphibian able to do small utility tasks from fields or rivers. With Osoaviakhim funds designed Sh-1 and built prototype in own small flat in Leningrad. Great success, and derived Sh-2 made in large numbers. In 35 appointed to USP (NTK) under L. V. Kurchyevskii to build Sh-3. In 36 assigned to build flying boat for Far East and amphibian for Glavsevmorput, but increasingly involved in administration, teaching and writing. Design consultant to ESKA-1 (see *Ekranoplans*). By far most important historian of Soviet aircraft, published articles in aviation press and two hardback books, latter dealing with periods up to 38 and 1938-50. Manuscript covering 1950-60 was well advanced when he died 23 Dec. 76.

MDR-7 (Samsonov)

Sh-1 Shavrov gradually formulated outline design for this simple light amphibian soon after beginning work at OMOS but failed to get it adopted as official OKB project. Bold decision to try and build it himself and in 27 approached Osoaviakhim and eventually obtained financial backing on extremely modest scale. Managed to obtain enthusiastic support of Viktor L'vovich Korvin (see MK-1 under Mikhelson) and mechanic N. N. Funtikov, and drawings mainly completed by time actual construction began on 16 April 28. Major airframe parts sized to fit into one-room flat and go downstairs, one result being division of wing into three equal-span sections. Simple wooden construction, with two-spar wing carried on four pairs of mild-steel tube struts forming two inverted pyramids all meeting at two points at upper sides of wood-planked hull, with two added bracing struts to rear. Outer wings cantilever. Small lower wings also cantilevered and carrying stabilizing floats. Landing gear with spoked main wheels cranked up around sides of hull. Simple wood/fabric tail, and engine mounted on front of wing with no pretence at cowl or nacelle, with fuel tank in wing. Cockpit with side-by-side seats and small space at rear for third occupant. Whole design rather compromised by limited funds, and limited objectives matched to adverse environment. Parts assembled at Grebno Port and, without landing gear, launched 4 June 29 and tested on water by L. I. Giks (Hicks). First flight 21 June, and subjected to serious measured testing by B. V. Glagolyev 1 to 8 July. Landing gear then installed and on 6 August took off from water and landed on Leningrad airfield. On 31 Aug flew from water at Grebno Port to Moscow Central Aerodrome where on instructions of P. I. Baranov of VVS Sh-1 was put through official NII testing by Glagolyev, M. A. Korovkin and A. V. Chekaryev from Central Aerodrome and Moscow reservoir. As result Shavrov was asked to build improved version with more power (original engine was all he could manage, not what he wanted). Original machine, first amphibian in the Soviet Union and also first to

Sh-1

have wooden structure protected by layer of varnished fabric, continued to be popular at aero club at Leningrad until no less a pilot than V. P. Chkalov demolished it (without injury to himself or mechanic Ivanov) flying in severe snowstorm on 26 Feb 30.

DIMENSIONS Span 10,7 m (35 ft 1¼ in); length 7,7 m (25 ft 3⅛ in); wing area 20,2 m² (217 ft²).
ENGINE One 85 hp Walter 7-cylinder radial.
WEIGHTS Empty 535 kg (1,179 lb); fuel/oil 60 kg+7 kg; loaded 790 kg (1,742 lb).
PERFORMANCE Max speed 126 km/h (78 mph); climb to 1 km, 17·5 min; ceiling 2470 m (8,100 ft); range 400 km (249 miles); take-off 200 m/20 s; landing 100 m/12 s/65 km/h.

Sh-2 Shavrov's simple homebuilt was right vehicle at right time. Sh-2 was similar to original machine except for different centre section, proper engine nacelle, increased dimensions and redesigned wing/hull struts and horizontal tail. Hull of pine, with some parts ash, with 25 frames (four being bulkheads) and numerous stringers, cross sections milled (planed) to 12 mm×15 mm up to 15 mm×35 mm. Ply skin 2,5 to 3 mm, with 6 mm at single step of simple planing bottom. Completed hull covering with casein-glued fabric, with top coat of varnish (then novel, but found to give long-term protection). Small lower wing of same construction and protective covering with two spars and watertight ply skin and incorporating two streamlined-box stabilizing floats. Wing redesigned with smaller centre-section, made of duralumin to avoid danger from fire and manually folding outer panels. Two built-up box spars with overlapping ply webs with grain at 45°. Built-up truss ribs, with ply skin over leading edge to front spar (dural on centre section) and fabric elsewhere. Slotted ailerons 'slanting construction'. Aft of rear spar centre-section hinged upwards to allow outer panels to fold to rear, trailing edges fitted with simple latch to link folded wings. Welded mild-steel tube engine mount with dural nacelle and crankcase fairing. Two-blade wooden propeller, often 2,35-mm U-2 type but occasionally with spinner and sometimes metal blades. Mild-steel wing/hull struts rearranged in conventional diagonal pairs aligned with spars, lower front attachment being disconnected to fold wing. Two extra near-vertical struts, with streamlined fairings, from nacelle to top of main landing gears and incorporating lower front wing-strut anchors at lower ends. Landing gear with light disc wheels and 700-mm tyres, or pivoted skis, carried on twin steel tubes pivoted to trunnion bolted to upper longeron (several arrangement of rubber shock-absorption) and retracted along rail guide by cockpit crank and cable. Tailplane with two bracing struts each side, elevators with previous horn balances deleted. Fuel tank in centre section for 330 lit (72·6 gal), and 30-lit oil tank in nacelle. Pilot and passenger or mechanic side by side, rear seat for third (at a pinch, fourth) occupant.

First Sh-2 test-flown from airfield by Glagolyev 11 Nov 30, and four days later (with Shavrov) made first test from water, successfully landing in 0·8-m waves in blinding snowstorm. Subsequently tested NTK UVVS at Sevastopol and 12-17 June 31 by NII GVF in Leningrad. Result was production 32-34 by GAZ-31 totalling 270. Standard utility transport with many Aeroflot directorates (Shavrov mentions Northern, Siberia and Far East), often on skis and including 16 **Sh-2S** (*Sanitarnyi*, sanitary ie, casevac) carrying one or two stretcher patients in open rear compartment with attachment notches. Duties of Sh-2 included frontier and fisheries patrol, reconnaissance for icebreakers and Arctic expeditions, civil utility transport, military liaison and above all training. In 39 Aeroflot shops began assembling additional Sh-2 s from spare parts, and by 40 were building 'several hundred' from scratch. Post-war designation **Sh-2bis** reflected mods such as compressed-air starting, ring cowl and glazed canopy. Survivors active to 64.

DIMENSIONS Span 13,0 m (42 ft 7¾ in); length 8,2 m (26 ft 10¾ in); wing area 24,7 m² (266 ft²).
ENGINE Originally one M-11; later M-11L or other version.
WEIGHTS Empty 660 kg (1,455 lb), (no landing gear) 620 kg, (Sh-2S) 680 kg; fuel/oil 87 kg+15 kg; loaded 937 kg (2,066 lb), (Sh-2S) 1 t (2,205 lb).
PERFORMANCE Max speed 139 km/h (86 mph); cruise 120 km/h (75 mph); climb to 1 km, 8·3 min; service ceiling 3850 m (12,600 ft); range up to 1300 km (800 miles); endurance up to 11 h; take-off 100 m/10 s; alighting 100 m/10 s/60 km/h (37 mph).

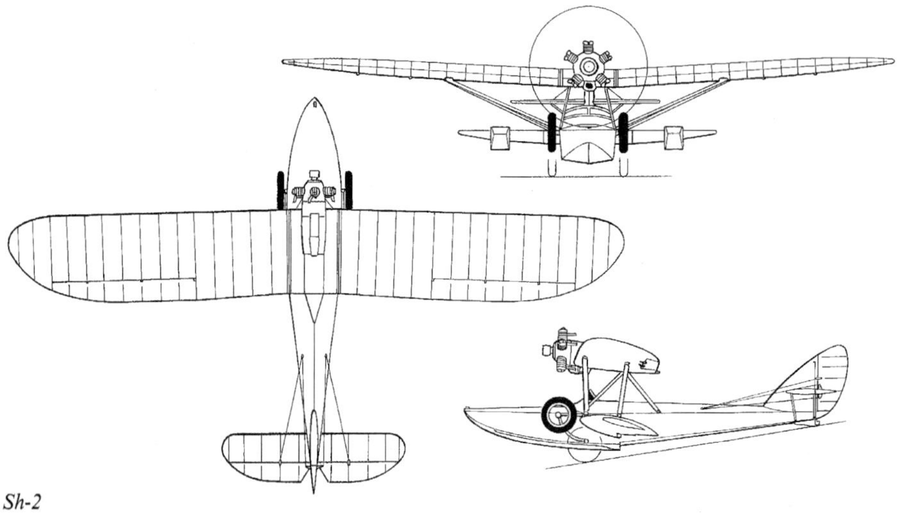

Sh-2

Sh-5 In 28 head of Soviet cartographic and geophysical committee made enquiry about use of specially designed aircraft for photography and aerial surveying. Early 29 special committee organized under Academician A. Ye. Fersman to study possibilities, and quickly concluded that special aircraft could photograph large areas and speed up mapping of vast regions of Soviet Union. Shavrov recommended large enough cabin for full photographic equipment (including developing and printing on board). Emphasized need for large (144°) field of view, need to keep landing gear out of field of view of cameras, absence of good airfields and probable desirability of amphibian. Eventually, early 30, decision to built aircraft. Further committees April/June 30 decided two versions, **FS-1** landplane and **FS-2** flying boat (FS, *Fotosamolyet*, photo aircraft). Committees by this time included NII AFS and NTK UVVS, while aircraft itself, Sh-5, assigned to Richard OKB, and on that bureau closing, March 31, to CCB. Project ostensibly handled by Shavrov, assisted by K. A. Vigand for calculations, but too many people involved for fast progress. June 31 development assigned to Marine Section at GAZ-28 (former Richard OKB) and construction to ZOK NII GVF, already overloaded with nine prototypes including DAR and several for Grigorovich and AGOS. Single Sh-5 slowly built, Shavrov having chosen amphibian from outset but having to increase

Sh-5

engine power from 300 hp each to meet progressively greater demands. Capacious fuselage and vertical tail made of gas-welded KhMA steel tube with fabric covering. Planing bottom, scaled from Sh-2, and underwing floats wood with glued fabric, varnished. Wing of MOS-27 profile, constant 16% thickness, with two widely spaced spars, made like horizontal tail from D1 dural with fabric covering. Struts steel tube with D1 fairings. Engines in Townend-ring cowls on D1 pylons/nacelles above wing. Metal propellers with ground adjustment of pitch. Main landing gears with air/oil shock struts pivoted to KhMA tubular struts between hull, floats and wing, main legs drawn up by cables from cockpit windlass to horizontal position for alighting on water. First flight eventually accomplished 19 March 34, on skis (like wheels, retractable for alighting on water). Flight testing satisfactory, but by this time Sh-5 was outdated and no longer excited original sponsors. At end of 34 one landing-gear strut broke because of manufacturing defect and aircraft never repaired. Surveying continued to be done mainly by R-6.

DIMENSIONS Span 24,0 m (78 ft $8\frac{7}{8}$ in); length 15,0 m (49 ft $2\frac{1}{2}$ in); wing area 73,1 m^2 (787 ft^2).
ENGINES Two M-22.
WEIGHTS Empty 3470 kg (7,650 lb), (as FS-1 landplane 3150 kg); fuel/oil 500 kg (FS-1, 800); loaded 5 t (11,020 lb).
PERFORMANCE Max speed 213 km/h (132 mph), (FS-1, 225 km/h); climb to 1 km, 5 min; service ceiling 4,9 km (16,000 ft); range 800 km (500 miles), (FS-1, 1200 km); take-off 250 m/16 s; landing 240 m/16 s/80 km/h.

Sh-3 In 35 Shavrov accepted invitation of L. V. Kurchyevskii to join NTK, and was permitted to form his own small team to build aircraft of his choice. Sh-3 unusual light transport amphibian with attractive lines. Cantilever low-wing monoplane with four landing wheels on oleo struts retracting into twin floats. Mixed construction, fuselage all-dural monocoque with oval section and almost headroom to walk upright at centreline. Dural centre-section integral with fuselage, dural struts and floats, wooden outer wings. Floats replaceable by wheels or skis. Enclosed cockpit at front for single pilot, space behind for two passengers or light cargo or mail. Windshield and door quickly jettisonable. Intention later to produce **Sh-3S** for stretcher patients and other variants such as **Sh-3F** for photo-survey. Seen as fast modern successor to Sh-2, though with efficient slotted flaps to keep down landing speed and thus increase utility in undeveloped regions. Whole project had to be dropped when NTK closed Feb 36, with prototype in early stage of construction.

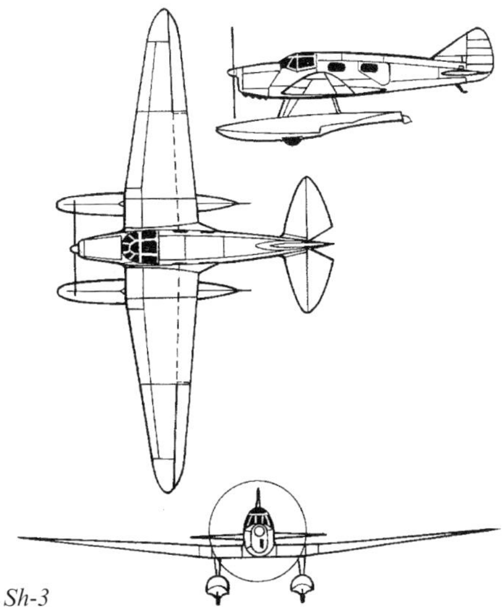
Sh-3

DIMENSIONS Span 14,0 m (45 ft $11\frac{3}{16}$ in); length 7,8 m (25 ft 7 in); wing area 24,0 m^2 (258 ft^2).
ENGINE Originally one 120 hp Cirrus Hermes inverted 4-inline, later to be M-11.
WEIGHTS Empty 720 kg (1,587 lb); fuel/oil 150 kg; loaded 1070 kg (2,359 lb).
PERFORMANCE Designed for 5-h endurance; never tested.

MDR-7 On termination of NTK work Shavrov was asked by Siberian Far East district of MA to design long-range reconnaissance flying boat. Agreement reached to use maximum (possibly 60%) of parts of Ilyushin DB-3B bomber, including wing, powerplant installation, tail, control system, equipment and armament. Visited Ilyushin OKB and design of aircraft begun Jan 37, sponsored by V. K. Blyukher and F. A. Ingaunis of Far East and GAZ director K. D. Kuznetsov. Terminated end 37 because of existence of MDR-6, same designation then used by Samsonov. Strange that designer has no record of data. Only illustration, retouched photo of model tested in CAHI towing tank. No data, though span and engines as DB-3B.

Sh-7 In late 38 Shavrov began design of this neat light transport amphibian for Glavsevmorput and Aeroflot. Cantilever high wing of NACA-230 series, 16% root and 12% tip, two spars with elliptical-tube booms of 30KhGSA in centre section and dural outer panels, and double spar webs in D1-0,8 sheet. Ribs mainly dural truss construction from channels 12 mm×10 mm×0,5 mm. Fabric covering, with slotted ailerons drooping 15° when slotted flaps driven to 50°. Tail mainly from dural tube and rolled sections, with fabric covering. Flight controls dual in side-by-side flight deck with cable operation. All dural hull with full-length one-step planing bottom, with pedal-controlled rear water rudder and tailwheel. Skin mainly 1 mm, 1,5 mm at step, 0,8 mm at rear, not flush-riveted. Dural wire-braced underwing floats. Main landing gears with oleo shock struts carrying wheels or skis (latter I-15bis type) with breaking

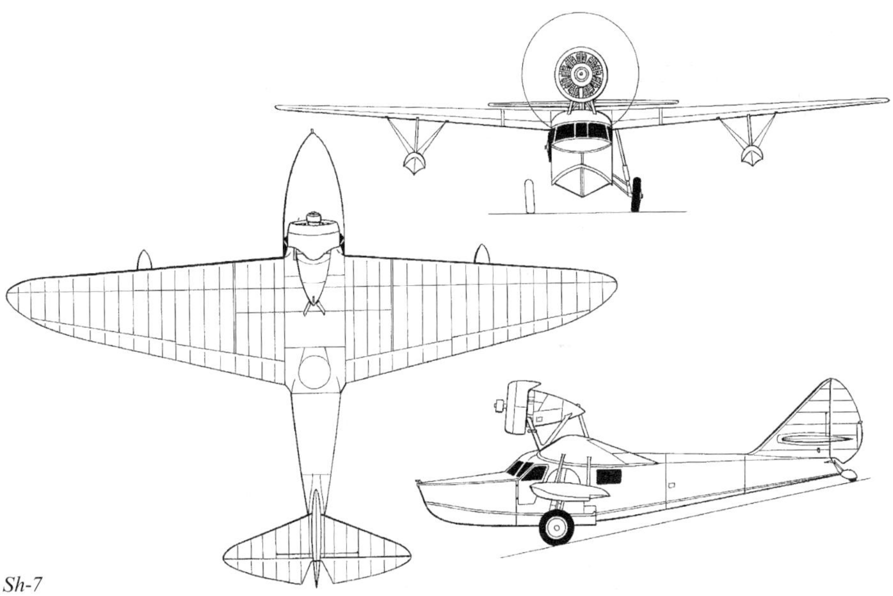
Sh-7

MDR-7 artwork

strut cranked up by hand until wheel housed in recess in side of hull. Two seats in cockpit and four in cabin, all using left/right cockpit doors. Circular roof hatch at rear of cabin at trailing edge for gun position, emergency escape or loading cargo. Two welded AMTs aluminium fuel tanks in centre section, oil tank in pylon-mounted nacelle. Originally designed for MV-6 but more powerful engine substituted during prototype construction in 39. Original propeller from Stal-2, later a three-blade VISh fitted. Fully equipped for open-sea operation, night flying and radio navigation, and military version with TT-1 ring mount for ShKAS with 300 rounds. Launched spring 40 for water tests and flown by Ye. O. Fyedorenko 16 June 40. Completed factory tests Sept and then SNII GVF testing, with Dec 40 recommendation for production with MG-31F and small increase in wing area. Production prevented by war, but prototype (No 359) busy throughout war on urgent cargo along Volga linking Saratov, Stalingrad, Astrakhan etc.

DIMENSIONS Span 13,0 m (42 ft 7¾ in); length 9,4 m (30 ft 10 in); wing area 23,3 m² (251 ft²).

ENGINE One MG-31, later MG-31F.

WEIGHTS Empty 1230 kg (2,712 lb); fuel/oil 320 kg; loaded 1,9 t (4,189 lb).

PERFORMANCE Max speed 218 km/h (135·5 mph) at SL, 184 km/h at height; climb to 1 km, 5 min; service ceiling 2960 m (9,700 ft); range 920 km (572 miles); take-off 280 m/25 s; alighting 210 m/20 s/90 km/h (56 mph).

Sh-7

Shchyerbakov

Aleksei Yakovlyevich Shchyerbakov was Ukrainian, born 02, admitted Kharkov Tech (not Aviation) Institute 24, two years later additionally joining design team at Kalinin OKB, graduated Kharkov 29, remained with Kalinin until 35 when to OSK in Moscow to specialize in high-altitude flight. Became national, if not world, leader in pressure-cabin technology, flexible GK (*Germetichyeskoi Kabinyi*) designed for Gribovskii G-14 sailplane, I-15 series, I-16 series and two unbuilt prototypes, and in addition welded aluminium ('hard') type for I-153 in design of which he played major role. In 38 led group under Korolyev in research/test of RP-318 and in flight test of PVRD under wings of I-15bis. In 41 managed design and mass-production of underwing *kasset* people-containers for U-2 and R-2, 42 designed floatplane variant of U-2 (never previously successful), 43 own OKB at Chkalov to build Shchye-2, whilst also working at Myasishchyev OKB on GKs for Pe-2, Yak-7B and La-5. Last project VSI VTOL fighter.

IVS Designation from *Isrebitel Vyisokoskorostnyi*, high-speed fighter. Project of 37, eventually received positive VVS decision 29 Sept 40, but prototype discontinued at invasion. Mid-mounted M-120 engine with shaft drive under floor of front cockpit to 0·666 gearbox for tractor propeller, plus auxiliary boost PVRD at tail. Glycol radiator under fuselage ducted to main PVRD duct. Hard-type pressure cabin, tricycle landing gear, nose armament one ShVAK and two BS. Design speed 700 km/h, rising to 825 km/h with PVRD; corresponding ceilings 12 and 14 km.

SP Few details survive of *Stratosfernyi Planer*, stratospheric glider, second project of 37. Supported by VVS rather than civilian funds, possibly intended clandestine recon. Span in neighbourhood of 30 m, believed single-seat GK (pressure cabin) and intended to be towed by high-performance aircraft to max height and then soar to still-greater altitudes. No mention of auxiliary rocket or other propulsion; intended operating height 15 km (49,200 ft).

Shche-2, TS-1 Often written Shche-2; alternative designation from *Transportnyi Samolyet*, transport aeroplane. Design undertaken to meet urgent need recognized about July 42 for utility transport to carry items in front-line areas and between factories and repair shops. Shchyerbakov was chief engineer and GAZ director, and also head of central office at NKAP for aircraft repair. U-2 and R-5 used in large numbers to carry many items, latter being able to carry fighter wings slung under lower wing, but neither could carry engines. Desperate need to fly engines such as ASh-82 and AM-37, weighing up to 1 t when packed, to front-line repair units whose location moved almost daily in regions devoid of usable roads. Such aircraft also useful for general front-line supplies, support of partisans and regular duties such as passenger and casevac missions. Financial cover obtained from MA naval aviation Sept 42, design completed in six weeks and prototype TS-1 started before end of year. As far as possible non-strategic materials, and airframe designed for lightness even at cost of labour-intensive construction. Intended engine MG-31 no longer available, and much less powerful M-11 accepted despite poor performance (judged not dangerous, and still adequate for need).

Wing R-II profile, 10% root to 6% tip, with single box spar and built-up truss ribs, glued and pinned joints. Slotted ailerons and flaps. Streamlined oval-section fuselage with built-up frames and even stringers made as light box sections with booms 4 mm×8 mm and webs of 1-mm ply made up to box with section 10 mm×30 mm. Wooden twin-finned tail, whole airframe fabric covered and spray-doped. Aluminium sheet floor to accept high point loads, steel-tube strut bracing high wing on each side, braced fixed landing gear with La-5 shock struts, tyres 600 mm×180 mm, alternative skis 2400 mm×620 mm. Castoring tailwheel with 300 mm×125 mm tyre or 800 mm×380 mm ski. Prototype and nearly all later aircraft initially built with spats, often later removed. Engines on welded mild-steel frame with rubber mounts. Prototype uncowled, most production with cowling. U-2 type propeller. Fuel in four welded AMTs tanks, total 850 lit, in centre section. Single door in left side 1,43 m×1,64 m with metal-clad frame. Design structural load factor 5·85.

First flight 17 April 43. Immediately proceeded to NII testing, and decision to produce in series taken Aug. Production begun in OKB with designation Shchye-2. Series aircraft flown from 1 Aug 44 and total of 550 built by 46. First production aircraft made 20 full-load flights on route Chkalov/Kuibyshyev/Moscow and demonstrated service life of 1,000 hours. Series aircraft widely used as 16-passenger transport, ambulance (11 stretchers), assault transport (nine paratroops) or as 5-pupil navigator trainer. Several experimental models including light (3,4 t max) variant with same tail but wing cut down to 55 m², and another flown with General Motors diesels from US armoured vehicle. OKB deputy chief, M. V. Lyapin, suggested twin-boom version with third M-11D at rear of cargo nacelle; not built. Final (45) model **Shche-2SKh** (ag) with main-gear legs 450 mm longer and tyres 800 mm×260 mm. Not accepted because of extremely poor performance and time 85 s to make 360° turn. By Western standards basic type grossly underpowered, but it was easy to fly and at 3 t (ie, with little payload) could even remain above stall with one engine stopped at about 80 km/h.

DIMENSIONS Span (TS-1, bulk production) 20,54 m (67 ft 4⅔ in), (45 and SKh) 20,484 m (67 ft 2½ in); length 14,27 m (46 ft 9¾ in); wing area 64,0 m² (689 ft²), (45, Skh) 63,88 m² (687·6 ft²).

ENGINES Two M-11D.

WEIGHTS Empty (TS-1) 2210 kg, later 2235 kg (4,927 lb), (SKh) 2,5 t; fuel/oil (TS) 370 kg+40 kg, (43-45) 300 kg+40 kg, (SKh) 100 kg+30 kg; loaded (TS) 3,3 t, 3,4 t (7,496 lb); max overload (not SKh) 3,7 t (8,157 lb).

PERFORMANCE Max speed (TS) 168 km/h, (43-45) 155 km/h (96 mph); cruise 140 km/h (87 mph); climb to 1 km, 13-17 min, (SKh) 24 min;

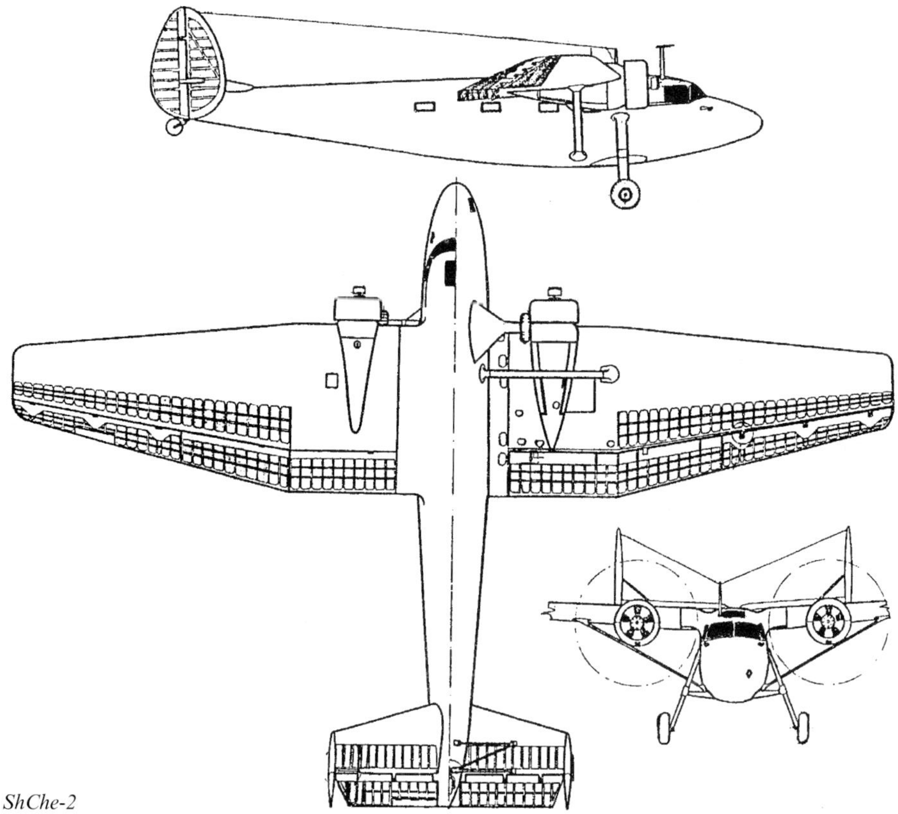

ShChe-2

ShChe-2

service ceiling 2-3 km; range 980 km (600 miles) with 1 t load; take-off 350 m/19 s; landing 170 m/15 s/70 km/h (43·5 mph).

VSI Conceptually years ahead of its time, this was world's first project for vectored-thrust jet VTOL. Designation as for same designer's earlier IVS but words rearranged: high-speed fighter. Streamlined fuselage accommodating retracted tricycle landing gear, broad but low T-tail, high wing of low aspect-ratio carrying tip pods housing engines. Each pod rotated through up to 120° to provide lift, thrust or braking force, with 'gas rudder' in each jet to give control in hover and in forward flight. Scheme submitted in 46 brochure using two Nene (RD-45) in fighter weighing 4 to 5 t. More detailed design 47 following closure of OKB. Favourable appraisal by B. N. Yuryev and V. S. Puishnov of VVIA (VVS engineering academy) and also from VVS central command. Decision to build flight test rig, but designer no longer in MAP system so construction begun at VVIA. Large rig with hovering vehicle of welded KhMA steel tube part-skinned in stainless sheet, supported on cables from four large towers and with underground tunnels for efflux gas. Only available engines German BMW 003; hovering begun 48 but work soon halted, conflicting reasons given by different sources.

Shishmaryev

Mikhail Mikhailovich Shishmaryev was one of earliest Russian designers with major programmes at Shchyetinin factory during the First World War. After Revolution he became major-general and professor at VVA.

GASN Designation from *Gidroaeroplan Spetsialno Naznachyeniya* (seaplane for special duties), incorrectly claimed (Shavrov and others) as world's first torpedo seaplane. Alternative designation **SON** (aeroplane for individual duties). Large three-bay biplane with twin floats and three-finned biplane tail, mixed construction with three-spar wings, steel-tube struts throughout (many with fabric fairings) and ply-skinned forward fuselage for nose observer/gunner, two pilots side-by-side and second gunner behind wings. Rubber shock-absorbers in float struts. Engines between wings on inclined groups of four struts aligned with float attachments, upper LE cut away near props. Torpedo hung under centreline, provision for several bombs under central lower wing. Ten ordered but only one completed at Shchyetinin works 1916-17, first flown 24 Aug 17 by Lt A. Ye. Gruzinov. Outstanding on water; in air CG too far aft. Rectified but damaged 24 Sept when float broke. Derelict until 20, then restored but 4 Nov 20 forced landing and frozen-in 2 km from shore.

DIMENSIONS Span 28,0 m (91 ft 10⅓ in); length 14,2 m (46 ft 7 in); wing area 150 m^2 (1,615 ft^2).
ENGINES Two 220 hp Renault.
WEIGHTS Not known but weapon load 1450 kg (3,197 lb).
PERFORMANCE Max speed 110 km/h (68 mph).

R-III First flown 25, this was a minimum-change rebuild of R-1 (ie, neé D.H.9a) with fuselage attached between upper wings instead of lower, and different cabane struts joining bottom of fuselage to lower wings. Involved exceptionally long-strut main gears and also slightly increased ground angle. Idea was to improve field of view in reconnaissance role. Some accounts suggest more than one R-III built, but R-1 remained sole recon aircraft in service units. No data (very similar to R-1 since regular wing panels, engine and fuselage retained).

Silvanskii

A. V. Silvanskii is a blurred figure, hardly taken seriously by Shavrov and other historians yet responsible in 1938-39 for small fighter aircraft.

I-220, IS *Istrebitel Silvanskii*, designed at OKB formed at GAZ where Silvanskii worked, no indication of how funded. Partners V. D. Yarovitskii and Yu. B. Sturtsel. Aircraft described as similar to I-16 but with slimmer M-88 engine and short single-strut landing gear, criticisms including inadequate recess for retracted wheel which accordingly ruined wing profile, and inadequate ground clearance for propeller. Eventually Silvanskii cropped 100 mm off each propeller blade. Thus, aircraft could hardly take-off and could not climb, but undeterred designer (refused flight-test facilities at factory) brought IS to Moscow and prevailed upon LII to carry out test programme. Unnamed pilot struggled into air and at full power found aircraft on point of stall at 300 m; managed to effect landing, claiming IS would not fly. Silvanskii bankrupted, and industry delegation banned him from further design. No data.

IS mock-up

Sitora

TAO Sitora ('Star') is aviation industry of Tadjik Republic, at capital Dushanbe. General Constructor Dr F. Mukhamedov, 30 years at RKIIGA. Biggest projects Eurasia-700 and Eurasia 18-50 (both see Aviaprom).

Sitora Same name as company, simple side-by-side 2-seat low-wing aircraft now being built. Prototype imported US engine (probably O-200A), later choice of CIS engines. Data not finalised.

Stal. See OOS, Bartini.

Sukhoi

Pavel Osipovich Sukhoi was one of greatest Soviet designers, but his early work never carried his name and his first 20 years as independent designer were heartbreakingly unrewarding, successive excellent designs for various reasons failing to be selected for production. Today his name is commemorated in about half of CIS tactical airpower. He was born 10 (22 in old calendar) July 95 at Glubokoye, Byelorussia. Studied from 05 Gomel Gymnasium, Moscow University (14, briefly) and at Moscow Highter Tech School MVTU under Zhukovskii (later Prof at VVA). Commissioned in the First World War, then managed to survive revolution to join Red Army. In 20 joined CAHI, initially in capacity of student but, after demonstrating to Tupolev considerable design ability, as junior member of AGOS. First assignment, Dec 24, design ANT-4 tailskid. Subsequently led entire design of ANT-5 (I-4) and made major contribution to ANT-6/-9 and -10 before in 32 heading KOSOS Brigade No 3, leading design of ANT-25, 29, 36, 37 and 51 and six types of torpedo boat; Red Star award 22 Dec 33; Order of Badge of Honour 13 Aug 36; 36-7 lead designer at ZOK until opened own OKB 39 producing not only Su designs but also UTB and Yer-2ON. Closed by Stalin 49, Su became deputy to Tupolev until after Stalin's death 53 he reopened OKB, title General Constructor 56, many awards, died at 81 on 15 Sept 75. Succeeded by Ye. A. Ivanov Nov 75 to 83 when succeeded by Mikhail Petrovich Simonov. Based at original address 23A Polikarpov St, Moscow, OKB became ANPK 'OKB Sukhoi'. See Ekranoplans.

ShB Prototype built at ZOK 38, *Shturmovik bombardirovshchik* differed from ANT-51 *Ivanov* in having M-88A engine in improved cowl driving different prop with spinner, and in main landing gears retracting backwards, wheels turning 90°. Bombload doubled to 600 kg, much better armour protection, max wt 4500. Flown early 40. Led to Su-2, but also projected with M-71 engine, abandoned late 40.

Su-2, BB-1 Very similar to final form of ANT-51, *Blizhniii Bombardirovshchik* (short-range bomber) replaced M-87A by M-88 and had armour and bomb capacity of ShB, but retained original inward-retracting main gears. Wing restressed for increased weights, profile CAHI B-series, 15·25% root and 8% tip. Centre section 3940 mm, outer panels with taper on both edges, dihedral uniform from roots. Two spars with D16 sheet webs and 30KhGSA booms, smooth flush-riveted dural skin with screwed fairing strip over centre/outer wing joint. All-dural split flaps in four sections driven pneumatically to 55°. All three units of landing gear retractable, main gears with 750 mm×250 mm tyres, pneu brakes, single shock struts on outer side of wheels, fully housed between wing spars pulled up by pneu cylinder 300 mm×125 mm diam, no standby, tailwheel semi-retracted to rear with own actuator, steering ±42°. Fuselage wooden semi-monocoque with 20 frames and bakelite-ply skin, with integral fin. Control surfaces dural with fabric covering, tailplane all-dural set at –5°, trimmers all-dural, elevators with inset hinges. Advanced engine installation in long-chord cowling with variable gills, opening in three petals (lower incorporating carb-air duct) with oil cooler immediately to rear and driving 3,25-m VISh-23-7 hydraulic propeller. Fuselage tank 425-lit, outer-wing tanks 140-lit each. Crew of two in tandem under larged glazed canopy, pilot high above bomb bay and radio/gunner in manual turret with RSB radio, AFA-13 camera installation and single ShKAS with 650 rounds. Ventral position with single ShKAS retained, but fixed wing ShKAS reduced from four to two, each with 650 rounds. Internal bay with left/right doors housing four FAB-100 (or 30×8, 20×15, 20×20, 12×25 etc.); wing racks for two bombs up to 250 kg (overload) or 14 containers or ten RS-82 or eight RS-130. Both cockpits well protected with 9-mm armour.

ShB

Su-2

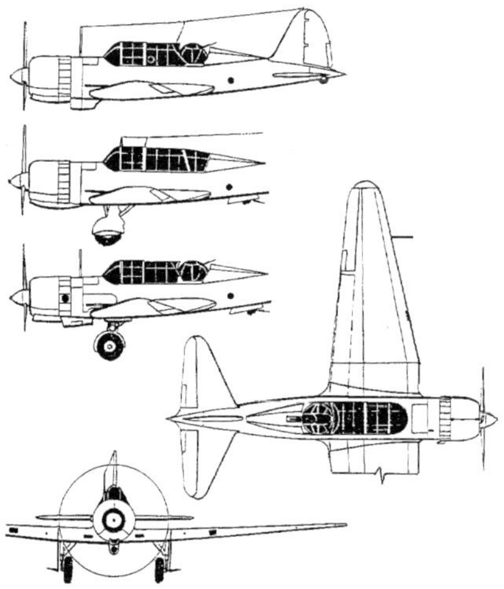

Su-2 (middle side view, variant with no turret but ventral gun; bottom, ShB)

Su-4

BB-1 first flown mid-40 with M-88, soon afterwards replaced by M-88B. State acceptance trials 2 Dec 40/17 Mar 41, mainly by Maj Dolgov. Result, order for 800, with new designation Su-2 under 41 scheme, minor changes including deletion of ventral gun. Numerous field mods including some with four wing guns (provision always there) used as emergency fighters. All units severe attrition and most re-equipped by end 42. Restored example (mainly replica) at Monino.

DIMENSIONS Span 14,3 m (46 ft 11 in); length 10,25 m (33 ft 7½ in); wing area 29 m² (312 ft²).

ENGINE One M-88B.

WEIGHTS Empty 2930 kg (6,459 lb); loaded 4345 kg (9,579 lb).

PERFORMANCE Max speed (SL) 378 km/h (235 mph), (5 km) 455 km/h (283 mph); climb 4 km in 8 min; service ceiling 8,9 km (29,200 ft); range (normal) 850 km (528 miles); TO 650 m; landing 130 km/h, 450 m.

Su-4 Further attempt to improve *Ivanov* design, mainly by fitting much more powerful engine, driving prop of ViSh-105 series, wing modified to reduce light-alloy content, armour further improved, ventral gun restored, two fixed BS, two ShKAS in turret (drawings show one), 400 kg bombs or 10 RS-132. Service tests by Capt Korobov 25 Feb/23 April 42, resulting in immediate decision to build in series, following Su-2 on Komsomolsk line. Quantity believed 1,000, single example with AM-37. Led to Su-6.

DIMENSIONS As Su-2 except length 10,79 m (35 ft 4¾ in).

ENGINE One M-82.

WEIGHTS Empty 3192 kg (7,037 lb); fuel/oil 525 kg; loaded (normal) 4,7 t, (max) 4,9 t (10,802 lb).

PERFORMANCE Max speed (SL) 459 km/h (285 mph), (5 km) 486 km/h (302 mph); climb 10·5 min to 5 km; service ceiling 8,4 km (27,560 ft); range 974 km (605 miles); TO 450 m; landing 135 km/h, 320 m.

Su-1, I-330 Sukhoi invited to Kremlin early 39 and agreed to produce short- and long-span versions of hi-alt fighter with TKs. Studied with GK pressure cabin but built without it. Main feature was twin TK (turbochargers) and large cannon firing through propeller hub.

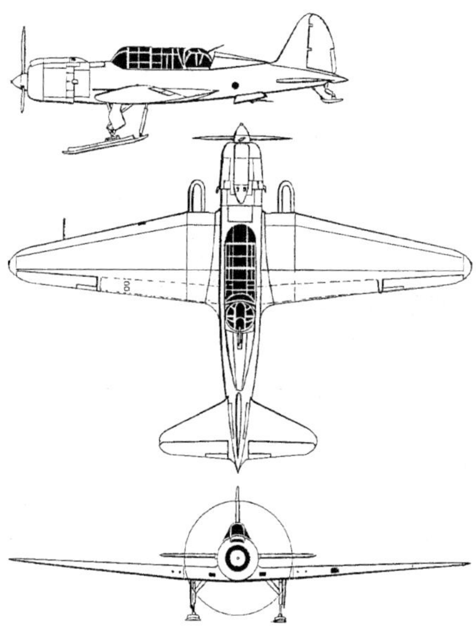

Su-4

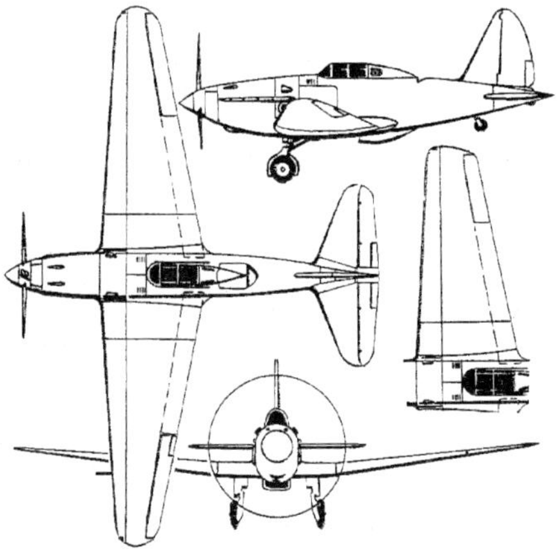

Su-1 (inset, Su-3)

Su-3

Straightforward low-wing aircraft with mixed construction similar to Su-2 family. Fuselage wood semi-monocoque, skinned in bakelite-ply 5/2,5 mm with integral fin. Neat sliding canopy, with rear fairing of teardrop shape but opaque. Engine cowl dural with Dzus fasteners, TK flush on each side of fuselage surrounded by square of stainless-steel skin. Single-spar centre-section and outer wings, profile not known, dihedral c/s 0°, outer 5°. Thick sheet spar web, heavy machined D16 T-section booms, pressed-sheet ribs, numerous angle-section stringers, flush-riveted D1 skin. Single-strut main gears retracted to rear, wheels pivoting 90° on legs. Four-section split flaps. Tailwheel retracting to rear with two doors, small portion of tyre showing. Duralumin tailplane with stressed skin, duralumin flight control surfaces with fabric covering (but all-dural trim tabs on rudder, both elevators and left aileron). Push/pull rods for ailerons and elevators, cables for rudder. Engine on welded 30KhGSA truss, glycol header tank behind V ISh-61P prop, carb inlets each side above TK-2 turbos, rad in duct under rear fuselage. Armament (not fitted) one ShVAK (200 rds) and two ShKAS (400 ea) above engine. Authorised spring 39, first flight Aug 40; prolonged TK trouble but performance outstanding. Damaged in evacuation from Moscow Oct 41, led to Su-3.

DIMENSIONS Span 11,5 m (37 ft 8¾ in); length 8,42 m (27 ft 7½ in); wing area 19,0 m² (204·5 ft²).

ENGINE One 1100 hp M-105P with twin TK-2.

WEIGHTS Empty 2495 kg (5,500 lb); fuel/oil 255 kg; loaded 2875 kg (6,338 lb).

PERFORMANCE Max speed 500 km/h at SL, 641 km/h (398 mph) at 10 km; climb to 5 km, 4·9 min; service ceiling 12,5 km (41,000 ft); range 720 km (447 miles); take-off 220 m; landing speed 111 km/h (69 mph).

Su 3, I-360 With such low landing speed Su clipped wings, also increasing design factor to 13·5 and removing dive limit. Minor changes incl small reduction fuel capacity. Almost complete at evacuation, flown late 41, LII test 42, NII retained as TK devt tool. Expected speed gain not realised.

DIMENSIONS, ENGINE As Su-1 except span 10,1 m (33 ft 1⅝ in), wing area 17,0 m² (183 ft²).

WEIGHTS Empty 2496 kg (5,503 lb); fuel/oil 225 kg+26 kg; loaded 2992 kg (6,596 lb).

PERFORMANCE Max speed 500 km/h at SL, 638 km/h (396 mph) at ht; climb 5·5 min to 5 km; service ceiling 11,9 km (39,000 ft); range (est) 700 km (435 miles); TO 220 m; ldg 122 km/h.

Su-6 Though it had misfortune to have to compete with Il-2 at every stage of development, this armoured *Shturmovik* was not a parallel design but much later project begun at Sukhoi OKB late 41. Authorization as Aircraft **81**, also called **OBSh** (experimental armoured attacker). OBSh made first flight early 42 with designation Su-6. Closely followed by almost identical second machine called A, or **Su-6(A)**. Having carefully studied earlier Ilyushin, Sukhoi adhered to single-seat formula but adopted most powerful available aircooled engine and designed improved scheme of armour. Basic airframe typical Sukhoi with wings and tail derived from Su-4 (wings mod with slats), fuselage scaled Su-1 and landing gear of ShB. Considerable detail in *ASU*. Internal bay for up to 400 kg bombs, armament two VYa-23 each with 180 (max 230) rounds, retaining four ShKAS further outboard, plus underwing 10×RS-82 or RS-132. NII testing by Dolgov Feb/May 42; in most respects superior to Il-2 but no intention of disrupting latter's production. OKB instructed to prepare two-seat version, **Su-6(2A)** with extended mid-fuselage with radio/gunner with UBT with four magazines each of 49 rounds, standing between outward-canted noses of four FAB-100 in revised bay divided along outer sides of fuselage but with same doors (relocated) and unchanged sighting for dive or level bombing. (Note: on Su-2 bombs were below pilot, not aft of gunner.) Heavier forward-firing armament of two 37-mm OKB-16 or 11-P-37 each with 45 rounds, plus two ShKAS with 1400 rounds used for sighting. Considerably increased fuel capacity, stronger main gears with larger tyres and bay doors, 643 kg armour. OKB and NII testing 29 June-30 Aug 43 by Cols Kabanov and Stefanovskii and Maj Dolgov, all highly impressed but engine production could not be arranged.

Final Su-6 launched together with Il-8 and Il-10 at 42 meeting to decide next-generation *Shturmovik* to replace Il-2. Sukhoi went ahead with totally different Su-8, but also pulled out one more stop with Su-6 programme. Adopting liquid-cooled engine (also chosen by rival) he at last increased wing area and produced best aircraft of series, called **2A(M-42)**. New wing with trailing edge at 90° to aircraft axis, slight increase in span, tip t/c 6·25%, additional 2,6 m² area (all to rear of spar, with increased-chord flap and aileron) but still less than three-quarters area of Il-2. Big engine carried in left/right trusses of welded 30KhGSA with four pin-jointed but undamped pick-ups on main longerons. Underslung radiator, carb-air inlet projecting at top, oil coolers in outer centre section as before but with discharge from lower surface. Four-blade 3,4 m AV-9L-172 prop, Hucks dogs retained, internal fuel again increased. OKB states three ShKAS, believed error. NII testing by Kabanov, Dolgov and Maj Sinelnikov 28 April-2 July 44, but Il-10 chosen.

Su-6(A)

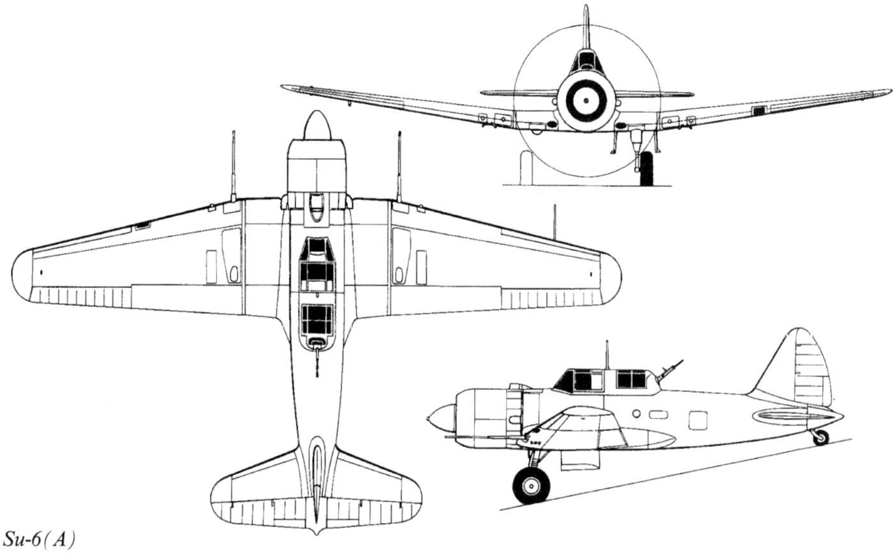

Su-6(A)

DIMENSIONS Span (except AM-42) 13,5 m (44 ft 3½ in), (AM-42) 13,58 m (44 ft 6⅔ in); length (A) 9,133 m (29 ft 11½ in); wing area (except AM-42) 26,0 m² (280 ft²), (AM-42) 28,6 m² (308 ft²).

ENGINE (A) one M-71, (2A) ASh-71F, (2A/M-42) AM-42.

WEIGHTS Empty (A) 3727 kg (8,216 lb), (2A) 4137 kg (9,120 lb), (AM-42) 4370 kg (9,634 lb); fuel/oil (A) 480 kg, (2A) 570 kg, (M-42) believed greater, despite reduced range; loaded (A) 5250 kg (11,574 lb), (2A) 5534 kg (12,200 lb), (M-42) 6,2 t (13,668 lb).

PERFORMANCE Max speed (A) 496 km/h (308 mph) at SL, 527 km/h (327 mph) at 2,5 km, (2A) 480 km/h (298 mph) at SL, 514 km/h (319 mph) at 2,5 km, (M-42) 492 km/h (306 mph) at SL, 521 km/h (324 mph) at 1,5 km; climb to 3 km, (A) 7·3 min, (2A) 8·2 min, (M-42) 8·5 min; service ceiling (A) 7,6 km, (2A) 8,1 km (26,575 ft), (M-42) 8 km (26,250 ft); range (A) 450, (2A) 973 km (605 miles), (M-42) 790 km (490 miles); take-off (A) 520 m, (2A) 410 m, (M-42) 540 m; landing (A) 440 m, (2A) 730 m/146 km/h, (M-42) 660 m/150 km/h (93 mph).

Su-6(2A)

Su-7 Authorization about Oct 42 to develop Su-6(A) into long-range single-seat fighter, very few changes except engine changed from M-71 to ASh-71F with long carb inlet above and TK-3 turbo each side (not fitted on first flight), armour reduced, internal bomb bay replaced by additional tank, three ShVAK each with magazine of 370 rounds. Aircraft completed late 43, but soon refitted with ASh-82FN without turbos. These added 44, plus RD-1 rocket engine in extended tailcone with nitric acid tank behind cockpit giving c4 min burn time at single (max) feed rate. Tested by G. Komarov 31 Jan-20 Dec 45, positive assessment but no production considered (RD-1 not built in series).

DIMENSIONS As Su-6(A) except length with RD-1 10,03 m (32 ft 10¾ in).

ENGINES One M-71F/2TK-3; one ASh-82FN; 82FN/2TK-3 + RD-1.

WEIGHTS (final standard) Empty c3250 kg; fuel/oil 480+50+180 kg acid; loaded (with RD-1) 4360 kg (9,612 lb).

PERFORMANCE Max speed (M-71F/2TK-3)

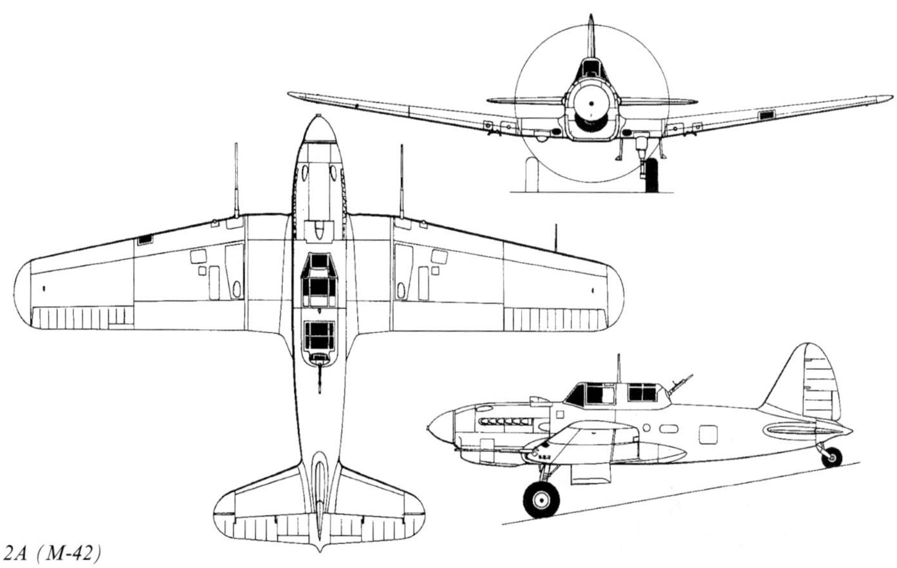

2A (M-42)

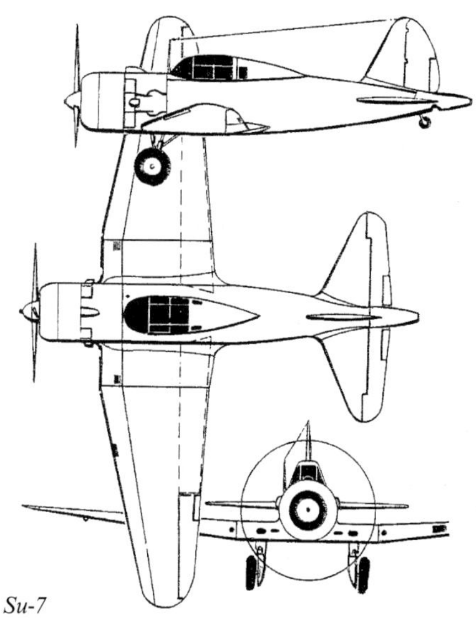

Su-7

520 km/h SL, 635 at height, (ASh-82FN) 500/600, (ASh-82FN/2TK-3+RD-1) 680 km/h at 7,5 km, 705 km/h (438 mph) at 12 km; service ceiling (M) 12 km, (A) 10,5 km, (R) 12 750 m (41,830 ft); range (M, max) 1240 km, (R) 880 km; TO (M) 525 m, (R) 300 m; landing 125 km/h, 450 m.

Su-8

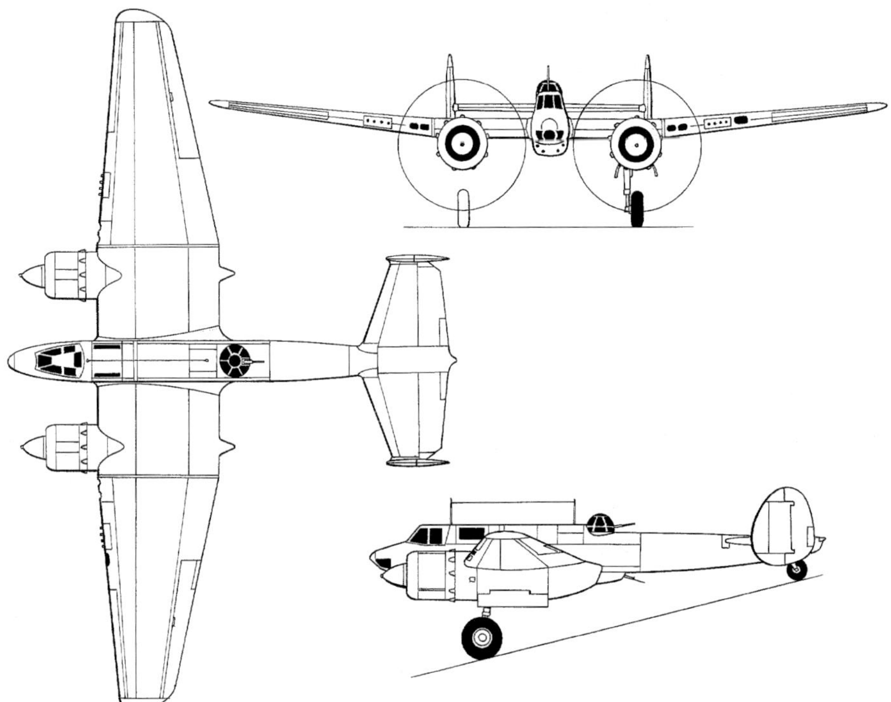
Su-8

Su-8 Jan 42 Kremlin conference on Shturmoviks also demanded next-generation machines. AM-42 Su-6(2A) was regarded by Sukhoi as interim, and it was his idea to build large long-range aircraft. Little evident need at time; it was only from late 44 that Nazi retreat became so fast that close-support air units could not keep up, and short range/endurance of Il-2 became a problem. Sukhoi said he predicted this situation and that Kremlin meeting concurred, authorizing two prototypes designated **DDBSh** (*Dvukhmotornyi Dalnii Bronirovannyi Shturmovik*, twin-engined long-range armoured attacker; according to Shavrov, second D = *Dvukhmyestnyi*, two-seat). OKB called new project aircraft B. Exceedingly impressive aircraft, with two of most powerful available engines, capacious tankage and devastating armament. Slim fuselage with pilot and radio/gunner separated by large central fuel tank, low wing and twin-finned tail. Mixed construction, with complex fuselage having entire front-end *bronyekorpus* (armour-body) with five thicknesses up to 15 mm. Detailed structure in *ASU*. Primary armament large ventral box housing (first aircraft) four 11P-37, (second) four OKB-16-45, each with 50 rounds in clips loaded by radio/gunner (OKB-16 had rig with auto feed, not installed in aircraft). In addition four ShKAS in outer wings firing ahead, used air/air and for air/ground sighting. Glazed nose used by pilot to sight forward-firing guns, all angled slightly downwards and with sight allowing for this. Glazed nose and same sight system, also used by pilot to aim bombs. Normal load four FAB-150 in four cells in wing aft of main spar between engines and ShKAS batteries. Overload 1400 kg including external. Upper rear defence single UBT aimed by radio/gunner. Lower rear ShKAS at rear of ventral gun box also aimed by backseater. Total 5900 rounds weighing 232 kg for all guns. Heaviest forward-firing firepower of any Second World War aircraft, with over 1 t/min for big guns alone. Total mass of armour on first aircraft 1680 kg. Designated Su-8 before completion, first aircraft on factory test late 43, second joined early 44. Both outstanding with superb handling even at full load, but prevailing mood was war could be won with existing equipment. Su-8 range and endurance would have been useful in final year of war. Land armies advanced over 1500 km and were often 150-200 km ahead of nearest close-support air unit.

DIMENSIONS Span 20,448 m (67 ft 1 in), length 13,58 m (44 ft 6⅔ in); wing area 60,0 m^2 (646 ft^2).

ENGINES Two ASh-71F.

WEIGHTS Empty (1st) 9180 kg, (2nd) 9208 kg (20,300 lb); fuel/oil 2370 kg; loaded (1) 12 425 kg, (2) 12 413 kg (27,366 lb); max 13,3 t (29,321 lb).

PERFORMANCE Max speed 500 km/h (311 mph) at SL, 550 km/h (342 mph) at 4,6 km; climb to 3 km, 7·3 min, to 5 km in 9 min; service ceiling 9 km; range with full weapons 600 km (373 miles), (max, no bombs) 1500 km (932 miles); TO 400 m; landing 470 m/140 km/h (87 mph).

Su-5 Second of OKB's experimental interceptors of 44, Su-5 was rival to MiG N (I-250) as mixed-power aircraft able to dogfight German jet and rocket aircraft. Prototype authorised as **I-107** early 44, using same combination of engines as I-250: modified VK-107A with step-up (13:21) geared drive to compressor in separate large fuselage duct for CIAM (Kholshchyevnikov) VRDK boost propulsion burning same fuel from seven ducted combustors and discharging through stainless-steel welded pipe and variable-area nozzle with hydromechanical actuation. Main-engine exhaust discharged straight overboard, but carb-air supplied via left/right pipes taken from high-pressure area downstream of VRDK compressor. Airframe entirely duralumin, including flight-control surface skins. Wing with single spar with heavy plate web, profile CAHI 1B10 inboard 16·5% at root, changing to NACA-230 outboard mainly 11%. Structurally in three sections each side, one bolted joint outboard of main gears and other at flap/aileron junction. Main gears with track 3,15 m, tyres 650×200 mm, retracting inwards hydraulically ahead of spar. Fully balanced Frise ailerons. Hydraulic split flaps driven to 54°. Fully retractable 300×125 mm tailwheel with asbestos insulated well. Four-blade 2890 mm propeller without Hucks dogs, 646 lit fuel in wing aft of spar (gravity fillers near outer ends of centre section), oil cooler in left centre section immediately behind spar fed by large shallow duct from leading edge, curved over retracted main leg and exhausting via guide-vanes in bottom skin ahead of flap. Spring-tab flight controls, with trimmers using same tabs by biassing neutral position. Armament one NS-23 with 100 rounds firing through propeller hub plus two

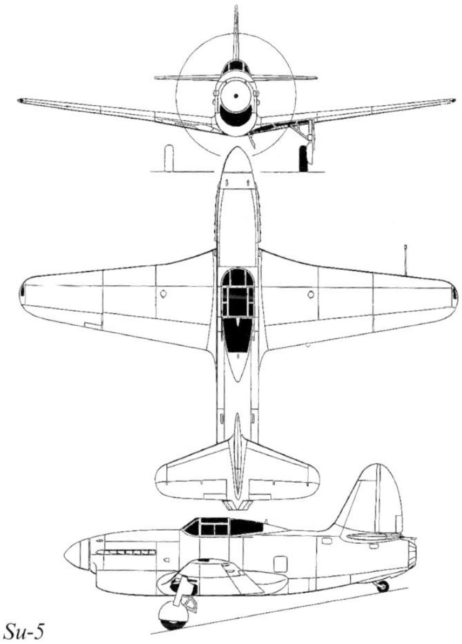

Su-5

Su-5

UBS with 400 rounds in top of fuselage. Pilot seat 10 mm armoured, canopy sliding back over glazed rear fairing, gyro sight, HF radio with wire aerials to fin including IFF. Su-5 with matt-painted finish on factory test, believed Novosibirsk, using wheeled landing gear from 6 April 45. Almost all flying by G. Komarov. Satisfactory aircraft, but no production requirement in view of unquestioned superiority of turbojet. Did not fly after 15 June 45.

DIMENSIONS Span 10,562 m (34 ft 7⁴⁄₅ in); length 8,21 m (26 ft 11¼ in); wing area 17,0 m^2 (183 ft^2).

ENGINES One VK-107A (modified) plus VRDK giving high-alt boost equivalent to 900 hp for max 10 min.

WEIGHTS Empty 2954 kg (6,512 lb); fuel/oil 465 kg; loaded 3804 kg (8,386 lb).

PERFORMANCE Max speed (SL) 645 km/h, (5 km) 768 km/h, (7,8 km) 810 km/h (503 mph); climb to 5 km 5·7 min; service ceiling 12,05 km (39,535 ft); range 600 km (373 miles); TO 345 m; landing 140 km/h, 600 m.

UTB This low-powered (700 hp ASh-21) trainer derivative of Tu-2 was developed by P. O. Sukhoi's OKB. See under Tupolev, page 412.

Su-9 OKB's first jet was straightforward fighter/bomber authorized as two flight prototypes plus static-test airframe May 44. OKB designation Aircraft **K**. General appearance strikingly similar to Me 262, and underwing engine nacelles almost identical apart from having longer inlet ducts making engine nose bullet less visible. All-dural airframe with flush riveting or bonding throughout. Wing of S-1-12 series, no attempt at sweepback, ruling t/c 10%, single main spar, fuel both inboard and outboard of nacelles with two gravity fillers. Fixed leading edge but trailing edge slotted flaps inboard of nacelles and novel airbrakes between nacelles and ailerons divided into upper (large) and lower (smaller) surfaces pulled apart to 88° and stressed to 900 km/h IAS. Fin integral with oval-section monocoque fuselage, carrying fixed tailplane with dihedral, rudder hinge behind ele-

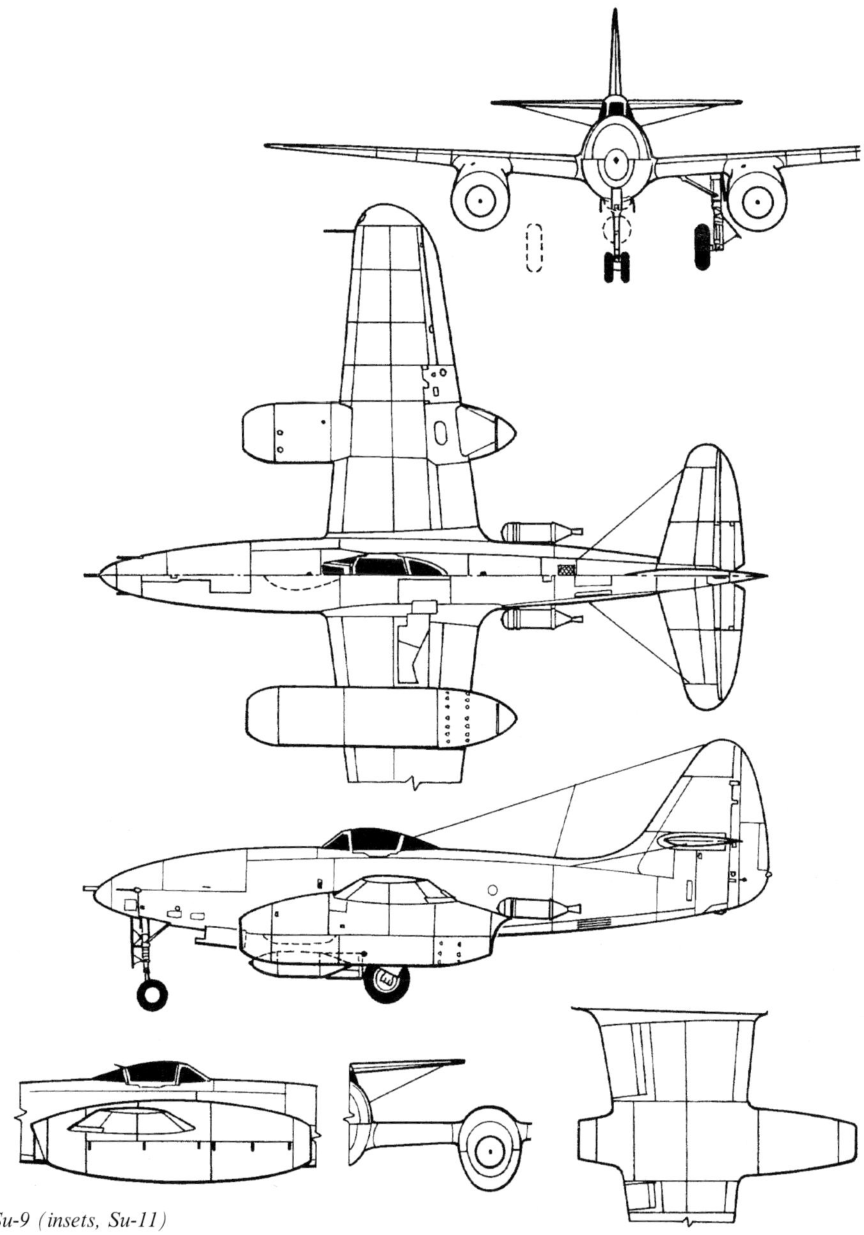

Su-9 (insets, Su-11)

vators. Flight-control surfaces metal-skinned and hydraulically boosted, with 12:1 gear ratio manual reversion (judged unacceptable for combat but adequate for landing). Airbrakes, flaps and landing gear also hydraulic, with single system but pump on each engine. Other new features included pressurized cockpit with completely transparent enclosure (with frameless blown Plexiglas main canopy sliding to rear), provision for two U-5 solid a.t.o. rockets on lower sides of lower rear fuselage (each 1150 kg for 8 s), cordite-fired Ye-24 ejection seat, toe-operated wheel-brakes, and braking parachute streamed from compartment immediately ahead of rudder. Nose gear with twin wheels, retracting to rear into bay with twin doors, main gear with wide track (3,51 m) and large low-pressure tyres retracting inwards behind spar face, visible. Three protected flexible fuel cells in fuselage supplementing those in wing, toal 2430 lit (535 gal); no provision for drop tanks. Armament one N-37 with 30 rounds plus two NS-23 each with 200 rounds; belly racks for two FAB-250. Flown by G. Komarov Aug 46, NII testing by Anokhin, Kochyetkov and Shiyanov, all confirmed excellent handling, fully up to standard of Me 262 (which was superb) but lacking German type's dangerous character at low speeds or with engine out. Testing included gun-firing and steep dive bombing using airbrakes. Rate of roll higher than MiG-9 (I-300) despite wing-mounted engines, and much better range, endurance (1 hr 44 min with fuel for spare circuit at base) and ammo capacity. Despite stigma of similarity to Me 262, so good that NII recommended series-production but no GAZ capacity available. One prototype flew publicly in Tushino parade 3 Aug 47.

DIMENSIONS Span 11,20 m (36 ft 9 in); length 10,55 m (34 ft 7⅓ in); wing area 20,2 m² (217 ft²).

ENGINES Two RD-10.

WEIGHTS Empty 4060 kg (8,951 lb); fuel 1750 kg max; loaded (normal) 6,1 t, (max) 6380 kg (14,065 lb).

PERFORMANCE Max speed (5 km) 885 km/h (550 mph); climb to 5 km 4·2 min; service ceiling 12,8 km (42,000 ft); range 1200 km (746 miles); TO 910 m, (with a.t.o.) 475 m; landing 150 km/h, 960 m (with drag chute 660 m).

Su-9 with FAB-500

Su-11

Su-11, KL Improved Su-9 with axial engines of greater thrust. Larger engines on steel-tube truss ahead of front spar, jetpipe close under wing, whole enclosed in large nacelle curved across wing to enhance lift. Wing made slightly larger, fuel reduced (2335 lit, 514 gal). NII testing by Anokhin and Shiyanov 28 May 47 to 15 April 48, during which engines replaced by uprated version, though with little change in Mach-limited speed. A. S. Yakovlev told Stalin 'Copy of Me 262 and dangerous to fly'.

DIMENSIONS As Su-9 except span 11,8 m (38 ft 8½ in), wing area 21,4 m² (230 ft²).

ENGINES Two TR-1, later TR-1A.

WEIGHTS Empty 4496 kg (9,910 lb); fuel 1682 kg; loaded 6350 kg (14,000 lb), later (1A engines) 6877 kg (15,161 lb).

PERFORMANCE Max speed (3 km) 910 km/h (565 mph); climb to 5 km, 3·2 min; service ceiling 13,2 km (43,300 ft); range 900 km (559 miles); TO 780 m.

Su-13, KD Final mod of Su-9 design to increase performance. New centrifugal engines, wing as Su-11 but thickness reduced from 11% to 9%, horizontal tail considerably enlarged and swept 35°, modified nose landing gear, armament changed to three N-37 each with 50 rounds. Provision for two underwing drop tanks. First flight early 48, but performance still inadequate.

DIMENSIONS As Su-11.

ENGINES Two RD-500.

WEIGHTS Empty 4550 kg (10,031 lb); loaded 6436 kg (14,189 lb), (max, with tanks) 7036 kg (15,511 lb).

PERFORMANCE Max speed (SL) 970 km/h (603 mph), (5 km) 960 km/h; climb to 5 km 4·0 min; range 1550 km (963 miles); (with tanks) 2000 km (1,243 miles); TO/ldg both 500 m.

Su-10 First large bomber designed under Sukhoi supervision since 34, this four-jet day bomber was authorized 46 as Aircraft **Ye**, actual construction begun 47. Large finely stream-

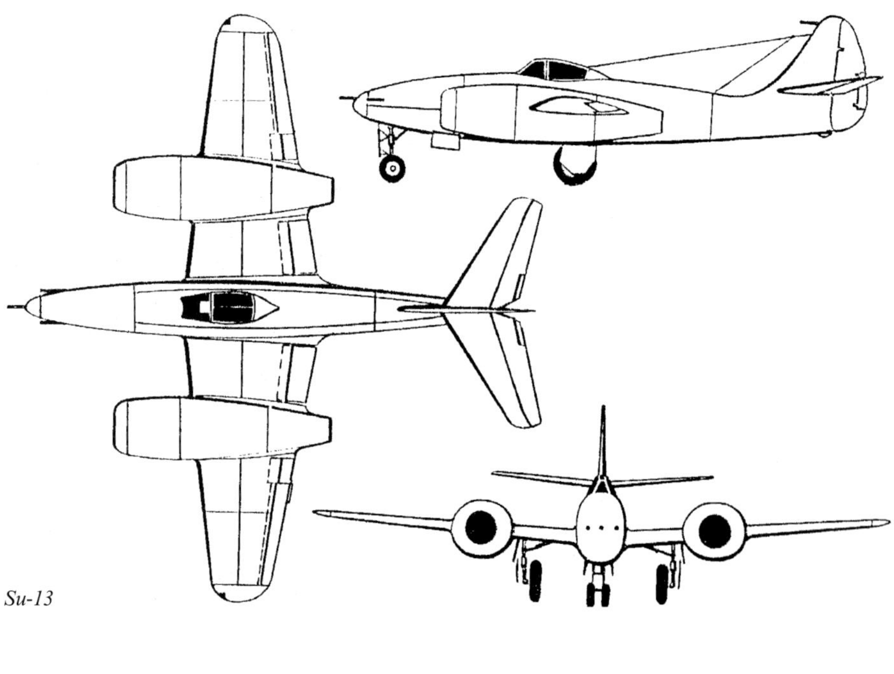
Su-13

lined fuselage of circular section with pressurized nose compartment for nav/bomb in glazed nose, pilot on upper centreline under large canopy, radio/gunner back-to-back with pilot, and observer/gunner in pressurized compartment at tail. All-dural construction with flush riveting or (new techniques) bonding. Unswept wing with main box enclosed by two spars with heavy plate webs and machined stretched skins forming integral tank over most of span, except between engines and fuselage where large cut-outs for retracted main gears. These required so much space engines had to be well outboard, and four turbojets installed in superimposed staggered pairs, each engine on welded 30KhGSA spaceframe carried off main box and with large dural nacelle of removable panels. Fixed leading edge, large Frise ailerons and large one-piece Fowler flaps driven hydraulically on internal rails to 32°. Main gears retracted inwards, single main wheels. Short twin-wheel nose gear retracting forwards under pressure floor. Large bomb bay occupying full cross-section of lower half fuselage, max capacity 4 t (8,818 lb), doors opened hydraulically inside fuselage. All flight controls hydraulically boosted including upper/lower rudders on fin swept at 45° on LE carrying high horizontal tail as far to rear as possible. Four flexible fuel cells in fuselage bringing total capacity to 8350 lit (1,837 gal) at which bombload not normally above 1500 kg. Fixed B-20E in nose, twin B-20E in dorsal turret and twin B-20E in tail turret; no inflight access between tail gunner and forward compartment and no control authority by either gunner over other pair of movable guns. Provision for four U-5 solid a.t.o. rockets on lower sides of rear fuselage, retractable tail bumper and tail braking parachute. Ejection seats for three forward crew, rear gunner with single-lever escape via ventral slide complete with seat. Programme abandoned 48; prototype airframe passed (unflown) to MAI for instructional purpose.

DIMENSIONS Span 20,6 m (67 ft 7 in); length 19,55 m (64 ft 1⅔ in); wing area 71,3 m^2 (767·5 ft^2).

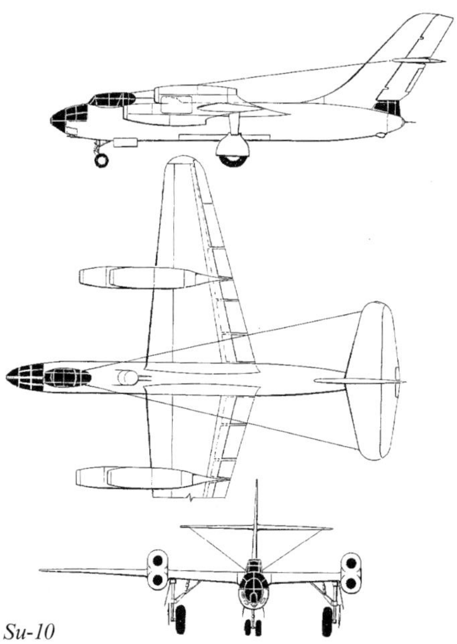

Su-10

Su-12

ENGINES Four TR-1.

WEIGHTS Empty 13 346 kg (29,621 lb); loaded 18 950 kg (41,777 lb), (max) 21 230 kg (46,803 lb).

PERFORMANCE Max speed (SL) 810 km/h (503 mph), (8 km) 850 km/h (528 mph); service ceiling 12 km (39,370 ft); range (10 km cruise) 1500 km (932 miles); TO 1 km. All estimates.

Su-12 This multi-role tactical aircraft was inspired by Fw 189 but on much more powerful scale. Unfortunate for OKB, such aircraft might have been useful in the Second World War and could have proved valuable in many post-war situations, but it was launched June 46 just as such heavy piston-engined combat aircraft were appearing outmoded. Project called Aircraft RK, from *Razvyedchik Korrektirovshchik* (reconnaissance and artillery spotting), but Su-12 design grew to have crew of four and appreciable offensive power. Modern all-metal stressed-skin structure almost entirely D1 and D16 and with flush-riveted exterior skins throughout. Unswept wing built as rectangular centre section and tapered outer panels with 6·5° dihedral, NACA-230 section 14% t/c at root and 9·25% at tip. Single straight spar at 40% chord with heavy plate web and extruded/machined booms. Pressed-sheet ribs, L-stringers and 0,6 mm or 1,0 mm skins. Hydraulic slotted flaps pivoted on brackets well below lower surface in four sections inboard and outboard of tail booms, outer sections stressed as airbrakes. Manual flight controls with D1 skin over leading edge back to spar, fabric to rear; servo tab on right aileron, elec trimmer on left. Central crew nacelle with wing in low/mid position, housing pilot on left, nav on right in small seat with tall vertical armoured back, radio/gunner at rear of compartment with seat and radio behind pilot and vertical pillar/saddle on centreline for aiming electric-powered mid-upper turret with twin B-20E with 200 rounds each, observer/gunner seated on centreline at rear of nacelle with control of rear B-20E with 200 rounds, recon camera in floor and various army liaison subsystems. Fourth B-20E fixed firing ahead with 100 rounds. Eight FAB-100 or various alternative weapon loads carried in bays with twin doors in tail booms at trailing edge, elec

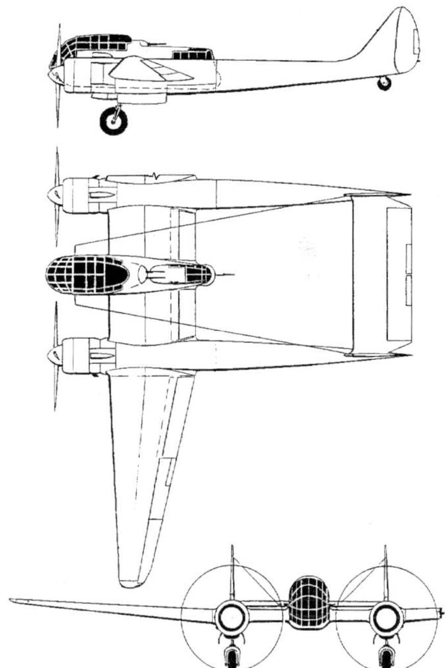

Su-12

door operation and release. Neat engine installations, filtered/de-iced carb-air inlets above, ducted oil coolers below, four-blade AV-9VF-21K propellers 3,6 m diam. Two 440-lit wing tanks and two 180-lit tanks between engines and front spar, total 1240 lit (273 gal). Nacelle protected by 28 pieces of armour from 2 mm to 12 mm, weighing 450 kg. Armour-glass 15 mm to 90 mm, thick over most glazed areas. All three units of landing gear retracting hydraulically, main unit with straight single oleo legs with fork-ends and 900 mm × 300 mm tyres folding to rear and tailwheel with levered suspension and 420 mm × 185 mm tyre retracting sideways to right into slightly bulged tailplane bay, all bays fully closed by doors. Su-12 flew Oct 47 (took part in Tushino flypast), NII testing by Kabanov, Stefanovskii, Zhdanov, Kubshkin and Tinyakov from 25 Dec 47 to 25 May 48. Excellent aircraft but judged outmoded.

DIMENSIONS Span 21,576 m (70 ft 9½ in);

length 13,053 m (42 ft 9·9 in); wing area 52,44 m^2 (564·5 ft^2).
ENGINES Two ASh-82FN (designed for ASh-82M).
WEIGHTS Empty 7552 kg (16,649 lb); fuel/oil 1820 kg; loaded 9510 kg (20,966 lb).
PERFORMANCE Max speed 460 km/h (286 mph) at SL, 531 km/h (330 mph) at 5,3 km; time to 5 km, 5·3 min; service ceiling 11 km (36,000 ft); range at 1 km alt, 1140 km (708 miles); take-off 220 m; landing 320 m/125 km/h (78 mph).

Su-15 OKB schemed radar-equipped interceptor 47, ahead of VVS demand, so having achieved funds for prototype, called **P**, this beat official rivals (La-250, I-320, Yak-50) first two of which followed Su engine layout. Wing S7s-12 built around main two-spar box incorporating integral tankage, four full-chord fences, outboard balanced hydraulically boosted ailerons, inboard hydraulic Fowler flaps, 0° dihedral. Plain nose inlet divided by vertical centreline splitter acting as structural support to underside of pressurized upper compartments for radar and (left of centreline) pilot. Right-hand air duct for second engine carried behind wing centre-section on main fuselage frames in sealed plenum box, removable upwards. Fixed rear fuselage carrying twin airbrakes and swept, hydraulically boosted tail surfaces, fin carrying electrically trimmed tailplane well above fuselage. Front engine at bottom of fuselage, fed by left duct, immediately behind box for rearward-retracting levered-suspension nose gear and attached to fuselage frames ahead of wing with surrounding plenum box and short jetpipe under wing. Four bag tanks, total 2875 lit (632 gal). Single-strut main gears retracting inwards hydraulically with 850 mm × 250 mm tyre housed not quite upright in fuselage. Short levered-suspension nose gear with 520 mm × 230 mm tyre retracting backwards. Pressurized cockpit slightly left of centre with heated windscreen, canopy sliding electrically to rear, access by ladder, cordite seat with faceblind, central display from *Toryii* radar with scanner above nose inlet. N-37 gun low on each side of nose, with 55-round magazine beside nosewheel bay. Attachments on rear fuselage for two U-5 a.t.o. rockets. Fin cap dielectric antenna. Prolonged ground testing delayed first flight to 11 Jan 49, NII testing by Anokhin and Shiyanov. Viewed favourably until suffered violent flutter during full-power run at 2 km, Anokhin ejecting.
DIMENSIONS Span 12,87 m (42 ft 2⅝ in); length 15,44 m (50 ft 7⅞ in); wing area 36,0 m^2 (387·5 ft^2).
ENGINES Two RD-45.
WEIGHTS Empty 7409 kg (16,334 lb); fuel 2760 kg; loaded 10 437 kg (23,009 lb) [OKB states normal 9050 kg, max 9900 kg, unexplained].
PERFORMANCE Max speed (SL) 1050 km/h (652 mph), (5 km) 1000 km/h; climb 5 km in 2·5 min; service ceiling 15 km (49,200 ft); range 1600 km (994 miles), (max, unexplained, no drop tanks) 2000 km; TO 450 m; landing 565 m.

Su-15 (1948)

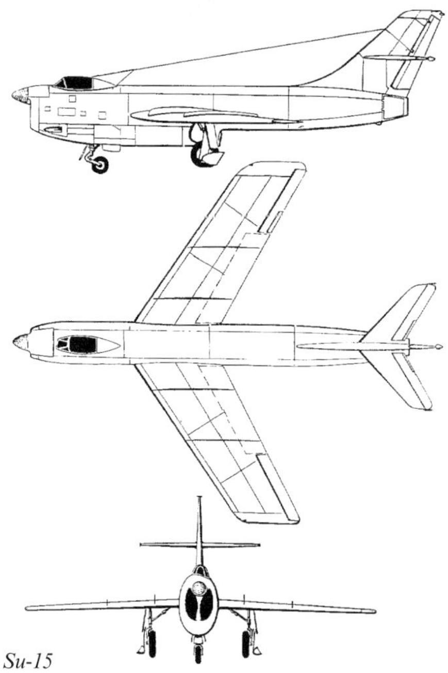
Su-15

Su-17, R Experimental aircraft to exceed Mach 1 and serve as prototype of front-line fighter. Generally rival to Yak-1000. Conventional layout, wing CAHI-9030 profile at root, SR-3-12 at tip, ¼-chord sweep 50°, two full-chord fences each side, third fence to aileron, large fin carrying fixed horizontal tailplane with 50° LE more than half-way up; all control surfaces hyd boosted (believed first in USSR), duplicated hydraulics also operating Fowler flaps, airbrakes and landing gear; levered main legs pivoted under wing root near TE rotating forwards into fuselage; levered nose unit retracting rearwards between inlet ducts split to pass cockpit. Provision for two N-37 each with 40 rounds. Non-afterburning engine exposed by removing rear fuselage; fuel in fuselage ahead of joint. Most radical feature, jettisonable nose, housing pressurized cockpit, guns and magazines. Separated at sloping bulkhead, automatically at 18 g or fired by pilot; drogue streamed 10 sec later, pilot taking to his own chute at lower altitude. Alternatively, pilot had ejection seat. Two flight articles and static test airframe, but CAHI said wing torsionally weak (aileron reversal) and refused to issue flight clearance. OKB closed on Stalin's order 1 Nov 49. First Su-17 near completion, used as gunfire target.
DIMENSIONS Span 9,95 m (32 ft 7¾ in); length 15,25 m (50 ft 0⅜ in); wing area 27,5 m^2 (296 ft^2).
ENGINE One TR-3.
WEIGHTS Empty 6240 kg (13,757 lb); fuel 1 t; loaded 7890 kg (17,394 lb).
PERFORMANCE (Est) Max speed (SL) 1209 km/h (751 mph, M 0·985), (5 km) 1207 km/h (M 1·05), (10 km) 1156 km/h (718 mph, M 1·085); climb to 10 km 3·5 min; service ceiling 14,5 km (47,570 ft); range 855 km (531 miles) [another document, 1080 km]; TO 460 m; land 755 m.

S-1 From Dec 49 Sukhoi and senior design engineers were assigned to Tupolev OKB, initially to assist Tu-91 but Su himself allowed almost complete freedom and chose to continue fighter aerodynamics in collaboration with CAHI. Played central role in establishing two configurations, 60° (or 62°) S (swept wing) and 60° (or 57°) T (delta). Immediately on Stalin's death Su applied for permission to reopen OKB and this resumed work at original premises 10 May 53. MAP signed contracts for design phase of **S-1** as prototype frontal fighter and **T-1** interceptor.

S-1 wing LE swept 62°, sized to large Lyulka engine and in many ways derived from Su-17 though bigger and completely new design. Wing below mid-position, SR-3S profile, thickness 5·9% root, 4·7 tip, described by OKB as single spar but actually two, joined by strong transverse beam providing main bending input to heavy forged root rib and trunnion mount for main landing gear. Tracked area-increasing slotted flaps, powered outboard ailerons, enormous fence ahead of flap/aileron junction (extending round LE to underside) and surrounding flat plate just inboard of tip. Root rib distributing

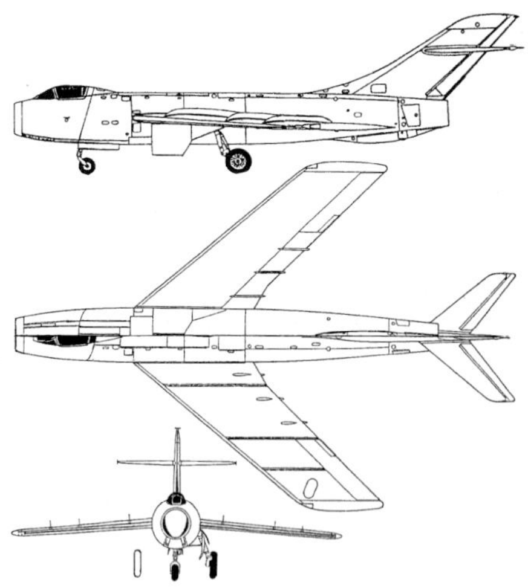
Su-17 (also see Addenda)

Su-17 (1948)

loads into three heavy forged frames (there being no spars across fuselage). Circular nose inlet with conical centrebody later made to slide axially according to flight Mach number (first in USSR). Duct immediately bifurcated past pressurized and air-conditioned cockpit and levered-suspension forward-retracted nose gear to compressor face behind rear-spar frame. Four frames further aft, large louvred compressor blow-off exit above fuselage each side adjacent to bolted double bulkhead enabling rear fuselage to be removed. Levered-suspension main gears, leg shortening and wheel and door rotating 50° to lie flat in wing in bay bounded by root rib, front spar and transverse beam. Wheelbase 5114 mm, track 3830 mm. Swept fin and powered rudder, horizontal tail just below mid-point of fuselage, anti-flutter masses inboard of tips, fully powered 'slab' surface, LE 60°; 12,95 m² ribbon parachute streamed from box under rear fuselage, with tail bumper. Door-type hydraulic airbrake high on each side of rear fuselage. Retractable landing/taxi lights under nose and left wing, pitot booms centred above nose and inboard of left wingtip. Provision for three NR-30 guns under nose and in wing roots and for two UV-8-57 rocket launchers on wing pylons inboard of fences.

Mock-up accepted Jan 54, S-1 completed Sept 55 and inadvertently made hop on 8 Sept when A. G. Kochyetkov (loaned from La OKB) found taxi test had exceeded liftoff V. Su appointed V. N. Makhalin c.t.p., who reached 2170 km/h April 56. On state acceptance S-1 crashed 21 Nov 56, killing NII pilot I. N. Sokolov, though S already accepted for production, see next. Major advances were perfection of inlet system and auto trim of slab tail for all flight conditions including use of flaps.

DIMENSIONS Span 9,309 m (30 ft 6½ in); length overall 17,45 m (57 ft 3 in), (excl probe) 15,7 m (51 ft 6⅛ in); wing area 34,0 m² (366 ft²).

ENGINE One AL-7F.

WEIGHTS Empty 7890 kg (17,394 lb); loaded 9423 kg (20,774 lb), (max) 10 859 kg (23,940 lb).

PERFORMANCE Max speed (SL) 1140 km/h (708 mph, M 0·932), (10 km) 2170 km/h (1,348 mph, M 2·04); ceiling 19,1 km (62,664 ft); range 1370 km (851 miles), (max, though no mention of drop tanks) 1950 km; TO/landing both 800 m.

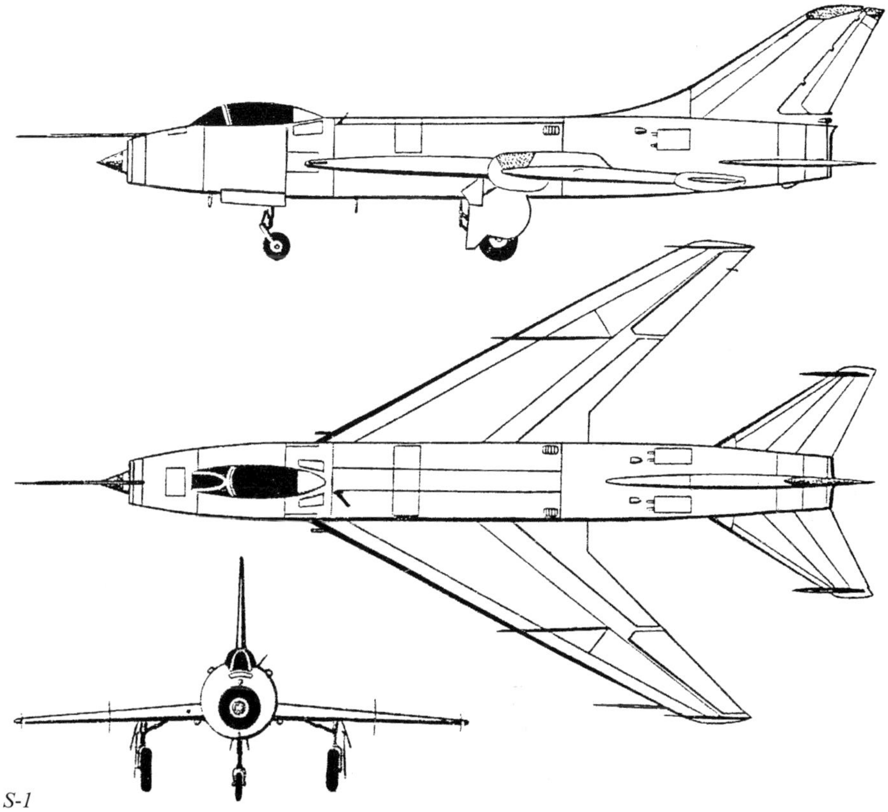

S-1

S-1

S-2, Su-7 Second prototype, stretched and area-ruled fuselage, improved inlet with rescheduled centrebody control and two suction-relief doors covering side of each duct, two additional airbrakes hinged to same frame as before but low on each side, fuel system with increased capacity and provision for two 600-lit drop tanks side-by-side under fuselage (endurance on internal fuel in full afterburner at SL eight minutes); additional cooling-air inlets and revised jetpipe fairing. Fuselage gun removed, leaving wing-root guns, and pop-out box of 32 mm×57 mm rockets added between fuselage pylons. KS-4 seat.

Assigned to N. I. Korovushkin, flown by him April 56, and to over 19 km (62,340 ft) in 57. Testing of pre-production batch, assisted by Makhalin and Ye. Kukushyev, followed by NII testing headed by V. Pronyarkin, cleared aircraft for production as **Su-7** frontal fighter (designation re-used), 100 produced at Komsomolsk 58-59 and assigned to local VVS regiment. ASCC name 'Fitter-A'.

DIMENSIONS As S-1 except length 18,055 m (59 ft 2⅞ in).

ENGINE One AL-7F.

WEIGHTS Empty 8200 kg (18,078 lb); loaded 12,0 t (26,455 lb), (max) 12,95 t (28,549 lb).

PERFORMANCE Max speed (SL) 1135 km/h (705 mph), (10 km) 2120 km/h (1,317 mph, M 2·0); service ceiling 19 km (62,340 ft); range with drop tanks 1875 km (1,165 miles); TO 1300 m; landing (drag chute) 900 m.

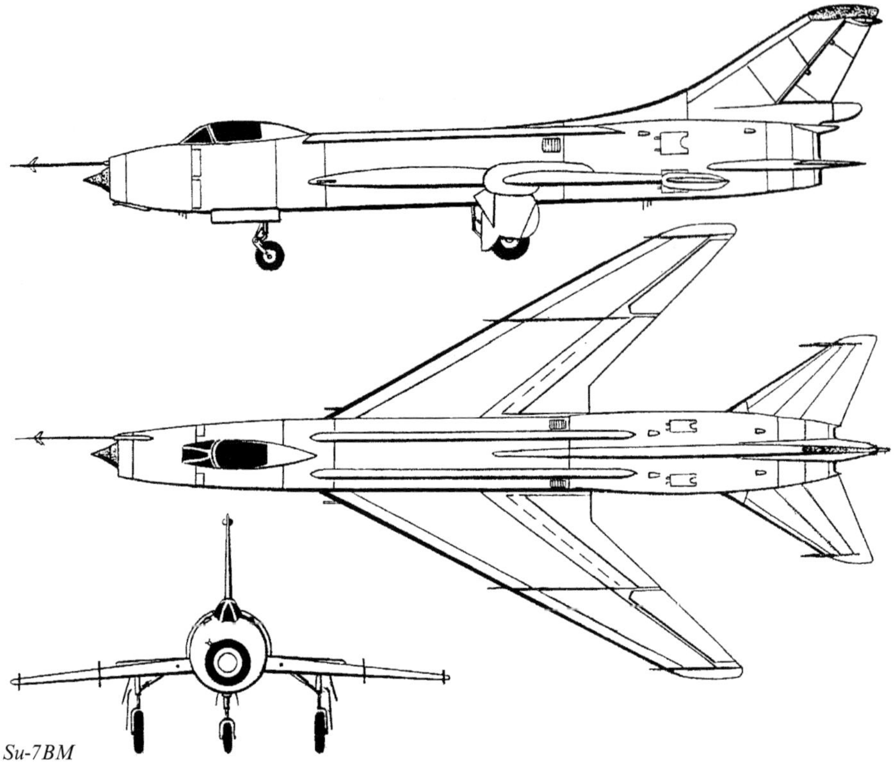

Su-7BM

S-22, Su-7B Change of policy (influenced by F-105 and MiG-21) resulted May 58 in decision to use Su-7 as ground-attack fighter. Basic aircraft unchanged, still equipped to carry four pylons, two side-by-side under fuselage and two just inboard of fences, but now wing pylons cleared to 500 kg loads and also plumbed for 600-lit (132 gal) tanks. Internal fuel 2940 lit (647 gal) in integral wing tanks and between and around inlet ducts with most system components and filler behind canopy. Rocket box removed but guns unchanged, installed in wing roots with right gun sufficiently far back for 73-round magazines round fuselage between frames to be in tandem; access doors in wing upper skin and ejectors underneath; fuselage skin near muzzles doubled, outer steel. Pitot boom moved to upper right side of nose, and q-feel pressure sensors linked to ARZ-1 flight-control system. Augmented avionics including expected VHF/UHF and RSIU radio, SRO-2 IFF, Sirena-2 warning, RV-2 radar alt, ARK-5 radio compass, MPR-48P beacon receiver, ATC/SIF, OSP-48 ILS, ASP-3VM ranging radar and autopilot. First flight by Ye. S. Solovyov April 59; NII testing by Fuhalov, Shatalov (later Cosmonaut) and Kutakhov (later Marshal). First full production version, delivered to FA from 61. Still very limited capability but loved for its strength and handling. 'Fitter-A'.

Su-7BM Uprated AL-7F-1 engine, with anti-FOD inlet, improved engine-bay cooling and nozzle fairing. Pitch and yaw vanes added to pitot. Improved wing and fuselage tanks, prominent pipe fairings along each side above centre fuselage. Braked nosewheel, canopy detonating cord and rear-view mirror, KS-3 seat, ASP-5ND ranging sight, *Sirena-3* warning, SRO-2M IFF, provision for vert or oblique cameral behind nose gear and chaff/flare/signal

Su-7BKL

dispenser under R wing root, pylons uprated to 750 kg (1,653 lb) and addition of two 500-kg pylons outboard of fences, all standard on later versions. Production 61-64, -200 engine from 62. One rebuilt as S-22I. 'Fitter-A'. Production at GAZ-153 Novosibirsk.

Su-7BKL Developed for dispersed operation from short unpaved airstrips and snow. Main-gear (tested on S-22-4) trailing arm extended aft to carry two parallel extensible shock struts on which are mounted steel-shod ski outboard of wheel; when extended can bear almost all weight of aircraft; modified doors. Nose-gear yoke modified for wheel with brake and larger low-pressure tyre; bulged doors. Attachments on rear fuselage for two SPRD-110 a.t.o. rockets. Twin ribbon drag chutes relocated in large box under rudder; latter reduced in area and tail nav light moved to fin tip. Small changes to increase effectiveness of flaps: small split flap added under inboard end, depressing to greater angle (50° instead of 30°) and sharp strip to extend TE chord of outer section. Plumbed for four drop tanks, KS-4 seat. Prodn from 65. 'Fitter-A'.

Su-7BMK Final fixed wing production 'Fitter'. Primarily for export, but also used by FA. Many features retrofitted to earlier variants. Similar to 7BKL without steel skis. SP-50 'Swift Rod' ILS at top of fin, facing aft. Rear-view mirror in fairing above canopy. Sirena 3 RWR, KM-1 rocket seat. Extra underwing pylon outboard of fence. Su-7 operators include Afghanistan, Algeria, Cuba, Czechoslovakia (Su-7BM), Egypt, India (Su-7BM), Iraq, North Korea, Poland, Romania and Syria, mostly to Su-7BMK standards.

Su-7UM Tandem dual trainer version of BM, Dash-200 engine. Instructor cockpit behind original, almost at same level, retractable periscope available below 600 km/h with gear extended in top of upward-hinged canopy faired into fin by dorsal spine (pipe fairings retained). Pupil cockpit upward-hinged canopy, mounted on strong frame between cockpits serving as crash pylon, with side windows giving instructor limited view ahead. This frame part of 300 mm extra section of fuselage (one additional frame). Rear cockpit required No 1 tank to be deleted. Each cockpit with own instrument venturi and RV-2 antenna relocated ahead of SRO-2M

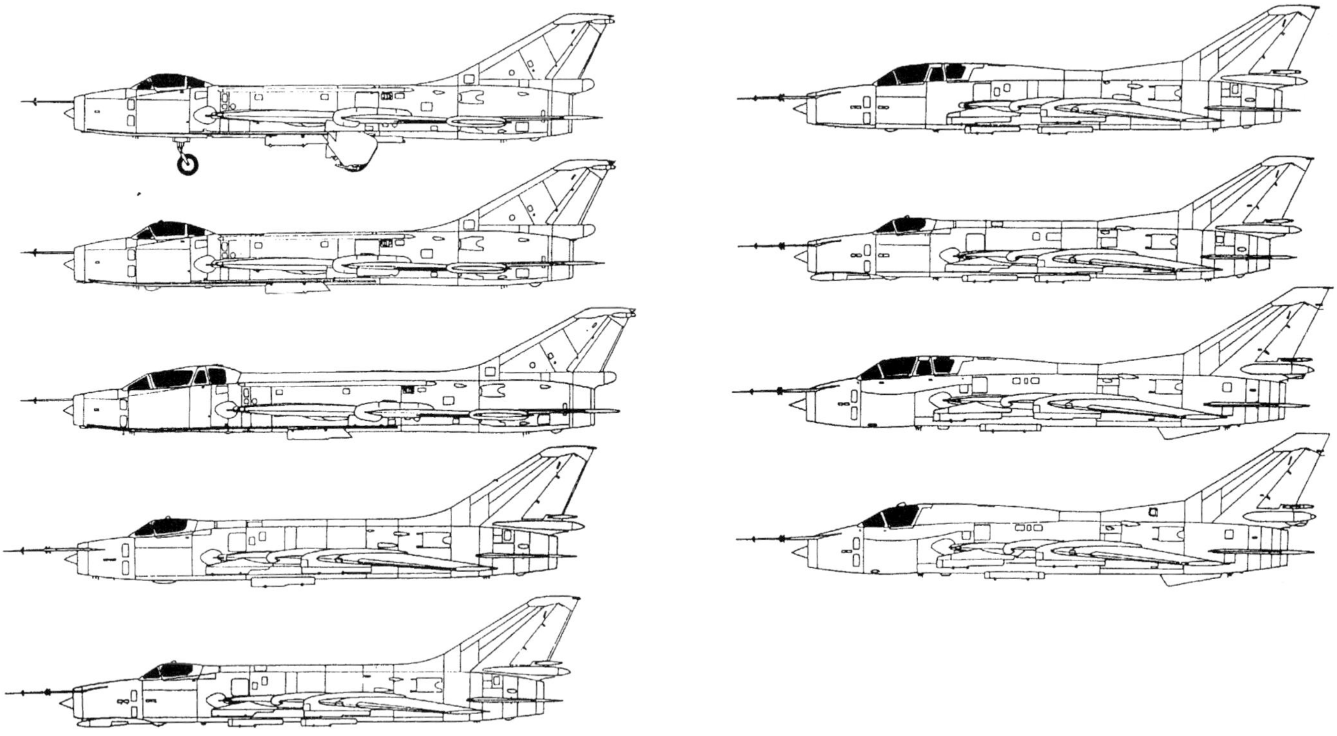

From top left: Su-7BKL, Su-7BMK, Su-7UM, Su-17, Su-17M, Su-17U, Su-20, Su-22UM, Su-22M-1/M-3

under nose. KS-4U-22 seats. Armament usually retained (bombload 500 kg), radar sometimes removed. Prototypes **U-22-1**, **-2** tested 65. **Su-7UMK** Export trainer, weapon system as BMK, twin drag chutes below rudder. Both in prodn 65-71 ASCC 'Moujik'.

S-22PDS, 100 LDU Su-7U fitted 69 with powered canards (greater span than on exptl MiGs, anti-flutter masses longer and nearer tips) to support T-4/100. Another 7U used to test K-36 seat.

DIMENSIONS Span 9,309 m (30 ft 6½ in); length 18,055 m (59 ft 2⅞ in), (U, UM) 18,36 m (60 ft 2⅞ in); wing area 34,0 m^2 (366 ft^2).

ENGINE (7B) one AL-7F, (BM) 7F1-150, then -200, (BKL) 7F1-250, (U,UMK) 7F1-200.

WEIGHTS (BKL) Empty 8370 kg (18,452 lb); fuel 3180 kg; loaded 13,6 t (29,982 lb), (max) 13,83 kg (30,489 lb).

PERFORMANCE (BKL) Max speed, SL (max dry) 850 km/h, (a/b) 1140 km/h (708 mph), (10 km) 2150 km/h (1,336 mph, M 2·02); climb 10 km in 90 s; service ceiling 18,5 km (60,500 ft); range (hi, internal fuel) 1280 km, (2 tanks, 1 t bombload) 1875 km (1,165 miles); combat radius (2 t bombs, no tanks, hi-lo-hi) 440 km (273 miles); TO 1450 m; landing (chutes) 990 m.

S-22I, Su-7IG Experimental swing wing Su-7 derivative intended not as a mere generic VG testbed but as an updated operational version. N. G. Zyrin managed design team led by V. Krylov who positioned pivot close to the main undercarriage pivots, giving swinging sections 4·5 m long and allowing the same basic structure to be retained. Wing sweep controlled by hydraulic pistons allowing two positions, fore and aft. Slats added as a result of wind tunnel testing. Prototype flown by V. I. Ilyushin 2 Aug

S-22PDS

S-22I

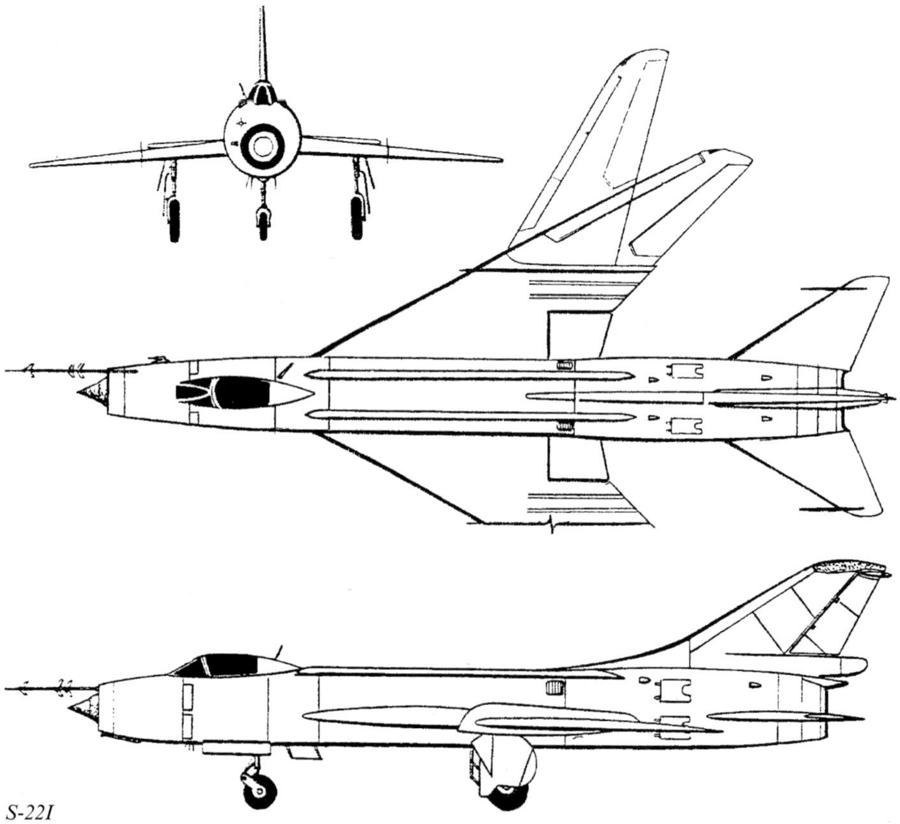

S-22I

66 and demonstrated by I. K. Kukushev at 67 Domodyedovo show.

S-32, Su-17 IG (var-geometry) rebuild resulted in ability to carry doubled bombload from airstrip half as long and extend combat radius c30%, so late 67 decision to develop series versions for FA and AV-MF for ground attack, anti-ship strike, multisensor recon, EW and training. All versions initially allotted OKB designation **S-32** and customer designation **Su-17**, in each case with sub-type suffix. By 68 build-standard of pre-series Su-17 settled, first flight mid-68. Wing refined internally, three-section slat, inboard flap refined (swept LE, higher C_L max), fence extended downwards to form pylon (plumbed for tank), second pylon added at Rib 2 near root to make total six; NR-30 guns in wing root as Su-7. Ldg gears strengthened, track 3,83 mm, wheelbase 5,38 mm, main tyres 800 mm×230 mm, nose gear as 7BKL, tyre 660 mm×220 mm, improved door bulges. Cockpit improved, K-36 seat, canopy with miniature detonating cord in left/right transparencies redesigned to hinge up with jack in front of spine (based on Su-7UM) housing air-cond, fuel tank and pipes, control runs and avionics. Retained external cable ducts of Su-7BM, Su-7BKL and Su-7BMK, and had only single fence and large Su-7BMK twin brake chute fairing up to 4th production batch. Internal fuel 2800 kg. One regiment for evaluation 70. ASCC reporting name 'Fitter-B Mod'.

S-32M, Su-17M First series version. AL-21F-3 engine with completely altered aux cooling-air inlets, compressor blow-off apertures not needed, twin pipe fairings above fuselage omitted because of spine, forward fuselage lengthened 0,23 m, rear-view mirror added to canopy, fin redesigned with three spars and increased height, braking parachutes replaced by single cruciform type in tubular container with twin doors, fuel capacity increased to 4550 lit (3630 kg) and both forward fuselage and outboard wing pylons equipped for PTB-800 or -600 drop tanks, extra fence round LE extending above wing to 60% chord in line with TE kink, nine pylons (centreline, front/rear pairs under fuselage, inboard/outboard under wing). Final avionic fit: RSB-70/R-831/RSIU-5 com, SOD-57M ATC/SIF, SP-50 ILS, ARK-10 radio compass, RSBN-2S nav receiver, NI-50BM doppler, RV-UM rad alt, MRP-56P beacon receiver, SRO-2M IFF (under nose and tail), Sirena 3M RWR (new forward receivers in LE, aft moved from top to bottom of fin with rudder cut-out to make room), ARL-S data link, SRD-5M ranging radar in ASP-5ND fire-control/sight system fed by instrument boom (as before at 10 o'clock on nose) with pressure head and triple pitch/yaw vanes, static head at 2 o'clock, total-temp probe near windscreen on left and AOA vane behind nose on left. Weapons included all tactical dropped stores and rockets plus Kh-23M, all three variants Kh-25 and self-defence R-3 or R-60. First flight 71, delivered to FA (Far East) 72 and to GSFG, Czech and Polish AF. Export version **S-32MK, Su-20** to Afghanistan, Algeria, Angola, Egypt, Iraq, North Korea and Vietnam. 'Fitter-C'.

Su-17R Few internal changes but with wiring and plumbing for choice of three multisensor recon pods carried on centreline, one with forward radar and Elint receivers, one with forward camera, SLAR and Elint modules, and one with four cameras, IRLS and battery of photoflash cartridges. Export version **Su-20R**. 'Fitter-C'.

S-32M2, Su-17M-2D First flight 74, pilot view improved by forward fuselage tilted down 3° while retaining inlet face vertical and instru-

Su-17M-3

mentation horizontal; nose also lengthened 0,38 m. KN-23 nav complex with ranging radar replaced by *Fon* laser ranger in fixed centrebody, second AOA sensor added to rear of first. Underside of nose almost horizontal, in S-**32M2D**, **Su-17M-2D** fitted with undernose fairing housing doppler; forward SRO-2M moved aft under LE. Usually camouflaged, delivered 76-80, 'Fitter-D'.

S-32M2K, Su-22M-2K Export version of S-32M2, R-29BS-300 engine in bulged, slightly shorter rear fuselage with different cooling inlets. Usually fitted with undernose fairing and with added straight dorsal fin, but avionic fit downgraded with Sirena 2, different doppler, no laser and fewer navaids. 'Fitter-F'. To Angola, Libya, Peru.

S-52U, Su-17UM-2D First of redesigned S-52 series, conversion trainer with rear instructor cockpit added, with different canopies similar to Su-15UM. Rear cockpit just possible without altering air ducts but displaced air-conditioning and avionics to new location in greatly enlarged dorsal spine joined to fin by straight dorsal fillet. Internal fuel as Su-17M3. Undernose fairing and left gun omitted, attachments on left wing root and fuselage for walkway between detachable ladder and rear cockpit. In production 75. 'Fitter-E'.

S-52, Su-17M-3 First single-seat S-52. Entire fuselage ahead of structural break for engine removal redesigned with increased depth, cockpit 200 mm forward giving view 15° down over nose instead of 9°, deeper canopy, large spine and dorsal fin now leading into new fin with increased height and square tip, internal fuel increased to 3890 kg, strengthened landing gear, landing/taxi lights increased to max five (retrac under nose and on each gear), *Klen*-PS laser in centrebody, doppler in deeper nose (no external fairing), gun magazines extended by left/right boxes in deeper spine, extra self-defence underwing pylons added midway between previous pair with rail for R-3 or R-60, various new weapon options such as four SPPU-22 or SPPU-6 pods all with depressed barrels and two firing to rear, all aircraft wired and plumbed for recon pod (as Su-17R but adding KKR-ITA/2-54 housing day/night cameras, flash cartridges and new Elint modules) on centreline and ECM jammer pod on left inboard

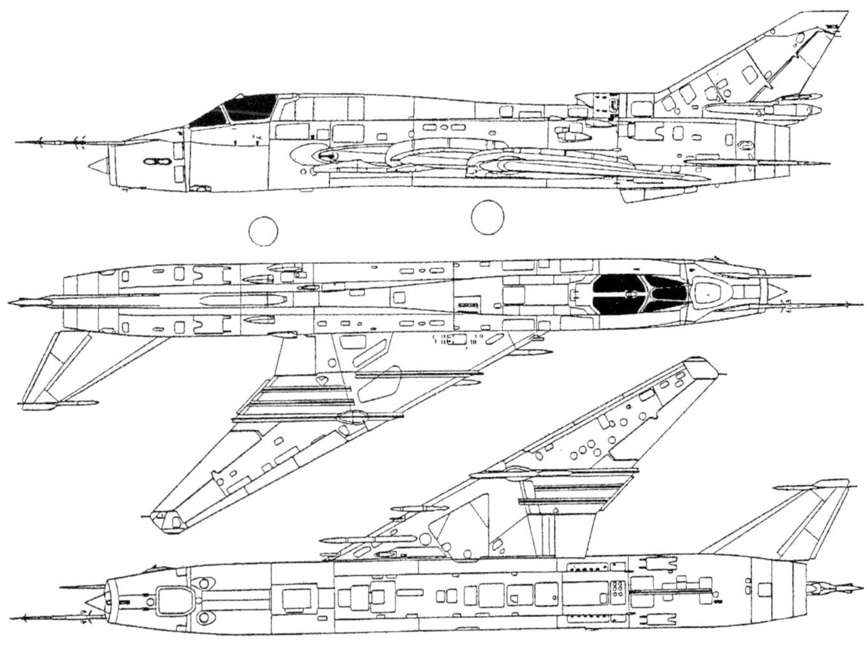

Su-17M-4 or 22M-4

pylon, provision for four 64-tube ASO chaff/flare dispensers scabbed above fuselage with manual or *Sirena-3* control. In production 76-77. 'Fitter-H'.

S-52M (still 17M3), **S-52UM** (still 17UM) modified during prodn with large removable ventral fin to improve gun aiming at extreme AOA. Some, including Hungarian AF, with extra AAM pylon underwing. In prodn 77-81. 'Fitter-H'.

S-52K (Su-22M), S-52UMK (Su-22UM) Export versions of 52M/UM, R-29, in production 77-82. 'Fitter-J', 'Fitter-G'. To Angola, Libya, Peru, Syria, Vietnam and both Yemens.

S-52UM3 (Su-17UM3) Prodn trainer from 78 with all systems/weapons/avionics unified with single-seaters. Afghan war led to refit with ECM dispenser blocks (four, above fuselage). 'Fitter-G'.

S-52MK (Su-22M3), S-52UM3K (Su-22UM3) Corresponding unified export versions, delivered from 82. From 83 Su-22UM3 was also delivered as **UM3K** with AL-21F-3 engine. 'Fitter-J/-G'.

S-54, Su-17M-4 Final new-build version, PRNK-54 (TsVM) nav/sighting complex, ram inlet above spine leading to heat exchanger in dorsal fin. Weapon options include several new attack or self-defence missiles; Afghan war led 81 to retrofit two pairs ASO-series ECM dispensers (max 256 payloads) above fuselage. Mach limit reduced from 2·09 to 1·75. First delivery 78. 'Fitter-K'.

S-54K, Su-22M-4 Export S-54, all Warsaw Pact so with AL-21F-3. Delivered from 83. 'Fitter-K'. To Poland, Czechoslovakia, East Germany.

After death of Sukhoi (75) programme led by N. G. Zyrin, later by A. A. Slezev. Total all versions 2,000+, completed 91. Several hundred in CIS storage, many transferred to former AV-MF. Export customers included: Afghanistan, Algeria, Angola, Czechoslovakia, Egypt, E Germany, Hungary, Iraq, N Korea, Libya, Peru,

Su-17M-4

Poland, Syria, Vietnam, N and S Yemen. Data for M-4.
DIMENSIONS Span (30°) 13,68 m (44 ft 10⅝ in), (63°) 10,03 m (32 ft 10⅞ in); length overall 19,03 m (62 ft 5¼ in), (excl probe) 17,34 m (56 ft 10⅝ in); wing area (30°) 38,49 m^2 (414·3 ft^2), (63°) 34,45 m^2 (370·8 ft^2).
ENGINE One AL-21F-3.
WEIGHTS Empty 10 667 kg (22,516 lb); fuel 3890 kg internal, 2860 kg ext (2×600 lit+2×1150 lit); max weapons 4250 kg; loaded 15,23 t (33,576 lb), (max) 19,43 t (42,835 lb).
PERFORMANCE Max speed (clean) 1400 km/h (870 mph, M 1·18) at SL, 1860 km/h (1,156 mph, M 1·74) at 11 km; max climb (clean) 225 m/sec (44,290 ft/min); service ceiling 15,2 km (49,900 ft); combat radius (4 t bombs, lo-lo) 495 km (308 miles); ferry range 2500 km (1,553 miles); TO 850 m clean, 1,5 km max wt; ldg 285 km/h, 1050 m (750 m with chute).
Data for UM3K: DIMS, ENG, as M-4.
WEIGHTS Empty 10,8 t (23,810 lb); loaded 15 t (33,068 lb); max 18,9 t (41,667 lb).
PERFORMANCE Max speed 1700 km/h (1,056 mph) at SL, 1900 (1,181 mph, M 1·79) at 11 km; climb to 10 km in 1·0 min; service ceiling 15,5 km (50,850 ft); range (4×840-lit) 2000 km (1,242 miles); TO 875 m, ldg 285 km/h, 1,2 km (950 m with chute).

T-3 In parallel with Mikoyan reopened Su bureau received order 54 for essentially identical aircraft using common components wherever possible with S and T wings (see S-1). T (*Treugolnyi*, triangular = delta) series all studied at first as perfect triangle, LE angle 60°, dihedral 2°. Su built only larger-size prototypes, and while rival was crippled by engines Su picked Lyul'ka and reached production with both configurations. T-1 with simple nose inlet mod during construction to T-3, first of several prototypes intended to explore supersonic inlets with radar. Radar *Almaz-3*, needing two antennas. As in MiG-17PF, ranging antenna put inside inlet and search scanner at 12 o'clock. Difference was that T-3 was to be highly supersonic, so main antenna was behind large

T-3

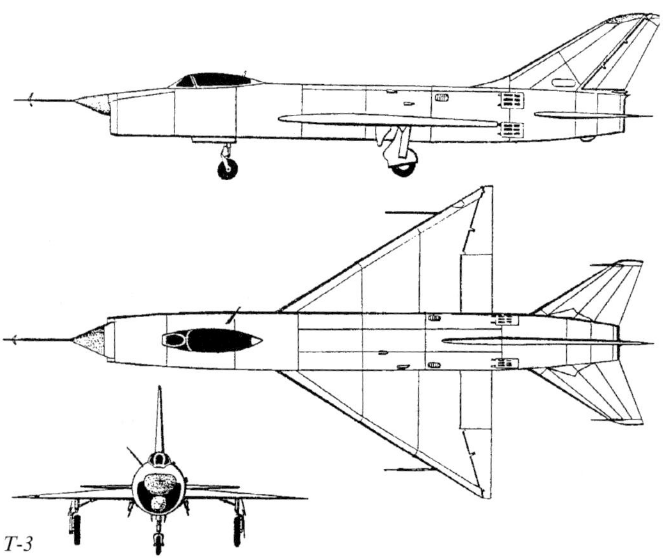
T-3

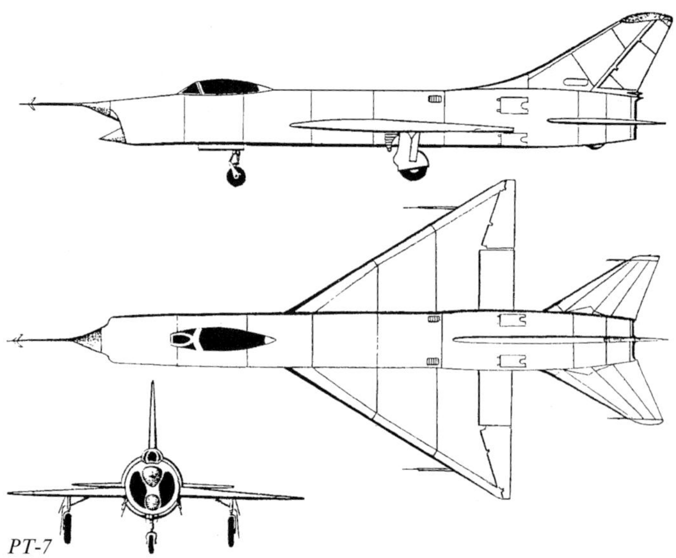
PT-7

radome of quasi-conical form, carrying PVD-7 instrument boom (with pitch/yaw vanes) on its tip. Compared with S-1, inlet appeared enormous, nose taper being almost non-existent. Remainder of aircraft similar to S-1 except for three-spar wing with rectangular slotted flaps (max 25°) and sharply tapered ailerons. Wing structure resulted in different main landing gears with track 4650 mm, wheelbase 5050 mm. Pitot tube near each tip. Fuselage not area-ruled, but fitted with four airbrakes, with large perforations. Tail as S-1 except tailplanes rotated about skewed axes on zig-zag fixed root portion. Pylons under outer wings for two K-6 missiles. No internal armament, but fuselage had steel-aluminium blast panels where gun muzzles would have been. Data indicate provision for drop tanks as well as AAMs. First flight by V. N. Makhalin 26 May 56. Flight 8 was as final aircraft in new-fighter fly-past at Tushino 24 June. Test team, Pronyarkin, Koznov, Kobishchan and future ctp Vladimir Ilyushin, all delighted. Dubbed *Balalaika*.

During devt LE extended with dogtooth, Su-7B style rear fuselage. Almaz 7 radar with small upper radome and conical ranging radome moved from centrebody to lower lip of intake. ASCC 'Fishpot-A'.

DIMENSIONS Span 8,7 m (28 ft 6½ in); length overall 18,82 m (61 ft 8⅞ in); wing area 24,9 m² (268·8 ft²).

ENGINE One AL-7F.

WEIGHTS Loaded 9060 kg (19,974 lb), (max) 11,2 t (24,691 lb).

PERFORMANCE Max speed (10 km) 2100 km/h (1,305 mph, M 1·98); ceiling 18 km (59,055 ft); range 1440 km (895 miles), (max) 1840 km (1,143 miles); TO/landing 1050 m/1200 m.

PT-7 Second interceptor delta (*Perekhvatchik Treugolnyi*), based on PT-3 but with area-ruled fuselage and more complex inlet. Antennas more widely separated and enclosed in conical radomes able to translate (move axially) in/out to match inlet to Mach as in simpler S-1. First flight 56.

DIMENSIONS, ENGINE As T-3.

WEIGHTS About 100 kg heavier than T-3.

PERFORMANCE Max speed (10 km) 2250 km/h.

PT-8 Designed 56 to explore inlet matched to single large antenna of Sokol-K, RP-1 and projected *Sapfir* radars. Extended circular nose with axi-symmetric inlet housing translating single-angle (two-shock) conical centrebody of nearly 1 m max diam, large suction-relief doors each side around upper part of duct. Wing as tested on modified PT-3, sweep 60° to dogtooth and 63° outboard, high-lift track-mounted flaps, aileron span reduced from 1,88 m to 1,45 m ending inboard of round tip of reduced-span wing. No pitot on right wing. Uprated engine, additional cooling inlets. Full combat avionics and weapons including two NR-30 guns. About 12 built, but proved too heavy.

DIMENSIONS Span 8,5 m (27 ft 10⅝ in); length 19,36 m (63 ft 7⅜ in); wing area 24,35 m² (262 ft²).

ENGINE One AL-7F1-100; no other data.

T-49 T-4 mod 9 in flight test from late 58. Left/right ducts around cockpit not joined at nose but taken to lateral inlets curved round fuselage, each forming 90° segment of circle. Sharp lips swept 60° in plan, inner wall variable in angle and throat area, single vertical suction-relief door in outer wall. Nose available for largest fighter radar, eg RP-25. Wing as PT-8 but LE 60° without dogtooth.

DIMENSIONS Span 8,6 m (28 ft 2½ in); length 19,8 m (64 ft 11½ in).

ENGINE One AL-7F1-100. No other data.

T-5 T-3 rebuilt early 58 with twin R-11F-300 engines in wide rear fuselage. Nose similar to S-1 but matched to greater airflow so larger and less-tapered, PVD on top and ILS underneath.

PT-8

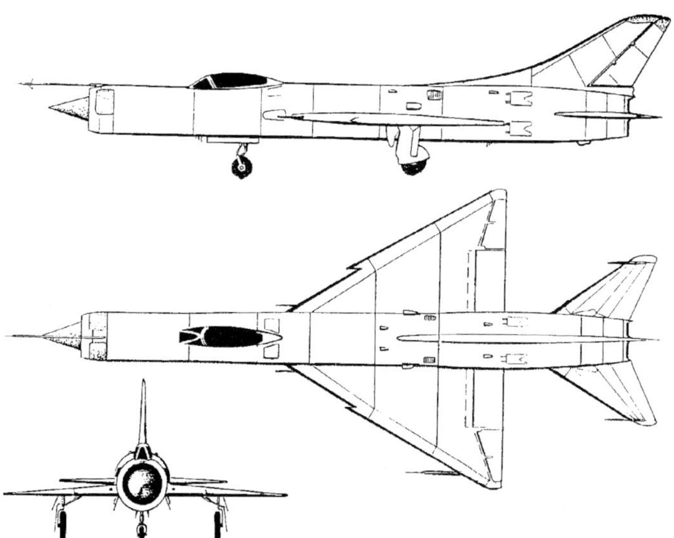

PT-8

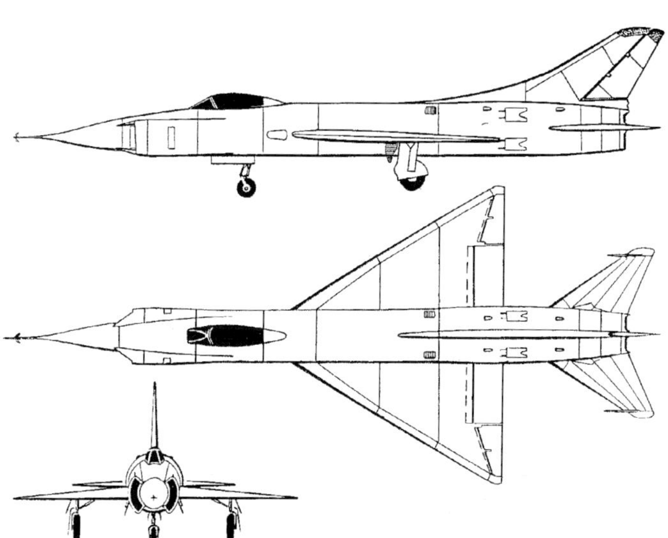

T-49

T-49

T-5

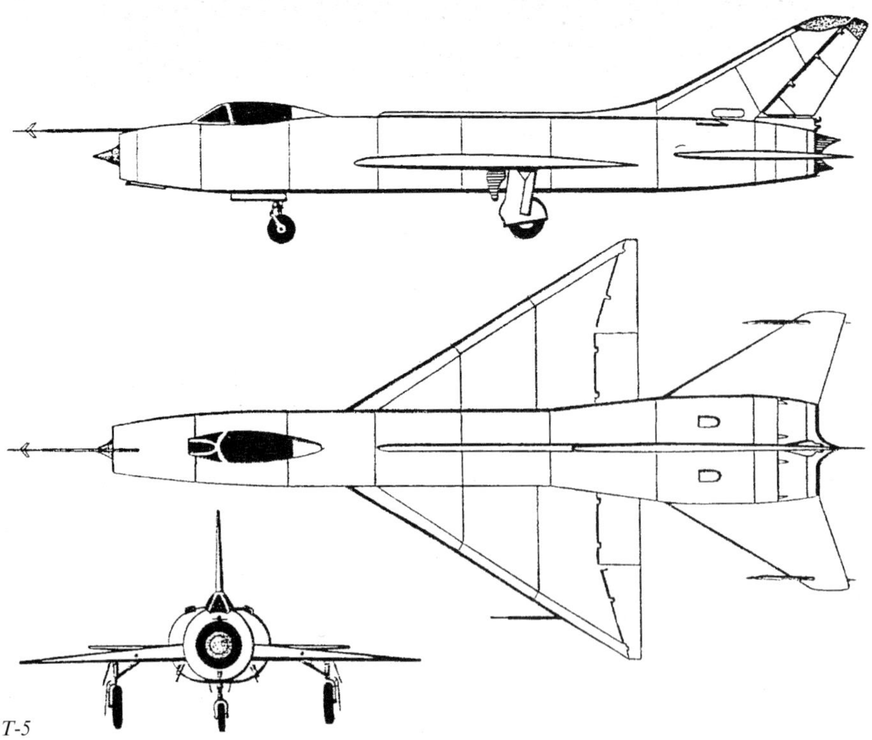

T-5

Track 4,5 m, wheelbase 5,3 m. Tailplanes enlarged nearly 60%. Long pipe fairing from No 1 tank to fin. No airbrakes or weapons. Flew 18 July 58, nicknamed 'Mother heroine' due to 'broad hips'.

DIMENSIONS Span 8,7 m; length 18,38 m (60 ft 3⅝ in).

ENGINES Two R-11F. No other data.

P-1 Two prototypes of two-seat heavy interceptor ordered 55 to meet PVO requirement, **P-1** intended to have AL-9 engine but actually powered by less powerful AL-7F1 and **P-2** with two VK-11. Wing scaled up from PT-8, but with kink reducing outboard sweep from 60° to 55°. Relatively small ailerons terminating well inboard. Large fuselage accommodating: PVD four-channel air-data probe; large erosion-resistant radome over prototype *Pantera* radar; armament bay wih two NR-30 guns and 50 NRS-70 spin-stabilized rockets fired from four bays with rapid-action front doors and rear flame doors; pressurized tandem cockpits with KS-1 seats and upward-hinged canopies for pilot and radar operator; unique circular inlets standing away from fuselage with sharp lip swept 45° in plan, small pointed splitter plate inboard, small cones centred in duct (thus at angle to fore/aft axis); max tankage between ducts with dorsal pipes in fairing from cockpit to fin. New tail, fin increased in height from 2,0 m to 2,5 m, tailplanes as T-5 but no anti-flutter masses and TE with greater sweep, reducing area. Twin parachutes under tail but no airbrakes. Provision for drop tanks under fuselage. Pylons under outer wings for two K-8 missiles. New nose gear retracting to rear, main gears inwards; track 4,6 m, wheelbase 7,6 m. Planned engine not ready, had to fly with lower thrust. OKB testing 12 July 57 to 22 Sept 58, by N. I. Korovushkin and Edward Elyan. Without engine, no point in going on, to Elyan's regret. Twin-engined second aircraft without dogteeth and different armament bay never completed.

DIMENSIONS Span 9,8 m (32 ft 1⅞ in); length 21,83 m (71 ft 7½ in); wing area c28,1 m^2 (302 ft^2).

ENGINE One AL-7F-200.

WEIGHTS Empty 8,4 t (18,520 lb); loaded 10,6 t (23,369 lb), (max) 11 550 kg (25,463 lb).

PERFORMANCE Max speed (15 km) 2050 km/h (1,274 mph, M 1·93); ceiling 19,5 km (64,000 ft); range (normal) 1400 km (870 miles), (max) 2000 km (1,242 miles).

T-37, RLD RLD = 'outstanding data provider'; fastest Su exptl interceptor, designed 58, built 60. Airframe based on PT-8 but with Tumanskii engine, nose shorter, downsloping and containing Oswatitsch centrebody with three cone angles for peak pressure recovery at over Mach 2 carrying PVD-7 air-data heads directly on tip, not on boom. Two suction-relief doors each side as on Su-7, but ducts larger to serve powerful engine also fitted to rival Ye-150. Engine accessed by unbolting double frame at trailing edge to remove rear fuselage. Upstream of this point fuselage was structurally remarkable in being almost pure monocoque, without stringers, 'a single piece of aluminium alloy including integral tanks and air ducts'. Downstream structure welded titanium and steel, with large ejector nozzle surrounding primary variable engine nozzle, diameter main-

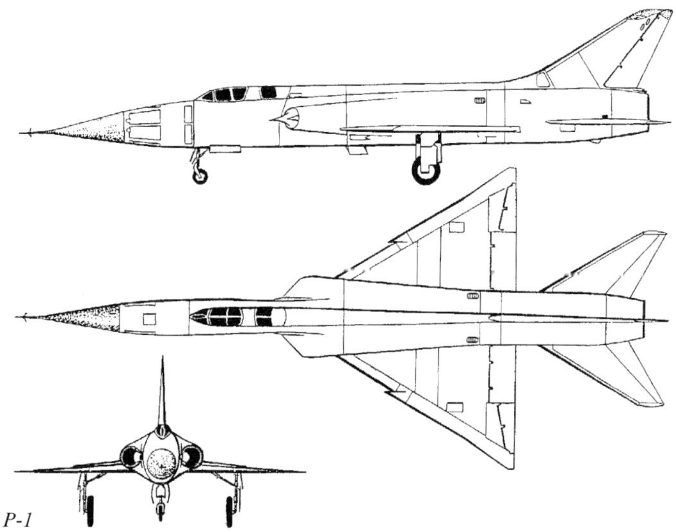

P-1

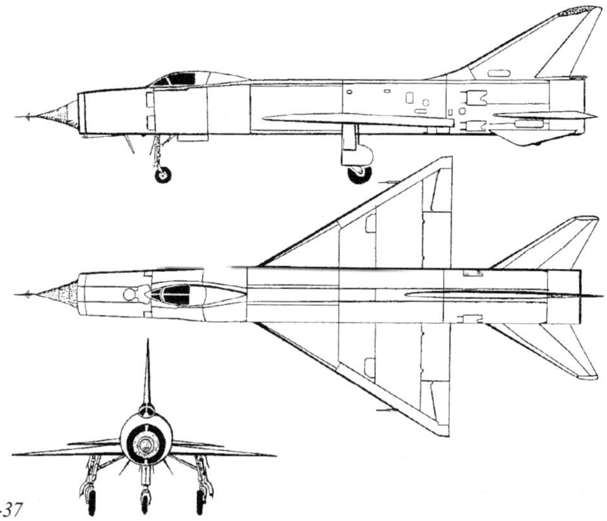

T-37

tained at maximum, meeting bottom edge of rudder; two oblique ventral fins, tailplanes without anti-flutter masses and with TE extended aft esp at tip. Wing without dogteeth, tips square but span unchanged. New main landing gears retracting directly inwards, longer nose gear retracting to rear, track 3,42 m, wheelbase increased to 6,35 m. Low-drag cockpit (not so extreme as Ye-150) with vee windscreen and left/right outward-hinged half-canopies. Four airbrakes, large drag chute. Provision for *Uragan*-5B radar in centrebody, two purpose-designed R-38 missiles under wings (pylon attachment doors visible in drawings, which also show IR sensor ahead of windscreen) and 1500-lit drop tank. Like Mikoyan, OKB crippled by engine unreliability and T-37 never flew, soon being scrapped.

DIMENSIONS Span 8,6 m (28 ft 2½ in); length 18,85 m (61 ft 10⅛ in); wing area c24,7 m^2 (266 ft^2).

ENGINE One R-15-300.

WEIGHTS Empty c7950 kg (17,525 lb); loaded 10 750 kg (23,700 lb), (max) 12 t (26,455 lb).

PERFORMANCE Design speed 3000 km/h at 15 km (1,864 mph, Mach 2·8); ceiling 25-27 km (c88,600 ft); range (est) 1500 km, (max) 2000 km.

Su-9

T-40 PT-8 served as basis for this family. First member, **T-43** flown April 57, introduced modified inlet with three-angle Oswatitsch translating centrebody lengthened to increase efficiency and provide room for RP-9 (SD-30) radar with parabolic antenna. Led to Su-9 (designation reused).

T-431

Su-9, T-43 By 58 PVO need for new interceptor was so pressing OKB was authorised to set sights lower. Almaz T-3-51 *kompleks* packaged simple RP-9U *Sapfir* radar and K-51 short-range missiles into aircraft based on proven parts. Most of airframe as T-3, combined with inlet similar to S-22 (Su-7B), with twin suck-in doors each side, but slightly lengthened to accommodate radar in translating single-angle 3-position centrebody. Duct bifurcated past cockpit to single central duct in mid-fuselage surrounded by three tanks plus injection water, but no dorsal pipe fairings; two tanks in each wing separated by main gears, twin fuselage pylons for side-by-side 600- or 950-lit drop tanks. Canopy with elec drive on rails to rear, top elec demisted, hot air over windscreen, KS-2 seat. Drag chute in tube below rudder. Some prototypes in T-40 series had NR-30 in each wing root, but series armament comprised four RS-2US on pylons cantilevered ahead of wing, each with command rod antenna projecting in front, or two RS-2US and two K-51 or two RS-2US and (replacing tanks) two UPK-23 gun pods.

First series a/c flown by Ilyushin 10 Sept 57. Production about 1,000, in service Moscow and Baku air-defence districts early 59. During prodn mods included: addition of dogteeth to wings; conversion of cannon bays to fuel tanks; conversion of fuselage tanks to integral, adding 11% to capacity; changed mid-fuselage construction from elimination of gun magazines and services; KS-1 seat replaced by KS-2A; and improved cooling around afterburner. On withdrawal 63-66 converted as **Su-9RM** radio-controlled targets. Data for initial series version. ASCC name 'Fishpot-B'.

Su-9U Tandem dual-pilot trainer with cockpits, canopies and spine as Su-7U, drop tanks retained but only two RS-2US. Prototype **U-43** July 60, prodn from 61. ASCC name 'Maiden'.

L02-10

DIMENSIONS Span 8,536 m (28 ft 0 in); length (incl PVD) 18 055 m (59 ft 2⅞ in); wing area 34,0 m² (366 ft²).
ENGINE One AL-7F1-100U.
WEIGHTS Empty 7675 kg (16,920 lb); loaded 11 422 kg (25,181 lb), (max) 12 515 kg (27,590 lb).
PERFORMANCE Max speed (SL) 1150 km/h, (11 km) 2120 km/h (1,317 mph, M 2·0); service ceiling 19,9 km (63,300 ft); range (2 tanks) 1800 km (1,118 miles); TO/landing both c1380 m.

Su-9 derivatives Several special mods. **T-405** (series a/c, 4+5=9) and **T-431** (T-43 #1) stripped of weapons/avionics for AL-7F-2 engine devt and record-breaking; 14 July 59, Ilyushin, zoom 28857 m; 2 Oct 60, B. M. Adrianov, 100-km circuit 2092 km/h; 7 Sept 62, Ilyushin, sustained 21270 m; 25 Sept 62, A. A. Koznov, 500-km circuit 2337 km/h. **Su-9UL** used for auto and semi-auto interception guidance and tests of ground-level seats. **100L** rebuilt 67 with new wing scaled from projected T-4/100, tested eight shapes [100L-1 thru -8] to 69. **L02-10** mod 78 for powerful direct side-force control with large powered vert fins (cropped delta pivoted at mid-chord with anti-flutter mass at 70% span) above and below nose, camera pod at top of fin to record lateral tracking of ground and air targets; flown 79, upper fin quickly removed and tested to 81.

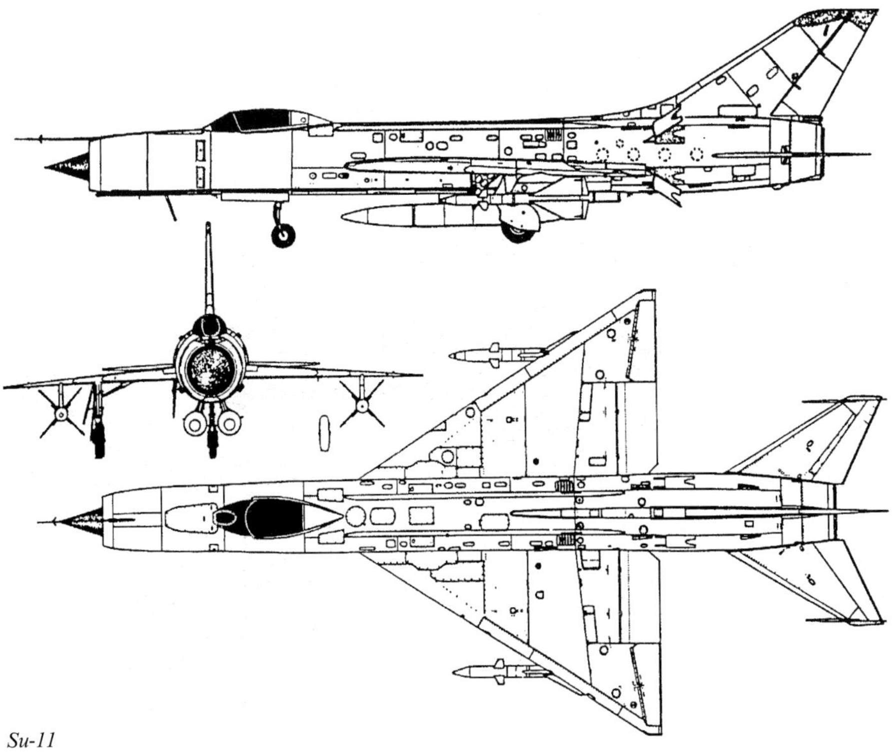

Su-11

Su-11, T-47 Designation reused. Interceptor with airframe based on Su-9 but uprated engine and *Uragan*-5B (T3-8M) complex for interception range 5-23 km. *Orel* radar in larger centre-body, thus almost untapered (and lengthened) fwd fuselage. Single pylons under each wing for R-8 missiles, usually one 8R and one 8T, later 8RM and 8TM. Pylons carry missiles further back than on Su-9, preserving directional stability without enlarging fin. Pipe fairing along each side of top of fuselage, air-data boom shortened to leave overall length unchanged. Radar 900 kg so nose gear strengthened, wheel/tyre larger. Mid-fuselage sealed against gas ingestion. Avionics upgraded including AP-39 autopilot, ILS, *Sirena*-3 warning and new beacon receiver. **T-47** flown by Evgenii Kukushyev late 58.

Su-11-8M accepted with T3-8M complex Feb 62, c100 built (3 regts), curtailed late 62 because of poor handling. ASCC 'Fishpot-C'. Most converted as **Su-11RM** targets. Trainer version not built in series.

DIMENSIONS Span 8,536 m (28 ft 0 in); length 18 225 (59 ft 9½ in); wing area 34 m² as before.
ENGINE One AL-7F-2.
WEIGHTS Empty 8562 kg (18,876 lb); loaded 12 674 (27,941 lb), max 13 986 kg (30,833 lb).
PERFORMANCE Max speed 1100 km/h at SL, 2340 km/h (1,454 mph, M 2·2) at 11 km; service ceiling 18,0 km (59,055 ft); range with weapons (2 tanks) 1710 km (1,063 miles); TO/ldg c1,4 km.

Su-11

Su-15, T-58 In 60 PVO issued reqt for interceptor radar offering enchanced detection range, matched to longer-range missiles, plus better resistance to ECM. This demanded antenna of greatest possible diameter, and OKB led by G. Kunyavskii produced *Orel-2* with 950-mm antenna. To accommodate this, aircraft needed lateral inlets (T-49 data) leading to twin engines (T-5 data). Flight performance had to be better than Su-11, which was deficient in range and ceiling. OKB designation **T-58D**, suffix=*Dva* (twin engines), leading engineer R. G. Yarmarkov. Design load factor 5 retained for all versions. Wing and other features based on Su-11, but with blown flaps (inoperative, see later, and reduced in span because increase in fuselage width greater than increase in wing span). Propulsion as T-5, but ducts fed from variable inlets of rectangular section, canted outwards and raked back at Mach angle in plan, with splitter plate and vertical boundary-

T-58D

T-58D-3

layer exit slits upstream of variable inner wall, removed flow escaping through gap between inlet and fuselage. Full-depth suction-relief door in outer wall mid-way between inlet and wing. Wing with LE at 60°, ducts passing over it making inward sweep to fit original narrow body. Fences added ahead of ailerons, four rear-fuselage airbrakes larger than T-5, upper pair scabbed on externally. Main landing gears uprated, water-cooled brakes, tyres 880 mm × 230 mm, wheels housed partly in wider fuselage, nose gear taller requiring longer bay with bulged doors for braked wheel with 660 mm × 200 mm tyre. Large cooling-air inlet above each afterburner, several other projecting inlets. Internal fuel increased to 4990 kg (6100 lit, 1,342 gal), provision for two 600-lit drop tanks under fuselage. Fin modified structurally, slab tailplanes 55° at ¼-chord moved to extreme tail (projecting behind nozzles) with 9° anhedral. Cruciform 25 m^2 parabrake under rear fuselage.

T-58D-1 First prototype built with inert mass inside nose painted black with sloping aft edge; OKB in parallel tested *Vikhr*-P and *Smerch*-AS radars. PVD boom on tip of nose. Flown by Ilyushin 30 May 62; fin then slightly enlarged and parabrake moved to tube below rudder. **T-58D-2** (No '32') flown 4 May 63, added RP-15 *Orel*-D58 radar with antenna behind longer nose with wider conical radome, autopilot, ILS and subsystem for two K-8M (R-98) missiles carried on PU-1-8 pylons well back under wing. These formed Su-15-98 *Kompleks*, derived from 11-8M except able to attack from front hemisphere and over widened height-band and to 35% greater range, and with security of duplicated aircraft systems. D-2 only: lengthened twin-wheel nose gear with blister in each door. **D-3** flown 2 Oct 63, corrected undesirably narrow centre fuselage by straightening sides; this increased fuel to 6860 lit (1,509 gal) plus 50 lit water. NII testing by S. Lavrentyev, L. Peterin, V. Petrov and PVO Marshal Ye. Savitskii.

Su-15 Series aircraft T-58-98 had to follow Yak-28P at GAZ-153 Novosibirsk, so first (#0015301) not flown until March 66. Each engine with two hyd pumps for four systems 215 kg/cm² for flight controls (BU-49 power units), gear, flaps, airbrakes, UVD-58M inlets, nozzles and radar antenna. Compressed air (280 kg/cm²) main/emergency brakes, emergency gear/flaps, hyd-tank pressn. ECS for cockpit and avionics, canopy demist, pressure-suit ventiln. Three fuselage and two wing tanks as before, plus two BDZ-59FK pylons for 600-lit drop tanks. Tyre press increased, KS-4 seat (H=0, V=140+ km/h), RSIU-5B (R-802V) radio, MRP-56P marker, RV-UM radar-alt, ARK-10 ADF, SOD-57M transponder, *Lazur* ARL-S ground guidance receiver, SRZO-2M IFF, *Sirena*-2 RWR, KSI-5 Tacan, RSBN-5S short-range/landing guidance, AGD-1 horizon. Weapon complex, *Orel*-D58, later replaced by D58M (better ECCM) two R-98RM or 98TM.

In PVO service April 67, certified using *Lazur* guidance for targets 500 km/h-3000 km/h at 0,5 km-23 km alt. Ten at airshow 9 July 67. Good weapon complex and handling, but inability of engine to deliver adequate air precluded use of flap-blowing, limiting depression to 25° and resulting in long TO/ldg. OKB developing new wing with extended outer panels, flown 66 on 0015301 which also tested probe/drogue refuelling. ASCC name 'Flagon-A'.

DIMENSIONS Span 8,616 m (28 ft 3¼ in); length (excl PVD) 20,54 m (67 ft 4⅝ in), (incl PVD) 21,44 m (70 ft 4 in); wing area 34,56 m² (372 ft²).

ENGINES Two R-11F2S-300.

WEIGHTS Empty 10,22 t (22,531 lb); normal loaded 16,52 t (36,420 lb), MTO 17 094 (37,685 lb).

PERFORMANCE Max speed (SL) 1200 km/h, (11 km) 2230 km/h (1,386 mph, M 2·1); service ceiling 18,5 km (61,000 ft); combat range 1260 km (783 miles), with drop tanks 1550 km (963 miles); TO 395 km/h/1300 m; ldg 320 km/h/1100 m.

T-58VD T-58D-1 rebuilt for STOL research to support T6-1. Wing redesigned outboard of fence with LE reduced to 45°, span increased with extended aileron ending inboard of broad square tip. Main-engine ducts straight, leaving room in centre for three RD-36-35 lift jets, one in forward bay and two in tandem to rear, each bay with louvred upward-hinged door at top and pilot-controlled cascade at bottom to vector thrust fore/aft. No military eqpt, telemetry antenna under forward fuselage. TO/ldg 240 km/h-290 km/h, 480 m-500 m. Flown 66, demo by Ye. Soloviev 9 July 67, later tested ogival radome and UPAZ IFR pod. 'Flagon-B'.

Su-15 Without change in designation, at 11th series block outer wing LE sweep reduced to 45°, extended in chord and drooped, linked by 0,5 m unswept section centred on fence with anhedral to join two different LE levels. Same radar and engines. 'Flagon-D'.

DIMENSIONS Span 9,34 m (30 ft 7¾ in); length unchanged; wing area 36,6 m² (394 ft²).

ENGINES Two R-11F2SU-300.

WEIGHTS Empty 10 350 kg (22,817 lb); normal

Su-15

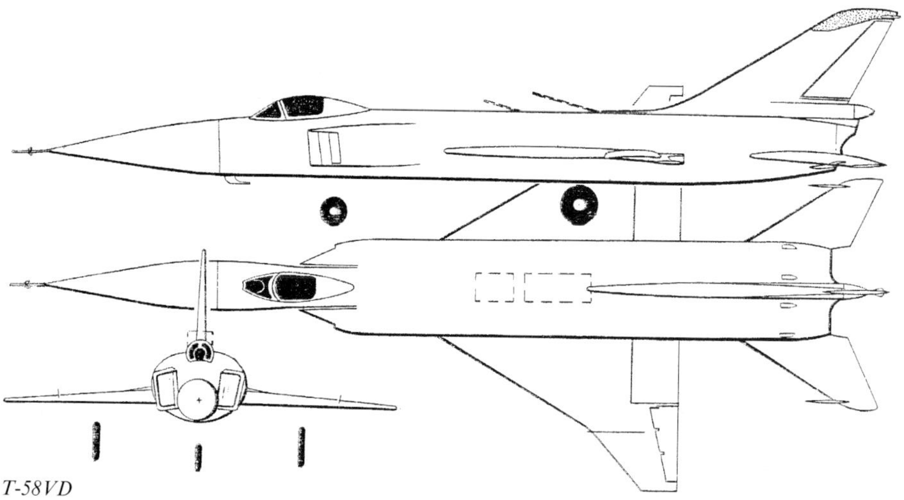

T-58VD

loaded 16 700 kg (36,817 lb).

PERFORMANCE As before except combat range 1305 km (811 miles); ferry 1600 km; TO 350 km/h/950 m; ldg 285 km/h/850 m.

Su-15L T-58D-2 rebuilt for research into operations from snow and surfaces softer than 8 kg/m². Large main skis, nose gear lengthened 350 mm to raise inlets and fitted with twin hyd-steered but unbraked KN-9 wheels with 620 mm × 180 mm tyres. New nose gear adopted.

Su-15T Prototype **Su-58T** flown 69, designation from RP-26 *Taifun* radar based on MiG-25 *Smerch* but in smaller radome. New engine reqd enlarged inlets and improved afterburner cooling eliminating large intakes each side of fin; provided air for blown flaps so angle cleared to 45°. Tailplane anhedral reduced. Stronger main gears with new tyres, heavy radar required new nose gear as tested on D-2. Radar antenna driven electrically so hyd systs

T-58VD

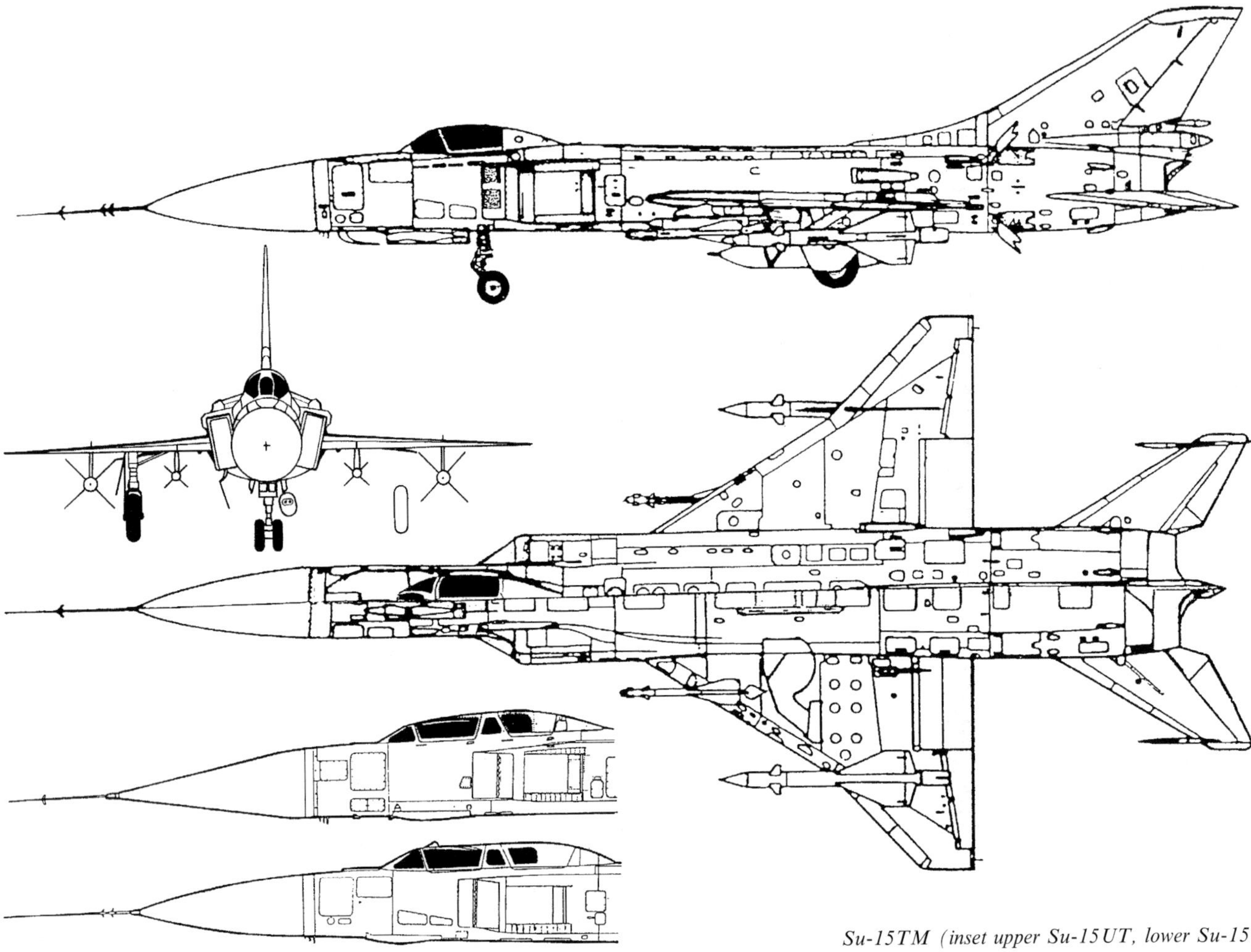

Su-15TM (inset upper Su-15UT, lower Su-15UM)

Su-15UT

reduced to three but each engine driving dual pumps and elec generators. Flight-control power units BU-49 replaced by BU-220 and BU-250; SAU-58 autopilot, radio R-832M (*Evkalipt*-SM), blind-landing RSBN-6S (*Iskra*-K), radar-alt RV-5, new ARL-SM (*Lazur*-SM), new warning SPO-10 (*Sirena*-3) and *Pion*-GT nose probe. Cleared to fire R-98M with revised guidance. Production begun 70 but soon halted because radar defective. Data little changed except engine R-13-300. 'Flagon-E'.

Su-15UT Tandem trainer able to convert pilot or backseater. Instructor cockpit added by extending fuselage 450 mm and reducing size of No 1 tank (total internal 5010 kg, 6250 lit), two upward-hinged canopies as Su-9U but smoother profile, pupil canopy longer, instructor canopy periscope. First block only, original wing and nose gear. Weapon complex deleted, but missile-firing simulation system added, plus weighted dummy missiles, jettisonable in emergency. Prototype **U-58T** Oct 69, prodn 70-72. 'Flagon-C'.

DIMENSIONS As Su-15 except length (excl PVD) 20,99 m (68 ft 10⅜ in).

ENGINES Two R-11F2S-300.

WEIGHTS Empty 10 750 kg (23,699 lb); loaded (2 dummy R-98) 16 690 kg (36,795 lb); max 17,2 t (37,919 lb).

PERFORMANCE Max speed 1200 km/h at SL, 1850 km/h at 15 km; service ceiling 16,7 km (54,790 ft); range 1700 km (1,056 miles); TO 400 km/h/1200 m; ldg 335 km/h/1200 m.

Su-15TM Major series version, with improved TM radar. To reduce drag and improve lateral control at low speeds, wing cleaned up with simple change in sweep on LE without droop. Kremlin decision 21 Jan 75 to adopt Su-15-

Su-15TM with UPK-23-250 pods

T-4 being built at Tushino

98M *kompleks*, linked via SAU-58 with *Vozdukh-1M* for fully auto ground guidance, using *Taifun*-M radar with greater transmitter power; internal reflections demanded (from 8th a/c) ogival radome which increased drag. Unable to 'look down', but from 78 SAU-58-2 linked with RV-5 made possible auto engagement of targets at 200 m; with *Iskra-K* cleared for blind landing to 50 m over threshold. PD-62 pylons added under inner wings with twin rail interface for R-60, max missile fit two R-98M (or R-98/R-8M-1/R-8) and two R-60T and/or R-60MK. All earlier versions in PVO service thus retrofitted. Firing parameters (front) 500 km/h-2500 km/h, 2 km-21 km; (rear) 500 km/h-1600 km/h, 0,5 km-24 km; max ht difference to target 9 km, or 4 km for collision course (1 km with R-60). Influenced by addition of gun to F-4 Phantom, tested internal GP-9 on left behind nose gear. Abandoned, replaced by modifying BDZ-59FK pylons to carry UPK-23-250 pods, aimed like R-60 by K-10T sight. Tested with two FAB-500, 4 × FAB-250, two UB-16 or 32, two S-24 or two ZB-500 tanks, though sight unsuitable for surface targets. In service 72, shot down KAL 707 20 April 78 and KAL 747 1 Sept 83. 'Flagon-F'.

DIMENSIONS, ENGINES As Su-15T.

WEIGHTS Empty 10 874 kg (23,973 lb); loaded (2 R-98M) 17 194 kg (37,906 lb); max (1st series) 17 660, (2 R-98, 2 R-60, 2 PTB) 17,9 t (39,462 lb).

PERFORMANCE Max speed 1300 km/h (808 mph, M 1·06) at SL, 2230 km/h at 13 km; service ceiling 18,1 m (conical radome 18,6 m); range 1380 km, ferry 1700 km (1,056 miles); TO 370 km/h/1000 m; ldg 290 km/h/900 m.

U-58UB Prototype dual trainer with TM radar and weapons; overloaded and abandoned.

Su-15UM Dual trainer version of Su-15TM, original-length fuselage, rear cockpit with modified canopy replacing No 1 tank, pupil with rear-view mirror. Combat trainer, no radar but R-98MT/R-60/UPK-23-250. Single example tested 76, small prodn run. 'Flagon-G'.

DIMENSIONS, ENGINES As TM.

WEIGHTS Empty 10 635 kg (23,446 lb); loaded (2 R-98) 17 200 kg (37,919 lb); max (2 PTB, 2 R-98, 2 R-60) 17,9 t (39,462 lb).

PERFORMANCE Max speed 1250 km/h at SL, 1875 km/h (1,165 mph) at 11,5 km; service ceiling 15,5 m (50,850 ft); range 1150 km (715 miles); TO/ldg as TM.

Despite poor serviceability, popular aircraft. Total for all variants 1,400; APVO regimental strength 550 to 88, 230 to 90. Then mass scrapping, few remain.

Su-15bis Prototype tested 73 with larger wing and new engines, ASK-T58 auto flight control, stressed for increased speed, higher ceiling, longer range for border defence. Recommended for prodn, but MiG-23/25 'look down' radars preferred so no production.

DIMENSIONS Span 10,52 m (34 ft 6⅛ in); length 21,95 m (72 ft 0 in); wing area 35,78 m² (385 ft²).

ENGINES Two R-25-300.

WEIGHTS Empty 12,7 t (27,998 lb); loaded 18 t (39,683 lb), (max) 19,3 t (42,550 lb).

PERFORMANCE Max speed (SL) 1400 km/h, (10 km) 2120 km/h (1,317 mph, M 2·0); max climb 175 m/sec (34,500 ft/min); service ceiling 18,2 m (59,700 ft); combat radius (hi) 1100 km (684 miles); field length 2 km.

T-58M Projected interceptor to replace Su-15, active 67-71, large delta, two AL-21F-3, *Smerch*-100M radar fire-control, four K-50 or K-100 missiles. Cancelled, neither missile placed in production, but wing basis of T-6.

T-4, 100 Challenging aircraft which pioneered Ti structure, FBW flight controls and was (Simonov) 'father of Su-27'. Dec 62 requirement was for Mach-3 interceptor to destroy B-70, 'Blackbird' and Hound Dog and Blue Steel missiles. At early stage primary mission changed to carriage of air-to-surface rockets. Gigantic technical challenge, not helped by political argument. Competition organised between Mya, Su and Yak, widespread view in MinAviaProm that result would justify abandoning project. Argument between cruise 2000 km/h/2300 km/h (aluminium) or 3000 km/h (ti/steel). Su proposed 3000 km/h, requiring structure to soak at 300°C, but higher propulsive efficiency and other advantages, plus highly innovative design, resulted in Su proposal being chosen. Further prolonged argument, finally won with support of V. Ryabikov, Gosplan 1st Deputy, but opponents got project allocated to GAZ named for La. These facilities were then given other work, and **T-4** (delta 4) reassigned to Buryevestnik design office and Tushino fac-

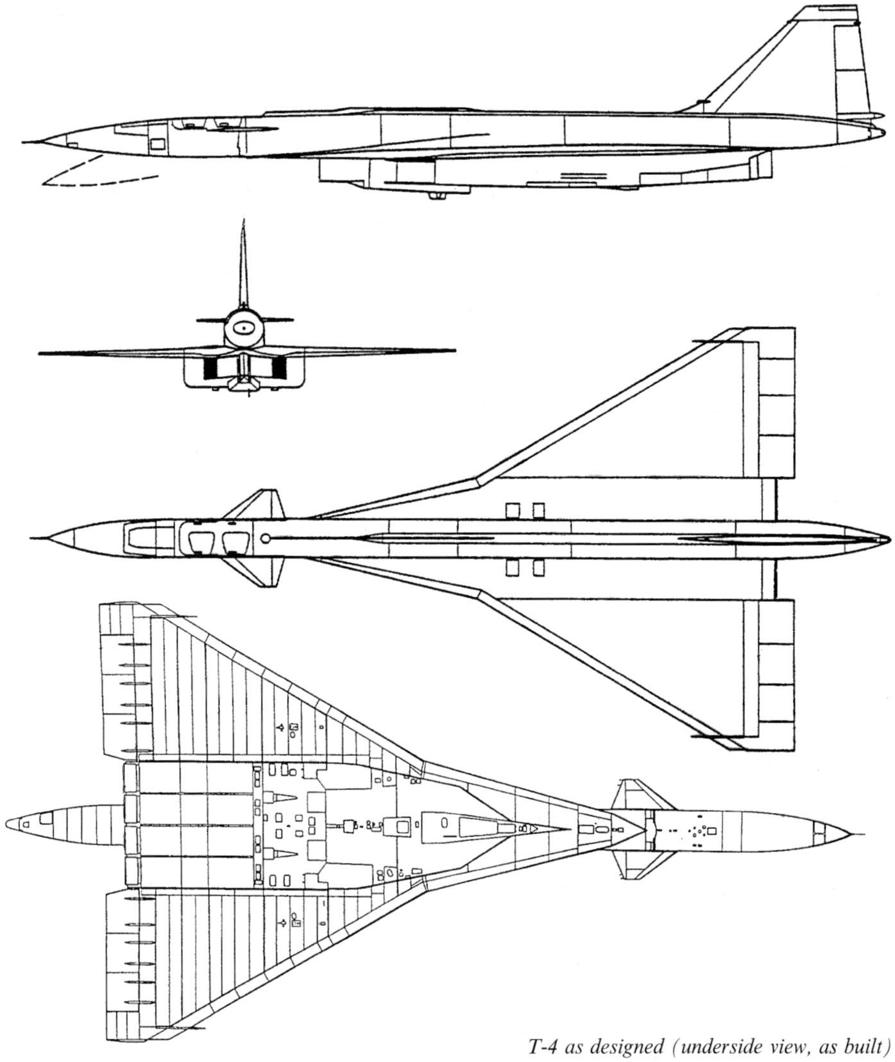

T-4 as designed (underside view, as built)

T-4 No 1

tory. Impossible situation with tenuous links between Su team and hostile partners who even went to Party Central Committee to get project abandoned. Gradually links improved and Tushino works realised technological benefits, in titanium fabrication, auto welding and electronics.

Chief designer N. Chernyakov recalled 'Required range and speed determined weight, and project dubbed *Sotka* (100) from estimate of 100 tonnes. Nine-year schedule reflected degree of innovation close to 100%'. About 20 configurations tunnel tested at CAHI, and explored with modified Su-7 and Su-9 testbeds, finally choosing same as XB-70 but with single fin. Wing initially pure delta, LE 77° inboard, 60° to tip, sharp LE t/c 3%, symmetric, dihedral 0°. Final design with true lifting section and TE elevons only outboard of engine box, four elevons on each wing, outermost with broad square tip forming giant horn balance. Longitudinal stability always positive, flapped foreplane increased stability 7% (to +2% subsonic and 3-5% supersonic) whilst reducing elevon hinge moments and obviating need for elevator deflection in cruise, drag further reduced by using fore/aft fuel shift to trim out CP migration. Minimal required yaw stability from fin, LE sweep 48°, and two-section powered rudder. For first time in USSR complete FBW (quadruplex, no mechanical backup) for flight control, engine and inlet control and nosewheel steering. FBW provided artificial stability augmentation. Major role played by c.t.p. Col (later Gen) S. V. Ilyushin, who recalled struggle to get stick adopted instead of wheel.

Three airframes funded, multispar wing akin to XB-70 but mainly welded Ti sheet, little machining and manufacture highly automated. Oval, then circular-section fuselage, diam 2 m, drooping nose. Pilot cockpit with large vertical front windows shut off by nose in cruise, all flying then on instruments. Engine matched to Mach-3 cruise, installation broadly similar to XB-70 but two engines each side of centreline. Sharp vertical central splitter with hinged duct walls to vary angle and throat area and extract boundary layer through perforations. Each duct splayed out to leave central nose-gear bay, then in again round main-gear bay before dividing at entry to stainless-steel engine compartments for four engines all parallel and close together. Innovative landing gears, high-strength steel struts, main unit with near-vertical twin-chamber shock absorbers with safety valve carrying pivoted truck with four twin-tyred wheels folding forward and rotating 90° to lie in vertical bay between engine duct and skin, with single ram for both retraction and locking; nose gear with two similar tyres with FBW steering and hydraulic retraction to rear into twin-door bay between ducts. Track 5,88 m, wheelbase 10,3 m. Twin drag chutes in tail compartment.

New standard set in avionics and computer control, among other things reducing crew to pilot and nav/systems operator, wearing pressure suits. Tandem upward ejection seats of special design incorporating windbreaks. Entry via ladder to upward-opening roof hatches on left side hinged near centreline. Small window of T2-55 glass in each hatch and on opposite wall. Difficulty in finding space for trad instruments and controls. ECS for cockpit and avionics by bleed air-cycle machines dumping heat by evaporation and by fuel/air radiators. Fuel occupying almost whole fuselage, special trim tanks at nose/tail with automatic Mach-controlled transfer. Innovative system filled with thermostable RG-1 (Natfil) fuel fed by air-turbine booster pumps, liquid N_2 purging of tanks and emergency high-rate jettison through large pipes in dorsal spine. Duplex twin-circuit hydraulics filled with thermostable KhS-1 (similar US Oronite 70) operating at 280 kg/cm² (3,980 lb/sq in), using pipes of VNS-2 alloy with soldered joints. This reduced size of power units, enabling twin drives to each rudder and elevon to be housed without bulges. Another Soviet first was generation of all electrical power by four constant-speed alternators giving stabilized frequency, DC being supplied by rectifiers. Powerful forward-looking radar for fire-control, targeting, nav and recon, plus recon SLAR and FLIR (another first) and optical sensors. Primary nav system astro-inertial. Intention to provide comprehensive EW systems. Armament in role finally adopted was to have been two ALBMs with solid-propellant rocket propulsion, preprogrammed dogleg flight and terminal homing. OKB said of project '200 new inventions, or c600 including manufacturing', but military eqpt never fitted.

First aircraft experienced cracks, which were inspected, drilled and left. No 2 had modified airframe with less-brittle alloys and composites, but this never flew. No 1 made first flight from Zhukovskii 22 Aug 72, pilot S. V. Ilyushin, observer A. N. Alferov. Deputy Project Manager Vladimir Yakovlev, 'I recall how the ground shook with the noise of its engines . . . it nodded its drooping snoot menacingly . . . it did not fly off but rather leaped into the clouds . . .' Ten successful flights, then in 82 to Monino, two others scrapped.

DIMENSIONS Span 22 m (72 ft 2⅛ in); length 44,5 m (146 ft 0 in); wing area 295,7 m² (3,183 ft²).

ENGINES Four RD-36-41.

WEIGHTS Empty 55,6 t (122,575 lb); internal fuel c62 t; loaded (no weapons) 114,4 t (252,205 lb), (intended max, plus external fuel) 135 t (297,620 lb).

PERFORMANCE Max speed 3200 km/h (1,988 mph, Mach 3·01), cruise (20 km) 3000 km/h (1,864 mph, Mach 2·82); range at M 2·82 4000 km (2,486 miles), (plus external fuel) 7000 km (4,350 miles); takeoff 1050 m; ldg (with chute) 900 m.

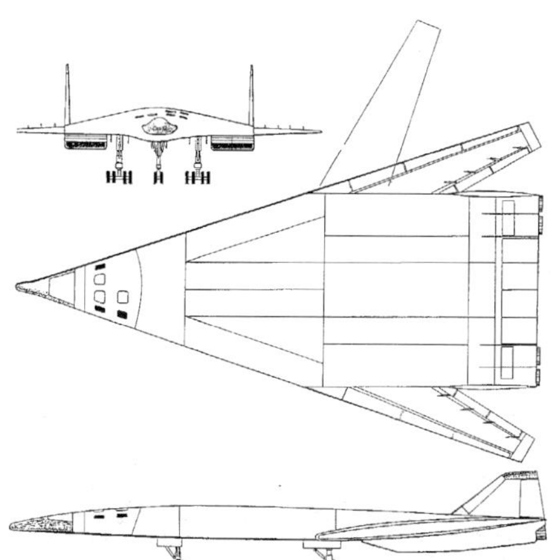
T-4MS/200

T-4MS, 200 In 70 KB won competition for even more advanced project for strike/recon aircraft. Strategic range cruising at M3+, ability to use existing runways. Answer was a totally integrated shape with no separate wing or fuselage, though small wings with T6-type hi-lift devices pivoted on sides. LE angles 72° and 30°. Cruise flight control by four elevons between underslung twin-engine nacelles with horizontal (Concorde-type) wedge inlets. Ceramic nose, three-seat cockpit, landing gears as 3-view. No money to construct, but influenced integrated body/wing of T-10.

DIMENSIONS Span (72°) 25 m (82 ft), (30°) 38,75 m (127 ft); length 39 m (128 ft).

ENGINES Four RD-36-41 or NK-32.

WEIGHTS Loaded (est) 140 t.

PERFORMANCE Range 7000 km at M 3·5.

T6-1 OKB picked 61 to design aircraft to replace Il-28 and Yak-28B, initially in lo attack mission, later in recon and EW. Requirements included: supersonic speed; ability to disperse away from airfields to hidden unpaved sites; maximum combat radius; day/night operation in all weather; max bombload (greater than predecessors); ability to make blind first-pass attack on point target; and max self-protection and invulnerability to ground fire including SAM warheads. OKB's largest-ever team formed under P. O. Sukhoi and deputy Yevgenii S. Felsner, with Vladimir Konokov flight-devt engineer, Mikhail P. Simonov deputy chief designer, Ye. Zazorin deputy chief engineer (eqpt) responsible for integrating 13 major avionics systems, and V. Yakovlev chief of scientific development.

Discussion with CAHI centred around two solutions: broad but short-span delta wing similar to British TSR.2 or VG wing similar to American TFX. In either case extreme STOL was a requirement; this was studied with central battery of lift jets. Final choice was TSR.2 layout but with four lift jets; MTO field length 300 m. **T6-1** lead engineer V. M. Torchinskii, lift-jet installn under test-pilot Yevgenii S. Solov'yev.

Wing aerodynamically close to TSR.2, with similar profile, fixed 60° LE, 4° anhedral, later modified with kink to reduce outboard sweep to 45°, ahead of conventional ailerons. Structurally three main spars at 90° to centreline, skins all machined panels, main interspar box integral tank. Wing mounted high (only spine projecting above). Forward fuselage of oval section housing radar (full equipment absent from T6-1) and many other avionics LRUs. After much discussion, side-by-side seating for pilot and right-seat navigator, main reason for choice ability to fit instructor flight controls into standard aircraft, with ideal crew communication, obviating need for trainer version. K-36D seats, left and right canopies individually hinged upwards from rear fairing into very broad spine. Tail controls and hydraulics routed far apart along sides of spine, and duplicated 210 kg/cm² hydraulics with each independent system energized by pumps on both engines. Downstream of inlets fuselage of broad box-like profile with sharp right-angled corners to give maximum lift in transonic attack. Tail again

T6-1 as built

influenced by TSR.2 with slab tailerons just below mid level serving as all-speed roll/pitch control, LE 55°, inboard TE cut back, driven by KAU-125 power units. Fin LE swept 60°, one-piece rudder driven by RP-280 power unit. No tabs or spoilers on aircraft.

Chosen engines R-27F-300, but devt delay forced switch to AL-21. Sharp-edged lateral inlets with vertical inner splitter plate standing 0,15 m away from fuselage wall, top/bottom lips swept 45°, inner wall variable in profile and throat area, tall but narrow rectangular duct changing to circle downstream of aux suck-in relief door ahead of wing. Aux overboard dump doors downstream in top of duct with projecting actuating arcs. Ducts parallel (though aircraft complied with Area Rule) leading to widely separated engines with nozzles inside rectangular box-like rear fuselage giving flat vert sides adjacent to tailerons and with cooling inlets across full width at top. Between ducts Nos 1 and 2 tanks and, in centre of aircraft, four lift jets in tandem fireproof bays each with upward-hinged louvred door; fore/aft thrust deflectors in underside. Levered-suspension main gears hinged to main spar body frame with twin wheels, LP tyres 900 mm × 230 mm, track 3,15 m, retracting forwards into bay covered by tandem doors hinged on outer side of fuselage, open on ground, wheels stowed upright outboard of duct. Steerable levered-suspension nose gear with twin wheels, tyres 600 mm × 200 mm with mudguard, retracting rearwards into bay under cockpit with L/R doors open on ground. Wheelbase 8,51 m, all wheels with anti-skid brakes. Cruciform braking parachute in rear fuselage on centreline. Four large wing pylons, outers projecting ahead of LE, two fuselage hardpoints inboard of main-gear bays, retractable landing light each side of forward fuselage, PVD-7/*Pion* GT nose antenna system, ASO-2 chaff/flare dispenser recessed above each inlet. Major effort to develop combat invulnerability and advanced avionics (for first time considered as PNK, attack and navigation complex) superimposed on basic problem of whether configuration was optimum.

First flight by Ilyushin June 67. Aircraft promising in full-throttle lo attack mode, but aerodynamic and control difficulties caused by lift engines, esp at low speeds. No bleed system for reaction-jet controls, and problems encountered by Ilyushin and Solov'yev (later killed in T-10). Mods included turning wingtips down 72° (as in TSR.2), with plan to terminate in

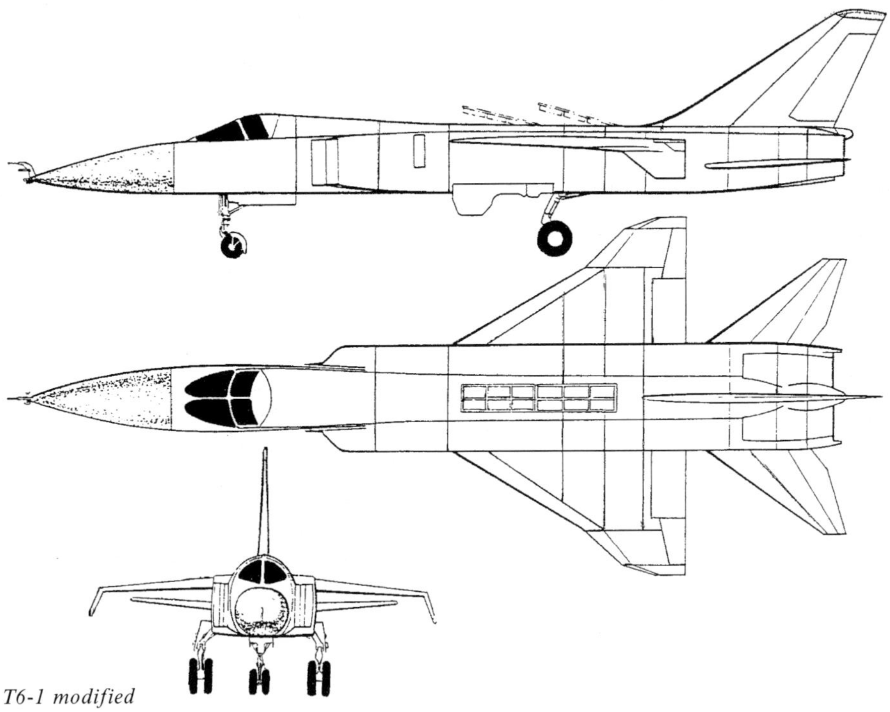

T6-1 modified

T6-1 modified

twin rails for self-defence missiles. Boundary-layer discharge jets added under inlets, with antenna fairings along outer sides of forward fuselage. Penalty of lift engines, and British replacement of TSR.2 by F-111, prompted change to F-111 arrangement of VG wings. T6-1 parked at Monino.
DIMENSIONS Span 9,2 m (30 ft 2¼ in); length 23,2 m (76 ft 1⅜ in); wing area 51 m^2 (550 ft^2).
ENGINES Two AL-21F, four RD-36-35.
WEIGHTS Not known, but max c28 t (62,000 lb).
PERFORMANCE Max speed (SL) M 1·2, (hi) M 1·9; field length c300 m.

T6-21, Su-24 Decision to redesign Dec 69. Decision rested primarily on impossibility of accommodating lift engines and still retaining desired low drag, internal fuel capacity and good handling at all speeds. Prime requirement was ability to attack at transonic speed at SL, and this was incompatible with large wing needed for STOL, because of unacceptable ride quality. Sukhoi/Felsner driven 68 to conclusion only poss answer was VG wing, with small fixed glove portion. Thus, span and area at max sweep would be minimum, essential for smooth ride during attack. Prolonged research at OKB and CAHI confirmed horizontal tails immediately aft of wing TE and at same level, requiring redesign of tailerons and rear fuselage, plus ventral fins. Rectangular cross-section retained on aft body, underfins as far apart as poss, set at 80°. Outer wings near-symmetric profile varying 8·5% to tip 10·5%, with 4°30' anhedral retained plus washout (negative twist towards tip). Four locked settings: TO/ldg, LE 16°; cruise, 35°; high-subsonic evasive action, 45°; Mach 0·9-1·35, 69°. At 16° high-lift from four sections of full-span LE slat (total 3,036 m^2) and three sections of double-slotted flap (total 10,21 m^2) on each wing, driven by hyd actuators, ballscrews and travelling nuts to 27° (slats) or 34° (flaps). Lateral control by delta tailerons (13,707 m^2 total), ¼-chord sweep 55° pivoted at top of fuselage and driven +11°/–25° by KAU-125, supplemented by two sections of spoiler (total 3,063 m^2) on each wing opened to 43° by screw/nut drives except with wing at 69°. Spoilers double as lift dumpers after landing.

Wing pivots and many other parts 30 KhGSNA or VNS-5 steel, others plastically deformed/diffusion bonded, integrally machined or composite. Small wing glove, LE 69°, ribs perp to LE. Fin 55°, 9,234 m^2, rudder 1,537 m^2 driven ±24°. Fuselage rectangular section for lift, edges now rounded. Canopies split wider on upward opening; T-6 escape required nav to fire seat, followed by pilot; simultaneous ejection from 72 with K-36DM zero/zero seats linked to canopy locking/jettison for ejection initiated by either crew member with very brief delay on one seat. Polycarbonate L/R windscreens meeting more severe birdstrike requirement. Downstream similar wide spine housing equipment as listed later. Variable inlets unchanged, ducts curving in to become circular tubes straight to engines. Fuel 11 860 lit (9850 kg, 2,609 gal) in No 1 tank between inlets, No 2 filling centre fuselage and (aft of MLG bays) around ducts, No 3 behind wing bridge and No 3A between engines. Piping down centreline for single 2000-lit (440-gal) drop tank and through each glove to fixed pylon centred on wing pivot, extended around LE as deep fence, for 3000-lit (660-gal)

T6-1 as built

T6-21

T6-21 on bomb trials

drop tank. Engines as before, improved dorsal cooling inlets, better grouping of accessories on underside, engines withdrawn on rails to rear. Landing gear strengthened for increased gross weight and redesigned with wheel truck rotating 90° to lie flat in bottom of fuselage in new main-gear bay closed by door hinged forwards with separately controlled ram to serve as air-brake. Track 3,31 m, wheelbase 8,51 m, main tyres 950 mm × 300 mm, nose 660 mm × 200 mm. Cruciform 50 m^2 braking parachute in tube under rudder with L/R doors, no hook.

T6-21 lead engineer R. G. Yarmarkov. Because interim aircraft, vulnerability, avionics, weapons discussed later. ASCC 'Fencer-A'.

First flight by Ilyushin 17 May 70. T6-21 fitted with all systems as at that time agreed, including terrain-avoidance radar, digital avionics including two main and various subsidiary computers, and integrated nav/weapon subsystems integrated by teams under Simonov, with cooling inlets plus an avionics heat-exchanger amidships in spine with air duct to right of centre.

Internal GSh-6-23 with 500 rounds, nine stores pylons, max load 8 t, bombs to 1500 kg, rockets 57 mm to 370 mm, all tac missiles plus air/air. Forward centreline equipped for 2000-lit tank or SSPU-6 gun pod (GSh-23/400 rds), glove pylons for 3000-lit tank or SSPU-6, outer wings (Nos 1 and 9) pivoted and linked mechanically to glove to remain aligned with slipstream. Ten devt aircraft made over 3,000 flights by Ilyushin, Solov'yev, V. Krechetov, S. Lavrentyev (like Solov'yev, killed in different aircraft) and N. Rukhlyadko, and navigators N. Alferov, V. Byelykh, L. Smyshlayev and L. Rudyenko. Not one ejection or serious damage, but numerous mods: redesign of wing pivots (225 diam in 55 mm thicknesses) and bridge (200 kg added), tight-fitting rear fuselage around afterburners giving better explosion protection and reduced base drag, full-width upper cooling inlets removed. All done at Komsomolsk, where full production authorized 72 as **Su-24**. ASCC name 'Fencer-A'. Addition of fairing for twin 25 m^2 brake chutes at base of fin, cooling intakes at base of fin and above spine led to allocation of 'Fencer-B' reporting name.

Special features for autonomous safe operation. Structure designed for 6·5 g at max weight with wings in any setting. Over 500 shots of calibre to 30 mm, including SAM warheads at proximity distance, fired at prototypes, components, mock-ups and production aircraft. Primary reason for redesign of wing bridge was to ensure strength after impact of directed splinters from modern SAM warhead. Wing skins and fuselage walls fabricated as large machined panels with integral stiffening and fewest joints. Outer wings with four spars, flight continued with any two severed. Ability of dense-packed avionics and equipment to arrest shells and warhead splinters reduced cockpit armour needed. Flight-control rods widely separated in spine and surrounded by dense equipment, with duplicate hyd piping widely separated and served by twin pumps on each engine; duplicate cylinders in each power unit with jamproof design, with pitch and roll backed up by elec signal channels and autonomous operation of separate spoiler sections. Two engines wide apart, separated by axial firewall, each bay divided by transverse fireproof bulkheads with own fire warning and extinguishing system; armoured inlet ducts and SUNA flameout prevention system with BPS-89P unit to maintain stable engine operation after heat/gas/blast waves from warhead. Fuel tanks withstand hydraulic shock from multiple splinter/shell hits, with inert-gas pressurization and gravity feed following pump failure. Fly-by-wire control of engines by either crew-member.

Series aircraft equipped with integrated nav/attack and self-protection systems, repeatedly upgraded as listed later. First major upgrades in 75, no change in designation. Four mods externally obvious: addition of internal active-jammer system (*Geran-F*), dual-wavelength targeting sensor (*Kaira-24*) and extended fin LE for Loran (A-711) antenna; in 76, *Berez* passive RWR (confused by West with *Geran-F*). Temporary change in early series block was to trans-

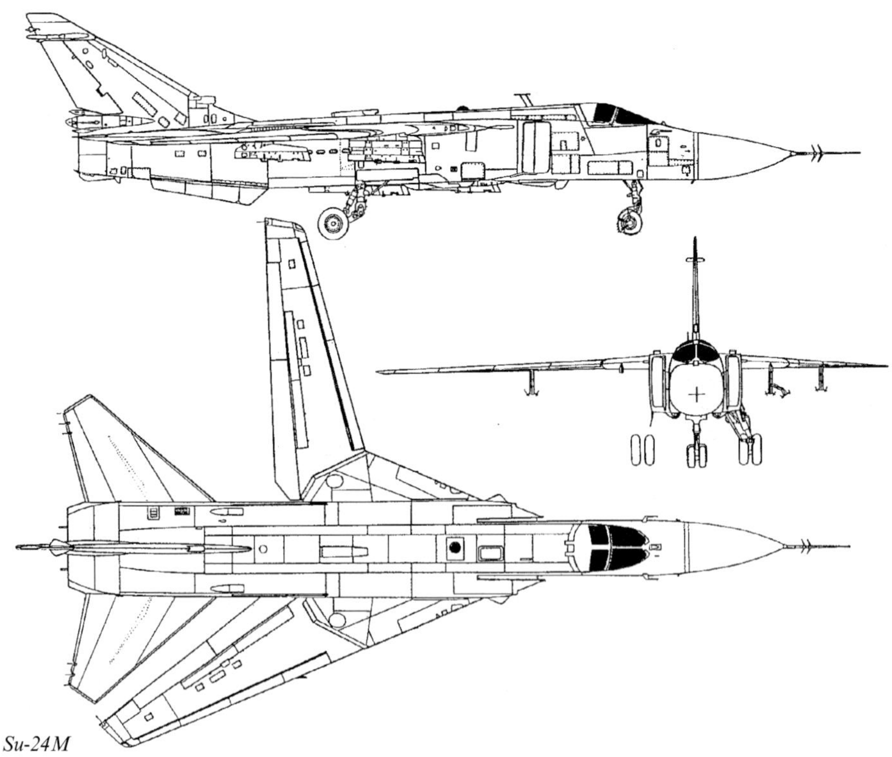

Su-24M

fer twin PVD-7 pitots to above nose probe, giving prong effect. Sustained lo-level operation, so inlets locked in TO position. Addition of triangular antennas on fins and intakes identified by ASCC as 'Fencer-C'.

Early 78 **Su-24M**, definitive series version. Inflight-refuelling probe at last available, mounted on telescopic pipe at 75° and raised by fuel pressure 0,8 m (32 in) from stowed position (no door) ahead of pilot windscreen. Some reports suggest retrofit on previous aircraft, all built with internal probe plumbing, and with pipe to centreline AKU hardpoint for UPAZ-A hose-drum pod converting aircraft to tanker transferring <15,000 lit. Probe only confirmed on Su-24M and subsequent sub-types. Following text outlines main features of 24M, many of which were on earlier production, notably unchanged armament (though stores carried have been improved repeatedly). Prominent blisters under fuselage for GSh-6-23 on right, with 260 rounds, and ASK-5 gun camera on left. Nine stores pylons, of four types with seven interface units, for 57 kinds of store including all tactical bombs and rockets, cluster dispensers, all laser and TV smart weapons (notably KAB-500L and -500KR), and all versions of Kh-23, -25, -29, -31, -58 and -59. Normal self-defence weapon R-60T carried in pair on APU-60-2 interface on No 1 pylon. PrNK-24M nav-attack complex housed in 53 LRUs, mainly in nose, behind and around cockpit and in spine, all designed for front-line test and replacement (high-power emitters to antenna test only, emission possible only after receipt of MLG retract signal or with laser door closed). Basic avionics: *communications*, managed by either crew-member, R-862 *Zhuravl*-K HF, R-864G *Zhuravl*-30 duplicate VHF/UHF, DNM-T6M command link and SPU-9 intercom; *monitoring*, MS-61, CVR and radio traffic recorder, *Tecon* (see offence), *Tester*-UZ Srs II flight-data recorder and MLP14-3 recorder (a sub-complex) recording entire parameters of NK-24M; *navigation*, NK-24M nav complex linking *Orion*-A search and mapping pulse-doppler main radar, *Relyef* terrain-avoidance radar, *Mis*-P Srs 2 inertial platform, *DISS*-7 doppler, RV-21-A1 radar altimeters and *Orbita* TSVU-10-058K central digital computer (all linked through SAU-6M1 auto flight-control

Su-24M

system), plus PPV twin-combiner HUD-sight, UVV *Binom*-TT keyboard (nav only) and VKU *Parus*-T switching unit (nav only); *radio-nav*, RSBN-6s Shoran and *Klystron* blind (50 m ceiling) landing system, with approach mode, ARK-15M auto radio compass (dipole antenna behind cockpit), SO-69 transponder, *Pion* GT-6M-9 (DL-5) antenna feeder (above brake-chute), *Skip*-2 A-711 Loran with A-713 digital co-ord converter using antenna added along LE of fin, and MRP-56P marker rcvr; *offence*, SUO-1-6M armament control system, attack modes with *Relyef* radar, boresighted *Kaira*-24M laser designator and target ranger and day/night TV sighting system, auto operation via NPP horizontal-situation, *Tecon*-1SM video monitor and SAU-6 flight-control or under pilot control via PPV sight; *defence*, BKO-2 *Karpat* integrated system comprising *Berez* (SPO-15S) passive detection and DF system (antennas each side near tip of fin, in spine behind canopies and each side of inlets ahead of LE), *Mak*-UL (LO-82) missile IR warning and DF system (antenna under dome on spine amidships), *Geran*-F (L-101G/102G) active jammer system (antennas under brake-chute, tip of fin above nav light and under nose), *Avtomat* (APP-50A) chaff/flare dispensers (in fairings over wing, extended upwards from pylons on some Su-24Ms or each side above rear fuselage) and Neon-F (L-167) jamming aids controller and energy management (nav only); 24M can also carry *Fantasmagoriya* LO-80/-81 passive receiver pod sensing virtually all hostile emissions in sector ahead, equally part of offensive nav/attack system; (*misc*) SRZO IFF (transponder only, stealth precluding an interrogator), SVS-PN-5-3A Srs III air-data system, PVD-18G-5M Srs II nose probe with duplicate 2-axis transducers, PVD-7 pitot tubes (each side ahead of windscreen) and UUAP-72M-13 AOA vane and g-meter. 'Fencer-D'.

Customer-tailored export versions with various avionics items eliminated (typically 48 LRUs) or replaced by simpler units already possessed by customer. Many enquiries but only sales (mid-94) Iraq 24, Libya 15, Syria 12. **Su-24MK** 'Fencer-D'. Total 24 and 24M/MK 720+.

Su-24MR Dedicated multisensor recon. Airframe as 24M except for removal of gun, magazine and gun camera and replacement by short blisters (see later), removal of three body store stations (hence stations numbered 1-6), replacement of large fence pylons by normal underwing pylons (usually AKU-470), removal of *Kaira*-24 laser/TV and increase in capacity of spine heat exchanger with projecting inlet/exit. Most comprehensive recon platform yet known, with five main internal systems supplemented by three types of external pod, M 1·35 attained

Su-24MR (OKB demo aircraft)

with even largest pod plus two 3000-lit tanks. Entire BKO-1 *Karpat* retained, as are option of R-60 missiles. NS-24MR radio-nav system normally tailored to manual, direct or auto flight modes around up to 12 waypoints with auto guidance to areas tasked for surveillance, with precision guidance from A-711/713, RSBN-6s, *Mis*-P and digital converter, and in manual mode waypoint crosses on PPV HUD-sight. IFF of new type (Parol-2D transponder). Internal sensors are: *Shtik* synthetic-aperture side-looking radar, with antenna occupying whole left side of nose; usable up to 1320 km/h at heights down to 100 m, linear resolution (with or without MTI, moving-target indication) down to 5 m, strip 4-28 km wide, 400 km long, with optional film record. *Zima* IRLS (infra-red linescan) in right side behind air inlet, giving 7 tones black to white (resolving 0·3°C) down to 200 m at 1320 km/h, scan width 3·4×height, again recording on film over 400-k, strip. *Aist-M TV*, in belly replacing gun/camera, single-line panoramic system usable to 1320 km/h at 200 m with resolution to 0,56 m with contrast 0·7 at illumination of 2,000 lux. A-100 framing camera looks forward obliquely outboard of L inlet duct, ahead of L airbrake, usable at 1320 km/h down to 50 m. A-402 panoramic camera replacing *Kaira*-24 behind nose-gear bay sweeps wide path (typically = 10×height) in daytime and drops on-board-processed film by KADR parachute container. Largest external pod (6 m, 20 ft) houses *Shpil*-2M laser system giving resolution to 0,25 m down to 150 m at 1200 km/h. *Tangazh* pod (4 m, 13ft) gives Elint 8-3,000 mm pulse wavelengths and 28-36 mm CW (continuous-wave). Third pod is *Efir*-1M, (3 m, 10 ft) usually on No 6 pylon; paper printout of radioactive-contam contours on ground or, above 0,5-km height, in air. All info from all sensors sent in real time to BKR-1 ground station by BRS-1 secure broadband link with antenna above heat-exchanger duct. Also *Chaika*-M telecode terminal, MS-61 tape recorder and MLP-14-3 recorder for all mission parameters, VKU *Parus*-T in/out switcher and *Sevan* common time system.

Sukhoi died 75, replaced by Simonov; Felsner died 89, replaced by Leonid Andreyevich Logvinov. MR deliveries from 84 to AV-MF and VVS, no export; OKB sent red/white/blue MR demo aircraft to UK 92. 'Fencer-E'.

Su-24MP (*Modification Pastanovchik* – to direct) dedicated EW, first of three planned by OKB with customer participation to increase effectiveness in selected role. Based on MR but devoted to Elint, Sigint, Comint and active jamming. Prominent dielectric fairing below nose radome, swept hockey stick antennas below intakes, flush antennas on sides of nose. Six main on-board subsystems for detecting, analysing, classifying (identifying), storing and if necessary countering every hostile emission of interest, together with record of location of surface emitters. On-board elec generating capacity and heat-exchanger power again increased, BKO-2 *Karpat* and twin R-60 defence unchanged. Pylons 2/5 for 3000-lit tanks, in exceptional circumstances external additional hi-power jamming or payload dispensing, but ability retained to precede or accompany friendly attack aircraft. Second version (no prodn) for radioactivity detection and mapping. 'Fencer-F'. 14 built, based with GSFG until withdrawn to Ukraine.

Su-24MP

DIMENSIONS Span (69°) 10 366 m (34 ft 0 in), (16°) 17 638 m (57 ft 10⅜ in); length (M, MR, MP, overall) 25 598 m (83 ft 11¾ in); wing area (16°) 55 168 m² (594 ft²), (69°) 51 024 m² (549 ft²).

ENGINES Two AL-21F-3A.

WEIGHTS Empty (M) 22,32 t (49,207 lb); fuel 9850 kg int/6590 kg ext; ext stores (M) 8 t; loaded (M) 32 260 kg (71,120 lb), (MK) 33 325 kg (73,468 lb), max 39,7 t (87,522 lb).

PERFORMANCE Max speed (SL) 1420 km/h (882 mph, M 1·16), (10 km) 1435 km/h (892 mph, M 1·35); combat radius (hi-lo-hi, 4 t bombs, no tanks) 420 km (4 t + 2 tanks) 650 km; ferry range 2500 km (4500 with one air refuel); TO 850 m-900 m (MR, 1100 m); ldg 800 m-850 m (MR, 1000 m).

T-8, Su-25, Su-28 Vietnam war suggested need for close-support aircraft able to operate from short unpaved front-line strips with minimal facilities and deliver max ordnance against ground targets wih max precision whilst surviving ground fire. OKB team under Zhurii V. Ivanshezkin began T-8 study 68, but official interest slow to grow. MAP requirement at last issued 73. Main team formed by chief designer Vladimir P. Babak including Ivanshezkin and D. N. Gorbazyev, U. M. Lyebedyev and O. S. Samoilovich. Configuration settled with CAHI, peak Mach 0·8, and **T-8-1** built at OKB. Wing laminar 11%, LE 20°, TE 0°, anhedral (upper surf) 5°, three spars, each wing 5 sections full-span slat with dogtooth at inboard end of middle section, two rectangular sections hyd double-slotted flap, single hyd-boosted aileron, two fences aft of slats (inboard smaller); wing shoulder-high above engine nacelles built integral with slab-sided fuselage. Small nose with laser ranger and SRO-2M IFF, single cockpit with K-36L seat, all-round armour, canopy hinged to right with mirror (otherwise poor rear view), simple nav by DISS-7 doppler and ARK-10 radio compass (dipole antenna on mast above fuselage), short HF wire antenna under tail, protected tanks 1-5 (total 4,270 lit) between cockpit and fin, simple tail with inset rudder driven manually by large spring tab, anhedral tailplane carrying spring-tab elevators. Two RD-9A engines with short fixed oval-section inlets hung under wing at slight nose-up angle with short jetpipes. Levered-suspension main landing gear mounted on two nacelle frames with LP tyre 840 mm×360 mm outboard of trailing arm, hyd retraction forwards about skewed axis turning wheel to lie flat under duct; steerable levered-suspension nose gear with tyre 660 mm×200 mm with mud-guard mounted left side of nose to retract forwards into bay with single door on L, open on ground. Prominent L/R air-data probes above nose with duplicate pitch/yaw vanes. AO-17A gun in gondola scabbed under R side of nose beside nose gear. Twin cruciform braking parachutes streamed from tube in tailcone. Basic airframe designed to load factor 6·5 with 1,5 t weapons, 5·2 with max (4,34 t).

T-8-1 trucked to Zhukovskii and flown by V. S. Ilyushin 22 Feb 75. Narrow (2480) track marginally adequate, handling generally good but aircraft would be underpowered with weapon load. Second, **T-8-2**, taller fin, small inset lower rudder with spring and anti-balance tabs, even smaller upper rudder driven automatically via yaw-damper circuit fed from transducers on PVD-7 nose boom; wing fences redesigned, larger fence now inboard; wingtip fitted with flat faired box containing avionics and with aft section split to form upper/lower airbrakes; four hardpoints under each wing rated 500 kg (1,102 lb) plus 200-kg outboard point, eight main points carrying any of five types of pylon for over 50 types of store, outers APU-68-1M rail for self-defence missile (initially K-13, later R-60); forward fuselage modified with improved titanium 'bath' 24 mm (0·95 in) thick surrounding cockpit, with retractable telescopic boarding ladder in L wall of outer structure; aft nacelle enlarged with jetpipe fairing leaving cooling-air gap fed by larger projecting intake on underside. T-8-2 flown by V. P. Valsilyev 26 Dec 75.

Change to available engine of adequate power; R-13-300 retrofitted to both prototypes, requiring deeper inlet. Tumanskii KB produced closely related R-95Sh cleared to run on kerosene, avgas, MT petrol or diesel oil (checks every 4 h on diesel) and designed to withstand ground fire. Non-flying **T-8-3** tested with R-95Sh against actual gun and SAM fire. This airframe also used for refinement of elec/hyd/fuel systems, with max duplication and physical separation, reticulated foam in tanks and inerting by Freon bottles, fuel lines and oil tanks armoured, flight-control rods titanium 40 mm diam, elevator rods duplicate. Standard eqpt included RV-5M radar alt, DISS-7 doppler (part of KN-23 nav complex based on Su-17M3), DUA-3M slip/AOA indicator, *Klen*-PS laser ranger and ASP-17BTs-8 sight.

After long delays NII testing passed 78, almost derelict Tbilisi plant given order for pre-production batch, while in April 80 both prototypes flown to Afghanistan where they proved far superior to other aircraft because of volume and accuracy of firepower. Prototype and pre-

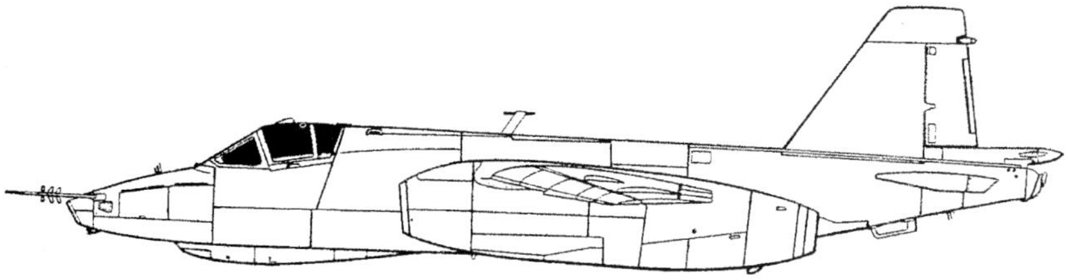

T-8-1

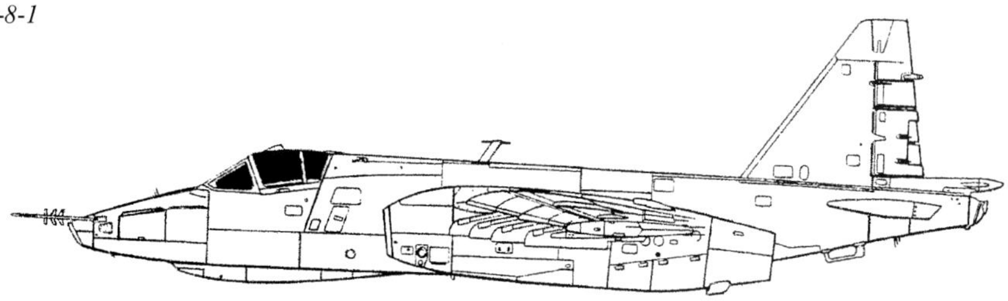

T-8-2

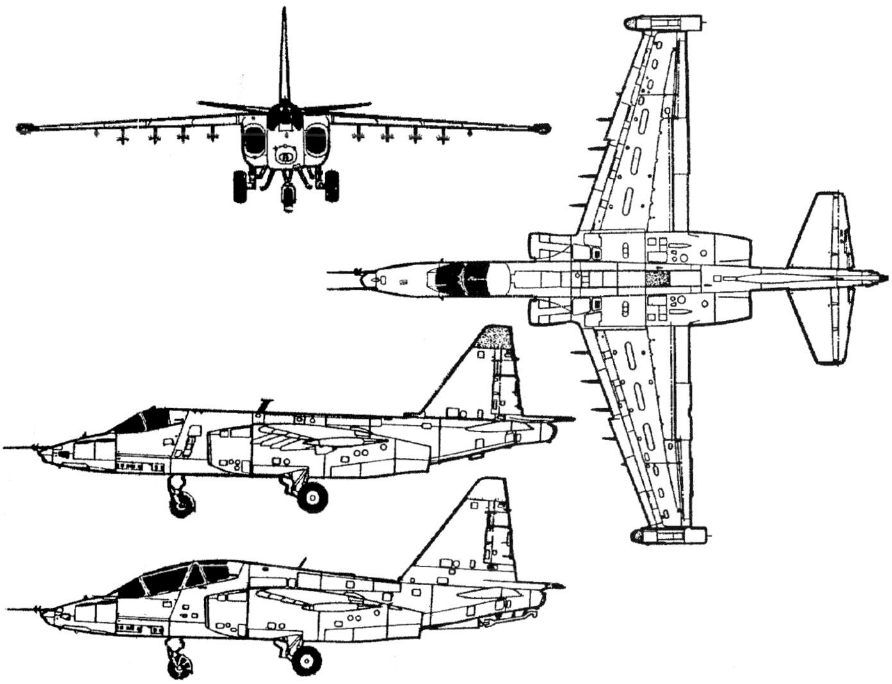

Su-25 (lower side view, Su-25UB)

production aircraft numbered (blue) from 81 (8-1); 86 destroyed during Kubinka demo 4 July 90. From No 10 '8' deleted. Blue 15 flew 600 h in action by Col (later Maj-Gen) A. V. Rutskoi (later Russian vice-president), hit by 62 shells and three missiles yet appeared immaculate at 89 Paris airshow! ASCC 'Frogfoot-A'.

Pre-production batch also known by Service designation **Su-25** (incorrect fighter odd number, just as ASCC name begins incorrectly with F). By Oct 80 200th Guards Attack Regt were at full strength, discovering need to prevent engine fire from spreading quickly to neighbour. P. Lyrshchykov fitted 5 mm steel armour between engines, after which every Su-25 hit by SAM returned to base. Babak: 'Combat experience shows Su-25 is world's most difficult aircraft to shoot down'. Pre-production aircraft had AO-17A installed at 25° tilt in left side of forward fuselage with 250-round box and row of bay gas vents below cockpit 'bathtub' on left (despite this, nose gear remained left of centreline). In hot/high conditions still need for greater engine power, but low priority. Tailplane anhedral changed to 5° dihedral, fences deleted, pylons usually lighter and lower drag, nose gear door small triangle. *Klen* laser window taller rectangle, Pion-GT front antennas on right (PVD-7) nose boom and rear on sides of tailcone above brake chute, left nose boom PVD-18G pitot, *Sirena-3* RWR installed and (initially as field mod) ram inlets above nacelles feeding pipe through jetpipe tailcone projecting beyond engine nozzle to reduce IR signature. All ground-support eqpt packaged into four pods carried on aircraft pylons. From 82 (again initially as field mod) ASO-2V chaff/flare dispensers scabbed on above rear of nacelles; later four-unit ASO-2V scabbed on above rear fuselage beside rudder. Tip pods modified with retrac landing lights at front (shielded from cockpit by anti-glare plate inboard), upper/lower *Sirena*-3 planar spiral antennas and redesigned airbrakes still opening 55° above/below but incorporating rear-hinged inset panels separately controlled for precise control of airspeed. Nacelle air inlets replaced by smaller inlet further forward and central inlet ahead of base of fin (jetpipe central air pipe deleted). Blade instead of rod antennas for SRO-2M, deletion of aileron tabs, increase in size of drag chutes, combined muzzle brake for

Su-25

Su-25UB

T-8(UTG)

Su-25T

both gun barrels and deletion of landing lights from nose and/or main gears. Stores carried never ceased to multiply, including versions of Kh-25 and -29 missiles, smart bombs (laser homers needing designation pod, *Klen* giving range only), SPPU-22-01 gun pods, active jammers (notably SPS-141) and many stores carried in tandem pairs on MBD-3 interfaces. Stations 3/5/6/8 plumbed for 1150-lit ferry tanks. Larger stores on Stations 3 and 8 required a cutout in TE of outboard flap. Normal external load 1340 kg, max 4340 kg. Prolonged pressure for better avionic fit. Near end of production 86 engine changed to more powerful R-195, designed to run after hits in eight places. Another late mod deletion of inbuilt boarding ladder. Total run c350, completed 91. Still 'Frogfoot-A'.

Su-25K export version, differing in detail (SRO-2M always rod antennas). Customers Bulgaria (20), Czechoslovakia (50), Hungary (50), Iraq (25), N Korea (20) and Syria (60). **Su-25BM** modified as target tug with TO-70 winch and *Kometa* target on No 5 pylon; gun deleted. **Su-25BMK** export version. All 'Frogfoot-A'. **Su-25UB** dual weapons trainer flown (T8UB-1) 85, produced in small series from 88. Retained gun and full combat capability, often with chaff/flare dispensers. New forward fuselage with tandem dual cockpits, rear (instructor) set higher, redesigned enclosure with new windscreen, two canopies hinged to right (front very long and with mirror repositioned internally on left, rear very short, with periscope) separated by intermediate frame and pressure-tight bulkhead and window to confine any pressure loss to one cockpit. Fin raised on to additional root section to increase area. Engine R-95Sh, fin-root inlet very small, aft nacelle inlets restored. **Su-25UBK** export version for Bulgaria, Czech, Iraq. All trainers built at Ulan-Ude. 'Frogfoot-B'.

Cutback in military market prompted OKB to develop trainer for civil use, targeting DOSAAF and foreign countries. Initially called **Su-28**, later changed to **Su-25UT**. All military equipment removed, reducing weight more than 2 t (4,409 lb), and structure restressed to +8 g/ –2 g. Nose gear moved to centreline and LE dogtooth replaced by gentle kink not generating vortex. Prototype was modified UB, flown 6 Aug 85, but no money for production. **Su-25UTG** prototype carrier-based trainer for AV-MF first flown 87, trials aboard *Tbilisi* (now *Adm. Kuznetsov*) 1 Nov 89, flown by I. V. Votintsev and LII's A. V. Krutov. Based on UB but HF antenna replaced by anchor for heavy tail hook. Gun, pylons and dispensers deleted, dogtooth retained Two blocks of five built for CIS navy. 'Frogfoot-B'.

Chief designer Vladimir P. Babak assigned to T-8M 83 to rectify shortcomings, prime mission anti-armour, secondary close support, in each case day/night in all weather. Basic need for greater endurance and more avionics led to airframe being based on capacious UB. Rear cockpit replaced by tankage (c1 t extra) and avionics. AO-17A replaced by GSh-30/II with 260 rounds in external package aft of nose gear on left side. Longer, more streamlined nose with *Merkuri* LLTV, other minor changes. First of three flown Ulan-Ude Aug 84. Led to initial series batch of 10 **Su-25T**, with operational eqpt. Completely new *Voskhod* navigation complex and *Shkval* EO attack system. Former includes inertial platform, autopilot and two digital computers. Latter has large nose window for combined boresighted optics for TV and laser (ranger and target designator) using stabilized mirror and ×23 magnification, *Merkuri* replaced by *Kinzhal* 8 mm radar and *Khod* FLIR with combined cockpit displays on HDD and HUD giving clear picture of tank in any weather at 3 km. Integrated defensive EW system under computer management with duplicated all-round RWR and IR warning with antennas in tip pods and behind base of rudder, internal active jammer in large tubular container above tailcone whose outer surface contains chaff/flare dispensers, with additional dispensers (to total 256) firing up from extreme tailcone. GSh-30/II replaced by original AO-17A but packaged with ammunition under right side of fuselage. Among new weapons on wing pylons are Kh-25ML, Kh-31P, Kh-58 and Kh-59, quad tubes for 9M114 *Skorpion* and APU-8 launcher for PTUR *Vikhr*. Latter scored 80% against moving targets in acceptance tests 92 but only eight to be bought. Sukhoi Stormovik consortium seeking customers for **Su-25TK** export version, now called **Su-39**.

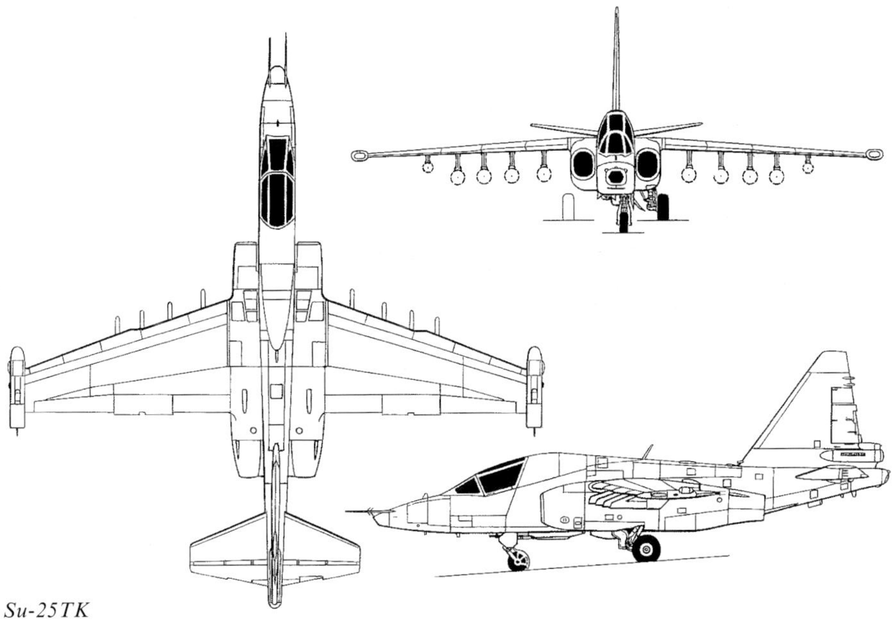
Su-25TK

DIMENSIONS Span 14,36 m (47 ft 1½ in); length (most) 15,53 m (50 ft 11½ in), (UB) 15,36 m (50 ft 4¾ in), (TK) 15,33 m; wing area 30,1 m² (324 ft²).

ENGINES (most 25 and UB) two R-95Sh, (late 25 and T) two R-195.

WEIGHTS Empty (25) 9430 kg, (R-195) 9505 kg, (UB) 9,19 t, (T) 9,62 t; fuel (25) 3,05 t, (UB) 2725 kg, (T) 3840 kg; normal landing (all) 12 t, max landing (25) 13,3 t; max TO (25) 17 530 kg (38,646 lb), (R-195) 17,6 t (38,800 lb), (UB) 17 222 kg (37,967 lb), (T) 19,5 t (42,989 lb).

PERFORMANCE Max speed, clean, SL to 2 km, (25, 95 Sh,T) 950 km/h (590 mph), (25/R-195) 975 km/h (606 mph), (UB) 980 km/h; Mach limit 0·82; climb (UB) 85 m/s (16,730 ft/min); service ceiling (25, UB) 7 km (22,970 ft); range (T) 400 km/700 km; ferry (25, ×4PTB-800) 1950 km (1,212 miles), (UB) 2150 km, (T) 2500 km; TO (25, max wt) 1150 m, (UB clean) 480 m, (T, max wt) 660 m; landing (25, T, with chute) 400 m, (UB, no chute) 600 m.

Su-26 Totally new venture funded 83, aerobatic competition aircraft, load factor +11/–9. Low/mid GFRP/CFRP wing symmetric profile 18% root 12% tip with large ground-tabbed ailerons each with two suspended balance triangles; pilot reclining in steel-tube fuselage with GFRP panels and metal cowl, jettisonable canopy hinged to R, windows sides and underneath; tank 91 lit, replaced in 85 by wing-root tanks 65 lit each (R tank not used in competition); landing gears titanium leaf springs, mainwheels hyd brakes and 350 mm × 135 mm tyres. First flight with 2,4 m V-530 prop June 84, second aircraft Hoffmann prop, four built. Followed by **Su-26M** (c/n 05) built 85 with wingtips squared off to carry horizon protractors, angular rudder, cowl with iris shutters, Nos 06-08 deeper but shorter windscreen, upward-hinged canopy and reduced glazing; dominated 86 World Champs. Fuel 60 lit in fuselage plus 140 lit in external ferry tank or two 83-lit LE tanks plus 100-lit ferry.

DIMENSIONS Span 7,8 m (25 ft 7 in); length 6820 mm (22 ft 4½ in); wing area (26) 10,85 m² (117 ft²), (26M) 11,9 m² (128 ft²).

ENGINE One M-14P.

WEIGHTS Empty 720 kg (1,587 lb); loaded 860 kg, reduced in 87 to 800 kg (1,764 lb).

PERFORMANCE Max speed 355 km/h, later 350 km/h (217 mph); climb 1 km, later 1080 m (3,540 ft)/min; ferry range 800 km (497 miles); TO 250 m.

Su-26MX Su-26M mod 89 with load factor +12/–10, 63 lit fuselage tank (competition) plus 200 lit in LEs; **Su-26MX** export version with Gerd Mühlbauer prop and other options. Data as before except:

DIMENSIONS Length 6845 mm (22 ft 5½ in).

WEIGHTS Empty 705 kg (1,554 lb); loaded (ferry) 1 t (2,205 lb).

PERFORMANCE Max speed 310 km/h (192 mph); TO 160 m.

Su-26

root and ending in fairing around single-contour nozzle. Centered above each engine, well forward on TE, closely spaced twin vertical fins. Box-section side beams (shelves) each carrying slab taileron, LE parallel to wing LE, square tip, driven by power unit faired above shelf. Inlet ducts and AL-21F-3 engines parallel so nozzles widely spaced, linked by aft end of fuselage resembling wing TE with small EW pod on centreline projecting beyond nozzle shrouds. Large integral fuel capacity aft of cockpit and in centre wing, reticulated-foam protection. Main gears pivoted on skewed axes under wing root to retract forwards and inwards into long bulged fairing, wheel stowed flat in root, bay closed by outward-hinged leg door and large forward door separately controlled to serve as airbrake. Tall nose gear with

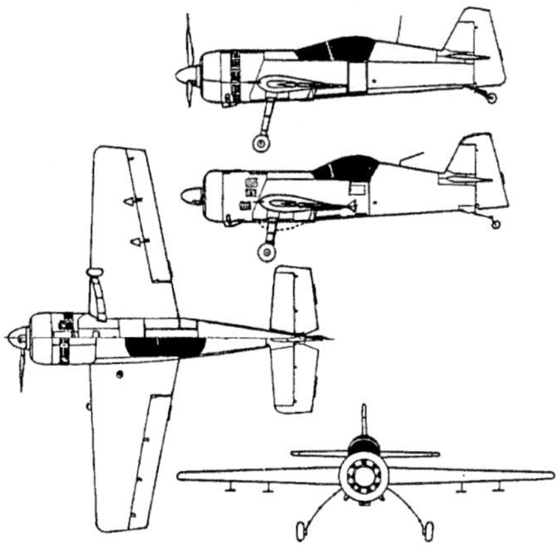
Su-26 (lower side view, Su-26M)

Su-26MX

T-10 In 69 OKB won design contract for IA-PVO interceptor to replace Tu-128, Yak-28P and Su-15bis. Requirement superiority over future foreign fighters (esp. USAF FX, later F-15), ability to dogfight and also shoot down aircraft and cruise missiles from long range. Thus aircraft large, for long range; high installed thrust (at low altitudes exceeding weight); blended wing/body layout; longitudinally unstable (first in USSR); fly-by-wire, without mechanical reversion; single seat. Wing tapered on LE rather than swept, so designation in T series. Design team led by Yevgeniy A. Ivanov, deputy Oleg Samolovich.

Wing symmetric profile 3·5%-5·0%, LE 79° at root, changing to constant 44° to curved (Küchemann) tip. Hinged LE, conventional ailerons and slotted flaps. Wing divided aircraft into two; above and ahead, oval-section fuselage with humped-back integral tankage; below and behind, twin engine nacelles tilted outwards and projecting above wing aft of 30% chord at single steerable wheel retrac backwards into bay under cockpit with one front and one side door, wheelbase 9,33 m!

Prototype **T-10-1** completed at OKB, flown by V. S. Ilyushin 20 May 77. Outstanding, but numerous snags and engines big, heavy and accessories on top where fins made access difficult. Modified with four small wing fences and anti-flutter rods ahead of wingtips and all tail surfs. **T-10-2** improvements included FBW flight controls based on T-4 with four channels.

T-10-1 as first built

Software error in T-10-2 led to unexplored resonance which overstressed airframe, killing Yevgeniy Solovyev; recorders showed cause.

In 79 aircraft accepted as **Su-27**, by which time five pre-production aircraft built at Komsomolsk-na-Amur (which had supplied outer wings and tails for first two). These were **T-10-5/6/9/10/11**. In 79 Komsomolsk built two further prototypes with new engine, AL-31F. These were **T-10-3** (flown 23 Aug 79) and **T-10-4** (31 Oct 79). New engines smaller and 0,5 t, 1,100 lb, lighter, with 12% greater thrust, improved fuel economy, variable con/di nozzle behind airframe and accessories on underside, transforming rear of aircraft. Among other changes, large inboard fences 5-70% chord at one-third semi-span, long fences above and below wing centered on tubular pod inboard of tip (these added to T-10-1), slats, modified landing gear and tail with widely spaced fins mounted outboard on shelves. Structure stressed for sustained 9 g. Problems continued, Aleksandr Komarov being killed by wing failure and Nikolai Sadovnikov (researching Komarov's crash) landing after losing major part of wing. Cause was sudden pitch-up in max-speed flight at low level resulting in destruction of slats (causing tail damage) and loss of outer wing panels. Another problem, airbrakes caused tail buffet.

Sukhoi succeeded by Simonov in 75, by which time info on F-15 available. OKB computer evaluation found T-10 was not meeting requirement. Simonov ordered what amounted to fresh start. Prototypes served later as testbeds. US DoD 'RAM-J', then ASCC 'Flanker-A'.

DIMENSIONS Span 15,75 m (51 ft 8 in); length (excl pitot) 20,79 m (68 ft 2½ in); wing area 59 m² (635 ft²).

ENGINES Two AL-21F-3, (Nos 3,4) AL-31F.

WEIGHTS Empty 18,2 t (40,123 lb); max 27,2 t (59,965 lb).

PERFORMANCE Not published.

T-10-1 preserved

T-10S, Su-27 Simonov 'We retained main wheels and ejection seat'. Wing redesigned with *centroplan* integral with fuselage with fixed ogive LE ending at axial discontinuity, outer panels with straight LE swept 42° carrying adaptive 4,6 m² flaps (called slats) powered to 30° in unison with inboard 4,9 m² flaperons driven +35°/–20° for high lift or roll up to 860 km/h IAS, ailerons omitted. Ineffective ogive tips replaced by missile rail projecting ahead as anti-flutter mass or EW receiver pod. Outer wing stiffened, 3-spar box forming integral tank with pylon hardpoint. *Centroplan* houses fuel

Su-30 (p. 367)

and EW boxes and, in upper part of right root, GSh-30/1 with 150 rounds loaded through left inboard LE.

Fuselage also redesigned with larger radome, forward cross-section reduced, aft of cockpit (Tank 1) with shorter untapered portion and reduced cross-section, then extended at rear far beyond nozzles to house main equipment group, aft tank, additional EW jammers and drag chute. Door airbrakes replaced by single 2,6 m² dorsal brake driven by jack to 54° during landing or balanced against air load to 1000 km/h IAS. Large canopy hyd opened upwards, K-36DM Srs 2 zero/zero seat with NAZ-8 surv pack, Komar-2M beacon and PSN-1 dinghy. Provision for retractable FR probe, used on later versions.

Prolonged testing with different tails. Final design, verticals 4,3 m apart, fin 15,5 m², LE 40°, six antennas, inset rudder 3,5 m², ±25°. Tailerons on redesigned shelves, 12,2 m², LE 45°, span 9,87 m, tip cropped at Mach angle, rod masses removed, driven ±10° for roll, otherwise +15°/–20°, by power unit in fairing under rudder. Vertical underfins added, 2,5 m², LE 28°. Centreline of forward fuselage tilted down 2·5°. Inlets redesigned for greater airflow, boundary-layer splitter separated by gap from underside of wing, multiple exit grille both sides and on top under wing, fixed lower lip but variable front/rear panels forming upper wall, aux-inlet louvre panel in underside, fine-grille (2 mm) screen normally recessed in bottom of duct hinges up when engines running on ground to prevent foreign-object damage. Main gear redesigned with strut vertical when extended and braced by latch to socket in blister on body, KT-156D wheel with fan-cooled disc brake and 1030 mm × 350 mm tyre, rotating as before to lie in wing in bay closed by door not used as airbrake. Tall nose leg with three lights with hyd steered levered-suspension single unbraked KN-27 wheel with debris guard and 620 mm × 260 mm tyre, retrac forwards to lie upright behind cockpit, wheelbase reduced to 5880 m; track 4330. Engine accessories now easily reached.

Systems also largely redesigned. Fuel 12 000 lit (2,640 gal) in three integral body/centrewing tanks and one in each outer wing. No provision for external tanks; FR probe not fitted. Two hyd systems each with pump on both engines at 280 kg/cm² (3,983 lb/in²) serving landing gear, slats, flaps, flight-control power units, airbrake, inlet screens, inlet ramp jacks, mainwheel brakes and steering. Pneumatic system drives nosewheel brakes, canopy and drag chute, pressurizes cockpit and avionics bays and can extend gear in emergency. Each engine drives constant-speed alternator supplying 115V at 400 Hz, single- and three-phase converters, DC supplies at 27V.

Avionics include SDU-10 quad FBW flight controls governed by pilot and by SAU-10 autopilot/computer which, with air-data system, warns of 'proximity of dangerous attitudes or flight regimes'. PNK-27 nav complex comprises IK-VSP velocity block, IK-VK course/altitude block, RV-10 radar altimeter and RSBN Shoran. *Radan* coherent TWS (track while scan) pulse-doppler radar with twisted-Cassegrain antenna used in conjunction with OEPS-27 opto-electronic (IR) sight ahead of windscreen on centreline (used whenever possible, for stealth) which together with PNK data is fed to ILS-15 HUD; all scanning heads and missile guidance can be cued by NSTs-27 helmet-mounted sight. Pilot combat switch can place flight-controls under guidance from GCI or A-50, SRO-2M IFF rod antennas under nose and above tailcone. *Berez* passive radar warning with receivers on sides of inlets and each side of tailcone, internal high-power jammers (option of *Sorbitsiya* pods on wingtips), flare/chaff dispensers (32 APP-50 triplet groups beside and above tailcone), SO-69 transponder, R-800 UHF, R-864 HF, cockpit-voice recorder, *Tester* flight-data recorder and *Ekran* data recorder. Underwing launchers for ten AAMs: two AKU tandem centreline (R-27R/T) or APU-470 (R-27ER/ET), one under each engine duct (same), one under each wing (same), one APU under each tip (R-73) and tip rails (R-73); underbody and wing AKU hardpoints can carry twinned launchers.

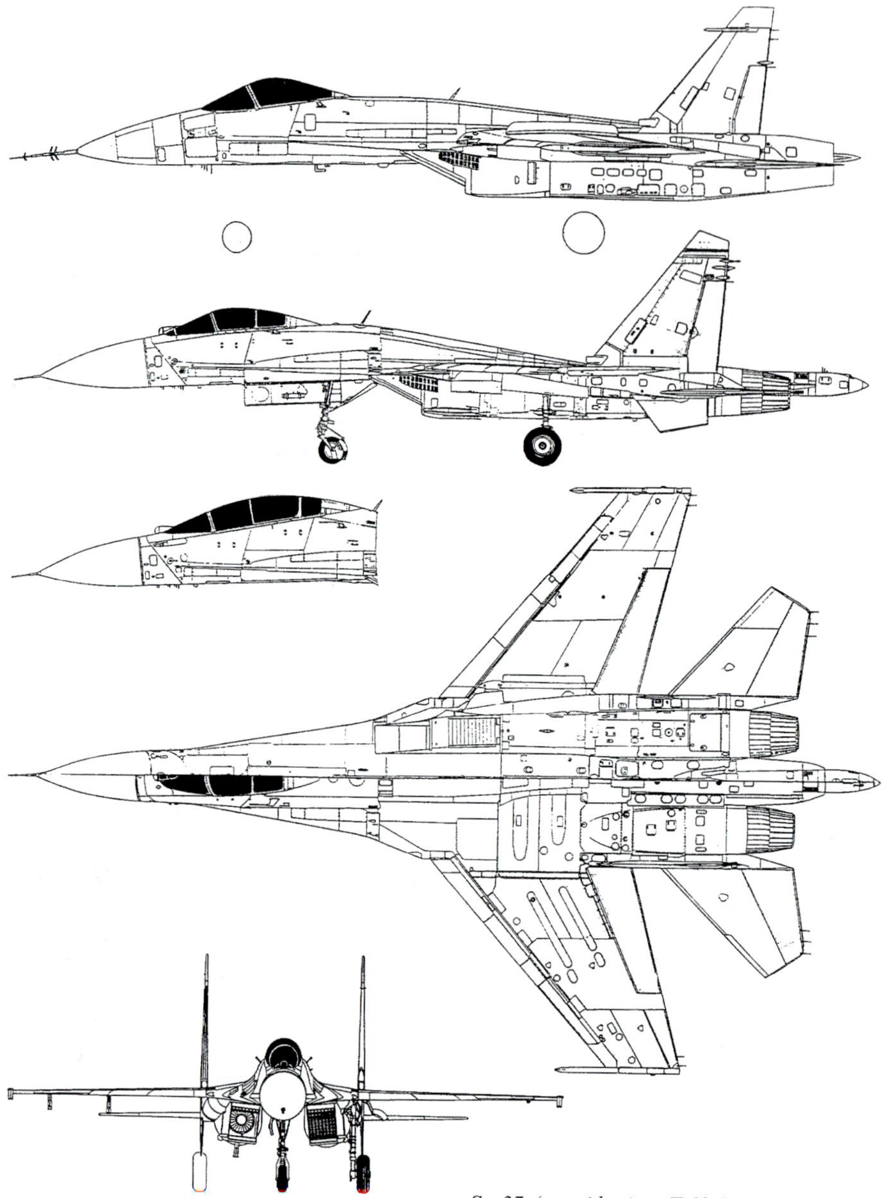

Su-27 (top side view, T-10-1; inset Su-27UB)

OKB designation **T-10S**. First flight of T-10S-1 (also called T-10-7) by V. S. Ilyushin 20 Ap 81. Enormous effort NII tests (T-10-17 and -22). First series rollout of **Su-27** from Komsomolsk Nov 82, but deliveries delayed to early 85 by radar problems. By this time anti-flutter weights and tank foam discarded; service limits set 28° AOA, +8·5/–2·5 g. Also in 85, well-used T-10 prototype modified under R. G. Martirosov as **P-42** for climb records; radar, armament, parabrake container, fin tops and ventral fins deleted, slats locked, light metal nose, unpainted finish, engines uprated to 13,6 t for thrust/weight ratio (at T-O wt 14,1 t) of 1·93. In 86-88 set 27 FAI records including 3/12/15 km in 25·4 sec, 57·4 sec and 75·7 sec, and sustained 19 335 m. T-10S-20 modified for cancelled attempt on distance/endurance record with extended tailcone, fuel tank in nose cone and cropped tailfins, T-10 type wingtips. Normal Su-27 outstanding; Ilyushin: 'Before Su-27 I

P-42

T-10-17

had flown 142 types, and came to know I was always 'more intelligent than the aircraft'. With Su-27 it was other way round; it always seemed more capable than I was. So to use its great potential a pilot must aim for perfection. When I flew first T-10S I understood this was the aircraft for which I had waited all my life . . .' One of many show-stoppers at Paris 89 was V. G. Pugachyev's Cobra: level at 425 km/h, switch off alpha-limit, pull nose up to AOA 120° (no gain of height), dynamic braking to 150 km/h in two sec, then recover to normal attitude in afterburner, autostab on. Delivered 100 mid-88, 200 by 91 inc first 8 of 24 for China. 'Flanker-B'.

DIMENSIONS Span 14,7 m (48 ft 2¾ in); length (excl pitot) 21 935 mm (71 ft 11½ in); wing area 62,0 m² (667·4 ft²).

ENGINES Two AL-31F.

WEIGHTS Empty 16 t (35,273 lb); fuel 9,4 t; normal loaded 22,5 t (49,600 lb); max 30 t (66,140 lb).

PERFORMANCE Mach limit 2·35; max speed (SL) 1400 km/h (870 mph, M 1·145), (11 km) 2500 km/h (1,553 mph); service ceiling 18,5 km (60,700 ft); range (SL) 1400 km (870 miles), (hi) 3900 km (2,425 miles); TO 650 m; landing 230 km/h, 620 m with chute.

Su-27 variants Tandem dual-pilot trainer prototype **T-10U** developed with no change in dimensions or fuel capacity. Instructor cockpit added behind and higher than original, with single aft-hinged canopy usually with one mirror for instr and two for pupil. Centre fuselage increased in depth to retain capacity of Nos 1 and 2 tank. Provision for FR probe on left of windscreen and back-up probe on left side. No change in avionics or weapons. Vertical tails raised on extra root sections, 18,6 m², airbrake longer and slightly wider. T-10U-1 flown by N. F. Sadovnikov 7 March 85. All 2-seat versions at Irkutsk, first delivery late 86 as **Su-27UB**. Tested with Il-78 tanker and UPAZ 'buddy pack'; Sadovnikov and I. Votintsev flew Moscow-Komsomolsk 13 440 km in 15 h 42 min with four refuelling contacts. Deliveries included 4 for China. 'Flanker-C'.

Su-27UB

DIMENSIONS, ENGINES As Su-27.

WEIGHTS Empty 17,53 t (38,646 lb); loaded 24 t (52,910 lb), max 30,5 t or 33,5 t (67,240 lb or 73,854 lb).

PERFORMANCE Mach limit 2·35; max speed (11 km) 2500 km/h (1,553 mph); service ceiling 17,2 km (56,430 ft); range (lo) 1300 km (808 miles), (hi) 3000 km (1,865 miles); takeoff run 550 m;

T-10-24

T-10K-1

landing 240 km/h, (with chute) 680 m.

Su-27K In some manoeuvres at extreme AOA tailerons are in wing wake, losing power. In 79 Simonov authorized research which culminated 82 into addition of PGO powered-canard system linked to flight-controls, tested from May 85 on **T-10-24**. Surfaces horizontal, pivoted to modified centreplane LE, increasing both stability (pitch and roll) and instability (pitch), and reduce trim drag; max lift coefft (30° AOA) increased from 1·75 to 2·1. Perfected widened manoeuvre envelope 86 in time to assist devt of **T-10K** (service **Su-27K**) for carrier *Tbilisi* (now *Admiral Kuznetsov*). This has foreplanes, strengthened gear with twin nosewheels, arrester hook under shortened tailcone, power-folding outer wings and tailerons, flaperons replaced by hi-lift slotted flaps in two sections extending to tips with inset inboard panel, OEPS sight moved to right to give clear view ahead, avionics changes including carrier precision landing system. FR probe from outset. **T-10K-1** (#37) flown by Pugachyev without wing fold 17 Aug 87. Same pilot made first Sov carrier landing with **K-2** (#39) 1 Nov 89. Other K devt aircraft 58/59/69/79, all slightly different and with AL-31FM engine. By Jan 93 ski-jump tests and arrested landings with max bombload (with #37) and bad-weather trials in Barents Sea. Production **Su-33** could have had avionics based on Su-35 (see later) and AL-35 engine with vectored nozzles but first batch ordered are merely navalised Su-27 with no multi-role capability. Circular vectored nozzle tested Jan 89 on left eng of T-10-28 (UB #8), later square nozzles flown. **Su-27PU** based on UB with special hand-off data links to ground and to accompanying radar-silent interceptors; two built 90. These devd into current production interceptor, displayed Paris 93 as **Su-30** (see separate entry).

Su-27IB for autonomous night/all-weather operation primarily against surface targets. Prototype (#42) flown 13 April 90 by Anatoly Ivanov, new forward fuselage, foreplanes as 10-24 extended as chine to shallow 'Platypus' nose; wide cockpit for pilot on left and WSO on right, left/right canopies and large left/right birdproof windscreens, no radar but advanced Su-27M type sensor displays, redesigned nose gear with twin wheels retracting to rear into bay with four doors and access ladder to cockpit and equipment bay, larger vertical tails but

T-10M-8 (Su-35)

ventral fins deleted. Evaluated to late 91 as Su-24 replacement, as **Su-34**, (which see) similar airframe proposed for **Su-27KU** carrier trainer. **Su-27M** see **Su-35**.

Su-28 See Su-25.

Su-29 Enlarged tandem dual version of 26MX, factors +11/–9, single upward-hinged canopy, MT prop, 24 built 92 for USA.
DIMENSIONS Span 8,2 m (26 ft 10⅞ in); length 7,32 m (24 ft 0⅛ in); wing area 12,24 m² (131·8 ft²).
WEIGHTS Empty 760 kg (1,675 lb); loaded 1,1 t (2,425 lb).
PERFORMANCE Max speed 340 km/h (211 mph); climb (dual) 960 m (3,150 ft)/min; ferry range 965 km; TO/ldg 120 m.

Su-30 Dedicated interceptor for PVO, derived from Su-27PU but meeting severe added demands, notably to fly 10-h missions. Airframe based on 27UB, cockpits almost identical, either pilot can command aircraft or command ejection, all systems certified to operate for 10 h with two refuellings (probe L of windscreen, EO sensor dome moved to right), hi-capacity O_2, toilet provisions. Integrated *Radan* fire-control with nine subsystems including Sh-101 radar, IRST, laser designator, laser ranger and HMS. *Anek* nav complex basically miniaturized airline type combining *Mars* Omega/GPS/INS/Loran, modern cockpit. First prototype T-10-05 flown Irkutsk 30 Dec 89 by Evgenii Ryavunov. Qual'd for prodn'n late 92, first two prodn aircraft stripped of combat eqpt, specially painted (#596/597) and sold (with a single-seater, #595) to Jupiter Insurance as aerobatic demo aircraft for Anatolii Kvotchur *Test Pilots* team, subsequent Su-30s operationally equipped. **Su-30M** multirole, adds air/ground capability, 12 hardpoints for 8 t (17,637 lb) weapons, including 6 R-27, 6 R-73, 6 R-77, 4 Kh-31R, 6 Kh-29T, 6 Kh-29L, 6 S-25LD, 2 'AGM-TVC', 6 GBU-500T, 16 FAB-500M54 bombs or 10 FAB-500M62 or 3B-500Sh or BETAB-500Sh or 36 FAB-100 or 6×B-8M (20×S-8 rockets) or B-13M (5S-13), retaining gun and 150 rounds. Will be carrier of aft-firing missile derived from R-73, using target-illuminating radar in enlarged tailcone. T-10-06 qual'd **Su-30MK** for export (initially China), foreign avionics/weapons at customer request.
DIMENSIONS, ENGINES As Su-27.
WEIGHTS Empty 17,7 t (39,021 lb); loaded 24,5 t (54,012 lb); max 33 t (72,751 lb).
PERFORMANCE As Su-27 except Mach limit 2·0; combat range 7000+ km (4,350+ miles) with 2 refuellings.

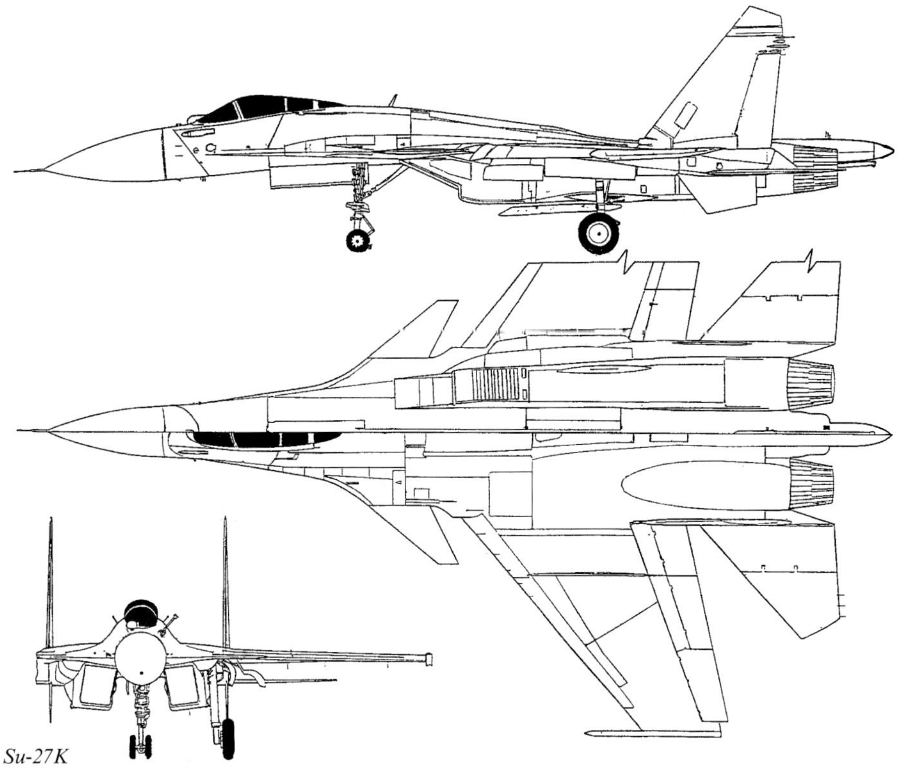
Su-27K

Su-31 Single-seat version of Su-29 with structure 70% GFRP/CFRP for +12/–10 manoeuvres while reducing mass; uprated engine; fuel 78 lit (17 gal), but can be 2×100 lit or 210

Su-27IB/27KU

Su-29

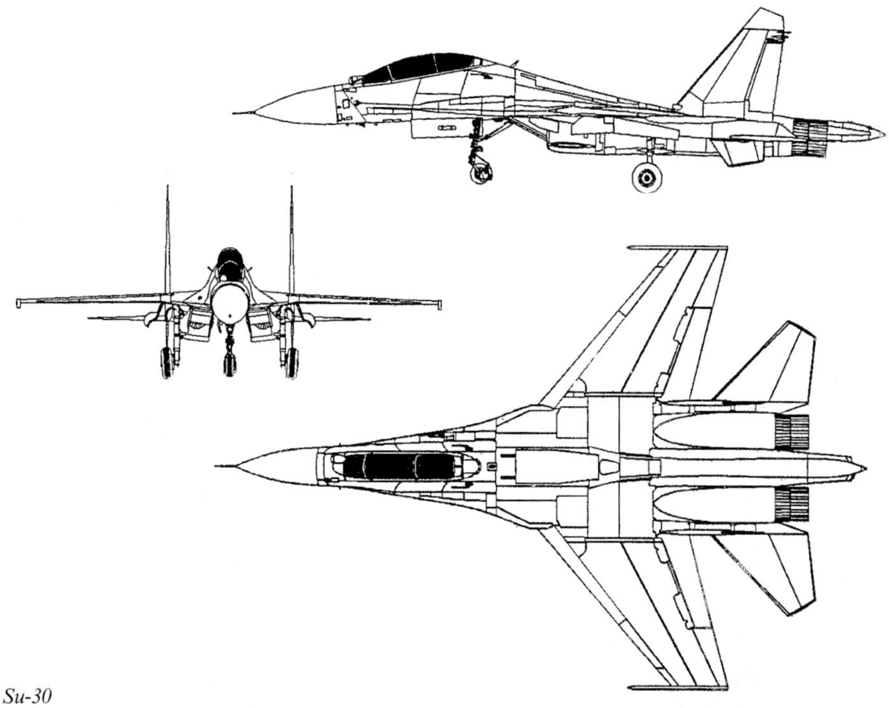

Su-30

scabbed underneath. **Su-31T** flew June 92, **Su-31U** with retrac gear in 94.
DIMENSIONS Span 7,8 m (25 ft 7 in); length 6,9 m (22 ft 7⅝ in); wing area 11,8 m^2 (127 ft^2).
ENGINE One M-14PF or NTK.
WEIGHTS Empty 672 kg (1,481 lb); loaded 968 kg (2,134 lb).
PERFORMANCE Max speed 340 km/h (211 mph); max climb 1080 m (3,543 ft)/min; range 290 km (180 miles), (ferry) 1200 km (745 miles).

Su-32 Trainer and light-attack aircraft. CFRP/Kevlar/GFRP airframe enlarged from Su-29, smaller ailerons, pneu-operated slotted flaps and tricycle gear (all 400×150) with hyd brakes, three tanks 310 lit (70·4 gal) internal, pressurized, tandem KS-38 seats thru hinged canopy on diverging trajectories, comprehensive radio/nav/mission avionics. Provision for gun, two 100-lit drop tanks or missiles. To fly 94-95, series from 96. Export **32X** with 435 hp TCM flat-six.

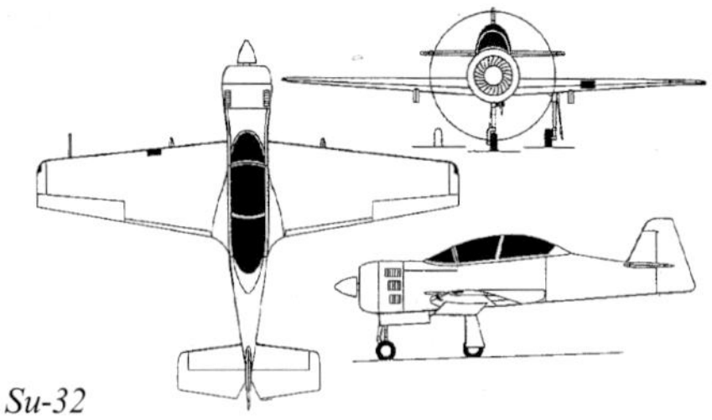

Su-32

DIMENSIONS Span 8,5 m (27 ft 10⅝ in); length 7285 (23 ft 10¾ in); wing area 12,2 m^2 (131 ft^2).
ENGINE One M-14NTK.
WEIGHTS Empty equipped 850 kg (1,874 lb); loaded 1500 kg (3,307 lb).
PERFORMANCE Max speed 370 km/h (230 mph); max climb 720 m (2,362 ft)/min; service ceiling 4 km (13,123 ft); range (max stores) 1200 km (746 miles), (max fuel) 2000 (1,242); TO 150 m; ldg 200.

Su-33, see Su-27K.

Su-34 Series all-weather attack aircraft to replace Su-24, chief designer Rollan Martirosov. Airframe based on Su-27IB, with following major changes: nose redesigned to house multi-mode navigation and terrain-following and avoidance radars; new tandem-wheel main landing gears (wheels in line, gear redesigned for greater weight, not to avoid rutting unpaved runways); vert tails restored to normal Su-27 height; enormous tailcone housing aft-facing radar for detecting targets and guilding aft-firing missiles; parabrake relocated in dorsal box ahead of tail; increased internal fuel capacity in fuselage, wings and fins; improved cockpit with new displays and toilet behind seats; and detail redesign throughout aircraft to minimise vulnerability to hostile fire (Su-25 experience), plus c1 t of armour including 17 mm-thick titanium bath enclosing cockpit; upgraded avionics including new central nav computer 'giving 1 m accuracy'. Heavy weapon load for use against targets up to 250 km from aircraft.

First of five preproduction aircraft came off Novosibirsk line, replacing Su-24M/MR, in Dec 93. Following OKB testing by Votintsyev and Ryavunov, ferried by Ryavunov and Igor Solovyev to Zhukovsky March 94 for LII testing. If funding permits, Su-24 to be completely replaced by 34 by 2002; Chief of Staff Pyotr Deniken: 'Crucial aircraft of CIS air forces in 21st century.' No known ASCC name.
DIMENSIONS Span 14,75 m (48 ft 4¾ in); length 23,95 m (78 ft 7 in); wing area 62,0 m^2 (667 ft^2).
ENGINES Two AL-31FM.
WEIGHTS Max 44,36 t (97,795 lb).

Su-34

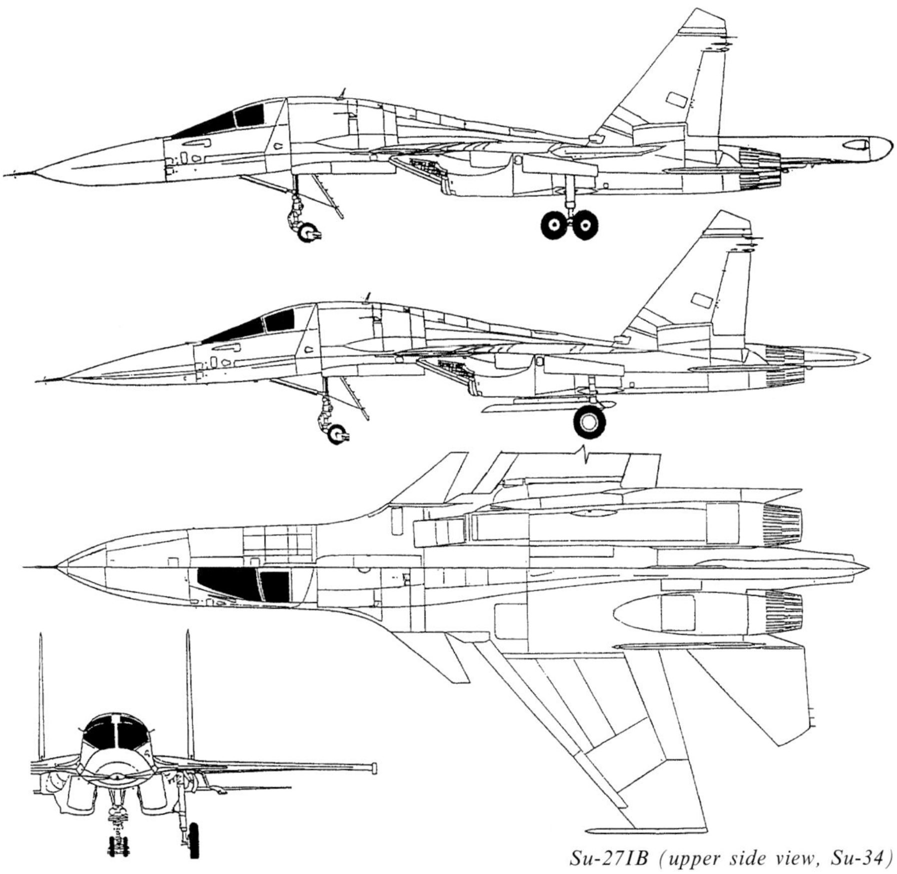

Su-27IB (upper side view, Su-34)

PERFORMANCE Max speed (SL) 1400 km/h (M1·15, 870 mph), (hi) 1915 km/h (M 1·8, 1,190 mph); range (internal fuel) 4000 km (2,485 miles).

T-60S Western artists have made several speculative – and so far very inaccurate – attempts to picture likely configuration of future long-range strike aircraft revealed by Simonov early 94 as 'intermediate-range bomber to replace Tu-16, Tu-22 and some Su-24 by turn of century'. Said to be stealth design for Mach-2 supercruise at over 18 km (59,000 ft), releasing weapons from internal bays at cruise height; combat radius c2000 km (1,242 miles). By mid-94 subject of intense speculation in West.

Su-35 Improved Su-27 developed with six T-10M prototypes (first flown by Oleg Tsoi 29 June 88) as '4th generation' multirole strike fighter. AL-31FM engine, canards, taller rudders and (except first aircraft) integral-tank square-top fins, high-power ECM in tip pods and tailcone (dorsal drag chute) and FR probe. Initially called **Su-27M**, now marketed as **Su-35** multirole fighter/strike/recon/EW platform. Digital avionics, four cockpit displays, NO-10 radar, OEPS with small external TV pod collimated with laser, duplicate reconfigurable computers and other features for hi-accuracy search by multi-channel tracking and fire-control against air/surface targets, automatic control of formation aircraft by commander, radar mapping, air-combat fire-control, auto selection and elimination of each most-dangerous threat, terrain following without external sensors whilst tracking air/surf targets, ECM suppression and IR decoy, passive radar warning, Elint ident and precision-location, autonomous nav including blind landing, and prevention of any 'extreme situation' caused by malfunction or pilot error. Main radar big advance over '2-target' Su-27 radar, with 400 km range and ability to track 15+ air targets and fire on any six. OEPS sight tracks ±60° to 50 km, with laser ranging for gun to 8 km, like attack radar feeding HUD and HDD and cued by HMS. Most comprehensive defensive avionics in history, with passive receiver covering all bands D thru J over 360° and vert ±30° and *Gardeniya* main jammer H/I bands and supplementary jammer G/J bands both ±45° azimuth directing emission at specific targets, with computer control to distort distance, angle, speed and apparent size of attacking force. External jammer pods can be used (eg G-beam tailored to Patriot). Elint library contains 512 threats, feeding auto selection of threats, crew warning (manual control poss) and inflight reprogramming. Com systems include UHF/VHF and HF to 1500 km plus data-link (up to 4800 bytes/sec) and voice scrambling. Armament: GSh-30/I, 150 rounds, 8 t (17,637 lb) on 12 hardpoints; missile types R-27T (2), R-27R (8), R-73 (6), R-77 (10), AAM-L (7) for air targets, Kh-29T (6), Kh-31P (6), KAB-500T (6), GBU-1500T (4), Kh-58A (6) and Kh-59 (4). Production Su-35 would have AL-35, aft-facing radar for 'over shoulder'

Su-27K

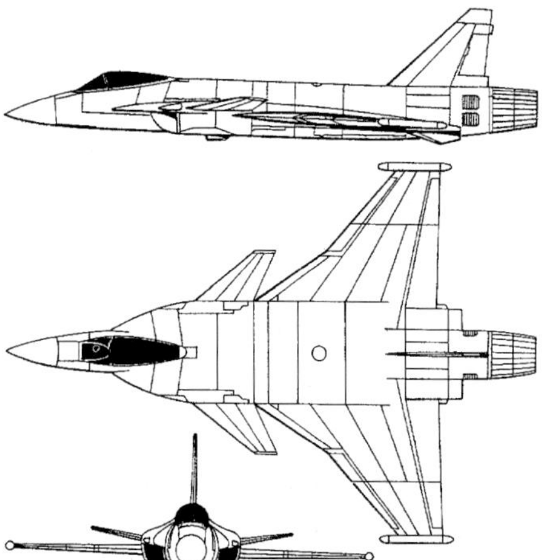

Su-37 project

weapon guidance and further protective features including part-multilayer structure, improved fuel-syst protection and additional redundancy.

Data for proposed series Su-35:

DIMENSIONS Span 15,16 m (49 ft 8⅞ in); length (excl PVD) 22 183 m (72 ft 9⅜ in); wing area 64,2 m² (691 ft²).

ENGINES Two AL-35F.

WEIGHTS Empty 17 t (37,478 lb); loaded (air combat) 26 t (57,320 lb), max 34 t (74,955 lb).

PERFORMANCE As Su-27 except speed at SL 1450 km/h (900 mph, M 1,19), service ceiling 18 km (59,055 ft); range (hi) 4200 km (2,610 miles).

Su-37 Projected multirole combat aircraft. Configuration as 3-view, blended wing/body with quarter-round inlets under LERX carrying foreplanes driven +10/–70°, other FBW controls slats/elevons/rudder. Incorporated Su-25 Afghan lessons to maximize survivability. Terrain-following radar (10-target capability), thermal imager, laser ranger/designator (front/rear), EW/IRCM (jammers on wingtips), retrac FR probe, gun in right wingroot, up to 8 t (17,637 lb) weapons on 18 attachments. Two-seat and naval versions intended, but reported cancelled early 94.

DIMENSIONS Span 11,8 m (38 ft 8½ in); length 17,5 m (57 ft 5 in).

ENGINE One Soyuz augmented turbofan in 18,4 t (40,600 lb) class.

WEIGHTS Max fuel 8,3 t; max external stores 8 t; normal loaded 16-18 t (35,275 lb-39,680 lb); max 25 t (55,116 lb).

PERFORMANCE Max speed at SL 1500 km/h (932 mph, M 1·22), at ht Mach 2 class; acceleration at 1 km (3,280 ft) 14 s from 600 km/h to 1100 km/h, 7·2 s from 1100 km/h to 1300 km/h; service ceiling 17 km (55,775 ft); combat radius, hi-lo-hi 3 t weapons, 1500 km (932 miles); ldg 220 km/h.

Su-38 Agricultural sprayer/duster derived from Su-29. Wing constant-chord, extended to large winglets, plain flaps; enlarged tail; rough-field gear with 500 mm tyres; room for seat behind pilot; 100-lit tank each wing; 500 kg chemicals in rear hopper or (with second seat) faired tank under centreline fed to eight atomisers along spraybars. To fly 95; marketed at $100,000.

DIMENSIONS Span 10,0 m (32 ft 9¾ in); length 7285 (23 ft 10¾ in); wing area 17,5 m² (188 ft²).

ENGINE One M-14PT.

WEIGHTS Equipped empty 970 kg (2,138 lb); max 1650 (3,638).

PERFORMANCE Max speed 300 km/h (186 mph); econ cruise 120 (74·5); range (max load) 700 km (435 miles), (max fuel) 1000 (621); TO 120; ldg 280.

Su-39 Production Su-25T/TM/TK.

S-21 SSBJ (supersonic business jet) planned in partnership with Gulfstream Aerospace of USA. Configuration settled as drawing, MTO 51,8 t (114,198 lb), span 19 920 m (65 ft 4¼ in) length 36,8 m (120 ft 8⅞ in), 6/10 pax at Mach 2 for 7400 km (4,600 miles), same range at 0·95; prototype considered with two AL-31F, production intended to have three D-21A1. Gulfstream withdrew 92, project deferred to late 90s.

S-51 OKB study for small SST for up to 68 pax, variable-cycle engines to permit same fuel burn subsonic or Mach 2 (so can overfly anywhere), range 'one stop to anywhere'. No decision to build.

DIMENSIONS Span 27,12 m (88 ft 11¾ in); length 50,7 m (166 ft 4 in).

ENGINES Not yet produced.

WEIGHTS Max 90 t (198,413 lb).

PERFORMANCE Cruise, either 2125 km/h, M2, or 1000 km/h, 621 mph, M 0·9; cruise height 17-19 km (55,775-62,335 ft); range 9200 km (5,717 miles).

S-54 Market for subsonic trainers saturated, OKB propose supersonic, FBW programmable to make easier or more difficult to fly, with pushbutton recovery from spin or dangerous attitude at low level; design factors +9 g/–3 g; avionics and weapons for air combat or surface attack.

DIMENSIONS Span 9,08 m (29 ft 9½ in); length 12,3 m (40 ft 4¼ in).

ENGINES Two R-195FS suggested.

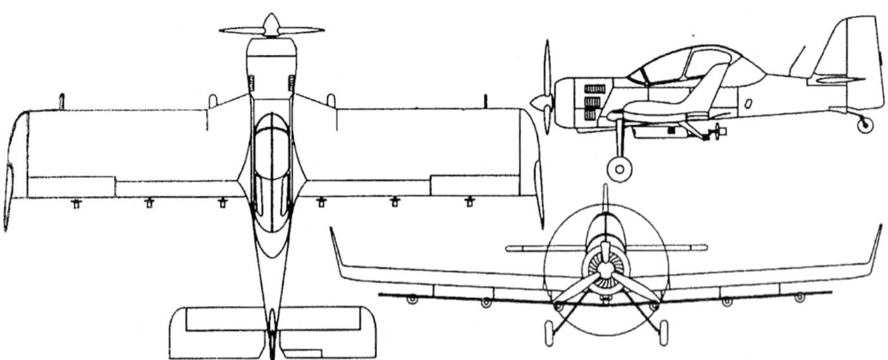

Su-38

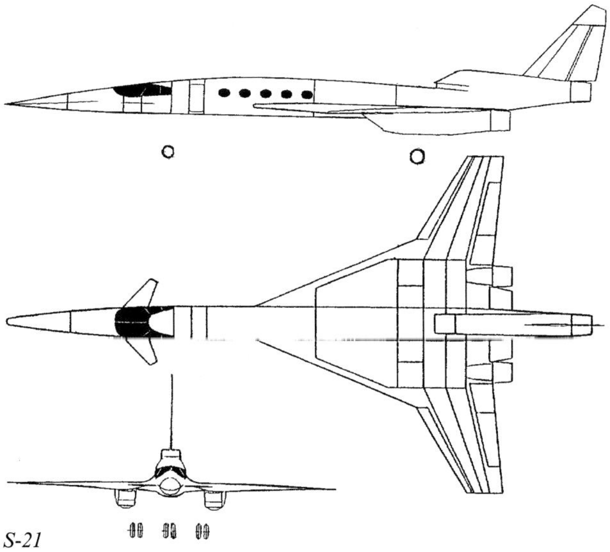

S-21

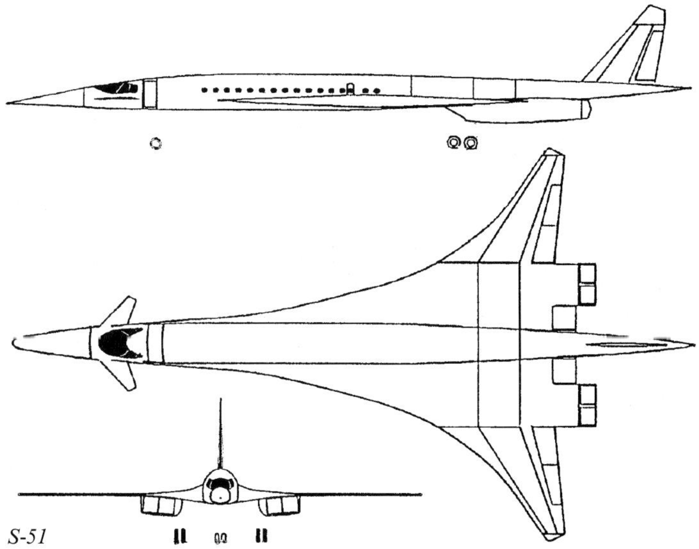

S-51

WEIGHTS Not published.
PERFORMANCE Max speed (SL) 1200 km/h (745 mph), (hi) 1650 km/h (1,025 mph, M 1·55); service ceiling 18 km (59,055 ft); range (SL) 820 km (510 miles), (hi) 2000 km (1,242 miles); TO 360, ldg 500.

S-80 Major programme for twin-turboprop multirole transport; config originally (91) tandem-wing, by 93 as drawing, advanced structure and STOL aerodynamics, cabin 6,3×2,13×1,9 m (20 ft 8 in×7 ft×6 ft 2¼ in) for 2,5 t (5,511 lb) cargo (full-section rear ramp door) or 23 pax or 10 stretchers or eqpt for mil assault (inc 21 paras), geological exploration, offshore patrol or other tasks; 6-blade Stupino props; eqpt (eg APU) for sustained ops from rough strips devoid of facilities except fuel. First of five to fly Komsomolsk 94, prodn of 2,000 planned in many versions.
DIMENSIONS Span 23,177 m (76 ft 0½ in); length 16,68 m^2 (54 ft 8⅝ in); wing area 44,0 m^2 (474 ft^2).
ENGINES Two TVD-1500S or GE CT7-9D.
WEIGHT OWE 6230 kg (13,735 lb); fuel 2370 (5,225); max 11,0 ft (24,250).
PERFORMANCE Max speed 500 km/h (311 mph); max climb 900 m (2,953 ft)/min; range (max payload) 1250 km (777 miles), (max fuel) 4500 km (2,800 miles); TO 360, ldg (reverse thrust) 180.

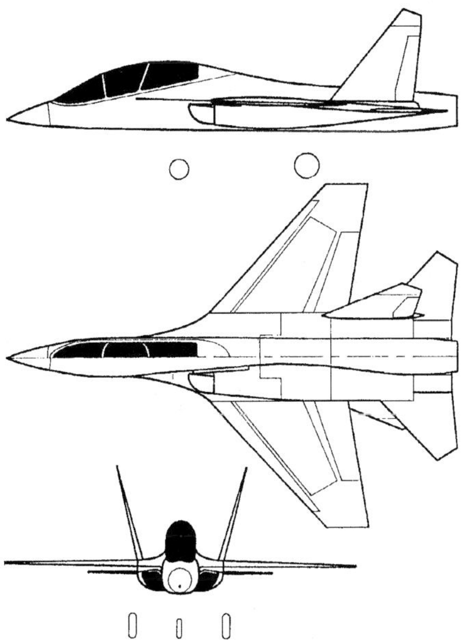

S-54

S-84 Utility aircraft, many design changes, 5-blade prop behind T-tail, all-composites, retrac nosewheel gear, cabin ahead of mid wing for up to 5 seats with provision for special missions, active flight controls and advanced avionics. Prototype building.
DIMENSIONS Span 12,55 m (41 ft 2¼ in); length 9,71 m (31 ft 10¼ in); wing area 16,83 m^2 (181 ft^2).
ENGINE One 350-hp TCM TSIO-550 flat-six.
WEIGHTS Empty 1105 kg (2,436 lb); fuel 430 (948); payload 500 (1,102); max 1900 (4,189).
PERFORMANCE Max cruise 380 km/h (236 mph); econ 220 (136); range (max fuel) 2910 km (1,808 miles), (max payload) 2540 (1,578); TO 340, ldg 180.

S-86 Multirole transport, repeatedly modified and (95) not yet committed to build; config as drawing; movable flaps on foreplane, pressurized cabin 6 m long and 1640 m wide for 540 kg (1,190 lb) cargo or 7 seats or furnished for spe-

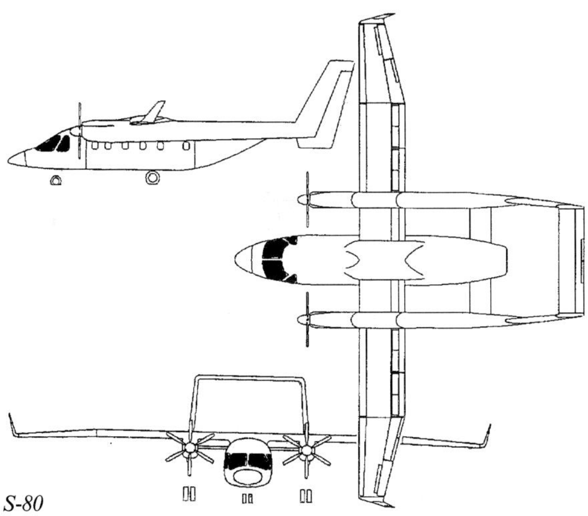

S-80

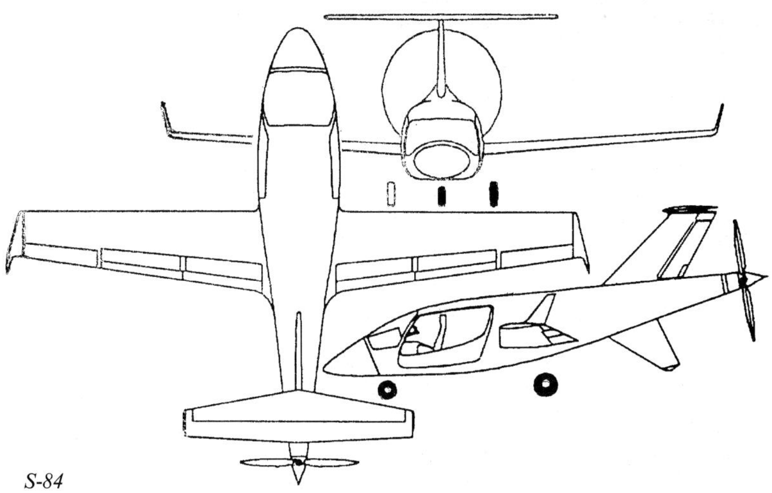

S-84

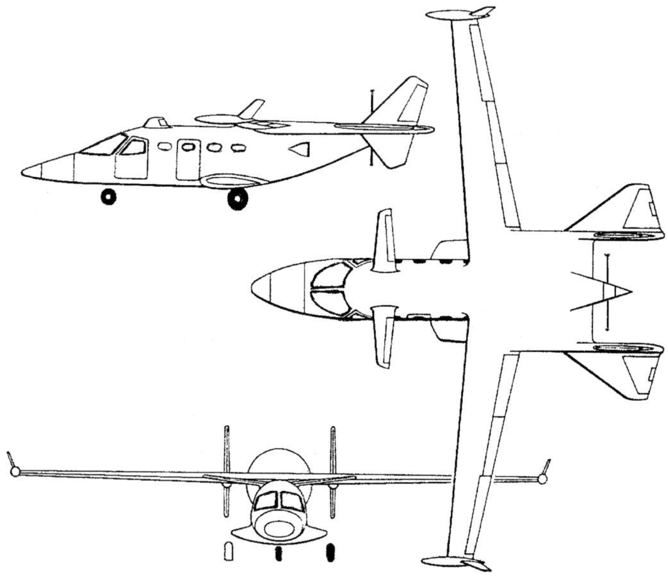

S-86

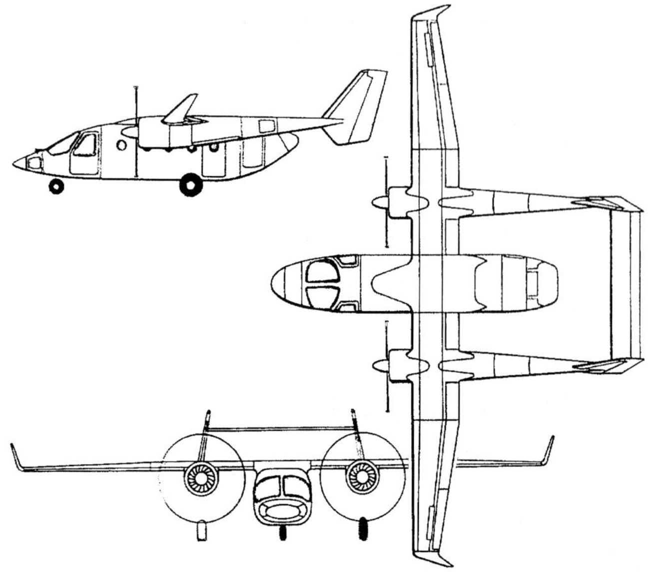

S-986

cial roles; 1350 kg fuel for twin engines driving pusher contraprop. Probably replaced by S-986.
DIMENSIONS Span 15,87 m (52 ft 0⅞ in); length 11,57 m (37 ft 11½ in); wing area 21,03 m^2 (226 ft^2).
ENGINE Two turboshaft (TVD-450 or AL-34) each rated at 404 kW (542 shp).
WEIGHTS Empty 2,9 t (6,393 lb); MTO 4,5 t (9,921 lb).
PERFORMANCE Max speed 600 km/h (373 mph); service ceiling 10,5 km (34,450 ft); range (max fuel) 3500 km (2,175 miles); TO/ldg 500 m.

S-986 Light utility transport, as drawing, cabin 1,68 m (66 in) wide and high, fuel 550 lit (121 gal), 9 pax or eqpt for rescue, firefighting etc; hyd flaps/spoilers/gear/rear ramp door; could fly 96. Data provisional.
DIMENSIONS Span 16 m (52 ft 6 in); length 11,6 m (38 ft); wing area 21,2 m^2 (228 ft^2).
ENGINES Two M-14P or NTK.
WEIGHTS Empty 2,3 t (5,070 lb); max 4 t (8,818 lb).
PERFORMANCE Max speed 350 (217); econ cruise 210 (130); range (max fuel) 1500 (932); TO 625; ldg 350.

Taifun

Moscow OKB formed 92 to build GA aircraft. Announced four projects at AeroSalon 93.

Duet Side-by-side, all-metal, fixed gear (floats and special role options), Czech engine and Avia-Hamilton prop.
DIMENSIONS Span 10 518 m (34 ft 6 in); length (land) 6370 m (20 ft 10⅞ in); wing area 11 055 m^2 (119 ft^2).
ENGINE One 140 hp M-332A.
WEIGHTS Empty 400 kg (882 lb); MTO 650 kg (1,433 lb).
PERFORMANCE (est) Max speed 250 km/h, cruise 215 km/h (134 mph); range 800 km (497 miles); TO 139 m, 103 km/h, ldg 92 km/h.

Selena 2-seat twin multirole, floats and special role options, engines from Austria.

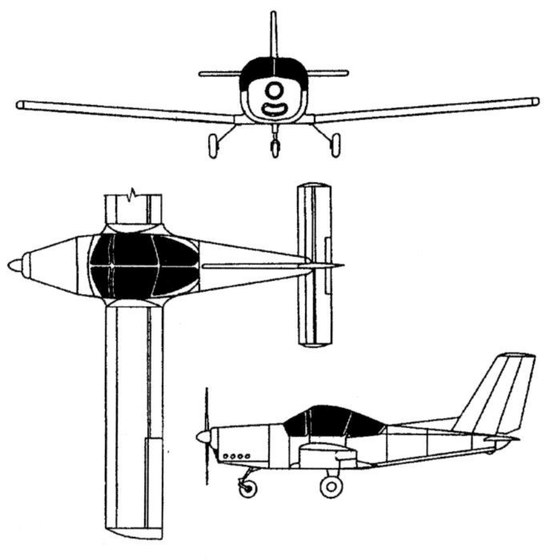

Duet

DIMENSIONS Span 10 m (32 ft 9⅝ in); length (land) 6,2 m (20 ft 4⅛ in); wing area 11 m^2 (118 ft^2).
ENGINES Two 60-76 hp Rotax.
WEIGHTS Empty 380 kg (838 lb); MTO 650 kg (1,433 lb).
PERFORMANCE (est) Cruise 180 km/h (119 mph); range 500 km (373 miles); TO 110 m, ldg 100 m.

Bekas Light twin, seating not announced, special role eqpt options; name = snipe.
DIMENSIONS Span 12 m (39 ft 4⅜ in); length 8 m (26 ft 2⅞ in); wing area 14 m^2 (151 ft^2).
ENGINES Two unspec. 140 hp.
WEIGHTS Empty 890 kg (1,962 lb); MTO 1,5 t (3,307 lb).
PERFORMANCE Max speed 300 km/h (186 mph); cruise 270 km/h (168 mph); climb 300 m (984 ft/min; range 800 km-1000 km; TO 250 m.

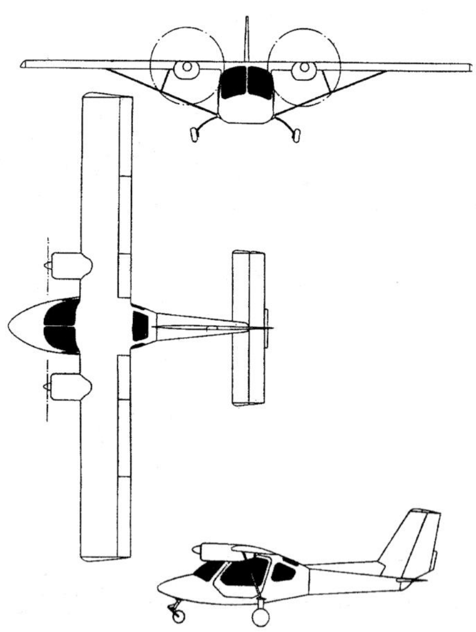

Selena

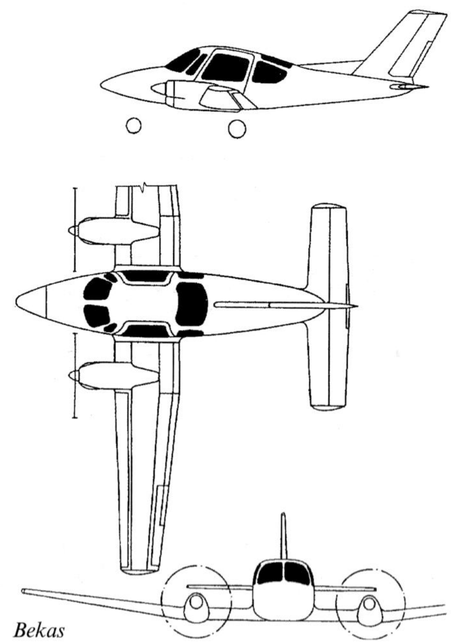

Bekas

Stayer Light 10-seat unpressurized multirole twin-propfan (previously unknown hi-b.p.r. engines), airways avionics.
DIMENSIONS Span 12,5 m (41 ft 0 in); length 11,5 m (37 ft 8⅞ in); wing area 40 m^2 (430·5 ft^2).
ENGINES Two Ryzhov 450 hp.
WEIGHTS Empty 2360 kg (5,203 lb); MTO 4,1 t (9,039 lb).
PERFORMANCE (est) Max speed 520 km/h (323 mph); cruise 475 km/h (295 mph); climb 480 m (1,575 ft)/min; operating ceiling 5 km (16,400 ft); range 1800 km (1,120 miles); TO 280 m/155 km/h, ldg 140 km/h.

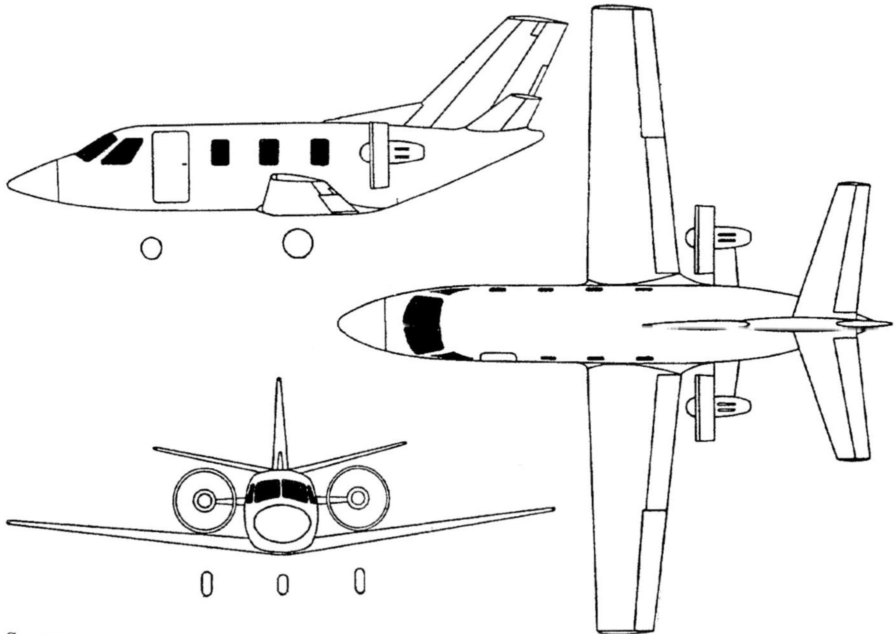
Stayer

Tikhonravov

Prof Mikhail Klavdiyevich Tikhonravov was member of staff at RNII.

302, I-302 This experimental mixed-power interceptor was advanced project authorized by NKAP 40 despite radical nature. Take-off by rocket, in tail of fuselage, and cruise on two VRDP (ramjets) under wings. Wooden construction, low wing of high aspect-ratio for fighter (7·3) with two spars and laminar profile. Wood monocoque fuselage with integral fin, four 20-mm ShVAK in nose, pilot in pressurized cockpit, hydraulically retractable tailwheel landing gear. Intended to have wing racks for bombs and RS-82 rockets. Ramjets not Merkulov but by brigade led by Vladimir Stepanovich Zuyev. Delays with propulsion led to construction of Aircraft 302P, glider for low-speed flight test, with VRDP mounts faired over. Two built 42 at VVA by authority of A. G. Kostikov of NKAP as stepping-stone to interceptor despite decision in that year not to pursue such aircraft. One airframe static-tested at CAHI, second flown spring 43 under tow by Pe-2; pilots included S. N. Anokhin and M. L. Gallai. Terminated by declining need and poor progress of chosen VRDP.
DIMENSIONS Span 9,55 m (31 ft 4 in); length 8708 mm (28 ft 6⅞ in); wing area 17,8 m^2 (192 ft^2).
ENGINES One 1,5 t (3307 lb at SL) NII-3 rocket engine, two Zuyev ramjets.

302P

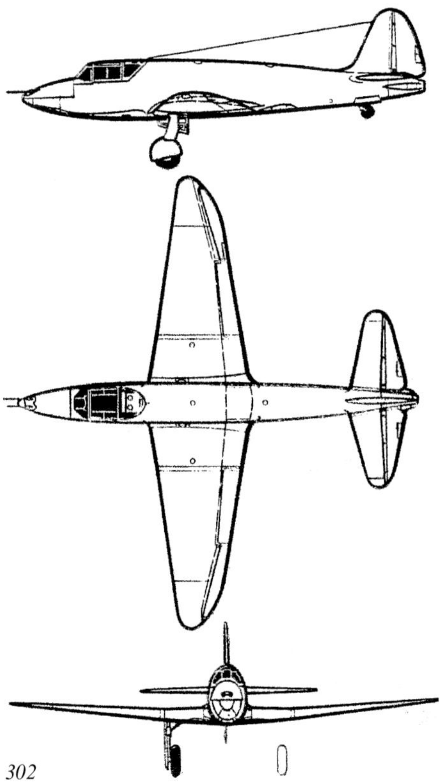
302

Tokayev

Utka research aircraft is described as MiG-8, see page 184.

Tomashyevich

Dmitrii Lyudvigovich Tomashyevich was born in Ukraine about 98, serving in the First World War and subsequent Red Army as engineer and finding work on RVZ-6 on its establishment in 23. Picked out by Kalinin as senior assistant and played active role in K-1 and K-2 whilst also attending Kiev Poly Inst for formal qualification. KPIR-5 lightplane was diploma project. Moved to CCB 31, and on Polikarpov's release from prison 33 joined his brigade as senior deputy constructor. Chief designer on I-180 programme, and on Chkalov's death in first I-180 immediately imprisoned. Assigned to

Special Prison KB-29 Jan 39 to lead Group 101, to create Fighter No 101. Abruptly terminated and assigned to assist Tupolev on No 103. In Oct 41 moved to Siberia and placed in charge of own semi-free KB at GAZ-266 in Siberia (Kulomzino, suburb of Omsk). Here produced Aircraft 110 and Pegas.

KPIR Successful firstborn of Tomashyevich was a neat cantilever monoplane of wood and fabric construction. Full designation was **KPIR-5**, though forerunners never built. Initials stood for Kiev Polytechnic Institute, where Tomashyevich built it as a diploma project in 1926-27.
DIMENSIONS Span 12,0 m (39 ft 4½ in); length 7,0 m (22 ft 11½ in); wing area 19 m^2 (204·5 ft^2).
ENGINE One 27/35 hp ABC Scorpion.
WEIGHTS Empty 233 kg (514 lb); fuel/oil 20 kg (44 lb); loaded 403 kg (888 lb).
PERFORMANCE Max speed 95 km/h (59 mph); landing speed 45 km/h (28 mph).

110 Often called **I-110** in West, this fundamentally conventional fighter was designed for simple field maintenance, as shown by removal of four bolts to release entire powerplant group complete with cowl, and provision of small hinged panels in top cowl for crane hooks. Mixed construction saving scarce alloys. Wing NACA-23 profile with tapered centre-section 3,6 m, dihedral 0° and outer panels dihedral 8°30'. Main spar D1 and 30KhGSA booms, front and rear secondary spars and truss/sheet ribs all-dural, but wing skin birch *shpon*. Fuselage steel-tube (mainly 30KhGSA) at front, with D1 cowl, but aft of main spar birch *shpon* monocoque. Wood fin and tailplane carrying control surfaces of D1 and fabric. Eight fuel tanks, four in centre section and four in outer wings, all claimed self-sealing against fire up to 12,7-mm calibre. Single-strut main gears, all three units fully retractable. Extremely large (29 m^2) radiator under engine in pressurized glycol circuit with integral oil radiator, all auto-controlled with elec-driven flaps at rear of duct and removed as part of complete powerplant. Large glazed areas over cockpit with central hood sliding to rear. Standardized instrument panel with quick elec-vacuum/pressure connections removable as one unit. Armament, one ShVAK firing through hub of 3,2-m three-blade propeller and two UBS (alternatively ShKAS) in top decking; belly rack for bomb or other store to max 500 kg. Previously thought completed 43, now known P. M. Stefanovskii made first flight Dec 42 and was generally satisfied. Handling fair, and performance adequate though severely compromised by extra weight of easy-maintenance and quick-build features such as making every dimension exact multiple of 10 mm. Major reasons for abandoning idea of recommended production was unavailability of engine and lack of production capacity for airframe.
DIMENSIONS Span 10,2 m (33 ft 5½ in); length 9,91 m (32 ft 6⅙ in); wing area 18,73 m^2 (202 ft^2).
ENGINE One VK-107.
WEIGHTS Empty 3285 kg (7,242 lb); fuel/oil 410 kg+35 kg; loaded 3980 kg (8,774 lb).
PERFORMANCE Max speed 508 km/h at SL, 610 km/h (379 mph) at 6,2 km; time to 3 km, 4 min, to 7 km, 7 min; service ceiling 10 km (32,800 ft); max range 1050 km (652 miles).

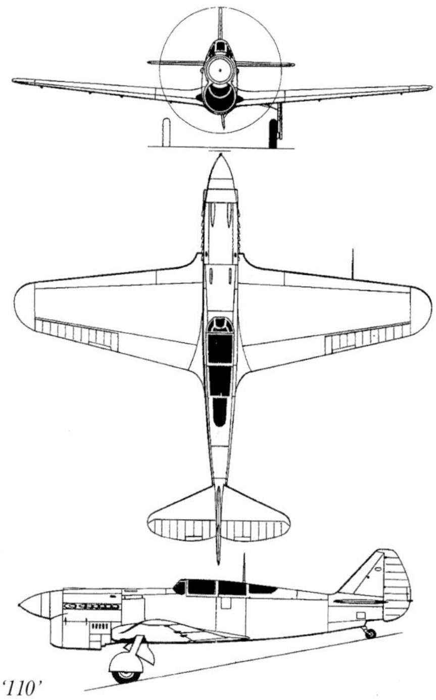
'110'

Pegas Successful use of U-2 (Po-2) variants as night bomber and close-support attack aircraft spurred launch of this much better light tactical attack aircraft making small demands on either construction workers or pilots, and sacrificing performance for simplicity and utility. Despite poor performance everything possible done to ensure survivability against ground fire (rifle-calibre for transparencies and 12,7-mm or 20-mm for vital areas) and ability to regain base and be quickly repaired. Authorized Aug 42 and first of five prototypes flying before end of year. Severely angular low-wing machine with two Po-2 engine/propeller groups. Aerodynamics ignored except to allow flight by unskilled pilot at angle of attack up to 20°. Construction almost entirely wood, mainly pine frames and birch ply skin. Two-spar wing with fixed slots over outer sections, large ailerons but no flaps. Option of removable or jettisonable upper plane. Fuselage built up on four pine longerons with ply skin except mild-steel armour over nose and cockpit. Armoured glass windscreen giving view 55° downwards directly ahead. Engines uncowled though with small dural shield above cylinders and exhausts piped around rear and over leading edge. Galvanized-iron fuel tanks forming nacelles jettisonable but leaving small emergency armoured tank giving several minutes' supply. Fixed tailwheel landing gear with rubber shock-absorption, first flight made on wheels but skis soon substituted. All-wood tail. Armament two VYa-23 and one UBS side-by-side in nose with breeches in cockpit. Alternative load, two 250-kg or one 500-kg bomb under fuselage, or fragmentation bombs. Original intention was mass-production at same airframe price as Po-2, using all locally available materials. First flight about same time as

110

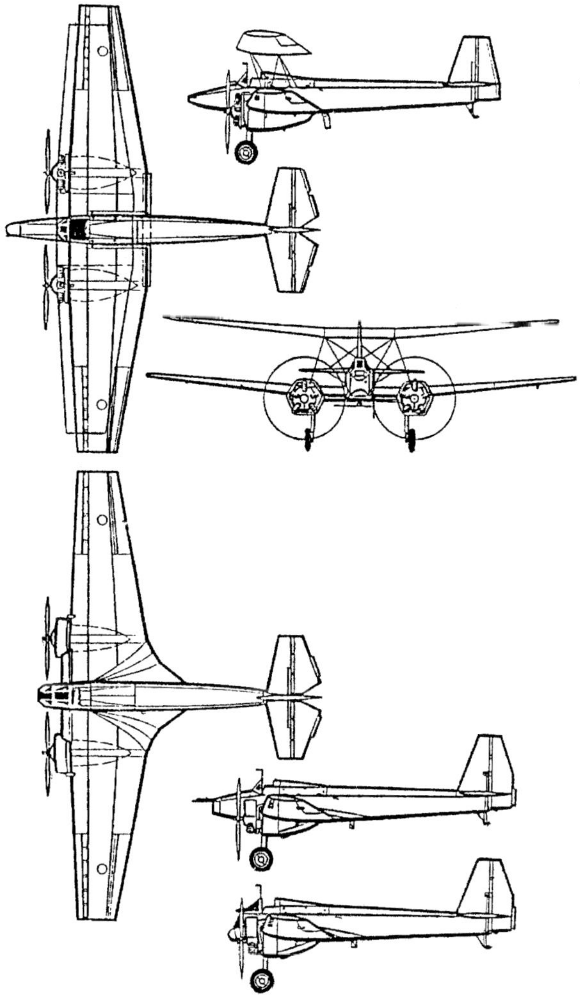

Pegas (3-view -01 with upper wing; middle side view -03; bottom and second plan view -04)

Pegas 04

Aircraft 110, at end 42. Stefanovskii and several other pilots carried out NII testing Feb 43 and found aircraft safe but overweight. Recommendation that examples be tested at Kursk front came to nothing, partly because of sheer distance of OKB. MG-31 engine was suggested but programme was dropped spring 43 because of declining need for such emergency solutions. Tomashyevich's design bureau was closed, but he continued to work as a designer. Now known that four prototypes were built, all different. Last two had larger vertical tail, ring cowl round top of engine and progressively shorter nose. Data are for -01, without upper wing.

DIMENSIONS Span 12,63 m (41 ft 5¼ in); length 8784 mm (28 ft 9⅞ in); wing area 26,6 m^2 (286 ft^2).

ENGINES Two M-11F.

WEIGHTS Empty 1800 kg (3,968 lb); loaded (normal) 2150 kg (4,740 lb), (max) 2320 kg (5,115 lb). Max wt does not allow for 500-kg bomb.

PERFORMANCE Max speed 172 km/h (107 mph) at SL, (max wt) 167 km/h (104 mph); initial climb 2 m/s (400 ft/min); service ceiling 2,62 km (8,600 ft); range 400 km (249 miles); landing speed 80 km/h (50 mph).

Tsybin

Pavel Vladimirovich Tsybin was important engineer working at NII GVF and at LII in period 1940-60. Collaborated with Kolyesnikov on KTs-20 assault glider 40, subsequently designing at OKB-256. Worked with S. P. Korolyev from 58 on recoverable manned aerospace glider; joined Korolyev OKB 61 to direct pilotless versions of Vostok, Molniya and Soyuz, from 74 deputy chief designer of Buran (see Molniya). Died 4 Feb 92.

Ts-25 Assault glider built to meet VVS specification for Aviation of Airborne Troops (AVDV) issued early 44 calling for 2,5-t payload. No direct rival known. Simple high-wing transport of wooden construction with wing tapered from root and braced by steel-tube strut at main spar. Slotted flaps and ailerons carried on brackets beneath wing. Except for full-section fabric-covered nose, manually hinged to right, entire fuselage and fin ply-skinned with four circular windows each side and door on right. Side-by-side cockpit for two pilots above unobstructed hold, access by internal steps. Tricycle landing gear, jettisoned on combat mission for landing on three skids. Original design load was anti-tank gun (57-mm M-1943) plus Jeep.

Produced in small series from 45; some used by AVDV for manoeuvres, at least one with 25 removable passenger seats tested as civil mixed-traffic glider on routes between Moscow, Novosibirsk, Gorkii and Kuybishyev, and in 52 two handed to Czech air force (where designated NK-25, Yak-14 being preferred). ASCC name 'Mist'.

DIMENSIONS Span 24,38 m (79 ft 11⅞ in); length 16,15 m (52 ft 11⅘ in); wing area 75,0 m^2 (807 ft^2).

WEIGHTS Empty 1787 kg (3,940 lb); loaded 4,2 t (9,259 lb).

PERFORMANCE Max tow speed 230 km/h (143 mph); landing speed 90 km/h (56 mph).

Ts-25M Powered version of Ts-25, believe single aircraft converted from production glider 46. Prototype used for NII trials but no production.

DIMENSIONS As Ts-25.

ENGINES Two M-11FR-1.

WEIGHTS Not known but empty about 2350 kg (5,181 lb); loaded probably 4,5 t (9,921 lb).

PERFORMANCE No data.

Ts-25

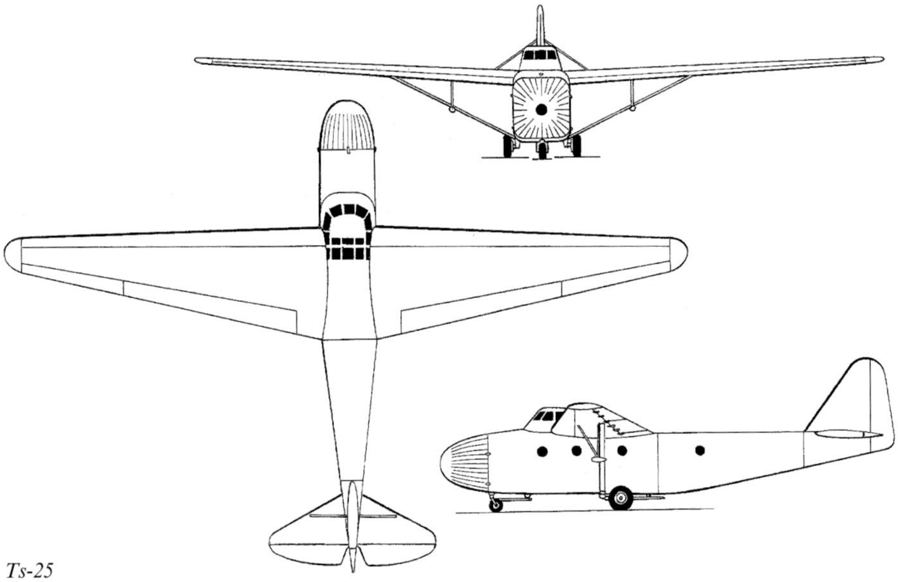

Ts-25

LL-1

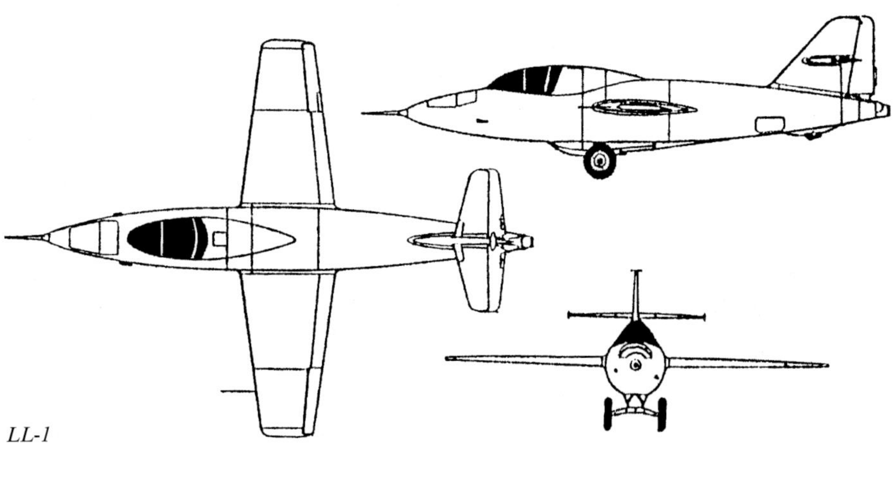

LL-1

Ts-1, LL In Sept 45 LII asked OKB-256 to build aircraft to test various wing shapes for transonic aircraft. Work began with **LL-1** (flying laboratory 1) wooden structure, two *Delta* spars in 5% straight wing, set at 2°, dihedral 0°, aileron 0,57 m^2, tailplane 2550 mm span and 1,97 m^2, fin 1,82 m^2, elev 0,67 m^2, rudder 0,54 m^2, flaps 1,36 m^2, all manual; ply-covered fuselage; short-duration rocket engine, tank for 1 t water ballast. Towed off two-wheel jettisoned dolly by Tu-2, cast off at 5-7 km, dived at 45°-60°, rocket fired, instrumentation on, camera photographing tufted wing. Ballast jettisoned, return as glider on central retrac skid. **LL-2** all-metal, wing swept back 30°, anhedral –4°, not completed (plenty of sweptback data). **LL-3** with wing swept forward 40°, dihedral 12°, completed with left wing tufted. LL-1 (30 flights) and -3 (100) flown by NII pilot M. Ivanov, subsequently by Amet-Khan Sultan, Anokhin and Rybko, very successful programme.

DIMENSIONS Span (LL-1, -2) 7,1 m (23 ft 3½ in), (LL-3) 7220 mm (23 ft 8¼ in); length 8,98 m (29 ft 5½ in); wing area (all) 10 m^2 (108 ft^2).
ENGINE One PRD-1500.
WEIGHTS (LL-1) Empty 1 t (2,205 lb); loaded 2039 kg (4,495 lb), landing 1,1 t (2,425 lb).
PERFORMANCE Max speed in 45° dive (-1) 1050 km/h (652 mph), (-3) 1200 km/h (M 0·97); landing 120 km/h.

RS On 4 March 54 Tsybin proposed a/c to fly 3000 km/h at 30 km with range 14 000 km, with 2·5% trapezoidal wing. On 23 May 55 OKB-256 received GUAP contract for design of RS (*Reaktivnyi Samolyet*, jet aircraft) to be released from Tu-95N at 9 km and thereafter accelerate on two rocket engines to 3000 km/h and then cruise for 13 000 km on ramjet power. Prelim design submitted 31 Aug 56 (vertical-launch *Burya* by Lavochkin and *Buran* by V. N. Chyelomey not accepted). Inability to carry thermonuclear bomb and success of S. P. Korolyev's giant R-7 ICBM (first flight 15 May 57) caused programme to be redirected, see next.

RSR Reqt changed to *RS Razvyedchik*, strategic day/night all-weather recon at 2500 km/h to 1700 km radius from 25,5 km without detectable emissions. D-16T, D-18 and D-20 structure, for extra rigidity outer wing, aileron and tail skins Al-Be (beryllium) alloy, design factor 2·5. Wing 2·5% thick, LE 58°, 5 spars changed to 16, conventional ailerons and plain flaps, low-BPR turbofans on 86 mm-thick tips. Circular 1,5 m-diam fuselage. One-piece tail surfs all 3·5%, vert ±18°, horiz +10°/–25°, all flight controls hyd power. Bicycle landing gear, main and steerable nose gears twin wheels, outrigger gears under engines, all units hyd retracting to rear. Braking parachute ahead of tail trim tank. Kerosene fuel, 7,3 t in D-20 integral tanks ahead of and behind wing, 4,4 t in two 650 diam drop tanks. Auto pump system to transfer fuel to tail tank during transonic accn, moving c.p. from 25% to 45% MAC, moved back to 26·4% for landing. Pilot in pressure suit in 30°C pressure cabin with ejection seat, APU and propane heater for instrument pallets and camera bay amidships, normally for two AFA-200 (mm focal length) and either one AFA-1000 or one AFA-1800, various mounts for

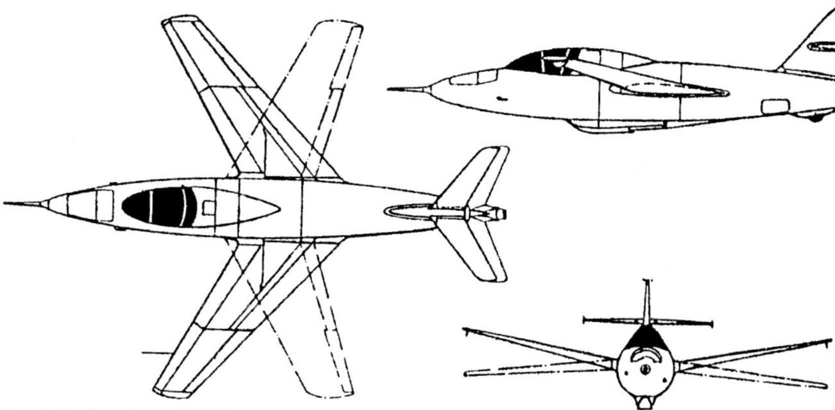

LL-3 (broken lines, LL-2)

vert/oblique/lateral cover, optical sights, autopilot, astro-inertial plus vert gyro, course system, panoramic radar, RWR, active/passive ECM. Operate from Class I runway, reach 8,5 km in 4 min and 20 km in 15 min, jettison tanks at 10,7 km at 420 m/s (940 mph), cruise M2+, avoid SAM by 2·5 g barrel roll to 42 km (137,800 ft) dynamic ceiling, return to Class II runway. Project issued 26 June 57, plans for initial five series aircraft to be built at Ulan-Ude, but project crippled by N. S. Khrushchyev's closure of OKB-256. Never built in this form, see **R-020**.

DIMENSIONS Span (over engines) 10,69 m (35 ft 0¾ in); length 27,4 m (89 ft 10¾ in); wheelbase 9,9 m (32 ft 5¾ in); wing area 58,4 m^2 (629 ft^2).

ENGINES Two D-21.

WEIGHTS Empty 8,9 t (19,621 lb); loaded 21 t (46,296 lb); landing 9,2 t (20,282 lb).

PERFORMANCE (Est) Max speed 3000 km/h (1,864 mph, M 2·82); service ceiling (M 2·65) 26,7 km (87,600 ft); range (20 km at M 2·0) 3760 km (2,336 miles); TO 330 km/h, 1,3 km; ldg 245 km/h, 1,2 km.

NM-1 *Naturnaya Model*, life (like) model, built 57 to assist design of RSR. Single-seater, five-spar 2·5% wing, powered flight controls including slab tail surfs. Tiny one-piece canopy (never flown with RS-4/01 canopy resembling RSR). Nose water tank to adjust CG to 25·5% MAC, two kerosene tanks plus hyd-oil tank. On CAHI advice small wing extensions outboard of engines. Takeoff from two-wheel trolley (as LL series), landing on retractable main skid, tip skids under nacelles and tail skid; braking parachute. Taxi runs by A-K Sultan from 1 Oct 58, first flight 7 April 59, later accompanied by Yak-25 taken to 500 km/h at 1,5 km. Stopped after 32 (one source says 13) flights by Sultan and Radiy Zakharov upon decision not to proceed with RSR.

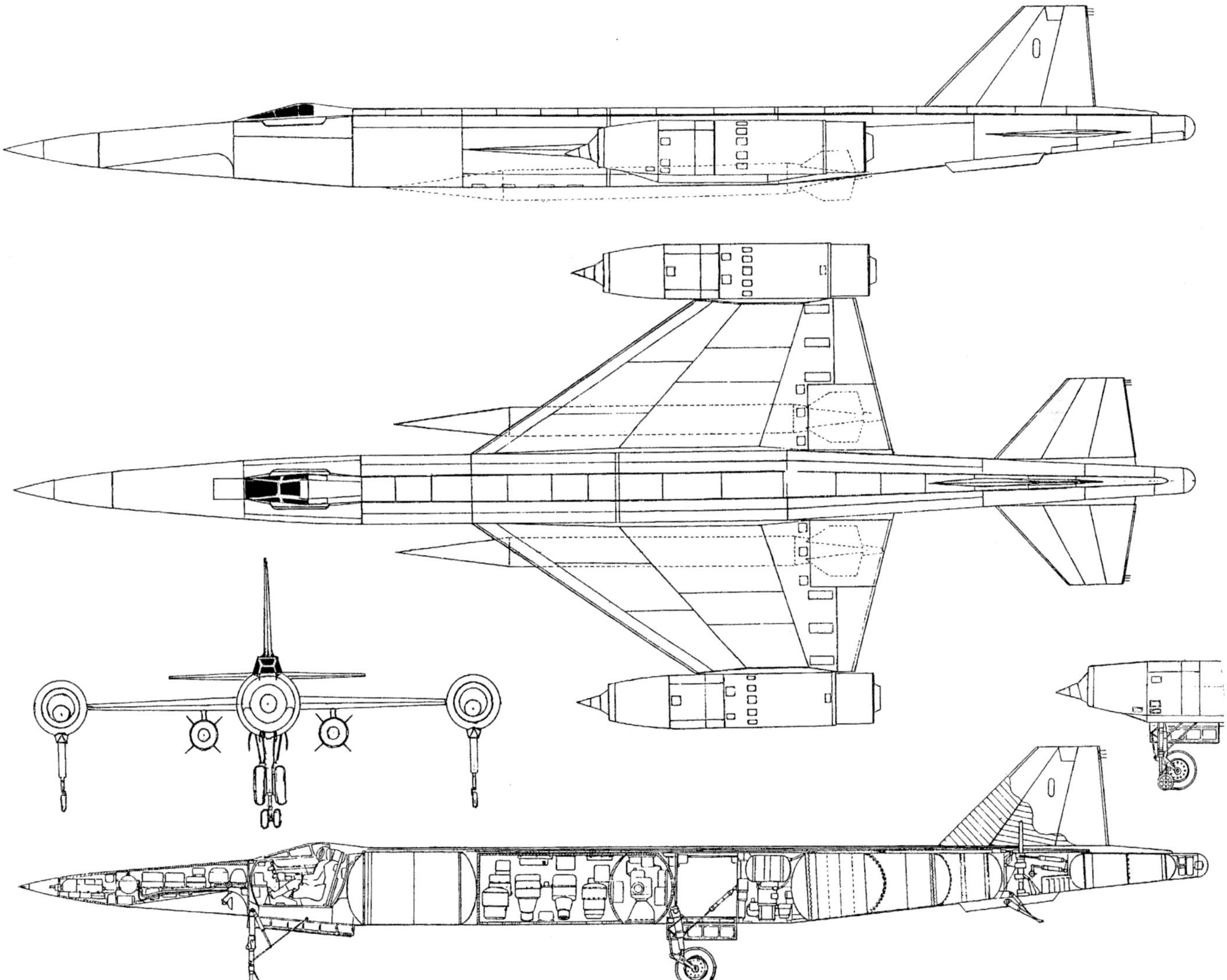

RSR as designed with twin-wheel main gear

NM-1

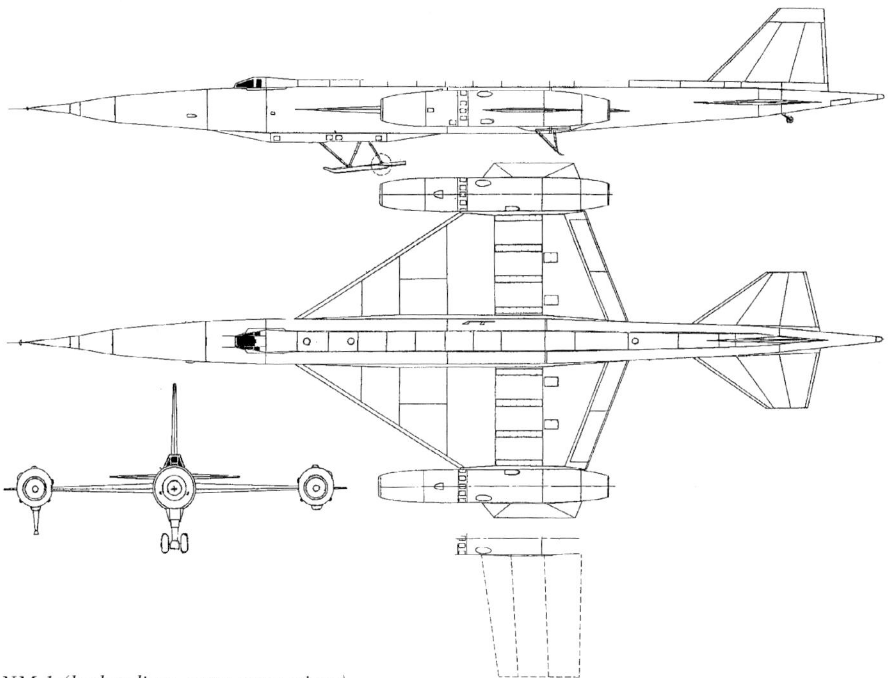

NM-1 (broken lines, extra outer wings)

DIMENSIONS Span (without outer wings) 10,48 m (34 ft 4⅝ in); length 26,57 m (87 ft 3¼ in).
ENGINES Two AM-5.
WEIGHTS Empty 6355 kg (14,010 lb); loaded 9 t (19,840 lb).
PERFORMANCE Max speed at low level 500 km/h (373 mph) but performance not explored; longest flight 15 min.

R-020/3 NM-1 data resulted in major changes to RSR. Redesign begun late 58, continued 59 upon receipt of NM-1 data. Major alterations: more powerful engines (turbojets instead of low-BPR turbofans) in longer but slimmer nacelles with axis of inlet spike horizontal; redesigned wing structure, area reduced by decreasing span and increasing TE taper, 'no movables except ailerons + trimmers', but drawings show flaps retained and LE flaps (max 10°) added, 'CAHI wing extensions did not produce desired effect and not included' but OKB drawings show them; 16-spar wing; longer fuselage with six thermally lagged tanks; nose gear with long levered suspension, 500 mm × 150 mm tyres, main leg raked 3° forward; main gear with four-wheel bogie, 750 mm × 250 mm tyres, changed leg/actuation geometry; outriggers changed to skids; tail surfs greater taper and reduced span, areas unchanged; tailplane power unit moved ahead of pivot so No 6 tank (for transonic trimming) moved forward and shortened; ventral strake removed, fuel-pipe fairing added; windscreen of acute V-type but less raked in side view, longer canopy fairing; mid-fuselage with hard lower chines tapered off at front and rear and flat underside with sliding door over camera bay; drop tanks only 5,8 m long but diam 700 mm; AOA sensor on nose probe and pitot on fin.

Prodn drawings issued to Ulan-Ude for initial five aircraft. On 1 Oct 59 Khrushchyev closed OKB-256, programme transferred to Myasishchyev at OKB-23, design team moved from Podberez'ye to Khrunichyev works, but on transfer of OKB-23 to spacecraft programme again moved to OKB-52 (V. N. Chyelomey) who found it interfered with OKB's main work. All five almost complete, waiting only for engines, when in April 61 order came to reduce to scrap. Workers resisted this; aircraft quietly stored for 3-4 years and then scrapped without protest. Parts of R-020/3 sent to MAI for instructional purposes.
DIMENSIONS Span 10 360 mm (33 ft 11⅞ in); length 29,0 m (95 ft 1⅞ in); wing area 57,9 m² (623 ft²).
ENGINES Two R-11F-300.
WEIGHTS, PERFORMANCE No data.

Tupolev

Andrei Nikolayevich Tupolev was responsible for more types of diverse aircraft than any other designer in history. From early stage his seniority put him more in position of administrator than designer, many of aircraft described below being created under Tupolev direction by teams led by Sukhoi, Petlyakov, Arkhangyelskii and Myasishchyev. As whole, however, combined ANT-Tu family has no parallel.

A. N. Tupolev born 10 Sept 88 at Pustomazov, near Kalinin, son of lawyer exiled for revolutionary activities. Mech engineering MVTU under Zhukovskii 08, arrested 11 for revolutionary activities but not exiled and returned MVTU 14. Engineer at Duks factory 15, remained there in Revolution and 18 co-founder with Zhukovskii of CAHI. With I. I. Sidorin chief technical member of 20 NTK VVF (state scientific commission on aircraft materials), member of KOMTA (which see) and founder and director of CAHI AGO (*Aviatsii i Gidrodinamiki Otdel*, aviation and hydrodynamics group) supervising various aero and sea designs including aerosleighs and hydroplanes at 16-17 Radio St, Moscow, all manufacturing in largest upstairs room. Head of 22 state committee on metal aircraft construction, strongly influenced by Junkers. Formed AGOS 24 as principal CAHI design bureau to manage challenge of ANT-4, brigades led by Arkhangelskii, A. A. Boykov, B. M. Kondorskii, N. S. Nekrasov, Petlyakov, N. I. Petrov, I. I. and Ye. I. Pogosskii, Putilov, Sidorin and Zimin. Appointed additionally 31 chief engineer of GUAP, but arrested 37 in strategic move to group major design strength in 'special prisons' as done earlier with Polikarpov and others; ridiculous charge was that he had passed design of Bf 110 to Germans. Lubyanka and Butyrkii prisons followed by GAZ-39 and -156, led KB-103 team to build Aircraft 103, later Tu-2, gaining freedom and first Stalin Prize 43. Managed Soviet copy of B-29 from 44 using derived technology in long and important succession of military and

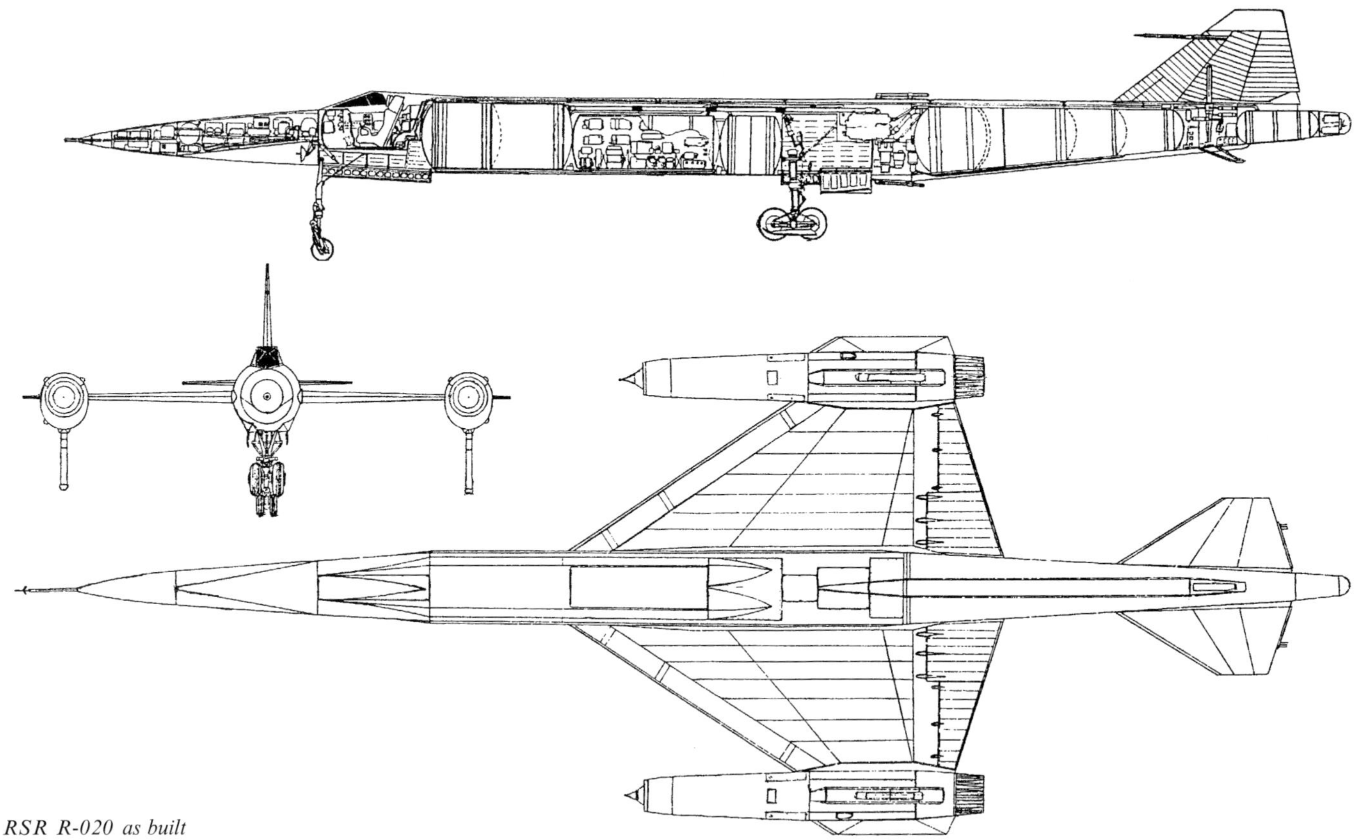

RSR R-020 as built

RSR R-020 mid-fuselage (4-wheel main gear)

civil piston, turboprop and jet aircraft. Followed Boeing 7-7 sequence in assigning OKB numbers ending in 4 to civil transports, resulting in out-of-sequence later types. Andrei Tupolev died 23 Dec 72; OKB continues to specialize in large aircraft, with chief designer for each programme, that for Tu-144 having been founder's son Dr Alexei Andreyevich Tupolev, until 92 General Constructor, deputy Andrei I. Kandalov. Today giant bureau is called ANTK Aviation Scientific/Technical Complex Imeni (named for) A. N. Tupolev, with links with groups within and outside Russian Federation, General Constructor Valentin Klimov.

ANT-1 Perfection of Kolchug duralumin-type alloy in early 22 led to pre-production fabrication of sheet and rolled sections such as angles, top-hats and Z in standard thicknesses and section sizes. To put specimens to use in actual airframe Tupolev himself designed small single-seater with minimum-risk structure using mainly wood, but with all ribs, control-surface frames and various other parts Kolchug. Part fabric covered, only complex feature differential ailerons. Designed from spring 22, built 17 Radio St (formerly house of fur merchant), flown from Cadet Sq nearby by Ye. I. Pogosskii 21 Oct 23. Flying (also by Petrov) curtailed by worn-out engine.

DIMENSIONS Span 7,2 m (23 ft 7½ in); length 5,4 m (17 ft 8⅝ in); wing area 10,0 m² (108 ft²).

ENGINE One 35 hp Anzani 6-cyl.

WEIGHTS Empty 229 kg (505 lb); fuel/oil 42 kg+9 kg; loaded 360 kg (794 lb).

PERFORMANCE Max speed 135 km/h (84 mph); greatest ht reached 400 m (1,300 ft); range about 450 km (280 miles); landing speed 70 km/h.

ANT-2 Success of ANT-1 prompted Tupolev to design this light transport with all-Kolchug airframe, first all-metal aircraft in Soviet Union. Design reminiscent of Junkers K 16 and skinned with same form of corrugated sheet (8 mm height, 40 mm pitch) in same gauges. Two-spar cantilever high wing with 13 truss ribs each side mounted on deep pot-bellied fuselage with almost flat sides coming together on lower centreline to give near-triangular section. Open pilot cockpit at leading edge, cabin with door on left and two seats facing each other. Main gears with wheels or skis, rubber springing inside fuselage and Kolchug skids under wingtips. First flight by Petrov 26 May 24, by end of year a further four built, most with larger fin and rudder. Recon version with rear gun-

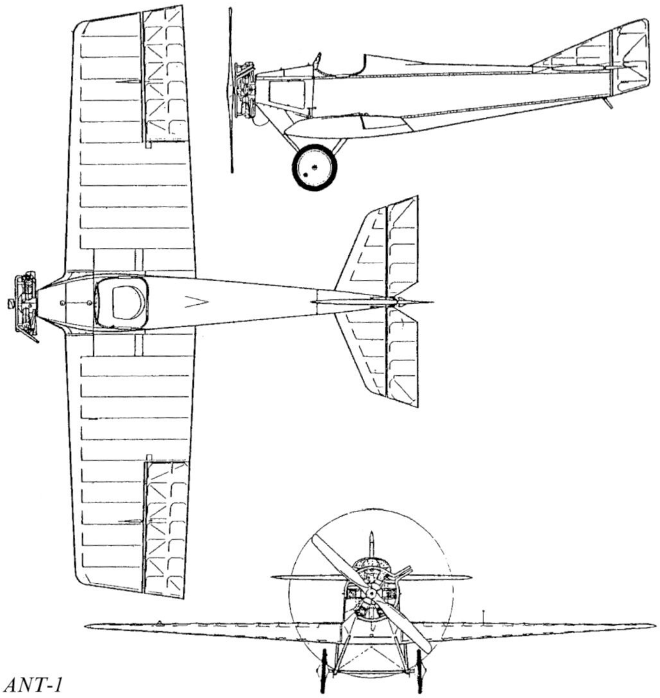

ANT-1

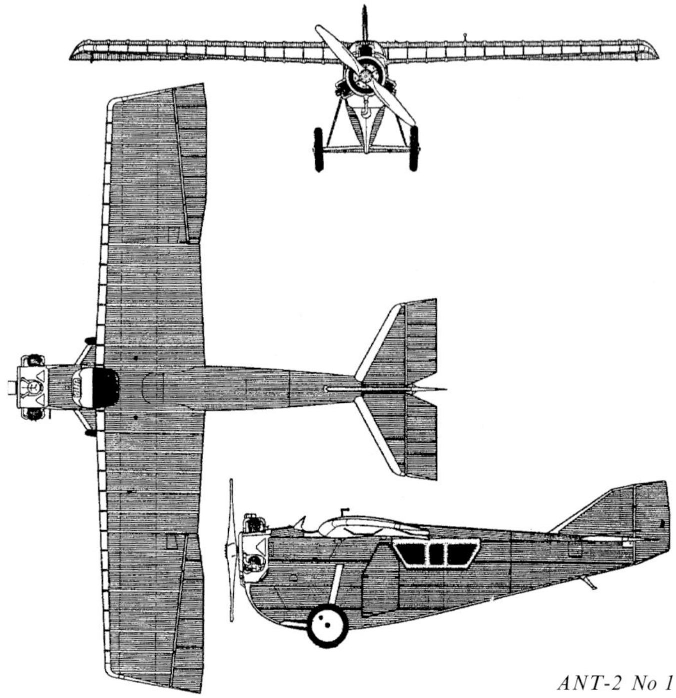

ANT-2 No 1

ANT-1

ANT-2

ner not built, but one aircraft modified as **ANT-2bis** for 30 All-Union contest with 200-hp Wright; no data for this. One at Monino.
DIMENSIONS Span 10,45 m (34 ft 3$\frac{3}{8}$ in); length 7,6 m (24 ft 11$\frac{1}{4}$ in); wing area 17,8 m^2 (192 ft^2).
ENGINE One 100 hp Bristol Lucifer.
WEIGHTS Empty 523 kg (1,153 lb); fuel 72 kg; loaded 837 kg (1,845 lb).
PERFORMANCE Max speed 170 km/h (106 mph) at SL; climb 8·5 min to 1 km; ceiling 3,3 km (10,800 ft); range (max) 700 km (435 miles); TO 350 m/12s; ldg 150 m/78 km/h.

ANT-3, R-3 Proof of Kolchug construction having been gained with ANT-2, VVF took important decision July 24 to adopt this construction for major military aircraft. AGOS (Tupolev personally) charged with building single-engined reconnaissance aircraft and large twin-engined bomber in all-Kolchug structure. Former authorized as R-3 forthwith, with AGOS designation ANT-3. Conventional single-bay biplane with Liberty engine (later M-5) but after studying Shishmaryev R-III Tupolev put upper wing very close to fuselage. Deep fuselage with same near-triangular section as ANT-2, with depth almost sufficient to fill gap between wings. Sesquiplane layout with 24,5 m^2 upper wing with two spars and 12,5 m^2 lower wing with single spar. Y-type interplane struts of Kolchug sheet wrapped to streamline form linking all spars. Various forms of Kolchug used for primary structure, entire skin (even control surfaces), ailerons (upper wings only) being 40 mm × 8 mm or 20 mm × 5 mm corrugated sheet, tail control surfaces having large horn balances. Same control system as ANT-2, cables to rudder only. Upper wing on two Kolchug fairings linking spars to fuselage frames, widely spaced lift wires with vee anti-lift wires joined at foot of interplane strut each side. Pyramid-type main gears with tall main strut incorporating rubber shock-absorbers in large faired section. First aircraft completed as passenger machine with rear cockpit having forward-facing seat and windscreen for passenger or mechanic. Liberty engine cooled by two Lamblin radiators and driving Chauvière laminated-wood propeller. Registered RR-SOV and bearing slogan *Aviakhim SSSR Proletarii.* Rolled out at NOA (scientific experimental aerodrome) July 25 and flown by V. N. Filippov early Aug. Excellent aircraft, factory tests complete Oct, and in Dec (before NII tests) decision for series production at GAZ-5 of ANT-3 aircraft both for propaganda tours and as military R-3. GosNII tests by M. M. Gromov and V. S. Vakhmistrov April 26. First ANT-3 prepared for European tour and on Gromov's advice re-engined with Napier Lion for better performance with more fuel. With mechanic Ye. V. Radzevich left Moscow 30 Aug 26 and flew Königsberg/Berlin/Paris/Rome/Vienna/Prague/Warsaw/Moscow in 34 h 15 min (7150 km) returning 2 Sept. First Soviet aircraft seen in outside world. Napier aircraft (**R-3NL**) superb in all respects and 30 already ordered for VVF, but engine expensive and for various reasons replaced in first production series by Liberty (about 12 aircraft) and M-5 (18). Great difficulty in issuing drawings and training workforce, and first production aircraft eventually flown GAZ-5 June 27. This was civil ANT-3, prepared for even longer propaganda flight, registered RR-INT and inscribed *Osoaviakhim SSSR Nash Otvyet* (our answer). Flown by S. A. Shestakov with mechanic D. V. Fufayev Moscow/Tokyo in almost constant bad weather about 22 000 km in 153 h flown between 20 Aug and 1 Sept 27. Aircraft then flew back. M-5 outdated engine, and 26 deal with France for some 100 Lorraine-Dietrich engines resulted in last 79 military R-3 having this engine, known thus as **R-3LD**. Like most preceding Liberty and M-5 aircraft, flat frontal radiator, in this case almost circular instead of rectangular and with prop shaft near bottom. Minor airframe modifications and almost all armed with synchronized Vickers (later PV), twin movable DA on Tur-4 ring in rear cockpit, racks for 10 FAB-50 bombs aimed by AP-2 sight, hand-held camera. A. I. Putilov schemed *Shturmovik* (attack) version with 400 kg armour protecting engine and cockpits, not built. One aircraft converted on line at end of run March 29 with BMW VI, even better aircraft but no reserve strength in structure and so no production. One series R-3 converted as **PS-3** transport and used by Aeroflot in Yakutsk directorate.

DIMENSIONS Span 13,02 m (42 ft 8⅔ in); length (NL, M-5) 9,42 m (30 ft 10¾ in), (LD), 9885 mm (32 ft 5⅛ in); wing area (see text) 37 m^2 (398 ft^2).

ENGINE (Prototype) one 400 hp Liberty, then 450 hp Napier Lion, (production) Liberty or M-5, finally 450 hp Lorraine-Dietrich, one aircraft 730 hp BMW VIz.

WEIGHTS Empty (Liberty) 1335 kg, (Lion) 1390 kg, (M-5 series) 1377 kg (3,036 lb), (LD) 1340 kg (2,954 lb); fuel/oil, 343 kg+34 kg, then (Lion and M-5) 387 kg, (LD) 322 kg+30 kg; loaded (Liberty) 2085 kg, (Lion) 2144 kg (2400 kg max), (M-5) 2128 kg (4,691 lb), (LD) 2090 kg (4,608 lb).

PERFORMANCE Max speed (Liberty) 207 km/h, (Lion) 226 km/h, (M-5) 194 km/h (120·5 mph), (LD) 204 km/h (127 mph), (BMW) 229 km/h; climb to 1 km (Lib) 4·7 min, (Lion) 3·5 min, (M-5) 3·9 min, (LD) 4·2 min; service ceiling (LD, typical) 4920 m (16,140 ft); range (LD) 880 km (550 miles); 360° turn (NL, LD) 21 s; TO (M-5) 200 m/15 s, (NL) 160 m/10 s; landing 140-180 m/11-14 s/85 km/h (53 mph).

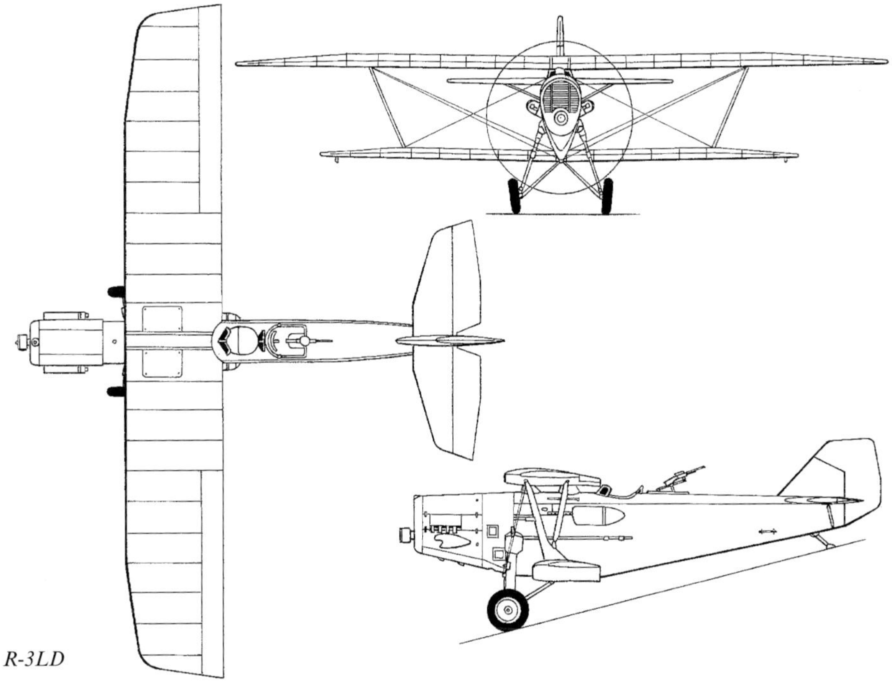
R-3LD

ANT-3, R-3LD

ANT-3, R-3/M-5

ANT-4, TB-1 Result of July 24 VVF demand for all-Kolchug heavy aircraft, this was one of most significant aircraft of its day and especially for Soviet Union. First large ANT, established basis for all subsequent large all-metal monoplanes up to ANT-42. Drew heavily upon Junkers structure, first large twin-engined all-metal monoplane bomber in world. Assigned chiefly to new AGOS brigade under Petlyakov, construction begun 11 Nov 24 in knowledge wall would have to be demolished to get aircraft out. Completed 11 Aug 25 in seven major portions, two months needed to get parts out of building, take to Moscow Central Aerodrome and assemble. Unpainted, two Napier Lion engines with British propellers, large Junkers ski landing gear with trolley under rear ski for ground manoeuvres, removed before take-off. Successful 7-min flight by A. I. Tomashyevskii 26 Nov 25. Skis heavy and weak, and no further flight possible until redesigned; then 35-min flight 15 Feb 26 followed by first properly planned GosNII state tests 26 March onwards with Tomashyevskii at Central Aerodrome and N. N. Morozov at Ostekhburo. Tomashyevskii made 25 test flights 11 June to 2 July 26 (always with second pilot or mechanic) expressing opinion '1st-class'. Decision to build in series primarily as VVF heavy bomber, using cheaper engine than Lion. Only possible production base Fili works, where Junkers operation closing down and 40 skilled light-alloy workers available. April 27 NTK UVVS decision to use BMW VI engine, subsequently substituting M-17.

Classic wing used with improvements and changes of scale in many subsequent ANTs. Profile Tupolev A° with ruling t/c 20%. Structurally 13,5-m centre section with five spars and 18 ribs, removable leading/trailing edges; removable 6825 m outer panels with 10 ribs each, held by new-type joints not copied from Junkers coupling nuts but using tapered conical bolts fitting in reamed holes. Spars Kolchug trusses riveted from tube and deep channel sections, booms being elliptical tubes 72,6 mm × 65,6 mm at root tapering to 30 mm × 25 mm at tip, all joints by four rows of five rivets; 2-mm sheet gusset at each joint between diag spar-web tubes and main booms. All ribs Kolchug trusses, A-section ribs being full wing profile with corrugated 40-mm × 8-mm skins (basic sheet thickness 0,33 mm but narrow walkway inboard of 0,8 mm) laid in strips between them, L-section ribs lying beneath skin with closed riveted joints. On prototype, extra skin stiffening by straps of 1,5-mm thickness and 100-mm or 150-mm wide arranged under skin at 45° (not used in later examples). Fuselage built up mainly of A-section upper longerons and 21 A-section frames, tubular lower longerons and sternframe, with 0,3 mm (40 mm × 8 mm) corrugated skin using open riveting. Structurally three parts, F-1 (nose), F-2 (integral with wing, from first to fifth spars), F-3 (aft of fifth spar). Trapezoidal cross-section with sloping sides as in previous ANTs but this time with width at bottom. All flight-control surfaces with Kolchug tube structure (plus sheet webs in tailplane) and corrug skin (20 mm × 5 mm) with large horn balances which in aileron projected beyond wingtip. Main gear struts imported Ni-Cr steel tubes (130 kg/mm^2) with machined ends and rubber shock-absorption, wire wheels with 1250 mm × 250 mm tyres, no brakes. Sukhoi did steel-shod tailskid. Galvanized-iron fuel tanks in wings, initially 2360 lit (519 gal); CG at 34·1%/34·3% MAC. Lion engines on welded steel-tube frames, each with twin Lamblin radiators under wings, short exhaust pipe from each cylinder. Fuselage equipped for side-by-side pilots with dual control by push/pull rods to ailerons and tailplane, cables to rudder. Glazed nose for nav, cockpit behind wing for observer. Tests soon resulted in additional tankage, Tomashyevskii making 12-h flight 10 July 26 and landing with 1075 kg load. Overload test 28 took off at 8790 kg with 3850-kg disposable load.

Second aircraft *dubler-etalon* (doubler standard) served as prototype of **TB-1** bomber, completed Feb 28 at CAHI and taken in pieces to Central Aerodrome where ready for test July. Flown by M. M. Gromov, S. A. Danilin and mechanic Kravtsov. State tests 15 Aug 28 to 26 March 29. Powered by BMW VI (7·3z) engines of 500 hp/730 hp, windmill elec generator on each inboard LE 0,7 m from root, new nose with glazed lower part and longer forward-sloping upper cockpit for gunner. Entry via ventral hatch between Nos 1-2 spars, ladder carried on board. Crew five, including three dorsal gunners each with Lewis. Provision for 730-kg bombload, or SBR-8 bomb installation or containers of small bombs in fuselage centre section F-2, plus complete radio and camera installations. Span reduced by clipping aileron horn balances, surprising because rudder horn increased and TB-1 always notorious for exhausting control forces, not easy to perform manoeuvres solo. TE of ailerons continued to curve behind TE of wing. Third aircraft, *dubler*

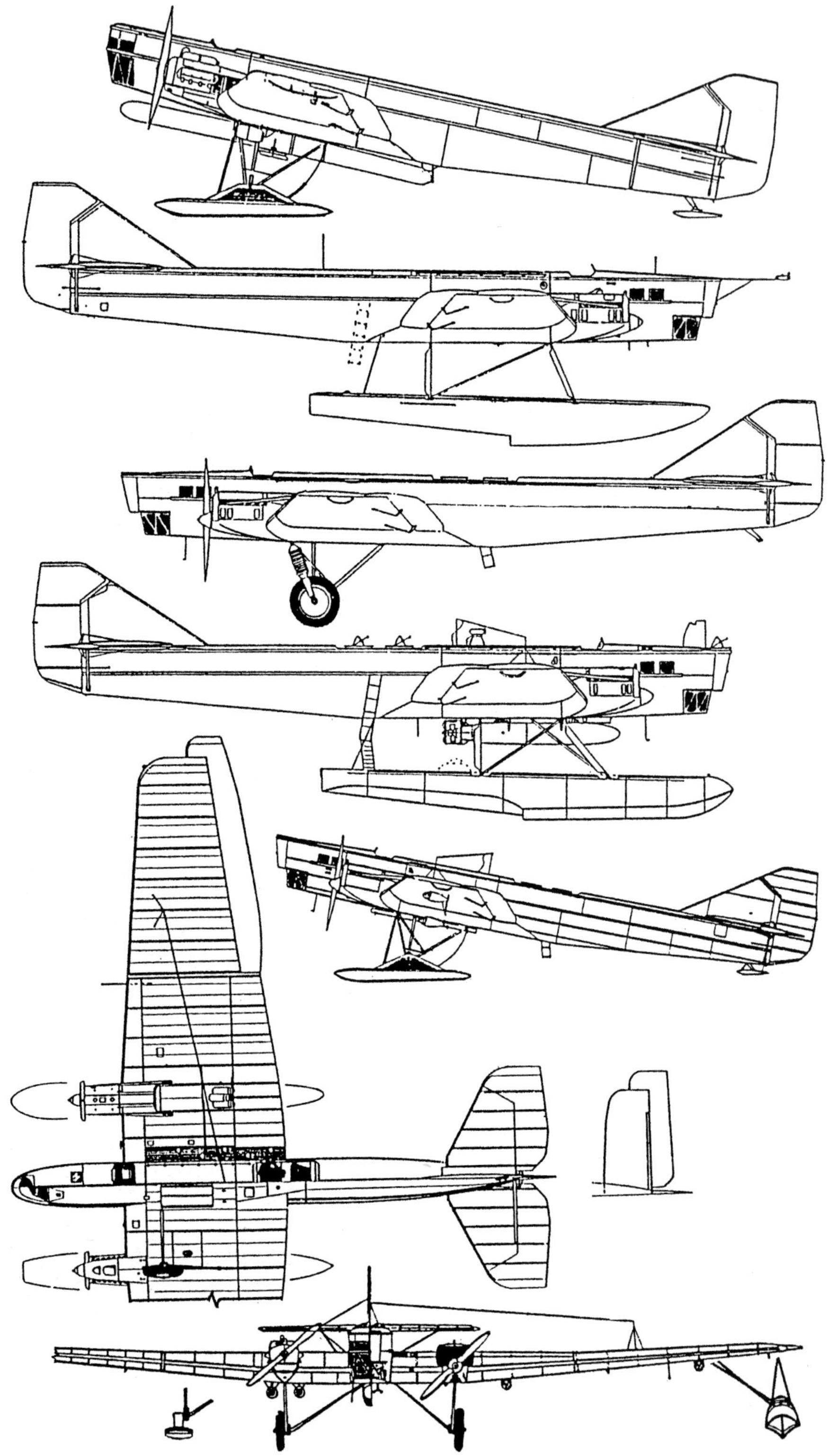

ANT-4, TB-1 (inset, 4th series tailplane); other side views from the top: ANT-4 prototype (also shown in stbd half of front view), Strana Sovietov *(on floats), Zvyeno-1, TB-1P*

TB-1 (**ANT-4bis**), first built at GAZ-22, former Junkers factory at Fili, subsequently centre for large ANT aircraft. Completed July 29, BMW VI engines in standardized installations with almost rectangular frontal radiators, crew of five (two pilots, engineer/gunner, nav/bomb, nose gunner and second dorsal gunner) often increased to six with full-time engineer, internal provision for bombload 1 t, armament three pairs DA on Scarff rings. GosNII state tests 1 Aug/19 Oct 29 and after fitting M-17 engines on operational trials winter 1929-30 flown by Gromov, P. M. Stefanovskii and A. B. Yumashyev. Series production under way mid-29 and first true production machine completed as unarmed civil propaganda aircraft URSS-300 named *Strana Sovyetov* (Land of Soviets) for flight eastwards to New York. Petlyakov in charge of land sectors, R. L. Bartini for water sectors including design and manufacture of floats and conversion to/from wheels. Bartini used JuG-1 floats lengthened 0,4 m to handle increased weight (see TB-1P). Crew: S. A. Shestakov and F. Ye. Bolotov (pilots), B. V. Sterligov (nav), D. V. Fufayev (eng). Left Moscow 8 Aug but damaged forced landing north of Chita. Crew made way back to Moscow, organized second aircraft, and with switch to floats and back completed mission 21 242 km in 137 h flying.

All prodn by GAZ-22 which delivered 2/66/146/2, total 216, in 29-32. Standard VVS bomber and used extensively for survey, airborne forces research and numerous experiments. Oct 33 trials by V. I. Dudakov with six a.t.o. rockets, two under and one above inner wing on each side, total weight 469 kg (later jettisonable), take-off time reduced to 'several seconds' (from typical 15), pilot N. P. Blagin. First major parent aircraft in *Zvyeno* (see Vakhmistrov). In 1933-35 I. Belozyorov and A. K. Zapanovannyi pioneered flight refuelling, piping fuel R-5 to TB-1, TB-1 to TB-1, and finally TB-1 to I-15 and I-16 simultaneously. P. I. Grokhovskii (which see) used TB-1 for initial VVS trials into supply dropping by parachute. From 1935-39 major research programme of remote radio control of TB-1, initially with pre-set bang/bang commands and finally with true analog remote piloting, claimed world first. Much used in Arctic, wheels/skis/floats; Aviaarktika aircraft large vee windshield for flight deck and raised fairing downstream; 5 March 34 TB-1 flown by A. V. Lyapidyevskii was first to succeed in landing on ice to rescue crew of *Chelyuskin*. Feb 37 F. B. Farikh flew civil N-120 Moscow/Sverdlovsk/Irkutsk/Anadyr/Cape Wellen/Arkhangel/Moscow (with many intermediate stops) in air temp –40° to –70°C.

TB-1P standard seaplane. Oct 28 UVVS bought floats from Short Brothers, dural with steel joints and brazier (lens)-head rivets, neatly made. Copied as Project Zh, final production form fuller nose, modified rivet (snap-head) and joints sealed by red-lead tape. Zh float 10,66 m long, 1,15 m wide and high, vol 7,25 m^3, weight with joints 299 kg, whole float gear with struts 816 kg. From 32 series floats, over 100 pairs, most used on 66 TB-1P of which 60 were MA torpedo droppers carrying weapon under fuselage. Skis studied by 15 Jan 26 meeting because those used for first ANT-4 inadequate. Five sets made, final choice was streamlined wooden form with cable bracing, 3700 mm × 900 mm, wt of pair 230 kg, 200 pairs made and used in winter by TB-1. On expiry of VVS life about 90 aircraft stripped for use as freighters by Aeroflot, designation **G-1** (*Gruzovoi* 1, cargo 1). Most tired airframes, many repairs, in some cases several hundred kg heavier than when new but removal of armament compensated. Many carried large ventral cargo container. Served nine years (1936-45) on such duties as carrying sulphur from Ashkhabad mines. Airframe time average 4800 h when scrapped.

DIMENSIONS Span (first) 29,6 m (rest) 28,7 m (94 ft 1 15/16 in); length (first) 17,3 m, (rest) 18,00 m or 18,01 m (59 ft 0 2/3 in), (TB-1P) 18,9 m (62 ft 0 in); wing area (first) 121,5 m^2, (rest) 120,0 m^2 (1,292 ft^2).

ENGINES (1) two 450 hp Napier Lion, (2) 500/730 hp BMW VI 7·3z, (series) 680 hp BMW VI or M-17.

ANT-4, TB-1P

PS-7 (p. 389)

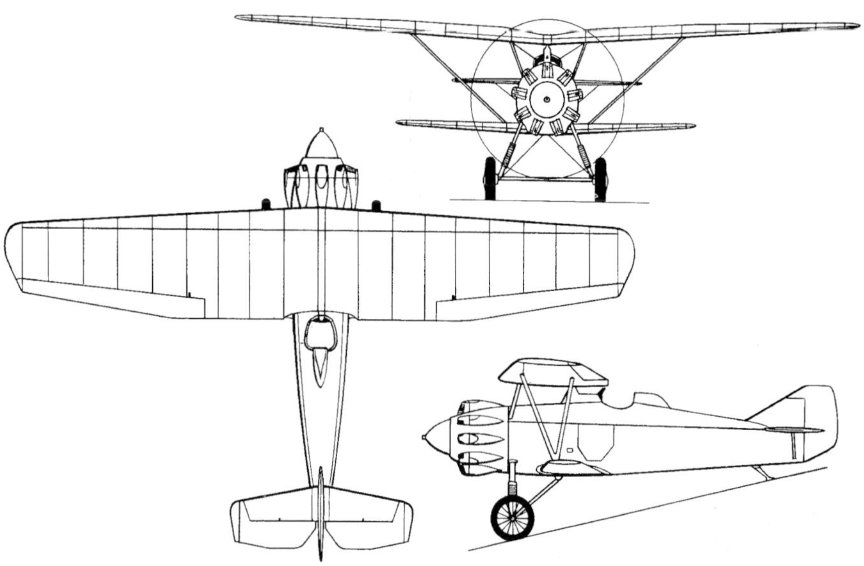

ANT-5, I-4

WEIGHTS Empty (1) 4014 kg, (series) 4520 kg (9,965 lb), (*Strana S*) 4630 kg, (TB-1P) 5016 kg (11,058 lb), (G-1) 4500 kg; fuel/oil (series) 1500 kg+110 kg, (*Strana S*) 2350 kg+155 kg, (TB-1P) 884 kg; loaded (1) 6·2 t, (series) 6810 kg (15,013 lb), (*Strana S*) 7928 kg, (TB-1P, G-1) 7,5 t (16,535 lb).

PERFORMANCE Max speed (1) 196 km/h, (series) 178 km/h (111 mph), (*Strana S*) 207 km/h, (1P) 186 km/h; cruise 156 km/h (97 mph); time to 1 km (typical) 5·5 min, (1P) 7·1 min, (G-1) 9·5 min; service ceiling (1) 4163 m, (series) 4830 m (15,850 ft), (1P) 3620 m, (G-1) 2850 m; endurance typically 6 h, LR tanks 12 h; range (series, M-17) 575 km (357 miles) with max bombload, 1000 km no bombs, (G-1) 950 km; time 360° turn (1) 47 s, (series) 26 s, (1P) 40 s; take-off (typical) 260 m/17 s, (1P) 660 m/26 s, (G-1) 18 s; landing (typical) 170 m/15 s/85 km/h (53 mph), (1P) 150 m/12 s/90 km/h.

ANT-5, I-4 Kolchug airframe (henceforth called dural) also spurred demand for fighter for comparison with trad structures. As part of 25 decree calling for speedy elimination of foreign combat types, meeting 21 Sept 25 decided to issue order for dural fighter. Assigned to P. O. Sukhoi as first major design under Tupolev supervision. Bold decision to build sesquiplane, almost parasol monoplane. Upper wing RAF profile t/c 16% max at outer extremity of parallel centre section and 12% on centreline and at tips, area 19,8 m²; small lower wing 4 m². Upper wing three spars, made left/right joined on centreline, typical truss structure with ten A-profile ribs in each half-wing with 40 mm × 8 mm corrugated skin in strips between them held by closed rivets to spars and intermediate L-section small internal ribs. Single spar lower wing with four A-section ribs and three intermediate inner ribs each side, vee-type interplane diagonal bracing strut linking first outboard A-rib of lower wing to fifth of upper. Rest of airframe typical ANT dural structure, imported engine on welded steel-tube mounts with smooth dural cowl continuing profile of large rounded spinner and leaving cylinders exposed, with two short exhaust pipes from each cylinder. Fine-pitch (20 mm × 5 mm) skin on tail, unbalanced ailerons upper wings only. Two synchronized PV in top of fuselage. First ANT-5 completed July 27, and after factory tests transferred to NII VVS September: M. M. Gromov, A. F. Anisimov, A. B. Yumashyev and I. F. Kozlov opinion 'no better fighter in world'. Production decision Dec 27, designation I-4 (though I-3 not then built). I-4 *dubler* flown with more powerful Jupiter VI Aug 28. First production I-4 completed 15 Oct 28; GAZ-1 built 170 that year, work then transferred to GAZ-22 which in 1929-31 delivered 2/163/12, total 347 not including two prototypes. Numerous modifications. From start of production, larger vertical tail with horn-balanced rudder, larger square-tipped horizontal tail, helmeted cowls on engine with front exhaust collector ring inside crankcase fairing, more pointed spinner, wire-spoked wheels left unskinned and improved tank/gun arrangement. **I-4Z** (*Zvyeno*, see Vakhmistrov) conversion of three 30 aircraft with very small (0,5 m each side) lower wings and attachment points on main-gear axle (left/right) and on ventral centreline under cockpit. Flown from TB-1 and TB-3 with both wheel and ski gear. I-4 with four guns, two in upper wing, tested on ground only. Many fitted with four 50-kg bomb racks. **I-4bis** major rebuild completed mid-31 (at least two aircraft). Lower wing removed completely, struts attached to fuselage, upper wing rebuilt with pilot-controlled slats over outer 44·5% span, radically different engine installation with combination of long helmets with Townend ring, pointed spinner and Bristol type front exhaust ring. Faster, better stability but poorer manoeuvrability; and decision to make future slats automatic. **I-4P** flown with AGOS floats August 31; no attempt at test programme because seaplane fighter thought useless. December 31 one I-4 tested GDL/Kurchyevskii DRP and APK in 76-mm calibre; one gun in each wing 1 m in from strut rib, later mounted under outer wing midway between strut rib and tip. Another I-4 used 35 for rocket a.t.o. trials, at first with two 5-kg gunpowder motors each giving 250-kg for 2 s, and finally with triple unit each side of fuselage at TE of lower wing each giving 500 kg for 2,5 s. I-4 remained in VVS service in fighter regts 1928-37, remarkable record.

ANT-5, I-4bis

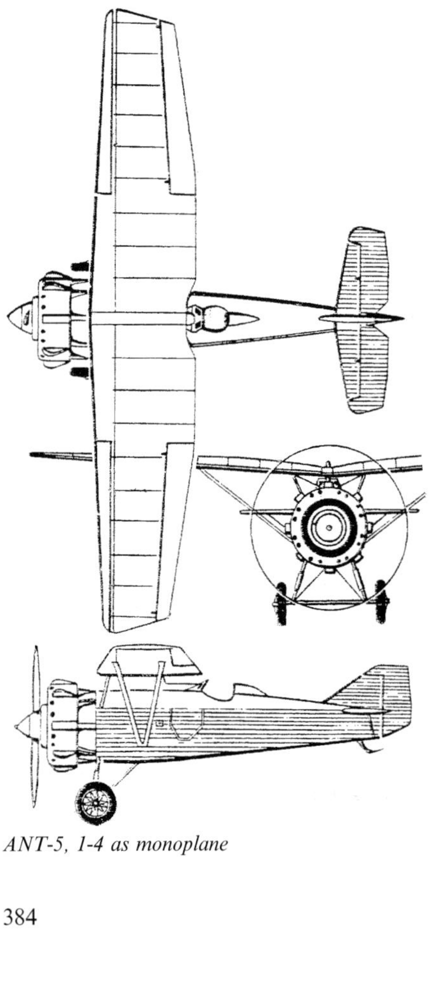
ANT-5, I-4 as monoplane

DIMENSIONS Span 11,4 m (37 ft 4¾ in), (lower wing 5,7 m); length 7,28 m (23 ft 10⅔ in); wing area 23,8 m² (256 ft²), (bis) 19,8 m².

ENGINE One Bristol Jupiter, (1) 420 hp GR9B (Jupiter VI, French-built), (2) 480 hp Jupiter IV, (series) M-22.

WEIGHTS Empty (1) 921 kg, (series) 978 kg (2,156 lb), (bis) 973 kg; fuel/oil 236 kg+25 kg; loaded (1) 1343 kg, (series) 1430 kg (3,153 lb), (bis) 1385 kg.

PERFORMANCE Max speed (1) 246 km/h at SL, 250 km/h at 3,5 km, (series) 220 km/h at SL, 231 km/h (144 mph) at 5 km, (bis) 268 km/h at SL (167 mph); climb to 1 km (typical) 1·8 min; service ceiling (1) 7,2 km, (series) 7 km (23,000 ft), (bis) 7 km; range 840 km (522 miles); time 360° turn (1) 11 s, (series) 13 s; take-off (series) 90 m/7 s; landing 210 m/19 s/95 km/h.

ANT-6, TB-3, G-2 Though natural next step beyond TB-1 this heavy bomber was one of greatest achievements in aviation. Allies in 41 compared it with their own designs of ten years later and drew wrong conclusions. Unlike other giant aircraft of 20s, designed to meet a real need, established configuration later universal with four engines, demonstrated national capability both to produce and also to develop over many years, and gave USSR greatest strategic capability in world. This capability dissipated from 36 by execution of senior officers who believed in it.

Genesis in 25 discussions on following TB-1 with four-engined 2,000-hp successor. Planning began early 26 as **ANT-6**. Lack of factory capacity held up work by Petlyakov (wing), Arkhangelskii (fuselage and controls), Nekrasov (tail unit), Putilov (landing gears) and Pogosskii brothers (power plant and calculations). VVS wished to operate TB-1 before taking final decisions, and design not settled until 29. Tunnel test began 3 Dec 29, mock-up review 21 Mar 30, rapid hand build of prototype at GAZ-22, assembly at Central Aerodrome complete 31 Oct 30.

Wing of Tupolev A° profile as in ANT-4, t/c 20% root and 10% tip, chord 8,0 m root and 2,95 m tip. Structurally, centre section (7 m) and outer wings each made in four easily unbolted parts of 4-m span for rail transport, two of joints in line with engines. Centre section integral with F-2 fuselage section as in ANT-4, leading and trailing edges removable to give chord 4 m for easy transport. Outer panels forward leading edge and whole trailing section removable. Four spars tip to tip, each dural truss using riveted elliptical tubes up to 100 mm×90 mm section with tapering booms and cone-bolts at steel joints. Truss ribs at 1,8-m spacing, 12 in each outer plane with 3-mm sheet gusset at all joints. A-section ribs to full wing profile with wing shape maintained by light stringers between spars each a trihedral girder formed from three strips with lightening holes riveted together. Skin larger corrugations than previously, 50 mm×13 mm, thickness 0,3 mm or 0,33 mm; 0,5 mm over upper surface of centre section (boots allowed, elsewhere soft shoes) and walkways near engines between spars 1 and 2 skinned with 0,8 mm. Leading edge each side of each engine hinged and containing folding ladder, eight in all. Fuselage scaled-up ANT-4 with modest increase in section with severe outlines softened by curved top decking. Structurally a scale of ANT-4 but with six of A-section frames formed into bulkheads with 1,5-m doors separating cabins for nav/bomb in glazed nose from open side-by-side cockpit and engineer and radio/gunner cabins to rear. Prototype provision for three dorsal Scarff rings as in TB-1, no guns fitted nor bombing equipment. Flight-control surfaces with 25 mm×8 mm corrugated skins, elevator with rigid (push/pull) system hinged to tailplane with 14° incidence adjustment by screwjack from cockpit, rudder with cable control and multiple-bungee tension device released by pilot to offset rudder following major asymmetric loss of power, ailerons with horn balances overhanging wingtips and cable circuits with two reduction gears and tension-compensators to give positive control with reduced pilot effort. Engines driving Curtiss aluminium propellers and mounted on welded M1 mild-steel trusses with three-point attachment (bolts to welded lugs, no vibration isolation) to front spar. Underslung water radiators at 45°, shutters at duct inlet. Inflight access to engines via wing is oft-quoted error. Four 1950 lit fuel tanks each divided into three compartments of dural sheet with special pan-head rivet and washer joints sealed by shellac-coated paper gaskets (tanks lasted life of aircraft).

Unpainted prototype mounted on special wooden skis basically scaled from ANT-4 type though more streamlined. Operationally basic aircraft with seats for five, including three/four engineers. First flight by M. M. Gromov 22 Dec 30, hair-raising take-off in which power rapidly lost and right wing only just missed hangar in unscheduled turn near stall. Power then came on again: transpired elevator loads had built up during run until both hands needed on wheel, while vibration caused throttle levers to close! After frantic signals by Gromov, mechanic kept holding levers. Flawless first flight in other respects, and 20 Feb 31 NII VVS recommendation for series construction as **TB-3**, but with M-17 engine. First ANT-6 returned to AGOS for change of engines and rectification of various deficiences. Elevator load removed by changed CG at 30·2% empty and 30·8% full load. Aileron horns removed (span reduced) tailplane span increased, balancing slots on ailerons and rudders, tailskid redesigned as scale of Sukhoi's ANT-4 type. New engine installation with better-cowled radiator and 3,5 m wooden props. Skis replaced by large (2 m×450 mm) tyres, soon replaced by tandem 1350×300 wire wheels in welded steel-tube frame with 12 rubber-plate shock absorbers.

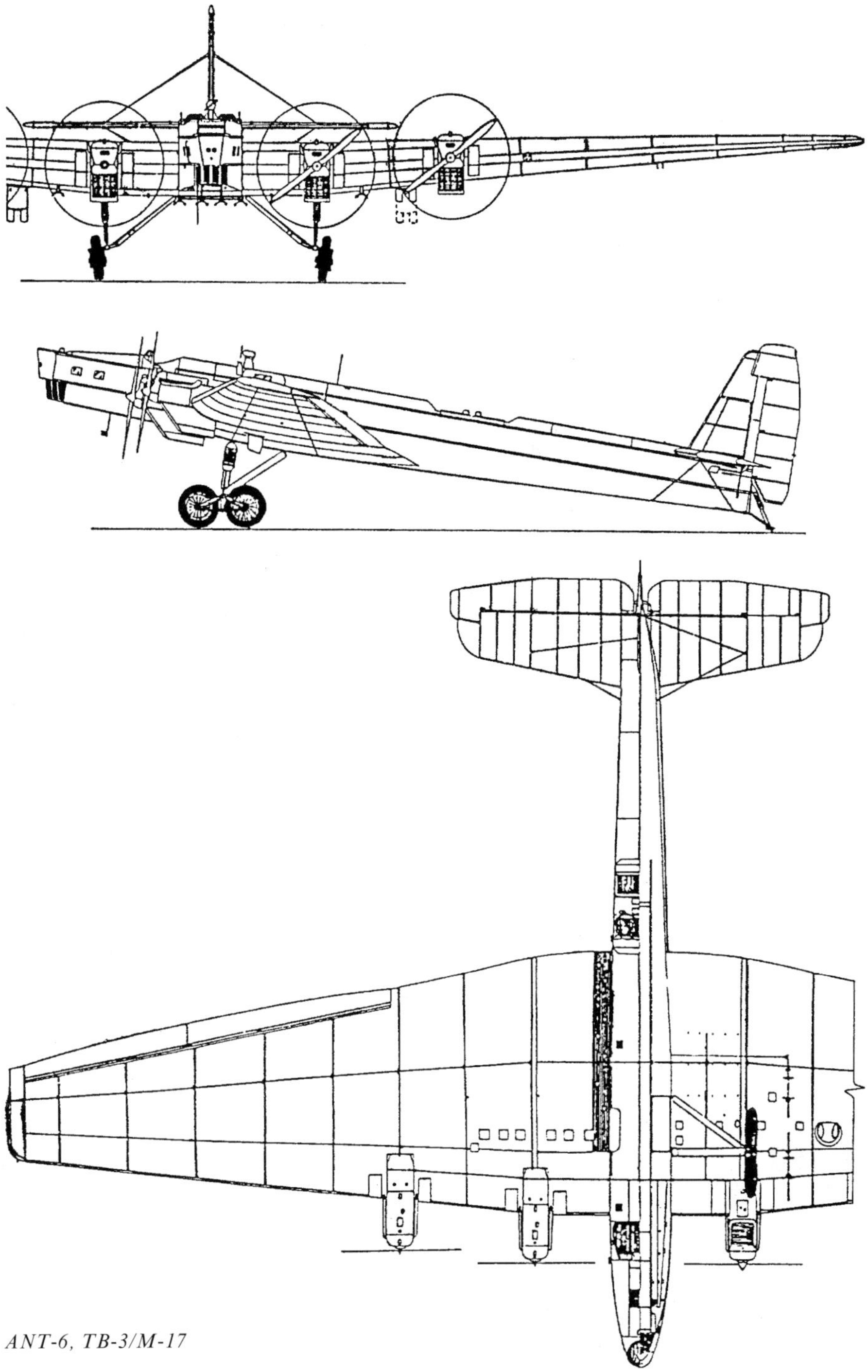
ANT-6, TB-3/M-17

Giant production plan organized at GAZ-22, GAZ-38 (rear fuse/tail) and GAZ-39, manufacture began at end 31. Prototype resumed testing 4 Jan 32 by A. B. Yumashyev and N. I.

Petrov; pronounced 'ready for service duty'. Fitted with armament. External racks of girder type under centre section and fuselage for 2-t (4,410 lb) bombload in Der-9, 13, 15 or 16 containers, Sbr-9 electro-mechanical release gear (weight with racks 437 kg) and 349 kg other military gear. Tur-6 (Scarff-derived) ring mount in nose with twin DA; two Tur-5 rolling-type rings at diag-opposite corners of large cockpit aft of wings each with twin DA; two B-6 turrets extended/retracted under wing near outer engines between spars 2 and 3 with handcrank traverse and hand-aimed DA. Total ammunition: 100 drums of 63 rounds. MF/HF radio with aerial wire strung between fuselage masts, tandem windmill DC generators on top of fuselage.

First production aircraft began NII test 10 Oct 32, mods continuing. Imported KhMA engine mounts and wing joints replaced by local KhNZA, but axles were still imported and weight-growth became severe problem. Weight-saving suggestions at GAZ level were rewarded at 100 rubles per kg saved in production. B-6 turrets were removed with second production block and main gears, elec cables, tanks, bombing system and fuselage lightened. Individual aircraft at first varied by 'tens or even hundreds of kg' and in some cases suffered vibration or structural cracks because of weight-reduction programme, but no inflight breakup and by mid-33 stabilized at about 10,97 t empty without armament or detachable equipment.

Single TB-3 *Zadrayennyi* (battened down) modified 33 with armament removed, all apertures and turrets faired, no aerials or other protrusions and long spats on bogies. Speed and thus range improved by only 4·5%, greater gain of 5·5% resulting from smooth fabric skin over wing from LE to rear spar on upper surface, tested on one aircraft 1 Jan/11 Feb 35. **TB-3/M-17** remained in service to 39, many later going to Aeroflot (see G-2). During 33 painstaking small improvements cleared overload weight to 19,5 t and by refinements to carburation and ignition improved range without bombs to 3120 km. In 34 introduction of stronger D6 and D16 to wing structure enabled tips to be extended, improving field length and ceiling. Engine-out handling improved by replacing bungee bias by rudder servo, and tail design refined in shape. Aileron gearing improved and from mid-33 production launched of improved skis tested previous winter with fine aerodynamic shape, size 5540 mm × 1460 mm (tail 1000 mm × 450 mm) with springing by ten 16-mm bungees in front and behind. One M-17 aircraft used by Grokhovskii (which see) in trials with heavy cannon. Test firings from 15 Dec 34 (in air from 17 Dec) with Model 1927 76 mm field gun slung under fuselage, cradle measuring recoil, fuselage skin reinforced near muzzle. Early 35 modified with two similar guns in wings outboard of outer props (skins locally reinforced) plus long-barrel 76 mm AA gun in shortened nose (nav station moved to mid-fuselage), all with manual loading, pilot with sight and fire button on control wheel to signal gunners to fire. Problem was poor accuracy.

In 31 discussion of M-34 engine led to prototype flying with M-34 in Feb 33 and to NII trials with two series aircraft Oct 33. After building about 400 TB-3/M-17F production switched to **TB-3/M-34** late 33, 'several tens' of this model. Visually distinguished by new radiator installation in neat duct further back under wing, with integral oil cooler, and gondola for bomb-aimer. Main 33 advance was flight of series TB-3 with M-34R with reduction gear driving 4,15-m propeller. NII tests completed 16 Oct 33, approved for production Feb 34. **TB-3/M-34R** introduced tail gun position, small curved cutout at base of rudder but initially without changing rear-fuselage shape and without communication with rest of crew. In production this position (twin DA) was refined and neatly faired in, intercom added and catwalk through rear fuselage. Main gears redesigned with oleo-pneumatic shock struts and hydraulic brakes on rear wheels; tailskid replaced by castoring wheel. In 35 new main gears with single wheels 2000 mm × 450 mm, hydraulic brakes, often retrofitted to M-17 aircraft. Further improvements in 35 TB-3/M-34R included CAHI-assisted fairings on nacelles, wing roots, fin and tailplane and fairings on main-gear legs (latter often removed). Both windmill generators made retractable. One modified aircraft set record flight 18·5 h. In 1933-34 nine special **ANT-6/M-34RD** built for overseas goodwill visits to Paris, Warsaw, Rome, Vienna and Prague. Extra tanks, refined cowlings, turrets faired over and bogie main gears with disc wheels with brakes on all four. In addition several (believed six) aircraft with single main wheels and three-blade metal propellers, demonstrating substantial performance gain. In 35 major test programme Aug-Oct of **TB-3/AM-34RN** with startling improvement in speed at alt and in ceiling. Inner engines four-blade wooden propellers, outers two-blade. Built with tandem gear, 2-m wheels substituted. New armament with Tur-8 (single ShKAS) in nose, dorsal, ventral (new rear-firing hatch) and tail; 2-t bombload using containers Der-19 or -20 or KD-2, release gear Sbr-9 and Esbr-2, bombsight SPB-2, OPB-1 and KB-5. Six supplied to China in 36.

Final production **TB-3/AM-34FRN** and FRNV, tested by Yumashyev 11 Sept 36 and set national records, eg, 8116 m with 5-t load, followed later by 8980 m with no load, 6605 m with 10-t load and 2700 m with 12-t. Extra tanks in outer wings, improved main gears with more powerful brakes (2-m wheels), Flettner tab on rudder, redesigned nose with enclosed turret, improved wing fairings, three-blade metal propellers, smaller radiators in improved ducts. Single **TB-3D** flown Jan 37 with AN-1A diesels, both to test engines and as possible prototype of long-range series version. To assist start of prodn of SB (ANT-40) output at GAZ-22 halted winter 34-35, so overall figures: at GAZ-22, 155 in 32, 270 in 33, 126 in 34, 74 in 35, 115 in 36, 22 in 37 and 1 in 38; at GAZ-18, 5 in 34 and 1 in 37; at GAZ-31, 5 in 32, 37 in 33 and 8 in 34, total 819. In exercises near Kiev 35 c3,700 paratroops and heavy eqpt dropped, 60 flew combat missions in Khasan Sea area spring 38, and larger numbers in Khalkin-Gol war of spring 39 and Winter War against Finland.

In 1939-41 nearly 50 tired M-17F bombers converted as unarmed **G-2** transports, while over 80 merely had armament removed, for glider towing, paratroop dropping and trials with heavy slung loads including light armoured vehicles. Among remarkable achievements of military G-2 were successful para-drops of Grokhovskii (which see) supply containers weighing 2,5 t, T-27 armoured car (1,7 t) and airborne assault of 1,200 paratroops 170 km behind front in 36 manoeuvres. In 1937-38 A. N. Tyagunin flew G-2 one metre above water on several occasions as 4-t T-37 amphibious tank was set free without parachute, on final test with crew inside. Like G-1 all G-2s had over 2,000 h and were tired and much-repaired but after conversion, with side and ventral doors, strong cargo floor with lashing points and complete structural audit they gave vital service throughout the Second World War. Final G-2 conversions had AM-34RN engines and at 22 t could carry over 7 t cargo.

Final variants were TB-3/AM-34RN modified on production line for use by *Aviaarktika* as chief aircraft supporting first permanent Soviet bases in Arctic. Five **ANT-6A**, also called **ANT-6/M-34R**, registered N-166 and N-169/172, modified with completely enclosed fuselage and flight deck, exhaust and combustion cabin heaters, limited (carb-air) de-icing, new nose packed with instrumentation, extra radio and navaids, quick-conversion wheel/ski main and tail gear, 14,0-m braking parachute, extensive emergency and survival eqpt and red-painted except undersurfs, assigned to Papanin/Schmidt expedition which finally left Central Aerodrome 21 April 37. Much heavier than other ANT-6s, packed with cargo and up to 22 men aboard, flagship N-170 made seven abortive take-off runs in flat calm and soft snow before getting away. On 21 May M. V. Vodopyanov radioed from N-170 he had landed on ice 20 km from North Pole and set up radio

TB-3 VAP-1000 poison-gas dispensers

TB-3/M-17F modified with twin Tur-8

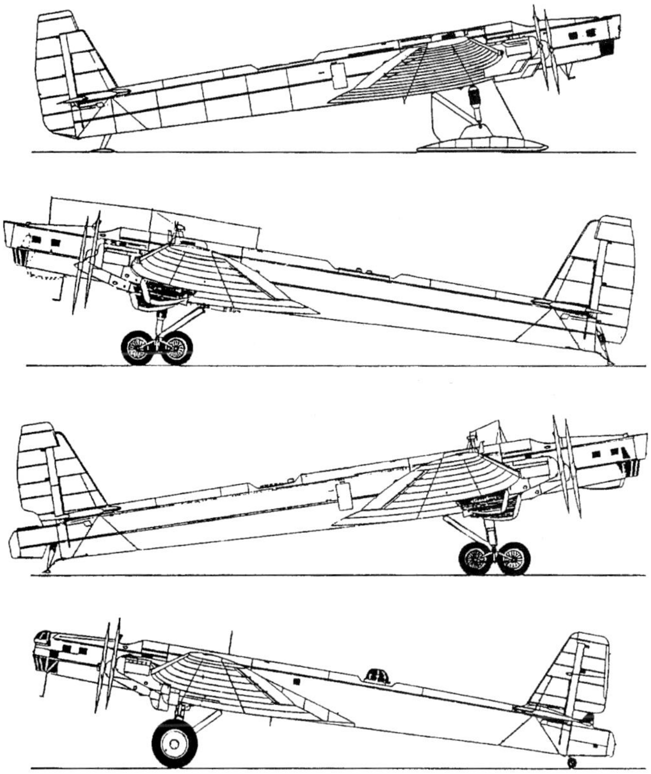

From top: ANT-6 prototype, TB-3/M-34, TB-3/M-34R, TB-3/M-34RN

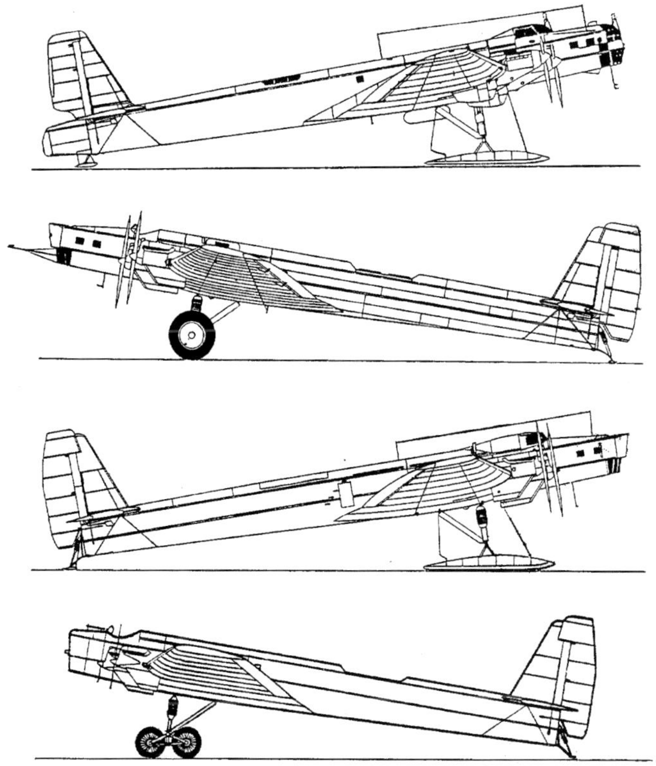

From top: ANT-6A, ANT-6/BMW VI, G-2, Grokhovskii 76 mm guns

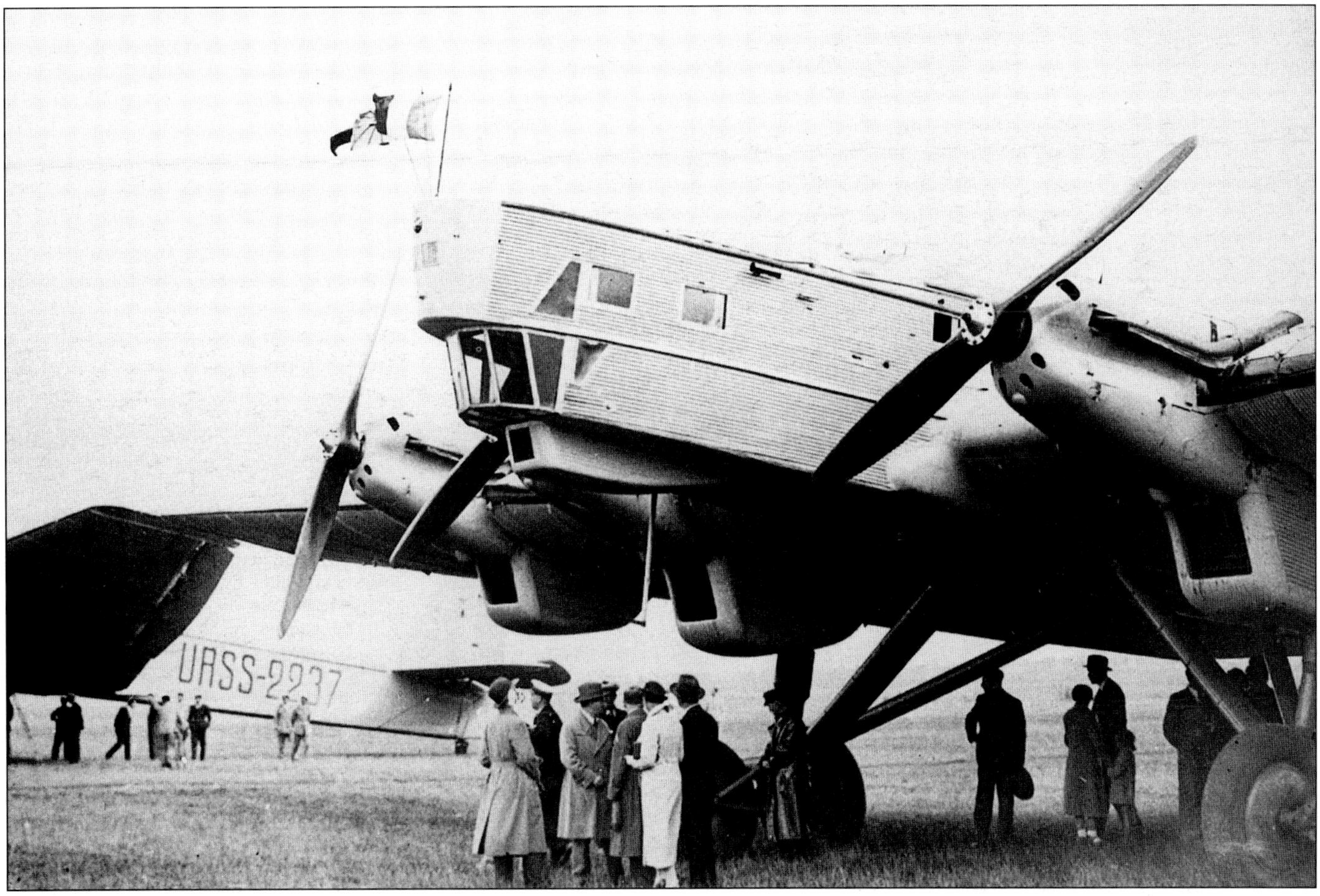

TB-3/M-34RD

ANT-6A

beacon. This was start of dogged air conquest of Sov Arctic, which from 39 used several improved AM-34FRNV transports, most of which survived war. One example being restored.

DIMENSIONS Span (1) 40,5 m (series M-17) 39,5 m (129 ft 7⅞ in), (series, 34 on) 41,8 m (137 ft 1⅔ in); length (1,2,3) 24,2 m, (series) 24,4 m (80 ft 0⅔ in), (35, M-34RN on) 25,1 m (82 ft 4¼ in); wing area (1) 231 m^2, (1932-33) 230 m^2 (2,476 ft^2), (34 on) 234,5 m^2 (2,524 ft^2).

ENGINES (1) four 600 hp Curtiss V-1760 Conqueror, (2,3) 730 hp BMW VI, (1932-34) M-17F, (34) M-34 or M-34R, (35) AM-34RN or FRN, (some late aircraft) AM-34FRNV.

WEIGHTS Empty (1) 9375 kg, (2) 10,08 t, (first series, GAZ-39) 11 207 kg (24,707 lb), (standard M-17F) 10 967 kg (24,178 lb), (M-34) 12,7 t (27,998 lb), (M-34R) 13 230 kg (29,167 lb), (G-2, M-17F) 11 179 kg, (G-2, M-34RN) 11 417 kg, (ANT-6A) 12,5 t; fuel/oil (1) 2610 kg, (34, M-34R) 2020 kg+300 kg, (RN) 2460 kg+360 kg, (G-2, M-17F, almost same as M-17F TB-3) 1600 kg+160 kg, (ANT-6A) 7280 kg+550 kg; loaded (1) 16 042 kg, (2) 16 387 kg, (M-17F standard) 17,2 t (37,919 lb), (M-34) 18,1 t, (M-34R) 18,6 t, (RN) 18 877 kg (41,616 lb), (G-2, M-17F) 17 t, (G-2, RN) 22 t (48,500 lb), (ANT-6A) 24,05 t (53,020 lb).

PERFORMANCE Max speed (1) 226 km/h at SL, (2) 213 km/h, (TB-3, M-17F) 197 km/h at SL (122 mph), 177 at 3 km, (M-34) 207 km/h at SL (129 mph), (RN) 245 km/h at SL, 288 km/h (179 mph) at 4 km, (FRN) 300 km/h (186 mph) at 5 km, (G-2, M-17F) 198 km/h at SL, (G-2, RN) 280 km/h at 3 km, (ANT-6A) 240 km/h at SL, 275 km/h (171 mph) at 3 km; time to 1 km, (TB-3, M-17F) 9·2 min, (M-34) 7, (R) 4·1, (G-2, M-17F) 24·0, (G-2, RN) 8·0, (ANT-6A) 8·8; service ceiling (1) 5,1 km, (TB-3, M-17F) 3,8 km (12,470 ft), (R) 4,5 km, (RN) 7740 m (25,400 ft); (FRN) 8 km (26,250 ft), (G-2, M-17F) 2,2 km (7,200 ft); range (M-17F) 1350 km (840 miles), (M-34, typical) 1000 km (620 miles), (ANT-6A) 2500 km (1,550 miles); time 360° circle (1) 54 s, (TB-3, typical) 36-40 s; take-off (1) 200 m/18 s, (M-17F) 300 m/28 s, (RN) 400 m/26 s, (G-2) often 625 m/40 s; landing (1) 150 m/13 s/106 km/h, (M-17F) 330 m/110 km/h; (M-34) 270-330 m/22 s/122-129 km/h (80 mph); (ANT-6A) over 400 m/159 km/h.

ANT-7, R-6, PS-7 Discussion in 24 had ranged over several possibilities, one being modern monoplane with two 500 hp engines expected to be as fast as fighter with one similar engine. Such machine seen as long-range escort, 'cruiser' fighter (able to penetrate deep into hostile airspace by itself), bomber, recon aircraft and torpedo carrier. Tupolev made proposal summer 25 for scaled-down TB-1 using same engines. Design started Oct 26 at low priority; VVS finally issued requirement for 'air fighting aircraft' 8 Jan 28. AGOS wished engines to be Hispano-Suiza V-12 or Jupiter, but design conference early 28 agreed BMW VI (licensed as M-17) and that, because of TB-1 experience, only one prototype need be built and that production planning be implemented forthwith. Prototype ANT-7 flown by M. M. Gromov 11 Sept 29.

Wing 80 m^2 compared with 120 for TB-1; A-section rib at end of centre section identical in both but no other common part. Centre section reduced span and rectangular in plan; outer panels t/c 18 to 14%, four spars throughout, eight A-section ribs in each panel. Ailerons constant chord, not projecting behind TE of wing and extending over entire span of outer panel. Improved low-friction control circuit, ball-bearing bell-cranks, small balance horn projecting beyond wingtip. Fuselage smaller than TB-1 and esp reduced cross-section, pure rectangle with round top decking, cockpit for two pilots side-by-side (in practice one often mechanic, ie, flight engineer, who assisted harsh manoeuvres as in TB-1 and TB-3), nose and dorsal cockpits for gunners or other crew members; unlike TB-1 not easy to crawl from one cockpit to another and in most ANT-7s impossible. Provision in prototype for fourth crew-member to use sliding turret (derived from B-6 of TB-3) in underside of centre fuselage. Simple pyramid main gears with rubber springing, no brakes but tailskid. Horizontal tail actually fractionally greater span than ANT-4 but much smaller chord; tailplane adjustable from cockpit ±5° and all tail skins finer (20 mm × 5 mm) corrugations than standard 40 mm × 8 mm used elsewhere. Honeycomb water radiators cranked in and out from underside of centre section by pilot. Engines on French AM type steel tube mountings, rigid dural pipe fuel system, exhausts gathered into pipe across wing. Prototype tail modified, and in early 30 given planned armament of twin DA in nose Tur-6, twin DA in dorsal Tur-5 and single DA in ventral 'dustbin'; up to 500 kg external bombs including 406-mm torpedo, Der-7 small-bomb containers and other options; Sbr-8 release system.

NII state tests were begun May 30, no factory tests having been made. GAZ-22 already tooled up and production of basic multi-role recon aircraft began as **R-6** mid-30. Production aircraft had several changes, most noticeable

ANT-7, MP-6

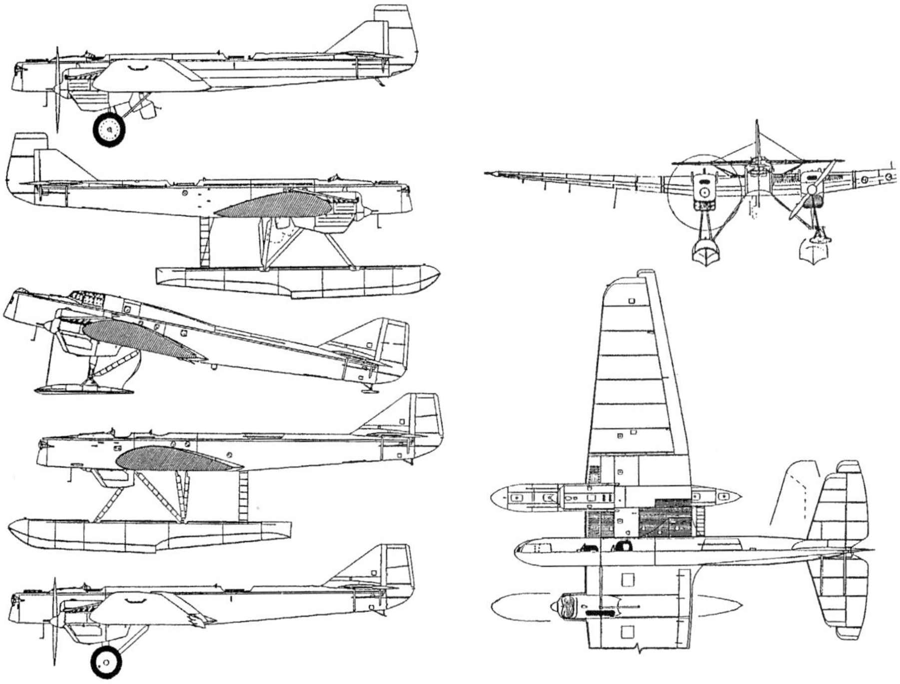

ANT-7, KR-6; other side views from the top: R-6, MP-6, PS-7, KR-6P (whose tailplane is shown in plan as scrap inset); some wings omitted for clarity

being installation of radiators in fixed 45° inclination in duct under engine, now M-17F. Remarkably, not until final state testing of first production R-6, 5 Oct 31, was it decided buffeting of tail unacceptable. Cure accepted was to rivet fixed flap at 29° angle right across centre-section TE, reducing speeds 7-8 km/h. Smooth flight ensued and R-6 production went ahead, terminating at 45th aircraft late 32. Whole programme transferred to GAZ-31 and R-6 then built at much higher rate there with wheel, ski and Zh-type float gear. Total programme closed June 34 at 435 in several versions. **KR-6** (*Kreiser Razvyedchik*, cruiser reconnaissance) carried no bombs and had crew two/three with dustbin removed, wing-root fillets, considerably greater fuel (tanks in outer wing as well as centre section), air/oil oleo legs, S. S. L'vov exptl wheel brakes and Zapp flaps copied from Northrop. **MR-6**, also designated **KR-6P**, was floatplane assembled at Taganrog, with remarkably little reduction in speed despite marked increase in weight. Several basic R-6 used in research and test programmes including use as engine test-bed, cooling-radiator research (eg, ducted radiators in wings, mainly by Myasishchyev's team, anti-vibration flaps being then removed, glider towing, armament trials (virtually every Soviet aircraft gun and RS projectile) and mapping/survey with two or even three large cameras.

Obsolescent even at end of production run, and by 35 transferred mainly to advanced training (until 38) and as utility transport, often with structural modifications to allow carriage of engines and other large loads (until late 44). From 35 at least 220 transferred to GVF, serving in all Aeroflot directorates, with Glavsevmorput and in almost all major Arctic expeditions. Basic civil transport designated **PS-7** or **P-6** (*Passazhirskii*). Almost all were down on performance due to age and local repairs, but lightened by removal of armament; normal PS-7 was little modified, original cockpit floors being used for up to 700 kg cargo loaded through gunner cockpits. One example, **R-6 Limuzin**, rebuilt in RVZ workshop with enlarged and enclosed centre fuselage seating nine passengers, flown July 33; crashed (no fault of aircraft) 5 Sept 33 on Moscow-Crimea flight killing P. I. Baranov, A. Z. Goltsman and others. Many civil machines were seaplanes, designation **MP-6**, with Zh-type floats, 22-31 km/h slower than military seaplanes and at first riddled with snags, three crashing on test. P. G. Golovin surveyed North Pole in R-6 prior to landing of 37 expedition led by I. D. Papanin (subsequently rescued nine months later, 1560 km away on ice-floe, by ship-based U-2). Production, GAZ-22 built 45 R-6, 170 KR-6 and 100 KR-6A, GAZ-31 71 R-6 and GAZ-126 20 R-6 in 36, total 406. A single airframe is today being restored at KIIGA.

DIMENSIONS Span 23,2 m (76 ft $1\frac{1}{3}$ in); length (1) 14,7 m, (R-6 and most land variants) 15,06 m (49 ft $4\frac{15}{16}$ in), (seaplanes) 15,71 m (51 ft $6\frac{1}{2}$ in); wing area 80,0 m^2 (861 ft^2).

ENGINES (prototype and *Limuzin*) two 730 hp BMW VI, (R-6) M-17F, (KR and civil) M-17.

WEIGHTS Empty (prototype) 3790 kg, (R-6) 3856 kg (8,500 lb), (MR) 4640 kg, (10,229 lb), (KR) 3870 kg (8,532 lb), (PS) 3520 kg (7,760 lb), (MP) 4457 kg (9,826 lb); fuel/oil (R) 1385 kg, (*Limuzin*) 1300 kg+120 kg, (KR) 2200 kg+200 kg; loaded (prototype) 5173 kg, (R) 6472 kg (14,268 lb), (MR) 7,5 t (16,535 lb), (KR) 5992 kg (13,209 lb), (PS) 6250 kg (13,779 lb), (MP) 6750 kg (14,881 lb).

PERFORMANCE Max speed (prototype) 222 km/h at SL, (R) 230 km/h (143 mph) at SL, 216 km/h at 3 km, (Limuzin) 248 km/h at SL, (MR) 234 km/h at SL, (KR) 226 km/h at SL, (MP) 211 km/h at SL; climb to 2 km (prototype) 7·1 min, (KR) 13·0 min, (MP) 34 min; service ceiling (R) 5620 m (18,440 ft), (MR) 3850 m; range (KR) 1480 km, (MP) 700 km; take-off (R) 160 m/11 s, (MR) 600 m/35 s; landing (R) 250 m/24 s/110 km/h.

ANT-8, MDR-2 This flying boat was yet another spin-off all-metal twin decided upon in 25 but AGOS design staff could not tackle everything at once and landplanes had priority. Eventually design went ahead, under direction of one of Pogoskii brothers (Ivan Ivanovich, head of KOSOS marine aircraft brigade). Wing based on that of ANT-7 and almost identical to ANT-9; horizontal tail almost identical to ANT-7 (not -9). Mock-up review 26 July 30; prototype completed 1 Dec 30 and flown 30 Jan 31.

Tupolev regarded ANT-8 as research tool; in-house designation not **MDR-2** (marine long-range recon) but **MER** (marine exptl recon). AGOS had little experience of marine aircraft and adopted ultra-conservative approach to prove hull design with minimal risk. Hull all-dural, deep and fairly narrow. Strong planing bottom and rigid upper hull with 27 main and 15 intermediate frames, heavy upper longerons, double chine stringers, twin keel from 3-mm sheet with 25 transverse watertight bulkheads 3-mm or 4-mm sheet. Planing bottom with central keel and two parallel auxiliary keels as pioneered by German Romar, finally curving down at chines; heavily riveted front step of 2,5-mm plate, second step 2,5-m aft of wing TE and planing bottom continued to tail. Hull mass 1395 kg, 42 kg/m^3 (25 being a common value). Stabilizing floats close to hull and took part of weight at rest; deep draft and nose-up angle on paired steel-tube N-struts with lateral brace. Exceptionally seaworthy, could safely operate in higher waves than any known contemporary aircraft. Engines on high paired N-struts with V lateral brace, pusher installation chosen to keep propellers remote from waves. Tanks in centre section, frontal radiators as TB-1 and same propellers. Wing similar to ANT-7 but shorter ailerons; same horizontal tail but braced by two (not parallel) struts each side without variable incidence near top of new-design vertical tail. Skins 40 mm×8 mm corrugations on wing, 20 mm×5 mm on tail. Floats smooth skin, and hull smooth but stiffened by four semicircular-section (40-mm rad) external stringers each side. Enclosed cockpit two pilots; Tur-6 bow cockpit with twin DA, Tur-5 retractable aft of wing with twin DA, bombs under wing roots, two 250 kg on DER-13 gear or four FAB-100 on DER-7.

Factory tests completed inside a week in 3 h 47 min flying. NII testing 15 Feb to 17 March 31 by S. T. Rybalchuk; generally satisfactory but required enlarged floats and rudder. Further NII tests 8 Oct to 14 Nov 31 by Gromov, Bukholts and N. G. Kastanayev; empty weight had grown and despite increase in gross weight 6665 kg to 6920 kg disposable load 2360 kg fell short of design figure 2,6 t. Takeoff and landing judged fast (by ancient flying-boat standards) and general conclusion was MDR-2 would soon be obsolete. Newer types were in prospect (MDR-3) and so MDR-2 was rejected; this proved major error, as it was sound and seaworthy and could have served long periods

ANT-8, MDR-2

while successors were failures. MRT-1 was proposal for civil regional transport.

DIMENSIONS Span 23,7 m (77 ft 9 in); length 17,03 m (55 ft 10½ in); wing area 84 m² (904 ft²).
ENGINES Two 680 hp BMW VI.
WEIGHTS Empty 4560 kg (10,053 lb); fuel/oil 1100 kg+90 kg; loaded 6920 kg (15,256 lb).
PERFORMANCE Max speed 203 km/h (126 mph) at SL; cruise 166 km/h (103 mph); time to 1 km, 7 min; service ceiling 3350 m (11,000 ft); range 1062 km (660 miles); endurance 5 h; time 360° turn, 34 s; take-off 29 s; landing 15 s/115 km/h.

ANT-9, PS-9 Though Kalinin provided indigenous light transports, lack of Soviet airliner for trunk routes caused question to be raised mid-27, and Tupolev called to Kremlin discussion 8 Oct 27. Order from GVF for large passenger aircraft, initially for Deruluft international services, placed Dec 27. Mock-up review Oct 28, prototype URSS-309 *Krylya Sovyetov* (Soviet Wings) rushed to completion in time for May Day 29; parts assembled in Red Square 28 April. First flight by Gromov believed 7 May; NII GVF trials from late May. Gromov, Mikheyev and Spirin flew Moscow/Odessa/Sevastopol/Odessa/Kiev/Moscow 6-12 June. Gromov with eight passengers then flew Moscow/Travemünde/Berlin/Paris/Rome/Marseilles/London/Paris/Berlin/Warsaw/Moscow 9037 km in 53 h between 10 July and 8 Aug 29. Basic aircraft satisfactory, but production had to use engine other than imported Titan, and M-26 selected.

Wing based on ANT-7, outer panels identical but centre-section 0,25-m greater span (one rib) on each side. Fuselage and tail new but typical AGOS dural construction. Main cabin of constant rectangular section with five strong built-up frames forming vert boundaries to four Celluloid windows each side. Nine passenger seats (5 right, 4 left), door aft on left side. Enclosed cockpit for two pilots ahead of high wing, access up two steps through cabin front bulkhead. Wing reduced cabin height but interior described as roomy. Bulkheads and frames aft of cabin cross-braced. Main gears with vertical shock struts pin-jointed to lower boom of front spar, multiple rubber shock-absorbers, V lateral braces to fuselage of streamline-section dural sheet. Wire wheels 1100 mm×250 mm or ANT-7 type skis. Horizontal tail much broader chord than ANT-8 or -7, tailplane derived from TB-1 but reduced span, ±5° incidence, large horns on all control surfaces. Five-cylinder Bristol (GR-built) radials completely uncowled, two-blade wooden propellers with spinners, 972 lit fuel in two centre-section tanks.

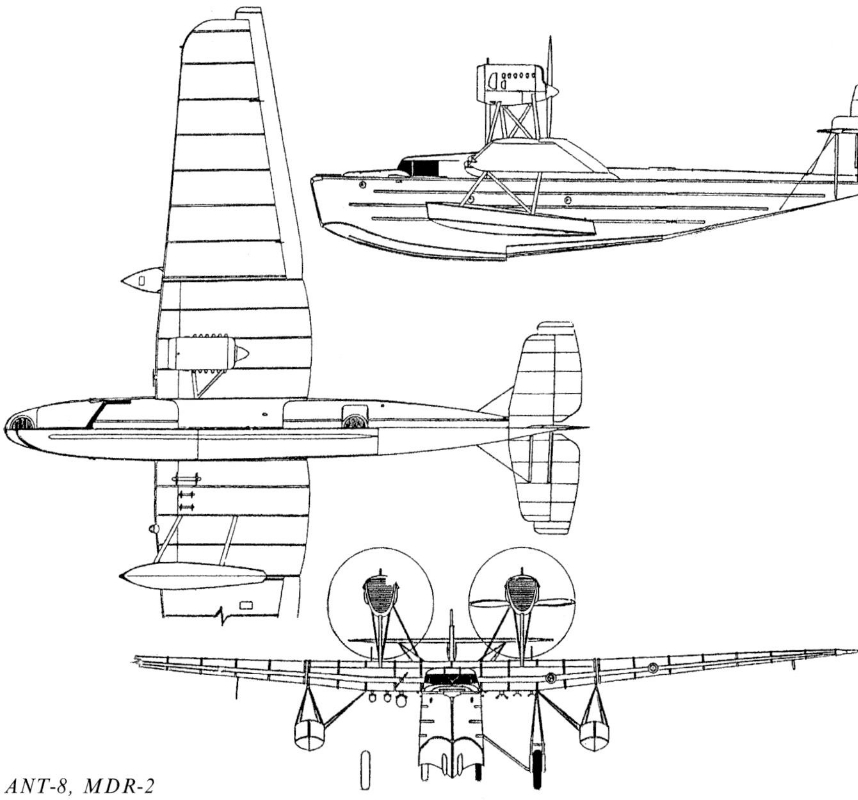

ANT-8, MDR-2

Prototype pleasant to fly, good ratio payload to empty weight, stable hands-off, could maintain 1200 m with nose engine shut down, 200 m with either wing engine shut down. Testing completed Sept 30. To avoid imported engine M-26 selected for production, uncowled except for cylinder baffles and driving Soviet aluminium AMTs two-blade propellers. Engine badly down on power and unreliable, and because of varied faults production of engine halted soon after first production ANT-9 flew early 31. Series aircraft equipped with radio, nav lights, landing light recessed inside left LE, windmill generator above fuselage. Four wing tanks instead of two. Usually toilet and small baggage hold at rear of cabin. Mid-31 decision to fit imported US engine with Townend-ring cowl and ground-adjustable Hamilton three-blade dural propeller, slightly reduced fuel, major

ANT-9, PS-9

alteration to tail area esp increased height of rudder, retrofitted on most aircraft already built. All first batch built with three engines, most eventually receiving US engine. For main series no Russian engine appeared suitable, and decision was major redesign to convert to twin, with M-17 engine in installation almost identical to ANT-7. This version designated **PS-9** in GVF service (AGOS designation believed ANT-9bis), economics penalized by heavier restressed centre section, most of cut in disposable load being reduced fuel though range claimed unaltered. This was main production version, 75 of this type delivered 1932-33 (130 in all). Three-engined ANT-9 saw small-scale use, two supplied 33 to Deruluft having modest increase in cruising speed from fabric covering over corrugated wing. Military bomber/transport with two gun turrets and six-stretcher ambulance versions not developed, but several (believed all three-engined) served with VVS as staff transports and in paratroop exercises. Basic PS-9 established high reputation. One paid for by *Krokodil* (satirical magazine) for *Maksim Gorkii* propaganda squadron had spatted main wheels with painted claws and long plywood nose with painted crocodile teeth. Many PS-9s logged over 5,000 h (amazing for this era), the record being 6,170 by 42. By this time survivors were on military cargo service, at least two surviving the war.

DIMENSIONS Span (prototype, M-26) 23,71 m, (series) 23,85 m (78 ft 10 in); length (prototype, M-26) 17,00 m, (J6) 16,65 m (54 ft 7½ in), (PS-9) 17,01 m (55 ft 9⅔ in) over pitot; wing area 84,0 m^2 (904 m^2).

ENGINES (prototype) three 230 hp GR Titan, (first series) three M-26, refitted with 300 hp Wright J6 Whirlwind; (PS-9) two M-17.

WEIGHTS Empty (prototype) 3353 kg, (M-26) 3950 kg, (J6) 3680 kg (8,113 lb), (PS-9) 4400 kg (9,700 lb); fuel/oil (prototype) 700 kg, (M-26) 1130 kg, (J6) 920 kg+80 kg, (PS) 720 kg+70 kg; loaded (prototype) 5043 kg, (M-26) 6 t, (J6) 5690 kg (12,544 lb), (PS) 6,2 t (13,668 lb).

PERFORMANCE Max speed (prototype) 209/km/h, (M-26) 185 km/h, (J6) 205 km/h (127 mph), (PS) 215 km/h (134 mph); cruise (J6) 170 km/h (105·6 mph), (PS) 180 km/h (112 mph); time to 1 km, (J6) 8·5 min, (PS) 6 min; service ceiling (prototype) 3810 m (M) 3400 m, (J) 4500 m (14,760 ft), (PS) 5100 m (16,730 ft); range (prototype, M) 1000 km, (J, PS) 700 km (435 miles); take-off (J) 170 m/16 s; landing (J) 150 m/18 s/93 km/h, (PS) 110 km/h.

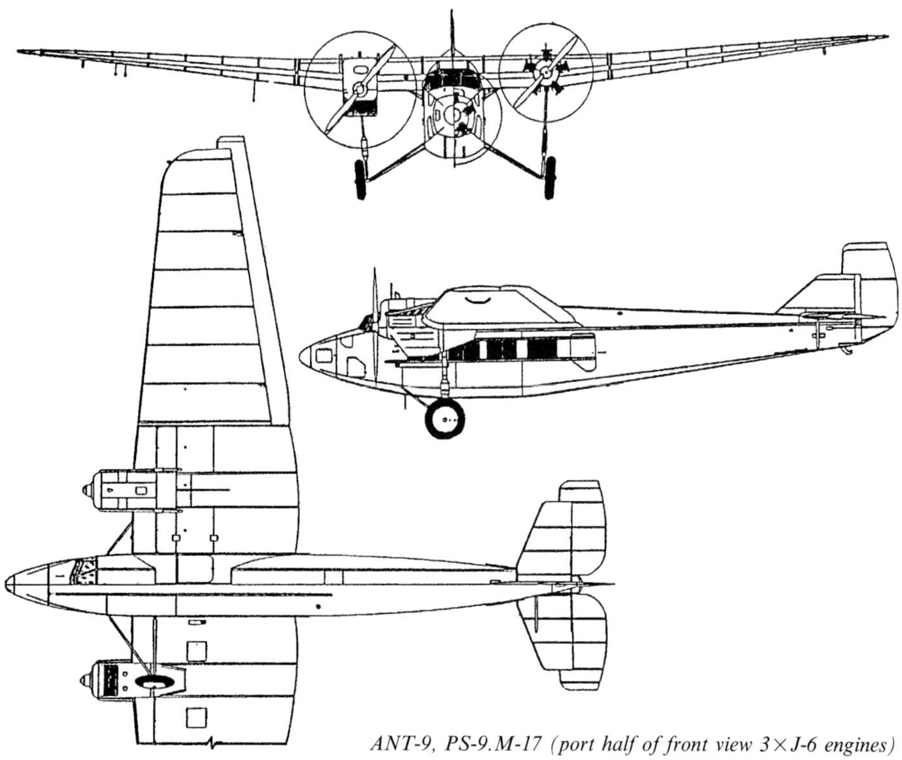

ANT-9, PS-9.M-17 (port half of front view 3×J-6 engines)

ANT-10, R-7 Original R-3 having failed to supplant R-1 as mass-produced VVS recon aircraft, emergence of R-5 triggered AGOS to design rival using dural construction. ANT-10 based on ANT-3 but fuselage of improved form with smaller frontal area, much larger wings (upper wing scaled up from ANT-5) with large gap between fuselage and upper wing, fuel tanks in upper wing (583 lit, much more than standard R-3), newer and more powerful engine (installation almost identical to TB-1 except for exhaust pipes on each side of fuselage), raised observer seat for better all-round view, nav lights and radio with battery charged by windmill generator on upper centre-section, two synchronized PV-1 and bomb load 500 kg in internal bay amidships. Project launched July 28, construction decision with VVS designation

ANT-10

R-7 on 28 March 29, aircraft on factory test 30 Jan 30. Satisfactory but performance surprisingly poor considering slightly smaller and lighter than R-5. NII tests March, returned to AGOS for installation of military equipment and rectification of faults. NII-tested again, late 30, but decided no advantage over R-5.

DIMENSIONS Span 15,2 m (49 ft 10½ in); length 10,9 m (35 ft 9⅛ in); wing area 49,0 m² (527 ft²).
ENGINE One 680 hp BMW VI.
WEIGHTS Empty 1720 kg (3,792 lb); fuel/oil 450 kg; loaded 2920 kg (6,437 lb).
PERFORMANCE Max speed 235 km/h (146 mph) at SL; cruise 184 km/h (114 mph); time to 1 km 3·1 min; service ceiling 5560 m (18,240 ft); range 1100 km (685 miles); endurance 5 h; time 360° turn, 25 s; take-off 150 m/10 s; landing 300 m/22 s/90 km/h.

ANT-11 Tupolev held discussion with GVF and Glavsevmorput early 29 on requirement for large civil transport flying boat, much more capable than ANT-8; same size aircraft also required for military MK (see ANT-22). ANT-11 project launched with scaled-up hull with same skin (smooth but with external bead stiffeners) and four M-17F or M-22 on pylons above wing (believed four separate tractor nacelles). Never built, but calculations and hull design greatly assisted ANT-22.

ANT-12, I-5 As part of first Five-Year Plan Tupolev was assigned task of designing new single-seat fighter by Aviatrust, acting as agency of government. Programme launched at meeting 22 June 28, stipulated engine being geared Jupiter VI for which Aviatrust was completing licence-negotiation. Conventional biplane, not all-dural but mixed construction, corrugated-skin wings but wood semi-monocoque fuselage, two synch PV. Service designation I-5, same timing as rival Polikarpov (later CCB) I-6. Prototype timed to fly Sept 29, but AGOS overloaded with all-dural large aircraft and received permission to drop ANT-12; service designation transferred to new design VT-11 by Polikarpov.

ANT-13, I-8 Another of projects overloading AGOS in late 20s was this small biplane fighter, conceived as exercise in using new steels studied by Tupolev on 28 visit to Germany. Fuselage welded truss using Cr-Mo hi-strength, wing spars Ni-Cr stainless, rest being dural with fabric covering. Extremely small, no bracing wires, aiming at very high performance with imported Curtiss engine. Received service designation I-8 and popularly called *Zhokei* (Jockey) after contemporary Vickers fighter. CG at 22%-23%, all flight controls by cable, direct-drive engine to metal propeller, spatted wheels, two synch PV-1. In early 29 AGOS management decided no possibility of continuing without dropping other programmes; but on initiative of engineer V. M. Rodionov decision privately to develop further and complete prototype; every worker donated 70 h and sponsorship freely given by UVVS in competition with OSS. Need formally affirmed 30 Dec 29, mock-up review 30 Jan 30 and first flight by Gromov 12 Dec 30. First time 300 km/h exceeded in Soviet Union, but decision not to license engine and existence of I-5 as major programme influenced adverse production decision – though handling good.

ANT-13

DIMENSIONS Span 9,03 m (29 ft 7½ in); length, 6,7 m (21 ft 11¾ in); wing area 20,3 m² (219 ft²).
ENGINE One 700 hp Curtiss Conqueror V-1570.
WEIGHTS Empty 960 kg (2116 lb); loaded 1424 kg (3139 lb).
PERFORMANCE Max speed 303 km/h at SL, 310 km/h (193 mph) at 3 km; time to 1 km, 1·0 min, to 5 km 10 min; service ceiling 8,5 km (27,890 ft); range 440 km (273 miles), 1·8 h; landing speed 118 km/h (73 mph).

ANT-14 This large passenger airliner was result of Tupolev's enthusiasm, coupled with fact AGOS had capability to build it, rather than having place in Five-Year Plan or order from GVF. Funding difficult to establish, but in fact costs remarkably low because major sections of wing, vertical tail and landing gear already designed for ANT-6. Completely new fuselage, most capacious of any landplane of era, usual all-dural structure with skin mainly 40 mm × 8 mm corrugated, detail structure same as ANT-9 but much larger square cross-section (3,2 m × 3,2 m externally) seating passengers 2×2 instead of 1×1; truss frames, cross-braced behind cabin, seven windows each side, door aft on left. Wing as ANT-6 except centre-section 0,8-m greater span (wider fuselage) and straight taper on TE root to mid-span each aileron. No horn balances on ailerons. Four tanks in wing, total capacity 2765 lit (608 gal). Imported engines (later licensed), direct drive 3,2-m two-blade wooden propellers, Bristol front exhaust collector ring. Nose engine uncowled, wing engines with unusual Townend rings with helmets. Two oil coolers in leading edge 1 m away on each side of each engine, four folding ladders in hinged LE sections between engines also serving as maintenance platforms. One of very few five-engined aircraft. Completed with large but thin (2000 mm × 450 mm) main wheels, with primitive brakes, all by Palmer Tyre, plus tailskid on steel-tube frame off aft face of stern frame under cut-away rudder. Bracing of horizontal tail (much smaller than ANT-6) arranged to allow ±5° variable incidence. At last moment AGOS built flight engineer observation blister above fuselage between wing spars 2 and 3; eight rows of 2×2 passengers seats.

Aircraft initially registered L1001, built rapidly and flown by Gromov 14 Aug 31. Virtually nothing wrong, but wheels soon replaced by TB-3 type tandem wheels with tyres 1350mm × 350mm. Extra row of seats installed, giving 36 passengers and crew of five. No quarrel with aircraft except far too large for any Aeroflot route. Languished for over a year, doing little flying but having various modifications including removal of ring cowls from wing engines, improved exhaust system fited, dural ground-adjustable propellers, elec system and top of fin extended to match rudder. In 1932 sponsored by newspaper Pravda and named thus; assigned to *Maksim Gorkii* propaganda squadron as flagship and intensively used in joyrides over Moscow as well as fund-raising sorties, twice to Kharkov and once to Leningrad, to collect money for ANT-20. Retired 41 after carrying 40,000 passengers (then record) without accident. Fuselage served during Second World War as cinema in Moscow Central Park of Culture.

ANT-14V bomber-transport with four M-34, two turrets, crew 11, never built.
DIMENSIONS Span 40,4 m (132 ft 6½ in); length

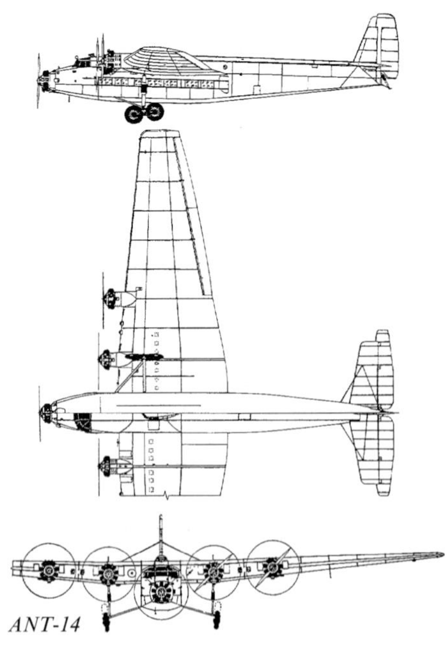

ANT-14

26 485 mm (86 ft 10¾ in); wing area 240 m² (2,583 ft²).
ENGINES Five 480 hp Gnome-Rhône Jupiter GR9Akx.
WEIGHTS Empty (as built) 10 828 kg (23,871 lb); fuel/oil 1990+166 kg; loaded 17 530 kg (38,646 lb).
PERFORMANCE Max speed 236 km/h (147 mph) at SL; cruise 195 km/h (121 mph); time to 1 km, 4·9 min; service ceiling 4220 m (13,850 ft); range 900 km (560 miles); take-off 250 m/18 s; landing 220 m/14 s/105 km/h.

ANT-15 Unbuilt design for all-dural two-seat biplane fighter with M-34 engine, intended to have smooth wing skins.

ANT-16, TB-4 Enormous bomber, launched as AGOS project March 30 with approval but no actual order from VVS. Subsequently Tupolev did obtain financial cover for single prototype, service designation TB-4, but no intention of order for bomber use until experience gained with TB-3. Same layout but scaled up 35% linear and almost 100% area and weight. Engine power, however, only 50% greater than TB-3/M-34 and even in design stage Tupolev had agonizing second thoughts over two tandem pairs of extra engines, one each side, instead of one pair on centreline. Eventually Tupolev chose to concentrate on load-carrying rather than performance. Thus wing built with 10,5-m centre section with t/c 16·6% root and 17·5% at steel structural joint to 21,75-m outer panel, where t/c tapered to 14% approx at tip. Except for direct short exhaust pipes from each cyl, engine installations similar to original M-17 with slanting water radiators directly below, exit under wing LE, two-blade wooden propellers. Extra pair on high steel/dural struts above frames 2 and 4 of F-2 fuselage section, water circuits between engines cooled by two radiators in dorsal box. Usual oval-section riveted-aluminium fuel tanks in wings, hinged LE sections with access ladders and thick (0,8-mm) sheet walkways above wing near engines; in addition novel feature of inflight access via crawlway between LE and front spar, and access via steps inside large pylon fairing to tandem centreline engines also considered but not pursued because of difficulty of sressing cowlings for opening. Fuselage equipped for ten of crew of 12, others being in underwing turrets (not fitted to prototype). Improved nose with large glazed underside and same ventral gondola for bomb-aimer as late models of TB-3. Large pilot cockpit with open roof but all-round windshield with glass windows. Main structure scaled from TB-3 but with five half-round external stiffening beads each side as introduced in ANT-8. Largest new design task was internal bomb bay, by far biggest ever attempted, housing up to 10 t (standard overloads 40 FAB-250 or 20 FAB-500). Bay comprised front section, by K. P. Sveshnikov's AGOS brigade and rear section by V. M. Myasishchyev's. Each bay 1,8-m wide and high and 2,5-m long, with powered doors, front between spars 1-2 and rear between spars 2-3, with sloping bracing walls and total five strong transverse frames. Rest of bombload (if any) hung on external beams under centre section and inboard outer wings. No details of defensive armament beyond total two hand-aimed 20-mm and ten DA. Cannon possibly in nose and tail positions and most 7,62-mm in pairs. Slender ailerons occupying most of trailing edge of outer panels, no horns; small horns on elevators, not projecting beyond tailplane which had the usual arrangement of variable incidence; rudder without balance but servo on TE. Large main gears each with twin Soviet-built 2000 mm × 450 mm wheels side-by-side in welded KhMA tube frame incorporating large fairings over rubber-disc shock-absorbers, track 10,645 m. Castoring tailwheel.

First flight M. M. Gromov 3 July 33. Unimpressive performance, and control forces, especially aileron, so high that only just possible with both pilots exerting full effort. Ailerons divided inner/outer to avoid pivots jamming and surface balance greatly increased; on second

ANT-14

ANT-16

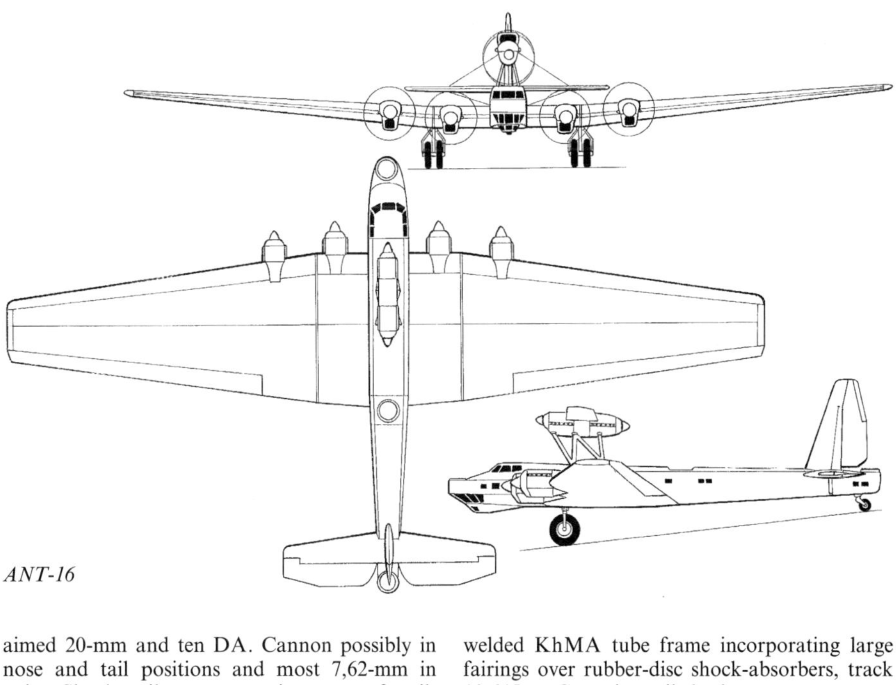

ANT-16

flight almost no effort needed and lack of 'feel'. On third flight aircraft reasonable to fly but performance still disappointing. Gromov suggested propellers blowing against wing 2-m thick were ineffective; Tupolev re-studied efficiency of tandem upper propellers. TB-4 subjected to full test programme by NII pilots Stefanovskii, Ryazanov, Nyukhtikov, but take-off, climb, altitude and manoeuvrability unacceptable. At normal gross could not equal TB-3 for range/bombload. Best achieved (at 37 t) was 8 t (17,637 lb) bombload over 940 km. Tupolev proposed fitting M-35 engines, never carried out. Largest landplane of day, never announced publicly; used in film *Bolshoi Krylya* (Big Wings) 35 in which simulated crash, scrapped 36.
DIMENSIONS Span 54,0 m (177 ft 2 in); length 32,0 m (105 ft 0 in); wing area 422 m² (4,542 ft²).
ENGINE Six M-34.
WEIGHTS Empty 21,4 t (47,178 lb); fuel/oil 4950 kg; loaded 33,28 t (73,369 lb), (overload) 37 t (81,570 lb).
PERFORMANCE Max speed 200 km/h (124 mph) at SL, 188 km/h (117 mph) at 5 km; cruise 159 km/h (99 mph); time to 1 km, 12·4 min; service ceiling 2750 m (9000 ft) at normal gross; range 1000 km (620 miles); range at 37 t with 2 t bombs, 2000 km (1,240 miles); take-off 800 m/36 s; landing 400 m/105 km/h.

ANT-17, TShB Four-seat armoured attack fighter, prototype building 30. Typical corrugated-metal sesquiplane with engines fully cowled above lower wings, water radiators projecting below. Fuselage filling space between wings housing pilot in open cockpit ahead of upper leading edge, front gunner with twin DA, rear gunner/radio with twin DA and probably fourth crew-member to navigate and load large guns. DPK and other large recoilless cannon rejected in favour of two conventional guns of 37-mm or 45-mm calibre in fuselage firing obliquely down below nose, backed up by reported four fixed (presumably inclined) machine guns. Bombload up to 600 kg, believed all external. Armour mass 1 t, of which 380 g to be structural. Project abandoned early 32. All published drawings are inaccurate.
DIMENSIONS Span c20,5 m (67 ft 3 in); length c13,5 m (44 ft 3½ in); wing area 67 m² (721 ft²).
ENGINES Two M-34F.
WEIGHTS Empty 3600 kg; loaded 5700 kg.
PERFORMANCE speed 255 km/h (158 mph).

ANT-18 Not one drawing or model appears to survive of this even larger armoured Shturmovik, parallel project to ANT-17, which was more powerful derivative of ANT-7 monoplane. Fuselage closely akin to ANT-17 and same crew, but different armament (DPK +4 PV firing ahead, two pairs movable DA, 1-t bombs) and armour; and same engine installations but on large monoplane wing almost same as ANT-7. Active project 1930 but prototype construction not believed started and no service designation known. Data are estimates.
DIMENSIONS Span 23,2 m; length 13,0 m; wing area 80,0 m².
ENGINES Two M-34F.
WEIGHTS Empty 4400 kg; loaded 7 t.
PERFORMANCE Max sped 275 km/h (171 mph).

ANT-19 Unknown.

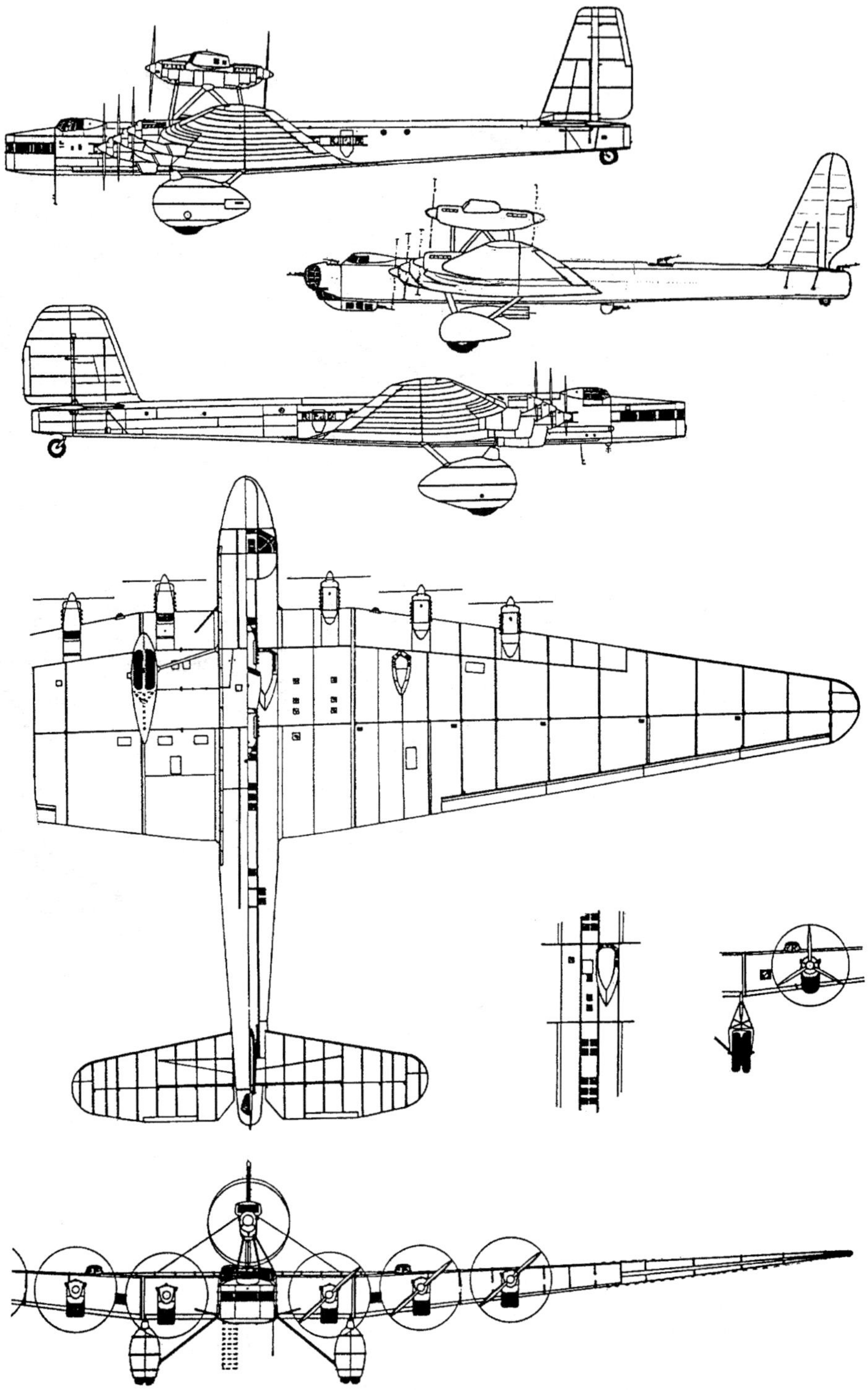

ANT-20, MG (inset, cabin roof); middle side view, ANT-20V project; bottom side view, ANT-20bis, PS-124 (inset, front view of ANT-20bis port inner and main gear with spat removed)

ANT-20, MG Initials are those of Maksim Gorkii, famed Soviet writer, whose career began 25 September 92, date celebrated annually in Soviet Union. For 40th anniversary president of Yurgaz (M. Ye. Koltsov) launched nationwide appeal for funds to build colossal aircraft to serve as flagship of newly formed *Maksim Gorkii Agiteskadril* (propaganda squadron). Gigantic project started, up to 6 million rubles (then £2 m, US$8 m) raised, and formidable 'technical Soviet' formed with representatives from some 100 institutions to assist design. Despite this Tupolev, inevitably chosen as chief designer, still managed to accomplish the task – and quickly. AGOS assigned formal contract July 33, but by that time design almost complete. In 31 AGOS had studied **ANT-20** for Aeroflot, 70-passenger devt of ANT-16, six AM-35R, extended span and other refinements, and this was modified to meet needs of **MG**.

ANT-20

(At this time it was expected 11 MG would be built.) In absence of M-35R two extra engines added on leading edge.

Team leaders Petlyakov/Arkhangelskii; Kondorski (general design), V. N. Belyaev (stressing), B. A. Saukke (ch.engineer), both Pogosskii (engines), A. A. Yengibaryan (eqpt) and Nekrasov. Wing profile as before CAHI-6, thickness 20% root, 10% tip, aspect ratio 8·2 (remarkable for 60 years ago). Traditional structures of three spars and built-up ribs retained, but because of great length of all members plain tube replaced by complex (usually closed-tube) sections riveted from wrapped sheet. Typical spar booms were 6-mm sheet wrapped into 160-mm diam or ellipse 160 mm×190 mm. As usual, A-type ribs projecting just outside wing profile, skin 0,3 mm to 0,8 mm thick with 50 mm×16 mm corrugations in strips between them. Centre section 10,645 m span, straight LE, same TE taper as outer panels, chord at outer joint 11,0 m exactly, two A-ribs each side of fuselage. Outer panels each 26,18 m span, tapering to 3,2 m chord at start of screwed-on semicircular tip, ten A-type ribs each side with skin stiffened by underlying L-section ribs and spanwise stringers. Ailerons divided into four sections each side to avoid TB-4 trouble of hinge-binding with wing-flexure, total 31,3 m². Tailplane symmetric section, 18,3-m span, 40×8 skin, two pairs bracing wires each side above and below to give inflight adjustment ±5° with new addition of elec servo/trim tabs on inset elevators. Static balance on rudder, servo rudder with elec positioning for trim, manual actuation to drive main surface. Dual flight controls almost entirely by steel cables with auto-tensioners to allow airframe flexure and gearing for ailerons. Main gears based on TB-4 but stronger KhMA struts with air/oil oleo strut resting at top in ball/socket and forked trunnion under front spar, similar large forked yoke rear brace privoted to third spar and single lateral strut to front spar at fuselage root. Specially made 2000 mm×450 mm wheels with four-shoe pneu brakes in side-by-side pairs. Tupolev proposed deep trousers over whole gear as in ANT-26 but accepted scheme was large aluminium spats stiffened by three exterior beads, removed after rollout but before first flight. Tailwheel 900 mm×200 mm on KhMA castoring sprung leg.

Tupolev disliked pylon tandem engines but could not achieve required power with six wing engines. Wing engines cantilevered far ahead of front spar on welded KhGSA trusses driving 4 m two-blade wooden propellers, water cooling radiators in short ducts under LE some resemblance TB-3/M-34. Upper centreline engine struts and cowls different from TB-4; Tupolev considered large dural pylon with ladder access in flight but accepted conventional arrangement with four pairs triangulated struts, single large ducted radiator above. Fuselage typical AGOS dural box 32,5 m long, 3,5 m wide and 2,5 m high but with flight deck projecting above at upper level. Structurally five parts, joined by steel eye/lug bolted joints at four longerons: F-1 nose saloon, four twin and three single glass windows, two seats in front, two pairs behind, with separate nav cabin to rear; F-1bis plywood roofed cockpit for two pilots, lower cabin for four pairs passengers, rear radio operator; F-2 between spars telephone (intercom) exchange with 16 lines, secretaries, toilet; F-3 buffet-bar, storeroom, film projector and photo library and processing lab, radio racks; F-4, cross-braced structure, unused. Inner wings offered spaces between spars and ribs roughly 2-m cube each, used for payload throughout. Usable floor area 109 m² (varied with mission). Normally four 2/3-berth sleeper cabins in centre-section on each side, small windows in roof and (three each side) in floor. Normal accommodation eight crew and 72 passengers, entry via large section of underside of F-2 hinged down from rear with integral stairs (when new, + red carpet). Roomy, access front/rear on centreline and into wing between spars 1/2 and 2/3 on each side. Wing compartments housed 30 hp APU/generator (pioneer use of 120V AC) and central electric station, audio/recording station with 'voice from sky' loudspeakers, four radio transmit/receive stations, printing press, pharmacy, laundry, leaflet dispenser and pneumatic-post tube centre. Outboard of centre section was cockpit between spars 1/2 each side for flight engineer with blister canopy above wing upper surface. Outboard of centre section all fuel in 28 riveted aluminium cylindrical tanks in groups of 8, 4 and 2 each side, total 9400 lit (2,067 gal). All engines started by compressed-air system.

Construction occupied nine months, and parts left GAZ-22 3 April 34 for assembly Central Aerodrome. Registration L-759 applied. Accepted by special committee 24 April, flown 35 min by M. M. Gromov 17 June 34. Second flight 19 June over Red Square during meeting of Chelyuskin expedition. After factory tests and installation of autopilot (first in USSR) and array of electric-light bulbs on underside of wings for displaying slogans, joined *Agiteskadril* as flagship 18 Aug 34. Established excellent record of serviceability and impressed all with size and majesty. Engines flight idling for air/ground broadcasting, established routine of two types of mission with crew eight (payload of passengers, mostly rewarded for public service) or crew 20 (using all on-board propaganda gear). Often escorted by small aircraft, I-5 or I-16, to show size; 18 May 35 escort N. P. Blagin in I-5 attempted unauthorized loop around MG wing and hit wing squarely from below. Aircraft crashed locked together, killing Blagin, CAHI pilot I. S. Zhurov, Sqn pilot I. V. Mikheyev, 10 crew and 33 pax.

Ten further MG never funded. Also unbuilt: **ANT-20V** bomber, with 12 t bombload, two 20 mm and 12 DA/ShKAS (model at Monino) and **ANT-20Sh** transport for Moscow and Volga HQ staff. Both terminated Jan 33.

DIMENSIONS Span 63,0 m (206 ft 8¼ in); length (on ground) 32,9 m (107 ft 11¼ in), (horiz) 32 476 m (106 ft 6⅝ in); height (on ground over centre prop) 10,6 m, (horiz, over fin) 11 253 m; wing area 488 m² (5,231 ft²).

ENGINES Eight M-34FRN.

WEIGHTS Empty 28,5 t (62,831 lb); fuel 7150 kg; loaded 42 t (92,593 lb), (max) 53 t (116,845 lb).

PERFORMANCE Max speed 220 km/h (137 mph) at SL, about 250 km/h (155 mph) at 5 km; service ceiling 4,5 km (14,760 ft); range 1200 km (750 miles); take-off/landing, both 400 m; landing speed 100 km/h (62 mph).

ANT-20bis, PS-124

ANT-21, MI-3

ANT-20bis, PS-124 Also known as MG bis and (from its civil registration) L-760, this was planned as first of several replacements for lost *Maksim Gorkii.* Fantastic emotion swept nation in aftermath of crash and 35 million rubles quickly collected to build whole fleet of giants. Authorization received to build mighty squadron of 16 MGs, first repeating name of lost L-759, others named for party leaders and revolutionaries. Intended these aircraft should be *dubler* ANT-20s, but several factors transpired to change plan. More powerful engines made it possible to dispense with tandem pair on centreline. Structural refinement was accompanied by various changes to systems, and 36 arrest of Tupolev at start of work caused upheaval and delay. Programme assigned to different GAZ outside Moscow where project placed under deputy director Boris Andreyevich Saukke, who had been on original AGOS ANT-20 team. Industry had no chance of building 16 giants, but work began on first, L-760, early 38. By this time feelings had changed; *Agiteskadril* was running down, and certainly did not want another MG. Decision to redesign as straightforward civil airliner, and no attempt made to build any further examples. Apart from using just six more-powerful engines, with three-blade 3,5-m constant-speed propellers, radiator ducts with inlet shutters and proper aerodynamic profile and exhaust stacks inside cowls terminating in single small pipe projecting above LE, ANT-20bis differed significantly in airframe (appreciably heavier), fuel system and many other parts. Nose lengthened, with landing light (in MG in LE between engines 2/3 and 4/5), larger nose cabin with nine wide windows extending to front of cockpit, improved cockpit with only four large glass windows and better roof profile, all crew amidships, vertical tail of reduced height but increased chord, spats on main gears retained. ANT-20bis test-flown by Gromov and E. I. Schwarz late 39; after NII tests passed to Aeroflot with designation PS-124, accommodation for 64 (max 85) day passengers in fuselage and centre section, with crew of nine. Put to work on route Moscow-Mineralnye Vody. Re-engined Dec 40, then after invasion assigned to heavy utility payload 14,4 t, transport in rear areas. Damaged beyond repair in heavy landing 14 Dec 42 after flying 969 h total-time. Last corrugated-skin aircraft in USSR.

DIMENSIONS Span 63,0 m (206 ft 8⅓ in); length 34,1 m (111 ft 10½ in); height (over fin, on ground) 7,7 m; wing area 486 m² (5,231 ft²).

ENGINES (as built) six AM-34FRNV, (41) AM-35.

WEIGHTS Empty 31,3 t (68,783 lb); fuel/oil 5830 kg+370 kg; loaded 44 t (97,000 lb), in wartime usually operated at 53 t (99,206 lb).

PERFORMANCE Max speed 235 km/h (146 mph) at SL, 275 km/h (171 mph) at 3,5 km; cruise at 3,5 km 242 km/h (150 mph); service ceiling 5,5 km (18,000 ft); range 1300 km (800 miles); take-off 500 m/26 s; landing 500 m/32 s/100 km/h.

ANT-21, MI-3 Launched by VVS prototype order 18 Jan 32, this multi-seat fighter, broadly successor to KR-6, was important in effecting transition from corrugated to smooth skin. Assigned to Arkhangyelskii's brigade, design used actual parts of KR-6 in wing (including spar booms) but with reduced t/c and span, bending moments coming out unchanged. Fuselage completely new, dural stressed-skin semi-monocoque with oval section, part flush-riveting, large wing root fillets. Engine installations almost exactly as KR-6 but with ducts for slanting ventral radiators continued to rear to provide compartments for retracting main gears, first in KB. Single KhMA air/oil shock strut with two welded diagonal side braces at top and fork at bottom for 1000 mm × 250 mm wheel retracting to rear by single hyd jack acting on rear breaker strut. Four wing tanks, capacities unknown. Twin-finned tail with corrug (mostly 25 mm × 10 mm) skin, elevators with tabs and internal mass-balance on fixed-incidence tailplane, rudders with large horns at top. Single pilot in open cockpit high above LE with two fixed PV-1 in nose, front gunner in KR-6 type nose with twin hand-

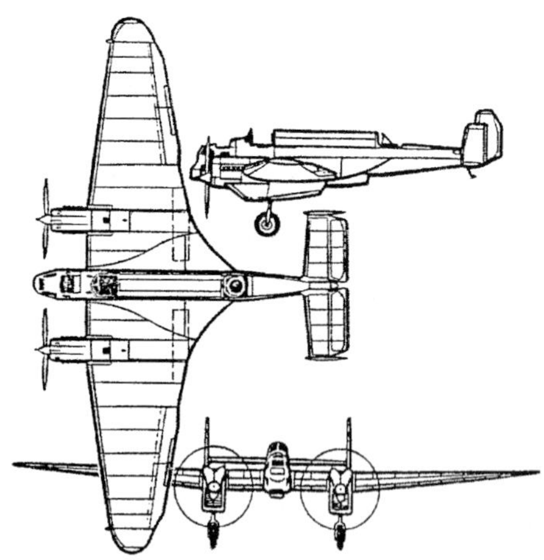

ANT-21, MI-3

aimed PV or DA and similar pair in rear dorsal position (believed Tur-6B); aircraft also built with lower rear position under wing TE for one DA manned by fourth crew-member, not known if gun fitted. This deep centre fuselage intended to house 600-kg bombload, never equipped for this.

Prototype completed as MI-3 (*Mnogomestnyi Istrebitel*, multi-seat fighter), flown Gromov 1 Aug 33 without problems. Popularly called *Mitrich*, liked by CAHI pilots until 14 Sept 33 when I. F. Kozlov made gentle dive to see how far beyond 350 km/h possible. Violent rudder flutter broke upper hinge of one rudder which then went hard-over and rotated aircraft 180° in yaw. This rudder separated but says much for strength of MI-3 that pilot made heavy landing. Aircraft replaced by **ANT-21bis, MI-3D** (*dubler*). Completely redesigned fuelage and tail with enclosed nose for gunner with five main windows, two lower front windows and one 20-mm (or UBS) in pivoted socket; fighter-type cockpit with sliding canopy and optical sight for fixed guns (twin PV); long 'greenhouse' with folding rear section for dorsal gunner (twin DA), lower rear gun position removed. New engines but same cowls. Smooth-skin tail with tapered tailplane braced half-way up single fin, mass-balanced rudder without horn but with Flettner tab. Wing span and aileron chord increased. Completed April 34, NII tested July-Dec but not accepted.

DIMENSIONS Span 19,11 m (62 ft 8⅜ in), (D) 20,76 m (68 ft 1¾ in); length 10,85 m (35 ft 7⅛ in), (D) 11,57 m (37 ft 11½ in); wing area 52,1 m^2 (561 ft^2), (D) 59,2 m^2 (637 ft^2).
ENGINES Two M-17B, (D) two M-34N.
WEIGHTS Empty 3412 kg (7,522 lb), (D) 4058 kg (8,946 lb); normal loaded 5088 kg (11,217 lb), (D) 5463 kg (12,044 lb).
PERFORMANCE Max speed 351 km/h (218 mph) at SL, (D) 285 km/h at SL, 350 km/h (217 mph) at 5 km; service ceiling 7885 m (25,870 ft), (D) 8,3 km (27,230 ft); no other data.

ANT-21bis, MI-3D

ANT-22, MK-1

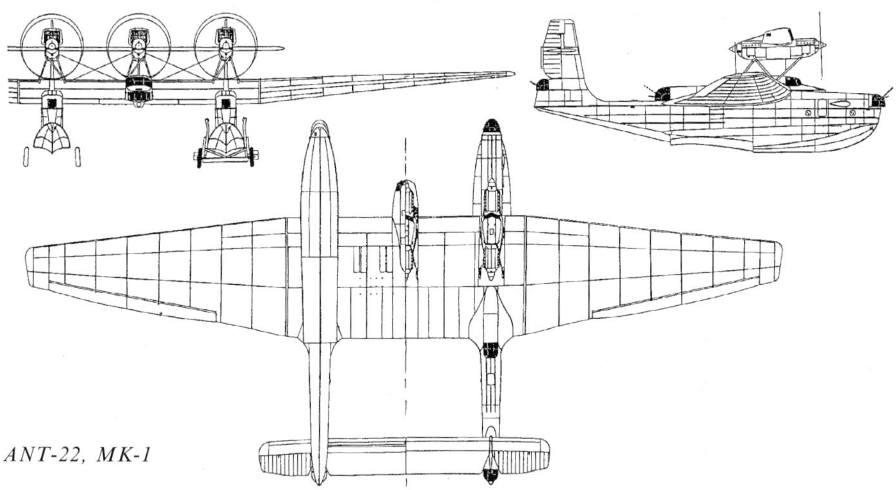
ANT-22, MK-1

ANT-22, MK-1 Twin-hulled flying boat, one of largest in history. Designation from *Morskom Kreiser* (sea cruiser), long-range aircraft for ocean recon and bombing; Tupolev also schemed civil transport. Preceded by unbuilt ANT-11 and by 30 project by CCB (Bartini), which were too large for various civil requirements but too small for continually upgraded military demand. Project grew out of ANT-11 and single prototype funded Dec 31; completed Aug 34.

All-metal flying boat with trad AGOS wing CAHI-6 profile, root thickness 20%, built as 16-m centre section of constant profile and 17,5-m tapering outer panels. Four spars, closed booms in centre section 125 mm×113 mm, drawings show no A-profile ribs in centre section but ten in each outer wing dividing corrugated skin (size/gauge not known) into 11 chordwise panels. Dural hulls 15,0-m apart on centrelines built as broad two-step planing bottoms (3,5-mm and 4,5-mm sheet at steps) with multiple keels and bulkheads and large hatches for inspection, superimposed by narrow upper portions with usable interior, relatively simple and light construction and smooth Alclad skin stiffened by two sets (front and rear, at different levels) of external half-round bead stiffeners. Tail on each hull with high-aspect-ratio upper 19,0-m tailplane projecting beyond fin on each side braced by wires and central inverted vee struts to narrow lower tailplane joining hulls and required because of short moment arm (only 2·2% MAC). DA-2 in each tail turret, Oerlikon 20 mm in left dorsal turret and right nose, and ShKAS in left nose and right dorsal. Bombload up to 6-t (13,120 lb) hung under centre section, wide variety of poss loads. Flight crew navigator/bomb-aimer, flight mechanic and side-by-side pilots in cockpit in small nacelle ahead of LE on centreline, emergency access via LE to hulls. Engines in tandem nacelles basically similar to TB-4 but larger radiator ducts on top between engines and, for CG reasons, engines on struts raked as far forward as poss. Same type 4-m propeller as ANT-20. Same 30-hp APU and generator set as ANT-20, compressed-air starting, full electrics, intercom and four radios.

Factory test began 8 Aug 34, no particular problem but prolonged until May 35. Outstanding on water, handling in air good though poor climb and ceiling. On paper, great capacity for overload, but when burdened with full military equipment performance sank to unacceptable level. NII tests 27 July-15 Aug 35 showed following figures with 6-t bombload; max speed 205 km/h; time to 1 km, 18 min; service ceiling 2250 m. These figures were at only 33,56 t, 9,44 t below max overload. VVS appreciated ability to operate in wind 12 m/s with waves to 1,5-m, but no production. Last two missions were world record 8 Dec 36 by T. V. Rybenko and D. N. Ilinskii; first carried 10,04-t useful load to 1942 m; second lifted 13-t load (just).

DIMENSIONS Span 51,0 m (167 ft 3⅞ in); length 24,1 m (79 ft 0¾ in); wing area 304,5 m^2 (3,278 ft^2).

ENGINES Six M-34R.

WEIGHTS Empty 21 663 kg (47,758 lb); fuel/oil 5,1 t; loaded (normal) 29 450 kg (64,925 lb), (overload) 33 560 kg (73,986 lb).

PERFORMANCE (normal gross) Max speed 233 km/h (145 mph) at SL; cruise 180 km/h (112 mph); time to 1 km, 10·3 min; service ceiling 3,5 km (11,480 ft); endurance 7 h; range 1350 km (840 miles) (at 4,36 kg fuel/oil per km ultimate range at overload wt was 4700 km, never approached in practice); time 360° turn, 85 s; take-off/landing both 600 m; alighting speed 110 km/h.

ANT-23, I-12 Unconventional fighter conceived by Viktor Nikolayevich Chernyshyov, AGOS brigade leader, to carry two APK-4 (see Recoilless Guns). Wing 18% thick, 5,4-m centre section, tapered outer panels with typical 'increased-chord' ailerons. Two D1 tubular spars, smooth skin 0,8 mm or 1,0 mm stiffened by external n-strips every 150 mm, underlying rib every 1,5 m; tail similar with fin braced to both tailplane spars. Fuselage welded KhMA truss with smooth D1 skin. Front engine helmeted ring cowl, rear with reverse helmets, 2,9-m wooden 2-blade props with metal spinners; riveted AMTs fuel tanks behind open cockpit with Aldis sight. Gun barrels formed tail booms, 3,7 m apart, seamless steel tube 170-mm diam, wall 1 mm to 3 mm, D1 fairing across wing. Tall main gears with outsize tyres and tall tailskids under tubes, all bungee-sprung. Flown by I. F. Kozlov from late 31, good handling but rear prop inefficient and endangered pilot-escape. Shell exploded in barrel 19 May 32, Kozlov awarded Order of Red Star for landing aircraft (when damaged barrel finally broke). **ANT-23bis** with jettisonable rear prop almost complete when programme abandoned.

DIMENSIONS Span 15,6 m (51 ft 2⅛ in); length 9,5 m (31 ft 2 in); wing area 30,0 m^2 (323 ft^2).

ENGINES Two GR9AK (licensed 525 hp Jupiter VI).

DIMENSIONS Span 15,67 m (51 ft 5 in); length 9,52 m (31 ft 2¾ in); wing area 33,0 m^2 (355 ft^2).

ENGINES Two 570 hp GR9Ak, later replaced by M-22.

WEIGHTS Empty 1818 kg (4,008 lb); loaded 2405 kg (5,302 lb).

PERFORMANCE Max speed 318 km/h (198 mph) at 5 km; climb to 5 km 7·7 min; service ceiling 9,32 km (30,580 ft); range 405 km (252 miles); ldg 100 km/h.

ANT-24 Little information on record about this twin-engined heavy fighter which came between MI-3D and ANT-29. One prototype authorized Nov 32, no evidence construction started. No data.

ANT-25, RD World records always impt to USSR, discussion at CAHI 20 May 31 examined options and decided to attempt absolute range. Tupolev showing existing (French) figure easily beaten by purpose-designed monoplane with greater aspect ratio and retrac gear. Series of meetings at Revolutionary War Council established special commission to build aircraft to Tupolev design and organise flight. Meeting 7 Dec 31 agreed to aim at 13,000 km using ANT-25, official designation RD (*Rekord Dal'nost*, record distance), flight set for summer 32; major task to provide improved equipment, train crew(s) and build paved runway. From start funding had one eye on importance of long range to future bombers. Further executive committee formed under K. Ye. Voroshilov to manage programme, including M-34 engine by Mikulin KB and building of runway. Design of RD often ascribed to Sukhoi but Tupolev adamant all his brigades played important role and refused to credit any one individually. Work took much longer than anticipated: design finished July 32 but first RD not complete until 11 months later, and then only with direct-drive M-34. Aug 32 work began on **RD** *Dubler* (double) with geared M-34R. Original direct-drive RD flown by M. M. Gromov 22 June 33; RD *dubler* flown by Gromov 10 Sept 33.

Design based on remarkable cantilever wing with aspect ratio 13·1. Modified CAHI-6 profile, t/c 20% root, 19·2% half semi-span, 18·5% tip. Remarkably, traditional corrugated structure adopted, with 3,75-m centre section and two 15,125-m outer panels, latter with 5° dihedral. Two main spars, plus auxiliary rear spar, all with usual elliptical-tube booms but this time KhMA heat-treated to 140 kg/mm^2, with truss web members and joint gussets all D6. Built-up truss ribs, 18 of A-type each side of fuselage projecting above wing profile with intermediate internal ribs; 40 mm × 8 mm skin riveted between each pair of A-ribs. Ailerons divided into four sections each side with slotted nose, inset hinges giving 100% mass balance and with servo drive tab on section 2 (numbered root to tip) on each side, sections 1 and 2 slightly greater chord. Oval-section fuselage in two parts, front integral with centre-section of wing, with strong truss internal structure and

ANT-23, I-12

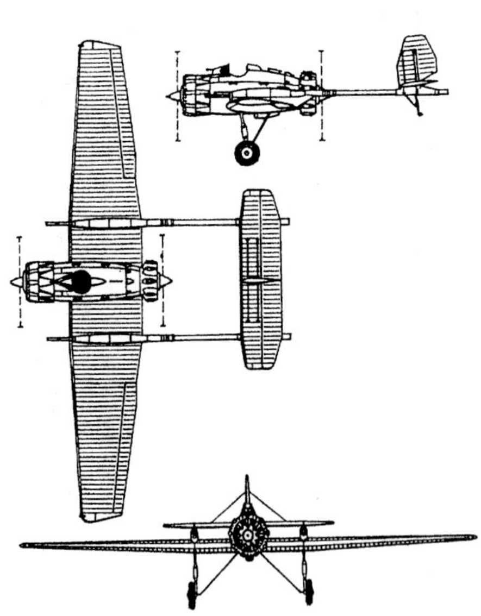

ANT-23, I-12

welded KhMA engine mount, lighter rear monocoque with L-section frames and stringers, smooth skin throughout but not flush-riveted, thickness 0,8 mm or 1,0 mm. Tail all-dural with corrugated skins, fixed tailplane with seven A-type ribs each side, elevators with 100% balance by recessed hinges, rudder part-balanced by recessed hinges on welded KhMA pyramids and part by small horn at top. Main gears with single air/oil shock strut of KhMA with twin side braces all hinged to front spar, rear bracing strut normally located at base of rear spar but on retraction pulled by elec motor and cable system along channel tracks almost to TE of wing, gear folding to rear; two 900 mm×200 mm tyres on each leg, half-projecting from wing ahead of box fairing (high drag, especially on take-off). Balloon tyre on tailwheel on fixed leg, half-projecting from fairing. Simple engine installation, large 4,5-m two-blade wooden propeller, water radiator cranked by pilot in vertical slides to extend beneath rear of cowling. Fuel relieved wing bending by being distributed in small tanks right across span between two main spars; centreline and inboard fuel used first to reduce wing loads. (Significant that, at time when fatigue ignored elsewhere, wing design factor was given as 3·0 to 4·8 depending on number of flying hours.) Fully enclosed accommodation, with separate tandem hinged canopies, for crew of three: pilot, radio/nav and engineer. Extremely comprehensive equipment including first Soviet gyro-compass, electro station generating 500 W at 12 V, various MF and HF radios including collapsible mast for emergency use after forced landing, sextant in hinged roof station and retractable ventral periscope.

Original RD was flown with M-34 with direct drive and compression ratio of 6, giving 750 hp. Though it showed good flying qualities it was soon clear max endurance could not exceed 48 h, putting any world record out of reach. This aircraft ceased to be of much interest, though it later flew with M-34 of compression ratio 7, giving 874 hp, and reaching 212 km/h. All hopes were focussed on **RD dubler** with geared M-34R engine distinguished by higher thrust-line and improved profile. Less obvious was vertical tail of increased chord, with servo-tab rudder. Original *dubler* engine had compression ratio 6·8 giving 900 hp driving two-blade prop. In Feb 34 engine of 6·6 c.r. was substituted, and later a 4,5 m propeller with three ground-adjustable light-alloy blades. This enabled full 6,1 t fuel load to be uplifted, though actual load never exceeded 5 t at this stage, giving 66-h endurance. Range still well short of design figure, though better than existing record, and Tupolev recognized a basic problem was corrugated skin. CAHI had established, for example, that airflow was seldom aligned with corrugations, and penalties much greater than previously believed. Entire wing and tail was thereupon covered with fabric, wrapped carefully around leading edges and sewn through many small holes with curved needles and tautened with dope. Propeller blades polished, fillet doped and sealed and attention paid to various other aerodynamically poor areas. Result was dramatic; drag at cruise was reduced by 36% and take-off weight increased. In mid-34 engine installation completely redesigned, with engine moved back 0,4 m and different radiator mounted in duct beneath engine with pilot-controlled entry shutters and variable flap at exit. Cowling sealed to reduce drag and exhaust stacks ducted to narrow fishtails near pilot canopy with surrounding ram-air pipe on each side providing heating for all three cockpits. Completely new vertical tail with increased height and rounded profile. Most of these changes effected by mid-34 when special paved runway at Shchyelkovo, with deliberate slope, was finished. Chosen crew, M. M. Gromov, A. I. Filin, I. T. Spirin, twice attempted max number circuits of triangle Moscow/Tula/Ryazan or Moscow/Ryazan/Kharkov only to be thwarted by engine trouble; 30 June 34 reached 4465 km in 27 h 21 min and three weeks later 6559 km in 39 h 1 min. On 10 Sept 34 same crew orbited this triangle 75 h 2 min covering 12 411 km, a new closed-circuit (even if multi-lap) record. Plan then made for what had always been main objective: over North Pole to USA. In midst of preparations. Gromov fell ill, and new crews organised under Levanevskii and Chkalov. In Aug 35 Levanevskii's crew flew towards Pole but suffered engine problems and struggled back to Leningrad. Decision to do Arctic testing: V. P. Chkalov, G. F. Baidukov and A. V. Belyakov detailed to use aircraft now painted with red wings and tailplane, with registration NO25-1, leaving Moscow after midnight 20 July 36 and flying Franz-Josef Land, Severnaya Zemlya, Petropavlosk Kamchatskii towards Nikolayevsk but had to land on island Udd (since named

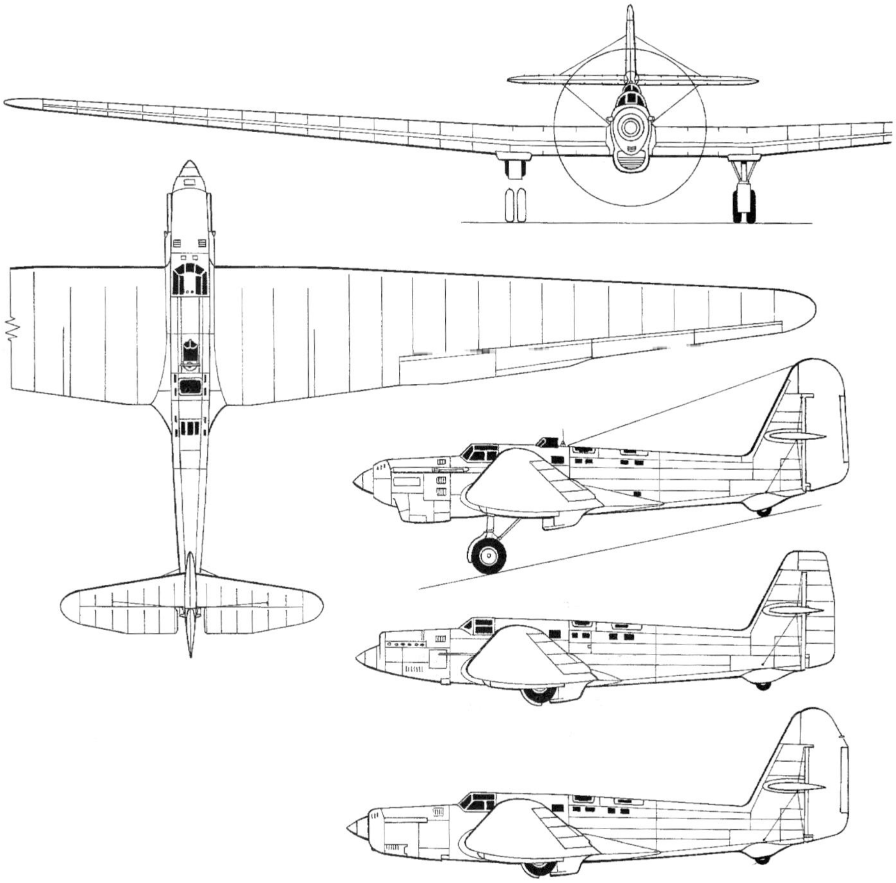

ANT-25 final configuration (centre side view, RD No 1; bottom RD dubler)

ANT-25, RD (third configuration)

Chkalov), 9374 km in 56 h 20 min. Using experience, same crew took off for USA 18 June 37; despite prolonged mechanical and weather adversities just managed to cross US frontier landing Portland (Washington, not Oregon) 9130 km in 63 h 25 min (FAI recognized great-circle 8504 km). Final and greatest RD flight was by M. M. Gromov, A. B. Yumashyev and S. A. Danilin, leaving Moscow 12 July 37 with record 6230 kg fuel and oil; easy flight, no bad weather, landed San Jacinto, Calif, 11 500 km (10 148 gt circle) in 62 h 17 min, with 1500 kg unused fuel.

In winter 1933-34 VVS ordered 50 slightly modified ANT-25 for use as long-range bomber and crew-training a/c. Not fulfilled, but GAZ-18 Voronezh did deliver 13 by early 36, with smooth skin on wings and tail and used for many research programmes including manufacturing techniques (including flush riveting and welding), engines (including Jumo 207 and, as **RDD**, RD *Dieselnyi,* Charomskii AN-1) and VVS bomber regt. Strangely, none made attempt on distance record nor on absolute altitude, for which RD would have been exceptional contender with different propulsion. Four military RDs were, however, rebuilt as BOK-1, BOK-7, BOK-11 and BOK-15.

DIMENSIONS Span 34,0 m (111 ft $6\frac{9}{16}$ in); length (1st) 13,4 m (43 ft 11½ in), (dubler) 13,08 m (42 ft 11 in); wing area 87,1 m^2 (937·5 ft^2).

ENGINE (1st) one M-34, (dub) M-34R; (mil RD) see text.

WEIGHTS Empty (1) 3,7 t (8,157 lb), (d, as built) 3784 kg, (d, 36) 4,2 t (9,259 lb); fuel (36) 5880 kg+350 kg; loaded (1) 8 t (17,637 lb), (d, 33) 10 t, (d, 36) 11,5 t (25,353 lb).

PERFORMANCE Max speed (874 hp) 212 km/h at SL, (d, 33) 244 km/h, (d, 36) 246 km/h (153 mph) at SL; service ceiling (d, full load) 2,1 km, (light) 7850 m (25,750 ft); endurance/range (1) 48 h/7200 km, (d, 33) 66 h/10 800 km, (36) 80 h/13 020 km (8090 miles); take-off (1) 1 km, (36) 1,5 km.

ANT-26, TB-6 Russian love of bigness led to same structural technology extrapolated from ANT-4 to ANT-6, -16 and -20, and by 30 Tupolev believed span 200 m (656 ft) possible. As next (intermediate) stage obtained 33 sanction for bomber of 70 t, span 95 m, as **ANT-26** (service designation **TB-6**), later adding ANT-28 transport.

ANT-26 was by far largest aircraft in world, and Tupolev was most anxious it should not be obsolescent when it appeared. Though underlying structure was trad diagonally-braced truss spars and ribs, all skins were smooth and there were no A-profile ribs to break contour of wing surfaces. Greatest problem was inevitable thickness of wing, which (as discovered with TB-4) could seriously reduce propulsive efficiency of propellers ahead of LE. CAHI studied five propulsion configurations and several wing forms before deciding on modified CAHI-6 profile (about 20% ruling t/c) with eight engines on LE and two tandem pairs above. It would have been possible to put all 12 on LE but this would have involved problems in resolving thrust outboard and appeared to offer poorer propulsive efficiency despite reduced performance of rear tandem propellers working in slipstream. More important, Tupolev hoped later to use engines of 1300 hp/1500 hp and eliminate pylon-mounted tandem units. Structurally wing was to have four main spars with 140 kg/mm^2 KhGSA booms, built as centre section and tapering outer panels with dihedral. Fuselage oval section at front changing to rectangular with rounded corners approx same size as MG but rather longer. Fixed tailplane on slightly raised platform on rear fuselage carrying large centreline and two smaller aux fins all braced by single inclined strut each side. Main landing gears with four-wheel bogies in tandem pairs (tyre size not known but smaller than MG) on single shock struts raised vertically into large fairings beneath wing at extremity of centre section. No crew in wing, and volume so large tanks for max fuel occupied only small proportion of space between front two spars. Normal crew 12, defensive armament one 37 mm (probably in tail turret), four 20 mm and four pairs ShKAS (two 20 mm probably in large turrets at rear of outer nacelles); bombload up to 20 t, normally 4 t for 2500 km. Most Soviet observers doubt that construction of TB-6 prototype ever started, though built and flew (test-pilot B. N. Kudrin) scale AGOS tandem-seat glider of some 20 m span which may have been ANT-33. By late 35 little was unknown about TB-6, resulting in 36 decision to discontinue project.

DIMENSIONS Span 95,0 m (311 ft 8⅛ in); length 39,0 m (127 ft 11½ in); wing area about 800 m^2 (8,600 ft^2).

ENGINES Twelve M-34FRN.

WEIGHTS Empty about 50 t (110,000 lb); loaded 70 t (154,000 lb), (max) 76 t (167,500 lb).

PERFORMANCE Max speed (est) 300 km/h (186 mph); range with 20 t bombload 1000 km, with 4 t about 2500 km.

ANT-27, MDR-4, MTB-1 Abysmal performance of MDR-3 (see under Chyetverikov) caused grave concern to MA and industry; lacking confidence in designer to produce solutions, whole programme (without Chyetverikov's participation) was transferred to KOSOS end-32 as **MDR-4**. Assigned to marine brigade under I. I. Pogosskii, who redesigned MDR-3 leaving nothing but basic hull structure. Wing was completely new, though derived from previous AGOS wings, with considerably greater span and area though tapering from root with unchanged root chord. Centre section with A-profile ribs, 40 mm×8 mm corrugated skin; outer panels fabric covered. New propulsion group comprising three more powerful water-cooled engines (in place of two tandem pairs), centre engine being pusher, all with circular radiator on front of nacelle and with geared drive to 4 m wooden two-blade props. Hull stiffened with AGOS (Pogoskii) style half-round external stiffeners, four each side. New single-fin tail with high braced tailplane and rear turret closely similar to MK-1. Armament 20 mm Oerlikon in Tur-9 dorsal turret twin ShKAS in nose cockpit and same in tail turret; bombload 500 kg, but armament never fitted. Prototype tested Taganrog, and taking off in choppy sea 16 April 34 centre nacelle crashed on to cockpit, killing Pogoskii and pilot Ivanov.

Urgent need for good large flying boat resulted in almost immediate order for second aircraft. This incorporated various changes, designation **ANT-27bis**, MTB-1 (*Morskoi Torpedonosyets-Bombardirovshchik*, sea torpedo-bomber). Basically similar but with much greater fuel capacity. Owing to sharp need for such aircraft MA decided to adopt MTB-1 prior to start of flight tests, and preparations made for series production of 15 at GAZ-31. First aircraft began factory testing May 35; on 23 Sept 35, prior to handing over at factory, suffered second fatal crash caused by separation of fabric from wing. Despite this, after urgent board of enquiry, decision to continue production, five MTB-1 being delivered 36 and ten 37 at which point production terminated. Despite sorry history and poor performance, service life satisfactory, used as ocean reconnaissance as well as torpedo-bombing, until 42.

DIMENSIONS Span 39,4 m (129 ft 3¼ in); length 21,9 m (71 ft 10¼ in); wing area 177,5 m^2 (1,911 ft^2).

ENGINES Three M-34R.

WEIGHTS Empty (4) 10,5 t (23,148 lb), (1) 10,521 kg (23,194 lb); fuel/oil (4) 2450 kg+196 kg, (1) 3746 kg+370 kg; loaded (4) 14,660 kg (32,319 lb), (1) 16,250 kg (35,825 lb).

PERFORMANCE Max speed (4) 232 km/h (144 mph) at SL, 211 km/h at 3 km, (1) 225 km/h at SL (140 mph), 200 km/h at 3 km; climb to 1 km (4) 4·6 min, (1) 5·4 min; service ceiling (4) 5450 m (17,880 ft), (1) 4470 m (14,665 ft); endurance (4) 7 h, (1) 11 h; range (4) 1230 km (often reported as 2130 km), (1) 2000 km; time 360° circle (4) 30 s, (1) 35 s; take-off (4) 30 s; alighting speed (4) 100 km/h (1) 105 km/h.

ANT-28 Transport version of the ANT-26 bomber, this enormous aircraft used a basically similar airframe but was quite different in detail. Major differences included long MG-like nose with row of large windows, two rows of five cabins on each side of fuselage inside wing, each providing accommodation at two levels, large single-finned tail, plain fixed landing gear and four integral stairways for access on ground. No provision for heavy or bulky freight except as underslung external load. In project state 33, so far as known purely military though no known defensive armament. Little published, but Tupolev said Petlyakov led design team and both he and Tupolev felt

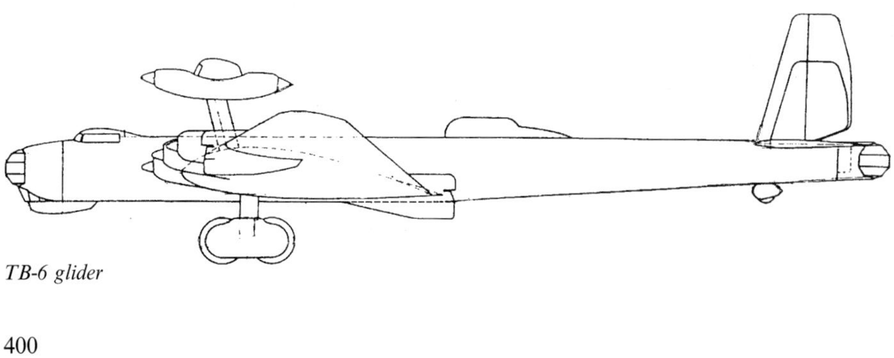

TB-6 glider

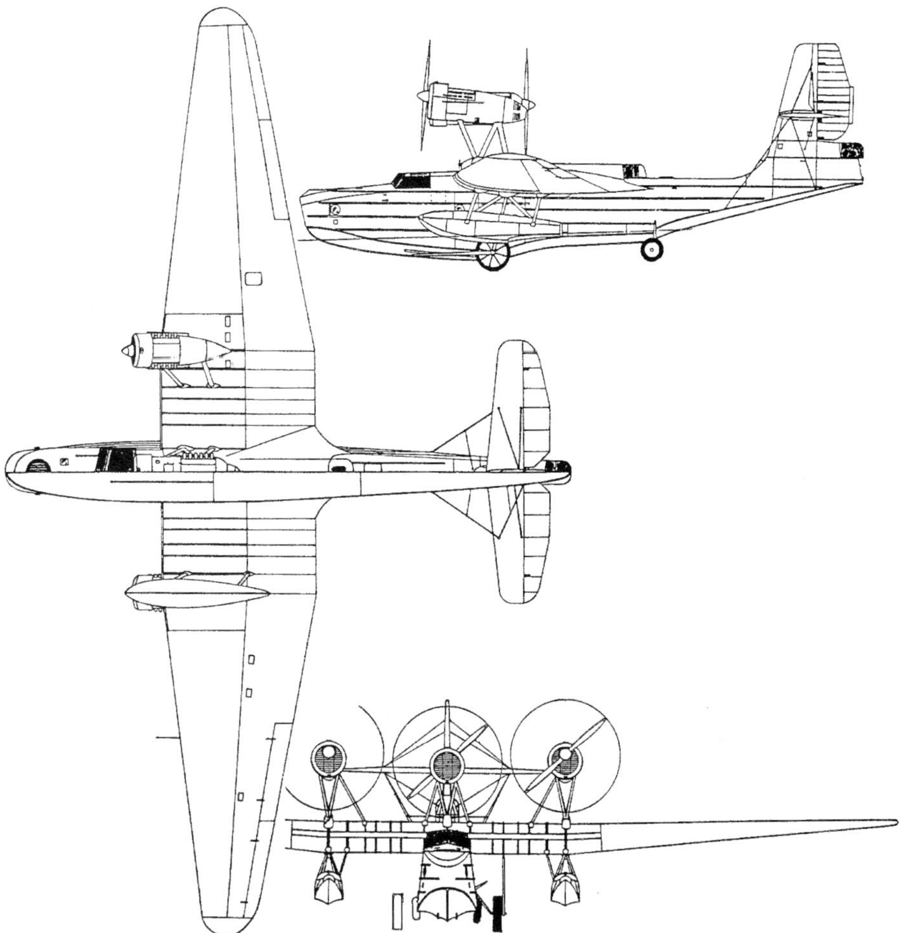

ANT-27, MDR-4

ANT-27, MDR-4

ANT-28, model

strongly ANT-28 should have been built, in contrast to TB-6 whose termination was generally considered wise. He said, from memory, max number of troops or other passengers about 250.

DIMENSIONS Span 95,0 m (311 ft 8⅛ in); length 40,0 m (131 ft 3 in); wing area about 800 m^2.

ENGINES Twelve M-34FRN.

WEIGHTS Empty about 47 t (103,600 lb); loaded 76 t (167,550 lb).

PERFORMANCE Max speed (est) 290 km/h (180 mph); range with 25 t payload 3600 km (2,250 miles).

ANT 29, DIP Natural successor to MI-3D, first AGOS smooth-skinned aircraft. Funded by VVS Sept 32 as heavy fighter with largest recoilless gun. Wing two spars at 16·6% and 66·6% chord, short centre section, 3-m horizontal inner sections with split flaps, 5,9-m tapered outer wings with dihedral, two-section balanced tabless ailerons. Stumpy fuselage seating pilot above LE under hinged/folding canopy and gun loader over TE with sliding canopy. High braced tailplane with mass-balanced tabbed elevators cut out for tabbed rudder. Main gears with 900 mm × 280 mm tyres on forked single legs retracted to rear into bay with twin doors. Engines in neat cowls with rads underneath with inlet shutters, driving French Chauvière 3,5-m dural 3-blade props. Fuel in four metal wing tanks. Two APK-8 about 4 m long in bottom of fuselage, shells loaded by observer, blast ejected behind tail. For sighting, two ShKAS guns in wing roots, loader with single ShKAS. Fixed optical sight ahead of windscreen.

Single DIP (*Dvukhmestnyi Istrebitel Pushechnyi*), 2-seat cannon fighter, completed Feb 35 but returned to ZOK for rectification (eg reskinning control surfaces) and did not fly until end-35. Failed NII testing.

DIMENSIONS Span 19,19 m (62 ft 11½ in); length 11,1 m (36 ft 5 in); wing area 56,8 m^2 (611 ft^2).

ENGINES Two 760-hp Hispano-Suiza 12Ybrs.

WEIGHTS Empty 3,9 t (8,598 lb); fuel/oil 720 kg+80 kg; loaded 4960 kg (10,935 lb), max 5,3 t (11,684 lb).

PERFORMANCE Max speed 296 km/h (184 mph) at SL, 352 km/h (219 mph) at 4 km; climb 5·6 min to 3 km, 9·6 min to 5 km.

ANT-30 SK-1 long-range fighter project

ANT-31, I-14 This fighter designed by Sukhoi's brigade 32 was first in world combining unbraced wing, retrac landing gear, cowled engine, enclosed cockpit and powerful armament. Negative feature was retention of corrugated skin on wings and horizontal tail. Wing of NACA 16% profile, two spars of KhGSA tube booms and riveted KhMA truss webs, ribs riveted from rolled sections, skin 1,0-mm and 0,8-mm, corrugation details unknown. Long-span slotted ailerons, no flaps, wing made as horizontal centre section and outer panels with considerable dihedral. Stumpy fuselage with modern structure of L and top-hat frames, L stringers and flush-riveted skin 0,8-mm throughout. Integral fin of remarkable height and small chord carrying braced horizontal tail (apparently 25 mm × 5 mm corrugation) well above fuselage. Large fuel tank on CG above

ANT-29, DIP

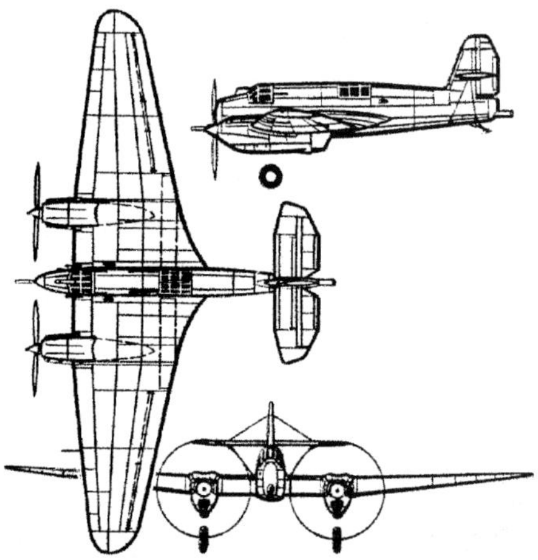
ANT-29, DIP

wing, cockpit behind trailing edge with glazed enclosure having roof hinged upwards at rear. Imported engine in locally designed installation with Bristol front collector ring around crankcase, large NACA cowl extending behind LE and direct drive to two-blade wooden propeller. Wide-track main gears with single oleo struts, inward and rear bracing struts, and retraction inwards by cockpit handwheel and cables. Armament one PV-1 upper right fuselage, four D-1 bomb containers and (not installed) two APK-37 under wings firing outboard of prop and with exhaust outboard of tailplane; alternative, two PV-1 and two underwing ShVAK. Sight as ANT-29. Flown on fixed skis by K. A. Popov 27 May 33, agile but tricky.

Second aircraft, **ANT-31bis, I-14bis,** redesigned with more powerful engine, new wing and horiz tail with smooth skin, narrow-track landing gear retracting outwards, narrow open cockpit and (intended) two PV-1 and two APK-37 (not installed at rollout). Completed March 34 and on factory and NII test to Oct. NII testing favourable and measured figures outstanding, but problems with spinning and with gear retraction mechanism. Pilots also had trouble with narrow track; original gear much better. Despite this, production order placed with more powerful licensed engine, revised armament two APK-11 and two ShKAS, modified horizontal tail. Of 55 ordered only 22 built (one ShVAK, one ShKAS) by GAZ-125, Nov 36/Feb 37.

DIMENSIONS Span 11,2 (36 ft 9 in); length (both) 6,1 m (20 ft 0 in); wing area (14) 16,8 m^2 (181 ft^2), (bis) 17,0 m^2 (183 ft^2).

ENGINE (14) One 580 hp Bristol Mercury VIS2 (not VS), (bis) 712 hp Wright Cyclone SGR-1820-F2, (production) M-25A.

WEIGHTS Empty (14) about 1088 kg, (bis) 1169 kg, (production) 1170 kg (2,579 lb); fuel/oil (production) 200 kg; loaded (14) 1455 kg, (bis) 1524 kg (production) 1540 kg (3395 lb).

PERFORMANCE Max speed (14) 316 km/h at SL, 384 km/h at 5 km, (bis) 323 km/h at SL, 414 km/h at 3 km, (production) 375 km/h (233 mph) at SL, 449 km/h (279 mph) at 3400 m; time to 1 km, not recorded, to 5 km (14) 8·2 min, (bis) 9·6 min, (production) 6·5 min; service ceiling (14) 9,4 km, (bis) 7,2 km, (production) 9430 m (30,940 ft); range, (14) 600 km (production) 615 km (382 miles); time 360° circle 16·5/14/14 s; take-off (14) 120 m, (production) 230 m; landing (14) 260 m, (production) 320 m/129 km/h.

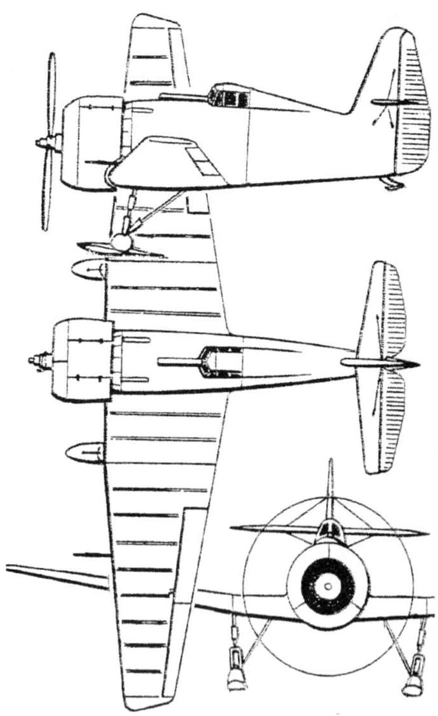
ANT-31, I-14

ANT-32 Projected single-seat fighter in KOSOS 34 programme, monoplane with conventional cannon armament (possibly *moteur-canon*), not built.

ANT-33 No information.

ANT-34 Projected development of ANT-29 with conventional cannon armament in fuselage, two seats, two Wright Cyclone F2 engines, smooth skin throughout, not built.

ANT-35, PS-35 In May 34 Aviavnito organised competition for two sizes of civil airliner. Arkhangekskii headed design of both civil and military versions of same basic design (see ANT-40 bomber), though no parts actually similar except landing gears. ANT-35 wing of typical KOSOS light-alloy construction with horizontal centre section extending to outer edges of nacelles and outer panels with 7° dihedral; straight taper from root on TE, two main and two secondary spars with riveted truss construction, ribs part pressed from sheet, skin 0,6 mm/1,0 mm on c/s, 0,5 mm/0,6 mm outer wings, riveting around LE to front spar being flush. Split flaps, three sections each side hydraulically driven to 60°. Modern tail unit of mainly pressed-sheet construction with flush riveted 0,5-mm skin on LE. Control surfaces with deeply inset hinges and with small horn balance on rudder, smooth LE skin but 20 mm×5 mm corrugated skin aft of spar. Trim tabs throughout, dual pilot control with electric autopilot. Main gears welded KhMA fork for 1000 mm×300 mm tyre, pneumatic brakes and hydraulic retraction rearwards into bay with twin doors leaving part tyre protruding. Fixed balloon tailwheel. French engines in NACA long-chord cowls with rear adjustable gills driving three-blade 3,2-m ground-adjustable dural props. Fuselage semi-monocoque of relatively small (1,5 m×2 m) section, five windows with adjacent seat each side, door rear on left with rear toilet and baggage, each passenger having individual hot-air outlet (exhaust heat-exchange system), light and ventilator. Payload 840 kg.

Construction of prototype NO35 swift and free from complication. First flight M. M. Gromov 20 Aug 36; on factory test showed good qualities, flew Moscow/Leningrad/Moscow (1266 km in 3 h 38 min) 15 Sept, and in Nov/Dec flew to Paris and back for static display. Judged easily best of Aviavnito twin-engined submissions, but cabin ceiling too low (1,68 m max on centreline) and serious obstruction by spars passing through fuselage at backs of 2nd and 3rd seat rows. Second aircraft therefore built. **ANT-35bis** with longer and deeper fuselage offering centreline interior height 1,83 m (6 ft) and only slight ridges at spars. Nine built 37, powered by licensed version of same engine.

These were followed 39 by 11 **PS-35** powered by M-62 (Cyclone) driving VISh-2 (Hamilton) 3-blade props, fuselage 0,45 m (18 in) longer, emergency hatch in cockpit roof with D/F loop behind, centre-section fuel increased to 990 lit. Used Moscow to Riga/Stockholm,

ANT-31bis, I-14bis

ANT-35bis, PS-35

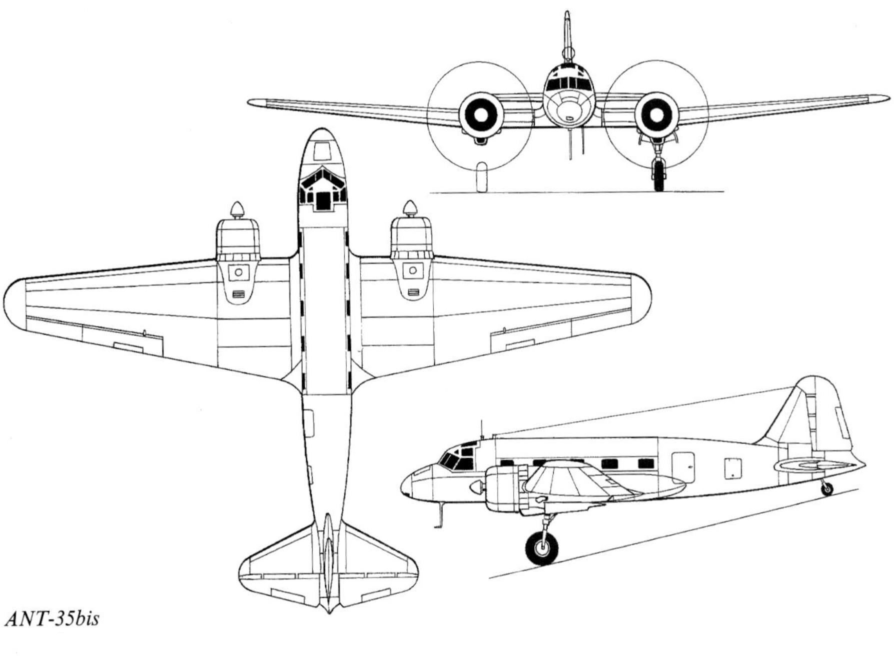

ANT-35bis

Prague, Leningrad, Lvov, Simferopol and Odessa to June 41. Never seen on skis but in winter spinners removed and perforated baffle added at front of engine. Engine installation identical to Li-2; speed higher but much less payload. Several continued to fly in rear areas during Second World War.

DIMENSIONS Span 20,8 m (68 ft 2⅞ in); length (ANT) 14,95 m (49 ft 0⅔ in), (PS) 15,4 m (50 ft 6⅓ in); wing area 57,8 m² (622 ft²).

ENGINES (ANT) two 800 hp GR 14K, later M-85, (PS) two M-62IR.

WEIGHTS Empty (ANT) 4710 kg (10,384 lb), (PS) 5012 kg (11,049 lb); fuel/oil (A) 690 kg, (P) 710 kg+90 kg; loaded (A) 6620 kg (14,594 lb), (P) 7 t (15,432 lb).

PERFORMANCE Max speed (A) 350 km/h at SL, 376 km/h (234 mph) at 4 km, (P) 350 km/h at SL, 372 km/h (231 mph) at 1,5 km; time to height (P) 6·1 min to 3 km, 13 min to 5 km; service ceiling (A) 8,5 km, (P) 7,2 km (23,620 ft); range with 10 pax (A) 920 km, (P) 1640 km (1,020 miles); take-off (P) 225 m/10 s; landing 300 m/19 s/105 km/h (65 mph).

ANT-36, RDD Long-range bomber *Rekord dalnost, diesel'nyi,* developed ANT-25 with Charomskii AN-1 diesel engine. Intended to fly 20-30 km/h faster and 20%-25% further, and related to BOK-1/7/11. Installation assigned to Moskalyev; aircraft flown spring 36 but no data.

ANT-37, DB-2 Recognizing futility of trying to make effective bomber from RD, Sukhoi's brigade assigned Dec 34 to creating new long-range twin-engined bomber using whatever RD parts were suitable. Believed to be Sukhoi's idea; much larger project than ANT-36 and whole Sukhoi brigade assigned, with direct Tupolev supervision. Sukhoi retained basic wing structure but completely altered rib peripheries to accept dural stressed skin, without corrugations, flush-riveted around LE as far as front spar. Span slightly reduced, ailerons increased in chord (projecting behind wing TE throughout instead of only two inner sections as on RD) and number of sections reduced to three. Almost entirely new fuselage, stressed-skin semi-monocoque, with accommodation for nose gunner in glazed turret filling whole depth of nose, pilot in enclosed cockpit with rearward-sliding hood above LE and mid-upper radio/gunner with retrac turret. Planned armament 20 mm in nose, twin ShKAS in dorsal turret, not fitted. Engines mounted on steel-tube frames at extremities of centre section in long-chord NACA cowls with cooling gills, 3,25 m two-blade wooden fixed-pitch props. Bomb bay in fuselage between wing spars increased to carry four FAB-250 or other loads totalling 1 t with powered doors. Tail based on ANT-25 but greater chord and reduced height, smooth skin and horizontal tail only just above fuselage. Landing gear with twin doors closing over legs, twin-wheel units being housed as before upstream of fixed rear fairing. Simple fixed tailwheel.

Prototype ANT-37, service designation **DB-2** (*Dal'nii Bombardirovshchik*, long-range bomber), first flown 16 June 35. Generally satisfactory but onset of tail buffet as speed increased and 20 July 35 in shallow dive violent vertical oscillation of horizontal tail caused structural failure of rear fuselage; pilot K. K. Popov and leading

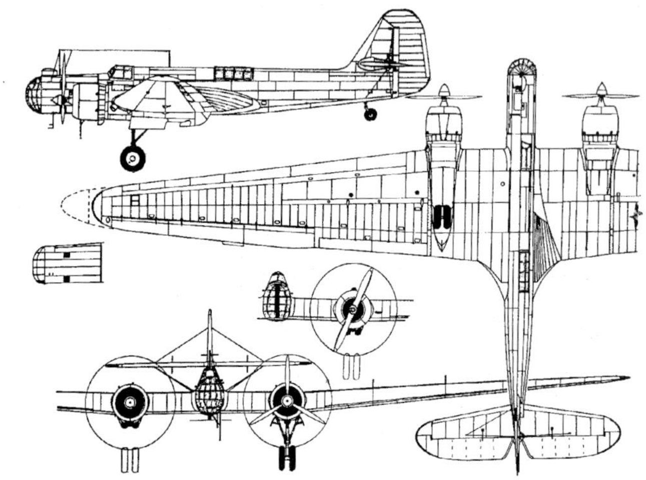

ANT-37bis, DB-2B Rodina, *inset, nose, head-on and wing tip of ANT-37*

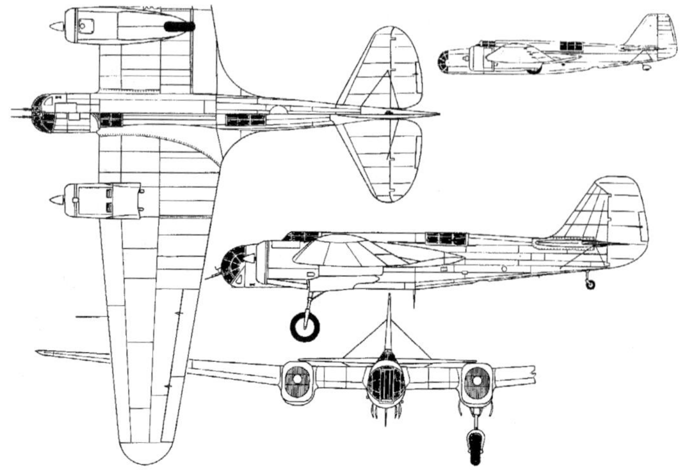

SB M/100A (inset SB, lengthened)

ANT-37, DB-2

engineer M. M. Yegorov escaped by parachute but electrical engineer killed. Major research programme begun into flight control surfaces and systems, aerodynamic buffet and induced flutter, published in *Tekhnika Vozdushnogo Flota* (air fleet engineering) in late 35. Sukhoi quickly completed *dubler*, **DB-2D**, with many modifications including improved wing/body fairing, strengthened rear fuselage and completely redesigned tail with defects corrected, with insignificant weight penalty. Three-blade Hamilton (VISh-2) propellers, extra tankage. Tested Feb 36 and in NII trials 20 Aug 36 demonstrated 1-t bombload carried Moscow/Omsk/Moscow 4995 km at 213 km/h average. After evaluation against Ilyushin possibilities ANT-37 ordered into production as DB-2, but order switched to CCB-30 (became DB-3).

On direct government order KOSOS also built third aircraft, **ANT-37bis (DB-2B)** as unarmed record-breaker. Airframe refined in further small details, with hemispherical glazed nose with curved panes of moulded Plexiglas, flush rear cockpit and retractable tailwheel. Fully equipped nav station in nose and radio station in rear cockpit. Fully retractable landing gears with bays closed by doors with no part of tyre projecting, fully retractable tailwheel, electrical operation by pushbuttons, first in Soviet Union. Considerably greater fuel capacity in 12 wing tanks, matched by greater engine power; engine installations as DB-2 dubler except addition of spinners. Rapid production of three DB-2B, a fourth with even greater outer-wing tankage not being completed. First aircraft flown Feb 36 and named *Rodina* (motherland). After long wait same aircraft set world women's long-distance record 24/25 Sept 38 in hands of V. S. Grizondubov, P. D. Osipenko and M. M. Raskov, flying Moscow 5908 km to wheels-up landing in flooded valley of Amur near Kerbi in 26 h 29 min. Subsequently long career with Aeroflot and, durng Second World War, at Moscow factory as research and trials aircraft.

DIMENSIONS Span (37) 33,2 m (108 ft 11 in), (rest) 31,0 m (101 ft 8½ in); length (both, approx) 15,0 m (49 ft 2½ in); wing area 84,9 m^2 (914 ft^2).

ENGINES (2) Two 800 hp GR K14, (2D) M-85, (bis) M-86.

WEIGHTS Empty (2) 5,8 t (12,787 lb), (bis) 5855 kg (12,908 lb); fuel/oil (2) 4050 kg+380 kg, (bis) 5525 kg+430 kg; loaded (2) normal 9450 kg (20,833 lb), overload 11,5 t (25,353 lb), (bis) 12,5 t (27,557 lb).

PERFORMANCE Max speed (2, normal weight) 301 km/h at SL, 342 km/h (213 mph) at 4 km, (bis) 300 km/h at SL, 340 km/h (211 mph) at 4,2 km; endurance (bis) 30 h; range (2) 5000 km, (bis) 7300 km (4550 miles); take-off (bis, full load) 1 km.

ANT-38 Projected high-speed bomber of 1934, believed to be standard bomber version of ANT-41 torpedo carrier.

ANT-39 Not known.

ANT-40 From these prototypes stemmed most important bomber in world of late 30s, thousands seeing action in five wars. VVS requirement for SB (*Skorostnoi Bombardirovshchik*, fast bomber) issued by NII October 33 and KOSOS took up challenge, basing **ANT-40** submissions on ANT-29 and other existing designs. Assigned by Tupolev to Arkhangyelskii brigade, work beginning January 34. Clean stressed-skin aircraft with smooth skin, wing raised to provide room for internal bomb bay, otherwise similar to ANT-29 wing but of simpler construction. Centre section 5 m wide tapered on TE only, no dihedral on upper surface, outer wings 5° dihedral, 6mod profile t/c 16% tapering to 12,5% at tip. Two spars throughout with 30-KhGSA tube lower booms, other members wrapped sheet and L-section diag bracing. Ribs riveted truss con-

SB rebuilt with longer fuselage

struction with U (top hat) periphery and tubular/diag bracing, ruling spacing 200 mm/250 mm. Centre-section skin 0,6 mm/1,0 mm, outer wings 0,5 mm/0,6 mm, flush-riveted throughout. Stumpy fuselage, modern structure with U-section frames and stringers, main frames (eg, at spars/bomb bay) pressed from sheet (no closed tubes or complex sections). Skin applied mainly in long strips arranged axially, ruling gauge 0,5 mm flush-riveted. Broken into three sections, F-1 for nav/bomb with glazed nose (Celluloid panels) with vert slits for two ShKAS; F-2 with integral wing c/s set at +2° and including pilot cockpit with sliding canopy and bomb bay with two doors; F-3 with radio/gunner able to aim upper or lower ShKAS through sliding canopy and small ventral hatch. Welded-KhMA landing gear with single leg carrying fork for braked wheel with 950 mm × 250 mm tyre, pulled to rear by hyd jack acting on rear breaker strut, doors closed over legs leaving part wheel exposed; spatted tailwheel. Engines in short NACA cowls without gills driving 3,2-m Hamilton 3-blade props, four wing tanks total 1670 lit but only two 360-lit inner tanks normally used (see data). Flight controls by push-rod, unbalanced tail surfs, trimmers on rudder, elevs and right aileron. Hydraulic split flaps to 60°.

First aircraft, called simply **SB**, first flown by K. K. Popov assisted by I. S. Zhurov on fixed skis 2800 mm × 820 mm (tail 800 mm × 320 mm) 7 Oct 34, cut short by landing accident 31 Oct. Rebuilt with much longer fuselage and more tapered outer wings (span unchanged) and M-85 engines and resumed testing 5 Feb to 31 July 35. Not submitted to NII because ANT-40$_1$ superior.

SB 2HS, built for comparison with water-cooled engines; completed with wing of greater span and with outboard LE taper 9° instead of 4·5°, and improved fin/rudder. Completed late 34, factory test from Feb 35. Flat frontal radiators, twin exhaust pipes to open rear of nacelle above wing, 3,3-m two-blade metal props, wire-braced tailplane, fixed tailwheel, landing light in left outer LE. Twin ShKAS in vert slits in nose, one upper rear and one lower rear, total 4,420 rounds; bomb bay with four (overload six) FAB-100 hung nose-up. Aircraft obviously outstanding, so massive production organized, but many faults remained, including unreliable oxygen and gear extension; some (eg inability of nav/bomb to escape quickly in belly landing) never rectified. Last prototype, **ANT-40$_2$**, on test Sept 35 to April 36, by which time production standard decided. Changes included licence-built engines moved 100 mm forward, driving VFSh fixed-pitch dural props with Hucks starter dogs, larger tail with inset hinges and 50%-80% mass balance, and bomb bay for six vert FAB-100 or two horiz FAB-250. Aircraft later used for trials with RS-132 underwing rockets and four ShVAK with 520 rounds in tray under fuselage.

DIMENSIONS Span (SB) 19,0 m (62 ft 4 in), (40$_{1,2}$) 20,3 m (66 ft 7¼ in); length (SB) 12,3 m (40 ft 4¼ in), (40$_{1,2}$) 12,27 m (40 ft 3 in); wing area (SB) 47,6 m^2 (512 ft^2), (SB rebuilt) 46,3 m^2 (498 ft^2), (40$_{1,2}$) 51,95 m^2 (559 ft^2).

ENGINES (SB) two 730-hp Wright Cyclone FF2, (SB rebuilt) M-85, (40$_{1,2}$) 760-hp HS 12Ybrs.

WEIGHTS Empty (SB) 3132 kg (6,905 lb), (SB rebuilt) 3210 kg (7,077 lb), (40$_1$) 3464 kg (7,637 lb), (40$_2$) 3900 kg (8,598 lb); fuel/oil (all) 530 kg + 60 kg; loaded (SB) 4717 kg (10,399 lb), 4850 kg (10,692 lb), (40$_2$) 5350 kg (11,794 lb).

PERFORMANCE Max speed (SB) 325 km/h (202 mph) at 4 km, (40$_1$) 332 km/h at SL, 404 km/h (251 mph) at 5 km; climb (40$_1$) 9·4 min to 5 km; service ceiling (SB) 6,8 km (22,310 ft), (40$_1$) 9,4 km (30,840 ft); range (40$_1$) 1850 km (1,150 miles); TO 300 m; ldg 350 m.

SB series From 34 Arkhangelskii organized production at GAZ-22 and GAZ-125, simplifying manufacture (flush riveting abandoned except LE of wing and tail) and dividing structure into 7 major parts, sheet thicknesses standardized (c/s 0,6 mm-1,0 mm, outer wings 0,5 mm-0,6 mm), standard fuel 1670 lit reduced to 1520 lit (2 × 360 lit + 2 × 400 lit) in 37 with protective tanks. Cleared for use with fixed skis (as on first flight), one tested retrac skis. Production **SB 2M-100** delivered from early 36. Served in Spain, initially excellent results (in action from 26 Oct 36, outpaced all opposing fighters), popular name *Katyuska* after character in musical. Only serious problem, propensity to catching fire in combat.

SB 2M-100A (ie, SB with 2M-100A)

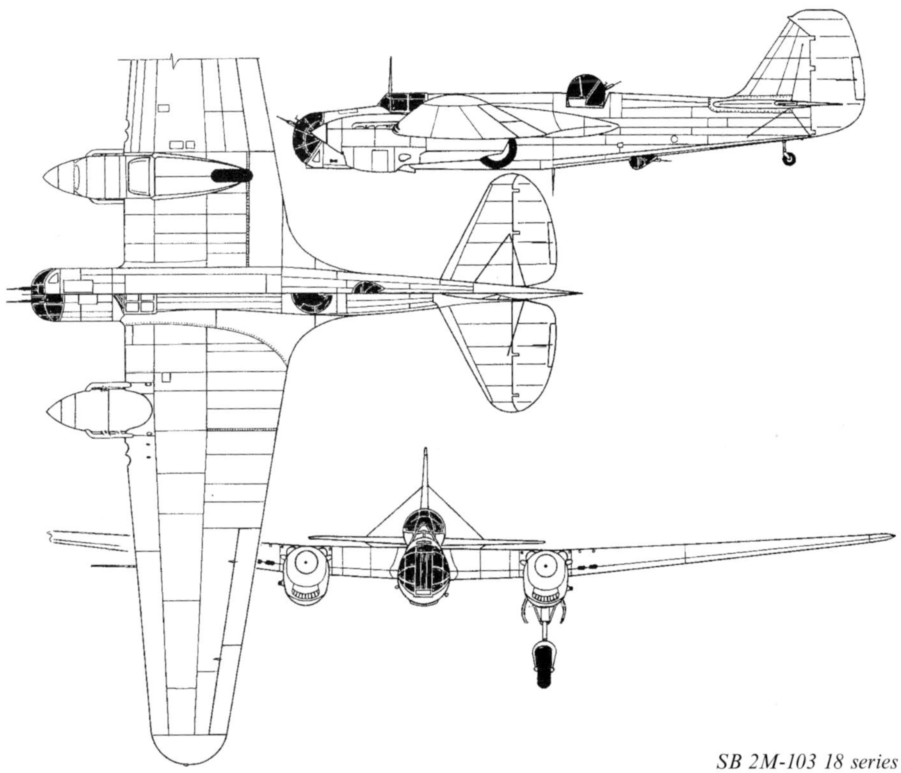

SB 2M-103 18 series

SB 2M-103 18th series

PS-41

SB 2M-100A More powerful engine supplied from Nov 36, in slightly improved cowl, from early 37 with elec starter and driving 3-blade VISh-2 v-p prop (Hucks dogs retained). Bomb bay able to carry horiz FAB-500, rack under each inboard wing for bomb or mine to 300 kg. Licensed to Czechoslovakia March 37, Avia HS12Ydrs engines and Czech eqpt (single nose gun) as **B 71**, 60 exported, 45 later built by Aero and 66 by Avia, all for Luftwaffe or puppet AFs.

SB 2M-100A modernizirovannyi Modernised version tested May 37, main difference MV-3 dorsal turret; recommended production not then actioned.

USB Frequent accidents at last resulted in decision to issue dual trainer version. Prototype Sept 37 with simple nose with open instructor cockpit on elevating seat. In high position instructor could do TO/ldg, looking ahead over small windscreen; with seat lowered he could get some outside view while pupil could see ahead. Total production 120+.

PS-40 One aircraft modified 36 as transport, with interior gutted and attachments provided for three aluminium cargo boxes total capacity 2,58 m^3 (91 ft^3); cleared to 6,4 t max, 7 t overload, speed 341 km/h. Several similar supplied Aeroflot 38.

SB 2M-103 10th series introduced with M-103 engine Oct 36; first on 1 Nov 36 reached 12 695 m (41,650 ft). First flight of definitive M-103 bomber by M. Yu. Alekseyev 2 Sept 37. Apart from new engine, major changes to nav compartment including providing emergency set of flight controls. Pre-series aircraft tested 27 July to 19 Sept 38, strengthened structure, provision for two external 368-lit tanks, max bombload increased to 1,6 t. Also with this version retrac skis developed, some retrofitted to previous, swinging aft on tandem parallel links.

SB bis 2 Polished wing surface, tested Mar 38 at 5905 kg (on skis), reaching 428 km/h with 1240 kg fuel. Overtaken by next.

SB bis 3 14th-series version distinguished by redesigned engine installn with modern ducted rad underneath, exhaust pipes still taken above upper wing skin, twin oil coolers in each LE outboard of engine. On test 1 Nov 37 to 17 Jan 38, by spring 38 was standard series version. During 38 simple aft ventral hatch began to be replaced by glazed gondola with ShKAS aimed from gunner's seat via periscope.

PS-41 Aeroflot transports based on bis 3, all armament removed, payload as PS-40, built with provision for retrac skis (main 2800 mm × 910 mm, tail 800 mm × 320 mm); most with 103U engine, **PS-41bis** added two external tanks (270 kg, hence c340 lit).

SB 2M-103 18-series, 39, based on bis 3, plus MV-3 turret (one ShKAS), VISh-22 3-blade v-p props and further improved radiator installn. All leading edges polished. In highest-rate production, reduced from late 40.

SB 2M-104, M-106 In 39 a few (c30) completed with M-104 engine, and at least two with M-106, but not successful and engines not in production. Last came off line early 41. All versions, production by years, GAZ-22 268 in 36, 853 in 37, 1,250 in 38, 1,435 in 39, 1,820 in 40 and 69 in 41; GAZ-125, 73 in 37, 177 in 38, 343 in 39, 375 in 40 and 168 in 41, total 6,831. Of these 292 (various) supplied to China 37-41. Special trials included rebuild by I. P. Tolstikh 40 with fixed nosewheel landing gear. With Tupolev imprisoned, Arkhangelskii continued devt (see his section).

DIMENSIONS Span (all) 20,33 m (66 ft $8\frac{3}{8}$ in); length (1936 m) 12,24 m (40 ft $1\frac{7}{8}$ in), (USB)

SB 2M-103 18th series

12,52 m (41 ft 0⅞ in), (rest) 12,273 m (40 ft 3¼ in); wing area (all) 56,7 m^2 (610 ft^2).
ENGINES (36) two M-100, then M-100A (also PS-40), (38) M-103 (also PS-41).
WEIGHTS Empty (36) 4138 kg (9,123 lb), (38) 4427 kg (9,760 lb), (38) 4768 kg (10,511 lb), (PS-40) 4222 kg, (PS-41) 4380 kg (skis 4550 kg); fuel (36) 530 kg+60 kg, (38) 1250 kg max, (PS-41) 1200 kg, (41 bis) 1730 kg; loaded (36) 5628 kg (12,407 lb), (37) 6013 kg, (38) 6175 kg normal, 7750 kg max (17,086 lb), (39) 6380 normal, 7880 kg max (17,372 lb), (PS-40) 6400 kg, (PS-41, 41bis) 7000 kg.
PERFORMANCE Max speed at SL (36) 326 km/h (203 mph), (M-100A) 372 km/h, (39) 375 km/h; max at 4 km (36) 393 km/h (244 mph), (M-100A) 423 km/h (263 mph), (39) 450 km/h (280 mph), (PS-40 skis) 341 km/h (212 mph), (PS-41) 428 km/h (266 mph); climb to 5 km (36) 11·7 min, (M-100A) 7·4 min, (39) 9·5 min; service ceiling (36) 9 km (29,530 ft), (M-100A) 9,56 km (31,365 ft), (39) 9,3 km (30,500 ft); range (36) 1250 km (777 miles), (38) 1800 km (1,118 miles), (39) 1900 km (1,181 miles); TO (36) 300 m, (39) 370 m, (PS-41) 660 m; landing (36) 300 m, (39) 400 m.

ANT-41, T-1 Torpedo bomber, two ordered March 34, assigned to KOSOS Myasishchyev, generally enlarged ANT-40. Two-spar wing of increased span, two-section balanced untabbed ailerons, split flaps, straight-edged tail with smaller tabs. Fuselage strengthened, many heavy box frames and stringers, all external skin flush-riveted. Streamlined engines driving 4 m 3-blade VISh props, ducted rads in inboard LE with aft-facing exits in upper surface. Crew as in SB though canopies and fittings quite different and no aft ventral gun. Landing gear stronger SB type, all units retracting fully. Twin-float version not built. Internal stowage for two torpedoes or two FAB-500N in tandem.

First flight June 36 by A. P. Chernavskii and two crew, better all-round performance than SB and good handling, every expectation of very successful aircraft. On 14th flight T-1 suddenly broke up in air; crew escaped. Cause traced to inadequate aileron mass-balance (80 instead of 105%) resulting in flutter violent enough to cause failure of wing. Plans for T-1 production immediately stopped; later manufacture of second aircraft, ANT-41bis, also discontinued.

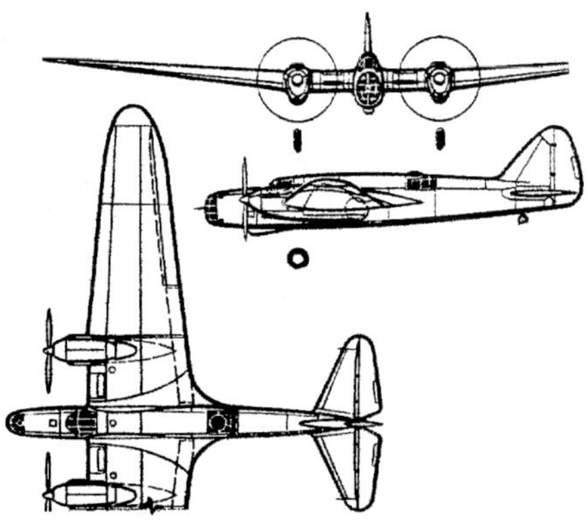

ANT-41, T-1

DIMENSIONS Span 25,73 m (84 ft 5 in); length 13,8 m (45 ft 3⅓ in); wing area 88,94 m^2 (957 ft^2).
ENGINES Two M-34FRN.
WEIGHTS Empty 5846 kg (12,888 lb); fuel/oil 1900 kg+150 kg; loaded 8925 kg (19,676 lb).
PERFORMANCE Max speed 435 km/h (270 mph) at SL, greater at height; service ceiling 9,5 km (31,170 ft); range (Shavrov) 4200 km; take-off 370 m; no other data.

ANT-42, Pe-8 Described under Petlyakov as Pe-8.

ANT-43 Not to be confused with PS-43 (no PS number was assigned), this KOSOS high-speed transport was designed 36. Clean low-wing monoplane with flush-riveted skin, 800 hp GR14Krsd engine, seven seats (this is believed to include pilot), construction described as similar to I-14 (suggesting Sukhoi's brigade), main gears retracting inwards. Radical manufacturing method, commonly used other countries, of photographing from templates direct to metal sheet. With Tupolev imprisoned nobody took responsibility for permitting flight test.

ANT-44, MTB-2 Reconnaissance/bombing flying boat or amphibian, more modern design than previous AGOS and KOSOS marine aircraft and one of first in world to have four engines on LE of wing mounted direct to hull. Wing 6mod profile, 15% t/c over constant-profile with sharp centre section dihedral, outer wings evenly tapered with upper surface horizontal. Two main spars, tubular truss structure but some outer ribs pressed from sheet; two-section balanced ailerons, slotted flaps on centre section and inboard outer panels. Deep but well-profiled hull, broad V-section planing bottom with main step at 90° and rear bottom tapering to aft knife-edge, narrower upper hull accommodating crew 7/8. Single fin carrying braced fixed-incidence tailplane, all control surfaces mass balance and tabs. First aircraft with French engines, KOSOS installations, wing LE hinged down each side to form work platforms, fuel between spars of wing, fixed underwing floats, provision for beaching chassis, armament three ShKAS in nose and tail turrets and sliding roof over rear dorsal cockpit (unusual change in level along top of rear hull); 2,5 t load of bombs, mines under centre section.

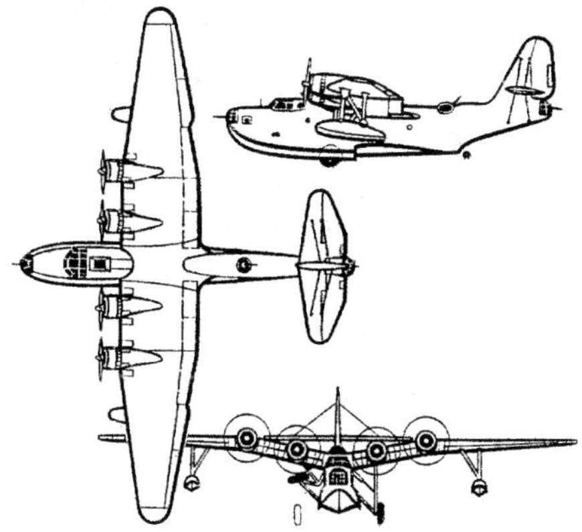

ANT-44D, MTB-2

First aircraft completed Sevastopol and test flown by T. V. Ryabenko and Il'inskii 19 April 37; successful factory testing followed by NII tests late 37. Second aircraft **ANT-44bis, ANT-44D**, built as amphibian with large single-wheel main gears pulled up (not retracted) for water operations by hydraulic jack acting on hinge of main oleo leg; castoring tailwheel aft of stern knife-edge. ANT-44D also had more powerful Russian-built engines, VISh-3 propellers of 3,5 m diameter, and MV-series dorsal turret in place of sliding hatch, larger vertical tail with rudder inset hinges but no horn balance, similar change to elevators. First flight Ryabenko 7 June 38; subsequently tests from Moscow reservoir and NII tests Sevastopol. No real deficiencies but decision not to build in series and all development ceased Jan 40. Later I. M. Sukhomlin set five amphibian class records: 17 June 40, 1-t load to 7134 m; 19 June two flights, 6284 m with 2 t and 5219 m with 5 t; 28 Sept, with underwing floats jettisoned, 1000-km circuit (Kerch/Kerson/Taganrog) with 1 t payload, 277,45 km/h; 7 Oct, same circuit with 2 t at 241,9 km/h. During the Second World War both aircraft flew transport and other missions, often in command of Sukhomlin, 1941-43 in Black Sea area.
DIMENSIONS Span 36,45 m (119 ft 7 in); length 22,42 m (73 ft 6⅔ in); wing area 144,7 m^2 (1,558 ft^2).
ENGINES (44) four 810 hp GR14Krsd; (44D) M-87.

ANT-44D

WEIGHTS Empty (44) 12 t, (44D) 13 t (28,660 lb); loaded (44) 18,5 t, 21,5 t overload (47,400 lb), (44D) 19 t (41,887 lb).
PERFORMANCE Max speed (44) 330 km/h at SL, (D) 355 km/h (221 mph) at SL, more at height; time to 1 km, (44) 3·5 min, (D) 3 min; service ceiling (44) 6,6 km (21,650 ft), (D) 7,1 km (23,300 ft); range (44) 4500 km (2,796 miles); endurance (44) 16 h, (D) 14 h; alighting speed (44) 125 km/h, (D) 130 km/h.

ANT-45 Unbuilt KOSOS design for 36 low-wing two-seat fighter.

ANT-46, DI-8 Another heavy fighter in same family as ANT-29 and SB series, ANT-46 was ordered as single prototype Nov 34 with VVS designation DI-8 (*Dvukhmestnyi Istrebitel*, two-seat fighter). Assigned to Arkhangyelskii on condition it did not interfere with SB, but troublefree development and first flight 9 Aug 35. Project thus ran only weeks behind ANT-29 (DIP) but later and better aircraft with basically $ANT-40_2$ airframe with pilot above LE and radio operator with ShKAS at rear. French engines driving two-blade VFSh props. Main armament two APK-11 buried in wings immediately inboard of ailerons; four ShKAS in nose each 500 rounds. On factory test to June 36, but Tupolev's arrest and closure of Kurchyevskii KB made project pointless.
DIMENSIONS Span 20,3 m (66 ft 7¼ in); length 12,24 m (40 ft 1⅞ in); wing area 55,7 m^2 (600 ft^2).
ENGINES Two 800-hp GR14Krsd.
WEIGHTS Empty 3487 kg (7,687 lb); loaded 5291 kg normal, 5553 kg (12,242 lb) overload.
PERFORMANCE Max speed 344 km/h (214 mph) at SL, 388 km/h (241 mph) at 4,25 km; climb 6·8 min to 3 km, 11·4 min to 5 km; service ceiling 8570 m (28,120 ft); range 1780 km (1,100 miles).

ANT-47 Unbuilt fighter project of 37, design in Tupolev's absence.

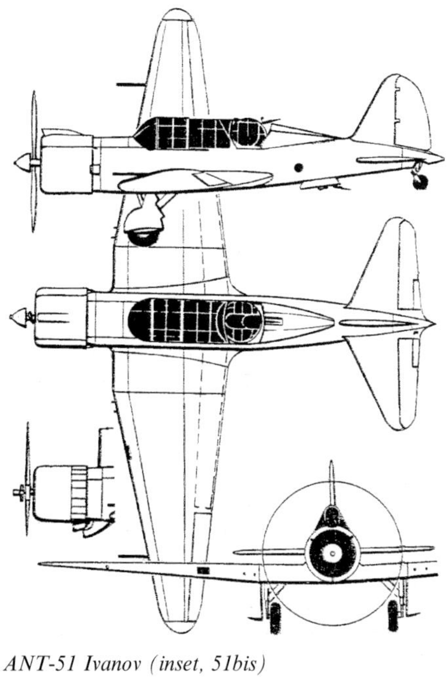

ANT-51 Ivanov (inset, 51bis)

ANT-46, DI-8

ANT-51

ANT-48 Unbuilt high-speed bomber project of 36.

ANT-49 Unbuilt project for reconnaissance version of SB 36, M-100A engines, three cameras in heated bay in place of original bomb bay, augmented fuel tankage.

ANT-50 High-speed passenger transport project 37 based on ANT-43, two AM-34, not built.

ANT-51 Single-engined tac-recon and attack aircraft begun 36 in Sukhoi's brigade under Tupolev supervision; after latter's arrest managed by chief engineer GUAP (position from which Tupolev dismissed) and prototype built at ZOK as contender in *Ivanov* programme: low wing, inward-retrac main gears, fixed tailwheel, greenhouse canopy, four ShKAS in wings, one in rear cockpit, internal bay for 300 kg bombload. Flown by Gromov 25 Aug 37, 51 bis flown 39, later see under Sukhoi.
DIMENSIONS (all) Span 14,3 m (46 ft 11 in); length (37) 9,92 m, (39) 10,25 m (33 ft 7½ in); wing area 29,0 m^2 (312 ft^2).
ENGINE (37) One M-62, (39) M-87A, later M-87B.
WEIGHTS Empty (37) 2604 kg, (39) 2816 kg (6,208 lb); fuel/oil (39) 550 kg; loaded (37) 3937 kg (8,679 lb), (39) 4030 kg/4080 kg (8,995 lb).
PERFORMANCE Max speed (37) 360 km/h at SL, 403 km/h (250 mph) at 4,7 km, (39) 375 km/h at SL, 470 km/h (292 mph) at 4 km; time to 5 km (37) 16·6 min, (39) 11·5 min; service ceiling (37) 7440 m, (39) 8800 m (28,900 ft); range (37) 1200 km, (39) 1160 km (720 miles); take-off (37) 380 m/20 s; landing 240 m/16 s/120 km/h.

ANT-52 No information.

ANT-53 Unbuilt 36 project in Petlyakov brigade for bomber and passenger aircraft with four M-34FRNV engines.

ANT-54 to **ANT-56** Last of these was assigned to SRB (*Skorostnoi Razvyedchik Bombardirovshchik*) described under A. P. Golubkov.

57, DPB Projected long-range dive bomber, four M-105, drawn while in prison 37.

58, Aeroplane 103, FB, Tu-58 This important prototype led to Tu-2, programme exceptional for scope and complexity even for Soviet Union. Main technical description appears under No 61, basic production Tu-2S; postwar variants also described separately but all consecutive and as far as possible in chronological sequence even where this disarranges ANT number sequence.

When Tupolev was arrested 27 Oct 37 much time wasted because plans for use of such design leaders had not been worked out; about 18 months spent in Lubyanka and Butyrkii prisons, occupying cell with his wife and drawing board. Vague command to design aircraft to beat Ju 88, and proceeded to scheme 58 (number by chance same as cell at Butyrkii); no organization until in Aug 38 NKVD formed CCB-29 at GAZ-156 with OTBs (special tech bureaux) numbered from 100, Tupolev being No 103. Aircraft thus called **Samolyet 103**; later led to type numbers 58-69 inclusive and Tu numbers 1 to 10. Design based for horizontal and dive bombing but also later adapted to torpedo, reconnaissance, air-combat, LR interception, *Shturmovik*, transport and training roles. Clean mid-wing monoplane with large bomb bay beneath wing, twin-finned tail and crew of

ANT-51bis

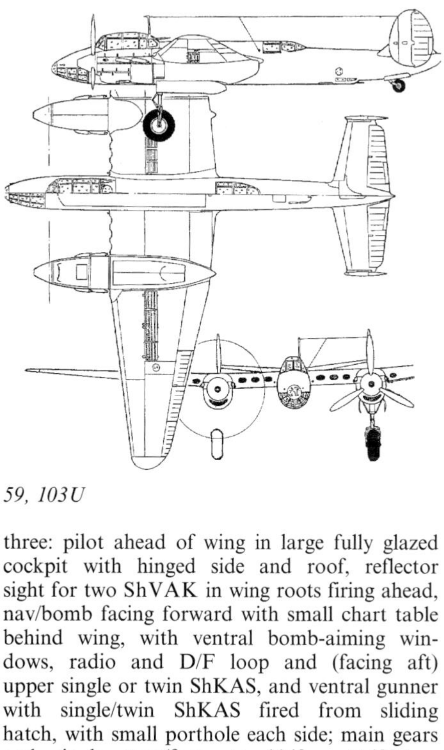

59, 103U

three: pilot ahead of wing in large fully glazed cockpit with hinged side and roof, reflector sight for two ShVAK in wing roots firing ahead, nav/bomb facing forward with small chart table behind wing, with ventral bomb-aiming windows, radio and D/F loop and (facing aft) upper single or twin ShKAS, and ventral gunner with single/twin ShKAS fired from sliding hatch, with small porthole each side; main gears and single strut/form to 1142 mm × 432 tyre retrac to rear, tailwheel 470 mm × 210 also retrac to rear; main and oil rads between centtre-section spars with three oval inlets in each LE and exit louvres above wing near TE, carb inlet under spinner; four sections split flap, steel Venetian-blind dive brakes under wings; 3,4-m VISh-61T props, fuel in self-sealing wing tanks with nitrogen protection; internal bay with twin doors for 2 t (max 3 t) load, size up to FAB-1000; gear/flaps/dive brakes/bomb doors hydraulic.

Design approved and prototype authorized 1 Mar 40, built under Bartini and Korolyev, complete 3 Oct except for engines, at last flown 29 Jan 41 by M. A. Nyukhtikov and engineer V. A. Miruts. Outstanding aircraft, assigned service tag FB (*Frontovoi Bombardirovshchik*, frontal bomber) and also called Tu-58. Factory test at GAZ-156 complete 28 April 41, CG demo 25·6%/30·6%, still lacking much equipment and both sets rear guns not fully developed. NII tests June 41; later (date not known) lost when back at KB (evacuated to GAZ-166) following fire in right engine, Nyukhtikov escaping but engineer A. Akopyan killed.

DIMENSIONS Span 18,86 m (61 ft 10½ in); length 13,2 m (43 ft 3⅝ in); wing area 48,8 m² (525 ft²).

ENGINES Two AM-37.

WEIGHTS Empty 7626 kg (16,812 lb); fuel/oil 2147 kg; loaded 9950 kg normal, 10 992 kg (24,233 lb) max.

PERFORMANCE Max speed 482 km/h at SL, 635 km/h (395 mph) at 8 km; time to 5 km 8·6 min; service ceiling 10,6 km (34,780 ft); range 2500 km (1,550 miles); take-off 440 m; landing 730 m/155 km/h.

59, 103U Second prototype with redesigned longer fuselage providing for fourth crew-member to fire single ShKAS from lower rear hatch; raised canopy over pilot and nav and latter seated facing aft with twin ShKAS augmenting rear dorsal pair; ten-RS rockets under outer wings. Taller fins/rudders, oil coolers in wings with LE inlets, props VISh-61P, then VISh-61Ye, finally AV-5-67, all 3,8-m with 150-mm cooling hole in spinner. Remarkable CG range 16·3%/32·25%. First flight 18 May 41 by Nyukhtikov and Miruts; NII testing complete autumn, and strong recommendation immediate series production, agreed Sept 41. Great problem with evacuation of OKB and abandonment of snag-ridden AM-37; after studying AM-39F decision to fit M-82 (ASh-82), task needing 1500 new drawings, done under pressure during evacuation and Aircraft 103U flew again 1 Nov 41 with M-82 engines driving AV-5-167 props (same 3,8 m) and other minor changes such as longer nacelles. Speed reduced at height but increased at low level, and floatless (injection) carbs greatly improved negative-g behaviour.

DIMENSIONS Span 18,8 m (61 ft 8¼ in); length 13,8 m (45 ft 3⅓ in); wing area 48,52 m² (522 ft²).

ENGINES As 103, then two M-82.

WEIGHTS (AM-37) empty 7823 kg (17,246 lb); fuel/oil 2456 kg; loaded 10 435 kg normal, 11 477 kg (25,302 lb) max.

PERFORMANCE (AM-37) max speed 469 km/h SL, 610 km/h (379 mph) 7,8 km, (M-82, 484 SL, 530 at 3,2 km); time to 5 km 9·5 min; service ceiling 10,5 km; range 1900 km (1,180 miles); take-off 435 m; landing 765 m/155 km/h.

60, 103V Production prototype with effort made throughout to reduce number of parts and man-hours, and planned for mass-production with greater emphasis than ever before on accurate tooling, control desks, bench-made wiring looms and pipe-runs, and widest range of subcontracted accessories. Two main assembly lines established at GAZ-166 and (later) 156, and major parts made at many others. Aircraft 103V very similar to re-engined 103U except in minor details of eqpt and in using saw-tooth profile corrugated sheet in wings instead of rectangular top-hat, saving wt; two ShVAK, five ShKAS, 3 t bombs plus 10 RS-82. Built 1 Aug-13 Nov 41, flown by M. P. Vasyarkin 15 Dec and after some rectification of engine faults completed NII testing 22 Aug 42. After further testing and rectification began VVS service as instructor trainer.

DIMENSIONS As before except length 13,71 m (44 ft 11¾ in).

ENGINES Two ASh-82.

WEIGHTS Empty 7335 kg (16,171 lb); fuel/oil 2411 kg; loaded 10 343 kg normal, 11 773 kg (25,955 lb) max.

PERFORMANCE Max speed 460 SL, 528 km/h (328 mph) at 3,8 km; time to 5 km 10 min; service ceiling 9 km (29,530 ft); range 2000 km (1,242 miles); take-off 516 m; landing 640 m/152 km/h.

103VS This aircraft No 308 (100308) regarded as first series example. Eqpt changes included all three movable ShKAS replaced by UBT, lower rear gunner 3 portholes each side, vert tails more pointed top, dive brakes removed, systems simplified. Flown Mar 42, NII test 13 Sept-28 Oct, by which time GAZ-166 in prodn; April 42 this a/c and next two sent to Kalinin front to 3rd Air Army (Gen M. M. Gromov),

61, Tu-2L (testbed for RD-10 and TR-1) at LII-VVS

v.successful front-line trials.
DIMENSIONS Span 18,86 m; length 13,8 m; area 48,8 m.
ENGINES Two ASh-82.
WEIGHTS Empty about 7,4 t; loaded 10 538 kg normal, 11 768 kg (25,944 lb) max.
PERFORMANCE Max speed 521 km/h (324 mph) at 3,2 km; time to 5 km, 10·2 min; service ceiling 9 km; range 2020 km (1,255 miles); take-off 450 m; landing 545 m/152 km/h.

61, Tu-2S First true production aircraft was No 716, with direct-injection engines and AV-5-167A propellers, and rocket rails removed. Other small changes included simpler nose glazing, longer detachable wingtips, further revision of electric, hydraulic and fuel systems, and refined armour. Aircraft designated **Tu-2** by NKAP early 42, and by end-42 GAZ-166 had equipped two Polk (30 each), first commanded by A. Perelyot who later joined OKB. Tupolev, released from captivity before evacuation, later awarded first Order of Lenin, Hero Socialist Labour and Hammer/Sickle medal.

Aerodynamically outstanding, particular attention to low drag with stiff sandwich wing skins, flush riveting on all external surfaces and painstaking attention to detail on original Aircraft 103, as witness speed higher than any Soviet fighter then in service. Wing CAHI-40 profile 13·75% root, 9·9% tip, MAC 2885 mm, main structural box formed by LE web at 6·3% chord and single main spar at 35·4% joined by double skins with inner corrugated layer (corrugations spanwise) with thickness 1,5 mm upper and 1,0 mm lower surface. Horizontal centre section tapered only on TE, span 6,56 m, joined by 19 bolts each side to 6,15 m outer panels with taper and 5° dihedral. Third light rear spar at 77% to carry Schrenk-type split flaps on piano hinges, set 15° take-off and 45° for landing. Three section balanced Frise ailerons with tabbed No 1 section and Nos 2 and 3 increased chord (projecting aft of TE on most series aircraft). Ruling structural material D16-T except for 30KhGSA at landing gear and engine mounts. Fuselage semi-monocoque with 44 ring frames pressed from sheet and four strong but open-section longerons, some L and top-hat frames and stringers but stringers later almost eliminated from mid-fuselage, skin 1,5 mm, thinner at ends (in first two production blocks of 500 each, forward fuselage redesigned in glued *shpon*, with some steel reinforcement).

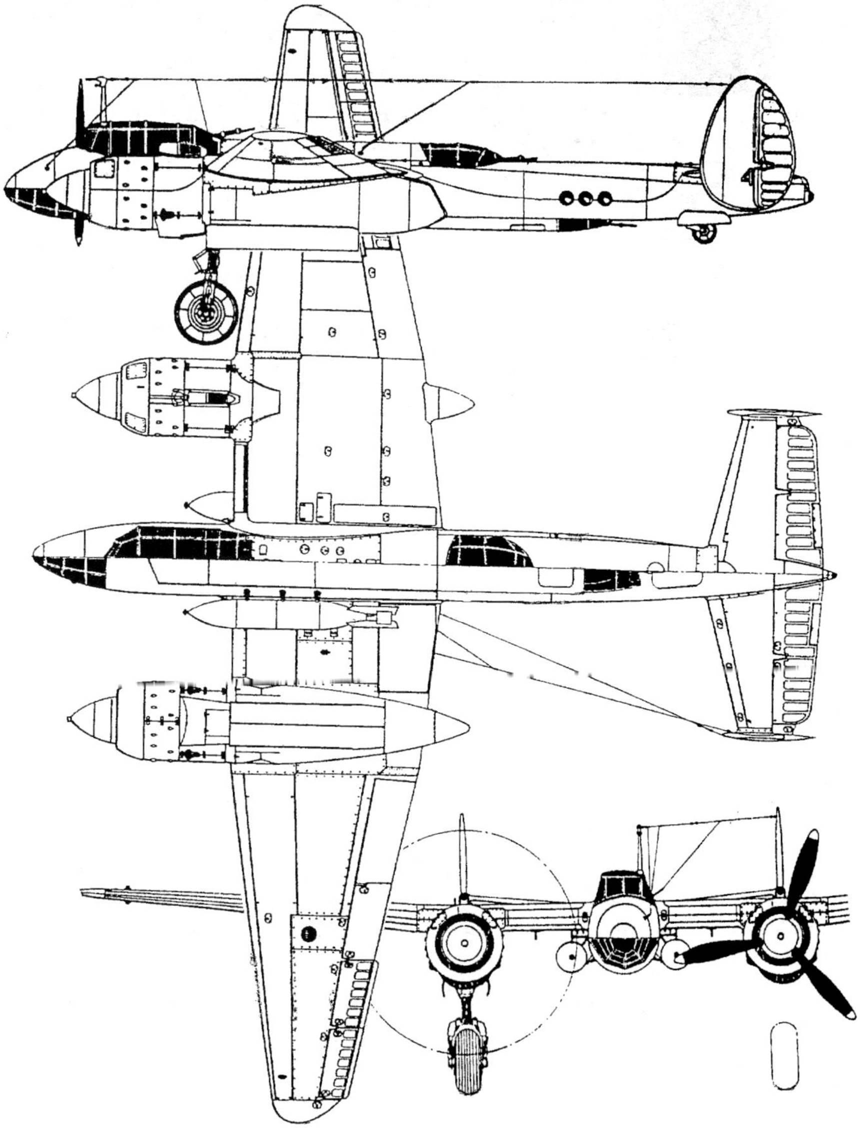

61, Tu-2S (external FAB-1000)

High fixed tailplane with 8° dihedral with inset hinges carrying balanced tabbed elevators and rudders with tabs or ground-adjustable TE strips (in first two production blocks fins often wood with bakelite-ply skins; by 44 D16-T returned with simple pressed ribs and 0,8 mm skin). Engines on KhGSA welded tube mounts in tight cowls with rear exhausts grouped in cluster each side (various arrangements), carb-air in at top, oil cooler in duct at bottom. Fuel

62T, Tu-2D

Tu-2 Paravan

in four inner and four outer tanks between spars, self-sealing and with NG (neutral gas) protection, total 2800 lit (616 gal). One (sometimes two) landing light hinged down below each outer wing. Hydraulic pump on both engines serving landing gear, flaps, bomb doors and wheel brakes (in some aircraft landing gear pneu). Main gears 1142 mm × 432 mm on forks on single struts retracting to rear; tailwheel 470 mm × 210 mm also retracting fully to rear, in all cases with twin doors. Access to nose via glazed ventral hatch. Pilot in armoured seat with canopy folding up at roof and down at sides, reflector sight for twin ShVAK in wing roots (usually 100 rounds each), ram's horn flight control by push/pull rods with provision for dual. Nav/bomb with aft-firing UBT with 200 or 250 rounds and prone nose position with stabilized sight for level bombing. Radio op at upper rear with tip-up seat and UBT with 250 rounds. Lower rear gunner with UBT with 250 rounds sighted by ventral periscope. Bombload 1-3 t internal (typical 4 × 500 kg) but overload 4 t with two FAB-1000 hung under wing roots.

During course of production, chiefly post-war, series Tu-2 grew in structure mass and equipment until by 47 empty weight had risen by roughly 1 t, from 7,4 t to 8,4 t. Engines did not change but installation varied through numerous mods affecting cowl panels (reduced diameter, with 28 small blisters for valve gear), exhaust (short pipes no longer visible), cooling-air outlet (in sides of nacelle, contoured to give positive thrust at high power), carb-air (inlets reprofiled and lengthened, with filter and anticing provision) and oil cooler (better radiator in reprofiled duct). Most late batches returned to single porthole each side (larger than in prototypes) for lower rear gunner. There were also variations in wheel size; no Tu-2 seen on skis. Because of delayed start only about 800 Tu-2S accepted by end of war, but post-war GAZ-125 built 218 to replace wartime attrition, bringing total (excl variants) to 2,527. Used post-war to test turbojets, radars, guns, ejection seats and looped-hose air refuelling. Exported to Bulgaria, China, Hungary, N. Korea, Poland and Romania. ASCC name 'Bat'.

DIMENSIONS Span 18,86 m (61 ft 10½ in); length 13,8 m (45 ft 3⅓ in); wing area 48,8 m² (525 ft²).

ENGINES Two ASh-82FN.

WEIGHTS Empty (43) 7474 kg (16,477 lb), (48) 8404 kg (18,527 lb); fuel/oil 2016 kg+300 kg; loaded (43) 10 360 kg/11 360 kg (max 25,044 lb), (48) 11 450 kg (25,243 lb).

PERFORMANCE Max speed (43) 482 km/h at SL (300 mph), 547 km/h (340 mph) at 5,4 km, (48) 550 km/h (342 mph) at 5,7 km; time to 5 km (43) 9·5 min, (48) 10·8 min; service ceiling (43) 9,5 km (31,170 ft), (48) 9 km (29,530 ft); range (43) 2100 km (1300 miles), (48) 2180 km; take-off (43) 485 m, (48) 540 m; landing (43) 675 m/158 km/h, (48) 500 m.

Tu-2/ASh-83 Also known as **Tu-2M**, series aircraft mod early 45 with more powerful engines driving AV-5V props with four broad hollow-steel blades.

DIMENSIONS As Tu-2.

ENGINES Two ASh-83.

WEIGHTS Loaded 10 585/11 575 kg (max 25,518 lb).

PERFORMANCE Max speed 605 km/h (376 mph) at 8,8 km; time to 5 km, 8·5 min; service ceiling 10,4 km (34,120 ft); range 1950 km; take-off 480 m.

62, Tu-2D From 41 Tupolev had schemed **103D** long-range bomber version with longer outer wings and revised forward fuselage for dual pilots. In 44 time could be spared to build this, and No 718 using original airframe flown by Perelyot 17 July; second (No 714) flown 20 Oct with wide-chord ailerons on extended outer wings housing two additional tanks each side, nav moved to nose 0,6 m longer and much wider, cockpit for two pilots side-by-side, tail span increased from 5,4 m to 5,7 m with larger fins/rudders, 4 t bombload normal maximum, main gears strengthened. GAZ/NII testing 20 Oct 44 to 31 Oct 45. Tested Jan-April 47 as **62T** with four-blade AV-5V props.

DIMENSIONS Span 22,06 m (72 ft 4½ in); length 14,42 m (47 ft 3¾ in); wing area 59,05 m² (636 ft²).

ENGINES Two ASh-82FN.

WEIGHTS Empty 8316 kg (18,333 lb); fuel/oil 2820 kg; loaded 12 290 kg/13 340 kg (max 29,409 lb).

PERFORMANCE Max speed 465 km/h at SL (289 mph), 531 km/h (330 mph) at 5,6 km; time to 5 km, 11·8 min; service ceiling 9,9 km; range 2790 km (1,734 miles); take-off 480 m; landing 610 m/149 km/h

104, Tu-2/104 Like Pe-2, attempts were made to turn Tu-2 into bomber-destroying interceptor, in this case with radar. Redesigned nose housing large radar and gunsight system (team led by A. L. Mints) with two VYa-23 in underside of forward fuselage. Normal crew 2. First flight 18 July 44 (Perelyot, with leading engineer L. L. Kerber). Radar operative 45, first such trials with fighter in the Soviet Union.

63, Tu-2SDB Two aircraft mod 43-44:

No 1, **SDB** (*Skorostnoi Dnyevnoi Bombardirovshchik*, fast day bomber), originally an early Type 103 rebuilt with modified airframe, more powerful liquid-cooled engines driving 3,6 m AV-5LV-22A three-blade props, dive brakes removed, fuselage rearranged for crew of two (pilot and nav/bomb), all guns removed except wing-root ShVAK, bombs as before and most of details as Tu-2S. First flight 21 May 44 (Perelyot) and joint GAZ/NII tests 5 June to 6 July.

No 2, **SDB-2**, with AM-39F engines (same props), totally new main gears with single oleo legs of increased length passing straight down to inner end of axle, larger tailwheel tyre (480 mm × 200 mm), hydraulic system with all tubing dural instead of steel, same 5,7 m horizontal tail as Tu-2D, crew 3 (as No 1 but radio/gunner at rear) in armoured cockpits, rear guns ShKAS, extra fuel in wing. First flight 14 Oct 44; combined test programme by M. A. Nyukhtikov and leading engineer V. A. Shubralov complete 30 June 45. Neither SDB had adequate nav vision, but No 2 was judged superior aircraft and would have gone into production had it not been for Type 68 (Tu-10).

DIMENSIONS Span (1) 18,86 m, (2) 18,8 m; length (1) 13,2 m (43 ft 3½ in), (2) 13,6 m (44 ft 7½ in); wing area (1) 48,52 m², (2) 48,8 m².

ENGINES (1) TWO AM-39, (2) AM-39F.

WEIGHTS Empty (1) 7787 kg (17,167 lb), (2) 8280 kg (18,254 lb); fuel/oil (1) 1767 kg, (2) 1750; loaded (1) 10,36 t (22,840 lb), (2) 10,925 kg (24,085 lb), (max) (1) 11,8 t (26,014 lb), (2) 11,85 t (26,124 lb).

PERFORMANCE Max speed (1) 527 km/h at SL, 645 km/h (401 mph) at 6,6 km, (2) 547/640 (398 mph) at 6,8 km; climb to 5 km (1) 7·45 min, (2) 8,7; service ceiling (1) 10,0 km, (2) 10,1 km; range (1) 1830 km (1,137 miles), (2) 1530 km (951 miles); TO (1) 470 m, (2) 535 m; ldg (1) 550 m, (2) 156 km/h, 650 m.

Tu-2 Paravan Two production aircraft modified for trials with deflectors for barrage-balloon cables. Steel deflector cable attached to tip of 6 m dural monocoque cone cantilevered ahead of aircraft nose and extending 13,5 m each side to cable cutter at tip of wing. First aircraft on test Sept 44 with 150 kg ballast in tail and most combat equipment removed.

DIMENSIONS Length 19,8 m (64 ft $11\frac{1}{2}$ in).
WEIGHT Empty 9150 kg (20,172 lb); loaded about 11 t.
PERFORMANCE Max speed 537 km/h (334 mph) at 5450 m; time to 5 km, 11 min; service ceiling 9150 m (30,000 ft).

Tu-2Sh Three armoured *Shturmovik* versions. First, 44, proposed by A. D. Nadashkevich (OKB a/t head): bomb bay filled by pallet carrying 88 modified PPSh-41 machine carbines (sub-machine-guns) with barrels parallel firing ahead at 30° depression. Second, also 44, single 75 mm gun under centreline reloaded by nav, successfully tested. Third, 46, 2-seater with poss record frontal firepower; two ShVAK, two NS-37 and two NS-45, plus UBT at rear. Data for this version.
DIMENSIONS, ENGINES As Tu-2.
WEIGHTS Not known.
PERFORMANCE Max speed 575 km/h (357 mph) at 5,8 km; time to 5 km, 9·0 min; service ceiling 10 065 m (33,022 ft); range 2500 km (1,553 miles); landing speed 160 km/h (99 mph).

Tu-2K At least two production aircraft modified 44 and 45 for ejection-seat trials; designation from *Katapult*.

Tu-2G Designation probably from *Gruzovoi*, freight; several production aircraft modified 1944-50 for carrying cargo internally and externally and especially for parachuting of bulky loads, 49 example of latter being GAZ-67b scout car (378 km/h, 6 km service ceiling). This particular aircraft one of several Tu-2s known with 3,6 m AV-5V prop hubs with four experimental hollow steel paddle blades with square tips.

Tu-2N Series a/c used to test RR Nene turbojet slung below fuselage July 47; at least two other a/c used to test different turbojets, one with afterburner.

Tu-10, 68 Series a/c completed 45 with liquid-cooled engines driving 3,8 m 3-blade AV-5LV-22A props, 4 seats, 3 ShVAK and 2 UB, 4 t bombload , strengthened main gears. Flown 19 May 45 by Perelyot, outstanding; fitted with AV-5LV-166B props, factory test to 8 July by F. F. Opadchii; then dihedral reduced to 1·5° and fitted new engines, 3,6 m 4-blade AV-9K-22A props, larger vert tails; prolonged test to 20 Nov 46. Small series (10) built, designated **Tu-4**.
DIMENSIONS As Tu-2S.
ENGINES Two AM-39FNV, later AM-39FN-2.
WEIGHTS Empty 8870 kg (19,555 lb); fuel/oil 1630 kg; loaded 11 650 kg/12 735 kg (max 28,075).
PERFORMANCE Max speed 520 km/h at SL, 635 km/h (395 mph) at 7,1 km; climb to 5 km 11·1 min; service ceiling 9,8 km (32,150 ft); range 1660 km (1,030 miles); TO 525 m; ldg 190 km/h.

Tu-2T A torpedo-carrying Tu-2 had been in original scheme for Aircraft 103, but not realized until design effort could be spared 44. Tu-2T (*Torpedonosyets*, torp-carrier) stemmed from two prototypes taken off 44 production line. First, **NT** (*Nizkii*, low), few changes apart from addition of TD-44 pylon and Der-4-44-U safety system under each wing root for 45-36-AN air torpedo. Combined factory/NII tests February/March 45, good results (data for this aircraft). Second Tu-2T had supplementary fuel tanks (1020 lit, 224 gal) in sealed bomb bay, strengthened landing gear and other changes.

68 (re-engined)

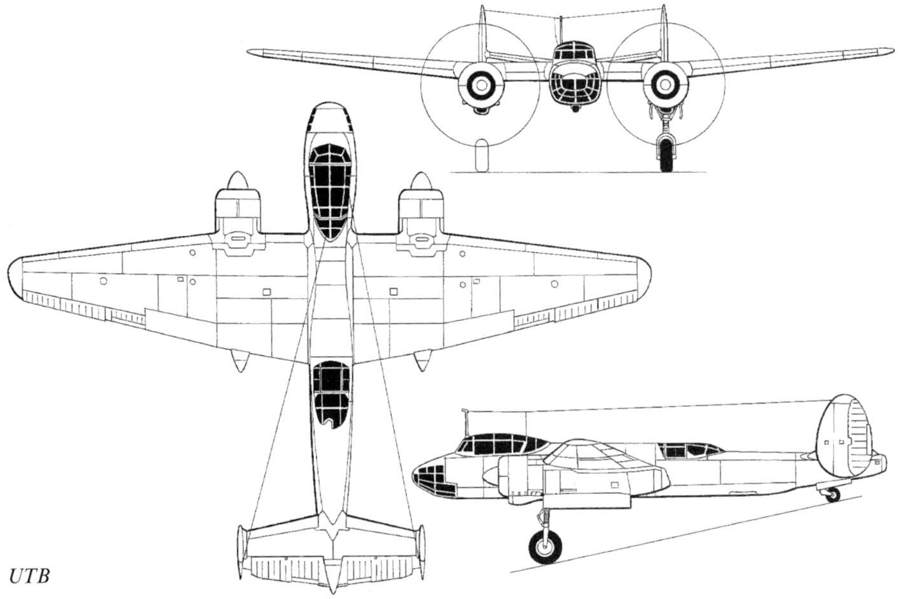

UTB

UTB-2

First flight by Opadchii and V. P. Marunov 2 Aug 46; remarkable ability to fly 3800 km; speed with two torpedoes 490 km/h, or just over 500 with one. Also flown wih three 800 kg torpedoes, but not cleared as normal load. Built in series 47 for AV-MF, serving in all three main fleet theatres until mid-50s.
DIMENSIONS, ENGINES As Tu-2.
WEIGHTS (No 1, loaded) 11 423/12 389 kg (max 27,313 lb).
PERFORMANCE Max speed 505 km/h (314 mph) with one torp, 493 with two; service ceiling 7,5 km (24,600 ft); range 2075 km (1,289 miles); take-off 580 m; landing 480 m/159 km/h.

UTB Low-powered trainer designed by P. O. Sukhoi. Sometimes called **UTB-2** (UTB, *Uchyebno-Trenirovochnyi Bombardirovshchik*, training bomber), this used simplified Tu-2 airframe restressed to lower factors, low-powered engines in short cowls with VISh-111V two-blade v-p metal props, large flight deck with full roof glazing and room for one or two pilot seats at front, with dual control, and one or two nav seats at rear, with new access tunnel through where bomb bay had been to rear rad/gunner cockpit with VUB-68 mount for UBT and 60 rounds. External belly racks for four FAB-50 or FAB-100 bombs aimed by

67

trainee bomb/nav through glazed nose. Main tyres 900 mm×300 mm, tail 440 mm×210 mm. No lower rear position. Totally different fuel and accessory systems, ailerons, flaps and many other parts. First flight summer 46; production UTB-2 (at least 500, about 100 for Poland) delivered 47.

DIMENSIONS Span 18,86 m (as Tu-2); length 13,985 m (45 ft 10½ in); wing area 48,8 m² (as Tu-2).

ENGINES Two ASh-21.

WEIGHTS Empty 5020 kg (11,067 lb); loaded 6550 kg (14,440 lb).

PERFORMANCE Max speed 391 km/h (243 mph) at 2,1 km; climb to 3 km 8·0 min; service ceiling 7,0 km (23,000 ft); range 950 km (590 miles); TO 460 m; ldg 125 km/h, 345 m.

Tu-2D, 67 Long-span long-range prototype with diesels; 5 seats, 1260 mm main tyres (bulged doors), 3,8 m AV-5L props. First flight by Perelyot 12 Feb 46; good aircraft but engine troubles caused abandonment 47.

DIMENSIONS As Tu-2D 62 except wing area 59,12 m² (636 ft²).

ENGINES Two ACh-30BF.

WEIGHTS Empty 8323 kg (18,349 lb); loaded 13 626 kg/15 215 kg (max 33,543 lb).

PERFORMANCE Max speed 509 km/h (316 mph) at 6,2 km; time to 5 km, 13·0 min; service ceiling 8850 m (29,000 ft); range 5000 km (3,100 miles); take-off 530 m; landing 700 m.

62T Torpedo version of long-span 62 (Tu-2D), two torpedoes external as Tu-2T, fuel in bomb bay, main gears as 2nd Tu-2T but 1260 tyres hence bulged doors. First flight 2 Jan 47, data as 62 except max speed with two 45-36-AN 501 km/h (311 mph) at SL, service ceiling 7,7 km (25,260 ft), range 3800 km (2,360 miles).

63P, Tu-1 Three-seat long-range escort fighter and interceptor, intended eventually to carry radar. Basically a modified Tu-10 (68) with AM-43V engines driving four-blade AV-9K-22A props, no lower rear crew position, main cockpit with pilot on centreline and nav/radar observer at rear facing forward. Armament two NS-45 in lower part of nose, two NS-23 in wing roots and twin UBT, bombload 1 t. Nose above large guns configured for PNB-1 'Gneiss-7' radar, based on German FuG 220 with tail-warning. Factory test 22 March to 3 Nov 47, superb but engines not in prodn.

DIMENSIONS As Tu-2 except length (excl guns) 13,6 m (44 ft 7⅓ in); with guns 13,72.

ENGINES Two AM-43V.

WEIGHTS Empty 9460 kg (20,855 lb); loaded 12 755 kg/14 460 kg (max 31,878 lb).

PERFORMANCE Max speed 479 km/h (298 mph) at SL, 641 km/h (398 mph) at 8,6 km; time to 5 km, 11·6 min; service ceiling 11 km (36,090 ft); range 2250 km (1,400 miles), take-off 605 m; landing 560 m.

Tu-2R, Tu-2F, Tu-6 Several factory and field conversions of Tu-2 bombers as reconnaissance aircraft made from 43, with three/four cameras with individual ventral doors, but not until 46 did purpose-designed variants appear: Tu-2R (*Razvyedchik*) with standard airframe and subsequent Tu-2F (*Fotorazvyedchik*) with long-span wing for high-alt operation, and new forward fuselage with nav station in nose. It was expected Tu-2R would be built in series, with service designation Tu-6 assigned. Four-seat aircraft with long-range tanks and provision for three or four reconnaissance cameras (usually AFA-33, AFA-3c/50 and AFA-33/50 or 33/100) in fuselage bays with bulge under TE. Two ShVAK (100 rds) and two UBT (250 rds). Flown Oct 46, NII test to 9 Ap 47. Later fitted with *radarom* chin radar.

DIMENSIONS, ENGINES As Tu-2S.

WEIGHTS Empty 8205 kg (18,089 lb); loaded 10 740 kg/12 755 kg (max 28,119 lb).

PERFORMANCE Max speed 509 km/h (316 mph) at SL, 545 km/h (339 mph) at 5,5 km; time to 5 km, 10·3 min; service ceiling 9050 m (29,700 ft); range 2780 km (1,727 miles); no other data.

RShR, Tu-2RShR Dedicated anti-armour aircraft with 57-mm RShR gun mounted on lower centreline of nose, with barrel and muzzle brake projecting only about 0,5 m. Basic aircraft very like Tu-2Sh with crew of two: pilot and nav/radio operator who reloaded gun breech in bomb bay. Main legs inclined to place wheels 125 mm further forward. NII tests completed 28 Feb 47; no production. Data as Tu-2.

RShR

65, Tu-2DB

65, Tu-2DB Penultimate DB (*Dalnii Bombardirovshchik*, long-range bomber) variant, this was based on long-span wing and twin-pilot cockpit, with bombload and weapons as Tu-2 but advanced liquid-cooled engines supercharged by TK-1B (TK-300) exhaust turbos, driven by exhaust from right-hand bank of cylinders on each engine (left bank had plain ejector exhausts); AV-5LV-166B 3,8 m three-blade props with mech control. Main wheels 125 mm further forward as previous. Five seats. Single example mod from a/c 714, flown by Opadchii 1 July 46, dropped mainly because engine snags.

DIMENSIONS As Tu-2D Type 67.

ENGINES Two AM-44TK.

WEIGHTS Empty 9696 kg (21,376 lb); loaded 13 006 kg/15 962 kg (max 35,190 lb).

PERFORMANCE (est) max speed 579 km/h (360 mph) at 9,3 km; climb 9·0 min to 5 km; service ceiling 11 km (36,090 ft); range 2670 km (1,659 miles); TO 480 km, ldg 490 km.

69, Tu-8 Last of basic line, this DB (long-range bomber) was heaviest variant and a most refined and impressive aircraft, but it arrived into OKB dominated by Tu-4 and a world turning swiftly to jets. Airframe based on 62 (Tu-2D) but with further increase in chord to give greatest area of series. Main wheels 125 mm further forward as in RShR, larger tyres to match gross weight (main 1170 mm × 435 mm, tail 580 mm × 240 mm). Five seats, completely revised defensive armament with B-20 on right side of fuselage fired by pilot, B-20 with 190 rounds fired from rear of main cockpit by nav/bomb or, usually, second pilot on swivel seat, third B-20 with 250 rounds in MV-11 dorsal turret and fourth/fifth B-20 in pair in ventral turret (type unknown) aimed by fifth crew-member via remote sight/control system using diagonal beam portholes upstream of recessed sides of rear fuselage. Fuel capacity as ANT-62, bombload increased to 4,5 t max. Single example flown briefly May 47. Versions with AM-42 (Tu-8B) and ACh-39BF (Tu-8S) never completed.

DIMENSIONS Span 22,06 m (72 ft 4½ in); length 14,61 m (47 ft 11⅛ in); wing area 61,26 m² (659 ft²).

ENGINES Two ASh-82FN.

WEIGHTS Empty not found; loaded 14 029 kg/16 663 kg (36,734 lb).

PERFORMANCE Max speed 507 km/h (315 mph) at 5,7 km; climb to 5 km 17·0 m; service ceiling 7650 m (25,100 ft); range 4100 km (2,548 miles); TO 860 m; ldg 632 m.

71 Tu-2 project 46 with new nose and M-93 engines.

72 Two Tu-2 derived projects: 46, med-range bomber with turbocharged ASh radial engines (believed ASh-90); 47, airframe based on 69 but with two RR Nene turbojets.

Launched 46 as medium bomber with two 3,300-hp ASh-2TK, terminated Jan 47 and restarted as OKB's first jet. Airframe based on Type 69, centre section modified to accept two long underslung nacelles for British turbojets. Split flaps outboard of nacelles. Main gears retracted forwards into nacelles, nose gear backwards. Fuselage stretched, nose more pointed for navigator/bomb aimer, pilot with sight for aiming single fixed B-20 with 125 rounds, radio/gunner behind pilot in UST-K2 powered turret with twin B-20 with 250 rounds, aft compartment with gunner sighting through beam windows as in 69 and remote control of inverted UST-K2 again with 250 rounds. Four wing tanks plus large upper-fuselage tank. Bomb bay for up to 4t, largest size FAB-1500. Prototype never completed.

DIMENSIONS As 69 except length 16,11 m (52 ft 10¼ in).

ENGINES Two 5,000-lb Rolls-Royce Nene I.

WEIGHTS Normal loaded 16 100 kg (35,494 lb).

PERFORMANCE Max speed est 725-750 km/h (max 466 mph); service ceiling 12 km; range (1t bombload, 7500m) 2000 km (1,242 miles).

Project 64 In Gt Patriotic War strategic bombers had no priority and lagged behind Allies, Stalin then waking up and making repeated requests for B-29s. By mid-44 decision taken to build aircraft in this class and protoypes ordered of Mya DVB-202 and Tupolev 64. Former a close copy of B-29, but '64' similar only in size, power and general technology; far less fuel. Outstanding wing, aspect ratio 11·9, NACA-2330 profile, rect c-s including inner engines and long outer panels tapered on LE. Fuselage same 2,9-m diam as B-29, same three GK (pressure cabins), but pilots with 'bug-eye' canopies and gunner compartment further aft. Completely different twin-fin tail and main gears with large single wheels. Long bomb bays in front of and behind spars of low/mid wing, ten B-20 guns in five turrets (lower front in chin posn ahead of twin-wheel nose gear). Liquid-cooled engines with turbos, driving 4,5-m 4-blade props. Secondary power systems hydraulic. Crew 7: two pilots, nav/bomb, engineer, radio/radar op, gunner plus tail gunner. Drawings issued early 46 but halted before prototype assembly started because of B-29 captures. A. N. Tupolev regretted this decision.

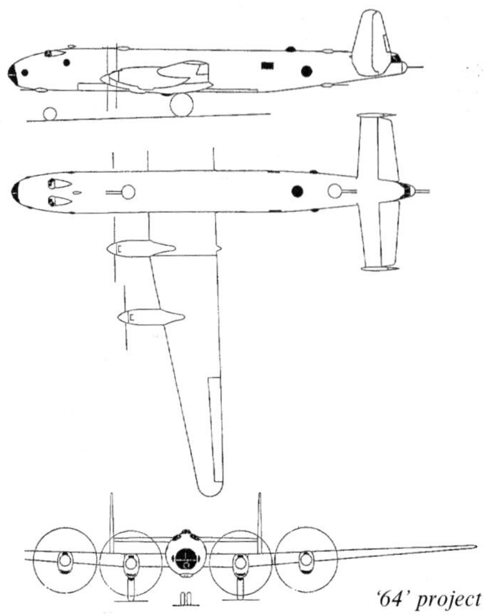
'64' project

DIMENSIONS Span 39 m, later 42,8 m (140 ft 5 in); length 26,75, later 29,0 m (95 ft 1¾ in); wing area 152 m² (1,636 ft²).

ENGINES Four AM-44TK.

WEIGHTS Empty c 22 t; loaded (5 t bombs) 36 t (79,365 lb).

PERFORMANCE Max speed 600 km/h (373 mph) at 6 km; range with 5 t bombload 3000 km (1,864 miles).

Project 66 Transport version of '64'; fully pressurized unobstructed fuselage, low-mounted wing, nose resembling B-29, tailcone carrying tail with smaller fins. No drawings issued to shops.

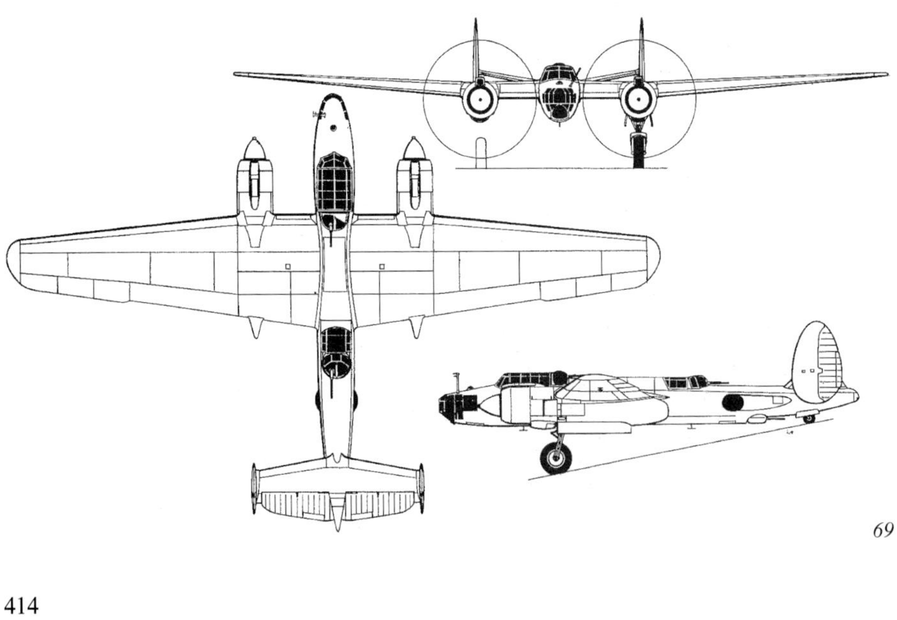
69

Tu-4 Sov leaders had little doubt they would soon get hands on actual B-29, and windfall happened sooner than expected. Though property of ally, three B-29s which force-landed in Sov Far East were instantly appropriated: 29 July 44, B-29-5-BW 42-6256; 20 Aug 44, B-29A-1-BN 42-93829; 21 Nov 44, B-29-5-BW 42-6358. All restored airworthy and flown by test pilots Reydel and Marunov to Moscow; -93829 used as model (wing used in Tu-70), -6256 parent of '346' and -6358 retained as crew trainer. Task of dismantling and analysis without precedent: Tupolev said 105,000 items checked for material spec, function, manufacturing processes, tolerances and fits, and translated into Sov drawings. Many parts new to Sov industry or alien to established Sov practice. Full-scale programme authorized end-44 to create copy for DA; original designation **B-4** (Bombardirovshchik, not Boeing), later changed to **Tu-4** to reflect immense 'copying' effort. Pre-production batch of 20 ordered mid-45, with tooling at four GAZ. Tu (miffed at cancellation of '64') said 'three-year effort', I. V. Stalin said 'You have two years'.

Few parts emerged identical to B-29, metric gauges generally overthick (first Tu-4 airframe 15 196 kg) and with compromises in piping and cables. Wing profile RAF-34 20%/10%, difficulty with 75-micron bolt interference fit and in stress-relief and precision-machining of new D16-ABTN, nose and tail being D16-AT. Tyres 1450 mm × 520 mm (5-5,8-at pressure) on mainwheels and 950 mm × 350 mm (3,6-4) on nosewheels. Bomb bays redesigned to carry up to 8 t Sov bombs. Major redesign of powerplant group, TK-19 turbos and 5056-mm VZ-A3, VZ-A5 or VZV-A5 props. Integral tankage abandoned, fuel only 3480 kg in first three Tu-4, later 11,3 t in 22 protected flexible cells in wing and option of two AMTs tanks in fuselage above bomb bay. Major tasks complete redesign of defensive gun system, fitting VHF radio and Sov IFF.

First three aircraft assigned to N. S. Rybko, M. L. Gallai and A. G. Vasilchyenko. First flown 6 months after Tu-70 on 19 May 47. Predictably severe problems with many systems, but success never in doubt. Total number built about 900, ending c51. First batch lacked many items and carried twin B-20E guns in five turrets. From 48 new turrets fitted with twin NS-23 guns, and proportion of deliveries had retractable *Kobalt* blind bombing radar between bomb bays. Production included **Tu-4R** strategic-recon aircraft with forward bomb bay permanently housing tanks and aft bay carrying camera groups. From 52 increasing numbers converted as **Tu-4N** air-refuelling tankers, testing all known techniques. A further group used for research and as carry-trials aircraft (though supersonic '346' carried by B-29), and biggest conversions were **Tu-4LI** flying laboratories. Crucial LI had No 3 engine replaced by TV-12 turboprop (5 × power of other engines), chief engineer D. I. Kantor and pilot M. A. Nyukhtikov to clear engine for prototype 95/II. Others tested AL-5, AL-7, AM-3, AM-5, NK-4, AI-20 (both above and below wing as in Il-18 and An-10) and VD-7. Three used for cruise-missile trials, initially with captured Fi 103 ('V 1') and later La-17 and Mikoyan KS-1. Several converted as transports, including single **Tu-4T** assault vehicle with 28 paratroops. As interim bomber OKB proposed **Tu-94** turboprop conversions with either TV-2 or NK-4 engines, but not accepted. Several (24+) supplied to China, where later re-engined with WJ-6 (AI-20) turboprops, reinvention of NK-4 installation of 20 years previously. ASCC name for all 'Bull'.

DIMENSIONS Span 43,05 m (141 ft 2⅞ in); length (47) 30,18 m, (50) 30,8 m (101 ft 0⅝ in); wing area 161,7 m^2 (1,740 ft^2).

ENGINES Four ASh-73TK.

WEIGHTS Empty (48) 35 270 kg (77,756 lb); fuel/oil 11,3 t max + 800 kg; loaded (47) 47,6 t, (48 max) 54,5 t (120,150 lb), (51 max) 66 t (145,500 lb).

PERFORMANCE Max speed 420 km/h (261 mph) at SL, 558 km/h (347 mph) at 10 km; climb to 5 km 18·2 min; service ceiling 11,2 km (36,750 ft); range with 2 t bombload 5100 km (3,170 miles); TO 960 m-2210 m depending on wt; landing 1070 km/h/172 km/h.

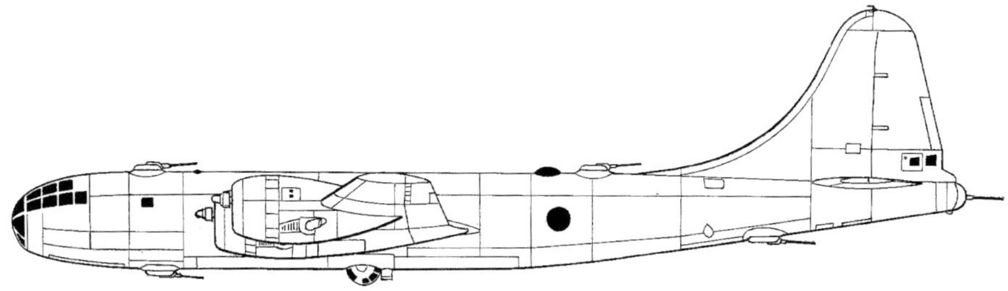

Tu-4

Tu-4

Tu-4/TV-12

Tu-70

Tu-70 Passenger transport, produced much quicker than prototype Tu-4 by using parts of B-29A-1 including wing and engines. Completely new pressurized fuselage, 3,5 m diam, nosecap as bomber, but new cockpit with air-line-type windscreens, then front, centre and rear compartments, centre (overwing) being small galley. Main cabins with wide rectangular windows, aft section 7 portholes each side. Built for crew 8 and 48 VIP seats, later crew 6 and 72 seats. Wider c-s increased span, nose gear redesigned and other changes. First flight by Opadchii 27 Nov 46, retained as VVS staff transport with service designation (on aircraft at roll-out) of **Tu-12**. ASCC 'Cart'.

DIMENSIONS Span 44,25 m (145 ft 2⅛ in); length 35,4 m (116 ft 1¾ in); wing area 166,1 m^2 (1,788 ft^2).

ENGINES Four R-3350-57.

WEIGHTS Empty 33 979 kg (74,910 lb); fuel 7430 kg normal, 16 030 kg max; loaded 51,4 t normal (113,316 lb), 60 t (132,275 lb) max.

Tu-75

PERFORMANCE Max speed 424 km/h (263 mph) at SL, 568 km/h (353 mph) at 9 km; climb 21·2 min to 5 km; service ceiling 11 km (36,090 ft); range 5000 km (3,730 miles) max; TO 670 m; ldg 600 m.

Tu-75 Second transport derivative of B-29, entirely Sov-built, cleared for higher weights with 50% greater fuel capacity. Fuselage generally as Tu-70, unpressurized, with large aft loading ramp for small vehicles fitted with roller conveyor for other cargo. Purely military, equipped for 10 t cargo or 120 assault troops and their weapons with small folding seats along walls. Designed to have Tu-4 armament (one dorsal, one ventral, plus tail turret) but built with only provision for this. First flight by Marunov July 60. ASCC 'Cart'.
DIMENSIONS Span 44,25 m (145 ft 2⅛ in); length 35,61 m (116 ft 10 in); wing area 162,7 m² (1,800 ft²).
ENGINES Four ASh-73TKNV.
WEIGHTS Empty 37 810 kg (83,355 lb); fuel/oil 10 540 kg normal, 24,9 t max; loaded 56 660 kg (124,912 lb) normal, 65,4 t (144,180 lb) max.
PERFORMANCE Max speed 545 km/h (339 mph) at c9 km; service ceiling 9,5 km (31,170 ft), range (normal fuel) 4140 km (2,573 miles).

Tu-80 First major attempt at improving basic B-29. Range was increased by redesigning wings with integral tanks, as in B-29 but using different materials, giving 15% more fuel in only slightly greater span. Section unchanged but wing structurally re-planned with longer centre section and slightly shorter outer panels with dihedral reduced to 0°, integral tankage throughout. New engines, installations redesigned with higher thrustlines, circular cowlings, separate ventral oil-cooler ducts and aerodynamics similar to US R-4360 in B-50. Redesigned forward fuselage establishing arrangement for crew repeated in Tupolev bombers for almost 20 years, with glazed nose for front gunner and nav/bomb, separate cockpit for two pilots with conventional windscreens (pilots had difficulty with internal reflections and optical distortion in early Tu-4s) and radar/nav seated behind facing to rear with large mapping/bombing radar in pressurized underfloor compartment. Communication tunnel restored to rear-fuselage compartment with several changes including lateral observation windows upstream of recesses in fuselage sides as in Type 69 (Tu-8). Extra nose length balanced by enlarged vertical tail with no structural part common to Tu-4. Modified gun turrets, tail turret sloping aft window (armament not fitted in only known photos); bomb bay redesigned for max 12 t (26,455 lb). First flight 1 Dec 49, but testing discontinued because of Tu-85.

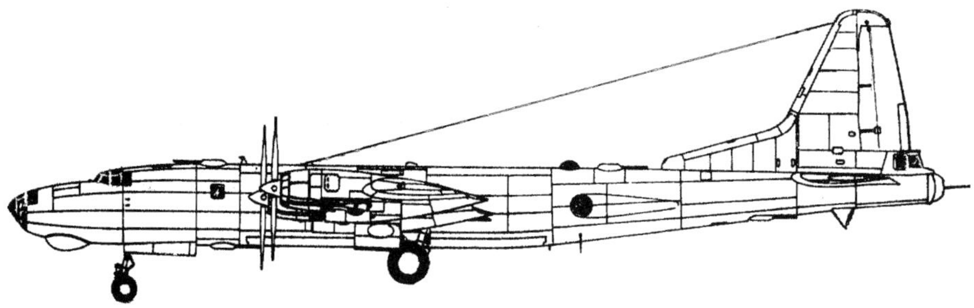

Tu-80

DIMENSIONS Span 43,45 m (142 ft 6⅝ in); length 34,32 m (112 ft 7¼ in) wing area 167,0 m² (1,798 ft²).
ENGINES Four ASh-73FN.
WEIGHTS Empty, 37 850 kg (83,444 lb); fuel 28,6 t; normal loaded 60,6 t (133,598 lb); max figure higher.
PERFORMANCE Max speed 428 km/h (266 mph) at SL, 545 km/h (339 mph) at 10 km; range (est) 7000 km (4,350 miles); TO 1200 m; ldg 505 m.

Tu-85 Even this ultimate extrapolation of B-29 was abandoned despite its excellence because of dramatic potential of Tu-95. Made possible by existence of ASh-2 and VD-4K engines, which enabled fuel capacity to be almost doubled. Success achieved with completely new wing. Same RAF-34 profile but rect c-s, dihedral only on tapered outer panels, aspect ratio 11·4. Structurally advanced with fewer ribs despite greater span, skins up to 10 mm at root and secured by precision bolts. Fuselage extended by 14 extra frames (6 ahead of wing, 8 aft) giving six main compartments, three pressurized for crew 11 or 12 (nav, bomb-aimer, two pilots, engineer, radio, radar and technician in front, three gunners rear and tail gunner); three light bunks for off-duty crew numbering up to 16 on longest missions. Bomb bays improved and lengthened to carry normal 5 t, max 20 t including FAB-9000 or nuclear. Defensive sight/control system as before but guns changed to ten NR-23. OKB-designed twin-wheel landing gears with pneu operation. Prototypes ordered with each engine, plus static-test airframe, but both completed with V. A. Dobrynin's engine. Distance between centrelines, inners 9,1 m, outers 20,38 m. Cooling air entering at front with forward gills around oil cooler ring, two-stage supercharger fed by large dorsal inlet, and driving 4,5-m reversing props with four solid-dural blades. Fuel in 48 flexible tanks total 63 600 lit (13,990 gal), exceeded at time only by B-36.

Construction took two years, first flight 9 Jan 51, crew Perelyot plus leading engineer N. A. Genov, nav S. S. Kirichyenko and engineer A. F. Chyernov. Tested to Oct 51, generally excellent aircraft, 12 Sept 51 No 2 flew 9020 km in 20·5 h and later with max fuel made flight of 13 018 km (8,089 miles). In view of this, decision to rely on jet/turboprop bombers may have been hasty. ASCC reporting name 'Barge'.
DIMENSIONS Span 55 939 m (183 ft 6⅜ in); length 39 905 m (130 ft 11 in); wing area 273,59 m² (2,945 ft²).
ENGINES Four VD-4K.
WEIGHTS Empty 54 711 kg (120,615 lb) (No 2, 55,4 t); fuel/oil 20 129 kg normal, 48 600 kg max; loaded 76 t (167,549 lb) normal, 107 292 kg (236,534 lb) max.
PERFORMANCE Max speed 459 km/h (285 mph) at SL, 638 km/h (396 mph) at 10 km; service ceiling 11,7 km (38,390 ft); range (5 t bombload) 12 000 km (7,457 miles); TO 1640 m; ldg 1500 m.

77, Tu-12 Twin-jet bomber based on airframe of Tu-2S but restressed with heavier-gauge materials for greater weight and higher IAS, with two extra tanks at wingtips. Tail likewise strengthened. Fuselage generally retained but

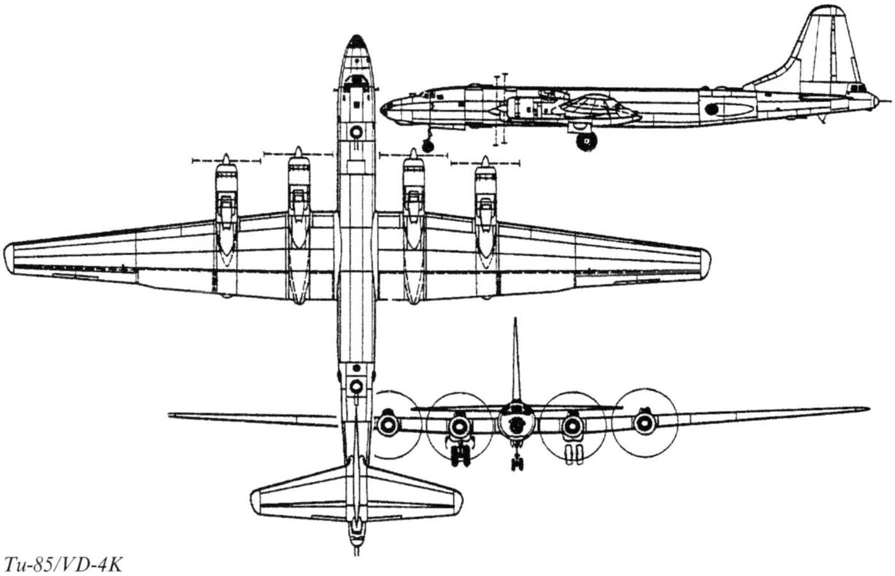

Tu-85/VD-4K

Tu-85

Tu-12

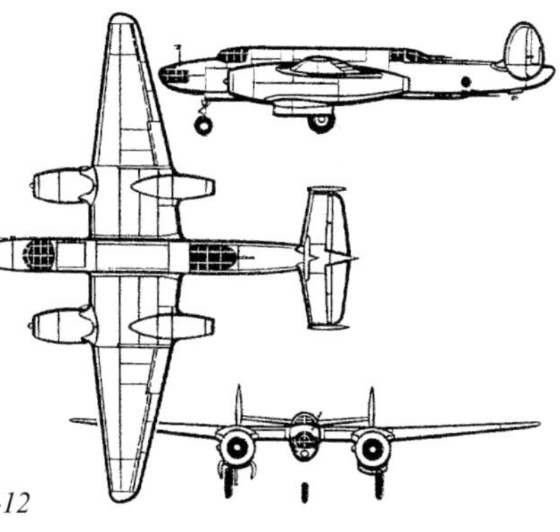

Tu-12

increased in depth 0,5 m to provide for additional fuel and equipment and with redesigned nose with greater length and complete glazing (not same as long-nosed Tu-2 variants). No pressurization. Engines mounted further apart than in Tu-2: 5,3 m for centrelines, 5,4 m track and span of tail unit (fin centrelines) 5,3 m. Moving engines out increased apparent size of centre section but in fact wing size overall unchanged. Long (6,2 m) nacelles fully underslung and designed by OKB, unlike those of Il-28 and with short main gears with large wheels rotating 90° on single legs to lie flat in wide bulge in underside of nacelle, leg retracting forwards. Single nosewheel retracting to rear, all bays having twin doors. New pilot cockpit on centreline with deep hinged side windows and glazed roof but no rear vision. One fixed NR-23 on left side of nose, two rear gunners each with UBT, bomb bay for up to 3 t bombload. First Type 77 built at Tu-2 production plant and flown by Perelyot 27 July 47. Two took part in Tushino show on 3 Aug. Total of five built by Sept 47, but never operational.

DIMENSIONS Span 18,86 m (61 ft 10½ in); length 15,75 m (51 ft 8 in); wing area 48,8 m^2 (525 ft^2).

ENGINES Built with two Nene, re-engined with similar RD-45.

WEIGHTS Empty 8993 kg (19,826 lb); fuel/oil 6727 kg max; loaded 14 700 kg (32,407 lb) normal, 15 720 kg (34,656 lb) max.

PERFORMANCE Max speed 778 km/h at SL, 783 km/h (487 mph) at 5 km; climb to 5 km in 8.0 min; service ceiling 11 360 m (37,270 ft); range 2200 km (1,367 miles); takeoff 1030 m; landing 163 km/h, 885 m.

73 In Jan 47 Tupolev launched this larger three-jet bomber to meet demand for aircraft to carry 3t bombload and exceed 800 km/h. Airframe scaled up from Type 69 (in one document Type 73 called Tu-8, as was 69, later crossed through as incorrect). Wing with plate spars and machined skins as in B-29, with four integral tanks, split flaps inboard and outboard of nacelled. Circular-section fuiselage, glazed nose seating nav/bomb-aimer further back with side portholes; pilot and radio-operator/gunner back-to-back under large glazed canopy; full-section tank; next section occupied by dorsal PS-23 turret and magazine; mid fuselage with twin-door bay able to carry FAB-3000, above this two tanks separated by wing-box tank; next bay for camera, typically AFA-33/75; pressurized rear compartment for gunner with bulged side windows as in 69 and remote control of inverted PS-23 behind pressure bulkhead with twin NR-23 fed by magazine in roof of pressure cabin. Behind fireproof bulkhead, plenum chamber for centre engine fed by duct from front of long dorsal fin. Fixed tailplane halfway up single fin, all surfaces metal-skinned and tabbed.

Main engines installed as in 72, ground-power socket on right of each cowl, twin-door bay underneath to house main gears with wheels horizontal above leg. Long-stroke castoring nose gear retrating to rear. Provision under centre section for four PSR-1500-15 takeoff rockets. Sprung retractable tailskid. Access hatches under both pressure cabins and under centre engine. Radio antenna strung from fin to mast inclined from right of canopy, flush antenna in fin.

First 73 completed without turrets and with retractable fairing so that centre-engine inlet could be blanked off in cruising flight. First flown by Opadchii 20 Dec 47. Second 73 with full armament and simple fixed inlet to centr engine, and with Tu logo and 'Tu-14' painted on fin. NII testing to 31 May 49.

DIMENSIONS Span 21,71 m (71 ft 2¾ in); length 20,32 m (66 ft 8 in); wing area 67,38 m^2 (725 ft^2).

ENGINES Two Nene I, one 3,500-lb Derwent 5.

WEIGHTS Empty 14 340 kg (31,614 lb); loaded 21,1 t (46,517 lb) normal, 24,2 (53,351 lb) max.

PERFORMANCE Max speed 840 km/h at SL, 872 (542 mph) at 5 km; climb to 5 km in 9.5 min; service ceiling 11,5 km (37,730 ft); range 2810 km (1,746 miles); takeoff 740 m; landing 173 km/h, 1170 m.

78 This prototype differed from 73 in minor

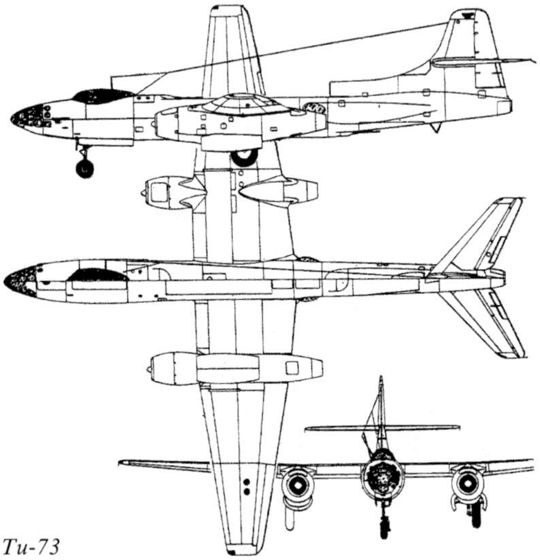

Tu-73

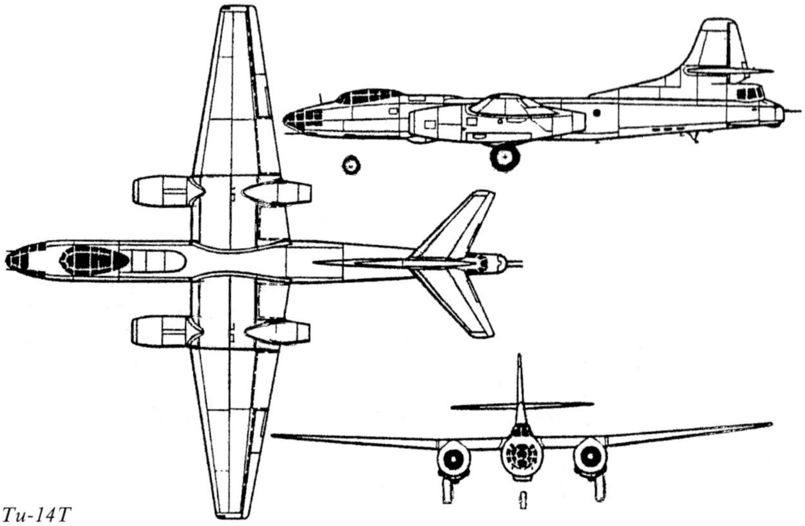

Tu-14T

Tu-73 with a.t.o. rockets

respects, such as increased tankage, Soviet engines, 0,3-m extension of fuselage, increase in chord instead of round navigator windows and restoration of tail-engine inlet fairing. OKB logo and 'Tu-14' on nose. First flight 7 May 48. In late 48 Marshal Vershinin ordered three VVS crews picked at random to evaluate against Il-28; all picked Il-28. Tu-78R recon aircraft not flown, devt continued for navy.

DIMENSIONS Span 21,71 m (71 ft 2¾ in); length 20,62 m (67 ft 7¾ in); wing area 67,36 m² (725 ft²).

ENGINES Two RD-45, one RD-500.

WEIGHTS Empty 14 290 kg (31,505 lb); fuel/oil max 8350; loaded 23 790 kg (52,447 lb) max.

PERFORMANCE Max speed 840 at SL, 875 (544 mph) at 5 km; climb to 5 km in 7.7 min; service ceiling 11,5 km (37,730 ft); range 3100 km (1,926 miles); takeoff 1000 m; landing 1300 m.

Tu-73

Tu-14, Tu-79, Tu-81, Tu-89 Well before competitive fly-off against Il-28 Arkhangyelskii had learned from Klimov of greater thrust possible from VK-1. Decision taken summer 48 to revert to twin-engined aircraft with reduced basic weight and increased fuel; MA had never liked third engine and OKB had never established best cruise technique with two or with three engines. First of new series **81** with few changes except removal of tail engine, rearward shift in installation of main engines to balance loss of tail mass, increase in fuselage fuel (small, no data) and increase in defensive armament to six NR-23 (two pairs in turrets and two fixed in lower part of nose fired by pilot). Not known if this aircraft was built; no record of flight test.

Further study showed possibly better answer was to fit tail turret similar to Tu-80/Tu-85 with radio racking at tail and remove other movable guns, reducing crew to three. Shift in CG enabled existing nacelles and engine installation to be retained and substantially reduced basic weight (though empty weight naturally grew by at least 1 t as result of combat equipment).

Ilyushin OKB assisted turret installation, but in May 50 limited traverse ±50° was one of criticisms voiced at MA evaluation. Fuselage again rearranged internally to make better use of extra length, new turret fitted offering ±70°, with NR-23 guns replaced by B-20E. Gunner's windows enlarged, and bottom of rudder cut off diagonally. At end 50 improved Type 81 accepted as **Tu-14T**, suffix indicating primary role as torpedo carrier. Prototype 81T established standard armament as two fixed B-20 and two B-20E in turret, with normal load of two torpedoes (any of four types) or 3 t (6,614 lb) bombs, including nuclear.

Total of 89 built by GAZ-125 Irkutsk, where Sukhoi sent as director of programme. Factory pleased (42 delivered a year from start), and aircraft popular in service, 52-62. Updates included improved autopilot and eng-monitoring, nav ejection seat and ribbon drag chute. Planned more economical replacement Tu-91 cancelled. Batch of 50 supplied 58-59 to China. Single **Tu-89** tested as prototype of Tu-14R recon version, with aux fuel, AFA-33-20, 30-50/75 and 33-100 cameras, drift sight, flares and provision for later side-looking radar. USAF designation 'Type 35', ASCC name 'Bosun'.

DIMENSIONS Span 21 686 m (71 ft 1¾ in); length (81) 21,945 m (72 ft 0 in), (89) 21,69 m (71 ft 2 in); wing area 67,36 m² (725 ft²).

ENGINES Two VK-1.

WEIGHTS Empty (81) 14,43 t (31,812 lb), (89) 14,49 t; fuel (89) 8,7 t; loaded (81) 21 t (normal, 24,6 t (54,233 lb) max, (89) 21 t normal, 25,35 t (55,886 lb) max.

PERFORMANCE Max speed (81) 800 km/h at SL, 861 km/h (535 mph) at 5 km, (89, poss with external torps) 774 km/h at SL, 845 km/h (525 mph) at 5 km; time to 5 km (both) 9·5 min; service ceiling (81) 11,5 km (37,730 ft), (89) 11,2 km; range (both) 3010 km (1,870 miles); takeoff (81) 1250 m, (89) 1200 m; landing (81) 1120 m/175 km/h, (89) 1100 m/176 km/h.

82, Tu-22 At height of Tu-14 effort Arkhangyelski suggested exptl bomber with swept wings, using max number of existing parts. During design of **Type 82**, early 48, mission changed to frontal bomber for VVS, given designation **Tu-22**. Major parts of horizontal tail, landing gear and nacelles as Tu-14, but fuselage cross-section changed to oval, reducing fuel capacity and weight, and whole aircraft much smaller, though crew as before. Nose extra roof glazing, pilot's canopy fighter type, offset to left. Tail turret with superimposed guns (two B-20E) never fitted. Wing with ¼-chord sweep 35°, high aspect ratio (6·9), fixed LE, conventional flaps/ailerons, four fences each side. Internal bay for 1 t various bombs (not nuclear). NR-23 forward-firing gun. First

Tu-14LL

Tu-22 (Type 82)

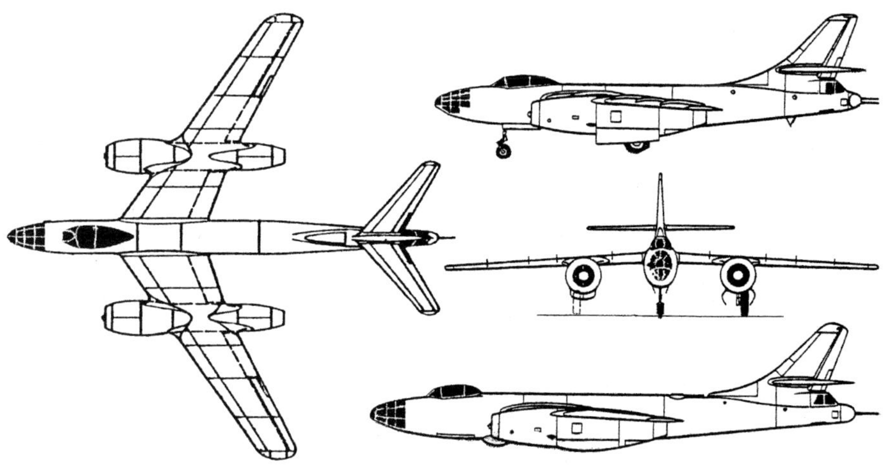

Tu-82 (bottom side view, '83')

flight by A. D. Perelyot 24 March 49. By this time Tu-88 was dominant, and Type 82 dropped, VVS number Tu-22 being used again for No 105.

DIMENSIONS Span 17,81 m (58 ft 5⅛ in); length 17,57 m (57 ft 7¾ in); wing area 46,24 m^2 (497·7 ft^2).

ENGINES Two RD-45F.

WEIGHTS Empty 11 226 kg (34,749 lb); fuel/oil 2250 kg (max, aux tanks) 5670 kg; loaded 14 919 kg (32,890 lb), (max) 18 339 kg (40,430 lb).

PERFORMANCE Max speed (SL) 870 km/h (541 mph), (4 km) 931 km/h (579 mph); service ceiling 11,4 km (37,400 ft); range (normal) 2395 km (1,488 miles); TO 1100 m; ldg 550 m.

83, 86, 87 Unbuilt projects for twin-jet bombers derived from 82. **No 83** stretched to 19 925 m (65 ft 4½ in) to permit addition of remotely sighted dorsal turret ahead of fin, radar and longer bomb bay. **No 86** further stretched (span 25,5 m, length 24,15 m) and powered by AM-02 for 1000 km/h; range 4000 km with 2 t to 6 t bombs; **No 87** same but TR-3 engines. VVS preferred Il-28 + Tu-16.

84 Bomber project of 49, 2 VK-2+ 1 VK-1.

Tu-88, Tu-16 Development by Mikulin engine KB of M-209 (later AM-3, RD-3) was key to this outstanding jet bomber. Tupolev had long sought to combine jet speed with range and bombload of Tu-4, and was ordered to do this 48. Initially studied projects with six VK-1. Project **90/88** was far better, with two TR-3F in wing roots and two more in underwing pods, and this nearly went ahead with first *Rubin* radar and OPB-11r sight. Breakthrough was availability of AM-3, air-tested under Tu-4LL. Late 48 OKB received order for two **Tu-88** prototypes plus static-test specimen. Project called **N**, design team led by Dmitri S. Markov (later a Tu chief designer). Tupolev himself decided location of engines, A. A. Judin then designing novel main landing gears which became OKB 'trademark'.

Basic requirement: carry 5 t (11,020 lb) bombs 5000 km, and also carry largest bomb, FAB-9000. This dictated width of bomb bay; to reduce drag Tupolev discarded circular section in centre fuselage and drew engines close beside FAB-9000, 1,9 m (75 in) apart, curving in inlets and curving out jetpipes. This gave Area Rule shape, and splayed-out jetpipes avoided scrubbing on rear fuselage.

Wing aerofoil PR-1-10S-9, theoretical t/c 15·7% on centreline and 12% on outer panels. LE sweep 41° from root to structural joint at 6,5 m, then 37° to tip. Mean chord 5021 mm (198 in), anhedral –3°. Main structural box thick machined skins, forming integral tanks throughout, hot-air LE anti-icing with exhaust from tip slits. TE tracked slotted flaps inboard and outboard of main landing gears, total each wing 25,17 m^2, driven by electric ball-screws to 35° for landing. Outboard balanced Frise ailerons each 14,77 m^2. Fences 200 mm deep (extended to LE during production) added at structural break and at flap/aileron junction. Tail scaled-up from Tu-82, fin 23,3 m^2, LE sweep 40°, rudder 5,21 m^2. Fixed tailplane 34,45 m^2, LE 42°, elevators each 8,65 m^2. All control surfaces tabbed and balanced for manual control. Production aircraft added hydraulic boost.

Fuselage mainly circular section, diameter 2,5 m (98·4 in), made in five sections. First, glazed nose, with radar under cockpit, and ventral crew door with telescopic ladder under avionics/systems compartment, bounded by convex rear pressure bulkhead. Next section housing nose landing gear, forward tanks, dorsal turret, cameras and landing lights. Next, long centre section with protected tanks and 6,5 m (256 in) bomb bay with two hydraulic doors, followed by bay for target indicators (or other loads in later versions). Next, Nos 5 and 6 tanks and ventral turret, followed by rear pressurized compartment with ventral door, side blister windows, retractable bumper, PTK-16 braking parachute and tail turret. Total of 36 main frames, those in centre section being KhGSA with extra side rings for engine or duct picking up wing spars at top, large engine doors under wing. Nav/bomb station in nose on armoured seat, side-by-side pilots, then *radist* to operate RBP *Rubin* nav/bombing radar, and dorsal gunner who managed electrical system and signals. In tail a ventral gunner with optical sights at lateral blisters and tail gunner with turret, defensive system described later. Pilots could blow off roof hatches and eject upwards, other crew-members ejecting downwards. Fuel in 27 tanks with inert-gas protection, self-sealing in fuselage and integral in wings, total capacity 43 800 lit (9,635 gal), fed via flow proportioner.

First Soviet bogie main landing gear; track 9775 mm (32 ft), tyres 1100 mm × 330 mm, pressure 6,4 at (94·3 lb/in^2) compatible with unpaved runways. Multi-disc anti-skid brakes, hyd retraction backwards, truck somersaulting to lie inverted in box fairing with twin doors behind wing. Strikingly unusual, this avoided cutting into highly stressed wing skins. Twin-wheel nose gear with tyres 900 mm × 275 mm, hyd steering and retracting to rear into bay with twin doors; wheelbase 10,91 m (35 ft 10 in).

Hyd systems 3,000 lb/in^2, completely duplex to ensure stand-by power. Electric power generated as DC at 28 V for flaps and turrets, and for starting main engines and (in later versions) S-300 gas-turbine APU. Raw AC supplied for deicing tail and windscreens. Air-cycle environmental system max dP 0,05 kg/cm^2 (7·1 lb/sq in).

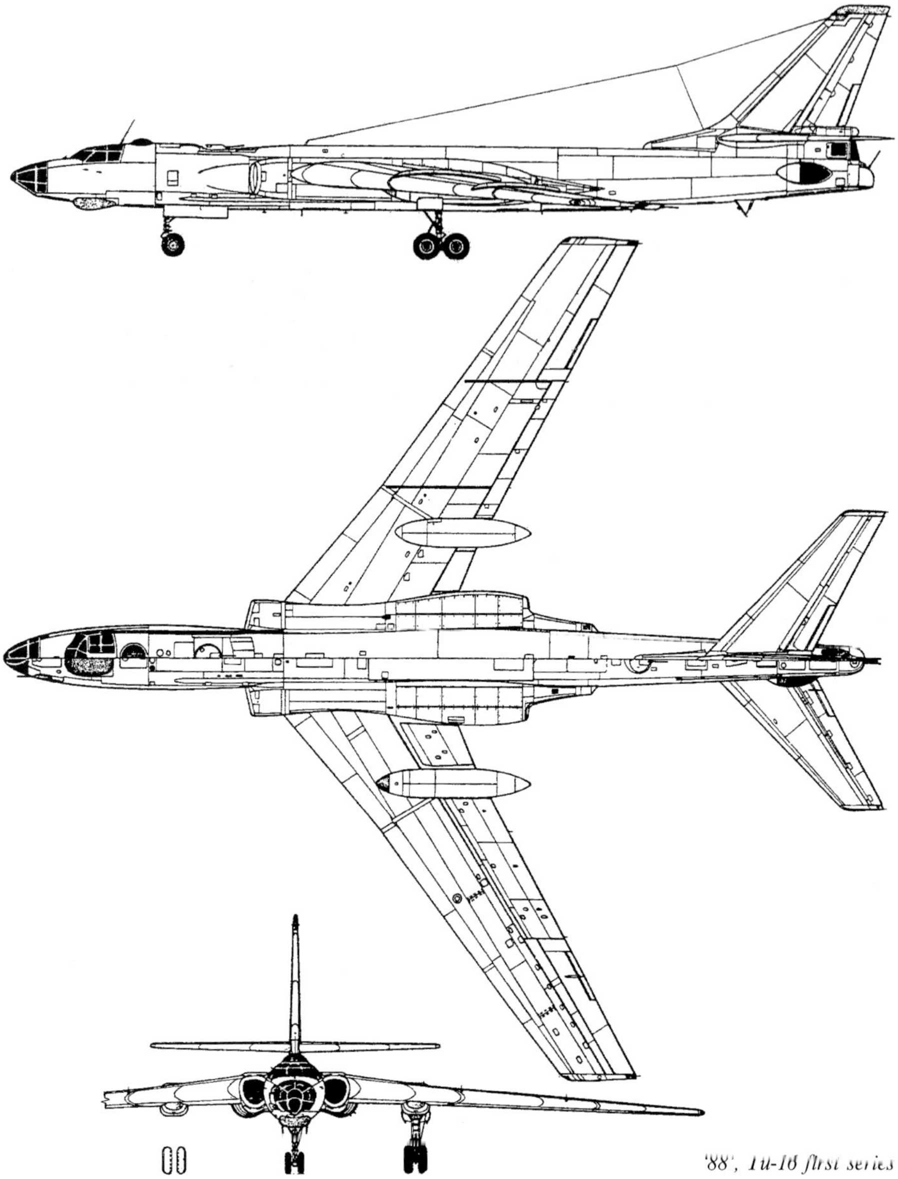

'88', Tu-16 first series

Tu-16A preserved

Tu-16KS

KP-23 gaseous-oxygen system and two IAS-5M five-place dinghies.

Tupolev was distressed by weight growth to over 80 t. He contested requirement for Mach 0·9 to be demonstrated at low level, and got this downgraded to 700 km/h, enabling airframe weight to be reduced by 5,5 t. First prototype flown by test crew headed by N. S. Rybko 27 April 52. Reached 1015 km/h but heavy structure prevented attainment of design range. Despite this, and arguably better handling of rival Il-46, potential of big Tu-88 was greater, and production was ordered Dec 52. OKB was permitted to delay programme while lighter No 2 was built, flown early 53. This reduced MTO to 71 560 kg and exceeded specified range (but Markov reprimanded for not getting it right first time).

Production with RD-3 engine began at Kazan late 53, followed in 54 by Kuibyshyev. Nine series aircraft flew over Red Square on May Day 54. Service designation Tu-16. Small number supplied to DA and also to AV-MF, most retained for test and training. USAF designation 'Type 39', ASCC name 'Badger'.

Tu-16A Late 54, *Atomnyi*, first operational version, powered by RD-3M, weapon bay configured for five types of nuclear bomb or FAB-9000 or other loads to 9 t (19,840 lb). PV-23 fire control installed, governing all defensive guns, linked to PRS-1 *Argon* tail radar under rudder. At rear of main crew compartment was DT-V7 dorsal with twin short-barrel AM-23 each with 500 rounds, 360° traverse and elev +90°/–3° safety cutout to prevent firing on own tail. Ventral turret DT-N7S, same guns but 700 rounds, slewing ±95°, elev +2°/–90°. Tail turret DK-7, twin long-barrel AM-23 with 1,000 rounds, slewing ±70°, and elev +60°/–40°. PU-88 installation on right of nose, single long-barrel AM-23 (100 rounds), aimed with PKI pilot reflector sight. PV-23 added S-13 (forward) and PAU-457-1 and -2 (turret) ciné cameras.

Basic avionics SPU-10 intercom, R-807 and (pilot) -808 HF, RSIU-3 UHF, RSIU-4 VHF, RBP *Rubin* main radar, RV-17 (high-alt) and RV-2 (low-alt) radar altimeters, AP-28 autopilot linked to NAS-1 nav system including DISS-*Trassa* doppler, MRP-48P ADF, KRP-F VOR and SD-1M DME linked via GRP-2 glide-slope receiver into SP-50 ILS. SRO-2 IFF with three rod antennas, plus Sirena-2 RWR (radar warning receiver). Standard cameras AFA-33M (day) and NAFA-8S (night). During production upgrades included AP-6E autopilot, RV-5 and -18 radar alts, SOD series ATC/SIF, ARK-15 ADF, ARK-5 tanker homing and *Khrom-Nikel* IFF. In 65 ILS was upgraded to SP-50M for Cat III landing, almost blind conditions. Usually RSDN *Chaika* Loran and, from 83, *Glonass* navsat receiver added. Production of Tu-16A c700, of total for all versions exceeding 1,520. Most Tu-16A rebuilt for other purposes after Jan 60. Egypt received 20, destroyed on ground in 67 war. ASCC name 'Badger-A'.

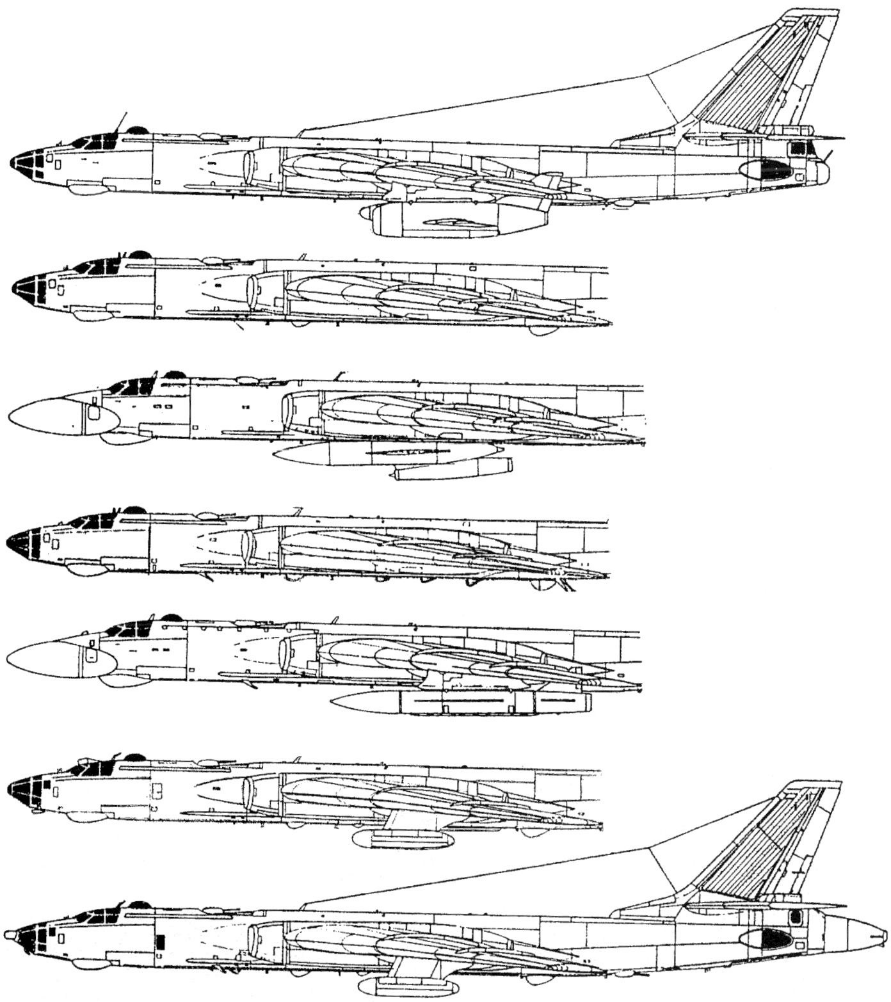

Tu-16 variants, from the top: KS, R, K-10, Ye, K-10-26, P, PM

Tu-16KS From outset AV-MF played prominent role in programme. In 48 it funded devt of KS-1 *Komet* cruise missile for stand-off attacks on ships. Tu-16KS, Aug 54, carried one KS-1 under each wing on pylon 7,75 m (25 ft 5 in) from centreline, plus *Kobalt-N* guidance transmitter from Tu-4K. Over 100 built 54-57 plus 25 of export version for AURI 41/42 Sqns (Indonesia) mid-61. AV-MF all survivors converted to later versions. ASCC 'Badger-B'.

Tu-16R, Type 92 Recon version, 135 produced

from 55 for AV-MF in distinct day and night photo versions. Later some conversions, eqpt standards varying. All included bomb-bay pallets for 5-9 AFA-series cameras, from 61 adding other sensors including EW comint/ Elint.'Badger-E'.

Tu-16Z Evaluated 55, tanker version using looped hose from right wingtip of tanker to left of receiver, as devd with Tu-4. Additional transfer fuel carried, but bombing capability retained. 'Badger-A'.

Tu-16T AV-MF torpedo bomber, prototype early 56, four 533 mm AN torpedoes, and/or bombs or sea mines. All built new, later converted to Tu-16S. 'Badger-A'.

Tu-16K-10 AV-MF missile carrier with K-10S cruise missile recessed into weapon bay with multipin guidance link and fuel connection. K-10S complex included A-329Z guidance system replacing nav station by YeN search/tracking radar with 2,0 m (80 in) antenna scanning through limited forward arc to provide offset to missile autopilot out to 250 km; guidance radar below cockpit. Bomb bay housed recessed missile, its top-up fuel tank and pressure cabin for YeN operator. Prototype 58, 200+ built from mid-59, but Fleet trials of K-10S not complete until Oct 61 when production of missile began. Most rebuilt as EW platforms. 'Badger-C'.

Tu-16Ye First AV-MF electronic-warfare version, 61; usually rebuilt Tu-16KS deleting bombing capability but retaining radar and adding row of three steerable receiver antennas under bomb bay occupied by multi-waveband (A thru I) analysis/classification and threat libraries and bulk chaff cutter/dispenser. 'Badger-D'.

Tu-16P Second AV-MF EW version, Tu-16KS converted from 61, 12 receiver/jammer antennas on nose, on small pods under centre fuselage and (usually) larger pods on underwing pylons. Tail turret replaced by long tailcone for tail-on reception/DF/jamming and other tasks. About 60 served to 90. 'Badger-F', 'Badger-K' or 'Badger-L' some with tailcone and all with thimble nose.

Tu-16K-11-16 From 59 A and KS refitted with *Rubin-1* radar, and late same year all unconverted Z and KS rebuilt for AV-MF with KR-11-16 *Kompleks* to interface with two new types of cruise missile carried on underwing pylons: KSR-11 (K-11 system) or KSR-2 (K-16). Theoretically retained free-fall bomb capability. Simplified K-16 used by Egypt in 73 war. AV-MF aircraft later rebuilt for further roles. ASCC 'Badger-G'.

Tu-16K-10-26 AV-MF missile carrier, 62, rebuilt K-10 to launch centreline K-10S and two KSR-5 (option of KSR-2); replaced by K-26. 'Badger-C Mod'.

Tu-16N Air-refuelling tanker conversions of DA and AV-MF, 63, mainly to serve Tu-22; bomb bay occupied by hose-reel for probe/drogue method, usual fuel capacity 54 000 lit (11,880 gal), transfer rate 1000 lit/min. Over 70 still airworthy with CIS Navy. 'Badger-A'.

Tu-16Sh Rebuilds fitted with radar, navigation and bombing complex of Tu-22 for use as crew trainers.

Tu-16PM Various further rebuilds of P Elint platforms with self-protection active jammer antenna in nose 'thimble' and additional receiver antennas mainly for VHF; all have tail turret replaced by ECM tailcone. 'Badger-L'.

Tu-16S AV-MF SAR (search and rescue) version, rebuilds from 65 of all remaining Tu-16T with *Fregat* radio-controlled rescue boat para-dropped from under fuselage, guided to survivors by *Rech'* transmitter. Still in service. 'Badger-A'.

Tu-16K-26 AV-MF missile carrier, 65, able to launch KSR-5P/K-26 from either right wing or both wings. Large steerable antenna for guidance radar ahead of bomb bay. ASCC 'Badger-G Mod'.

Tu-16PP A or KS rebuilt as stand-off jammer; tandem Elint receivers in front of and behind bomb bay rebuilt to dispense up to 8 t (17,640 lb) of chaff cut to length matching receiving wavelengths. 'Badger-H'.

Tu-16RM AV-MF EW platform, second rebuilds of K-10 from 69 primarily as active jammers with interlinked groups of palletized receivers/transmitters mainly in bomb bay but including wingtip plate antennas usually covering Bands A or B to J, most also having *Chaika-M* rail, 'Badger-J'.

M-16 Remotely piloted targets.

Tu-16KRM Basically RM transferred to PVO and rebuilt to launch and guide rocket target drones; small number remained through 1980s.

Tu-16 Tsyklon Weather-recon conversions, c77 onwards, for research, measurement and cloud-seeding by various chemical dispensers.

Tu-104G Demilitarized prototype/devt aircraft used as civil crew trainers and transports.

Tu-103 Transoceanic Tu-16 with four engines, not built.

Testbeds As early as 54 one aircraft fitted with air-refuelling hose and drogue extended from tube at extreme tail for expts with fighters. From 58 c12 Tu-16 used for various LL (flying laboratory) duties including engines in pods extended under bomb bay (eg, AL-7F-1, VD-7 and -7F and D-36); one LL flew with zero-track 'bicycle' landing gear.

In Sept 57 China was granted licence to manufacture Tu-16A. In 59 two pattern aircraft delivered, one assembled Harbin and flown 27 Sept 59. In 61 Chinese decided to build as H-6 (Hongzhaji 6, bomber 6) at what became XAC, Xian A/c Mfg Co. Engine produced as WP-8. First H-6A flew 24 Dec 68, delivered to PLA air force and navy by 87, plus four Iraq and spare parts to Egypt. Current version **H-6D** (**B-6D** in Westernized notation). Most have large Chinese radar and computer providing targeting information for two C-601 anti-ship cruise missiles. Unlike 6A, 6D has no fixed nose gun. Many other variants studied, including tanker.

DIMENSIONS Span (88) 35,5 m (109 ft 11 in), (prodn excl tip antennas on some) 33,4 m or (most) 32 989 (108 ft 2¾ in); length (basic) 34,8 m (114 ft 2 in), (mod) up to 37 902 (124 ft 4 in); wing area 164,65 m^2 (1,772 ft^2), (H-6D) 167,55 m^2.

ENGINES (88) Two AM-3, (16) RD-3M, (from 57) RD-3M-500.

WEIGHTS Empty (88 #1) 37,2 t, (16A) 36,6 t (80,688 lb), (max EW versions) 38 t, (H-6D) 38,53 t (84,944 lb); fuel (16A) 34,36 t (equiv to 41 400 lit T-1 or 43 750 lit TS-1), (H-6D) 33 t; normal loaded (88 #1) 80 t, (prodn, all) 72 t (158,730 lb); (max) 75,8 t (167,108 lb); landing (paved) 50 t, (unpaved) 48 t, (H-6D) 55 t.

PERFORMANCE Max IAS (exc 88 #1) 700 km/h; max at 6 km (88 #1) 945 km/h, (prodn) 1050 km/h (652 mph), (10 km) 1010 km/h (628 mph, Mach 0 95); max cruise H-6D with C-601 786 km/h (488 mph); service ceiling (88 #1) 11 km, (A) 15 km (49,200 ft), (KS) 12,8 t, (H-6D) 12 km (39,370 ft); range with 3 t bombload 5760

Tu-16K-26

Tu-16LL (NK-6)

M-16

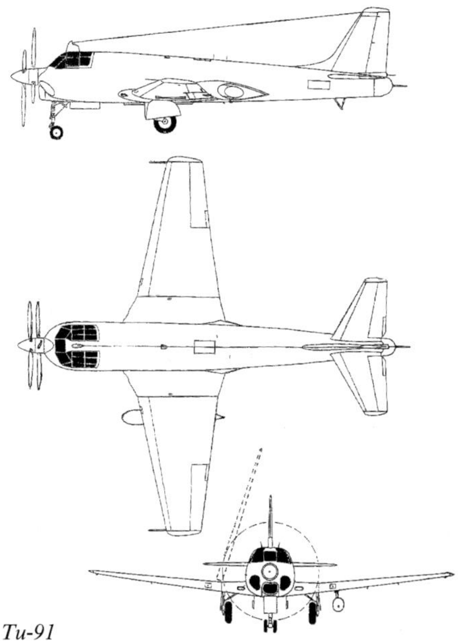

Tu-91

km (3,580 miles), with 9 t 4400 km; radius (KS, H-6D with missiles) 1800 km; TO 1250 m; ldg 223 km/h, 1100 m.

Tu-91 To equip force of giant carriers planned from 47, AV-MF carrier-based torpedo, bomber, minelayer, attack, recon, even fighter. Launched as projects **507** and **509** (year 50, #7 and #9) to fly Tu-14 missions with better fuel economy, esp at low level. Sukhoi recalled from Tu-14 line at Irkutsk as director but later left to re-form own OKB. Studies by B. M. Kondorsky and A. A. Yudin. Twin-jet 509 inadequate endurance; 507 single turboprop met demands (most powerful single-prop aircraft in history). Design went ahead under V. A. Chizhyevskii with V. I. Bogdanov (airframe), A. M. Shumov (propulsion) and M. G. Pynyegin (equipment). Several forms, one with NK-12 and swept wing.

Conventional D16-T structure with low wing, three spars, sharply tapered centre section housing tanks and main gears, outer panels with fixed LE, tracked slotted flaps, manual tabbed ailerons. Horizontal tailplane 10,2 m^2 on fuselage, manual elevators/rudder. Twin nosewheels (tyres 570 mm × 140 mm) retrac to rear, single mainwheels (HP tyres 1050 mm × 300 mm) retrac inwards, track 3370 mm (133 in), all wheels on trailing arms, disc brakes. Cockpit in nose, pilot on left and observer on right, 8-18 mm ANBA-1 light-alloy armour, bulletproof glazing except hinged overhead panels. Ejection seats could be fired simultaneously. Three nose inlets to ducts between/beside crew to plenum chamber feeding turboprop mounted on wing spars driving through shaft between seats to nose gearbox and six-blade reverse-pitch contraprop used as brake in dive attacks. Jetpipe bifurcated to exit both sides, gas passing below tailplane. Fuel in four centre-section cells plus two between jetpipes. Doors beside and above engine, accessories and gas-turbine APU/starter above engine, oil cooler in each wing root with inlet on LE and louvred exit above wing. Two NR-23 with 100 rounds in root of each outer wing. DK-15 tail turret with two NS-23, each 150 rounds, aimed by observer periscope. No internal weapon bay but provision for TAN-53, RAT-52 or 45-36 torpedo under centreline; alternatively FAB-1500 or 12 FAB-100 or three mines or rockets (8 TRS-212, 36 TRS-132, 120 TRS-85) on four wing pylons (two only fitted, beneath upper-surf fences) inboard of power fold. Sting hook at rear, camera in rear fuselage with ventral doors, LAS-3M dinghy in dorsal box, retrac landing lights under outer wings, mast above cockpit for HF wire, VHF antenna above rear fuselage. Order for one static-test and one flight article. Dubbed *Golavl'* (chub).

Propulsion system tested on Tu-4LL. Tu-91 flown 17 May 1955 by D. V. Zyuzin and observer K. I. Malkhasyan. Highly manoeuvrable, though barrel roll forbidden. AV-MF lost requirement with cancellation of carrier, and Tu-91 redesigned with non-folding wing, cat hooks replaced by a.t.o. rockets, carrier eqpt removed, LP tyres and VVS radio. Single pre-production a/c built with defects eliminated,

Tu-91 No 2

lighter structure, better cockpit view. NII testing by Alekseyev/Sizov favourable, and production authorized Jan 55 including versions for ASW, ECM jamming and dual training. Scornful (ignorant) opposition of N. S. Khrushchyev resulted in termination of programme. Failure of Il-40 was seen as confirming obsolescence of such aircraft. Shown to Western delegation at Kubinka June 56. ASCC name 'Boot'.

DIMENSIONS Span 16,4 m (53 ft 9⅝ in); length 17,7 m (58 ft 0⅞ in); wing area 47,48 m² (511 ft²).

ENGINE One TV-2M.

WEIGHTS Empty c8 t (17,640 lb); loaded normal 12,85 t (28,329 lb), max 14,4 t (31,746 lb).

PERFORMANCE Max speed (clean) 800 km/h (497 mph); econ cruise 250 km/h-300 km/h (155 mph-186 mph); service ceiling 11 km (36,000 ft); range 2350 km (1,460 miles); TO (MTO wt, rockets) 518 m, ldg 652 m (with reverse 438 m).

92 OKB *Izdelye* number of Tu-16R.

93 Variant of Tu-14 with 'new Klimov engines', probably VK-5.

94 Turboprop variants of Tu-4 (52-53) with TV-2, VK-2 or NK-4 engines. None built.

Tu-95, Tu-142 Termination of Tu-85 was based on vulnerability of B-29 in Korea and premise that gas-turbine engines would in acceptable timescale make possible intercontinental bomber meeting Kremlin 24 March 51 objective of 5 t bombload at 900 km/h for 12 000 km. CAHI helped define wing swept 35° with aspect ratio 7. Studies with various engine combinations finally led to two: six AM-3 turbojets (superimposed pairs in wing roots plus two under rear fuselage) or turboprops totalling c40,000 hp. Jet 'looked exotic but could not meet reqd range.' Turboprop rested on two Kuznetsov engines, TV-2 in 6,000-hp class and TV-12 with potential twice as great.

Type 95 authorized 11 July 51 under N. Bazyenkov. Project in two forms, one with four TV-12, other with eight TV-2, both c150 t. Study of long push/pull nacelles and other arrangements of eight engines led Jan 51 to devt of double 2TV-2F, built as one unit with left power section driving front 4-blade unit of 5,6-m (220 in) AV-60 contraprop via shaft passing thru partner power section. Propulsion posed high risk, whereas airframe was extrapolation of OKB practice. Wing SR-5S profile, 12·5% at root, three spars root to tip plus fourth ending at structural break 90° to rear spar outboard of outer engines; further joint 90° to rear spar between flap/aileron. Fixed LE swept 37° to first break, thence 35°. Flaps inboard/outboard of landing gear of electric tracked slotted type, aileron in three sections, inner section with tab, with hyd boost drive. TE straight, static anhedral −1°. Three 200-mm fences from behind LE to TE, one between prop wakes, one in flap/aileron gap and third at outer/middle aileron joint. Duplex hyd ballscrew flap drives; single hyd spoiler ahead of inner aileron as all-speed airbrake and roll assister.

Fuselage almost same 2,9 m-diam circular section as Tu-85. Two dorsal, two ventral and tail turrets, each with two NR-23, pressurized forward compartment for six crew entering via

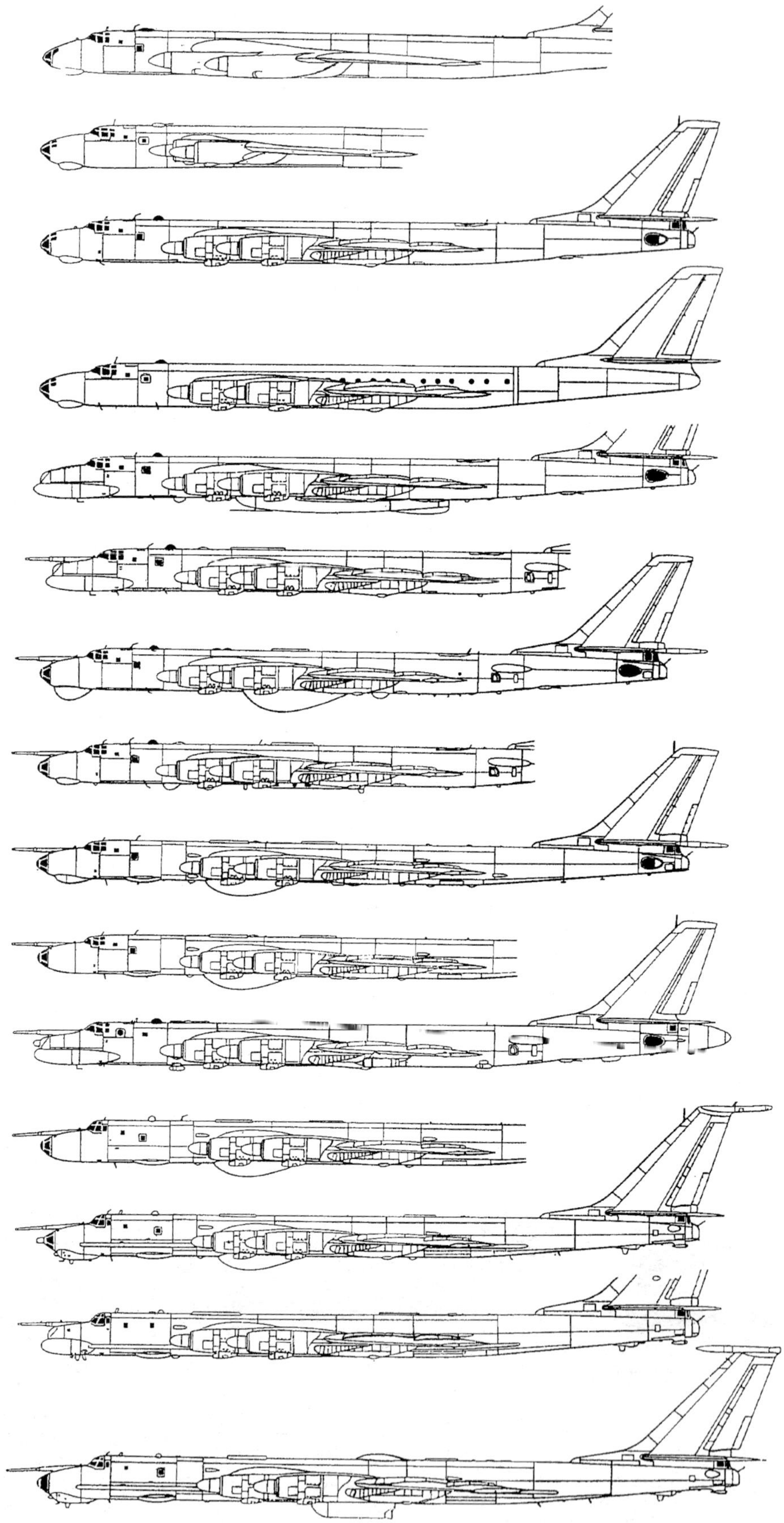

Tu-95 variants, from the top: 95/I, 96, 95 1st series, 116, 95K-20, 95KD, 95RTs, 95MR, 142, 142A, 95K-22, 142M, 142MZ, 95MS-6, 142MR

nose-gear bay ladder and tail section (two) with linking tunnel. No ejection seats, emergency escape for main crew assisted by powered belt 'conveyor' to open nose-gear bay, floor hatches for tail gunners. Sweeping wing put spars ahead of large bomb bay centred on CG able to take TN bomb or FAB-9000, max bombload 12 t (26,455 lb). Swept vertical tail scaled up from Tu-88, hyd boosted tabbed elevators and rudder. Landing gears again similar to Tu-88 but stronger, main units 4 tyres 1500 mm×400 mm, hyd internal expanding brakes, hyd retrac to rear, bogie somersaulting to lie inverted in rear of inboard nacelles. Tall steerable nose gear inclined forwards, twin tyres 1180 mm× 350 mm, hyd retraction back into bay with left/right doors. Tailskid bay with twin doors, further back compartment for twin brake chutes.

Prototype **95/I** increasingly became research vehicle to provide data for 95/II, as well as insurance against failure of TV-12. Eight TV-2 power sections in four twinned installations, wide cowls with chin inlets to feed left and right engine power sections driving shafts to intermediate gearbox and separate coaxial propellers. Successful 50-min flight 12 Nov 52 by A. D. Perelyot and test crew, but on 17th flight No 3 engine inter-gearbox failed and fire resulted in loss of aircraft and Perelyot, navigator Kirichyenko and NISO engineer Bolshakov, six crew escaping.

Airframe 95/2 completed June 54, but painstaking devt of TV-12 engine, unique reduction gear and prop by K. Zhdanov delayed first flight to 16 Feb 55; crew headed by M. Nyukhtikov and I. Sukhomlin. Engines on long steel trusses to put AV-60N propellers in same posn as in 95/1, far ahead of wing, inners set at +3°. Key feature extremely coarse pitch in cruise, giving high thrust at 750 rpm to Mach 0·835 but requiring new solutions in absorbing powerful low-freq vibration into airframe. Annular inlet behind spinner, jetpipes both sides of nacelle, gas passing under wing. Heat exchangers in jetpipes for wing-deice air. Tail deiced by TA-12 APU in dorsal fin used for independent engine start, dorsal fin also housing two IAS-5M liferafts. Total 72 980 lit (16,055 gal) in wing box tanks. Engine rpm constant at 8,300 but power variable by fuel flow to 12,000 hp. From this point project never looked back, production organised at GAZ-18 Kuibyshyev.

DIMENSIONS Span 49,8 m (163 ft 4⅝ in); length 44,35 m (145 ft 6 in); track 9,4 m; wing area 280 m² (3,014 ft²).
ENGINES Four 2TV-2F.
WEIGHTS Loaded 140 t (308,640 lb).
PERFORMANCE Not fully recorded.

Tu-95 First two series Tu-95 flew Oct 55, NII trials with these plus 95/II completed 20 Dec. Tu-95 cert operational with DA Aug 57 though speed/ceiling did not satisfy. Many used for trials or rebuilt as Tu-95U trainers. USAF designation 'Type 40', name 'Bear'. Data for 95/2.
DIMENSIONS Span 50,0 m (164 ft 0 in); length 47,3 m (155 ft 2 in); wing area c284 m² (3,057 ft²).
ENGINES Four NK-12.
WEIGHTS Empty 77 480 kg (170,811 lb); max (95/2 t) 155 t (341,710 lb), (series) 172 t (379,189 lb).
PERFORMANCE Max speed (95/2) 920 km/h (572 mph) at 7 km; service ceiling 11,3 km (37,075 ft); range with 5 t bombload 13 460 km (8,364 miles).

Tu-95M TV-12, by 55 called NK-12, devd into NK-12M, fitted to 2nd series a/c onwards, giving performance satisfying customer. AV-60N props with braking capability, drag chute(s) deleted. Tu OKB had already designed four extra tanks (filling wing box), total 95,000 lit (20,900 gal), and together with new engine this was main feature of Tu-95M (modernized). Forward dorsal and forward ventral turrets removed, rear dorsal turret made retractable with optical sight in dome at rear of main cabin, aft PV-23 fire control as Tu-16 with side blisters and PRS-1 *Argon* radar at base of rudder, total six AM-23 guns. Weapon bay with two doors able to carry 12 t bombload including FAB-9000; provision for further 6 t external (normal max load 12 t). Flight controls improved and tailplane hinged and driven +3°/–1° by elec ballscrew actuator for trim. Tyre pressures increased, retractable twin tailwheels. Basic avionics: SPU-10 intercom, RSIU-3 and -4, R-803/807/808 com radio, RBP-4 *Rubin* main nav/bombing radar plus nose optical sight, integrated nav system with autopilot linked to doppler, ADF, VOR/DME (plus later RSBN Tacan) and SP-50 ILS. High and low-range radar altimeters, SOD/ATC transponder and IFF. *Chaika* Loran rail in some aircraft. Chaff dispenser in unpressurized rear fuselage, various groups of ASO chaff/flare dispensers in main-gear fairings. Normal crew 8: two pilots, nav (astro dome), engineer, nav-operator (managing radar/optical bombsight), radio/gunner (forward cabin) and two gunners (tail).

Loss of aircraft March 57 led to NK-12MV engine with rapid autofeather backed by positive manual feather capability. NII testing completed Oct 57, replaced Tu-95, total for both 150 by end-59 all by GAZ-18 for ADD. Survivors at repair centres over many years for rework and role conversion, some transferred to AV-MF. ASCC 'Bear-A'.
DIMENSIONS Span 50,03 m (164 ft 1⅝ in); length 46,17 m (151 ft 5¾ in); wing area 283,7 m² (3,054 ft²).
ENGINES Four NK-12MV.
WEIGHTS Empty 79,6 t (175,485 lb); max 182 t (401,234 lb).
PERFORMANCE Max speed 910 km/h (565 mph) at 7 km; service ceiling 12,5 km (41,000 ft); range with 5 t bombload 14 960 km (9,296 miles).

Tu-95K One aircraft modified 55 to carry Mikoyan SM-20 recessed into belly and launch it. This piloted aircraft tested guidance for Kh-20 cruise missile, and carrier aircraft had instrumentation and com systems to monitor its flight after release; designation **Tu-95SM-20** changed to **Tu-95K**. Flown from 1 Jan 56.

Tu-96 Discussion mid-52 regarding tradeoff of reduced range in order to get still higher alt and speed resulted in NK-16 engine, leading to Tu-96 designed to carry 5 t bombload 10 000 km at 850 km/h cruise at heights to 17 km (55,775 ft). Slightly larger dimensions, new centre wing, forward fuselage, cowls and props. GAZ test 56 but taken no further because SAM defences made hi-alt capability pointless.

Tu-116, Tu-114 Described separately.

Tu-119 One Tu-95M modified from 57 as testbed for reactor for nuclear propulsion system, flown briefly 'early 60s'.

Tu-95K-20 From 56 altitude was replaced by supersonic cruise missiles as best way to deliver, hence variant based on Tu-95K with K-20 *kompleks* to carry and launch Kh-20 AS-3 'Kangaroo' cruise missile recessed into belly with nose air inlet sealed against curved fairing which rotated into fuselage on weapon drop. Glazed nose replaced by A-336Z guidance installation with giant 'duckbill' 'Crown Drum' I-band radar scanning over small forward arc to illuminate target. Many innovations, not cert operational until autumn 59, c50 built, later rebuilt for other roles. ASCC 'Bear-B'.

Tu-95N One aircraft modified as carrier for Tysbin NM-1/RSR programme 59.

Tu-95M

Tu-95RTs with SPS tailcone

Tu-95KD Tu-95K missile and guidance radar degraded range and led May 60 to probe/drogue FR tests, resulting in retrofit to aircraft in service. Resulting Tu-95KD featured large fixed probe at top of nose on centreline with external pipe on right of pressure cabin to No 1 tank behind, standard on most subsequent versions together with crew rest area for missions up to 36 h. Produced alongside K 61-65. 'Bear-B' or '-C'.

Tu-95MR AV-MF strategic-recon version, small batch in 61. FR probe, Elint blister both sides of rear fuselage and usually camera both sides also. Long HF strake antenna above fuselage, other flush or blade antennas depending on Elint fit. Dorsal turret often removed because of rear-fuselage electronics. Some with different aft fire-control radar, and a few with turret replaced by extended tailcone housing further Elint and defensive ECM. Ventral turret and lateral sighting blisters retained. Post-75 mods include chaff-dispensing pod under each outer wing. 'Bear-E'.

Tu-95RTs AV-MF version with bomb/missile capability deleted. Primary mission recon and target indication for submarines and surface warships armed with cruise missiles. Two large radars, one under nose (c14·5 GHz, four PRF modes, for missile guidance) and huge radar in former weapon bay (c9 GHz) for targeting ships on or over horizon. Twin dorsal antennas for link with Soviet com and surveillance satellites. Upgraded Elint and recon installation with receiver blister and camera on both sides of rear fuselage and 'farm' of up to seven blade antennas under rear fuselage. Streamlined antenna pod on each tailplane tip, HF strake antenna above fuselage. Dorsal turret often deleted. Intensive testing took two years before cleared for oceanic service 64. Operated from Cuba, Libya, Angola, Mozambique, Guinea, Ethiopia and Vietnam; 50 on strength in 80s excluded from SALT/START treaties. 'Bear-D'.

Tu-126 Described separately.

Tu-95U From 65 several Tu-95M gutted and refurbished with weapon bay sealed and interior furnished for various forms of crew training; red band around rear fuselage. 'Bear-A' or unofficially 'Bear-T'.

Tu-95KM Modernized K-20/KD aircraft, reissued from 67 with IFR probe, Elint antennas, upgraded radio and nav systems for precision flying including Polar region. 'Bear-C'.

Tu-142 New AV-MF family, planned from 63 as best available platform to meet threat of Polaris/Poseidon submarines. Funded as new design (different *Izdelye* number) though max use of Tu-95 experience. Wing redesigned with improved aerodynamics, double-slotted flaps and long-life structure, increasing lift at all speeds, reducing cruise drag and increasing fuel capacity. Forward fuselage extended by 1,78 m (70 in) plug ahead of rear pressure dome giving better working and rest facilities for main and relief crew. Rudder chord increased progressively towards top. To reduce footprint pressure main gears redesigned with six twin-tyred wheels requiring wider and longer fairings, nose gear tyres enlarged requiring bulged doors. Tailwheels omitted. Fuselage rearranged with glazed nose compartment, J-band nav radar with slim radome under nose and large overwater search radar (c21 GHz) ahead of weapon bay with several blisters and ram-air cooling inlets upstream near leading edge. Weapon bay configured for ASW torpedoes and nuclear/conventional depth charges. Rear-fuselage Elint/camera installation and dorsal and ventral turrets all deleted and replaced by sonobuoy bay with twin doors and new electronics with row of ventral blisters. Ram-air inlet on left of rear fuselage. Twin dorsal HF strakes, satcom antennas and tailplane tip fairings as RT. Prototype flew July 68, certn for service 72. 'Bear-F'.

Tu-142A After producing c15, production switched to simpler standard with no nav radar, single cooling inlet each side for main radar and, during production, return to four-wheel main gears (higher tyre pressure) with smaller fairings. 'Bear-F Mod I'.

Tu-95M-5 To meet threat of B-52 with cruise missiles, late-70 decision to integrate *kompleks* 95K-26 based on **Tu-16K-26**, including wing

Tu-95K-22

pylons for two KSR-5 rockets. OKB **Tu-95K-26**, changed to **Tu-95M-5**, but decision in 71 not to produce in quantity, leaving this missile to Tu-16 and 22M.

Tu-95K-22 To replace M-5, further major rebuild of 95K to carry weapon system already used by Tu-16 and 22/22M. Tu-95KM nose retained, but radar with improved performance and ECCM qualities in radome 3,5-m wide, operating in various modes for nav and, especially, with targeting of Kh-22 missile, one carried internally, or one on pylon under each wing root. ECM thimble on nose, left/right pairs of small pylon-mounted pod antennas under forward and rear fuselage, Elint fairing and camera each side of rear fuselage, dorsal turret deleted and tail turret replaced by SPS fairing housing high-power tail-on ECM. External elec/fuel pipe fairing past pressure cabin. Ventral turret and sighting blisters retained. Prototype flown Oct 75, most K-20/KD/KM rebuilt late 70s. Over 45 with Irkutsk Air Army in 93. 'Bear-G'.

Tu-142M OKB (now led by Dr. A. A. Tupolev) began late 74 to meet threat of low-noise submarines. Upgraded avionics and sensors (especially sonar system), with precise triple inertial navigation and automated com systems. Forward fuselage redesigned to improve crew efficiency still further. Nose to Frame 3 revised with FR probe angled down 4°; between Frames 3-6 cockpit roof 0,36 m (14 in) higher, deeper windows, new cockpit displays; aft of Frame 6, structurally altered with rearranged interior, available length increased 0,23 m, comsat antennas replaced by steerable antenna in dome. Modified MA engines, small main gears, square access doors along underside of wing LE as far as structural break beyond outer engines. Tu-142M prototype flown by I. Vedernikov and test crew 4 Nov 75; major task, not operational until 80. 'Bear-F Mod 2'.

Tu-142M2/Tu-142ML Interim ASW platform with *Ladoga* MAD sensors projecting aft from tip of taller fin, which also carries swept VHF antenna. Aft (sonobuoy) bay doubled in length, but doors slightly narrower, and ram air inlet above this bay removed. Five simplified exports serving Indian Navy 312 Sqn, Dabolim. 'Bear-F Mod 3'.

Tu-142MZ In 76 Bazyenkov died and leadership passed to N. Kirsanov. For next two years further major revision of sensors, with tailplane tip MAD system replaced by *Ladoga* MAD spike projecting aft from fin tip. ECM thimble radome on nose, large sensor group including FLIR, com antennas and extra radar altimeter in chin compartment linked by external cooling-air and multi-cable duct on left side of fuselage to generating and air-conditioning group in rear fuselage (ram inlet each side) and ESM passive receiver under tail with twin aft-facing passive antennas in small pylon-mounted pods under tailcone. Rear fuselage sighting blisters deleted. In 93 c59 to this standard. 'Bear-F Mod 4'.

Tu-95MS-16 Final new production for CIS Air Armies, long-range missile carrier first flown at Kuibyshyev Sept 79. Based on airframe/engines of Navy Tu-142M, but with original shorter forward fuselage (though new cockpit adds the 0,23 m) with diam exactly 2,9 m (114 in), new main radar with deeper antenna scanning 20° left/right, reinforced horizontal FR probe and shorter fin. Same nose thimble, left-side duct fairing and some ECM/ESM as Tu-142M, two HF dorsal strakes, satcom dome on rear fuselage. Extremely comprehensive defensive avionics, jammers and main-gear chaff/flare dispensers, various antennas under nose and tail, SO-type multi-facet IR warning receivers under nose and above rear fuselage and aft-facing spiral antenna in fin tip. Ventral turret deleted and from 88 tail turret fitted with two GSh-23. Weapon bay modified to contain rotary launcher for six RKV-500A missiles and four wing pylons: under each wing root for two RKV-500A and between engines for three (total 16). To comply with SALT/START all mod to MS-6. 'Bear-H16'.

Tu-95MS-6 Final version, as MS-16 but wing pylons deleted. Production complete after 38 years early 92. About 110 with Air Armies. 'Bear-H6'.

Tu-142MR Perhaps final rebuild version, to provide secure com link between CIS national command and naval units, esp missile submarines submerged on station around world. Airframe generally as 142M, with large fuselage areas occupied by powerful VLF relay station transmitting via suspended wire antennas, main wire c8 km long unreeled from external box under centre fuselage. Dorsal antennas: HF rails, satcom dome and large Glonass navsat blister. Unglazed nose with pimple radome, FR probe, chin sensors, cable duct, rear fuselage and vertical tail unchanged (as late M versions) but extended fin carries forward-facing HF antenna, terminated at rear by circular spiral passive receiver antenna as on 95MS. About five each with Northern and Pacific Fleets. 'Bear-J'.

Tu-95LL At least nine aircraft modified for special test purposes; many projects to launch missiles or piloted aircraft were never tested.

Tu-95I Improved version with NK-20 engines in 20,000 shp class, not built.

DIMENSIONS Span (142) 50,0 m (164 ft 0 in), (95MS) 50,04 m (164 ft 2 in); length overall (142) 48,17 m (158 ft 0½ in), (95MS) 49,13 m (161 ft 2¼ in), (K-22) 51,5 m (168 ft 11½ in), (probed 142MZ) 53,9 m (176 ft 10 in); length of 142 fuselage 45,0 m, with probe 46,9 m; wing area 289,9 m^2 (3,121 ft^2).

ENGINES Four NK-12MV, (95MS) -12MA.

WEIGHTS Empty (142) 82,1 t (181,000 lb), (95MS) 84 t (185,185 lb); fuel (K) 76 t, (142, 95MS) 77 t; MTO (K) 184 t (405,644 lb), (142) 182 t (401,234 lb), (142M, 95MS) 185 t (407,850 lb), (after air refuel) 190 t (418,871 lb); MLW (all) 135 t (297,619 lb).

PERFORMANCE Max speed (K) 880 km/h (547 mph), (142, 142M) 850 km/h (528 mph), (95MS) 845 km/h (525 mph); service ceiling (M) 12,4 km (40,700 ft), (142) 11,2 km (36,745 ft), (142M) 10,7 km (35,100 ft), (95MS) 10,5 km (34,450 ft); practical range with max load (K) 13 000 km (8,080 miles), (142, 142M) 12 000 km (7,460 miles), (95MS) 10 900 km (6,775 miles); TO 300 km/h, 2450 m; landing 270 km/h, 1400 m.

Tu-98 Challenge posed by 53 (post-Stalin) reqt

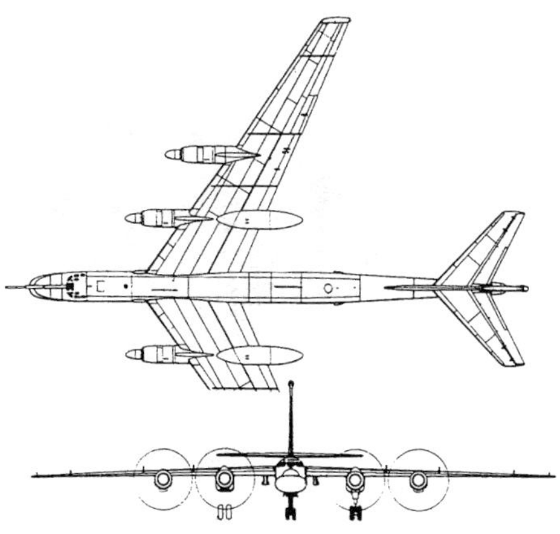

Tu-95MS

Tu-142M

Tu-95MS-6

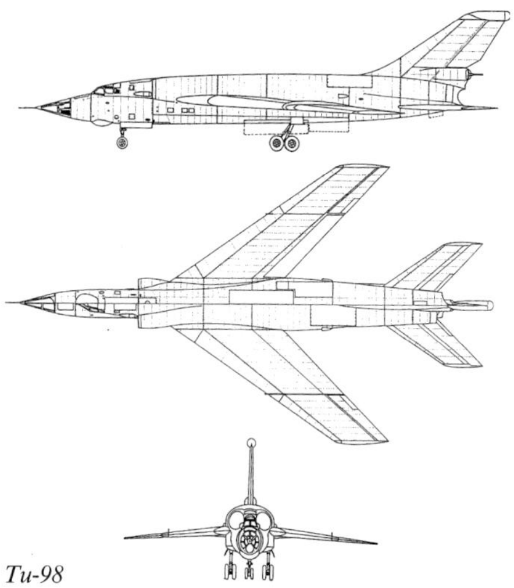

Tu-98

for bomber with supersonic speed in target area. First in OKB to need supersonic aerodynamics (wing thinner than Tu-16, all leading edges 58° and much more highly stressed) and afterburning engines. Mid wing with thick machined skins, shallow full-chord fence between tabbed aileron and track-mounted slotted flap, anhedral 5°. Small pointed glazed nose for nav/bombardier, pilot on centreline with V-windscreen and hinged glazed roof, radio/gunner behind with small window each side, all on forward-facing KT-1 seats with roof hatches; unpressurized under floor for *Initsiativa* radar, fixed AM-23 gun and nose gear. Rear pressure bulkhead flat and vertical, camera bay behind with No 1 tank above between ducts and No 3 to rear. Engines close together in area-ruled rear fuselage fed by ducts curved inwards over wing from high oval inlets with small part-cone blisters to focus oblique shockwaves on sharp lips. Steerable twin-wheel nose gear (660 mm × 200 mm) retracting backwards, four-wheel bogies (900 mm × 275 mm) on vertical main legs, with front and rear dampers to control truck beam attitude, retracting to rear to lie inverted in bay with two doors. Track only 2,5 m (98 in) but wheelbase (10 901 m (35 ft 9 in), max ground angle 11°30'. Low slab horizontal tails with tabbed elevators, span 7,7 m (25 ft 3 in), fin with tabbed rudder and tip fairing for *Argon* gun-direction radar. Five tanks filling fuselage ahead of and behind bomb bay and between engine ducts. Short (3,68-m) bomb bay behind wing box, max load 4 × FAB-1500 or two nuclear. Twin remotely sighted AM-23 guns in tail. Twin braking parachutes in box under rear fuselage. Stressed for a.t.o. rockets.

Single prototype flown early 56 by V. Kovalyov, who completed OKB testing same year. Terminated by Khrushchyev prior to start of NII testing. Shown to US delegation Kubinka June 56; misidentified as 'Il-140' or 'Yak-42'. ASCC name 'Backfin'.

DIMENSIONS Span 17 274 m (56 ft 8 in); length 32 055 m (105 ft 2 in); height 8063 m (26 ft 5½ in); wing area 87,5 m² (942 ft²).
ENGINES Two AL-7F.
WEIGHTS Max 49,5 t (109,127 lb).
PERFORMANCE Max speed 1365 km/h (848 mph, M 1·29); service ceiling 12,75 km (41,830 ft); range (max bombload) 2440 km (1,516 miles).

99 High-alt 4-jet strategic recon-bomber project, late 55.
101 Twin-turboprop military transport project, 54-54.
102 Pressurized passenger version of 101, neither built.

Tu-104 No large modern passenger aircraft has been produced as cheaply or effortlessly as this re-fuselaged derivative of Tu-88 (Tu-16) bomber. Tupolev showed drawing to Stalin and gained latter's approval a few weeks before his death in Feb 53. Headed S. M. Yeger. Main design task was new fuselage and uprated environmental system handling more than three times pressurized volume. Minor changes design of new wing centre section between engines, absent in Tu-88, longer nose gear and bag-type wing tankage. Resulting aircraft had no appeal to Western operators, despite initial selling price £425,000, because by any normal DOC formula relatively uneconomic. But in Soviet context Tu-104 wrought transformation of all long-distance routes with much greater carrying capacity than existing eqpt (Il-12 or Il-14 usually with 18/21 seats) and 2½ times speed. Typical example, flight time Moscow/Irkutsk fell on introduction of Tu-104 from 13 h 50 min to 5 h 30 min. Restored OKB to position of pre-eminence in GVF passenger aircraft sustained for next 25 years.

Wing closely similar to Tu-88 in early variants, except for new wide centre section with three heavy plate spars with KhGSA booms tying banjos encircling engines. Latter in almost straight ducts with circular inlet well outboard of fuselage, unlike Tu-88, and requiring no boundary-layer bleed. Rest of wing almost unchanged. Fuselage diameter increased from 2,9 m to 3,4 m, pressurized at original 0,5 km/cm² level from nose to bulkhead at LE of tailplane, with normally closed pressure bulkhead (flat) at Frame 11 between flight deck and 142,3 m³ main cabin. Flight deck similar to Tu-

Tu-104

Tu-98

88 though military equipment deleted, different nose glazing, smaller mapping radar in chin location and with provision for flight crew three (two pilots, nav) to five (add radio, eng). Main passenger cabin 16,11 m long, 3,2 m wide and 1,95 m high. Divided into small forward section with four circular windows each side and large rear section by inconvenient raised floor over wing centre section. In prototype and early aircraft raised area used for galley, with three roof windows left of centreline and two right-side windows higher than remainder. Main rear cabin with seven windows each side; toilets each side at rear and another forward on right; total of seven windows removable inwards as emergency exits. Except in prototype, six small underfloor baggage/cargo holds with plug doors on ventral centreline. Rest of aircraft generally as Tu-88, though tailplane lowered to upper part of fuselage and elevators separated from roots by fixed inboard sections. Nose gear moved much further forward and lengthened to reach higher fuselage.

Prototype L-5400 flown by Yu. T. Alasheyev 17 June 55. Production Tu-104 established as 50-seater. Two small compartments ahead of spars with 6 and 8 seats and tables, then 8-seat cabin aft of spars and rear cabin for 28 in forward-facing pairs each side of aisle. Heavy brass-finish baggage racks along sides with numerous hooked-on coat hangers, emergency oxygen at knee level along walls, Plexiglas windows in non-structural dividing bulkhead partitions, tables mahogany and antimacassars lace. Fuel 33 150 lit (7,292 gal), derated engines with elec starting, operating pressurization and anti-icing. Avionics included British VOR and ILS. Crew training under A. K. Starikov started with Tu-104G (demilitarized Tu-16) summer 55 and continued with NII trials of L-5400 Oct 55. Aircraft passed all test programmes winter 1955-56 and L-5400 flew to London Heathrow 22 March 56. Line service begun to Irkutsk 15 Sept 56, followed by Prague 12 Oct. Subsequently Tu-104 variants used by every Aeroflot directorate and leading Soviet passenger aircraft until mid-60s.

Robust engine enabled TBO to rise slowly from 300 to 2000 h, and to be fully rated for civil purposes as AM-3M, in turn enabling OKB to increase gross weight. Resulting **Tu-104A**, flown winter 1956-57, increased seating to 70: forward cabin with four rows 2+2 and

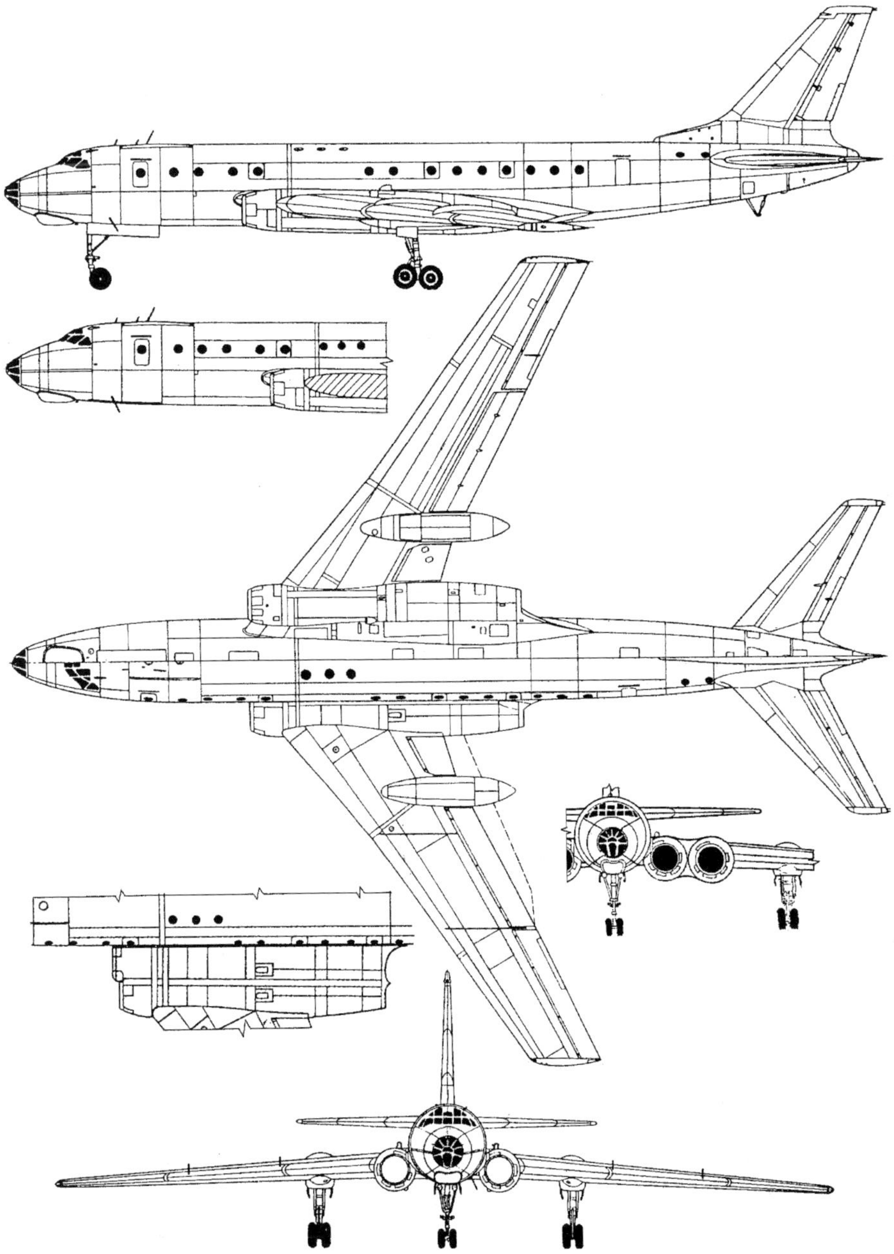

Tu-104 (broken line, 104B); insets, nose of 104B and plan/front views of Tu-110

rear cabin with aft-facing 2+2 and ten rows forward-facing 3+2, generally modernized furnishings, small changes elsewhere. Class records: 6 Sept 57, 11 221 m with 20 t load; 11 Sept, 847,498 km/h over 2000-km circuit with 2-t load; 24 Sept, 970,821 km/h over 1000-km circuit with 10-t load. Replaced Tu-104 on production line at Kharkov at about 11th aircraft and large number built.

By 58 first Tu-104 (L-42399) introduced 1210 mm (3 ft 11½ in) stretch, dP increased to 0,57 kg/cm^2, AM-3M-500 engine, three windows above wing on each side (none in roof) at slightly higher level, revised underfloor holds with greater capacity (lowered hold floors and raised passenger floor) and more sensible doors on right side, increased flap chord to maintain field length despite greater weight, and normal interior for 100 passengers, seated 3+2, in cabins seating 30 (front), 15 (over wing, galley being aft of flight deck) and 55 (rear cabin). Entered service as **Tu-104B** on Moscow/Leningrad 15 April 59, production being completed 60 at about 200 of all variants. Tu-104B records: 1 Aug 59, 1015,86 km/h over 2000-km circuit with 15 t load; 4 Aug, 12 799 m with 25 t.

Tu-104V designation applied to Tu-104A modified internally to carry 100 passengers (cabin for 25, 15 and 60) by rearranging toilets and wardrobes. **Tu-104D** in-service mod raising accommodation to 85 only, later also increased to 100, but with 3M-500 engines. **Tu-104Sh** Tu-22 crew trainers. Most B refurnished in GVF shops for 104 or 115 pax from 67. Only foreign customer CSA, five plus a replacement supplied from Nov 57, basically 104A with 81 seats. Small number used by VVS for staff transport and cosmonaut training in weightless flight. In 64 first line aircraft, 5412 (re-registered 42318) intensively flown as structural fleet leader, in 67 transferred to tank testing. By 81 this type withdrawn from line duty after circa 90 million passengers carried nearly 20 bkm with reliability and safety better than any other Soviet aircraft. One or more used by Gidrometsovcentr as met platform with pointed nose radome and wing pylons for cloud-seeding rockets. **Tu-104G**, see Tu-16. Military variant, see Tu-107. ASCC name 'Camel'.

DIMENSIONS Span 34,54 m (113 ft 3¾ in); length (104, A, V) 38,85 m (127 ft 5½ in), (B, D) 40,05 m (131 ft 4¾ in); wing area (104, A, V) 174,4 m^2 (1,877 ft^2), (B, D) 183,5 m^2 (1,975 ft^2).

ENGINES (104) two AM-3, (A, V) AM-3M, (B) first few as A, most AM-3M-500, (D) AM-3M-500.

WEIGHTS Empty (5400) 39,5 t, (104) 41,6 t, (B, V, D) 44,02 t (97,046 lb); fuel (standard) 26,5 t; max TO (5400) 72,5 t, (104, V) 75,5 t, (B, D) 78,1 t (172,178 lb).

PERFORMANCE Max speed (all) 950 km/h (590 mph) at 10 km; cruise M 0·72, 770 km/h (478 mph); service ceiling (typical) 11,5 km (37,730 ft); range with max payload against 50 km/h wind with 1 h reserve (104) 2650 km (1,647 miles) with 5,2 t (11,464 lb), (A) same with 9 t (19,841 lb), (B) 3600 km (2,237 miles) with 12 t (26,455 lb); TO (typical) 2,2 km (7,220 ft), ldg (61 t) 1,85 km (6,070 ft)/240 km/h, or 1,3 km (4,265 ft) with parachute.

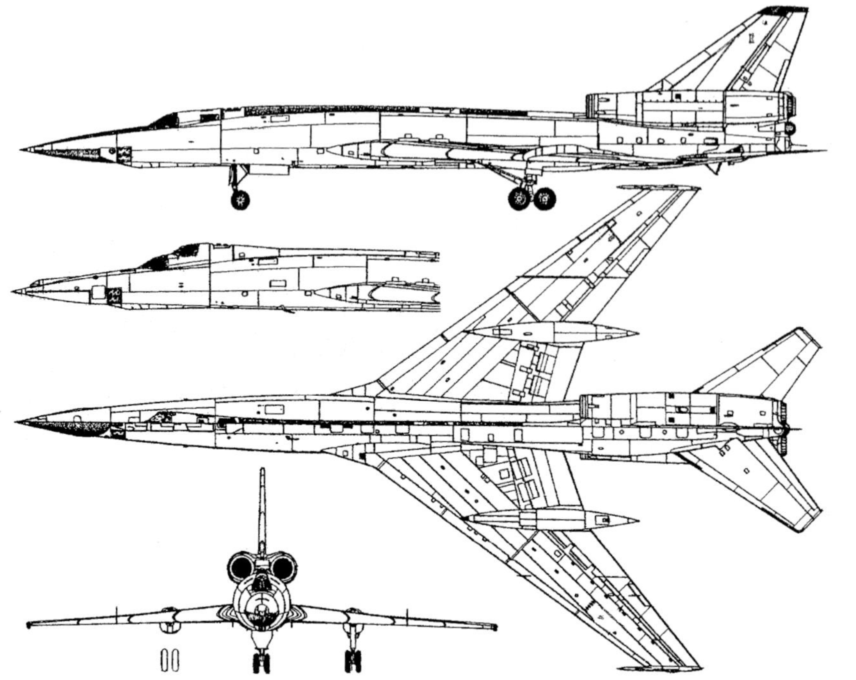

Tu-22 (inset, 22U)

Tu-22, Tu-105, Yu Major programme launched Dec 55 for theatre-type bomber and missile carrier to fly Tu-16 missions but at supersonic speed over target. Required to operate from existing or planned VVS and AV-MF bases. Accepted that, compared with Tu-16, radius would be less and costs greater. Basic design aerodynamically and structurally assisted by Tu-98. OKB code **Aircraft Yu**, no **105**. Type 105 initially schemed with Type 88 fuselage and 45° wing carrying four VD-5 or VD-7 engines. Redesigned from Aug 55.

Wing modified SR-5s profile, thickness 9% at root, 7% at tip, LE sweep 70° at root, quickly changing to 52° to tip. First Tu design with machined skins enclosing integral tankage bounded by spars at 10% and 60% chord. Fixed LE, tracked slotted flaps driven hydraulically to 50° inboard of main gears and to 30° outboard, fully powered inner/outer ailerons moving as one unit. Single fence at flap/aileron junction. No spoilers. Tips extended forwards carrying anti-flutter mass and pitot head. Circular-section fuselage of area-ruled profile tapering from cockpit to tail. Wing remarkably low, making weapon bays bounded by spar bulkheads. Pointed nose occupied by radar, then pressurized crew capsule with navigator behind radar with two superimposed windows each side giving undistorted view down and to side. Pilot at upper level on centreline with radio/gunner behind. Pilot seat ejects upward, others downward through ventral entrance hatches in gondola with bombsight window at front, faired at rear into aft-retracting steerable nose gear. Seats serve as lifts to raise crew from ground-level. Type 105 had main gears retrac inwards into thickened wing roots with TE fillets. To improve area-ruling this arrangement was changed to gear folding aft into pods, the TE being swept back with no fillet.

Main gears similar to Tu-134 but stronger, bogie beam inclined so that front wheels hit runway first; beam rotates horizontal, shock strut compresses and whole unit then swings aft restrained by long diagonal strut at front. This strut incorporates jack which pushes gear backwards, linkage rotating beam so that bogie is stowed inverted in faired box with twin doors. Rear fuselage protected by retractable steel-shod skid, behind which is compartment for twin braking parachutes anchored to tailplane frame. Tailplanes for pitch control only, single slabs with LE 59°, tips cut off at angle appropriate to Mach 1·5, spars at 10 and 70% chord, steel pivot trunnion at 58% in bearings at bottom of fuselage, dihedral 7°. Sharply tapered vertical tail incorporating powered tabbed rudder with inset hinges. Distinctive mounting of engines above rear fuselage attached to fin-spar frames. Thus most of fuselage aft of pressure cabin available for fuel. Variable vee inlet rejected in favour of two plain inlets, with front section able to translate forward to admit extra air all round on takeoff. Bomb bay for 3 t normal, 9 t max; tail guns two AM-23. Pipe fairing from cockpit to fin past tanks 51,830 lit, 42,5 t. Prototypes at OKB, first flown by Yu. T. Alasheyev 21 June 58.

DIMENSIONS Span 23,745 m; length 41,921 m; wing area 166,6 m^2.

ENGINES Two VD-7M-1.

WEIGHTS Empty 37,9 t; MTO c80 t.

PERFORMANCE Max speed at ht 1450 m; service ceiling 13,6 km; range 5800 km.

Tu-22B At this time intention was to replace with T-tail and either two VD-7M or two NK-6. This continued to May 60 as **106**, span 23,89 m, length 40,2 m, MTO 99 t, two NK-6 for 2000 km/h, but cancelled. Instead **105A** modified with VD-7M and accepted for series manufacture at Kazan. Mod bomb bay max 12 t, production tail turret with single AM-23 (600 rounds) directed by RPS-3 *Argon* fire-control radar and TP-1A TV sight. *Rubin*-2 or -3 main radar optimised for anti-ship attack, bombsight, RV-25 radar-alt, DISS A-322Z doppler, twin ARK-11 ADF, NI-50BM nav computer, DAK-DB-5 astrocompass, HF/VHF, SP-50M ILS, SRO-2 or 2M warning receiver and Elint systems, including antennas in front of tip pods (pitots on fuselage). Provision for overload takeoff with four solid a.t.o. rockets. From 64 FR probe above nose, external portion detachable (later plus deflector guard below), all plumbing internal. Weapon bay with two double-hinged doors for one FAB-9000 or: two nuclear bombs, 3 × FAB-3000, 8 × FAB-1500, 24 × FAB 500, 42 × FAB-250 or -100, or 8 or 18 mines. First 105A (second Tu-22 prototype) flown by Alasheyev, I. Gavrilyenko and K. Shchyerbakov 7 Sept 59. Series aircraft **Tu-22B**, about 12 built, VVS-DA acceptance Sept 62. Nice to fly but popularity marred by excessive pilot workload and poor view through sharp wedge windscreen. ASCC name 'Bullshot', then 'Beauty', then 'Blinder-A'.

Tu-22K Main production (believed 250) missile carrier. First flight early 61. Later *Rubin* radar with wider antenna for targeting and initial guidance of Kh-22 cruise missile recessed by removing parts of bay doors which could be replaced to enable original loads to be carried instead. Provision to carry three Kh-22 not used. Aft bay of main-gear pods housing chaff/flare dispensers and strike camera. FR probe compatible with Tu-16N. In service with VMF late 64, mod to **22KD** with RD-7M-2 engines, 25 in action Afghanistan; 17 repainted and sold to Libya (one shot down by SAM over Chad), others to Iraq. About 70 with upgraded avionics in Smolyensk and Irkutsk air armies until 92. ASCC 'Blinder-B'.

DIMENSIONS Span 23,646 m (77 ft 6⅞ in); length (excl probe) 41,6 m (136 ft 5¾ in); wing area 162,25 m² (1,746·5 ft²).

ENGINES Two VD-7M.

WEIGHTS Empty 38,3 t (84,436 lb); max 84,0 t (185,185 lb), (a.t.o.) 94 t (207,231 lb); ldg 60 t (132,275 lb).

PERFORMANCE Max speed at 12 km 1510 km/h (938 mph, M 1·42); service ceiling 14,7 km (48,230 ft); range 5850 km (3,635 miles); TO 2200 m; ldg 310 km/h, 2400 m.

Tu-22P Dedicated missile director (*Postanovshchik*); few built, all re-engined as **PD**.

Tu-22R Dedicated reconnaissance for AV-MF. About 60 with weapon bay reconfigured for up to seven AFA cameras and, from 79, mod to **Tu-22RDK** with *Kub*-4 SLAR or IR linescan, on 5,55-m (18 ft 2 in) pallet with aft section magazine ejecting photoflash cartridges; all with eqpt for visual recon and most for Comint/Elint. 'Blinder-C'.

DIMENSIONS Wing area only 162 m².

ENGINES Two RD-7M-2.

WEIGHTS Empty 40,9 t; max 92 t (202,822 lb).

PERFORMANCE Max speed (SL) 890 km/h (550 mph), (12 km) 1610 km/h (1,000 mph, M 1·52); service ceiling 13,3 km (43,635 ft); combat radius (400-km full-power dash, hi-lo-hi) 2200 km (1,367 miles); max range (normal fuel) 4900 km (3,050 miles); ferry range (bomb-bay fuel) 5650 km (3,511 miles); TO 2250 m; landing speed/run 310 km/h, 2170 m (with chute 1650 m).

Tu-22U Pilot-conversion trainer with second (instructor) cockpit above and behind pupil in place of usual rear cockpit. Unchanged fuel.

Tu-22K

Tu-22U

Radar replaced by simpler mapping/weather radar, tail turret replaced by fairing to preserve CG posn. 'Blinder-D'.

Tu-22RM, RDM Former 22K aircraft converted for Elint (AV/MF) and ECM jamming (CIS AF) duties. Former has REB-K complex with 'farm' of blade antennas (14 wavelengths from metric to cl cm) and twin 6,2-m (20 ft 4 in) canoes with four (one each end) Chipthorn 'hockeysticks'. Latter has large Elint pod scabbed under belly and receiver/jammer antennas each side of cockpit, nose-gear door, wing roots, tip pods, landing-gear pods, engine nacelles and rear fuselage, and 2,4-m SPS receiver/jammer fairing replacing tail turret. Over 20 still airworthy with each service late 92. 'Blinder-E'.

About nine Tu-22 converted as testbeds for wide range of items from supersonic propulsion to flight-control/nav systems. In 67, when 'Tu-22M' design complete, 22 was mod via 106 into **145**, which see.

Tu-107 Military variants of Tu-104, distinguished by DK-7 tail turret (twin AM-23), side cargo door and special avionics and furnishing (at least one VIP). Data as 104B.

Tu-110 OKB assigned brigade under Markov to prepare four-engined version of Tu-104. Major redesign with extended centre section incorporating staggered engines at same level, though smaller ducts and different installn. Rear-spar KhGSA banjo forgings round engines. Two aircraft, first (#5600) with original fuselage and wing flown by D. V. Zyuzin 11 March 57 (red star on fin). Second aircraft with stretched body and extended-chord wing flown April 59. ASCC name 'Cooker'.

DIMENSIONS Span 37,5 m (123 ft 0⅜ in); length (first) 38,3 m (125 ft 7⅞ in); wing area 182 m² (1,959 ft²).

ENGINES Four AL-7.

WEIGHTS Empty 44,25 t (97,553 lb); max 79,3 t (174,824 lb).

PERFORMANCE Max speed 1000 km/h (620 mph); cruise 890 km/h (553 mph); service ceiling 12 km (39,370 ft); range 3450 km (2,144 miles) TO 1,6 km; ldg 1,2 m.

111 Project 55 for 24-passenger twin-turboprop.

Tu-114 Like Tu-104 this transport derived from Tu-95 was logical development, main problem being sheer size and weight of aircraft for which GVF runways in mid-50s totally unprepared, and inability of routes to fill 220 seats. Like Tu-95 and Tu-116 assigned to Arkhangyelskii, to fulfil GVF 53 requirement.

Turning existing Tu-95 into passenger transport almost same task as done by Tupolev himself with Tu-88/104. Bomber already had a wing centre section but despite moving wing from mid to low position this had to be extended, slightly increasing span, though structure outboard of fuselage virtually unchanged, except major increase in chord of secondary upper-surface and underlying flap at TE. Same thing done by OKB several times previously, most recently with Tu-104B. Flaps often called Fowler but differ in having only small overlying portion of fixed upper surface and resemble track-mounted plain flaps. Operation hydraulic, major break with all B-29 successors. Horizontal tail lowered to fuselage, trimming by variable-incidence tailplane, all flight controls hyd boosted. Main landing gear as Tu-95, nose gear set new record height, twin left/right inclined 30KhGSA struts meeting at steerable lower twin-wheel unit, single rear strut pulling unit up rearwards into unpressurized box. Fuselage circular section 4,2-m diam, pressurized 0,59 kg/cm^2 (8·4 lb/in^2) back to bulkhead aft of tailplane LE, with floor just below horiz diam leaving underfloor height adequate for kitchen and walk-in service compartments as well as cargo holds front 24 m^3 and rear 46 m^3; fresh air for cabin air-cycle machines via ram inlet under fuselage. Wing LE deicing by jetpipe heat exchangers, tailplane/props/windscreens raw AC, fin rubber boot. Heavy protective fuselage skin in plane of props. Engine installations as Tu-95, fuel very similar, capacity 71618 lit (15,754 gal).

Prototype L5611 *Rossiya* first flown by A. P. Yakimov 15 Nov 57; few major problems, displayed Paris 59, GVF NII testing completed July 60 but line service did not begin until 24 April 61 (Moscow/Khabarovsk, 6800 km in 8 h 15 min schedule). Wide range of interior layouts for upper deck, though all aircraft with same window and door arrangement. Normal crew nav (nose), two pilots, eng, rad, two in underfloor galley and three or more cabin crew. High-density 220-seat interior first in world with seats 4+4; more common arrangements were 170 seats (main cabins for 42, 48 and 54) or 145, or four small compartments just behind wing with two divans each or six seats and folding bunk. Most configurations provided middle cabin, and restaurant with tables linked by elec lift to underfloor kitchen. While on GVF test, I. M. Sukhomlin and crew established 32 records including: 24 March 60, 1000 km circuit with 25 t payload, 871,38 km/h; 1 April, 2000 km circuit with 25 t, 857,277 km/h; 9 April, 5000 km circuit with 25 ft 877,212 km/h; 21 April, 10 000 km circuit (Moscow/Sverdlovsk/Sevastopol/Moscow) with 10 t, 737,352 km/h; 12 July 61, height 12 073 m with 30 t payload. Another notable flight non-stop Moscow/New York with N. Khrushchyev (then prime minister) 15 Sept 59. Special ultra-long-range version (designated **Tu-114D**) for route to Havana, 80,000+ lit fuel, two crews and seating reduced to 120, service from Jan 63 with refuel at Murmansk outbound, return 10 900 km non-stop. Other routes Delhi, Montreal, Tokyo (Japan Air Lines titles) and Conakry/Accra. Kuibyshyev/Samara built 31, last delivered 65; in 15 years carried 6+ million pax, replaced mainly by Il-62 in 71-75. Much used for evaluation of new systems. Largest/heaviest airliner prior to 747. ASCC name 'Cleat'.

DIMENSIONS Span 51,4 m (168 ft 7⅝ in); length 54,1 m (177 ft 6 in); wing area 311,1 m^2 (3,348 ft^2).

ENGINES (L-5611) four NK-12M, (series) NK-12MV.

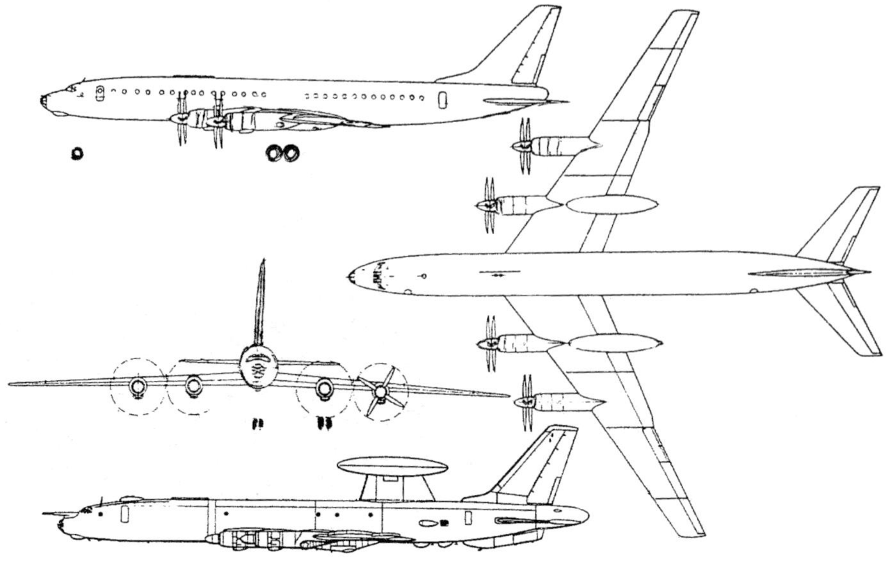

Tu-114 (lower side view Tu-126)

Tu-114 prototype

WEIGHTS Empty c91 t (200,600 lb); fuel/oil 57,7 t; payload 22,5 t; MTO 179 t (394,621 lb); MLW 128 t.

PERFORMANCE Max speed 870 km/h (541 mph) at 8 km; cruise 770 km/h (478 mph); service ceiling 12 km (39,400 ft); range (15 t load, 1 h reserve) 9720 km (6,040 miles); TO 1700 m; ldg 205 km/h, 1550 m.

115 Military transport version of Tu-114, provision for bulky loads, tail turret.

Tu-116 As parallel project to support Tu-114 passenger aircraft OKB assigned Arkhangyelskii to build direct civil variant of Tu-95 (Tu-20). This was relatively simple, and Tu-116 first flew 23 April 57, well before prototype 114. Few changes compared with VVS aircraft. Forward crew compartment virtually unaltered though smaller mapping-only radar same as Tu-104) and duplicated VOR/ILS. Weapon bay skinned over. Rear fuselage furnished for payload, 24/30 passengers in 2+2 seats, galley, two toilets, or various cargo. Rear pressurized compartment and turret replaced by tailcone, twin braking parachutes; wing, flaps and tail as bomber. Access tunnel linking front and main cabins, separate ventral door with stairway from ground to rear cabin with emergency side door. Purpose of Tu-116 to prove propulsion system in civil operation, prove routes and airfields (latter had virtually no means of handling aircraft in 57) and if possible set records and gain publicity. First aircraft completed NII GVF testing from early 58. Two examples only constructed, as military 7801 and 7802. In 58 the second was re-registered as civil No 76462 and assigned to Aeroflot. First used for crew training for Tu-114, first major flight to be revealed Moscow/Irkutsk/Moscow April 58, but biggest publicity task was series of three zig-zag tours of the Soviet Union later in 58 with 24 officials and journalists covering 34 400 km (landings at Vladivostok, Tashkent, Minsk for night stops) in 48 h 30 min.

DIMENSIONS As Tu-95M.

ENGINES Originally four NK-12M; later NK-12MV.

WEIGHTS Empty 79 t; lo aded 124,1 t, MTO 143,6 t.

PERFORMANCE Max cruising speed 870 km/h (541 mph) at 6300 m; range 11 190 km (6,953 miles).

117 Military transport version of Tu-110, special cargo provisions.

121 Prototype strategic unmanned bomber.

123 *Yastreb* (Hawk) cruise missile programme.

Tu-124 Tu-104 scaled down with new engines to meet 57 spec for short-hauler to replace Il-14. Wing with same LE kink, 35° LE outboard but redesigned movables and TE inboard of landing gears at 90°, outer TE straight. Hyd double-slotted flaps, manual spring-tabbed ailerons and spoilers ahead of outer flaps driven asymmetrically to 52° in air to assist roll and flicked open on touchdown as lift-dumpers; large hydraulic airbrake under fuselage driven to 40° and normally open on approach. Twin upper-surface fences each side, fixed LE, about 3° anhedral on ground, retracted flaps increased wing chord but much less than on Tu-104. Manual elevators and rudder, again with spring tabs, and tailplane with elec irreversible trim. Wing anticing by bleed air as in Tu-104, but fin and tailplane electrothermal, using raw AC in glassfibre elements beneath skin. Fuselage circular section, same 2,9 m diam as B-29 and jet bombers, pressurized to dP 0,57 km/cm² by system energized by engine bleed, all doors plug-type. Short landing gears with hyd retraction to rear, main tyres 865 mm × 280 mm (6,5 kg/cm²), bogie stowed inverted, nose tyres 660 mm × 200 mm (same), anti-skid multi-disc brakes, optional mudguards on nose gear, emergency braking parachute. Normal fuel capacity 13 500 lit (2,970 gal) in integral centre-section and outer-wing tanks and 4 flex cells in inboard wings, pressure-fuelling under each wing. Engines in straight-through nacelles with top level with wing upper surface and jetpipe inclined downward. Each engine carries DC starter/generator; emergency ram-air turbine in each wing for immediate purpose of driving tank booster pumps. Minimum flight crew two pilots and nav in glazed nose; optional seat for radio operator or second nav. Usual GVF mapping radar and all expected avionics, retrac landing lights each side of nose, towing point on nose leg after steering disconnected.

Prototype L-45000 flown by A. D. Kalina 24 March 60. Entered service Moscow/Tallinn 2 Oct 62 as 44-seater, with all seating 2+2, forward cabin 12, small raised mid-cabin over wing eight seats and rear cabin 24; insufficient underfloor depth for holds, cargo and baggage accommodated on main floor on right side behind flight deck and at extreme rear. Furnishing more modern than Tu-104 fleet with fluorescent lights and synthetic fabrics in cream, pale green or grey. Toilet and coat space aft and galley forward on left. Production replaced Tu-104 at Kharkov and main run was of **Tu-124V** with minor changes and 56-seat interior arranged 12/12/32. Surprising that no other airline (except three for CSA and two for Interflug) bought this cheap ($1·45 m in 65) and reliable aircraft, with published DOC 22c/short-ton mile. One ditched in Neva (Leningrad) without passenger injury and re-entered service. About 100 built, incl 20+ **Tu-124K** (*Konvertibelnyi*, convertible) cargo/pax versions, some with VIP interior, **K2** exports to Interflug, LSK and Indian and Iraqi AF. **Tu-124Sh** navigator trainer for Tu-128 radar operators. ASCC reporting name 'Cookpot'.

DIMENSIONS Span 25,55 m (83 ft 9⅞ in); length 30,58 m (100 ft 4 in); wing area 119,37 m² (1,285 ft²).

ENGINES Two D-20P.

Tu-116 (side elevation, page 424)

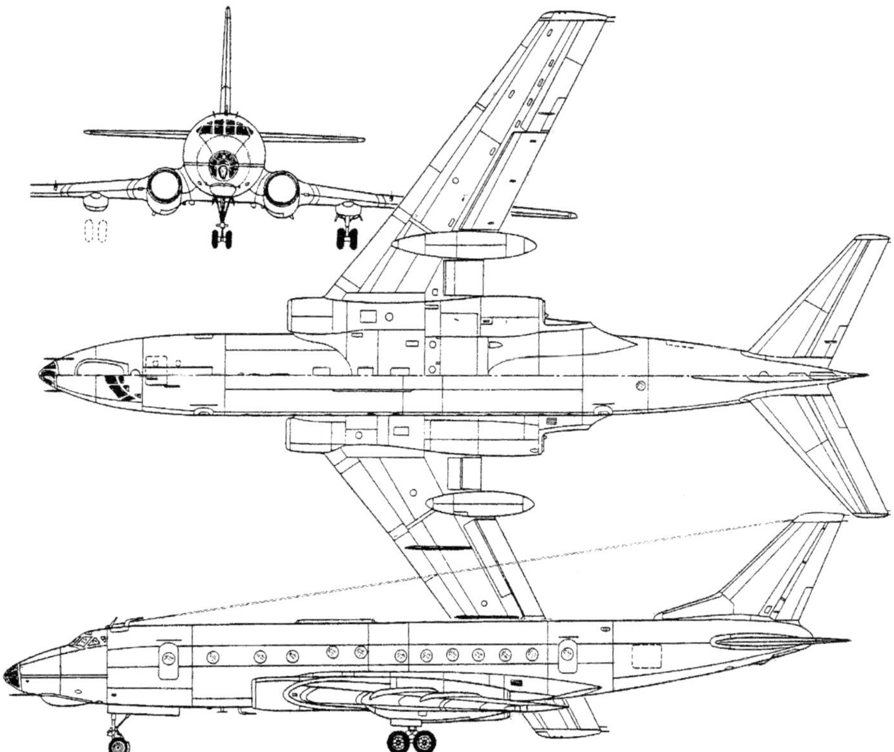
Tu-124

Tu-124

Tu-126 prototype

WEIGHTS Empty 22,9 t (50,486 lb); fuel 10,8 t; payload 6 t; loaded 36,5 t, MTO 37,5 t (82,673 lb).
PERFORMANCE Max speed 970 km/h (603 mph), cruise 800 km/h (497 mph); max climb 12,0 m/s (2,360 ft/min); service ceiling 11,5 km (37,730 ft); range (max payload) 1500 km (932 miles), (max fuel) 2200 km (1,367 miles); TO 1200 m; ldg 190 km/h, 900 m.

Tu-126 Airborne early-warning platform based on modified airframe of Tu-114, after Tu-107 found inadequate. Work began May 60 on a/c and on *Liana* pulse-doppler radar by what became NP Vega. Beam antenna 10,5 m (34 ft 5 in) wide, electronic scanning in elevation, faired into deep rotodome carried on single broad pylon above rear fuselage. No armament but purpose-designed nav and EW complexes, latter including SPS tailcone and rear ventral gondola. Two crews each of 12, one off-duty. Aerody prototype Dec 61, eight series a/c from Kuibyshyev 65-67. Despite operational shortcomings, aircraft popular because of spacious interior and good facilities for air-refuelled missions up to 20 h. One detached to operate with Indian AF in 71 war. All nine served A-PVO to 84, when replaced by A-50. ASCC 'Moss'.
DIMENSIONS Span 51,4 m (168 ft 7⅝ in); length with probe 56,5 m (185 ft 4⅜ in); wing area 311·1 m^2 (3,348 ft^2).
ENGINES Four NK-12MV.
WEIGHT Empty 103 t (227,000 lb); fuel 60,8 t; loaded 171 t (376,984 lb).
PERFORMANCE Max speed 790 km/h (491 mph); operating regime 520 km/h at 9 km (323 mph at 29,500 ft); service ceiling 10,7 km; endurance (1 refuelling) 20 h; normal range (internal fuel) 7000 km (4,350 miles).

127 Military derivative of Tu-124, 59 project.

Tu-128, 28-80 Growing demands on IA-PVO rapidly made Yak-25 inadequate to defend Soviet airspace, and by 58 it was clear La-250 would also fail. Main threats Mach 2 Vigilante from carriers in Arctic, Snark and Hound Dog missiles and possibly B-70. Assumed target 2000 km/h at 21 km; interceptor also had to perform 2-hour escort function. Task assigned Tupolev OKB; similar in size/shape/weight/power to Tu-98.

A. N. Tupolev personally managed refined aerodynamics of wing and inlets; project brigade led by S. M. Yeger, chief designer **Tu-128** I. F. Nyezval. OKB shorthand **Type 28**, first prototype being **Tu-28-80** from -80 complex and K-80 missile. No requirement for internal carriage of weapons so wing moved to low/mid position. Main wing box bounded by heavy spars at 5% and 40% chord with machined skins forming integral tankage. Structurally made as centre section integral with fuselage, with TE at 90°, and fully swept outer panels. Fixed LE with hot-air deicing, constant sweep (56°22'24". Large-chord tracked slotted flap inboard and outboard of main gear, outboard aileron, spoiler ahead of outer flap. Horiz tails pivoted low/mid on fuselage, with elevators increasing camber and acting as back-up, LE 58°. Fin LE 60° extended fwd to large ram inlet, inset tabbed rudder, tip cropped at Mach angle to increase flutter-critical speed, all tail LE deiced by 5th-stage bleed. All surfaces fully powered with manual reversion; transfer from manual to power only on ailerons and rudder, which have trimmers. Spoilers prevent excursion beyond crit AOA and not as lift dumpers after landing. Area-ruled fuselage wide but shallow, rounded underside but broad flat top, becoming like letter B (flatter bottom, twin-bulge top) at engine bays. Close engines fed by long ducts from D-inlets standing well away from forward fuselage, this time with translating central half-cone. Eight fuel tanks in centre fuselage, with wing tanks total 14 850 kg (32,738 lb, 18,200 lit, 4,005 gal). No FR probe or drop tanks. Steerable nose gear, HP tyres 660 mm×200 mm, hyd retraction back into bay with 4 doors. Main gears based on Tu-98, four 900 mm×275 mm tyres, anti-skid brakes, hyd retrac backwards whilst bogie somersaults to lie inverted in large fairing. Track 6300 m, wheelbase 10,65 m. Braking parachute 50 m^2 (538 ft^2) in thermally insulated container under rear fuselage. Axis of forward fuselage inclined 3° down, large radome a challenge to V. Volkov KB who produced RP-5 radar as part of *Smerch* fire-control system. Pressurized cockpits for pilot and nav/radar operator with KT-1 seats, red/white lighting, V-windscreen and upward-hinged metal canopies, boarding via 11-rung ladder.

Aircraft designed for VD-19 engines, delayed delivery caused urgent redesign for AL-7F. First flight 18 March 61 by Mikhail V. Kozlov and observer K. I. Malkhasyan. This aircraft

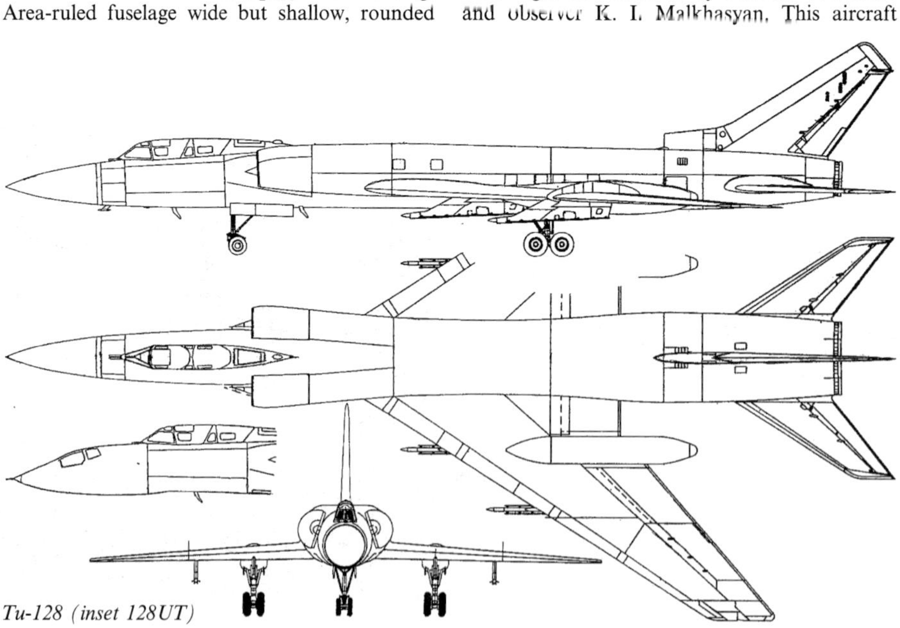
Tu-128 (inset 128UT)

Tu-128UT prototype (series UT had enlarged fin with horizontal top)

had aileron chord extending behind wing TE, two underwing pylons for missiles (prototype Biesnovat K-80R, K-80T) and giant ventral fairing for receiver antenna to test reception of target echoes, which required two inclined rear ventral fins. One of only two types in 61 Tushino Aviation Day to go into production; identified by Western analysts as Yakovlev. Prolonged difficulties with radar fire control, and once NII testing had started (20 March 62) criticisms voiced by NII pilots V. Ivanov and future Cosmonaut G. Beregovoi. Poor subsonic yaw stability, unacceptable with missiles, precluding desired armament of two R-4R and two R-4T (prodn versions of K-80) and fact that at over M 1·45 pedal pressure resulted in yaw reversal (left rudder = yaw to right). Ventral fairing and fins removed, vertical tail modified. Internal fuel standardized at 18 600 lit (4,085 gal), 14 850 kg. Fences increased in depth from under LE to mid-chord, stopping ahead of flap, but tricky handling demanded **Tu-128UT** trainer (devd in 66) with radar replaced by cockpit for instructor with hinged V-canopy flush with upper surface. Backseaters trained with Tu-124Sh. AP-7P autopilot, NVU-B1 nav complex and Put'-4P flight-control system giving semi-automatic guidance in level flight with altitude/heading hold, airfield homing, runway approach and, esp, auto return to pre-programmed position and missile lock-on to targets. Tu-128 damaged by first air firing against target (Il-28M), Sept 62.

Series production delayed to 66 as **Tu-128**, complex 128S-4 including *Smerch* radar and two R-4P and two R-4T missiles, cleared to fire at 40 km. **Tu-128Ch** added *Chaika* Loran, small porthole window between canopies and roof windows for backseater (though smaller than pilot's). Engines repeatedly upgraded to improve performance (data), **Tu-128A** introducing 7F-2 and RP-SA (*Smerch-A*) radar, aft canopy no roof window, and **Tu-128M** first flown 15 Oct 70 with definitive 128S-4M fire-control (RP-5M *Smerch*-M radar) and two R-4PM and two R-4TM missiles. NII testing complete spring 74, PVO service 79; some earlier aircraft retrofitted. **Tu-128B** bomber with VD-19 engines, 4,5 t bombload, *Initsiativa-2* radar and aft-hemisphere search, rejected in favour of Su-24. Tu-128 in PVO service 68-88. ASCC reporting name 'Fiddler'.

DIMENSIONS Span (128) 17,53 m, (M) 17,67 m (57 ft 11⅝ in); length (128) 30,06 m, (M) 30,49 m (100 ft 0⅜ in); wing area 96,94 m² (1,043·5 ft²).
ENGINES (28-80) two AL-7F, (128) AL-7F-1, (first series) AL-7F-2, (third series and retrofit) AL-7F-4, (B)VD-19.
WEIGHTS (first series) empty 25 960 kg (57,231 lb); max 43,0 t (94,797 lb).
PERFORMANCE Max speed at 11 km, first series, clean, 1915 km/h (1,190 mph), M 1·8), with missiles, 1665 km/h (1,035 mph, M 1·57), (F-4 engines) 2085 km/h (1,296 mph, M 1·96); loiter speed M 0·85; service ceiling (F-1) 15,6 km (51,180 ft); range (combat allowances) 2565 km (1,595 miles); PVO interception line 600-965 km; TO 1350 m; landing 1050 m.

130 Small unmanned winged (shuttle-type) space launcher, prototype built 61-66.

131 Long-range air-defence missile, 59 project.

Tu-134 Though Tu-124 was excellent short-haul jet its outdated layout inherited from Tu-104 resulted in high structure weight, reduced engine efficiency, impaired wing/flap geometry and high interior noise and vibration. Fashionable rear-engined T-tail configuration solved most of these problems, though doing little for structure weight, and about 62, as Tu-124 entering service, GVF agreed with OKB to fund Tu-124 derivative with this configuration. Designation **Tu-124A**, assigned to brigade under Leonid Selyakov with Arkhangyelskii (far senior in age and experience) as deputy.

Wing differs from 124 in having increased-span centre section (without engine installations) and outer-wing panels, two section ailerons with spring and trim tabs, increased-travel flaps, all sections coming down together to 20° for take-off, 38° for landing, and slightly less static anhedral (1° 30'). Fuselage same diam but 1,6 m longer and with wing box at last accommodated under level floor throughout giving constant 1,96 m headroom along aisle. First aircraft D-20P-125 engines, rest D-30 of greater power and efficiency, hung on short stubs on two frames at rear of fuselage within pressurized portion. Considerably larger vertical tail carrying variable-incidence trimming tailplane and manual geared-tab elevators. Airframe anti-icing as 124; bleed for engines, wing and fin, electric for tailplane. Three fuel tanks each wing, system capacity 16 500 lit (3,629 gal)

Tu-128A

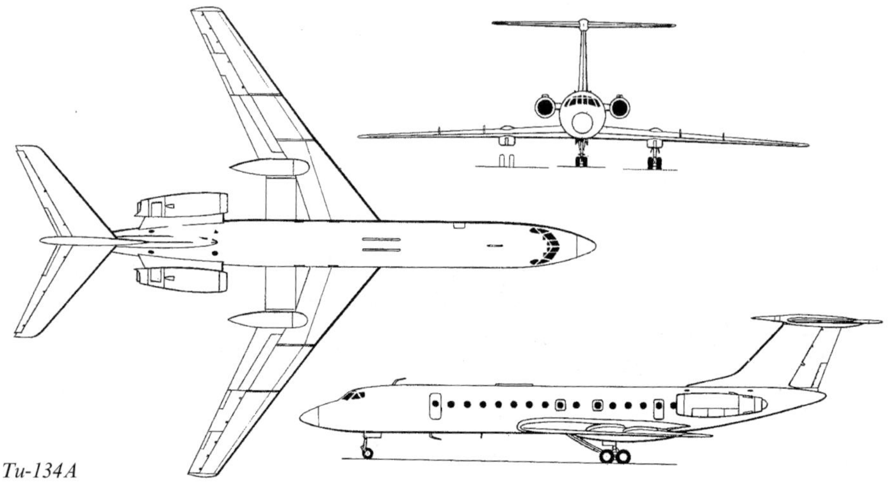

Tu-134A

Tu-134UBL prototype

Tu-134A-3

with pressure coupling at right LE. Landing gears as 124 but strengthened, and main tyres enlarged to 930 mm×305 mm and pressure reduced to 6 kg/cm² (85 lb/in²) for poor surfaces, increased-capacity brakes, no braking parachute or thrust reversers.

This outline accepted by GVF Aug 62, by which time two aircraft on 124 production line (45075, 45076) had been set aside for completion as 124A. First flown by A. D. Kalina 29 July 63. Designation changed to **Tu-134** and aircraft unveiled on 100th test flight 29 Sept 64. By this time several other 134s had flown, but like many modern Soviet civil transports development was protracted and despite use of large number of production aircraft no passenger service flown until Sept 67 (Moscow/Stockholm), after several aircraft had flown unscheduled domestic cargo services. Attractive aircraft, improvement on Tu-124 and contrary to original idea a different and more capable vehicle. Original Tu-134 flight crew two pilots plus nav, one cabin crew, 64 passengers 2+2 (16 first class in forward cabin, 20 tourist mid and 28 tourist rear). No underfloor holds, cargo/baggage being as before at extreme rear and behind flight deck, moved to left side with galley on right. Seating soon increased to 72, with 44 in front cabin and 28 in rear. In 67 main electrical power system settled as frequency-wild AC from inverters, and in 69 engines fitted with constant-speed drives (English Electric) to alternators, first in Soviet line service. Later same year reversers of twin-clamshell target type introduced, engine designated Series II. Final absentee, APU, introduced 70 for main-engine start (electric), and ground electric power and air-conditioning.

Production switched 70 to **134A**, stretched 2,1 m (83 in) to seat 76; strengthened main gears with wheels/brakes/tyres as Il-18 (but tyre pressure held at original level), Srs II engine standard, locally strengthened wing, VHF probe forming long pointed nose to tail/fin bullet, and improved avionics (Arkhangyelskii claimed Cat III). All known subsequent airline deliveries with 28-seat rear cabin; front cabin various (inc 12 first-class) up to total capacity 84, all 2+2, high-density reducing baggage space, normally 4,0 m³ or 6,0 m³ forward and 8,5 m³ aft. First Soviet transport to sell to several foreign airlines, including CSA, Interflug, Malev, Balkan Bulgarian, LOT and Aviogenex. Third Aviogenex (71) was first to do away with visual nav station in nose and replace it by *Groza* weather radar.

In 80 production switched to **134B** with advanced cockpit, direct-lift spoilers and other changes, seating 80, or 90 with galley removed (**134B-1**) or up to 96 on lightweight seats with D-III engines (**134B-3**). First Soviet transport to win major export orders (to 21 airlines), total production 853, over 360 m passengers carried.

Withdrawals from 88, 27+ converted as: **Tu-134UBL** to train Tu-160 pilots, bomber's radar increases length to 41,92; **Tu-134BSh** for radar/bombardiers, wing bomb racks, like UBL numerous cabin consoles, Tu-22M radar gives length 41,87; **Tu-134SKh** multisensor strategic survey for land use/ice patrol/pollution/disaster monitoring etc; **Tu-134LL** various testbeds including CAHI deep-stall research with giant tail chute. ASCC name 'Crusty'.

DIMENSIONS Span 29,01 m (95 ft 2 in); length 34,95 m (114 ft 8 in), (A, B) 37,32 m (122 ft 5¼ in); wing area 127,3 m² (1,370 ft²).

ENGINES Two D-30, (A) D-30-II, (B) D-30-III.

WEIGHTS Empty 26,5 t (58,422 lb), (A, B) 28,6 t (63,052 lb); fuel 10,53 t, (A, B) 14,58 t; loaded 44 t (97,000 lb), (A) 44,5 t (98,104 lb); MTO 45 t (99,200 lb), (A) 47,5 t (104,720 lb), (B) 47,6 t (104,940 lb).

PERFORMANCE Max speed 870 km/h (540 mph), (A, B) 900 km/h (559 mph); cruise (all) 800 km/h (497 mph); max climb 14,5 m/s (2,855 ft/min); service ceiling 11,5 km (37,730 ft); range (max fuel) 2000 km (1,242 miles); TO 1,2-2 km; ldg 240 km/h, 1,6 km.

135 Long-range strategic bomber project of 61, configuration as B-70, rival of Sukhoi T-4, four NK-6 engines.

139 Reconnaissance version of 123, prototype flown 68.

143 Major *Reis* (Voyage) pilotless recon aircraft produced in series.

Tu-144 SST was first in world to fly and was expected to lead to early use of at least 75 to transform travel times throughout world's largest country. Instead Tu-144 designation has

Tu-144 68001

identified succession of completely redesigned aircraft.

Original GVF requirement dated 63, based on premise Aeroflot already then saved average 24·9 h journey-time per passenger and with widespread use of SST this could rise to over 36 h. Programme cost expected to be three times as great as estimate for Il-62 (previous highest) but immediate decision to undertake largest-ever research programme at CAHI, CIAM and elsewhere to underpin slender-delta aircraft made mainly of light alloy to cruise at Mach 2·35 for 6500 km with 121 passengers. Design authorized at OKB under personal supervision of A. N. Tupolev but with his son Alexei Andreyevich chief designer, Yu. N. Popov deputy, Yu. N. Kashtanov chief engineer. Propulsion assigned N. D. Kuznetsov. Two flight prototypes and static/fatigue specimen ordered early 64, and project disclosed with model at Paris Salon 65. Gross weight then 130 t (286,600 lb). Mikoyan charged with building Analog-144 to fly wing scaled down to fit MiG-21, but this severely delayed by inability to settle details of wing and elevons. Only one aircraft seen by UK delegation to Zhukovskii factory 67 but told two other airframes well advanced and first flight early 68. First flight article, 68001, made maiden flight 31 Dec 68, accompanied by Analog-144; crew Edward Elyan, copilot Mikhail Kozlov, flight-test director V. N. Benderov and engineer Yu. Seliverstov.

First prototype (68001) with 2·5% wing, LE swept at two distinct angles (76° and 57°), slight camber on inboard LE, whole wing made by Antonov OKB. Rectilinear structure of multiple spars attached to ten fuselage frames, mainly machined forgings, with skins chemically-milled. Ruling material VAD-23 aluminium alloy with titanium LE and titanium or steel in region of engines, landing-gear rib and other highly stressed or hot areas. Trailing edge on each side formed by four approx-square elevons each driven by two power units faired into underside of wing. Three separate hydraulic systems 210 kg/cm² with majority-vote feature to give quad reliability. Likewise rudder divided into upper half with two power units on right side and lower half with two power units on left, each power unit in different system. Fuselage of basically circular section about 3,4 m diam, with forged frames and thick chemically-milled skin,

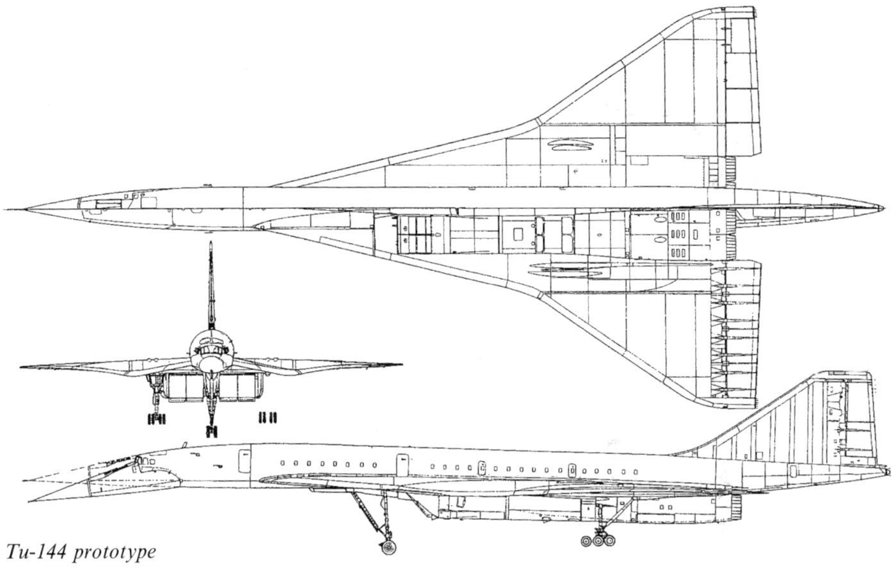

Tu-144 prototype

integrally stiffened with crack-stoppers along row of small rectangular windows, 25 each side. Underside rear fuselage titanium/steel. Interior width 3,05 m (120 in), height along aisle 2,16 m (85 in) over passenger section 26,5 m (87 ft) long, 120 seats (max 126) mainly 3+2, all baggage/cargo on same deck front and rear, two main doors each side 1680 mm × 762 mm (66 in × 30 in), small Class 2 emergency door each side at rear. Flight crew 3, pilots side-by-side with nosecone drooping 12° for take-off/landing. Engines in pairs, originally to have been in one box of four but modified 66 with four circular cowls fed by two paired ducts with wide centreline gap for nose gear. Nacelles fixed-geom inlets overall length 23 m (75 ft), titanium except for light-alloy access doors. No reversers. Remarkable main landing gear with 12-wheel bogie (4 tyres on each of three fixed axles) retracting forwards, bogie rotating 180° to lie inverted in wing; despite 13 kg/cm² (185 lb/in²) tyres of minimum size, long fairings necessary in wing upper surface and bay doors in line with each wheel. Tall nose gear almost identical with Tu-114 retracts rearwards into ventral fairing beneath pressurized fuselage. Brakes on all 26 wheels, main gears having quad steel-plate brakes each side of beam on each axle. Braking parachute in rear fuselage. Fuel in integral tanks in outer wing, LE inboard and bottom of fuselage, with trim tanks nose/tail, announced capacity 87 500 lit (19,247 gal, 70 t). Pipes to tail trim tank external on 68001.

Early flying devoted not only to performance/handing but also to getting duplicated environmental systems working, installing variable inlets and control system, inertial/doppler nav and autopilot. All four test crew with upward ejection seats. Mach 1 exceeded 5 June 69, Mach 2 on 26 May 70, later reached 2·4. Publicly unveiled Sheremetyevo 21 May 70. Preproduction **77101**, flown July 71 was virtually a different design. New fuselage with length increased by 6,3 m (20 ft 8 in), greatly modified structure and materials, 34 windows each side and Class 1 exit at rear each side. New wing with span increased 1,15 m (4 ft) with camber across entire chord root to tip and marked downward curvature at TE, skins integrally stiffened by machining from slab and completed panels then attached by welding instead of riveting. More extensive honeycomb structure, most movable surfaces titanium alloy (extensive use Ti/4V and similar). Redesigned engine nacelles resembling Concorde with oblique rectangular inlets separated by deep central splitter, electrically de-iced inlets, variable upper/lower profile (not fully developed spring 73) and four improved variable nozzles separated by width of fuselage and extending aft of wing TE. Redesigned landing gears. Main gear with eight-wheel bogies (four tyres on each of two axles), tyres 950 mm×400 mm, attached to forged frames in engine nacelles and retracting forwards between each pair of engine ducts; during retraction bogie first rotates 90° inwards about axis parallel with fuselage to lie upright in narrow thermally insulated and cooled bay between ducts. Nose gear moved forwards 9,6 m (31 ft 6 in) and of necessity made even longer; redesigned to retract forwards into unpressurized underfloor box within fuselage. All legs ultra-high-strength steel, nose gear carrying six (previously four) landing/taxi lights. Addition of retractable canards of 6,1 m (20 ft) span to top of fuselage just aft of flight deck. Each surface almost rectangular, with double LE slat and double-slotted TE flap, swinging open forwards to zero sweep but sharp anhedral; this powerful lift at nose enabled elevons to deflect down instead of up to give greatly improved lift overall at low speeds, improve low-speed handling and agility and reduce field length. Ejection seats replaced by hydraulically opened entry door on left of nose. Nosecone with greater glazed area. Fuel capacity increased to 118 750 lit (26,121 gal) with greater capability for rapid transfer nose-tail (Elyan said at this time transfer aft at M 0·7 and back when decelerating thru 1·5.

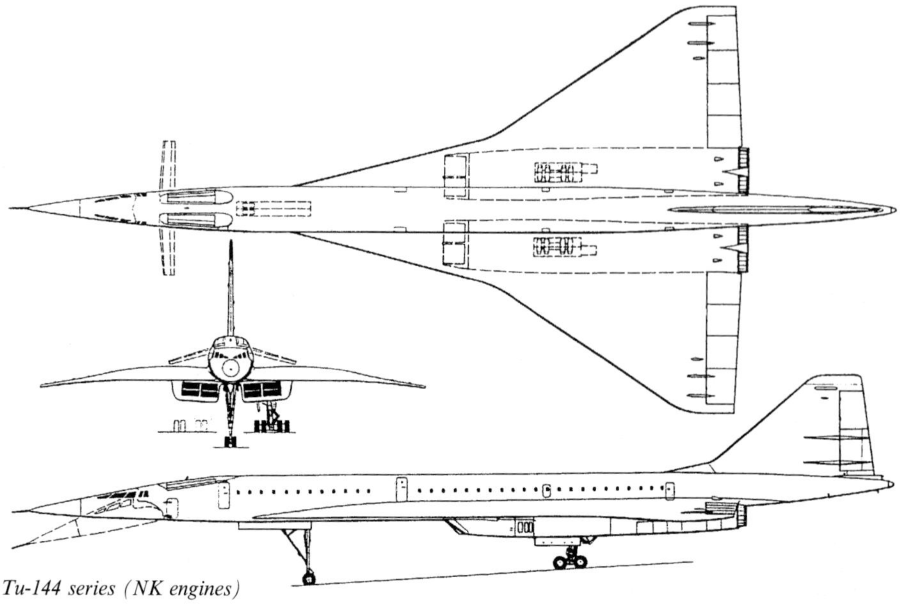

Tu-144 series (NK engines)

Eight production aircraft seen at Voronezh during French visit Dec 72; second aircraft 77102 flew at 73 Paris Salon and suffered inflight breakup during unplanned violent pitch manoeuvre 3 June 73, killing Kozlov and Benderov among others. No 77106 began proving flights Moscow/Alma Ata with cargo 26 Dec 75, flying 3260 km sector regularly in under 2 h block time (77106 retired to Monino 80). Normal passenger accommodation 140: 11 in first-class front cabin seated 2+1, 30 in mid cabin 2+3 and 75 in rear cabin 2+3, final 24 seated 2+2. Rear baggage/cargo accessed via door at TE on right. From 22 Feb 77 series of 50 proving flights Moscow/Khabarovsk. At last, about five years behind planned schedule, passenger service began Moscow/Alma Ata 1 Nov 77; five of next six flights cancelled but 102 revenue flights made when service suddenly terminated 1 June 78 following fatal accident 28 May to aircraft testing new engines. Cause traced to fuel system, but passenger service never resumed.

P.A. Kolesov engine KB devd uprated engines (see data) which resulted in **Tu-144D**, able to cruise without afterburner, but this never entered Aeroflot service. Visually distinguished by large nozzles with centre plugs. Aircraft 08-2 (77114) set records in 83 as **Type 101**. All early aircraft scrapped or used for ground instruction at KAI and KuAI. Better examples are dwindling fleet at Zhukovskii (three types of engine) used to study ozone layer and to assist Buran pilot training, five cannibalised to keep several able to undertake many important tests and quasi-operational missions. ASCC name (all) 'Charger'. Proposed to lease one flyer to NASA.

DIMENSIONS Span (prototype) 27,65 m (90 ft 8½ in), (production) 28,80 m (94 ft 5⅞ in); length (prototype) 59,40 m (194 ft 10½ in), (production 65,7 m (215 ft 6⅔ in); wing area (prototype) 469 m², (production) 506 m² (5,447 ft²).

ENGINES (prototype) four NK-144, 17,5 t (38,580 lb); (series) NK-144A, 20 t (44,092 lb); (144D) NK-144F, 22 t (48,501 lb) and RD-36-51A, 20 t.

WEIGHTS OWE (prototype) 79 t, (series) 85 t; fuel (prototype) 70 t, (series) 95 t; payload 12 t; MTO (prototype) 150 t, (series) 180 t, (D) c190 t.

PERFORMANCE Max speed (all), also max cruise, 2500 km/h (1,550 mph, M 2·35); normal cruise 2100 km/h (1,305 mph, M 1·98) whilst drifting up from 16 km to 18 km (max 59,000 ft); max climb from max take-off (production) 50 m/s (9,843 ft/min); range (prototype) no meaningful figure but far short of 6500 km objective (est generally around 3200 km), (production) 6500 km with 140 pax (4,300 miles), (144D) 7000 km (4,350 miles) with max payload; take-off (prototype) 2300 m, (production) 1980 m; landing (prototype) 1490 m/305 km/h.

Tu-22M, Tu-145 This devt of Tu-22 was undertaken from 64 without govt decree, so same *Izdelye* number retained, though it became swing-wing bomber and rocket carrier with few parts unchanged. Dr Alexei A. Tupolev said 'Disguise as a mere modification was political'. Pivoted high-lift wings were used to reduce field length and increase weapon load rather than combat radius.

In 66 metal cut on prototype **Tu-22M**, also called **Article 45-1**. Two NK-144-22 engines

Tu-144 77102

Tu-22M-0

inside modified rear fuselage, fed by long ducts from fully variable inlets beside forward fuselage. Inboard wing as Tu-22 aerodynamically but structurally new. At one-third semi-span pivots far back at almost 50% chord; plain steel bearings in titanium/steel central box permitting outer wings to pivot 20°/50°/60°. Wings driven by duplex hyd motors in separate systems. Large fence around LE at tip of fixed wing. Shallow aerodynamic fairing on underside in line with hinge, protruding behind TE. Max dash Mach at hi-alt set at 1·89; structural load factor 2·5, same for subsequent versions. Fixed inboard LE constant 55° to root, TE 90° to fuselage. Five spars, Nos 3 and 4 diverging to leave bay for main gears by Hydromash, based on Tu-22 bogie but with two extra wheels added closer together at rear and centre pair splayed apart on gear extension to give three sets of ruts on soft ground. Top of main leg curved sharply out towards wingtip; thus, as breaker strut pulled gear inwards, whole assembly rose into wing with bogie housed under wing inside fuselage; bay doors hinged near centreline, closed on ground, legs having own door. New gear tested on Tu-22. Tyres all 1030 mm × 350 mm. Steerable levered-suspension twin-wheel nose gear retracted backwards into bay aft of crew compartment, front door on leg and left/right bay doors open on ground. All three gears with anti-skid brakes. Track 7,3 m, wheelbase 13,67 m.

Inlets with large gap (min 150 mm) to divert boundary layer through aft-facing ducts above and below. Splitter plate almost vertical, unperforated but with vert boundary-layer slits in inner wall positioned by multiple rams to vary profile and throat area according to flight Mach number. Lips sharp, top and bottom swept 60° in plan. At low airspeeds extra air admitted through suck-in door in outer wall just behind wing LE. Engines installed at angle to bring nozzles close together, with afterburner petals beyond fuselage. Access doors under engines but engines removed to rear. Rear fuselage stressed for quad a.t.o. rockets (not used). Drag chute in large container below rudder.

Aircraft area-ruled without waisting fuselage, so inlet ducts straight, inner wall curved out when rising over wing. Crew compartment pressurized 0,59 kg/cm^2 (8·39 lb/in^2) for pilot (often a/c commander) on left, copilot on right, nav/bomb behind on left, *radist* electronic-systems officer on right, each on forward-facing KT-1 seat (limits 250 m and at low level 130 km/h, or 300 m for forced simultaneous ejection) under individual gull-wing roof hatch with anti-flash blind. For medium/high level bombing nav/bomb could lie at ventral blister with oblique forward window, but normally seated at radar/video displays. LAS-5M dinghy behind pressure cabin.

Outer wings t/c 3·5% at root and 2·9% at tip, resulting in pronounced inflight flexure; three sections powered LE slat and three sections double-slotted TE flap each running out on two tracks and preceded by hinged cove strip. Outboard sections flaperons, backed up by spoilers ahead of flaps operating differentially in conjunction with differential tailplanes, all six spoilers also used as airbrakes/lift dumpers. Tailplanes larger than Tu-22 but set at same low level, reflex (upturned LE) profile, LE sweep 60°, tips aligned with airflow, +25°/–40°.

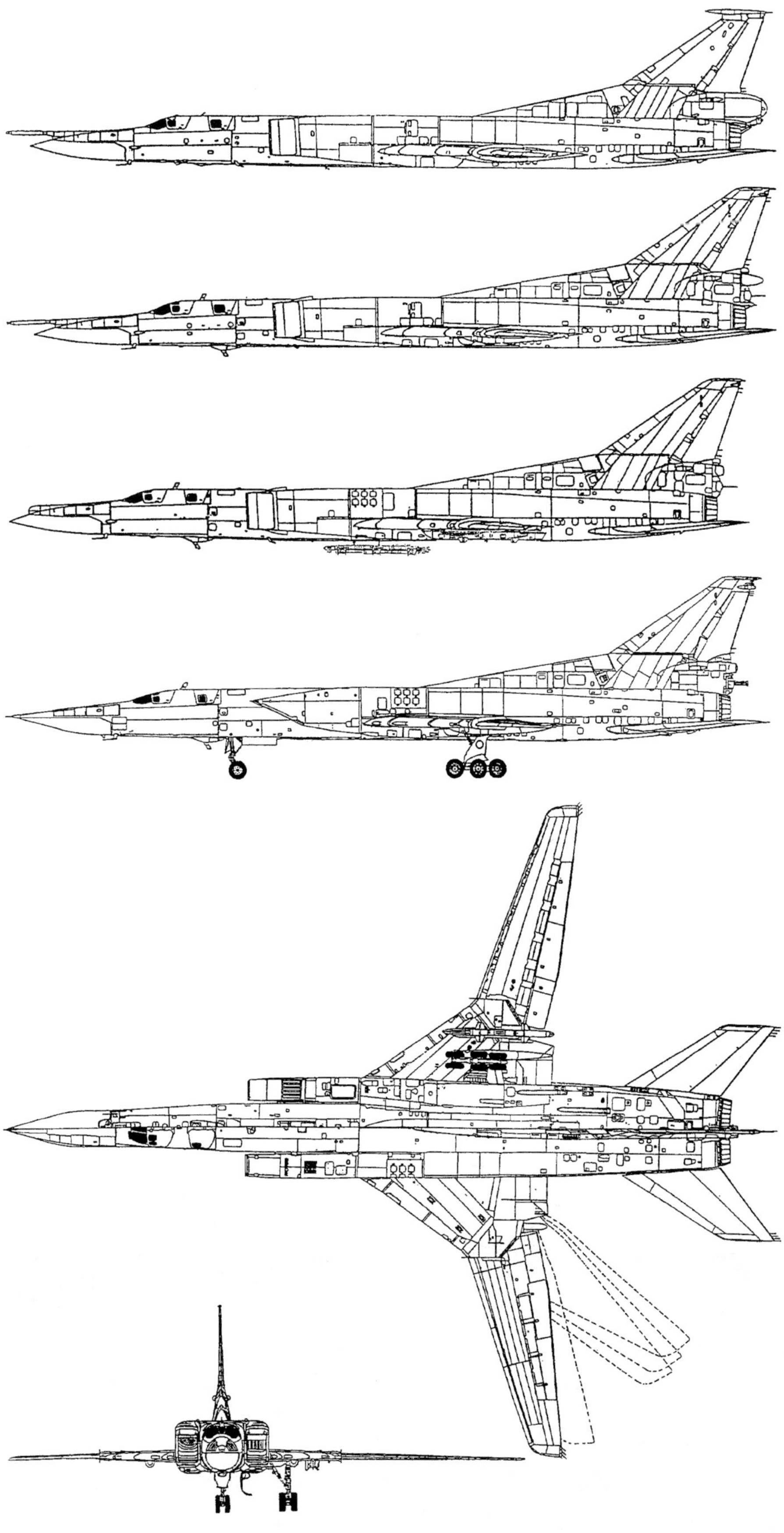

Tu-22M3 (upper side views, Tu-22M-0, Tu-22M-1 and Tu-22M-2)

Vert tail derived from Tu-22, LE 80° on dorsal fin and 60° on fin proper, all control surfaces fully powered. Slim antenna fairing at tip of fin. No defensive armament, RBP-series main radar, upper half of nose access hatches, light each side to assist night use of FR probe. First 22M (in OKB documents **Tu-22M-0**) flown at GAZ-22 Kazan by test crew headed by V. P. Borisov 30 Aug 69. Immediate conclusion: overcame problems of predecessor by having two pilots with good forward and downward view. Total M-0 one static and nine flight prototypes, last example having newly designed NK-22 engines, underwing fairings removed and tail barbette with GSh-23 cannon, drag chute being relocated under rear fuselage. One (#33) at Monino, another (#156) at Kiev VVS Inst. Seen by US satellite July 70. ASCC name 'Backfire', later (first eight) 'Backfire-A'.

Tu-22M-1, 45-01 Ten pre-production aircraft, all for service test. First early 71 very similar to ninth M-0, last almost identical to first M-2, all with NK-22 engines. Description see M-2.

Tu-22M-2 First series version, developed as **Article 45-02** at Moscow and Kazan from 71 by team led by Boris Levanovich. Extended outer-wing panels, slats/flaps remaining as before thus terminating inboard of tips, wings able to oversweep to 65°, shallow aerodynamic fairings removed. Inlets given large aft-facing boundary-layer ejectors above and below. TA-series gas-turbine APU in dorsal fin, inlet on right, exhaust on left. Improved paralleled AC and 27V 500A DC electrics, with four engine-driven generators and two on APU, and 210 kg/cm^2 (3,000 lb/in^2) hydraulics. Fuel system augmented, removable probe above nose with internal plumbing. New landing gears with mod pivots requiring underwing fairings, tyre pressure increased, centre bogie wheels no longer splayed apart. Electronics complex, 13 subsystems in 80 LRUs, including updated inertial plus A322Z doppler (retaining RSBN-2S Shoran and twin ARK-15 with blades above and below cabin) and new automatic flight-control system for smooth low-level ride, manoeuvre limiting and bad-weather landing. *Rubin* nav/bombing radar, plus autopilot link to dual RV-18 radar altimeters for sustained flight at 150 m (492 ft), other sensors including fairing ahead of nose gear for EO and visual sight. Braking parachute (cruciform) relocated under fuselage to permit installation of UKU-9K-502 turret (twin GSh-23, 600 rds each) with PRS-4 fire control using TP-1 EO (TV) sight between aft radar and rudder. Fin-tip fairing deleted. Great effort integrating computers controlling navigation and weapon-delivery system using OPB-15 sight. Defensive avionics with 18 antennas mainly in inboard LE, above/below fuselage and on vertical tail, C-VU-10-022 computer for energy management. Main RWR passive receivers on each side of nose and centre fuselage, along inner LE (6), front and rear near bottom of fin, facing aft inboard of tailplanes and at top of fin (raised above rudder to house four antennas). LO-82 IR warning above and below forward fuselage. Ejectors for 192 chaff/flare cartridges in strips of eight triplets along each side of rear fuselage firing up and down. Normal free-fall bomb load 12 t, max 24 t, carried partly internally and partly on four bolt-on external MBDZ-U9M racks under inlet ducts each with nominal capacity up to three triplets of FAB-250 or single Kh-31 missile. Free-fall conventional loads include 2 FAB-3000, 8 FAV-1500, 42 FAB-500 (25 external), or 69 FAB-250 (36 external). Internal bay can carry combinations of N or TN bombs or rotary launcher for six Kh-15P or 12 RKV-500B missiles. AV-MF option is up to three Kh-22 cruise missiles, one recessed on centreline (special doors), others on D2M (AERT-150) pylons just outboard of landing gears, each pylon with missile conditioning system, aircraft with A359Z missile guidance transmitter.

Tu-22M-0

Certified at 122 t (268,959 lb), service test from Jun 71 with 185th Guards Regt of 13th TBAD. Total c250, split equally between Air Armies and AV-MF, crews trained at Tambov but initially based Ryazan. During prodn at Kazan, inlet ducts redesigned with aux suck-in side door replaced by nine smaller doors (3×3); tail radar changed, avionics updated, FR probe removed to comply with SALT-2. Small force Afghan ops from Poltava and Maryi from Dec 87. 'Backfire-B'.

Tu-22M-3 Devd as **45-03** in 80 to take new engine with greater airflow and pressure ratio, increased thrusts and better fuel economy. This

Tu-22M-2

enabled fuel capacity to be slightly increased, whilst reducing field length and raising combat ceiling. As duct had to be enlarged inlets redesigned for higher efficiency at peak Mach (Tu-144 experience) changing to wedge type with variable upper wall and hinged lower lip; mid fuselage refined eliminating orig aux-inlet frame and moving 9 aux inlets further aft, mod secondary airflows and larger nozzles required reprofiling rear fuselage. Fence geometry (shorter, equal above/below wing) and other changes perfected 81. Advanced multimode radar with DBS (doppler beam sharpening) in upturned nose and new functions tailored to maritime use. FR probe not fitted but associated lights and plumbing retained. KT-1M seats, ht limit 60 m (one a/c with K-36D seats). Fuel again increased, Class 1 runways only. New tail turret with single GSh-23M with superimposed barrels and ejection chutes. Though AV-MF aircraft equipped for external free-fall bombs, officially referred to as rocket launcher, qualified on five missile types (Kh-22, RKV-500B, Kh-15P, Kh-31A/-31P, Kh-35). Prolonged service test by 185th Guards complete 86; joined M2 over Afghanistan Oct 88. About 130 delivered, to Dec 92. 'Backfire-C'.

Predictably said by US Dept of Defense to pose direct threat to USA and thus to justify massive expenditure on counter systems (unrefuelled combat radius said to be '3,000 nautical miles', 5470 km). SALT-2 reqd removal of FR probes and prodn limited to 30 annually. Alarm now expressed at CIS wish to find export customers for new and used versions.

DIMENSIONS Span (M-0) 31,5 m at 20°, 22,3 m at 60°, (M-2, -3) 34,28 m (112 ft 5⅝ in) at 20°, 23,3 m (76 ft 5⅜ in) at 65°; length (excl probe) 42,46 m (139 ft 3⅝ in); wing area (M-3, 20°) 183,58 m^2 (1,976 ft^2), (65°) 175,8 m^2 (1,892 ft^2).

ENGINES (M-0) NK-144-22, (M-1, -2) two NK-22, (M-3) two NK-25.

WEIGHTS (M-3) Empty 54 t (119,050 lb); max fuel 53,55 t (118,060 lb); max weapons 24 t (52,910 lb); max takeoff (M-0) 105 t, (M-2) 122 t, (M-3) 124 t (273,370 lb), (with a.t.o.) 126,4 t (278,660 lb); landing (normal) 78 t, (max) 88 t (194,000 lb).

PERFORMANCE (M-3) Max speed (SL) 1050 km/h (652 mph), (11 km) 1900 km/h (1,181 mph, M 1·79), (at light weight) 2300 km/h (1,429 mph, M 2·16); cruising speed (hi) 900 km/h (559 mph); mission radius (all lo, max bombload) 1500 km (932 miles), (hi, max bombload or 3×Kh-22) 2000 km (1,242 miles), (normal bombload) 2200 km (1,367 miles); service ceiling (max wt) 13,3 km (43,650 ft), (light, not under M 1·3) 14 km (45,930 ft); TO 370 km/h, 2050 m (with a.t.o., 1920 m); landing 285 km/h, 1450 m (1200 m with chute).

Tu-148 Advanced interceptor to replace Tu-128, designed for Mach 3, variable-sweep wing, two RD-36-41 engines. *Zaslon* radar, internal bay for six or eight missiles, design PVO interception line at 1200 km. Launched 62, funding stopped 69.

Tu-154 Large trijet closely similar to stretched 727 in capability but with larger airframe and much larger and more powerful engines because of demanded ability to use short unpaved airports. Designed initially by team led by S. M. Yeger but 70 Dmitri Markov appointed project chief designer. Major GVF requirement, about 64, for jet to replace Tu-104, Il-18 and An-10 offering dramatic advance in fleet mean cruise speeds without requiring significant airfield improvement. Range specified was severe, stages to 6000 km though with reduced payload.

Tu-22M-3

Configuration essentially scaled-up Tu-134 with three engines and triplexed hydraulic and flight-control systems. Wing profile 12·2/9%, sweep 40° to structural break (90° to rear spar) at main gear, 38° to tip, 35° mean at 25% chord, dihedral 0°, incidence +2·5° decreasing to tip; five sections elec powered slat on outer 80% each wing, hydraulic triple-slotted flaps inbd and outbd of pods, powered ailerons from fence to tip, four section hyd spoiler above each wing, outermost differential for low-IAS roll, mid two sections symmetric airbrakes and innermost (inboard of pods) airbrake and also lift dumper. Tail almost exact scale of 134 though tailplane sweep increased to 45° (40° at 25%) and flight-surfaces honeycomb skins and fully powered by triplexed system without manual reversion, tailplane becoming primary control as well as trimming surface. Large fuselage with circular section 3,8 m (149·6 in) diameter, almost same as B-727 (except for Trident all major Western jetliners by this time no longer used circle) with straightforward fail-safe structure and level floor unobstructed by wing box. Max dP 0,63 kg/cm^2 (9 lb/in^2). Basic interior width 3580 m (141 in), usable length 27,45 m (90 ft), rectangular windows in crack-stop panels, two doors ahead of wing on left, two service doors opposite, all 1650 mm × 762 mm (65 in × 30 in) opening outward. Unlike 124 and 134 underfloor cargo/baggage holds, two forward of wing total 38 m^3 and small unpressurized compartment aft. Main gears with six-wheel bogies (left/right wheel on each of three fixed axles) with 930 mm × 305 mm tyres at 8 kg/cm^2 (114 lb/in^2) retracting hydraulically rearwards to stow bogie inverted in wing pod. Steerable nose unit, 800 mm × 225 mm tyres, retracting backwards. Side engines hung on rear fuselage with R-R (Greatrex) type reversers, centre engine fed via S-duct, no reverser. Two integral tanks each wing, service tank in centre section, total 41 140 lit (9,050 gal); optional four c-s flex cells raising total to 46 825 lit. Airframe de-icing by bleed air except elec heating of slats and windscreens. Elec power from three CSD alternators, plus alt and starter/generator on TA-8 APU. Avionics include moving-map GPI, VOR, DME, 3-channel autopilot with hybrid flt-control computer, cleared to Cat II ILS. Two pilots and engineer, jump seats for nav or others. Passenger seats 128-167 mainly 3+3.

Prototype 85000 flown by Yu. V. Sukhov 3 Oct 68. Prodn authorized Kuibyshyev mid-70, wings supplied by Kumertao (now Avn Prodn Assoc); six devt a/c delivered Aeroflot Aug 70, cargo flights to 71, sched cargo from July 71, irregular passenger service (Tbilisi) July 71, sched passenger (Mineralnye Vody) 9 Feb 72, int'l (Prague) 1 Aug 72. In 74 first delivery of **Tu-154A**, more powerful engine, centre-section tank 8250 lit (6,6 t) contents transferable to main system only on ground, auto connect flaps/tailplane to maintain trim (pilot over-ride if +3° tailplane movement demanded), two emergency exits at rear, improved systems and augmented avionics.

Production switched 77 to **Tu-154B**, French (Thomson-CSF/SFIM) flight control and ILS for Cat II, weights increased, spoilers improved,

Tu-22M-3

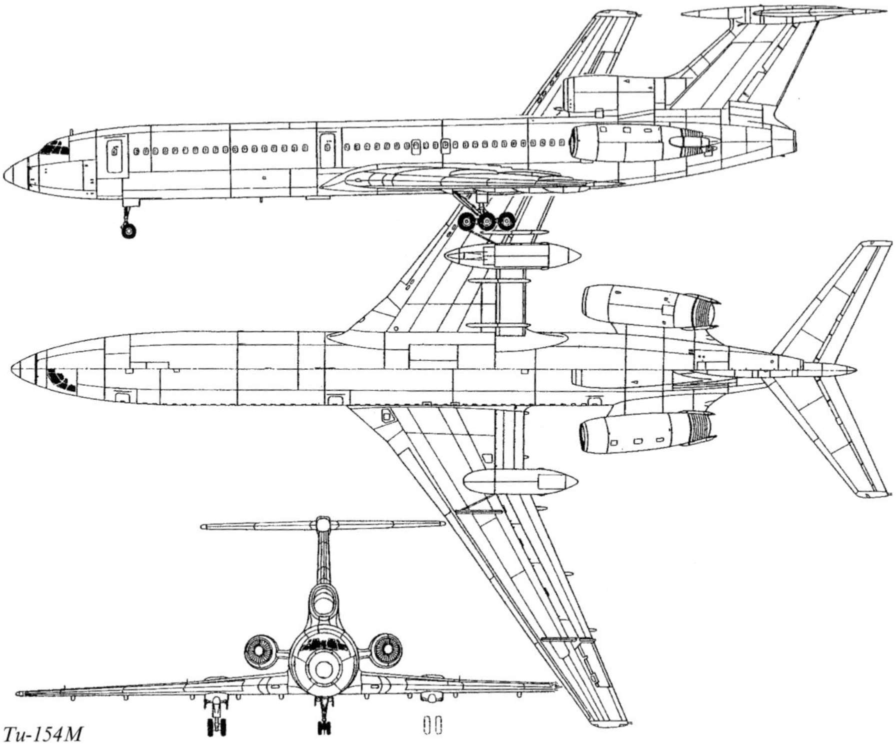

Tu-154M

Tu-154B-2

cabin dP reduced to 5,9 kg/cm^2 (8·4 lb/in^2), two extra doors at extreme rear and cabin extended aft for max 180 seats. **B-2** followed 80 with new radar, centre-section made part of flight fuel system, main-bogie front axles equipped to align with landing runway in crosswinds. In 82 **154S** available with cargo door at front on left 2,8 m × 1,87 m (110·25 in × 73·5 in), ball/roller floor, max nine ISO (88 in × 108 in) pallets. From 84 standard aircraft **154M** (original B-2 conversion called **164**) with different engines, with IL-62M clamshell reversers, TA-92 APU moved from below rudder to forward in fuselage; production M has smaller slats, larger spoilers, triplex INS, non-flam furnishing, enclosed overhead baggage bins. Produced to 92, 996 all versions delivered, service at ±50°C, export customers Afghanistan, Bulgaria, China, Cuba, Czech, Egypt (returned), E Germany, Hungary, Iran, Iraq, Latvia, Mongolia, N Korea, Poland, Romania and Syria. Single aircraft converted as **Tu-155** to research problems of using LNG (liquefied natural gas) and LH$_2$ (liquid hydrogen) as left-engine fuel. ASCC name for all "Careless."

DIMENSIONS Span 37,55 m (123 ft 2½ in); length 47,9 m (157 ft 1¾ in); wing area 201,45 m^2 (2,168 ft^2).

ENGINES (154) three NK-8-2, (A,B,B-2,S) NK-8-2U, (155) two NK-8-2U and one NK-88, (M) three D-30KU-154-II.

WEIGHTS Empty (154) 43,5 t (95,900 lb), (A,B) 50 775 kg (111,938 lb), (M) 55,3 t (121,915 lb); fuel (154) 33,15 t, (A, B, M) 39,75 t; max (154) 90 t (198,416 lb), (A) 94 t, (B) 96 t, (M) 100 t (220,460 lb).

PERFORMANCE Max speed 575 km/h IAS or M 0,9; max cruise (exc. M) 900 km/h (M); 950 km/h (590 mph); normal cruise (all) 850 km/h; max alt 11,9 km (39,000 ft); range with max payload (154) 2520 km, (B) 3200 km, (S) 2900 km, (M) 3900 km (2,425 miles); balanced field length (154) 2100 m, (B) 2200 m, (M) 2500 m.

155 See 154.

156 Airborne warning and control aircraft to replace Tu-126, config as E-3, D-30KP engines, not built.

Tu-160 Heaviest and most powerful combat aircraft of all time, programme launched 73 solely to counter USAF B-1; VVS Col Evgeni Vlasov called it 'An expensive countermeasure'. Uniquely, has OKB designation '**Aircraft 70**' (out of sequence) and VVS designation which seems more like that of OKB. Authorization to create optimised aircraft regardless of cost, and continued after cancellation of B-1 on 30 June 77; indeed this is when OKB received order for prototypes. Team leader Vladimir I. Bliznuk, prototype assembly at Zhukovskii, series planned at Kazan.

According to Vlasov 'Looked at from outside, Tu-160 and B-1B are similar. This explained by similar objectives: long radius of action despite heavy equipment/stores load, capability of deceiving enemy defence systems at low or high altitudes, and having minimum radar, IR, optical and acoustic signatures. However, they bear resemblance from outside only; differences are fundamental.' Biggest single difference is that Tu-160 has 79% more installed engine power. Last to bear initials of KB director Kuznetsov, NK-321 is most powerful military engine in history (see engines section). Second difference is that, though significantly larger than USAF bomber, Tu-160 has smaller radar cross-sections and lower aerodynamic drag. Dr Alexei A. Tupolev 'We believe, most aerodynamically efficient supersonic aircraft ever built'. Col-Gen Boris F. Korolkov comments there is marked difference between radar cross-sections even ignoring B-1B's external carriage of missiles and (claimed) much better performance of Tu-160 EW systems. Finally, partly thanks to idealised inlets, Tu-160 is faster than B-1B at SL and almost twice as fast at hi-alt.

Four engines installed in paired nacelles under enormous fixed inner wing which has LE of large radius (for max power and efficiency of internal EW antennas) which in plan curves continuously to sweep angle 90° where it blends imperceptibly into fuselage near cockpit. Two giant beams link pivots for outer wings, 19,2 m, apart, with sweep angles selected by buttons 20°, 35°, 65°. Each outer wing straight-tapered root to tip, full-span four-section hydraulically driven slotted LE flaps (essentially slats) and full-span four-section double-slotted TE flaps. Outboard of TE flaps, but stopping 3,5 m short of tip, are powered ailerons. These are all-speed primary roll-control surfaces, but backed by spoilers and tailerons; with flaps extended they droop 20°. After endless juggling decision taken to retain optimum engine installation, leaving nowhere for wing TE to penetrate at 65°, and to solve problem by making inboard TE hinge progressively up to form vertical fence at max sweep. Large vertical lower fin carrying bearings for left/right one-piece tailerons and one-piece unswept upper rudder (no fin above tailerons). All surfaces driven by electro-hydraulic power units with dual FBW and mechanical signalling. Fuselage cross-section minimum necessary for crew, fuel and payload, significantly less than B-1B or Tu-22M. Drag and radar cross-section further reduced by angle of conical nose, and by use of special computer routines to achieve optimum shape and control machine tools in pro-

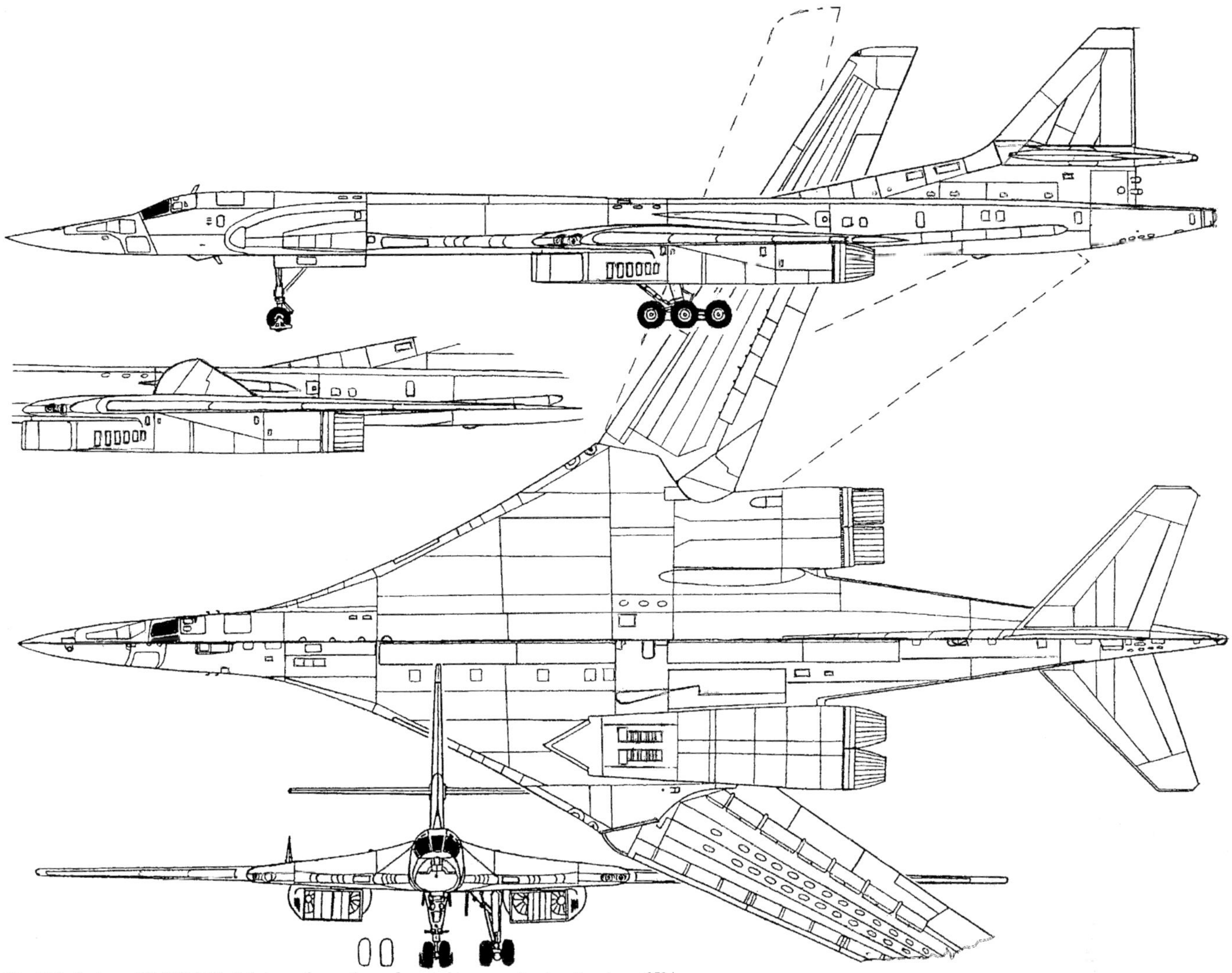

Tu-160 (wings 20°/35°/65°, inlets and nozzles subsonic/supersonic; inset, wings 65°)

duction. Of over 45 antennas, only three project as blades or spikes. Apart from hinged 'flap fence', no fence or vortex generator anywhere. Extensive use of honeycomb sandwich skin and precision-controlled RAM (radar-absorbent material) covering. Engine pairs in rectangular-section nacelles hung under inboard wing with rear upper portions projecting above as separate jetpipe fairings. Despite experience with Tu-144/Tu-22M-3, for minimum radar cross-section inlets similar to B-1A (not B-1B) with vertical splitter leading back to inner wall variable in profile and throat area (claimed to achieve higher pressure recovery and lower drag from 0 to M 1·9 than any other). Outer wall also vertical, incorporating five inward suction-relief doors of progressively reduced height matching profile of duct which at throat is tall and narrow. Consideration of modifying these doors with zigzag edges. Complete variable nozzle of engine projects behind nacelle, engine withdrawn on rails to rear. Nose landing gear behind, not under, crew compartment for min cross-section; carries landing/taxi lights and spray/slush deflector, hyd steerable ±55° and retracted to rear into bay with left/right doors open when gear down. Main gears carry six-

Tu-160

Tu-160

wheel bogies, unlike Tu-22M last pair of wheels in line with preceding, mounted on massive oleo leg made by Hydromash, installed inboard of engines thus restricting track to 5,4 m. Main drag strut incorporates retraction jack which pushes gear to rear, leg pivoting back and lower portions also inwards while bogie somersaults to lie inverted in box which projects upwards to cause canoe blisters between engines and fuselage. These blisters taper at rear into pipe/cable-loom fairings carried externally on each side of rear-fuselage integral tanks to closure bulkhead at aft end. Large braking parachute streamed from fairing between horiz/vert tails. Internal fuel (see data) 50% greater than B-1B, integral tankage throughout, main refuel doors in front face of inboard LE, unused provision for retractable probe above nose.

Crew enter via pull-down ladder to forward weapon bay and walk forward through main avionics compartment. Crew of four as Tu-22M in forward-facing K-36T seats under jettisonable roof hatches. Front windows same birdstrike strength as Su-24, all windows nuclear flash blinds and cockpit side windows openable on ground. FBW flight control, fighter-type sticks, despite severe flight limitations so-called 'carefree' manoeuvres possible under protection of avionics. Nav/bomb has use of multimode radar with terrain-following and optical sight plus video camera looking ahead from ventral blister. Usual com/ADF/triple INS/Glonass/ILS suite, while *radist*, having no gun to manage, is full-time defensive systems operator. According to Vlasov 'EW systems fully meet purpose. Easily removable and maintainable, and modules quickly replaceable. Nav's panels provided with eight computers, complex bomb/missile systems, radar displays and long-range nav systems linked with Glonass satellites. A computer generates actual flight route on topographic chart. More than 100 computers aboard Tu-160. Automatic air-combat and defence systems can interrogate many targets simultaneously'. Passive radar and IR warning systems are new, and through computers serve most comprehensive ECM/IRCM active defence systems ever created. Main batteries of chaff/flare dispensers in flush triplet groups surrounding rear fuselage. Systems officer has complete tactical display and can at any time take over manual control in situations where auto response might betray aircraft's presence. All weapons carried internally in two enormous bays extending 10 m from nose gear to engine nozzles, one ahead of wing bridge and second behind. Simple outward-hinged doors. Nominal capacity 30 t (66,139 lb), and this was used setting various speed/alt/load records but would seldom be approached in normal missions. Forward bay normally equipped with rotary launcher for six Kh-15P; rear bay with two rotary launchers each for 12 smaller RKV-500B. ANTK Tupolev state no problem in carrying loads heavier than 30 t, but no requirement. Effort applied to radius and radar/IR cross-section.

Test team headed by L. V. Kozlov and B. I. Veremey, first flight at Ramenskoye (later called Zhukovskii) 19 Dec 81. Development fleet of ten aircraft, all slightly different. Production of 100 at Kazan authorized 85. Aircraft No 12 presented for inspection by US delegation 2 Aug 88, unblemished white finish being admired. No 14, also called **Aircraft 70-03**, used for 1000 and 2000 km circuit records with 25 t load at M 1·58 and 1·63, and for heights to 13 894 m (45,584 ft) with 30 t. Instructor training carried out from 85 at Dolon AB, Siberia, with development and early production aircraft. Later crews qualified with 16 converted Aeroflot aircraft comprising Tu-134UBL pilot trainers and Tu-134BSh bomb/nav trainers, both with Tu-160 radar and cleared for low terrain-following. BSh, with underwing pylons for practice bombs, serves at Tambov school which also trains Tu-22M-3 crews. UBL serves alongside first operational unit of Strategic Aviation, 184th Regiment at Priluki, Ukraine. Unit operational April 87, later qualifying with RK-55 and by early 88 with aircraft on 24-h nuclear alert, this not required since 91. Unit still in Ukraine 93 because construction work at Engels not yet completed. Regt has two squadrons of ten available aircraft each, others in reserve; nuclear weapons under control of special sub-unit as usual. One pre-production aircraft exhibited by 184th 92, showing cockpit with single display for each pilot, rest of area dial instruments. Major problem no two aircraft alike, even differing in inlets, quite apart from avionics standards. CO, Col Valerii Gorgol, said 'Aircraft is complex and costly to maintain, and crews have complained of cockpit environment, escape system and nav system'! Also severe shortages (incl helmets and ground-crew ear defenders). Also in 92 Ukraine laid claim to aircraft. Production halted at about No 38 in late 92, though intention to resume. In Aug 92 seven of development fleet still at Zhukovskii, five considered not worth bringing up to operational standard and cannibalized. Reporting name 'Blackjack'.

DIMENSIONS Span (20°) 55,7 m (182 ft 9 in), (35°) 50,7 m (166 ft 4 in), (65°) 35,6 m (116 ft 9¾ in); length 54,1 m (177 ft 6 in); wing area 340 m^2 (3,660 ft^2).

ENGINES Four NK-321.

WEIGHTS Empty c118 t (259,000 lb); fuel 130 t; loaded 275 t (606,260 lb).

PERFORMANCE Max speed (SL) 1250 km/h (777 mph, M 1·02), (11 km) 2000 km/h (1,243 mph, M 1·88); cruising speed 850 km/h (528 mph); service ceiling 18 km (59,000 ft); max unrefuelled combat radius 7300 km (4,535 miles).

164 Project derived from 134 (not version of 154 as reported), 71-72.

Tu-204 Aeroflot mainliner to replace Tu-154. Launched 82, design led by Lev Aronovich Lanovskii settled 86. Close resemblance unbuilt Hawker Siddeley 134, designed for 60,000 h and 45,0000 cycles in 20 years. Low wing with two underslung engines, low tailplane. Wing supercritical profile 14% root 9·5% tip, LE sweep 28°, three-spar box with one-piece 22 m (72 ft) skins forming integral tank each wing with Al-Li and composite structure. Each wing four sections hyd-powered LE slat, two sections inboard double-slotted TE flap (outer part deflection reduced on landing to avoid jet) at 90° to fuselage, each running on one track, preceded by two airbrake/dumpers, one-piece sweptback outboard double-slotted flap, running on two tracks, preceded by five spoilers, tabless ailerons not extending to washed-out tip with winglets. Skin aft of wing box and large wing/body fairing all glassfibre honeycomb, all movables carbon-fibre composite. Fuselage constant oval section 3,8×4,1, load-bearing parts include Al-Li and Ti, left/right doors front and rear, Class 1 exits both sides ahead of and behind wing, ruling max cabin width 3570, normal dP 0,6 kg/cm^2 (8·53 lb/in^2). Tail LE and one-piece movables carbon-fibre composite, fin 2820-lit integral trim tank. Flight controls conventional yokes via triplex digital FBW with analog standby. Engines hung in full-length pods with carbon-fibre cowl panels, original (PS-90A) engines incorporating fan-duct blocker doors and all-round reverser cascades. Fuel 30 000 lit (52,090 lb) in wing and fin, single 3000 lit/min socket under fuselage. Main landing gears bogies with four tyres 1070 mm×390 mm, nose 840 × 290 steerable ±70°, retracting forwards, twin doors open on ground. TA-12-60 APU in tailcone for engine-start, ground ECS and emergency elec. Main elec 400-Hz AC, with rectified 27V DC; triple hyd systems at 210 kg/cm^2 (3,000 lb/in^2) including APU-elec pump for flight controls. 'Dark' type cockpit with six multifunction displays, two pilots and engineer plus jump seat. Aeroflot avionics including triple INS (Honeywell laser gyros), VOR/DME and ILS to Cat IIIA. Basic one-class seating for up to 214 pax 3+3, or (typical) 12 4+4 and 184 tourist. Underfloor hold limits 3625 kg forward, 5075 kg aft, total 1,123 ft^3.

First prototype (64001) flown by A. I. Talalakin 2 Jan 89. Three more, plus static and fatigue airframes, former Aeroflot announced order in 88 for 350 of requirement for 500, renegotiated by successor airlines. Length increased during prototype construction, and now several variants all with basically same airframe:

Tu-204 Prototypes, MTO 93,5 t (206,125 lb), fuel 24 t, max payload 21 t, MLW 86 t, PS-90AT engines, range (M 0·78, max fuel, 196 pax = 19 t) 3850 km (2,392 miles).

Tu-204-100 Basic production by Aviastar at Ulyanovsk, MTO 99,5 t (219,355 lb), payload

Tu-204-120

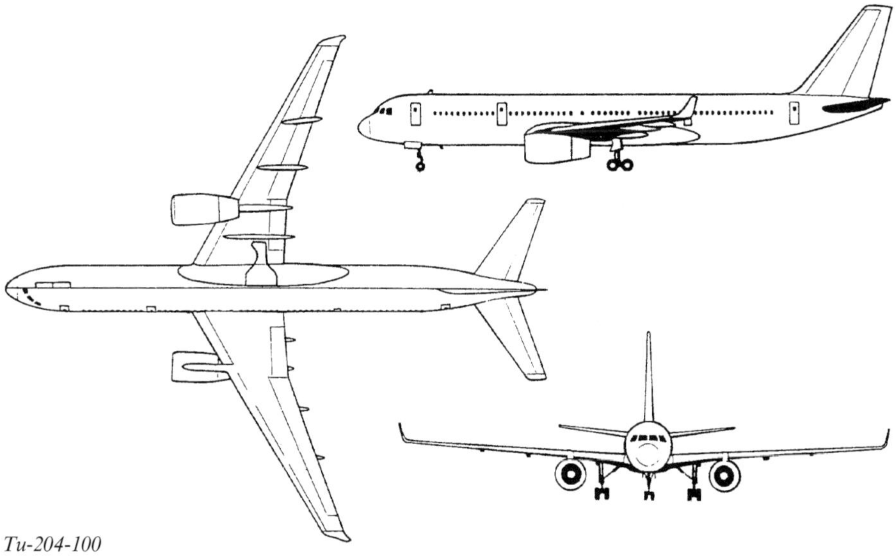

Tu-204-100

24 t, range with full tanks and max pax, 5300 km (3,293 miles).

Tu-204-120 Initial production with RR 535E4 engines; first = No 6 off line, first flight 14 Aug 92.

Tu-204-122 RR engines and Collins avionics.

Tu-204-220 Extended-range, RR 535-F5 engines; data below.

Tu-204-200 Extended-range, version, MTO 110,75 t (244,175 lb), fuel 32,7 t, payload 25,2 t, MLW 87,5; range (M 0·78 with reserves, max payload) 6330 km (3,933 miles).

Tu-204-222 As -220 but Collins avionics.

Tu-204-230 Proposed version with PW2240 engines.

Tu-204C Cargo version with 2,19×3405 side door and interior for eight LD-3 or 12 88×107 pallets; first example No 10 off line.

Tu-204-300 Destretched version, length 40,2 m, PS-90P engines.

Tu-204-400 Stretched version, about 52 m.

Deliveries begun 15 Sept 93, cost c$38 m (757 c$57 m); all-CIS cheapest, Bravia tradeoff price to customer wish to use RR engines, Collins avionics or other Western parts. Conscious cockpit commonality with Il-96M. All-CIS certificated June 93, RR eng March 94, JAA cert with Collins early 96.

DIMENSIONS Span 42,0 m (137 ft 9½ in); length 46,22 m (151 ft 7¾ in); wing area 168,62 m^2 (1,815 ft^2).

ENGINES (See above).

WEIGHTS OWE 59,0 t (130,071 lb); wt-limit payload 25,2 t (55,559 lb); MTO 110,75 t (244,160 lb); ZFW 185,639 lb; MLW 197,310 lb.

PERFORMANCE Cruising speed at MTO wt 810-850 km/h (503-528 mph); cruise height 10 650-12 200 m (34,950-40,000 ft) reached in 22-25 min; range, see above; TO 269 km/h, 1230 m; landing 245 km/h, 850 m.

Tu-214 Projected stretched and aerodynamically improved Tu-204 with LNG-fuelled NK-94 engines.

Tu-24 Light STOL utility aircraft with metal/composite airframe, hi-lift flaperon/fixed slat/drooped-tip wing, separate fixed main gears, cabin for 1-6 persons or cargo. Mock-up of **SKh** ag-version shown 93 with hopper in rear of cabin. Data provisional.

DIMENSIONS Span 13,0 m (42 ft 8 in); length 9,25 m (30 ft 4 in); wing area 28 m^2 (301 ft^2).

ENGINE One M-14P.

WEIGHTS Empty 990 kg (2,183 lb); payload 900 kg (1,984 lb); max 2100 kg (4,630 lb).

PERFORMANCE Cruise 200 km/h (124 mph); spraying speed 130 km/h; ferry range 2000 km (1,242 miles); TO 180 m (590 ft); ldg 100 m (330 ft).

Tu-244 ANTK Tupolev has longer SST experience than any other group, and in recent years has spent large sums with CAHI flying modified Tu-144LL aircraft in aerodynamic, noise, ecological and systems tests. Results underpin design of 2nd-generation aircraft (SST-II), L/D of wing 15 at M 0·9, 10·0 at M 2. Mainly VT-6Ch titanium. Design refined 79-93 when first published.

Wing broadly symmetric but drooped LE, inboard 75° (drawing shows 80°) and fixed, outboard 35° and hinged. TE eight elevon/flaps separated by individual nacelles for advanced turbofans derived from NK-321 with inlets devd from Tu-160. All landing gears retrac forwards, main 12-tyre (950×400) bogies flat in wing and on centreline, nose gear (1050×400) derived from Tu-144. Twin rudders, all flt controls FBW. Normal passengers 300, total 11 above-floor doors and emerg exits. ANTK prepared to merge design with those of Western partners (all of which are smaller).

DIMENSIONS Span 54,47 m (178 ft 8½ in); length 88,7 m (291 ft 0 in); wing area 1200 m^2 (12,917 ft^2).

ENGINES Four TVVRD each 33 t (72,750 lb).

WEIGHTS Empty 172 t (379,189 lb); fuel 178 t (392,416 lb); MTO 350 t (771,605 lb).

PERFORMANCE Cruise Mach 2·05 at 18-20 km (1,355 mph at <65,600 ft); range 9200 km (5,717 miles); cert to Cat IIIA.

Tu-304 Projected wide-body transport, 300 seats, 2-class, two NK-44.

Tu-306 Projected wide-body transport, 450 seats 2-class, two NK-46.

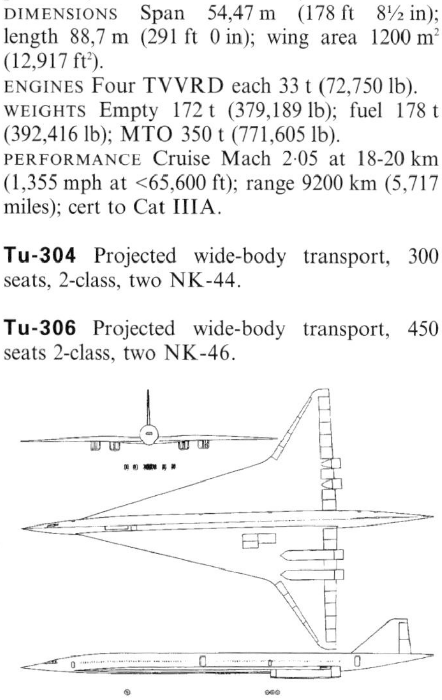

Tu-244 project

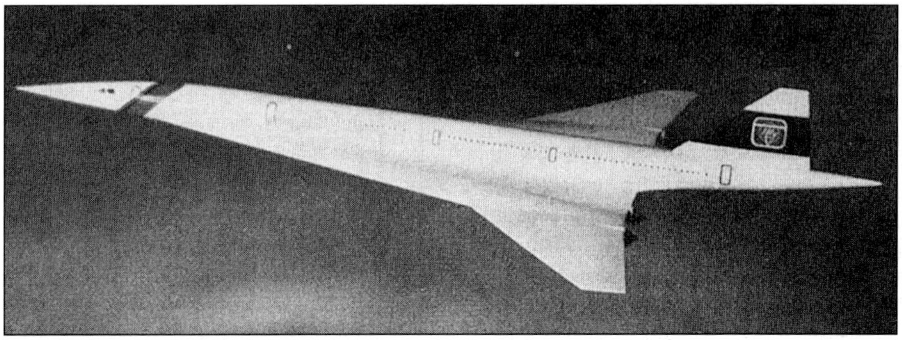

Tu-244 artwork

Tu-130 Multirole twin-turboprop; high wing with lift dumpers, pressurized fuselage with cabin 8,5×2,8×2,2 high for 5 t (11,023 lb) cargo or 53 pax 3+2, twin nosewheels and tandem main wheels, engines cleared for LNG or propane/butane. Prototype intended for 96. Data provisional.
DIMENSIONS Span 26,54 m (87 ft 1 in); length 22,75 m (74 ft 7⅝ in; wing area 60 m^2 (646 ft^2).
ENGINES Two TV7-117S.
WEIGHTS Max 21 t (46,295 lb).
PERFORMANCE Cruise 500-520 km/h (311-323 mph); range (3 t) 4000 km (2,485 miles), (5) 2000 km (1,242 miles); field length 1,8 km (5,900 ft).

Tu-330 Major programme for airlifter. High wing and pylons derived from Tu-204, pressurized fuselage 5,35 m diam, cargo hold 19,5×4×3,55 high, rear ramp door but fixed nose, triple tandem main gears; first of 10 to fly at Kazan 96 with PS-90AT engines, export versions planned with GE, PWA or RR engines.
DIMENSIONS Span 43,5 m (142 ft 8⅝ in); length 42,0 m (137 ft 9½ in); wing area 220 m^2 (2,369 ft^2).
ENGINES See above.
WEIGHTS Empty c46 t (101,400 lb); payload 35 t (77,160 lb); max 103 t (227,072 lb).
PERFORMANCE Cruising speed 800 km/h (497 mph); range (30 t) 3000 km (1,864 miles), (20 t (5600 km (3,480).

Tu-334 Completely new and uncompromised transport to replace Tu-134 (task originally assigned Yak-42). To save time/cost forward fuselage and cockpit same as Tu-204 apart from short nose gear (tyres 620 mm×180 mm) raised into small bay. Wing almost exact scale of Tu-204, though inboard flap one piece on single track and outboard spoiler has four sections (both as originally schemed for 204). T-tail with sharply swept, untapered fin carrying tailplane with anhedral and same modest sweep (28°) as wing. Inward retracting main gears with short levered-suspension legs and twin wheels (tyres 1070 mm×390 mm). Engines hung on rear-fuselage fin-spar frames. Full-length cowl, blocker doors close fan duct and translating cowl opens all-round cascade reverser. Door front/rear on left with service doors opposite, buffet/toilet at each end, basic 3+3 seating for 102 at 810 mm pitch, 110 at 780 mm or 130 at 750 mm; typical two-class 8+84. Systems/avionics based on Tu-204.

334-100 Basic version, prototype should have flown mid-92 and been delivered Aeroflot 93; actual delivery to successor airlines now unlikely before late 96. Data are for prototype, intended now to fly spring 95. Later in 95 No 2 is to fly, followed in 96 by No 3, intended eventually to have 18,000-20,000-lb BMW Rolls-Royce 715 engines.

Originally ANTK Tupolev planned a later TU-334 with D-27 propfans and a stretched cabin, but this was discontinued in 94.
DIMENSIONS Span 29,77 m (97 ft 8 in); length 31,26 m (102 ft 6½ in); wing area 83,2 m^2 (895·6 ft^2).
ENGINES Two D-436T.
WEIGHTS Empty 41,05 t (90,498 lb); max payload 11 t; MTO 46,1 t (101,631 lb).
PERFORMANCE (est) Cruising speed 800-820 km/h (497-510 mph); range with 9251 kg (102 pax) payload 2000 km (1,242 miles); balanced field length 2200 m.

Tu-336 Projected stretched Tu-334, 120 seats at 810 mm pitch, NK-112 engines.

Tu-338 Projected stretched Tu-330, NK-94 engines.

Tu-34 First light aircraft from Tupolev since ANT-1, multirole utility of sophisticated design for STOL from unprepared strips. Wing of high aspect ratio with structural box behind cabin, LE slats, TE track-mounted flaps, ailerons and spoilers, streamlined fuselage with cabin 2750 mm×1450 mm×1300 mm with large side doors, T-tail, pusher engines with 3-blade c/s props, ldg gear retrac into fuselage. To be available from 96 carrying 4 passengers, or 450 kg (992 lb) or 2 stretchers or patrol sensors. Many versions projected.

Tu-24 mock-up

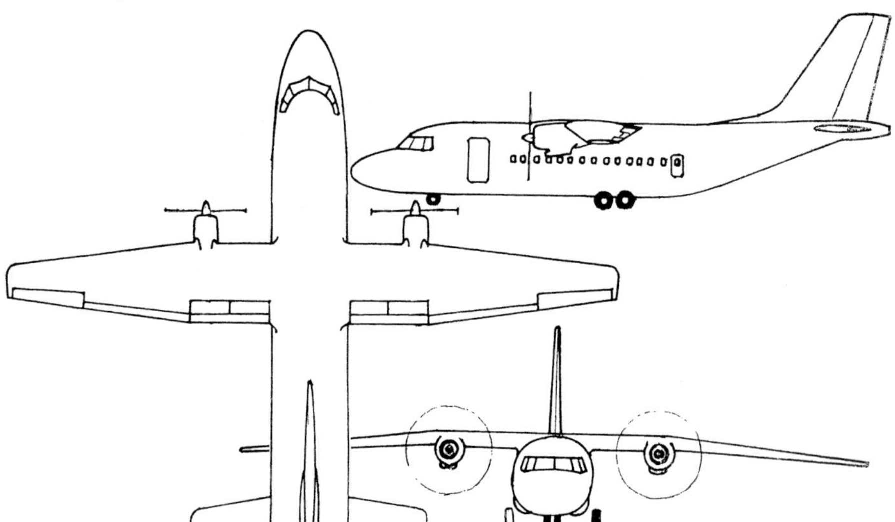

Tu-130

Tu-330 model

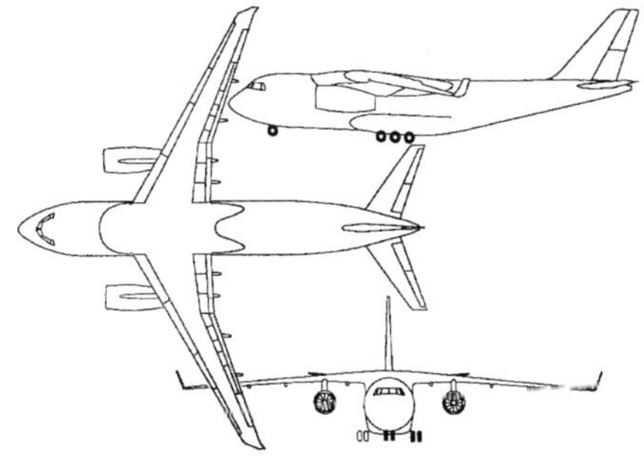

Tu-330 project

Tu-334 artwork

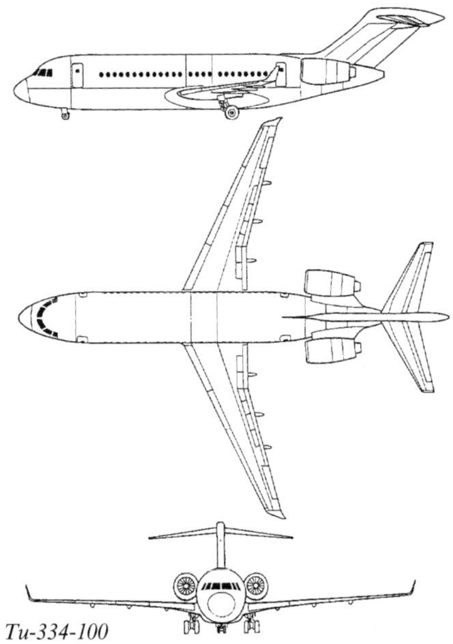

Tu-334-100

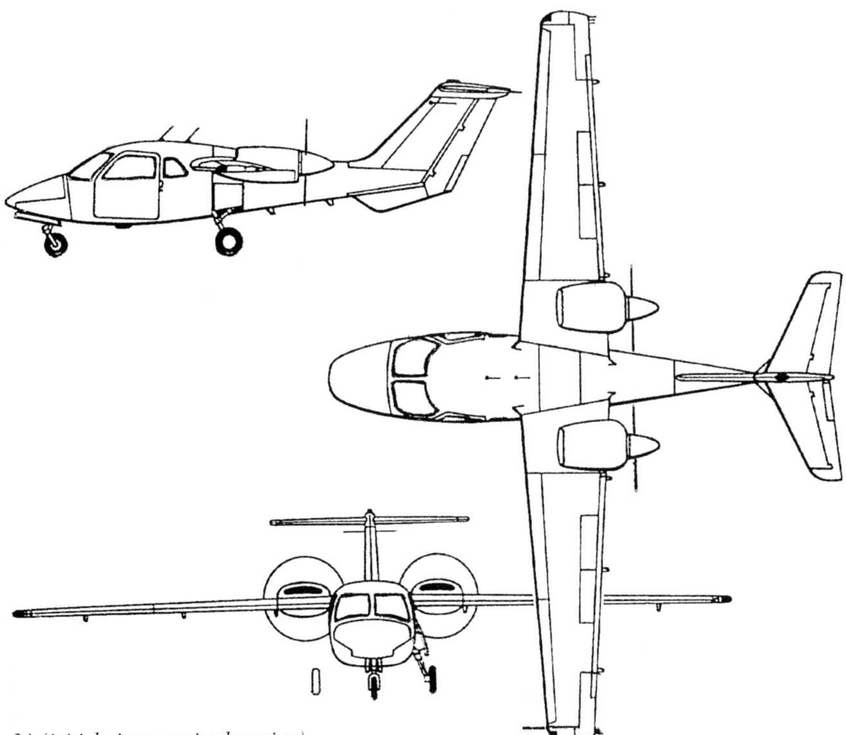

Tu-34 (initial piston-engined version)

DIMENSIONS Span 13,2 m (43 ft 3⅝ in); length 9440 m (30 ft 11⅝ in); wing area 19,8 m^2 (213 ft^2).
ENGINES Two D-200 or US types in 220 hp class.
WEIGHT MTO 1,9 t (4,189 lb).
PERFORMANCE Max speed 400 km/h (249 mph); cruise 280-340 km/h (174-211 mph); service ceiling 8,2 km (26,900 ft); range 2100 km (1,300 miles); field length 400 m.

Tu-404 Ambitious programme to build UHCA (ultra-high-capacity aircraft). Began with conventional configuration with 4 underwing engines, but various fuselages all posed severe problems and Ch.Designer Yurii V. Vorobyov was led to superior layout with conventional wings attached to edges of very wide lifting fuselage. Interior for 700-750 pax 'like theatre or cinema, floor and roof joined by slender ties to bear pressurization loads.' Model shown Paris 93 is favoured configuration with projecting 2-pilot cockpit at nose, widely spaced oblique tails with ventral fences beneath, and six advanced turbofans along trailing edge (model showed propfans). LNG is envisaged as principal fuel. A major task is precise stress analysis. Discussions with poss partners, eg Aérospatiale and DASA.
DIMENSIONS Span c100 m (328 ft); length c52,5 m (172 ft); wing area c700 m^2 (7,535 ft^2).
ENGINES Six NK-62 assumed.

Tu-404 model

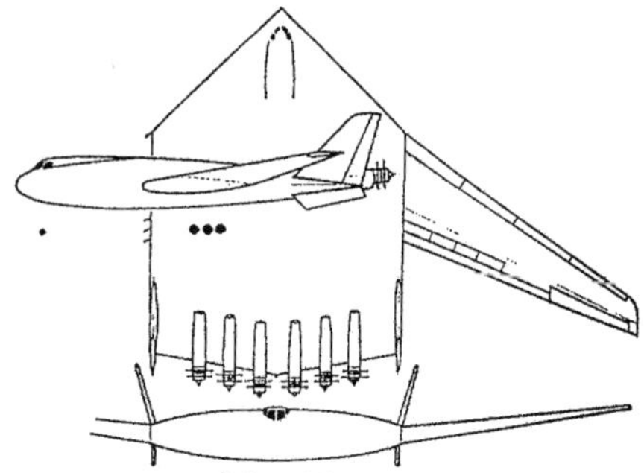

Tu-404 project

Tu-414 model

WEIGHT MTO c550 t (1,212,500 lb).
PERFORMANCE Cruise c840 km/h (520 mph); range 13000-15000 km (<9,320 miles).

Tu-414 Long-range bizjet and regional transport with supercritical wing, cabin for up to 30 pax. Data provisional.
DIMENSIONS Span 28,8 m (94 ft 6 in); length 28,2 m (92 ft 6 in).
ENGINES Two 14,900 lb BR710.
WEIGHTS Max payload 3,3 t (7,275 lb); max 40 t (88,183 lb).
PERFORMANCE Cruise 860 km/h (534 mph); range (8 pax) 10 560 km (6,560 miles), (30) 8500 km (5,282 miles); field length 2040 m (6,690 ft).

Tu-2000 Project for exptl aerospaceplane, single stage to orbit. To research propulsion, materials and other aspects to underpin commercial payload carrier with short response time, versatile capabilities and independence of ground support. Main initial tasks devt of powerful computational aerothermodynamics, followed by materials choices and devt to efficient air-breathing powerplant usable from rest to M 25. Circular-section body riding on low wing with 70° drooping LE, elevons on TE, vert tail, FBW controls to aerodynamic surfs and reaction-jet controls for thin atmos. Crew two, tankage for 35-50 t LO_2/LH_2, conventional TO/ldg. Start of construction possibly 2000.
DIMENSIONS Span c14 m (45 ft 11 in); length c60 m (197 ft); wing area 160 m² (1,722 ft²).
WEIGHT MTO 90 t (198,400 lb).
PERFORMANCE Max speed M6 (6400 km/h, 3,965 mph) at 30 km (98,425 ft); with 2nd stage M15 to orbit, with M25 in more distant future.

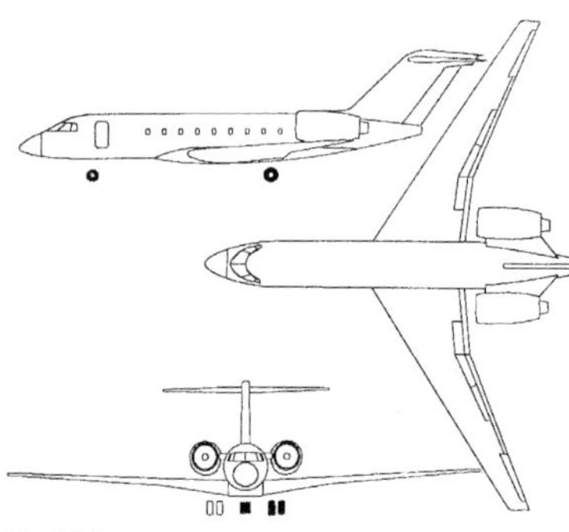

Tu-414

Triton Projected multirole amphibian by Kazan design bureau of Tupolev Avn Corp. See KKB.

Tyrov

This is best English spelling of name, but Russian rendering Tairov (which English person would pronounce differently) leads to aircraft designations Ta. Vsyevolod Konstantinovich Tyrov was lecturer at MAI circa 1930-34, and in 1934-35 first deputy in Polikarpov OKB. Late 35 helped form OKO at Kiev and became chief constructor. OKO managed several major projects until Tyrov was killed flying Moscow/Kuibyshyev Dec 41.

OKO-1 Straightforward passenger transport funded in part by Ukrainian regional government. Clean low-wing monoplane with trousered landing gear. All-wood construction except for dural control surfaces with fabric covering and dural skin on underside of split flaps. Two-spar wing made in one piece, with two metal fuel tanks inboard, ply skin. Monocoque fuselage basically circular section, enclosed cockpit for two pilots side-by-side, main cabin with three seats each side or 550 kg cargo. Engine in long-chord metal cowl with louvred front baffle plate (poss based on I-16) and driving Hamilton-type two-blade VISh-6 propeller. Split flaps elec drive, pneu wheel brakes, trimmers on all tail control surfaces, lightng and instruments for blind or night flying, cabin heated and soundproofed (lagged with Viamiz material), full GVF service eqpt. First flight Oct 37 and testing completed June 38. Performance regarded as outstanding; no reason known for failure to win order.
DIMENSIONS Span 15,4 m (50 ft 6⅓ in); length 11,6 m (38 ft 0¾ in); wing area 35,1 m² (378 ft²)
ENGINE One M-25A.
WEIGHTS Empty 2370 kg (5,225 lb); fuel/oil 364 kg+42 kg; loaded 3,5 t (7,716 lb).
PERFORMANCE Max speed 305 km/h at SL, 347 km/h (216 mph) at 2,6 km; time to 1 km, 3 min; service ceiling 6740 m (22,100 ft); range 1700 km (1060 miles) with full payload, 2300 km max fuel; take-off 420 m/20 s; landing 430 m/22 s/95 km/h.

OKO-4 Small single-seat sesquiplane fighter/attack (Shavrov: *Shturmovik*, implying armoured) aircraft powered by M-88. Armament two BS plus 100 kg bomb. Prototype built from mid-38 but when completed following year clearly unsuccessful and removed from factory programme without being flown.

Ta-1, OKO-6 This twin-engined escort fighter was built to meet VVS requirement of early 38 in parallel with several other similar aircraft. Though large project for OKO, **No 6** was carried through rapidly and successfully, and air-

OKO-1

OKO-4 tunnel model

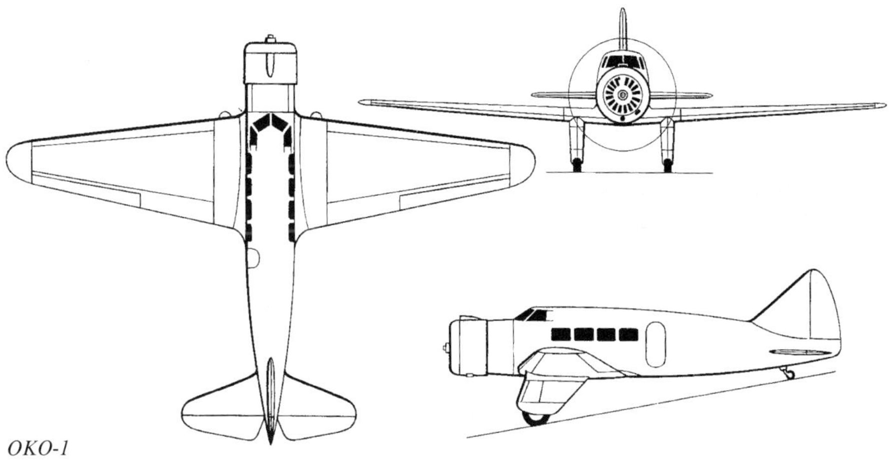

OKO-1

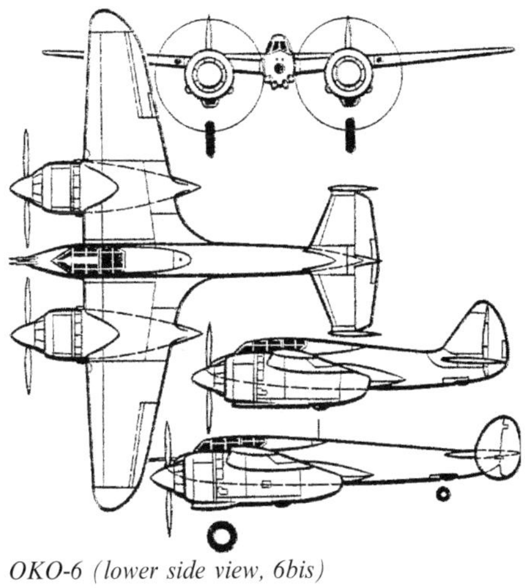

OKO-6 (lower side view, 6bis)

Ta-3

craft had potential for combining exceptional speed and firepower, though of course with basic lack of agility of relatively large fighter. Single-seat mid-wing, mixed construction. Wing with two spars each 30KhGSA T-section booms and D1 plate webs, rubber-pressed D1 sheet ribs, flush-riveted D1 skin except LE skinned with thick Elektron, split flaps. Slim fuselage semi-monocoque, mainly D1 but with all-wooden tail section. Conventional single-finned tail with stressed-skin fin/tailplane and fabric-covered control surfaces with trimmers. Engines in Elektron cowls, driving VISh-23 propellers, handed to eliminate torque. II landing gear units folding to rear with twin doors, main gears twin legs with actuator to mid-point of rear bracing strut. Heavy armament of four ShVAK-20 in underside of forward fuselage, two BS in top of nose ahead of cockpit. Compact cockpit with sliding canopy, bulletproof front windscreen and armour ahead and behind. First flight 31 Dec 39 by Yu. K. Stankevich. Tested by LII until summer 40 when con-rod broke. Directional stability unsatisfactory and second prototype, **OKO-6bis**, built with tailplane of increased span with added tip fins. Later original vertical tail removed and replaced by larger twin fins and rudders. More powerful engines with reduction gears, both LH built with longer fuselage, twin-finned tail, more powerful geared engines, both LH rotation, improved KPD props. Four ShVAK and two ShKAS. Handling much improved. (Factory drawing shows dihedral from root, photo shows horiz c-s). Third aircraft, initially OKO-6bis, later called **Ta-3**, flown by Yu. K. Stankevich May 41. M-89 engines, one AM-37 and two ShVAK. **Ta-3bis** with increased span (intended for M-82 engines) left incomplete by invasion. Designer's death halted further work.

DIMENSIONS Span 12 658 m (41 ft 6 3/8 in), (3bis) 14,0 m; length (6) 8430 m (27 ft 7 7/8 in), (remainder) 9827 m (32 ft 2 7/8 in); wing area 26,9 m^2 (290 ft^2), (3bis) 33,5 m^2.

ENGINES (6) two M-88, (6bis) M-88R, (3bis) M-89.

WEIGHTS Empty (6) 4 t, (3bis) 4,5 t; loaded (6) 5,8 t, (3bis) 6626 kg.

PERFORMANCE Max speed (6) 488 km/h at SL, 567 km/h (352 mph) at 7,5 km, (6bis) 477/595, (3bis) 448/595 ; service ceiling (6) 10,3 km, (3bis) 11 km; range (6, 6bis) 1060 km (660 miles), (3bis) 2065 km (1,283 miles).

OKO-7 Tyrov's last known design, heavy high-speed fighter to 40 VVS requirement, two M-90 or AM-37, three ShVAK and two ShKAS.

Vakhmistrov

Vladimir Sergeyevich Vakhmistrov was a senior LII research engineer. One of his projects of 30 was development of glider for use as air/air gunnery target, and he eventually perfected method of carrying it on upper wing of R-1 and releasing in flight. This suggested one of first 'parasite' schemes: fighters carried by bombers far over hostile territory and if necessary released for their protection, thereafter hooking back on. Vakhmistrov did preliminary calculations and then obtained VVS approval for flight test. First combination, called *Zvyeno* (link) 1 or **Z-1**, chosen to be TB-1 carrying I-4 above each wing. Aircraft modified (see ANT-5) for purpose. Each fighter positioned behind wing of bomber and hauled up wooden ramp by ropes from towing crew from front. Secured by hold-down link on maingear axle and rear fixture on tripod attached to bomber wing holding fighter beneath cockpit. First flight Monino 3 Dec 31, fighters flown by V. Chkalov and A. F. Anisimov, bomber by A. I. Zalevskii and A. R. Sharapov with Vakhmistrov as observer. First take-off made with fighter engines full power. Co-pilot, presumably Sharapov, detailed to release fighters; improperly briefed and released Chkalov's axle without waiting for fighter pilot's signal he had released rear attachment: I-4 reared up but Chkalov was quick enough to release rear hold-down before disaster. Release of other I-4 some seconds later according to plan. Original plan had been to let both go together, and this incident showed TB-1 rock-steady with fighter on one wing only.

Next combination, **Z-1a**, comprised TB-1 with two I-5. This was first flown Sept 33 by pilots I. F. Grozd and V. K. Kokkinaki, with TB-1 flown by P. M. Stefanovskii. Testing then progressed to much larger TB-3, starting with Zvyeno **Z-2** comprising TB-3 carrying I-5 on each wing and a third above fuselage. Flown Aug 34 by Zalevskii (TB-3), Suzi, S. P. Suprun and T. T. Alt'nov. **Z-3** would have hung I-Z under each wing of TB-3. **Z-4** unknown. **Z-5** first attempt at hooking back on; combination chosen to be single I-Z with large suspension structure of steel tubes with curved top guide terminating at rear in sprung hook releasable by wire to cockpit (exact copy of F9C Sparrowhawk). TB-3 had large steel-tube trapeze under fuselage, folded up for take-off and landing. Prolonged tests with TB-3 flown by Stefanovskii and I-Z by V. A. Stepanchenok, latter practising by breaking strings holding rows of coloured flags one by one. Final hook-on successfully accomplished 23 March 35, first under-fuselage hook-on in world. Final combination of original series **Z-6** was TB-3 with I-16 under each wing, suspended between two large V-strut links of aluminium streamline tubing picking up sliding horiz spigots above main spar. First flight Aug 35, TB-3/AM-34 flown by Stefanovskii, fighters by K. K. Budakov and Nikashin.

Culmination of concept was grotesque *Aviamatka* (mother aircraft) in which in Nov 35 TB-3/AM-34 took off carrying two I-5 above wings, two I-16 below wings and with under-fuselage trapeze. While flying over Monino trapeze was lowered and Stepanchenok hooked on in I-Z, making combination of six aircraft of

I-Z hooking on Aviamatka

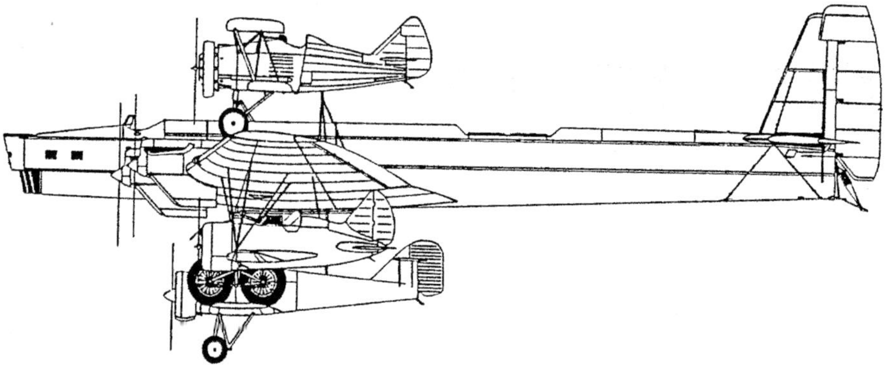

Aviamatka

Z-7

four types. After several passes all fighters released simultaneously. By this time Vakhmistrov had scheme for specially designed ultra-fast fighters (see I-Ze entry) carried in large number (6 or 8) under/over large tailless bomber (full-scale derivative of Kalinin VS-2). When purges began this withered through elimination of its supporters, but Vakhmistrov did manage to fly one final combination. SPB, described under I-16, replaced I-16 underwing fighters by slightly modified I-16s carrying two FAB-250 bombs each. First flown July 37, TB-3/AM-34RN flown by Stefanovskii, dive bombers by A. S. Nikolayev and I. A. Taborovskii.

Unexpectedly, Vakhmistrov managed to fly one more combination; **Z-7** with I-16 under each wing and third latched on fuselage trapeze in air was tested Nov 39, pilots Stefanovskii, Suprun and Nyukhtikov, though severe difficulty. Final outcome of all *Zvyeno* experiments was decision early 40 to form combat unit with SPB. Equipped with six modified TB-3/AM-34RN and 12 SPB, based Yevpatoriya. Made one famous combat mission 25 Aug 41 against Danube bridge at Chyernovod, Romania (main rail link to Constanta). This bridge was destroyed. Other missions, from Crimean bases.

I-Z, I-Ze Project for simple fighter drawn by Vakhmistrov 1934-35: low-wing monoplane with 850 hp GRKs engine, 1910 kg but only 7,75 m span and 10 m^2 wing, speed 518 km/h. No landing gear except centreline skid. Prototype started but discontinued 36 at start of purges.

Yakovlev

Aleksandr Sergeyevich Yakovlev rose from humble start to become top Russian fighter designer during the Second World War and deputy aviation minister. His OKB has produced exceptional diversity of aircraft, though he personally liked small sporting types. Most exceptionally for his generation he was never imprisoned. For many years he was chief father-figure invited to sit in at all top decision-making, especially concerning aircraft design.

Born 19 March 06 to prosperous family (father worked for Nobel Oil in Moscow). Teenage aviation enthusiast, managed via K. K. Artseulov to be appointed helper to V. Anoshchyenko building glider for Koktebel meeting 23. Following year, with technical help from Ilyushin, built his own glider, successful, but lacked army service necessary to enter VVA. With Ilyushin's help got job March 24 as menial labourer at VVA workshops, and gradually learned workshop processes, aircraft design and entire technology from strictly practical shopfloor basis. In 26 with help from V. S. Pyshnov designed powered aeroplane, VVA-3; excellent design and not only set world class record but gained entry for Yakovlev to VVA 27. Other designs followed, designated AIR for A. I. Rykov, then Lenin's successor as Chairman of Council of People's Commissars. Graduated 31 and joined Polikarpov KB not as designer but as engineering supervisor, giving freedom to create own sporting prototypes with distinctive red/white colour scheme. This frowned on, and when AIR-7 lost aileron in flight Yakovlev expelled from plant. After much lobbying awarded derelict bed factory on Moscow Leningradskii Prospekt, gathered own team and July 34 set up OKB which is still on same site with Yak museum. Rykov victim of purge 37, Yakovlev changed to Ya designations. Amongst many lightplanes, powerful Ya-22 put him on military and political map, showered with honours. On 1 Jan 40 first Yak-1 (then called I-26) rolled out, first of 37,000 derived Yak fighters outnumbering all other fighters of the Second World War. On 9 Jan 40 appointed Deputy Commissar of Aviation Industry and head of Dept of Experimental Aircraft Construction and Research, throughout war spending morning at OKB and rest of day to about 2 am at NKAP. Organized industry evacuation from Moscow area Sept/Oct 41. Helicopter dept 46, terminated 60. Post-war great variety of aircraft, notably trainers/aerobatic, tactical jets and

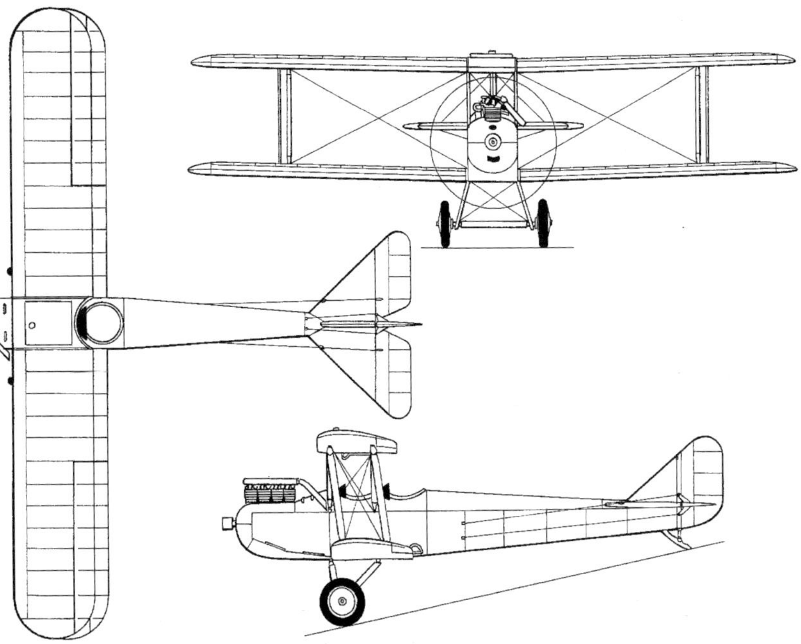
AIR-1

AIR-1

jet transports. Founder died 22 Aug 89, today Yakovlev Corp and A. S. Yakovlev design bureau headed by Aleksandr S. Dondukov, with founder's son a deputy general designer (Yak-40 and -77).

AIR-1, VVA-3 Success with gliders naturally led to wish to design aeroplane, conceived as 18 hp *Aviette* (ultralight). Discussed with senior VVA students, especially Pyshnov and Ilyushin. Former stressed two-seat dual trainer with more powerful engine would be more useful, and offered essential technical help. A year from late summer 25 on design and drawings, approved Osoaviakhim 26 Aug 26 and funds allotted (some added by Pioneers youth movement) and aircraft built in VVA lab by designer and his friends 5 to 11 pm daily. Generally positive attitude by VVA, but some jealousy (Yak a mere manual worker) and one final-year student sent report of faulty detail design, causing ban on work and prolonged interrogations of designer until Pyshnov bravely submitted personal guarantee of VVA-3 airworthiness. Completed on wire wheels 1 May 27; after fitting skis flown by Yak's squadron commander, Yuri I. Piontkovskii, 12 May 27.

Single-bay tandem dual biplane, constant-section (Göttingen 387) wings with two wood box spars and 13 ribs in each of four identical wings, ply LE, fabric elsewhere, upper/lower ailerons again ply LE and fabric with aluminium-tube tie-rod links, upper centre-section same width as fuselage with galvanized iron tank 65-lit of deeper aerofoil section inserted. Wooden frame fuselage with wire-braced trusses, fabric except for aluminium front cowl (engine cylinders exposed) and ply skin to rear of front cockpit and ply decking back to tail. Wooden tail, ply-skinned except fabric aft of spar on movable surfs. Control wires emerged from fuselage behind rear cockpit, rudder wires needing slots cut in tailplane LE. Bungee-sprung landing gear and tailskid. Many aluminium details (eg, top of fuselage above tailplane LE), padded leather edges to cockpit cut-outs, aluminium celluloid windscreens. Ply skin varnished, rest doped white with red trim (eg, fin, prop and most of each aileron). Wire wheels later given thin aluminium discs. Simple and positive to fly. With Osoaviakhim permission, long-range flight began 12 June 27, Moscow-Kharkov/Sevastopol. Long-range tank then put in passenger seat (in place of designer, who returned by train) and faired over. Piontkovskii returned non-stop 1240 km in 15 h 30 min, world class record. By this time called AIR-1, and civil registration R-RAIR, all to please commissar Rykov. Autumn 27 took part in Odessa Mil District manoeuvres. Preserved at OKB.

DIMENSIONS Span 8,85 m (29 ft 0⅓ in); length 6,99 m (22 ft 11⅛ in); wing area 18,7 m² (201 ft²).
ENGINE One 60 hp ADC Cirrus.
WEIGHTS Empty 335 kg (739 lb); fuel/oil (normal) 50 kg; loaded 535 kg (1,179 lb).
PERFORMANCE Max speed 140 km/h (87 mph) (recent account by Zasypkin, 150 km/h); cruise 120 km/h; time to 1 km, 8 min (3 km, 35); service ceiling 3850 m (12,630 ft); range (normal) 500 km (311 miles); take-off 80 m/8 s; landing 60 m/6 s/60 km/h.

AIR-2 In 28 Yakovlev tinkered with AIR-1 design and assisted by awards for long-range record built second aircraft called *Pioner* (pioneer). Eventually six built in four versions. First flew with Cirrus in better aluminium-sheet cowl. Larger fuel tank, cabane braced by diagonal forward struts, slightly modified wings with full-span ailerons on lower planes only, rounded LE on tailplane, rudder cables internal emerging close beside fin. Flown late 28. By late 29 second aircraft, **AIR-2s** (or **AIR-2S**), built with Siemens engine and several other changes, and with twin floats designed by V. B. Shavrov on struts of wire-braced mild-steel tube with al fairings. Red trim, with red/silver striped vertical tail which became Yak trademark. AIR-2s taken to Gorkii Park and flown 18 May 31 from Moscow River by B. L. Bukholts and Shavrov. Later flown by Piontkovskii, Yakovlev and others with complete success; Bukholts flew it under Old Crimea Bridge on Moscow River. Two similar aircraft built with interchangeable float/wheel gear, latter having bracing strut not behind but in front, pinned to bottom of engine-mount structure and with foot-step to assist work on top cylinder of engine. Final one (maybe two) aircraft had enclosed cockpits with side-folding individual Celluloid hoods, slats on upper wing (set at increased incidence) and powerful slotted ailerons on lower wing. Three-cylinder Soviet engine with aluminium-sheet front cowl, simple electric light system for night flying. First flown July 31; instant response to controls during spin recovery but hoods resonated with engine vibration.

DIMENSIONS Span 8,9 m (29 ft 2½ in); length 7,0 m (22 ft 11½ in), (2s seaplane) 7,7 m (25 ft 3 in); wing area 18,7 m² (201 ft²).
ENGINE One 60 hp Cirrus, 85 hp Siemens Sh 11 or 65 hp M-23.
WEIGHTS Empty (Siemens 2s) about 470 kg, (land) 420 kg (926 lb), (M-23) 403 kg (888 lb); fuel/oil 72+11 kg loaded (2s) about 710 kg (1,565 lb), (land) 660 kg (1,455 lb), (M-23) 646 kg (1,424 lb).
PERFORMANCE Max speed (2s) 140 km/h (87 mph), (land) 150 km/h (93 mph), (M-23) 141 km/h; time to 1 km (M-23) 8·7 min (46 min to 3 km); service ceiling (M-23) 3534 m (11,594 ft); range (M-23) 540 km (336 miles); take-off (M-23) 80 m/7 s; landing 140 m/9 s/60 km/h.

AIR-3 Designer's glider experience used in this parasol monoplane, again using almost same fuselage and tail as AIR-2. Wing again Göttingen 387 section, rectangular centre section of constant profile and tapered outer panels, dihedral 0° on upper surface. Two box spars with spruce booms basically 60 mm × 40 mm, twin webs 2 mm ply (3 mm at outer-panel joint), rib trusses glued and pinned from 6 mm × 6 mm with 1 mm ply. Neat ply skin from front spar on underside to rear spar above, giving better aerodynamic efficiency. Three metal fuel tanks, total 176-lit, recessed flush with aerofoil in centre section between spars. Centre section on same cabane struts as AIR-2; main wing struts parallel each side from bottom of fuselage to spars ½-way between last two ribs of c-s, wrapped mild-steel sheet with streamline section 64 mm × 32 mm. Fuselage as before, longerons 27 mm × 27 mm mid-section, 20 mm × 20 mm at tail. Engine on welded steel-tube mount, designed for Czech Walter but provision for M-23 and fitted later. Al-sheet nose, oil tank in front of front instrument panel. Landing gear as AIR-1 but stronger with steel tubes (most 27 mm OD by 1,5 mm) faired by wood, axle 44 OD × 1,5 mm.

Built at GAZ-39 from April 29, completed end June and tested for VVA by A. I. Filin, later with Piontkovskii, D. A. Koshits and A. B. Yumashyev. Later based at October Field, painted red fuselage with SSSR-310 white on left and *Pionerskaya Pravda* (young communist paper) on right. From 26 Aug 29 used for intensive trials between Moscow and Mineralnye Vody, 1750 km/1830 km, non-stop, usually flown by Filin and A. F. Korolkov, best time 10 h 23 min on 6 Sept, flew 12 000 km in three months. In 38 made 'invisible', see Kozlov EI.

DIMENSIONS Span 11,1 m (36 ft 5 in); length 7,1 m (23 ft 3½ in); wing area 16,5 m² 178 ft²).
ENGINE Originally one Walter NZ-60 rated 60 hp; by Feb 31 had M-23.
WEIGHTS Empty 392 kg (864 lb); fuel/oil 176 kg+20 kg; loaded 762 kg (1,680 lb).
PERFORMANCE Max speed 146 km/h (91 mph)

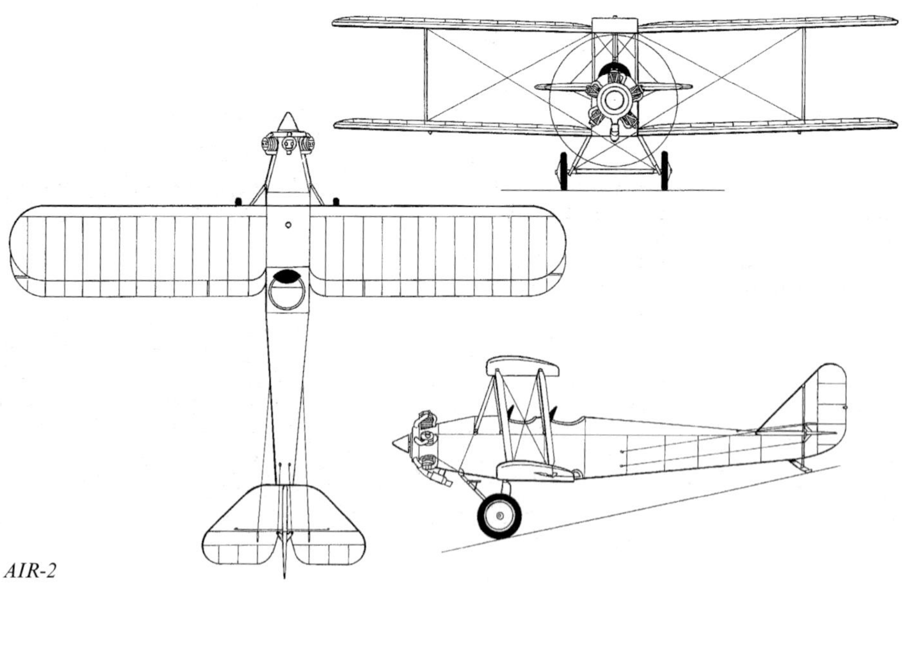
AIR-2

AIR-2

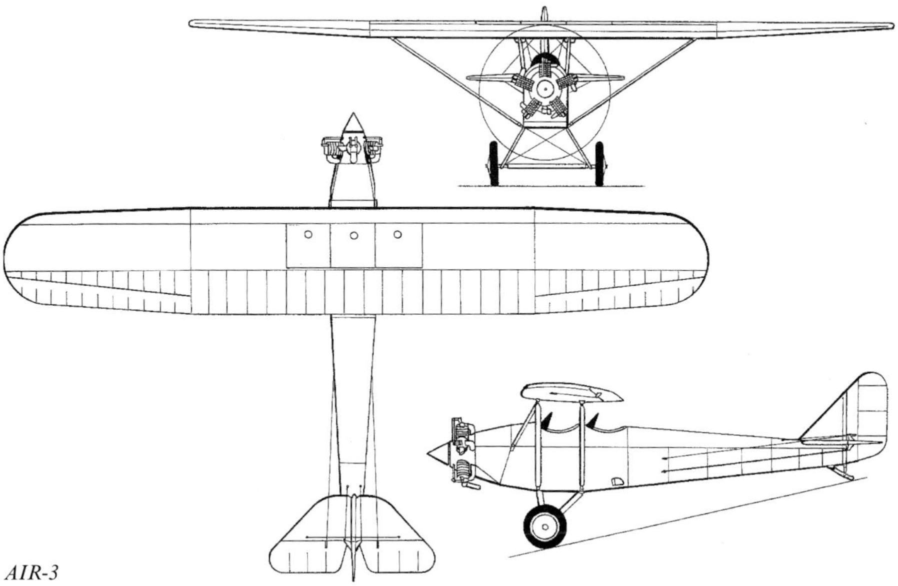

AIR-3

though with favourable wind averaged 166,8 on best trip, see above; time to 1 km, 6 min (34 min to 3,2 km); service ceiling 4,2 km (13,780 ft); range 1835 km (1,140 miles); take-off 60 m; landing 66 km/h (41 mph).

AIR-4 Essentially AIR-3 with new main landing gear, divided (pyramid) type with rubber-sprung shock strut from wheels to pin-joint linking front main wing strut to V-struts attached to fuselage. Cabane struts sloped to pick up wing further outboard, fuel capacity 110 lit, more comfortable cockpits with complete dual insts, other detail changes. Two built at GAZ-39, SSSR-311 and 312 (both striped tail, red fuselage with black-lined silver side panel), tested Sept 30. Following month circuit patterns between Moscow and Sevastopol.

Late 30 re-engined with M-23. Early 33 selected by NII GVF as vehicle for major attack on achieving maximum lift coefficient. First AIR-4, still with Walter engine, fitted with new wing by B. N. Zalivatskii and L. M. Shekhter. Work apparently done at NII; aircraft redesigned **E-31** and given this registration. Wing similar structure and profile but made in one piece to square tips beyond which were then-novel ailerons forming complete rounded tips, pivoted about steel-tube aileron spar 25% chord held in last three ribs of wing, movement ±15° (Zasypkin), +30°/–60° (works drawing), long tube ahead of aileron root with streamlined lead weight on tip. Entire trailing edge of fixed wing occupied by four sections large Schrenk flap driven by worm gear from cockpit. Span/area of wing increased by ailerons, but flaps produced mainly high drag at full deflection 60°. Alternative designation of E-31 was *Mekhanizirovannyi Krylo*, mechanized wing. Flown about July 33, outstanding results; lowest min-flight and landing speeds then known for conventional aeroplane with modest penalties in other performance.

DIMENSIONS Span (4) 11,1 m (36 ft 5 in), (E) 12,55 m (41 ft 2⅛ in); wing area (4) 16,5 m^2 (178 ft^2); (E) 18,05 m^2 (194 ft^2).

ENGINE One 60 hp Walter NZ-60, (second, 30) M-23.

WEIGHTS Empty (4) 395 kg (871 lb), (E) 440 kg (970 lb); fuel/oil 65 kg; loaded (4) 630 kg (1,389 lb), (E) 670 kg (1,477 lb).

PERFORMANCE Max speed (4) 150 km/h (93 mph), (E) 145 km/h (90 mph); time to 1 km (4) 6 min (30 min to 3 km); service ceiling (4) 4 km, (E) still climbing at 5,2 km; range (4) 500 km (310 miles), (E) 450 km; take-off (4) no data, (E) about 50 m/5 s; landing (4) 100 m/10 s/66 km/h, (E) 40 m/4 s/34 km/h (21 mph) (Zasypkin and Kondratyev both claim E-31 flown at 35 km/h and could land at 30 TAS).

AIR-5 Logical next stage beyond original AIR-4, this light transport incorporated designer's lately acquired knowledge of welded-steel tube structures. Enclosed cabin fuselage made of welded MS (mild-steel) tube using designer's own accurate jigging. D1 dural skin over engine crankcase and back to cabin, fabric to rear. Two-spar wooden wing with ply skin to front spar, fabric elsewhere. Ailerons Frise type with large slots, wood/fabric but ply skin around LE to spar. Tail D1 dural and fabric, tailplane with

AIR-5

AIR-6

pilot-adjusted incidence and asymmetric duo-convex section. Pyramid-type divided landing gear with welded steel tubes and rubber-sprung shock strut. Main wing struts riveted MS streamline section with added vertical braces from struts to spars. Comfortably furnished cabin with 4 to 5 total seats, single pilot, door each side. Single AIR-5 built 31, red fuselage, striped tail, registered No 38. Excellent aircraft but no available production engine (proposal to license Whirlwind as M-48 never implemented).
DIMENSIONS Span 12,8 m (42 ft 0 in); length 8,0 m (26 ft 3 in); wing area 23,0 m² (248 ft²).
ENGINE One 200 hp Wright Whirlwind J-4.
WEIGHTS Empty (Shavrov) 812 kg (1,790 lb); fuel/oil 200 kg+20 kg; loaded 1390 kg (3,064 lb).
PERFORMANCE Max speed 193 km/h (120 mph); cruise 159 km/h; time to 1 km 6·5 min (27 min to 3 km); service ceiling 4275 m (14,000 ft); range 1000 km (620 miles); take-off 100 m; landing 100 m/75 km/h.

AIR-6 Yakovlev next designed similar small transport to reduced scale matched to readily available 100 hp engine. Resulting AIR-6 identical in detail design to AIR-5, probably over-strength as same sections and detail fittings, but scaled down with two wing tanks of 75 lit each and fuselage tailored to two (max three) seats. Prototype two seats in tandem, door each side, painted as AIR-5. Ideal for GVF and in 36 GAZ-23 built 53, including 20 ambulance variant (possibly designated **AIR-6S**) with larger triangular door on left and provision for attaching standard GVF stretcher and with seat for medic. At least one aircraft with spats and Townend-ring cowl, first in world formally called Executive Aircraft; at least one used by GVF central direction committee. Flight of (believed three) AIR-6 flew Moscow/Irkutsk/Moscow in stages 34. In 33 one fitted twin floats designed by V. B. Shavrov (not same as AIR-2 floats, larger and difficult to make with frames of curved glued pine sections 4 mm × 12 mm). Eventually several float examples, not all identical but all called **AIR-6A**. Rather sluggish (floats weighed 120 kg, compared with 35 kg for land gear) but one set class record 583 km in 6 h 5 min and 23 May 37 Ya. B. Piss'menyi flew one with extra tanks Kiev/Batum 1297 km non-stop 10 h 25 min.
DIMENSIONS Span 12,08 m (39 ft 7⅝ in); length 7,8 m (25 ft 7 in); wing area 19,8 m² (213 ft²).
ENGINE One M-11.
WEIGHTS Empty (prototype) 584 kg (1,287 lb), (series) 616 kg (1,358 lb); fuel/oil 110 kg+12 kg; loaded (p) 843 kg, (s) 961 kg (2,119 lb).
PERFORMANCE Max speed (p) 166 km/h, (series) 169 km/h, (6S) 156 km/h, (6A) 150 km/h (93 mph); climb (s) 6·6 min to 1 km, 29·3 min to 1 km; service ceiling (s) 4 km; range (s) 715 km (444 miles), (6A) 600 km; TO 85 m; ldg 120 m, (6S) 165 m, (6A) 150 m, all 68-70 km/h.

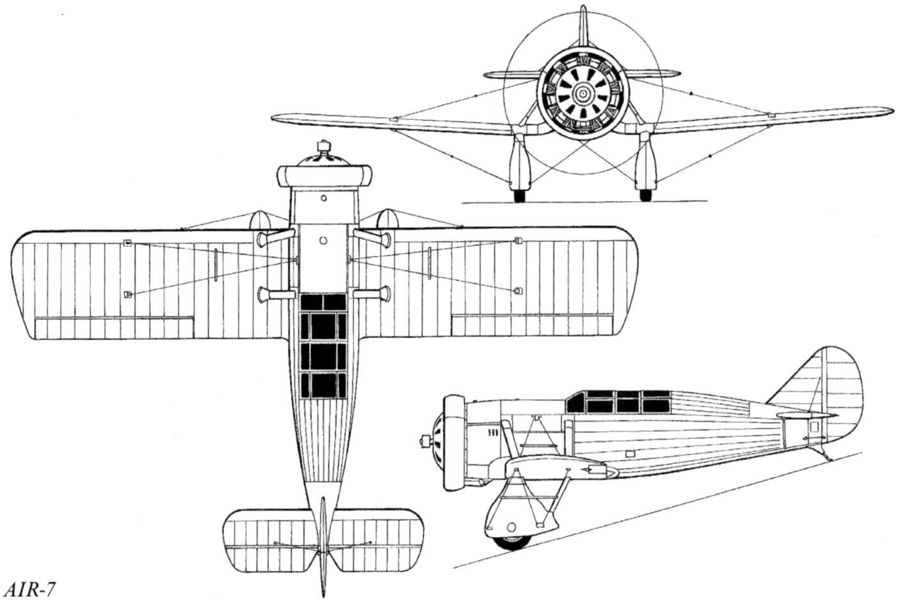

AIR-7

AIR-7

AIR-7 Main occupation of GAZ-39 design office was I-5 fighter, and Yakovlev became obsessed with belief he could design much faster aircraft using same engine. Winter 1931-32 spent in project study which settled on tandem-seat sporting machine with braced low wing. Wing 8% Göttingen 436 profile, wood structure with two spars bolted to centre section integral with fuselage, ribs glued/pinned from pine sections and ply skin over LE, fabric aft of front spar. Fuselage rectilinear truss of welded mild-

AIR-8

steel tube built up with wood/D1 secondary structure to more rounded shape, dural panels to front of cockpit, fabric thence to tail. Main fuselage bay ahead of cockpit welded to centre-section spars, span 2,3 m between outer bolt-holes carrying wings and main landing gear, with two parallel struts (steel tube of streamline section) from tips of c-s spars to upper longerons of fuselage. Main gears mainly steel tube with rear V-struts providing pivot for horizontal fork locating wheel (710 mm × 135 mm tyre) fore/aft and vertical fork and single leg forming shock strut with 12 rubber and 13 steel interleaved washers. Surrounding trouser fairing aluminium, wire bracing streamline profile 18 mm × 4 mm. Tail and ailerons D1 and fabric, fin offset 1·5° to counter torque, slotted ailerons in inner/outer sections. Engine in Townend-ring cowl driving 2700 mm prop with two polished dural blades. Oil tank top of first fuselage bay, oil cooler between bottom cylinders, two fuel tanks in second fuselage bay, total 400 lit. Tandem cockpits, passenger in rear with three flight instruments, folding celluloid hoods, pitot tube under outer right wing, venturi for gyro turn/slip (Kollsman).

Basically simple aircraft, construction started April 32 and though unpopular with designer's employers caused intense interest among Chkalov and other VVS test pilots. Completed Nov 32 in usual red/silver livery, no registration. First flight 19 Nov by Yu. I. Piontkovskii with 80 kg ballast on rear seat. Exceeded designer's estimate of 320 km/h and fastest aircraft in country. In 33 flown by many pilots with generally glowing opinions, 25 Sept 33 flown by Piontkovskii to national speed record of 332 km/h, engine overspeed 80 rpm. Many suggestions for AIR-7 to be produced in series as fighter or as transport for newspaper matrices to other cities. Early 34 special show for VVS officers and other VIPs, with L. Malinovskii, Osoaviakhim deputy chairman passenger. Bad weather, aileron came off at high speed at low level, Piontkovskii made masterly forced landing in railway yard. Discovered design error; Shavrov states failure due to flutter. Commission (which heard no evidence from Yakovlev) concluded 'to forbid Yakovlev to carry on with design work and to notify government he is unworthy to receive award' (he had been recommended for decoration). Yak and associates eventually evicted from GAZ-39 and forbidden to build any more aircraft.

After great difficulties Yak's party membership enabled him to win limited support from Ya. E. Rudzutak, chairman of party central control commission, who sampled 'executive' AIR-6 and eventually got NKAP to offer Yak old bed factory on Leningradskii Prospekt, near Moscow Central Aerodrome. This was converted to aircraft production autumn 34 and home of OKB-115 ever since.

DIMENSIONS Span 11,0 m (36 ft 1 in); length 7,8 m (25 ft 7 in); wing area 19,4 m² (209 ft²).

ENGINE One M-22.

WEIGHTS Empty 900 kg; fuel/oil 300 kg; loaded 1400 kg (3,086 lb).

PERFORMANCE Max speed 332 km/h (206 mph); time to 1 km 3·6 min (12 to 3); service ceiling 5,8 km (19,000 ft); range 1300 km (808 miles); landing 150 m/110 km/h.

AIR-8 Single AIR-4 type aircraft built at GAZ-39 mid-34 with small modifications requested by VVS. Wing increased area but same span, low-pressure balloon tyres and modified shock-strut extending up to wing, for rough-field or snow operation. First flown with M-23 but incurable vibration resulted in substitution of 60 hp Walter and finally 85 hp Siemens (data for this). Passed NII GVF testing 34 but never put to use.

DIMENSIONS As AIR-4 but wing area 18,0 m² (194 ft²).

ENGINE One 85 hp Siemens Sh 11.

WEIGHTS Empty 430 kg (948 lb); fuel/oil 85 kg; loaded 675 kg (1,488 lb).

PERFORMANCE Max speed 150 km/h (93 mph) landing 65 km/h. No other data.

AIR-9 One of least-known Soviet aircraft, this simple sports machine was first in series of Yak low-wing sporting and training aircraft made in vast quantities to present day. Designed 33 as contender in Aviavnito contest for tandem cabin tourer/trainer powered by M-11. Structurally similar to AIR-7: welded mild-steel tube fuselage with light secondary fairing structure, wooden two-spar wing with ply skin over upper surface and on underside back to front spar, dural tail with fabric, fabric-skinned fuselage and fabric rest of wing except dural/ply flaps, latter innovation for Yak. Tandem cockpits with dual control and folding Celluloid hoods. Wire-braced spatted main wheels on single rubber shock struts. Engine in helmeted cowl, prop with spinner. Contest laid stress on safe flight, hence auto slats over 60% span. Yak began prototype at GAZ-39 but soon evicted. Construction continued in wooden hut, then no progress possible until OKB established July 34. Two eventually completed April and Oct 35, first, AIR-9, with inverse-slope windscreen and second (**9bis**, sometimes called **9A**) with normal windscreen and Townend-ring cowl. Second (9bis) used by Irina Vishnyevskii and Yekaterina Mednikova 3 May 37 for women's class record climb to 6518 m.

DIMENSIONS Span 10,2 m (33 ft 5⅝ in); length 6,97 m (22 ft 10⅜ in); wing area 16,87 m² (182 ft²).

ENGINE One M-11.

WEIGHTS Empty (bis) 495 kg (1,091 lb); fuel/oil 63,5 kg+17,5 kg; loaded (9) 799 kg, (bis) 768 kg (1,693 lb).

PERFORMANCE Max speed 215 km/h (134

AIR-9

mph); climb 4·8 min to 1 km, 16·4 min to 3 km; service ceiling 6080 m; range (bis) 695 km (432 miles); TO 6·5 s/80 m; ldg 65 km/h/90.

AIR-10 Derived from AIR-9 but with open cockpits, built to VVS standards with mods to fit it to tough life as primary and intermediate pilot trainer. Airframe closely similar to AIR-9 except locally strengthened to give design factor increased from 8 to 10, with wire bracing removed except strut fin/tailplane. No Townend ring or spinner, tandem cockpits with mil eqpt and frameless Plexiglas windshields, no slats but flaps retained and trimmers added to meet VVS requirement. Single aircraft in OKB red/silver livery, took part in prolonged U-2 replacement contest 1935-36 and won 500 km trial (Piontkovskii/Demeshkyevich). No production but development continued via Ya-20.

DIMENSIONS Span 10,2 m (33 ft 5⅝ in); length 6,8 m (22 ft 3¾ in), (floats) 7,65 m (25 ft 1¼ in); wing area 16,87 m² (182 ft²).

ENGINE One M-11.

WEIGHTS Empty 510 kg (1,124 lb), (floats) 624 kg (1,376 lb); fuel/oil 113,5 kg+19,5 kg (floats 100 kg+12 kg); loaded 820 kg (1,808 lb) (floats 896 kg/1,975 lb).

PERFORMANCE Max speed 217 km/h (135 mph) (floats 200 km/h); climb 4·2 min to 1 km, 16 min to 3 km; service ceiling 5,7 km (floats 3,2 km); range 950 km (590 miles) (floats 700 km); TO 10 s/100 m; ldg 70 km/h, 120 m.

AIR-10

AIR-11

AIR-11 Using almost same airframe as immediate predecessors, this neat cabin aircraft still looks modern today. Designed first quarter 36, intended as three-seat tourer or single-seat mailplane with 200-kg mail load. Wider than predecessors with room for two seats side-by-side at front, third at rear. Fitted with flaps but no slats. Tailplane braced below only. Full blind-flying panel, even shelf for radio (not installed), exceptional soundproofing, exhaust-muff heating. Completed late Nov 36, successful GOS NII testing. Later won Sevastopol/Moscow contest, then after much pleading given to S. V. Ilyushin as his personal mount commuting from OKB to GAZ at Voronezh. About a year later Ilyushin scarred across face for life in night forced landing; mechanic had forgotten to replenish Gipsy's oil.

DIMENSIONS Span 10,2 m (33 ft 5⅝ in); length 7,32 m (24 ft 0¼ in); wing area 16,8 m² (181 ft²).

ENGINE One 120 hp Gipsy Major.

WEIGHTS Empty 566 kg (1,248 lb); fuel/oil 80 kg+14 kg; loaded 891 kg (1,964 lb).

PERFORMANCE Max speed 209 km/h (130 mph); climb 5·2 min to 1 km, 21·6 min to 3 km; service ceiling 4480 m; range 720 km (447 miles); TO 16 s/200 m; ldg 82 km/h, 340 m.

AIR-12

AIR-12 36 racer incorporated unusual features, most notable being fully retractable main landing gear. Structure as before though aluminium trunnions in wing to take main gears with vertical rubber-disc shock struts and pinned rear braces, retracting inwards under cable from cockpit crank-handle. Wing of increased span with straight TE and all taper on LE (even then CG not quite far enough aft with heavy pilot). Aluminium skin over nose as usual, no attempt to fair cylinder heads. Pilot in open cockpit well aft of wing with narrow Plexiglas windshield extending over 1 m ahead of pilot's face and left cockpit wall hinging down for access. Large fuselage bay above wing either for passenger, with transparent roof, or aux fuel tank for long-range flight. Though intended for competitive sport, use of regular M-11 ensured mundane performance. First flight about end-June 36 and on factory test by Piontkovskii solo with aux tank flew Moscow/Kharkov/Sevastopol/Kharkov about 2000 km in 10 h 45 min flying time on 21 Sept. Later fitted with M-11Ye and gained women's class record 24 Oct 37 with Moscow/Aktyubinsk (1444 km straight-line) flown by V. S. Grizodubov and M. M. Raskova.

DIMENSIONS Span 11,0 m (36 ft 1 in); length 7,17 m (23 ft 6¼ in); wing area 15,6 m² (168 ft²).

ENGINE One M-11.

WEIGHTS Empty 558 kg (1,230 lb); fuel/oil 430+40; loaded 1204 kg (2,654 lb).

PERFORMANCE Max speed 235 km/h (146 mph); range 2990 km (1,860 miles); TO 220 m; ldg 93 km/h.

AIR-13 Not recorded.

AIR-14 First of Yak's smash hits, this followed exactly same aerodynamics and structural principles but was scaled down to single-seat size and stressed to factor of 10 for unlimited aerobatics. It began as another of designer's fun aircraft, but it was accepted by VVS as standard advanced trainer for fighter pilots.

Fuselage truss welded MS tube braced with piano wire tightened by turnbuckles. Engine on separate welded frame attached at front by four bolted fittings. Wooden one-piece wing with front and rear spars at 13·5% and 56% chord with sharp taper mainly on TE. Ply skin except aft of rear spar inboard of ailerons (no flaps). Unbalanced control surfaces of D1 with fabric aft of D1 LE, ailerons hung on three brackets below wing with large slots in neutral position. Usual braced fork-fittings for main wheels with 500 mm × 125 mm tyres and rubber-disc shock absorbers, faired inside large aluminium spat and leg trouser moving vertically with wheel. Steel-leaf sprung tailskid. Cylinders uncowled, top three with short exhausts and two bottom cylinders discharged via carb-air heater duct. Two metal fuel tanks ahead of cockpit. Venturi for gyro turn/slip.

Prototype completed early 36 (preserved by Yak). Piontkovskii did 300 landings in one day and 1,000 over next several days; state tests completed 29 March 36. Not simple to fly and need for care and precision likened to I-16 and sparked off VVS interest, formally expressed April 36. Fitted M-11Ye 37, set various class records including fitted with floats 218 km/h 2 Oct 37 and 1174 km Moscow-Ufa 21 Oct.

DIMENSIONS Span 7,3 m (23 ft 11⅜ in); length 5,75 m (18 ft 10⅜ in) (floats 6,65 m/21 ft 9¾ in); wing area 9,58 m² (103 ft²).
ENGINE One M-11, later 11Ye.
WEIGHTS Empty 423 kg (933 lb), (37, 449; floats 505); fuel/oil 68 kg (37, 62,5) + 13,5 kg (Ye/floats 79); loaded 590 kg (1,300 lb), (37, 618 kg, floats 673 kg).
PERFORMANCE Max speed 249 km/h (155 mph), (37, 245; floats 218); climb 2·6 min to 1 km, 10 min to 3 km (37, 3/13; floats not recorded); service ceiling 6180 m (37, 5 km); range 670 km (416 miles); TO 8 s/120 m; ldg 80 km/h, 160 km (37, 180).

UT-1 Standard fighter trainer, spec agreed Aug 36, but failure to weld accurate fuselages resulted in OKB team taking up residence at GAZ, practice soon universal in USSR. GAZ-47 built 152/384/445/1 in 37-40 while GAZ-150 built 3/150/106, total 1,241. Spade grip standard, popular and used for many purposes, many local mods incl early 41 L. I. Sutugin fitted M-12 engine, a gun and oleo legs, while late 41 A. I. Volkov fitted two underwing ShKAS (200 rds ea) and K. A. Moskatov fitted two ShKAS and two RS-82; others fitted floating ailerons, improved gears/props and wing with experimental VVS profile.

AIR-14, UT-1

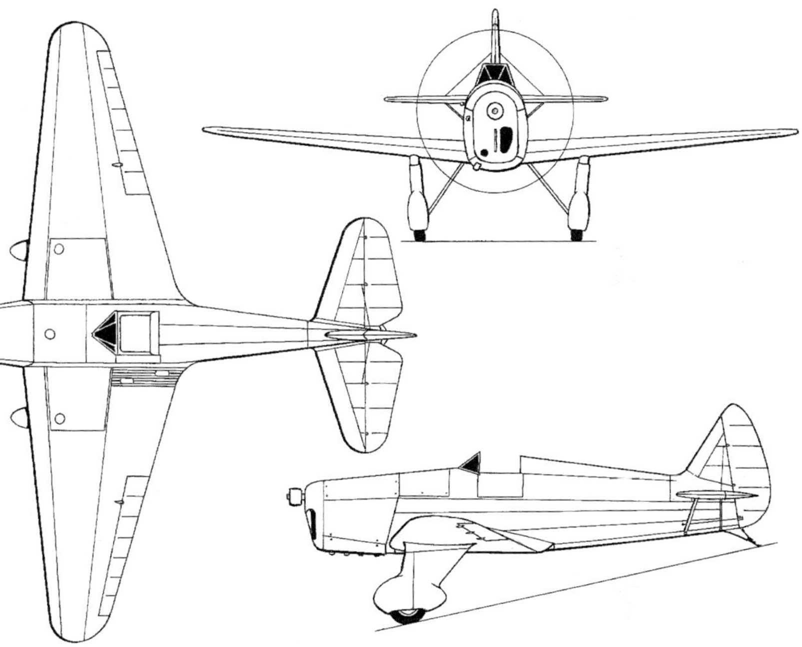
AIR-14

DIMENSIONS As AIR-14.
ENGINE One M-11Ye.
WEIGHTS Empty 429 kg (946 lb); fuel/oil 79 kg; loaded 597,5 kg (1,317 lb).
PERFORMANCE Max speed 257 km/h (160 mph); climb 2·45 min to 1 km, 8·7 min to 3 km; service ceiling 7120 m (23,360 ft); range 670 m (416 miles); TO 5·5 s/90 m; ldg 80 km/h, 190 m.

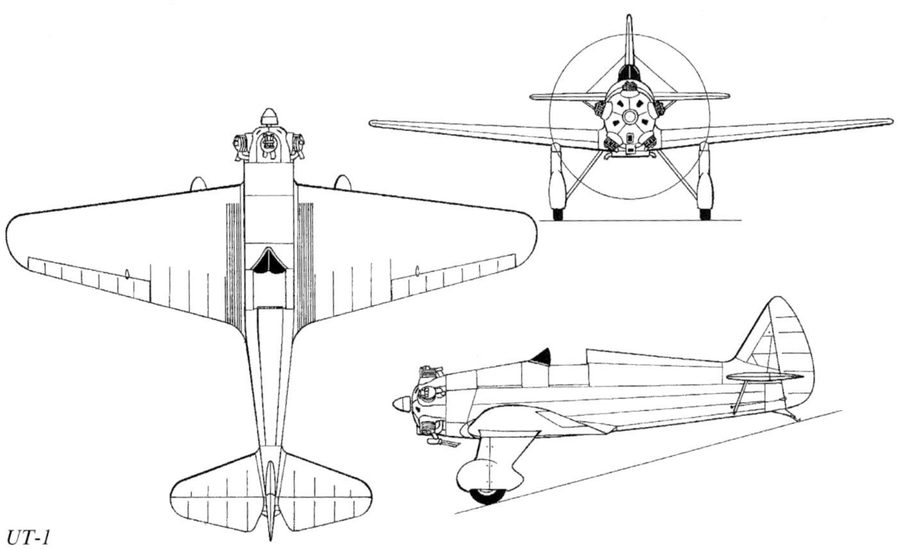
UT-1

AIR-15 One of aircraft actually built for which few details can be found. Almost certainly derivative of AIR-14, built for racing 36 with experimental wing to formula (profile) of F. G. Glass (so-called zero-moment section). No details of flight results.

AIR-16 Another aircraft completed but almost lost without trace. Superficially similar to AIR-11, with four-seat enclosed cabin, but much more powerful engine. AIR-16 was on Central Airfield 37 but never released for flight. OKB has drawing showing retractable gear, uncertain if this was fitted.
DIMENSIONS Presumed as AIR-11.
ENGINE One 220 hp Renault MV-6.
WEIGHTS, PERFORMANCE No data.

AIR-17 This twin-engined machine was complete break with small sporting types and also believed first to be completed after purge of A. I. Rykov and Yak's decision to abandon AIR designations. Designed early 37 as trainer for pilots of large civil and military aircraft with added capability of training other members of crew if necessary. Structure similar to previous designs. Fuselage welded MS tube with D1 and ply secondary structure (fairings and floors) and fabric covering, wooden one-piece two-spar wing but this time Clark YHC section, D1/fabric control surfaces and hydraulic system driving Schrenk flaps and rearwards-retracting single-leg main gears copied from SB. Prototype built as military crew trainer with pilot above LE with folding canopy, glazed nose for nav/bomb with small table and ShKAS in pivoted nose mount, and rear radio-gunner with ShKAS (Shavrov 'twin ShKAS in rotary nose mount, none in rear'). Four FAB-50 hung under fuselage. First AIR-17, French engines, Ratier v-p props, completed 31 Dec 37, flown on skis. Weight 400 kg over est, but still good perf/handling and adopted for production at two GAZ, but as two-seat pilot trainer, see next. Fourth prototype was fitted with 205-hp DH Gipsy Sixes with fixed-pitch DH props.

DIMENSIONS Span 15,0 m (49 ft 2½ in); length 10,7 m (35 ft 1¼ in); wing area 33,42 m^2 (360 ft^2).
ENGINES Two 220-hp Renault Bengali 6Q.
WEIGHTS Empty 2040 kg (4,497 lb); fuel/oil 285 kg+42 kg; loaded 3108 kg (6,852 lb).
PERFORMANCE Max speed 273 km/h (170 mph); climb 5·4 min to 1 km, 22·6 min to 3 km; service ceiling 4 km; range 1000 km; TO 19·9 s/319 m; ldg 90 km/h, 245 m.

UT-3 Series aircraft two-seater with streamlined ply nose, dual pilot controls, no armament (but structural provision for fixed guns and light bombs), licensed engine and AV-3 prop. Long argument, standard not settled until Feb 41, and only small batch, believed to number 30, built at one GAZ. In war VVS had no light twin trainer.

DIMENSIONS As AIR-17 except length 10,83 m (35 ft 6⅞ in).
ENGINES Two MV-6A.
WEIGHTS Empty 2042 kg (4,502 lb); fuel/oil 350 kg+32 kg; loaded 2627 kg (5,791 lb).
PERFORMANCE Max speed 260 km/h (162 mph); climb 3·8 min to 1 km, 11·4 min to 3 km; service ceiling 6,2 km; range 1050 kg (652 miles); TO 15·5 s/230 m; ldg 95 km/h, 115 m.

AIR-17, UT-3

AIR-18 Series UT-1 modified as racer with French engine, tightly enclosed cockpit (sliding hood and folding side panels), inwards retracting gear, fuel in two wing tanks; flown mid-37, later briefly fitted with floats.

DIMENSIONS As UT-1 except length 5,99 m (19 ft 7⅞ in).
ENGINE One 140 hp Renault Bengali 4.
WEIGHTS Empty 475 kg (1,047 lb); fuel/oil 68,5 kg+12,5 kg; loaded 645 kg (1,422 lb).
PERFORMANCE Max speed 310 km/h (193 mph); service ceiling 6,5 km; range 600 km (373 miles); TO 8 s/120 m; ldg 85 km/h, 160 m.

No 19, Ya-19 Rykov purged in 37 terror, so Yakovlev changed via 'No' to 'Ya' and later 'Yak'. No 19 civil version of 17, lead design engineer Oleg K. Antonov, first use of stainless by OKB, fuselage truss of welded 30KhGSA tube with piano-wire bracing. Skin dural flush-riveted over nose, ply to rear of cabin and fabric from there to tail. Wing identical to UT-3 except no bomb racks, same slotted ailerons and Schrenk flaps. Main gears as for pre-production UT-3 with SB type geometry and fork fitting for wheel with tyre 650 mm×250 mm. Comfortable cockpit with full glazing, usually single pilot only. Main cabin with five passenger seats, or three stretchers and attendant seat, or (drawings prepared) reconnaissance/survey cameras, cargo and other duties. Prototype flown late 38 and factory tested in white/striped-rudder livery on wheels and skis. Passed GOS NII tests spring 39 (pilots G. A. Muratov, E. I. Schwartz); at aft CG (31·3%/32·5%) fin considered too small and rudder large, but overall excellent, production being thwarted by termination of UT-3 programme.

DIMENSIONS Span 15,0 m (49 ft 2½ in); length 10,02 m (32 ft 10½ in); wing area 33,42 m^2 (360 ft^2).
ENGINES Two MV-6.
WEIGHTS Empty 2134 kg (4,705 lb); fuel/oil 280 kg+26 kg; loaded 2950 kg (6,504 lb).
PERFORMANCE Max speed 271 km/h (168 mph); climb 5·3 min to 1 km, 17 min to 3 km; service ceiling 5,6 km; range 783 km (487 miles); TO 17 s/410 m; ldg 86 km/h, 365 m.

Ya-20 Outstanding basic design, leading to standard pilot trainer of Gt Patriotic War. Derived from AIR-10 but structurally complete break with past in all-wood airframe, using MS or D1 only for simple pressings and sheet laminates at major joints. Fuselage spruce/pine built up on wire-braced truss integral with engine mounts. Main frame at front of each tandem cockpit integral with 2,24 m centre section spars. Ply top decking, fabric elsewhere. Outer wing panels, Göttingen 387 section, 5° dihedral on LE from straight c/s with joints at two spars reinforced by three MS plate laminates bolted each side of spar booms to give ultimate load factor 10 (common to all variants); ply ribs, multiple stringers and ply skin. Slotted ailerons with fabric covering over ply nose, mounted on brackets beneath lower surface and driven via rods/bellcranks. Entire tail D1 (only metal structure) with fabric covering wrapped over

AIR-18

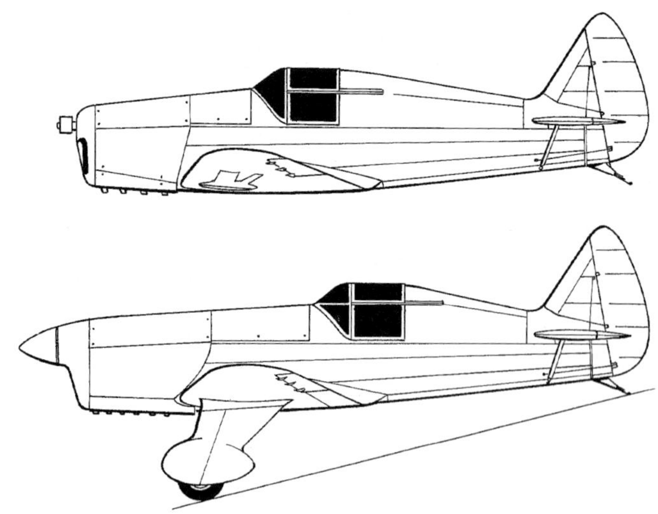

AIR-18 (lower, Ya-21)

Ya-19

Ya-20 Gidro (1st UT-2)

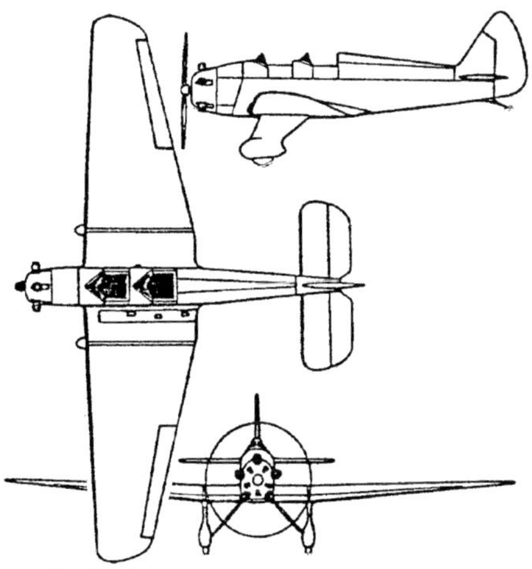

Ya-20, UT-2

flush-riveted D1 on all LE including large horn on rudder, cable actuation. Usual welded MS landing gear with fork fittings on rear bracing strut and main leg with stack of 16 rubber shock-absorbers; tyre 500 mm × 125 mm, surrounded by large spat/leg-trouser of aluminium in front/rear sections. Tailskid multi-leaf steel spring, each laminate 50 mm × 5 mm. Engine in aluminium cowl driving 1,96 m wooden propeller. Welded 90 lit fuel tank in each side of c-s between spars outboard of fuselage; welded oil drum ahead of front instrument panel. Open cockpits with padded leather edges and folding sidewall on left. Comfortable padded seats on aluminium-tube basis made at OKB. Baggage hatch in decking behind rear seat.

First aircraft fitted with French engine and flown on floats from AIR-6 early 37. Second with M-11Ye and only 83 lit fuel; third with 200 lit. NII testing summer 37 and decision Sept 37 to produce in large series at four GAZ as standard pilot trainer.

DIMENSIONS Span 10,2 m (33 ft 5⅝ in); length (Renault) 7,65 m, (No 2) 7,11 m, rest 7,0 m (22 ft 11⅝ in); wing area 17,2 m^2 (185 ft^2).
ENGINE (1) one 140 hp Renault Bengali 4, (rest) one M-11Ye.
WEIGHTS Empty (1) 569 kg, (floats) 677 kg, (rest) 574 kg; fuel/oil (1) 128 kg+15 kg, (2) 60 kg, (rest) 143 kg+14,5 kg; loaded (1) 996 kg, (landplane) 888 kg, (2) 804 kg, (rest) 900 kg (1,984 lb).
PERFORMANCE Max speed (1) 210 km/h, (land) 240 km/h, (2) 230 km/h, (rest) 210 km/h; climb 3·3-4 min to 1 km; 13·2-15·3 min to 3 km; service ceiling (1) 3267 m, (landplanes) 6·1 km-6·5 km; range (1) 834 km, (2) 450 km, (rest) 1000 km; TO (land) 12-13 s/150-230 m; ldg 75-85 km/h, 200-263 m.

UT-2 Standard decided 38 with M-11 (not Ye) and only 89 lit, but before prodn started decision to fit two wing tanks of 90 lit and upper (fuselage) of 20, for 7 h endurance; oil 16 lit. Bucket seats for seat-type pack, compass and turn/slip, from early 40 adding blind-flying hood over rear cockpit. First few spatted wheels, rest bare gear or skis. Prodn begun GAZ-301, delivering 110 in 38, 276 in 39; GAZ-23 35 in 38, then 254/385/5, ending Jan 41. Wartime mass-prod by GAZ-47 and 116, countless local mods including fixed guns, bombs (eg four FAB-50 as single-seater), glider hook, supply dropping and psy-war loudspeaker. K. A. Moskatov fitted Nos 23/24 with MV-6 engine, another two ShKAS and four RS-82 and another with crash arches/canopy and aileron/elevator trimmers. **UT-2M** described separately, total prodn by end 44 (incl 2M) 7,243, training over 100,000 pilots. ASCC name 'Mink'.
DIMENSIONS As Ya-20 except length 7,0 m (22 ft 11⅝ in).
ENGINE One M-11, but several blocks had M-11G or 11D.
WEIGHTS Empty 616 kg (1,358 lb); fuel/oil 146 kg+13 kg; loaded 938-940 kg (2,068 lb).
PERFORMANCE Max speed 205 km/h (127 mph); climb 5·0 min to 1 km, 31 min to 3 km; service ceiling 3,1 km; range 1130 km (702 miles); TO 12·5 s/175 m; ldg 95 km/h, 200 m.

UT-2M New outer wing panels with all taper on LE and 7° dihedral, wing area as before, but ailerons smaller. Modified tail and other minor changes, generally giving better handling esp spin recovery. First flight June 41, began replacing UT-2 in prodn May 42. Remaining shortcoming basic simplicity, causing pupil problems on next step, eg Yak-7UTI; so in 43 designed **UT-2L** with enclosed cockpit, blind-flight insts, split flaps driven 50° by hand worm gear, wheelbrakes and tailwheel, later with spats and helmeted cowling. Flown early 44 with M-11D, became known as **UT-2/1944** or **Etalon 44**, led to Yak-5. **UT-2MV** modified 2M as single-seat tactical bomber, four FAB-50 (or more, limit not recorded). None of 2M derivatives built in series. Data for 2MV, dimensions as previous.
ENGINE One M-11D.
WEIGHTS Empty 716 kg; fuel/oil 146 kg+20 kg; loaded 1150 kg.
PERFORMANCE Max speed at SL 177 km/h (110 mph); climb 8 min to 1 km; service ceiling 3350 m; range 480 km.

Ya-21 Mew Gull and Caudrons inspired Yak to build racer based on Ya-18 but with big 6-cyl engine and fixed spatted gear. Göttingen 387 wing, D1 manual split flaps, welded MS fuselage with 170-lit fuel drum filling X-section between spars, fabric control surfs, 2120-mm 2-posn prop, rubber shock struts to 500 mm × 125 mm tyres, no brakes. Flown late 38, tricky to fly and considerably altered as **Ya-25** early 39 but never completed thus because of warplane priority.
DIMENSIONS Span 7,3 m (23 ft 11⅜ in); length 6,4 m (21 ft 0 in); wing area 9,58 m^2 (103 ft^2).
ENGINE One 220 hp MV-6.
WEIGHTS Empty 611 kg (1,347 lb); fuel/oil 120 kg+12 kg; loaded 831 kg (1,832 lb).
PERFORMANCE Max speed 290 km/h at SL, 322 km/h (200 mph) at 2 km; climb 2·7 min to 1 km, 7·5 min to 3 km; service ceiling 9,1 km; range 715 km; TO 7 s/100 m; ldg 85 km/h, 189 m.

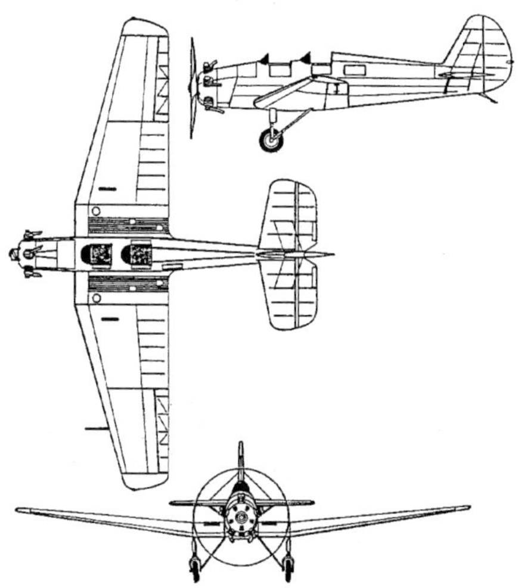

UT-2M

Ya-21

Ya-22, BB-22, Yak-2 Despite lack of experience Yak invited May 38 to build hi-speed light bomber. Lo-wing 2/3-seater, mixed structure using SB technology (eg ldg gear). No 22 drawings showed provision for two ShVAK under fuselage, not then fitted.

Wing of Göttingen 387 profile modified with flat underside, made as centre section integral with fuselage mid-section, wholly wood construction, usual two box spars with ply ribs glued and pinned and stressed for factor of 8 (in this context means 8 g ultimate strength). Dihedral 6° from roots. Ply skin throughout, graded thickness. Schrenk flaps of D1 structure inboard/outboard of nacelles, hyd drive to 55°. Ailerons of D1 with fabric covering over D1 LE skin, hung on hinges below wing in usual way. Fixed tail surfaces D1 with stressed skin, movable surfaces as ailerons but with trimmers. Main mid-portion of fuselage wood with ply skin, integral with c/s of wing. Nose D1 with stressed skin flush-riveted. Rear fuselage welded 30KhGSA tube with wood upper decking and fabric covering. Landing gear strengthened version of Ya-17 with welded MS fork to single wheels with 550 mm × 250 mm tyres, hydraulic retraction rearwards into nacelles with twin doors enclosing bay completely. Fully retractable tailwheel. Fuel in four wing tanks between spars and two large tanks in mid fuselage, total 1232 lit. Pilot ahead of LE with remarkably broad cockpit enclosed by sliding hood, nav aft of TE with hinged hood. Wing stressed for bombs hung under spars, four FAB-100 on drawings but no racks fitted. M-103 engines neatly cowled with carb-air inlets on underside, oil-cooler ram inlets on outer sides, glycol radiators inside rear of nacelle behind wheel bay with vertical slit inlet each side and adjustable vertical exit doors aft of TE. VISh-22 three-blade c/s 3,1 m props.

First **No 22** in OKB red livery flown by Piontkovskii early 39 (one report, 22 Feb). Outstanding aircraft, fastest in Soviet Union. Handling generally good and rather precipitate Stalin command 15 March 39 to put into immediate production at two GAZ with mission short-range bomber (*Blizhnii Bombardirovshchik*, hence designation **BB-22**. Yakovlev had already organized special group called KB-70 to speed production by photocopying over 3,300 drawings. Personally called to Kremlin 27 April 39 and interrogated; then awarded Order of Lenin, ZIS car and 100,000 rubles. But getting BB-22 into production was major task, complicated by 100+ engineering change orders which included redesign of centre fuselage so that navigator was seated directly behind pilot with continuous glazing, pilot canopy now hinged, backseater's able to rise up bodily for using aft-facing ShKAS, given field of fire by entire upper deck of rear fuselage hinged at tail to pivot down in combat. Four FAB-100 in internal bay plus two external under fuselage, single fixed ShKAS in nose, four large wing tanks, oil tanks outer LE.

First **BB-22** overtaken by **R-12** recon 2-seater with original layout with widely separated pilot and nav with aft-facing ShKAS firing through large aperture at rear of hinged hood kept closed in flight; three cameras (usually AFA-33) in bottom of fuselage, bay behind

Ya-22

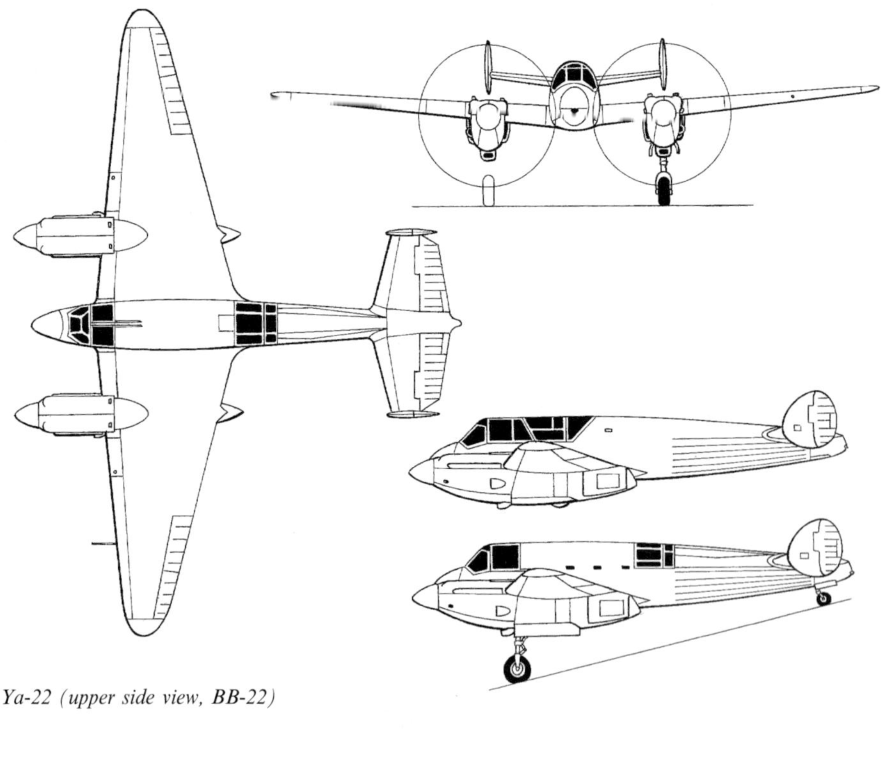

Ya-22 (upper side view, BB-22)

pilot for eight FAB-20 bombs, fixed ShKAS in nose. Also built to original layout **I-29** (also called **BB-22IS**) escort fighter with restored fuselage fuel and two ShVAK underneath, but this not flown (P. M. Stefanovskii) until early 41. Both remained prototypes; no data unearthed. First **BB-22** flown on skis 20 Feb 40. GAZ-1 produced 81 in 40, and GAZ-81 57 in 40 and 63 in 41, total 201. Redesignated **Yak-2** in 41, seeing action with two regts.

DIMENSIONS Span 14,0 m (45 ft 11⅛ in); length (all) 9,34 m (30 ft 7¾ in); wing area 29,4 m^2 (316 ft^2).

ENGINES Two M-103.

WEIGHTS Empty (22) 3796 kg (8,369 lb), (Yak-2) 4043 kg (8,913 lb); fuel/oil (22) 678 kg, (2) 600 kg+62 kg; loaded (22) 5123 kg (11,294 lb), (2) 5380 kg (11,861 lb).

PERFORMANCE Max speed (22) 567 km/h (352 mph) at 4,9 km, (2) 515 km/h at 5,2 km; climb to 5 km (22) 5·75 min, (2) 7·7 min; service ceiling (22) 10,8 min, (2) 8,9 min; range (22) c1000 km, (2) 800 km (497 miles); TO (22) 375 m, (2) 500 m; ldg (22) 160 km/h, 855 m, (2) 155 km/h, 500 m.

Yak-4 Designation originally applied to BB-22/Yak-2 fitted with M-105 engines, but in winter 39-40 new aircraft emerged with new fuselage, crew generally as Yak-2 but different glazing, pilot with windows low on each side, nav with bulged roof with rear closure pivoting up and forwards for firing ShKAS across lowered rear fuselage no longer needing to fold down. New main gears with twin narrow wheels (700 mm × 150 mm) partly protruding for belly landing. Cooling system as before but increased capacity. All fuel in four large wing tanks, oil tanks outer LE, attachments under outer wings for drop tank (c250 lit) or FAB-100; same fuselage bay (4 × FAB-100) as Yak-2; same guns (two ShKAS) as R-12. Series aircraft with radio mast above right nacelle with wire to tail, and new cowling with oil rad in duct under engine. In prodn late 40, c400 built by GAZ-1 and -81. Brief front-line career, mainly structural problems.

DIMENSIONS Span 14,0 m (45 ft 11⅛ in); length 10,18 (33 ft 4¾ in); wing area 29,4 m^2 (316 ft^2).

ENGINES Two M-105.

WEIGHTS Empty (1st) 4251 kg, (series) 4560 kg (10,053 lb); fuel/oil 800 kg+70 kg; loaded (1st) 5845 kg, (series) 6115 kg (13,481 lb).

PERFORMANCE Max speed (1st) 573 km/h at 5,3 km, (series) 533 km/h (331 mph) at 5,0 km; climb to 5 km (1st) 5·45 min, (series) 6·5 min; service ceiling (1st) 10 km, (series) 9,7 km (31,824 ft); range (1st) 960 km, (series) 925 km (548 miles); TO (1st) 12 s/300 m, (series) 415 m; ldg (1st) 160 km/h, 550 m, (series) 496 m.

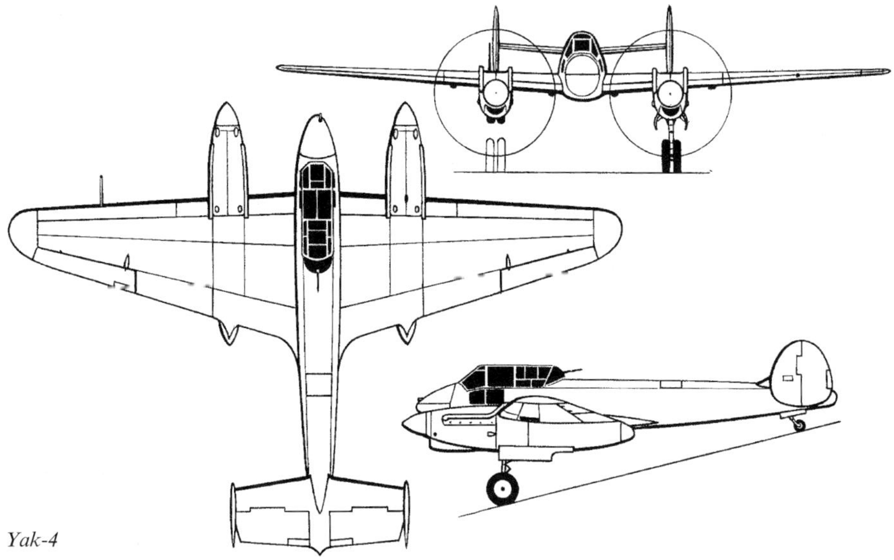
Yak-4

Yak-4 prototype

Yak-4

Ya-26, I-26 Yakovlev's chief ambition was to build a small, high-speed fighter. As soon as design effort on No 22 was tailing off he obtained NKAP authority to begin design of 'frontal' fighter; this was in Nov 38, not subsequent to Jan 39 Kremlin meeting of OKB heads. Yak already on good terms with Klimov and planned **No 26**, or Ya-26, around M-106. From outset determined to keep aircraft small and put all armament in fuselage. Studied stressed-skin structures, but adhered to OKB traditional mixed construction, which was known quantity, did not impose severe penalties, was easy to repair and used materials not in short supply.

One-piece wing Clark-YH profile, t/c 14% root and 10% tip, dihedral 5°30' from root, incidence 0°30'. All-wood structure with usual pair of widely spaced box spars, ply/pine strip ribs with cut-outs and birch-ply skin with varnished fabric overlayer. Schrenk flaps (not following TE of wing but at 90° to fuselage, following hinge line along kinked rear spar) driven by single pneu ram to 50°, wholly D1 construction. Ailerons D1 with fabric covering, Frise type with slots and hinged behind rear spar (not to brackets below wing). Fuselage built around truss of welded 30KhGSA with diagonal (Warren) bracing, oval section ahead of instrument panel achieved by D1 secondary structure carrying Dzus-fastened D1 cowl and access panels. Aft of cockpit plywood formers above and below carried light pine laths with thin birch-ply upper decking and overall covering of fabric. Tail D1 with fabric covering, tailplane symmetric, fixed, span 3,4 m, tail control surfaces driven by cables (ailerons by push-rods) on plain hinges, elevators with simple extended mid-section LE for balance, all surfaces with trim tabs. Main landing gear with air/oil shock struts and single curved fitting holding inboard end of axle, tyres 650 mm × 200 mm, pneumatic retraction inwards ahead of spar, track 3,25 m. Fully retractable tailwheel 300 mm × 120 mm with two doors. Engine, second-best choice of M-105P because 106 not available, mounted direct on forward fuselage tubes without vibration

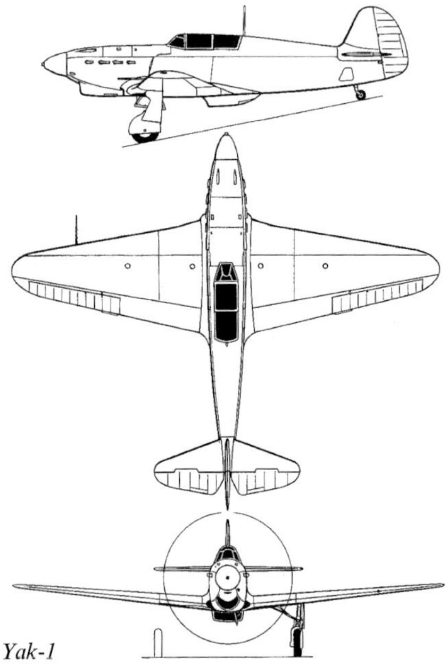
Yak-1

dampers, with glycol radiator and oil cooler in controllable duct under wing TE, carb-air inlet under wing LE, plain exhaust stubs. Three-blade 3,0 m (139 kg) VISh-52P prop with manually variable pitch. Four flexible fuel cells between wing spars outboard of fuselage, inners 130-lit and outers 74. Cockpit with bucket seat above rear spar, fully glazed Plexiglas windscreen (unarmoured), sliding hood (spring-latch either open or closed) and fixed Plexiglas rear section; small 8-mm armour plate behind pilot's head. Planned armament one ShVAK mounted on engine firing through spinner (130 rounds) and four ShKAS (420 rounds each) in top decking and in fuselage sides above wing; PAN-23 optical sight. No radio.

I-26-1 completed in OKB red livery first week of 40 dubbed *Krasavits* (beauty). Flown without armament by Yu. I. Piontkovskii 13 Jan 40. Generally excellent, no problem with wheeled gear on ice/snow, GAZ testing yielded 586 km/h but because of manufacturing defect crashed 27 April, killing Piontkovskii. By this time complete confidence, no need for modification apart from various planned changes, and even at time of crash decision taken to produce in large series at two GAZ. **I-26-2** completed with strengthened wing, armour from 8 to 9 mm, upper fuselage behind canopy widened, ejector exhausts, VISh-61P prop, carburettor fed by ducts from inlets in wing roots, oil cooler moved forward under engine, lower pair of ShKAS removed, upper magazines altered to 380 rounds, ShVAK reduced to 120; PBP-1 reflector sight. Flown by P. Ya. Fedrovi 14 April 40, then by S. A. Korzinshchikov. **I-26-3** flown 18 Sept, minor changes including 135 rounds ShVAK and 750 each ShKAS. NII testing from 13 Oct by Stefanovskii and A. S. Nikolayev.

DIMENSIONS Span 10,0 m (32 ft 9¾ in); length 8,48 m (27 ft 9⅞ in); wing area 17,15 m^2 (184·6 ft^2).

ENGINE M-105P.

WEIGHTS Empty 2206 kg/2318 kg/2401 kg (max 5,293 lb); fuel 305 kg; loaded 2600 kg/2700 kg/2801 kg (max 6,175 lb).

PERFORMANCE Max speed (1) 580 km/h (360 mph) at 5 km, (2) 490 km/h at SL, 585,5 km/h (364 mph) at 4,8 km, (3) not cited; climb to 5 km (1) 5·2 min, (2) 6·0 min; service ceiling (1) 11,15 km, (2) 10,2 km; range (2) 700 km (435 miles); time 360° circle (2) 24 s, (3) 20-21 s; TO (2) 300 m; ldg (2) 135 km/h, 540 m.

UTI-26 Third and fourth prototypes tandem dual-control two-seaters, instructor in rear, based closely on I-26-2 but lightened; VISh-61P prop, two ShKAS each 500 rounds, wing moved 100 mm aft, main radiator moved forward under wing, two sliding canopies, step to rear cockpit, front with blind-flying screen, doors on fuselage to cover retracted wheels, leg/wheel doors separated, no radio. **UTI-26-1** flown by Fedrovi 23 July 40, NII testing from 28 Aug. Like I-26, many defects. Data as I-26 except:

WEIGHTS Empty 2181 kg (4,808 lb); loaded 2750 kg (6,063 lb).

PERFORMANCE Max speed 500 km/h at SL, 586 km/h at 4,5 km; 5·5 min to 5 km; service ceiling 9,4 km.

I-28 Also called **Yak-5**, exptl high-alt interceptor for PVO; new wooden wing reduced span but same area with large auto slats, pitot on mast under left wing; Plexiglas area behind cockpit, PD engine with V. A. Dollezhal's E-100 2-stage supercharger, 61P prop, armament as I-26-2, RSI-3 radio (not fitted during factory tests begun by Fedrovi 1 Dec 40).

DIMENSIONS Span 9,74 m (31 ft 11½ in); length 8,5 m (27 ft 10⅝ in).

ENGINE One M-105PD.

WEIGHTS Empty 2450 kg (5,401 lb); fuel 310 kg; loaded 2928 kg (6,455 lb).

PERFORMANCE Max speed 515 km/h at SL, 650

UTI-26 No 1

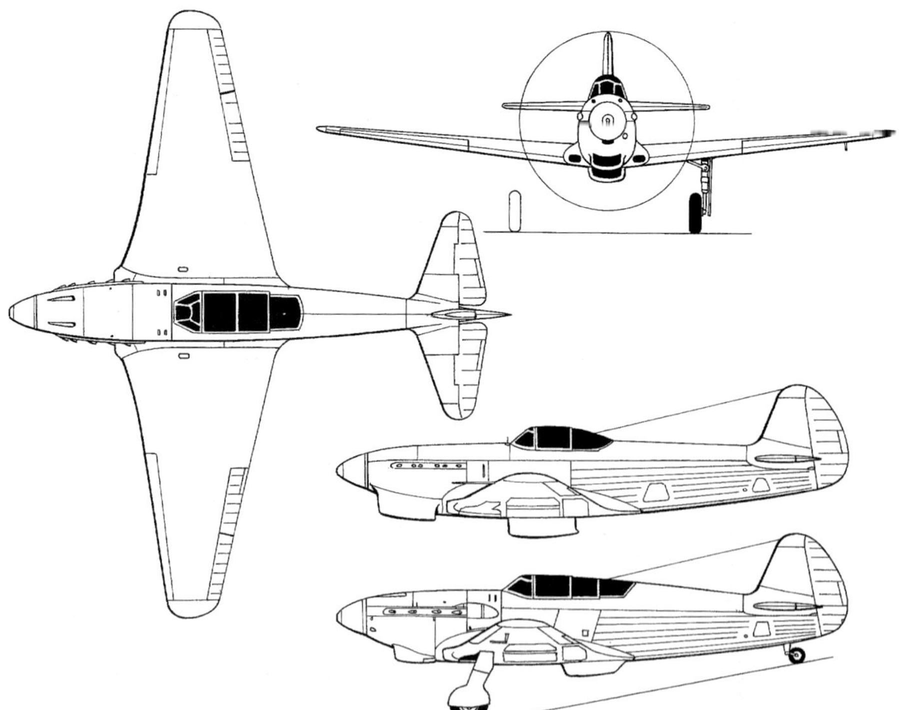
I-28 (upper side view, Yak-9PD)

km/h (404 mph) at 9,0 km; time to 5 km 5·2 min; service ceiling 12 km; 360° circle 17·6 s; ldg 132,5 km/h.

I-30 Also called **Yak-3** (repeated 44 for different aircraft), I-28 wing redesigned in light alloy with plate spars, machined booms and pressed-sheet ribs, made as centre section (0° dihedral) and outer panels (5·5° dihedral) with landing light inboard of L slat, fabric ailerons with tab on L, PD engine as I-28, main leg faired by single door and oleo pressure up from 7 to 10 at, tailwheel tyre 125 mm wide, cockpit canopy sliding over transparent rear section with cut-down rear fuselage, mast/wire antenna for RSI-4 radio, RPK-10 radio compass, two extra 50-lit wing tanks, two ShVAK cannon added at roots of outer wing each 120 rounds. Factory test of **Yak-3-1** 5 April-17 June 41, then rebuilt with unslatted wooden outer wings and production 105P engine. **Yak-3-2** (*Dubler*) on test from May 41 with oil cooler under engine replaced by two rads in centre-section LE; conventional canopy as on I-26. Decision to build Yak-3-1 at GAZ-81 Moscow, 83 Khabarovsk and 292 Saratov overtaken by German invasion. Later 3-1 mod by removing wing guns and c/s tanks, reducing weight to 2725 kg. Data for 3-1 July 41.

DIMENSIONS As I-28; ENGINE see above.

WEIGHTS Empty 2550 kg (5,622 lb); fuel 383 kg; loaded 3130 kg (6,900 lb).

PERFORMANCE Max speed 476 km/h at SL, 571 km/h (355 mph) at 4,9 km; time 7 min to 5 km; service ceiling 9 km (29,530 ft); range 975 km (606 miles); TO 303 m; ldg 142 km/h, 525 m.

Yak-1 No 38-55 (1st with PA engine, Nov 41)

Yak-1 (Black Sea Fleet)

Yak-1 Urgent need so production overlapped prototype devt, 64 delivered 40 (48 GAZ-301 Moscow, 16 GAZ-292 Saratov), 105P engine oval inlet L wing root, oil cooler under engine. Poor firepower, hood jamming, pipe breakage from vibration, unreliability of pneu gear/flaps and armament problems. Subsequent devt: 41, skis, 105PA engine, two BI-42 bomb containers (each one FAB-100 or 4×FAB-25), six rails for RS-82 rockets; armament changes (1×20 mm + 1/2×12,7 mm), more ammo (from 120 to 140 20 mm, from 200 to 240 12,7 mm), grille to protect rad from shell cases (later deleted), 6-ply tyres, ejector exhausts, and painstaking drag reduction. RSI-4 receiver standard, every tenth a/c with transmitter also. In Sept 41 to Feb 42 c830 built with semi-retrac skis; degraded perf, so instead snow cleared from runways. In March 42 No 45-95 first of lightened batch, removal of many items reduced wt from c2917 to 2780, better agility. From June 42 engine 105PF, except one a/c at GAZ-153 with M-106 in Jan 43; 47 more by GAZ-292 never delivered. July 42 No 35-60 first **1B**, better view as I-30 from lower aft fuselage and all-round canopy, 2 ShKAS replaced by 1 UBS (200, later 240, rounds). Sept 42 new control stick with firing button/trigger. **Yak-1MPVO**, 385 built 43-44 improved night eqpt and DF receiver. Total prodn (almost all GAZ-292) 8,734 ending July 44, of which 5,672 PF engine, 4,188 1B version. Data 43-44 versions, same dimensions, PF engine:

WEIGHTS Empty 2316 kg (5,106 lb); loaded 2884 kg (6,358 lb).

PERFORMANCE Max speed 531 km/h (330 mph) at SL, 592 km/h (368 mph) at 4,1 km; climb 5·4 min to 5 km; service ceiling 10,05 km; range 700 km (435 miles); 360° in 19 s; TO 340 m; ldg 560 m.

Yak-7 Acute shortage of fighters after German invasion (22 June 41) led to urgent conversion of UTI-26 trainer into single-seat fighter. Unexpectedly became major type, 18 versions. First version **Yak-7UTI** trainer, one-piece main leg fairing, fixed tailwheel, engine governed 2,350 rpm (normal 2,700 rpm), left ShKAS only (500 rds), in prodn GAZ-301, first flight 18 May 41, 186 built, evac to GAZ-153 Sept 41. Also used for recon and artillery correction. Late July 41 OKB (K. V. Sinelshchikov) proposed fighter version, PA engine, VISh-61P prop, armour, self-sealing inerted tanks, ShVAK (120 rds) and two ShKAS (750 each), RS-82 and ESBR-3 bomb racks; canopy and tailwheel unchanged. Surprisingly good handling, so OKB team produced **Yak-7M** with PA/VISh-61P, armour, 9,74 m span, slats, larger flaps, ShVAK (120) plus two wing ShVAK (110) reducing fuel 23 lit each wing made up by 80 added in rear cockpit; FS-155 landing light. NII test from 5 Oct 41 but not adopted except rear-c'pit tank. Instead **Yak-7A** with semi-retrac tailwheel, T/R radio, rear canopy replaced by hinged hatch, ShVAK + 2 ShKAS (500 ea) with case/link collectors. In prodn GAZ-153 Jan 42 (skis).

DIMENSIONS As Yak-1 except length 8,5 m (27 ft 10⅝ in).

ENGINE One M-105PA.

WEIGHTS Empty (skis) 2523 kg (5,562 lb), (wheels) 2450 kg; loaded 3008 kg (6,631 lb), (wheels) 2935 kg.

PERFORMANCE (skis) Max speed 451 km/h (280 mph) at SL, 555 km/h (345 mph) at 5 km [wheels, 495 mm/571 mm]; climb 6·8 min to 5 km; service ceiling 9,25 km; range 590 km; 360°, 22 s; TO 460 m; ldg 445 m.

Yak-7B ShVAK (120) and two UBS (L, 260; R 140) [one tested with RS-82, not adopted]. Fully retrac tailwheel, PA engine cleared 2,700 rpm, numerous drag reductions and better elec/pneu/radio (transmitter every tenth a/c). Produced GAZ-153 April-July 42. Drawbacks, poor climb, long field length, bad transparencies, high-drag rads and crashes caused by separation of wing skin. Following tests with a 7UTI in Aug 41 about 350 Yak-7B produced as recon fighters with UBSs removed and an AFA-1M camera in rear cockpit, RSI-4 transceiver and extra armour. Detail changes mid-42 reduced weight, and PF engine fitted. Few built as **Yak-7MPVO** day/night interceptors, RPK-10 D/F, FS-155 light and elec ldg-gear indicator. In April/May 42 single **Yak-7-37** tested with MPSh-37 gun (20 rds) and two UBS, cockpit moved 400 mm to rear (no rear cockpit), 9,74 m slatted wing, larger tailwheel; 22 built Aug 42, to 42 IAP. Single **Yak-7-82A** tested Jan-May 42 with M-82A engine and AV-5L prop, 9,74 m wing, 2 ShVAK outboard plus 1 UBS, two extra 40-lit tanks; poor results. **Yak-7PD** lightened interceptor, continuing I-28 programme, PD engine with E-100 supercharger, one ShVAK (120) only, NII tests Sept 42 halted by engine problems.

Definitive trainer **Yak-7V** tested Feb-Mar 42, PA (later PF) engine with rectangular inlet L root, VISh-61P, simplified airframe, no armament/self-sealing tanks/oxygen/radio, fixed gear, full dual controls; 510 built plus 87 converted 7B. DM-4S ramjets tested on UTI-26-2 in Jan 42, more successfully by S. N. Anokhin March-Dec 44 with a Yak-7B but never adopted. **Yak-7R** with M-105 replaced by cockpit and two UBS, D-1A rocket in tail and DM-4S under wings never completed (est 800 km/h). **Yak-7D** long-range prototype tested June 42; new 9,74 m wing with bakelite-ply skins on light-alloy spars/ribs, 11 tanks totalling 925 lit, single MPSh-20 with 60 rds, 2-h oxygen, AFA-B camera; poor perf except range 2785 km. **Yak-7DI** long-range fighter tested June-Aug 42, wing similar to 7D, four Dural tanks total 673 lit in balanced fuel system controllable by pilot, ShVAK + one UBS, rear deck lower and all-round vision canopy with hatch over aft compartment for passenger, bombs, camera or tank; Hucks dogs but no ejector exhausts. Outstanding versatile aircraft, redesignated **Yak-9**. Final exptl version **Yak-7P**, tested Nov-Dec 43 with both UBS replaced by two ShVAK; perf/handling not impaired but lash-up guns hit prop, cases hit tail etc. Total production of Yak-7 6,399, of which 4,888 at GAZ-153. ASCC name 'Mark'. Data for 42 7V with wheeled ldg gear:

DIMENSIONS, PA engine, as 7A/7B (10,0 m, 8,5 m).

WEIGHTS Empty 2210 kg (4,872 lb); loaded 2725 kg (6,007 lb).

PERFORMANCE Max speed 410 km/h at SL, 472 km/h (293 mph) at 5 km; 6·6 min to 5 km; service ceiling 9,9 km; range 615 km; 360° in 18 s; TO 295 m, ldg 360 m.

Yak-7-82A

Yak-7V

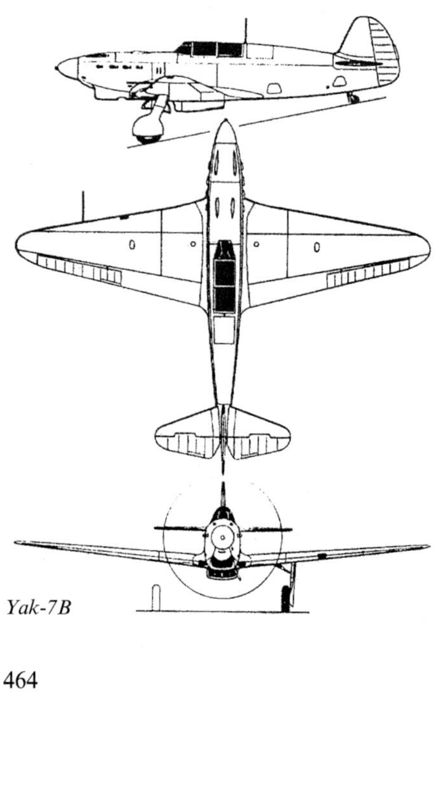
Yak-7B

Yak-9 Experience with Yak-1 and -7 and increased supply of light alloys made possible most numerous Sov fighter; GAZ-153 produced 20 per day in mid-44, when 9/9T/9D outnumbered all other fighters combined. Basic aircraft as 7DI but with only two Dural tanks (320 kg instead of 500 kg), oil reduced from 50 kg to 26 kg; M-105PF/VISh-61P standard, ShVAK (120) and left UBS only (200), no bomb racks; 459 built at GAZ-153 and 166, first use Stalingrad Dec 42, later suffered wing skin failures, teams sent to front-line units to rectify joints. Back in Nov 42 OKB modified much-used Yak-7 as **Yak-9/M-106** with all-round canopy, fully retrac tailwheel, sealed gaps and improved rads, but high perf marred by unreliability of M-106 engine.

First major variant **Yak-9T** anti-tank, 105PF and 3 m VISh-105SV prop, one 37 mm gun (OKB-16 11P-37, later replaced by NS-37) with 30-32 rounds (see data for increased length), and one UBS (200 rds); cockpit moved 400 mm to rear, VV-1 ring/bead sight using UBS to sight main gun. Max two (at most three) rounds possible before 5500-kg recoil depressed nose off target, but single hit usually destroyed target. Batch of 34 tested in action July 43, average one kill per 31 rounds (cf 147 of 20 mm). Total 9T delivery from GAZ-153 2,748 by June 45. Data:

DIMENSIONS Span 9,74 m (31 ft 11½ in); length 8,66 m (28 ft 5 in); wing area as before 17,15 m².

ENGINE One M-105PF.

WEIGHTS Empty 2298 kg (5,066 lb); loaded 3025 kg (6,669 lb).

PERFORMANCE Max speed 533 km/h (331 mph) at SL, 597 km/h (371 mph) at 4 km; climb 5·5 min to 5 km; service ceiling 10 km; range 735 km (457 miles); 360° 19 s; TO 380 m; ldg 500 m.

Yak-9D Standard long-range, similar to 9T but four tanks 650 lit (480 kg) and 48 instead of 25 kg oil; armament as 9. NII test Jan-Feb 43, snags high TO wt, no horizon or D/F, short range of radio. Prodn 3,058, end June 46. Single **Yak-9P** tested April 43, differing from 9/PF only in having UBS replaced by one ShVAK (SP-20) with 175 rds. **Yak-9PD** hi-alt with PD engine, separate non-ejector exhaust each cyl, rad deeper and moved forward, oil cooler larger, ShVAK only (120 rds), five built April 43, severe cooling problems. One (01-29) given wing 10,74 m span, 17,8 m² area, improved cooling, N_2 inerting, wt cut to 2845, reached 12,5 km but engine problems. Oct 43 reached 13,1 km after fitting M-106PV. In April 44 improved PD/M-106PV (6 ejectors each side with fairings above/below) reached 13,5 km; altogether 35 PD built, all lacking specified pressure cabin. Sept 43 GAZ-166 built 35 **Yak-9R**, with either 324 kg or 480 kg fuel, ShVAK alone or with UBS and rear-fuselage AFA-1M or 3S/50 for vert photography 300 m-3000 m; all had AGP-2 horizon, RPK-10 DF and RSI-4 transceiver. Supplemented Pe-2 in high-risk areas. Single **Yak-9TK** Oct 43 equipped for ShVAK or VYa-23, NS-37 or NS-45 (a/c length 8,66); NS-37 preferred single shots at 300+ km/h; NS-45 single shots only at max airspeed. Anti-tank **Yak-9K** based on 9T

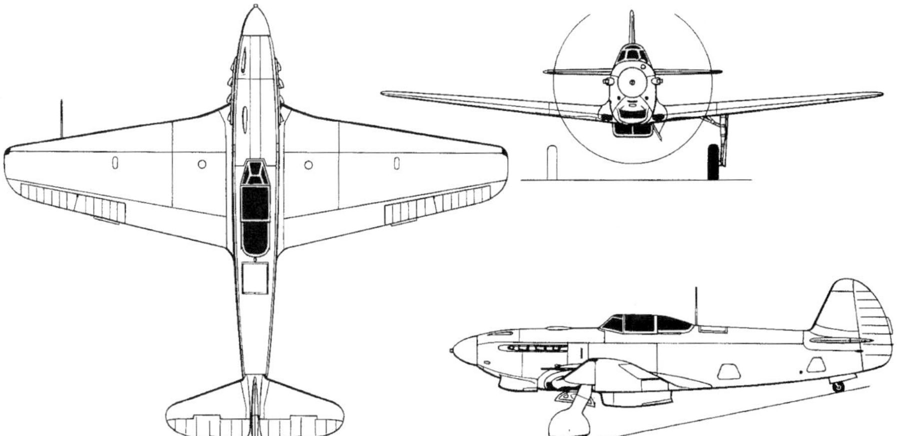

Yak-9D

but VISh-16P 3 m prop, 475 kg fuel; NS-45 with large muzzle brake (a/c length 8,87 m) and 29 rds, plus UBS (200 rds), PBP-1A optical sight. Tested Jan 44, followed by 53 service eval with 350 kg fuel (outer tanks deleted). **Yak-9B** fighter/bomber (OKB 9L) based on 9D plus 4 tubes behind seat at 75° each for FAB-100 nose-up or container for 32 PTAB-2,5-1,5 bomblets, front tubes normally used, all four = overload (3556 kg). Bombs aimed by various procedures in level flight; good results but lack of proper sight restricted 9B to eval batch of 109 Feb-Mar 44. Extended-range **Yak-9DD** for escorting bombers, eight metal tanks (each wing 210 lit/120 lit/50 lit almost to tip, plus three fuselage) total 845 lit; 70 lit oil; ShVAK only; cockpit 400 mm to rear as in 9T, 8-lit O_2, dual long-range radio, DF, horizon. One DD April 44, 299 built by Sept 45; 12 flew 1300 km to Bari Aug 44 impressed Allies, but at max wt underpowered and could not properly escort Tu-2. Underwing drop tanks never tested. Fuselage standardized with aft cockpit by **Yak-9M**, basically 9T/9K fuselage on 9D plus stronger wing (no skin failures), many minor improvements, total 4,239 by June 45. Few produced as **9M/PVO** with multichannel radio, DF, FS-155 light, SCh-3 IFF and illuminated gear indicator, fuel reduced to 420 kg. Date for 9M as before (9,74/8,5 m, PF engine), except:

WEIGHTS Empty 2428 kg (5,353 lb); fuel/oil 480 kg; loaded 3095 kg (6,823 lb).

PERFORMANCE Max speed 518 km/h at SL, 573 km/h (356 mph) at 3,75 km; 6·1 min to 5 km; service ceiling 9,5 km; range 950 km (590 miles); TO 420 m; ldg 550 m.

Yak-9 Kur'yer OKB match of DD wing (9 tanks, 846 lit) with unarmed fuselage with rear passenger cockpit with maps/baggage stowage but no inst panel, both cockpits aft-sliding jettisonable hood and urinal. Flown July 44, not submitted NII test. Similar fuselage mod from Yak-9T to produce **Yak-9V** trainer, full dual, transceiver, DF and intercom, one ShVAK (90 rds), no oxygen, armour or tank protection, smaller oil tank. OKB built 44, NII test April 45, 456 built by Aug 47 plus 337 conversions from 9M. Two **Yak-9S** mod May 45 from 9M with 105SV-01 prop and heavy armament: NS-23 (60 rds) plus two B-20S (120 rds each). Poor speed/climb compared with Yak-9U and Yak-3, so discontinued.

Yak-9U major upgrade, prototype Nov 43: PF2 driving VISh-105SV-01, improved ejectors, no dorsal inlet, improved oil rad moved to centre section, improved main rad moved aft with ART-41 auto control, fuel 320/oil 25 but improved protection, fuselage fabric replaced by 2 mm ply skin, radio button on throttle and mast deleted, O_2 doubled to 4 lit, better armour, lower rear deck and all-round canopy, auto tailwheel lock, VYa-23 (60 rds) and two UBS (170 rds each); provision for ShVAK/B-20/NS-37. Tested Nov 43, best fighter with PF2 but further mods, including VK-107A engine and 3,1 m VISh-107LO prop first flown Dec 42. Engine unreliability delayed 9U/VK-107A to Dec 43, prototype again new rads, fuel/oil 355 kg/35 kg, wing moved 100 mm forward, slightly smaller elevators, inner collector taking exhaust from each bank and discharging through 6 ejectors, inlet above cowl to cool inner collectors, plugs and generator, reached 700 km/h at 5,6 km. Series delivery with ShVAK + 2 UBS from April 44, engine gradually improved, bigger rads permitting full 3,200 rpm and drag redn, 3,921 delivered ending Aug 45. Outstanding combat results. Data for 45 series:

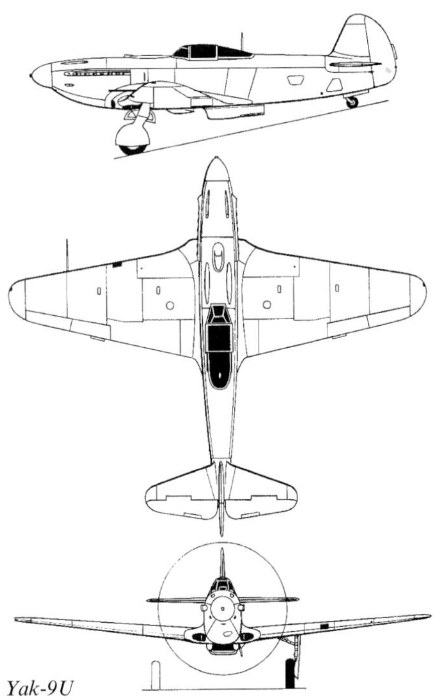

Yak-9U

DIMENSIONS Span 9,74 m; length 8,55 m (28 ft 6⅛ in); wing area 17,15 m^2.

ENGINE One VK-107A.

WEIGHTS Empty 2512 kg (5,538 lb); fuel 380 kg; loaded 3204 kg (7,063 lb).

PERFORMANCE Max speed 575 km/h at SL, 672 km/h (418 mph) at 5 km; 5 min to 5 km; service ceiling 10,65 km (34,950 ft); range 675 km; TO 375 m; ldg 530 m.

Yak-9UV Improved trainer, VK-107A, one B-20M (100 rds), otherwise as 9V/9U, flown June 45, no prodn because imminent jets. **Yak-9UT** built Feb 45, as U/VK-107A but air inlet to cool plugs and valve gear; one N-37 (lighter, less recoil than NS-37) with long muzzle brake (30 rds) plus two B-20S (120 rds); good but high stick forces, 282 produced Feb-May 45 with N-37 replaced by NS-23.

Yak-9P final variant for post-war service, much higher standards for extended life; all-metal wing including skin and ailerons, wingtips elliptic, armoured headrest, flap jack ball lock, improved tankage (Dyurite), wide wing-root inlets with filters and splitters separating ducts to supercharger and oil cooler. Designed for 8 g, 720 km/h dive limit. Two tested July 46, approved with two outboard tanks 350 kg normal and 515 kg overload, RPKO-10M DF, SCh-3 IFF, FS-155 light, UV panel lights, PAU-22 gun camera, and choice of armament; two B-20S (120 rds each) and either one B-20M (120 rds), NS-23 (80 rds), N-37 (30 rds) or N-45 (30 rds) or one B-20S (120 rds) and one 57 mm (20 rds). Latter discontinued and ammo for central gun changed after test to 115/75/28/25. Retested April-July 47 but no production because of jets. In 46 GAZ-153 produced 29 with ply fuselage skin and 10 Yak-9P all-metal. Further test Oct-Dec 47 with all-metal fuselage, two further tanks (682 kg total), rpm restricted to 3,000, one tested with VISh-107R reverse-pitch (shorter landing). At last 801 Yak-9P delivered Jan-Dec 48, of which 29 all-metal wing only and 772 metal fuselage, some supplied Albania, China, Hungary, Jugoslavia, N Korea, Poland, ending piston-fighter prodn; Yak-9 total 16,769. ASCC name 'Frank'.

DIMENSIONS as before 9,74 m/8,55 m/17,15 m^2.

ENGINE One VK-107A.

WEIGHTS (3×B-20) empty 2708 kg; fuel 516 kg; loaded 3550 kg.

PERFORMANCE Max speed 590 km/h at SL, 660 km/h (410 mph) at 5 km; 5·8 s to 5 km; service ceiling 10,5 km (34,450 ft); range 1130 km (702 miles); TO 540 m; ldg 146 km/h, 582 m.

Yak-3 Smallest, most refined and most agile piston-engined Yak, built in two main versions 'frontal' and hi-alt, but actually 18 variants many without designations. Began with **Yak-1M**, refined offshoot of final Yak-1 with almost same fuselage (low rear deck and all-round canopy) and landing gear (three separate doors each unit) but wing of mixed Yak-9 construction, reduced in span/area and in L/R halves for field replacement, tail as lightened Yak-1 but smaller, tailwheel unlocked by moving control column from neutral, no aileron mass balance, initially PF engine and VISh-61P prop, OP-492 rad better submerged in

Yak-3/VK-105PF-2

Yak-3U/ASh-82FN

fuselage in longer duct further aft, two 9 in (measured in inches) oil coolers parallel in wing roots under cockpit floor, supercharger inlet between them, two nitrogen-protected 120-kg wing tanks and one unprotected 30-kg centre tank, ejector exhausts, one ShVAK (120) and one UBS (200), radio transceiver without mast. First prototype 1M, with oil cooler under engine, flown 28 Feb 43; later cleared for PF2 engine with 1100-mm boost in low gear, above this height held to 1050-mm by poor 78-oct fuel. Consensus: best piston-engined fighter; marred by poor oil cooling and loss of oil and bad radio perf. Second (*Dubler*) improved aerodynamics, PF2 engine, six even-spaced ejectors each side, VISh-105SV with lighter spinner and reprofiled blade roots, wing-root oil coolers with bigger area, fuselage skin 2-mm ply, jettisonable canopy, elevator mass balance inside fuselage, lighter main legs, centre tank protected and better pilot protection, OP-554 main rad with ART-41 thermostat, ShVAK replaced by lighter ShA-20M with 110 rds and second UBS added (now 150 rds each), ring/bead replaced by PBP-1a sight, RSI-4 radio with twin antenna wires, rear-view mirror; first flown 20 Sept 43, much faster, satisfactory cooling and better radio. To be produced as Yak-3.

Yak-3/VK-105PF2 As 1M-*Dubler*, except inner wing strengthened for 700 km/h dive, main legs taller and stronger, fuel gauges moved outboard (seen better), rads again improved finish, but poor mass-prod quality control resulted in higher weights and reduced perf. First prodn Yak-3 flown GAZ-292 Saratov 8 March 44. To batch 13 (197 a/c) one ShVAK + one UBS; thence extra UBS added. From batch 16 fuel reduced from 372 lit to 350 lit-360 lit. Most (4,797 of 4,848) standard with PF2 engine; selected by Normandie-Niemen and flown in France by them to 56.

DIMENSIONS Span 9,2 m (30 ft 2¼ in); length 8,5 m (27 ft 10⅝ in); wing area 14,85 m^2 (160 ft^2).

ENGINE One VK-105PF2.

WEIGHTS Empty 2123 kg (4,680 lb); fuel 270 kg; loaded 2692 kg (5,935 lb).

PERFORMANCE Max speed 567 km/h at SL, 646 km/h (401 mph) at 4,1 km; 4·5 min to 5 km; service ceiling 10,4 km; range 648 km (403 miles); TO 280 m; ldg 137 km/h, 455 m.

Yak-3PD Hi-alt version with PD engine, water-methanol, VISh-105L-2 prop, polished blades, supercharger inlet under centreline, span 9,8 m, wing area 15,35 m^2, intended one central and one synch ShA-20M, 150 rds each, but actual a/t one NS-23 (60 rds). Flown 20 May 45, many engine problems. Second aircraft with VK-105PV, various props, reduced wt, flown 29 Sept 45, later reached 710 km/h and 13,3 km but no prodn. **Yak-3T**, flown 19 Jan 45, standard PF2 but L-28 prop, armed by N-37 (reduced wt and barrel shortened 150 mm) with 25 rds and two synch B-20S (100 ea); cockpit 400 mm to rear, armour reduced and inerting eliminated, exptl prop and radiator, pneu-system bottles moved to cowling, cockpit floor vented, fuel reduced to 355 lit. Only good feature, good platform for heavy firepower. Single **Yak-3RD** with RD-1 rocket in tail with kerosene (50 kg) and HNO_3 (200 kg) tanks in wings; rudder chord increased to make up for lost lower portion, elevs mod inboard; PF2 eng driving rocket pump, main/oil rads combined and supercharger inlet L root only, 200 kg petrol; single NS-23 (60 rds). Flown 22 Dec 44, elec rocket ign replaced by RD-1KhZ with chemical, many rocket problems but reached 782 km/h (486 mph); destroyed by mystery dive into ground 16 Aug 45, killing V. L. Rastorguyev. **Yak-3P** mod series PF2 aircraft with B-20M (120 rds) and two synch B-20S (130 rds ea), lighter than standard a/t and new electro-pneu firing and pneu charging of central gun; minor mods incl RSI-4A replaced by RSI-6M receiver. Tested 23 March 45, produced in series April 45 to mid-46, total 596. Data little changed (weights 2150 kg/2708 kg, speeds 572 km/h/646 km/h, range 610 km).

Yak-3/VK-107A Prototype complete 6 Jan 44, flown 15 Ap 44, new engine driving 107LO 3,1 m prop, cockpit 400 mm to rear, stronger main gears, 518 lit (390 kg) fuel in 4 wing + distrib tank, 60 lit oil, 6 evenly spaced ejectors with inner pipes and plugs in ram-air duct, wing-root inlets extended forward. Reached 720 km/h at 5750 m but repeated cooling problems, small ground angle limited braking (longer field length), inadequate struc rigidity. *Dubler* tested earlier (from 25 Jan), this rectified defects from 13 July but VK-107 still unreliable and mixed structure inadequate. GAZ-292 in March 46 built three VK-107A a/c with all-metal wing and tail, stronger fuselage and other parts,

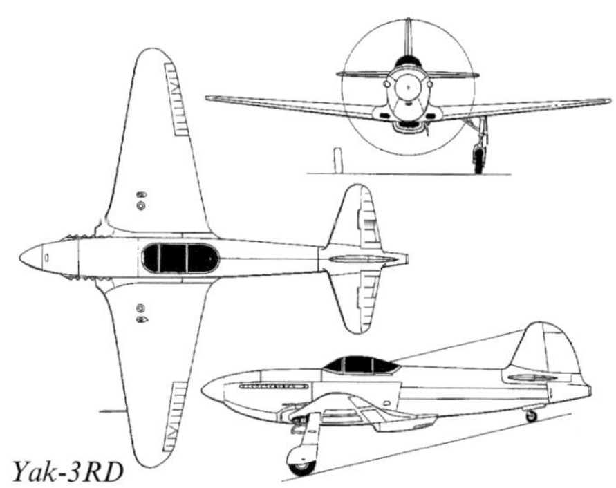

Yak-3RD

improved cooling; one with two synch B-20S (120), other two with one B-20M and two B-20S (100 ea). All three NII-tested from 10 April, many defects. In March 45 VK-107A a/c from GAZ-31 rebuilt as all-metal, bigger oil rad, ball-bearing elevators, Elektron (Mg alloy) fuselage and control-surf skin, cockpit 400 mm aft, fuel/oil reduced (to 320 kg/32 kg); a/t remained one B-20M and 2 B-20S. Wt reduced from 3059 kg to 2935 kg, but poor metal finish reduced speeds (eg max from 720 km/h to 706 km/h) and range from 1060 km to 777 km, but handling better. Total production 48, all needing rework; engines instead to Yak-9U/9P.

Yak-3/VK-108 Single mixed-constn aircraft fitted with VK-108 driving VISh-107LT-5,

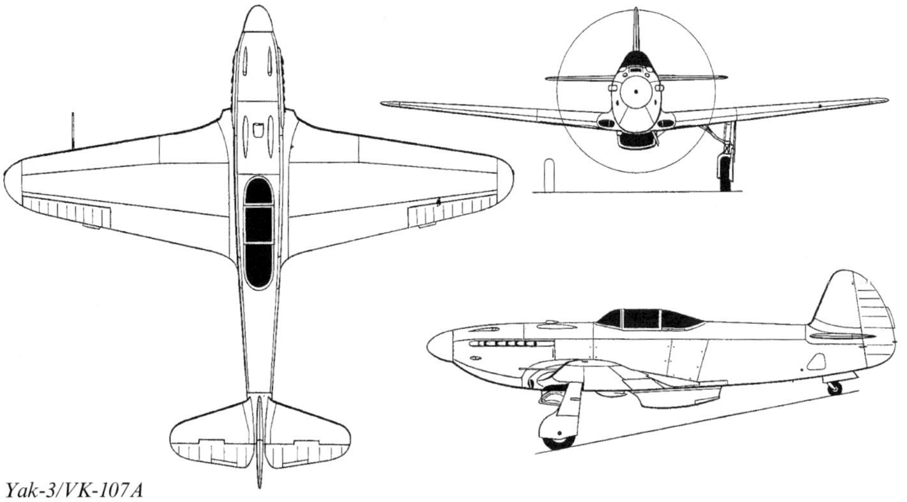

Yak-3/VK-107A

OP-624 and 622 rads, metal aileron skins, no exhaust collector (individual pipes), supercharger inlet under engine, one NS-23 (60 rds). First flown 7 Oct 44, outstanding perf but engine problems. Second conversion Nov 45 with two synch B-20S, but abandoned because of jets. Data for first:
DIMENSIONS as before 9,2 m/8,55 m/14,85 m².
ENGINE One VK-108.
WEIGHTS Empty 2510 kg; fuel/oil 350 kg/50 kg; loaded 3000 kg.
PERFORMANCE Max speed 639 km/h at SL, 745 km/h (463 mph) at 6290 m; 3·5 min to 5 km; service ceiling 10,4 km; range not recorded; TO 250 m; ldg 145 km/h, 560 m.
Yak-3U/ASh-82FN Standard PF2 a/c rebuilt with powerful radial engine, wing increased in span and moved forward 219 mm, cockpit 84 mm higher, ply fuselage skin, metal tail and ailerons, careful wt and drag reduction, two B-20S (120 ea). Flown 29 April 45, outstanding; returned to OKB 15 June for all-metal wing and 80 mm longer legs for better ground angle; even better, but languished unwanted. Provided basis for **Yak-U** trainer which became Yak-11. Total prodn of Yak-3 4,848 (291 Saratov 3,840, 31 Tbilisi 1,008).
DIMENSIONS Span 9,4 m (30 ft 10 in); length 8,17 m (26 ft 9⅝ in); wing area 17,15 m² (185 ft²).
ENGINE One ASh-82FN.
WEIGHTS Empty 2273 kg; fuel/oil 340 kg+35 kg; loaded 2792 kg, (metal wing) 2740 kg (6,041 lb).
PERFORMANCE Max speed 705 km/h (438 mph) at 6,1 km; 3·9 min to 5 km; service ceiling 11,25 km; range 778 km; TO 470 m; ldg 135 km/h, 430 m.
Yak-3UA In partnership with Museum of Flying, Santa Monica, Yak A/c Corpn is producing 20 repro Yak-3U from original drawings/tooling. Modern instruments/radio/materials, slightly more metal skin, US engine requiring carb-air inlet above cowl. All 20 have customers, c$500,000 with standard US avionics, paint to customer spec. FAA xptl category.
DATA Brochure figs as original except engine reconditioned Allison 2L, 1,240 hp.

Yak-5 Single-seat advanced trainer based on UT-2MV but with many new features. Structure mixed as before, but new wing with equi-taper LE/TE. Front cockpit removed, sliding canopy over rear cockpit, engine in helmeted cowl as before but driving VISh-237 v-p prop, semi-retractable gears with oleo legs, tyres 500 mm×125 mm, brakes, tailwheel (200 mm×80 mm), split flaps. One synchronized ShKAS with 300 rds above fuselage, PBP reflector sight. Flown Sept 44, rated highly by V. L. Rastorguev but crashed, suspected wood failure. Led to all-metal Yak-18.
DIMENSIONS Span 10,5 m (34 ft 5⅜ in); length 7,3 m (23 ft 11⅜ in); wing area 17,0 m² (183 ft²).
ENGINE One M-11D.
WEIGHTS Empty 770 kg (1,698 lb); fuel/oil 62 kg+8 kg; loaded 940 kg (2,072 lb).
PERFORMANCE Max speed 250 km/h (155 mph) at SL; no data on climb or ceiling; range 450 km (280 miles); landing 85 km/h (53 mph).

Yak-6, NBB This aircraft was designed by Yak himself in second half 41 as simple utility aircraft, making minimal demands on contractors or users, to serve as NBB (*Nochnoi Blizhnii Bombardirovshchik*, night short-range bomber) and multi-role transport, an improved successor to Po-2. Hoped to carry 1 t bombs using two 190 hp M-12 engines but unavailability of these resulted in fall back on lower-power M-11F. Despite this, simplicity of aircraft resulted in production decision.

S. A. Yakovlev with first Yak-3UA in 1993

Basic structure wood with some fabric covering. Wing of Clark-YH section, max t/c (at outer end centre section) 16·8%. Centre section tapered on TE only, dihedral 0°; outer panels fully tapered, dihedral 8°. Two box spars, ply ribs, ply skin over LE back to front spar, fabric elsewhere. Slotted wood/fabric ailerons each hung on three brackets well below surface, driven by bellcranks and levers pivoted to top of aileron spar. Single dural flap across centre section, perforated Schrenk type, manually actuated by transverse rods and pivoted links driven from cockpit handwheel. Fuselage a rounded box with four longerons. Wing had light stringers (10 forward of wing TE, 22 aft) and 10 transverse frames, with glued *shpon* nose, ply skin forming semi-monocoque ahead of TE and fabric aft. Tail wood/fabric with ply skin LE to front spar, tailplane with variable incidence on prototype only, pilot-actuated trimmers on rudder and elevators. Welded steel-tube nacelle trusses with dural skin extended forward to enclose crankcase and form small baffles round cylinder barrels; installation made from same drawings as UT-2 with same carb-air heating by two lower cylinder exhausts. Drum oil tank in nacelle, 195 lit (42·9 gal) dural fuel tank in each side of centre section between spars. Two-blade fixed-pitch 2,2 m wooden props. Prototype built with main landing gears each with single air/oil shock strut with pinned side braces, welded to inner end of axle carrying pneu-braked wheel with 600 mm×180 mm tyre, retracted to rear by crank-arm to sliding trunnion driven by cables to handwheel in cockpit. Sprung castoring tailwheel, 255 mm×110 mm.

Prototype completed as NBB with enclosed cockpit for two with dual pedals and control-column fittings but spectacles yoke not fitted on right side. Rear cabin with roof hatch (left/right halves folding inwards) and pintle for ShKAS

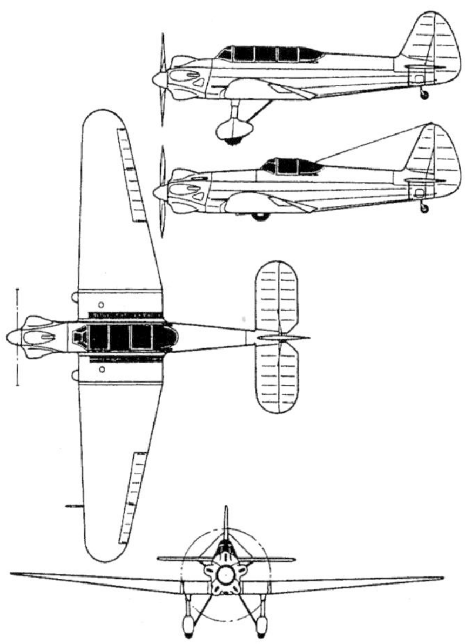

UT-2L [p. 459] (lower side view, Yak-5)

Yak-6, NBB, with bombload

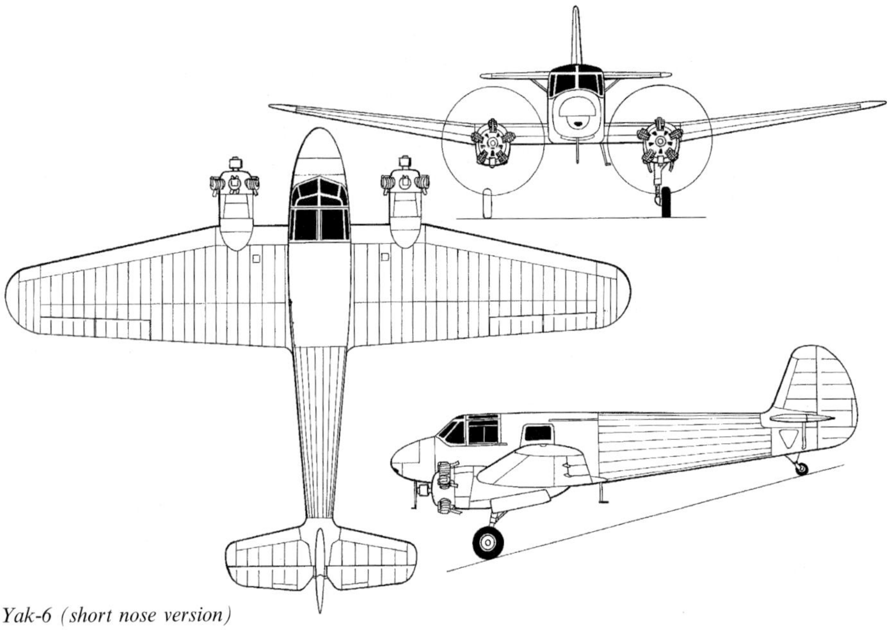

Yak-6 (short nose version)

with satchels for eight ammo drums on walls. Five racks under centre section for FAB-100 or Der containers from 50 kg to 250 kg, max total load 500 kg (1,102 lb). Windmill generator on right side of fuselage, 12A-30 battery, RSI-3/4 radio and D/F loop under nose. First flight about June 42, completed NII tests Sept 42 and cleared for production with modifications including improved outer panels, reduced dihedral, wing-root fillets, larger ailerons, main cabin equipped for transport with seats for four passengers (door on left) or 500 kg (1102 lb) cargo load (loaded through large hatch on right). Except for first few, production Yak-6 had simple fixed main gear with bay sealed by detachable dural doors. Though of low performance, about 1,000 built and used with great effect in front line, both as transport and as close-support and night attack aircraft with various arrangements of pilot-aimed bombs and up to ten RS-82 rockets under outer wings. Said to have acquired popular name *Dyerevyannyi Duglas* (wooden Douglas, ie, small partner of Li-2). Standard utility transport of VVS regiments 1944-50. Known with glider hook, skis, photo-survey gear and ambulance conversion; not seen on floats. One aircraft modified by CAHI with wings almost scaled from Li-2, retrac gear, helmeted cowls, VISh-327 props, pivoted tailplane and increased tankage.

DIMENSIONS Span 14,0 m (45 ft 11¼ in); length 10,35 m (33 ft 11½ in); wing area 29,6 m² (319 ft²).

ENGINES Two M-11F.

WEIGHTS Empty (42) 1368 kg, (43) 1415 kg (3,119 lb); fuel/oil 280 kg+24 kg; loaded 2,3 t (5,071 lb).

PERFORMANCE Max speed 187 km/h (116 mph); climb 5·4 min to 1 km; service ceiling 3380 m (11,090 ft); range 900 km (559 miles); TO 285 m; ldg 93 km/h, 265 m.

Yak-8 Still expecting M-12 engine to become available, small team under Antonov developed Yak-6 *Modifikatsirovannyi* into this larger aircraft in late 43, prototype flying in VVS markings early 44. Basic design similar and wing believed identical to predecessor. Fuselage considerably larger, because Yak-8 was dedicated transport, with no secondary roles. Max body width remained 1240 mm but external depth of fuselage increased from 1500 mm to 1760 mm, giving much better headroom internally, and length increased. From nose to wing TE structure strengthened with main frames built-up boxes 80 mm deep, with 2 mm ply skin throughout. Extreme nose and entire rear fuselage identical to Yak-6 Mod. Tail construction unchanged but surfaces enlarged with greater height/span and almost straight edges. Engine installations and main gears same as Yak-6 Mod, though engine slightly more powerful. Engines drove air compressor, hydraulic pump and generator, gear retraction being hydraulic. Assumed flap likewise. Brakes remained pneumatic. Tailwheel smaller, tyre 220 mm × 110 mm. Tankage same as Yak-6 Mod (greater than series Yak-6). Cockpit as Yak-6 Mod, main cabin with seats for six passengers with three Yak-6 size windows each side.

NII test satisfactory but no real requirement and judged future aircraft should be all-metal, so no order. Despite this, ASCC name 'Crib'.

DIMENSIONS Span 14,8 m (48 ft 6⅔ in); length 11,35 m (37 ft 2⅞ in); wing area 30,0 m² (323 ft²).

ENGINES Two M-11FM (M-11M).

WEIGHTS Empty 1750 kg (3,850 lb); fuel/oil 340 kg; loaded 2,7 t (5,952 lb).

PERFORMANCE Max speed 248 km/h (154 mph) at SL; cruise 190 km/h (118 mph); time to 1 km, 6·4 min; service ceiling 3,9 m (12,795 ft); range 890 km (553 miles); TO 380 m; landing 260 m/100 km/h (62 mph).

Yak-10 (Yak-14) One of two light aircraft designed 44 to permit comparison between high wing (this) and Yak-12 (later called Yak-13, which see). Direct extrapolation of AIR-6, and like earlier type was of very mixed construction. Four-seat cabin fuselage built around welded truss of mild-steel tube with D1 sheet over nose, door and rear hatch, fabric elsewhere. Wing of 11% Clark-YH profile, wooden two-spar structure with ply LE and fabric elsewhere, braced by vee-struts. Tail and all control surfaces D1/fabric. Rubber-sprung main gears with neat faired legs and spats over 500 mm × 150 mm wheels with brakes; tailwheel 200 mm × 80 mm. No slats, flaps or tabs; essence of concept was simplicity for easy handling, low cost and good reliability. Yak-14 flew late 44 and immediately showed bad qualities. Considerable tinkering with several aircraft (believed all military) ensued, designation being changed to Yak-10 early 45. Eventually passed NII tests June 45.

Yak-8

Uninspired design, but 40 built and used in several versions, without spats and with improved engine installation and VISh-327 propeller. **Yak-10V** (*Vyvoznoi*, carrier) had dual control; **Yak-10S** (*Sanitarnyi*) had long hatch for loading stretcher on left side; **Yak-10G** (*Gidro*) had two AIR-6 type floats and, though it passed factory tests on completion in 46, its performance was unacceptable. One 46 series machine had M-11FR of 160 hp and specially designed skis (main 1930 mm×340 mm, tail 460 mm×120 mm; data appear under (L) for *Lyizhi*, skis. Yak-10 did not form basis of later Yak-12.

DIMENSIONS Span 12,0 m (39 ft 4½ in); length 8,45 m (27 ft 8⅝ in); wing area 22,0 m² (237 ft²).
ENGINE (prototype) one M-11FM, (rest) M-11FR.
WEIGHTS Empty (p) 792 kg, (series) 820 kg (1,808 lb); (L) 812 kg; fuel/oil 94 kg+14 kg; loaded (p) 1150 kg, (s) max 1250 kg (2,756 lb), (L) 1170 kg.
PERFORMANCE Max speed 200 km/h/206 km/h/195 km/h; climb to 1 km 5·5 min/5·5 min/9·5 min; service ceiling 3,4 km/3,5 km/2550 m; range 576 km/605 km/350 km; TO 260 m/228 m/350 m; ldg 79 km/h/73 km/h/-, 280 m/193 m/-.

Yak-10

Yak-11

Yak-11 OKB discussed advanced fighter/trainer variant of Yak-3 with VVS mid-44 and design went ahead at once; though low priority, OKB grown so large plenty of spare design capacity. Prototype Yak-3UTI (Yak-UTI) flown 45. Many parts common to Yak-3 (wing span as Yak-3U, 9,4 m) and designed to exceptional factor 15·4; first installation of ASh-21 engine and VISh-111V-20 prop, and fitted with UBS and 100 rounds plus wing racks for two FAB-100 or other loads. No evidence of major shortcoming but no production. Used as basis for slightly refined aircraft designated Yak-11 first flown 46; precise differences and first-flight date not known but evidence Yak-3UTI had short-chord cowl. Yak-11 no longer used identical Yak-3 parts, though differences generally small. Fuselage modified shape with welded 30KhGSA truss, ply covered forward and fabric aft, on wood stringers with ply top decking. Tandem cockpits (neither in original position) each with sliding canopy in long glasshouse. Well-equipped cockpits; pilot's notes key 53 items (Yak-3, 38 items). All-metal two-spar wing with Clark-YH profile, 5° dihedral from root, D1 flush-riveted skin 2 mm over LE to front spar, 1 mm elsewhere. Tail, ailerons and control surfaces D1 with fabric. Neat engine installation on rubber anti-vibration mounts in long-chord cowl faired into fuselage with exhaust collector ring discharging below; side exits for cooling air with adjustable flaps. VISh-111V-20, two blades, constant-speed, 3 m diameter, large spinner. Oil tank 38 lit with dorsal vent, cooler in wing served by ram inlet left wing root, exit chute in undersurface. Split flaps 0° and 43°. Main landing gear with 600 mm×180 mm tyres retracting inwards ahead of front spar; tailwheel with 255 mm×110 mm tyre retracting aft with two doors. All these services, flaps, gear, brakes and cooling-air flaps actuated from pneumatic system with main and emergency bottles charged by engine-driven piston compressor. Fuel in 173 lit protected tank in each wing feeding small 13,5 lit tank behind engine. UBS on left of decking firing through cowl lip, 100 rounds; same variety of underwing loads up to 100 kg each side; provision for various role fits including PAU-22 combat ciné camera in fairing above windscreen.

NII testing completed Oct 46, generally excellent aircraft though really agile only in roll; soon cleared for production with deliveries from mid-47. Series aircraft slightly heavier with extra equipment, RSIU-3M/-4M radio with mast behind rear cockpit, fixed tailwheel, often ShKAS instead of UBS and landing light in left LE. Total 3,859, plus 707 (called **Le-10**, later **C-11**) Czech-built by LET, all with ShKAS.

In 51 OKB produced **Yak-11U** with tricycle gear for training jet fighter pilots. Like later series aircraft, no spinner, RSI-6 radio, S-2V gun camera and tow hook. Modest numbers,

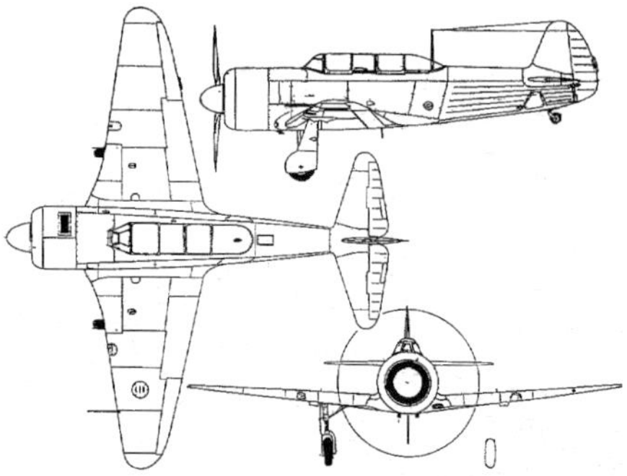

Yak-11

including Czech **C-11U** version. Original version exported 18 countries, many preserved, several airworthy. Numerous C-1d records, incl 100-km circuit at 479,97 km/h. ASCC name 'Moose'.

DIMENSIONS Span 9,4 m (30 ft 10 in); length 8,2 m (26 ft 10⅞ in), (11U, 8,66 m, 28 ft 4⅞ in); wing area 15,4 m² (166 ft²).

ENGINE One ASh-21.

WEIGHTS Empty 1,9 t (4,189 lb), (U) 2066 kg; fuel/oil 230 kg+24 kg; loaded 2440 kg (5,379 lb), (U) 2,5 t.

PERFORMANCE Max speed 465 km/h (289 mph), (U) 460 km/h; climb to 1 km 1 min; service ceiling 7950 m (26,080 ft), (U) 7,2 m; range 1250 km (777 miles); TO 395 m, ldg 89 km/h, 500 m.

Yak-12

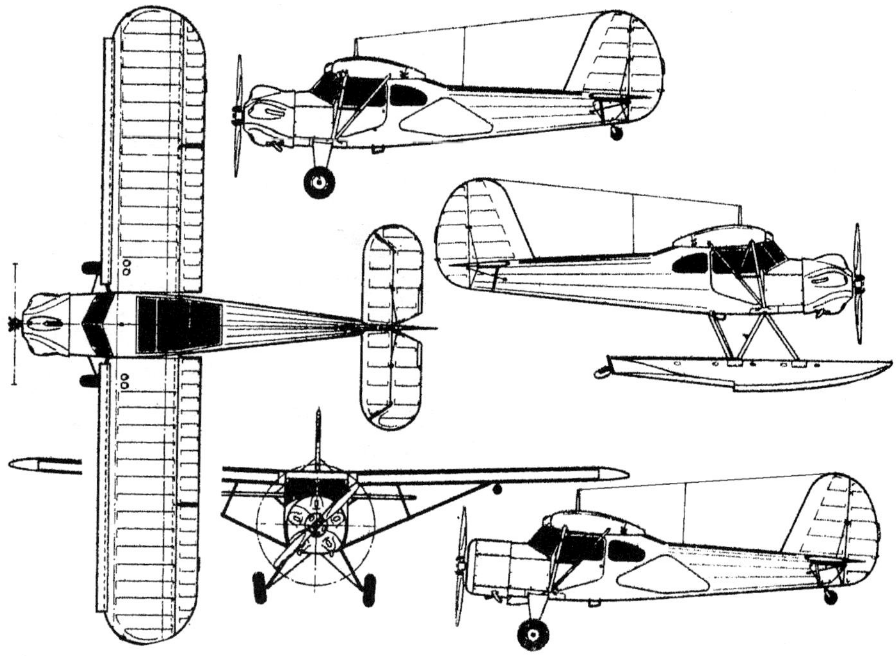

Yak-12 (middle side view 12GR, bottom 12M ambulance)

Yak-12 In early 47 Yak decided to build a better high-wing utility cabin machine to replace Yak-10, and resulting Yak-12 (no relation to previous Yak-12) flew same year. Wing Clark-YH rectangle with two spars, wood construction, ply skin over LE as far back as front spar, glued fabric over whole wing. Dihedral 3° 30', braced by drawn D1 V-struts pinned to both spars and meeting at bottom of fuselage. Triangle of sub-struts bracing midpoints of main struts together and to front spar of wing. Fixed D1 slats along entire LE, D1/fabric slotted ailerons with inset hinges at mid depth (unusual for OKB) and plain manual flaps inboard. Fuselage completely reprofiled compared with Yak-10, basis welded truss of KhNZA tube with some secondary wood formers and stringers and fabric covering. Tail entirely D1/fabric with wire bracing. Fuel in two 100 lit aluminium tanks between spars just outboard of fuselage. Engine installation same as Yak-18 with helmeted cylinders and fixed-pitch aluminium prop. Improved low-drag main gears with pin-jointed legs sprung by rubber in fuselage attached to axles by steel tapes, low-pressure tyres 600 mm×180 mm. Fixed tailwheel 200 mm×110 mm. Standard ski installation, interchangeable with Yak-18. Side-by-side bucket seats with flight controls on left, cable operation except push/pull in wings. Normal missions liaison and artillery spotting (*korrektirovshchik*) with RSIU-16 radio. Small number built, believed all for VVS, and a few with third seat at rear. ASCC name 'Creek'.

Yak-12S (*Sanitarnyi*, ambulance) flown 48 with mounting for single standard stretcher loaded through large triangular hatch in left side. Standard agreed late 48 with seat for attendant and stowage for 22 kg medical kit. Replaced U-2S progressively. Second 48 version **Yak-12SKh** (agricultural) equipped with glassfibre or metal tank for chemical dispensed through aluminium-sheet fabricated spreader below fuselage in almost same installation as Po-2AP. Several different variations but built only in small numbers because underpowered. From start Yak had suffered from absence of M-12 engine and waited for forthcoming AI-14. Last of original M-11FR series was single **Yak-12GR** (*Gidro*) with AIR-6 type twin floats. In this case lack of power made performance marginal.

AI-14R engine flight tested in Yak-12 June 50. In late 51 resulted in production **Yak-12R**. New engine of almost doubled power in long-chord cowl with collector ring to ventral exhaust outlet, radial shutters (usually 28) to regulate airflow, side cooling-air exits, two-blade 2,75 m VISh-530L-11 bracket-type two-pitch prop. Redesigned wing with greater span and area, entirely D1 structure and flush-riveted D1 over LE, fabric elsewhere. Similar D1 skin for top of fuselage and much of area previously ply-skinned, including all-D1 stretcher hatch. Smaller fuel tanks but surprisingly little difference in range. Larger wing was one of several changes to enable 12R to use smallest and most confined fields and clearings. Special dozer blade (*soshnik*, spade) smoothed surface and reduced take-off run by up to 50 m. First mass-produced variant (at least 1,000) initially for VVS as two-seat liaison, with occasional third seat, and from 1953-54 in several related civil transport forms for GVF with normal upholstered seats and in some models provision for small cargo load or stretcher. 'Creek-B'.

In 55 OKB produced **Yak-12M** with better overall distribution of masses and areas, balancing heavier engine by longer rear fuselage and enlarged tail, vertical tail area increased 41% and horizontal 36% with two bracing struts. For first time a true four-seater, with substantial increase in allowable weights. Produced in ag and ambulance models, former with alleged 2 h spray/dust between reloads; another model Dec 55 configured for sport parachuting, door on right equipped with runningboard. 'Creek-C'.

Final production variant, of 57, was **Yak-12A** with new wing and many other improvements. Wing of Cessna profile/plan, left/right wings each one piece but with discontinuity at 55% semi-span. Inboard untapered, two spars widely separated at root but meeting at 34% chord at discontinuity; outboard single spar at 34% chord with taper on TE. Manual plain flaps on inboard section, slotted ailerons outboard. All D1 structure with riveted D1 over LE and back to rear spar inboard and single spar outboard; fabric to rear. Ailerons/flaps D1 over LE, fabric to rear. Wing braced by single pin-jointed aerofoil-profile strut meeting Y-junction of spars, with secondary sub-strut joining mid-length of strut to wing. Fixed slats replaced by retractable slats on outboard section only. Vertical tail as 12R but new horizontal tail with bracing (wire) above and below, D1 skin back to front spar and improved controls and tabs. Larger fuel tanks in tapering space between wing spars. Landing light in left LE. Completely revised cabin interior with US-style panel, wheel instead of stick, and more comfortable seats for pilot (left) and passenger and two-seat divan at rear. Despite increased weight, considerably faster and longer-range. Over 1,000 built for local services. 'Creek-D'.

Single STOL **Yak-12B** biplane built with AI-14RF and rectangular lower wing of 10 m span with full-span slotted flap (four flaps in all, pneu actuation) and single interplane struts. Several versions widely exported, military and civil. Yak-12, Yak-12R and Yak-12M licence-built in Poland by WSK-Swidnik, later developed into PZL-101 Gawron. Also basis for several Chinese aircraft in this class.

DIMENSIONS Span (12, S) 12,0 m (39 ft 4½ in), (R, M, A) 12,6 m (41 ft 4 in); length (12, S) 8,36 m (27 ft 5⅛ in), (R) 8,49 m (27 ft 10¼ in), (M, A) 9,0 m (29 ft 6⅓ in); wing area (12) 21,6 m^2 (232·5 ft^2), (S) 22,0 m^2 (237 ft^2), (R, M) 23,8 m^2 (265 ft^2), (A) 22,66 m^2 (244 ft^2).

ENGINE (12,S) one M-11FR, (others) one AI-14R.

WEIGHTS Empty (12) 830 kg (1,830 lb), (S) 852 kg (1,878 lb), (R) 912 kg (2,011 lb), (M) 1026 kg (2,262 lb), (A) 1059 kg (2,335 lb); fuel/oil (12/S) 145 kg+14 kg, (R, M) 135 kg, (A) 165 kg; loaded (12) 1185 kg (2,612 lb); (S) 1232 kg (2,716 kg), (R) 1172 kg (2,584 lb), (M) 1450 kg (3,197 lb), (A) 1588 kg (3,501 lb).

PERFORMANCE Max speed at SL (12) 169 km/h (105 mph), (S) 172 km/h (107 mph), (R) 184 km/h (114 mph), (M) 182 km/h (113 mph), (A) 220 km/h (143 mph); cruise (except A) about 150 km/h (93 mph), (A) 190 km/h (118 mph); time to 1 km (typical) 8 min; service ceiling (12) 3 km, (R) 5,8 km, (M) 4160 m, (A) 4050 m; range (12) 810 km, (S) 620 km, (R) 510 km, (M)

Yak-12A

PZL-101A (Yak-12M)

Yak-12B

Yak-13

765 km, (A) 1070 km (665 miles); TO typically 130 m (B, 50); ldg 60-70 km/h, 80-130 m.

Yak-13 First flown Nov 44, this neat low-wing cabin aircraft was designed to afford direct comparison with braced high-wing Yak-10. Original designation, **Yak-12** (unrelated to later aircraft just described). Yak-10 and Yak-13 had as nearly as possible same engine installation, cabin, rear fuselage and tail, though Yak-13 fuselage was shallower. Apart from welded-tube fuselage truss construction virtually all-wood, with fabric covering. New wing designed for this machine, with horizontal but tapered centre section (Clark-YH, 15%) with ply covering, and tapering outer panels to 9% tip, two spars throughout, ply skin as far back as front spar. Ailerons and tail D1/fabric. All-D1 split landing flap under centre section, manually driven to 40°. Main gears with neat single shock struts to outer ends of axles for 500 mm × 150 mm braked wheels, manual cable/winch retraction inwards ahead of front spar. Castoring tailwheel 200 mm × 80 mm. Fuel in two 75 lit tanks in centre section; engine helmeted with short ventral stub exhaust, VISh-327 prop. Comfortable cabin with fully glazed canopy (sliding, later left/right halves hinged upwards at centreline of roof). Four seats, but (Shavrov) for practical purposes a three-seater. Aerobatic, fitted booster coil and starter but no instrumentation for night flying or a radio. Completed NII trials 45; much better than Yak-10 but not as good as Yak-12, so no order. Despite this, ASCC name 'Crow'.

DIMENSIONS Span 11,5 m (37 ft 8¾ in); length 8,45 m (27 ft 8⅝ in); wing area 22,0 m² (237 ft²).

ENGINE One M-11FM.

WEIGHTS Empty 868 kg; fuel/oil 120 kg+14 kg; loaded 1230 kg (2,712 lb).

PERFORMANCE Max speed 245 km/h (152 mph); climb to 1 km in 4·2 min; service ceiling 4 km; range 815 km (506 miles); TO 330 m; ldg 89 km/h, 375 m.

Yak-14 Designation repeated for submission to meet the final *desant* (assault) glider specification issued mid-47. Called for 3,5 t payload and made other important demands to meet A-VDV needs, with some interest by GVF which was still studying civil cargo gliders and glider trains. Ilyushin assigned metal glider (Il-32), Yak a wooden one. Yak-14 designed second half 47, neat and straightforward. Hold cross-section 2 × 2 m (79 in × 79 in) with max values ahead of wing 2,3 m × 2,25 m; usable length 8 m (26 ft 3 in). Entirely wood structure with steel fittings. Nose hinged to open vertically (gas bottle or hand pump with stand-by winch/cable) for full-section access including small vehicles and artillery. Fabric-covered rear fuselage, detached for unloading on combat mission. Two-spar high wing, rectangular centre section and tapered outer panels with strong ribs at interconnection incorporating anchor for strut. Full-span slotted flaps driven by gas bottles with manual (80:1 ratio) crank emergency, outer sections carrying inserted slotted ailerons. Wire-braced tail with rudder trimmer. Steel-tube steps up fuselage exterior to flight deck mounted above left side of fuselage seating two pilots side-by-side with optional third jump seat at rear. Door on left of cockpit, all-round glazing, no known armament apart from infantry weapons but at least one glider with radio. Canvas seats for 35 troops, normal door at rear on left. Tricycle landing gear jettisoned on combat mission with skid landing. Satisfactory NII tests 48, immediate production at Rostov of 413 including one of mixed (part-metal) construction. Three to Czech AF, designated NK-14. ASCC name 'Mare'.

DIMENSIONS Span 26,17 m (85 ft 10¼ in); length 18,44 m (60 ft 6 in); wing area 89,0 m² (958 ft²).

WEIGHTS Empty 3095 kg (6,823 lb); loaded 6750 kg (14,881 lb).

PERFORMANCE Tow speed 300 km/h (186 mph); sink after release (no flap) –2,6 m/s; landing 380 m/93 km/h.

Yak-15 First jet (as distinct from rocket) aircraft completed in Soviet Union, this simple fighter was jet derivative of Yak-3, sole example apart from Swedish J21R of successful production jet fighter based on piston-engined design. This was possible because at late-Feb 45 Kremlin meeting Stalin had instructed Yak to build jet fighter using single German engine, matching size of existing piston fighters (La had same instruction but made fresh start with La-150). Yak explained his intentions in a positive way, not suggesting that compromised result would be second-rate, and gaining support from promise of fast timing; in prevailing mood, even inferior jet fighter in numbers would have been preferred to better one later. Crash programme headed by Ye. Adler and Leon Shekhter, basic 3-view layout complete in three days and first drawing issued late May.

Existing Yak-3 production drawings retained for most of tail, wing and landing gear, and much of centre and rear fuselage also retained apart from using metal skin as Yak-3U. New forward fuselage of chiefly D1 construction though with steel firewall at front spar and welded-steel-tube fuselage truss extended forward to two main frames carrying engine. Latter rearranged with accessories on top to suit unusual installation. Axis of engine inclined slightly nose-up, with rear half under wing and nozzle as close as possible under TE to minimise base drag. This required that D1 front spar be arched over engine; both spars also

Yak-14

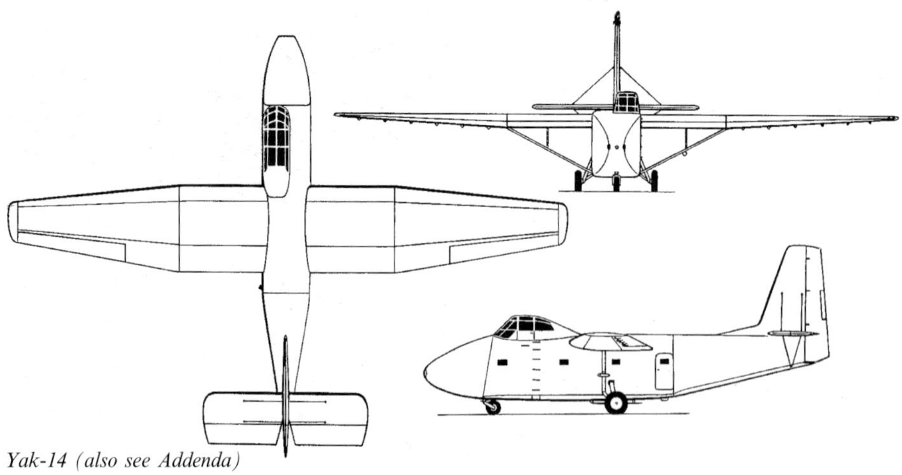

Yak-14 (also see Addenda)

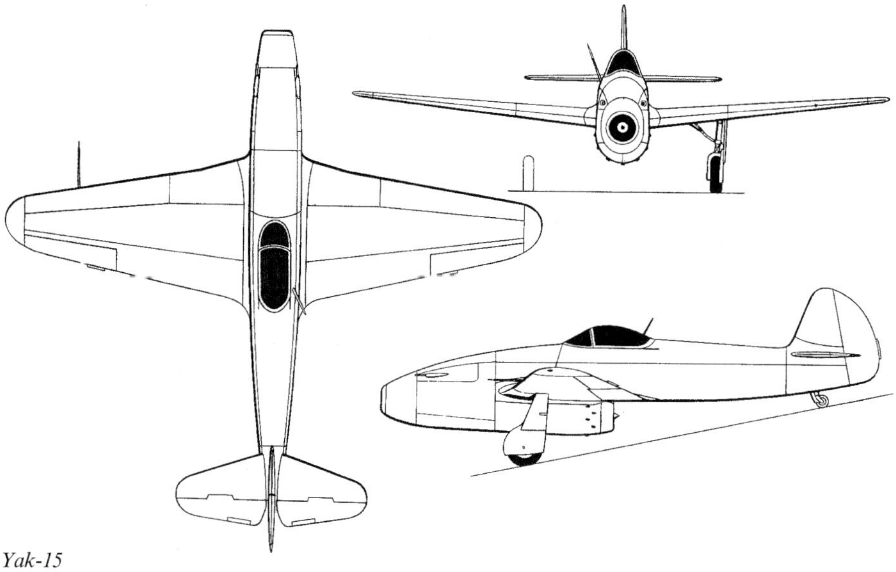

Yak-15

Yak-15

slightly strengthened by using thicker material for booms. Skin of forward fuselage all D1, almost all detachable for access to engine, accessories and armament. Design armament one ShVAK and two UBS, changed during prototype construction to two B-20 and finally to two NS-23 each with 60 rounds. This enabled larger fuel tank to be accommodated in rear fuselage, though for reasons of weight not all 720 lit (590 kg) was normally used. Eventually fitted with thin frameless windscreen, aft-sliding canopy, HF antenna wire from fin to inclined mast, KhGSA underskin behind nozzle, small steel guard in front of retractable tailwheel and D1 skin over lower part of rudder.

Prototype hand-built with utmost urgency, fitted with Jumo 004B, then subjected to testing including CAHI tunnel Oct 45 to April 46. Suggestion remains could have flown months before MiG rival, but at Zhukovskii on 24 April toss of coin put I-300 in air first. Three hours later Yak-15 made successful flight in hands of Mikhail I. Ivanov. Both prototypes made public debut at Tushino Aviation Day 18 Aug 46. Stalin then decreed that 12 of each should participate in October parade, 80 days hence! Yak himself was sent, with Deputy Minister A. Kuznetsov, to chosen GAZ with instructions not to leave until 12th aircraft despatched. Task completed using inefficient hand-building, initially with no production jigging. First pre-series Yak-15 crated and sent by rail 20 Sept, next on 5 Oct and last of required dozen on 21 Oct. (Tushino fly-past cancelled by fog). First pre-series aircraft flown 5 Oct 46 and NII testing by Stefanovskii generally successful. Lightest jet fighter and possibly cheapest. Total of all versions believed 280. USAF designation 'Type 2', ASCC name 'Feather'.

Single example built of **Yak-21T** tandem dual trainer, modified from series Yak-15, first flown May 47. Pupil cockpit added in front, streamlined side-hinged canopies, painted dark green with number 'white 101'. Another modified as **Yak-15U** with nosewheel landing gear.

DIMENSIONS Span 9,2 m (30 ft 2¼ in); length 8,7 m (28 ft 6½ in); wing area 14,85 m² (160 ft²).
ENGINE (first) one Jumo 004B, (rest) one RD-10.
WEIGHTS Empty 1918 kg (4,228 lb); fuel/oil 590 kg+10 kg; loaded 2634 kg (5,807 lb).
PERFORMANCE Max speed 700 km/h at SL (435 mph), 805 km/h (500 mph) at 5 km; climb to 5 km in 4·8 min, to 10 km in 13·8 min; service ceiling 13,35 km (43,800 ft); range 510 km (317 miles); TO 600 m; ldg 135 km/h, 720 m.

Yak-16 Rejection of Yak-8 left Yak in no doubt GVF needed rather larger passenger transport of modern design with all-metal construction, to partner Li-2 and Il-12 on lower-traffic routes. Yak-16 was designed by largely new stressed-skin engineering team and built under direction of former flying-boat designer P. D. Samsonov. Simple airframe with Clark-YH wing, two spars, horizontal centre section and tapered outer panels with dihedral, oval-section fuselage and clean unbraced tail. Slotted ailerons, rudder and elevators fabric-covered, all with tabs. All-D1 split flaps on centre section and outer panels with pneumatic actuation to 40°. Main gears with single oleo legs to inner ends of axles for wheels with 900 mm × 300 mm tyres, retracting forwards to leave wheel part-

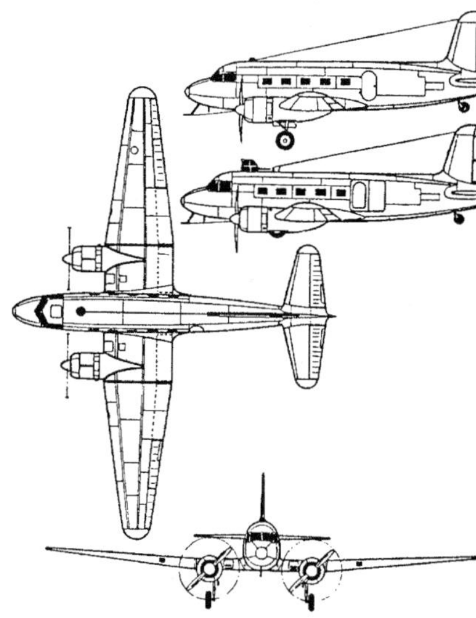

Yak-16-I (lower side view 16-II)

Yak-16-I

exposed, brakes and retraction pneumatic. Castoring fixed tailwheel 470×210mm. Four wing tanks total 1800 lit (396 gal); two 35 lit oil tanks. Engines in neat cowls with gills and ventral exhausts, oil coolers in wing inboard of nacelle with LE inlet, VISh-111Va two-blade constant-speed props with large spinners. Flight deck for crew of three, five passengers each side, small toilet and 100 kg baggage/cargo at rear.

Civil **Yak-16-I** tested by F. L. Abramov from Sept 47, later passed NII testing by Anokhin and Gallai but no production (An-2 thought more appropriate); made foreign sales tour. Single example flown of **Yak-16-II** assault transport, large cargo door, dorsal turret with UBT, glider tow hook, 7 troops plus weapons or 6 stretchers, three bombs (eg FAB-100) and para-container bags. ASCC reporting name 'Cork'.

DIMENSIONS Span 21,5 m (70 ft 6½ in); length 16,0 m (52 ft 6 in); wing area 56,2 m^2 (605 ft^2).

ENGINES Two ASh-21.

WEIGHTS Empty 4465 kg (9,843 lb), (II) 4993 kg (11,007 lb); fuel/oil 480 kg+45 kg; loaded 6021 kg (13,274 lb), (II) 6615 kg (14,583 lb).

PERFORMANCE Max speed 370 km/h (230 mph) at SL, (II) 348 km/h (216 mph); climb to 1 km 2·3 min; service ceiling 7,55 km (24,800 ft); range 800 km (497 miles); TO/ldg 250 m.

Yak-17 (under T-4)

Yak-17 This was a logical modest next step beyond Yak-15, retaining mixed construction yet with far more detail redesign than immediately apparent. Obvious advantage of nose-wheel landing gear, tested on Yak-15U, led to more extensive redesign to achieve optimum fighter with underslung RD-10. Accommodation of retracted nose gear had proved impossible in 15U, and presence of low-slung engine still posed insuperable problem. In Yak-17 nose gear continued to be largely external, with 700 mm×190 mm wheel on levered-suspension oleo strut carrying large fairing forming front half of ventral bulge. Wing restressed for increased weights and airspeeds, design factor 12, main strengthening in wing spars, inboard ribs and fuselage truss. D1 split flaps and landing gears pneumatic, main legs being same levered-suspension forgings as on 15U and pulled inwards by Y-type breaker strut to lie between spars, with D1 bay door hinged at wing root. Reduction in wing tankage now countered by adding 300 lit jettisonable external tanks under wingtips, requiring only local structural revision but restricting aerobatic capability when attached. Tail redesigned, partly to match greater side area of nose. Two NS-23 each 60 rounds, landing light in left LE, flat bulletproof windscreen (most with ciné camera at top). Single examples presented Czech AF (called S-100) and Poland.

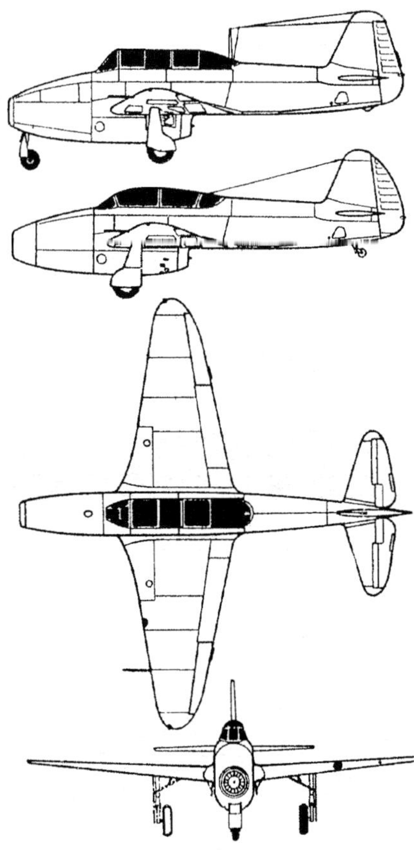

Yak-17UTI (lower side view Yak-21T, p. 473)

Prototype Yak-17 flown mid-47, passed NII testing and put into production, almost all being **Yak-17UTI** tandem dual trainers. Tall less-sloping windscreen, widely spaced separate flat-topped canopies for pupil and instructor, fixed aft section with vert mast for RSI-6 HF radio. Prototype single UBS, production UTI unarmed. Combined total both versions 430, completed late 48. Yak-17 USAF 'Type 16', ASCC 'Feather'; Yak-17UTI 'Type 26', ASCC 'Magnet'.

DIMENSIONS Span 9,2 m (30 ft 2¼ in); length (both) 8,78 m (28 ft 9⅝ in); wing area 14,85 m^2 (160 ft^2).

ENGINE One RD-10A.

WEIGHTS Empty (17) 2081 kg (4,588 lb), (UTI) 2148 kg (4,735 lb); fuel (17) 553 kg or 875 kg, (UTI) 464 kg; loaded (17) 2890 kg or 3240 kg (max 7,143 lb), (UTI) 2806 kg (6,186 lb).

PERFORMANCE Max speed at SL (both) 702 km/h, at 5 km (17) 751 km/h (467 mph), (UTI) 724 km/h (450 mph); climb to 5 km (both) 5·8 min; service ceiling (both) 12,6 km (41,340 ft); range (17) 717 km (446 miles), (UTI) 330 km (205 miles); TO (both) 640 m; ldg (17) 160 km/h, 560 m, (UTI) 164 km/h, 700 m.

Yak-18 Descended from UT-2, this family has grown to encompass many single-seat aerobatic aircraft as well as totally redesigned four-seat cabin machine, in addition to tandem trainers standard throughout WP air forces since the Second World War. Origin was UT-2MV of 43, from which OKB developed more powerful tandem trainer with retractable main gear (as pupil-pilot exercise, not to improve performance) and part metal skinning over fuselage and wings. This prototype, Yak-18, flown late 45.

Wing Clark-YH, 14·5% at root and 9·3% at tip, rectangular centre section with 0° dihedral, tapered outer panels with 6° 30' dihedral, two-spar D1 construction with pressed ribs and riveted D1 skin over LE back to front spar, fabric remainder. D1/fabric slotted ailerons, split flap over centre section with manual operation to 55° for landing. Fuselage welded-MS tube truss with wood formers and stringers, D1 panels (many detachable) from nose to rear cockpit, fabric aft. Tail D1/fabric, wire-braced. Tandem cockpits with bucket seats (padded seats for civilian pilots with back-type parachute) and dual controls. First prototype with fixed landing gear with single side-braced oleo struts carrying braked 500 mm×150 mm wheels. All subsequent aircraft with legs folding rearwards mechanically, with pilot/cable operation, legs lying externally beneath wing and half wheel exposed. Fixed castoring tailwheel. Engine on steel-tube mounts, initially without rubber vibration absorbers fitted later, with pressed D1 helmeted cowl and driving VISh-327 prop. Two 65 lit aluminium tanks in centre section, drum-type oil tank behind engine. No armament, initially no radio.

Prolonged factory and NII testing completed late 47, and series production authorized with HF radio (usually RSIU-3M/-4M), blind-flying instruments, engine-driven generator, nav lights, cockpit lighting and, later in production, V-501 prop. Replaced UT-2 and Po-2 in all military and civilian training units and Dosaaf clubs with clearance for unrestricted aerobatics including flick manoeuvres. Engine cowl improved early 50s, spinner usually omitted, ARK-5 radio compass installed and pneumatic system added with engine/compressor-charged bottle to operate landing gear retraction and wheelbrakes, flap remaining manual. Civil pilots gained eight FAI class records including 100 km circuit at 262,771 km/h (16 Sept 49), 2000 km circuit (in two flights) at 209,664 km/h (16 Sept 49), non-stop flight of 2004,62 km (25 Sept 54) and height of 6311 m (18 June 54). Aircraft widely exported outside Warsaw Pact, including barter with Austria for steel sheet.

Yak-18

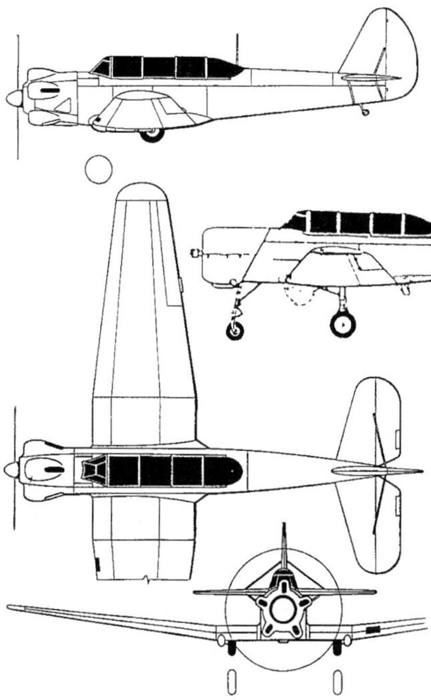

Yak-18 (lower side view 18A)

Yak-18U (*Uluchshyennyi*, improved) flown 54, entered production 55 though not replacing original model. Tricycle landing gear with different main units hinged to rear spar with legs on inside of wheels, retracting forwards with upper part of wheel in small recess in wing ahead of front spar. Nose gear with long single steel-tube leg pivoted to bottom of engine mount frame, on front of lengthened forward fuselage, with rear breaker strut pulled up to fold unit to rear with half wheel (400 mm × 130 mm) exposed. As in Yak-18, provision for belly landing on retracted gear with no damage except to prop, and bottom of cowl. Tailwheel replaced by fixed bumper, dihedral of outer wings increased to 7° 20', landing light(s) in left LE, cowl further improved with large open inner front to improve cooling in climb (or in glider towing which was task of many aircraft with modified lower rudder and hook), welded-tube access step added on left at TE of wing. Performance, hardly sprightly in Yak-18, deteriorated with greater weight.

Yak-20W rectified problem with much more powerful AI-14R as used in Yak-12 since 50. New engine in smooth long-chord cowl with oblique trailing edge, made in detachable or hinged upper/lower halves. Rear collector ring with twin ventral exhaust pipes, front of cowl open but fitted with 28 pilot-controlled radial shutters to regulate cooling airflow. V-530-D35 controllable (counterweight-type) 2,3 m prop, oil cooler in duct in right wing root. Airframe locally strengthened, span slightly increased, tail redesigned similar to Yak-12A, fuel tanks enlarged to 65 lit each, pneumatic system (0,49 MPa on all subsequent models) extended with main/emergency and actuator added to drive flaps via push/pull spanwise rods, canopy made deeper to increase pilot headroom, avionics greatly improved with HF/VHF radio, telephone-type intercom in place of Gosport tube, improved blind-flying instruments, marker receiver, radio compass and better nav lights. NII tested spring 57 and cleared for production same year, replacing all previous variants at Arsen'yev plant (East Siberia) where Yak general-aviation types by this time centred. In production redesignated **Yak-18A**. Example Monino with two rocket launchers.

Belated recognition of importance of international aerobatic contest (then Lockheed, UK, Trophy) prompted development of dedicated aerobatic single-seat variant. Competitions 1955-59 missed whilst exploring front and rear positions of single-seat cockpit and numerous other argued features. In early 59 several production aircraft built as **Yak-18P** (*Pilotazhnyii*) in two forms. All had longer-span ailerons and modified fuel system permitting 5 minutes inverted flight at any power setting. Most promising machine had front cockpit retained and rear cockpit eliminated, and neater main gears retracting inwards to lie wholly within wing between spars. An alternative scheme, also tested, eliminated front cockpit and retained rear, this being a converted Yak-18A and retaining original landing gear. B. N. Vasyakin gained 5th place in 60 championships at Bratislava with rear-cockpit Yak-18P. In 62 championships at Budapest four front-cockpit machines participated, gaining 2nd (Lochikov), 7th, 9th and 10th places. In 64 at Bilbao slightly modified Yak-18Ps with fin/rudder of increased chord and stressed to +9 g/ −6 g gained 4th, 5th, 6th, 7th places, and women gained 1st, 2nd, 3rd, 4th. Aerobatic development continued with **Yak-18PM** of 65, with more powerful AI-14RF driving V-530D-25 prop, rear cockpit moved 120 mm further back than original rear cockpit, dihedral reduced from 7° 20' to 2°, flap replaced by instant-

Yak-18U

Yak-18PM

action 3 m² airbrake under belly, and various minor changes. Result was great improvement in performance (eg, see climb data) and CG moved from 20% to 26% chord giving near-unstable agility. Yak-18PM decisively won 66 championships at Moscow Tushino. At Hullavington (UK) in 70 joined by single **Yak-18PS**, similar to PM but with tailwheel gear using rearward-retracting main legs pivoted to front spar as in original Yak-18, legs passing outside wheels, and other changes including oil cooler in duct under nose. Aerobatic series continued with Yak-50, Yak-52, Yak-53, and Yak-55 described in following entries.

Last variant, **Yak-18T**, was completely different aircraft, first announced by OKB as work of son Sergei Yakovlev, though designer later named as Yuri Yankyevich. Planned late 64 for GVF in five variants: primary trainer, advanced trainer (instrument flight and all-weather radio), transport for three passengers and baggage, mail/cargo for 250 kg, and ambulance with one stretcher and attendant. Totally new and much larger fuselage, rounded box semi-monocoque wholly of stressed-skin without welded-tube truss (rear fuselage spot-welded). Two-spar wing with stressed-skin centre section extended in span from 2500 mm to 2706 mm, still 14·5% Clark-YH, outer panels basically as Yak-18A with 7° 20' dihedral, slotted ailerons and pneumatic split flap. AI-14RF in prototype as in Yak-18PM but new broad oil cooler in duct in right inboard wing, and fuel tanks enlarged to 104 lit size and repositioned between spars in outer wings outboard of main gears. Improved pneumatically retracted landing gear with smoother differential brakes, tyres 500 mm × 150 mm; 400 mm × 130 mm nose gear almost completely housed in fuselage when retracted. Comfortable and well-equipped cabin with large door both sides; roof window only at extreme front. Normally side-by-side dual controls with lateral control by wheels. Removable divan for two at rear as alternative to cargo, plus large rear baggage compartment accessed via door on left also used to load stretcher. Comprehensive avionics fit including VHF/UHF and intercom, ILS, VOR/DME, radio altimeter and flight recorder. Capable electrical system includes starter and full night

Yak-18T

equipment with rotating beacon. Prototype in white OKB livery, unregistered, shown at Paris 67; two tested in 450 flights 1968-69 at Sasov by G. A. Taran leading to numerous mods, chief being installation of new Vedeneyev engine driving V-530TA-D35 prop. Prolonged further trials by pre-production aircraft culminated in intensive testing throughout 73 of four series machines from plant at Smolyensk. Whole 74 intake (100 pupils) at Sasov trained on Yak-18T and by 81 over 1,000 in use including first ambulance, Aeroflot line-service and liaison examples, with skis cleared and floats on test. Various military, fire patrol and photo versions.

Production of original Yak-18 variants terminated end-67 with 6760 produced. Including Yak-18T total 8000. ASCC name 'Max'.

DIMENSIONS Span (18,U) 10,3 m (33 ft 9½ in), (A,P,PM,PS) 10,6 m (34 ft 9⅓ in), (T) 11,16 m (36 ft 7¼ in); length (18) 8,07 m (26 ft 5¾ in), (U) 8,53 m (27 ft 11¾ in), (A,P,PM,PS,T) 8,35 m (27 ft 4¾ in); wing area (18,U) 17,00 m² (183 ft²), (A,P) 17,80 (191·6), (PM,PS) 17,9 (192·7), (T) 18,75 (201·8).

ENGINE (18,U) one M-11FR, (A,P) AI-14R, (PM,PS) AI-14RF, (T) M-14P.

WEIGHTS Empty (18) 816 kg (1,799 lb), (U) 970 kg (2,138 lb), (A) 1025 kg (2,260 lb), (P) 818 kg (1,803 lb), (PM) 825 kg (1,819 lb), (T) 1200 kg (2,646 lb); fuel/oil (18) max 112 kg, (U,A) 94 kg+14 kg, (P,PM,PS) 50 kg, (T) 150 kg+25 kg; loaded (18) 1120 kg (2,469 lb), (U) 1300 kg (2,866 lb), (A) 1316 kg (2,901 lb), (P) 1065 kg (2,348 lb), (PM) 1100 kg (2,425 lb), (T) 1650 kg (3,637 lb).

PERFORMANCE Max speed (18) 248 km/h (154 mph), (U) 235 km/h, (A) 260 km/h (162 mph), (P) 275 km/h (171 mph), (PM) 315 km/h (196 mph), (T) 295 km/h (183 mph); cruise (18) 195 km/h, (U) 186 km/h, (A) 224 km/h, (P) 256 km/h, (PM) 282 km/h, (T) 250 km/h; initial climb (18) 3,5 m/s, (U) 2,4 m/s, (A) 5,4 m/s, (P) 8,6 m/s, (PM) 10,0 m/s, (T) 5,0 m/s; service ceiling (18) 4 km (13,123 ft), (U) 3,4 km, (A) 5,06 km, (P) 6,5 km, (PM) 7,0 km, (T) 5,5 km; range (18) 1050 km (652 miles), (U) 780 km, (A) 750 km, (P) 420 km, (PM, PS) 400 km, (T) max fuel, 900 km (560 miles); take-off (18) 205 m, (U) 295 m, (A) 215 m, (P, PM) 140 m, (T) 400 m; landing (18) 270 m/89 km/h, (U) not known, (A) 250 m, (P, PM) 130 m, (T) 500 m/100 km/h.

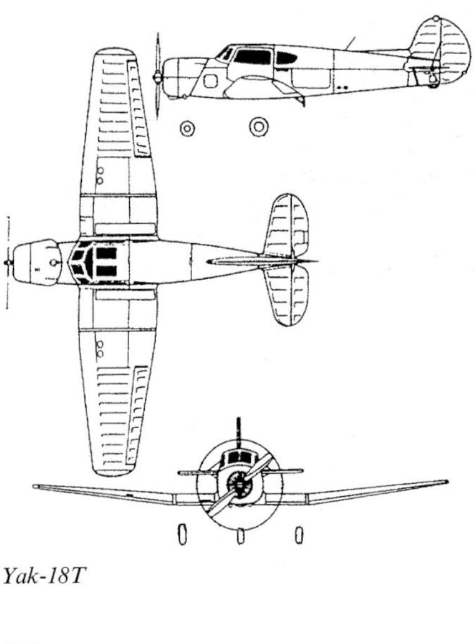
Yak-18T

Yak-19 First stressed-skin aircraft designed as such from outset by OKB, this uninspired jet fighter was completely fresh design tailored to more powerful version of RD-10. Midmounted wing, CAHI S-1-12 laminar section 12% thick at root, tapering to KV-3-12 (still 12%) near tip, equal taper (taper ratio 2·5, aspect ratio 5·6), dihedral 3·5°. Two spars, pressed ribs, skin 1,2 mm/1,8 mm, flush-riveted (first time at OKB). Frise ailerons, manual, metal-skinned; split flaps, hydraulic. Barrel fuselage with nose inlet to left/right ducts past unpressurized cockpit and over wing spars to engine in rear fuselage with reheat nozzle at extreme tail; flush-riveted skin 0,8 mm/1,8 mm. Rear fuselage detachable, with engine, latter then withdrawn forwards. Fixed tailplane mounted half-way up tall fin, rudder/elevators manual, metal-skinned and divided into upper/lower and left/right sections with fixed cruciform at centre (no bullet). Tricycle landing gear with levered-suspension main legs and 570 mm × 135 mm tyres, brakes and retraction pneumatic with leg housed in wing between spars and wheel in fuselage. Nose gear 375 mm × 135 mm, castoring, levered suspension, pneumatic retraction forwards with twin doors to narrow compartment between ducts. Primitive ejection seat, flat bulletproof windscreen, sliding canopy. Total 815 lit fuel in two fuselage tanks, plus 11 lit petrol for two-stroke starter engine and 14 kg oil. Later fitted with 200 lit tip tanks. Electrically-driven pump supplying supplementary fuel to crude after-burner. Two Sh-Z 23 mm cannon with 150 rounds. Flown early 47 and factory testing completed 21 Aug 47; not submitted for NII testing as engine outdated.

DIMENSIONS Span 8,70 m (28 ft 6½ in); length 8,11 m (26 ft 7⅓ in); wing area 13,5 m² (145 ft²).

ENGINE One RD-10F.

WEIGHTS Empty 2192 kg (4,832 lb); fuel/oil 658 kg+14 kg; loaded 3 t (6,614 lb).

PERFORMANCE Max speed with reheat 875 km/h at SL and at 10 km, 904 km/h (562 mph) at 5 km, normal cold thrust 760 km/h at SL, 818 km/h at 5 km and 788 km/h at 10 km; initial climb (reheat) 25,8 m/s (5,080 ft/min), (cold thrust) 16,4 m/s (3,228 ft/min); time to 5 km and 10 km (cold) 3·9 min and 10·5 min; service ceiling 15 km (49,200 ft); range, 700 km; take-off 550 m; landing 520 m/180 km/h (112 mph).

Yak-20 Side-by-side trainer and sporting aircraft designed in three months summer 50 and test-flown at end of year. Fuselage retained welded steel tube truss with secondary wood stringers and fabric covering, with width for

Yak-19 No 2 with tip tanks

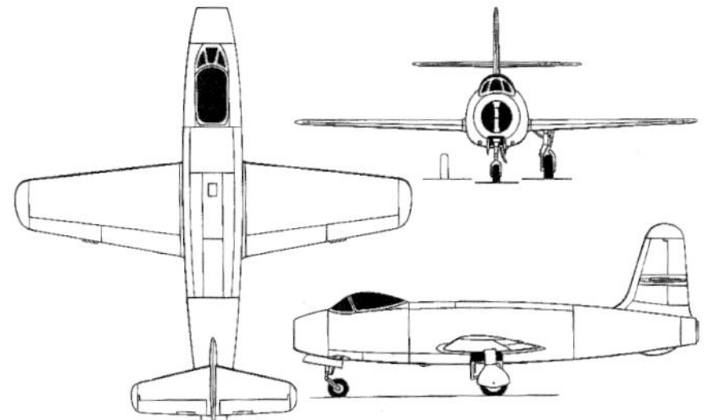

Yak-19

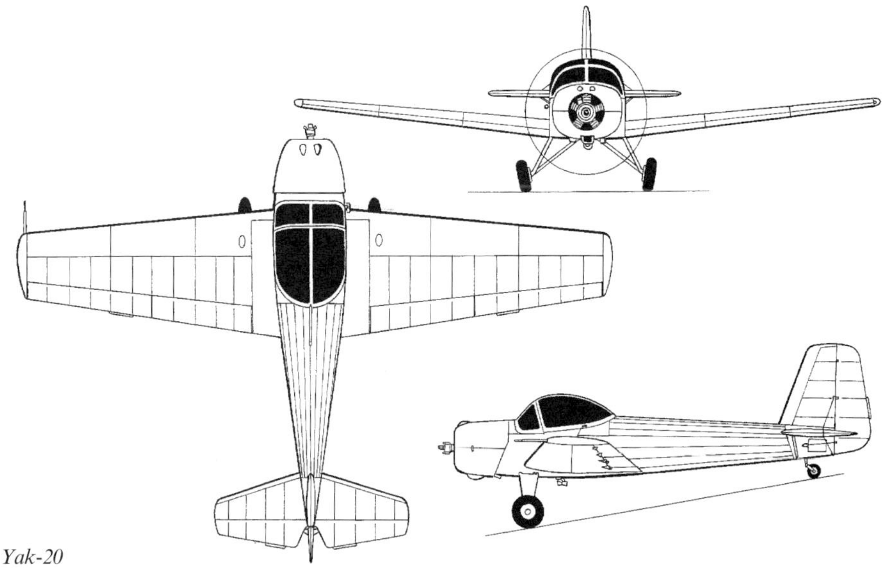

Yak-20

two pilots side-by-side. New simple single-spar wing, Clark-YH 14·5%/9% thick, D1 construction with riveted skin over LE to front spar, fabric to rear. Slotted Frise ailerons worked by push-rods, 17° down and 20° up, slotted flaps inboard set by pilot linkage to 18° for take-off and 32° for landing. New Ivchyenko engine on rubber mounts at front of tubular truss, D1 cowl in hinged upper/lower halves, 2 m V-515 controllable-pitch (counterweight type) prop. Fuel in two 35 lit tanks in wing roots ahead of spar. Tail D1/fabric but with D1 skin over fin LE. Simple main gears with 500 mm × 150 mm wheels on welded V-type legs (joined by fairing) hinged to deflect outwards, inner bracing strut having wrapped rubber shock-absorber bungee on transverse fuselage member. Tailwheel 200 mm × 80 mm sprung by loop of rubber. Sliding canopy moulded in left/right halves in D1 frame, dual stick-type controls, venturis ventral (large) and on right side, no radio. Prototype painted in VVS insignia, tested by Anokhin and Georgii Shiyanov and from 1 Jan 51 flown by numerous Dosaaf members. Proved capable of performing all required aerobatics with two pilots, but no series aircraft built.

DIMENSIONS Span 9,56 m (31 ft $4\frac{1}{3}$ in); length 7,06 m (23 ft 2 in); wing area 15,0 m^2 (161·5 ft^2).

ENGINE One AI-10.

WEIGHTS Empty 470 kg (1,036 lb); fuel/oil 50 kg+10 kg; loaded 700 kg (1,543 lb).

PERFORMANCE Max speed at SL 160 km/h (99 mph); cruise 142 km/h (88 mph); climb, not known; service ceiling 3 km (9,850 ft); range 400 km (249 miles); take-off 295 m; landing 180 m/60 km/h.

Yak-20

Yak-21 Dual tandem Yak-15, described in that entry.

Yak-23 To meet Oct 46 demand for next-generation fighter able to use existing airfields and facilities and not incorporating swept aerodynamic surfaces. Extraordinarily quick design possible because same drawings used for wing (with tip tanks) and landing gear as Yak-19.

New fuselage of stressed-skin construction identical in material, frames, stringers and technique to Yak-19, without welded tube truss. Engine in sealed plenum chamber ahead of front spar (latter arched over jetpipe) with entire nose section ahead of diagonal frame removable for access or for engine-change. Internal fuel 920 lit in protected tanks immediately ahead of and behind unpressurized cockpit; tip tanks each 200 lit (44 gal). Improved ejection seat, but still simple and driven by single cordite charge. Sliding teardrop canopy, 57 mm bulletproof windscreen and 8 mm back armour. RSIU-6 hf radio with mast and wires. Push/pull rods to tabbed elevators on tailplane fixed at 0·5° part-way up fin; rudder with ground-adjustable tab driven by cables. Tail all stressed skin, horizontal tail reduced from 3 m^2 in Yak-19 to 2,96 m^2 and vertical from 2,03 to 1,64. Two NR-23 each with 90 rounds.

Prototype flown mid (believed 17 June) 47 by Ivanov; Anokhin, not yet fully back in harness, also flew it. Successful and attractive aircraft, no problem with jet noise or heating on rear fuselage and only visible mods were to reduce height of rudder, add 5° dihedral to tailplane and tinker with landing gear and nozzle fairing. Combat camera and landing light added in inlet bifurcation above nosewheel bay, tip-tank attachments modified to accept alternative 60 kg bombs and radio added. Factory testing

Yak-23

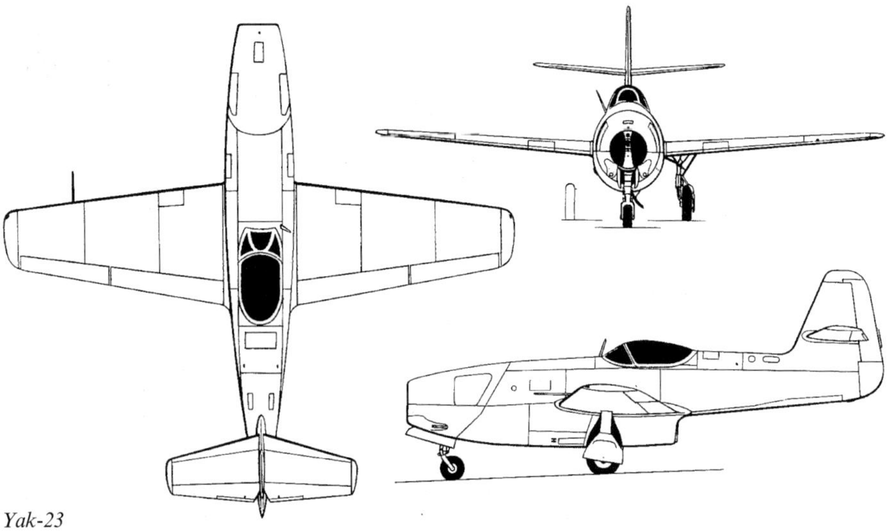

Yak-23

completed 12 September 47 and following good NII report cleared for production early 48. Intended batch curtailed at 310 because of success of MiG-15. RPKO-10M homer/radio compass and slightly larger stainless skin on underside at rear. Service with VVS brief, but popular with Warsaw Pact air forces (Czech designation S-101). Could outclimb MiG-15, and Polish civil-registered Yak-23 set record 21 Nov 57 by reaching 3 km in 119 s and 6 km in 197 s. ASCC name 'Flora'.

OKB built one **Yak-23UTI** in first half 49, tested by G. S. Klimushkin. Rear cockpit retained but given different bulged canopy with fixed rear portion. Forward cockpit for pupil severely reduced fuselage tankage (660 lit total) and had additional combat camera above windscreen. Height of vertical tail increased. Armament, one UB on left side of nose.

DIMENSIONS Span 8,37 m (28 ft 7¾ in); length 8,12 m (26 ft 7⅔ in),; wing area 13,5 m^2 (145 ft^2).

ENGINE One RD-500.

WEIGHTS Empty (prototype) 1980 kg, (production) 2 t (4,410 lb), (UTI) 2220 kg (4,894 lb); fuel/oil 760 kg (1085 kg with tip tanks)+12 kg; (UTI) 530 (850 kg with tip tanks +12 kg; loaded (production) 3036 kg (6,693 lb) or 3384 kg (7,460 lb) with tip tanks, (UTI) 2950 kg (6,504 lb) or 3300 kg (7,275 lb) with tip tanks.

PERFORMANCE Max speed (production) 923 km/h (574 mph) at SL, 868 km/h (539 mph) at 5 km, initial climb 47 m/s; time to 5 km and 10 km (production) 2·3 min and 6·2 min; service ceiling (production) 14,8 km (48,600 ft), (UTI) 14 km; range (production) 755 km (469 miles) internal fuel, 1030 km (640 miles) with tip tanks; take-off 440 m (UTI, 400 m); landing 485 m (UTI, 500 m)/157 km/h (95·6 mph).

Yak EG OKB formed helicopter group in 46, and assigned design management to Igor A. Yerlikh. Adopted coaxial configuration with pair of two-blade rotors with max vertical separation. Maximum help from CAHI, and OKB engineers also visited Breguet which had flown Dorand coaxial machine before war and was completing prototype G-11E. No direct liaison with Kamov but GUAP provided funds and oversaw both teams, which (except for pilot controls) did come to similar conclusions. Rejecting TE outrigger tabs considered at start, rotor fully articulated with friction dampers and controlled by modern cyclic/collective levers acting via oil dashpots on swash-plates, with yaw control by pedals giving differential collective to upper/lower rotors. Blades built up laminated hardwood and pine with ply covering overlain by glued fabric. Simple fuselage based on welded steel truss with D1 sheet from nose to rear of engine compartment, fabric at tail. Two seats side-by-side in nose (not known if dual control) with door each side. Radial engine in normal attitude with forward drive via cooling fan and centrifugal clutch to bevel box and rotor shafts, with hydraulic coupling and additional oil-depressed spring which on failure of drive forced blades into autorotative pitch. Tricycle landing gear with vertical oleo struts; ground resonance avoided by pure chance. Completed early 47 with long fabric-covered rear fuselage carrying twin-finned tail and tailskid. Tested by V. V. Tezavrovskii who made 40 tethered hovers followed by 75 flights for total airborne time of 20 h. EG (Eksperimentalnyi Gelikopter) flew, but CG too far aft and tail removed; oil tank moved from engine compartment to rear cockpit bulkhead, skid removed and rear fuselage re-faired. Normal rotor rpm 233, severe vibration, and though positive control never lost, stick force and phugoid instability worsened beyond 30 km/h forward speed, severely limiting practical value of EG. After prolonged investigation Yak decided to leave this configuration to Kamov.

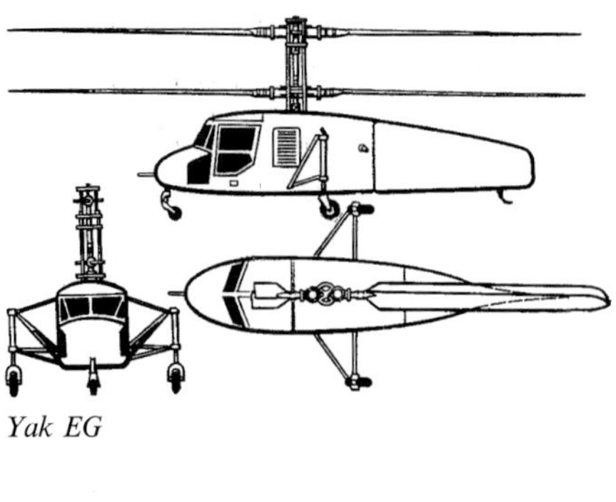

Yak EG

DIMENSIONS Diameter of rotors 10,0 m (32 ft 9⅔ in); length of fuselage 6,35 m (21 ft 5 in).

ENGINE One M-11FR-1.

WEIGHTS Empty 878 kg (1,936 lb); loaded 1020 kg (2,249 lb).

PERFORMANCE Max speed did not exceed 70 km/h and at this speed stick force so great as to be impractical; ceiling given as 250 m hover and 2,7 km in forward flight, but in fact greatest height actually reached was 180 m; range, in theory 235 km claimed, but again in practice shorter.

Yak-100 This helicopter looked like exact copy of S-51, though Yak insisted OKB team, under Yerlikh, merely reached similar design conclusions. Complete reversal of original scheme, this single-rotor machine did retain similar dynamic parts, drive and hub mechanism with damped flapping, feathering and drag hinges and hydraulic-dashpot control via swashplate. Three main blades of hardwood and pine as before, with ply skin and glued fabric overall. Root of each blade in upper/lower halves of bolted D1, steel hub. Main change in engine installation was vertical engine mount with several changes to enable engine to run in this attitude. Cooling fan, clutch/freewheel and reduction gear and angle box for tail rotor were more Sikorsky than scaled-up EG, though pilot controls and control system retained OKB belief in spring and oil vibration dampers (this time with pilot adjustment) and in spring-loaded actuation of pitch into autorotation following loss of drive torque. Pedal control of tail-rotor pitch. Air-

Yak EG as built with tail

Yak-100

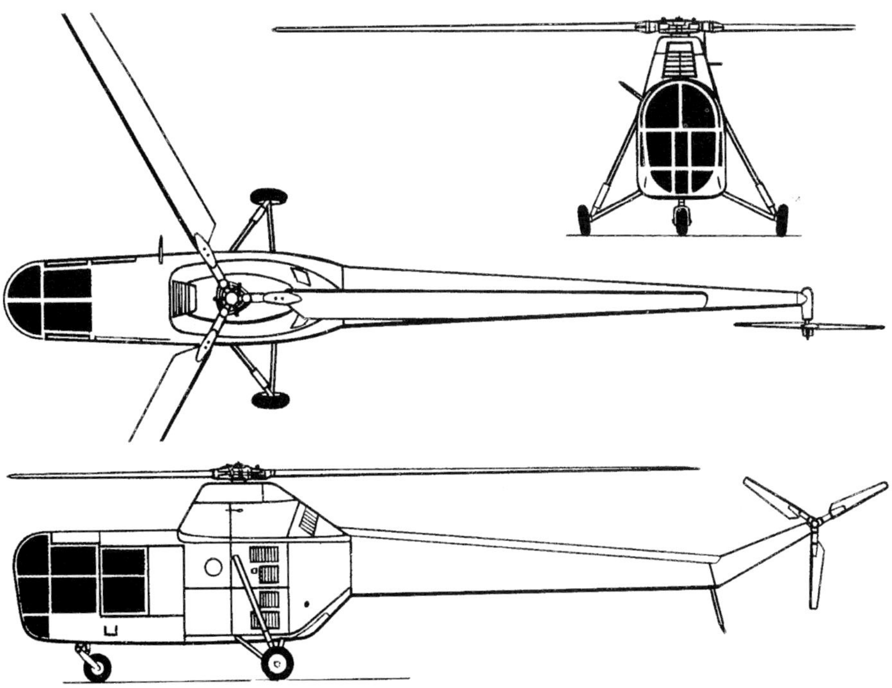

Yak-100 (machine as built differed slightly)

frame based on welded steel-tube frame carrying landing gear, seats, rotor hub and engine. Skin riveted D1, and cranked tailboom D1 monocoque. Basic design completed late 47 as tandem trainer or as transport with single pilot in front and two-seat divan or other payload behind. Two prototypes built, first with VVS funds and bearing VVS insignia. Began factory test Nov 48, initially suffered from severe vibration and apparent blade flutter. Blades given ground-adjustable TE tabs and eventually modified with CG further aft, behind flexural axis. These blades first fitted to No 2 Yak-100, which began test July 49. Factory test complete June 50 and NII tests successfully accomplished later same year, but Mi-1 already adopted and Yak-100 lost by being later.

DIMENSIONS Diameter of main rotor 14,5 m (47 ft 6⅞ in); length of fuselage 13,9 m (45 ft 7¼ in); tail rotor 2,6 m (102·36 in).

ENGINE One AI-26GRFL.

WEIGHTS Empty 1805 kg (3,979 lb); fuel/oil 80 kg+14 kg; loaded 2180 kg (4,806 lb).

PERFORMANCE Max speed claimed 170 km/h (106 mph); ceiling (hover) 2720 m, (forward flight) 5,250 m (17,200 ft); range, claimed 325 km (202 miles).

Yak-24 This large and powerful transport helicopter was created in response to order of Stalin at Kremlin meeting autumn 51. Instruction was for Mil to build single-engine machine with 1200 kg military load, Yak a machine with twice this load, prototypes to be ready in a year. Mil had already prepared suitable design, and Yak gained permission to use essentially same main rotor and drive from similar engine, merely doubling up to use two engine/rotor systems at ends of boxcar fuselage. Yak awed at size of task and short timescale, assembled large team including Yerlikh, veteran helicopter man N. Skrzhinskii, P. D. Samsonov (famed flying-boat designer who had long managed Yak prototype dept), L. Shekhter, L. S. Vil'dgrub and many other well-known engineers. Plan was to build four Yak-24, already called **LV** (*Letayushchaya Vagon*, flying wagon), two for static and resonance test and two for flight. Promised 'unlimited support' in crash programme.

Basic engine/rotor design described under Mi-4. Fuselage functional container based on welded KhGSA truss with minimal secondary stringer/fairing formers of D1 or wood. Unstressed D1 sheet covering over front and rear engine bays and large fin, fabric elsewhere. Aluminium plank cargo floor with full-section access via rear ramp/door; passenger door forward on left side. Rear rotor mounted on top of vertical fin (TE curved to right to give side-thrust to left in flight) with drive from engine installed in normal horizontal attitude at base of fin, with open cooling-air inlets each side of fin and clearance under engine for vehicles and other cargo on ramp. High-speed connecting shaft to front rotor, mirror-image with rotation anti-clockwise seen from above, driven by engine at 60° angle between cockpit and cabin. Nose cockpit for two pilots, radio-operator and engineer, entirely glazed with aft-sliding door each side and sliding door(s) at rear giving restricted access past engine to main compartment. Latter measured 2 m×2 m×10 m (78¾ in square by 32 ft 9⅔ in long) with intended

Yak-24

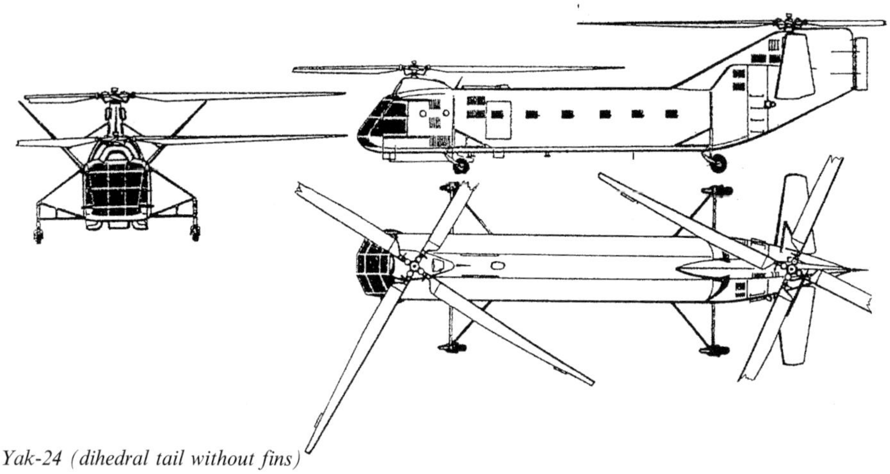

Yak-24 (dihedral tail without fins)

accommodation for up to 40 troops on canvas wall seats or light vehicles or 4 t (8,818 lb) cargo, with crane operation using central hook on underside of fuselage. Four similar levered-suspension landing gears, each normally castoring ±30°, on rigid welded steel-tube outriggers.

While numerous establishments tested complete engine/rotor rigs, blade fatigue and truss structure of fuselage, first flight article readied spring 52 and began 300 hr endurance test with wheels tied down. Vibration in evidence from start, and usually severe. With greater experience OKB might have recognized a fundamental N_1 main-rotor mode and altered critical dimension. As it was, at 178th hour, rear engine tore free from fatigued mounts, machine being destroyed by fire. Second flight article, ie, 4th airframe, finally began tethered flight piloted by Sergei Brovtsyev and Yegor Milyutchyev 3 July 52. Hops at partial power were followed by full-power flights, when vibration reared its head dangerously. Five months by every available expert found no cure; then Yak claims he personally ordered 0,5 m (19·7 in) cut off each main-rotor blade. This effected immediate great improvement. No 4 aircraft delivered for NII test Oct 53, but destroyed when tethers snapped during ground running. OKB delivered improved aircraft with numerous mods including modified tail with no fins but braced tailplanes with dihedral 45°. This finally passed NII April 55 and production began at GAZ in Leningrad. First four pre-series Yak-24 (visibly not all identical) flew at Tushino, Aug 55. Series version had strengthened floor with tracks for vehicles, tie-down rings, attachments for pillars carrying 18 stretchers, full radio and night equipment and facilities for field servicing. Normal max load 20 armed troops or 3 t (6,614 lb). Only 36 built, most with tailplane dihedral only 20° and with fixed endplate fins canted 3°30' to give side-thrust to left. On 17 Dec 55 Milyutchyev took payload 2 t to 5082 m and Tinyakov lifted 4 t to 2902 m.

OKB produced improved versions but none built in series. **Yak-24U** (*Uluchshyennyi*, improved) flew Dec 57 with numerous mods resulting from prolonged research. Rotor blade length unchanged but diameter restored by adding long tubular tie at root. Side-thrust at tail reduced by canting axes of rotors 2°30' (front to right, rear to left), so curved rear of fin removed. Fuselage frame strengthened, metal skinned throughout and cabin increased in width 0,4 m (15¾ in). Flight-control system fitted with two-axis autostab and autopilot of limited authority, developed within OKB. External slung load attached to winch in roof of cabin with large door in floor. Rear landing gear oleos changed in rate to eliminate last vestiges of ground resonance, and other minor changes including revised fuel system. In production GAZ-33 early 59, though halted at No 40. This variant could at last lift 40 troops or 3,5 t (7,716 lb) and at least some production machines had tailplane dihedral 0°.

One example built by 60 of **Yak-24A** (designation from *Aerolinyi*, airline) similar to late Yak-24U with horizontal tailplane and latest avionics but with comfortable civil interior for 30 passengers seated 2+1. Continuous glazing down sides of fuselage, compartment for 300 kg baggage and fold-down steps at door; rear freight door eliminated. Appeared in Aeroflot markings though never in service. A further example built by 60 of VIP model, **Yak-24K.** Fuselage shorter, electrically extended airstairs and luxurious accommodation for (usually) nine passengers with four large windows each side. The 30-pax **Yak-24P** was never built.

Yak-24 posed immense problems, and though it took much longer than Stalin's year, development was eventually completed to point at which this Flying Wagon could be put into military service. It was used for various purposes including crane role and for special photo missions, but remained a slow and rather unpopular machine. Had OKB persevered it might have produced more satisfactory machines with turbine engines, but it was glad to leave helicopter field to others. ASCC name 'Horse'.

DIMENSIONS Diameter of four-blade main rotors, (prototype, 24U, A, K and P) 21,0 m (68 ft 10¾ in), (production 24) 20,0 m (65 ft 7⅓ in); length of fuselage (most) 21,3 m or 21,34 m (70 ft 0⅛ in), (A) 22,1 m, (K) not known but about 19 m.

ENGINES Two ASh-82V.

WEIGHTS Empty (pre-production) 10 607 kg (23,384 lb), (U) 11 t (24,250 lb); fuel/oil, not known; loaded (24) 14,24 t or 14,27 t (31,459 lb), (U) 15,83 t (34,899 lb), (A) 16,0 t (35,273 lb), (P) 18 t (39,683 lb).

PERFORMANCE Max speed (24, U, A, K) 175 km/h (109 mph), (P) 210 km/h (130·5 mph); cruise 156 km/h (97 mph); ceiling (hover) (24) 2 km, (U) 1,5 km, (forward flight) (24) 4,2 km, (U) 2,7 km; range (24) 265 km (165 miles), (U) 255 km (158 miles), (A) 200 km (124 miles); take-off and/landing, within 50 m (164 ft).

Yak-25 First aircraft to bear this designation was small further step beyond Yak-19 and based on similar airframe though with swept tail. Wing similar family to Yak-19 and Yak-23 though improved all-metal construction and with later high-speed 'laminar' section, S-9S-9 at root tapering to KV-4-9 at tip, 9% thickness throughout, with aspect ratio 5·64 and taper ratio 2·5. OKB's first wing with welded 30KhGSA T-booms on two main spars and V-95 webs, tip to tip. Short centre section with auxiliary spar carrying landing gear (again similar to Yak-19 and Yak-23). Fuselage based on Yak-19 but longer and with plenum chamber amidships for centrifugal engine of greater diameter (though frontal area actually less than either Yak-19 or Yak-23). OKB's first pressurized-cockpit jet, with aft-sliding canopy and 57 mm bulletproof windscreen. Cordite ejection seat with leg restraints and 8 mm back armour. Internal fuel capacity 875 lit (192·5 gal). Swept tail with fixed tailplane swept at 35° on LE, 3 m^2 area, mounted 1 m above fuselage on fin swept at 40° on LE. Total vertical tail area 2,12 m^2. Complete rear fuselage and tail removed for engine access. Hydraulic system in place of many power functions previously pneumatic, and hydraulic boosted flight conrols (initially ailerons). New feature in powered speed brakes

Yak-25E

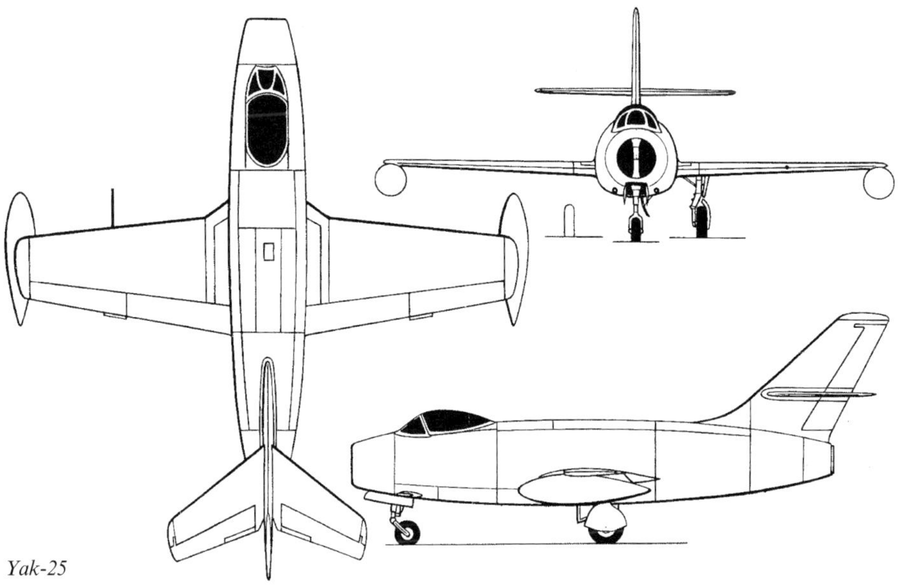

Yak-25

in place of lower section of rudder, opened automatically by pitot/static system to restrict airspeed to safe level. Armament three NR-23 each with 75 rounds. Same 200 lit (44 gal) tip tanks as Yak-23. First of two prototypes (number 15) began flight-test programme under Anokhin 31 Oct 47. Great improvement on any previous Yak fighter and demonstrated excellent manoeuvrability with rapid roll and performance seldom attained by any straight-wing aircraft. Factory test continued to 3 July 48, by which time higher performance of MiG-15 was evident. OKB never again recovered its former ascendancy in fighters. Second, mod as **Yak-25E** with refuelling probe.

DIMENSIONS Span 8,8 m (28 ft 10⅓ in); length 8,66 m (28 ft 5 in); wing area 14,0 m^2 (150·7 ft^2).

ENGINE One RD-500.

WEIGHTS Empty 2285 kg (5,037·5 lb); fuel/oil 700 kg, 1025 kg with tip tanks; loaded 3185 kg, with tip tanks filled 3535 kg (7,793 lb).

PERFORMANCE Max speed 982 km/h (610 mph) at SL, 953 km/h (592 mph) at 5 km; time to climb to 5 km and 10 km, 2·5 min and 6·3 min; service ceiling 14 km (45,932 ft); range 1100 km, or 1600 km (994 miles) with tip tanks; take-off 510 m; landing 825 m/175 km/h.

Yak-30 First aircraft to carry this designation, which curiously is even and thus not normally assigned to a fighter, was OKB's answer to March 46 requirement for Mach 0·9 interceptor for use from existing unpaved airfields. Design based upon Yak-25 but with entirely new swept wing mounted in mid position, in turn calling for different main landing gear. Wing of S-9S series but swept 35° at LE, with dihedral –2°; aspect ratio 5, taper ratio only 1·5. Two spars throughout with 30KhGSA booms and V-95 webs, skin flush-riveted D16 of ruling thickness 2 mm. Boosted tabbed ailerons, slotted split flaps and four full-chord fences; TE kinked to short unswept inboard portion. Fuselage basically same as Yak-25 but longer, though internal fuel unchanged. Identical cockpit, canopy and seat as Yak-25 and same armament of three NR-23 with 225 total rounds. Split speed brake at tail retained, though no longer any Mach limitation. Fuselage slightly greater cross-section as perfect circle (1,36 m) instead of oval 1,36 m wide and 1,34 m high. Fin essentially as Yak-25 but tailplane slightly enlarged to 3,05 m^2, still 35° (not 45°). Drop tanks slightly modified to fit as slippers under wing at mid-span, same 200 lit capacity. First prototype (number 42) flown by Anokhin 4 Sept 48, too late to rival MiG-15. Second Yak-30, number 54, had various minor changes, most obvious being revised main-gear doors forming large section of fuselage skin. Mach 0·935 often demonstrated and handling generally good, but out classed by MiG-15. Factory testing concluded 16 Dec 48 without submission to NII. Anokhin noted unsuccessful wing profile and dissatisfaction with aileron balance.

DIMENSIONS Span 8,65 m (28 ft 4½ in); length 8964 mm (29 ft 5 in); wing area 15 m^2 (161·5 ft^2).

ENGINE One RD-500.

WEIGHTS Empty 2415 kg (5,324 lb); fuel/oil 700 +12 kg; loaded 3305 kg (7,286 lb), or with drop tanks 3630 kg (8,003 lb).

PERFORMANCE Max speed 1060 km/h (659

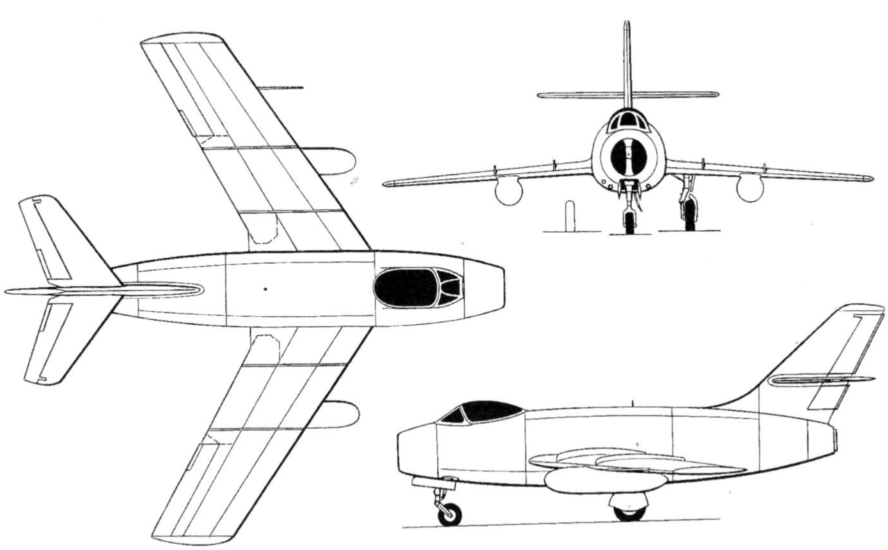

Yak-30

Yak-30 (first aircraft)

Yak-50-II with central underfin

mph) at SL, 1025 km/h (637 mph) at 5 km; time to climb to 5 km and 10 km, 2·6 min and 6·6 min; service ceiling 15 km (49,200 ft); range 1000 km (621 miles), or 1500 km (932 miles) with drop tanks; take-off 510 m; landing 610 m/188 km/h.

Yak-50 First aircraft to carry this designation, which for second time was a non-fighter even number, was dramatic attempt to equal performance of MiG OKB by using same VK-1 engine and increasing sweep throughout. Described as light fighter/interceptor, and intended to meet requirement for single-seat all-weather aircraft, but real value was in providing basis of later Yak-25. Wing of unknown section (believed SR-9 series) with constant chord and generally unvarying profile outboard of short centre section with TE at 90°. Ruling sweep angle 45° throughout, dihedral –5°, skin stretch-formed and 3 mm thick between spars, three fences each side (outers only as far as aileron LE), aspect ratio 4. Fuselage enlarged to 1,4 m diameter and lengthened to 9465 mm (excluding tail). Dummy radome at top of inlet in nose, ducts enlarged to handle mass flow increased from 29 kg/s to 45 kg/s, pressurized cockpit redesigned with lower-drag canopy, wing in true mid-position, rear fuselage and tail detachable for engine access. Sweep of vertical tail increased to 45°, increased in area to 3,0 m^2, horizontal tail at last swept 45° and reduced in size to 2,86 m^2, tailplane still fixed. Top of fin dielectric aerial, rudder in upper/lower parts, airbrakes relocated on sides of rear fuselage, ventral strake/fins of 0,344 m^2 added to increase high-Mach yaw stability. Completely new area-increasing tracked flaps of basically Fowler type moving parallel to fuselage (unlike Yak-30). Considerably greater fuselage fuel, 1065 lit (234 gal). Completely new tandem (*velosipedno* tipa, bicycle type) landing gears, rear twin 600 mm×150 mm wheels taking 85% of total weight. New nose gear retracting to rear, plus outrigger tip-protection gears with extremely small solid-tyred wheels retracting to rear housed in fixed tip fairings downstream of large

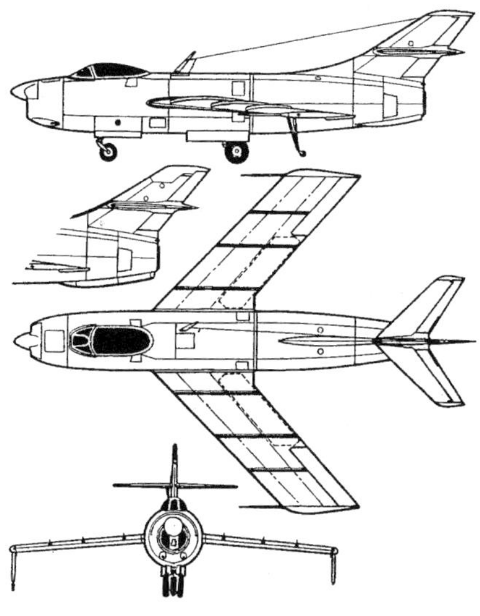

Yak-50-I (inset 50-II)

pitot heads. Armament reduced to two NR-23 each with 80 rounds. Braking parachute added to first prototype No 20; **Yak-50-II** numbered 35. Flight test programme by Anokhin opened 15 July 49. Soon proved to have tremendous performance; few fighters of era could equal speed and climb, and manoeuvrability also outstanding with rate of roll better than MiG-15 and tighter turning circle. VVS interest focused on low gross weight, well below MiG-15 and conferring great high-altitude performance and good handling despite wing area 16 compared with 22,6 m^2. In general Yak-50 outperformed predicted figures for new MiG SI, later called MiG-17, which was to fly a year later. Yak almost made it with this fighter, but ultimately giant established programme with existing MiG-15 proved unshakeable. Eventually factory test programme was halted 30 May 50. Only real fault was poor directional stability at high speed on ground and Anokhin said uncontrollable on wet surface. Led to Yak-140.

DIMENSIONS Span 8010 mm (26 ft 3⅓ in); length 11 185 mm (36 ft 8⅓ in); wing area 16,0 m^2 (172 ft^2).
ENGINE One VK-1.
WEIGHTS Empty 3085 kg (6,901 lb); fuel/oil 850 kg+10 kg; loaded (½ fuel) 3650 kg, (max fuel) 4100 kg (9,039 lb).
PERFORMANCE Max speed, 1170 km/h (727 mph) at SL, 1135 km/h (705 mph) at 5 km, 1065 km/h (662 mph) at 10 km, M-max 1·03; initial climb 68 m/s (13,390 ft/min); time to 5 km/10 km/15 km, 1·5 min, 3·5 min and 7 min; service ceiling 16,6 km (54,462 ft); range 1100 km (684 miles); take-off 587 m; landing 965 m/200 km/h (124 mph).

Yak-1000 Another inexplicable OKB number, this was assigned to a remarkable experimental fighter designed 49 to explore configuration for supersonic interceptor. Again *velosipedno* landing gear was adopted, because at this time problems with Yak-50 were not yet manifest. Configuration fairly like Douglas D-558-I Skystreak, with long barrel-like fuselage of minimum cross-section tailored to slim axial engine of highest possible thrust. Pressurized cockpit near nose with rudder pedals right in bifurcation of duct, pilot being in semi-reclining posture in modified ejection seat to reduce frontal area. Low-drag canopy without any fighter attributes such as flat bulletproof windscreen. Air ducts passed only above three spars of wing inside fuselage (fuel and gear below). Engine installed immediately aft of rearmost spar with short jetpipe in removable rear fuselage and tail unit. Wing of low aspect ratio and sharp taper, entire surface aft of rear spar hinged to form slotted flap incorporating powered aileron. Landing-gear as Yak-50, but outriggers moved inboard to point at which chord of wing equalled length of leg. Low aspect ratio tail with fixed tailplane carried well aft on fin so that powered elevators were downstream of rudder. Airbrakes on sides of rear fuselage, braking parachute unknown location. Limited fuel capacity ahead of and behind main-gear bay; space for two cannon in lower part of

Yak-1000

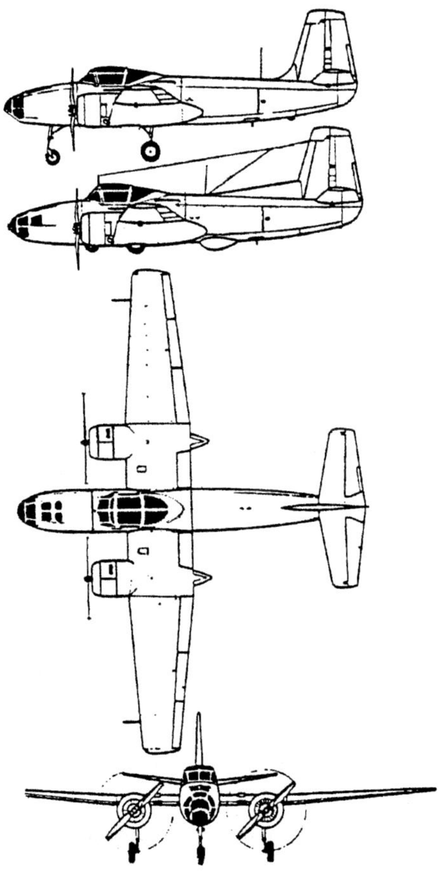

Yak-200 (lower side view 210)

nose, not installed. Taxi tests in 51 showed such dangerous instability whole programme eventually abandoned. Design speed 1750 km/h, M 1·65, or 2000 km/h (M 1·88) with afterburner.
DIMENSIONS Span 4,52 m (14 ft 10 in); length, excl instrument boom, 11,69 m (38 ft 4⅛ in).
ENGINE One AL-5.
No other data.

Yak-200, 210 Prototypes only of stressed-skin trainer for bomber pilots (**Yak-200**) and navigator/bomb-aimers (**Yak-210**). Mid-mounted wing, rectangular centre section 15% thick, dihedral and tapered outer panels (8%), slotted flaps inboard/outboard of nacelles, two-part drooping ailerons, conventional tail with dihedral tailplane, all controls manual, fabric-skinned. Fuel 760 litres (167 gal) in centre section, piston engines driving 2-blade VISh-111-B20A props. Simple nosewheel-type landing gear, like flaps pneumatic operation. Side-by-side cockpit amidships for instructor/pupil pilots, sliding canopy; glazed nav station in nose (in 210, with bombsight); provision for nav/bomb instructor also.

RSIU-3 and -5 radio, ARK-5 ADF, RV-2 (lo) and RV-10 (hi) radar alts, marker receiver, provision for SRO-2 IFF; Yak-210 in addition RSBN-M nav/bombing radar under rear fuselage, AFA-BA-40 camera and provision for 300 kg of practice bombs. First flight of 200 March 53, followed by 210 in June. NII tested, but no reqt, Il-28U being considered worth extra costs.
DIMENSIONS Span 17 455 m (57 ft 3¼ in); length 12 950 m (42 ft 5⅞ in); wing area 36 m^2 (387·5 ft^2).
ENGINES Two ASh-21.
WEIGHTS Empty (200) 3910 kg (8,620 lb), (210) 4542 kg (10,013 lb); loaded (200) 4715 kg (10,395 lb), (210) 5422 kg (11,953 lb).
PERFORMANCE Max speed (200) 400 km/h (249 mph); service ceiling (200) 7,16 km (23,490 ft), (210) 6,7 km (22,000 ft); range (200) 1280 km (795 miles); TO/ldg (200) 360 m/130 km/h, 430 m.

Yak-120, Yak-25 OKB did not respond to 49 call for radar-equipped interceptor, but in Jan 51 decided to scale up Yak-50 with two engines and radar. Unlike rivals chose to put engines under wings, leaving fuselage free for radar, weapons and maximum fuel. Project authorised summer 51, OKB No **120**.

Mid-position wing scaled up to 28,94 m^2, retaining same SR-9s profile, 45° sweep and

Yak-200

Yak-25 No 03

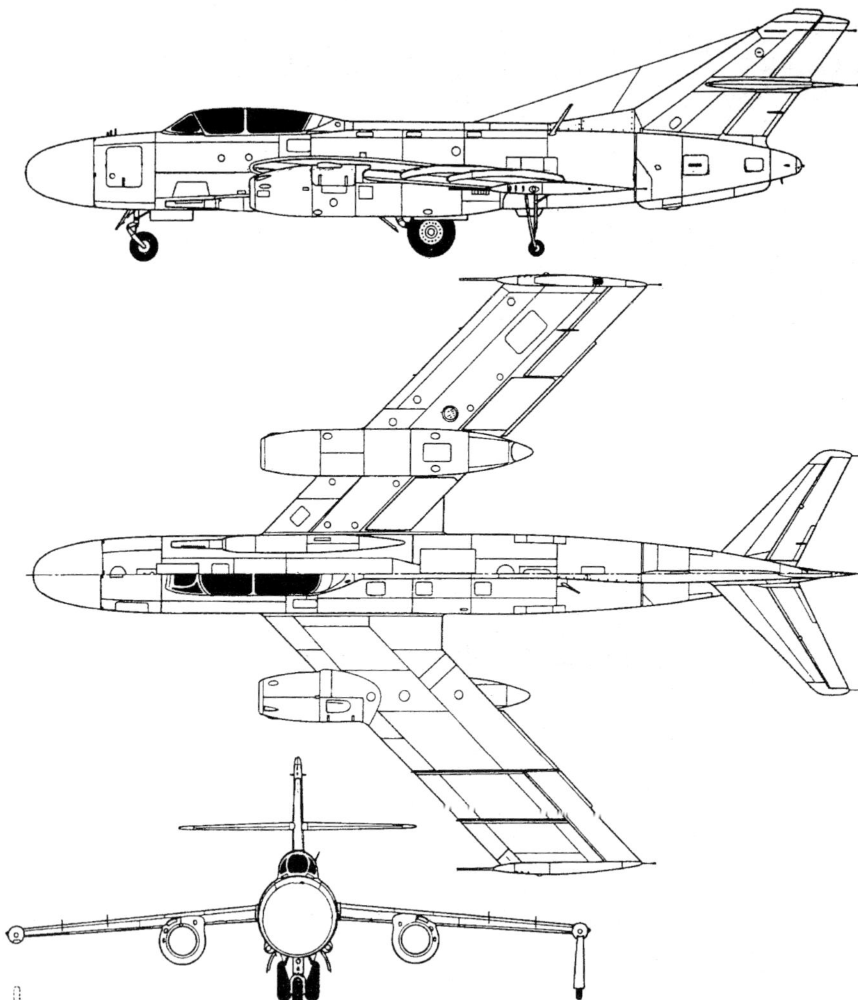

Yak-25M

constant chord outboard of kinked inboard TE, but anhedral reduced from –5° to –3° and underslung nacelles enabled innermost fence to be omitted. As before, two parallel spars, slotted flaps (now divided by nacelle) and boosted aileron. Fuselage same circular section but much longer, with pilot and radar operator in tandem (same seats as before) under single large canopy sliding 2,2 m to rear. Bulletproof (57 mm) windscreen and armoured headrests, front bulkhead and side panels. Airbrakes moved forward, midway between wing and tail. Tail also scaled Yak-50 but with dorsal fin faired into spine to canopy taking tail control rods past tanks. Landing gear changed only in detail, main leg strengthened for higher weights and castoring nose leg lengthened. Outrigger gears modified to retract backwards into oval-section pods forming wingtips, carrying long pitot tubes as before. Engines slung below wing in simple nacelles with nozzle faired into small nib projecting behind TE. Plain inlets with three projecting aux inlets to cool generator, oil cooler and ventilate bay. Late start meant radar could be *Korshun*, rather than *Torii*. Sole armament two N-37L on lower sides of fuselage each with 50 rounds, cases/links ejected.

This aircraft flown 19 June 52. Despite marginal handling at high AOA overall assessment by OKB and NII favourable, esp on grounds of combat radius. Selected May 53 for PVO service as **Yak-25**, subject to further improvements. Most impt of these, completely new RP-6 *Sokol* radar, in bluff radome with good electronic perf but initially prone to icing. Further devt led to **Yak-25M**, fitted with AM-9A engine with elec-heated inlet grille, hot-air wing/tail deicing and provision for non-conformal 685-lit drop tank. Total 480 built 55-57, serving to 65. From 55 Yak-25M used for various test programmes, esp on air refuelling, air/air missiles, ejection seats and avionics, and in support of later devts of same family, those at OKB having red or grey nose with circular border each side. Reporting name 'Flashlight', later 'Flashlight-A'.

DIMENSIONS Span 11,0 m (36 ft 1 in); length 15 665 m (51 ft $4\frac{3}{4}$ in); wing area 28,94 m^2 (311·5 ft^2).

ENGINES (25) two AM-5A, then AM-5B, (25M) AM-9A.

WEIGHTS Empty (25) c6,35 t (14,000 lb); loaded (25) 9220 kg (20,326 lb), (25M) 10 045 kg (22,145 lb).

PERFORMANCE Max speed (25) 1090 km/h (677 mph), (25M) 1040 km/h (646 mph) at 9 km (M 0·954); climb to 10 km (25M) 6·4 min; service ceiling (25M) 13,9 km (45,600 ft); range (25M) 2730 km (1,696 miles).

Yak-25 variants Single example built 53 of **Yak-121**, 25M reconfigured for fighter/attack. At least one **Yak-122** reconnaissance conversion. **Yak-123** two-seat bomber, at least one flown. **Yak-125** recon version. Rear cockpit removed, pilot with short upward-hinged canopy faired into spine, and radar replaced by navigator station in glazed nose with seat able to eject through aperture of jettisoned roof hatch. Pallet of cameras in belly, single NR-23 on right side. **Yak-125B** bomber version of 125, prototype #25, blunt glazed nose, twin-wheel nose gear with radar immediately to rear. **Yak-25L** ejection-seat testbed. Front and rear cockpits separated by pressure bulkhead, with individual canopies, though photo shows jettisoning of single long canopy with metal rear portion. **Yak-25K-75** one of several testbeds for missiles (mainly K-5, K-8) and guidance radar. **Yak-25N** unbuilt variant for tactical nuclear bomb.

Yak-25R Frontal Aviation recon aircraft, based on Yak-125. Built in small series. ASCC name 'Flashlight-B'.

WEIGHTS Loaded 9840 kg (21,693 lb).

PERFORMANCE Max speed 1080 km/h (671 mph).

Yak-25RV Single-seat ultra-high-alt recon aircraft derived from Yak-25R in 59. Completely new unswept wing mounted shoulder-high, completely different flaps and ailerons. New engines in nacelles of increased diam, extending well behind TE. Fin and dorsal fin extended in chord and height, but ventral fin removed. Fuselage broadly as 25R except for metal-skinned nose with pitot and air-data booms. New outrigger gears retracting forwards into pods extending far ahead of wingtip. Armament removed, fuel increased to c5000 lit. Special provisions for crew comfort on sorties up to 6 h. **RV-I** pilotless version flying mission on preprogrammed autopilot. **RV-II** similar but RPV with control receiver to feed autopilot commands from operator flying in modified Yak-30. Total of all versions 165, initially with

Yak-125B

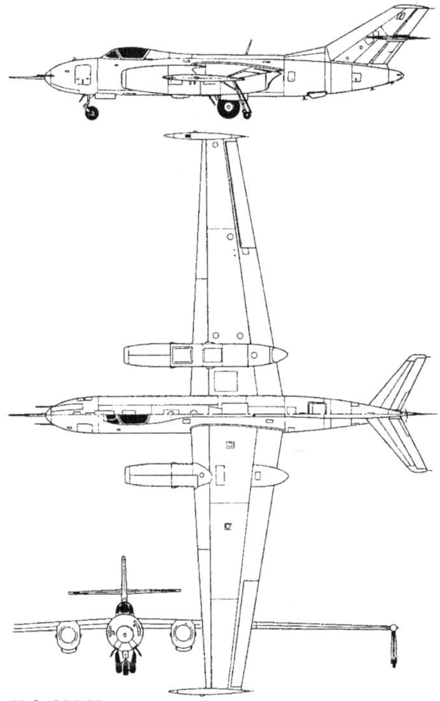

Yak-25RV

Yak-25K-75 (two K-8)

Yak-25RV

simple camera fits, from 66 with SLAR and various other sensors, though withdrawn from front-line use early 70s. OKB pilot Vladimir Smirnov sets records July 63 including 20 456 m with 1 t; Marina Popovich set women's circuit speed records 65-67. Smirnov once reached 22 km, where 'unable to descend' until he lowered landing gear. ASCC name 'Mandrake'.

DIMENSIONS Span 23,4 m (76 ft 9¼ in); length 18,45 m (60 ft 6 in); wing area 51,5 m^2 (554 ft^2).

ENGINES Two R-11V-300.

WEIGHTS Normal loaded 9012 kg (19,868 lb), max 9950 kg (21,936 lb).

PERFORMANCE Max speed (above 11 km) 870 km/h (541 mph); service ceiling 21,0 km (68,900 ft); range 3500 km (2,175 miles).

Yak-26 First supersonic member of family, derived from Yak-123. Aerodynamically improved, thinner wing (5·5%) with extra sweep at LE root, longer fuselage, pointed nose, more powerful engines in different nacelles, main landing gears strengthened and nose gear as Yak-125B, steerable with twin wheels, retracting forwards. Unchanged: mid posn of wing, two outboard fences, rear-retrac tip gears. Bomb bay in centre fuselage, racks under outer wings for rocket (not missile) launchers. Prototype flown 55 (#54, metal nose with various square windows); followed by eight further prototypes with different noses, most being all-glazed. Aircraft unacceptable because of instability at hi-AOA which drooping outer LE failed to cure. No ASCC name.

DIMENSIONS As 25 except length c20 m (65 ft 7 in).

ENGINES Two RD-9AK.

WEIGHTS Loaded 11,2 t (24,691 lb).

PERFORMANCE Max speed (10 km) 1235 km/h (767 mph, M 1·16).

Yak-26 prototype (modified nose and B-20E in tail)

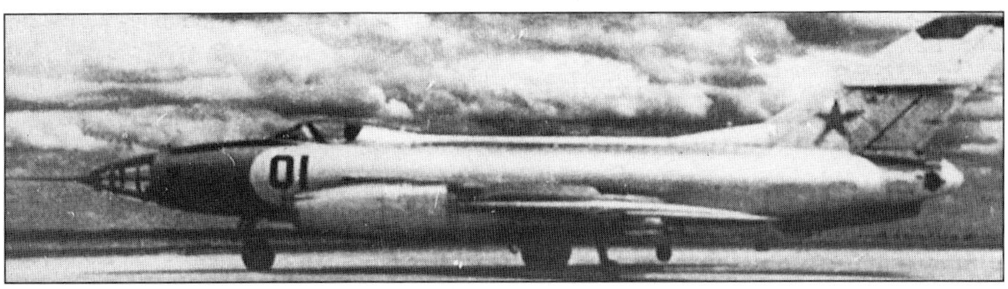

Yak-27R prototype

Yak-27 Small group of interceptors derived from Yak-121. Wing of modified profile (same thickness as 26) with LE extended in chord and drooped from inboard fence to tip-gear fairing, beyond which tip slightly extended. Tip gear (retrac aft as before) housed in improved fairing tapering at front into long pointed anti-flutter mass; outer fence deleted. Nose gear as 25. Radar (TsD-30) in nose with dish housed in pointed radome (lower drag, less erosion and less icing), some with pitot boom on tip. Cockpits almost as 25, tail, armament unchanged; engines RD-9AK. **Yak-27I** #55 gun armament. Several **Yak-27K** with one R-98 (K-8) missile under each inboard wing. Failed NII testing through instability at hi AOA.

Yak-27V Special hi-alt interceptor with rocket engine in rear fuselage. Installation (esp fuel feed) claimed 'far more effective' than on rival MiG Ye-50. Total internal fuel considerably increased, though duration under rocket power 160s at max thrust, 264 at cruise. Stringent weight-reduction eliminated rear cockpit, brake chute and various eqpt. First flown May 57 by Valentin Mukhin, who later reached 23 km (75,460 ft). Failed NII 59 on grounds of hi-AOA behaviour.

Yak-27R Two-seat recon aircraft with improved afterburning engines in 27-type airframe and 25R type accommodation. Nav in fully glazed nose coming to sharp point with instrument boom. Tankage increased to c3800 lit (3040 kg), camera pallet usually as 25R. First flight 58, passed NII testing and 180 delivered, serving 60-71. Data for 27R. ASCC name 'Mangrove'.

DIMENSIONS Span 11,7 m (38 ft $4\frac{5}{8}$ in); length c20,5 m (67 ft 3 in).

ENGINES Two RD-9AF.

WEIGHTS Empty c8,25 t (18,187 lb); loaded 12,5 t (27,560 lb).

PERFORMANCE Max speed (11 km) 1285 km/h (798 mph, M 1·2); range 2250 km (1,400 miles).

Yak-27R (1st series)

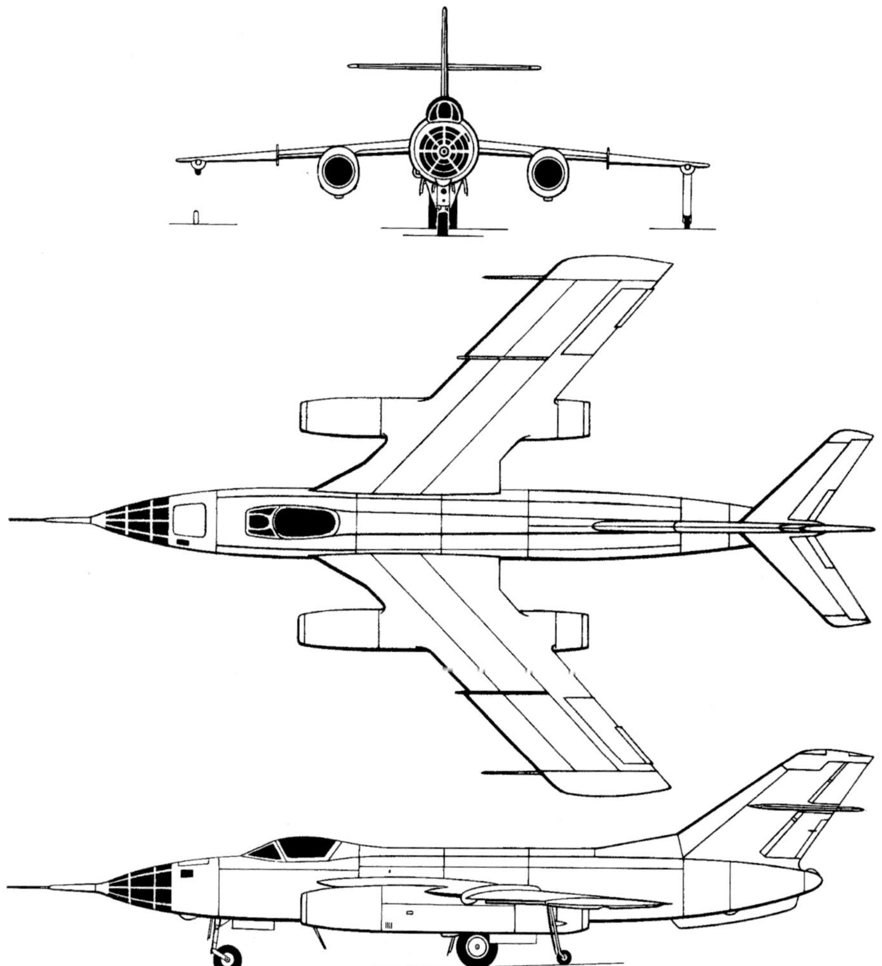

Yak-27R

Yak-129, Yak-28 Prototype **129** of 58 led to prolific new generation to fly many missions. Prototype was 'front-line bomber' derived from Yak 26, with almost identical crew layout and landing gear. Major differences: new engines in redesigned nacelles; totally revised structure for increased gross weight; greater wing chord; wing moved up to shoulder height; and increase in area and authority of tail. Wing raised for aerodynamic reasons and to enable larger stores to be carried in bomb bay (though usual load still typically three pairs FAB-500). Wing thickness retained while chord increased, inboard LE swept 60°, inboard TE 90° with rectangular slotted flap hinged without tracks with large fence upstream of mid flap span, outboard fence removed and LE sharply extended and drooped, aileron chord increased and fixed strips added behind TE of fixed outer wing, tips outboard of gear fairings as 27R but extended chord. Tail increased span and area, tailplane pivoted and fully powered and elevators cut away at root to enable one-piece rudder to be used. Forward main landing gear as Yak-26 but strengthened, powered steering; aft main gear strengthened, both gears with anti-skid multi-disc brakes. Fuselage strengthened for increased weights and airspeeds, though Mach 1 not to be exceeded under 8 km. Larger but shorter afterburning engines in deeper nacelles with vertical oval inlets fitted with translating conical centrebodies. Variable nozzle exposed behind nacelle. First flight 5 March 58, OKB testing to Aug 58, NII testing thence to May 59. Accepted as frontal bomber, pre-production batch designated **Yak-28**.

Yak-28B First full production version. Two-seat bomber with avionics for night and bad-weather attack. One NR-23 with 200 rounds, normal bombload 3 t (6,614 lb), internal fuel 4315 lit (949 gal), external fuel (two tanks scabbed under outer wings, jettisonable in unison) 2000 lit (440 gal). RBP-3 nav/bombing radar with scanner in prominent radome behind nose gear. ARK-2A, MRP-56A, SOD-2M, SRO-2, RV-17 and RV-UM. ASCC name 'Brewer-A'.

DIMENSIONS Span 11,7 m (38 ft $4\frac{5}{8}$ in); length 21,52 m (70 ft $7\frac{1}{4}$ in); wing area 35,25 m^2 (379 ft^2).

Yak-28B

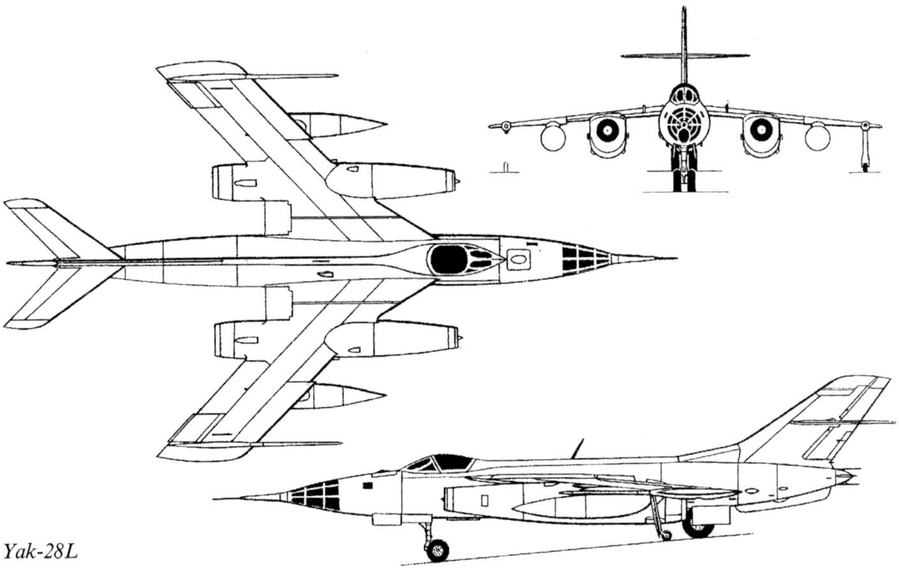

Yak-28L

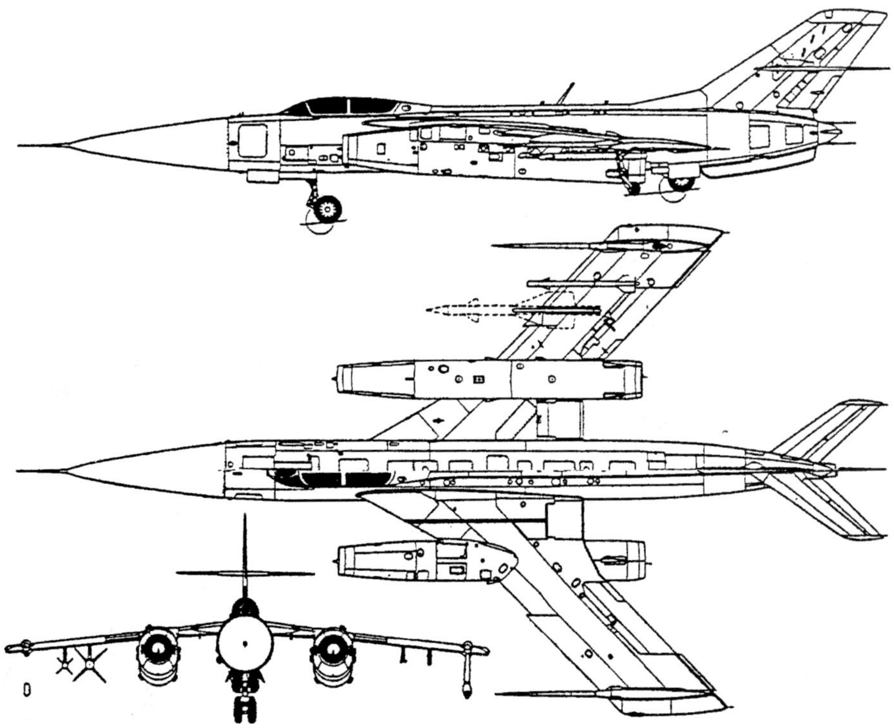

Yak-28PM

ENGINES Two R-11AF-300.

WEIGHTS Empty 7220 kg (15,920 lb); loaded 13,63 t (30,050 lb), max 16 t (35,273 lb).

PERFORMANCE Max speed 980 km/h (609 mph) at SL, 1900 km/h (1,181 mph, M 1·79) at 12 km; service ceiling 16,2 km (53,150 ft); range 1950 km (1,212 miles), (with drop tanks) 2630 km (1,635 miles); TO 1 km; landing 225 km/h, 1,3 km.

Yak-28L While autonomous capability conferred by radar was valuable, location of radome over scanner made it essential to raise entire aircraft off ground in order for bomb trolleys to reach bomb bay, making rapid turnround between sorties impossible (normal time for bomb loading 90 min). Yak-28L replaced radar by *Lotos* precision guidance from groups of ground stations. This enabled high sortie rate to be maintained, at cost of inability to undertake independent target search, or operate beyond line-of-sight range of guidance stations, or bomb with equal precision except visually. Engines as 28B. Production from 62, total 111. ASCC name 'Brewer-B'.

DIMENSIONS, ENGINES As 28B.

WEIGHTS Empty 7,2 t (15,873 lb); loaded 15 545 kg (34,270 lb).

PERFORMANCE Max speed at 12 km 1945 km/h (1,209 mph, M 1·83); service ceiling 16,2 km (53,150 ft); range (ext fuel) 2420 km (1,504 miles).

Yak-28I Main production bomber version, first flown 60, designation from *Initsiativa* nav/bombing radar replacing RBP-3 and giving improved capability for independent operation in bad weather. During development engines replaced by AF2-300 version in redesigned nacelles with long tapering inlets of circular section with better pressure recovery at all Mach numbers. Forward fuselage lengthened 0,78 m (31 in), nose gear moved forward with it but not pilot cockpit, giving more room for navigator and radar/sight system. NR-23 replaced by GSh-23Ya with 350 rounds. Modified glazing to nose, forked twin-sensor nose PVD instrument boom. Drop tanks slightly increased in size (1100 lit) and redesigned with pointed conical nose. In production 63, total 223. ASCC 'Brewer-C'.

DIMENSIONS As 28B except length 22,3 m (73 ft 2 in).

ENGINES Two R-11AF2-300.

WEIGHTS Empty 7660 kg (16,887 lb); loaded 16,16 t (35,626 lb).

PERFORMANCE Max speed at 12 km 1805 km/h (1,122 mph, M 1·7); service ceiling 14,5 km (47,575 ft); range (ext fuel) 2070 km (1,286 miles).

Yak-28BI All Yak-28B aircraft in FA service were retrofitted with longer nose and *Initsiativa* radar, becoming BI versions. Engines, nacelles and gun unchanged. 'Brewer-C'.

Yak-28P Interceptor version to replace Yak-25 was obvious, and prototype flew 60. Designed for all-weather interception at low and medium alt. Airframe generally as Yak-28, except design load factor increased from 4 to 6, bomb bay occupied by fuel bringing total (all internal) to 6350 lit, tandem cockpits for pilot and radar operator with bulletproof front windscreen and one-piece four-pane canopy sliding (elec) to rear, prominent superimposed static discharge wicks projecting aft from tailcone, *Orel*-D radar occupying nose, pylon under each outer wing

Yak-28P

for one K-8R, one K-8T, STOL capability conferred by twin SPRD a.t.o. rockets and braking parachute triggered by long bar (hinged down from ventral strake by extending gear) hitting runway. Prototypes with AF-300 engines and original nacelles, production from Dec 61 with new propulsion systems (as 28I).

Yak-28PM Improved interceptor introduced to production 64, *Orel*-DM radar with improved power and discrimination, longer pointed radome giving higher flight performance, two additional underwing pylons for R-3R/-3S missiles, upgrading avionics. Delivered Irkutsk 65-68, total both versions 435. ASCC name (both) 'Firebar'.

Yak-28PD Despite inboard ailerons lack of wing stiffness caused control reversal at highest speeds at low level; PD cured this by adding delta ailerons at front of wingtips, but no production. Official PVO carrier of same radars/missiles was Su-15; Yak-28P/PM never officially in service but in fact Arctic regts used over 200 to 80.

DIMENSIONS Span 11,7 m (38 ft 4⅝ in); length (both radomes) 21,47 m (70 ft 5¼ in); wing area 35,25 m² (379·4 ft²).

ENGINES Two R-11AF2-300.

WEIGHTS Empty 7,75 t (17,085 lb); loaded (P) 15,7 t (34,612 lb), (PM) 15,9 t (35,050 lb).

PERFORMANCE Max speed (13 km) (P) 2060 km/h, (PM) 2110 km/h (1,311 mph, M 1,99) [service limit with missiles 1840 km/h, 1,143 mph, M 1·73]; service ceiling 16 km (52,500 ft); range 2600 km (1,615 miles); TO (a.t.o.) 400 m; landing (chute) 620 m.

Yak-28U Conversion trainer developed 62, armament deleted, instructor cockpit as bombers, pupil cockpit replacing nav/bomb station in nose with fixed windscreen and canopy hinged to right, short nacelles, short forward fuselage, fuel tanks in bomb bay but no drop tanks. Total production 183. ASCC name 'Maestro'.

DIMENSIONS As 28B except length 20,2 m (66 ft 3¼ in).

ENGINES Two R-11AF-300.

WEIGHTS About 1 t lighter than operational versions.

PERFORMANCE Max speed at 12 km 1900 km/h; range c2500 km.

Yak-28R Reconnaissance version first flown 63. Airframe as 28I except for modified nose

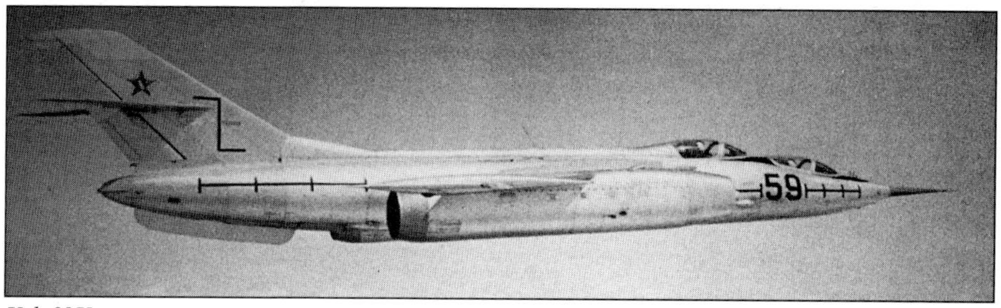

Yak-28U

glazing with diagonal rear edge and pilot windscreen comprising left/right elliptical panes meeting at knife-edge. Fuel in bomb bay (total internal 6200 lit) and optional drop tanks, choice of several payload pallets carrying optical cameras, IR linescan or side-looking radar (often cameras only). Prototype (#45) later served as prototype **Yak-28RL** with radar replaced by *Lotos* guidance. In production 70, total 183. 'Brewer-D'.

DIMENSIONS Span 11,64 m (38 ft 2¼ in); length 22,3 m (73 ft 2 in); wing area 35,2 m² (379 ft²).

ENGINES Two R-11AF2-300.

WEIGHTS Loaded 15 725 kg (34,667 lb).

PERFORMANCE Max speed at SL 1100 km/h as before, at 12 km 1880 km/h (1,168 mph); service ceiling 15,65 km (51,345 ft); range 2450 km, (ext fuel) 3000 km (1,864 miles).

Yak-28PP Final series version, based on 28I but with windscreen of 28R. Dedicated EW (electronic warfare) platform, primarily as active jammer to precede or escort friendly attack aircraft. SG-1 series RWR installation with receiver antennas (5) around nose, at base of large-diam conical PVD probe, and on each side of forward fuselage. Direction-finding blade antennas mounted diagonally on nacelles. High-power directional jammer and chaff-dispenser installations aft of nose gear and completely filling previous bomb bay, latter being preceded by high-capacity heat exchangers cooled by ram air from three inlets around underside of fuselage. Pylons under outer wings (as outboard on 28PM) for 32×57 rockets with chaff warheads. Internal fuel 4,55 t (10,031 lb). In production 67, in FA service to 94 and beyond because Su-24MP absorbed by Ukraine. ASCC name 'Brewer-E'.

DIMENSIONS, ENGINES As 28R.

WEIGHTS Empty 7,92 t (17,460 lb); loaded 15 955 kg (35,174 lb), max with drop tanks and rockets 17 865 kg (39,385 lb).

PERFORMANCE Max speed at 11 km (no tanks but with rockets) 1570 km/h (976 mph), (clean) 1725 km/h (1,072 mph, M 1·62); service ceiling 14,9 km (48,900 ft); range (with tanks and rockets) 1980 km (1,230 miles).

Yak-28-64 Major redesign to compete with Su T-58. R-13-300 engines close together in rear fuselage, fed by vertical rectangular var-geom inlets on sides of forward fuselage. Wing part-redesigned and restressed, pylons for two K-8 and two K-13. Larger square-cornered ventral fins under tail.

Yak-33 Projected long-range Mach 2·5 interceptor to replace Yak-28M, terminated c73.

Yak-140 Major attempt to get back in single-seat fighter business, project begun 52 with order for two prototypes. First with two NR-30 guns, later to have missile armament (probably two K-5 under wing roots). Compared with Yak-50, similar body cross-section but considerably longer, circular nose inlet with gunsight radar in conical centrebody, cockpit similar but canopy faired into spine leading to more sharply swept vertical tail, powered tailplanes with camber-increasing elevators centred on fuselage, two door airbrakes low on rear fuselage, wing larger but same depth (so t/c ratio reduced), four fences, slotted flaps (planned later to be blown). See Addenda.

First prototype with two NR-30 guns and interim engine did taxi tests and short hops in 55. Second with later engine and three guns never complete, because Ministry refused to clear Yak-140 to fly.

Yak-28-64 with drop tanks

Yak-140 (No 1)

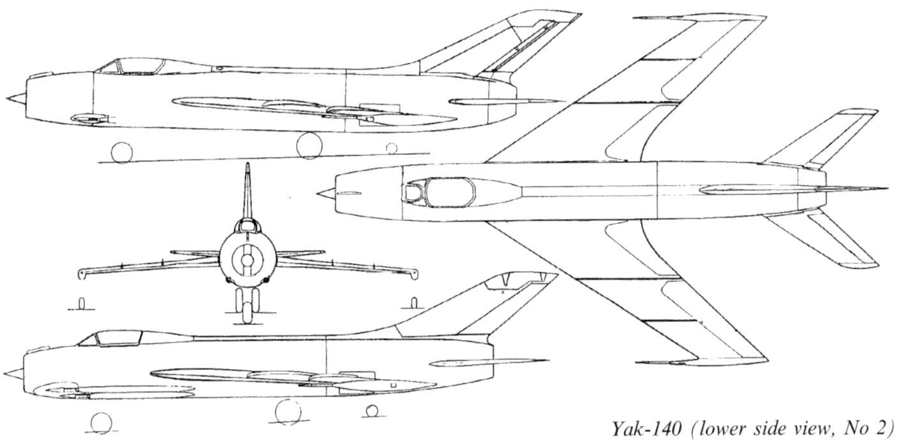

Yak-140 (lower side view, No 2)

DIMENSIONS Span (No 1) 7935 (26 ft 0½ in), (2) 8,0 m (26 ft 3 in); length (1) 13 340 (43 ft 9¼ in), (2) 12 950 (42 ft 5⅞ in); wing area 19 m^2 (205 ft^2).
ENGINE (1) one AM-9D, (2) one AM-11.
WEIGHTS Loaded est 7 t (15,430 lb).
PERFORMANCE Max speed est 2000 km/h (1,243 mph).

Yak-30, OKB number **Yak-104**, also called **No 104**, built 59 to meet VVS requirement for efficient jet trainer, designed for job and not modified from fighter. Stress laid on low cost, reliability, long life. Tumanskii KB charged with producing new small turbojet; stressed-skin construction with tandem ejection seats. Wing SR-9S series, two spars, pressed ribs, flush-riveted skin, pneumatic split flaps, manual ailerons with spring tabs, pitot on left LE. Well-streamlined fuselage with long nose with landing/taxi light, upper half nose hinged up for access. Tandem unpressurized cockpits (no radio), rear instructor seat slighter higher than pupil but cill of single aft-sliding canopy horizontal, dorsal spine to swept vertical tail carrying unswept fixed tailplane and spring-tab elevators with cut-outs for rudder. Engine in bottom of fuselage aft of rear spar of low-mounted wing fed by shallow root inlets. Complete left/right cowls hinged and removable to expose engine; no jetpipe. Fuel in 600 lit (132 gal) tank above wing in fuselage. Levered-suspension legs, retraction inwards (main) between spars or forwards into nose. Provision for armament but none fitted. Six prototypes or static-test aircraft built 1960-61. Competed against Polish TS-11 and Czech L 29 in Moscow Aug 61; L 29 adopted. Immediately after choice, one Yak-30 set 15/25 km speed record 767,31 km/h. Three days later, 25 Sept 61, another set C-1d class record altitude 16 128 m.

Yak-32 was single-seat sporting derivative, rear cockpit deleted but no change in fuel capacity. Two built in parallel with Yak-30, funded by Dosaaf and both gave aerobatic display 61 Aviation Day. Height (C-1d class record) 14 283 m set 22 Feb 61; data included above-average weight (2137 kg) and below-normal thrust (800 kg). In 65 Yak-32 set 15/25 km speed 775 km/h. Yak-30 No 90 and Yak-32 No 70 parked on field at 66 World Aerobatic Championships. ASCC names, 30, 'Magnum'; 32 'Mantis'.

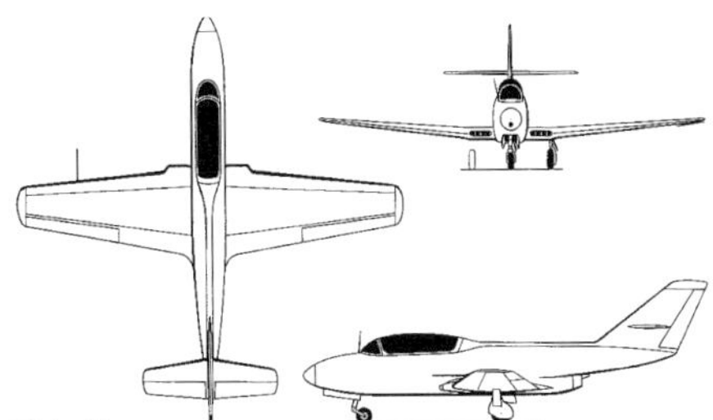

Yak-30

DIMENSIONS Span 9,38 m (30 ft 9¼ in); length 10,14 m (33 ft 3¼ in); wing area 14,3 m^2 (154 ft^2).
ENGINE One RU-19.
WEIGHTS Empty (30) 1555 kg (3,428 lb), (32) 1435 kg (3,164 lb); fuel/oil 480 kg+20 kg; loaded

Yak-30

Yak-36

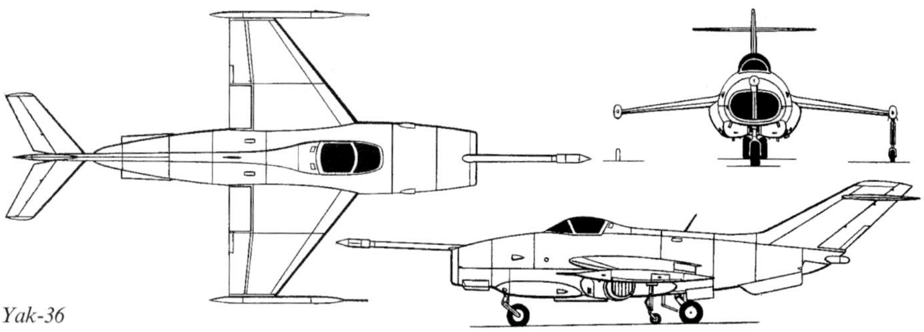

Yak-36

(30) 2250 kg (4,960 lb), (32) 1930 kg (4,255 lb).
PERFORMANCE Max speed (30) 660 km/h (410 mph) at low level, (32) 700 km/h (435 mph); initial climb (30) 18 m/s (3,540 ft/min); service ceiling (30, official figure) 11,5 km (37,730 ft), but see record climb figs; range (30) 965 km (600 miles); take-off 425 m; landing 450 m/140 km/h.

Yak-36 Yak OKB selected 62 to build first Soviet jet V/STOL aircraft, Prof Matveyev and A. N. Rafaelyants transferred from Turbolyet and new Yak team formed under S. G. Mordovin. Full NII funding for experimental aircraft and required engines. Decision not to fit so-called composite scheme of cruise plus lift jets, but to use lift/cruise engine(s) with vectored thrust in small single-seat airframe designed for high-subsonic performance.

All-metal stressed-skin, mid-mounted wing, powerplant in lower forward fuselage with nozzles below wing roots, cockpit at upper level, many features (tail, landing gear) reminiscent of Yak-25. Wing apparently SR-9S profile, 40° cropped delta with –5° or –6° anhedral and fixed LE, TE occupied by tabbed ailerons and area-increasing slotted flaps. Large plain nose inlet immediately bifurcated into left/right ducts to engines occupying lower part of forward fuselage, with all efflux discharged through lateral rotary cascade deflector identical in geometry with Rolls-type patented 61. Left/right nozzles driven in unison aft for thrust and down through at least 90°; not seen at Domodyedovo in forwards braking position but assumed capable. Reaction control jets in wingtip pods, at tail and on long pipe ahead of nose; no provision for eliminating uncontrollable roll-couple on single-engine failure. Pilot in ejection seat in pressurized cockpit with thin moulded windscreen and sideway-hinged canopy. Tail based on Yak-25/28 but smaller (tailplane span 3500 mm instead of 4500 mm) and tailplane mounted higher up fin. Tailplane fixed. Fuel tank behind cockpit, dorsal spine linking cockpit to fin. Landing gear of *velosipedno* type similar to Yak-50 but with nose unit and outriggers all retracting forwards and main unit to rear. Nose gear retracts to occupy front of inlet bifurcation with twin doors hinged to large left/right flap hinged down from rear to improve pressure/flow under fuselage and minimize reingestion of hot gas. Edges of these doors carry angle strakes. Further strakes on each side of centre fuselage, between which is second (smaller) rear-hinged door lowered in jet-lift mode. Double-hinged door ahead of twin-wheel main gear stressed to act as hot-gas deflector, bay/leg door and airbrake. Large left/right canted ventral strakes on rear fuselage. Outriggers fold forward into oval-section tip fairings extended far ahead of LE; large freedom in roll, outrigger often 15 cm off ground when parked.

Two years static testing to perfect engines, controls and autostab, followed by first flight 65 by Yu. A. Garnayev (killed in Mi-6 April 66), flying continued by Valentin G. Mukhin. Ten airframes, several for ground tests, #37 and 38 demonstrated Aviation Day 67, latter carrying 36-57 rocket launchers. One fitted with gun, another did deck trials aboard *Moskva*, but emphasis was on solving control problems, esp severe pitch stability problem in accelerating transition. Led 68 to different aircraft with separate lift and lift/cruise engines (Yak-36M). ASCC name 'Freehand'. See Addenda.
DIMENSIONS Span 8,25 m (27 ft 0 in); length (excl probe) 12,5 m (41 ft 0 in), (overall) 17,5 m (57 ft 5 in); wing area 16 m^2 (172 ft^2).
ENGINES Two R-11V.
WEIGHTS Empty 5300 kg (11,684 lb); max 8 t (17,640 lb).
PERFORMANCE Max speed 1120 km/h (696 mph); range (max) 1300 km (807 miles).

Yak-36M, Yak-38 Second jet-lift type, Yak-36M, designed under Mordovin's team by chief designer Arkadi Borisovich Zvyagintsev. Intended to serve with AV-MF in training and operational role, though mainly to learn techniques and problems. Structure conventional light alloy. Wing loosely based on Yak-36 but smaller and with hydraulic fold near mid-span. Profile SR-9S, LE 42°, anhedral 6°, two spars, powered ailerons with setback hinges and trim tabs on outer panels, tracked slotted flaps with central actuator inboard, fixed LE. Oval-section fuselage, with slim nose, pressurized cockpit with armoured front windscreen, canopy hinged to right and K-36V seat, bay for two lift jets, mid-section occupied by large lift/cruise engine with fuel bladders in front, above and to rear, trimming anhedral tailplane, powered elevators, rudder with trim tab, fin faired into spine downstream of ECS cooling-air inlet. Main engine simple round-lipped inlets with inboard boundary-layer splitter plate, five auxiliary suck-in inlets immediately to rear; jetpipe bifurcated to discharge through twin 90° nozzles driven through 95° range by dual hydraulic motors and chains. Two lift engines in tandem fed by inlet (roof) panel with 24 spring-loaded doors hinged up at rear and discharging through left/right hinged exit doors. Lift engines installed at 87° with nozzles fitted with deflector rings to vector thrust over 30° range for accelerating and decelerating transitions; nozzles undeflected in hover, with main nozzles at 95°, three jets meeting at 50 m. Jet reaction controls bled from main engine comprise up/down roll valves in wingtips and left/right yaw valves near tail; pitch governed by differential throttling of engines. Single electronic control of all engine functions, linked to powered mechanically-signalled flight controls (aerodynamic and jet). System fed by air-data and inertial sensor and linked to seat to eject pilot automatically if combined height and descent rate exceed safe limit. Levered MLG hyd retrac fwd, nose gear aft, track 2760 (9 ft $0\frac{5}{8}$ in), wheelbase 6060 (19 ft $10\frac{5}{8}$ in), no drag chute.

Prototype (see Addenda) flown by Mukhin 28 May 70. Avionics developed in two stages. Phase I tailored to opeval and refinement of techniques using simple ranging radar, laser system for ship guidance, twin radar altimeters, ADF, doppler, twin-gyro platform, IFF and VHF/UHF. One 36M prototype visited *Moskva* in 72 during year-long service trials. From 74 trials aboard *Kiev*, followed by summer 76 Med cruise of test squadron.

Aircraft adopted by AV-MF as **Yak-38**. ASCC name 'Forger-A'. Production of 100 plus (believed 20) tandem dual **Yak-36U** (AV-MF **38U**) trainer, with no weapons or ranging radar, pupil in nose, extra plug in rear fuselage. ASCC 'Forger-B'. Production aircraft with

Yak-38

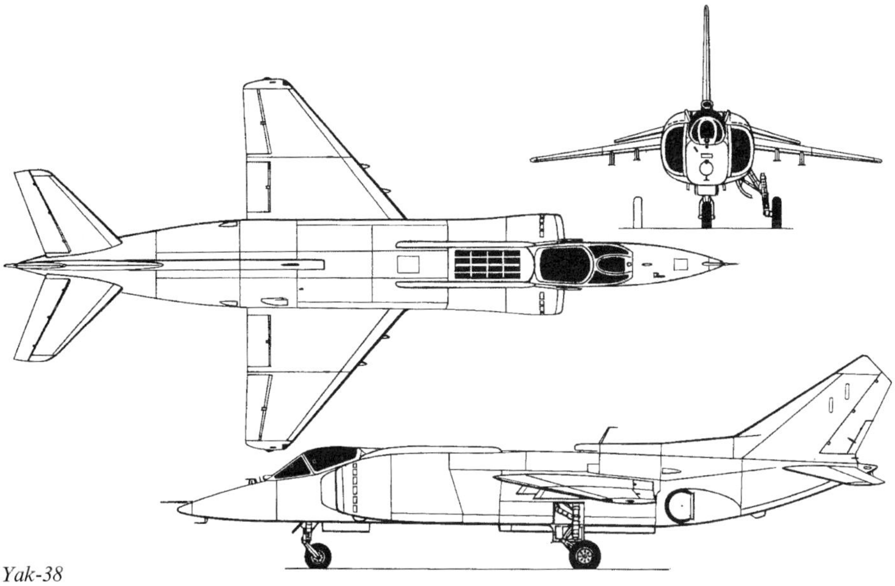

Yak-38

Yak-38U

revised inlets, strakes beside lift-engine inlets, two pylons added under inboard wing for various loads to 500 kg on each, no internal gun. Prolonged evaluation emphasized sensitivity to high ambient temp. Most obvious route to practical endurance/radius with useful load was to adopt STOL takeoff, which enabled disposable load (fuel or weapons) to be increased by 1000 kg. OKB regretted aircraft lacked Phase II avionics, but in any case installed thrust inadequate. LII pilot Oleg Kononyenko killed 79 (Indian Ocean, *Minsk*), and many critical situations during hot/high opeval (by FA) in Afghanistan 80. All fitted with K-36LV seats with auto system triggered by dangerous combination horiz/vert speed, height, airspeed; of 20 ejections one pilot killed through omission to arm auto system.

Yak-38M Initial step to alleviating problem. Increase of 1000 kg in available thrust, permitting significant increase in internal fuel, addition of underwing drop tanks and steerable nosewheel. Production ordered, and AV-MF withdrew all Yak-38 and also some 38M not equipped with underwing tanks. FR probe not adopted, and operation still limited on hot day. 'Forger-A'.

Yak-39 Variant combining larger wing of 38M with further increase in installed thrust to carry Phase II avionics multimode radar, related to N-O19, and laser seeker/guidance unit, giving capability in air combat and surface attack. MAP and AV-MF showed little interest, and aircraft not tested. Refurbished 38M aircraft available for sale.

DIMENSIONS Span 7,022 m (23 ft 0½ in), span folded 4,88 m (16 ft 0 in); length (excl PVD) 15,43 m (50 ft 7⅞ in), (38U) 17,65 m (57 ft 10½ in); wing area 18,69 m² (201 ft²).

ENGINES Two R-27V-300 plus two RD-36-35; (38M) R-27VM-300 plus two RD-36-35FVR.

WEIGHTS Empty (38) 7370 kg (16,240 lb); max (38) 11,7 t (25,794 lb).

PERFORMANCE Max speed (SL) 978 km/h (608 mph), (11 km) 1009 km/h (627 mph); max SL climb 4,5 km (14,750 ft)/min; service ceiling 12 km (39,370 ft); combat radius (38) max 200 km (124 miles), (38M) 390 km (242 miles).

Yak-40 Agreed autumn 64 as replacement for all major short-haul GVF (Li-2, Il-12 and Il-14 and to small extent An-2), Yak himself being instrumental in securing adoption of STOL jet, then a new concept. Required new engine from Ivchyenko, sized also for trainer and executive

Yak-38M

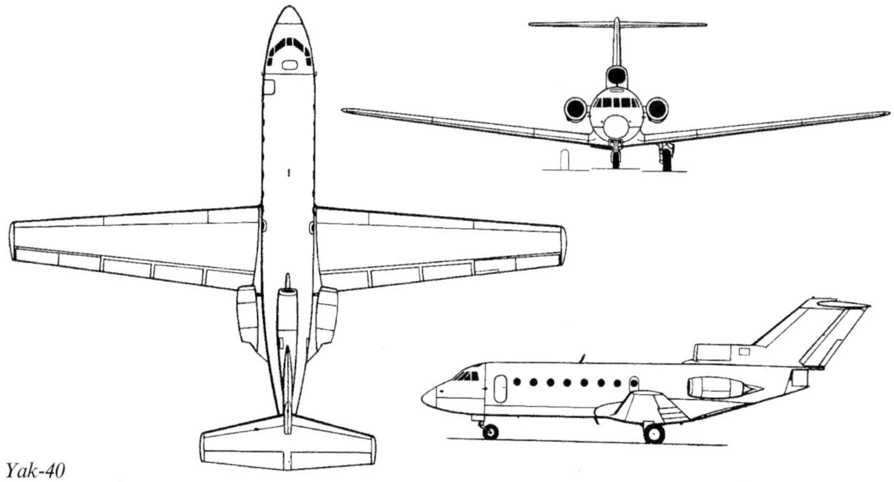

Yak-40

Yak-40K

aircraft and with two giving sufficient cruise thrust, third for take-off boost (plan dropped during flight test in favour of all three being used in cruise). Yak led team, with Adler chief engineer and son Sergei chief project engineer. Careful compromise between demand for economy and STOL performance (Class 5 grass fields 700 m max), all-round modernity for long service life and export success, need to meet international certification including 30,000 h airframe life, reliability and independence of ground support.

Three engines resulted in rear-engine T-tail layout with low wing unusual in aspect ratio 8·93 and 0° sweep. Section SR-9 series, 15/10%, dihedral 6° (later reduced, as noted) structurally in left/right halves joined on centreline with straight main spar tip-to-tip, auxiliary front/rear spars, chemically-milled skins, assembly by spot-weld/bonding/riveting with bolts and steel forgings at main joints (eg, centreline). Manual ailerons, tab on right only, split inner/outer halves for wing flexure. Plain flaps split into three each side for flexure, hydraulic drive to 15° take-off, 35° landing. Spoiler/dumpers omitted at late stage. Fuselage circular section 2,4 m diameter, again D1 structure assembled by spot- welding/bonding/riveting and with one-level floor of low-density foam sandwich. Fin sweep 50° at LE; fixed bullet at top upstream of hinge for electrohydraulic trimming tailplane, unswept, 13,03 m² with manual elevators, one-piece manual tabbed rudder. Levered-suspension nitrogen-oleo landing gear, main 1120 mm × 450 mm 393 kPa (circa 57 lb/sq in), nose 720 mm × 310 mm, steerable. Nose unit forwards, main inwards with no doors, hydraulic brakes and retraction. Bogies omitted at late stage, though initially offered as alternative standard. Engines 1 and 3 hung on rear frames and 2 inside tail of fuselage fed by S-duct with APU above it to provide air for engine start and limited cabin heat on ground. Bleed-air LE heating of wings and tail, electrically heated windscreen. Cabin dP 29 kPa (c4¼ lb/sq in). Hydraulic system 15,2 MPa (c2200 lb/sq in). Fuel 3800 lit (839·9 gal) in integral tanks forward of main spar. Flight deck for two-crew, autopilot, Grosa-40 weather radar, ILS for 500/50 m (later Cat 2). Three DC generators, Ni/Cd batteries. Cabin length 6,7 m normally seating 24 (2+1) aligned with eight circular windows each side, hydraulic rear ventral airstair and service door forward on left.

First of five prototypes, No 1966, flown 21 Oct 66. Rapid flight-test and NII clearance 68 for series at GAZ-292 Saratov. Dihedral reduced to 5° 30', No 2 engine inlet raked backwards, improved engine-bay fire protection, minor avionics changes. Passenger service from 30 Sept 68. Second series, extra fuel in outer wings (total 3900 lit, 858 gal), main gears strengthened, clamshell reverser on No 2 eng. Third series, minor changes including removal of fin/tailplane bullet. Max payload 2720 kg (5,996 lb), standard seating 27 (2+1) or 32 (2+2), but numerous other interiors and **Yak-40K** fitted wide upward-hinged cargo door on left.

Total production by July 76 1,011. Stretched **40M** not built, nor proposed military **40B** or export **40V**. Conversions included **40REO** with giant ventral canoe and observation blister(s), **Liros** with nose pylon and wingtips carrying choice of long probes and various cantilevers carrying dipoles, **Shtorm** with long nose probe and scabbed-on dispensers along forward fuselage, **Kalibrovshchik** with 'farm' of antennas including large blades above fin and under rear fuselage, **40-25** with nose replaced by that of MiG-25R, and testbed with nose mounting Czech M-602 turboprop. **Yak-40TL** announced 91 would be joint programme by Skorost (see Yak intro) and Textron Lycoming. Redesigned rear fuselage, added dorsal fin, original three engines replaced by two LF 507-1N each rated 7,000 lb, no reversers, accommodation/payload unchanged. MTO 16 575 kg (36,541 lb). In 93 still evaluating different engines. ASCC name 'Codling'.

DIMENSIONS Span 25,0 m (82 ft 0¼ in); length 20,36 m (66 ft 9½ in); wing area 70 m² (753·5 ft²).
ENGINES Three AI-25, (40B) AI-25T.
WEIGHTS Empty (early) 9 t (19,685 lb), (series) 9,4 t (20,725 lb); fuel 4 t (8,818 lb) max; loaded (early) 13,75 t (30,313 lb), (series) 16,2 t (35,714 lb).
PERFORMANCE Max speed 600 km/h (373 mph) at SL; max cruise 550 km/h (342 mph) at 7 km; initial climb 8 m/s (1,575 ft/min); service ceiling, 8 km (same as max cruise height); range (initial, at 550 km/h) 600 km (373 mph), (series, max payload, 470 km/h at 8 km) 1800 km (1,118 miles); take-off (initial) 550 m, (series) 700 m (2,297 ft); landing 500 m, 320 m (1,050 ft) with reverser/110 km/h.

Yak-42 OKB's longest development task, demonstrating Soviet Union's long timescales with modern civil aircraft. Yak accepted job about 72 as potentially largest-ever GVF programme, for standard regional transport; planned 2,000 to be bought, replacing not only Tu-134 and other jets but even An-2 in areas where traffic has risen and longer runways built. First HBPR (high bypass ratio) turbofan in country ordered from Lotarev for this aircraft, design generally scaled-up Yak-40 but with some sweep and reliance on high-lift

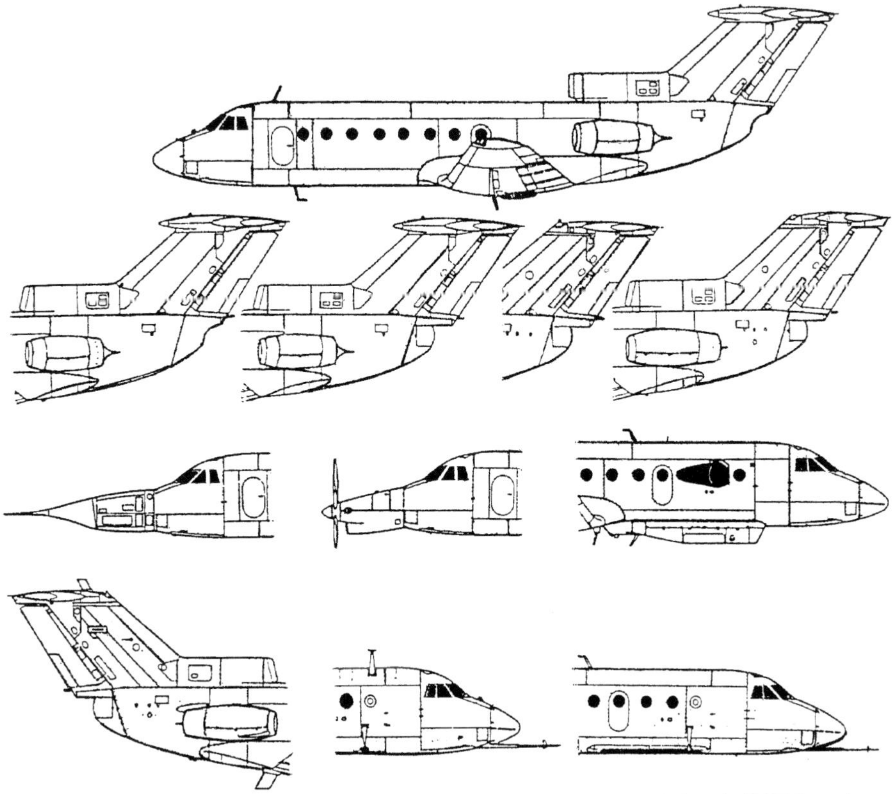

Yak-40 variants, from the top: prototype; 1,2 and 3 and 40K; 40-25, 40/M602 and 40RZO; and Kalibrovshchik, Liros *and* Shtorm

devices for field length of 1,8 km (compared with 0,7 km for predecessor). Model and mock-up revealed June 73.

Wing t/c about 8·5/6·5%, dihedral 0°; despite much tunnel testing unique decision to fly two sweep angles, first 11° and subsequently 25° (later changed to 23°). Structurally 13,05 m centre section with 90° TE, and left/right outer wings, all with chemically-milled skins on two-spar torsion box with secondary leading/trailing structure including honeycomb. Wing movables subject to prolonged change and research. Ailerons eventually fully powered, and after flight test changed in design and split into inner/outer halves in separate circuits. Hydraulic LE slats extensively tunnel tested and flown on first Yak-42 but omitted 76. Final standard of flap area-increasing tracked slotted type in two sections with hydraulic drive. Three flight/ground, roll/airbrake spoilers ahead of outer flaps, lift dumpers inboard. Circular fuselage, 3,8 m (149·6 in) diam, dP 48·9 kPa (7·16 lb/sq in), with engines hung exactly as Yak-40, APU above S-duct with exhaust on right. Rear fuselage and tail reshaped about 74 to become higher and less-swept. Acorn bullet survived in this aircraft, powered tailplane primary surface +4°/–8° with trim tabbed elevators, spring servo tabbed one-piece powered rudder. Landing gear with nitrogen oleos, main 1300 mm × 480 mm twin wheels retracting inward with doors over legs only; hydraulic brakes and retraction. Nose unit, twin wheels, steerable, 930 mm × 305 mm, hydraulic retraction forwards with auto-brake to arrest wheelspin (not used on landing). No reversers or parachute, complete reliance on lift dumpers and multi-disc main brakes. Fuel (1976-80) 15 795 lit (3,474·4 gal) in main wing box integral cells; in 81 increased. Bleed-air used for 10 wing/tail de-icing and environmental systems, electrically heated windscreens. Two-crew flight deck with third jump seat. SAU-42 flight control linked with autopilot and area-nav system, weather radar, ILS Cat 12.

First prototype, No 1974, flown 7 March 75. Wing swept 11°, heavy hydraulically-powered crew door forward on left serving as escape door at all IAS; main hydraulically-powered ventral airstair at rear. Apart from test gear, equipped as 100-seater (basically 3+3) with large carry-on baggage/coat compartments front and rear. Second aircraft, No 1975, with many changes including longer fuselage with 19 instead of 17 windows each side, 120 seats and no carry-on baggage (though two coat areas). Two shallow underfloor holds, forward for six 2,2 m³ bins and rear for three bins, total vol 26,2 m³ (924·5 ft³), all loaded through door ahead of wing on right and positioned by chain conveyor. Third aircraft, No 1976 (later in line-service as 42303) with de-iced tail, improved main-gear fairings and emergency exits (two each side) moved forward closer to LE. Max payload intended 14,5 t, reduced in production to 12,8 t; by 76 final standard 23° wing, increased weights (as data) and four-wheel bogie main gears by Hydromash with steerable twin-wheel nose gear with brakes only to stop rotation during retraction forwards, main gears fold inwards to flattened belly, all tyres 930 mm × 305 mm, all brakes and retraction hyd, nitrogen oleos, track 5,63 m, wheelbase 14,78 m. Fuel capacity increased to 23 175 lit (5,100 gal) but no reversers. Rear airstair retained, second airstair door 1,81 mm × 0,83 mm at front on left. Normal seating 120 (3+3); or 96 plus 8 First Class with carry-on baggage; chain-drive loading of bulk cargo or small (2·2 m³) containers under floor. ASCC name 'Clobber'.

Yak-42AM Minor changes to give standard 144-seat interior.

Yak-42A Convertible interior with left front doors 2025 mm × 3230 mm wide. Production Smolyensk from 81, fewer than 100 by 93 inc 6 China and 2 Cubana.

Yak-42D Small changes in fuel capacity (range with 120 pax 2200 km, 1,365 miles) and furnishing, replaced 42 in production 88, intention to deliver 100 by 97. Demonstrator with Bendix-King avionics.

Yak-42E-LL Experimental testbed (No 42525) with right engine replaced by D-236 driving 4,2 m Stupino SV-36 tractor contraprop (front 8 blades, rear 6), first flight 15 March 91.

Yak-42F Research/survey aircraft (No 42644) with Aeroflot titles used for Earth-resource and

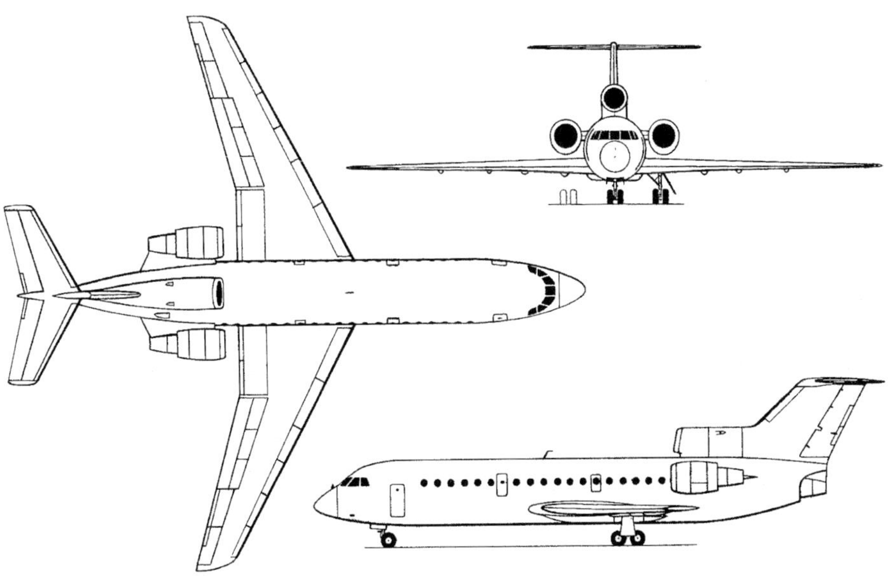

Yak-42 series

Yak-42E-LL (D-236 propfan)

Yak-42F

geophysical survey with EO sensors in large cylinders with rotating front section hung on pylons ahead of inner wing.

Yak-42M Described below.

Yak-142 Designation of latest series version announced Sept 93. Numerous minor improvements, plus almost wholly new avionics, flight-deployable spoilers, wider range of flap settings, TA-12 APU. Data generally as below except MTO 57 t (125,661 lb) giving ranges 1380 km (max payload) and 2000 km (120 pax).

DIMENSIONS Span (1st) 35,0 m (114 ft 10 in), (1st 23°) 34,2 m, (series) 34,88 m (114 ft 5¼ in); length (1st) 35,0 m, (series) 36,38 m (119 ft 4¼ in); wing area (all) 150 m² (1,615 ft²).

ENGINES Three D-36 (in 1st, derated to 6320 kg).

WEIGHTS Empty (2nd) 28,96 t, (series) 34 515 kg (76,092 lb); fuel (1st) 12,65 t, (series) 18,5 t, (42D) 19,2 t; max payload (2nd) 14,5 t, (series) 12,8 t (28,218 lb); MTO (1st) 50 t, (2nd) 52 t, (initial series) 53,5 t, (standard) 56,5 t (124,560 lb).

PERFORMANCE Max cruise 810 km/h (503 mph), (normal) 740 km/h (460 mph); max cruise ht 9,6 km (31,500 ft); range (series) 1300 km (808 miles) with max payload, 1900 km (1,180 miles) with 120 pax; landing from 15 m (50 ft) at 210 km/h, 1100 m; balanced field length 2200.

Yak-42M New aircraft retaining little except 3,8 m diameter of fuselage and (modified) landing gear. Wing supercritical profile with more span, less area, winglets, double-slotted flaps, greater fuel capacity and new materials. Stretched fuselage for two pilots (optional engineer) and 156 passengers 3+3, other configurations max 168, doors as 42. Redesigned tail with tailplane pivoted to fuselage. New engines, arranged as DC-10, with reversers. FBW flight controls, CPNK-42M digital avionics, five-screen 'glass cockpit', Cat IIIA. Fuel burn (est) 21 g/pax-km. Design begun 1987 with expectation of prodn for Aeroflot; prototype to be completed as Yak-242.

DIMENSIONS Span 36,25 m (118 ft 11⅛ in); length 38,0 m (124 ft 8 in); wing area 120 m² (1,292 ft²).

ENGINES Three D-436M.

WEIGHTS Empty 37,4 t (82,451 lb); max payload 16,5 t (36,376 lb); MTO 63,6 t (140,212 lb).

PERFORMANCE (est) Max cruise 830 km/h (516 mph); econ cruise 800 km/h (497 mph); range (max payload) 1950 km (1,212 miles), (156 pax) 4500 km (2,796 miles); balanced runway 2,2 km (7,218 ft).

Yak-44 Twin-turboprop for service aboard CIS navy carriers. Basic aircraft is **44E** similar in configuration and role to Grumman E-2C selected mid-90 over An-71DRLD as 'radio-electronic support aircraft . . . for observation, target indication and homing'. Abandoned at mock-up stage early 92 for financial reasons, but OKB resumed low-budget work 93 in absence of external funding. Lighter and more compact than E-2C, aiming at greater patrol ht at expense of flight endurance (air-refuelling by probe intended). Features include antennas in rotodome of 7,3 m (23 ft 11 in) diam retrac to reduce height from 7,0 m to 5,7 m (same as twin-fin tail); high-lift outer wings with full-span slats, double-slotted flaps, spoilers, two-section ailerons and winglets on skewed hinges folding up ahead of rotodome (R over L, see drawing); radar radiator above forward fuselage; four sections elevator and eight of double-hinged rudders; contra-rotating props; and twin-wheel landing gears of track 7929 and wheelbase 9373. Project exists for derived transport version. Data are intended values.

DIMENSIONS Span 25,7 m (84 ft 3¾ in); length 20,39 m (66 ft 10¾ in).

ENGINES Two turboprops in 3500 kW (4,700 shp) class.

WEIGHT Max in 30 t (66,140-lb) class.

Yak-46-1 Proposed variant with high-bypass turbofans hung under wing. This requires wing to be moved forward. Shorter fuselage with conventional tail similar to 42M. No funding 92.

DIMENSIONS As 42M except length 38,8 m (127 ft 3½ in).

ENGINES Probably two NK-92.

WEIGHTS (est) Empty 34 840 kg (76,808 lb); max payload 17,5 t; MTO 60,2 t (132,716 lb).

PERFORMANCE As TVDD except range 3450 km (2,143 miles), (max payload) 2200 km.

Yak-46-2 Project for Yak-42M with further stretched fuselage and new engines driving pusher 3,8-m propfans. Accommodation 3+3 for 168 one-class or 114+12, rear airstairs deleted. Fuel burn (est) 14 g/pax-km.

DIMENSIONS As 42M except length 41,0 m (134 ft 6 in).

ENGINES Two D-27M.

WEIGHTS Empty 37,3 t (82,230 lb); max payload 17,5 t; MTO 61,3 t (135,141 lb).

PERFORMANCE Normal cruise 830 km/h (515 mph) at 11,1 km; range (168 pax) 3500 km (2,175 miles), (max payload) 1800 km (1,118 miles); balanced field 2,1 km (6,890 ft).

Yak-48 Project begun late 89 as totally fresh design for regional 30-pax jet or large bizjet; replaced by Yak-77 and Yak-Astra Galaxy.

Yak-50, Yak-52 Continuing Yak-18 family development, led by Sergei Yakovlev and Yu. Yankyevich, resulted in new aircraft for 76 world aerobatic championships at Kiev which

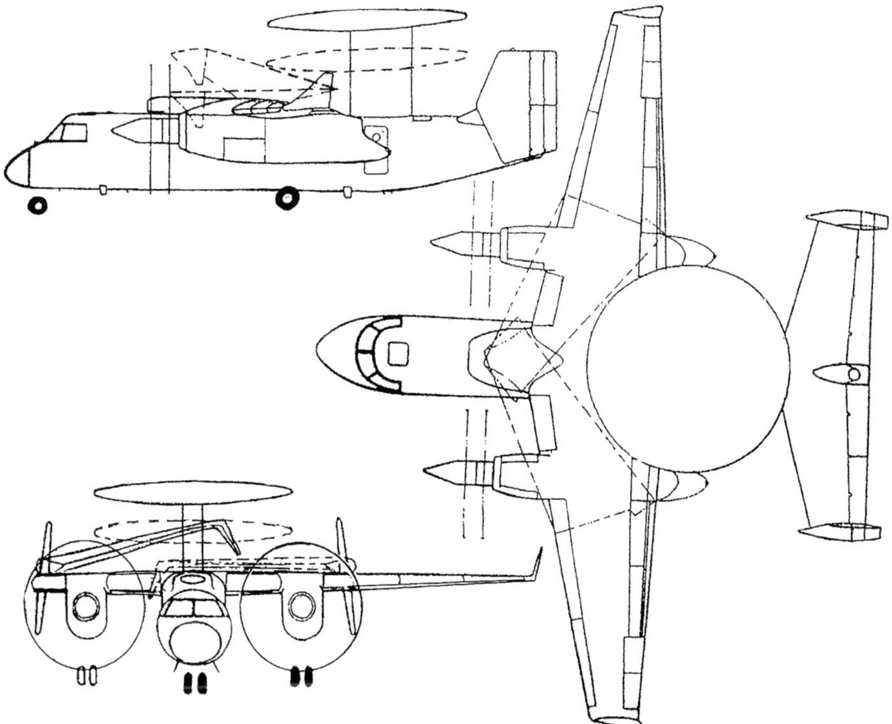

Yak-44E

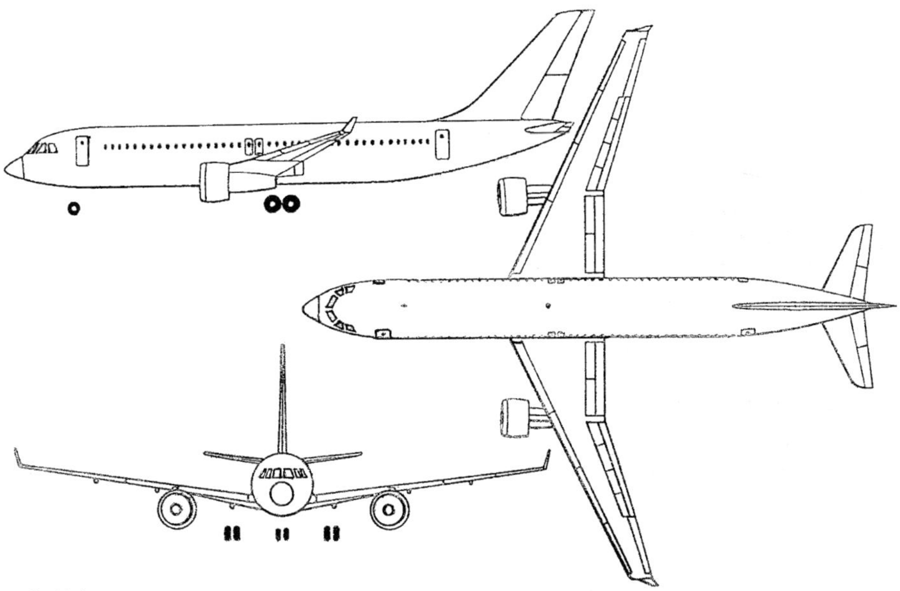

Yak-46-1

Yak-50

repeated Yak-50 designation previously used for fighter. Based on tailwheel, aft-cockpit Yak-18PS but new airframe. Left/right wings (based on Yak-20 of 26 years earlier) simple straight-tapered panels attached to fuselage. Clark-YH, 14·5/9%, dihedral 2°, D1 stressed-skin structure with main spar at 37·5% and auxiliary spars inboard at LE and TE to carry movable surfaces, five L-section stringers above and five below, 15 pressed ribs each wing, riveted skin. Fabric-covered slotted ailerons each mounted on two forged brackets from below, push-rod actuation, ground-adjustable tabs, D1 split flaps inboard operated by links from push/pull rods to pneumatic rams in fuselage. All-D1 stressed-skin semi-monocoque fuselage (no steel-tube truss) with fabric covering aft of cockpit. D1 stressed skin fin and tailplane, fabric-covered rudder with ground-adjustable tab and elevators with pilot-controlled tabs. Engine installation generally as Yak-18T but prototype fitted with large spinner on V-530TA-D35 prop of 2,4 m diameter (not 2,0 as often reported), fuel in single 55 lit tank in forward fuselage, and oil cooler in short duct under nose behind cowling. Prototype also fitted with teardrop canopy and fixed, spatted main gears with 500 mm × 150 mm tyres; fixed 200 mm × 80 mm tailwheel. Standard vhf radio with whip aerial ahead of fin, 27-V battery behind seat. Prototype completed at Progress Works, Arsen'yev, and tested by Anatoli Sergeyev mid-75, cleared to unrestricted +9/–6 g manoeuvres. Aircraft 02 built with Yak-18PS type main gear folding to rear with wheels part-exposed Yak-18PS canopy with fixed rear portion, and spinner omitted. This became series type, six of which swept board in 76 world championships. Substantial number built for Dosaaf and used throughout Warsaw Pact. Three-bladed Hoffman prop fitted 82.

Yak-52 is primary trainer variant. Based on stressed-skin airframe of Yak-50 but important differences. Tandem cockpits (rear seat in unchanged position) with usual pair of sliding canopies, front (pupil) with some additional instruments but instructor with all light and other electric switches. Tricycle landing gear with single oleo legs, main tyres 500 mm × 150 mm, nose 400 mm × 150 mm, all retracted by pneumatic rams but entire gear remaining exposed outside aircraft and designed for inadvertent gear-up landing. Fuel in two 65 lit tanks in wing roots ahead of spar feeding small (5,5 lit) inverted-flight fuselage tank. Oil cooler in broad duct under root of right wing behind spar with controllable exit flap. Electric battery in left wing. Designed by Komsomol members under OKB guidance at Arsen'yev and prototype in OKB livery built in 12 months 1978-79. Demonstrated Tushino 8 May 79 and completed NII testing same year. Production assigned to IRAvB, Bacau, Romania, (now called Aerostar), 1,800 by late 94. No ASCC name.

DIMENSIONS Span 9,5 m (31 ft 2 in); length 7676 mm (25 ft 2¼ in); wing area 15,0 m^2 (161·45 ft^2).

ENGINE One M-14P.

WEIGHTS Empty (50) 780 kg (1,720 lb), (52) 1 t (2,205 lb); fuel/oil (50) 40 kg+14 kg, (52) 180 kg+25 kg; loaded (50) 915 kg (2,017 lb), (52) 1290 kg (2,844 lb).

PERFORMANCE Max speed (50) 320 km/h (199 mph), (52) 285 km/h (177 mph); limit in dive

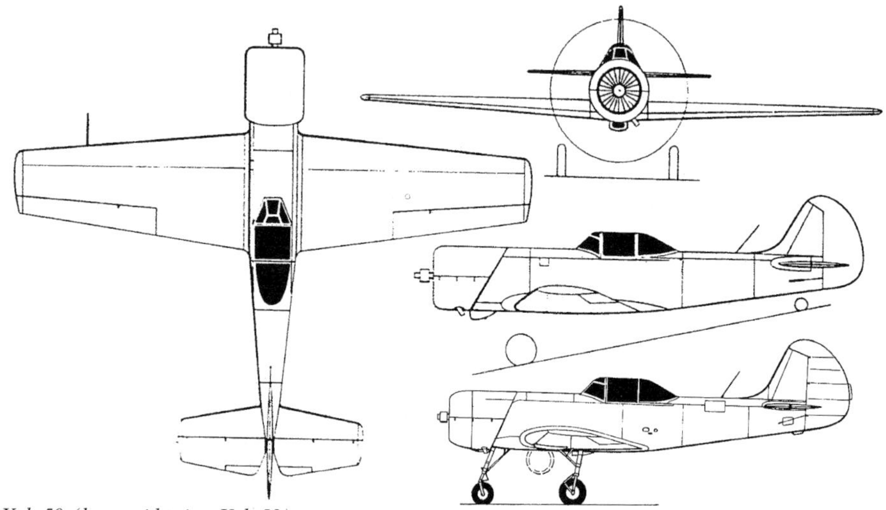

Yak-50 (lower side view Yak-53)

Yak-52

Yak-55

470 km/h (292 mph); initial climb (50) 15 m/s (2,950 ft/min), (52) 10 m/s; service ceiling (50) 5,5 km, (52) 6,0 km; range (50) not published, but 495 km with 120 lit auxiliary tank, (52) 500 km; take-off (50) 200 m, (52) 160 m; landing (50) 250 m, 110 km/h, (52) 260 m, 110 km/h.

Yak-53 Yak-50 with Yak-52 landing gear. Simplified flight controls without spring bias, all avionics removed except VHF com.

DIMENSIONS As Yak-52.

WEIGHTS Empty 900 kg (1,984 lb), loaded 1060 kg (2,337 lb).

PERFORMANCE As Yak-52 except max speed 300 km/h (186 mph), max climb 900 m (2,953 ft/min), take-off 150 m (493 ft), stalling speed 115 km/h (71·5 mph).

Yak-54 Tandem-seat aerobatic trainer replacing Yak-56, based on 55M, similar all-metal but even higher factor +6/-7, fuselage shorter. In prodn 93.

DIMENSIONS Span 8,16 m (26 ft 9¼ in); length 6,91 m (22 ft 8 in); wing area 12,89 m² (139 ft²).

ENGINE One M-14NTK.

WEIGHTS Loaded (competition) 850 kg (1,874 lb), (training) 990 kg (2,183 lb).

PERFORMANCE Max speed 450 km/h (280 mph); stall 110 km/h (68 mph); max climb 15 m/s (2,953 ft/min); roll 6 rad (510°)/s; ferry range 700 km (435 miles).

Yak-55 Made surprise debut at 82 World Aerobatic Championships. Marked departure from preceding Yak aerobatic series, only common feature being M-14P radial. Construction all-metal, with mid-mounted cantilever wing of thick section and low aspect ratio; incidence and dihedral nil. Control surfaces horn-balanced, tailwheel steerable, spring-leaf landing gear, and bubble canopy. Stressed to ±9 g.

DIMENSIONS Span 9,81 m; length 7,48 m (24 ft 6½ in); wing area 16 m².

ENGINE One M-14P.

WEIGHTS Max take-off 840 kg (1,852 lb); Empty 705 kg.

PERFORMANCE Max speed 290 km/h (199 mph), initial climb 960 m/s (3150 ft/min), take-off 120 m (492 ft), landing 250 m (656 ft).

Yak-55M Further refined version with smaller and thinner wings of symmetric profile, normal tips, almost full-span ailerons with large suspended balance tabs. Soviet team 89, production 91.

DIMENSIONS Span 8,1 m (26 ft 6¾ in); length 7,5 m (24 ft 7¼ in); wing area 12,8 m² (137·8 ft²).

ENGINE One M-14P.

WEIGHTS Empty 640 kg (1,411 lb); max 855 kg (1,885 lb).

PERFORMANCE Max speed 450 km/h (280 mph); SL climb 930 m (3,050 ft)/min; roll 345°/sec.

Yak-56 Projected aerobatic trainer, see Yak-54.

Yak-58 Completely new multirole aircraft; untapered high-lift wing with downcurved tips, manual ailerons, pneumatic flaps, central nacelle for pilot and 5 passengers (2+2) or 450 kg cargo, engine in annular duct driving pusher 3-blade v-p propeller, high tailplane carried on twin fins mounted on short booms, pneumatic tricycle landing gear for rough surfaces. Claimed choice of Western avionics. Model

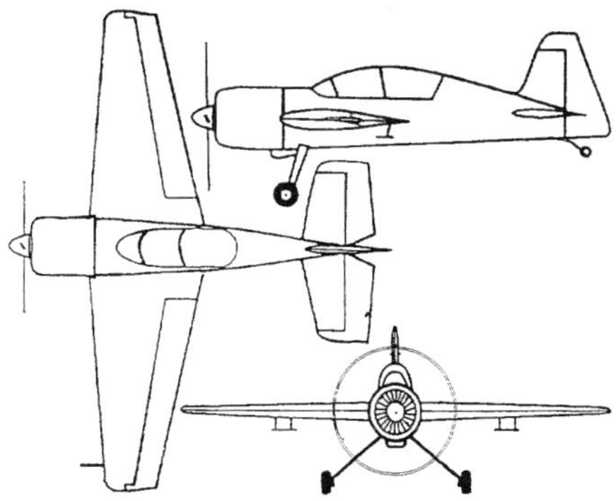

Yak-54

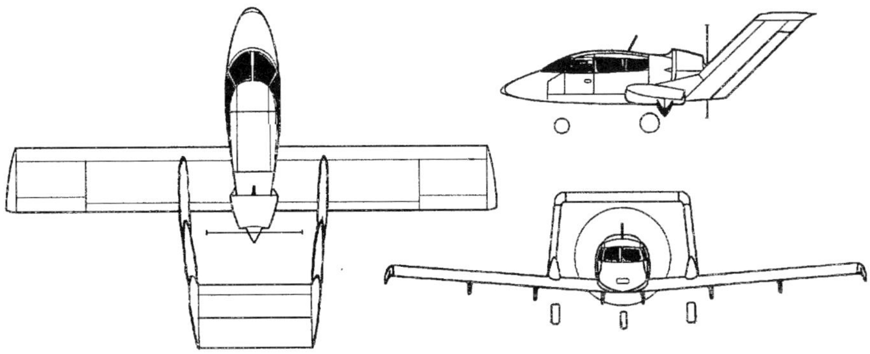

Yak-58

shown 90, enlarged 92 from 14 m² wing to 20 without change in seats or payload. In prodn Tbilisi 93 against claimed 250 letters of intent, first flight 26 Dec 93.
DIMENSIONS Scan 12,7 m (41 ft 8 in); length 8,55 m (28 ft 0½ in); wing area 20,0 m² (215 ft²).
ENGINE One M-14PT; intend to replace by M-16 later.
WEIGHTS Empty 1270 kg (2,800 lb); max 2100 kg (4,630 lb).
PERFORMANCE Max speed 300 km/h; max cruise 285 km/h (177 mph); econ 190 km/h (118 mph); range (max payload, 45-min reserve) 1000+km (620+ miles); TO/ldg 125 km/h, 610 m.

Yak-77 Long-range bizjet/regional transport. CAHI supercritical wing, cabin for up to 32 passengers (2+2), Collins ProLine 4 avionics, to fly 1996.
DIMENSIONS Span 21,55 m (70 ft 8½ in); length 20,45 m (67 ft 1¼ in).
ENGINES Two Allison AE 3007 turbofans of 7,200 lb thrust; alternative (unstated) engines intended for Russian Federation.
WEIGHTS Empty not stated; max payload 3,5 t (7,716 lb); max 25,2 t (55,555 lb).
PERFORMANCE Max cruise M 0·8; econ cruise M 0·75 (797 km/h, 496 mph) at 12·2 km/40,000 ft; range (8 seat) 10 000 km (6,214 miles), (32) 6000 km (3,725); field length 2,2 km (7,220 ft).

Yak-112 Won 88 competition for two-seat trainer/glider tug, developed into multirole light four-seater, designer Vladimir Mitkin. Extensive composites, large transparent area, downturned wingtips, MIKBO-3 integrated avionics for air data/nav/display. First flight with US IO-360-ES engine 20 Oct 92; production versions from Irkutsk for CIS to have M-17 or DN-200.
DIMENSIONS Span 10,25 m (33 ft 7½ in); length 6,96 m (22 ft 10 in); wing area 16,96 m² (183 ft²).
ENGINE See text.
WEIGHTS Empty c775 kg (1,709 lb); max payload 270 kg (595 lb); MTO 1,290 kg (2,844 lb).
PERFORMANCE Max cruise 250 km/h (155 mph); econ 190 km/h (118 mph); range with max payload (45-min reserve) 850 km (528 miles), max fuel 1200 km (746 miles); TO/landing 125 km/h, 500 m.

Yak-141 Supersonic V/STOL interceptor, secondary surface attack. Begun under Mordovin and Zvyagintsev in 75, later under Konstantin Popovich. Yak-36M layout retained but for supersonic flight single afterburning main engine with 'straight through' flow. Problem of placing deflected nozzle only slightly behind CG solved by mounting tail on deep booms. Wing similar to 36M but larger and thinner; sharper 40° LE with full-span droop flaps and large curved root extension, plain flaps and ailerons divided by power fold. Fuselage merged into almost square inlet ducts from sharp-edged fixed wedge intakes with four suck-in aux doors above and four on outer side,

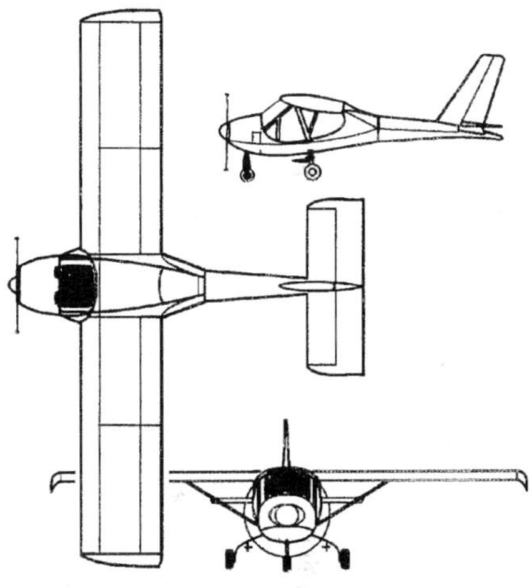

Yak-112

Yak-112

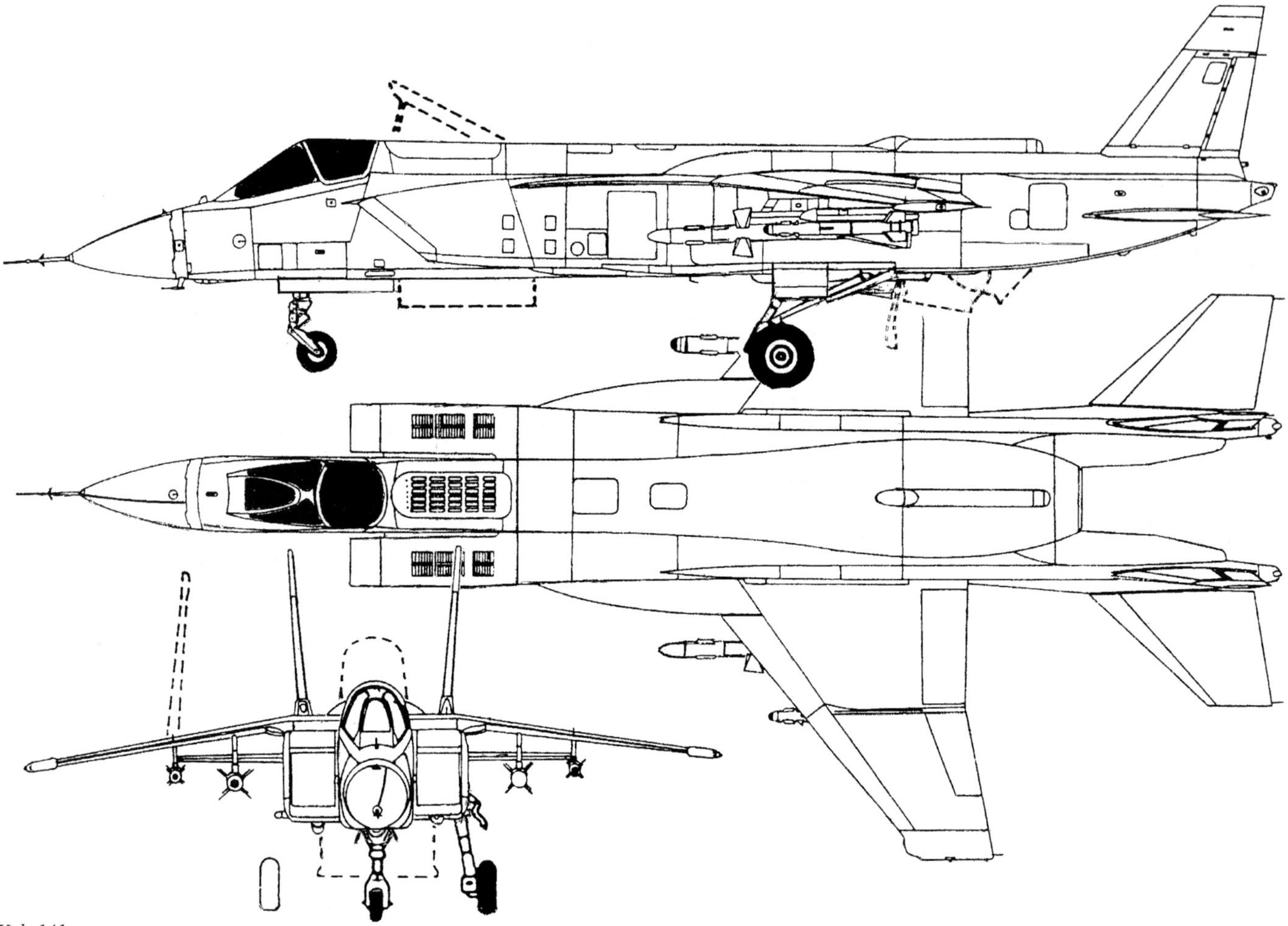
Yak-141

inner wall 150 mm from fuselage. Fin, LE 40°, carried on each boom, canted out 6°, with inset rudder. Slab taileron, LE 50°.

Main engine with duplicated hyd motor and chain drive to three wedge pipe sections which rotate afterburner nozzle to 0° (forward flight), 63° (STOL) or 95° (VL and hover). Triplex electronic control for main-engine fuel flow and bleed air to start lift engines and control aircraft in hover via twin tandem wingtip roll nozzles and circular yaw valve in tip of each tail boom. Two lift engines in tandem under upward-hinged louvred inlet door and twin outlet doors as on 36M, inclined at 85° with powered nozzle deflectors to vector from 2° forward to 24° aft. Main afterburner on all takeoffs. STO with lift engines at 24° and main at 63° under auto control. VTO possible from steel deck. In all jet-lift modes lower skin of fuselage under main engine hinges as enormous airbrake door; recirculation reduced by left/right flaps hinged down to form dam across mid-fuselage at upstream end of wide (350 mm) horizontal strakes along square lower chine, rear strake section formed by main-gear door. Protected tanks ahead of, above and behind main engine total 5500 lit (1,210 gal); 2000-lit non-conformal drop tank on centreline. FR probe above nose not installed on prototypes. All landing gears with single leg carrying single wheel on levered suspension, main units with anti-skid brakes, tyres 880 mm × 230 mm, retracting forwards beside duct, nose unit steerable, tyre 500 mm × 150 mm, retracting to rear. Track 3 m, wheelbase 6945 mm. Cruciform 17 m² drag chute deployed from tube above main nozzle. Dial-instrument cockpit with bird-proof windscreen and canopy hinged to right, K-36V seat armed whenever main nozzle exceeds 30° and fired if safe parameters exceeded under 300 km/h.

Airframe designed to 7 g, mainly Al-Li alloy, but 26% composites (carbon and glass) including movable surfaces, fins and LE root extensions. Underside of rear fuselage coated heat-resistant resin putty, inner walls of booms titanium, MLG brake pipes lagged and, to avoid tyre damage, TO/landings are rolling. Avionics based on Yak-36M Phase II including same radar, laser/TV targeting, HUD, HDD, HMS, radar alt, IFF and EW system. Triplex digital FBW flight/engine controls linked through main and flight-control computer to INS/ADF, air-data and aerodynamic limiters for 'round the clock manual, director and auto flight control from TO to landing in any weather worldwide'. Fins extended as spines above sides of fuselage available for chaff/flare dispensers, right spine extended forward to ECS ram-air inlet. Single GSh-30/I under left side with 120 rounds, four wing pylons rated 500 kg each for R-27/R-73/Kh-25/Kh-31/KMGU-2/UPK-23-250. Total external load 2600 kg (5,732 lb), leaving 900 kg for weapons when tank carried.

Funding covered two flight articles (#48, 77), static and fatigue airframes. First flight by #48 by Andrei Sinitsin March 89. #77 completed to higher standard with EW suite, gun, weapons and later low-speed controls including tail yaw nozzles replaced by single left/right nose reaction-control valve for greater authority. Over 150 h flown in first year, but CIS Navy terminated funding Sept 91. On 5 Oct 91 #77 severely damaged by impact /fire on anchored *Adm Gorshkov*, pilot safely ejecting. This aircraft rebuilt by OKB with mods, but #48 removed from storage, renumbered 141 and flown at Farnborough 92 by CTP Vladimir A. Yakimov. OKB regards devt as 85% complete, including partial completion of **Yak-141U** trainer. Partner sought to share (est) $400 m needed to complete to **Yak-141M** production standard with MTO 21,5 t (47,399 lb), giving weapon load 4,2 t (9,259 lb) and radius with 2 t weapons 900 km. This would require two more static and three more flight articles. OKB working on new STOVL for 21st Cent. ASCC 'Freestyle'. Data for prototypes:

DIMENSIONS Span 10,105 m (33 ft 1½ in), (folded) 5,9 m (19 ft 4¼ in); length 18,36 m (60 ft 2¼ in); height 5,5 m (18 ft 0½ in); wing area 31,7 m² (341·6 ft²).

Yak-141

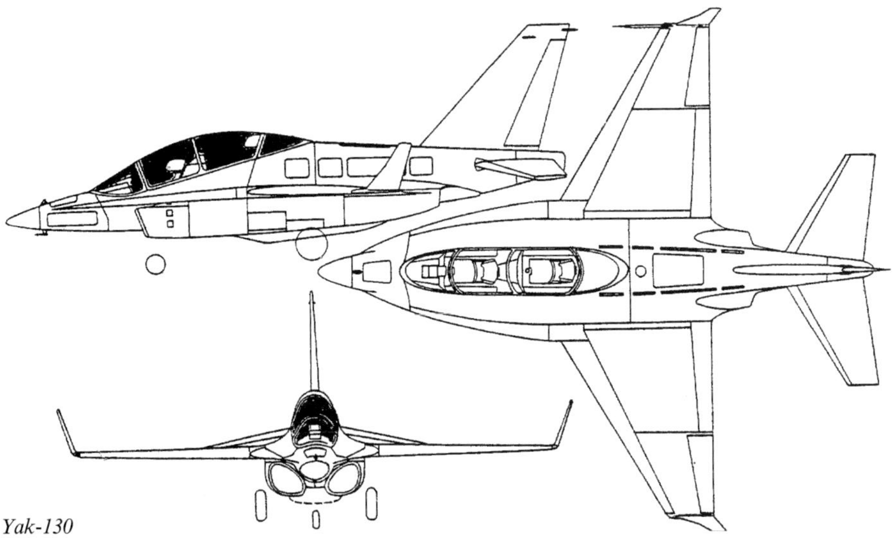
Yak-130

Yak-130 model

ENGINES One R-79V-300, two RD-41.
WEIGHTS Empty 11 650 kg (25,683 lb); fuel 4,4 t (internal), 1,75 ext; loaded (VTO) 15 800 kg (34,832 lb), (STO) 19 500 kg (42,989 lb).
PERFORMANCE Max speed (SL) 1250 km/h (777 mph, M 1·02), (11 km) 1850 km/h (1,150 mph, M 1·74); dive M limit 1·8; max climb 250 m/s (49,213 ft/min); service ceiling 15 km (49,200 ft+); ferry range 2100 km (1,305 miles); combat radius (STO, 2 t weapons, hi-lo-hi) 690 km (429 miles); normal TO 120 m; landing 240 m.

Yak-130 Originally Yak-UTS, contender for advanced trainer to replace L-39 in CISAF, part of integrated computer-based training complex. Programmed stability, design factor +8/–3; wing tapered on LE with large winglets, full-span slats for controlled AOA to 35° using vortices from sharp wing roots. Engines under wing roots, stepped tandem seats, dorsal airbrake, low slab tailplane, variable FBW flight control, rough-field gear track 2500, wheelbase 4000, 7 weapon hardpoints, digital avionics. Internal fuel 1800 lit (396 gal), 700-lit conformal ventral tank option. First flight cMarch 94, expected to enter service 95; export version with any desired (more costly) Western content marketed by Aermacchi. Intention to build one navalised prototype.
DIMENSIONS Span 10,64 m (34 ft 11 in); length 11,9 m (40 ft 8¼ in); gross wing area 44,0 m^2 (474 ft^2).
ENGINES Two DV-2S (R-35); export may be different.
WEIGHTS Empty 4150 kg (9,150 lb); fuel 1600 kg (2200 max); loaded 6 t (13,228 lb), (max) 8,5 t (18,740 lb).
PERFORMANCE (Est) Max speed (SL) 850 km/h (530 mph), (9 km) 975 km/h (606 mph); service ceiling 12+km (39,370+ ft); max range 2200 km (1,370 miles); TO 200 km/h, 380 m; ldg 195 km/h, 670 m.

Yak-242 Potentially giant programme to replace all CIS medium passenger aircraft (Tu-134/154, Yak-42). Basically twin derived from Yak-42M. Latest possible aerodynamics, structure and digital FBW avionics with 2-pilot cockpit with 5 multifunction displays. Configuration as 3-view. Normal seating 3+3, 162 pax at 780 mm pitch; other schemes 132-180, max payload 18 t (39,683 lb). Chief designer Vladimir G. Dmitriev 'We expect to fly by Jan 95 and enter service with Cat IIIA clearance early 97, with fuel burn 18,9 g/pax-km. IAE have asked us to build a 242 with V2500 engines. In 98 we hope to fly with high-b.p.r. engines; there is underwing clearance for 25 b.p.r., giving c14,8 g/pax-km. By 2000 we hope to get to 12 g with pusher propfans.' Biggest civil project of Aviation Ministry.
DIMENSIONS Span 36,25 m (118 ft 11 in); length 38,0 m (124 ft 8 in); wing area 120 m^2 (1,292 ft^2).
ENGINES Two PS-90A12.
WEIGHTS OWE 38,4 t (84,656 lb); MTO 64,6 t (142,416 lb).
PERFORMANCE Cruise 800 km/h-850 km/h (497 mph-528 mph) at 11,1-11,6 km (36,400-38,000 ft); range with CAR reserves (max payload) 1600 km (994 miles), (162 pax) 2700 km (1,678 miles), (max fuel) 5000 km (3,107 miles); field length (hot/hi) 2200 m.

Yak-58 and Yak-242 models

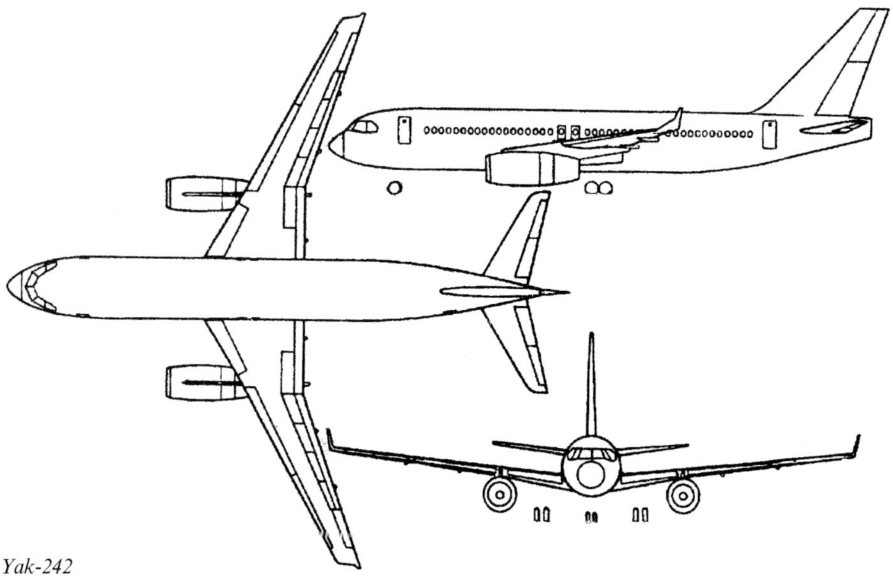

Yak-242

Yatsyenko

Vladimir Panfilovich Yatsyenko (1892-1970) was worker in pre-17 aero factory, qualified as engineer 24, on drawing board at CCB on foundation as member of Polikarpov's stressing team, rose to replace S. A. Kochyerigin in managing release of DI-6 through NII tests and entire production programme. Following closure of his KB Jan 41 assigned to Il-2 production; after evacuation worked for Ilyushin and Mikoyan, retiring on pension 66.

I-28 While still in charge of DI-6 production Yatsyenko completed project design of new fighter and submitted to NII VVS. Aug 38 design accepted as I-28, and before 1 Oct 38 personally assigned to build two prototypes of **I-287** (7th mod). Planned engine M-90 not available and M-87 adopted with only slight changes from Il-4 programme, driving slightly modified Il-4 prop, 3 m VISh-23Ye. Wing of inverted-gull form to reduce landing-leg length, two spars of wooden box construction, simple ply ribs, ply skin built up from glued *shpon* with thickness 12 mm at root and 4 mm at tip, polished to mirror finish. Balanced push-rod ailerons 40% semi-span D1/fabric. All-D1 Schrenk flaps with pneumatic actuation to 45°. Forward fuselage simple welded KhMA steel tube frame carrying engine, guns and tank; rear *shpon* semi-monocoque, all with mirror finish. Long (record says NACA-type) constant-diameter D1 engine cowl covering accessories, oil tank, part fuel tank, and gun barrels. Ventral oil cooler, unusual lateral exhausts (two pipes, 7 cylinders each) projecting through sides of cowl, large spinner. Wooden fixed tail integral with fuselage, D1/fabric control surfaces with inset hinges (unlike triple mass balances on ailerons) and tab on each elevator, push-rod actuation. Main gears with rear-braced oleo legs, 600 mm × 170 mm tyres, pneumatic brakes and pneumatic inwards retraction, doors on legs but not on wheel bay. Retractable tailwheel. Cockpit well aft of wings, with rear-sliding hood. Armament two ShVAK-20 with 300 rounds and two ShKAS with 1700 rounds.

Completed 30 Ap 39, factory test 1 June to 4 July. NII testing by Stefanovsky and Kub'shkin, excellent but former dived to 725 km/h and a/c broke up, pilot escaping. Steel casting in tail fractured, but this probably caused by impact of disintegrating cowling. Second aircraft with open cockpit, later engine and one ShVAK, two BS, four RS-82 or four FAB-100 (overload). Series production ordered, following R-10/PS-5 at GAZ-292, for 30 **I-28Sh** *shturmoviks* (2911 kg gross). Order rescinded Feb 40, Yatsyenko joined MiG team, later moving to assist Ilyushin, but five series a/c flown.

DIMENSIONS Span 9,6 m (31 ft 6 in); length 8,54 m (28 ft 0 in); wing area 16,5 m² (178 ft²).

ENGINE (1) one M-87A, (2) M-88.

WEIGHTS Empty (1) 2257 kg (4,976 lb), (2) -; fuel 275 kg; loaded (1) 2660 kg (5,864 lb), (2) 2730 kg (6,019 lb).

PERFORMANCE Max speed (1) 412 km/h at SL, 545 km/h (339 mph) at 6 km, (2) 439 km/h/566 km/h at 7 km (352 mph); climb to 5 km (1) 6·3

I-28

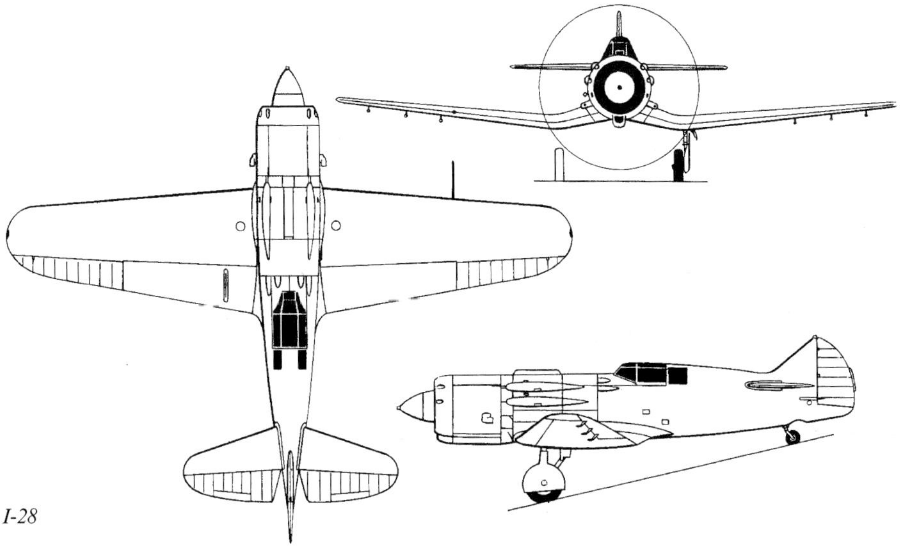

I-28

min, (2) 6·1 min; service ceiling (1) 10,4 km, (2) 10,8 km (35,430 ft); range (2) 450 km (280 miles); TO (2) 325 m, ldg 140 km/h.

Yermolayev

Vladimir Grigor'yevich Yermolayev was born 08 and graduated 30 from physics faculty of Moscow state university. Joined Bartini's design group following year as estimator, working on Stal-6 and Stal-7. A political activist, he saw his chance when Bartini was arrested Jan 38, and became prominent member of small KB formed under Zakhar Borisovich Tsentsiper to continue Stal-7 programme and design a bomber version. All Stal-7 record flights organized by this group, funded by Aeroflot deputy head M. F. Kartushyev, but VVS (and Stalin) interest in bomber meant priority for bomber version from Jan 39, before 5068 km flight. Large OKB immediately established within Aeroflot based on existing GAZ-240 of D. S. Maksimov, and Kartushyev appointed Yermolayev head and M. V. Orlov his deputy. Effective organization swiftly created outstanding long-range bomber, flown June 40 and ordered into production October. OKB at this time transferred to NKAP and aircraft redesignated Yer-2 for Yermolayev (number 2 because bomber, ie, even number). Potentially great programme hampered by lack of suitable engine and need to move entire programme to Siberia just as prodn in full swing. At height of effort 31 Dec 44 Yermolayev died of typhoid; OKB taken over Mar 45 by P. O. Sukhoi.

DB-240 Because of GAZ number allotted VVS designation DB-240 (*Dalnyi Bombardirovshchik*, long-range bomber), little remained but general shape of Stal-7. Wing redesigned as modern stressed-skin structure in D16, extensive rubber-presswork in simple ribs and details but heavy plate spars and extruded/machined spar booms and simple open top-hat or angle stringers, all suited to rapid mass-production. Centre section little changed, still electro-welded steel-tube space-frame, but because of increased aircraft weight KhMA replaced by 30KhNZA and changed in detail. Two main spars 2,5 m (98·4 in) apart at root left room for capacious bomb bay in redesigned semi-monocoque fuselage, which retained slightly triangular oval section of Stal-7 but longer and much stronger with heavy pressed frames, multiple stringers and skin 0,8 mm/1,0 mm. Flush riveting extensively used, a bold step forward. Accommodation for nav/bomb in fully glazed streamlined nose with hand-aimed UB, pilot in fully glazed blister cockpit on left of centreline, radio alongside on right with side windows (in combat to go aft and man lower rear ShKAS firing through sliding floor panel) and gunner in dorsal power turret with UBT. Stressed-skin twin-fin tail 5,8 m span, fabric movables with inset hinges, mass balance and trim tabs. Desired M-106 engine never materialised, replaced by less-powerful M-105. Neat cowlings, radiators outboard in wing with LE inlet and cascade outlet in upper surf, carb and oil cooler inlet below with latter exhausting on top of nacelle. AV/5L 3 m 3-blade c/s props. Twin-oleo main gears for greater weight than Stal-7, tyres 1000 mm × 350 mm. Retractable tailwheel. Two-section slotted tabbed ailerons, inbd/outbd split flaps, max 45°, operation of gear/flaps/bomb doors hyd, turret elec. All fuel between spars, normal inbd of engines 1900 lit each, max adding two outer-wing tanks total 5500 lit. Normal bombload 2 t (4 × FAB-500 or 1 × FAB-2000) or 1 t for max range/ overload internal/external 4 t (8,818 lb).

Prototype authorised Aug 39 and flown at Moscow Central by N. P. Shyebanov 14 May 40. NII testing begun 27 Sept 40, outstanding, order for series production as Yer-2 at Voronezh (probably GAZ-64) under A. B. Shyenkman. Tooling swift and aircraft available from spring 41, with around 50 delivered by 22 June 41. Second *dubler* prototype complete Sept 40 handled VVS TTT evaluation. Believed it was this aircraft (not commonly reported first) which at start of 41 flew round-trip Moscow/Omsk/Moscow loaded to represent dropping of 1 t bomb at Omsk.

DIMENSIONS Span 23,0 m (75 ft 5½ in); length 16,3 m (53 ft 5¾ in); wing area 72,1 m² (776 ft²).

ENGINES Two M-105.

WEIGHTS Empty 7076 kg (15,600 lb); fuel/oil 2726 kg (max 4650 kg); loaded 11,3 t, max 13 556 kg (29,885 lb).

PERFORMANCE Max speed 395 km/h at SL,

DB-240

Yer-2/ACh-30B

445 km/h (277 mph) at 4,2 km; climb 16·5 min to 5 km; service ceiling 7,7 km (25,260 ft); range (max fuel) 4100 km (2,548 miles); TO 580 m.

Yer-2 Series aircraft designated thus (in Cyrillic, Er-2) June 41. Deliveries from Oct 41, two regts hastily formed Smolyensk from GVF line crews and other civil personnel; 747th Polk operational 1 Oct 41, regts named *Novodranov* and *Gusyev*. Initial missions from Rzhev with 1 t loads against Berlin, Königsberg, but soon severe attrition from unsuitable close-support sorties. GAZ-64 evacuated Oct 41 at 128th a/c, relocating at Irkutsk.

DIMENSIONS, ENGINES As before except length 16,4 m (53 ft 9¾ in).

WEIGHTS Empty 7,5 t (16,535 lb); fuel/oil 2726 kg normal, 4,6 t max; loaded 12 570 kg (27,712 lb) normal, 14 150 kg (31,195 lb) max.

PERFORMANCE Max speed 387 km/h at SL, 437 km/h (272 mph) at 4 km; climb 9·0 min to 3 km, 16·5 min to 5 km; service ceiling 7,5 km (24,600 ft); range (max fuel) 4000 km (2,486 miles); TO 650 m.

Yer-2/AM-37 Deep penetration of Nazi armies accentuated need for greater range, and Dec 40 studies into more powerful engines (AM-35, 37 and 40, ASh-71 and 82, M-30, 40 and 120 and MB-100) led to DB-240 being fitted with AM-37s Nov 41. Outer wings redesigned with less LE taper. This was fastest of family.

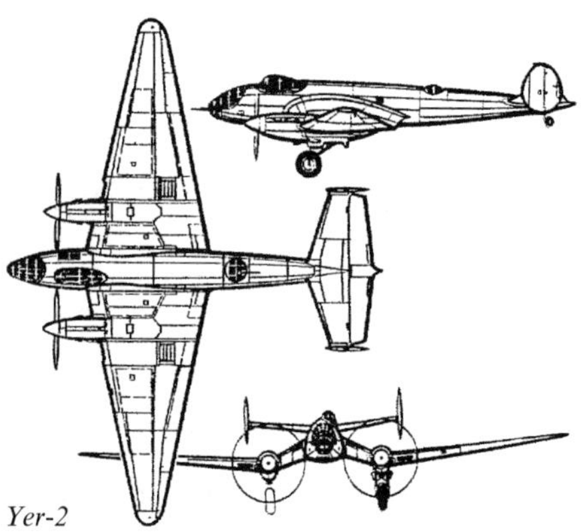

Yer-2

DIMENSIONS As DB-240 except wing area 73,1 m² (787 ft²).

ENGINES Two AM-37.

WEIGHTS Empty 7775 kg (17,141 lb); fuel/oil 2730 kg normal, 4650 kg max; loaded 13 t normal, 15 850 kg (34,943 lb) max.

PERFORMANCE Max speed 407 km/h at SL, 519 km/h (323 mph) at 6950 m; climb 10·2 min to 3 km, 17·3 min to 5 km; service ceiling 9150 km (30,000 ft); range 2930 km normal, 3500 km (2,175 miles) max; TO 760 m.

Yer-2/M-40F Early 42 OKB received orders to fit M-120, followed Dec 42 by order to fit MB-100. Single aircraft, believed one of the DB-240s, did fly 45 with single example of this massive 3200-hp engine, but only replacement for M-105 was Charomskii diesel. Fourth (pre-series) a/c had been retained as OKB testbed and this converted after evacuation with first flight-cleared diesels. This aircraft had completely redesigned outer wings with slightly reduced span but greater chord and revised tankage for oil fuel. Flown Dec. 41 On this a/c struc clearance obtained for max bombload (int/ext) of 5 t (11,023 lb).

DIMENSIONS Span 22,3 m (73 ft 2 in); length 16,3 m; wing area 75,3 m² (811 ft²).

ENGINES Two M-40F.

WEIGHTS Empty 10 t (22,050 lb); fuel/oil 4780 kg+390 kg; loaded 13,5 t normal, 15,5 t (34,171 lb) max.

PERFORMANCE Max speed c360 km/h at SL, 440 km/h (273 mph) at 6 km; climb no data; range 5300 km (3,293 miles) normal fuel; TO 780 m.

Yer-2/ACh-30B Same a/c re-engined 43 with prodn diesels in new installn which proved to be excellent, despite imperfect running of engine, with slim cowl, AV-5LV-116 prop and completely redesigned cooling system with main radiators transferred to centre section (usual LE slit inlet and flush cascade discharge in upper surface) and oil coolers in prominent circular ducts projecting ahead of LE outboard of cowl. Nacelle reprofiled to extend to TE providing enlarged bay for 1200 mm×600 mm tyres to support further-increased weights due to denser oil fuel, clearance for overload 5 t bombs and improved a/t with UB in nose and aft ventral and new elec-driven turret with UBT or ShVAK. Outer wings again redesigned with considerably greater area. Prolonged tests marred only by engine faults, NII tests completed Dec 43, and of first series a/c Feb 44. Irkutsk scientific institutes helped solve problems of making large parts in KhGSA and c312 built before termination late 44, forming core of newly created ADD force and winning major competitive evaluation against Il-6. Production aircraft had added AMTs protected tanks in outer wings, modified bomb bay able to take three FAB-1000 or three 980 kg torpedoes (or wide variety other stores), improved radiator discharge giving positive thrust in cruise, GS-1000 generators, TN-12 fuel pumps, new 15° flap setting for take-off, and usual defensive

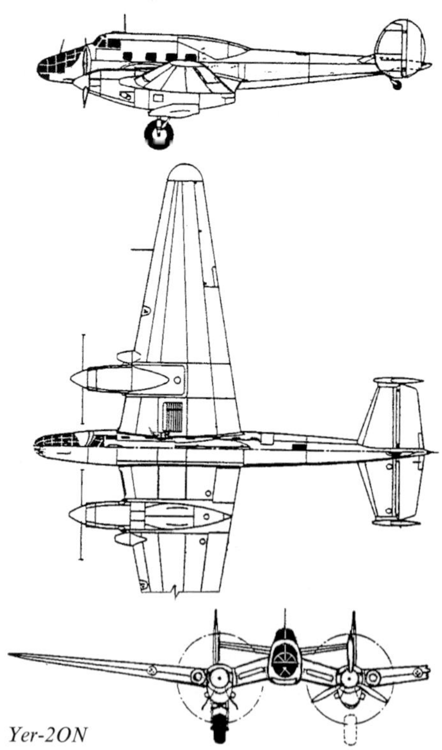

Yer-2ON

ARK-Z-1

armament one UBT in nose, one UBT in lower rear and one ShVAK in TUM-5 turret. Modifications introduced during production included addition of co-pilot (both tandem and side-by-side cockpits flown, latter giving aircraft symmetry and being accepted despite increased drag). Key role in Battle for Berlin played by 45 a/c of 16th Air Army.

DIMENSIONS Span 23,0 m (75 ft 5½ in); length 16,42 m (53 ft 10½ in); wing area 79,0 m^2 (850 ft^2).

WEIGHTS Empty 10 455 kg (23,049 lb); fuel 5460 kg; loaded 14 850 kg (32,738 lb) normal, 18 580 kg (40,961 lb) max.

PERFORMANCE Max speed 360 km/h at SL, 420 km/h (261 mph) at 6 km; climb 18·6 min to 3 km, 30·4 min to 5 km; service ceiling 7,2 km (23,620 ft); range (1 t bombload) 5500 km (3,418 miles); TO 810 m, ldg 140 km/h.

Yer-2 variants In man-hours, most impt was **Yer-2ON** (*Osobogo Naznachyeniya*) 'special assignment'; two airframes taken off production and in Jan 44 under Sukhoi completed as transports for Soviet leaders. Armament and mil eqpt eliminated, dural fuel tanks in bomb bay and fuselage struc rebuilt and furnished to carry up to 12 passengers in comfort on flights lasting up to 15 h. Data as bomber except cleared to 19 t (41,887 lb) max. A third aircraft, built 41, mod as OKB shuttle between Irkutsk/Moscow. **Yer-2N** (*Nocitel*, carrier or bearer of load) used late 45 to fly As 014 resonant-duct engines from V-1 missiles; five flights showing 'poor power, high drag'. Final variant **Yer-4** was series aircraft No 11 (originally M-105s) rebuilt 45 with ACh-30BF engines and redesigned coolant system with wing ducts on each side of nacelle (not just on one side) further developed to give better positive thrust in cruise.

Zlokazov

Aleksandr Ivanovich Zlokazov was senior airframe engineer at GVF repair plant at Irkutsk in 30s. About end 33 he formed small KB to create new transport for Arctic regions.

ARK-Z-1 Cast in Junkers mould, this low-wing monoplane was of substantial size for single engine, and embodied such advanced features as air/oil oleo legs and powered landing flaps. Much of airframe based on PS-4, and wing used corrugated skin of identical rolled profile. Enclosed and heated cockpit, main cabin for ten passengers or 880 kg cargo with double door on left, fuel tanks in inner wings, fixed-pitch wooden prop, large low-pressure tyres replaceable by skis. Completed spring 35, satisfactory test programme and then flown in stages to Moscow for NII GVF testing. Completed programme, including long-range flight, during Oct/Nov 35 but decision not to produce in series because of obsolescent structure.

DIMENSIONS Span 21,8 m (71 ft 6¼ in); length 15,0 m (49 ft 2½ in); wing area 70,2 m^2 (755·6 ft^2).

ENGINE One M-34R.

WEIGHTS Empty 3,2 t (7,055 lb); fuel/oil 900 kg total; loaded 5150 kg (11,354 lb).

PERFORMANCE Max speed 240 km/h (149 mph) at SL, 180 km/h (112 mph) at 2,5 km; time to 1/2/3 km, 4·3/10·5/20 min; service ceiling 3,8 km (12,500 ft); range, not recorded; take-off 280 m/15 s; landing 450 m/25 s/95 km/h.

Avietka

In this section are grouped light aircraft produced by designers not part of a recognised OKB. They are listed in chronological order rather than by alphabetical order of designer's name, aircraft name or designation. Some entries are classed as *motor planer* (powered gliders), but gliders and sailplanes are not included. A few entries were the result of nationwide competitions, organised for example by Dosaaf, but the majority were the result of individuals and small groups, and organised student bodies in aviation institutes, managing to obtain official authority or scrounging materials and building to their own design, usually without state backing or test facilities. Before the Second World War, M-11 engines were available on loan or at a reduced price to individual constructors, on the understanding that the state had free use of any successful design powered by it.

Shyukov Little is known of this triplane, begun in 19 but not completed. Clearly a single seater.
DIMENSIONS Span 7,8 m (25 ft 7 in); length 5,8 m (19 ft 0 in); wing area 16 m² (172 ft²).
ENGINE One 120 hp Le Rhône 9-cylinder rotary.
WEIGHTS Empty 360 kg (794 lb); fuel 70 kg (154 lb); loaded 557 kg (1,228 lb).
PERFORMANCE Speed at sea level 190 km/h (118 mph).

Vigulya Pyetra Antonovich Vigulya was behind a new design built at the Aviarabotnik (aviation-worker) repair factory in Moscow in 20. Based on the Farman 30 pusher, it had the original wings but tandem 80 hp Le Rhône engines, one at each end of the two-seat nacelle. It suffered rudder failure and crashed. No data.

ShM No details are available of this biplane built in 21, beyond the following data.
DIMENSIONS Span 11,5 m (37 ft 9 in); length 8,9 m (29 ft 2½ in); wing area 35 m² (376·7 ft²).
ENGINE One 100 hp Gnome Monosoupape 9-cylinder rotary.
WEIGHTS Empty 500 kg (1,100 lb); fuel/oil 98 kg/31 kg (216 lb/68 lb); loaded 800 kg (1,764 lb).
PERFORMANCE Max speed 120 km/h (75 mph) at sea level.

Kasyanenko No 6 Known as the Cavalry Aeroplane, described as a diminutive but neat design. Intended for army reconnaissance. Construction begun 21 at No 6 Aircraft Repair Factory in Kiev. Never completed; no data.

SP The *Sinyaya Ptitsa* (Bluebird) was almost too large to be included here, for it carried three passengers in a cabin behind the open cockpit. Designed 22 and built 23 at the Aviarabotnik factory; the work of Ignatii Aleksandrovich Valentei and Nikolai Efimovich Shvaryev, the latter being the designer, the basis being a captured Schneider reconnaissance aircraft. Fuselage completely rebuilt, aircraft doped blue overall, and flown in summer of 23 by I. G. Savin. Rear fuselage too large and aircraft unstable.
DIMENSIONS Span 13,0 m (42 ft 7¾ in); length 8,0 m (26 ft 3 in).
ENGINE One 220 hp Benz Bz III 6-cylinder inline.
WEIGHTS Empty 980 kg (2,160 lb); loaded 1490 kg (3,285 lb).
PERFORMANCE Max speed 180 km/h (112 mph).

VOP-1 First of two attractive sporting machines created by one of the most famed aviators, Victor Osipovich Pissarenko. Single-seat low-wing cantilever monoplane, with open cockpit and discontinuity in fuselage underside where deep cockpit joined slim tail boom. Structure wood, with birch veneer glued and pinned to form skin. Pissarenko flew it on 27 Nov 23; ferried to Moscow and flown by over 100 pilots.
DIMENSIONS Span 7,5 m (24 ft 7¼ in); length 5,0 m (16 ft 4½ in); wing area 10 m² (107·6 ft²).
ENGINE One 35 hp Anzani.
WEIGHTS Empty 222 kg (489 lb); fuel/oil 20 kg (44 lb); loaded 322 kg (710 lb).
PERFORMANCE Max speed 120 km/h (75 mph); practical ceiling 1,2 km (3,940 ft).

Savelyeva This tandem-seat quadruplane was a failure. Handling poor, and attempts to improve manoeuvrability made things worse, landing being especially difficult. Photographs show it on skis. Designer V. F. Savelyeva; flown 23, by which time generally considered obsolete.
DIMENSIONS Span 5,6 m (18 ft 4½ in); length 6,4 m (21 ft 0 in); wing area 20,7 m² (224·8 ft²).
ENGINE One 120 hp Le Rhône 9-cylinder rotary.
WEIGHTS Empty (wheels, not skis) 506 kg (1,116 lb); fuel/oil 106 kg/40 kg (234 lb/88 lb); loaded 802 kg (1,768 lb).
PERFORMANCE Max speed 164 km/h (102 mph); time to 1 km alt 4 min; service ceiling 3500 m (11,480 ft).

LM Despite its trivial power this ultralight monoplane was a complete success. Designed and built by V.A. Likoshin and N.G. Mikhelson (whose later work is described under his own OKB). Built at Red Airman plant in Leningrad and flown about Feb 24.
DIMENSIONS Span 8,4 m (27 ft 6¾ in); length 5,2 m (17 ft 0 in); wing area 12,7 m² (136·7 ft²).
ENGINE One 7½ hp Indian (motorcycle).
WEIGHTS Empty 89 kg (196 lb); fuel/oil 5 kg (11 lb); loaded 174 kg (386 lb).
PERFORMANCE Not published.

Buryevestnik This name (Stormy Petrel) identified ultralight monoplanes created in 20s by Vyacheslav Pavlovich Nyevdachin. He had been a glider pioneer at the meetings at Koktebel. His **P-5** (*Planer* No 5) was flown by Jungmeister to a national record of 532 m in Nov 23, 18 months after it appeared. In 24 Nyevdachin fitted a P-5 with a Harley-Davidson engine, the first of numerous motor-cycle engines that offered power at minimal cost. Tested on 3 Aug 24 the resulting Buryevestnik **S-2** managed to take-off but the power was too low. Construction mainly pine, with glued and pinned veneer on the fuselage and leading part of the wing. Main gear, an arc of ash laminates carrying two wheels. Several made, with different engines and reduced dimensions.
DIMENSIONS (No 1) Span 10,0 m (32 ft 9½ in); length 6,0 m (19 ft 8¼ in); wing area 15 m² (161·5 ft²).
ENGINE One motorcycle type, originally 5/7 hp Harley-Davidson V-twin.
WEIGHTS (No 1) Empty 120 kg (264·5 lb); fuel/oil 5 kg (11 lb); loaded 220 kg (485 lb).
PERFORMANCE (No 1) Max speed 70 km/h (43·5 mph).

Buryevestnik S-3 Despite its more powerful engine, this 26 machine never managed to fly further than 300 m. Tested with wheels and skis.
DIMENSIONS Span 9,8 m (32 ft 1¾ in); length 5,8 m (19 ft 0 in); wing area 12,5 m² (134·5 ft²).
ENGINE One 12 hp Harley-Davidson V-twin.
WEIGHTS Empty 135 kg (298 lb); fuel/oil 7 kg (15·4 lb); loaded 220 kg (485 lb).
PERFORMANCE Poor.

Buryevestnik S-4 This at last performed like a real aeroplane, the key lying in the engine. On 29 July 27 A. I. Zhyukov set ultralight height record. Take-off and landing run 30 m, and 360° turn made in 14 seconds. Fame followed a

flight in five stages from Moscow to Odessa. Numerous pilots flew S-4, most performing aerobatics.
DIMENSIONS Span 9,0 m (29 ft 6¼ in); length 5,8 m (19 ft 0 in); wing area 9,6 m² (103 ft²).
ENGINE One 18 hp Blackburn Tomtit.
WEIGHTS Empty 130 kg (287 lb); fuel/oil 20 kg (44 lb); loaded 230 kg (507 lb).
PERFORMANCE Max speed 140 km/h (87 mph); landing speed 60 km/h (37 mph).

Buryevestnik S-5 Last of Nyevdachin's ultralights, this was the most powerful, the State Aviatrust having imported a few Bristol Cherubs along with its licence for the Jupiter. This made the S-5 a first-class performer and it was exhibited at 29 International Aviation show in Berlin, after eight months of successful flying. The same aircraft was later given to the Osoaviakhim club of Irkutsk.
DIMENSIONS As S-4.
ENGINE One 32 hp Bristol Cherub flat-twin.
WEIGHTS Empty 145 kg (320 lb); fuel/oil 25 kg (55 lb); loaded 250 kg (551 lb).
PERFORMANCE Max speed 160 km/h (100 mph).

Pissarenko T A small monoplane like its predecessor, the T (for Trainer) was a different machine, for it outperformed most aircraft of its day. It flew in 25, engine being a 150 hp Hispano. Single-seat parasol monoplane, of wooden construction, with ply covering on the wing leading edge and fuselage back to the cockpit, which was behind the trailing edge. Wide spacing of wing bracing struts suggested spars at about 5 and 80% chord. Maximum speed approached 300 km/h (186 mph); studied as a *Strelbom* (fighter/bomber), considered novel.

Alekseyev Igor G. Alekseyev was son of the painter Georgiya Alekseyev. He produced two ultralights when civil war had not completely stopped. His **IgA-1** *Blokhi* (Flea) was one of the simplest machines imaginable, being a parasol monoplane of backyard appearance. In 24 K. K. Artseulov offered to test it, but it is doubtful that it flew.
DIMENSIONS Span 4,5 m (14 ft 9 in); length 3,0 m (9 ft 10 in); wing area 5 m² (53·8 ft²).
ENGINE One 8 hp JAP V-twin.
WEIGHTS Empty 65 kg (143 lb); fuel/oil 5 kg (11 lb); loaded 130 kg (287 lb).

IgA-2 This 26 machine used same engine but was larger. It broke after making several attempts to fly.
DIMENSIONS Span 6,0 m (19 ft 8 in); length 4,5 m (14 ft 9 in).
No other data.

OSO-1 This ultralight was a diploma project by Sergeya Sergei Ivanovich Kamenyev, supervised by the head of OSO CAHI, A. A. Baikov. Designed 25, flown in 26 by AGOS engineer N. I. Petrov. Considered to be underpowered.
DIMENSIONS Span 10,9 m (35 ft 9 in); length 5,8 m (19 ft ⅓ in); wing area 15 m² (161·5 ft²).
ENGINE One 18 hp Blackburn Tomtit.
WEIGHTS Empty 180 kg (397 lb); fuel/oil 20 kg (44 lb); loaded 267 kg (589 lb).
PERFORMANCE Max speed 120 km/h (74·5 mph); landing speed 55 km/h (34 mph).

SK Designation derived from A. N. Sedyelnikov and V. L. Korvin (who collaborated with Mikhelson on the MK-1 described earlier). Well-braced biplane, unusual in having interplane struts just outboard of the fuselage; further struts linked upper and lower ailerons on each side. Fuselage completely skinned with glued ply.
DIMENSIONS Span 9,0 m (29 ft 6¼ in); length 5,3 m (17 ft 4½ in); wing area 19,8 m² (213 ft²).
ENGINE One 12 hp Harley-Davidson tuned to 16 hp.
WEIGHTS No data.
PERFORMANCE Max speed 100 km/h (62 mph); landing speed 50 km/h (32 mph).

VEK Named *Strekoza* (Dragonfly), neat monoplane built at Kharkov OAVUK by student A. A. Bromberg, supervised by professor G. F. Proskur. Made several flights, powered by 27/35 hp ABC Scorpion engine. No other data.

IT-2 One of three powered machines designed by Igor Pavlovich Tolstikh, his other designs being gliders. Fuselage ply-skinned, the seat being ingeniously incorporated into the structure. Bad workmanship in the engine prevented take-off. Date, late 25.
DIMENSIONS Span 6,5 m (21 ft 4 in); length 3,2 m (10 ft 6 in); wing area 10,5 m² (113 ft²).
ENGINE One IEL motorcycle, nominal 9/12 hp.
WEIGHTS Empty 85 kg (187 lb); fuel/oil 10 kg (22 lb); loaded 170 kg (375 lb).

Ivanov Ultralight created by Viktor Appolonovich Ivanov and pilot N. D. Anoshchenko. Made numerous flights in 25-26.
DIMENSIONS Span 7,5 m (24 ft 7¼ in); length 4,5 m (14 ft 9 in); wing area 11 m² (118 ft²).
ENGINE One 8/12 hp JAP.
WEIGHTS Empty 150 kg (331 lb); fuel/oil 20 kg (44 lb); loaded 250 kg (551 lb).
PERFORMANCE Max speed 100 km/h (62 mph); landing speed 63 km/h (39 mph).

MB This small biplane few in 26; designer not reported.
DIMENSIONS Span 6,2 m (20 ft 4 in); length 5,2 m (17 ft 0¾ in); wing area 9,8 m² (105·5 ft²).
ENGINE One 45 hp Anzani.
WEIGHTS Empty 160 kg (353 lb); fuel/oil 20 kg (44 lb); loaded 260 kg (573 lb).
PERFORMANCE Max speed 145 km/h (90 mph).

Dzerzhinski Biplane also known as **SCh**, from builder Sergei Dmitryevich Chernikhovsky. Fuselage and tail welded steel tube, wings being wood. Work began 25; from 27 numerous flights in Moscow and Odessa.
DIMENSIONS Span 8,8 m (28 ft 10½ in); length 5,3 m (17 ft 4¾ in); wing area 12,8 m² (138 ft²).
ENGINE 27/35 hp ABC Scorpion.
WEIGHTS Empty 170 kg (375 lb); fuel/oil 20 kg (44 lb); loaded 270 kg (595 lb).
PERFORMANCE No data.

Pishchalnikov N. N. Pishchalnikov, an engineer at OMOS, made his lightplane alone at home; then took it by rail to a Finnish beach where he made the first flight in late 26.
DIMENSIONS Span 8,0 m (26 ft 3 in); length 5,1 m (16 ft 9 in); wing area 12,5 m² (135 ft²).
ENGINE 15 hp Reading-Standard.
No other data.

Mars A rebuild of a glider of the same name, both by Kharkov constructor S. N. Ruiltsyev. Parasol monoplane, flown 27 wheels and skis.
DIMENSIONS Span 12,0 m (39 ft 4½ in); length 5,5 m (18 ft 0½ in); wing area 13 m² (140 ft²).
ENGINE One 27/35 hp ABC Scorpion.
WEIGHTS Empty 190 kg (419 lb); fuel/oil 20 kg (44 lb); loaded 290 kg (639 lb).
PERFORMANCE Max speed 110 km/h (68 mph); landing speed 55 km/h (34 mph).

STI Two-seater designed and built by students at Siberian Technical Institute, Tomsk, under professors G. V. Trapeznikov (airframe) and A. V. Kvasnikov (engine), 1925-27. Engine, designated for five students who built it, used parts from a Le Rhône rotary; airframe had braced low wing and incorporated tail of Nieuport and landing gear from Morane. First flight 27; later flew as far as Kansk.
DIMENSIONS Span 10,0 m (32 ft 9½ in); length 7,0 m (22 ft 11½ in); wing area 15 m² (161·4 ft²).
ENGINE One Tuzhkut flat twin, 24 hp.
WEIGHTS Empty 300 kg (661 lb); fuel/oil 20 kg (44 lb); loaded 400 kg (882 lb).
PERFORMANCE Max speed 100 km/h (62 mph).

Tri Druga Name of this attractive parasol monoplane (Three Friends) reflected those who built it: S. A. Semyenov, L. I. Sutugin and S. N. Gorelov – and motto of 23 Aviakhim contest 'strength, simplicity, cheapness, three friends of aviation'. In fact less simple than most, because it was a refined design with monocoque fuselage of glued veneer, with cantilever main gear of original design. Pilot ahead of front spar; two-seater carried passenger behind under the wing. Flew 28 and exhibited in Berlin.
DIMENSIONS Span 12,0 m (39 ft 4½ in); length 6,9 m (22 ft 8 in); wing area 18,7 m² (201 ft²).
ENGINE Bristol Cherub 30 hp.
WEIGHTS Empty 245 kg (540 lb); fuel/oil 22 kg (48·5 lb); loaded 417 kg (919 lb).
PERFORMANCE Max speed 127 km/h (79 mph); landing speed 45 km/h (28 mph); ceiling 3,2 km (10,500 ft) (single-seat, 4,3 km, 14,100 ft).

Budilnik No data on this ultralight built by group of railwaymen at Krasnoyarsk and flown 27. Had motorcycle engine, described as 'weak'. Name, however, means 'alarm-clock'.

LAKM Designation from Leningrad Aero Club-Museum. Possibly only machine to fly with neat flat-twin engine designed by L. Ya. Palmen. Low-wing single-seater, designed with Osoaviakhim support by M. V. Smirnov (famed *Ilya Mouromets* pilot) and Ya. L. Zarkhi. First flight 27 Nov 28; still flying 31.
DIMENSIONS Span 10,4 m (34 ft 1½ in); length 5,6 m (18 ft 4½ in); wing area 12 m² (129 ft²).
ENGINE One 20 hp Palmen flat-twin.
WEIGHTS Empty 175 kg (386 lb); fuel/oil 27 kg/13 kg (59·5 lb/28·7 lb); loaded 285 kg (628 lb).
PERFORMANCE Max speed 115 km/h (71·5 mph); landing speed 50 km/h (31 mph); ceiling 1620 m (5,300 ft).

S-1 Little known about this parasol single-seater designed by test-pilot Vasili Andreevich Stepanchenok. Despite poor engine performance (bad workmanship) the S-1 managed to

fly moderately well. First flight 28, demonstrations in Moscow 29.
DIMENSIONS Span 7,7 m (25 ft 3 in); length 5,0 m (16 ft 4¾ in); wing area 9,2 m^2 (99 ft^2).
ENGINE One 40 hp Anzani.
WEIGHTS Empty 207 kg (456 lb); fuel/oil 27 kg (59·5 lb); loaded 314 kg (692 lb).
PERFORMANCE Max speed 150 km/h (93 mph); landing speed 75 km/h (46·6 mph).

Pavlov Built by Aleksei Nikolayevich Pavlov, famous instructor at Orenburg flying school, this was another high-wing single-seater. Flown spring 29; made several flights, including one from Orenburg to Moscow, but crashed fatally late 29.
DIMENSIONS Span 7,5 m (24 ft 7¼ in); length 5,5 m (18 ft 0½ in); wing area 10 m^2 (107·6 ft^2).
ENGINE 100 hp Bristol Lucifer.
WEIGHTS Empty 480 kg (1,058 lb); fuel/oil 90 kg (198 lbs); loaded 650 kg (1,433 lb).
PERFORMANCE Max speed 190 km/h (118 mph); landing speed 85 km/h (53 mph).

Ivensen Paul Albertovich Ivensen (Yevyensyen) was famous for his gliders, but his single-seat aeroplane, flown 29, was a new design. Parasol wing, a single wooden structure, held by four short struts. Flew with wheels and skis. In 30s Ivensen was deputy to Isacco.
DIMENSIONS Span 9,0 m (29 ft 6¼ in); wing area 13,5 m^2 (145 ft^2).
ENGINE One 30 hp Bristol Cherub.
No other data.

IT-6 Next in IT family of I. P. Tolstikh to be powered. Parasol two-seat tandem, named *Komakademiya* or *Krupskoi*. Again a refined veneer-skinned aircraft which made many flights 1929-30.
DIMENSIONS Span 10,6 m (34 ft 9¼ in); length 6,7 m (22 ft 0 in); wing area 16 m^2 (172 ft^2).
ENGINE One 60 hp Cirrus Hermes.
WEIGHTS Empty 430 kg (948 lb); fuel/oil 60 kg (132 lb); loaded 650 kg (1,433 lb).
PERFORMANCE Max speed 105 km/h (65 mph).

RG-1 Designed by Sergei Nikolayevich Goryelov, unusual in being STOL sesquiplane with slatted main wing. First Soviet lightplane powered by Czech Walter radial, which though expensive was reliable. This machine had the more powerful seven-cylinder model. Wooden, with monocoque fuselage, and engine fully cowled. Tandem-seater, said to be 'easily understood'.
DIMENSIONS Span 10,0 m (32 ft 9½ in); length not recorded; wing area 20 m^2 (215 ft^2).
ENGINE One 85 hp Walter.
WEIGHTS Empty not recorded; loaded 900 kg (1,984 lb).
PERFORMANCE Max 135 km/h (84 mph); landing speed 45 km/h (28 mph).

Prokopenko A bus-driver in Krivoi Rog, Ukraine, G. I. Prokopenko built a parasol monoplane that flew on 24 Feb 30 with complete success. Data lacking, but wing was upper plane from Hanriot 14 and engine incorporated parts from 80 hp Le Rhône. Wing collapsed during later flight, constructor being killed.

Omega Attractive low-wing cabin monoplane built at Kharkov 31 by Aleksei Nikolayevich Gratsianski, later Hero of the Soviet Union; tested by none other than B. N. Kudrin. Efficient wooden single-spar wing with aspect ratio of 9; welded steel-tube fuselage. Photograph shows tandem; dual control. Original 60 hp Walter ran with severe vibration, and was replaced. This type (a number were built) served as trainer at Poltava Osoaviakhim school.
DIMENSIONS Span 11,0 m (36 ft 1 in); length 7,5 m (24 ft 7¼ in); wing area 13,4 m^2 (144 ft^2).
ENGINE One M-23 (NAMI-65).
WEIGHTS Empty 500 kg (1,100 lb); fuel/oil 110 kg (243 lb); loaded 760 kg (1,675 lb).
PERFORMANCE Max speed 170 km/h (106 mph); to 1 km, 6 min; to 2 km, 13 min; service ceiling 4,5 km (14,750 ft); landing speed 60 km/h (37 mph).

Vinogradov 3B/M Named *Igrado* (plaything), neat biplane designed by Ivan Nikolayevich Vinogradov and constructed at FZU plant at Frunze 1929-30. Construction mixed, forward fuselage being corrugated duralumin and rear wood, with veneer skin. Wings wood, with fabric covering, with streamlined I interplane struts. Test flying 31. In final form ailerons were full-span, but on lower wings only.
DIMENSIONS Span 10,2 m (33 ft 5½ in); length 5,4 m (17 ft 9 fin); wing area 22,4 m^2 (241 ft^2).
ENGINE 60 hp Walter 5-cylinder.
WEIGHTS Empty 500 kg (1,100 lb); loaded 675 kg (1,488 lb).
PERFORMANCE Max speed 120 km/h (74·5 mph); cruising speed 92 km/h (57 mph).

Gup-1 Designation of this two-seat tourer intended to be surname of German designer, Friedrich Guep. He was invited to build by ZOK-NII-GVF; first flight Oct 33. Tandem open cockpits in steel-tube truss fuselage; wooden wings of constant section braced by four compression struts to top of fuselage. Control surfaces skinned with glued ply; stability and handling 'put it in a class of its own'.
DIMENSIONS Span, 10,5 m.
ENGINE One M-11, with ring cowl.
WEIGHTS Empty 615 kg (1,356 lb); fuel/oil 190 kg (419 lb); loaded 880 kg (1,940 lb).
PERFORMANCE Max speed 160 km/h (100 mph); climb to 1 km, 8 min; service ceiling 3,5 km (11,500 ft); landing speed 70 km/h (43·5 mph).

IT-9 Last and best aeroplane of I. P. Tolstikh. Tandem two-seat parasol, demonstrated outstanding performance and handling. Fuselage welded steel-tube truss; folding wings were wooden, of constant (R-IIS) section, with slats ahead of ailerons and manually lowered flaps inboard. Metal tail, with tight fabric covering. Bungee shock-absorbers on long-stroke landing gears, fitted with either wheels or skis. Test-flying early 34; IT-9 was so good it was considered for mass-production, but familiarity and experience of U-2 (Po-2) rendered the biplane unassailable.
DIMENSIONS Span 12,9 m (42 ft 4 in); length 7,7 m (25 ft 3 in); wing area 28 m^2 (301 ft^2).
ENGINE One M-11 with ring cowl.
WEIGHTS Empty 625 kg (1,378 lb); fuel/oil 65 kg/5 kg (143 lb/11 lb); loaded 1000 kg (2,205 lb).
PERFORMANCE Max speed 140 km/h (87 mph); take-off 80 m (262 ft) in 7 s; range 400 km (248 miles); landing speed 55 km/h (34 mph).

VVA-1 In 34 Osoaviakhim and the Aviavnito (aviation technical society) sponsored a competition for a superior touring aircraft. It was hoped to create 'a flying Ford'. Many entries submitted and several of designs built, including the winner, VVA-1. Though one of the 'official' entries, it was rejected as being structurally difficult. Designer one of chief VVA professors, Vladimir Sergeyevich Puishnov, who played a part in many aircraft prior to 41. Tandem-seat cabin sesquiplane, with N-Warren interplane bracing of rear-folding wings, and neat landing gears using copies of Dowty internally sprung wheels. What the committee found too much was the beautiful monocoque wooden fuselage, of streamlined profile and circular section. Tailplane was ahead of fin, on top of rear fuselage.
DIMENSIONS Span 10,9 m (35 ft 9 in); length 7,8 m (25 ft 7 in); wing area 24 m^2 (258 ft^2).
ENGINE One M-11 in long-chord cowling.
WEIGHTS Empty 800 kg (1,764 lb); loaded 1120 kg (2,470 lb).
PERFORMANCE (Est) Max speed 170 km/h (106 mph); range 550 km (342 miles); landing speed 60 km/h (37 mph).

Anito-1 Another attractive contest entry, name being abbreviation of sponsoring society. Low-drag low-wing monoplane, seating pilot and two passengers one behind the other in enclosed (seemingly rather cramped) cabin. Wooden, elliptical-section monocoque fuselage, two-spar wing with split flaps, veneer-covered balanced control surfaces, and helmeted engine cowl. Designer/builders Nikolai Georgievich Nurov and Suren Alekseyevich Elibekyan. Flight testing 2-9 Sept 35.
DIMENSIONS Span 12,6 m (41 ft 4 in); length 8,4 m (27 ft 6¾ in); wing area 24 m^2 (258 ft^2).
ENGINE One M-11.
WEIGHTS Empty 815 kg (1,797 lb); fuel/oil 86 kg/14 kg (186 lb/31 lb); loaded 1160 kg (2,557 lb).
PERFORMANCE Max speed 204 km/h (127 mph); take-off 190 m in 15 s; service ceiling 5040 m (16,535 ft); range 666 km (414 miles); landing speed 60 km/h (37 mph).

Komarov No details of Aviavnito contest entry by A. A. Komarov built at Novocherkassk and flown 35, beyond the fact it had an M-11. It crashed, killing the designer.

Chyerednichenko Ultralight built from parts of a glider and another powered aircraft by engineer/pilot Vladimir Grigoryevich Chyerednichenko in 35. Wooden low-wing monoplane, with main wheels protruding from underside of wing. Powered by locally created engine, made several ground runs and one flight.
DIMENSIONS Span 11,0 m (36 ft 1 in); length not recorded; wing area 24 m^2 (258 ft^2).
ENGINE One 18 hp Stulov flat-twin.
WEIGHTS Empty 270 kg (595 lb); fuel/oil 20 kg (44 lb); loaded 370 kg (816 lb).
PERFORMANCE Not recorded.

Pleskov Another of bumper 35 crop was simple parasol monoplane seating two in tandem built by famed glider pilot A. I. Pleskov, assisted by colleagues Kovalyenko and Mazalov, and also Sevastyanov, mechanic at Saratov aero club. Of wood/fabric construction, it managed to fly despite its car engine (56 hp GAZ-M-1 as described in KSM-1 entry), adopted to save expense. No data.

Vasilyev A *Motoplaner* (powered glider) constructed 1935-36 by G. S. Vasilyev who mounted a 12 hp outboard motor from a boat on parasol wing, driving tractor propeller above cockpit canopy. Wing said to have 'freely floating ailerons', which may explain why it slid one wingtip along ground and crashed. No data.

Blokhi family News of Henri Mignet's *Flying Flea* triggered a rash of Fleas in the Soviet Union, and as elsewhere most never flew. Of those that were completed, many were greatly (and probably advantageously) modified by their builders. One of the first, flown in 35, was the *Blokha* (Flea) built at GAZ-1. Main constructor A. A. Shteiner. Became a prototype for further construction, examples of which went to newspapers *Pravda* and *Izvestya* and adopted their names. Another placed in Polytechnic Museum in Moscow. Data typical.
DIMENSIONS Span 5,6 m (18 ft 4½ in); length 3,5 m (11 ft 5¾ in); wing area 10 m^2 (108 ft^2).
ENGINE One 18 hp Aubier-Dunne.
WEIGHTS Empty 100 kg (220 lb); fuel/oil 20 kg (44 lb); loaded 200 kg (441 lb).
PERFORMANCE Max speed 105 km/h (65 mph); ceiling 1800 m (5,900 ft).

Sheremetyev One of the more attractive Fleas was flown 35 by glider factory engineer Boris Nikolayevich Sheremetyev. Also known as **ShBM** and **ZAOR**. Rear wing pivoted at 25% chord and left/right halves could function as ailerons and as elevators. This was an important improvement; described as having flown quite well.
DIMENSIONS Span 5,8 m/6,1 m (19 ft 0 in/20 ft 0 in); length 3,9 m (12 ft 9½ in); wing area (total) 12,5 m^2 (135 ft^2).
ENGINE One 18 hp Aubier-Dunne.
WEIGHTS Empty 150 kg (331 lb); fuel/oil 18 kg (40 lb); loaded 253 kg (558 lb).
PERFORMANCE Max speed 115 km/h (71·5 mph); ceiling 2 km (6,500 ft); range 290 km (180 miles); landing speed 48 km/h (30 mph).

Moskit Built by S. V. Konstantinov, assistant professor at Novocherkassk, assisted by V. V. Belyaninuim, this was not quite a Mignet type but almost a Flea-sized regular aeroplane with full-span ailerons and large slab tailplanes pivoted at 25% chord. All these surfaces were operated by the stick with rudder interlinked, there being no pedals. Test-flown by head of experimental KB, G. M. Zhuravlyev.
DIMENSIONS Span 6,0 m/4,0 m (19 ft 7 in/13 ft 1¼ in); length 4,0 m (13 ft 1½ in); wing area 7,5 m^2/4,8 m^2 (81/51·6 ft^2).
ENGINE One 27 hp/35 hp ABC Scorpion.
WEIGHTS Empty 120 kg (265 lb); fuel/oil 30 kg (66 lb); loaded 230 kg (507 lb).
PERFORMANCE Max speed 100 km/h (62 mph); landing speed 45 km/h (28 mph).

Katsur VO-4 Katsur was an engineer in Kharkov, whose *Avietka-Blokha* had a rear wing pivoted to function as an elevator and also carrying ailerons; front wing fixed and without movable surfaces. First flight early 36.
DIMENSIONS Span 6,0 m (19 ft 7 in); length 3,9 m (12 ft 9½ in); wing area 13,8 m^2 (148·5 ft^2).
ENGINE One 27 hp, believed ABC Scorpion.
WEIGHTS Empty 120 kg (265 lb); fuel/oil 21 kg (46 lb); loaded 238 kg (524 5 lb).
PERFORMANCE Max speed 115 km/h (71·5 mph); ceiling 3 km (9,850 ft); landing speed 47 km/h (29 mph).

Ruibchinsko In 1935-36 three Flea-type machines were built at GAZ-1 in Moscow by Ruibchinsko, assistant professor (reader) at *Dniepropetrovsk* metallurgical institute and associates from Chardzhou and Gomel. No data.

Oktyabrnok A. I. Mikoyan's first creation is best covered here because it was designed and built with K. Samarin and N. A. Pavlov in 36 as their diploma project at the VVA. Neat pusher, with shoulder wing and open cockpit offering almost perfect view ahead of long-stroke main gears.
DIMENSIONS Span 8,0 m (26 ft 3 in); length 6,2 m (20 ft 4 in); wing area 11,4 m^2 (123 ft^2).
ENGINE One 25 hp flat-twin by P. Labur.
WEIGHTS Empty 150 kg (331 lb); fuel/oil 20 kg (44 lb); loaded 264 kg (582 lb).
PERFORMANCE Max speed 126 km/h (78 mph); take-off/landing 85 m/50 m; ceiling 3 km (9,850 ft); landing speed 45 km/h (28 mph).

Sidoryenko Another tandem-wing ultralight, built in Tashkent by this constructor, flying 36.
DIMENSIONS Span 9,5 m (31 ft 2 in); length 5,4 m (17 ft 8½ in); wing area 12 m^2 (129 ft^2).
ENGINE One 17 hp Aubier-Dunne.
WEIGHTS Empty 154 kg (340 lb); fuel/oil 6 kg (13 lb); loaded 240 kg (529 lb).
PERFORMANCE Max speed 110 km/h (68 mph); ceiling 1,5 km (4,900 ft); landing speed 42 km/h (26 mph).

KSM-1 Shortage of good yet cheap light aircraft engines led in the early 30s to attempts to use car engines or engines using car components. Most important Soviet attempt, under direction of E. V. Agitov, was GAZ-M-1 four-in-line, with water cooling and electric starter. KSM-1 *Komsomolyets*-1 was first aircraft designed for this engine, though beaten into the air by simpler Pleskov. Low-wing tandem two-seater of wooden construction, monocoque fuselage and leading edge having skin of *shpon* (birch ply) and wing fabric behind front spar. Split flaps fitted, and engine cowled in circular casing with radial nose apertures for radiator air. Designer Aleksei Andreyevich Smolin. Built under expert supervision at GVF; experimental shop at Gorki car works made one-piece propeller. Construction complete 35; first flight 36. Performance unimpressive because of heavy engine installation. At least two built. Attempts continued to 39 to improve matters. Gribovskii assisted, and used same type of engine in G-23.
DIMENSIONS Span 12,0 m (39 ft 4½ in); length 7,0 m (22 ft 11½ in); wing area 18 m^2 (194 ft^2).
ENGINE One 56 hp GAZ-M-1.
WEIGHTS Empty 605 kg (1,334 lb); loaded 860 kg (1,896 lb).
PERFORMANCE Max speed 121 km/h (75 mph); take-off 400 m/20 s; ceiling 1620 m (5,315 ft); landing speed 57 km/h (35 mph).

MIIT Student at MIIT, Sergei Vasilyevich Popov, built neat single-seater in 36 as diploma project. Named *Miitovyets*, said to grow more beautiful in flight, and to have passed factory tests with commendation.
DIMENSIONS Span 10,6 m (34 ft 9 in); length 6,0 m (19 ft 9 in); wing area, not recorded.
ENGINE One 25 hp (possibly ABC).
WEIGHTS Empty 175 kg (386 lb); fuel/oil 30 kg (66 lb); loaded 285 kg (628 lb).
PERFORMANCE Max speed 120 km/h (75 mph); landing speed 50 km/h (31 mph).

MB-1 In 36 M. P. Beschastnov, skilled aviation worker in far east of Siberia, built single-seat trainer looking very much like an I-16.
ENGINE One M-11; trials successful, speed being 230 km/h (143 mph). No other information.

Vlasov Another *motor planer*, built and flown at Kuibyshev 37 by A. Vlasov. Based on Antonov PS-2, plus 12 hp LM-4 flat-twin (outboard boat engine).

Caudron/Dubrovin In 1936-39 a major effort was made to develop the Caudron-Renault racers and their engines. Main types involved were the C.690 and C.713, the former a main competitor in Coupe Deutsch de la Meurthe races in 1934-35, and the latter related to light fighters of 1937-40. The Renault engines adopted in 4, 6 and 12-cylinder forms, MV-6 used in Soviet aircraft 1938-41 period. Airframe bureau chief A. A. Dubrovin, with A. G. Brunov as chief deputy; others were Z. I. Itskovich (of KAI) and Ye. G. Adler (of Yak). Work discontinued 39.

Sh-13 Another design of Boris Nikolayevich Sheremetyev actually built, this was intended to be a super-efficient sporting single-seater to take the world absolute long-range record in its class. Construction described as 'one-piece wooden' and its outstanding feature was the wing whose aspect ratio had not been approached before (in the Soviet Union, at least) except for two sailplanes. Well streamlined, with single-seat cockpit having moulded plastics panels in hinged canopy. Test flown by P. G. Golovin 39; pleasant at light weights, but had no chance of lifting design fuel load which would have given endurance of 17 hours.
DIMENSIONS Span 13,0 m (42 ft 8 in); length 5,92 m (19 ft 5 in); wing area 10,56 m^2 (113·7 ft^2).
ENGINE One 40 hp Salmson radial.
WEIGHTS Empty 397 kg (875 lb); fuel/oil 176 kg (388 lb); loaded 663 kg (1,462 lb).
PERFORMANCE Max speed 180 km/h (112 mph); range 2500 km (1,553 miles) with normal tankage; landing speed 90 km/h (56 mph).

G-5 From 37 purges and terrifying environment almost halted light aircraft construction, until in 40 a refined *motor planer* was built at Kharkov with this designation. Constructor was Stepan Vasilyevich Grizodubov. Wooden, built for low-drag penetration, with single open cockpit and pylon-mounted wing. Engine on steel-tube framework above wing, driving pusher propeller. Flights successful.
DIMENSIONS Span 12,15 m (39 ft 10 in); length 6,0 m (19 ft 8¾ in); wing area 17,65 m^2 (190 ft^2).
ENGINE One 35 hp ADG-4 flat-twin.
WEIGHTS Empty 178 kg (392 lb); fuel/oil 32 kg (70· 5 lb); loaded 300 kg (661 lb).

UPO-22 Having achieved indifferent results with KSM-1, A. A. Smolin persisted with low-wing two-seaters with car engines and by 41 completed this refined and more powerful machine. Assumed similar to KSM in construction, and engine was 80 hp GAZ-M-4. All documents lost in War.

Dosav During its two-year life, in 50, this organization sponsored the first (and almost only) nationwide design competition since the Second World War. Objective: a training and utility machine worthy of succeeding Po-2. Numerous entries received, from which five designs listed below were selected for prototype construction. Contest then abandoned.

Elibekyan S. A. Elibekyan, one of the designers of *Anito* of 35, produced entrant with simple (mixed steel tube, wood and fabric) construction and adequate performance. Side-by-side dual trainer, high-wing cabin machine with slats, slotted flaps and high tailplane.
DIMENSIONS Span 11,6 m (38 ft 0¾ in); length 7,6 m (24 ft 11½ in); wing area, not recorded.
ENGINE One M-11.
WEIGHTS Empty 560 kg (1,235 lb); loaded 830 kg (1,830 lb).
PERFORMANCE Max speed 190 km/h (118 mph); landing speed 60 km/h (37 mph).

Golayev Pure sporting machine, entry by A. Golayev was small wooden racer with wings of 'laminar' section and high aspect ratio.
DIMENSIONS Span 7,0 m (22 ft 11¾ in); length 5,7 m (18 ft 8¾ in); wing area 7 m² (75 ft²).
ENGINE One 90 hp radial.
WEIGHTS Empty not recorded, loaded 500 kg (1,100 lb).
PERFORMANCE (Est) Max speed 306 km/h (190 mph); landing speed 110 km/h (68 mph).

Gribovskii This experienced designer proposed stressed-skin low-wing two-seater, with side-by-side seating in enclosed cockpit with dual control. Wings high aspect-ratio, elevator one piece behind rudder, fixed tricycle landing gear.
DIMENSIONS Span 11,4 m (37 ft 5 in); length 7,05 m (23 ft 2 in); wing area 12,2 m² (131 ft²).
ENGINE One 90 hp radial.
WEIGHTS Empty 502 kg (1,107 lb); loaded 750 kg (1,653 lb).
PERFORMANCE Max speed 195 km/h (121 mph); landing speed 65 km/h (40 mph).

Nikitin Another experienced builder, V. V. Nikitin adopted stressed-skin for small ultralight single-seater. Braced high wing, tricycle gear and liberal slats and flaps.
DIMENSIONS Span 9,3 m (30 ft 6½ in); length 6,4 m (21 ft 0 in); wing area 13 m² (140 ft²).
ENGINE One 50 hp flat-twin.
WEIGHTS Empty 272 kg (600 lb); loaded 400 kg (882 lb).
PERFORMANCE Max speed 145 km/h (90 mph); landing speed 55 km/h (34 mph).

Ollo/Shprangel only successful entry by a new team, low-wing single-seater with light-alloy framework and fabric covering. Wing resembled Blanik in being swept slightly forward, with flaps.
DIMENSIONS Span 8,5 m (27 ft 11 in); length 7,1 m (23 ft 3¾ in); wing area 12 m² (129 ft²).
ENGINE One 225 hp radial.
WEIGHTS Empty 430 kg (948 lb); loaded 830 kg (1,830 lb).
PERFORMANCE Max speed 318 km/h (198 mph); landing speed 88 km/h (55 mph).

Zherebtsov One of several post-war ultralight tip-drive helicopters. Designed from 47 by B. Ya. Zherebtsov, Yu. S. Braginski and Yu. L. Starinin, mainly to test proposed pulsejet drive; the three engineers led group formed for this project. Rotor had two blades, of 7,0 m (23 ft 0 in) diameter, intention being to substitute 9 m unit later. Pulsejets of 250 mm (10 in) length gave up to 17 kg (37·5 lb) thrust. In 50 prototype flown (pilot Smirnov) with Zis-150 truck engine on ground driving rotor up to speed.

Ilyin Soviet Union has persistently produced ideas for flapping-wing aircraft (ornithopters). In 57 reported that small machine built by Dmitri V. Ilyin had flown, using engine of 5 hp. Said to be based on widely published principles of I. N. Vinogradov (lead designer of 3B/M described earlier in this section); official Soviet view is that it flew 3 km non-stop.

Diskoplan Engineer Sukhanov led a small team trying to perfect aircraft with circular wing. Glider (250 kg loaded) with true circular wing of 10 m², small nacelle under LE and tail above TE, flown at Tushino 58; improved glider flown 60, third in 62 (preserved Monino) having near-rectangular wing with elevons (no separate horiz tail) and canopy above LE. Intention was to build jet.

Leningradets Built and flown 62, one of world's smallest two-seaters. Designers Lyev Sekinin, Valentin Tatsiturnov and Lyev Kostin, in Leningrad. Tandem cabin parasol monoplane of traditional form.
DIMENSIONS Span 7,0 m (23 ft 0 in); length 4,8 m (15 ft 9 in).
ENGINE One 51 hp Zündapp Z9 inverted 4-inline.
PERFORMANCE Max speed 150 km/h (93 mph). No other data.

Malysh Meaning *Little One*, only designation of primitive but creditable ultralight by aeromodelling section of Young Pioneers of Zlatoust, Urals. Wood/fabric single-seat parasol monoplane designed and built under direction of Lyev Aleksandrovich Komarov. Flown on skis, 12 April 64.
DIMENSIONS Span 6,9 m (22 ft 7¾ in), length 4,74 m (15 ft 6¾ in); wing area 7,8 m² (84 ft²).
ENGINE One 30 hp flat-twin motorcycle.
WEIGHTS Empty 110 kg (242·5 lb); loaded 200 kg (441 lb).
PERFORMANCE Max speed 130 km/h (81 mph); landing speed (min) 50 km/h (31 mph).

Riga-1 Formed 62, Riga RIIGA (institute of civil aero engineers) immediately launched this ultralight, diploma project of students G. Ivanov and F. Mukhamedov, assisted by others. Strongly reminiscent of Tipsy Nipper, though high aspect-ratio wings were from BRO glider. Steel-tube/wood/fabric fuselage carried balanced rudder but no fin. Riga-1 flew 65.
DIMENSIONS Span 9,0 m (29 ft 6¼ in); length 5,2 m (17 ft 1 in); wing area 9,0 m² (97 ft²).
ENGINES Various 26 hp/30 hp (eg, M-61/M-62/K-750.
WEIGHTS Empty about 190 kg (419 lb); loaded about 300 kg (661 lb).
PERFORMANCE (Est) max speed 140 km/h (87 mph); landing speed 65 km/h (40 mph).

Riga-50 Following at least two other sailplane-derived ultralights RIIGA built this autogyro mid-60s, completed 67. Designed by diploma student V. Ustinov, assisted by D. Osokin and V. Prishlyuk. Difficult parts – rotor blades, hub and dynamics – based on Bensen philosophy, though Ka-18 served as pattern for blade design. Duralumin keel provided foundation for nacelle, with glassfibre casings. No spin-up auxiliary drive provided. Later Riga-50M with nacelle of welded steel tube and plywood.
DIMENSIONS Rotor diameter 6,1 m (20 ft 0 in); length (discounting rotor) 3,4 m (11 ft 2 in); disc area 29,2 m² (314 ft²).
ENGINE One 45 hp M-77 flat-twin.
WEIGHTS Empty 140 kg (309 lb); loaded 225 kg (496 lb).
PERFORMANCE Max speed 85 km/h (53 mph); take-off run about 100 m (330 ft).

Chaika Using same design of main rotor and control rods as on Riga-50, this rotor-kite, also designated **Ch-1**, was designed for towing behind car or boat. First flown by V. Tseitlin in Aug 70 (listed here to be near its predecessor). Structure built from standard thin sheet-steel U-sections resembling Dexion, empty weight under 50 kg (110 lb). Normal towing speed 30 km/h to 45 km/h (19 mph-28 mph).

VIGR-1 Another ultralight rotorcraft, early student project at KuAI (Kuibyshev Aviation Institute). Two-blade rotor driven by tip pulsejets similar to Zherebtsov's. First hovering flight Oct 65. No spin-up engine fitted.
DIMENSIONS Rotor diameter 8,5 m (27 ft 11 in); disc area 56,75 m² (611 ft²).
WEIGHTS Empty 95 kg (209 lb); loaded 190 kg (419 lb).
PERFORMANCE Not recorded.

Sverchok-1 Meaning *Cricket*, KuAI's second light rotorcraft. Autogyro, diploma project by student group led by S. Pyatinitskii and tested 71. Simple structure of light alloy and glassfibre, with veneer-skinned blades on steel spar.
DIMENSIONS Rotor diameter 6,4 m (21 ft 0 in); length (discounting rotor) 3,75 m (12 ft 3½ in); disc area 32,2 m² (346 ft²).
ENGINE One 38 hp M-61.
WEIGHTS Empty 110 kg/125 kg (242 lb/278 lb); loaded 180 kg/196 kg (397 lb/432 lb).
PERFORMANCE Max speed 100 km/h (62 mph); take-off speed 30 km/h/40 km/h (19 mph/25 mph); landing speed 20 km/h/25 km/h (12 mph/16 mph).

Chkalov Ch-1 Also designated **C-12** and **VAT** (Voronezh Aero-Technical college); qualified for inclusion in 'Flea' section for same layout. Metal construction, fabric covering. Fixed tricycle gear with twin nosewheels; mainwheels, Yak-12 tailwheels. Full-span ailerons on front wing, enormous horn-balanced elevators at rear. Motorcycle saddle and handlebars, though conventional flight controls with pedals for rudder. First flight 70.
DIMENSIONS Span (wing) 5,4 m (17 ft 8½ in), (tail) 4,4 m (14 ft 5 in); length 3,7 m (12 ft 1½ in); height 1,64 m (5 ft 4½ in); areas (wing) 6,72 m² (72·3 ft²), (tail) 5,58 m² (60·1 ft²).
ENGINE One 28 hp K-750.
WEIGHTS Empty 230 kg (507 lb); loaded 330 kg (726 lb).

PERFORMANCE Max speed 105 km/h (65 mph); take-off 15 m (49 ft); landing speed 31 km/h (20 mph); range up to 620 km (358 miles).

Tourist In Feb 70 Kiev engineers, Demchyenko, Khitry and Gusyev, said to have completed nine years' development of personal helicopter foldable into 1 m (39 in) suitcase. Coaxial rotors 'inflated hydraulically' and driven by 'kerosene turbine of kind used to start turboprops of large aircraft' (NK-12MV). Unfolding said to take 3 min. Empty weight 25,5 kg (56 lb) with two-way radio; max speed 160 km/h (99 mph).

AT-1 Mriya Name (as An-225) Ukrainian for 'dream', lightplane by Mikhail Artyumov and Viktor Timofeyev early 70s.

Bekshta Romuald Bekshta of Lithuanian SSR built series of model canard (tail-first) monoplanes, notably RB-4 in 47, followed by RB-7 and culminating in full-scale RB-15 of 74. Single-seat enclosed cockpit in steel-tube fuselage with braced wing at rear with ailerons and Scheibe tip fins/rudders, and controllable foreplane pivoted to nose; fixed tandem wheels, pusher engine at back. Rebuilt 79 as RB-17 with Scheibe tips replaced by large central fin ahead of prop and rudder behind prop, and with normal pair of main gears.
DIMENSIONS Span 8,05 m (26 ft 4⅞ in), foreplane 3,6 m; length (15) 4,1 m (13 ft 5⅜ in), (17) 5,0 m (16 ft 5 in); wing area 10,5 m^2 (113 ft^2).
ENGINE Two-cylinder Belkoshapkin, 25 hp.
WEIGHTS Empty 205 kg (452 lb); max 320 kg (705 lb).
PERFORMANCE Max speed 110 km/h (68 mph).

Omega Artyumov and Timofeyev then designed their own aircraft; Artyumov's was an attempt to use channel-wing idea unsuccessfully worked on by Custer in USA in 1948-60. Small metal/fabric amphibian with gull wings forming channel ahead of propeller. Tricycle landing gear with main wheels non-retractable but raised clear of water; nose wheel replaced by ski for first flight from ice on 8 Jan 75.
DIMENSIONS Span 8,22 m (26 ft 11½ in); length 5,4 m (17 ft 8½ in).
ENGINE One 32 hp IZ-56 flat-twin driving glassfibre-skinned pusher propeller.
WEIGHT Max loaded 270 kg (595 lb).
PERFORMANCE Max speed 100 km/h (62 mph); landing 65 km/h (40 mph).

T-1 Mustang Timofeyev built a conventional and attractive high-wing pusher with configuration almost indistinguishable from Polish Janowski J-1 of same period. Mixed construction, cantilever main legs in line with front wing-struts, tailskid, streamlined nose cockpit open at sides. Flew early 75.
DIMENSIONS Span 8,0 m (26 ft 3 in); length 5,6 m (18 ft 4½ in).
ENGINE One 32 hp IZ-56 flat-twin driving pusher propeller.
WEIGHTS Empty 150 kg (331 lb); max 270 kg (595 lb).
PERFORMANCE Max speed 105 km/h (65 mph); take-off 65 km/h (40 mph) in 100 m (328 ft).

RKIIGA-74 Also called Riga-74, and named *Experiment*, this novel amphibian used hull of a Progress speedboat, wings and tail of a Primoryets glider and powerful Czech aero engine. Windshield moved 0,15 m (6 in) back to avoid propeller; hull strengthened at attachments of the wings, landing gear and tail. Side-by-side seats with 90 lit (19·8 gal) fuel tank behind. First flight by V. Abramov and Riga professor V. Cezhtlin on 17 Sept 74, as flying boat. Water rudders and retracting landing gear then added.
DIMENSIONS Span 13,2 m (43 ft 4 in); length 8,1 m (26 ft 7 in); height 2,4 m (7 ft 10½ in) on wheels; wing area 20,2 m^2 (217 ft^2).
ENGINE One 140 hp Avia M332 4-inline driving tractor propeller.
WEIGHTS Empty 600 kg (1,323 lb); max 900 kg (1,984 lb).
PERFORMANCE Max speed 160 km/h (99 mph); climb at 90 km/h, 6 m/s (1,180 ft/min); range 500 km (311 miles).

Mikrosamolyet Low-wing cabin single-seater (named Micro-aeroplane) designed and built in Frunze by team led by V. Dmitriyev. Mainly wood and fabric, disclosed 75, no details except 40 hp engine, empty weight 250 kg (551 lb) and (optimistic?) design speed 270 km/h (168 mph).

Enthusiast Streamlined low-wing single-seater with spatted landing gear built by RIIGA (Riga) to attack national long-distance class record. Completed 75 and exhibited in Moscow 76, possibly without having flown.
DIMENSIONS Span 8,0 m (26 ft 3 in); length 7,6 m (24 ft 11¼ in).
ENGINE One 140 hp Avia M 332 inverted 4-inline.
WEIGHT Max 750 kg (1,653 lb).
PERFORMANCE Max speed 265 km/h (165 mph); service ceiling 6 km (19,700 ft); range 3000 km (1,864 miles).

MVTU Curious reverse-delta (straight leading edge, trailing edges meeting at extreme tail) with 25 hp pylon-mounted engine described as 'trainer and for testing future lunar vehicles' when displayed at 6th Scientific and Technical Youth Exhibition 76.

Strekoza Second avietka with this name (dragonfly), built 77 by SKB at Kuibyshev Aviation Institute, marking their return to fixed wings. Materials steel tube, light alloy, fabric and glassfibre. Junkers-type double-wing of R-P-14 profile with flaps inboard and ailerons outboard. Engine mounted far ahead of leading edge on centreline, with pilot gondola underneath hung from spars and carrying tail on lattice boom of inverted-triangle section. Two bracing struts to wing, and two linking main gears to rear of engine mount.
DIMENSIONS Span 7,2 m (23 ft 7½ in); length 6,07 m (19 ft 11 in); height 2,0 m (6 ft 7 in); wing area 9,8 m^2 (105·5 ft^2).
ENGINE One 25 hp VIKhR-25 flat twin.
WEIGHTS Empty 150 kg (331 lb); max 230 kg (507 lb).
PERFORMANCE Max speed 100 km/h (62 mph); cruise 90 km/h (56 mph); landing 45 km/h (28 mph).

KuAI Szmiel Built in parallel with *Strekoza*, this has more conventional layout, with engine on nose and wings attached each side above single-seat cabin. Triangular truss fuselage has apex along underside instead of top, and is fabric-covered throughout; tailplane part-way up fin. Name means bumblebee.
DIMENSIONS Span 7,6 m (24 ft 11¼ in); length 5,42 m (17 ft 9½ in); wing area 10,24 m^2 (110·2 ft^2).
ENGINE One 38 hp M-73 flat-twin aircooled.
WEIGHTS Empty 220 kg (485 lb); loaded 310 kg (683 lb).
PERFORMANCE Maximum speed 120 km/h (74 mph); take-off 400 m/55 km/h; landing 300 m/50 km/h.

Semurg Meaning bird of happiness, name applied to small passenger jet exhibited as model 80 and stated to have flown 81. Design group under Professor A. Badyagin, with assistance from Tashkent Poly and local aviation plants. Pilot plus 5/7 passengers or two stretchers or 800 kg (1,323 lb) cargo. Clean high wing, T-tail, engine located in rear fuselage and fed by dorsal inlet. Small APU turbine provides thrust for emergencies.
ENGINE One AI-25.
WEIGHTS Max 3200 kg (7,055 lb).
PERFORMANCE Cruise 360 km/h (223 mph); take-off 150 m (492 ft).
No other data.

Kh-12S Second homebuilt by Viktor Dmitriyev, driver at Kirghiz University, Frunze, flew early 81. Ultralight built by KB formed with various friends.
DIMENSIONS Span 5,54 m (18 ft 2 in); length 3,35 m (11 ft 0 in); height 1,26 m (4 ft 1⅔ in).
ENGINE One 20 hp by Nikolai Kitz.
WEIGHT Empty 51 kg (112 lb).
PERFORMANCE Speed 105 km/h (65 mph); take-off 30 m (98 ft).

Gamlet Built 82 by Aeroprakt syndicate, designation A-11M, low glassfibre wing Wortmann 02-196 profile, wood fuselage with upward-hinged canopy, tricycle gear with leaf-spring legs, 24-lit tank. Flown 83 with 30-hp Vikhr engine, then with 40-hp Ch-Z (high prop axis because of reduction drive). Max 150 km/h (93 mph).

Troika Tandem-seat pusher biplane built 84 by Alekseyevich Khobutovskii. Mainly light alloy, wings fabric aft of main spar, twin-fin tail carried on booms, fixed nosewheel gear, open cockpits, tanks in upper centre section.
DIMENSIONS Span (upper) 7,3 m (23 ft 11½ in); length (excl probe) 6,9 m (22 ft 7⅝ in); wing area 10,8 m^2+9,7 m^2 (221 ft^2).
ENGINE One Czech M-331 Mikron 4, 115 hp.
WEIGHTS Empty 470 kg (1,036 lb); max 670 kg (1,477 lb).
PERFORMANCE Max speed 150 km/h (93 mph); range 160 km (100 miles); TO 120 m, ldg 80 m.

Oleg Antonov Conventional low-wing design, named for famed Ukrainian designer, built Pyatigorsk 85 by L. F. Mayentyako. Fixed tailwheel gear, sliding canopies over tandem cockpit, 1,8 m c/s prop.
DIMENSIONS Span 8,74 m (28 ft 8 in); length 6,25 m (20 ft 6 in); wing area 13,6 m^2 (146 ft^2).
ENGINE One 105 hp Czech Walter Mikron.
WEIGHT Loaded 705 kg (1,554 lb).
PERFORMANCE Max speed 170 km/h (106 mph); SL climb 3,5 m/s (690 ft/min).

Duet Conventional low-wing monoplane built 87 at Kalinin by A. Einovkin. Untapered 15% wing with ply to main spar, fabric aft, outer panels dihedral, split flaps. Steel-tube fuselage, fixed tailwheel gear, side-by-side seats, dual control, hinged canopy.
DIMENSIONS Span 9,0 m (29 ft 6⅜ in); length 6,3 m (20 ft 8 in); wing area 11,25 m^2 (121 ft^2).
ENGINE One 105 hp Walter Mikron.
WEIGHTS Empty 525 kg (1,157 lb); max 750 kg (1,653 lb).
PERFORMANCE Max speed 150 km/h (93 mph); TO 150 m, ldg 100 m.

Baikal Unconventional powered sailplane built 87 by Vladimir Fedorov. Streamlined, tandem seats under hinged canopies ahead of slender wing with flaps and airbrakes, tandem centre-line wheels, two oblique pylons carrying engines driving tractor 1,4 m props, retractable into dorsal fairings for soaring.
DIMENSIONS Span 18,0 m (59 ft 0⅝ in); length 8550 m (28 ft 0⅝ in); wing area 18,9 m^2 (203 ft^2).
ENGINES Two 40 hp.
WEIGHT Loaded 817 kg (1,800 lb).
PERFORMANCE Max speed 150 km/h (93 mph).

Argo-02 Single-seater built 87 at Kalinin by Ye. Ignatyev, Yu. Gulakov and A. Abramov. Untapered low wing, mainly wood/ply/fabric, steel-leaf main legs and skid. Belt reduction to 1430 mm prop.
DIMENSIONS Span 6,3 m (20 ft 8 in); length 4550 m (14 ft 11 in); wing area 6,3 m^2 (67·8 ft^2).
ENGINE One RMZ-640.
WEIGHTS Empty 145 kg (320 lb); max 235 kg (518 lb).
PERFORMANCE Max speed 130 km/h (81 mph).

M-5 Feniks Ultralight twin built 87 at Kuibyshyev by N. P. Masterov. Metal stressed-skin, untapered low wing, T-tail, fixed nose-wheel gear, engines on horiz pylons above wing, 1·5 m between centrelines of 850-diam pusher props, fighter-type canopy, underfloor 25 lit A-70 fuel.
DIMENSIONS Span 7,0 m (22 ft 11½ in); length 4,9 m (16 ft 0⅞ in); wing area 5,6 m^2 (60 ft^2).
ENGINES Two 25 hp Vikhr.
WEIGHTS Empty 160 kg (353 lb); max 254 kg (560 lb).
PERFORMANCE Max speed 195 km/h (121 mph); min speed 70 km/h (43·5 mph); duration 80 min; TO 110 m, ldg 100 m.

Debut Cabin biplane built 89 at Voronezh, fourth design by V. P. Pivovarov and S. and P. Goltsyov. Mainly light alloy and steel tube, cockpit with side-by-side dual and aft-facing rear seat all covered by canopy sliding over rear fuselage, fixed tailwheel gear.
DIMENSIONS Span 10,95 m (35 ft 11 in); length 6,7 m (21 ft 9¾ in); wing area 24,0 m^2 (259 ft^2).
ENGINE One 140 hp Czech M-332.
WEIGHTS Empty 602 kg (1,327 lb); max 800 kg (1,764 lb).
PERFORMANCE Max speed 150 km/h (93 mph).

VK-8 Aushza Another Lithuanian, Vladas Kensgaila, won SLA-89 (5th national light-aviation convention) with this low-wing multirole aircraft; hi-lift slats/flaps, fixed tailwheel gear, 2-seat enclosed cockpit, c/s prop, payload 1 t.
DIMENSIONS Span 15 m (49 ft 3 in); length 9,57 m (31 ft 4¾ in); wing area 28,4 m^2 (306 ft^2).
ENGINE One M-14P.
WEIGHTS Empty 1140 kg (2,513 lb); max 2,3 t (5,070 lb).
PERFORMANCE Max speed 250 km/h (155 mph); econ cruise 160 km/h (99 mph); range (max payload) 450 km (280 miles); TO 35 m, ldg 60.

I-2 Impuls Light twin built at Kuibyshyev 89 by Vladimir Gaslov, low wing, mainly wood and glassfibre, large ailerons but small flaps, nosewheel gear with elec retraction, side-by-side seats under gull-wing canopies, 1,32-m fixed-pitch props.
DIMENSIONS Span 7,2 m (23 ft 7½ in); length 4860 m (15 ft 11⅜ in); wing area 7,48 m^2 (80·5 ft^2).
ENGINES Two RMZ-640 with belt reduction drive.
WEIGHT Max 600 kg (1,323 lb).
PERFORMANCE Max speed 180 km/h (112 mph).

Optimist Hi-wing single-seat pusher built 89 at Sverdlovsk by Viktor Babov. Light-alloy/fabric, tail carried on drawn tube, enclosed cockpit, levered nosewheel, steel-leaf main legs, tank above low-mounted engine with long belt drive to propeller.
DIMENSIONS Span 8450 (27 ft 8⅝ in); length 5150 (16 ft 10¾ in); wing area 9,0 m^2 (96·9 ft^2).
ENGINE One 36-hp DVS.
WEIGHTS Empty 210 kg (463 lb); max 301 kg (664 lb).
PERFORMANCE Max speed 130 km/h (81 mph).

A-15 Neat single-seater built 89 Kuibyshyev by Igor Vakrushyov. Mainly wood, untapered wing with almost full-span aileron/flap under TE, steel-leaf nosewheel gear, upward-hinged canopy.
DIMENSIONS Span 6,8 m (22 ft 3¾ in); length 4,5 m (14 ft 9⅛ in); wing area 5,2 m^2 (56 ft^2).
ENGINE One RMZ-640.
WEIGHTS Empty 135 kg (298 lb); max 220 kg (485 lb).
PERFORMANCE Max speed 140 km/h (87 mph).

Chibis Tandem-seat pusher built 89 at Sverdlovsk by V. Kalyut and A. Sukhov. All-wood, levered nosewheel, steel-leaf main legs, direct drive to 1,7 m prop.
DIMENSIONS Span 10,3 m (33 ft 9½ in); length 6,1 m (20 ft 0 in); wing area 13,4 m^2 (144 ft^2).
ENGINE One flat-4 45 hp.
WEIGHTS Max 400 kg (882 lb).
PERFORMANCE Max speed 120 km/h (75 mph).

M-12 Kasatik High-wing 3-seat pusher twin by N. P. Masterov group, now based at Samara, details (but not aircraft) exhibited Moscow aero show 93.
DIMENSIONS Span 10 m (32 ft 9¾ in); length 6,2 m (20 ft 4 in); wing area 12,2 m^2 (131 ft^2).
ENGINES Two 35 hp Vikhr-30.
WEIGHTS Empty 300 kg (661 lb); loaded 520 kg (1,146 lb).
PERFORMANCE Max speed 130 km/h (81 mph); min speed 75 km/h (46·6 mph); climb 180 m (590 ft)/min; range 400 km (249 miles); TO/ldg 150 m.

Ekranoplans

Absent from 83 volume, this section can hardly be omitted. Even aviation enthusiasts are often unaware that former Soviet aerospace industry created a remarkable, and often awesome, new species of vehicle which combines modern airframes with jet power on a scale previously seen only in shipbuilding.

Ekranoplan means 'screen plan', in sense that a screen covering an area sustains a cushion of air at slightly greater than atmospheric pressure, enabling a vehicle to fly near Earth's surface. Unlike ACV (hovercraft) an Ekranoplan generates its cushion in cruising flight by virtue of its forward motion, uses aerodynamic controls, and banks in turns. Unlike an aeroplane it generates most lift by underside of wing(s), not upper surface.

Thus, Ekranoplans are vehicles which, by avoiding contact with surface, can travel much faster than traditional ships or land vehicles. At rest they need powered lift in order to rise from surface (in most cases, water); then they generate lift by virtue of their speed and proximity to Earth's surface, using wings or suitably profiled bodies to obtain greater lift than they could at high altitude. Most are not intended to fly in normal sense (but, eg, see S-90-200). Rather should they be considered as ships which can sustain themselves in a medium (air) whose resistance is only 1/800th that of water.

Past inventors worked on air-lubricated ships, hydrofoil boats, air-cushion vehicles (hovercraft), surface-effect ships, ground-effect machines and surface-effect craft. English term for Ekranoplan was originally WIG (wing in ground-effect) craft and today wingship. Imagine emotions aroused by enormous aircraft-like structures moving with jet speed close above the sea, or even flat Tundra! Mikhail Simonov, hard-bitten head of Sukhoi, said 'I was involved in many tests . . . land, water and airborne . . . but I never experienced such delight and excitement as on wing-effect machines'.

Nevertheless, maybe some in Soviet Union tried to run before they could really walk. Building a jet-propelled merchant ship is fine, so long as you know you are always in control. Supertankers take miles to slow down from 15 knots. You need a bit more real estate to slow down something almost as big from 250 knots. But the prospects are exciting. Previous military commanders or commercial operators had a choice of three types of vehicle: displacement ships, which despite anti-roll systems respond in a sick-making way to severe sea states, and in any case are limited to very slow speeds; hydrofoils, which cannot be made very large and whose problem of cavitation limits speed to about 120 km/h (65 kt); and air-cushion vehicles, which can be made slightly bigger than hydrofoils and travel slightly faster, but can still make passengers feel ill and have to slow down in severe seas. Ekranoplan promises vehicles with no obvious limit in size, which can travel five times as fast as previous large marine craft, hardly respond to rough seas, are untroubled by baulks of timber and other floating obstructions, and need no power to maintain height (except to overcome backward component of resultant wing force), riding at low altitude automatically maintaining a mean flying height.

Workers in related fields over past 200 years numbered hundreds. A few were ostensibly aircraft constructors (Do X flying boat of 29 used surface effect whenever it could). One of first to build a wing-effect vehicle was T. Kaario (Finland); other pioneers included G. Yurg and Alex Lippisch of Germany, Lippisch testing small vehicles in USA and Germany in 61-80. Soviet workers included B. N. Yur'ev (CAHI) and P. I. Grokhovskii.

CCB-SPK

Rostislav Yevgenievich Alekseyev was born Dec 16, no relation to aircraft designer of same surname. He was General Constructor at NPO 'CCB po SPK' (Central Hydrofoil Design Bureau), at Gorkii (Nizhni Novgorod). He died 80, but bureau is named for him, today led by Viktor V. Sokolov. Commercial production and sales are handled by Amfikon (see below), Oboroneksport and Transal (*Transport Alekseyev*). Serious funding began 61, when bureau

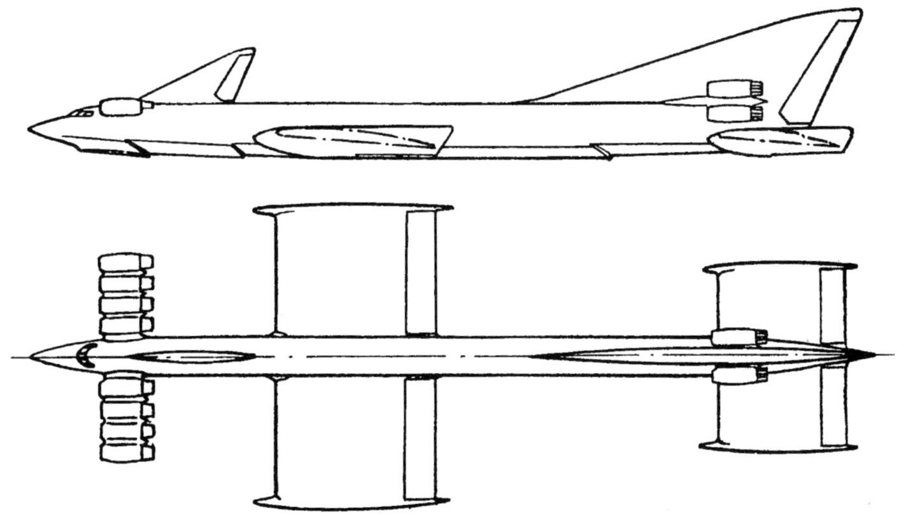

Typical 62-63 tandem-wing design

SM-1

had established hydrofoils in production and was considering how to overcome limitations. Ekranoplan offered independence of water or terrain beneath (subject to limitations of gradient or obstructions) and ability to reach very high speeds. Subsequently absence of wavemaking or damage to vegetation were seen as other attributes. Initial studies involved lightly loaded hydrofoils running just under (almost along) water surface to establish optimum design for longitudinal (pitch) stability. Models were used to explore free-flight configurations. Tandem-wing layouts were soon replaced by wing/tailplane.

SM-1 Three-seat full-scale vehicle, unswept central wing with tip skegs (vertical plates) to contain dynamically generated cushion, trailing-edge flaps, low tailplane and elevator, fixed fin amidships, external turbojet ahead of sharply swept canted fins.

SM-2 Three-seater, ship bow, central keel, deeper wing skegs, aft fixed spray deflector, large single fin/rudder carrying swept tailplane/elevators, dorsal inlet to internal turbojet.

SM-2P7 Small single-seater to explore nose inlet to feed cushion retained by twin skegs projecting ahead of nose under fuselage.

SM-3 Wing of enormous chord, bluff 'ski' nose with one main and eight aux inlets.

SM-4 Two-seater, nose and multiple dorsal inlets for lift, aft dorsal inlet for propulsion, unique dorsal nose rudder.

SM-5 Powered lift from turbojet with dorsal inlet and two vectored nozzles blowing under wing; spray wall ahead of aft inlet to propulsion engines.

SM-8 Research vehicle scaled up from SM-5, starting engine fed from 'chimney' inlet discharging through eight pivoted nozzles simulating KM system; cruise engine retained at rear, tailplane with dihedral.

KM Designation from *Korabl' Maket* (ship model) but popularly interpreted as *Kaspian Monster*. Boldest of all Ekranoplans, leaping from 3,5 t to 500 t! Aerodynamics/control scaled from SM-8, but eight starting engines on fixed stub wings at top of fuselage behind flight deck. Cruise propulsion: two similar engines with inlet spray deflectors high on each side of fin. Controls: jet deflectors to vector down starting jets under wing, two powered flaps forming each wing TE, eight sections of powered elevator and three powered air rudders. Hull with planing bottom under forward part, with succession of steps to assist unstick. Construction Al/Ti stressed skin. First of eight completed 65; hull needed strengthening for bad-weather operation. According to *Kryl'ya Rodiny* 'The great ship – which was not a ship nor an aeroplane – always appeared in early morning and roared across unruffled surface of Caspian. At times, as if tired, it settled; then, after a short run, it rose into air and at low level vanished into distance. Then one day, in dense fog, it disappeared'. This was in 69; captain lost visual

SM-2

SM-2P7

SM-3

SM-4

SM-5

SM-8

KM-04 in spring 74

SM-6 being refuelled by KM-08 in 78

horizon and entered uncontrolled pitching which resulted in a crash and fire. Another lost in 80, again 'pilot error'.
DIMENSIONS Span 32 m, later 40 m (131 ft); length 92 m to 106 m (348 ft).
ENGINES Ten (8+2) VD-7.
WEIGHT Loaded, normal 495 t (1,091,000 lb), max 540 t (1,190,500 lb).
PERFORMANCE Max cruise 500 km/h (311 mph); range, up to 3000 km (1,864 miles) at economical 400 km/h (248 mph).

Lun' (Lune), Spasatel' Slightly smaller successors to KM, chief designer V. Kirillovyikh. Names mean Harrier and Lifesaver, former a multirole craft with MF (Sov Navy). Differences compared with KM: greater span, eight turbofans at front on pivots to blow under wing at start, no separate cruise engines, shorter fuselage of modified profile; taller less-swept fin carrying large swept tailplane without dihedral, four flaps on each wing, three rudders but six elevators, hydraulically-damped ventral ski. Radar on nose, fore/aft sensor pod above tail, and aft control station with forward-facing windows projecting ahead from top of fin. First Lun' completed 87 as attack version; aft control station oversees loading and operation of three pairs cruise missiles (3M80 or 80M Moskit [NATO 'SS-N-22 Sunburn']), fired from retractable launchers. *Konversiya* led to *Spasatel'* rescue version; aft station oversees use of three pairs rescue capsules for 500 persons replacing missiles, second ship to this standard for marketing by GED/Oboronexport.
DIMENSIONS Span 44,0 m (144 ft 4 in); length 73,8 m (242 ft 1½ in); hullborne draught 2,5 m (98 in).
ENGINES Eight NK-87.
WEIGHTS Loaded 400 t (882,000 lb).
PERFORMANCE Speed (3 external rescue capsules) 450 km/h (280 mph); max 550 km/h (342 mph); range 3000 km (1,864 miles); cruise height 1-4 m (max 13·1 ft); operable in 3,5 m waves.

A-90 Orlyonok First large wingship regarded as commercially practical and operated in substantial numbers. Name = Eaglet, designed 79 under Sokolov as VMF assault craft. Operable in Sea State 5, able to travel up beach and over unobstructed terrain. Conventional stressed-skin construction, aircraft-type systems. Wing

Lun' *of AV-MF Fleet*

Lun' *firing 3M80*

Spasatel' *model*

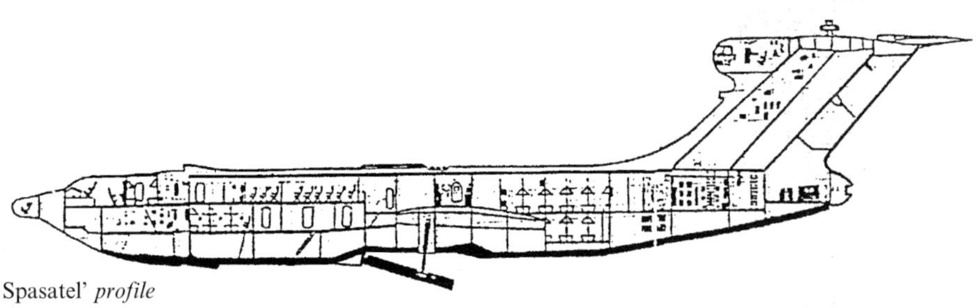

Spasatel' *profile*

mounted lower, thus smaller skegs at tips, each TE formed by five flaps. Tall swept fin with two balanced rudders carrying tapered and swept tailplane with zero dihedral and four elevators each side. Improved planing bottom, amphibious with two or four front wheels and transverse row of 10 mainwheels. Starting by two turbofans in nose fed by dorsal inlets with doors closed in cruise, jetpipes vectored to blow under wing to liftoff speed and then accelerate with jetpipes horiz blowing across wing; cruise propulsion by large turboprop with AV-68N propellers; vectored jets used again for alighting. APU in tail, fuel in wings, capacity with allowances 15 t for each 1000 km (621 miles) of operating range. Normal crew 5. Cargo version with hinged nose and ramps for loading 30 t cargo into hold 21 m × 3,2 m × 3 m. Passenger version with cabin 25 m × 3,3 m × 3 m for 75 (luxury), 150 normal or 300 (two decks), crew 9. Naval versions with pylon-mounted surv radar and other avionics, some with missile launchers and/or 76 mm gun turret. Arctic prospecting with *Progress* seismic sensor, Grad-AM system, MVS MAD, INEI-P analog recorder, ES-5017 collector, sonar, ADSR sea-bed seismic and GTMD gravity sensor, PSK-6 seismic sensor, proton magnetometer and other items; 2 diesel compressors, 2 diesel generators and ZKD12NL-520 for search propulsion at 5 or 10 km/h. SAR version with 10 rescue/medical crew and 150 rescuees, much special eqpt; schemed as rapid-response system carried to scene riding on An-225. In Sept 92 captain of Orlyonok-21 felt nose pitch down slightly, overcorrected and crashed from over 40 m, one fatality, vehicle a write-off. CIS navy received c30, marketing of civil versions now by LIKO, General Director B.V. Litvinenko.

DIMENSIONS Span 30,6 m or 31,5 m (103 ft 4 in); length 58,0 m (190 ft 3½ in); wing area 304,6 m^2 (3,279 ft^2).

ENGINES Starting, two NK-8-4K; cruise, one NK-12MK.

WEIGHTS Empty c83 t, normal fuel 28 t; typical payload 20 t; loaded 125 t-140 t (308,640 lb).

PERFORMANCE Cruise speed 350 km/h-400 km/h (217 mph-249 mph); range (cargo) usually 1100 km (684 miles), other versions up to 3000 km (1,864 miles).

Orlyonok II Projected civil transport with wing redesigned to have skegs at ends of untapered centre section and narrow-chord swept outer wings with downturned tips. Tailplane smaller and unswept.

SM-6 A-90 scaled down to match AI-25 starting engines and AI-24 cruise turboprop. Several examples built of this experimental design with main object improving ability to climb over obstacles. Track-mounted flaps, slightly swept tailplane, single elevator each side, single rudder, glazed cockpit, deflector strakes each side of nose so no doors over starting engines.

UT Small trainer with hyd-damped flaps under fuselage and tip skigs, pylon-mounted Czech piston engine, no starting engine.

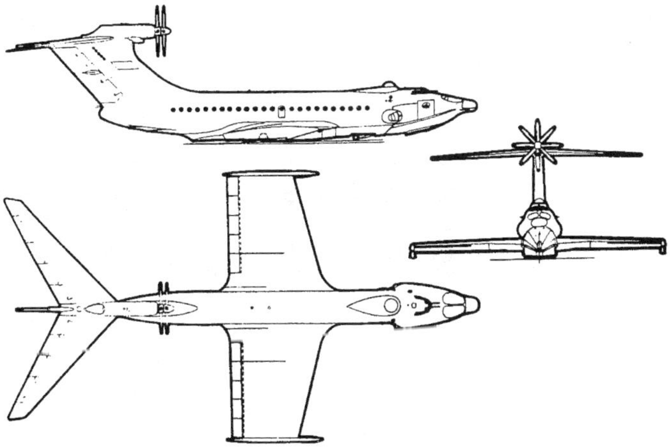

A-90 civil passenger version

A-90 (CIS Navy No 26) over Caspian

A-90 naval transport version

Orlyonok II *model (Ekranoflot titles)*

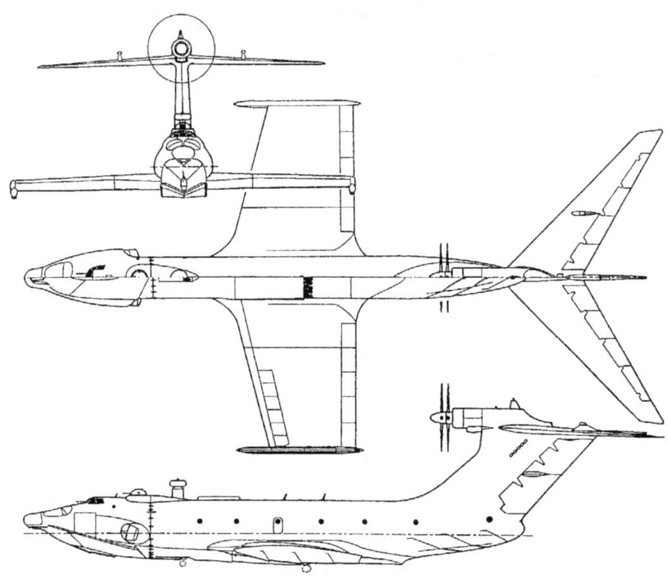

A-90 naval transport version with two radars and gun turret

SM-6 No 6M79

SM-6 No 6M80

UT (p. 514)

Vikhr-2 Projected open-sea 250-seater with SM-6 layout except sole propulsion two turbofans on pivoting mounts each side of bow. Name = Whirlwind.

DIMENSIONS Span 28 m (91 ft 10⅜ in); length 54 m (177 ft 2 in).

ENGINES Two D-36.

WEIGHTS Loaded 105 t (231,480 lb).

PERFORMANCE Cruise 280 km/h (174 mph); range 1500 km (932 miles).

Amfikon

Nizhni Novgorod enterprise *Amfibiinye Konstruktsii* with grandiose plans for range of wing-ships characterized by deep but short-span aerofoil wing and twin rear booms merging into fins carrying tailplane; multiple wing/fin/tailplane control surfs. NVA-30G twin turboprops on tailplane, 60G twin turboprops above centre, 60P twin turbofans above centre, 120P twin jets at base of fins, 120GP and 600GP twin turbofans above each tail boom. Artwork/models labelled 'design version of A. Sukhov'.

Beriev

In 59 R. L. Bartini (see main section of book) proposed giant vehicle called **M** able to rise from sea and fly at high speed to destroy Polaris submarines. Advantages, ability to operate in severe sea states, use ground effect to reduce fuel burn, and transit to operational area at high alt at jet speed. His own OKB being shut, he approached G. M. Beriev and was permitted to form design cell No 14 at GAZ-31, Taganrog. Test vehicle designed 61 by Bartini and Viktor Biryulin, later replaced by Nikolai Pogorielov, built by Beriev. Bartini lived Moscow, seldom visited Taganrog.

Be-1 Test vehicle to explore stability, control and performance. Central nacelle with single-seat cockpit and instrumentation projecting from deep high-lift wing of short span with TE flap. On each side at rear (TE in line) short conventional wings with flaps and endplates. Under outer edges of central wing, two floats with V planing bottoms under forepart plus V-type surface-piercing hydrofoils at front and

NVA-30G

Be-1

NVA-60G

Be-1 on test 62

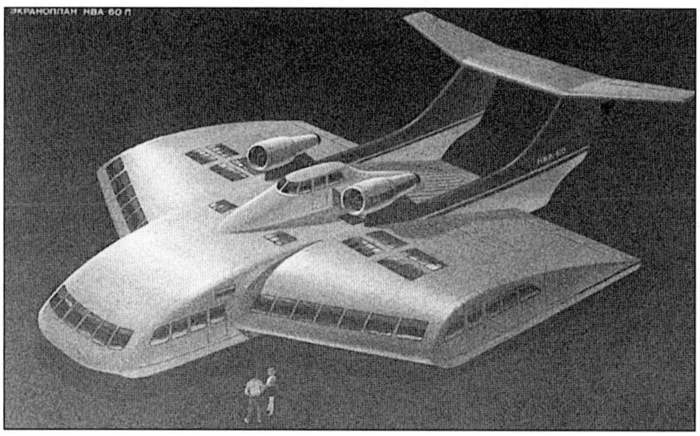
NVA-120GP

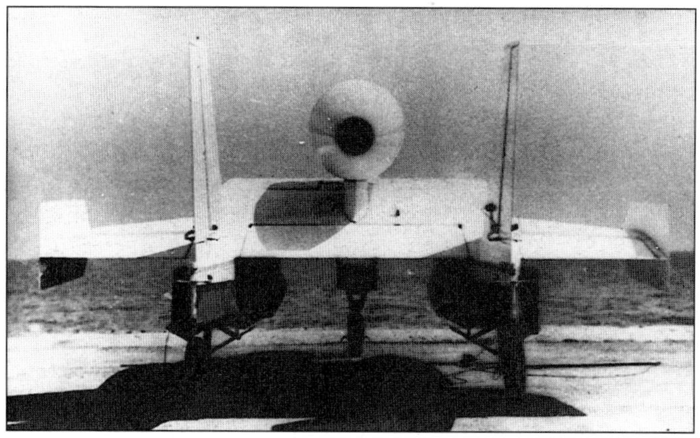
Be-1

rear. Above rear corners of central wing, swept vertical fin/rudder. RU-19 turbojet in dorsal nacelle, retractable tricycle land gear. No lift engines, cushion generated by ram air contained between floats; prolonged exploration of wing height, attitude, flap angles and other variables.

VVA-14 Research to support 2500-tonne ASW search/attack platform, which by 62 was called **M-62**. Be-1 scaled up with changes: central wing even lower aspect ratio, outer wings increased span with LE slats, TE flaps (rotating aft about 3 hinges 1 m below wing) and ailerons inboard of downturned tips, swept horiz tailplanes/elevators added. Pressurized nose for pilot and copilot on K-36L seats, mid compartment for 12 RD-36-35PR lift turbojets, two behind wing box and rest in left and right rows in front, two turbofans above rear of lifting body for cruise, two TA-6 turbogenerators, 15,5 t fuel in 12 cells and two integral tanks, SAU-M autopilot, powered controls and jet reaction controls. Designation = 'vertical take-off amphibian', three built. First, VVA-1M, No 19172, built with tricycle landing gear using Tu-22 nose gear, Tu-22 central bogie and complete 3M outrigger-gear pods. Weapon bays for two torpedoes or 16 PLAB-250, or 8 IGDM-500 mines or many other loads. First flight from Taganrog runway by Yuriy M. Kupriyanov and nav L.F. Kuznetsov 4 Sept 72. Flown to AOA beyond 30°. Later fitted with giant inflatable pontoons for VTO and STO from open sea. Converted to ekranoplan, see next.

DIMENSIONS Span over tip pods 30,0 m (98 ft 5 in); length (excl PVD) 25,97 m (85 ft 2½ in); lifting area 280 m² (3,014 ft²).
ENGINES Two D-30M, 12 RD-36-35PR.
WEIGHT Empty 35356 kg (77,945 lb); loaded 52 t (114,638 lb).
PERFORMANCE Max speed 760 km/h (472 mph), patrol speed (also min cruise speed) 360 km/h (224 mph); service ceiling 10 km (32,800 ft); range 2450 km (1,522 miles).

14M1P In 76 No 19172 was converted into an Ekranoplan. Pontoons replaced by rigid floats with no step but deep inclined skegs along sides to contain cushion at start. Two additional D-30M start engines added on sides of nose to

VVA-14 (initial tests from runway)

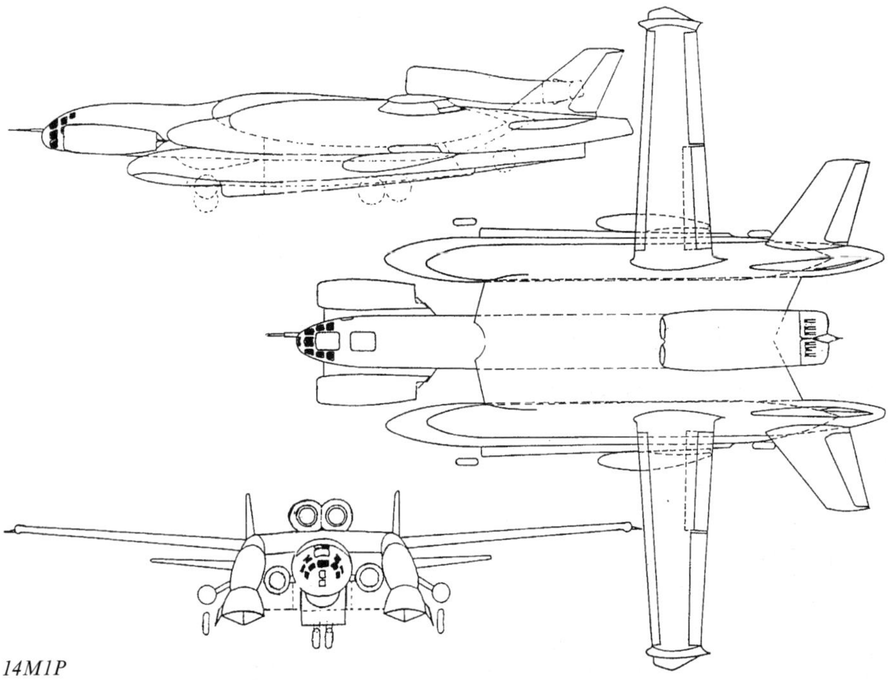

14M1P

14M1P with forward starting engines

VVA-14 testing inflatable pontoons

blow under wing between floats. Outrigger wheels attached to short anhedral wings, later removed during sea trials by test-pilot V. Demyanovskii, nav Kuznetsov and engineer I. Vinokurov. Bartini died in 74 and pressure of TANTK work on A-40 and A-50 forced abandonment. Third VVA-14, No 10687 (Aeroflot titles), suffered fire damage, dumped at Monino. Three-view drawings, see Addenda.

Studies Plan views show two of numerous studies by OKB into Ekranoplans ranging up to 1000 t, 700 km/h and range 8000 km.

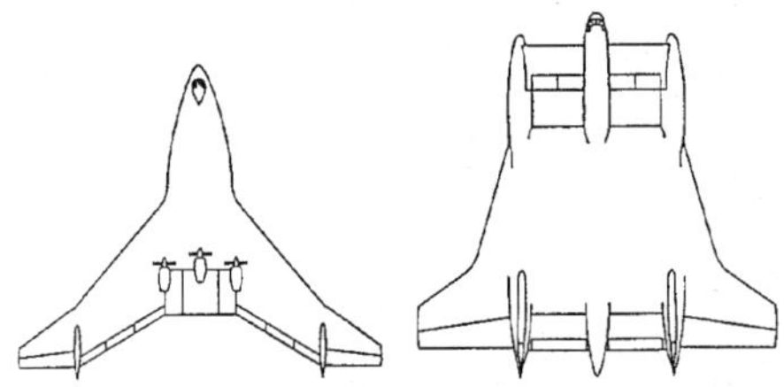

Beriev studies

CLST

Central Lab of Lifesaving Technology, Moscow, produced family of small machines inspired by Lippisch aerofoil boats. Cushion created under broad anhedral wing inboard of stab floats beyond which are small wings with sharp dihedral.

ESKA Prototype (data refer) 2 seats plus cargo, tested 29 Aug 73, sustained 30 m (100 ft) but design cruise height up to 1,5 m (5 ft). ESKA = Ekranoplan amphibious lifeboat. Four built by 65, many larger successors.

DIMENSIONS Span 6,9 m (22 ft 7⅝ in); length 7,55 m (24 ft 9¼ in); wing area 13,85 m^2 (149 $ft)^2$).
ENGINE One 30 hp M-63 motorcycle flat-twin.
WEIGHTS Empty 230 kg (507 lb); loaded 450 kg (992 lb).
PERFORMANCE Cruise 110 km/h (68 mph); TO distance 50 m snow, 90 m water; range 350 km (217 miles).

NGAZ Sokol

NizhnyeNovgorodskii State Avn-building works *Sokol* (Falcon) in production from late 93 with economical small vehicle. Though at same city, not part of CCB-SPK.

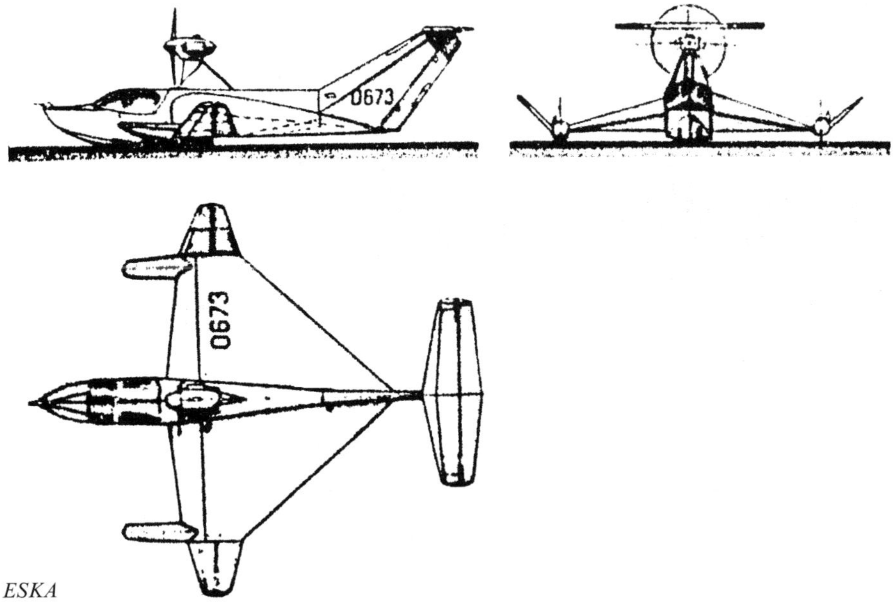

ESKA

ESKA

Volga 2 *test vehicle*

Sever-4 Utility *Kater-amfibiya* able to run over terrain with obstructions up to 300 mm (12 in) high. Name = North.
DIMENSIONS Span 2,5 m (8 ft 2½ in); length 5,8 m (19 ft 0½ in).
ENGINES Two RMZ-640.
WEIGHTS Empty 360 kg (794 lb); fuel 60 lit 92-oct; payload 4 adults or 430 kg (948 lb).
PERFORMANCE Speed (rough soil) 60 km/h, (snow) 80, (water) 70; endurance c3·5 h.

SDPP

Suda na Dynamicheskom Printsipye Podderzhaniya, dynamic-support craft; producing small passenger vehicles.

Volga-2 First of family of simple low-tech vehicles (Russian term, *Kater* = cutter) for lakes, rivers and flat land including ice/snow. Crew 2 and 8 passengers, rectangular wing blown by two 4-blade propellers in ring cowls with 5 horiz deflector vanes, twin inflatable floats with side skegs, T-tail. Autostable with simple controls, waves to 0,5 m (20 in), land gradient 10%. Production by Sukhoi group.
DIMENSIONS Span 7,63 m (25 ft 0½ in); length 11,43 m (37 ft 6 in); draft at rest 0,25 m (10 in).
ENGINES Two VAZ-413.
WEIGHT Empty 1,6 t (3,527 lb); loaded 2,5 t (5,511 lb).
PERFORMANCE Cruise 120 km/h (75 mph); range 500 km (311 miles).

Strizh Name 'swift', two-seat trainer by chief designer V. Bulanov. Configuration as Volga-2, but large tip floats replaced by skegs. Mainly Al-Mg alloy, fixed-pitch props, fuel 0,35 lit/km (8 mpg). Model shown 93 of 8-seater with dihedral winglets outboard of skegs.
DIMENSIONS Span 6,6 m (21 ft 7⅞ in); length 11,4 m (37 ft 4⅞ in).
ENGINES Two 160 hp VAZ-4133.
WEIGHT Loaded 1630 kg (3,593 lb).
PERFORMANCE Max speed 200 km/h (124 mph), cruise 175 km/h (109 mph); TO 500 m, alight 300 m; range 500 km (373 miles); riding height 0,3-1,0 m.

Raketa-2-2 Projected 90-passenger craft of different form with twin turboprops at front blowing under small canard wing with downturned tips ahead of rectangular main wing with tip skegs; cruise turboprop on fin carrying large lifting tailplane; 4 wing flaps, 4 elevs, 2 rudders; amphibious. Name = Rocket.
DIMENSIONS Span 19,5 m (63 ft 11¾ in); length 33,5 m (109 ft 10⅞ in).
ENGINES Three TV3-117Ye.
WEIGHT Loaded 31 t (68,342 lb).
PERFORMANCE Cruise 180 km/h (112 mph); range 500 km (311 miles).

Sukhoi

This large OKB has for many years been a partner in Ekranoplan devt with CCB-SPK 'Alekseyev', CLST and Krylov Shipbuilding Research Inst., and was responsible for construction of many large vehicles. Sukhoi Marine Systems is large independent company, but

Volga 2

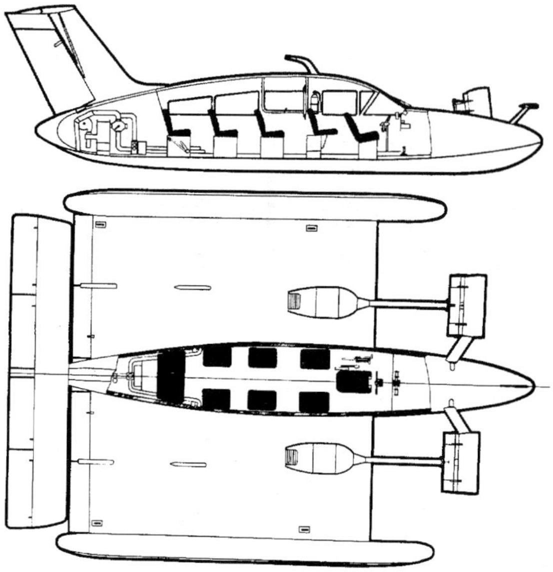

Volga 2

Strizh

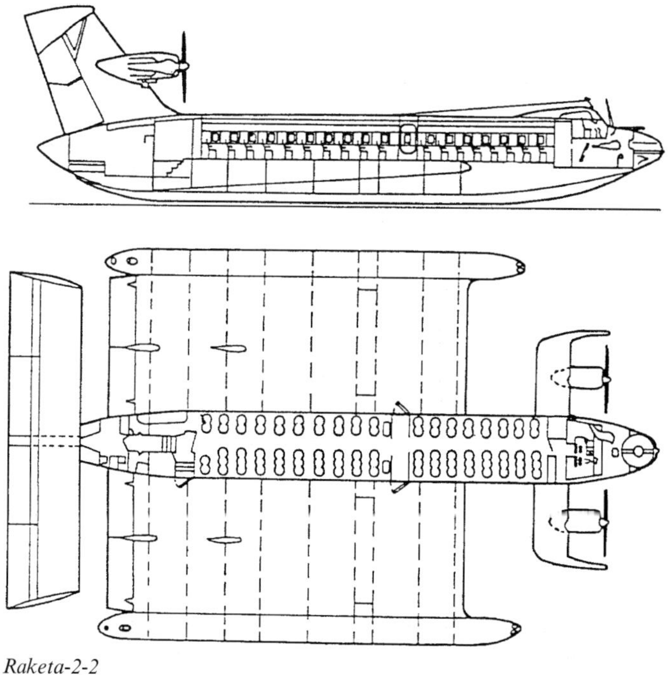

Raketa-2-2

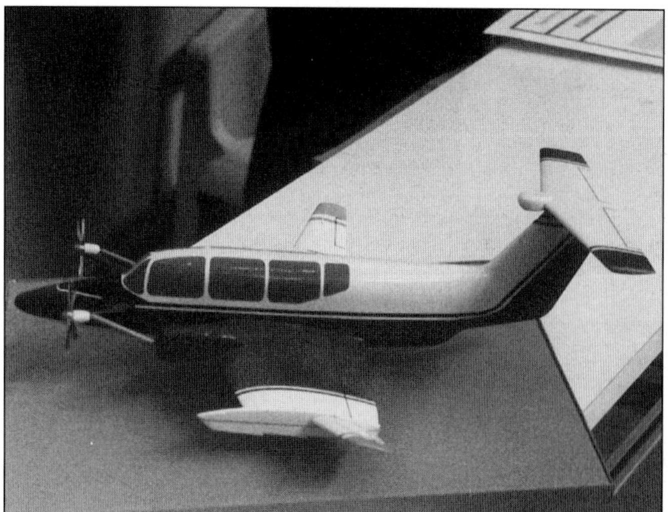

Model of 8-seat Strizh development

Raketa-2-2 artwork

S-90-200 model

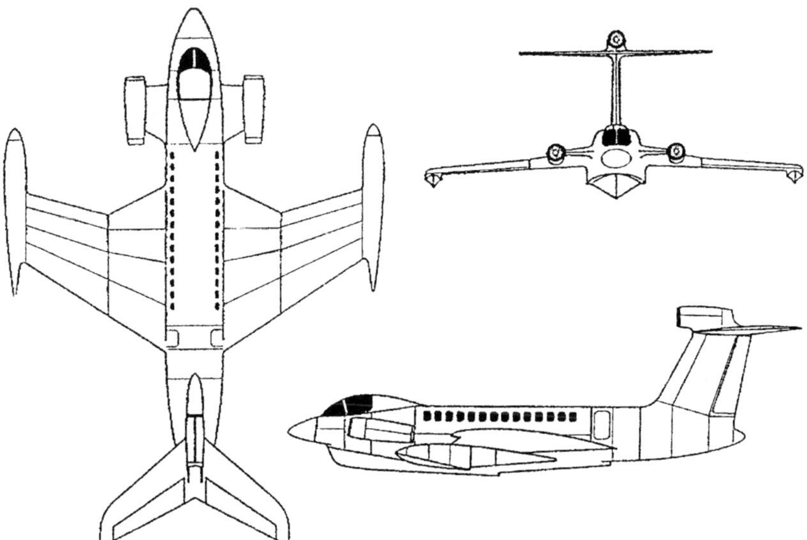

S-90

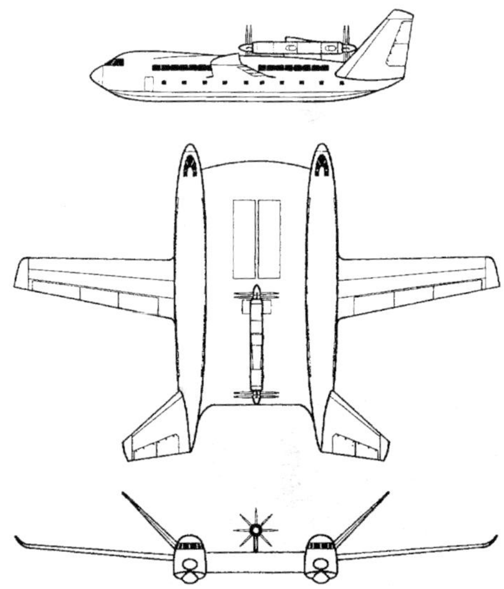

S-90-200

Ekranoplan design is at OKB, headed by Rollan G. Martirosov.

S-90 Intended to have true flight capability though normally in ground effect. Configuration as 3-view. Fuel 6 t, payload 5 t (typically 40 passengers). Not built.
DIMENSIONS Span 20,6 m (67 ft 7 in); length 28,0 m (91 ft 10$\frac{3}{8}$ in); wing area 120 m^2 (1,292 ft^2).
ENGINES Three HBPR turbofans.
WEIGHTS Empty 25,6 t (56,435 lb); loaded 35,6 t (78,483 lb).
PERFORMANCE Cruise 400 km/h (249 mph); altitude range 2,5 m-4000 ml; range 2000 km (1,243 miles).

S-90-200 Unrelated to S-90; amphibious open-sea transport, configuration as 3-view with twin planing-bottom hulls with two-decks for <400 pax or various other loads, crew 14. Seeking funds for prototype jointly with Aero Marine Singapore Pty and other possible partners. Design is being updated.
DIMENSIONS Span 60,9 m (199 ft 9½ in); length 40,6 m (133 ft 2½ in).
ENGINES Two NK-12MK (push/pull).
WEIGHT Max payload 25 t (55,116 lb); loaded 132 t (291,000 lb).
PERFORMANCE Max cruise 470 km/h (292 mph); cruise height 3 m (10 ft) with ability to climb if necessary to 2,5 km (8,200 ft); range 1200 km (746 miles).

S-90-8 Proof of concept, S-90-200 scaled down to use single Czech M 601 turboprop driving Avia Hamilton 5-blade prop. Eight seats, structure special VILS al-alloy. Su can afford to build this; big enough to be a viable commercial vehicle.

ADDENDUM

Most of the items on this and the following four pages were discovered too late for inclusion in the body of the book. Except for the tables of production numbers, each is keyed into a relevant page in the preceding text.

Alfa (p.18). This photograph of the prototype Finist was taken soon after its first flight in mid-94. Chief designer of Alfa is Slava Kondratiev.

Alfa Finist

Beriev (p.45). Previously known only for his seaplanes and flying boats, G.M. Beriev decided in 40 to try to add high-speed land-based aircraft. On 15 Feb he took *Eskiznyi* (preliminary drawing) of his B-10 project to A. S. Yakovlev, deputy head of Aviation Ministry, and urged that it be included in 40-41 plan. An interceptor and dive bomber, it was to be powered by two coupled tandem M-107 (VK-107) engines driving a single contra-rotating pusher propeller. Drawing shows scale of metres. With an NACA laminar wing, speed was calculated to reach 818 km/h (508 mph). General Direction of war material said "Realistic, but Beriev is busy with KOR-2 and Bolkhovitinov has *sparka* (twin) experience with Aircraft S, so project is transferred to him". It became Bolkhovitinov Aircraft I, suffering many problems. Power of two M-107 too much for structure, shaft torque and vibration. Engine plant manager Lavrentyev "We are overloaded with VK-105, VK-107 and M-120, pick a different engine". Bolkhovitinov chose twin M-40 diesels to save fuel weight, but this demanded complete redesign, incomplete at German invasion.

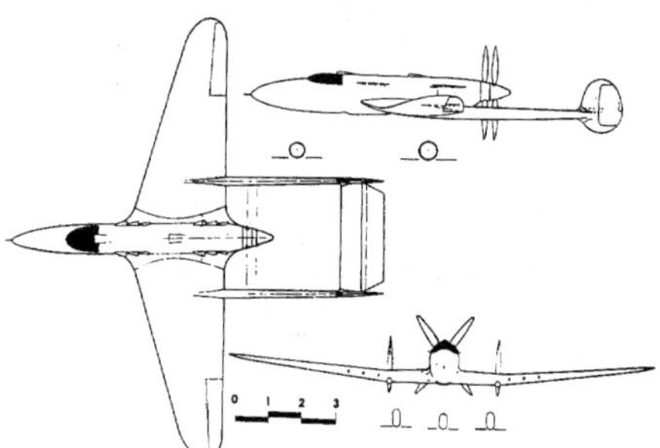

Beriev B-10

Bolkhovitinov (p.60). Seldom-illustrated second DB-A, with turbocharged M-34FRN engines and modified inboard nacelles housing retracted main landing gears and aft-facing turrets, as later used on ANT-42.

Bolkhovitinov DB-A

Borovkov-Florov (p.62). Recent copy of OKB drawing of IS-207 (unrelated to I-207), 41 fighter project with pusher M-71 and two Merkulov DM ramjets (in tail booms).

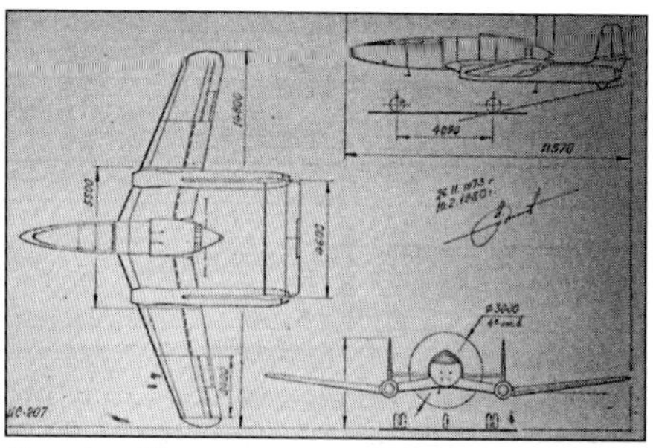

Borovkov-Florov

Chyetverikov (p.71). Sketch of 45-47 project for PTI (carrier-based torpedo fighter).

Chyetverikov PTI

Chyetverikov (p.71). Model of LK, of which 3-view appears earlier.

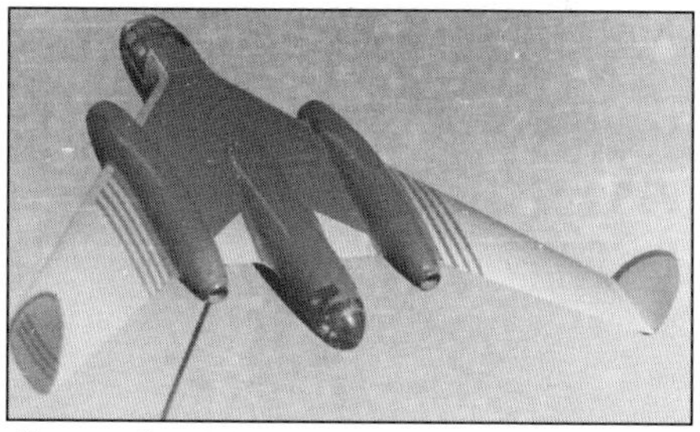

Chyetverikov LK

Grigorovich (p.82). Recently discovered *Eskiznyi* drawing of MM-1 (two M-5 engines) of 24-25; designation used again in 28 for monoplane minelayer.

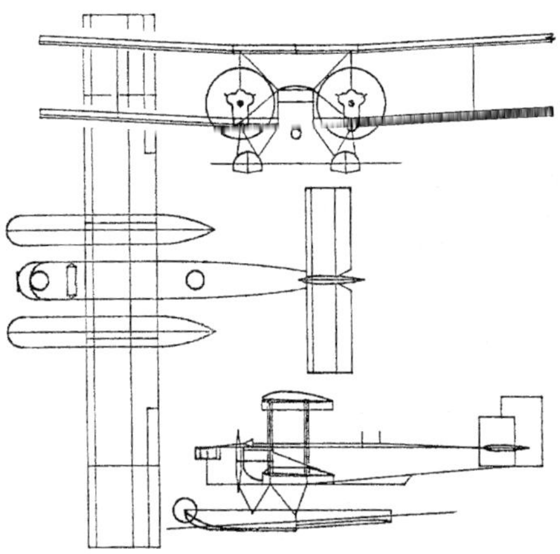

Grigorovich MM-1

Grigorovich (p.82). Until recently only known illustrations of Stal-MAI were of a model. Yu. V. Makarov's book *Letatel'nye Apparatye MAI*, published 94, contains this 3-view, with dimensions, previously unknown. Wing area 40 m^2 (430.6 ft^2), 830-hp M-34R, max speed 320 km/h (199 mph), landing 100 km/h; weights not yet discovered.

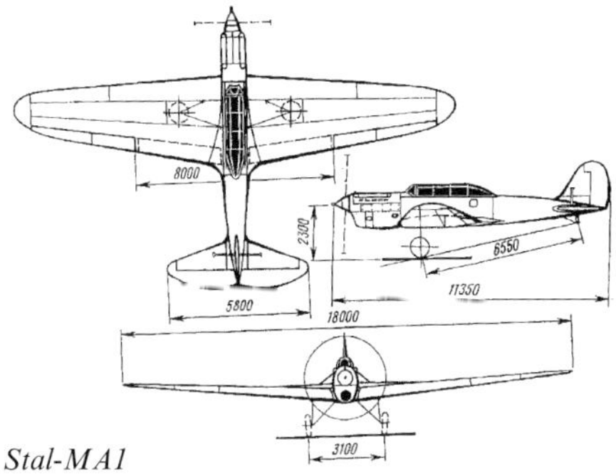

Stal-MA1

Kalinin (p.132). A comprehensive book about Kalinin's work – Vyacheslav Savin's *Planeta "Konstantin"*, published 94 – contains drawings of several of his unbuilt projects:

K-11 utility transport, 33 project, span 18,5 m (60 ft 8⅜ in), length 12,0 m (39ft 4½ in).

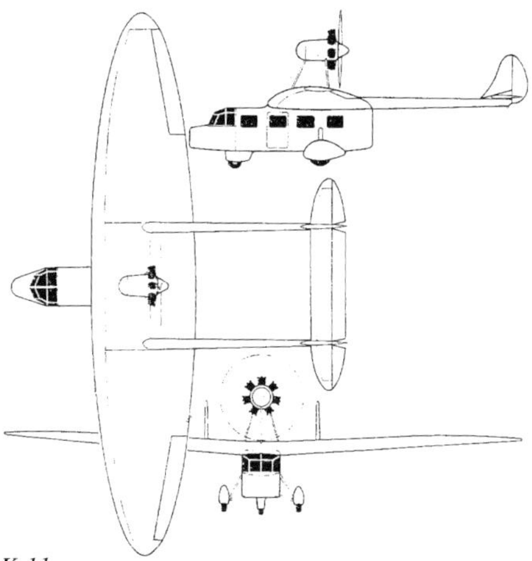

K-11

K-12 glider to test aerodynamics of K-12 bomber, span 10,45 m (34 ft 3⅜ in), length 5,21 m (17 ft 1⅛ in); flown by V. O. Borisov 33.

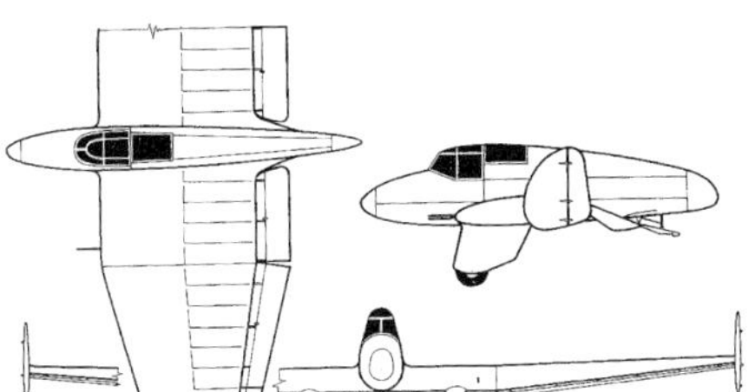

K-12 glider

K-12 bomber, initial project of 1933, with (lower side view) passenger version.

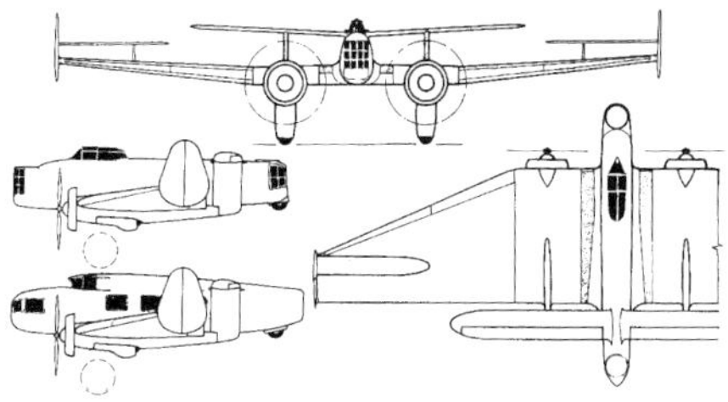

K-12

K-13 (SB) fast bomber, two AM-34F, span 23,0 m (75 ft 5½ in), length 13415 mm (44 ft 1½ in), with (lower side view) **K-14/1** passenger version, length 13,0 m (42 ft 7⅞ in).

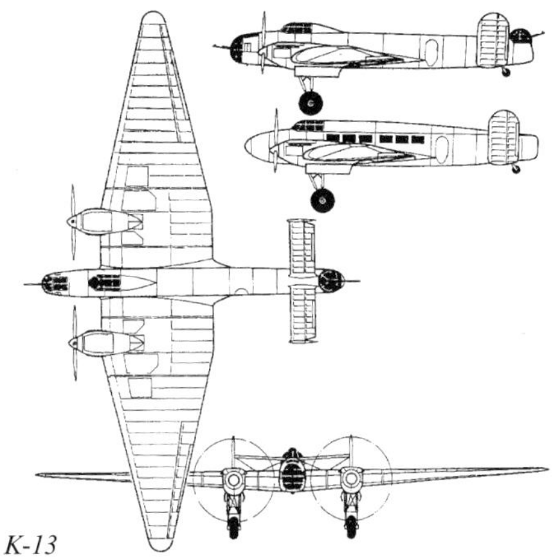

K-13

K-15 rocket aircraft, 36 project, span 9,5 m (31 ft 2 in), length 10,0 m (32 ft 9¾ in).

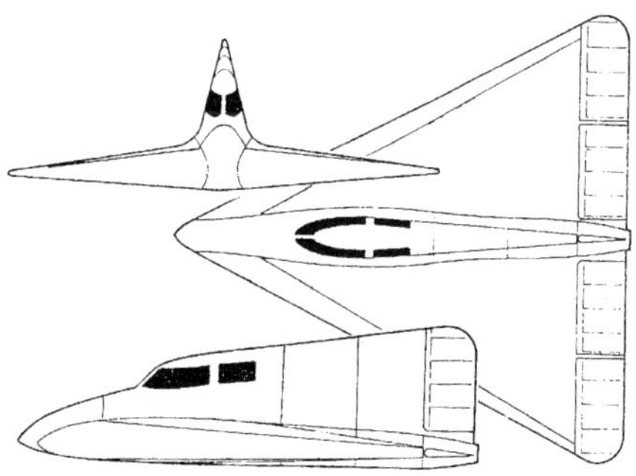

K-15

K-16, VS bomber-reconnaissance project of 36, remarkably small: span 10,0 m (32 ft 9¾ in), length 7,6 m (24 ft 11¼ in); 3200 kg (7,055 lb), speed 513 km/h (319 mph); engines "M-14 MARS", unknown to author.

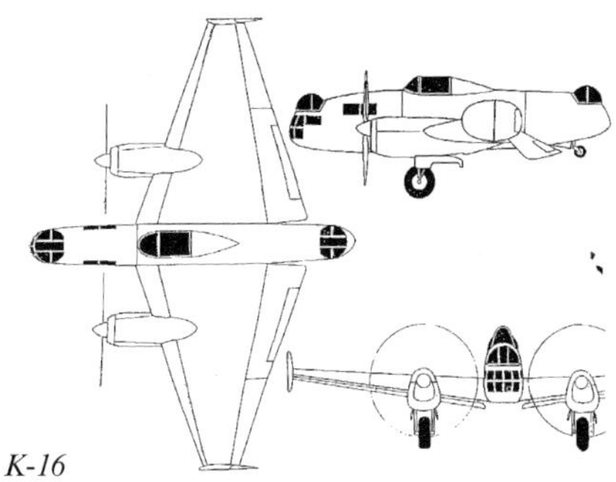

K-16

K-17 long-range bomber, Kalinin's last project, late 36, four M-85, span 35,0 m (114 ft 10 in), length 13,5 m (44 ft 3½ in), wing area 200 m^2 (2,153 ft^2), 30 t (66,140 lb), 365 km/h (227 mph), claimed 8000 km (4,970 miles).

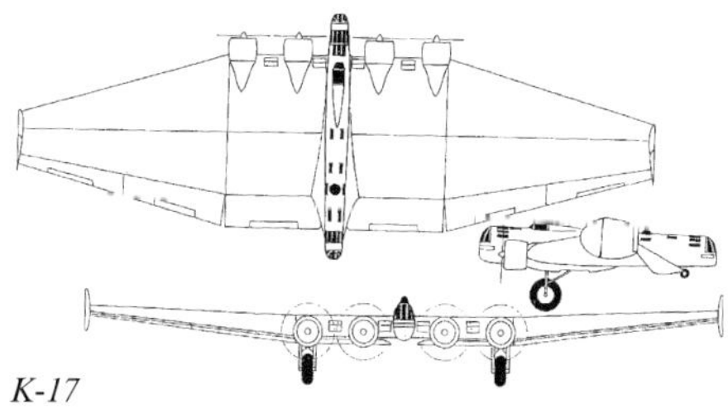

K-17

Kochyerigin (p.155). The first sketch of the OPB, quite different from the aircraft as built.

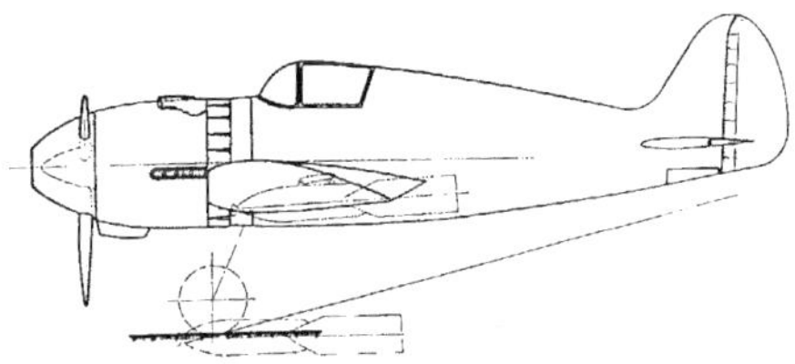

OPB

Kochyerigin (p.156). This is part of a recently unearthed three-view of a high-speed torpedo-bomber of 1940.

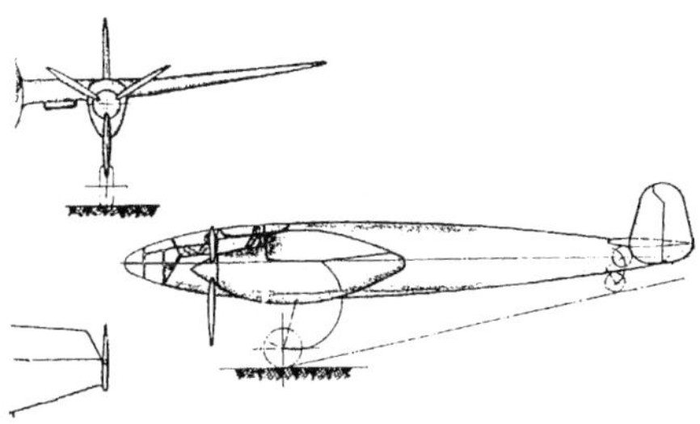

Kochyerigin torpedo-bomber

Kurchyevskii (p.161). A three-view has now been discovered of the BICh-17 tailless fighter, of basic design by Chyeranovskii.

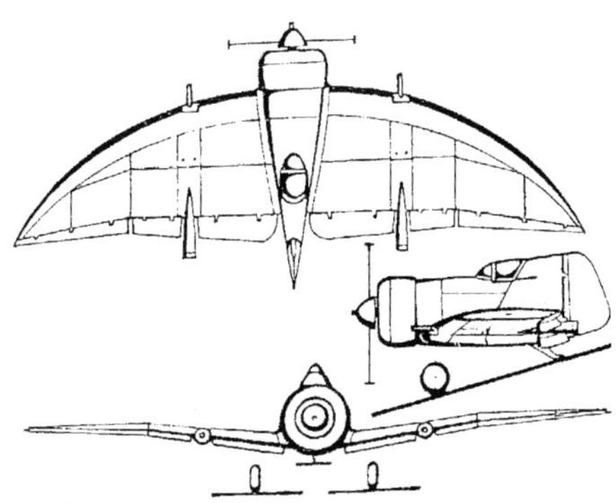

BICh-17

MAI (p.174). MAI book previously mentioned contains this 3-view of E-MAI, or EMAI, showing length was actually not 7,0 but 7,03 m (23 ft 0¾ in).

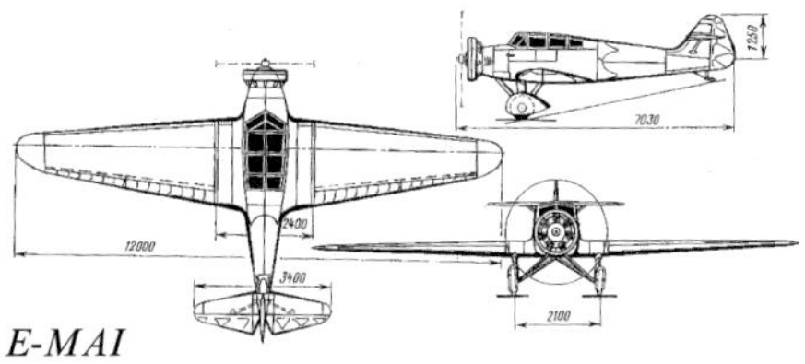

E-MAI

MiG (p.208). All MiG-21s, except possibly Ye-8, suffered from compressor stall in yawed flight or under other severe combat conditions. LII-VVS (air force flight research institute) at Kiev designed this improved inlet, extended forwards with a ring of large power-driven auxiliary doors. Above nose are two control pipe fairings. Radar was removed from this aircraft (Red 01 outlined in black). Results were excellent, but scheme was not adopted.

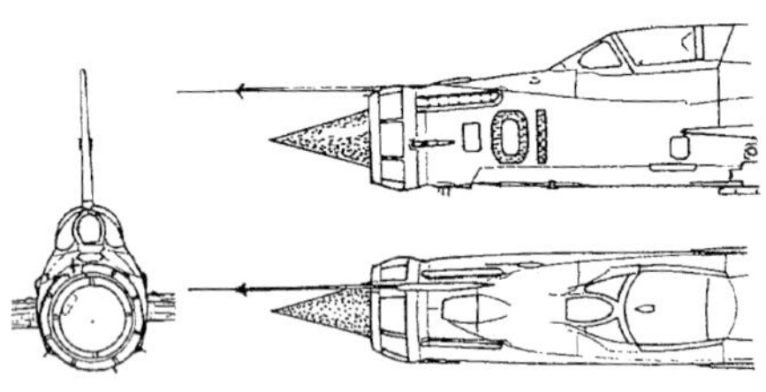

MiG-21 Kiev

MiG (p.191). A three-view of the I-320 (R-2).

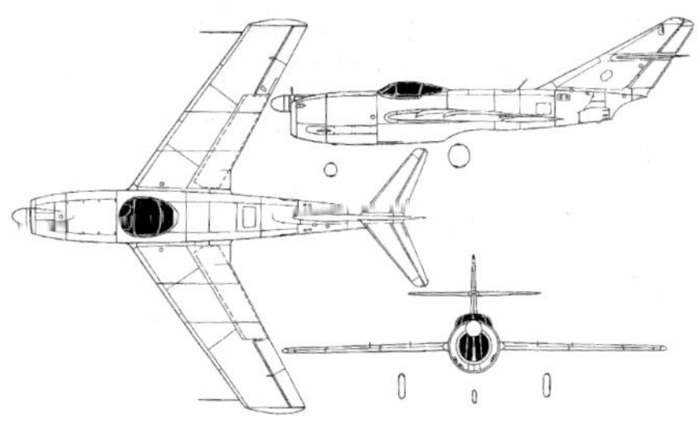

MiG I-320 (R2)

MiG (p.229). A three-view of the Type 110.

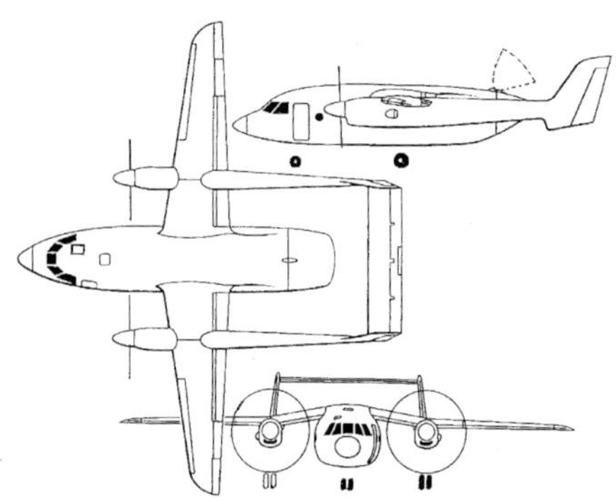

MiG Type 110

Mikhelson (p.231). A better indication of MP general arrangement is afforded by this recently discovered freehand *Eskiznyi* sketch. Side elevation shows hinged engine and wing of (TB-3?) carrier aircraft.

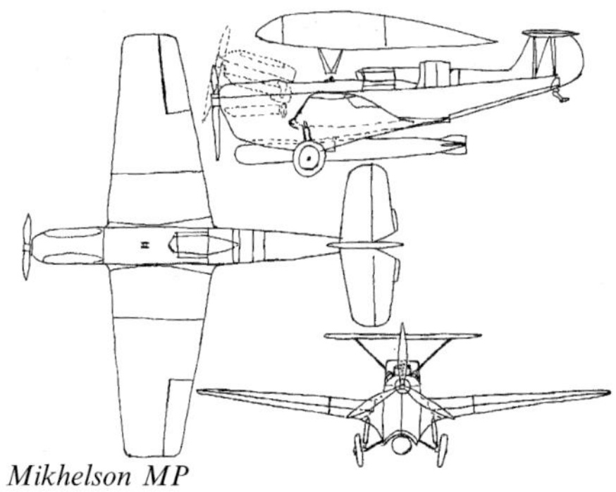

Mikhelson MP

Moskalyev (p.250). A further variant of this designer's Sigma series was this fighter of 33-34 period, powered by two 860-hp Hispano-Suiza 12Ybrs piston engines in tandem driving propellers turning in opposite directions (not a contraprop; either engine could be shut down in flight). Span 5,63 m (18 ft 5⅝ in); wing area 32,5 m² (350 ft²); loaded weight 3080 kg (6,790 lb); max speed 1000 km/h (621 mph, wildly optimistic); landing speed 125 km/h (78 mph).

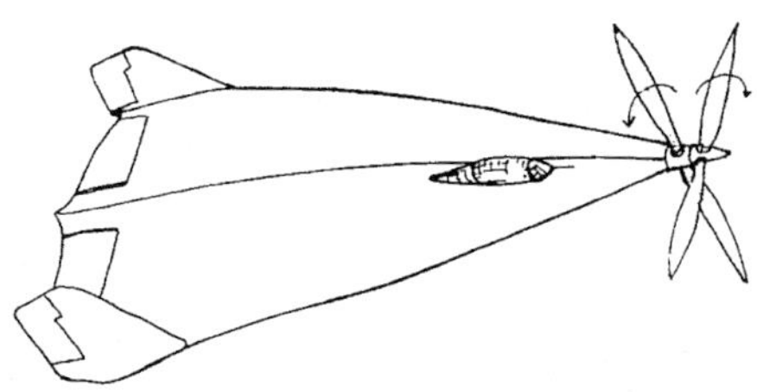

Moskalyev Sigma

Myasishchev (p.251). Recent copy of OKB drawing of DSB-17 (VM-24).

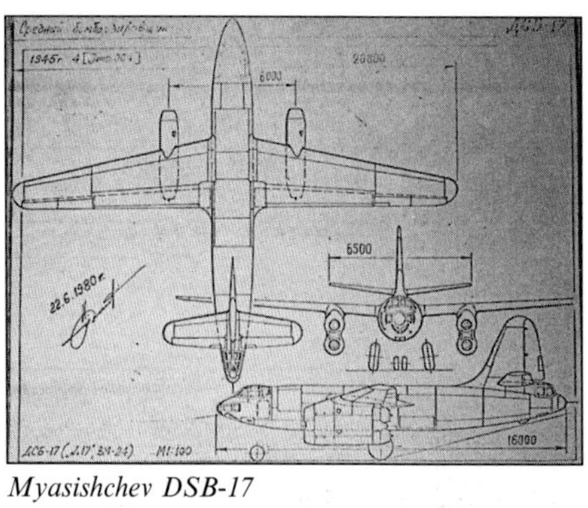

Myasishchev DSB-17

Myasishchev (p.258). Artwork of high-altitude Project 28, also designated 2M, showing changed method of mounting engines. Myasishchev had previously had to redesign wing roots in order to change type of engine fitted. Yefim Gordon archive.

Myasishchev 28

Nikitin (p.271). A three-view has been found of the MU-5 training amphibian. Basic design by Mikhelson.

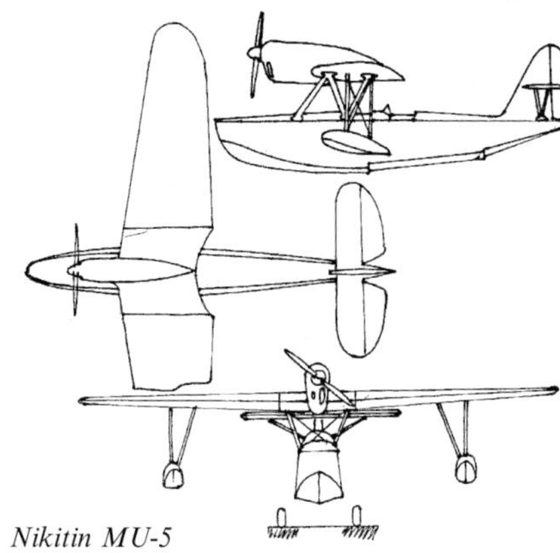

Nikitin MU-5

Petlyakov (p.282). Figure in main text for production of Pe-8 (93 = 91 plus two prototypes) is widely accepted, but 79 were built at GAZ Nos 22 and 124 after 18 delivered 1939-40. Research to verify total continues.

Polikarpov (p.309). Photograph of I-190 was discovered in early 94. It shows beautifully cowled M-88B engine and retractable skis. Photograph of I-190GK being sought.

Polikarpov I-190

Richard (p. 315). This drawing of TOM, discovered in 94, was also called Model 71, and dated 5 Mar 29. It included a scale of metres showing span at this time to have been about 31,8 m (104 ft 4 in), slightly less than TOM-1 as built. It also makes it obvious that photograph on p.315 is in fact a TB-1P, so photograph of TOM-1 is still being sought.

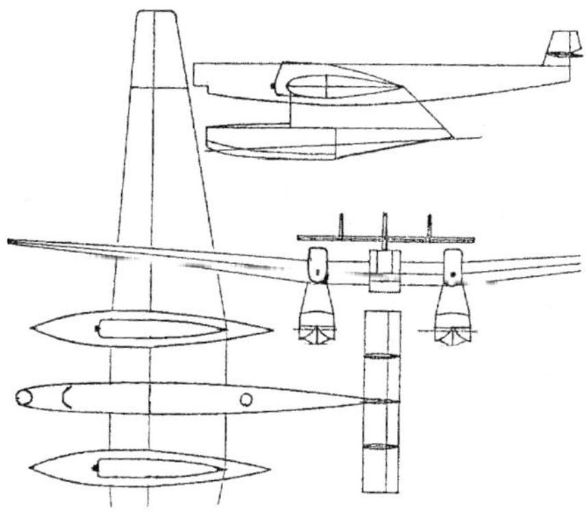

Richard TOM

Samsonov (p.321). Earliest known Samsonov design, this OKB drawing of MBR-3 discovered in 94 shows spatted wheels for alighting on land; stabilizing floats appear in side elevation only. To meet Aviatsiya VMF spec for bombs, mines or other stores and be defended by guns. Accompanying 10-m scale suggests span about 20,5 m (67 ft 3 in). Engines would have been c500 hp, possibly M-25. No evidence aircraft completed.

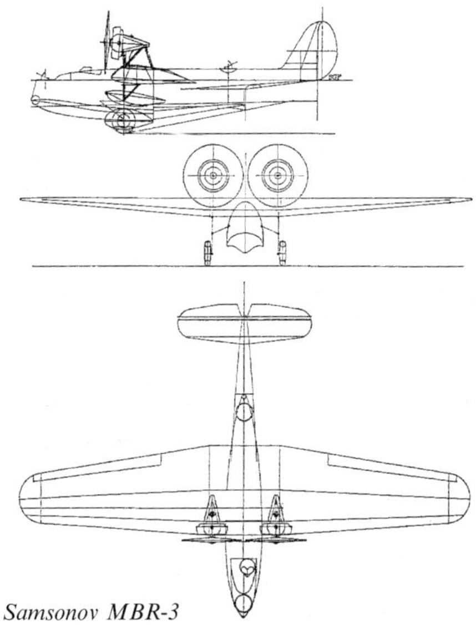

Samsonov MBR-3

Sukhoi (p.336). Because little-known, these drawings of Su-17 of 49 are of interest. Cross-section 5-5 shows position of two N-37 guns. Lyul'ka engine looks enormous.

Sukhoi (p.368). Improved three-view drawing of Su-34, based in part on one appearing in new magazine *A-K AS*.

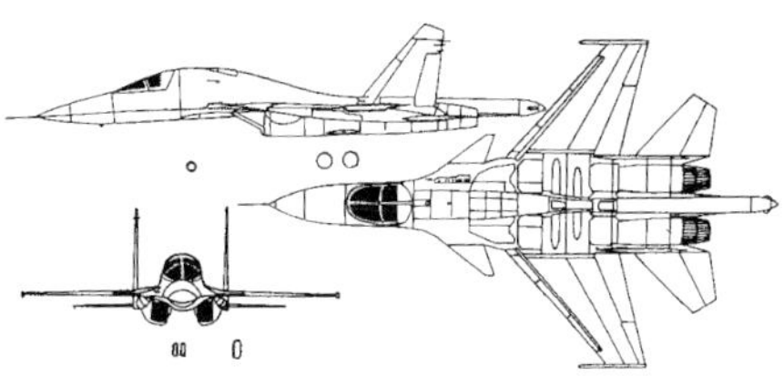

Su-34

Tupolev (p.430). *Eskisnyi* drawing of 105, immediate predecessor of 105A and Tu-22.

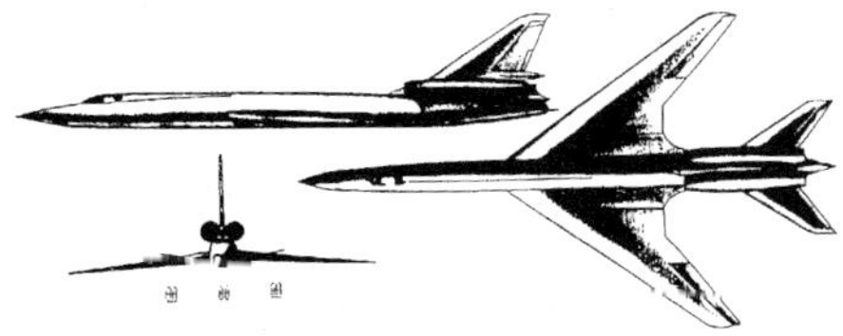

Tu-105

Tupolev (p.436). *Eskiznyi* drawing of Variant No 1 of 135, one of nine unbuilt supersonic projects; four 23-t NK-6, span 28,0 m (91 ft $10\frac{3}{8}$ in); length 44,8 m (146 ft $11\frac{3}{4}$ in); area 380 m^2 (4,090 ft^2); MTO 190 t (418,870 lb); range 7950 km (4,940 miles) at 2350 km/h (1,460 mph) at 22,3 km (73,165 ft); drawing shows Kh-22 missile.

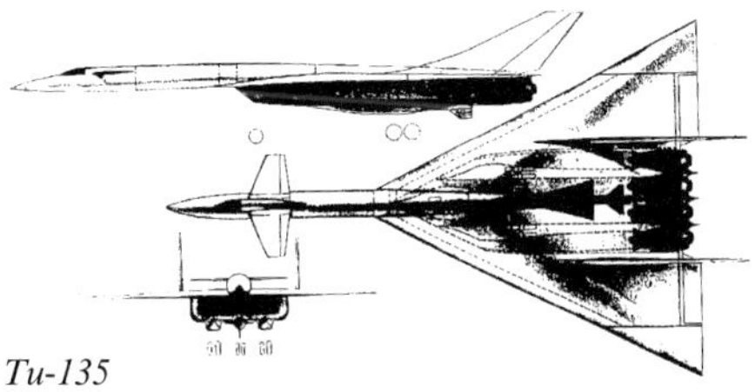

Tu-135

Tupolev (p.442). Desk-top model of 156, powered by D-30KP turbofans, intended to replace Tu-126. Kremlin decided timescale too long, asked Ilyushin to modify Il-76, but Novozhilov didn't like risk, so A. Konstantinov (Beriev) got job with A-50. Yefim Gordon archive.

Tupolev (p.447). Updated 95 drawing of Tu-34, now projected with choice of 230-hp piston engines (max speed 360 km/h, payload 4 pax or 450 kg for 1800 km) or 420-hp turboprops (max speed 450 km/h, payload 5 pax or 700 kg cargo for 1340 km).

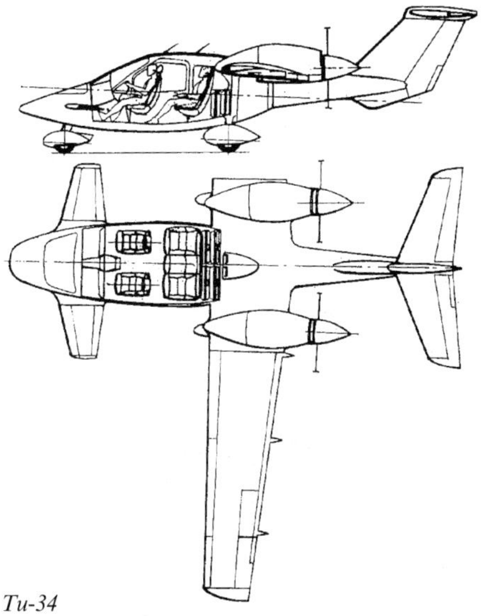

Tu-34

Tu-156

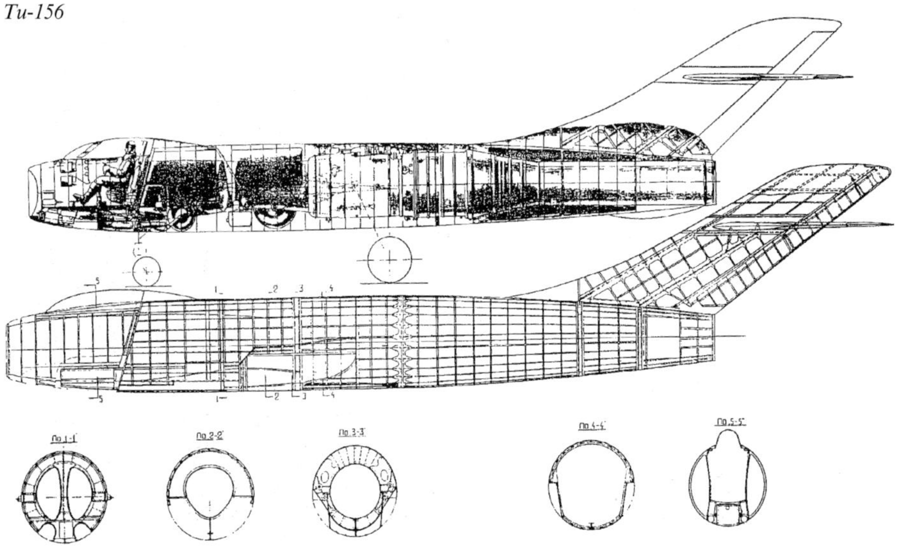

Su-17

Tyrov (p.448). Copy of OKB drawing of Ta-2 (OKO-4).

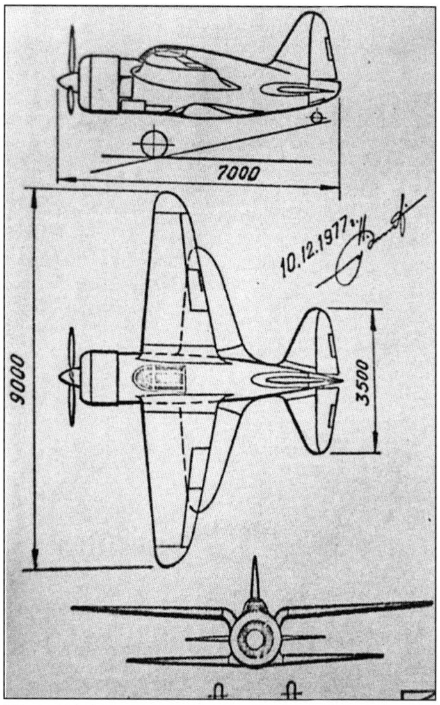

Tyrov Ta2

Tyrov (p.449). Just discovered, a three-view of the Ta-3bis, showing forward-swept wings.

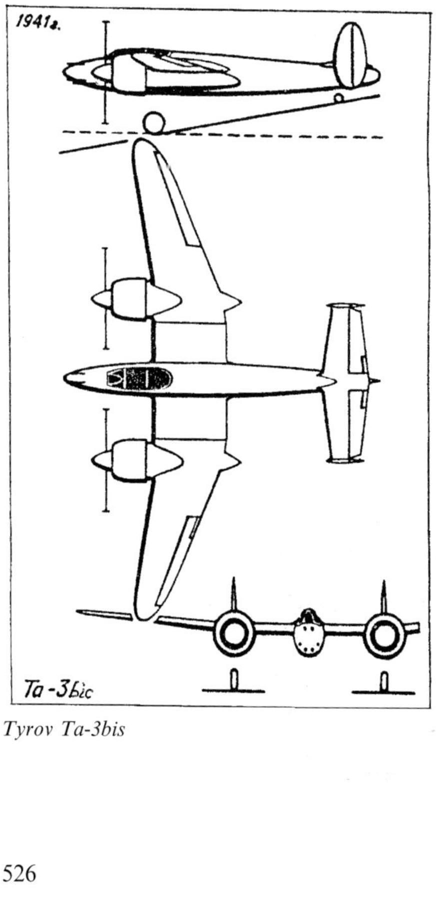

Tyrov Ta-3bis

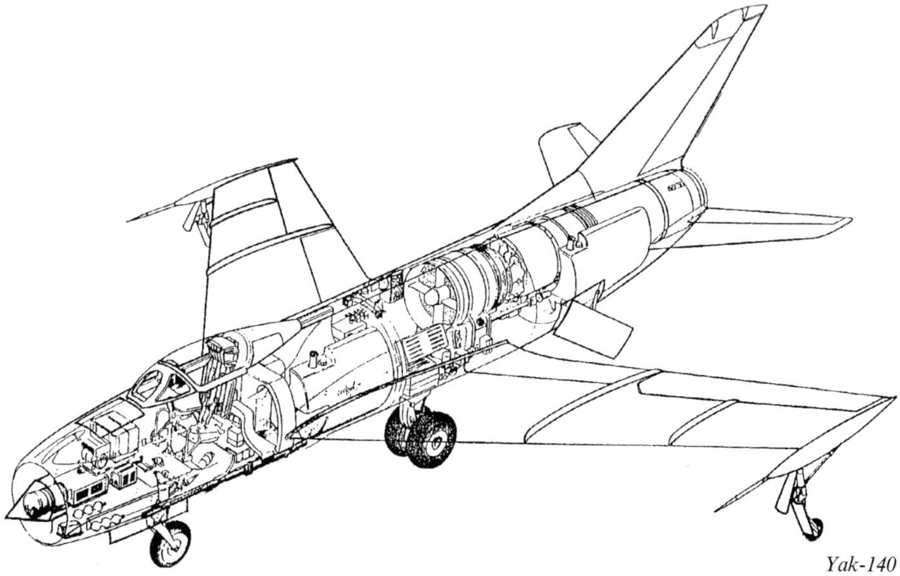

Yak-140

Yakovlev (p.488). Because it is so little known, this drawing of Yak-140 (No 1) equipment will be of interest; engine was an AM-9D and radar RP-21 *Sapfir.*

Yakovlev (p.489). Another view of Yak-28-64, with K-8 and K-13 missiles but no tanks.

Yak-28-64

Yakovlev (p.490). As originally built, Yak-36 No 38 differed from original No 36 in having a larger inlet to R-27V-300 engines. Later inlet was again enlarged and auxiliary door with downturned edges added under nose. Yefim Gordon archive.

Yak-36

Yakovlev (p.490). Grey-painted VM-02, first prototype of Yak-36M, seen here with Kh-23 (H-23) attack missiles. Yefim Gordon archive.

Yak-36M

Beriev (p.518). Comparative plan views of VVA-14 and 14M1P.

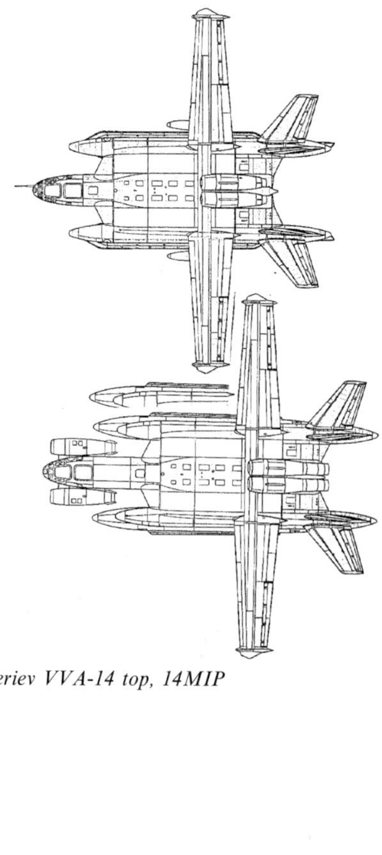

Beriev VVA-14 top, 14MIP